ENAM 98

ENAM98 — Exotic Nuclei and Atomic Masses

23-27 June, 1998   Bellaire, Michigan USA

ENAM 98

Exotic Nuclei and Atomic Masses

Bellaire, Michigan June 1998

EDITORS
Bradley M. Sherrill
David J. Morrissey
National Superconducting Cyclotron Laboratory
Michigan State University

Cary N. Davids
Argonne National Laboratory

AIP CONFERENCE
PROCEEDINGS 455

American Institute of Physics   Woodbury, New York

Editors:

Bradley M. Sherrill
Department of Physics and
National Superconducting Cyclotron Laboratory
Michigan State University
East Lansing, MI 48824-1321

E-mail: sherrill@nscl.msu.edu

David J. Morrissey
Department of Chemistry and
National Superconducting Cyclotron Laboratory
Michigan State University
East Lansing, MI 48824-1321

E-mail: morrissey@nscl.msu.edu

Cary N. Davids
PHY203
Argonne National Laboratory
9700 S. Cass Avenue
Argonne, IL 60439-4843

E-mail: davids@anl.gov

Articles on pp. 90–93, pp. 367–370, pp. 415–421, pp. 486–489, pp. 682–685, and pp. 952–959 were authored by U. S. Government employees and are not covered by the below mentioned copyright.

Authorization to photocopy items for internal or personal use, beyond the free copying permitted under the 1978 U.S. Copyright Law (see statement below), is granted by the American Institute of Physics for users registered with the Copyright Clearance Center (CCC) Transactional Reporting Service, provided that the base fee of $15.00 per copy is paid directly to CCC, 222 Rosewood Drive, Danvers, MA 01923. For those organizations that have been granted a photocopy license by CCC, a separate system of payment has been arranged. The fee code for users of the Transactional Reporting Service is: 1-56396-804-5/ 98 /$15.00.

© 1998 American Institute of Physics

Individual readers of this volume and nonprofit libraries, acting for them, are permitted to make fair use of the material in it, such as copying an article for use in teaching or research. Permission is granted to quote from this volume in scientific work with the customary acknowledgment of the source. To reprint a figure, table, or other excerpt requires the consent of one of the original authors and notification to AIP. Republication or systematic or multiple reproduction of any material in this volume is permitted only under license from AIP. Address inquiries to Office of Rights and Permissions, 500 Sunnyside Boulevard, Woodbury, NY 11797-2999; phone: 516-576-2268; fax: 516-576-2499; e-mail: rights@aip.org.

L.C. Catalog Card No. 98-88781
ISBN 1-56396-804-5
ISSN 0094-243X
DOE CONF- 980672

Printed in the United States of America

Contents

Preface xxiii
Committees and Sponsors xxv

MASSES, MOMENTS, AND RADII

Mass Measurements with a Penning Trap Mass Spectrometer at ISOLDE 3
    G. Bollen, F. Ames, G. Audi, D. Beck, F. Herfurth, H.-J. Kluge, A. Kohl,
    D. Lunney, R. B. Moore, M. de Saint Simon, E. Schark, S. Schwarz,
    J. Szerypo, and the ISOLDE Collaboration

Experiments with Stored Relativistic Exotic Nuclei 11
    H. Geissel, T. Radon, F. Attallah, K. Beckert, F. Bosch, A. Dolinskiy,
    H. Eickhoff, M. Falch, B. Franczak, B. Franzke, Y. Fujita, M. Hausmann,
    M. Hellström, F. Herfurth, Th. Kerscher, O. Klepper, H.-J. Kluge,
    C. Kozhuharov, K. E. G. Löbner, G. Münzenberg, F. Nolden, Yu. Novikov,
    Z. Patyk, W. Quint, H. Reich, C. Scheidenberger, B. Schlitt, J. Stadlmann,
    M. Steck, K. Sümmerer, L. Vermeeren, M. Winkler, Th. Winkler,
    and H. Wollnik

Self-Consistent Microscopic Approaches to Nuclear Masses 22
    P. Ring

Atomic Mass Knowledge: 1999 Status 30
    G. Audi and A. H. Wapstra

Single Particle Behaviour at (and away from) Stability 38
    J. Duflo and A. P. Zuker

Persistence of the N=28 Shell Closure Far from Stability 44
    F. Sarazin, H. Savajols, W. Mittig, G. Auger, D. Baiborodin,
    A. V. Belozyorov, C. Borcea, A. Gillibert, A. S. Lalleman, M. Lewitowicz,
    S. M. Lukyanov, F. de Oliveira, N. Orr, Y. E. Penionzhkevich, D. Ridikas,
    P. Roussel-Chomaz, H. Sakurai, and A. de Vismes

Recent Measurements of Nuclear Moments Far from Stability 50
    M. Keim

Nuclear Moments of Exotic Nuclear States Studied with Level
Mixing Techniques 58
    G. Neyens

Isospin Nonconserving Effects in Light, $T_z=-3/2$ Nuclei 66
    P. F. Mantica, D. W. Anthony, B. A. Brown, B. Davids, G. Georgiev,
    M. Huhta, R. W. Ibbotson, P. A. Lofy, J. I. Prisciandaro, and M. Steiner

On-line Laser Spectroscopy of Refractory Radioisotopes at the
JYFL IGISOL Facility 72
    P. Campbell, D. M. Benton, J. Billowes, P. Dendooven, D. E. Evans,
    D. H. Forest, I. S. Grant, J. A. R. Griffith, A. Honkanen, A. Jokinen,
    J. M. G. Levins, M. Oinonen, H. Penttilä, K. Peräjärvi, D. S. Richardson,
    G. Tungate, G. Yeandle, and J. Äystö

Large Odd-Even Staggering in the Very Light Platinum Isotopes from Laser Spectroscopy.. 78
    F. Le Blanc, J. Pinard, L. Cabaret, J. E. Crawford, H. T. Duong, J. Genevey,
    M. Girod, G. Huber, M. Krieg, J. K. P. Lee, J. Lettry, D. Lunney, J. Obert,
    J. Oms, S. Péru, J. C. Putaux, B. Roussière, J. Sauvage, V. Sebastian,
    S. Zemlyanoi, and the ISOLDE Collaboration

Nuclear Radii of ^{14}Be and ^{17,19}B.. 84
    T. Suzuki, K. Sümmerer, O. Bochkarev, L. Chulkov, D. Cortina,
    H. Geissel, M. Hellström, M. Ivanov, R. Janik, K. Kimura, T. Kobayashi,
    A. A. Korsheninnikov, G. Münzenberg, F. Nickel, A. A. Ogloblin,
    A. Ozawa, M. Pfützner, V. Pribora, H. Simon, B. Sitár, P. Strmeň,
    I. Tanihata, M. Winkler, and K. Yoshida

Mass Measurement in the fp-Shell Using the TOFI Spectrometer............ 90
    Y. Bai, D. J. Vieira, H. L. Seifert, and J. M. Wouters

Application of the Laser Ion Source for Isotope Shift and Hyperfine Structure Investigations... 94
    A. E. Barzakh, I. Ya. Chubukov, D. V. Fedorov, V. N. Panteleev,
    M. D. Seliverstov, and Yu. M. Volkov

Mass Extrapolations in the Region of Deformed Rare Earth Nuclei.......... 98
    C. Borcea and G. Audi

Mass Measurements of Proton-Rich Medium Mass Nuclides................ 102
    D. S. Brenner, B. D. Foy, C. J. Barton, C. N. Davids, D. Seweryniak,
    D. Blumenthal, R. L. Gill, N. V. Zamfir, and D. D. Warner

^{208}Pb Neutron Density: A Mean Field Problem?.......................... 106
    S. Gmuca

Measurement of the Magnetic Moments of ^{7}Be and of the Halo Nucleus ^{11}Be... 110
    S. Kappertz, W. Geithner, G. Katko, M. Keim, G. Kotrotsios, P. Lievens,
    R. Neugart, L. Vermeeren, S. Wilbert, and the ISOLDE Collaboration

Nuclear Mass Formula with the Shell Energies Obtained by a New Method... 114
    H. Koura, M. Uno, T. Tachibana, and M. Yamada

Interaction Radii of Proton-Rich Radioactive Nuclei at A=60−80............ 118
    G. F. Lima, A. Lépine-Szily, A. C. C. Villari, W. Mittig, M. Chartier,
    R. Lichtenthaler, D. Hirata, J. C. Angelique, G. Audi, J. M. Casandjian,
    A. Cunsolo, C. Donzaud, A. Foti, A. Gillibert, M. Lewitowicz,
    S. Lukyanov, M. MacCormick, D. J. Morrissey, N. A. Orr,
    A. N. Ostrowski, B. M. Sherrill, C. Stephan, T. Suomijärvi,
    L. Tassan-Got, D. J. Vieira, and J. M. Wouters

Recent Mass Predictions in the Infinite Nuclear Matter Model.............. 122
    L. Satpathy and R. C. Nayak

Spectroscopic Applications of the ISOLDE Laser Ion Source................ 126
    V. Sebastian, R. Catherall, V. N. Fedoseyev, U. Georg, G. Huber,
    Y. Jading, O. Jonsson, U. Köster, M. Koizumi, K.-L. Kratz, E. Kugler,
    J. Lettry, V. I. Mishin, H. L. Ravn, C. Tamburella, A. Wöhr,
    and the ISOLDE Collaboration

Status of the Canadian Penning Trap Mass Spectrometer at the
Argonne National Laboratories 130
    K. S. Sharma, R. C. Barber, F. Buchinger, J. E. Crawford, X. Feng,
    H. Fukutani, S. Gulick, G. Hackman, J. C. Hardy, D. Hofman,
    J. K. P. Lee, P. Martinez, R. B. Moore, G. Savard, D. Seweryniak,
    and J. Uusitalo

Correlation Parameters in Nuclear Binding Energies. 134
    S. I. Sukhoruchkin and D. S. Sukhoruchkin

The Magnetic and Quadrupole Moment of Oriented Nuclei Measured
with the β-LMR-NMR. 138
    S. Teughels, G. Neyens, N. Coulier, G. Georgiev, S. Ternier, K. Vyvey,
    D. L. Balabanski, R. Coussement, W. F. Rogers, D. Cortina-Gil,
    F. de Oliveira, M. Lewitowicz, M. Mittig, P. Roussel-Chomaz,
    and A. Lépine-Szily

β-NMR Measurement of the Nuclear Quadrupole Moments of $^{20,26-31}$Na 142
    S. Wilbert, B. A. Brown, W. Geithner, U. Georg, S. Kappertz,
    M. Keim, P. Lievens, R. Neugart, M. Neuroth, L. Vermeeren,
    and the ISOLDE Collaboration

The Wigner Term in Heavy Nuclei 146
    N. Zeldes

Isotope Shifts of Optical Transitions of Kr I Using Collinear Fast
Beam Laser Spectroscopy 150
    J. Lassen and H. A. Schuessler

NUCLEI AT THE DRIP LINES

Exploration Beyond the Proton Drip-Line 157
    P. J. Woods

Status of Neutron Drip-Line Nuclei. 166
    B. Jonson

Probing Halo Structures with Breakup Reactions. 174
    I. J. Thompson

Studies of Excited States in ^{11}Li. Spectroscopy of ^{7}He 182
    A. A. Korsheninnikov, M. S. Golovkov, A. Ozawa, E. A. Kuzmin,
    E. Yu. Nikolskii, K. Yoshida, B. G. Novatskii, A. A. Ogloblin, I. Tanihata,
    Z. Fulop, K. Kusaka, K. Morimoto, H. Otsu, H. Petrascu, and F. Tokanai

Study of Exotic Nuclei by Proton Scattering in Inverse Kinematics 188
    G. D. Alkhazov

Energy and Angular Correlations in Peripheral Fragmentation
of Halo Nuclei. 196
    T. Aumann and L. V. Chulkov

Coulomb Excitation of Exotic Nuclei and Segmented Germanium Detectors... 204
    T. Glasmacher

Spectroscopy of ^{25}Al and ^{26,27,28}P Using High-Energy Stripping Reactions 209
    N. Alahari, D. Bazin, B. A. Brown, B. Davids, G. Gervais, T. Glasmacher,
    K. Govaert, P. G. Hansen, M. Hellström, R. W. Ibbotson, V. Maddalena,
    B. Pritychenko, H. Scheit, B. M. Sherrill, M. Steiner, J. A. Tostevin,
    and J. Yurkon

Coulomb Dissociation of ^{19}C 215
    T. Nakamura, N. Fukuda, T. Kobayashi, N. Aoi, H. Iwasaki, T. Kubo,
    A. Mengoni, M. Notani, H. Otsu, H. Sakurai, S. Shimoura, T. Teranishi,
    Y. Watanabe, K. Yoneda, and M. Ishihara

New Evidence for Parity Inversion in ^{10}Li from ^{9}Li and γ-Ray Coincidences 221
    M. Chartier, J. R. Beene, B. Blank, L. Chen, A. Galonsky, N. Gan,
    K. Govaert, P. G. Hansen, J. Kruse, V. Maddalena, M. Thoennessen,
    and R. L. Varner

Sizes of the He Isotopes Deduced from Proton Elastic Scattering Measurements 227
    J. A. Tostevin and J. S. Al-Khalili

New Neutron-Rich Isotope ^{31}F and Particle Instability of ^{25}N and ^{28}O 233
    H. Sakurai, S. M. Lukyanov, M. Notani, N. Aoi, D. Beaumel, N. Fukuda,
    M. Hirai, E. Ideguchi, N. Imai, M. Ishihara, H. Iwasaki, T. Kubo, K. Kusaka,
    H. Kumagai, T. Nakamura, H. Ogawa, Yu. E. Penionzhkevich, T. Teranishi,
    Y. X. Watanabe, K. Yoneda, and A. Yoshida

Study of ^{11}Be Structure via the p(^{11}Be, ^{10}Be)d Reaction 239
    S. Fortier, J. S. Winfield, S. Pita, W. N. Catford, N. A. Orr, Y. Blumenfeld,
    R. Chapman, S. P. G. Chappell, N. M. Clarke, N. Curtis, M. Freer, S. Galès,
    K. L. Jones, H. Langevin-Joliot, H. Laurent, I. Lhenry, J. M. Maison,
    P. Roussel-Chomaz, M. Shawcross, M. Smith, K. Spohr, T. Suomijärvi,
    and A. de Vismes

Measurement of the E1 Strength Function of ^{11}Be 245
    R. L. Varner, N. Gan, J. R. Beene, M. L. Halbert, D. W. Stracener,
    A. Azhari, E. Ramakrishnan, P. Thirolf, M. R. Thoennessen,
    and S. Yokoyama

Do the Excited States in the System of Two Neutrons Exist? 252
    D. V. Aleksandrov, E. Yu. Nikolskii, B. G. Novatskii, D. N. Stepanov,
    and R. Wolski

Continuum Excitations in Neutron-Rich Oxygen Isotopes 256
    T. Aumann, A. Leistenschneider, K. Boretzky, D. Cortina, J. Cub, W. Dostal,
    B. Eberlein, Th. W. Elze, H. Emling, H. Geissel, A. Grünschloß,
    M. Hellström, J. Holeczek, R. Holzmann, S. Ilievski, N. Iwasa, M. Kaspar,
    A. Kleinböhl, J. V. Kratz, R. Kulessa, Y. Leifels, E. Lubkiewicz,
    G. Münzenberg, P. Reiter, M. Rejmund, C. Scheidenberger,
    Ch. Schlegel, H. Simon, J. Stroth, K. Sümmerer, E. Wajda, W. Walus,
    and S. Wan

Visualizing ^{11}N by Resonance Reactions 260
    L. Axelsson, K. Markenroth, M. J. G. Borge, S. Fayans, V. Z. Goldberg,
    S. Grévy, D. Guillemaud-Mueller, B. Jonson, K.-M. Källman, T. Lönnroth,
    M. Lewitowicz, P. Manngård, I. Martel, A. C. Mueller, I. Mukha, T. Nilsson,
    G. Nyman, N. A. Orr, K. Riisager, G. V. Rogatchev, M.-G. Saint-Laurent,
    I. N. Serikov, O. Sorlin, O. Tengblad, F. Wenander, J. S. Winfield,
    and R. Wolski

Proton Decay Studies at HRIBF 264
    J. C. Batchelder, C. R. Bingham, K. Rykaczewski, K. S. Toth, T. Davinson,
    T. N. Ginter, C. J. Gross, R. Grzywacz, Z. Janas, M. Karny, S. H. Kim,
    B. D. MacDonald, J. F. Mas, J. W. McConnell, A. Piechaczek, J. J. Ressler,
    R. C. Slinger, J. Szerypo, W. B. Walters, W. Weintraub, P. J. Woods,
    C.-H. Yu, and E. F. Zganjar

Probing the Nuclear Structure of ^{8}B and ^{19}C 268
    T. Baumann, H. Geissel, H. Lenske, K. Markenroth, T. Aumann,
    L. Axelsson, U. Bergmann, D. Cortina-Gil, L. Fraile, M. Hellström,
    M. Ivanov, N. Iwasa, R. Janik, B. Jonson, G. Münzenberg, F. Nickel,
    T. Nilsson, A. Ozawa, A. Richter, K. Riisager, C. Scheidenberger,
    G. Schrieder, W. Schwab, H. Simon, B. Sitar, M. H. Smedberg, P. Strmen,
    K. Sümmerer, T. Suzuki, M. Winkler, and M. V. Zhukov

Measurement of Parentage in Stripping Reactions of Halo Nuclei........... 272
    D. Bazin, N. Alahari, B. Blank, J. E. Bush, J. A. Caggiano, L. Chen,
    B. Davids, T. Glasmacher, K. Govaert, V. Guimares, P. G. Hansen,
    R. W. Ibbotson, D. Karnes, J. J. Kolata, V. Maddalena, B. Pritychenko,
    H. Scheit, and B. M. Sherrill

On the Road to Doubly-Magic ^{48}Ni... 276
    B. Blank, J. Benlliure, F. Boué, R. Collatz, S. Czajkowski, F. Davi,
    R. Del Moral, J. P. Dufour, A. Fleury, A. Heinz, M. Lewitowicz,
    C. Marchand, M. Hellström, Z. Hu, Z. Janas, M. Karny, M. Pfützner,
    E. Roeckl, M. S. Pravikoff, M. Shibata, and K. Sümmerer

Toward a Complete Picture of ^{11}Be Interaction with Silicon 278
    C. Borcea, F. Carstoiu, F. Negoita, M. Lewitowicz, M. G. Saint-Laurent,
    R. Anne, D. Guillemaud-Mueller, A. C. Mueller, F. Pougheon, O. Sorlin,
    A. Fomitchev, S. Lukyanov, Yu. Penionzhkevich, N. Skobelev, and Z. Dlouhy

Deformation Dependence of Proton Decay Rates and Angular Distributions in a Time-Dependent Approach 282
    N. Carjan, P. Talou, and D. Strottman

Evidence for 2p-Radioactivity in ^{17}Ne... 286
    M. J. Chromik, P. G. Thirolf, M. Thoennessen, M. Fauerbach,
    T. Glasmacher, R. Ibbotson, R. A. Kryger, H. Scheit, and P. J. Woods

Cross-Section Studies of Light Exotic Nuclei............................... 290
    D. Cortina-Gil, K. Sümmerer, T. Baumann, and H. Geissel

New Insight in Halo Fragmentation 294
    B. V. Danilin, S. N. Ershov, T. Rogde, and J. S. Vaagen

Spacial Probability Distribution of Nucleons and Nuclear Binding-Energy Losses in ^{6}He and ^{8}He Associated with Halo Formation...... 298
    F. Everling

Density-Dependent Pairing in Nuclei Far from Stability 302
    S. A. Fayans, S. V. Tolokonnikov, E. L. Trykov, and D. Zawischa

Semimicroscopic Calculations of Total Reaction Cross Sections for Light Exotic Nuclei.. 306
    S. A. Fayans, D. V. Bolotov, O. M. Knyazkov, and I. N. Kuchtina

Nolen-Schiffer Anomaly and Atomic Masses 310
    S. A. Fayans

Exotic Molecular and Halo States in ^{12,14}Be............................314
    M. Freer, N. A. Orr, M. Labiche, F. M. Marqués, J. C. Angélique,
    L. Axelsson, B. Benoit, U. Bergmann, M. J. G. Borge, W. N. Catford,
    S. P. G. Chappell, N. M. Clarke, G. Costa, N. Curtis, A. D'Arrigo,
    F. de Oliveira, E. de Goes Brennard, O. Dorvaux, B. R. Fulton,
    G. Gardina, C. Gregori, S. Grévy, D. Guillemaud-Mueller, F. Hanappe,
    B. Heusch, B. Jonson, G. Kelly, C. Le Brun, S. Leenhardt, M. Lewitowicz,
    K. Markenroth, M. Motta, A. C. Mueller, J. T. Murgatroyd, T. Nilsson,
    A. Ninane, G. Nyman, I. Piqueras, K. Riisager, M. G. Saint Laurent,
    F. Sarazin, S. Singer, O. Sorlin, L. Stuttgé, and D. L. Watson

Resonance Scattering to Study Nuclei at the Borders of Nuclear Stability.....319
    V. Z. Goldberg

Final-State Interactions in the System ^{8}He+n.............................323
    L. Chen, K. Govaert, B. Blank, M. Chartier, A. Galonsky,
    P. G. Hansen, J. Kruse, V. Maddalena, M. Thoennessen, K. Ieki,
    Y. Iwata, Y. Higurashi, S. Takeuchi, F. Deak, A. Horvath,
    A. Kiss, and Z. Seres

Neutron Angular Distributions from the Core Break-Up Reactions
of the ^{11}Be and ^{11}Li Halo Nuclei 327
    S. Grévy, L. Axelsson, J. C. Angélique, R. Anne, D. Guillemaud-Mueller,
    P. G. Hansen, P. Hornshoj, B. Jonson, M. Lewitowicz, A. C. Mueller,
    T. Nilsson, G. Nyman, N. Orr, F. Pougheon, K. Riisager,
    M. G. Saint-Laurent, M. Smedberg, and O. Sorlin

The Description of Exotic Nuclei Elastic Scattering in the
Framework of Glauber Model with Non-Eikonal Corrections 331
    K. A. Gridnev and T. V. Taroutina

Two-Neutron Removal Reactions for Three-Body Halo Nuclei...............335
    E. Garrido, D. V. Fedorov, and A. S. Jensen

Phase Equivalent Potentials for Three-Body Halos........................339
    E. Garrido, D. V. Fedorov, and A. S. Jensen

Angular Correlation in Breakup of Three-Body Halo Nuclei 343
    E. Garrido, D. V. Fedorov, and A. S. Jensen

Study of the Unbound Nucleus ^{11}N by the ^{12}C(^{14}N, ^{15}C) ^{11}N Transfer
Reaction...347
    A. Lépine-Szily, J. M. Oliveira, Jr., A. N. Ostrowski, H. G. Bohlen,
    R. Lichtenthaler, A. Blazevic, C. Borcea, V. Guimarães, R. Kalpakchieva,
    V. Lapoux, M. MacCormick, F. Oliveira, W. von Oertzen, N. A. Orr,
    P. Roussel-Chomaz, Th. Stolla, and J. S. Winfield

Survey of the Beta-Strength of the Halo Nucleus ^{11}Li 351
    I. Mukha, M. J. G. Borge, D. Guillemaud-Mueller, P. Hornshøj, B. Jonson,
    H. Fynbo, T. Leth, T. Nilsson, G. Nyman, K. Riisager, G. Schrieder,
    M. H. Smedberg, and O. Tengblad

Shell-Quenching in the Infinite Nuclear Matter Model 355
    R. C. Nayak

Half-Life Measurements of ^{31,32}Ne.......................................359
    M. Notani, N. Aoi, N. Fukuda, H. Iwasaki, K. Yoneda, H. Ogawa,
    T. Teranishi, S. M. Lukyanov, Yu. E. Penionzhkevich, T. Nakamura,
    H. Sakurai, E. Ideguchi, A. Yoshida, Y. Watanabe, T. Kubo,
    and M. Ishihara

Elastic Transfer in ^{4}He (^{6}He, ^{6}He) ^{4}He 363
  A. Piechaczek, R. Raabe, A. Andreyev, D. Baye, W. Bradfield-Smith,
  T. Davinson, M. Gaelens, W. Galster, M. Huyse, J. McKenzie, A. Ninane,
  A. C. Shotter, G. Vancraeynest, P. Van Duppen, and A. Wöhr

Proton Radioactivity from Highly Deformed Nuclei 367
  A. A. Sonzogni, C. N. Davids, P. J. Woods, D. Seweryniak,
  J. C. Batchelder, C. R. Bingham, T. Davinson, D. J. Henderson,
  R. J. Irvine, G. L. Poli, J. Uusitalo, and W. B. Walters

Quasifree Scattering of Neutron in ^{11}Li from Deuteron 371
  S. Takeuchi, S. Shimoura, T. Teranishi, Y. Ando, M. Hirai,
  N. Iwasa, T. Kikuchi, S. Moriya, T. Motobayashi, H. Murakami,
  T. Nakamura, T. Nishio, H. Sakurai, T. Uchibori, Y. Watanabe,
  Y. Yanagisawa, and M. Ishihara

Proton Decay of the Closed Neutron Shell Nucleus ^{155}Ta 375
  J. Uusitalo, C. N. Davids, P. J. Woods, D. Seweryniak, A. A. Sonzogni,
  J. C. Batchelder, C. R. Bingham, T. Davinson, J. DeBoer, D. J. Henderson,
  H. J. Maier, J. Ressler, R. Slinger, and W. B. Walters

NUCLEAR STRUCTURE AND SHAPES

Exotic Nuclei from a Theoretical Perspective 381
  W. Nazarewicz

Monte Carlo Shell Model Calculations for Exotic Nuclei 391
  T. Otsuka, T. Mizusaki, and Y. Utsuno

Quantum Monte Carlo Calculations of Light Nuclei 399
  S. C. Pieper

Superdeformation and Smooth Band Termination in A~60 Nuclei 407
  C. E. Svensson

In-Beam γ-Ray Spectroscopy in the Vicinity of ^{100}Sn 415
  D. Seweryniak

Nuclear Structure Information from Recoil Decay Tagging Experiments 422
  M. Leino, R. Julin, J. F. C. Cocks, P. A. Butler, O. Dorvaux, K. Eskola,
  P. T. Greenlees, P. Jones, S. Juutinen, K. Helariutta, H. Kankaanpää,
  H. Kettunen, P. Kuusiniemi, M. Muikku, R. D. Page, P. Rahkila,
  A. Savelius, W. H. Trzaska, and J. Uusitalo

Microsecond Isomers Studies 430
  R. Grzywacz

Shell-Model Monte Carlo Studies of Nuclei Far from Stability 438
  D. J. Dean

In-Beam γ-Ray Spectroscopy in the Ground-State Proton Emitter ^{113}Cs 444
  C. J. Gross, Y. A. Akovali, C. Baktash, J. C. Batchelder, C. R. Bingham,
  M. P. Carpenter, C. N. Davids, T. Davinson, D. Ellis, A. Galindo-Uribarri,
  T. N. Ginter, R. Grzywacz, R. V. F. Janssens, J. W. Johnson, J. F. Liang,
  C. J. Lister, J. Mas, B. D. MacDonald, S. D. Paul, A. Piechaczek,
  D. C. Radford, W. Reviol, K. Rykaczewski, W. Satula, D. Seweryniak,
  D. Shapira, K. S. Toth, W. Weintraub, P. J. Woods, C.-H. Yu,
  E. F. Zganjar, and J. Uusitalo

Probing the Structure of the N=Z=31 Nucleus $^{62}_{31}$Ga 450
   S. M. Vincent, P. H. Regan, D. D. Warner, W. Gelletly, J. Simpson,
   R. Bark, D. Blumenthal, M. P. Carpenter, C. N. Davids, D. J. Henderson,
   R. V. F. Janssens, C. J. Lister, D. Nisius, C. D. O'Leary, C. J. Pearson,
   T. Saitoh, J. Schwartz, D. Seweryniak, and S. Törmänen

Decay of Very Neutron-Rich Mn Nuclides and Vanishing of the N=40 Subshell Closure in ^{66}Fe 456
   A. Wöhr, M. Hannawald, W. B. Walters, T. Kautzsch, B. Pfeiffer,
   K.-L. Kratz, V. N. Fedoseyev, V. I. Mishin, D. Forkel-Wirth, V. Sebastian,
   M. Koizumi, U. Köster, J. Lettry, H. L. Ravn, and the ISOLDE Collaboration

Fine Structure in the Alpha-Decay of the Neutron-Deficient ^{191}Po Isotope 462
   A. N. Andreyev, N. Bijnens, J. F. Cocks, K. Eskola, K. Helariutta,
   M. Huyse, H. Kettunen, P. Kuusiniemi, M. Leino, W. H. Trzaska,
   P. Van Duppen, and R. Wyss

Total Coulomb Excitation Cross Section Measurements of Radioactive Nuclear Beams in Low Energy Inverse Kinematics 466
   C. J. Barton, D. S. Brenner, R. F. Casten, N. V. Zamfir, R. L. Gill,
   and D. Shapira

Spectroscopy of N=82,83 ^{136,137}Xe Isotopes from ^{248}Cm Fission 470
   P. Bhattacharyya, C. T. Zhang, P. J. Daly, Z. W. Grabowski, R. Broda,
   B. Fornal, I. Ahmad, T. Lauritsen, L. R. Morss, W. R. Phillips, J. L. Durell,
   M. J. Leddy, A. G. Smith, W. Urban, B. J. Varley, N. Schulz,
   E. Lubkiewicz, M. Bentaleb, and J. Blomqvist

Alpha Decay of the $h_{9/2}$ Ground and $s_{1/2}$ Intruder States in Light Bi and At Isotopes 474
   C. R. Bingham, J. C. Batchelder, J. A. Cizewski, C. N. Davids,
   R. J. Irvine, W. Reviol, D. Seweryniak, K. S. Toth, W. B. Walters,
   J. Wauters, J. L. Wood, X. J. Xu, J. Uusitalo, and E. F. Zganjar

Elastic and Inelastic Proton Scattering on the Unstable ^{20}O Nucleus Measured with the "Must" Detector Array 478
   E. Khan, Y. Blumenfeld, T. Suomijärvi, N. Alamanos, F. Auger, N. Frascaria,
   A. Gillibert, T. Glasmacher, M. Godwin, V. Lapoux, I. Lhenry, F. Maréchal,
   D. J. Morrissey, A. Musumara, N. Orr, S. Ottini, P. Piattelli, E. C. Pollacco,
   P. Roussel-Chomaz, J. C. Roynette, D. Santonocito, J. E. Sauvestre,
   and J. A. Scarpaci

The Isomeric Cross Section Ratios (ICSR) Method at the Investigation of the Role of Isospin at the Population Compound States Cloused Yrast Band in the Nuclear Reactions with Neutron Heavy Ions 482
   T. V. Chuvilskay, Yu. G. Seleznev, A. A. Shirokova, and M. Herman

Evolution of Collective Motion in Light Polonium Nuclei 486
   J. A. Cizewski, K. Y. Ding, N. Fotiades, D. P. McNabb, W. Younes,
   R. Julin, M. Leino, J. Cocks, P. Greenlees, K. Helariutta, P. Jones,
   S. Juutinen, A. Kankaanpää, H. Kettunen, P. Kuusiniemi, M. Muikku,
   P. Rahkila, A. Savelius, C. N. Davids, R. V. F. Janssens, D. Seweryniak,
   M. P. Carpenter, H. Amro, P. Decrock, P. Reiter, D. Nisius, L. T. Brown,
   S. Fischer, T. Lauritsen, J. Wauters, C. R. Bingham, M. Huyse,
   and A. Andreyev

**Inelastic Proton Scattering on the Radioactive Nuclei ^{18}Ne and ^{20}O
in Inverse Kinematics** 490
    P. D. Cottle, L. A. Riley, J. K. Jewell, T. Glasmacher, K. W. Kemper,
    Y. Blumenfeld, M. Chromik, S. E. Hirzebruch, R. W. Ibbotson,
    F. Maréchal, D. J. Morrissey, H. Scheit, and T. Soumijärvi

**Interplay Between Nuclear Structure and Reaction Mechanism in the
Production of Projectile-like Short-lived Isomers** 494
    J. M. Daugas, M. Lewitowicz, R. Anne, J. C. Angélique, L. Axelsson,
    R. Béraud, C. Borcea, E. Chabannat, Th. Ethvignot, S. Franchoo,
    M. Glogowski, R. Grzywacz, H. Grawe, D. Guillemaud-Mueller, M. Huyse,
    Z. Janas, M. Karny, C. Longour, M. J. Lopez-Jimenez, A. C. Mueller,
    A. Nowak, F. de Oliveira-Santos, N. A. Orr, A. Płochocki, M. Pfützner,
    K. Rykaczewski, M. G. Saint-Laurent, J. E. Sauvestre, O. Sorlin,
    P. Van Duppen, and J. S. Winfield

Mass Dependence of the Effective Isovector Charge in the sd-Shell 498
    M. Fauerbach, P. D. Cottle, T. Glasmacher, R. W. Ibbotson, K. W. Kemper,
    B. Pritychenko, H. Scheit, and M. Steiner

On the Q_β-Puzzle Near ^{132}Sn 502
    B. Fogelberg, K. A. Mezilev, H. Mach, and V. I. Isakov

High-Spin States in ^{71}As, ^{72}Se, and ^{72}Br 506
    N. Fotiades, J. A. Cizewski, C. J. Lister, C. N. Davids, R. V. F. Janssens,
    D. Seweryniak, M. P. Carpenter, T. L. Khoo, T. Lauritsen, D. Nisius,
    P. Reiter, J. Uusitalo, I. Wiedenhover, A. O. Macchiavelli, and R. W. McLeod

**Analysis of the Elastic $^6Li+^{12}C$ Scattering: Energy Dependence,
"Abnormal Dispersion" and Dynamic Polarization Potential**................ 510
    S. A. Goncharov, A. S. Dem'yanova, and A. A. Ogloblin

Symmetry Structure in Neutron Deficient Xenon Nuclei 515
    I. M. Govil

Fragmentation of Alpha-Cluster States in ^{32}S. Bozonization 519
    K. A. Gridnev, M. Brenner, A. E. Antropov, S. E. Belov, B. Z. Taibin,
    K. N. Ershov, D. K. Gridnev, M. P. Kartamishev, I. V. Krouglov,
    and T. V. Taroutina

Identification of ^{162}Gd and a New Type of Identical Bands................. 523
    E. F. Jones, P. M. Gore, J. H. Hamilton, A. V. Ramayya, R. S. Dodder,
    C. J. Beyer, J. K. Hwang, X. Q. Zhang, S. J. Zhu, A. P. de Lima,
    J. Kormicki, J. D. Cole, R. Aryaeinejad, W. C. Ma, G. M. Ter-Akopian,
    Yu. Ts. Oganessian, A. V. Daniel, J. O. Rasmussen, S. J. Asztalos,
    I. Y. Lee, A. O. Macchiavelli, M. A. Stoyer, R. W. Lougheed,
    S. G. Prussin, and R. Donangelo

A Fully Relativistic Hartree-Bogoliubov Approach for Deformed Nuclei 527
    D. Hirata and B. V. Carlson

**Decay Studies of Neutron-Rich $A \approx 70$ Nuclei Produced by Fragmentation
of 70 MeV/A ^{76}Ge Projectiles** 532
    J. I. Prisciandaro, P. F. Mantica, D. W. Anthony, M. Huhta, P. A. Lofy,
    R. M. Ronningen, M. Steiner, and W. B. Walters

**In-Beam Coulomb-Excitation Studies of Odd-A $\nu(fp)$- and $\pi(sd)$-
Shell Nuclei** 536
    R. W. Ibbotson, T. Glasmacher, P. F. Mantica, and H. Scheit

β-Decay Half-Lives of New Neutron-Rich Lanthanide Isotopes 540
    S. Ichikawa, K. Tsukada, M. Asai, A. Osa, M. Sakama, Y. Kojima,
    M. Shibata, I. Nishinaka, Y. Nagame, Y. Oura, and K. Kawade

Conversion Electron Measurements in ^{125,127}Ba 544
    H. Iimura, S. Ichikawa, T. Sekine, M. Oshima, and M. Miyaji

Fusion Reactions of Deformed Nuclei Near Coulomb Barriers 548
    H. Ikezoe, T. Ikuta, S. Mitsuoka, T. Kuzumaki, J. Lu, Y. Nagame,
    I. Nishinaka, K. Tsukada, and T. Ohtsuki

Properties of Two Neutron-Hole ^{130}Sn and Two Neutron-Particle ^{134}Sn 552
    V. I. Isakov, K. I. Erokhina, B. Fogelberg, and H. Mach

Study of the First Excited $K^\pi=0^+$ Band in ^{162}Dy 556
    B. Liu, N. V. Zamfir, D. S. Brenner, R. F. Casten, G. Cata-Danil,
    C. W. Beausang, R. Krücken, J. R. Cooper, J. R. Novak, C. J. Barton,
    and R. L. Gill

Shape Coexistence in the Light Po Isotopes 560
    A. M. Oros, K. Heyde, C. De Coster, B. Decroix, R. Wyss, B. Barrett,
    and P. Navratil

Fine Structure in ^{192}Po α-Decay and Shape Coexistence in ^{188}Pb 564
    R. D. Page, R. G. Allatt, T. Enqvist, K. Eskola, P. T. Greenlees,
    P. Jones, R. Julin, P. Kuusiniemi, M. Leino, and J. Uusitalo

Some Regularities in the Production of Isotopes in ^{32,34,36}S-Induced Reactions in the Energy Range 6-75 A MeV 568
    O. B. Tarasov, Yu. E. Penionzhkevich, R. Anne, D. S. Baiborodin,
    A. V. Belozyorov, C. Borcea, Z. Dlouhy, D. Guillemaud-Mueller,
    R. Kalpakchieva, M. Lewitowicz, S. M. Lukyanov, V. Z. Maidikov,
    A. C. Mueller, Yu. Ts. Oganessian, M. G. Saint-Laurent, N. K. Skobelev,
    O. Sorlin, V. D. Toneev, and W. Trinder

Study of Neutron-Rich Nuclei Near the N=20 Neutron Closed Shell 570
    R. Allatt, J. C. Angélique, R. Anne, C. Borcea, Z. Dlouhy, C. Donzaud,
    S. Grévy, D. Guillemaud-Mueller, M. Lewitowicz, S. Lukyanov,
    A. C. Mueller, F. Nowacki, N. A. Orr, Yu. E. Penionzhkevich, R. D. Page,
    F. Pougheon, A. Reed, M. G. Saint-Laurent, W. Schwab, E. Sokol,
    O. Tarasov, W. Trinder, and J. S. Winfield

Spectroscopy of Neutron Rich Nuclei in the Vicinity of N=20 572
    A. T. Reed, R. D. Page, R. G. Allatt, P. J. Nolan, D. Guillemaud-Mueller,
    C. Donzaud, S. Grévy, A. C. Mueller, F. Pougheon, O. Sorlin,
    Yu. Penionzhkevich, S. Lukyanov, E. Sokol, O. Tarasov, J. C. Angélique,
    F. M. Marques, N. A. Orr, R. Anne, M. Lewitowicz, G. Martinez,
    M. G. Saint-Laurent, W. Trinder, C. Borcea, V. Burjan, Z. Dlouhý,
    J. Novák, W. N. Catford, P. H. Regan, and S. M. Vincent

Single-Particle and Cluster Levels in ^{107,108,109}Sn Populated in the Decay of ^{107,108,109}Sb 575
    J. J. Ressler, D. Seweryniak, L. Conticchio, D. Ciurczak, J. Swider,
    H. Penttilä, W. B. Walters, J. Wauters, C. Bingham, R. de Haan, B. Foy,
    and C. N. Davids

Production and Identification of New, Neutron-Rich Nuclei in the ^{208}Pb Region 581
    K. Rykaczewski, J. Kurpeta, A. Płochocki, M. Karny, J. Szerypo,
    A.-H. Evensen, E. Kugler, J. Lettry, H. Ravn, P. VanDuppen, A. Andreyev,
    M. Huyse, A. Wöhr, A. Jokinen, J. Äystö, A. Nieminen, M. Huhta,
    M. Ramdhane, G. Walter, P. Hoff, and the ISOLDE Collaboration

Deformation Change Between Isomeric and Ground States in the ^{184}Au and ^{183}Pt Isotones. 585
    J. Sauvage, N. Boos, L. Cabaret, J. Crawford, H. T. Duong, J. Genevey,
    M. Girod, G. Huber, F. Ibrahim, M. Krieg, F. Le Blanc, J. K. P. Lee,
    J. Libert, J. Obert, J. Oms, J. Pinard, J. C. Putaux, B. Roussière,
    V. Sebastian, S. Zemlyanoi, and the ISOLDE Collaboration

Coulomb Excitation of a ^{78}Rb Radioactive Beam 589
    J. Schwartz, C. J. Lister, D. H. Henderson, S. M. Fischer, P. Reiter,
    A. Aprahamian, J. A. Cizewski, C. N. Davids, R. deHaan, R. V. F. Janssens,
    D. Nisius, D. Seweryniak, and S. M. Vincent

Magnetic Dipole Bands in ^{82}Rb, ^{83}Rb and ^{84}Rb. 594
    R. Schwengner, H. Schnare, S. Frauendorf, F. Dönau, L. Käubler, H. Prade,
    E. Grosse, A. Jungclaus, K. P. Lieb, C. Lingk, S. Skoda, J. Eberth,
    G. deAngelis, A. Gadea, E. Farnea, D. R. Napoli, C. A. Ur, and G. Lo Bianco

Proton Scattering on ^{40}S 598
    F. Maréchal, T. Suomijärvi, Y. Blumenfeld, A. Azhari, D. Bazin,
    J. A. Brown, P. D. Cottle, M. Fauerbach, T. Glasmacher, S. E. Hirzebruch,
    J. K. Jewell, K. W. Kemper, P. F. Mantica, D. J. Morrissey, L. A. Riley,
    J. A. Scarpaci, and M. Steiner

Decay Properties of Ground-State and Isomer of ^{103}In 602
    J. Szerypo, R. Grzywacz, Z. Janas, M. Karny, M. Pfützner, A. Płochocki,
    K. Rykaczewski, J. Zylicz, M. Huyse, G. Reusen, J. Schwarzenberg,
    P. Van Duppen, A. Wöhr, H. Keller, R. Kirchner, O. Klepper, A. Piechaczek,
    E. Roeckl, K. Schmidt, L. Batist, A. Bykov, V. Wittmann, and B. A. Brown

Low-Energy Structure of Neutron-Rich S, Cl and Ar Nuclides Through β Decay 606
    J. A. Winger, H. H. Yousif, W. C. Ma, V. Ravikumar, W. Lui,
    S. K. Phillips, R. B. Piercey, P. F. Mantica, B. Pritychenko,
    R. M. Ronningen, and M. Steiner

Coulomb Excitation of ^{56}Ni 610
    Y. Yanagisawa, T. Motobayashi, S. Shimoura, Y. Ando, H. Fujiwara,
    I. Hisanaga, H. Iwasaki, Y. Iwata, H. Murakami, T. Minemura, T. Nakamura,
    T. Nishio, M. Notani, H. Sakurai, S. Takeuchi, T. Teranishi, Y. X. Watanabe,
    and M. Ishihara

New Signatures of Phase Transitional Behavior in Nuclei 614
    N. V. Zamfir, R. F. Casten, R. Krücken, C. W. Beausang, G. Cata-Danil,
    J. R. Cooper, B. Liu, J. R. Novak, and C. J. Barton

Extraction of Cluster Spectroscopic Factors from Anomalous Large-Angle Scattering of ^{3}He and α-Particles 618
    N. Burtebaev, B. A. Duisebaev, A. Duisebaev, G. N. Ivanov, and S. B. Sakuta

HEAVY ELEMENTS, FISSION, CLUSTER RADIOACTIVITY

New Elements Produced at GSI 625
    S. Hofmann

The FLNR (JINR) Experiments on Synthesis of Superheavy Nuclei with ^{48}Ca Beam 633
    Yu. Ts. Oganessian, A. V. Yeremin, M. G. Itkis, G. G. Gulbekian, and V. B. Kutner

Stability of the Heaviest Elements 639
    A. Sobiczewski

Fission Studies of Nuclei Far from Stability 647
    K.-H. Schmidt, J. Benlliure, C. Böckstiegel, H.-G. Clerc, A. Grewe, A. Heinz, M. de Jong, A. R. Junghans, J. Müller, M. Pfützner, and S. Steinhäuser

Structure of Neutron-Rich Nuclei and Rare Processes in Spontaneous Fission of ^{252}Cf 655
    A. V. Ramayya, J. H. Hamilton, J. K. Hwang, and GANDS95 Collaboration

Fission Processes in ^{238}U-Collisions on Pb and Be Targets at Relativistic Energies 664
    M. Bernas

Proton and Cluster Radioactivity and Nucleus Shapes 672
    S. G. Kadmensky

Fission of Nuclei with Z=102-112 Produced in Reactions with ^{22}Ne and ^{48}Ca Ions 678
    M. G. Itkis, Yu. Ts. Oganessian, E. M. Kozulin, N. A. Kondratiev, L. Krupa, I. V. Pokrovsky, A. N. Polyakov, V. A. Ponomarenko, E. V. Prokhorova, B. I. Pustylnik, A. Ya. Rusanov, and V. I. Vakatov

Study of the High-j States in ^{249}Cm 682
    I. Ahmad, B. B. Back, A. Bacher, G. P. A. Berg, R. R. Chasman, C. C. Foster, J. P. Greene, T. Ishii, W. R. Lozowski, L. R. Morss, W. Schmitt, E. J. Stephenson, and T. Yamanaka

Experiments on Projectile Fission and Fragmentation Relevant for Accelerator-Driven Systems 686
    T. Enqvist, J. Bennlliure, F. Farget, J. Taieb, K.-H. Schmidt, P. Armbruster, C. Böckstiegel, M. de Jong, M. Bernas, B. Mustapha, C. Stéphan, L. Tassan-Got, A. Boudard, S. Leray, R. Legrain, C. Volant, W. Wlazlo, S. Czajkowski, M. Pravikoff, and J. P. Dufour

Neutron-Rich Nuclei Produced in Intermediate Energy Fission 690
    P. Dendooven, S. Hankonen, A. Honkanen, M. Huhta, A. Jokinen, G. Lhersonneau, M. Oinonen, H. Penttilä, V. A. Rubchenya, J. C. Wang, and J. Äystö

Identification of μs Isomers in Fission Products 694
    J. Genevey, J. A. Pinston, H. Faust, T. Friedrichs, M. Gross, F. Ibrahim, T. Larqué, and S. Oberstedt

Systematics of Calculated Cold-Fusion Barriers for Reactions Leading to Compound Systems from Z=104 to Z=126 698
    P. Möller, P. Armbruster, S. Hofmann, and G. Münzenberg

The Berkeley Gas-Filled Separator 704
    V. Ninov, K. E. Gregorich, and C. A. McGrath

Fragment Angular Momentum and Descent Dynamics in ^{252}Cf
Spontaneous Fission 708
    G. S. Popeko, G. M. Ter-Akopian, J. H. Hamilton, J. Kormicki,
    A. V. Daniel, Yu. Ts. Oganessian, A. V. Ramayya, J. K. Hwang,
    A. Sandulescu, A. Florescu, W. Greiner, J. Kliman, M. Morhac,
    J. O. Rasmussen, M. A. Stoyer, J. D. Cole, and GANDS95 Collaboration

Angular Momentum Effects in Multimodal Fission of ^{226}Th 711
    G. G. Chubarian, B. J. Hurst, D. O'Kelly, R. P. Schmitt, M. G. Itkis,
    N. A. Kondratiev, E. M. Kozulin, Yu. Ts. Oganessian, V. V. Pashkevich,
    I. V. Pokrovsky, V. S. Salamatin, A. Ya. Rusanov, L. Calabretta,
    C. Maiolino, K. Lukashin, C. Agodi, G. Bellia, F. Hanappe, E. Liatard,
    A. Huck, and L. Stuttgé

Early and Current Stability Predictions for Nuclei Near Z=110
and N=162 715
    P. Möller

BETA DECAY AND FUNDAMENTAL MEASUREMENTS

Limits on Physics Beyond the Standard Model Using Exotic Beams 719
    A. García

Beta Strength Distribution in Neutron-Deficient Nuclei 725
    Z. Janas, J. Agramunt, A. Algora, L. Batist, B. A. Brown, D. Cano-Ott,
    R. Collatz, A. Gadea, M. Gierlik, M. Górska, H. Grawe, A. Gulielmetti,
    M. Hellström, Z. Hu, M. Karny, R. Kirchner, F. Moroz, A. Piechaczek,
    A. Płochocki, M. Rejmund, E. Roeckl, B. Rubio, K. Rykaczewski, M. Shibata,
    J. Szerypo, J. L. Tain, V. Wittmann, and A. Wöhr

Superallowed Fermi Beta Decay 733
    J. C. Hardy and I. S. Towner

Indication for Superallowed Fermi Decay from the N=Z Nuclei ^{78}Y,
^{82}Nb, ^{86}Tc 739
    Ph. Dessagne, C. Longour, J. Garcés Narro, D. Applebe, L. Axelsson,
    B. Blank, A. M. Bruce, W. N. Catford, C. Chandler, R. Clark, D. Cullen,
    S. Czajkowski, J. M. Daugas, A. Fleury, L. Frankland, W. Gelletly,
    J. Giovinazzo, B. Greenhalgh, R. Grzywacz, M. Harder, K. L. Jones,
    N. Kelsall, T. Kszczot, M. Lewitowicz, Ch. Miéhé, R. D. Page, C. J. Pearson,
    A. T. Reed, P. H. Regan, O. Sorlin, and R. Wadsworth

Beta-Decay Strength and Isospin Mixing Studies in the sd and fp-Shells 745
    A. Jokinen, J. Äystö, P. Dendooven, A. Honkanen, P. Lipas, K. Peräjärvi,
    M. Oinonen, and T. Siiskonen

The Mechanism of β-Delayed Two-Proton Emission in ^{31}Ar 749
    M. J. G. Borge, L. Axelsson, J. Äystö, L. M. Fraile, H. O. U. Fynbo,
    A. Honkanen, P. Hornshøj, A. Jokinen, B. Jonson, I. Martel, I. Mukha,
    T. Nilsson, G. Nyman, M. Oinonen, B. Petersen, K. Riisager,
    M. H. Smedberg, O. Tengblad, and the ISOLDE Collaboration

Spectroscopy of ^{22,23,24}Si and ^{22}Al 753
    S. Czajkowski, S. Andriamonje, B. Blank, F. Boué, R. Del Moral,
    J. P. Dufour, A. Fleury, E. Hanelt, N. A. Orr, P. Pourre, M. S. Pravikoff,
    and K.-H. Schmidt

Beta Decay of Neutron-Rich Cobalt and Nickel Isotopes 757
    S. Franchoo, B. Bruyneel, M. Huyse, U. Köster, K.-L. Kratz, K. Kruglov,
    Y. A. Kudryavtsev, W. F. Mueller, B. Pfeiffer, R. Raabe, I. Reusen,
    P. Thirolf, P. Van Duppen, J. Van Roosbroeck, L. Vermeeren, W. B. Walters,
    L. Weissman, and A. Wöhr

Is There a β3p Branch in the Decay of ^{31}Ar? 761
    H. O. U. Fynbo, J. Äystö, M. J. G. Borge, L. M. Fraile, A. Honkanen,
    P. Hornshøj, Y. Jading, A. Jokinen, B. Jonson, I. Martel, I. Mukha,
    G. Nyman, M. Oinonen, K. Riisager, T. Siiskonen, O. Tengblad,
    F. Wenander, and the ISOLDE Collaboration

Progresses in Statistical Analysis of β-Delayed Proton Emission 765
    J. Giovinazzo, Ph. Dessagne, Ch. Miehé, and the ISOLDE Collaboration

Beta-Decay of ^{97}Ag: Evidence for the Gamow-Teller Resonance Near ^{100}Sn ... 769
    Z. Hu, L. Batist, J. Agramunt, A. Algora, B. A. Brown, D. Cano-Ott,
    R. Collatz, A. Gadea, M. Gierlik, M. Górska, H. Grawe, M. Hellström,
    Z. Janas, M. Karny, R. Kirchner, F. Moroz, A. Płochocki, M. Rejmund,
    E. Roeckl, B. Rubio, M. Shibata, J. Szerypo, J. L. Tain, and V. Wittmann

Charge-Exchange Reactions with a Radioactive Triton Beam 773
    J. Jänecke

Beta-Decay of ^{103}In Studied by Using a Total Absorption Spectrometer 777
    M. Karny, L. Batist, B. A. Brown, D. Cano-Ott, R. Collatz, A. Gadea,
    R. Grzywacz, A. Guglielmetti, M. Hellström, Z. Hu, Z. Janas, R. Kirchner,
    F. Moroz, A. Piechaczek, A. Płochocki, E. Roeckl, B. Rubio,
    K. Rykaczewski, M. Shibata, J. Szerypo, J. L. Tain, V. Wittmann,
    and A. Wöhr

Decays of Very Neutron-Rich Fission Products ^{113}Ru and ^{113}Rh 781
    J. Kurpeta, A. Płochocki, G. Lhersonneau, J. C. Wang, P. Dendooven,
    A. Hokanen, M. Huhta, M. Oinonen, H. Penttilä, K. Peräjärvi, J. R. Persson,
    and J. Äystö

Structure Studies of Nuclear Systems Close to the Doubly-Magic ^{132}Sn
Using Advanced β^--Spectroscopy 785
    H. Mach, J. Blomqvist, B. Fogelberg, V. I. Isakov, L. Jacobsson,
    A. Lindroth, K. A. Mezilev, M. Sanchez-Vega, and R. B. E. Taylor

Deformation Signature from the Gamow-Teller Decay of N=Z Nuclei 789
    Ch. Miehé, J. Giovinazzo, Ph. Dessagne, A. Huck, A. Knipper, G. Marguier,
    C. Longour, V. Rauch, M. J. G. Borge, I. Piqueras, O. Tengblad, A. Jokinen,
    M. Ramdhane, and the ISOLDE Collaboration

Gamow-Teller Strength in the $f_{7/2}$-Nuclei ^{54}Co and ^{42}Sc Studied
Through the Beta Decay of ^{54}Ni and ^{42}Ti 793
    I. Reusen, A. Andreyev, J. Andrzejewski, N. Bijnens, B. Bruyneel,
    S. Franchoo, M. Huyse, Y. A. Kudryavtsev, K. Kruglov, W. F. Mueller,
    A. Piechaczek, R. Raabe, K. Rykaczewski, J. Szerypo, P. Van Duppen,
    J. Van Roosbroeck, L. Vermeeren, J. Wauters, L. Weissman, and A. Wöhr

Statistical Deliberations for Exotic Nuclei 797
    K. Riisager

The Beta-Delayed Proton Decay of ^{23}Al 801
    M. W. Rowe, D. M. Moltz, T. J. Ognibene, J. Powell, and J. Cerny

Delayed Neutron Emission in the Semi-Gross Theory of Nuclear β-Decay..... 805
    T. Tachibana and M. Yamada

The GT Resonance Revealed in β^+-Decay Using New Experimental
Techniques. 809
    J. Agramunt, A. Algora, L. Batist, R. Borcea, D. Cano-Ott, R. Collatz,
    A. Gadea, J. Gerl, M. Gierlik, M. Górska, O. Guilbaud, H. Grawe,
    M. Hellström, Z. Hu, Z. Janas, M. Karny, R. Kirchner, P. Kleinheinz,
    W. Liu, T. Martinez, F. Moroz, A. Płochocki, M. Rejmund, E. Roeckl,
    B. Rubio, K. Rykaczewski, M. Shibata, J. Szerypo, J. L. Tain, V. Wittmann,
    and the German Euroball Col.

Gamow-Teller Decay of Even Isotopes ^{68}Ni to ^{78}Ni........................ 813
    J. Zylicz, J. Dobaczewski, and Z. Szymański

NUCLEAR ASTROPHYSICS

The rp-Process in X-Ray Bursts.. 819
    M. Wiescher, A. Aprahamian, J. Döring, J. Görres, and H. Schatz

New Information on r-Process Nuclei 827
    K.-L. Kratz

Explosive Nucleosynthesis and the Astrophysical r-Process.................. 837
    F.-K. Thielemann, C. Freiburghaus, T. Rauscher, E. Kolbe, B. Pfeiffer,
    K.-L. Kratz, and J. J. Cowan

Review of Radioactive Beam Nuclear Astrophysics Experiments............. 849
    J. Vervier

Direct Measurements of the ^{7}Be (p,γ) ^{8}B Reaction Cross Section 858
    G. Bogaert

Measurement of the ^{7}Be (p,γ) ^{8}B Cross-Section with an Implanted
^{7}Be Target 864
    M. Hass, C. Broude, V. Fedoseev, G. Goldring, G. Huber, J. Lettry,
    V. Mishin, H. L. Ravn, V. Sebastian, and L. Weissman

Determination of $S_{17}(0)$ from Transfer Reactions 868
    R. E. Tribble, A. Azhari, H. L. Clark, C. A. Gagliardi, Y.-W. Lui,
    A. M. Mukhamedzhanov, A. Sattarov, L. Trache, V. Burjan,
    J. Cejpek, V. Kroha, Š. Piskoř, and J. Vincour

Determining the Astrophysical S_{17} with Transfer Reactions 876
    J. C. Fernandes, R. Crespo, F. M. Nunes, and I. J. Thompson

High Energy Coulomb Breakup Experiments for Nuclear Astrophysics....... 882
    T. Motobayashi

Measurement of E2 Transitions in the Coulomb Dissociation of ^{8}B........... 890
    B. Davids, D. W. Anthony, S. M. Austin, D. Bazin, B. Blank, J. A. Caggiano,
    M. Chartier, H. Esbensen, P. Hui, C. F. Powell, H. Scheit, B. M. Sherrill,
    M. Steiner, P. Thirolf, J. Yurkon, and A. Zeller

Asymptotic Normalization Coefficients for $^{14}N \to {}^{13}C+p$ and $^{10}B \to {}^9Be+p$ 896
    A. Azhari, H. L. Clark, C. A. Gagliardi, Y.-W. Lui, A. M. Mukhamedzhanov,
    L. Trache, R. E. Tribble, H. M. Xu, X. G. Zhou, V. Burjan, J. Cejpek,
    V. Kroha, and F. Carstoiu

Extrapolation of the Astrophysical S Factor for ^{7}Be (p,γ) ^{8}B to Solar
Energies 900
    S. Karataglidis, B. K. Jennings, and T. D. Shoppa

Beta-Decay of ^{40}Ti and Its Implication for Solar-Neutrino Detection 904
    W. Liu, M. Hellström, R. Collatz, J. Benlliure, L. Chulkov, D. Cortina Gil,
    F. Farget, H. Grawe, Z. Hu, N. Iwasa, M. Pfützner, A. Piechaczek, R. Raabe,
    I. Reusen, E. Roeckl, G. Vancraeynest, and A. Wöhr

Study of the d (^{7}Be, ^{8}B)n Reaction 908
    C. F. Powell, D. J. Morrissey, D. W. Anthony, B. Davids, M. Fauerbach,
    P. F. Mantica, B. M. Sherrill, and M. Steiner

Status of the LOREX: Geochemical ^{205}Tl Solar Neutrino Experiment 912
    K. M. Subotic and M. K. Pavicevic

EXPERIMENTAL DEVELOPMENTS AND RADIOACTIVE BEAMS

New Results from Advances in ISOL Techniques......................... 919
    P. Van Duppen

REXTRAP, an Ion Buncher for REX-ISOLDE............................. 927
    F. Ames, G. Bollen, G. Huber, P. Schmidt, and the REX-ISOLDE Collaboration

Status of RNB Facilities in Europe 933
    A. C. Mueller

Status of RIB Facilities in Asia 943
    I. Tanihata

Status of RNB Facilities in North America 952
    J. A. Nolen

Possibilities for the Production of Neutron-Rich Isotopes 960
    J. Benlliure, F. Farget, A. R. Junghans, and K.-H. Schmidt

A Radio Frequency Quadrupole Ion Beam Buncher for ISOLTRAP 965
    G. Bollen, J. Dilling, A. M. Ghalambor Dezfuli, S. Henry, F. Herfurth,
    A. Kellerbauer, T. Kim, H.-J. Kluge, A. Kohl, E. Lamour, D. Lunney,
    R. B. Moore, W. Quint, S. Schwarz, P. Varfalvy, and L. Vermeeren

New Approach to the Analysis of Total Absorption Spectra................. 969
    D. Cano-Ott, A. Gadea, B. Rubio, J. L. Tain, M. Karny, Z. Janas,
    K. Rykaczewski, R. Kirchner, E. Roeckl, L. Batist, F. Moroz,
    and V. Wittmann

The ISAC Radioactive Beams Facility in Canada: Progress and Plans........ 973
    G. Ball, R. Bartmann, J. Behr, P. Bricault, L. Buchmann, J. M. D'Auria,
    P. Delhaij, M. Dombsky, G. Dutto, R. Kiefl, K. P. Jackson, R. Laxdal,
    J. M. Poutissou, P. Schmor, and G. Stanford

EXOGAM: A γ-Ray Spectrometer for Exotic Beams...................... 977
    G. de FRANCE for the EXOGAM Collaboration

Cooling, Bunching and Isobar Separation of Radioactive Ion Beams
at IGISOL 981
    A. Jokinen, J. Äystö, P. Dendooven, V. S. Kolhinen, J. Huikari,
    A. Nieminen, and K. Peräjärvi

Enhancement of the Radiative Transitions Between the Ground and the 3.5-eV Isomer States in the Hydrogen-Like ^{229}Th^{89+} Ion 985
    F. F. Karpeshin, S. Wycech, I. M. Band, M. B. Trzhaskovskaya,
    M. Pfützner, and J. Zylicz

On-line Separation of Short-Lived Beryllium Isotopes 989
    U. Köster, J. Barker, R. Catherall, V. N. Fedoseyev, U. Georg, G. Huber,
    Y. Jading, O. Jonsson, M. Koizumi, K.-L. Kratz, E. Kugler, J. Lettry,
    V. I. Mishin, H. Ravn, V. Sebastian, C. Tamburella, A. Wöhr,
    and the ISOLDE Collaboration

MISTRAL:* The Beginning of a New Mass Measurement Program at *ISOLDE . 995
    D. Lunney, C. Toader, M. de Saint Simon, G. Audi, C. Borcea, H. Doubre,
    M. Duma, M. Jacotin, S. Henry, J.-F. Képinski, G. Lebée, G. Le Scornet,
    C. Monsanglant, C. Thibault, and the ISOLDE Collaboration

BEARS: Radioactive Ion Beams at LBNL 999
    J. Powell, F. Q. Guo, P. E. Haustein, R. Joosten, R.-M. Larimer, C. Lyneis,
    D. M. Moltz, E. B. Norman, J. P. O'Neil, M. W. Rowe, H. F. VanBrocklin,
    Z. Q. Xie, X. J. Xu, and J. Cerny

The Use of (d,xn) Reactions: RIB Production and Energy Generation 1003
    D. Ridikas and W. Mittig

CONCLUDING REMARKS

Concluding Remarks 1009
    D. Guillemaud-Mueller

ENAM98 Scientific Program 1015
List of Participants 1023
Author Index 1043

Preface

The ENAM98 conference was held from June 23-27, 1998 at the Shanty Creek Resort in Northern Michigan, USA. This conference has a long and distinguished history and follows the successful ENAM95 conference held in Arles, in southern France. ENAM stands for Exotic Nuclei and Atomic Masses and represents the merger of the Nuclei Far From Stability (NFFS) series of conferences that began in 1966 in Lysekil, Sweden and the Atomic Masses and Fundamental Constants (AMCO) series that traces its roots to Mainz in 1956. While ENAM98 was the second merged conference, it actually should be called the 16th in a series of conferences that include the past NFFS and AMCO meetings.

There were 260 participating scientists and 23 accompanying persons who attended ENAM98. These numbers included a very large number of students and young scientists. The content of the program was selected based on advice from the International Advisory and the Local Organizing committees. The committee members read and considered over 200 abstracts in selecting oral contributions. The program consisted of 67 invited talks and 134 poster presentations. Many thanks are due to all the speakers and poster presenters for the very high quality of their excellent contributions. Although the program was very full, the attendance for all talks and the poster session was exceptionally high. The active participation of the attendees contributed to the success of the conference. Of particular note, Dominique Guillemaud Mueller provided a wonderful and insightful conference summary.

The conference site, Shanty Creek Resort, was beautiful and the hospitality of the Shanty Creek employees was outstanding. Brenda McLellan and Jeanne Belanger from the resort staff worked very hard before and during the conference to make everything run smoothly. Their flexibility and "can do" attitude were a great help. Among the many notable events, which included a rain shower during the conference photo and a great swing band at the conference banquet, was the Schussy Cats show that was interrupted about half way through by one of the most spectacular thunderstorms any of us has experienced. We were also treated to outings at Mackinac Island, Sleeping Bear Dunes National Lakeshore and the Legend and Shanty Creek golf courses.

Argonne National Laboratory, The National Superconducting Laboratory, the Department of Energy, the National Science Foundation, and the International Union of Pure and Applied Physics sponsored the conference. We also received financial support from various industries listed on the following pages. Finally, the real success and smooth running of the conference was due to the conference secretaries Barbara Weller and Donna Nelson from Argonne National Laboratory. They were tireless in their efforts and spent many long hours at the conference desk and nearby lobby.

We all look forward to ENAM2001 which will be hosted in Finland and wish the organizers the best.

Bradley M. Sherrill, David J. Morrissey and Cary N. Davids
Editors

International Organizing Committee:

J. H. Äystö (Finland)
W. Benenson (USA)
J. D'Auria (Canada)
C.N. Davids (USA) Chairperson
D. Guillemaud-Mueller (France)
J.H. Hamilton (USA)
J.C. Hardy (USA)
K. Heyde (Belgium)
M. Ishihara (Japan)
B. Jonson (Sweden)
S. Kubono (Japan)
A.A. Ogloblin (Russia)
E.W. Otten (Germany)
Ph. Quentin (France)
E. Roeckl (Germany)
I. Thompson (Great Britain)
J.S. Vaagen (Norway)
J. Vervier (Belgium)
D.J. Vieira (USA)
A.H. Wapstra (Netherlands)
N. Zeldes (Israel)
J. Zylicz (Poland)

National Organizing Committee:

C.N. Davids (ANL)
J.H. Hamilton (Vanderbilt)
C.J. Lister (ANL)
D.J. Morrissey (NSCL)
K. Rykaczewski (ORNL)
B.M. Sherrill (NSCL) Co-Chair
W.B. Walters (Maryland)
E. Zganjar (LSU)

The following institutions and companies provided financial support for the ENAM98 conference. Their help is gratefully acknowledged.

Institutions:

International Union of Pure and Applied Physics (IUPAP)

US Department of Energy

US National Science Foundation

Argonne National Laboratory (Physics Division)

Michigan State University (NSCL, Department of Physics, Department of Chemistry, Graduate School, College of Natural Science)

Companies:

Micron Semiconductor Ltd., Lancing, UK

Midwest Vacuum, Hinsdale, IL 60521, representing VAT, Norcal,
EVAC International, Spectra International, Alcatel,
Surface Interface, Plein & Baus Corp.
Springfield, OH 45505

Shanty Creek Resort, Bellaire, MI 49615

EG&G Ortec, Oak Ridge, TN 37380

GMW Associates/Danfysik A/S, Redwood City, CA 94064

LeCroy Research Systems, Chestnut Ridge, NY 10977

Bruker Analytische Messtechnik GMBH, Karlsruhe, Germany

MASSES, MOMENTS, AND RADII

Mass measurements with a Penning trap mass spectrometer at ISOLDE

G. Bollen[1], F. Ames[2], G. Audi[3], D. Beck[4], F. Herfurth[4], H.-J. Kluge[4], A. Kohl[4], D. Lunney[3], R. B. Moore[5], M. de Saint Simon[3], E. Schark[2], S. Schwarz[4], J. Szerypo[7] and the ISOLDE Collaboration[1]

[1]*CERN, Geneva, Switzerland,* [2]*Institut für Physik, Universität Mainz, Germany,* [3]*CSNSM-IN2P3-CNRS, Orsay, France,* [4]*Gesellschaft für Schwerionenforschung GSI, Darmstadt, Germany,* [5]*McGill University, Montreal, Canada and* [6]*Institute of Experimental Physics, Warsaw University, Warsaw, Poland*

Abstract. Penning trap mass measurements on radioactive isotopes are performed with the ISOLTRAP mass spectrometer at ISOLDE/CERN. In the last years the applicability of the spectrometer has been considerably extended. The most recent measurements were carried out on isotopes of rare earth elements and on isotopes with Z = 80 - 85. An accuracy of $\delta m/m \approx 1\cdot 10^{-7}$ was achieved.

INTRODUCTION

The binding energy of the atomic nucleus is one of the most fundamental properties of such a many-body system. Accurate mass data serve as testing grounds for nuclear models and stimulate their further improvement. Furthermore, systematic investigation of the binding energy as a function of proton and neutron number allows the direct observation of nuclear properties like pairing, shell and sub-shell closures, as well as deformation effects, and leads to a deeper understanding of nuclear structure. In addition, very precise mass differences are for example required in the context of precision weak interaction studies in nuclear β-decay. Therefore, large efforts are presently devoted to the application of classical as well as new mass spectrometric techniques, such as time-of-flight, Smith-RF or Schottky mass spectrometry, for the accurate mass determination of short-lived isotopes far from the valley of beta stability (1,2).

Penning traps have proven to be very accurate mass spectrometers (3). A large variety of mass measurements with highest accuracy have been performed on stable, mostly light particles. ISOLTRAP is so far the only spectrometer operational for the investigation of short-lived radioactive isotopes. Another project, the CPT trap system (4) installed at ANL will start to deliver first results very soon.

ISOLTRAP has very successfully continued its mass measurement program since last ENAM. More than 70 mass values of rare earth isotopes in the vicinity of ^{146}Gd, of neutron-deficient mercury isotopes and of isotopes with Z = 82 - 85 have been determined with an accuracy typically better than 20 keV. In parallel with these measurements, the Penning trap mass spectrometer has undergone a number of important modifications and improvements.

THE ISOLTRAP SPECTROMETER

The basic principle of mass measurements with ISOLTRAP is the determination of the cyclotron frequency $\omega_c = q/m \cdot B$ of ions with a charge-over-mass ratio q/m stored in a Penning trap with known magnetic field B. Fig. 1 shows the present layout of the ISOLTRAP spectrometer (5, 7). The first section of the spectrometer has the task to stop the 60 keV ISOLDE beam and to prepare it for an efficient transfer into the cooler trap. In the past a stopping/re-ionization technique was applied which limited the applicability of ISOLTRAP to surface ionizable elements. Recently the system was considerably improved by the installation of an RFQ trap ion beam buncher (6), which allows to capture the continuous ISOLDE beam in flight. The lower Penning trap (7) has the task to accumulate, cool, and mass separate the ions delivered from the ion

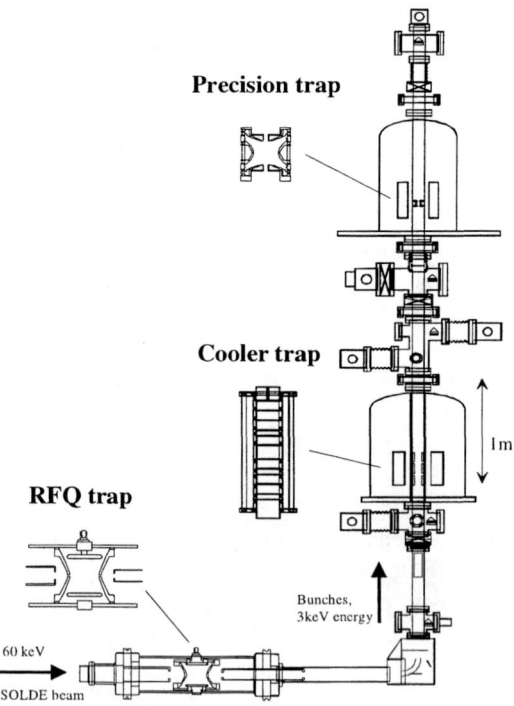

FIGURE 1. Experimental set-up of the ISOLTRAP mass spectrometer at ISOLDE/CERN

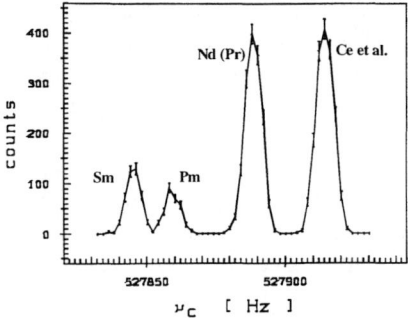

FIGURE 2. 'Mass scan' with the cooler trap for A=138 ions delivered from a Ta-foil target with surface ionizer. Shown is the number of ions extracted from the trap as a function of the applied radio frequency.

preparation section and to bunch them for an efficient delivery to a second Penning trap. This precision trap is the actual mass spectrometer where the cyclotron frequency of the captured ions is determined. Important aspects of the performance of ISOLTRAP will be discussed in the following.

Isobar and isomer separation

Penning trap mass measurements require rather clean beams in order to avoid systematic errors in the mass determination arising from Coulomb interaction of different ion species in the trap. The resolving power of the ISOLDE general-purpose separator is by far not high enough to deliver such isobarically pure beams. A mass selective cooling technique (89, 10), based on the simultaneous application of a buffer gas cooling and radio frequency excitation of the ion motion, is employed in the first "cooler" Penning trap. This trap system has been optimized for a mass selectivity high enough to resolve isobars and for the delivery of clean and cooled ion bunches to the precision trap, an essential ingredient for highly accurate mass measurements. As an example,

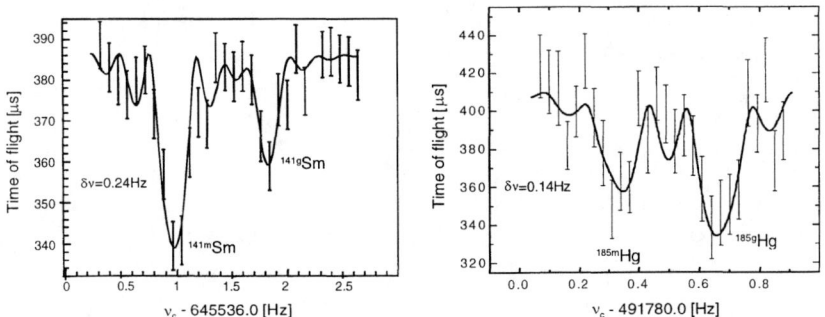

FIGURE 3. Resolved isomeric and ground states for ^{141}Sm ($\Delta E = 175$ keV) and ^{185}Hg ($\Delta E = 118$ keV).

Fig. 2 shows a 'mass scan' performed with the cooler trap for an A=138 ion beam delivered by ISOLDE from a Ta-foil target. Shown is the number of ions extracted from the trap as a function of the applied radio frequency. The mass resolving power achieved here is about $R=10^5$, which is sufficient to resolve and separate isobars even close to stability.

The precision trap in which the cyclotron frequency determination of the ions takes place is in operation without major modification since several years and performs excellently (11). This trap is normally operated with a resolving power R close to one million corresponding to rf excitation times of $T_{rf} = 1$ s for A = 100 ions. If required, the resolving power R ~ T_{rf} can be considerably increased by increasing T_{rf}. The maximum resolving power that has been realized in off-line tests with ^{133}Cs ions is R = 8 million using T_{rf} = 12s. This corresponds to a mass resolution of Δm_{fwhm} = 15 keV.

For the investigation of trends in nuclear binding energies an accuracy of $\delta m/m \approx 10^{-7}$ is normally sufficient and is already achieved with modest resolving powers (R < 10^6). Higher resolving powers become important in the case of long-lived isomers produced simultaneously with isotopes in their ground state. Over the nuclide chart nearly one third of the isotopes have long-lived isomeric states with (in many cases unknown) excitation energies down to < 100 keV. Only in a few cases information about the production ratio exists which may vary drastically depending on the half-lives and release times from the targets. Therefore, the resolution of isotopes in their ground or isomeric state is essential for an unambiguous determination of the mass of the isotope in one or the other state. That this can be achieved with ISOLTRAP has now been demonstrated several times. Two recent examples are shown in Fig. 3 for ^{141}Sm and ^{185}Hg.

MASS MEASUREMENTS

In total 76 isotopes and states were investigated since 1994, which are listed in table 1. An accuracy in the mass determination of $\delta m/m = 1 \cdot 10^{-7}$ was achieved for most of these isotopes. The measurements concentrated on rare earth isotopes, isotopes of mercury and of heavier elements, which became only possible by the recent improvements of the spectrometer.

Rare earth isotopes

So far direct mass measurements in this region were hampered by the fact that many isobars are delivered simultaneously by ISOLDE. Since the cooler trap can be operated as an isobar separator, clean ion samples can be prepared and sent to the precision trap. In several beam times it was possible to investigate more than 50 isotopes in the vicinity of ^{146}Gd, most of them with N ≤ 82 and Z < 64. For most of the isotopes the detailed analysis of the data is finished (12) and an atomic mass evaluation similar to the work

TABLE 1. List of isotopes investigated with ISOLTRAP since 1995.

Element	Mass Number	Element	Mass Number
Ba	123, 125, 127, 131	Ho	150
Cs	133 (reference isotope)	Tm	165
Ce	132, 133, 134	Yb	158, 159, 160, 161, 162, 163, 164
Pr	133-137	Hg	184, 185g+m, 186-190, 191m,
Nd	130, 132, 134-138		192, 193g+m, 194-196, 197g
Pm	136-141, 143	Pb	196, 198, 208 (reference isotope)
Sm	136-140, 141m, 141g, 142, 143	Bi	197
Eu	139, 141-149, 151, 153	Po	198
Dy	148, 149, 154	At	203

by G. Audi et al. (13) has been performed. The evaluation shows that the ISOLTRAP measurements have a large impact on this mass region. This is illustrated in Fig. 4, which shows the trend of the two neutron separation energies. The upper and lower part of the figure show the situation before and after the ISOLTRAP data have been included.

Prior to the ISOLTRAP measurements strong discontinuities were observed in the S_{2n} trends derived from estimated but also from experimental mass values as can be seen in the upper figure. Above the N=82 shell closure, for Z=67 and Z=78, these discontinuities are now removed and the separation energies follow the regular trend observed in the neighboring isotopic chains. Most of the isotopes investigated by ISOLTRAP are in the region with N < 82 and Z < 64. Also here trends are now more clearly established. Systematic deviations from a linear trend for Z > 56 around N=76, 77 are now visible up to Z=60. They might be related to the eradication of the proton sub-shell gap at $Z \approx 64$ as one departs from N=82 and be accompanied by a change in nuclear deformation.

Since mass values of many isotopes are linked via known Q-values to other isotopes, accurate mass measurements of a few key isotopes can have a large impact on the knowledge of masses over a whole mass region. The case of ^{150}Ho will be discussed as an example. Mass differences between 19 isotopes linked to ^{150}Ho, some of them beyond the proton drip-line around $Z \approx 80$, are already known via experimental Q-values. No link existed between these nuclei and the backbone of stability, since a doubtful experimental Q-value for ^{150}Ho was rejected in the 1995 atomic mass evaluation (14). This unsatisfactory situation is now resolved by the ISOLTRAP measurement on ^{150}Ho, which justifies the early rejection of the old experimental datum, which is 810 keV away from the ISOLTRAP value. The ISOLTRAP measurement therefore not only gives an accurate experimental mass value for ^{150}Ho but also anchors the masses for all 19 isotopes linked to it.

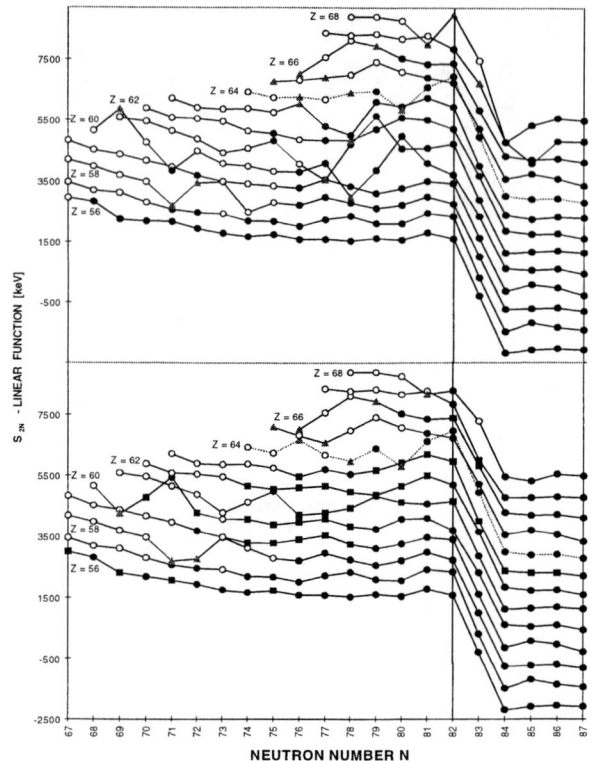

FIGURE 4. Two-neutron separation energies as a function of neutron number. Shown are S_{2n}-values excluding (top) and including (bottom) ISOLTRAP data in the atomic mass evaluation. The isotopes are marked by squares (ISOLTRAP data), filled circles (other experimental data), open circles (estimates from systematic trends), and triangles (doubtful experimental value).

Neutron-deficient mercury isotopes

The interest for nuclear structure investigations and mass measurements in this region arises from the appearance of shape coexistence at low excitation energies in the region around the shell closure at Z=82. The onset of rotational bands built on low-lying 0^+ states has been found (15) in even-even Pt, Hg, Pb and Po isotopes mid-shell between N = 82 and N = 126. A large staggering in the $\delta\langle r^2\rangle$ values determined from isotopic shift measurements was observed for A $\leq$ 185 for the ground-states of the light Hg isotopes, a jump from small to strong deformation in the neighboring Au isotopes at A $\leq$ 186 and a smooth transition in the Pt isotopes (16, 17, 18). However, until recently no mass values were known in this mass region. Today, precise information is still lacking for A $\leq$ 185, where the strongest structural changes happen.

The neutron-deficient isotopes of elements around Z = 82 are all members of long α-decay chains with well-known Q-values. Therefore, an accurate determination of

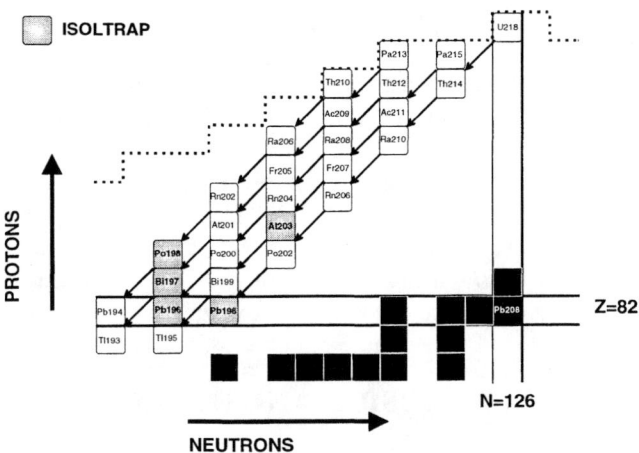

FIGURE 5. Q_α-value decay chains with isotopes investigated by ISOLTRAP (shaded). The dashed line indicates the borderline of known nuclei.

such isotopes allows to fix these chains, making a large impact on a whole mass area starting at the upper part of the rare earth region and reaching to the border of known proton-rich isotopes.

With ISOLTRAP, a first series of mass measurements on the neutron-deficient mercury isotopes $^{185-197}$Hg was carried out in December 1996 after the installation of the RFQ trap ion beam buncher. In the case of the even isotopes where no isomeric states exist, the evaluation was straightforward and an accuracy of $\delta m \approx 20$ keV can be assigned to all mass values. However, in the case of the odd isotopes long-lived isomers exist and are produced at ISOLDE. The excitation energies of typically 100 - 150 keV of these isotopes are very low. In the first measurements in December 1996 the spectrometer was operated with a resolving power of $R \approx 500000$, which corresponds to a mass resolution of 300 keV in this mass range. Therefore it was not possible to resolve isomeric and ground states.

Therefore, in a second run the attempt was made to verify the production of the isomers and to resolve them and the corresponding ground states. For this purpose the spectrometer was operated with resolving powers up to $R = 5$ million with which a mass resolution of $\delta m \approx 30$ keV was achieved. Using such a scenario it was possible to resolve isomeric and ground state in the cases of ^{185}Hg (see Fig. 3) and ^{193}Hg. Furthermore, it was verified that the ground state is dominantly produced in the case of ^{197}Hg, while for ^{195}Hg only the isomeric state has been seen, for which the excitation energy is known. Therefore for all these isotopes the ground state masses are now identified and determined with an accuracy of 20 keV. In addition, during this run it was possible to extend the measurements in the mercury chain out to ^{184}Hg.

Isotopes with Z = 82 - 85

Using the Paul trap ion beam buncher, the investigation of a new region of isotopes with $Z \geq 82$ was started very recently. In a first experiment isotopes were selected which are members of long alpha decay chains, those either not linked to an isotope with known mass or to one with a large mass uncertainty. Fig. 5 shows the decay chains and the isotopes investigated by ISOLTRAP. Due to the high accuracy of the ISOLTRAP data together with the availability of the Q_α-values accurate information on nuclear binding energies is now available even for very heavy proton-rich isotopes like ^{210}Th, ^{213}Pa, or ^{218}U, situated at the borderline of known nuclei.

CONCLUSION AND OUTLOOK

Penning trap mass spectrometry provides high accuracy mass data far from stability. The most recent investigations by ISOLTRAP have contributed significantly to our knowledge about nuclear binding of neutron-deficient rare earth isotopes and of isotopes with Z = 80 - 85. On-going measures (19) to increase the efficiency of ISOLTRAP will allow to extend the studies even farther from stability and to explore new mass regions. One example is the hardly explored region of neutron-rich isotopes at and above the magic proton number $Z = 82$, which is of importance in the context of nuclear astrophysics but also for the theoretical prediction of properties of super-heavy elements.

---

1. Bollen, G., Nucl. Phys. A626, 297c (1997).
2. Mittig, W. et al., Annu.Rev. Nucl. Sci. 47, 22 (1997).
3. Proc. of the Nobel Symposium 91 on Trapped Charged Particles and Related Fundamental Physics, Lysekil, Sweden, August 19-26,1994, Physics Scripta T59 (1995).
4. Savard, G., et al., Nucl Phys A626, 353c (1997).
5. Bollen, G., et al., Nucl. Instr. Meth. A368, 675 (1996).
6. Moore, R.B, et al., J. Mod. Optics 39, 361 (1992).
7. Raimbault-Hartmann, H., et al., Nucl. Instr. Meth. B126, 374 (1997).
8. Bollen, G., et al., J. Appl. Phys. 68, 4355 (1990).
9. Savard, G. et al., Phys. Lett. A158, 247 (1991).
10. König, M., et al., Int. J. Mass Spec. Ion. Proc. 142, 95 (1995).
11. Beck, D., et al., Nucl. Instr. Meth. B126,378 (1997).
12. Beck, D., et al., Nucl. Phys. A626, 343c (1997).
13. Audi G., et al., Nucl. Phys. A565, 1 (1993).
14. Audi G., et al., Nucl. Phys, A595, 409 (1995).
15. Wood, J.L., Phys.Rep. 215, 101 (1992).
16. Ulm, G., et al., Z. Phys. A325, 2471 (1986).
17. Passler, G., et al., Nucl. Phys. A580, 173 (1994).
18. Hilberath, T., et al., Z. Phys. A342, 1 (1992).
19. Bollen G., et al, A Radio Frequency Quadrupole Ion Beam Buncher for ISOLTRAP, these proceedings.

Experiments with Stored Relativistic Exotic Nuclei

H. Geissel[1], T. Radon[1], F. Attallah[1], K. Beckert[1], F. Bosch[1],
A. Dolinskiy[1], H. Eickhoff[1], M. Falch[2], B. Franczak[1], B. Franzke[1],
Y. Fujita[3], M. Hausmann[1], M. Hellström[1], F. Herfurth[1],
Th. Kerscher[2], O. Klepper[1], H.-J. Kluge[1], C. Kozhuharov[1],
K.E.G. Löbner[2], G. Münzenberg[1], F. Nolden[1], Yu. Novikov[4],
Z. Patyk[5], W. Quint[1], H. Reich[1], C. Scheidenberger[1], B. Schlitt[1],
J. Stadlmann[6] M. Steck[1], K. Sümmerer[1], L. Vermeeren[1],
M. Winkler[1], Th. Winkler[1], H. Wollnik[6]

[1] *Gesellschaft für Schwerionenforschung mbH, Planckstraße 1, D-64291 Darmstadt, Germany*
[2] *Sektion für Physik, LMU München, Am Coulombwall, D-85748 Garching, Germany*
[3] *Department of Physics, Osaka University, Toyonaka, Osaka 560, Japan*
[4] *St. Petersburg Nuclear Physics Institute, Gatchina 188350, Russia*
[5] *Soltan Institute for Nuclear Studies, 00-681 Warsaw, Poland*
[6] *II. Physikalisches Institut, JLU Gießen, Heinrich-Buff-Ring 16, D-35392 Gießen, Germany*

Abstract. Beams of relativistic exotic nuclei were produced, separated and investigated with the combination of the fragment separator FRS and the storage ring ESR. The following experiments are presented: 1) Direct mass measurements of relativistic nickel and bismuth projectile fragments were performed using Schottky spectrometry. Applying electron cooling, the relative velocity spread of the circulating secondary nuclear beams of low intensity was reduced to below 10^{-6}. The achieved mass resolving power of $m/\Delta m = 6.5 \cdot 10^5$ (FWHM) in recent measurements represents an improvement by a factor of two compared to our previous experiments. The previously unknown masses of more than 100 proton-rich isotopes have been measured in the range of $54 \leq Z \leq 84$. The results are compared with mass models and estimated values based on extrapolations of experimental values. 2) Exotic nuclei with half-lives shorter than the time required for electron cooling can be investigated by time-of-flight measurements with the ESR being operated in the isochronous mode. This novel experimental technique has been successfully applied in a first measurement with nickel fragments. A mass resolving power of $m/\Delta m = 1.5 \cdot 10^5$ (FWHM) was achieved in this mode of operation. 3) Nuclear half-lives of stored and cooled bare projectile fragments have been measured to study the influence of the ionic charge state on the beta-decay probability.

INTRODUCTION

The study of exotic nuclei is an important challenge in nuclear physics and has revealed exciting new nuclear properties which are not present close to the valley of beta stability [1,2]. Research of nuclear structure is extended by new secondary beam facilities, where recently special effort has been devoted to experiments using radioactive nuclei at energies above the Coulomb barrier [3].

An advantage of exotic nuclear beams at relativistic energies is that the reaction products are bare or populate only few-electron states. Indeed, the selection of the projectile energy can be used to prepare the fragments in desired charge-state distributions [4]. This situation allows experiments under conditions which prevail in stellar plasmas and are therefore relevant for basic astrophysical studies [5,6].

In addition to the ion-optical performance of a separator system, the atomic interactions (slowing-down characteristics and charge-state population) of the ions determine the isotopic separation quality and particle identification [7,8]. Advantages of relativistic secondary beams have been demonstrated at the fragment separator FRS with the identification of the two doubly magic nuclei ^{78}Ni [9], ^{100}Sn [10], and of 117 new fission fragments. New nuclear structure properties of halo nuclei have been discovered using the FRS as a high-resolution spectrometer [11-13].

In this contribution we will concentrate on mass and half-life measurements of relativistic projectile fragments performed with the unique combination of the FRS

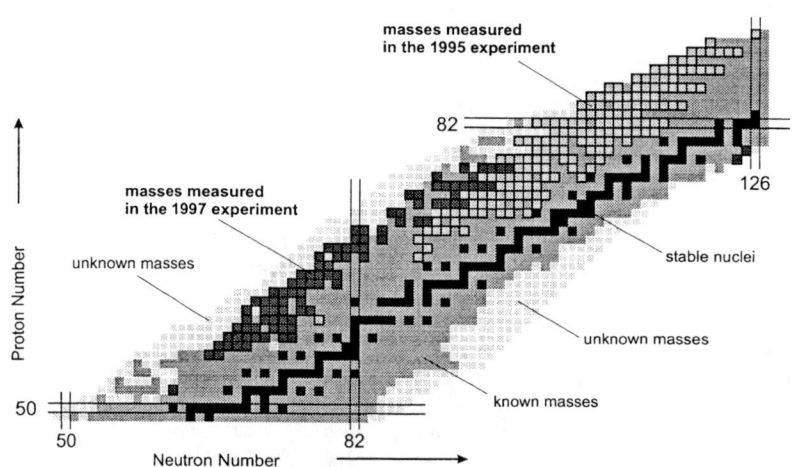

FIGURE 1. Range of FRS-ESR mass measurements covered in experiments with ^{209}Bi projectile fragments in 1995 and 1997. In the latter experiment we extended the measured mass surface down to cesium and also to neutron-rich fragments near the shell closures at Z=82 and N=126. The marked areas of our studies represent the previously unknown masses only.

and the experimental storage ring ESR [14,15]. The progress of mass measurements are presented in some recent review articles [16–18].

Our mass measurements with stored fragments in the ESR covered a large area of proton-rich nuclei 54≤Z≤84 which includes also members of α-chains linked by precise Q_α values but not yet connected to the backbone of known masses [19]. An overview of the mass surface covered by our mass measurements with ^{209}Bi projectiles is shown in fig. 1. The isotopes with masses known before our experiments as well as those where new mass measurements were performed in 1995 and 1997 are indicated in a section of the chart of nuclides. In our 1997 experiment we extended the measured mass surface to smaller Z values down to cesium (Z=55) and also to neutron-rich fragments near the shell closures at Z=82 and N=126.

MASS SPECTROMETRY OF STORED RELATIVISTIC IONS AT THE SIS-FRS-ESR FACILITY

Stable beams of relativistic heavy ions provided by the synchrotron SIS [20] with a maximum magnetic rigidity (Bρ) of 18 Tm are converted into exotic nuclei by nuclear collisions in the production target at the entrance of the spectrometer FRS. Projectile fragmentation and projectile fission combined with in-flight separation are versatile tools for experiments with relativistic exotic nuclei [3]. The FRS separates the fragments in flight and injects them into the ESR [14] for precise mass determination, performed by measuring the revolution frequency of the stored ions. The ESR is equipped with an electron cooler [24] and can store ions in the range of (0.5≤ Bρ ≤10)Tm . The storage time of the nuclei (τ_{st}) is limited by atomic collisions with atoms of the residual gas (pressure ≤ 10^{-10} mbar) and with the electrons of the cooler. τ_{st} can range from hours up to days depending on the velocity and the charge state of the stored ions. The phase-space density of the stored ions can be drastically reduced by electron cooling, e.g., the relative velocity spread of a low-intensity cooled beam can be less than 10^{-6}.

Two methods are employed to perform precise mass measurements of stored ions circulating in the ESR: 1) Mass spectrometry using cooled ion beams. 2) Mass spectrometry of hot fragments operating the ESR in the isochronous mode. Both principles can be easily understood by

$$\frac{\Delta m}{m} = -\gamma_t^2 \frac{\Delta f}{f} + \left(\gamma_t^2 - \gamma^2\right) \frac{\Delta v}{v}, \qquad (1)$$

the first-order relation between the mass m, the revolution frequency f, and the velocity v. The mass resolution is given by the precision achieved for the frequency and velocity determination. The formation of the exotic nuclei always causes a relatively large velocity spread, which is inherent due to the nuclear reaction dynamics and depends on the selected fragments. γ is the relativistic Lorentz-factor and γ_t represents the reduced transition energy which characterizes the ion-optical mode of the ring. From this formula it is obvious that either cooling $\frac{\Delta v}{v} \to 0$ or the isochronous condition, $\gamma \to \gamma_t$, are the basis of precise mass measurements.

Schottky Mass Measurements with Cooled Fragments

Schottky spectroscopy is widely used for beam diagnosis in circular accelerators and storage rings. The induced signals of the stored circulating ions in non-destructive probes are recorded and analyzed. Already in our pilot experiments with cooled projectile fragments [15,21] we have applied Schottky diagnostics and since then we are gradually developing this technique for the requirements of precision mass spectrometry [19,22,23]. The stored and cooled ions circulate in the ESR with revolution frequencies of about 1.9 MHz. In the experiment we used the Schottky-noise signal of the 32th harmonic of the revolution frequencies. This high-frequency band has to be reduced by mixing with an external oscillator to match the working range of 1 kHz to 300 kHz of the Fourier-analyzer system. An improved data-acquisition system digitizes the Schottky signals with a sampling rate of 625 kHz. The data are sequentially recorded on tape with the time correlation of the events. The frequency bandwidth of 200 kHz covers most of the Bρ acceptance of the stored beam in the ESR. Fast Fourier transformation of the time-correlated data can be performed off-line in order to obtain the revolution frequencies of the stored ions as well as the half-life information of the radioactive species. Furthermore, undesirable drifts of the experimental conditions can be corrected in the off-line analysis. In the 1997 experiment we used ^{58}Ni and ^{209}Bi projectiles and set the magnetic fields of the FRS and ESR and the cooler voltage corresponding to a constant Bρ value of 6.5 Tm. This means for example, that the kinetic energy of ions with a mass-over-charge ratio of 2.35 is 315.6 A·MeV. This energy range allowed an optimum performance of the electron cooler and, in addition, presented the opportunity to measure the masses of bismuth fragments in bare, H-like, and He-like charge states. The mass measurement in different charge states yields not only redundant data for the same isotope, but is also advantageous for the calibra-

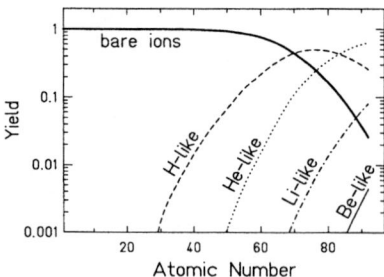

FIGURE 2. Calculated ionic charge state population of heavy ions [4] emerging from a beryllium target with velocities corresponding to a fixed magnetic rigidity of 6.5 Tm. Masses of stable nuclei are used in these calculations.

tion. Reference masses in different ionic charge states are automatically included in the corresponding $B\rho$ settings and appear in this way in the neighborhood of the isotopes with unknown masses in frequency spectra. The rest mass of the electron and its binding energies are known with an accuracy of the order of 10^{-10} for masses in the lead region. This uncertainty is completely negligible compared to our present errors for the nuclear masses [19] being in the order of 10^{-6}. The incident energy of the projectiles, focused on the production targets at the entrance of the FRS, was varied to select different fragments for injection into the ESR. The production targets were 4 and 8 g/cm^2 beryllium for the ^{58}Ni and ^{209}Bi projectiles, respectively. The ionic charge state population of fragments emerging from the targets with velocities corresponding to the fixed magnetic rigidity of 6.5 Tm is shown in fig. 2. The populations were calculated by computer codes which are based by heavy ion charge state measurements at GSI and LBL [4]. The fragments were separated with the FRS by pure $B\rho$ analysis during the major part of the mass measurements. In this case up to 60 different fragments were injected into the ESR by a single bunch of projectiles from SIS. However, we also successfully applied the $B\rho$-ΔE-$B\rho$ separation method [7] and stored and cooled a monoisotopic beam in the ESR, as discussed below. A number of important improvements could be achieved between the 1995 and the 1997 mass measurements: A new data acquisition system was taken into operation (see above). The performance of the ESR cooler was considerable improved yielding a much stronger cooling force which led to shorter cooling times and, therefore, gave access to isotopes with shorter half-lives. Furthermore, the better stabilization of the power supplies and the improved field homogeneity of the ESR magnets helped to significantly improve the resolving power in the mass spectra. In fig. 3 we present Schottky frequency spectra for bismuth and nickel fragments characterized by a mass resolving power of 6.5·10^5 which is about a factor of two better than in our 1995 experiment [19]. It should be mentioned that the half-life of neutral ^{151m}Er atoms is 0.58 s, however, in our case with bare nuclei the half-life is prolonged by roughly a factor of 21, see below.

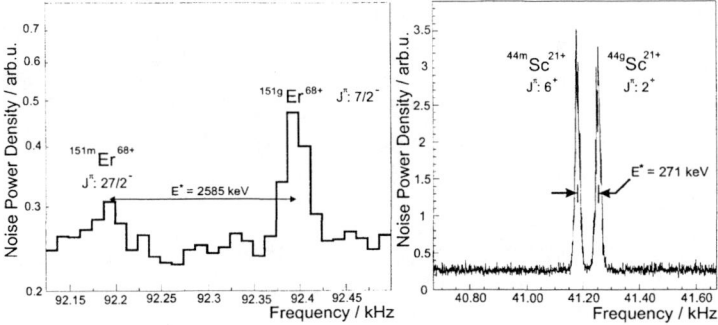

FIGURE 3. Mass resolved Schottky frequency spectra of bare ^{151g,m}Er ions (ground and isomeric states separated by 2585 keV, left panel) and of bare ^{44g,m}Sc ions (ground and isomeric states separated by 271 keV, right panel)

Isochronous Mass Measurements with Hot Fragments

Exotic nuclei with half-lives shorter than the cooling time can be investigated by the time-of-flight techniques where the ESR is operated in the isochronous mode [25,26]. In this case, the magnetic fields of the ESR quadrupole and hexapole magnets are set such that the revolution frequency of an ion species becomes independent of its velocity, see equation (1). This novel experimental technique has been successfully applied in first measurements with nickel fragments [26]. The ESR lattice was tuned to $\gamma_t=1.37$ corresponding to a kinetic energy of 345 A·MeV for the stored ions. γ_t is mainly determined by the dispersion function inside the dipole magnets. The first frequency spectra recorded in the isochronous mode are shown

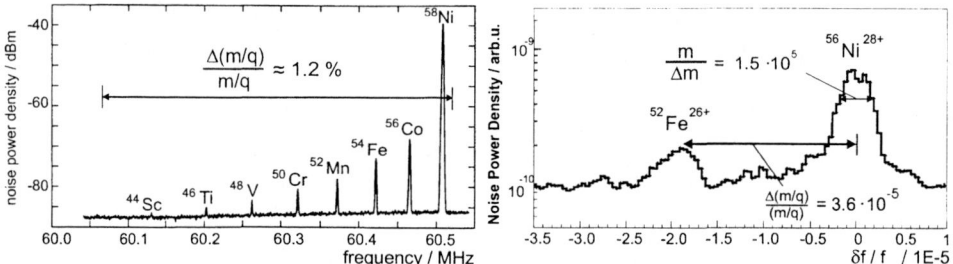

FIGURE 4. First frequency spectra of hot fragments recorded in the isochronous mode of the ESR. Left panel: Nickel fragments and the primary beam (^{58}Ni) simultaneously injected within the same window of the Bρ acceptance of the ESR. Right panel: Frequency sprectrum of ^{52}Fe and of the doubly magic nucleus ^{56}Ni measured in the isochronous mode.

in fig. 4. In the left panel the primary beam (^{58}Ni) and the fragments in the same window of the Bρ acceptance of the ESR injection are shown. The relative velocity spread of these stored ions was about 10^{-3}. Although the momentum acceptance inside the ESR is strongly reduced in the isochronous mode, as compared to the standard operation, the measured m/q acceptance still reached 1.2%. Already in this pilot experiment we achieved a remarkable mass resolving power of m/Δm = $1.5 \cdot 10^5$ (FWHM) as demonstrated by resolving ^{56}Ni and ^{52}Fe fragments, see the right panel of fig. 4. In the off-line analysis the measured dependence of the frequency on the velocity could be well understood and was reproduced by ion-optical model calculations. New ion-optical calculations suggest strong improvements for the isochronous mode which will be tested in future experiments [27].

RESULTS OF THE MASS MEASUREMENTS

After having demonstrated that the results obtained with our novel experimental methods for direct mass measurements agree well with already known light masses, a large number of isotopes with unknown masses has been measured (fig. 1). The interesting nuclides have been produced via fragmentation of bismuth projectiles.

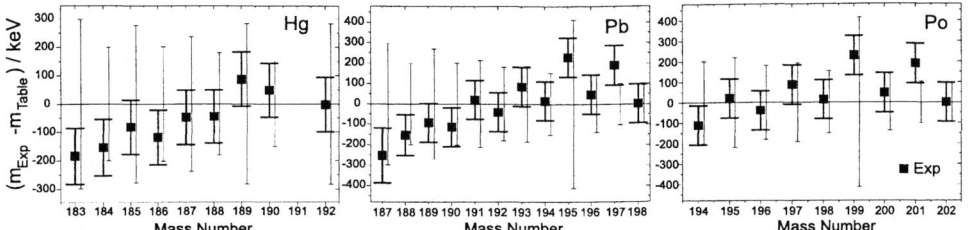

FIGURE 5. Differences of masses from the 1995 experiment obtained by Schottky spectrometry and from the Audi-Wapstra table [28] for Z=80, 82, 84. The bold bars represent the errors of the experimental data, whereas the thin bars indicate the uncertainties given in the mass table.

As already mentioned, the population of different charge states served as an excellent tool for unambiguous identification and redundant determination of ionic masses from frequency spectra. In particular, the H- and He-like proton-rich nuclei appeared in the spectra in the neighborhood of bare nuclei which are positioned closer to the valley of beta stability and have very well known masses used for calibration of the frequency scale. The masses were evaluated by application of the maximum likelihood method. The relation of the mass-over-charge ratio as a function of the revolution frequency was fitted by a first-order polynomial connecting the known masses as references [28]. The unknown masses were determined by searching the maximum of the probability density for all data. The total error of the new masses is in the range of 100 to 200 keV, where a systematic error of about 95 keV has been included. The systematic error has been evaluated by a comparison with previously well-known masses. In the following, some of the data measured in the 1995 experiment are compared with the tabulated values derived from systematics [28]. The comparison presented in fig. 5 demonstrates that, in general, our data are in good agreement with the values of the mass table [28]. The good accuracy of our measured data is a solid basis for a better understanding of the nuclear forces. The comparison of the data with microscopic theories yields new insight for the different models based on the nucleon-nucleon interactions parametrized as nuclear forces (e.g. Skyrme, Gogny) [29,30]. However, at the present status of

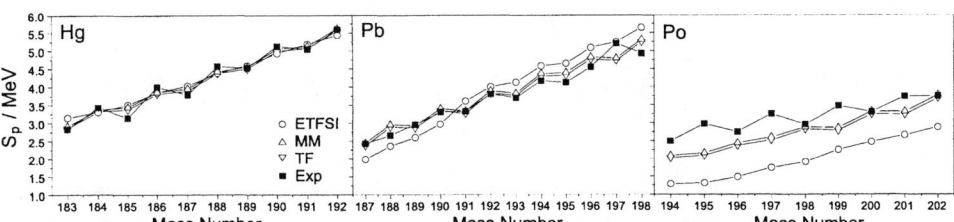

FIGURE 6. Measured 1-proton separation energies (Exp) for mercury, lead, and polonium isotopic chains compared with model predictions. The abbreviations for the different theoretical models are explained in the text. The experimental errors are smaller than the symbols.

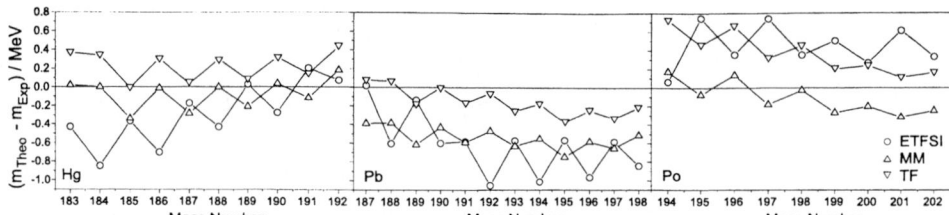

FIGURE 7. Differences of the measured masses for mercury, lead, and polonium isotopic chains and the corresponding model predictions [32–34]. The experimental errors are those displayed in fig. 5.

the microscopic theories the deviations from the experimental results are still up to several MeV. Progress in this field might be expected from modern Hartree-Fock-Bogoliubov calculations and Relativistic Mean Field theory which yield also reliable microscopic descriptions of other nuclear properties, like radii, of course using the same nucleon-nucleon interactions as employed for the mass predictions. This subject is for example reviewed in the theoretical contribution by P. Ring [31] at this conference. Therefore, we restrict ourselves here to comparisons with the Macroscopic-Microscopic (MM) [32] approach, with predictions based on the Thomas-Fermi model (TF) [33], and with the Extended Thomas-Fermi model with Strutinski Integral (ETFSI) [34]. In fig. 6 our experimental 1-proton separation energies, are compared to the MM, TF, and ETFSI predictions. In general, a stronger staggering effect between even and odd masses is observed in the experimental data, in particular for the most proton-rich isotopes, as compared to the mass model predictions. This observation clearly indicates that the description of the pairing force has to be improved. The need for improvements of the theoretical models is even more obvious from the comparison of absolute mass differences, as depicted in fig. 7. In the future direct mass measurements will be extended to shorter-lived nuclei using the isochronous method. However, there are also regions on the chart of nuclei where the Schottky mass spectrometry can still significantly contribute to an improved knowledge of the mass surface. The better mass resolution as well as the successful development of the stochastic cooling in the ESR [35], decreasing the cooling time for hot fragments, will considerably enhance the potential of Schottky mass spectrometry of cooled exotic nuclei.

LIFETIME MEASUREMENTS OF STORED RELATIVISTIC IONS

Mass and half-life measurements both address fundamental properties of atomic nuclei. In the case of half-life measurements at relativistic energies, one has a unique possibility to study the nuclear decay as a function of the ionic charge state [36]. Decay studies of bare and few-electron radioactive ions are of fundamental interest and are also relevant for the understanding of nuclear decay and reactions in stellar plasmas. The possibility to access bare nuclei in the laboratory is experimentally also quite advantageous as demonstrated with our example with ^{151m}Er nuclei where the half-life prolongation allowed to apply Schottky mass spectrometry.

Half-lives can be measured in the ESR via the decreasing intensity of stored mother nuclei [21], by directly detecting the daughter nuclides, or even by both. The decay of stored nuclei can be measured in several scenarios: 1) If the magnetic rigidity difference between mother and daughter nuclei is within the 3.6 % acceptance of the ESR, then the Schottky pickups can simultaneously monitor both species. This represents a unique possibility. 2) The ionic charge state remains the same, however, the Q-value of the decay is large enough to allow the separation of both frequency signals. 3) The daughter products are detected by particle detectors at positions where they are going to leave the closed orbit of the mother nuclei [15]. In the case of β decay into bound states, a large Bρ change can be introduced via stripping off the decay electron in the internal gas target of the ESR [5,6]. Half-life measurements of stored ions using the time-dependent intensities of both the mother and daughter nuclei require that the daughter is not already dominantly present in the circulating beam or that it cannot be formed by the decay of other stored nuclei. Therefore, it is desirable to have a monoisotopic beam stored in the ESR.

The feasibility of this method at the FRS-ESR is demonstrated by applying the Bρ-ΔE-Bρ method [7] for nickel fragments with the goal to measure the half-life of the ^{52m}Mn isomer. In fig. 8 Schottky spectra are shown for stored fragments without (left panel) and with a degrader (right panel) placed in the midplane of the FRS. In the latter case, the FRS-ESR system was set to transmit and accept only bare ^{52g,m}Mn^{25+} ions which are separated only by 378 keV. The experimental results successfully illustrate the injection and storage of monoisotopic beams. The

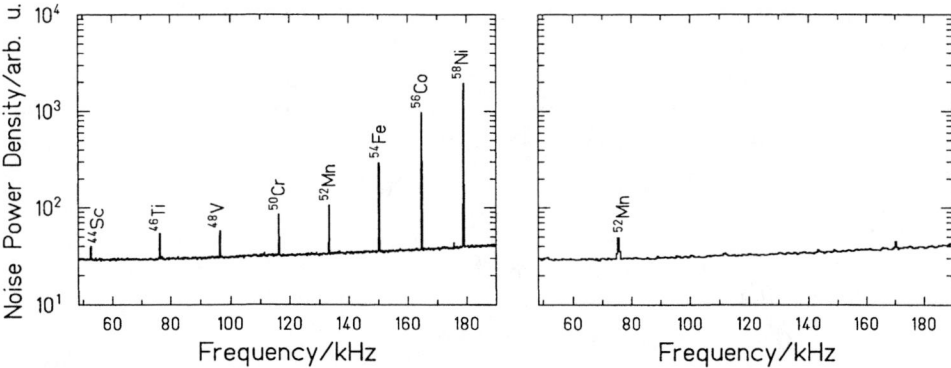

FIGURE 8. Measured cooled nickel fragments in the ESR after separation with the FRS by pure Bρ analysis (left panel) and by applying the Bρ-ΔE-Bρ method [7] with an intermediate degrader (right panel). In this broad-band spectrum the ground and isomeric states of ^{52}Mn nuclei are not resolved.

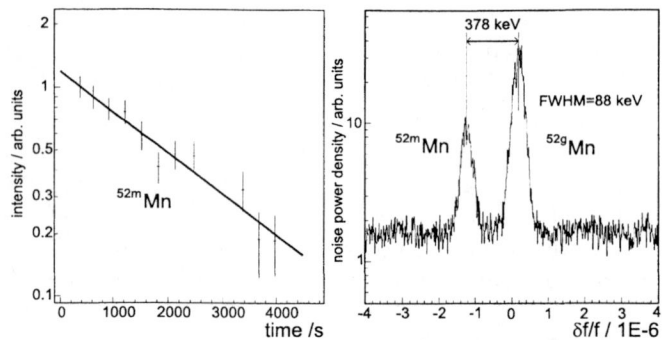

FIGURE 9. Half-life measurements of the mass resolved ^{52}Mn isomer (decay curve, left panel). The ground and isomeric states are well separated in the Schottky mass spectra (right panel). The excitation energy is 378 keV and the measured mass resolving power is $5.5 \cdot 10^6$.

improved mass resolving power in our recent experiments is clearly demonstrated in fig. 9 where the ground and isomeric states of ^{52}Mn nuclei are well resolved in the Schottky frequency spectrum. The decay of cooled fragments was monitored by the change of the peak area in the corresponding frequency spectrum. The basic assumption is that in this measurement the area of a Schottky frequency peak is strictly proportional to the number of stored ions. The absolute velocity of the stored ions was deduced from the terminal voltage of the electron cooler to transform the measured half-life from the laboratory frame to the rest frame of the ion. In the left panel of fig. 9 the decay curve for the bare isomer ^{52m}Mn ($J^\pi=2^+$) is shown. The measured half-life of ^{52m}Mn is 22.7 ± 3.0 min where the particle losses due to atomic interaction were taken into account from the storage time of the much longer-lived ground state $T_{1/2}(^{52g}\text{Mn}) = 5.59$ d. The result is in good agreement with the experimental half-life for neutral atoms because no major decay branch was suppressed in the bare state, and because the correction of the modified Fermi function is small in this case [21].

A goal for our half-life studies in the future is to systematically investigate the beta decay of heavier ions as a function of the ionic charge. The influence of screening and the branch of beta decay into bound states are further items which can be ideally investigated with the new data acquisition developed for the Schottky mass spectrometry.

The authors Z. P. and Y. N. would like to acknowledge the support of the Polish Committee for Scientific Research (KBN), grant no. p03b 117 15 and the WTZ grant RUS-654/96.

REFERENCES

1. Proceedings of ENAM95 and references therein,
   Ed. by D. Guillemaud-Mueller, Editions Frontieres, ISBN 2-86332-186-2.

2. P.G. Hansen, A.S. Jensen, B. Jonson, Ann. Rev. Nucl. Part. Sci. 45 (1995) 591.
3. H. Geissel, G. Münzenberg, K. Riisager, Ann. Rev. Nucl. Part. Sci. 45 (1995) 163.
4. C. Scheidenberger, Th. Stöhlker, W.E. Meyerhof, et al., GSI-Preprint 98-03 (1998).
5. M. Jung, F. Bosch, K. Beckert, et al., Phys. Rev. Lett. 69 (1992) 2164.
6. F. Bosch, T. Faestermann, J. Friese, et al., Phys. Rev. Lett. 77 (1996) 5190
7. H. Geissel, P. Armbruster, K.H. Behr, et al., Nucl. Instr. and Meth. B70 (1992) 286.
8. M. Pfützner, H. Geissel, G. Münzenberg, et al., Nucl. Instr. and Meth. B86 (1994) 213.
9. Ch. Engelmann, P. Armbruster, M. Bernas, et al., Z. Phys. A352 (1995) 351.
10. R. Schneider, J. Friese, J. Reinhold, et al., Z. Phys. A348 (1994) 241.
11. W. Schwab, H. Geissel, H. Lenske, et al., Z. Phys. A350 (1995)283.
12. B. Jonson, contribution to this conference.
13. T. Baumann, H. Geissel, H. Lenske, et al., contribution to this conference.
14. B. Franzke et al., Nucl. Instr. and Meth. B24/25 (1987) 18.
15. H. Geissel, K. Beckert, F. Bosch, et al., Phys. Rev. Lett. 68, (1992) 3412.
16. H.-J. Kluge, in reference [1] page 3.
17. G. Audi, O. Bersillon, J. Blachot, A.H. Wapstra, Nucl. Phys. A624 (1997) 1.
18. W. Mittig, A. Lepine-Szily, N.A. Orr, Ann. Rev. Nucl. Sci. 47 (1997) 27.
19. T. Radon, Th. Kerscher, B. Schlitt, et al., Phys. Rev. Lett. 78 (1997) 4701.
20. K. Blasche, B. Franczak, Proc. of the third European Part. Acc. Conf., Berlin, 9, eds. H Henke, H Homeyer, Ch. Petit-Jean-Genaz. Gif-sur-Yvette: Editions Fronitière (1992).
21. H. Irnich, H. Geissel, F. Nolden, et al., Phys. Rev. Lett. 75 (1995) 4182.
22. B. Franzke K. Beckert, T. Beha *et al.*, Physica Scripta T59 (1995) 176.
23. B. Schlitt *et al.*, Hyp. Int. 99 (1996) 117, and Nucl. Phys. A626 (1997) 315c
24. M. Steck, K. Beckert, H. Eickhoff, et al., Phys. Rev. Lett. 77 (1996) 3803.
25. H. Wollnik, et al., GSI-Report 86-1 (1986) 372.
26. M. Hausmann, K. Beckert, H. Eickhoff, et al., GSI-Report 98-1 (1998) 170.
27. M. Hausmann, K. Beckert, H. Eickhoff, et al., Proc. of EPAC conference 1998.
28. G. Audi, A.H. Wapstra, Nucl. Phys A595 (1995) 409, and Nucl. Phys. A624 (1997) 1.
29. Z. Patyk, A. Baran, J.F. Berger et al., GSI Preprint, GSI-97-40 (1997), submitted to Phys. Rev. C.
30. H. Geissel, G. Bollen, B. Franzke et al., Nucl. Instr. and Meth. B126 (1997) 351.
31. P. Ring, contribution to this conference.
32. P. Möller, J.R. Nix, W.D. Myers and W.J. Swiatecki, *At. Data Nucl. Data Tables* **59** (1995) 185.
33. W.D. Myers and W.J. Swiatecki, *Nucl. Phys.* **A601** (1996) 141.
34. Y. Aboussir, J.M. Pearson, A.K. Dutta and F. Tondeur, *At. Data Nucl. Data Tables* **61** (1995) 127.
35. F. Nolden, B. Franzke, A. Schwinn, GSI-Report 98-1 (1998) 171.
36. O. Klepper, Nucl. Phys. A626 (1996) 199c.

Self-consistent microscopic approaches to nuclear masses

P. Ring

*Physikdepartment der Technischen Universität München
D-85748 Garching, Germany*

Abstract. In this talk we present the status of self-consistent theoretical methods, to describe the ground state properties and in particular nuclear masses over the entire range of the periodic table. We discuss their advantages and their shortcomings for the description of nuclear masses as compared to earlier more phenomenological methods based on the macroscopic-microscopic approach. In particular we emphasize, that in regions close to the drip-lines one expects a softening of the nuclear surface and a reduction of the spin-orbit splitting, which leads the quenching of some shell effects. This has to be described self-consistently. As compared the non-relativistic methods, relativistic models have the advantage, that they are assumed to be more reliable in taking into account the proper isospin dependence of the spin-orbit splitting.

INTRODUCTION

Exotic nuclei far form the line of beta-stability have gained considerable interest in recent years both on the experimental as on the theoretical side. New phenomena are expected such as dramatic changes in the nuclear shell structure. In large areas of the periodic table experimental data on such nuclei are not yet available. Therefore one needs a reliable theoretical description to make predictions. There are many models well adjusted to presently known data, which have been used for this purpose. However, in practice it turns out, that the predictions of the different models deviate strongly from one another in the unknown region. This is seen already in the prediction of nuclear masses.

At present our understand on nuclear structure is far from being able to predict nuclear masses over large regions of the periodic table based on a many-body theory using only the nucleon-nucleon interaction. One is bound to phenomenological descriptions. However one should use as much theoretical input as possible. Hartree-Fock calculations based on density dependent energy functionals and Relativistic Mean Field theory are the modern tools, which are used to give a rather reliable microscopic description of many nuclear data all over the periodic table.

PHENOMENOLOGICAL APPROACHES

The experimentally known nuclear masses form very large set of extremely accurate data, which have been used extensively to derive mass formulas in order to predict new masses. There are many such mass formulas, which differ in their phenomenological ansatz. Some of them use no input of physics and modern techniques of neural networks, where the system learns by adding more and more date. They have turned out not to be very successful in predicting nuclear masses. Others use general physical ideas based on properties like saturation, surface energy or Coulomb repulsion or pairing in order to make an ansatz with contains a number of parameters fitted to the experimental data. The most famous is the Bethe-Weizsäcker formula and its extensions, as the Liquid Drop Model (LDM) of Myers and Swiatecki.

MICROSCOPIC-MICROSCOPIC MODELS

Nuclear masses depend on the mass number A, the neutron number N and the proton number Z only on a first glance in a smooth way. Shell effects cause deviations from a smooth behavior. Therefore, in order to get a reliable description of nuclear masses by simple analytical mass formulas, one first has to remove shell effects. This can be done with great success by the method of Strutinski. Within this method the shell effects are calculated in a shell model, as for instance in the generalized oscillator model presented by Nilsson, in a Saxon-Woods model or a folded Yukawa model. In all these models the nuclear binding energy is calculated by summing over the binding energies of the individual particles. Because of many-body correlations the binding energies obtained in this way deviate dramatically from experimental values even if the parameter of the corresponding single particle potentials are fitted carefully. The advantage of the shell model binding energies is the fact that, in contrary to the analytical mass formulas, they show shell effects, which depends only on the shell structure in the neighborhood of the Fermi surface. This quantity can be properly described in the single particle shell model by a careful adjustment of the parameters of the single particle potential to experimental single particle spectra. The merit of Strutinski was to find a method to extract from the shell model binding energies the smooth part, which is not described reliably in this model. This has to be done with great accuracy. The method proposed by Strutinski is the one, which is mostly used nowadays for this purpose. However, it requires discrete single particle levels and it fails for cases, where the Fermi level approaches the continuum limit, i.e. for exotic nuclei close to the neutron drip line. If one removes the smooth part of the binding energies from the shell model values, one is left with the shell corrections. The smooth part of the shell model energies is then replaced by the macroscopic formulas.

The basic idea of the macroscopic-microscopic models is to divide the nuclear properties in gross properties containing many-body correlations, which behave smooth as functions of the particle numbers A, N and Z, and shell effects. Both parts are fitted to experimental data independently. This method allows to reproduce the experimental masses in a very accurate way. Pairing effects present correlations depending very strongly on the single particle properties as for instance the level density in the vicinity of the Fermi surface, have to be taken into account by specific methods. Therefore these macroscopic-microscopic models need a rather large number of parameters, which are directly fitted to minimize the rms-deviations of nuclear binding energies over nearly 1500 nuclei.

The predictive power of these macroscopic-microscopic models seems to be rather high. However, it is predicted to fail in cases, where either the decoupling of bulk properties and shell effects breaks down. This is certainly the case in light nuclei and in most of the applications the regions of very light nuclei is therefore excluded in the fits. The method is also expected to fail in cases, where the single particle potential is changing so dramatically that it cannot be described so easily by a small number of parameters. Exotic nuclei in are such a case, where one expects neutron skins and neutron halo's connected with changes of the surface diffuseness. A third case, where the macroscopic methods have difficulties, are the regions where the Fermi surface of the neutrons come close to the continuum limit. In such cases there are not enough bound single particle states available for the Strutinski method to work.

NON-RELATIVISTIC ENERGY FUNCTIONALS

Starting from in a microscopic framework one tried to derive nuclear properties from the nucleon-nucleon interaction determined from scattering experiments. This interaction is much too strong for mean field approximations of the Hartree-Fock type. Many-body correlations not taken into account in such a framework cannot be neglected. In fact it is hard to understand on a first glance, why a mean field description as the single particle shell model works at all in nuclei. It has been found, that is works only because of the Pauli principle, which prevents the nucleons from coming to close in the interior of the nucleus. In fact Pauli effects change the effective interaction in the interior of the nucleus considerably. Brückner theory takes into account such effects on the ladder level and it yields in the interior of the nucleus a relatively weak residual interaction, which is strongly density dependent. This approximate method gives also relatively good results for the kinetic and for the potential energy, which are both large and with opposite sign. However, the accuracy is not large enough to reproduce the binding energies properly, which is only a small difference between these two large numbers. Non-relativistic Brückner calculations and more recently also exact calculations based on two-body interactions obtained from scattering data fail to reproduce even the binding energy and the density of the simplest case, of symmetric nuclear matter.

In any case Brückner calculations have shown us, that self-consistent mean field calculations with effective density dependent forces are possible and meaningful in nuclei. Skyrme, Vautherin and Brink therefore developed a phenomenological force for Hartree-Fock calculations, which depends on the density. It is relatively simple, mostly zero range and containing only a few (roughly 10) parameters, which are adjusted to nuclear matter data and to data of a few closed shell doubly magic nuclei like ^{16}O, ^{40}Ca or ^{208}Pb. This procedure is equivalent to the variation of a phenomenologically determined energy functional depending on the nuclear density. This method is therefore equivalent to the Kohn-Sham method, which is used with great success in other areas of Physics based on the Coulomb force, such as molecular physics or condensed matter physics.

Gogny has later-on extended this idea by using in addition to the density dependent zero range term in the Skyrme force finite range forces, which allow also a self-consistent treatment of pairing correlations in the framework of density dependent Hartree-Fock-Bogoliubov theory.

RELATIVISTIC MEAN FIELD THEORY

In the RMF theory, the saturation and the density dependence of the nuclear interaction is obtained by a balance between a large attractive scalar i σ-meson field and a large repulsive vector ω-meson field. The asymmetry component is provided by the isovector ρ meson. The nuclear interaction is hence generated by the exchange of various mesons between nucleons in the framework of the mean field. The spin-orbit interaction arises naturally in the RMF theory as a result of the Dirac structure of nucleons.

In the RMF theory, the nucleons are described as relativistic particles moving independently in average potentials determined in a self-consistent way by the exchange of mesons. The relativistic single particle equation is the Dirac equation. In contrast to the non-relativistic Schroedinger equation, which contains one average potential of Saxon-Woods shape, the Lorentz structure of the Dirac equation allows in principle several types of fields:

a) The vector field $(V_0(\mathbf{r}), \mathbf{V}(\mathbf{r}))$, which is four-dimensional in space-time and behaves like a four-vector under Lorentz transformations, is similar in structure as the electromagnetic potentials $(A_0(\mathbf{r}), \mathbf{A}(\mathbf{r}))$ of Maxwell, well known from the Dirac equation in atomic physics. This vector field contains a time-like component $V_0(\mathbf{r})$ corresponding to the Coulomb field $A_0(\mathbf{r})$ and three space-like components $\mathbf{V}(\mathbf{r})$ equivalent to the magnetic potential $\mathbf{A}(\mathbf{r})$ in electrodynamics. Assuming time-reversal invariance we can neglect currents and the corresponding space-like parts $\mathbf{V}(\mathbf{r})$. We are then left with the time-like part $V(\mathbf{r})$ (for simplicity we neglect in the following the index 0). The Lorentz structure of the theory implies that this time-like part of the vector fields is repulsive. As we will see the essential part of this field is determined by the short range repulsion of the nucleon-nucleon interaction caused by the exchange of vector mesons.

b) In addition to vector fields familiar from atomic physics, the Dirac equation in nuclear physics contains a scalar field $S(\mathbf{r})$, which behaves like the rest mass and stays invariant under Lorentz transformations. The Lorentz structure of the theory implies that scalar fields are attractive. The origin of this field is the attractive part of the nucleon-nucleon interaction at intermediate distances. It is caused to a large part by correlated two-pion exchange and by two-pion exchange with a Δ-particle in the intermediate state. Both processes lead to a parity conserving mean field. In principle the Lorentz structure we could allow a pseudo-scalar field caused by the one-pion exchange. However, this part has to vanish, because it is well know that the nuclear mean field is parity conserving. Therefore the pion contributes on the Hartree level only via the two-pion exchange.

Neglecting nuclear magnetism, i.e. assuming time reversal invariance of the mean field, we than have the stationary Dirac equation containing only the time-like part of the vector V and the scalar potential S:

$$\{\alpha \mathbf{p} + \mathbf{V}(\mathbf{r}) + \beta[\mathbf{m} - \mathbf{S}(\mathbf{r})]\}\psi_i = \epsilon_i \psi_i. \tag{1}$$

It contains the four-dimensional Dirac matrices α and β. m is the rest mass of the nucleon. Expressed in terms of the Pauli spin matrices this Dirac equation reads

$$\begin{pmatrix} m - S + V & \sigma \mathbf{p} \\ \sigma \mathbf{p} & -m + S + V \end{pmatrix} \begin{pmatrix} f \\ g \end{pmatrix}_i = \varepsilon_i \begin{pmatrix} f \\ g \end{pmatrix}_i \tag{2}$$

The single particle wave-functions ψ_i are four-dimensional spinors, which describe stationary states of the nucleons with the index i and the single particle energy ε_i By summing over the occupied orbitals we can use this wave-functions in order to calculate two types of densities, the usual density

$$\rho(\mathbf{r}) = \sum_{i=1}^{A} \psi_i^+ \psi_i = \sum_{i=1}^{A} \mathbf{f}_i^+(\mathbf{r})\mathbf{f}_i(\mathbf{r}) + \mathbf{g}_i^+(\mathbf{r})\mathbf{g}_i(\mathbf{r}), \tag{3}$$

which is the zero component of the four-dimensional relativistic current vector, and the scalar density.

$$\rho_s(\mathbf{r}) = \sum_{i=1}^{A} \overline{\psi}_i^+ \psi_i = \sum_{i=1}^{A} \mathbf{f}_i^+(\mathbf{r})\mathbf{f}_i(\mathbf{r}) - \mathbf{g}_i^+(\mathbf{r})\mathbf{g}_i(\mathbf{r}). \tag{4}$$

In the *no-sea* approximation these sums do not include negative energy solutions of Eq.. (2). The fields $V(\mathbf{r})$ and $S(\mathbf{r})$ are obtained by averaging over the interactions induced by the exchange of vector and scalar mesons with the corresponding densities.

$$V(\mathbf{r}) = \int v_v(\mathbf{r},\mathbf{r}')\rho(\mathbf{r}')d^3\mathbf{r}, \qquad (5)$$

$$S(\mathbf{r}) = \int v_s(\mathbf{r},\mathbf{r}')\rho_s(\mathbf{r}')d^3\mathbf{r}. \qquad (6)$$

The two-body interactions $v_v(\mathbf{r},\mathbf{r}')$ and $v_s(\mathbf{r},\mathbf{r}')$ are of Yukawa type, because they correspond to the exchange of scalar mesons σ (isoscalar) and vector mesons ω (isoscalar) and $\vec{\rho}$ (isovector), where the fields are defined as

$$S(\mathbf{r}) = -g_\sigma \sigma(\mathbf{r}), \qquad (7)$$

$$V(\mathbf{r}) = g_\omega \omega(\mathbf{r}) + g_\rho \rho_3(\mathbf{r}) + A_0(\mathbf{r}), \qquad (8)$$

where $\sigma(\mathbf{r})$, $\omega(\mathbf{r})$, $\rho_3(\mathbf{r})$ are the classical mesons fields and $A_0(\mathbf{r})$ is the Coulomb field having its origin in the exchange of photons is the. These equations of motions for the mesons fields are the Klein-Gordon equations

$$(-\Delta + m_\sigma)\sigma(\mathbf{r}) = -g_\sigma \rho_s(\mathbf{r}), \qquad (9)$$

$$(-\Delta + m_\omega)\omega(\mathbf{r}) = g_\omega \rho(\mathbf{r}), \qquad (10)$$

$$(-\Delta + m_\rho)\rho_3(\mathbf{r}) = g_\rho(\rho_n(\mathbf{r}) - \rho_\mathbf{p}(\mathbf{r})), \qquad (11)$$

$$-\Delta A_0(\mathbf{r}) = e^2 \rho_c(\mathbf{r}), \qquad (12)$$

where $\rho_c(\mathbf{r})$ is the charge density. Having in mind, that the Greens functions of these equations are of Yukawa and Coulomb type we obtain for the interactions

$$v_s(\mathbf{r},\mathbf{r}') = -\frac{g_\sigma^2}{4\pi}\frac{e^{-m_\sigma|\mathbf{r}-\mathbf{r}'|}}{|\mathbf{r}-\mathbf{r}'|}, \qquad (13)$$

$$v_v(\mathbf{r},\mathbf{r}') = \frac{g_\omega^2}{4\pi}\frac{e^{-m_\omega|\mathbf{r}-\mathbf{r}'|}}{|\mathbf{r}-\mathbf{r}'|} + \vec{\tau}\vec{\tau}'\frac{g_\rho^2}{4\pi}\frac{e^{-m_\rho|\mathbf{r}-\mathbf{r}'|}}{|\mathbf{r}-\mathbf{r}'|} + \frac{e^2}{4\pi}\frac{1}{|\mathbf{r}-\mathbf{r}'|}, \qquad (14)$$

where $\vec{\tau}$ are the isospin matrices.

This set of coupled equations for relativistic nucleons moving in classical meson fields are Euler equations obtained from Hamilton's variational principle based on the following relativistic Lagrangian density of the Walecka model

$$\begin{aligned}\mathcal{L} = &\bar{\psi}\left(i\gamma\cdot\partial - m\right)\psi + \frac{1}{2}(\partial\sigma)^2 - \frac{1}{2}m_\sigma\sigma^2 \\ &-\frac{1}{4}\Omega_{\mu\nu}\Omega^{\mu\nu} + \frac{1}{2}m_\omega^2\omega^2 - \frac{1}{4}\vec{R}_{\mu\nu}\vec{R}^{\mu\nu} + \frac{1}{2}m_\rho^2\vec{\rho}^2 - \frac{1}{4}F_{\mu\nu}F^{\mu\nu} \\ &- g_\sigma\bar{\psi}\sigma\psi - g_\omega\bar{\psi}\gamma\cdot\omega\psi - g_\rho\bar{\psi}\gamma\cdot\vec{\rho}\vec{\tau}\psi - e\bar{\psi}\gamma\cdot A\frac{(1-\tau_3)}{2}\psi,\end{aligned} \qquad (15)$$

where $\Omega^{\mu\nu}$, $\vec{R}^{\mu\nu}$, and $F^{\mu\nu}$ are field tensors and the dots abbreviate a scalar product in Minkowski space ($\gamma\cdot\omega = \gamma^\mu\omega_\mu = \gamma_0\omega_0 - \vec{\gamma}\vec{\omega}$).

Using the experimental masses m, m_ω and m_ρ for the nucleons and the ω- and ρ-mesons we are left with only four parameters, m_σ, g_σ, g_ω and g_ρ, which are adjusted to experimental data in a few spherical nuclei. Already very early it has been recognized, that this simple model is not flexible enough to describe quantitatively the properties of real nuclei. An effective density dependence has been introduced in replacing the quadratic σ-potential $\frac{1}{2}m_\sigma \sigma^2$ in the Lagrangian by a quartic potential $U(\sigma)$ including a the nonlinear σ self-interaction.

Having in mind, that we want to calculate here only ground state properties of nuclei, the question arises, why it is necessary to use a relativistic formulation. In fact the kinetic energies and the Fermi momenta are relatively small as compared to the rest mass of the nucleons. We therefore certainly can neglect relativistic kinematics. However the Dirac equation contains more. In contrast to an equivalent Schödinger equation with a potential of a depth of roughly 50 MeV, which is also small as compared to the rest mass, the Dirac equation contains two potentials $V(\mathbf{r})$ and $S(\mathbf{r})$, which are both very large (roughly 350 and 400 MeV). They cannot be neglected as compared to the rest mass of 938 MeV and one needs relativistic dynamics in order to describe the interplay of these two strong potentials properly. From Eq. (2) we sea that in the upper equation of the large components only the difference $V - S$ enters, which is in fact small as compared to the rest mass and this leads to the relatively small Fermi momenta in this system . In the second equation for the small components, however, the very large sum of both potentials $V + S$ enters. It cannot be neglected and it is well known that this leads to the strong spin-orbit term in nuclear physics. In fact it is the advantage of this theory that the strength and the shape of the spin-orbit term are determined here in a fully self-consistent way.

Finally we want to emphasize, the fact, that $V - S$ is small leads to relatively small Fermi momenta and allows in principle a non-relativistic reduction of the Dirac equation to a Schroedinger equation with momentum dependent potentials. Therefore, a non-relativistic theory with additional spin- and momentum-dependent terms and adjustable parameters can also provide a reliable description of nuclei. However, in general it requires more parameters and its predictive power is probably reduced as compared to that of a fully relativistic theory.

RESULTS AND DISCUSSION

In Table 1 we show a comparison between the different models. With respect to the masses is is very clear that the macroscopic-microscopic models yield much smaller rms deviations for the nuclear masses. We have to keep in mind, however, that these model contain a very large number of adjustable parameters, which are fitted in such as way as to minimize the rms-deviations for the masses. The self-consistent models of Skyrme- and Gogny-type as well as the RMF-models contain much less parameters, which are adjusted to only a few spherical nuclei. In particular the relativistic models contain only 6 or 7 parameters, which is roughly half of the number of parameters used in the non-relativistic models. It is astonishing, that there is practically no essential difference between relativistic and non-relativistic

TABLE 1. Mass and charge radius rms deviations

MM(TF):	0.57	
MM(FRDM):	0.65	
MM(FRLDM):	0.76	
ETFSI:	0.80	0.028
SIII	4.74	0.059
SIII$^\delta$	3.07	0.057
SIII$^{\delta\rho}$	2.26	0.065
SkP	2.37	0.040
SkP$^\delta$	2.53	0.033
SkP$^{\delta\rho}$	2.32	0.043
SkM*	6.32	0.022
SkM*$^\delta$	5.36	0.021
SkM*$^{\delta\rho}$	4.74	0.023
Gogny	2.07	0.031
RMF(NL1)	3.94	0.026
RMF(NL2)	11.24	0.031
RMF(NL3)	2.48	0.028

models here, in some cases the relativistic models are even superior. When one considers masses together with radii the relativistic models are definitely superior to the non-relativistic models

CONCLUSIONS

Taking into account only the rms deviations for nuclear masses, at present the phenomenological macroscopic-microscopic models are still superior to the self-consistent models. However, since this result has been obtained by fitting this number with a very large number of parameters, it is not clear whether the predictive power of these models stays at the same level if one moves further away from the valley of beta-stability and approaches the nuclear drip-lines. Concerning the radii the self-consistent models are in most cases superior, in particular the relativistic versions.

Of course one should extent the relativistic models, taking into account further mesons, as for instance a scalar isovector meson and carry out a fit-program over many nuclei like it was possible for the macroscopic-microscopic models. Present computer facilities do not allow that. It will certainly be possible in near future.

I am grateful to G. Lalazissis, Z. Patyk, A. Baran, and A. Sobiczewski, for the important contributions to this investigations. We also acknowledge support from the Bundesministerium für Bildung und Wissenschaft under the Project 06 TM 784.

Atomic Mass Knowledge: 1999 Status

G. Audi* and A. H. Wapstra†

*Centre de Spectrométry Nucléaire et de Spectrométry de Masse CSNSM, IN2P3-CNRS
Bâtiment 108, F-91405 Orsay Campus, France
† NIKHEF, POBox 41882, 1009DB, Amsterdam, The Netherlands

Abstract. An overview is given of developments of interest for the determination of atomic masses of atoms around nuclear ground-states.

ABOUT THE BACK-BONE

In this report we discuss data new since our 1995 "update" [1] (which we will denote here AME95) of our 1993 mass table [2]. For several cases, we will mention work in progress, of which we assume that pertinent results will be available before we finish next mass evaluation. For this reason our title refers not to today but to the near future.

Not only for this Conference but also for nuclear physics in general, the most important new facts are concerned with nuclides far removed from the line of stability against β-decay. Yet, some new data for nuclides along that line deserve notice. We will therefore devote some attention to them. This will also serve to show several difficulties of the type we have to deal with in our evaluations.

We know people are working on the greatest problem along the line of β-stability: the masses of the stable Hg isotopes. For them, mass spectroscopic results from 1980 [3] deviate many keV, far outside the reported errors, from the results derived from reaction chains combined with spectroscopic results for other elements. Preliminary new mass spectroscopic results on one Hg isotope seem to agree with the earlier one. Though the precision is still rather less than that reported for the earlier result, a great problem would arise if it would be confirmed.

New results start to emerge about the most fundamental masses. The Stockholm SMILE group [4], working with a Penning trap, starts to get results of which they were so kind to send us provisional values. Their new value for H may be even somewhat more precise than the ones used in AME95 and agrees perfectly with them. But the new mass value for ^{4}He may deviate somewhat.

There are also developments that will affect the mass value for the neutron. It is derived essentially by combining the mass spectroscopic results for H and D

CP455, *ENAM98: Exotic Nuclei and Atomic Masses*
edited by B. M. Sherrill, D. J. Morrissey, and Cary N. Davids
© 1998 The American Institute of Physics   1-56396-804-5/98/$15.00

with the energy of the γ-rays emitted in the capture of thermal neutrons in hydrogen. The most precise value for the latter is derived from measurements of its wavelength, by diffraction in a silicon crystal [5]. Unfortunately, there was some uncertainty about the value of its lattice constant. It is hoped, that the current evaluation by Taylor and Cohen of data, important for determination of fundamental physical constants, will improve the situation. Also, it has been reported to us that a new measurement is in progress.

Somewhat less accurate γ-energy values are obtained by measurements in semiconductor diodes. These are calibrated, essentially, with the 411 keV γ-ray of ^{198}Au, which is again measured by the Deslattes group [6]. They may also be revised somewhat.

Van der Leun and Helmer are working on the general calibration of γ-rays. We are, in line with this, reconsidering the calibrations for many (n,γ) reactions, and so find that several old neutron binding energies can be improved. Following case presents an illustration. A value with a somewhat large error (650 eV) was reported [7] for the neutron binding energy in ^{54}Cr. Studying the paper taught, that this value was essentially the sum of the energies of two capture γ-rays of nearly the same enery. In recent work, a much improved value was given for one of them. And since the original paper gave a rather precise value for the difference in energy of the two rays, we can derive a much improved value for the resulting neutron binding energy.

The calibration for precision particle energies, e.g. for (p,γ) reactions and (p,n) thresholds we treat too. Unfortunately, new data [8] reportedly more precise (about 20 ppm) than old ones differ rather more than expected (more than 100 ppm) from older ones, causing yet unsolved difficulties.

The SMILE group also measured mass values for ^{22}Ne, ^{36}Ar and ^{133}Cs. The provisional results give rise to the following comments.

The ^{36}Ar result is some 1.2 keV lower than the AME95 value, to which an error of 0.3 keV was assigned. The latter value is, essentially, due to mass spectroscopic results for ^{35}Cl and ^{37}Cl, combined with reaction energies for 5 reactions. These data do agree quite well if combined in a least squares analysis: $\frac{R_e}{R_i} = 1.13$. But if the (provisional) new value for ^{36}Ar is added, $\frac{R_e}{R_i}$ is increased to 2.00. But this value is reduced to a reasonable 1.35 if, of the two available values for the ^{36}Ar(n,γ)^{37}Ar reaction energy, the oldest not well documented one is no longer used.

The ^{22}Ne result agrees quite well with the earlier ^{20}Ne value combined with the neutron capture ray energies in ^{20}Ne and ^{21}Ne mentioned in AME95 - but is over ten times more precise. Yet, the mass situation in this region is not perfect either. The mass of ^{28}Si is known with high precision. Its difference with that of ^{22}Ne also follows from a bridge of some 4 reaction energies. The so derived value was already in a not perfect agreement with the mass spectroscopic data in AME95. The new ^{22}Ne result makes it worse: the difference is over 1 keV, rather far outside the reported errors. This situation too requires a further analysis.

The SMILE ^{133}Cs result is important for the determination of masses many Cs and Ba isotopes: as discussed below, their relations with ^{133}Cs have been determined mass spectroscopically. The (provisional) SMILE value is about 5 keV higher than the AME95 one, to which an error of 3 keV had been assigned. The latter is mainly the result of a set of connections, through known Cs β^+ decay energies, with Xe nuclides, for which mass spectroscopic mass values were available (see the scheme fig. 1 in [2].) The nearest ones are those at mass numbers 124, 128, 129, 130 and 132. Analyzing them, we find that the connection with ^{132}Xe would make ^{133}Cs 15 [7] keV higher, that with ^{124}Xe 35 [20] keV lower. The first one, thus, is improved by the SMILE result. The other connections are not severely affected. In total, specifically, this analysis throws some doubt on the ^{125}Cs β^+ decay energy.

THE NUBASE EVALUATION

Already since long, we maintain a file of approximate mass values as input in our computer programs. (Mfile. Essentially, these programs calculate the differences with the input values.) In cases where isomers occur, one has to be careful to check which one is involved in reported experimental data, such as β-decay energies. Cases have occured where authors were not (yet) aware of isomeric complications. For that reason, our Mfile contained known data on such isomeric pairs (half-lives; excitation energies).

The matter of isomerism became even more important, when mass spectroscopic methods were developed to measure masses of exotic atoms, far from β-stability and therefore having small half-lives. The resolution was then limited, often insufficient to separate isomers. Then, one so obtains an average mass of the isomeric pair. A mass of the ground-state, our primary purpose, can then only be derived if one has information on the excitation energy and on the production rates of the two isomers. And in cases where e.g. the excitation energy was not known, it might be estimated by extrapolation of the values for isomers or isotones in the neighbourhood. We therefore decided, that it might be useful to make our Mfile as complete as possible. This turned out to be a major job. And since it was judged possible, that the result might be useful for others, it was published [9].

MASS SPECTROSCOPIC EXOTIC RESULTS

A group, originally from Mainz but working at ISOLDE in Geneva, developed a method of measuring masses of radio-active isotopes in a Penning trap. Mass values of many isotopes of Cs and some of Ba were already used in our 1993 mass evaluation, and will soon be published fully [10]. These new masses were derived from comparisons with the mass of ^{133}Cs and therefore may change somewhat in a new evaluation, as discussed above.

With this ISOLTRAP, masses [11] of several neutron-poor isotopes of lighter rare earth isotopes have also been determined, with a precision of about 20 keV. Especially mentioned may be their results for ^{148}Dy, ^{149}Dy and ^{150}Ho. These nuclides are the endpoints of α-decay chains, for which α-particle energies are known, starting with ^{180}Pb, ^{181}Pb and ^{174}Au respectively. No earlier value was known for ^{150}Ho; the value for ^{148}Dy is a decided improvement. Reversely, they found a value for ^{158}Dy which agreed quite well with the earlier value derived from its two member α-decay chain.

Finally, in a private communication we learned of their measurements on Hg isotopes with mass numbers from 184 to 200, with a precision of the order of 20 keV. They were obtained by comparison with ^{208}Pb; and thus are affected with the uncertainty about the mass of the stable Hg isotopes mentioned above. Especially usefull was, that in several cases they could separate isomers, at the odd mass numbers. Thus, in combination with α-decay data, good information is obtained for even-Z nuclei between ^{176}Pt and ^{210}Th. These data, combined with Pb(α) energies, allow a check on neutron pairing energies in proton-rich Hg and Pb isotopes. The Jensen-Hansen-Jonson [12] estimate is decidedly better than the earlier formula 12 MeV/$A_{\frac{1}{2}}$.

A quite exciting development occured at GSI, Darmstadt. They succeeded in storing radio-active ions in a storage ring and determining their masses [13] by, essentially, measuring their cyclotron frequencies. Those atoms were produced by bombarding targets with heavy ions, and caught in flight. As mentioned in two Muenich theses [14,15], masses were so determined for many dozens of proton-rich nuclides, roughly in the mass region A=140–200. Between many of the measured nuclides, connections exist due to chains of α-decays. Thus, checks are possible; and masses can be derived for many other nuclides.

MEASUREMENTS OF PROTON DECAYS

Our AME95 used a few results of measurements of energies of protons emitted in proton decay, made in Daresbury. Now we posses new measurements, first from the same place [16,17], but later at Argonne National Laboratory [18–20] and Oak Ridge National Laboratory [24] in collaboration with groups there. Thus, data were obtained for several very proton-rich nuclides, from ^{131}Eu to ^{195}Bi.

These data are quite important, mainly for two reasons. In the first place, we apply systematics of some quantities (among them proton separation energies) for estimating mass values for nuclides, for which no experimental mass data are available. For this purpose, knowledge of proton separation energies just beyond the proton drip line is quite valuable.

In the second place, the properties of proton decay allow in several cases to find proton decay energies from both members of an isomeric pair. Since, often, both are observed to decay to the ground-state of the daughter, one so derives the excitation energy of the isomer. And these studies even allow to get a fair estimate of the spin-parities of the separate members.

The advantage of this is especially valuable, if for both members α-decay is observed. In a particular case, even a succession of several such decays was found.

Their study showed that decays for some of these daughters observed in earlier work did not belong to ground-states, as assumed earlier; with evident consequences for the masses assigned to them.

THE α-DECAY CHAINS

Measured α-decay energies in such chains yields often quite precise information about differences in the masses of their members. It is therefore fortunate that new information on α decay is still regularly reported, by laboratories in Finland, Germany, Japan and the USA.

We may remind you that for even-even nuclides, like the ^{148}Dy and ^{158}Dy mentioned above, α-particle energies immediately determine their α-decay energies, since the strongest feeding occurs to the daughter's ground-state. Also, no complications are known to occur here due to isomerism. The only trouble that could occur is, that the nuclide involved is misassigned; but this is expected to occur only rarely.

Unfortunately, this is not true for other nuclides. There is a large number of cases, where no information is available about the levels, that are fed in the α-decay of nuclides with an odd number of neutrons and/or protons. Rather long ago, we tried to get some information about the average energy of the levels, fed by the highest energy α-particles, by comparing the values for a chain of successive ones with those of neighbouring even-even chains. They did not appear to differ much. Thus, we provisionally accepted them as due to transitions between ground-states. But in order to take the uncertainty into account, a value of 50 keV is (quadratically) added to the reported errors in the resulting α-decay energies. In principle, a better solution would be to accept the final level as an independent variable and to assign to its excitation energy some value with, say, an error 50. But we doubt that the extra complication due to the hundreds of new variables would be worth the trouble.

In the case that α-decay is observed for two isomers, it is rather unlikely that they preferentially feed the same level in the daughter nuclide. Often, the resulting problem can partially be solved by the fact that isomerism also occurs in the daughter. By measuring correlations, one can often decide which daughter-isomer is fed by which parent. And, as said, results for isomers decaying by proton emission can also be valuable in desentangling the complications due to the occurence of isomers.

The fact mentioned above, that mass specroscopic results now start to become available for several members of an α-decay chain is a help in alleviating the above problem. Yet, it does not eliminate it as long as the errors in those mass measurements are rather larger than those in α-ray energies.

MASSES AT THE HIGHEST MASS NUMBERS

Since AME95, a further new element, Z=112, was discovered at GSI [21], with mass number 277, the highest yet. Only two cases were found. A very remarkable fact was, that the two atoms of the daughter, ^{273}110, occured after delays that were a factor 1000 different. Also, the most delayed one had a very significantly (1.3 MeV) lower α-energy. This points to the influence of a semi-magic number of neutrons.

Other reports on α-decays in this neigbourhood also give very wellcome new information.

Not important for the mass work, but interesting to notice is, that the International Union of Pure and Applied Chemistry now accepted a set of names for the elements 103-109 [22]. No names have yet been proposed for the last three known elements.

In the high mass region too, long chains of α-decays are known. Again, the excitation energies of levels fed by the observed α-rays are often not known. A help is here, that the prominent decays are regularly those to states with the same Nilsson model quantum-numbers as the parent (favored α-decays). Differences between the positions of such particle-levels are often known for isotones or isotopes; and they do not change drastically as a function of N or Z. We made a study of them, which allows us to make decent estimates for corresponding excitation energies of states fed in favored α-decays.

In principle, the same could be done in the mass number region 160-200. We have started to analyze them too. The situation here, though, does not yet look promising.

MASSES FROM ISO-MULTIPLET MASS EQUATIONS

Recently, several cases have been studied of very proton-rich nuclides that β-decay to levels emitting protons, of which the energies are measured. Among them may be protons coming from the isobaric analogue of the mother nuclide. This may then permit to derive a decent mass value for that isobaric analogue. Authors then calculate the mass difference with the mother isotope from the (quadratic) Isobaric Multiplet Mass Equation (IMME). Before we discuss the consequences, the following is of interest.

A regular occurrence is, that an atomic mass value derived from a certain experiment disagrees rather severely with the value we would expect from extrapolation of masses of neighbouring nuclides (systematics). The difference may be outside the experimental error; but it may occur too that the reported error is compatible with the difference but (of course, then) quite large. In both cases, we had the habit to report both values. Earlier, the systematic value was given in the main table; but in our last table, AME95, we there give the experimental value but with a flag warning that a probably better value derived from systematics is given in a separate table.

We consider to use a similar method in cases where mass values for proton-rich nuclides can be derived from consideration of isobaric analogue states. Mass values may be known for three or more isobaric analogues, so that its mass value can be derived from the IMME. In some cases, an experimental value for that nucleus may be available too, but with a far larger error. Examples: ^{28}S and ^{40}Ti, where the experimental errors are 160 eV but the ones in the IMME derived values about 14 keV. We would wellcome reactions to the proposal, to give then in the main table the experimental value, but with a flag; and the one influenced by IMME in the separate table.

But consider now the cases mentioned above, where the mass of an isobaric analogue is known from proton decay. The mass of the "mirror" of the mother isotope is always known. If no mass value is known for one more analogue, one can not apply IMME. But then one can use the observation, that the constants in IMME are somewhat regular functions of the mass number [23]. Thus, using interpolated values, one can derive an estimate for the mass difference of the mother nuclide and its analogue.

The resulting mass of the mother may be compared with the one following from the kind of systematics mentioned above. And we have cases where they do not quite agree. Here again, we might mention one in the main table, with a flag; the other in the supplementary table. We would, in this case, value opinions, which of the two should appear in the main table ...

For completeness, it may be mentioned that we made an elaborate study of the applicability of IMME, be it restricted to analogues of ground-states. As a result, we advocate to increase the errors for proton-rich nuclides derived from IMME alone by certain amounts. Also, earlier considerations did not take into account the difference between proton- and neutron-pairing energies, which one of the present authors noticed to have a not negligible influence on the constants in the IMME. Last but not least: in the second situation mentioned above, we think that care should be taken to select the constants used in deriving the mass of the mother in a way to agree with the mass of its "mirror". Authors of such work sometimes do not take this into account.

One might say, that possibly the IMME is not exact anyhow. In cases where dependable mass values for four or more isobaric analogues are known, though, it seems to work quite well. We therefore feel that, in the cases mentioned, the mirror nuclide should indeed be taken into account.

ACKNOWLEDGMENT

One of the authors (AHW) thanks the management of NIKHEF, Amsterdam, for the permission to use the facilities of this Institute and help of some of its members even rather long after his retirement.

REFERENCES

1. Audi, G. and Wapstra, A. H., *Nucl. Phys.* **A595**, 409-480 (1995).
2. Audi, G. and Wapstra, A. H., *Nucl. Phys.* **A565**, 1-397 (1993).
3. Kozier, K. S. et al., *Can. J. Physics* **58**, 1311-1316 (1980).
4. Bergström, I. et al., *"An attempt to measure the proton mass .."* presented at ENAM95, Arles, June 1995, Proceedings pp. 787-797 (1996).
5. Greene, G. L. et al., *Phys. Rev. Lett.* **56**, 819-822 (1986).
6. Deslattes, R. D. et al., *Annals of Physics* **129**, 378-434 (1980).
7. White, D. H. et al., *Nucl. Instr. Meth.* **66**, 70-76 (1968).
8. Brindhaban, S. A. et al., *Nucl. Instr. Meth.* **A340**, 436-441 (1994).
9. Audi, G. et al., *Nucl.Phys.* **A624**, 1-124 (1997).
10. Ames, F. et al *Nucl.Phys.* to be published.
11. Beck, D., et al., *Nucl.Phys.* **A626**, 343c-352c (1997).
12. Jensen, A. S., Hansen, P. G. and Jonson, B., *Nucl. Phys* **431**, 393-418 (1984).
13. Radon, T. et al., *Phys. Rev. Lett.* **78**, 4701-4704 (1997).
14. Beha, T., *Thesis* München 1995.
15. Kerscher, T. F., *Thesis* München 1996.
16. Davids, C. N., et al., *Phys. Rev. Lett.* **76**, 592-595 (1996).
17. Page, R. D., et al., *Phys. Rev.* **C53**, 660-670 (1996).
18. Davids, C. N., et al, *Phys. Rev.* **C55**, 2255-2266 (1997).
19. Irvine, R. J., at al., *Phys. Rev.* **C55**, 1621-1624 (1997).
20. Davids, C. N., et al., *Phys. Rev. Lett.* **80**, 1849-1852 (1998).
21. Hofmann, S., et al., *Zs. Physik* **354**, 229-230 (1996).
22. Comm. Nomencl. Inorg. Chem., *Pure and Applied Chemistry* **69**, 2471-2473 (1997).
23. Antony, M. S., Pape, A., et al, *Atomic Data and Nuclear Data Tables* **33**, 447-478 (1985), **34**, 279-299 (1986), **40**, 5-56 (1988).
24. Batchelder, J., et al., *Phys. Rev.* **C57**, 1042-1046 (1998).

Single particle behaviour at (and away from) stability

J. Duflo[a] and A. P. Zuker[b]

(a) Centre de Spectrométrie Nucléaire et de Spectrométrie de Masse (IN2P3-CNRS) 91405 Orsay Campus, France

(b) Physique Théorique, Bât27, IRES-CNRS/Université Louis Pasteur BP 28, F-67037 Strasbourg Cedex 2, France

Abstract. Short of proposing a full microscopic mass formula, we outline the steps that should make it possible. The construction rests on defining the nuclear monopole Hamiltonian, H_m that has to be extracted phenomenologically because of the bad saturation properties of the realistic forces. We propose a preliminary form of H_m that makes clear the origin of shell effects, and with only three paraters, reproduces the known spectra of particle and hole states on doubly magic cores to within 300 keV. Predictions are made for the yet unobserved levels around ^{132}Sn, and those associated to ^{22}O, ^{34,42}Si, ^{68,78}Ni, ^{100}Sn and for the particle-hole gaps in these nuclei. Finally we analyse the single particle gaps for the N=50 and 82 isotones, to conclude that the major closures are likely to persist away from stability.

Text-books on many body theory usually start from the assumption that the Hamiltonian may be split as $H = H_0 + (H - H_0)$, where H_0, is some "conveniently" defined single-particle field. There a difficulty here: H is two-body, and it is not clear how to extract from it a one body part. The mathematically correct solution is to extract all quadratic (two body) forms in the scalar products $a_r^+ \cdot a_s$. The resulting object, the monopole Hamiltonian H_m is closed under Hartree Fock (HF) variation (i.e., unitary transformations of the a operators). The correct H_0 is then the diagonal part of H_m

$$H_m^d = \frac{1}{2}\frac{\hbar\omega}{\hbar\omega_0} \sum_{k,l} V_{kl} m_k (m_l - \delta_{kl}), \tag{1}$$

where $V_{kl} = \sum_J (2J+1) V_{klkl}^J / \sum_J (2J+1)$ is the centroid calculated at $\hbar\omega_0$, and m_k is either the number of neutrons n_k, or protons z_k in orbit k. The expectation value of H_m^d for any state is the average energy of the configuration to which it belongs (a configuration is a set of states with fixed m for each orbit). In particular H_m^d reproduces the exact energy of closed shells (cs) and single particle (or hole) states built on them ($(cs) \pm 1$), since for this set ($cs \pm 1$) each configuration contains a single member. To arrive at this form we assume that a nucleus of $A = N + Z$ particles, and isospin $T = (N - Z)/2$, has the correct radius. In other words, we assume that HF variation of H_m with respect to $\hbar\omega$ yields a minimum at the observed radius $<r^2> \approx 0.9 A^{2/3}(1 - (2T/A)^2)$ [1], so that [2]

$$\hbar\omega = \frac{34.6 A^{1/3}}{<r^2>} \approx 40/\rho, \quad \rho = (A^{1/3}(1 - (2T/A)^2). \tag{2}$$

As we know that a typical matrix element goes as [4,5]

$$V(\omega)_{klmn} \cong \frac{\omega}{\omega_0} V(\omega_0)_{klmn}, \tag{3}$$

and then eq. (1) follows after a bit of work.

Unfortunately, H_m, as extracted from potentials describing the NN scattering data does not have the correct properties. Recent *exact* results for $A \leq 8$ [3] indicate that a potential describing *perfectly* the NN phase-shifts, supplemented by a three body term, produces excellent spectroscopy, but still has problems with the absolute bindings, the symmetry energy and the spin-orbit splitting, terms essentially given by H_m. Very much the same happens in shell model calculations, where the damage due to bad saturation increases with particle number. The problem cannot be solved at present, but it can be bypassed cleanly by the separation $H = H_m + H_M$. All the trouble comes from H_m and can be phenomenologically corrected. Then, the multipole term H_M produces detailed agreement with the observables *provided it is derived from potentials consistent with the NN phase-shifts* [4,6,7,5]. The necessary condition that H_m must satisfy, is to reproduce properly the $(cs) \pm 1$ excitation spectra.

To understand the crucial importance of the quadratic effects in H_m, let us consider two shells h, p of degeneracies D_p, D_h on top of a core. Then

$$H_m = \varepsilon_p m_p + \varepsilon_h m_h + \frac{1}{2} V_{pp} m_p(m_p - 1) + \frac{1}{2} V_{hh} m_h(m_h - 1) + V_{ph} m_p m_h. \tag{4}$$

To make things more tranparent, let us introduce the operators

$$\Gamma_{ph} = \frac{m_p D_h - m_h D_p}{D_p + D_h}, \tag{5}$$

which produce unit splittings betwen orbits p, h: Then, we can recast eq. (4 as

$$H_m = \varepsilon_0 + \frac{1}{2} m(m-1) W_0 + \Gamma_{ph}[\varepsilon_1 + (m-1) W_1] + \Gamma_{ph}^2 W_2. \tag{6}$$

Take now a core of ^{12}C, call the $p_{1/2}$ orbit on top of it h, and the next two, $s_{1/2}$ and $d_{5/2}$, p. When 4 particles are allowed in this space, we have a simple model for ^{16}O, where one can infer from binding energies that the lowest ph states should come at around the value of the "gap" i.e., BE(^{17}O) +BE(^{15}O)−2BE(^{16}O)=11.5 MeV. Experimentally, there is a doublet near 6 MeV, consisting of a 3^- ph state and a 0^+ state of well established 4p-4h nature. The symmetry energy (i.e. a $T(T+1)$ term) brings easily down the 3^- from 11.5 to 6 Mev. What about the 0^+: If we neglect the quadratic term in Γ_{ph}^2 in eq. (6, it would come at about $4 \times 11.5 = 46$ Mev. By including it, and borrowing numbers from the calculation that first gave a consistent picture of the low lying states in the region [8], the average energies for $kp - kh$ excitations are roughly those of Fig. (1).

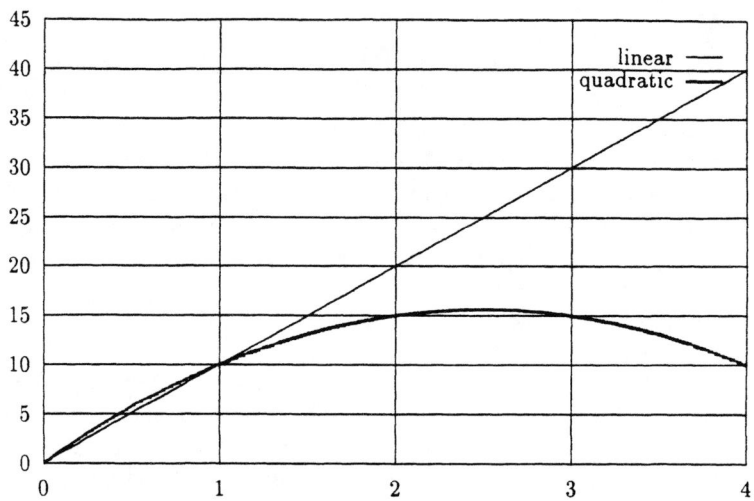

FIGURE 1. kp-kh excitation energies for two shells

Let us describe briefly the indispensable ingredients of the general H_m, bearing in mind this simple example that suggests we separate "constant" (bulk terms in total m) from linear an quadratic contributions. We shall certainly need the kinetic energy ho, compensated by an attraction which we take to go as the leading monopole term in the decomposition of the interaction (gm) [5,9]. The combination responsible for shell effects (p is the principal quantum number, an overall factor $\rho \equiv \omega/\omega_0$ is assumed),

$$gm - 4ho = (\sum_p \frac{n_p + z_p}{\sqrt{(p+2)(p+1)}})^2 - \sum_p \frac{n_p + z_p}{(p+2)(p+1)} - 4(p+3/2)(n_p + z_p), \quad (7)$$

has no bulk or sufuce contributions (we ignore them in this talk).

Next we need linear operators that we take to be either of "$l \cdot s$" type (i.e., they split spin-orbit (so) partners $l_<, l_>$), or "$l \cdot l$" type (i.e., they split the l centroids). Finally, we should decide on the quadratic operators, which produce basically modulations of the above ones. As the data demand a special treatment for the so partners with $l = p$, we introduce the following sets of orbits with degeneracies D_x:

$$\begin{aligned}
p, &\quad \text{the full shell,} &\quad D_p &= (p+1)(p+2) \\
p_>, &\quad \text{largest orbit,} &\quad D_{p_>} &= 2(p+1) \\
p_<, &\quad \text{its } l \cdot s \text{ partner,} &\quad D_{p_<} &= 2p \\
q, &\quad p_> + p_<, &\quad D_q &= 2(2p+1) \\
r, &\quad p - q, &\quad D_r &= p(p-1) \\
R, &\quad p - p_>, &\quad D_R &= p(p+1),
\end{aligned}$$

define the following set of operators, all simple variants of $l \cdot s$, and $l \cdot l$, (for the r-orbits we use $m_l = m_{l_<} + m_{l_>}$):

$$a_p = \frac{p}{p+2}\Gamma_{p_>p_<}, \quad b_p = \Gamma_{rq}, \quad s_p = \Gamma_{p_>R}, \quad v_p = \Gamma_{p_<r} \tag{8a}$$

$$d_p = \sum_{l \neq p} \frac{l}{p+2}\Gamma_{l_>l_<}, \quad e_p = \sum_{l \neq p} \frac{[2l(l+1) - (p-2)(p+1)]\, m_l}{2D_r} \tag{8b}$$

and select the ones we shall need

$$\mathcal{O} = \sum_p \mathcal{O}_p, \quad ff'g\mathcal{O} = \sum_{[pp'][xx']} \frac{b_p^x \mathcal{O}_{p'}^{x'}}{(D_p D_{p'})^{1/4}}, \quad x, x' \equiv n \text{ or } z \tag{9}$$

where $\mathcal{O} = a, b, v, d, e$, the denominators ensure splittings of order unity, and the sums are restricted so that

$$\begin{aligned}
ff'g\mathcal{O} &= ffi\mathcal{O}, \text{ for } x = x', \, p = p' \\
ff'g\mathcal{O} &= ffc\mathcal{O}, \text{ for } x = x', \, p \neq p' \\
ff'g\mathcal{O} &= zni\mathcal{O}, \text{ for } x \neq x', \, p = p' \\
ff'g\mathcal{O} &= znc\mathcal{O}, \text{ for } x \neq x', \, p \neq p'
\end{aligned}$$

The operators $\Gamma(ij)$ can act in either the same or different fluids (ff or nz) and either the same or different major shells (i or c). Eqs. (9) achieve maximum symmetry.

The quadratic operators in eq. (9) have the property of producing negligible splittings in hole states for HO closures. There is an exception that the data seem to demand. We call it simply ffi because it is the only representative of its class that will appear:

$$ffi = \sum_{xp} \frac{m_{p_>}^x}{D_{p_>}}\left(s_p^x - \frac{D_p - m_p^x}{D_p - 1}\right). \tag{10}$$

The guiding idea in writing these forms is that in the harmonic oscillator closures (HO), only the linear terms act. Since the major EI (extruder intruder) closures at N,Z=6(sic), 14, 28, etc. differ from HO ones only by the presence of a $p_>$ shell—whatever difference there is in the $(cs) \pm 1$ spectra with respect to the linear contributions must be due to a quadratic in which one of the factors contains a $p_>$ operator. (The "driver' b in the equation above.)

It proves convenient to regroup terms, and we have adopted the following, minimal, combination (using H'_m as a reminder that bulk terms have been omitted):

$$-\rho H'_m = c_1(d + znce + znie) + c_2(znib + ffi) + c_3\{W - 4K + 2[a + b(1 - 3/\rho)] + 2zncv + ffcv\}, \quad (11)$$

which yields $(c_1, c_2, c_3)=(12.90, 4.75, 8.03)$ with a deviation (rmsd) of 282 KeV, in a fit to the experimental excitations adopted in table I. The parameters change to $(12.91, 4.45, 8.17)$ with rmsd=244 KeV, if only the $A > 60$ data are kept, and to $(14.71, 5.18, 8.03)$ with rmsd=280 KeV for $A < 60$. These last parameters when used for $A > 60$ yield rmsd=324 KeV, and no important differences with the numbers in Table I, indicating a remarkable stability of the fit with respect to changes in input data.

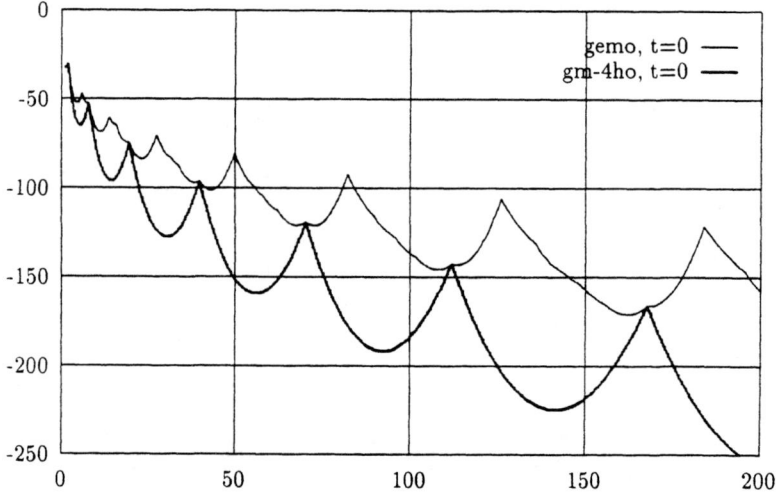

FIGURE 2. Shell formation mechanisms

In Fig 2 we can follow the energy for $T = 0$ states as a function of N. The $gm - 4ho$ in eq. (7) yields the shell effects associated with the HO closures. When all the contributions are added (*gemo*), the HO closures go (except at the very beginning) and are replaced by the EI ones.

Energies of the of the $(cs) \pm 1$ states around existing and possible new closures can be retrieved from ref. [10]. Because of lack of space we shall not comment on them, except to say that the agreement with observation is good enough to give plausibility to the predictions.

To close, the last figure compares the predictions for the single particle gaps around N=50, with the data from [11]. We need here the symmetry energy

$-32[4T(T+1)](1-1.6/\rho)/A$ taken from [9], and a perturbative estimate of multipole correlations, which amounts to a constant shift of about 1 MeV. However we have not estimated the correct occupancies of the proton orbits, which will smooth the residual subshell effects. The agreement is quite satisfactory, and there no sign of shell erosion in going away from stability. Very much the same happens for the closures at N,Z=82 and Z=50.

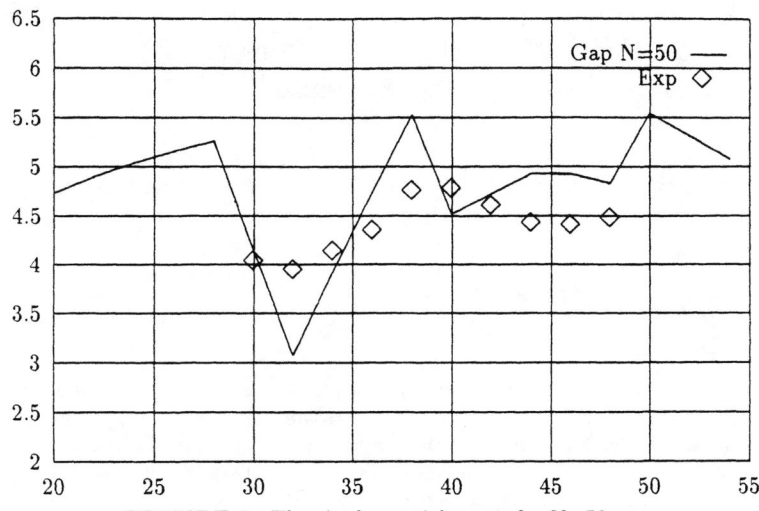

FIGURE 3. The single particle gaps for N=50

REFERENCES

1. J. Duflo Nuc. Phys. **A576**, 29 (1994)
2. A. Bohr and B. Mottelson, *Nuclear Structure* vol I (Benjamin, Reading, 1964).
3. See S. Pieper contribution to ENAM98 (this volume)
4. A. Abzouzi, E. Caurier et A. P. Zuker, Phys. Rev. Lett. **66**, 1134 (1991).
5. M. Dufour and A. P. Zuker, Phys. Rev. C **54**, 1641 (1996)
6. G. Martínez-Pinedo, A. P. Zuker, A. Poves and E. Caurier, Phys. Rev. C **55**, 187 (1997)
7. A. P. Zuker, in Contemporary Shell Model, edited by X. W. Pan (Springer Lecture Notes 482, 1997)
8. A. P. Zuker, B. Buck and J. B. Mc Grory Phys. Rev. Lett. **21**, 39 (1968).
9. J. Duflo and A. P. Zuker, Phys. Rev. C **52**, R23 (1995)
10. J. Duflo and A. P. Zuker, http://csnwww.in2p3.fr/amdc/ (theory file du_zu_ph.ps)
11. G. Audi and A. H. Wapstra, Nuc. Phys. **A565**, 1 (1993)

Persistence of the N=28 Shell Closure far from Stability

F. Sarazin [a], H. Savajols [a], W. Mittig [a], G. Auger [a], D. Baiborodin [c],
A.V. Belozyorov [c], C. Borcea [d], A. Gillibert [b], A.S. Lalleman [a],
M. Lewitowicz [a], S.M. Lukyanov [c], F. de Oliveira [a], N. Orr [f],
Y.E. Penionzhkevich [c], D. Ridikas [a], P. Roussel-Chomaz [a], H. Sakurai [e],
A. de Vismes [a]

a GANIL, B.P.5027, F-14076 Caen cedex 05, France
b CEA/DSM/DAPNIA/SPhN, CEN Saclay, F-91191 Gif-sur-Yvette, France
c LNR, JINR, Dubna, P.O. Box 79, 101 000 Moscow, Russia
d IAP, Bucharest, Roumania
e RIKEN Wako, Saitama 351-01 Japan
f LPC-ISMRA, Bd. Maréchal Juin, 14050 Caen cedex, France

Abstract. The masses of 16 neutron-rich nuclei in the mass range from 35 to 45 have been measured using a direct time of flight technique following the fragmentation of a ^{48}Ca beam at 60 MeV/nucleon. The masses of ^{35,36}Mg, ^{38}Al, ^{39,40}Si, ^{42,43}P and ^{43,44}S are reported for the first time. Preliminary analysis shows that the N=28 shell closure persists, even if weakened by the large neutron excess.

I. Introduction

One of the present fundamental question for nuclear structure is whether the magic numbers are universal or whether they change in certain regions far from stability or even disappear altogether. An example of such a magicity breaking is given by the N=20 neutron rich nuclei where a collapse of the standard N=20 shell closure have been observed [1-2]. Recently, there has been an increase of interest in the N=28 isotones far from stability, motivated by the possible existence of anomalies in the shell closures, as already found for N=20.

An experimental observable that may give a first answer to this question is the separation energy of the last two neutrons, S_{2n}, which can be deduced from nuclear masses.

Indeed, the nuclear masses are directly related to the binding energy of the nucleus as a consequence of the relation $E=mc^2$. One can see the evolution of the binding energy of isotopes by simply looking at its derivative which is more or less the S_{2n} observable, i.e.

$$S_{2n} = M(A-2,Z) - M(A,Z) + 2M_n .$$

CP455, *ENAM98: Exotic Nuclei and Atomic Masses*
edited by B. M. Sherrill, D. J. Morrissey, and Cary N. Davids
© 1998 The American Institute of Physics   1-56396-804-5/98/$15.00

We have then performed at GANIL a mass measurement experiment by using a direct time of flight technique which will be briefly described in the next section. Our goal was to investigate the N=20 and N=28 neutron shell closures for nuclei from Carbon (Z=6) to Calcium (Z=20). The production of these neutron-rich nuclei have been carried out by the fragmentation of a ^{48}Ca primary beam at 60 MeV/nucleon on a Ta target located in the SISSI device.

II. Direct mass measurements by the time of flight method

The technique consists of the use of the long flight path between a start detector located near the production target and a stop detector following a high resolution spectrometer [3-4]. In such a device, the mass can be deduced by the relation

$$B\rho = \frac{\gamma\, m\, v}{q}$$

where $B\rho$ is the magnetic rigidity of a particle of mass m, velocity v and charge q. The velocity is simply the ratio of the flight path over the time of flight measured between the start and stop detectors.

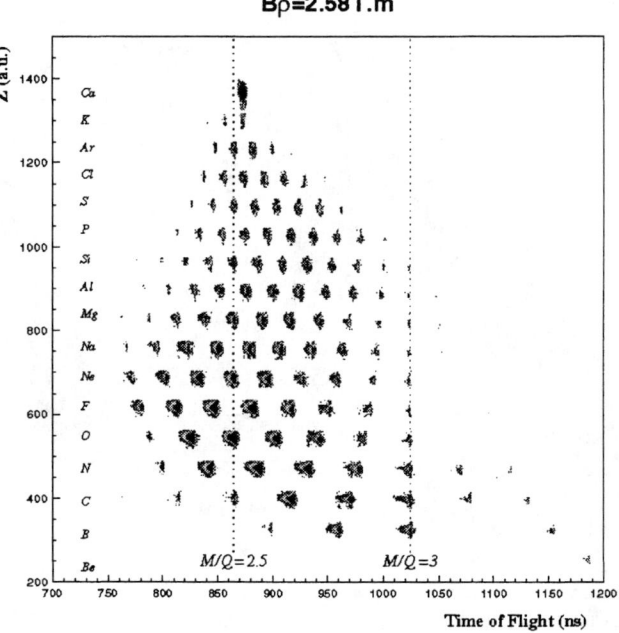

FIGURE 1. Transmitted nuclei for the $B\rho$=2.58T.m setting.

In the case of the SPEG [5] spectrometer, the flight path of 82m (i.e. time of flight is around 1 µs) allows a time of flight resolution to be of the order of typically $2,5.10^{-4}$. Combined with the intrinsic resolution of SPEG (10^{-4}), it leads to a mass resolution of $\sim 3.10^{-4}$. A large number of well-known nuclei are transmitted (Fig. 1) in a same setting of the beam line and spectrometer which provide a calibration from which the unknown masses are derived. The final uncertainties range from 100keV (thousands of events) to 1MeV for the most exotic nuclei (several tens of events), which should be sufficient to distinguish between different theoretical approaches.

III. Preliminary results and tendencies

During the experiment, two field settings of the spectrometer and beam line were tuned. The first, Bρ=2.58 T.m, provided broad element and isotopic distributions up to ^{46}Cl. The second setting of 2.73 T.m gave rise to more restricted distributions with higher yields for nuclei in the region of the neutron-rich sodium and magnesium isotopes. The preliminary results of the first setting (Bρ = 2.58 T.m) are presented in this paper.

We measured 8 masses with a precision better than before and 8 new masses with a precision of better than 1MeV (Fig. 2).

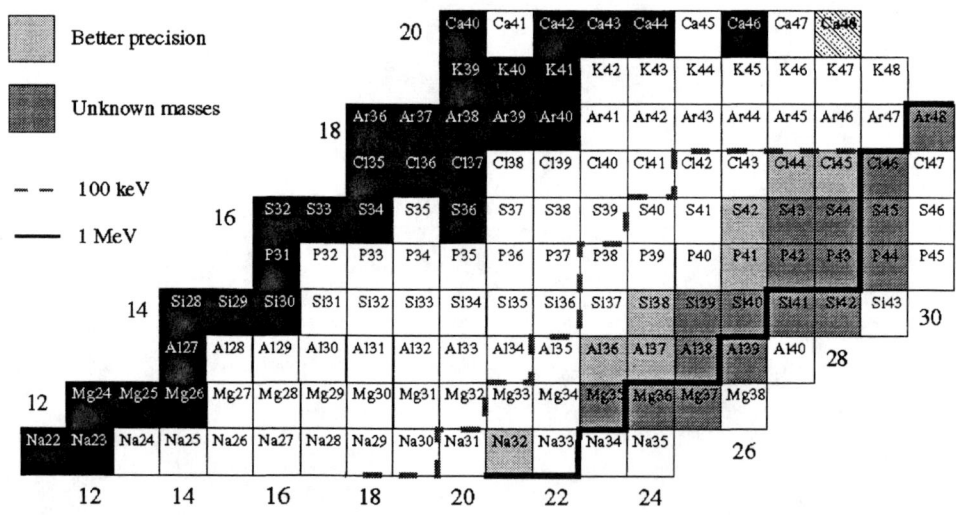

FIGURE 2. New measured masses. The dashed and solid lines delimit respectively masses of nuclei which are known with a precision better than 100 keV and 1 MeV.

Among them, the new measurements of neutron-rich nuclei around N=28 allow us to express some preliminary remarks concerning this shell closure.

The experimental two neutron separation energies, S_{2n}, in this region (Fig. 3) have some common features for different Z values. Nevertheless, if we assume that the S_{2n} for Ca isotopic chain is a good reference for a standard shell structure (dashed lines in Fig. 3), and if we compare this behavior to the other isotopes in this region, an excellent agreement is found for all Al, Si, Ar and K isotopes. However, overbinding energies are observed for several P, S and Cl isotopes between the N=20 and N=28 shell closures, which might be explained as deformation effects. These results are consistent with recent measurements of B(E2) values and excitation energies of the first 2^+ state in the S isotopes [6-8], where the evidence of moderate deformation have been pointed out.

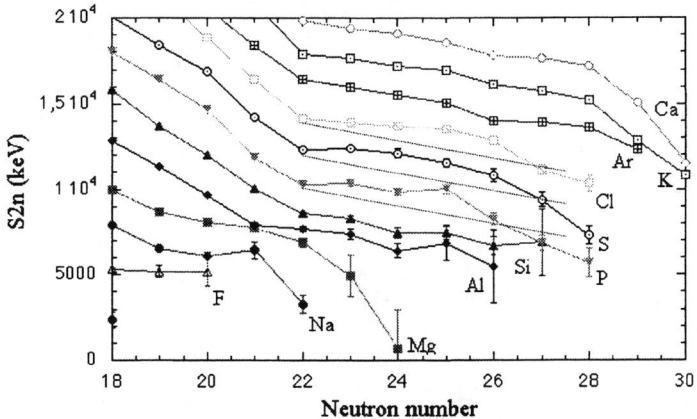

FIGURE 3. Experimental S_{2n} energies near the N=28 isotones. The dashed lines represent the Ca reference.

The present measurements of the masses of ^{43}P, ^{44}S and ^{45}Cl are of great interest since they show that the N=28 shell closure seems to persist.

In Fig. 4 a comparison is made between the two neutron separation energies measured in our present work for Si, P, S, Cl and Ar isotopes with shell model calculations [10-11] where the valence space includes the *sd* shell for protons and the *pf* shell for neutrons without any restriction. An excellent agreement is found between experimental data and this theoretical calculation, which might lead to a persistance of the N=28 shell. Moreover, these calculations predict also that ^{40}S and ^{42}S present some moderated deformation which is confirmed the S_{2n} values.

Finite-range liquid droplet macroscopic calculations (FRLDM) [12] have been performed and compared to the experimental values for Ca, S and Mg isotopes (Fig. 5). In the case of Ca isotopes, one can observe two minima centered at N=20 and N=28 and corresponding to a minimisation of microscopic effect due to this two shell closures. In contrary, for the Mg isotopes, an increase is observed near N=20 where a decrease would be expected according to this magic number. This increase corresponds to deformation effect observed in this region for the so called « island of inversion » [1-2-12]. The S isotope curve presents some common behavior with the Ca isotope near N=28. These results are consistent with a signature of a shell closure even if the N=28 shell gap is weakened for the ^{44}S.

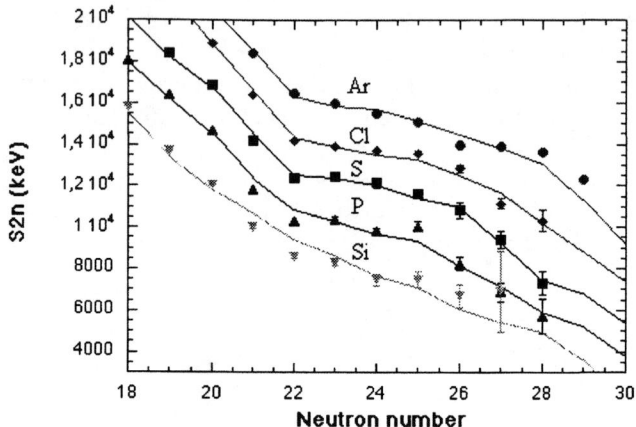

FIGURE 4. Experimental S_{2n} energies for Si, P, S Cl and Ar isotopes plotted together with shell model calculations.

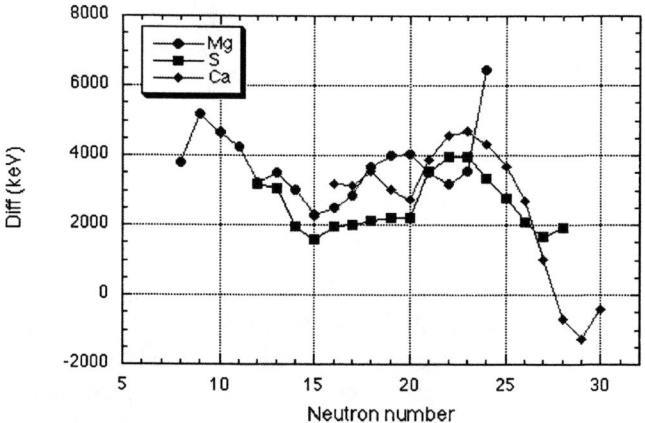

FIGURE 5. Comparison of experimental and FRLDM mass excess. The curves show the difference between these two quantities for Mg, S and Ca isotopes.

Such conclusion is opposite to that coming from mass systematic extrapolations, which lead almost constant S_{2n} near N=28, interpreted as a signature of a new region of deformation.

IV. SUMMARY

In summary, we have measured the mass of 16 nuclei, 8 masses with a precision better than it was obtained before and 8 new masses with a precision of better than 1 MeV. We used a direct time of flight technique following the fragmentation of a 48Ca beam at 60 MeV/nucleon. Results are in particular relevant in the region of β-unstable nuclei, near N=28, allowing us to study the persistence of the shell closure for large neutron excess (compared to the experimental evidence for an « island » of deformed nuclei located near the closed shell N=20 for Z < 14), and also because these neutron-rich nuclei play an important role in the nucleosynthesis of the heavy Ca-Ti-Cr isotopes. We observed an overbinding in the S_{2n} for P, S and Cl isotopes between the two magic numbers N=20 and N=28, indicating a new region of deformed nuclei near N=28. The measurement of the N=28 nuclei like ^{43}P, ^{44}S and ^{45}Cl and the comparison between shell model calculations and finite-range-liquid drop model allows us to suggest the persistence of this magic gap, even if eroded by the large neutron excess.

REFERENCES

[1] C. Detraz et al., Phys. Rev. **C 19**, 171 (1978).
[2] N.A. Orr et al., Phys. Lett. **B 258**, 29 (1991).
[3] W. Mittig et al., Nuclei far from stability, AIP Conf. Proc., Vol. **164** (AIP, New York, 1988) p.11.
[4] A. Gillibert et al., Phys. Lett. **B 176**, 317 (1986).
[5] L. Bianchi et al., Nucl. Inst. and Meth. **A 276**, 509 (1989).
[6] T.R. Werner et al., Phys. Lett. **B 335**, 259 (1994).
[7] H. Scheit et al., Phys. Rev. Lett. **77**, 3967 (1996).
[8] T. Glasmacher et al., Phys. Lett. **B 395**, 163 (1997).
[9] J. Retamoza et al., Phys. Rev. **C 55**, 1266 (1997).
[10] F. Nowacki, private communication
[11] P. Möller et al., Atomic Data and Nuclear Data Tables, Vol. **59**, 185 (1995).
[12] E.K. Warburton, J.A. Becker and B.A. Brown, Phys. Rev. **C 41**, 1147 (1990).

Recent Measurements of Nuclear Moments far from Stability

Matthias Keim

CERN, CH-1211 Geneva 22, Switzerland

Abstract. Measurements of nuclear moments far from stability have been performed mainly applying laser spectroscopy and β-NMR spectroscopy. A short overview of the experimental techniques are given. The β-NMR experiments on the quadrupole moments of sodium isotopes and of the magnetic moment of ^{11}Be, both using optically pumped radioactive beams, are discussed in more detail.

I INTRODUCTION

The investigation of nuclear moments has proven to be a powerful method to study the structure of the atomic nucleus. While magnetic moments are especially sensitive to the single-particle structure of the outer nucleon, the electric quadrupole moments reveal information about collective properties. Evidence for regions of strong deformation between the shell closures were actually discovered by systematic studies of nuclear quadrupole moments.

Even though the study of nuclear moments started more than 50 years ago, a quantitative analysis of the experimental data was often restricted by the limited theoretical models. Only in few cases where simplified single-particle approaches could be applied to nuclei close to the shell closures, an adequate description could be obtained.

The recent development of large-scale nuclear shell model calculations for light nuclei up to the fp-shell improved this situation considerably. It became possible to reproduce various observables such as nuclear moments with a high predictive power. The comparison with these calculations opens up new perspectives for the analysis of experimental data especially in the study of nuclei far from stability.

In parallel to the improvement of theoretical models several sensitive experimental techniques have been developed to measure nuclear moments of short-lived nuclei. Laser-spectroscopic methods have been successfully applied to heavy nuclei with production rates of even less than one per second. Furthermore, by the improvement of high resolution techniques optical spectroscopy could more and more be applied to light systems where the effects due to nuclear structure become tiny.

A whole group of experimental techniques is exclusively devoted to the investigation of exotic light nuclei. The methods employ polarized radioactive beams to perform β-ray detected nuclear magnetic resonance spectroscopy (β-NMR). They have reached a sensitivity sufficient to yield precise results for the nuclear moments of nuclei sometimes even situated at the drip lines.

II EXPERIMENTAL METHODS

A Optical Spectroscopy

In optical spectroscopy the nuclear moments are deduced from the hyperfine structure splitting of atomic resonances. The hyperfine structure can be parametrized in terms of the interval factors A and B which are essentially given by the product of the nuclear moments and the atomic hyperfine fields $H(0)$ and V_{ZZ}. In addition the differences of nuclear charge radii $\delta \langle r^2 \rangle$ can be derived from optical data. These are related to the isotopic field shifts by an isotope-independent electronic factor.

This electronic factor and the hyperfine field of the quadrupole interaction V_{ZZ} show a strong dependence on the nuclear charge and become very small for light nuclei. Therefore, optical spectroscopy is mainly used to investigate nuclei in the heavy or medium mass region where large effects are observed.

The experimental methods which have been developed during the long history of atomic spectroscopy extend from the simple irradiation of a gas cell with the light of a spectral lamp to elaborate on-line techniques at radioactive beam facilities employing a complicated experimental setup.

In the last years two concepts of laser spectroscopy have become particularly important for experiments on short lived and weakly produced isotopes.

1. Resonance ionization spectroscopy has been applied to neutron-deficient gold [1] and platinum [2] isotopes to study deformation properties and shape coexistence in the region of light mercury. Another experimental program was devoted to the investigation of fission isomers. In spite of extremely low production rates it became possible to measure the quadrupole moments and the radii of the super-deformed isomers ^{240f}Am and ^{242f}Am [3].

2. Collinear fast beam laser spectroscopy was used to measure moments and radii of a long series of lutetium isotopes from ^{162}Lu to ^{179}Lu [4]. The results extend considerably the systematics of data in the upper rare earth region, a region where extensive laser spectroscopy experiments were performed previously.

The high resolution of collinear laser spectroscopy makes it possible also to investigate lighter nuclei with sufficient precision. In experiments on short-lived argon isotopes [5] moments and charge radii have been derived to study the N=20 shell closure in the vicinity of the doubly magic ^{40}Ca.

A less common method based on laser spectroscopy in a buffer gas cell, was developed for measurements on radioactive bismuth isotopes [6]. With the results for the quadrupole moments and radii core polarization properties close to ^{208}Pb were studied systematically.

B β-NMR Spectroscopy with Polarized Beams

For the investigation of nuclear moments of light nuclei β-NMR methods using polarized radioactive beams have become more and more important in the last years. Results for nuclei in the p-shell such as boron, carbon, nitrogen and oxygen isotopes published by the two RIKEN groups [7] and the MSU group [8] were obtained with polarized beams from fragment separators. The experimental program at ISOLDE where a low-energy beam is polarized by optical pumping with laser light was concentrated on the investigation of the quadrupole moments of sodium isotopes and the magnetic moment of ^{11}Be. These two experiments will be described in section II B 2 and II B 3.

1 β-NMR technique

The spectroscopic technique is basically the same for all these experiments. In a typical setup (see figure 1) polarized nuclei are implanted into an appropriate crystal situated in a strong magnetic field of a few kG. The angular distribution of their Gamow Teller β-decay shows an asymmetry which is proportional to the degree of the nuclear polarization. This is detected by two scintillation-counters placed at 0° and 180° relative to the magnetic field.

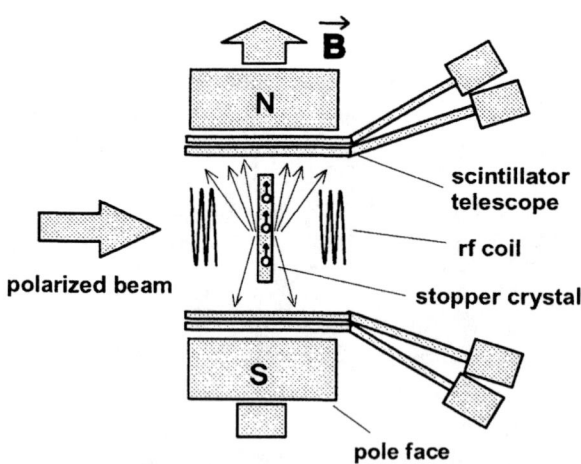

FIGURE 1. Experimental setup for β-NMR spectroscopy

The sample is now irradiated with a radio frequency field. If the frequency matches the Zeeman splitting of the nuclear spin system in the magnetic field, the levels are coupled and their population is equalized. This means that the polarization is reduced which is observed as a change in the β-asymmetry.

Without any additional interaction the equidistant Zeeman splitting is given by the Larmor frequency ν_L which yields with known magnetic field strength the nuclear g-factor $g_I = \mu_I/I$.

For the measurement of quadrupole moments the interaction of the implanted nuclei with the intrinsic electric field gradient of the host lattice is investigated. These field gradients are present at certain sites of crystal lattices with non-cubic symmetry. The interaction results in a shift of the Zeeman level energies and leads to a $2 \times I$ fold splitting of the NMR-signal, which is proportional to the product of the field gradient and the quadrupole moment.

2 The quadrupole moments of sodium isotopes

The isotopes of neon, sodium and magnesium around $N = 20$ show properties which are in contrast to the behaviour at a spherical shell closure. The unexpected large binding energies found in direct mass measurements [9] gave rise to the assumption that nuclear deformation and not the shell gap stabilizes these nuclei. This idea was further supported by the measurement of the energy of the first 2^+ level [10] and the B(E2) value [11] for the $N = 20$ isotope ^{31}Mg yielding a value for the quadrupole deformation of $\beta_2 \approx 0.5$.

In a microscopic picture these properties are related to configuration mixing with intruder states from the fp-shell. The population of these states before the sd-shell orbitals are completely occupied is driven by an enhanced proton-neutron interaction due to the large neutron excess [12].

This region of isotopes, the so called "island of inversion", which is centered around ^{31}Na, has been investigated at the ISOLDE on-line mass separator by the β-NMR measurement of the quadrupole moments of sodium isotopes. The nuclear polarization is produced by optical pumping with circularly polarized laser light [13].

Three different host lattices were used in the measurements. Quadrupole moments between 5 mbarn and 150 mbarn made it necessary to use lattices with strong (LiNbO$_3$) and weak (magnesium) electric field gradients. To obtain absolute values a measurement of ^{28}Na in a NaNO$_3$ single crystal with known intrinsic field gradient for implanted sodium was performed. Spectra of ^{28}Na in these three lattices are shown in figure 2.

The experiments yielded the quadrupole moments of $^{20,26-31}$Na. They are shown in figure 3 and compared to shell-model calculations [14]. The first set of theoretical values is based on a pure sd-shell model space using the USD interaction. The effective charges $e_p = 1.35$ and $e_n = 0.35$ are used, and the radial integrals are obtained with harmonic oscillator wave functions [14,15].

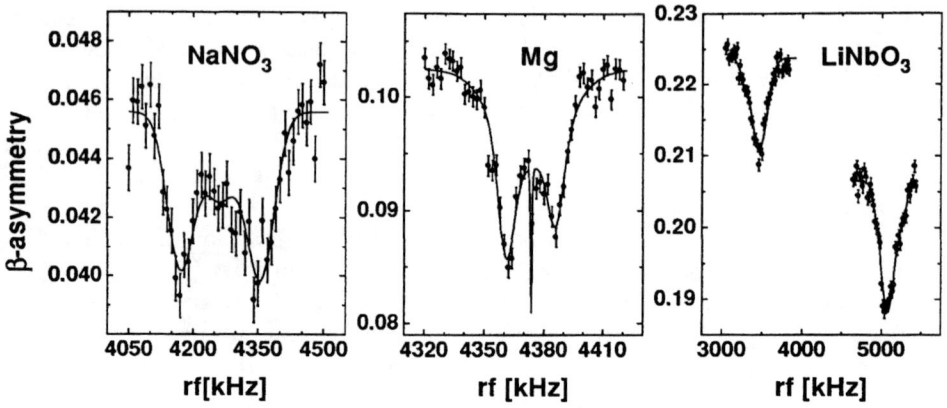

FIGURE 2. β-NMR spectra of ^{28}Na in single crystals of NaNO$_3$, Mg and LiNbO$_3$

The experimental results are well reproduced up to mass $A = 28$. For these isotopes a pure sd-shell structure can be assumed from β-decay and binding energy studies [16]. Going to heavier masses the description becomes less satisfactory. Especially for the isotope ^{30}Na the deviation of about 240 mbarn exceeds by far what is expected from the systematics of sd-shell nuclei.

The second set of theoretical results includes admixtures from the fp-shell. The calculations have been performed for the isotopes ^{29,30,31}Na where a considerable configuration mixing can be assumed. For ^{30}Na the description is considerably improved. This suggests that already in this isotope intruder states affect the ground state structure more than expected from former studies [16,17].

3 The magnetic moment of ^{11}Be

In the last years halo nuclei have been the subject of extensive experimental and theoretical investigations. The halo structure was observed for nuclei with low binding energy and an exotic N/Z ratio.

The isotope ^{11}Be is up to now the only known case of a one-neutron halo nucleus. First measurements of the high-energy interaction cross section yielded a large root-mean-square matter radius which was taken either as an indication for a halo structure or for deformation properties [18].

Later, the matter distribution was extracted from scattering experiments at different energies and with various targets [19]. An extended tail of low matter density which extends far beyond the radius of the nuclear core clearly confirmed the existence of a halo.

In contradiction to the standard shell model the position of the two bound states in ^{11}Be, a $1/2^+$ and a $1/2^-$ state, is inverted. Recent calculations [20] employing

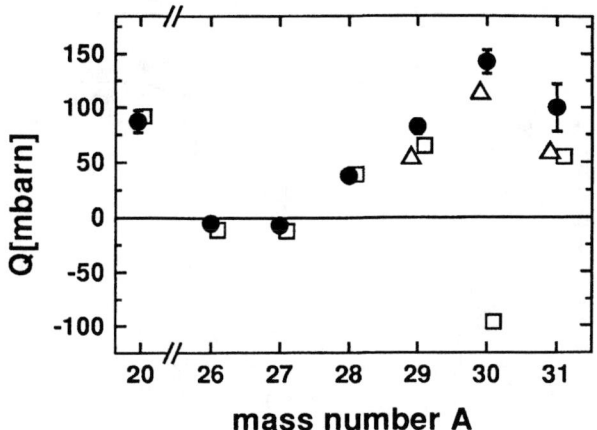

FIGURE 3. Electric quadrupole moments of sodium isotopes. Experimental values (solid dots) are compared to calculations on a pure sd-shell model space (rectangles) and including fp-shell admixtures (triangles).

a variational shell model approach were able to reproduce the $1/2^+$ ground state as well as the extended matter distribution. The admixture of a $[(^{10}\text{Be})\,2^+ \times d_{5/2}]$ component to the dominant $[(^{10}\text{Be})\,0^+ \times s_{1/2}]$ configuration and Pauli blocking due to the presence of the last neutron [21] were considered to be the underlying mechanisms for the inversion of levels.

The magnetic moment of the halo nucleus ^{11}Be has been measured by β-NMR spectroscopy applied to optically polarized nuclei. An intense beam of ^{11}Be was provided by a laser ion source [22] recently developed at ISOLDE. Other than in the sodium measurements the efficiency of the optical pumping was limited by the low available laser power in the ultra violet. The half-life of 13.8 s made it necessary to reduce the relaxation of nuclear polarization by cooling down the host lattice (a beryllium single crystal) to about 50 K. An experimental spectrum is shown in figure 4. The preliminary result of the magnetic moment is $\mu_I = -1.6814(13)\mu_N$.

The magnetic moment of ^{11}Be is in particular sensitive to the admixture of the $[(^{10}\text{Be})\,2^+ \times d_{5/2}]$ configuration. This was shown by Suzuki et al. [23] who calculated the magnetic moment including meson exchange current corrections for different amplitudes of this admixture. Their results obtained (a) without quenching ($g_s^{eff} = g_s$) and (b) with an effective g_s factor $g_s^{eff} = 0.85 g_s$ as deduced from the equivalent state in ^{15}C are shown in Figure 5. The experimental value is included in the figure.

The calculations yield an increasing absolute value for the magnetic moment for increasing $s_{1/2}$ contribution. Figure 5 shows that if no quenching is assumed the contribution of the $2^+ \times d_{5/2}$ configuration would be around 50%. The more realistic model including quenching reaches the value $\mu_I = -1.62\mu_N$ for a pure $s_{1/2}$ wave function. This is still too small to reproduce the experimental result of $\mu_I = -1.68\mu_N$.

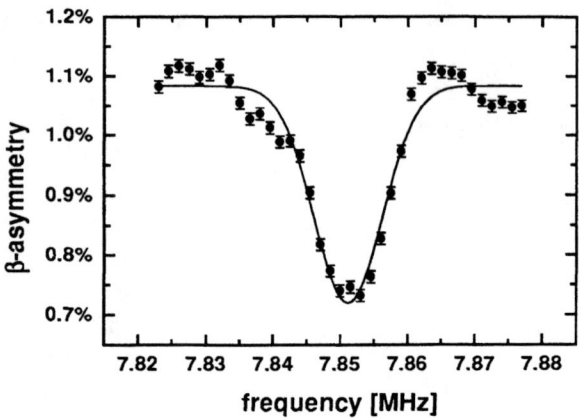

FIGURE 4. β-NMR spectrum of ^{11}Be measured in a single crystal of beryllium.

When comparing the experimental results to these calculations the uncertainties in the determination of the effective g_s factor have to be considered. Especially the influence of the halo structure on the quenching is not taken into account.

A reduced quenching compared to the case of ^{15}C due to the smaller overlap of the outer neutron with the core would result in a situation somewhere in between the cases shown in figure 5. This would suggest a ground state of ^{11}Be which is mainly $s_{1/2}$ with a correspondingly smaller $d_{5/2}$ contribution.

III CONCLUSION

In the last few years measurements of nuclear moments far from stability were performed with several experimental techniques. Methods of laser spectroscopy yielded important results in the heavy mass region. Furthermore, experimental techniques could be improved so that measurements even on light systems like argon became possible. The additional information about nuclear charge radii which can be extracted from laser spectroscopic data is an ideal complement to the nuclear multipole moments to analyze the nuclear shape.

A large progress on the experimental as well as on the theoretical side was achieved in the study of light nuclei in the p-shell and the sd-shell. The high predictive power of advanced theoretical approaches made it possible to obtain a detailed description of the nuclei by comparison with experimental results.

β-NMR techniques and methods to produce polarized beams of exotic isotopes have become an important tool to investigate moments of light nuclei. Experiments far from stability partly extending to drip line nuclei were performed. Here, the measurement of the magnetic moment of ^{11}Be is one of the outstanding results.

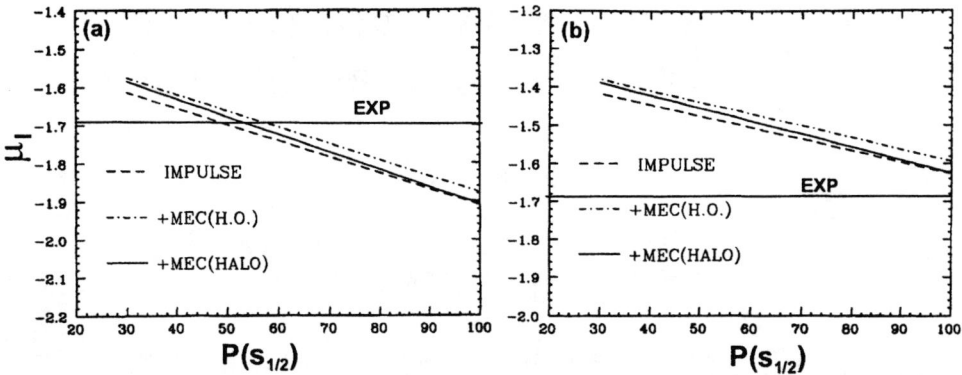

FIGURE 5. Magnetic moment of ^{11}Be as a function of the $s_{1/2}$ contribution to the ground state [23] compared to the experimental result. Left part: $g_s^{eff} = g_s$, right part $g_s^{eff} = 0.85 g_s$.

REFERENCES

1. Le Blanc F. et al., *Phys. Rev. Lett.* **79**, 2213 (1997).
2. Le Blanc F. et al., contribution to this conference.
3. Backe H. et al., *Phys. Rev. Lett.* **80**, 920 (1998).
4. Georg U. et al., submitted to *Europ. Phys. J. A*.
5. Klein A. et al., *Nucl. Phys. A* **607**, 1 (1996).
6. Kilgallon J. et al., *Phys. Lett. B* **405**, 31 (1997).
7. Minamisono T. et al., *Phys. Lett. B* **420**, 31 (1998) and references therein.
8. Huhta M. et al., *Phys. Rev. C* **57**, R2790 (1998).
9. Thibault C. et al., *Phys. Rev. C* **12**, 644 (1975).
10. Guillemaud-Mueller D. et al., *Nucl. Phys. A* **426**, 37 (1984).
11. Motobayashi T. et al., *Phys. Lett. B* **346**, 9 (1995).
12. Warburton E.K. et al., *Phys. Rev. C* **41**, 1147 (1990).
13. Keim M. et al., *Hyp. Int.* **97/98**, 543 (1995) and to be published.
14. Brown B.A., private communication.
15. Carchidi M. et al., *Phys. Rev. C* **34**, 2280 (1986).
16. Wildenthal B.H. et al., *Phys. Rev. C* **28**, 1343 (1983).
17. Poves A., and Retamosa J., *Phys. Lett. B* **184**, 311 (1987).
18. Tanihata I. et al., *Phys. Lett. B* **206**, 592 (1988).
19. Fukuda M. et al., *Phys. Lett. B* **268**, 339 (1991).
20. Otsuka T. et al., *Phys. Rev. Lett.* **70**, 1385 (1993).
21. Sagawa et al., *Phys. Lett. B* **309**, 1 (1993).
22. Köster U. et al., Contribution to this conference.
23. Suzuki T. et al., *Phys. Lett. B* **364**, 69 (1995).

Nuclear Moments of Exotic Nuclear States studied with Level Mixing techniques

Gerda Neyens[1]

*University of Leuven, Instituut voor Kern- en Stralingsfysica,
Celestijnenlaan 200 D, B-3001 Leuven, Belgium*

Abstract. Different applications of the nuclear level mixing principle have been used to study nuclear moments of nuclei far from stability and of high-spin isomers. The basic principles are explained and an overview of results obtained over the last years is given. We discuss in particular the first application of this technique to projectile fragmentation products, as well as the results obtained on quadrupole moments of high-spin isomers in the Pb-region and in the mass A=180 region.

INTRODUCTION

Many nuclear states can be labelled as "exotic". In this paper, we address two types of exotic nuclei : nuclei far from stability having a very unbalanced proton and neutron number, so high isospin. Another class of states is obtained when adding the excitation degree of freedom, giving rise to exotic isomeric states with very high spin. The first features that are studied, as soon as a new exotic state is discovered, are its binding (excitation) energy, its lifetime, its decay scheme and its spin and parity. These nuclear properties allow to define the nuclear structure of the exotic state in an indirect way. However, to get more insight into the details of the nuclear structure and to confirm what has been derived, one needs to measure the nuclear moments. Such measurements usually require a higher production yield (some 1000/s are needed) and will be performed mainly after the other parameters are known.

We present in this paper the "nuclear Level Mixing" principle (1) as a tool to study the static electric quadrupole and magnetic dipole moment of exotic nuclei in their ground state or excited isomeric state. The techniques based on this principle are complementary to other methods that can be used to study nuclear moments (see for example M. Keim, P. Campbell and F. Le Blanc in this conference proceedings). The level mixing techniques (LMR and LEMS) (2-5) are applicable to nuclear states with lifetimes between 50 ns and a few seconds, having a spin larger than $1/2\hbar$. For one of the techniques (LEMS), there is no upper limit on the spin : the technique remains

---

[1] post-doc researcher of the F.W.O.-Vlaanderen, Belgium

applicable up to spins as high as 35ℏ. To perform a level mixing experiment, initial orientation of the nuclear spins is needed (as for all Perturbed Angular Distribution measurements). This orientation can be either obtained in the nuclear reaction or by some other means, such as optical pumping (6) or tilted foil polarisation (7). Orientation from the production reaction itself is often high and anyhow for free, therefor all our experiments until now have been performed using like this. Spin-orientation in fusion-evaporation reactions is known and used already for a long time (8). Some years ago, it has been shown that also in projectile fragmentation reactions, spin-orientation of the reaction products is obtained. Spin-polarisation is measured for fragments selected under a small angle with respect to the primary beam (9), while spin-alignment is measured for fragments emitted in the forward direction (10).

LEVEL MIXING TECHNIQUES : BASIC PRINCIPLES

After production, the nuclei are recoil-implanted into a crystal with a static electric field gradient (EFG) V_{zz}, causing a quadrupole interaction with the nuclear quadrupole moment Q. The quadrupole interaction is proportional to m^2 (m = projection of the nuclear spin I on the principal axis of the EFG) and gives rise to a typical quadrupole splitting with basic frequency $\omega_Q = eQV_{zz}/4I(2I-1)\hbar$. The host crystal is mounted into a static magnetic field (with variable field strength B), inducing a magnetic dipole interaction with the magnetic moment μ (Larmor frequency $\omega_B = g\mu_N B/\hbar$). The combination of the two interactions gives rise to crossing of hyperfine levels at well-defined values for the magnetic field (fig. 1).

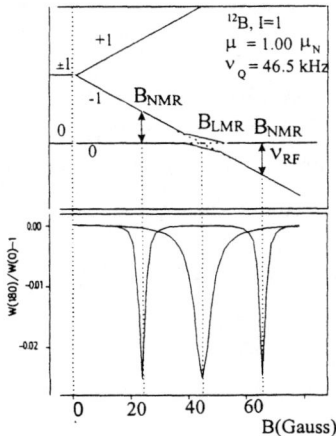

FIGURE 1 : Rabi-plot of the hyperfine levels of a nucleus with spin I=1, submitted to a non-collinear quadrupole and magnetic interaction. Mixing of the levels occurs at well-defined values for the ratio of the interaction frequencies, and thus a well-defined value of B. If an RF-field with constant frequency is added, this will induce also a local mixing of the level populations at B-values for which the level splitting matches the RF-frequency.

If the electric field gradient is slightly misaligned with respect to the magnetic field ($1° < \beta < 15°$), the breaking of the axial symmetry in the quantum system causes mixing of the population of two crossing m-states near the crossing fields. Due to this level mixing, the nuclear spins are re-oriented, giving rise to a resonant change of the angular distribution of the radioactive decay as a function of B (Level Mixing Resonances - LMR) (2,3,5). The position of the resonance field is defined by $\omega_B(LMR) = 3(m+m')\omega_Q/\cos\beta$.

For large misalignment angles β the perturbation is no longer small and the resonant behaviour is lost because all hyperfine levels are mixing. In that case, a decoupling curve is measured as a function of the magnetic field (Level Mixing Spectroscopy - LEMS) (4). The advantage of applying a large angle β is that the amplitude of the LEMS-curve (fig. 2) is not depending on the spin of the isomeric state. That means the level mixing technique is not restricted to low-spin states, but can be applied to high-spin isomers too. In both cases, LMR and LEMS, it is possible to deduce the ratio of the magnetic to the quadrupole moment of the nuclear state, provided the EFG is known.

Recently we have developed a third variant of the level mixing technique, adding an RF-field to the set-up (11). By applying a radio-frequent field with constant frequency ν_{RF} perpendicular to the static magnetic field, transitions between m-states are induced resonantly if the level splitting matches the RF-frequency. This occurs again at particular values for the magnetic field B (fig. 1), inducing resonances in the angular distribution as a function of B, as in the LMR-case. These resonances are similar to the classical NMR (Nuclear Magnetic Resonance), except that now the resonances are measured as a function of B, keeping ν_{RF} constant. From the position of two of this NMR-like resonances or an LMR and an NMR-like resonance, one can deduce unambiguously the magnetic and quadrupole moment of the nucleus. Furthermore, by comparing the sign of the measured beta-asymmetry for the magnetic field parallel or anti-parallel to a chosen z-axis, it is possible to deduce the sign of the magnetic to quadrupole moment ratio.

The advantage of this combined LMR-NMR technique as compared to classical NMR, is especially important for the study of the magnetic and quadrupole moments of β-decaying exotic ground states produced in projectile fragmentation. While the

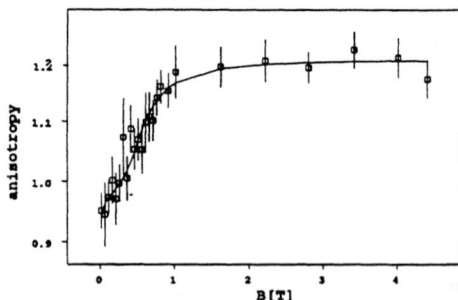

FIGURE 2 : LEMS-curve for the I=K=35/2 isomer in ^{179}W : the γ-anisotropy for the isomeric decay is measured as a function of the magnetic field strength. At high magnetic field the quadrupole interaction is decoupled, while the transition field gives information on the magnetic to quadrupole moment ratio.

classical β-NMR technique requires spin-polarisation (and thus has to be performed on a lower fragment yield), the β-LMR technique can be applied to any type of spin-oriented system, also on the aligned (forward selected) fragment beam. By applying a combined quadrupole and magnetic interaction, spin-polarisation is induced at resonance fields, allowing a β-asymmetry measurement.

OVERVIEW OF RESULTS

High-spin isomers from fusion-evaporation reactions : LEMS

In the late eighties and early nineties, we have investigated systematically the quadrupole moments of high-spin isomers in the trans lead region in Bi, At, Fr, and Ra isotopes (12-14). In this region near the magic proton Z=82 and neutron N=126 shell gaps, over 100 isomers occur, mainly due to decay via low-energy gamma-transitions of high multi-polarity and/or due to big changes in the single particle structure of the valence nucleons. The high spin of the isomers is obtained by alignment of the single particle spins of the valence protons and neutrons, giving rise to rather pure configurations. Isomers with spins up to $34\hbar$ have been identified. To obtain such high spin, the neutron core has to be broken, and particle-hole excitations across the neutron shell gap into high-j orbits are needed. The question rises what influence these core-excitations have on the softness of the Pb-core, a question that can be answered by investigating the quadrupole moment of these high-spin isomers.

The isomers on the neutron-deficient side of the shell closure are produced easily in heavy-ion fusion evaporation reactions on Tl, Pb or Bi-targets. In the reaction, angular momentum is transferred from the projectile to the compound nucleus, producing a spin-aligned ensemble of highly excited nuclei. As Tl and Bi are both crystals with an axially symmetric EFG, it is possible to use the target also as a host to perform the quadrupole moment measurement using the LEMS-technique. The use of the decoupling technique has the advantage that polycrystalline foils can be used as host. To decouple the quadrupole interaction, sometimes high magnetic fields are needed. The host crystal was mounted between the coils of a super-conduction split-coil magnet, with a maximal field of 5 Tesla. Gamma-spectra were recorded with high-purity n-type Ge detectors, placed at 0° and 90° with respect to the magnetic field (which is oriented parallel to the beam direction). In order to improve the peak-to-background ratio for the isomeric gamma-rays, a pulsed beam has been used in most experiments. The time integrated intensity is measured for different values of the magnetic field strength, giving rise to a typical decoupling curve (fig. 2).

The theoretical curve is a numerical calculation of the angular distribution, taking explicitly into account the perturbation due to the combined interaction. By diagonalising the combined interaction Hamiltonian $H = H_B + H_Q$, the eigenstates and eigenvalues of the quantum system are derived, allowing to calculate the time-integrated perturbation of the system. The parameters in the fit procedure are the initial orientation (a Gaussian distribution with width σ/I is assumed), a normalisation parameter depending on the relative detection efficiencies, and the ratio of the magnetic moment to quadrupole interaction frequency $v_Q = eQV_{zz}/h$. The initial orientation determines the amplitude of the LEMS-curve, the normalisation parameter is just a scaling factor, and the nuclear moment ratio determines the shape of the curve. So, if the magnetic moment and the electric field gradient are known from other experiments, the quadrupole moment can be deduced from the experimental data.

In figure 3, a compilation of all quadrupole moments of isomers in the N=126, 125 and 124 isotones is represented with their main proton configuration (also TDPAD-data are used (15)). No core-excited isomers are considered here, only isomers with a closed N=126 shell, or respectively 1 and 2 neutron holes in the $\nu 3p_{1/2}$ orbit. Note that a universal effective charge is found for each isotone, regardless the proton orbits that are occupied. It can be shown quantitatively that this neutron-dependent proton effective charge is related to the interaction between the collective quadrupole vibrational state in the underlying Pb-core and the valence nucleons (16,17). For the core-excited isomers, it could be shown that no extra core-polarisation charge has to be taken into account to reproduce the experimental quadrupole moments (12,13). They can be reproduced using the same proton effective charge as for the other isomers.

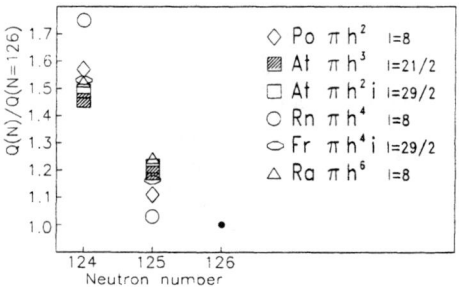

FIGURE 3 : Quadrupole moments of isomers are compared to their partner value for the N=126 isomer with the same proton configuration. A systematic trend as a function of neutron number is found.

Recently we have started a project to study the quadrupole moments of high-spin isomers in the deformed mass A=180 region. An abundance of multi-quasiparticle isomeric states has been observed in this mass region where many high-Ω orbits close to the Fermi surfaces are present in both neutrons (N ≈ 106) and protons (Z ≈ 74), leading to low-lying multi-quasi particle isomers with high K values. In many nuclei, rotational bands built on multi-quasi particle isomeric states were established, which seems to show that the axial symmetry of these deformed systems is preserved up to very high angular momentum and seniority. On the other hand, the moments of inertia of these bands cannot be described theoretically, accepting similar deformation to that of the ground state unless reduction of the pairing strength is taken into account. Thus the knowledge of the quadrupole moments is important for the understanding of the structure of these states. As a first case, we have considered the I=K=35/2⁻ 5-qp isomer in ^{179}W ($T_{1/2}$ = 750 ns), which is known to have an anomalous decay into the 1-qp ground state K=7/2 band (18). The isomer was produced in a ^{170}Er(^{13}C,4n) reaction with a beam energy of 63 MeV. A thin target (500 µg/cm^2) allowed 90% of the ^{179}W nuclei to recoil out of the target into a Tl-foil, heated up to 473 K in order to avoid radiation damage. From a fit to the data (fig. 2), using a preliminary experimental value µ=8.4 n.m. (19) for the magnetic moment, a quadrupole interaction frequency v_Q = 66.5(7.5) MHz has been derived (20). To extract the quadrupole moment, we need to know the EFG of W in Tl at the experimental temperature of 473 K. Theoretical band structure calculations using the LAPW method as implemented in the WIEN97 package (21) have been performed to determine the EFG for W(Tl) (22). The W impurity in hcp Tl is simulated by a 36 atom super cell approach, allowing full relaxation of 3 shells of neighbouring atoms. From the self consistent charge density in the cell (23) an EFG V_{zz} = 4.1 10^{21} V/m^2 at T=0 K is obtained. The EFG in Tl is known to be very temperature dependent, following an empirical T$^{3/2}$ law, $V_{zz}(T) = V_{zz}(0)$ (1-bT$^{3/2}$). For several impurities in Tl, b-values are ranging from 6 to 8.5 10^{-5} K$^{-3/2}$ (19,24). This range of values for the EFG, results in a spectroscopic quadrupole moment 1.8 eb < Q < 5.3 eb. In the strong coupling limit this gives an intrinsic quadrupole moment 2.8 eb < Q_0 < 6.5 eb. Potential energy surface calculations, taking into account pairing and the quasi particle configurations for the isomer, yield a value Q_0 = 7.2 eb (25). Although a detailed study on the EFG is needed (being analysed at the moment), it seems that the experimental quadrupole moment of the high-K 5 qp-isomer is lower than the value predicted by recent theoretical models.

Exotic ground states from projectile fragmentation : β-LMR-NMR

Since the development of intermediate energy particle accelerators followed by high-resolution in-flight mass separators, it has become possible to study nuclei far from stability in rather clean conditions. This allows nuclear moment measurements on exotic ground states, even with production rates as low as a few 100 per second. First experiments with the LMR-NMR method, have been performed at GANIL (Caen, France), on the neutron rich ^{18}N isotope. Both the magnetic and quadrupole moment of ^{18}N are unknown, but the implantation behaviour of N in a Mg single crystal has been studied extensively (26), as well as the electric field gradient (27). Spin-aligned ^{18}N projectile fragments were produced by selecting nuclei in the forward direction after the reaction of a ^{22}Ne (60.3 MeV/u) beam with a ^{12}C (350 mg/cm^2) target. A nearly pure secondary beam of ^{18}N-fragments (only 5.5% of ^{20}O) was obtained using the LISE spectrometer and the Wien-filter (28). The fragments were stopped in a Mg single crystal, which was mounted on the cold finger of a continuous flow cryostat into a vacuum chamber. The crystal was cooled to 40K, in order to reduce the influence of spin-lattice relaxation on the spin-orientation. A static magnetic field up to 2000 Gauss was induced by two coils mounted around the vacuum chamber. A small RF-coil was build around the crystal, to induce an RF-field with constant frequency. The induced spin-polarisation is measured by monitoring the β-decay asymmetry as a function of the magnetic field strength. The measured high-precision LMR (29) is in very good agreement with a first less-precise measurement (30). From the NMR-resonances, measured as a function of B (29), we derive a very small magnetic moment $\mu = 0.157(7)$ n.m.. Using this value together with the ratio μ/Q derived from the LMR-resonance field, we deduce a rather large quadrupole moment Q=32(3) mb for the ^{18}N, $I^{\pi}=1^-$ ground state (31). Comparison to shell model calculations using the universal sd-interaction (USD) of Brown and Wildenthal (32) shows that the magnetic moment is smaller than expected. However, modifying slightly the interaction parameters, such that the magnetic moment of the $3/2^+$ first excited state in ^{19}O is reproduced better, gives perfect agreement between experiment and theory for $\mu(^{18}$N) (33). No shell-model calculations for the quadrupole moment have been performed so far. Comparison to mean-field calculations (34), shows an experimental quadrupole moment which is about 60% larger that the theoretical value for the proton quadruole moment, using the rotational model relation (K=0) to relate the intrinsic theoretical and measured quadrupole moment ($Q_0 = -(2I+3)/I\ Q$).

CONCLUSIONS

Several applications of the Level Mixing principle, based on a combined non-collinear magnetic and quadrupole interaction, allow to study the magnetic and quadrupole moments of exotic nuclear states. In a first series of experiments, the level mixing technique has been applied for a systematic study of quadrupole moments of high-spin isomers, by investigating the change in the γ-anisotropy of spin-aligned fusion-evaporation products. A second application is the study of the magnetic and quadrupole moment of ground states far from stability, produced in projectile fragmentation reactions. By applying a combined interaction, it is possible to induce polarisation, starting from a spin-aligned system. This has the advantage that β-asymmetries can be investigated, starting from spin-aligned projectile fragments selected in the forward direction after the fragmentation process (and thus selecting the highest fragment yield). This will allow nuclear moment measurements on β-decaying nuclei, with production yields of the order of a few 100/s. By adding an RF-interaction, it is possible to extract both the magnetic and the quadrupole moment from the combined LMR-NMR data. In the future one might also consider to investigate γ-anisotropies of isomers produced in fragmentation (35), to improve our understanding of nuclear structure approaching the driplines, also in excited states.

ACKNOWLEDGMENTS

I would like to thank all my collaborators from Leuven University, especially R. Coussement who initiated this whole project and the Ph.D. students who were so enthousiastic for doing the experiments and performed most of the data analysis. Also thanks to all our collaborators for stimulating discussions, encouraging us to continue the investigation of the many possibilities of the level mixing techniques and of course their help in realising the experiments.

We are grateful to Bob Darlington, SERC Daresbury Laboratory, UK for producing the targets, to the engineers of the CYCLONE cyclotron at Louvain-la-Neuve, Belgium for the technical developments, as well as to the engineers and technical staff of the GANIL accelerators at Caen, France for the support during setting up the experiments.

REFERENCES

1. Tanihata I., Kogo S., Sugimoto K., *Physics Letters* **67B**, 392 (1977).
2. Coussement R. et al., *Hyperfine Interactions* **23**, 55 (1985).
3. Scheveneels G. et al., *Hyperfine Interactions* **52**, 273 (1989).
4. Hardeman F. et al., *Physical Review* **C43**, 130 (1991).
5. Neyens G. et al., *Nuclear Instruments and Methods in Physics Research,* **A 340**, 555 (1994).
6. Neugart R., *Zeitschrift fur Physic* **261**, 237 (1973).
7. Rogers W.F. et al., *Physics Letters* **177 B**, 293 (1986).
8. Morinaga H. and Yamazaki T., *In-beam γ-spectroscopy*, Amsterdam: North-Holand, 1976, p. 324-332.
9. Asahi K. et al., *Physics Letters* **251 B**, 488 (1990).
10. Asahi K. et al., *Physical Review* **C43**, 456 (1991).
11. Neyens G. et al., *Heavy Ion Physics* **7**, 101 (1998).
12. Scheveneels G. et al., *Physical Review* **C43**, 2560 and 2566 (1991).
13. Hardeman F. et al., *Physical Review* **C43**, 514 (1991).
14. Neyens G. et al., *Nuclear Physics* **A555**, 629 (1993).
15. Raghavan P., *Atomicr Data and Nuclear Data Tables* **42**, 189 (1989).
16. Heyde K.L.G., *The Nuclear Shell Model 2nd edition*, New York, Berlin, Heidelberg : Springer-Verlag, 1994, ch. 4, pp. 177.
17. Neyens G. et al., *Nuclear Physics* **A625**, 668 (1997).
18. Walker P.M. et al., *Physical Review Letters* **67**, 433 (1991).
19. Byrne A.P., *Private Communication*, 1998
20. Vyvey K. et al., "Study of the static quadrupole moment of the K=35/2 isomer in ^{179}W" *Journal of Physics G, Proceedings of the Int. Conf. on Nuclear Structure at the Extremes*, Lewes, 1998.
21. Blaha P. et al., WIEN97 (Vienna University of Technology), improved and updated Unix version of Blaha P. et al., *Computer Physics Communications* **B 59**, 399 (1990).
22. Blaha P., *Private Communication*, 1998
23. Schwarz K. and Blaha P., *Zeitschrift fur Naturforschung* **47a**, 197 (1992).
24. Neyens G., *Ph.D. Thesis*, 1993, unpublished
25. Xu F., *private communication*, 1998
26. Katigawa A. et al., *Hyperfine Interactions* **60**, 869 (1990)
27. Minamisono T. et al., *Physics Letters* **420 B**, 31 (1998) and references theirin.
28. Anne R. et al., *Nuclear Instruments and Methods in Physics Research,* **B 70**, 276 (1992).
29. Teughels S. et al., *this conference proceedings ENAM '98*
29. Neyens G. et al., *Physics Letters* **393 B**, vol. 1-2, 36 (1997).
30. Coulier N. et al., "Quadrupole moment of ^{18}N measured with the LMR method" *Nuovo Cimento , Proceedings of the XVI Nuclear Physics Divisional Conference - SNEC98*, Padova, 1998.
31. Brown B.A. and Wildenthal B. H., *Annual Review of Nuclear and Particle Sciences*. **38**, 29 (1988).
32. Brown B.A., *private communication*, 1998
33. Heenen P.H., *private communication*, 1998 and Patra S.K., *Nuclear Physics* **A559** (1992) 73
34. Grzywacz R. et al., *Physics Letters* **355 B**, 439 (1995).

Isospin nonconserving effects in light, $T_z = -3/2$ nuclei.

P.F. Mantica[1], D.W. Anthony[1], B.A. Brown[1], B. Davids[1],
G. Georgiev[2], M. Huhta[1], R.W. Ibbotson[1], P.A. Lofy[1],
J.I. Prisciandaro[1], and M. Steiner[1]

[1]*NSCL, Michigan State University, East Lansing, MI 48824 USA*
[2]*Katholieke Universiteit Leuven, Leuven 3030 BELGIUM*

Abstract. The ground state magnetic dipole moment of ^{9}C has been remeasured using the β-NMR technique and our new value of 1.396(3) μ_N is in agreement with the previously measured value for this nuclide. The quenched magnetic moment of ^{9}C and the anomalously large isoscalar spin expectation value for the ^{9}C - ^{9}Li $T = 3/2$ mirror pair compare favorably with the results of shell model calculations employing isospin nonconserving (INC) interactions. The ground state magnetic moment of ^{21}Mg was also investigated to determine the significance of INC effects in the heavier $A = 21$ $T = 3/2$ mirror system.

INTRODUCTION

Recently, the ground state magnetic dipole moments of the proton-drip line nuclei ^{9}C ($S_p = 1.3$ MeV) and ^{13}O ($S_p = 1.5$ MeV) have been obtained [1,2] using spin-polarized radioactive beams and the technique of nuclear magnetic resonance on β-emitting nuclei (β-NMR). The deduced ground state magnetic moments of ^{9}C, $(-)1.3914(5)\mu_N$, and ^{13}O, $(-)1.3891(3)\mu_N$, are significantly quenched with respect to the Schmidt limit value of -1.91 μ_N expected for a pure $\nu p_{3/2}$ ground state. While the magnetic moment of ^{13}O agrees with the results of shell model calculations employing Cohen-Kurath wavefunctions [2], the same calculations are unable to account for the small magnetic moment value of ^{9}C.

For isospin mirror nuclei the isoscalar and isovector combinations of known ground state magnetic moments provide an important and sensitive framework to identify deviations in the moment values from theoretical predictions [3]. The isoscalar spin expectation value, $<\sigma>$, can be calculated using the following relation assuming that isospin symmetry is conserved

$$<\sigma> = \frac{\mu(T_z = -1/2) + \mu(T_z = +1/2) - J}{(\mu_p + \mu_n - 1/2)}. \tag{1}$$

Experimentally, ground state magnetic dipole moments have been deduced for all the particle stable $T = 1/2$ mirror partners up to $A = 43$. The $<\sigma>$ values extracted for these mirror pairs are shown in Fig. 1, where the experimental magnetic moments are taken from Ref. [4] with the exception of ^{23}Mg [5], ^{41}Sc [6], and ^{43}Ti [7]. In general, the experimental $<\sigma>$ values for $T = 1/2$ pairs are bounded by the extreme single-particle estimates, and show a regular decrease with an increase in occupancy of a given single-particle orbital.

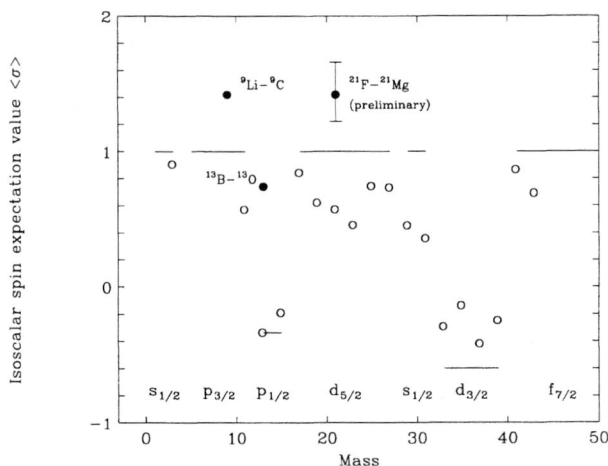

FIGURE 1. Isoscalar spin expectation values for the $T = 1/2$ nuclei (open circles) and $T = 3/2$ nuclei (filled circles) in the p, sd, and fp shells. The solid lines indicate the limits of $<\sigma>$ derived from the extreme single-particle model.

The measurements of the ground state magnetic dipole moments of ^{9}C and ^{13}O provide the first complete data set for $T = 3/2$ mirror partners for a comparison of the isoscalar and isovector moments with theoretical predictions. The $<\sigma>$ deduced from the ^{13}B – ^{13}O mirror pair, compares favorably with the systematic trend of $<\sigma>$ values for $T = 1/2$ nuclei. The $<\sigma>$ determined for ^{9}Li – ^{9}C using the known value [8] of $+3.4391(6)$ μ_N for the ground state magnetic moment of ^{9}Li and the small experimental magnetic moment deduced for ^{9}C is 1.44, a value that lies well outside the $T = 1/2$ systematics.

^{9}C REMEASUREMENT AND INTERPRETATION

We have remeasured the ground state g factor of ^{9}C to confirm the anomalous value of the magnetic dipole moment of this nuclide. A secondary beam of ^{9}C was produced at the National Superconducting Cyclotron Laboratory (NSCL) at Michigan State University by the fragmentation of an 80 MeV/nucleon ^{20}Ne beam in a 107 mg/cm^2 thick ^{93}Nb target. The beam, target and beam energy combination were chosen to enhance the polarization of ^{9}C fragments at the peak of the momentum yield distribution [9]. The spin-polarized ^{9}C fragments were collected at $+2.5°$ relative to the beam axis and separated from other reaction products using the A1200 fragment separator [10] with a 425 mg/cm^2 Al degrader placed at the second dispersive image of the device. The secondary beam was directed to the β-NMR apparatus, which consisted of two β telescopes, a 0.25 mm thick Pt implantation foil, and a set of radiofrequency (RF) coils, all placed between the pole faces of a large dipole magnet. Details on the β-NMR apparatus are given in Ref. [11].

The ^{9}C resonance curve we obtained is shown in Fig. 2. The multiple adiabatic fast passage technique with continuous beam implantation [11] was used to allow for a 100% duty cycle for data collection. An efficient data collection method was crucial for this measurement, since the ^{9}C implantation rate was ≈ 2 ions/s.

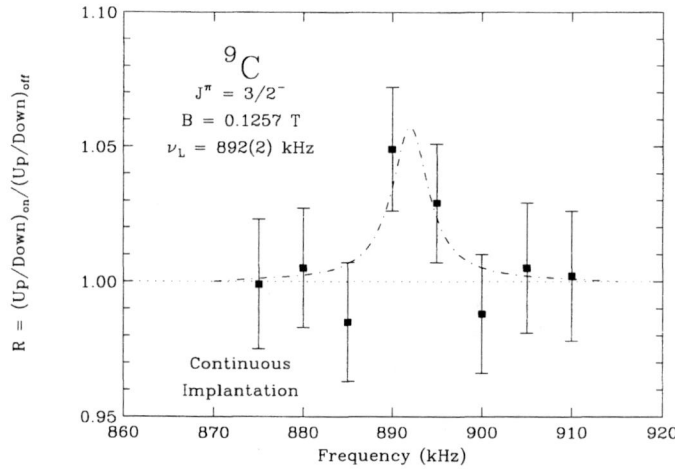

FIGURE 2. Resonance curve obtained for ^{9}C. The frequency modulation for each point is ± 10 kHz of the central frequency using a triangle waveform with a 500 Hz repetition rate.

The ^{9}C resonance data were fitted using a Lorentzian peak shape with a peak centroid of 892(2) kHz and a width of 5 kHz. We deduced a g factor of 0.931(2) for the ground state of ^{9}C using the measured holding field value of 0.1257 T. Since the ground state spin of ^{9}C is known to be $J = 3/2$, a value of 1.396(3) μ_N was obtained for the ^{9}C ground state magnetic dipole moment. The corrections for diamagnetic shielding and the Knight shift for the carbon in platinum system, given in Ref. [2], are smaller than the statistical error in our value for the magnetic moment of ^{9}C. Therefore, these corrections have only been included in the overall error.

Our new measurement for the magnetic dipole moment of ^{9}C is in agreement with the previous value [2] of $(-)1.3914(5)\mu_N$ for this nuclide, once again suggesting the unique character of this nucleus. We have performed shell model calculations in an attempt to reproduce the quenched g factor of ^{9}C and the large $<\sigma>$ extracted

TABLE 1. The isoscalar spin expectation values for $T = 3/2$ mirror pairs.

Mirror Pair	$<\sigma>$	shell model PTBME+INC	PTBME	Extreme S.P.
^{9}C – ^{9}Li	1.44	1.18	1.09	1.00
^{13}O – ^{13}B	0.76	0.71	0.71	1.00
^{21}Mg – ^{21}F	—	1.15	1.10	1.00

for the $A = 9$ $T = 3/2$ mirror pair. The PTBME interaction of Julies, Richter, and Brown [12], which includes a mass dependence for the two-body matrix elements, was chosen as it reproduces well the level energies and static electromagnetic moments of $0p$-shell nuclides. We employed the simpler, bare g-factors for calculating the magnetic dipole moments of the ^{9}Li – ^{9}C and ^{13}B – ^{13}O mirror pairs as the results with effective g-factors were only slightly different from those derived with the bare nucleon values. The results of the shell model calculations are compared with the experimental magnetic moments in Table 1. The present calculations reproduce the experimental magnetic moments and $<\sigma>$ for the $A = 13$, $T = 3/2$ mirror partners. The calculated magnetic dipole moment of ^{9}C was $-1.44\mu_N$, and contains a significant proton contribution, suggesting a breaking of paired proton spins as was observed in the antisymmetrized molecular dynamics (AMD) calculations of Kanada-En'yo and Horiuchi [13]. The derived value of $<\sigma> = 1.09$ for the $A = 9$ $T = 3/2$ partners, although greater than unity, is still well below the experimental value.

To explore further the large value of $<\sigma>$ for the ^{9}Li – ^{9}C mirror pair, we extended the shell model calculations described above to include the isospin-nonconserving (INC) processes outlined by Ormand and Brown [14]. The Coulomb interaction should play a significant role in the low-energy structure of loosely bound nuclei near the proton drip-line, and it is important to consider isospin mixing in these systems. The INC interaction of Ormand and Brown contains five parameters: (1) the isovector single-particle energies for the $p_{1/2}$ and $p_{3/2}$ orbitals (two parameters), (2) the Coulomb matrix elements scaled by a strength parameter (one parameter), and (3) isovector and isotensor contributions to the strong interaction each scaled to the isospin-conserving matrix elements by one parameter (two parameters). The five parameters are determined by a least squares fit to the 15 b and 7 c coefficients of the isobaric-mass-multiplet-equation (IMME) for p-shell nuclei in the mass $A = 9 - 15$ region. The INC wave functions are generated in a proton-neutron formalism and used to calculate the magnetic moments of the ground states of interest. For the $A = 9$ $T = 3/2$ states, the INC interaction allows mixing with the $T = 5/2$ states at higher energy.

The results of the calculations employing the PTBME interaction and INC interactions are also given in Table 1. Inclusion of the Coulomb interaction improved the agreement of $<\sigma>$ with the experimental value for the ^{9}Li – ^{9}C mirror pair. The scaling parameter for the isovector contribution to the strong interaction was observed to have the largest effect on the calculated $<\sigma>$ value. There was no change in the calculated $<\sigma>$ for the ^{13}B – ^{13}O mirror pair as these nuclei have three holes outside the shell closure at eight particles, and therefore have pure $T = 3/2$ isospin. Although we have demonstrated that INC effects are important for the interpretation of mirror moments, the present calculations do not give the full effect observed experimentally for the ^{9}C-^{9}Li mirror pair, but they go in the right direction. The Coulomb contribution to the INC is calculated using harmonic-oscillator radial wave functions, which may not be appropriate for the more loosely bound p-shell nuclei (including ^{9}C), and may be a source of isospin asymmetry beyond the present model.

^{21}Mg PRELIMINARY ANALYSIS

To extend the study of INC interactions in heavier $T = 3/2$ partners, we have completed an experiment to measure the ground state magnetic dipole moment of ^{21}Mg. The ground state magnetic moment of ^{21}F has previously been determined as $+3.9194(12)\mu_N$ [15]. The shell model calculations employing the PTBME interaction predict a $<\sigma>$ value greater than unity for the ^{21}Mg – ^{21}F partners (see Table 1). The inclusion of INC effects in the shell model hamiltonian results in $<\sigma> = 1.15$ for the $A = 21$ $T = 3/2$ mirror system. A ^{21}Mg secondary beam was produced by the fragmentation of a ^{24}Mg primary beam at 80 MeV/nucleon in a 642 mg/cm^2 ^{93}Nb target. Spin-polarized ^{21}Mg fragments were collected at +2.5° relative to the primary beam axis and were separated from other fragmentation products using a 233 mg/cm^2 plastic degrader placed at the second dispersive image of the A1200 separator.

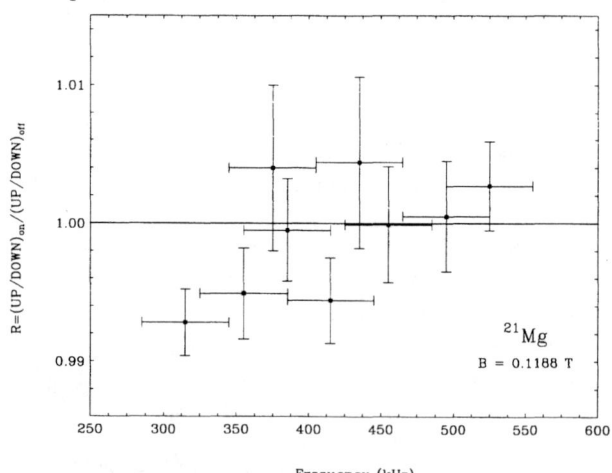

FIGURE 3. Resonance sweep obtained for ^{21}Mg. The frequency modulation ($\pm$ 30kHz) is indicated by the horizontal error bars. The applied radiofrequency was modulated using a triangle waveform with a 500 Hz repetition rate.

The polarized ^{21}Mg nuclei were implanted into an annealed Pt foil at the center of the β-NMR apparatus. Beam cycling implantation (170 ms collection, 250 ms decay) was used to minimize the contribution of ^{20}Na (the major secondary beam contaminant) β rays to the total β spectrum collected by the two β telescopes. A preliminary analysis of the resonance data for ^{21}Mg in the frequency range 285 to 550 kHz for a magnetic holding field of 0.1188 T is shown in Fig. 3. An NMR effect of -0.72(24)% has been detected at a RF field of 315(30) kHz, corresponding to a ground state magnetic dipole moment in the range 0.79 to 0.95 μ_N assuming $J = 5/2$ for the ground state of ^{21}Mg. Resonance data have been collected using a more narrow frequency modulation and will allow for a more precise determination of the ^{21}Mg magnetic moment. However, the preliminary analysis suggests that the $<\sigma>$ value for the $A = 21$ $T = 3/2$ mirror system is larger than unity (see Fig. 1) as predicted by the shell model calculations.

This work has been supported by the National Science Foundation under Contract No. PHY-9528844 and PHY-9605207.

REFERENCES

1. K. Matsuta et al., Hyperfine Int. **97/98**, 519-526 (1996).
2. K. Matsuta et al. Nucl. Phys. **A588**, 153c-156c (1995).
3. K. Sugimoto, J. Phys. Soc. Jap. Suppl. **34**, 197-200 (1973).
4. P. Raghavan, At. Data Nucl. Data Tables **42**, 1-291 (1989).
5. M. Fukuda, T. Izumikawa, T. Ohtsubo, M. Tanigaki, S. Fukuda, Y. Nakayama, K. Matsuta, Y. Nojiri, T. Minamisono, Phys. Lett. **307B**, 278-282 (1993).
6. T. Minamisono, Y. Nojiri, K. Matsuta, K. Takeyama, A. Kitagawa, T. Ohtsubo, A. Ozawa, M. Izumi, Nucl. Phys. **A516**, 365-384 (1990).
7. K. Matsuta, et al. Hyperfine Int. **78**, 123-126 (1993).
8. F. D. Correll, L. Madansky, R. A. Hardekopf, and J. W. Sunier, Phys. Rev. C **28**, 862-874 (1983).
9. H. Okuno et al., Phys. Lett. B **335**, 29-34 (1994).
10. B. M. Sherrill, D. J. Morrissey, J. A. Nolen, Jr., and J. A. Wigner, Nucl. Instr. Methods Phys. Res. B **56/57**, 1106-1110 (1991).
11. P. F. Mantica, R. W. Ibbotson, D. W. Anthony, M. Fauerbach, D. J. Morrissey, C. F. Powell, J. Rikovska, M. Steiner, N. J. Stone, and W. B. Walters, Phys. Rev. C **55**, 2501-2505 (1997).
12. R. E. Julies, W. A. Richter, and B. A. Brown, S. Afr. J. Phys. **15**, 35-55 (1992).
13. Y. Kanada-En'yo and H. Horiuchi, Phys. Rev. C **54**, R468-471 (1996).
14. W. E. Ormand and B. A. Brown, Nucl. Phys. **A491**, 1-23 (1989).
15. T. Onishi et al., Osaka University Laboratory for Nuclear Science Annual Report 1996 p. 45-47.

On-line Laser Spectroscopy of Refractory Radioisotopes at the JYFL IGISOL facility

P. Campbell[a], D.M. Benton[b], J. Billowes[a], P. Dendooven[c], D.E. Evans[b], D.H. Forest[b], I.S. Grant[a], J.A.R. Griffith[b], A. Honkanen[c], A. Jokinen[c], J.M.G. Levins[a], M. Oinonen[c], H. Penttilä[c], K. Peräjärvi[c], D.S. Richardson[b], G. Tungate[b], G. Yeandle[a] and J. Äystö[c]

[a] Schuster Laboratory, University of Manchester, Manchester M13 9PL.
[b] Department of Physics and Space Research, University of Birmingham, Birmingham B15 2TT.
[c] Accelerator Laboratory, University of Jyväskylä, Jyväskylä SF-403 51, Finland.

Abstract. A major objective of the laser–IGISOL programme has been realized with the first ever on–line observation of collinear laser induced fluorescence from an ion of a refractory element. The measurements demonstrate that the IGISOL can be operated in a mode that produces ion beams of good emittance with reasonable extraction efficiency. The technique has been used to study the neutron–deficient Hf isotopes.

Laser spectroscopic studies of short–lived radioactive nuclei has been very successfully performed at on–line isotope separators for over twenty years [1]. The majority of the work has used the collinear ion–laser beam technique (or hybrids thereof). Access to many nuclear systems has however been severely hindered by the release times and chemical selectivity of the ion sources used. In particular on–line laser spectroscopy of the refractory elements, for example Zr, Mo, Hf and W, has not be achieved because of the cripplingly low production efficiency for these metals. In this report we detail the first ever on–line measurements on refractory systems using an IGISOL (ion guide isotope separator on–line). The IGISOL, developed at the University of Jyväskylä, Finland, offers a fast, ~1ms, and chemically non–selective method for producing singly charged ionic species from the products of nuclear reactions [2]. The application of collinear laser spectroscopy to these ion beams has required considerable off–line development work with high current stable beams [3] used to investigate emittance properties as a function of ion guide parameters. The on–line experiments use a sensitive photon–ion coincidence method as a refinement of the standard collinear technique. Experimental results for zirconium and hafnium ions are presented. A previous test experiment on neutron–rich barium isotopes is summarized elsewhere [3].

Experimental Details

A full description of the IGISOL can be found in ref. [2]. In brief, recoiling products from nuclear reactions in a target foil are stopped and thermalized in He gas ($\sim$200mbar) flowing in a differentially pumped jet. Production reactions using light ion beams are favoured because of the low recoil velocities and low primary ionization of the buffer gas. In this report only work using the (p,xn) reaction is summarized. A test production of Hf using (α,xn) is reported in [4] and proton–induced fission production of Ba in [3]. A substantial fraction, $\sim$10%, of the thermalized products are singly charged ions when they enter the helium jet expansion region. During the expansion the ions are skimmed from the buffer gas by a shaped negative potential electrode and injected into the high vacuum of a conventional mass separator. The skimmer electrode has been found to be crucial in determining both the ion guide efficiency and the energy spread of the extracted beam. Laser–based investigations have revealed that a sharp local maximum in extraction efficiency can be found at a point where the ion beam has low energy spread ($\sim$5eV). Under these conditions, where the skimmer is operated close to -10V, up to 30% of the optimum ion guide efficiency can be recovered. This mode of guide operation was exclusively used during on–line laser investigations.

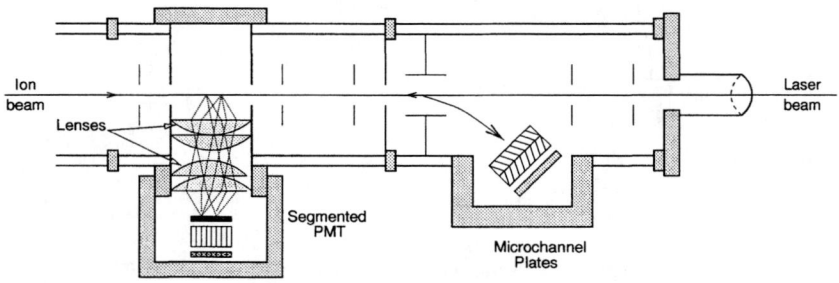

FIGURE 1. The laser–ion interaction region.

The ion beam is mass–analyzed by a 55° dipole magnet and transported 15m to the laser–ion beam interaction (figure 1). The ion beam is shaped and controlled by three electrostatic quadrupole elements in the transport line. The focusing can be optimized to deliver a narrow (10mm^2), low divergence (1 mrad half angle) ion beam to the interaction region. An 18mm length of the region where the laser overlaps the ion beam is imaged onto a photomultiplier tube (PMT). The entire interaction region is electrically insulated from the rest of the grounded beamline and held at a controlled potential. Adjusting this potential allows the ion beam to be Doppler tuned into resonance with a locked laser frequency. Downstream of the imaged region the ions are deflected onto a microchannel plate. Only those photons detected in correct time coincidence with an ion are accepted [5]. A recent improvement on background rejection has been gained by using a position sensitive PMT. Position information on the origin of the fluorescent ion can be used to narrow the photon–ion coincidence time window to a level that corresponds to the resolution limit of the apparatus. In the reported Hf work a 16 fold segmented Hamamatsu R5900P–03–L16 photomultiplier was employed and a spatial resolution of 2–3mm was achieved. Ionic transitions at 301nm (Hf) and 327nm (Zr) were studied in this work. The UV light was produced by intracavity frequency doubling with $LiIO_3$ in a Spectra Physics 380D dye laser. Laser stabilization giving a linewidth of ~3MHz is achieved by the Stabilok^{TM} system and active locking of the cavity to absorption lines of an iodine vapour cell.

Substantial metastable ion production is observed from the ion guide. Under on–line conditions the loss of population from the ground state, compared to room temperature production, is large, ~90–95%. Attempts to quench these populations using diatomic gases have been unsuccessful. Further investigation of quenching mechanism will be performed at a later stage with the on–line ion beam cooler currently under development.

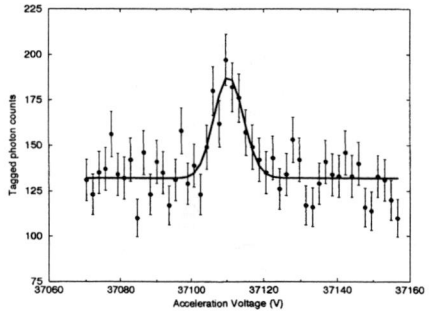 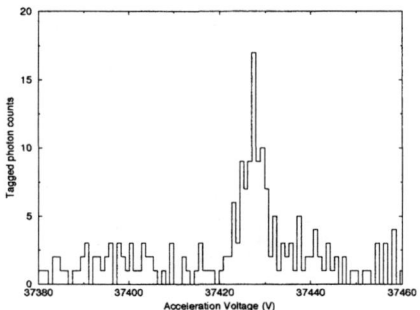

FIGURE 2. Resonance fluorescence spectra for ^{88}Zr (left) and ^{174}Hf (right).

Results for ^{88}Zr and ^{170,172,173,174}Hf

A beam of 8×10^3 s^{-1} ^{88}Zr was produced from the ^{89}Y(p,2n)^{88}Zr reaction at an energy of 25MeV. The primary beam, 50MeV H$_2^+$ was maintained at 8pμA and the yield of ^{88}Zr was calibrated using scaled known cross sections and directly measured (γ–counted) A=89 isomer production. The total intensity at A=88 was 12×10^3 ions s^{-1} of which 2×10^3 s^{-1} were ^{88}Y and a further 2×10^3 s^{-1} were non–radioactive contaminant background. Figure 2 shows the coincident photon counts using a non–segmented PMT in 5.25 hours as function of total acceleration voltage. The laser power (2mW) produced some power broadening and a resonance width (FWHM) of 10V was measured. The laser was locked to an I_2 resonance at

15261.20cm^{-1} to provide λ=327nm in the 2nd harmonic and a total acceleration of 37,110.3V was required to bring the system to resonance on the d^{2}s ^{4}F$_{3/2}$ – dsp ^{4}F$_{5/2}$ transition. The fluorescent efficiency was a factor of 20 less than for off–line production of an even Zr isotope. The on–line pressure, 220mbar, was selected to optimize production efficiency with respect to contaminant levels (the contaminant levels increase at lower pressure due to the reduction in gas cooling of the guide). Frequency calibration of single on–line measurements to off–line results are complicated by the 500V potential field within the off–line discharge source [3]. Either on–line calibration with a known isotope or an absolute frequency determination is required to eliminate the systematic error.

In the investigation of the near–stable neutron deficient Hf isotopes using the ^{175}Lu(p,xn)$^{176-x}$Hf reaction, no γ–counted yield calibration was possible due to the prohibitively long half–lives. The on–line period was instead immediately preceeded by an off–line optimization on the naturally occurring, radioactive ^{174}Hf isotope and initial on–line production was maximised for this mass. The mirror symmetry and construction of the ion guide allows two targets to be installed and for the guide to be rotated between them. Optimization of the reaction channel, and absolute IGISOL output, was achieved using reactions on a natural Ta target in this rotated position. Products of the ^{181}Ta(p,2n)^{180m}W reaction provided a swift method of optimisation and initial ^{174}Hf production (with the ion guide rotated back to the Lu target) commenced at 20MeV. Using 5μA of protons a beam flux of 3×10^3 s^{-1} was observed for ^{174}Hf, estimated from the change in the A=170–180 background beams upon rotation.

The hafnium ion has a ds^2 ground state configuration. The oscillator strength for ds^2 – dsp transitions is notably stronger than that of d^{2}s – dsp transitions (such as that studied in Zr) and a factor of four less laser power was required to provide optimum signal to background. The use of the position sensitive photomultiplier further provided a further five–fold increase in signal to background and a comparison of the Hf resonance to that of Zr, shown in fig. 2, clearly demonstrates the improvement in background rejection. Following the observation of ^{174}Hf fluorescence a variety of beam energy changes were made and the spectroscopy of A = 170, 172, 173 and 174 completed. The sample spectrum shows the acceleration voltage required to bring ^{174}Hf into resonance with a doubled laser frequency of fundamental 16579.2883 cm^{-1} exciting the ds^2 ^{2}D$_{3/2}$ – dsp ^{2}D$_{5/2}$ transition.

Proton beam energies of 20, 30, 40 and 55 MeV were used during the investigation. At 30 MeV isotopes of A = 172, 173 and 174 were simultaneously produced allowing their resonance positions to be measured under the same IGISOL conditions. The results are shown in table 1. The magnetic moment of the I$^\pi$=1/2$^-$ ground state of ^{173}Hf was determined to be +0.502(7)n.m. (using hyperfine parameters scaled from ^{177,179}Hf and assuming no hyperfine anomaly). At 55 MeV evidence of A = 171 structure was apparent with a large number of real counts forming a substantial peak in the timing spectrum. Insufficient statistics were obtained for a determination of hyperfine parameters or structure centroid. The A = 171 system is complicated by the production of a 30s isomer [4].

TABLE 1. Measured Hf isotope shifts and charge radii.

	170	172	173	174
$\delta\nu^{178,A}$ (MHz)	−10180(30)	−6632(10)	−5017(7)	−3704.3(31)
$\delta\langle r^2\rangle^{178,A}$ (fm^2)	−0.363(14)	−0.235(9)	−0.177(7)	−0.130(5)

Systematic shifts between the same resonance peaks were observed at different beam energies. The cause of these ∼3V shifts is unclear but appear to be related to different plasma conditions within the IGISOL. These systematic shifts will be removed by the decoupling provided by an operational ion beam cooler.

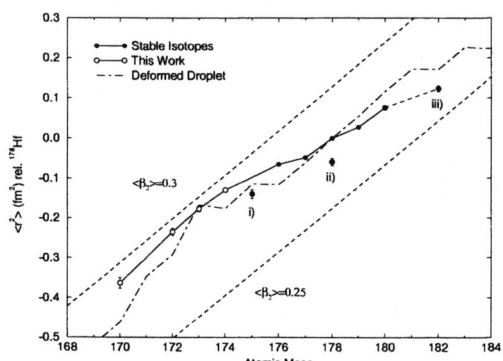

FIGURE 3. The mean square charge radii in the Hf isotope chain. Results are taken from this work and those of i) ref.[9], ii) ref.[6] and iii) ref.[11].

The extracted $\delta\langle r^2\rangle$ for Hf are included in table 1. The $\langle r^2\rangle$ extraction uses the atomic parameters of Boos et al. [6]. These measurements substantially extend the Hf chain. In figure 3 the measured $\langle r^2\rangle$ is compared with the droplet model estimates [7] including the predicted deformation corrections [8]. The trend in the charge radius closely corresponds to that of the predicted β_2 and the peaking of the deformation beyond the midshell, at A = 173, appears common to both the theoretical and measured deformation. The trend in the measured $\langle r^2\rangle$, which appears physically reasonable, gives greater confidence in the atomic calibration of the optical measurements (a discrepancy, ∼35%, exists between optical and non-optical measurements for the stable isotopes [6]). A number of interesting features are apparent in figure 3. The unusually large increase in deformation at A = 173 proves sufficient to invert the odd–even staggering (OES) at this point. The large normal OES reported for A = 175 [9] appears rather surprising as a sizeable increase in deformation is also expected for this isotope [8].

Investigations into the systems of Zr and Hf will continue at JYFL. Neutron–rich Zr are readily produced by the fission ion guide and the rapidly changing and varied nuclear structure in this region provides excellent motivation for spectroscopic investigations. In hafnium further extension to lighter masses are clearly required to investigate the course of the deformation. The OES must also be investigated in this extension and an immediate measurement of the A = 171, 171^m and 175 systems will be attempted at JYFL. Investigation of the high–K isomeric systems are also planned. The case of ^{178}Hf, for which the $I^\pi = 16^+$ isomeric state has been studied off–line, is of long standing interest [6]. Direct access to this system and neighbouring isomers (particularly ^{178m1}Hf) can now be achieved.

Prospects with Cooled Beams

The future programme will be greatly assisted by the successful installation of an on–line ion beam cooler which will be commissioned at the JYFL during late 1998. The device, a low pressure gas filled radio frequency quadrupole (RFQ) held a few volts below the acceleration potential, is intended to reduce the spatial extent and energy spread of the ion guide beam. Within the device the decelerating and expanding ion beam experiences transverse cooling provided by friction in the viscous gas and the ionic ensemble is cooled on to the axis of the RFQ. The positioning of the device beyond the mass analysing magnet ensures minimal space charge limitations and short term trapping or bunching of ionic ensembles can be investigated. At 100% transmission an operational cooler will provide realistically two orders of magnitude improvement in laser spectroscopic sensitivity. The large improvement results from i) the use of maximum IGISOL efficiency (instead of the compromised efficiency currently used to achieved low velocity spread), ii) lower laser power for the reduced spatial size and iii) the reduction in residual energy spread from the current 5eV to 1eV. The cooler may also be useful in relaxing metastable ionic populations. This has been successfully achieved [10] in the on–line gas cell technique with very low partial pressures, $\sim 10^{-2}$ mbar, of diatomic gases. Beyond this the cooler, irrespective of its efficiency, will also vitally decouple the spectroscopic investigation from the IGISOL parameters, removing any calibration uncertainty related to changing IGISOL conditions.

REFERENCES

1. Otten E. W. *Treatise on Heavy–Ion Science 8*, Plenum Press, New York (1988)
2. Dendooven P. et al. *Nucl. Inst. Meth.* **B126** 182 (1997)
3. Billowes J. et al. *Nucl. Inst. Meth.* **B126** 416 (1997)
4. Campbell P. et al. *J. Phys.* **G23** 1141 (1997)
5. Eastham D. A. et al. *Opt. Comm.* **60** 1583 (1986)
6. Boos et al. *Phys. Rev. Lett.* **72** 2689 (1994)
7. Myers W. D. and Schmidt K. H. *Nucl. Phys.* **A410** 61 (1983)
8. Möller P. et al. *At. Nuc. Data Tables* **59** 185 (1995)
9. Jin W. G. et al. *Phys. Rev.* **C55** 1545 (1997)
10. Schecker J. et al. *Phys. Rev.* **A46** 3730 (1992)
11. Anastassov A. et al. *Z. Phys.* **A348** 177 (1994)

Large Odd-Even Staggering in the very light Platinum Isotopes from Laser Spectroscopy

F. Le Blanc[1], J. Pinard[2], L. Cabaret[2], J.E. Crawford[3], H.T. Duong[2], J. Genevey[4], M. Girod[5], G. Huber[6], M. Krieg[6], J.K.P. Lee[3], J. Lettry[7], D. Lunney[8], J. Obert[1], J. Oms[1], S. Péru[5], J.C. Putaux[1], B. Roussière[1], J. Sauvage[1], V. Sebastian[6], S. Zemlyanoi[9] and ISOLDE collaboration[7]

[1] *Institut de Physique Nucléaire, IN2P3-CNRS, 91406 Orsay Cedex, France*
[2] *Laboratoire Aimé Cotton, 91405 Orsay Cedex, France*
[3] *Foster Radiation Laboratory, Mc Gill University, H3A2T8 Montréal, Canada*
[4] *Institut des Sciences Nucéaires, IN2P3-CNRS, 38026 Grenoble Cedex, France*
[5] *C.E.A, Service de Physique Nucléaire, BP 12, 91680 Bruyères-le-Châtel, France*
[6] *Institut für Physik der Universität Mainz, 55099 Mainz, Germany*
[7] *CERN, 1211 Genève 23, Switzerland*
[8] *C.S.N.S.M., IN2P3-CNRS, 91405 Orsay Cedex, France*
[9] *Flerov Laboratory of Nuclear Reaction, JINR, Dubna 141980, Moscow Region, Russia*

Abstract. Laser spectroscopy measurements have been carried out on very neutron-deficient platinum isotopes with the COMPLIS experimental set-up on line with the ISOLDE-Booster facility. For the first time, Hg α-decay was exploited to extend the very light platinum chain. Using the $5d^96s\ ^3D_3 \rightarrow 5d^96p\ ^3P_2$ optical transition, hyperfine spectra of ^{182,181,180,179,178}Pt and ^{183}Pt^m were recorded for the first time. The variation of the mean square charge radius between these nuclei, the magnetic moments of the odd isotopes and the quadrupole moment of ^{183}Pt^m were thus measured. A large deformation change between ^{183}Pt^g and ^{183}Pt^m, an odd-even staggering of the charge radius and a deformation drop from $A=179$ are clearly observed. All these results are discussed and compared with microscopic theoretical predictions using Hartree-Fock-Bogolyubov calculations using the Gogny force.

INTRODUCTION

The very neutron-deficient platinum isotopes belong to a mass region rich in shape instabilities. Along an isotopic chain, the crossing of the neutron mid-shell $N = 104$ is accompanied by a number of fluctuations of the nuclear shape. This is especially the case in the very neutron-deficient mercury nuclei where a huge odd-even staggering of the nuclear charge radius between $N = 106$ and $N = 101$ has been observed from laser spectroscopy [1,2]. This phenomenon has been interpreted as alternating oblate to prolate shape transitions between the even and the odd isotopes. Therefore, the important question to address is whether or not the lighter Pt nuclei (around $N = 104$) display the same behavior as their Hg isotones. In addition, laser spectroscopic studies have already been performed on platinum isotopes from $A = 193$ to $A = 183$ [3–5] and a large radius change was observed between $A = 186$ and $A = 185$ [4].

The nuclear structure of the very light platinum isotopes has been extensively studied. In particular, all the low lying energy levels in ^{181}Pt and ^{179}Pt have been associated with a prolate shape [6,7] while shape coexistence is suggested for the even ^{176,178}Pt isotopes [8]. It is therefore essential to measure the change in the mean square charge radius ($\delta < r_c^2 >$) in order to determine the deformation parameter of these exotic nuclei and the magnetic moment μ_I in order to confirm the neutron configuration of the odd isotopes. Moreover, if the nuclear spin is greater than 1/2, the measurement of the quadrupole moment is crucial to determine the sign of the β deformation for axially symmetric nuclei.

We present here high precision optical spectroscopy measurements on very neutron-deficient platinum isotopes. We recorded for the first time the hyperfine structures (HFS) of ^{182,181,180,179,178}Pt and ^{183}Pt^m. The HFS of ^{183}Pt^g and ^{185}Pt^{g,m} was also more precisely measured. We thus extracted the isomeric and isotope shift (IS) from which we can determine $\delta < r_c^2 >$ along this isotopic chain. In the odd isotopes we also extracted the magnetic A_i hyperfine constants from which we could deduce the μ_I values. In ^{183}Pt^m and ^{185}Pt^g, we measured the electrostatic B_i hyperfine constant to determine the spectroscopic quadrupole moment Q_s. Since platinum is very refractory, such measurements are impossible without a secondary beam obtained from radioactive mercury decay. We used the COMPLIS experimental set-up [9] which is especially well suited for the study of such elements.

EXPERIMENTAL METHOD

The experiment was performed on-line with the PS-Booster ISOLDE mass separator at CERN. The experimental procedure is described in Ref. [10] : the mercury ions are slowed from 60 kV to 1 kV, deposited on a graphite substrate and are laser desorbed as atoms after accumulation as grand-daughter isotopes. They are then ionized in 3 atomic steps by a set of 3 pulsed tunable dye laser beams where the first laser excitation step at 306.5 nm ($5d^96s\ ^3D_3 \rightarrow 5d^96p\ ^3P_2$) is obtained from frequency doubling. The ions are finally detected with time-of-flight mass identification.

^{185}Pt^{g,m}, ^{183}Pt^{g,m} and ^{182}Pt were obtained via β^+/EC decay of Hg nuclei. For the lighter isotopes, we have used for the first time the α decay mode to perform our measurements. The ISOLDE production yield is otherwise too low to produce them in sufficient quantities via successive β decay. For example, it drops to $4 \cdot 10^1$

atoms/s for ^{178}Hg compared to $8 \cdot 10^6$ atoms/s for ^{182}Hg. Even though the α branch carries only 15.2 % intensity in the decay of ^{182}Hg and the desorption efficiency is slightly lower (than that of β^+/EC decay) because the nucleus recoils isotropically at about 120 keV, we could gain a factor of 10^4 on the number of ^{178}Pt atoms.

RESULTS AND DISCUSSION

The HFS and IS of ^{185}Pt^{g,m}, ^{183}Pt^{g,m} and $^{182-178}$Pt are shown in Figure 1.

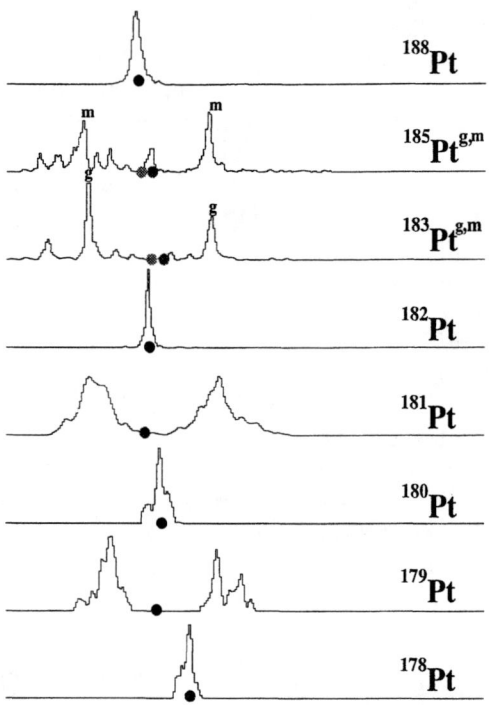

FIGURE 1. Hyperfine spectra of the light platinum isotopes. ^{g} and ^{m} represent the hyperfine lines of the ground and isomeric states respectively.

Extraction of the results

For ^{183}Pt^m and ^{185}Pt^g the B_i factors of the $^3D_3$ atomic ground state have been extracted from the hyperfine spectra. Q_s is related to B_i via:

$$Q_s(^xPt) = -0.685 \cdot B_i(^xPt)$$

which is obtained from the calculated electric field gradient as described in Ref. [5]. The Q_s values have then to be corrected for the Sternheimer shielding factor

R_{5d}. This factor, obtained for platinum from systematics on the 5d shell, lowers the measured Q_s by 9.1% [11]. The Q_s and Q_s^{corr} values are presented in Table 1.

The magnetic moments of the odd isotopes are obtained from A_i of the atomic ground state and the precisely known value of μ_I of ^{195}Pt [12,13]:

$$\mu_I(^xPt) = 0.214(2) \cdot A_i(^xPt) \cdot I(^xPt)$$

the error being due to 1% hyperfine anomaly uncertainty. The values of μ_I are presented in Table 1.

The experimental IS values relative to ^{194}Pt are given in Table 1. The IS consists of a mass shift $\delta\nu_{MS}^{A,A'}$ and a field shift $\delta\nu_{FS}^{A,A'}$ [14]. For a $ns \to np$ transition, it has been estimated to be $(1.3\pm0.9)\Delta\nu_{NMS}$ where $\Delta\nu_{NMS}$ is the easily calculable normal mass shift [14]. The nuclear parameter $\lambda^{A,A'}$ related to the change in the nuclear charge radius is extracted from the remaining $\delta\nu_{FS}^{A,A'}$ using $\delta\nu_{FS}^{A,A'} = F_{306} \cdot \lambda^{A,A'}$, where F_{306} is the electronic factor of the atomic transition. Since no relativistic multi-configuration Dirac-Fock (MCDF) calculation has been performed for this wavelength, we derived this factor from a King plot made on the stable isotopes, using the calculated F_{266} value of the 266 nm transition [5] and the experimental IS data of both transitions. This leads to $F_{306} = -18.5(10)$ GHz/fm^2. It is thus possible to extract $\delta < r_c^2 >^{194,A}$ following the method described in Ref. [15]. Table 1 and Figure 2 show the measured $\delta < r_c^2 >$ for the platinum isotopes. One can make the following remarks : i) there is a marked decrease of the radius and thus of the deformation from mass 179, ii) there is quite a strong inverted odd-even staggering (OES) below mass 186 and iii) there is a large difference in the radius of the ground and isomeric states for both ^{185}Pt and ^{183}Pt.

TABLE 1. Nuclear moments and $\delta < r_c^2 >^{194,A}$ in the light platinum isotopes. The B(E2)-value of ^{194}Pt [16] is used to extract the rms deformation parameter $< \beta^2 >^{1/2}$ obtained from the deformed part of $\delta < r_c^2 >^{194,A}$.

A	I^π	μ [μ_N]	Q_s [b]	Q_s^{corr} [b]	$\delta\nu^{194,A}$ [GHz]	$\delta < r_c^2 >^{194,A}$ [fm^2]	$< \beta^2 >^{1/2}$
185^g	9/2$^+$	-0.723(11)	+4.10(19)	+3.73(17)	1.582(35)	-0.093(7)	0.231(3)
185^m	1/2$^-$	+0.503(5)			3.611(7)	-0.212(6)	0.207(3)
183^g	1/2$^-$	+0.502(5)			3.67(5)	-0.216(8)	0.227(3)
183^m	7/2$^-$	+0.782(14)	+3.71(30)	+3.37(27)	1.81(4)	-0.106(8)	0.246(3)
182	0				4.75(8)	-0.279(10)	0.225(4)
181	1/2$^-$	+0.484(21)			4.27(20)	-0.251(15)	0.239(4)
180	0				6.13(5)	-0.360(11)	0.229(4)
179	1/2$^-$	+0.431(32)			5.70(30)	-0.335(21)	0.243(5)
178	0				8.99(20)	-0.529(16)	0.216(5)

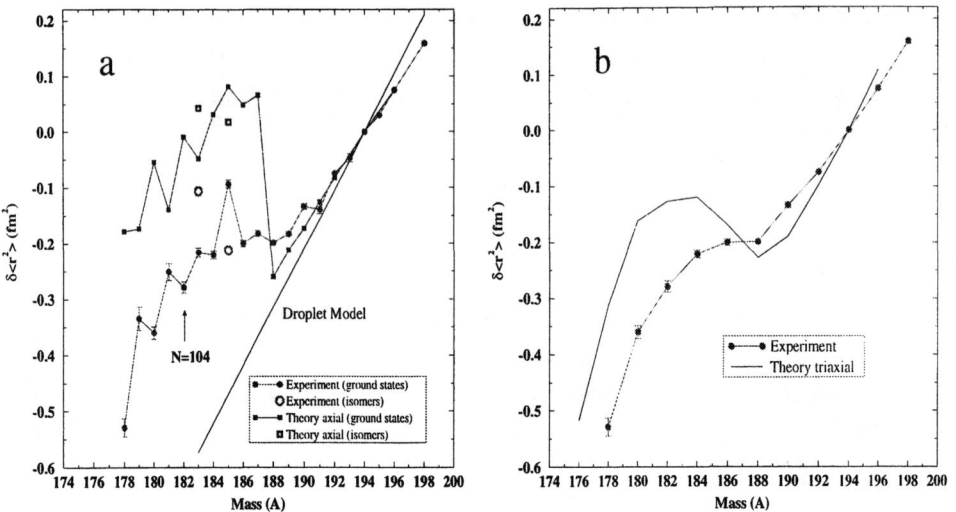

FIGURE 2. Comparison between the experimental $\delta < r_c^2 >$ with the microscopic Hartree-Fock-Bogolyubov calculations.

Discussion

If we assume a strong coupling scheme and axial symmetry, the β parameter can be extracted from the intrinsic quadrupole moment Q_0 deduced from Q_s^{corr}. This leads to $\beta(^{183}\text{Pt}^m)=+0.242(18)$ and $\beta(^{185}\text{Pt}^g)=+0.229(10)$. These values are in perfect agreement with the $<\beta^2>^{1/2}$ extracted from $\delta<r_c^2>$ (see Table 1) which shows that axial symmetry is fully justified for the two states. For the other odd nuclei, indications of axial symmetry are obtained from Total Routhian Surface calculations [5] that give $\gamma = 0°$ for ^{183g,181,179}Pt. Moreover the measured μ_I are in satisfactory agreement with the calculated ones [5] assuming axial symmetry.

To try to interpret this odd-even staggering, we have carried out microscopic Hartree-Fock-Bogolyubov calculations as described in Ref. [17,18]. Figure 2a shows the calculated values assuming axial symmetry. The differences of the radius between the ground states and the isomers in ^{185}Pt and ^{183}Pt are well reproduced, the $7/2^-$ and $9/2^+$ states being more deformed than the $1/2^-$ states. But in general the light even nuclei are predicted too deformed and the odd-even staggering is predicted to be normal. This normal OES is associated with pairing effects which increase $<r_c^2>$ for the even-even nuclei [19]. Recently Esser et al. extracted the γ and β deformation parameters from the analysis of B(E2) measurements and 2_1^+ and 2_2^+ energies, for the even isotopes $^{180-198}$Pt [20]. They found all those nuclei triaxial with γ varying from about $20°$ for ^{180}Pt to $33°$ for ^{190}Pt. Therefore, we have also performed microscopic HFB calculations on the even isotopes taking into account the triaxiality (see Figure 2b). Even though the γ-values are slightly lower and the β deformation a bit larger than the ones of Ref. [20], the general trend is well reproduced and particularly the kink appearing at mass 188. Moreover, the $<r_c^2>$ calculated using triaxiality are closer to the experimental values than those calculated with the axial symmetry. All this supports the triaxial shape in the even nuclei. The inverted OES results from this shape change between the odd and the even isotopes, provided that the odd nuclei are more deformed than the even ones.

CONCLUSION

In summary, the odd platinum isotopes can be interpreted as prolate shaped whereas the even ones are rather triaxial (with $\gamma \leq 20^o$ for $A < 186$). The inverted OES can be compared with the huge one occurring in the mercury isotopes below $N=106$, that is likely associated with alternating shape transitions, the odd isotopes being prolate and the even ones having a small oblate deformation [2]. In mercury, the large γ variation is associated with a large change in β. The phenomenon observed in the corresponding platinum and mercury isotones is thus of the same kind: in mercury, the shape transition is between triaxial-oblate and prolate while in platinum it is between triaxial-prolate and prolate. In both cases, the inverted OES is the signature of a shape change.

REFERENCES

1. J. Bonn et al., Z. Phys. A **276**, 203 (1976).
2. G. Ulm et al., Z. Phys. A **325**, 247 (1986).
3. J.K.P. Lee et al., Phys. Rev. **C38**, 2985 (1988).
4. H.T. Duong et al., Phys. Lett. B **217**, 401 (1989).
5. T. Hilberath et al., Z. Phys. A **342**, 1 (1992).
6. E. Hagberg et al., Phys. Lett. **78B**, 44 (1978).
7. J. Sauvage et al., Nucl. Phys. **A592**, 221 (1995).
8. G.D. Dracoulis et al., J. Phys. G **12**, L97 (1986).
9. F. Le Blanc et al., in *Proceedings of the 8th International Symposium on Capture Gamma-Ray Spectroscopy and Related Topics, Fribourg, Switzerland, 1993*, edited by J. Kern (World Scientific, Singapore, 1994) p. 1001.
10. F. Le Blanc et al., Phys. Rev. Lett. **79**, 2213 (1997).
11. W. Childs and K. Cheng, Phys. Rev. **A30**, 667 (1984).
12. S. Büttgenbach et al., Z. Phys. A **317**, 237 (1984).
13. P. Raghavan, At. Data Nucl. Data Tables **42**, 189 (1989).
14. K. Heilig and A. Steudel, At. Data Nucl. Data Tables **14**, 613 (1974).
15. E.C. Seltzer, Phys. Rev. **188**, 1916 (1969).
16. S. Raman et al., At. Data Nucl. Data Tables **42**, 1 (1989).
17. J. Dechargé and D. Gogny, Phys. Rev. **C21**, 1568 (1980).
18. M. Girod et al., Phys. Rev. **C37**, 2600 (1988).
19. D. Zawischa, Phys. Lett. B **155**, 309 (1985).
20. L. Esser et al., Phys. Rev. **C55**, 206 (1997).

Nuclear radii of ^{14}Be and ^{17,19}B

T. Suzuki[*1], K. Sümmerer[†], O. Bochkarev[‡], L. Chulkov[‡],
D. Cortina[†], H. Geissel[†], M. Hellström[†2], M. Ivanov[#], R. Janik[#],
K. Kimura[*], T. Kobayashi[*3], A. A. Korsheninnikov[*],
G. Münzenberg[†], F. Nickel[†], A. A. Ogloblin[‡],
A. Ozawa[*], M. Pfützner[†4], V. Pribora[‡], H. Simon[†],
B. Sitár[#], P. Strmeň[#], I. Tanihata[*], M. Winkler[†], K. Yoshida[*]

[*] *The Institute of Physical and Chemical Research (RIKEN), Wako, Saitama 351-0198, Japan*
[†] *Gesellschaft für Schwerionenforschung (GSI), D-64291 Darmstadt, Germany*
[‡] *Kurchatov Institute, Kurchatov sq. 1, 123182 Moscow, Russia*
[#] *Faculty of Mathematic and Physics, Comenius University, 84255 Bratislava, Slovak Republic*
[*] *Faculty of Engineering, Nagasaki Institute of Applied Science, Nagasaki 851-0193, Japan*

Abstract. The interaction cross-sections (σ_I) of light radioactive nuclei close to the neutron drip line (^{17,19}B, ^{14}Be) have been measured at around 800 A MeV. The effective mean–square–matter radii of these nuclei have been deduced from σ_I by a Glauber-type calculation. A large increase in the radius from ^{15}B to ^{17}B has been observed. A similar rate of increase is seen in ^{17}B to ^{19}B and in ^{12}Be to ^{14}Be. Questions concerning the neutron halo structure in ^{17,19}B are discussed.

INTRODUCTION

The neutron excess of ^{19}B ($T_z = \frac{9}{2}$) has one of the largest values currently known in the low-mass region of the nuclide chart. Since the first observation of ^{14}Be and ^{17}B in 1973 [1] and ^{19}B in 1984 [2], interest in these nuclei has greatly increased. Some properties, such as the mass and matter radius, have been studied. The two-neutron separation energy (S_{2n}) is known to be 1.39 ±0.14 MeV for ^{17}B and 1.34±0.11 MeV for ^{14}Be, and is estimated to be 0.46±0.43 MeV for ^{19}B [3]. The matter radius of ^{14}Be is reported to be 3.36±0.19 fm by Liatard *et al.* [4] and 3.11±0.38 fm by Tanihata *et al.* [5]. The authors of ref. [4] also reported

---

[1]) Present address: Department of Physics, Niigata University, Niigata 950-2181, Japan
[2]) Present address: Division of Cosmic and Subatomic Physics, Lund University, S-22100 Lund, Sweden
[3]) Present address: Department of Physics, Tohoku University, Miyagi 980-8578, Japan
[4]) Present address: Institute of Experimental Physics, Warsaw University, Hoza 69 00-691, Poland

a matter radius of ^{17}B as 4.10±0.46 fm, while Ozawa et al. reported 3.0±0.6 fm [6]. These anomalously large matter radii have been interpreted as being an indication of a neutron halo, which is a specific character seen in very neutron-rich and loosely bound nuclei, such as ^{11}Li and ^{11}Be. The uncertainty of the previously reported r.m.s. radius is rather large, so that a quantitative comparison between theoretical predictions and the experimental value was not possible. It should be noted here that no experimental data existed concerning the radius of ^{19}B. The narrow longitudinal momentum distribution of ^{12}Be from the breakup of ^{14}Be [7] suggests a halo structure in ^{14}Be. Thus, it is interesting to determine the effective root-mean-square (r.m.s.) matter radii ($\tilde{r}_m \equiv <r_m^2>^{1/2}$) of these nuclei.

In this paper we report on a new result obtained from an experiment at the FRS/GSI, to measure the interaction cross section (σ_I) of ^{17,19}B, ^{14}Be on a carbon target at 880, 740 and 850 A MeV, respectively.

EXPERIMENT AND ANALYSIS

Boron isotopes were produced as secondary beams through the projectile fragmentation of ^{40}Ar accelerated by the heavy–ion synchrotron SIS. A maximum beam intensity of 1×10^{10} particles per spill was used to produce ^{19}B. The experimental setup was essentially the same as that used in our previous studies [8–10], except for the use of newly constructed TPCs [11], replacing the MWPCs. The rigidity acceptance of ±0.5% allowed us to measure σ_I of ^{14}Be simultaneously under the magnet setting for ^{19}B. The isotopes produced in a production target of Be (4007 mg/cm^2) were separated by rigidity in the first half of the FRS. The rigidity-separated isotopes were identified before incidence on a reaction target by velocities [time–of–flight (TOF)] and by charges (pulse height in scintillation counters). No contamination of more than 0.1% was observed in any selected isotope beam. The σ_I value was measured by a transmission experiment using the second half of the FRS. The obtained σ_I values are listed in Table 1. The largest systematic uncertainty in σ_I, which amounts to about 0.8%, stems from the uncertainty in estimating the particle transmission through the second half of the FRS for non-interacting B/Be particles.

The $\tilde{r}_m$ values of the nucleon distribution were deduced from σ_I using Glauber–model calculations by assuming a harmonic–oscillator (HO) type density distribution including the contribution from the sd–shell:

$\rho_n(r) = 4\pi^{-3/2}\lambda^{-3}(1-1/A)^{-3/2}N/(N+8)exp(-x^2)(1+2x^2+(N-8)/15x^4)$,
$\rho_p(r) = 2\pi^{-3/2}\lambda^{-3}(1-1/A)^{-3/2}exp(-x^2)(1+(Z-2)/3x^2)$,

where $x = (r/\lambda)^2 A/(A-1)$ and λ denotes the width parameter. The details are described in ref. [12]. In ref. [12], the r.m.s. charge radii are said to agree well with those from electron–scattering experiments for stable light nuclei. The parameter λ was determined so as to fit to the experimental σ_I. Here, we assumed the same width for protons and neutrons, though mean–field models predict different potentials for protons and for neutrons in nuclei with a large neutron excess. This

TABLE 1. Interaction cross-sections (σ_I) in mb and the effective rms radii of the nucleon distributions in fm. $\tilde{r}_m$ is the effective $r.m.s.$ radii for the point nucleon distributions, deduced by assuming harmonic-oscillator distributions. The last three columns show the ratio of the valence radius to the core radius ($\tilde{r}_v/\tilde{r}_c$) from ref. [13] for $l = 0, 2$ and that from eq. (1) with the experimental data. See text.

Nuclei	σ_I	$\tilde{r}_m$	ref. [13]		eq. (1)
			$l=0$	$l=2$	
^{14}Be	1082 (34)	2.94 (9)	2.40	1.40	1.9 (0.2)
^{17}B	1118 (22)	2.90 (6)	2.35	1.45	1.9 (0.1)
^{19}B	1219 (83)	3.11 (13)			1.7 (0.3)

assumption is justified based on the insensitivity of the Glauber calculations to the shape of density distributions. The resultant $r.m.s.$ radii are listed in Table 1. The $\tilde{r}_m$ value of ^{17}B and of ^{14}Be in this work is consistent with the previous value [5,6], respectively, and has a higher accuracy.

Figure 1 shows the mass number dependence of $\tilde{r}_m$ for both B and Be isotopes. It can be seen that the $\tilde{r}_m$ value is almost constant during filling of the $p_{1/2}$ orbital, except for ^{11}Be. This is a quite general behavior for p-shell stable nuclei where all of them have nearly the same matter radius. Afterwards, it starts to gradually increase, reflecting the occupancy in the sd-shell. A large increase of the radius from ^{15}B to ^{17}B is observed, and a similar rate of increase is seen in ^{17}B to ^{19}B and in ^{12}Be to ^{14}Be.

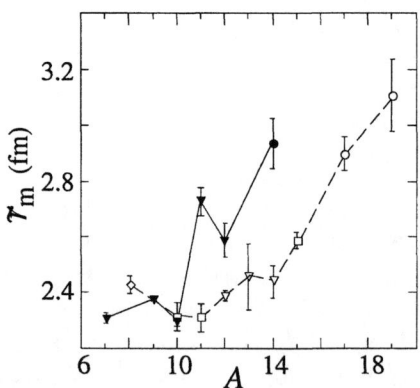

FIGURE 1. Effective $r.m.s.$ matter radii for Be and B isotopes. The closed and open symbols are for Be and B isotopes, respectively. The circles are data points obtained in this work. The squares, triangles and rhombus are from ref. [15], [5] and [14], respectively. The curves are to guide the eye.

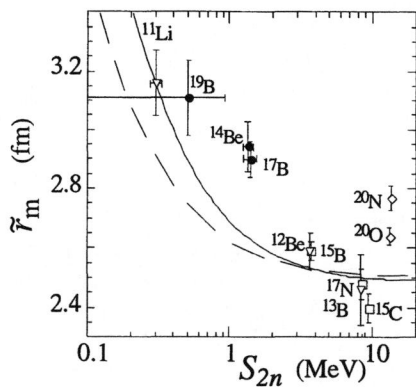

FIGURE 2. The r.m.s. matter radius versus the two-neutron separation energy. The closed circles are data points obtained in this work. The squares, triangles and rhombuses are from refs. [6,15], [5] and [19], respectively. The results of calculations from eq. (1) are shown by the dashed and solid curve for ^{19}B and ^{11}Li.

DISCUSSIONS

Could the large sizes be due to the large neutron halo in ^{17,19}B? In fig. 2, the $\tilde{r}_m$ values of odd-mass B isotopes and those of even-mass Be isotopes ($N \geq 8$) are plotted against S_{2n} in order to estimate their radii, since there is a large sensitivity of the spatial structure to the separation energy close to the threshold [16]. For a comparison, some p-shell nuclei with $N \geq 8$ are also plotted. It can be seen that the lower is S_{2n} the larger are $\tilde{r}_m$ values. The curves in the figure indicate the calculated matter radius for a given S_{2n} with the equation

$$\tilde{r}_m^2(A) = \frac{A-k}{A}[\tilde{r}_c^2 + \frac{k}{A}\tilde{r}_v^2] \qquad (1),$$

where k and $\tilde{r}_c$ represent the number of valence neutrons ($k=2$) and the core radius, respectively. This was introduced by Hansen and Jonson [17] in 1987. The effect of S_{2n} appears through the decay length, ρ ($=\hbar/(2\mu S_{2n})^{1/2}$), where μ is the reduced mass of the system. The $\tilde{r}_c$ value was chosen to be 2.5 fm, which is a typical value for the majority of the p-shell nuclei. The square-potential radius R was adjusted to reproduce $\tilde{r}_c$ for a large separation energy. Here we assumed the idea of a "core" plus "2n" picture in these nuclei ($k=2$). We also assumed a dineutron picture in the two halo neutrons for simplicity. The former assumption is justified in the case of ^{14}Be by the narrow longitudinal momentum distributions of ^{12}Be from the breakup of ^{14}Be [7], while the ^{17}B structure might be a tetraneutron around ^{13}B [18]; the same holds for ^{19}B. The curves reproduce the tendency of the data, although the uncertainty in data for ^{19}B is still large. Thus, one may expect a halo structure in

^{19}B under the assumptions mentioned above.

The $\tilde{r}_m$ value of ^{20}N deviates by ~0.2 fm from $\tilde{r}_c$, which can be understood to be due to the formation of a neutron skin in ^{20}N [19]. The $\tilde{r}_m$ values of ^{17}B and ^{14}Be are much larger, where the deviation from $\tilde{r}_c$ amounts up to ~0.4 fm. These facts suggest neutron–skin formation in these nuclei. Qualitatively, the term "neutron skin" describes an *excess* of neutrons at the nuclear surface, whereas "neutron halo" stands for such an *excess* plus a *tail* of the neutron-density distribution. The simple model apparently failed to reproduce the $\tilde{r}_m$ values of ^{17}B and ^{14}Be. As suggested by the authors of ref. [7], there is a strong correlation between the two halo neutrons, which accounts for the rather strong binding of the two neutrons in halo nuclei ^{14}Be. Both the $\tilde{r}_m$ value and the S_{2n} of ^{17}B are similar to those of ^{14}Be. Moreover, those quantities of its core ^{15}B are also similar to those of ^{12}Be. Thus, one may expect the halo structure in ^{17}B.

It is interesting to find the valence r.m.s. radius ($\tilde{r}_v$) using eq. (1). Taking $\tilde{r}_m$ of ^{15}B, ^{17}B, and ^{12}Be as the core radius in ^{17}B, ^{19}B, and ^{14}Be, the $\tilde{r}_v$ value for ^{17}B, ^{19}B, ^{14}Be becomes 4.9±0.4 fm 4.8±0.9 fm, and 4.9±0.5 fm, respectively. The $\tilde{r}_v$ values for all three nuclei are nearly the same within the experimental uncertainties. The ratio ($\tilde{r}_v/\tilde{r}_c$) becomes 1.9±0.1, 1.7±0.3 and 1.9±0.2 for ^{17}B, ^{19}B, and ^{14}Be, respectively. The ratio ($\tilde{r}_v/\tilde{r}_c$) can be compared with that calculated in ref. [13], where the valence radius is computed for various stable and unstable nuclei. In ref. [13], the criteria to identify the halo nuclide is simply given by the inequality ($\tilde{r}_v/\tilde{r}_c$) $\gtrsim$ 2 from known halo systems, whereas the ratio lies between 1.1 to 1.6 for most nuclei. Therefore, the ratio under the assumption of "core" plus "2n" supports the idea of a halo structure in ^{17}B as well as in ^{14}Be.

Next, we assume that the ^{19}B (^{17}B) structure is a tetraneutron around ^{15}B (^{13}B) (k=4). The ratio stays 1.9±0.2 for ^{17}B and increases to 2.0±0.2 for ^{19}B. In ^{8}He, the ratio increases from 1.4±0.1 (k=2) to 2.7±0.1 (k=4) using the known matter radii [20]. This suggests that the situation for ^{19}B is similar to that for ^{8}He. In He isotopes ^{6}He consists of an inert ^{4}He core plus two neutrons, whereas ^{8}He does not have an inert ^{6}He core [20].

SUMMARY

In summary, interaction cross sections for ^{17,19}B and ^{14}Be on a carbon target have been measured at 880, 740, and 850 A MeV, respectively. Nuclear r.m.s matter radii have been derived from the measured σ_I. The $\tilde{r}_m$ value of ^{17}B and of ^{14}Be in this work is consistent with the previous value, respectively, and has a higher accuracy. A large increase in the radius from ^{15}B to ^{17}B is observed, and a similar rate of increase is seen in ^{17}B to ^{19}B and in ^{12}Be to ^{14}Be. Assuming the "core" plus "1 dineutron" structure in these nuclei, the matter radius of ^{19}B could be understood from its two–neutron separation energy. However, the simple model failed to explain the matter radii of ^{17}B and ^{14}Be. Both of them were found to have a similar matter radius and separation energy. The existence of a neutron skin in these nuclei was suggested. The existence of a halo structure in ^{17}B was expected in analogy to the case of ^{14}Be. Then the possible neutron halo structure in ^{17,19}B was considered in terms of the valence radius. Assuming the "core" plus "2n" structure in these nuclei, the ratio of the valence radius to the core radius of ^{17}B as well as that of ^{14}Be supported the idea of a neutron halo. A similar estimation with the assumption of a "core" plus tetra–neutron suggested that the situation of ^{19}B is similar to that of ^{8}He. It would be useful to determine the width of the momentum distribution and/or the two (four) neutron–removal cross section after the breakup of ^{17}B and ^{19}B for a further discussion on this issue.

REFERENCES

1. J.D.Bowman et al., Phys. Rev. Lett. 31 (1973) 614.
2. J.A.Musser and J.D. Stevenson, Phys. Rev. Lett. 53 (1984) 958.
3. G. Audi, O. Bersillon, J. Blachot, and A. H. Wapstra, Nucl. Phys. A 624 (1997) 1.
4. E. Liatard et al., Europhys. Lett. 13 (1990) 401.
5. I. Tanihata et al., Phys. Lett. B 206 (1988) 592.
6. A. Ozawa et al., Phys. Lett. B 334 (1994) 18.
7. M. Zahar et al., Phys. Rev. C 48 (1993) R1484.
8. T. Suzuki et al., Phys. Rev. Lett. 75 (1995) 3241.
9. L. Chulkov et al., Nucl. Phys. A 603 (1996) 219.
10. T. Suzuki et al., Nucl. Phys. (1998) in press.
11. T. Baumann et al., Acta. Physica 37 (1996) 3.
12. I. Tanihata et al., Phys. Rev. Lett. 55 (1985) 2676.
13. R. Sherr, Phys. Rev. C 54 (1996) 1177.
14. M. Obuti et al., Nucl. Phys. A 609 (1996) 74.
15. A. Ozawa et al., Nucl. Phys. A 608 (1996) 63.
16. B. Jonson and K. Riisager, preprint, Göteborg University, CTHSP-97/08.
17. P. G. Hansen and B. Jonson, Europhys. Lett. 4 (1987) 4.
18. P. Descouvemont, Nucl. Phys. A 581 (1995) 61.
19. O. V. Bochkarev et al., Eur. Phys. J. A 1 (1998) 15.
20. I. Tanihata et al., Phys. Lett.B 289 (1992) 261.

MASS MEASUREMENT IN THE fp-SHELL USING THE TOFI SPECTROMETER

Y. Bai[1,2], D. J. Vieira[1], H. L. Seifert[1] and J. M. Wouters[1]

[1]*Los Alamos National Laboratory, Los Alamos, NM 87545*
[2]*Physics Department, Utah State University, Logan, UT 84322*

Abstract. The masses of 48 neutron-rich nuclei extending from ^{55}Sc to ^{75}Cu have been determined from the final set of data to be acquired with the time-of-flight-isochronous (TOFI) spectrometer. The masses of eight isotopes (^{68}Fe, ^{70,71}Co, ^{73}Ni, and $^{72-75}$Cu) are reported for the first time. The resulting masses now tie in neatly with the masses of previously measured neutron-rich Zn and Ga isotopes determined from fission product β-endpoint measurements. A careful evaluation of the calibration sensitivity is made with respect to inclusion or exclusion of these heavy known species and excellent calibration stability is found. Contrasting these results with previous TOFI measurements, we find that these new results fall between the results of Tu *et al.*(1) which trend to slightly less bound masses as one proceeds to the most neutron-rich species and Seifert *et al.*(2) which shows the opposite trend. Good agreement with the predictions of several mass models and Audi-Wapstra systematics are found.

TOFI MEASUREMENTS

This is the sixth and final set of mass measurements to be reported on using the time-of-flight isochronous (TOFI) recoil spectrometer at Los Alamos. Our long-term goal has been the systematic measurement of neutron-rich nuclei ranging from ^{11}Li as produced in proton-induced target fragmentation to the light-mass fission fragment region. Herein we achieve this goal by completing the measurement of many neutron-rich species of Fe, Co, Ni, and Cu such that they tie in with existing mass measurements of fission product Zn and Ga nuclei (see Fig. 1). A more detailed account of this work is forthcoming (3).

As in previous TOFI experiments, an 800-μA, 800-MeV proton beam induced fragmentation and fission reactions on a thin (~1.0 mg/cm^2) ^{nat}Th target. A fraction of the reaction products which recoil out of the target are captured by a beam transport line (located at a target angle of ~90°), separated in a low-resolution, separated-sector E x B mass-to-charge filter which is integrated into the transport line, and introduced into an isochronous, 4-dipole magnetic spectrometer - TOFI. A measurement of the recoil's time-of-flight through the spectrometer provides a high-precision measurement

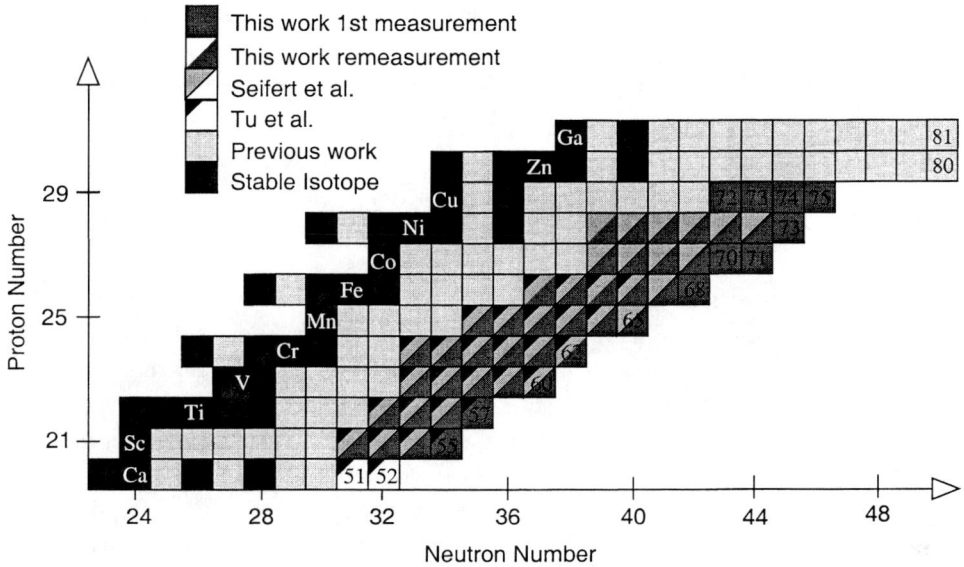

FIGURE 1. A section of the chart of the nuclides showing previous mass measurements and those of the TOFI group.

of the recoil's mass-to-charge ratio. Additional measurements of the recoil's velocity, stopping power, and total kinetic energy (using thin-foil, secondary-electron, fast-timing detectors and a Bragg-curve gas ionization counter) serve to uniquely identify each recoil according to its charge state (Q) and atomic number (Z). At one setting of the transport line and spectrometer, a relatively wide variety of nuclei (typically over a hundred different species in two or three charge states) are measured. Some have well-known masses and are used as calibrants; others are treated as unknowns. By accumulating sufficient statistics, direct mass measurements are extracted from the Z- and Q-gated time-of-flight (mass-to-charge) spectra.

The measurements that we report on here are performed in the same manner as those outlined by Seifert et al. (2). For brevity we will not go into details, but only mention that the settings of the transport line, spectrometer, and Bragg counter were optimized for slightly higher-energy, higher-Z reaction products than that of Seifert et al. Data for this experiment was acquired over a running period of two months.

After extensive analysis the mass measurement of forty-eight neutron-rich nuclei ranging from ^{52}Sc to ^{73}Cu were determined. Eight of the masses have been measured for the first time (see table 1) and forty are remeasurements of other neutron-rich nuclei previously measured by the TOFI group (1,2). Compared to these previous TOFI measurements, this data set contains heavier nuclei including several neutron-rich nuclei of zinc (Z=30) and gallium (Z=31) with well-known masses that we can use as calibrants. This enables the unknown region to be fully bracketed by both lighter and

TABLE 1. Mass excesses for previously unmeasured nuclei. 1σ errors are given in parentheses.

^{A}Z	Δm (MeV)	^{A}Z	Δm (MeV)
^{68}Fe	-43.1 (0.5)	^{72}Cu	-59.7 (0.5)
^{70}Co	-45.6 (0.6)	^{73}Cu	-58.45 (0.35)
^{71}Co	-43.9 (0.6)	^{74}Cu	-55.3 (0.4)
^{73}Ni	-48.9 (0.5)	^{75}Cu	-54.1 (0.7)

heavier mass calibrants which improves the reliability of the calibration. This improvement is shown in Figure 2 where we plot the deduced mass difference for the unknowns when we exclude or include the heavy zinc and gallium isotopes as calibrants. A small (less than 1σ), but steadily growing deviation is observed as one proceeds to the most neutron-rich species of cobalt, nickel, and copper. In this way we have increased confidence in this data over that of Seifert *et al.* (2), especially for the most neutron-rich species which they published (e.g., $^{65-67}$Fe, ^{68,69}Co and $^{70-72}$Ni). This effect appears to be verified when the two data sets are contrasted against one another. A small, but consistent trend towards increased binding is noted in the difference between Seifert *et al.* and these results as one progresses towards the most neutron-rich species. A comparison with Tu *et al.* (1) gives slightly better agreement, but deviations towards less binding are noted for ^{54,55}Sc, ^{59,60}V and ^{61,62}Cr. Consequently, the results of this work generally fall in between our two previous measurements with most masses agreeing within 2σ.

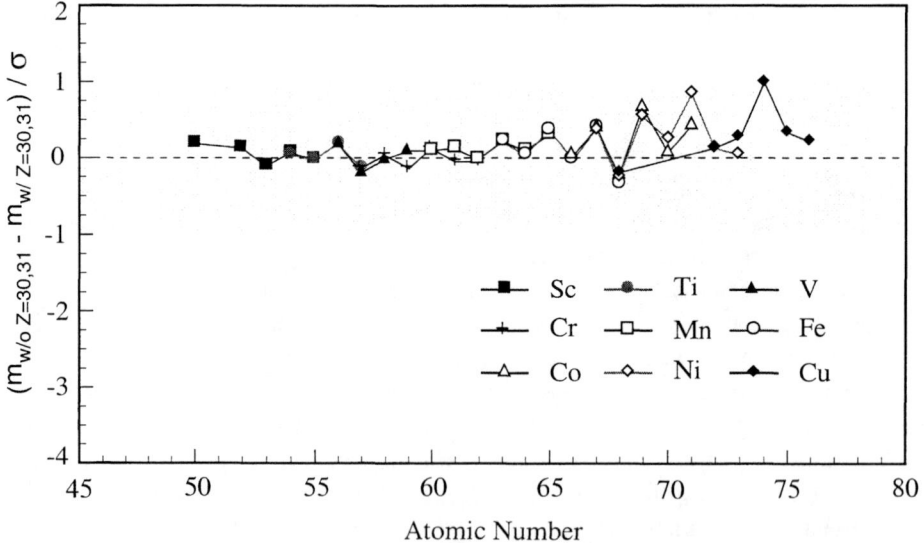

FIGURE 2. Mass difference for unknowns calculated without and with known mass Z=30 and 31 species divided by 1σ errors. The shaded area represents the 1σ variance region.

DISCUSSION

Since the mass resolution ($\Delta m/m \sim 2 \times 10^{-4}$ FWHM) of these measurements is insufficient to intrinsically resolve isomer and ground state masses should a sufficiently long-lived isomer exist, we have excluded from the calibration data set all nuclei which are known (4) to have isomers of excitation energy greater than 150 keV and lifetimes of 0.2 µs or longer. In addition we have also excluded $^{67-69}$Ni isotopes because of known or suspected isomers in these nuclei. Although such steps are appropriate in keeping the calibration set as pure as possible, it does not preclude the possibility of measuring a mixed isomer + ground state mass. In the present case, we can clearly substantiate our earlier work (2) where the population of a high-lying isomer in ^{68}Ni is indicated. However, without knowledge of the isomer excitation, lifetime, and isomer-to-ground state population ratio (e.g. isomer ratio) in 800 MeV protons on ^{nat}Th reactions, there is no way to correct our data. The reader should be aware of this potential problem.

Examining two-neutron separation energies, a smoothly decreasing trend is observed. No indication of a subshell closer is indicated at N=40 (i.e. with the completion of the fp shell) for neutron-rich Fe, Co, and Ni nuclei. A local, but small decrease in S_{2n} trend is noted at ^{66}Co suggesting the existence of an unknown isomer.

Comparing experiment to a variety of mass models, we find that the predictions of Möller and Nix (5) and Jänecke and Masson (6) are generally the best overall. The shell model predictions of Richter et al. (7) work well up to N=39,40 where in Mn and Fe nuclei a dramatic breakdown occurs due to basis space limitations. The extended Thomas-Fermi calculations of Aboussir et al. (8) tend to predict too much binding in the Sc and V isotopes, but obtain good agreement for Fe and heavier nuclei. Further discussion is contained in Ref. 3.

ACKNOWLEDGMENTS

We wish to thank K.E.G. Löbner for the continued use of the Bragg-curve gas ionization counter. This work was performed under the auspices of the U.S. Department of Energy.

REFERENCES

1. Tu, X. L. et al., Z. Phys. A **337**, 361 (1990).
2. H. L. Seifert et al., Z. Phys. A **349**, 25 (1994).
3. Bai, Y. et al. (to be published).
4. Audi, G. et al., Nucl. Phys. **A624**, 1 (1997).
5. Möller, P. and Nix, J. R., At. Data Nucl. Data Tables **59**, 185 (1995).
6. Jänecke, J. and Masson, P. J., At. Data Nucl. Data Tables **39**, 265 (1988).
7. Richter, W. A., Van der Merwe, M. G., Brown, B. A., Nucl. Phys. **A586**, 445 (1995).
8. Aboussir, Y. et al., At. Data Nucl. Data Tables **61**, 127 (1995).

Application of the Laser Ion Source for Isotope Shift and Hyperfine Structure Investigations

A.E. Barzakh, I.Ya. Chubukov, D.V. Fedorov, V.N.Panteleev, M.D. Seliverstov, Yu.M. Volkov

Petersburg Nuclear Physics Institute, 188350, Gatchina, Leningrad district, Russia

Abstract. A high-efficient method for measuring isotope shifts and hyperfine structures in optical transitions of radioactive atoms is presented. The method is based on application of laser resonance ionization in the mass-separator ion source. The sensitivity of the method is determined by a high efficiency of the laser ion source and low background of the detection system, making use of counting α-particles following the decay of the isotope under investigation. The possibilities of this method are shown in the experiment with ^{155}Yb and ^{154}Tm (I=9). The isotope shifts and electromagnetic moments have been measured.

INTRODUCTION

Studies of the isotopic dependencies of charge radii and electromagnetic moments for nuclei far from stability have been developed very efficiently during the last decade. Newly obtained information enables one to draw important conclusions about the structure of the nuclear ground states. It is of great value that new data allow to construct the two-dimensional picture of the nuclear properties (with the reference to N – neutron number and Z – proton number). The systematic trends in isotopic, isotonic and isobaric dependencies are caused by the fundamental properties of the nuclear forces. One of the most important characteristics of these dependencies is the slope of the corresponding curves, for instance, the value $\delta\langle r^2\rangle/\delta N$ for isotopic dependencies. This value prove to be remarkably constant (independently on Z) in the limits of the fixed region on the nuclide chart. This is the case, for example, for nuclei with N=82–88 and Z=55–68. Theoretical predictions for $\delta\langle r^2\rangle/\delta N$ prove to be very sensitive to the choice of an effective force in the framework of the HF method (1).

Recently the marked deviation from the established systematic behaviour was found for the Yb isotopes with N = 82–84 (2). It was shown that the radii of these isotopes increase essentially faster with the growth of neutron number than the radii in the other isotopic chains in this region. This effect has not found any satisfactory theoretical explanation yet. Thus it is of importance to continue the detailed experimental investigation of this region on the nuclide chart. In particular, it is necessary to determine relevant nuclear characteristics for odd Yb isotopes (A=153,155) and to look for the similar effect for adjacent elements, first of all – for Tm isotopes (Z=69) with N=82–87.

EXPERIMENTAL TECHNIQUE

To study nuclei very far from stability, an increase in sensitivity is needed. The conventional laser spectroscopic techniques failed for the isotopes with the production rate less than 10^3 atoms per second.

So we apply a new method – resonance ionization spectroscopy in the laser ion source (RIS/LIS). The method is based on the multistep resonance photoionization of atoms of the isotope under investigation immediately in the ion source of a mass-separator. The RIS/LIS method was tested in the trial experiment for ^{154,156}Yb (3). The isotope shifts for these isotopes have been measured. It was shown that this method allows to carry out measurements for isotopes with the production rate 10^2 atoms per second.

This method was applied for investigation of ^{155}Yb and ^{154}Tm (I=9). The following schemes of ionization were used:
1) for Yb atoms:

$6s^2\ ^1S_0 \xrightarrow{555.6\ nm} 6s6p\ ^3P_1 \xrightarrow{581.1\ nm} 4f^{13}(^2F_{7/2})6s^26p_{3/2} \xrightarrow{581.1\ nm}$ continuum,

2) for Tm atoms:

$4f^{13}6s^2 \xrightarrow{597.1\ nm} 4f^{13}6s6p \xrightarrow{600.3\ nm} 4f^{13}5d6s \xrightarrow{552.4\ nm}$ autoionizing state.

The laser frequency of the first excitation step was scanned. The optical spectrum represents the number of photoions detected at the exit of the mass-separator versus the frequency of the scanning laser radiation. The scheme of the experimental set-up is presented in Fig.1.

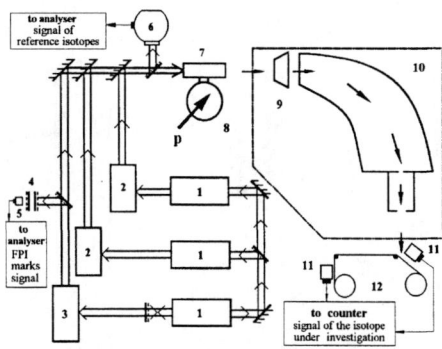

FIGURE 1. Experimental set-up for investigation of the α-radioactive isotopes. 1 – Cu-vapour lasers; 2 – broadband dye lasers; 3 – narrowband scanning dye laser; 4 – Fabry-Perot interferometer; 5 – photodiod; 6 – reference chamber; 7 – laser ion source; 8 – target; 9 – extraction electrod; 10 – mass-separator; 11 – α-detectors; 12 – tape-driving device.

The Cu-vapour lasers (average power 9 W, repetition rate 9 kHz) pumped the narrowband (bandwidth 1 GHz) and two broadband (30 GHz) dye lasers. The rays of three lasers were merged in the niobium tube of the laser ion source (diameter of the laser beam was 1 mm). Average power of the narrowband scanning dye laser was about 50 mW, power of the broadband dye lasers, tuned to the second and the third transition frequency, was about 550 mW. Part of the scanning laser radiation was splitted off and directed on the Fabry-Perot interferometer (free spectral range 5 GHz) to produce the frequency marks for frequency scale calibration. Another part of the laser beam was introduced into reference chamber in order to receive a reference spectrum of the stable isotopes. A new high temperature tantalum foil target inside a tungsten container has been developed for this experiment. Radioactive atoms were produced in this target of the mass-separator by 1 GeV protons. Yb and Tm atoms were ionized in the laser ion source and extracted by the extraction electrode of mass-separator.

Photoions were detected at the mass-separator exit by counting α-particles of the characteristic α-lines in the ^{155}Yb (E_α=5194 keV) and ^{154}Tm (I=9) (E_α=5031 keV) decay spectra. The optical spectra of radioactive isotopes are shown in Fig. 2.

EXPERIMENTAL RESULTS

The values of the isotopic shift $\delta\nu_{155,168}$, $\delta\nu_{154,169}$ and hyperfine structure constants a_{155}, b_{155}, a_{154}, b_{154} for ^{155}Yb and ^{154}Tm (I=9) have been extracted from the experimental spectra:

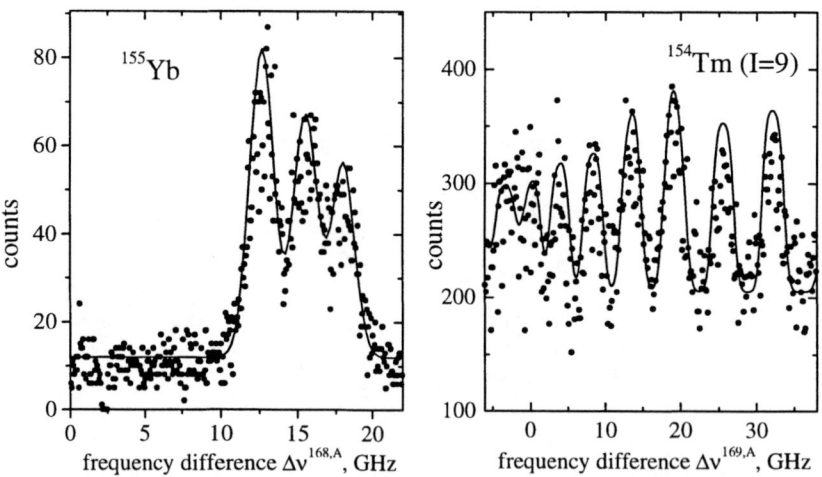

FIGURE 2. Hfs spectra for ^{154}Tm (I=9) and ^{155}Yb (I=7/2). Solid line – theoretical discription of these experimental spectra.

1) for ^{155}Yb: $\delta v_{155,168}= -15460(80)$ MHz, $a_{155}= -1050(25)$ MHz, $b_{155}= 160(80)$ MHz,
2) for ^{154}Tm (I=9): $\delta v_{154,169}= -15470(150)$ MHz, $a_{154}= 533(9)$ MHz, $b_{155}= 1020(500)$ MHz.

The magnetic dipole moments μ_{155}, μ_{154} and electric quadrupole moments Q_{155}, Q_{154} were evaluated with the aid of the well known scaling relation, which connected the values in question with the moments and hfs constants of the stable isotope (4). The results are:

$\mu_{155} = -0.913(22)$ nm , $Q_{155} = -0.5(3)$ b , $\mu_{154} = 5.92(10)$ nm , $Q_{154} = -0.75(3)$ b.
We neglected the hfs anomaly, which does not exceed 1% usually (4).

Isotopic change of the charge radius $\delta \langle r^2 \rangle_{AA'}$ can be determined through the simple formula (4):

$$\delta v_{A,A_0} = F \cdot \delta < r^2 >_{A,A_0} + M \frac{A - A_0}{A \cdot A_0}. \qquad (1)$$

where F – electronic factor, M – mass shift constant, $A_0=168$ for Yb and $A_0=169$ for Tm. With $F = -11.9$ GHz/fm^2 and $M = 296$ GHz from (2) for Yb and $F = -10.3$ GHz/fm^2 and $M = 268$ GHz (5), for Tm we obtain:

$\delta \langle r^2 \rangle_{155,168}$ (Yb) $=1.316$ fm^2, $\delta \langle r^2 \rangle_{154,169}$ (Tm)$=1.517$ fm^2.

The error of this value is determined by the uncertainty of the semiempirical calculation of the F and M values. It is supposed to be about 5% (4).

The smooth behaviour of isotopic dependency of $\delta \langle r^2 \rangle$ at ^{155}Yb points to the similarity of the structure of adjacent Yb nuclei with A=154–156. In particular, they must have the identical deformation, otherwise a sudden jump on the $\delta \langle r^2 \rangle$ curve would be observed. The data for Tm isotope may be compared with the analogous data for Ho isotope with the same neutron number and the same spin ^{152}Ho (5): $\mu_{152} = 5.92(5)$ nm , $Q_{152} =-1.3(8)$ b . The similarity of Ho and Tm data along with the smooth behaviour of Tm $\delta \langle r^2 \rangle$ curve evidences to the close conformity of the structure of the corresponding nuclei.

ACKNOWLEDGEMENTS

We would like to express gratitude to T.T. Fedorov, A.G. Poljakov and V.N. Fedoseev (IS RAS, Moscow) for helpful assistance.

REFERENCES

1. Barzakh, A.E., and Denisov, V.P., *Zs.Phys.* **A346,** 265 (1993).
2. Sprouse, G.D., et al., *Phys.Rev.Lett.* **63,** 1463 (1989).
3. Alkhazov, G.D., et al., *NIM* **B69,** 517 (1992).
4. Otten, E.W., *Treatise on heavy-ion science*, V.8, Plenum Publishing Corp, 1989, p. 517.
5. Alkhazov, G.D., et al., *Nucl. Phys.* **A477,** 37 (1988).

Mass Extrapolations in the Region of Deformed Rare Earth Nuclei

C.Borcea[a] and G.Audi[b]

[a]) IFIN-HH, P.O. Box MG-6, 76900 Bucharest-Magurele, Romania
[b]) CSNSM-Orsay, Bat.108, 91405 Orsay Campus, France

Abstract. A procedure based on the regularity property of the mass surface is proposed to make predictions for the masses of neutron rich deformed nuclei in the rare earth region. Tables are given for the estimated masses; they extend up to the presumed limit of the deformation region.

A striking aspect of the chart of nuclides in the rare earth region is the deep "gulf" present on the neutron rich side. In this "gulf" the last known isotopes are often only 4−6 neutrons away from the stable isotopes while model predictions for the neutron drip line lay much further away. The limit of nuclei for which masses are known comes even closer. The onset of important deformation effects starting above N=88−90 is another characteristic of the region. In addition, some isotopic chains contain a rather small number of measured masses, centered around the stable isotopes. All these facts make this region a difficult one for mass predictions. Indeed, a comparison between the values given by various models for the isotopes beyond the last measured one indicates a growing divergence when one goes away from the last experimental values.

Based on the global property of regularity of the surface of masses, a method has been developed [1] to extrapolate starting from the known masses into the adjacent regions. It starts from the observation that derivate quantities like S_{2n} or S_{2p} (which are not affected by the staggering effects due to pairing) align themselves on straight lines when displayed as a function of neutron or respectively proton number. That will suggest a quadratic dependence on n or p. This is valid only for a regular region in which neither shell (or subshell) closure appears, nor deformations in the ground state. At a closer look, these lines show a slight curvature. Consequently we tried a cubic (in n and z) local fit of the masses of nuclei comprised in between two magic numbers both for neutrons and protons. Perhaps the most convenient region to test such a procedure is that of nuclei having n and z in between magic numbers 28 and 50, as can be seen in [2]. Indeed, the result was quite encouraging: the rms deviation of the fitted values with respect to the data was 67 keV, while the same rms was higher for other model predictions [3]; e.g. 106 keV for the model of

TABLE 1. Mass excess predictions (in MeV) for the rare earth deformed nuclei situated between Xe and Ta. The predictions start after the last systematic value in the tables of Audi and Wapstra [6] and extend for each isotopic chain up to the expected end of the deformation region.

Nucleus	Mass excess (MeV)	Nucleus	Mass excess (MeV)	Nucleus	Mass excess (MeV)	Nucleus	Mass excess (MeV)
148Xe	−40.130	170Cs	84.240	174La	76.480	166Nd	−21.150
149Xe	−34.580	171Cs	91.290			167Nd	−15.270
150Xe	−30.630	172Cs	100.010	158Ce	−36.940	168Nd	−10.770
151Xe	−24.690			159Ce	−31.640	169Nd	−4.550
152Xe	−20.400	154Ba	−33.840	160Ce	−27.820	170Nd	.300
153Xe	−14.070	155Ba	−28.240	161Ce	−22.120	171Nd	6.810
154Xe	−9.400	156Ba	−24.180	162Ce	−17.980	172Nd	12.010
155Xe	−2.720	157Ba	−18.190	163Ce	−11.910	173Nd	18.780
156Xe	2.300	158Ba	−13.810	164Ce	−7.330	174Nd	24.350
157Xe	9.310	159Ba	−7.480	165Ce	−.920	175Nd	31.440
158Xe	14.680	160Ba	−2.720	166Ce	4.020	176Nd	37.380
159Xe	22.040	161Ba	3.960	167Ce	10.800	177Nd	44.700
160Xe	27.750	162Ba	9.110	168Ce	16.170		
161Xe	35.430	163Ba	16.150	169Ce	23.310	164Pm	−38.470
162Xe	41.480	164Ba	21.670	170Ce	29.050	165Pm	−34.930
163Xe	49.530	165Ba	29.020	171Ce	36.470	166Pm	−30.030
164Xe	55.880	166Ba	34.890	172Ce	42.580	167Pm	−26.030
165Xe	64.220	167Ba	42.570	173Ce	50.260	168Pm	−20.710
166Xe	70.730	168Ba	48.750	174Ce	56.710	169Pm	−16.240
167Xe	79.380	169Ba	56.760	175Ce	64.660	170Pm	−10.560
168Xe	86.210	170Ba	63.270			171Pm	−5.640
169Xe	95.200	171Ba	71.560	160Pr	−36.440	172Pm	.380
170Xe	102.360	172Ba	78.390	161Pr	−32.730	173Pm	5.610
171Xe	111.660	173Ba	87.000	162Pr	−27.590	174Pm	11.900
				163Pr	−23.480	175Pm	17.390
152Cs	−30.230	156La	−33.630	164Pr	−17.960	176Pm	24.000
153Cs	−26.040	157La	−29.690	165Pr	−13.430	177Pm	29.810
154Cs	−20.330	158La	−24.240	166Pr	−7.510	178Pm	36.570
155Cs	−15.750	159La	−19.950	167Pr	−2.600		
156Cs	−9.650	160La	−14.140	168Pr	3.700	166Sm	−40.660
157Cs	−4.710	161La	−9.440	169Pr	9.010	167Sm	−35.900
158Cs	1.720	162La	−3.310	170Pr	15.640	168Sm	−32.350
159Cs	7.000	163La	1.760	171Pr	21.360	169Sm	−27.190
160Cs	13.800	164La	8.260	172Pr	28.250	170Sm	−23.260
161Cs	19.430	165La	13.730	173Pr	34.310	171Sm	−17.710
162Cs	26.510	166La	20.540	174Pr	41.490	172Sm	−13.360
163Cs	32.500	167La	26.380	175Pr	47.960	173Sm	−7.500
164Cs	39.940	168La	33.520	176Pr	55.360	174Sm	−2.780
165Cs	46.300	169La	39.750			175Sm	3.370
166Cs	54.060	170La	47.210	162Nd	−39.370	176Sm	8.480
167Cs	60.770	171La	53.750	163Nd	−34.320	177Sm	14.920
168Cs	68.840	172La	61.510	164Nd	−30.710	178Sm	20.410
169Cs	75.850	173La	68.430	165Nd	−25.220	179Sm	27.060

TABLE 1. (continuation)

Nucleus	Mass excess (MeV)	Nucleus	Mass excess (MeV)	Nucleus	Mass excess (MeV)	Nucleus	Mass excess (MeV)
168Eu	−39.350	180Gd	5.700	183Dy	−1.290	183Tm	−27.900
169Eu	−35.920	181Gd	11.750			184Tm	−23.760
170Eu	−31.140			176Ho	−38.930	185Tm	−19.960
171Eu	−27.330	172Tb	−39.260	177Ho	−35.770	186Tm	−15.650
172Eu	−22.200	173Tb	−35.960	178Ho	−31.630		
173Eu	−18.000	174Tb	−31.440	179Ho	−28.120	182Yb	−38.580
174Eu	−12.530	175Tb	−27.760	180Ho	−23.690	183Yb	−34.830
175Eu	−7.990	176Tb	−22.990	181Ho	−19.880	184Yb	−32.070
176Eu	−2.250	177Tb	−19.020	182Ho	−15.080	185Yb	−28.000
177Eu	2.650	178Tb	−13.970	183Ho	−10.930	186Yb	−24.880
178Eu	8.630	179Tb	−9.620	184Ho	−5.990	187Yb	−20.630
179Eu	13.970	180Tb	−4.300				
180Eu	20.110	181Tb	.400	178Er	−40.020	185Lu	−33.580
		182Tb	5.920	179Er	−35.940	186Lu	−29.960
170Gd	−40.890			180Er	−33.100	187Lu	−26.810
171Gd	−36.310	174Dy	−40.590	181Er	−28.710	188Lu	−23.090
172Gd	−32.910	175Dy	−36.310	182Er	−25.430		
173Gd	−28.000	176Dy	−33.110	183Er	−20.690	187Hf	−32.950
174Gd	−24.260	177Dy	−28.530	184Er	−16.940	188Hf	−30.640
175Gd	−18.990	178Dy	−24.940	185Er	−12.100	189Hf	−26.970
176Gd	−14.900	179Dy	−20.080			190Hf	−25.140
177Gd	−9.330	180Dy	−16.180	180Tm	−38.010		
178Gd	−4.910	181Dy	−11.040	181Tm	−35.070	189Ta	−31.560
179Gd	.960	182Dy	−6.730	182Tm	−31.210	190Ta	−28.360

Duflo with 12 parameters [4], or 161 keV for the macroscopic-microscopic model of Moller [5]. In principle, the described method could provide reliable extrapolations for the next 4−5 masses, but in some particular cases its range of validity may extend further away. The procedure has been tested simply by excluding from the fit few (3−4) of the last known masses in each isotopic chain; the retrieved values agreed excellently with the real ones. For nuclei in the rare earth region the method encounters serious difficulties because here the regularity property is broken by the extra binding brought by the onset of deformation. However, one can still apply it to the region of masses with $50 \leq Z \leq 82$ and $82 \leq N \leq 126$ from which the deformed nuclei have been excluded. Though the number of nuclei left after this procedure is rather small, the fit is stable and leads to a hypothetical smooth mass surface for which the deformations are absent. By comparing to the real mass surface, the deformation region shows up prominently, presenting neat contours and a well developed symmetry The deformation sets in after N=88 and its amplitude grows gradually up to a maximum value; then it starts decreasing and disappears at N=116. The extension on Z ranges from Cs to Ir (with a small effect in both cases), having a maximum amplitude around Z=68. The position of maximum overbinding due to deformation along each isotopic chain varies from N=100 for small Z to N=106 for large Z. While for high Z the isotopic chains are almost complete from the point of view of measured masses, for lower Z the chains become ever shorter. Upper chains may therefore provide informations on the trends that can be used to complete the others. This operation is facilitated by the continuous comparison with the hypothetical undeformed mass surface where all isotopic chains should land at the end of the deformation region. The amplitude of the overbinding brought by the deformation could also be estimated and amounts to 5 MeV at the maximum of this effect (for ^{168}Dy). Interestingly, most of the systematic values given in the tables of Audi and Wapstra [6], lay very close or overlap the extrapolated values. Table 1 is a list of masses estimated by this procedure for nuclei supposed to belong to the deformation region and that are not yet measured, from Xe to Ta. Only the values placed after the systematic values of Audi and Wapstra are given.

New mass measurement in this region, the sole criterium of validity for extrapolations and mass models are therefore strongly advocated.

REFERENCES

1. C. Borcea and G. Audi, Rom. Jou. Phys. 38(1993), 455
2. C. Borcea et al., Nucl. Phys. A565(1993), 158
3. Web site: http://www-csnsm.in2p3.fr/amdc/
4. J. Duflo and A. P. Zuker, Phys. Rev. C52(1995), 23 and private communication
5. P. Moller et al., At. Data and Nucl. Data Tables 59(1995), 185
6. G. Audi and A. H. Wapstra, Nucl. Phys. A565(1993), 1

Mass Measurements of Proton-Rich Medium Mass Nuclides

D.S. Brenner,[1] B.D. Foy,[1] C.J. Barton,[1] C.N. Davids,[2] D. Seweryniak,[2] D. Blumenthal,[2] R.L. Gill,[3] N.V. Zamfir,[1,3,4,5] and D.D. Warner,[6]

[1]*Clark University, Worcester, MA 01610 USA*
[2]*Argonne National Laboratory, Argonne, IL 60439 USA*
[3]*Brookhaven National Laboratory, Upton, NY 11973 USA*
[4]*Institute of Atomic Physics, Bucharest Magurele, Romania*
[5]*WNSL Yale University, New Haven, CT 06520 USA*
[6]*CCLRC Daresbury Laboratory, Daresbury, Warrington WA4 4AD, United Kingdom*

Abstract. Beta-ray end-point energies for the decay of ^{150}Er and ^{151}Er were measured using a plastic scintillator - Ge, β^+ - γ coincidence spectrometer. Mass excess values are reported for these nuclides and for members of the α-decay chain which terminates at ^{151}Er. Our results are compared with extrapolations based on systematic properties for other nuclei.

EXPERIMENTAL METHODS

We have measured the decay of the 23.5 s ground state of ^{151}Er to ^{151}Ho using conventional $\beta-\gamma$ coincidence spectroscopy. ^{151}Er was produced by the ^{96}Mo(^{58}Ni,2pn) reaction at the ATLAS accelerator facility, Argonne National Laboratory. A 1 mg/cm^2 target of ^{96}Mo, enriched to an isotopic abundance of 97 %, was bombarded with a ~ 10 pnA beam of ^{58}Ni ions with incident energy of 250 MeV. Recoil products entered the Fragment Mass Analyzer (FMA) where they were separated according to their A/q values. FMA settings and slit positions were optimized for ^{151}Er such that the ion beam that was collected at the exit focal plane of the FMA on an aluminized mylar tape consisted primarily of ^{151}Er with a small amount of mass 150. At 48 s intervals the deposit spot was moved into a shielded environment where decay of the sample was measured using a Ge γ-ray detector and a cylindrical plastic scintillator β-ray detector positioned 180° relative to the Ge detector in close proximity to the activity spot. Singles and event-mode coincidence spectra were recorded for off-line analysis.

The Ge detector was calibrated using standard sources and well-known γ-rays produced in the experiment. The plastic scintillator was calibrated from β^+ spectrum endpoints for ^{65}Ga and ^{67}Ge, produced under similar experimental conditions by bombardment of ^{12}C with ^{58}Ni ions, and from decay of 35 s ^{151}Ho to the 527 keV level in ^{151}Dy, recorded during the course of the ^{151}Er experiment. The effects of summing in the plastic scintillator were carefully scrutinized and found to be minimal.

Positron spectra were generated by gating on intense γ-rays and time windows for each daughter nucleus, taking care to subtract the appropriate γ-ray and time backgrounds. In order to simulate a Fermi-Kurie analysis the positive square root of the counts was plotted as a function of energy bin. The known decay scheme was then used to guide the selection of the high energy region of the β^+ spectrum where only a single β^+ decay branch was thought to be present. The selected portion of the spectrum was quite linear

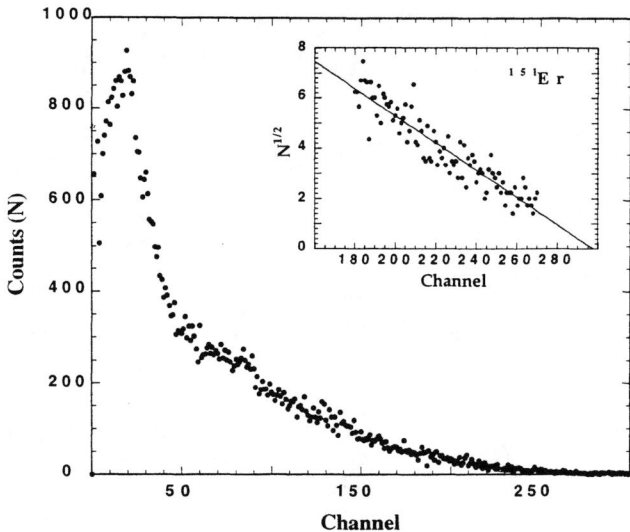

FIGURE 1. Positron spectrum of ^{151}Er gated by the 638 keV ^{151}Ho γ-ray. The inset shows a square root plot of the endpoint region. The linear fit to the data was made over the channel range 230 - 270 which corresponds to a β+ energy range of 2780 - 3240 keV. This excludes any contributions from feeding via the 1280 and 1833 keV levels.

and a least squares fit to the data was used to determine the endpoint channel. For a few cases a complete Fermi-Kurie analysis was done in order to check the adequacy of the simple procedure and in each instance the corrections to the high energy portion of the β+ spectrum were neglible. Therefore, we elected to use the simplified procedure.

RESULTS AND DISCUSSION

150Er

In our experiment a small amount of 18 s ^{150}Er was present in the ^{151}Er specta. This provided us with the opportunity to measure Q_{EC} values for both nuclei simultaneously under identical conditions. Since the Q_{EC} value for ^{150}Er is known to high precision from the work of Keller *et al.* (1) our measurement of this nuclide served as a test of our experimental set-up and data analysis method.

The determination of Q_{EC} values by measurement of β+ spectrum endpoints is complicated by summing effects in the β+ detector due to random as well as coincident radiations. We dealt with these effects as follows. We calibrated our β+ detector using sources produced and measured under similar conditions. In each case the calibration sources have á single strong β+ branch to a low-lying level which subsequently decays directly to the ground state by a single γ ray. In such cases coincidence summing due to cascade γ rays and other β+ branches will be minimal. Summing from annihilation radiation is systematically incorporated in the scintillator energy calibration curve generated from the calibration spectra endpoints. Thus, in those instances where an unknown nuclide and the calibrators have similar decay schemes, i.e., a single strong β+ branch whose endpoint falls within the region defined by the calibration curve and which

TABLE 1. β-endpoint Energies and Q_{EC} Values (keV).

Nucleus	Endpoint Energy	γ-ray gate	Q_{EC} This work	Others
^{150}Er	2646(100)	476	4144(100)	4110(15)[a]
^{151}Er	2788(188)	641	5090(188)	----
	3564(186)	638	5224(186)	----
	3661(184)	256	5081(184)	----
		(weighted mean)	5130(110)	----

[a] Ref. (1).

subsequently decays directly to the ground state, summing effects should not impact the extraction of an accurate β+ endpoint energy. This is the case for ^{150}Er where >99% of the β+ decay is to a single level in ^{150}Ho which depopulates via a 476 keV γ-ray to the ground state. The β+ spectrum of ^{150}Er coincident with the 476 keV γ-ray yielded a Q_{EC} value of 4144 ± 100 keV in excellent agreement with the result of Keller et al., 4110 ± 15 keV. This close agreement validated our analysis method and provided evidence that summation effects due to annihilation radiation in our β+ detector were properly handled by our calibration procedure.

151Er

Our primary focus was the determination of the mass of ^{151}Er. Based on the decay scheme of Toth et al. (2) significant β+ decay branches to levels at 398, 638, 667, and 1280 keV are expected. The experimental data for the β+ spectrum gated by the 638 keV ^{151}Ho γ-ray are shown in Fig. 1 and the β+ endpoints for this gate and two additional ones, the 641 and 256 keV γ-rays which depopulate levels at 1280 and 398 keV, respectively, are listed in Table 1. The weighted mean for these three γ-ray gates

TABLE 2. Mass Excess Values (keV)

Nucleus	J^π	Mass excess This work	Nubase[a]
^{150}Er	0+	-57940(100)	-57970(100)
^{151}Er	(7/2-)	-58500(110)	-58260(300)
^{155}Yb	(7/2-)	-50740(110)[b]	-50490(300)
^{159}Hf	[7/2-][c]	-43100(120)[b]	-42850(300)
^{163}W	[3/2-]	-35150(130)[b]	-34900(310)
^{167}Os	[3/2-]	-26750(140)[b]	-26500(310)
^{171}Pt	[3/2-]	-17710(150)[b]	-17470(310)
^{175}Hg	[5/2-]	-8250(160)[b]	-8000(320)

[a] Ref. (4). Values and uncertainties were estimated, in part, from systematic trends.
[b] Mass excess values computed using Q_α values and uncertainties from Ref. (5).
[c] Spin and parity values in square brackets are estimated from neighboring nuclides (4).

is $Q_{EC} = 5130 \pm 110$ keV and the values for individual gates all fall within one standard deviation. Since summing effects in the 256 keV gated β+ spectrum due to the 100 keV cascade γ-ray to the ground state are expected to be small, we take the agreement among the three gated spectra, which determine Q_{EC} through different pathways, to indicate that the effects of summing are smaller than the stated error of ±110 keV.

At the same time we also acquired a good quality β+ spectrum coincident with the 667 keV ^{151}Ho γ-ray which yielded an endpoint energy of 2666 ± 70 keV. According to the decay scheme of Toth *et al.* one expects substantial β+ decay directly to the 667 keV level in ^{151}Ho. Using this premise our β+ data yield a Q_{EC} of 4355 ± 70 keV, a result in clear disagreement with the results for the other gates. Puzzled by this inconsistency, we contacted the lead author and requested his assistance in assessing whether substantial direct feeding to the 667 keV level was established by their data or whether feeding somewhat higher in scheme, followed by deexcitation via unobserved γ rays to the 667 keV level, was a possibility. His conclusion was that the latter could neither be confirmed nor ruled out by their data (3). Because three of the four gated spectra gave results consistent among the set and in agreement with the estimated value based on systematic trends of 5379 ± 300 keV (4), we have not included the result for the 667 keV gate in our computation of Q_{EC} for ^{151}Er.

The ^{175}Hg → ^{151}Er alpha decay chain

As noted above the α-decay energies have been measured with good precision for the chain, ^{175}Hg → ^{151}Er. In principle, if the level schemes of all these nuclei were known in detail then the masses of all members of the chain could be determined since α decay down the chain is presumed to proceed through states of similar structure. This premise is valid for chains of even-even nuclei but it is not for odd-A or odd-odd chains. In these instances there is no foolproof method for establishing the relationship between the α–decay energies and the ground state masses. We have assumed a notional error due to this uncertainty which is somewhat larger than the stated error for the α–decay energies and projected the mass excess values for the chain ^{175}Hg → ^{151}Er. The Q_α values and uncertainties are taken from the Table of Isotopes (5). The results are tabulated in Table 2 where they are compared to estimates based mostly on systematic trends by Audi *et al.* (4). In all instances there is agreement within the stated errors.

ACKNOWLEDGMENTS

This work was supported by the Department of Energy under grants and contracts: DE-FG02-88ER40417, DE-FG02-91ER40609, DE-AC02-76CH00016, W-31-109-ENG-38. D.D.W. and D.S.B. also had support from NATO Research Grant No. 910214.

REFERENCES

1. Keller, H., *et al.*, Z. Phys. A - Hadrons and Nuclei **340**, 363-370 (1991).
2. Toth, K.S., *et al.*, Phys. Rev. C **44**, 1868-1877 (1991).
3. Toth, K.S., private communication.
4. Audi, G., *et al.*, Nucl. Phys. **A624**, 1-124 (1997).
5. Firestone, R.B., *et al.*,*Table of Isotopes*, 8th edition, John Wiley and Sons, New York (1996).

^{208}Pb Neutron Density: A Mean Field Problem?

Stefan Gmuca

Institute of Physics, Slovak Academy of Sciences,
Dúbravska cesta 9, SK-842 28 Bratislava, Slovakia

Abstract. The ground–state nuclear densities and radii of ^{208}Pb doubly–magic nucleus have been evaluated within the framework of the relativistic mean–field approach. It is pointed out that the neutron density and the neutron radius in the RMF approach are quite different from both, the empirical data and the predictions of the Skyrme-Hartree-Fock model.

Presently, the precise experimental charge density distributions in the nuclear ground states are known for many nuclei throughout the chart of nuclides. Due to its high precision, the charge densities are frequently employed in testing of nuclear models by comparison of model predictions with experimental values. However, the dominant contribution to the charge density comes from the proton density folded with the single proton electric form factor; thus any such comparison reveals the results sensitive mainly to the proton density distributions.

On the other hand, experimental information on neutron density distributions are rather scarce. Our current knowledge of neutron densities comes from non-relativistic impulse approximation analyses of the high quality 800 MeV polarized proton elastic scattering data on several closed shell nuclei [1]. Due to the strong absorption of protons and significant uncertainty of t matrix, the neutron densities obtained are characterized by large uncertainties in the nuclear interior. Mostly the neutron rms radii were, therefore, used in the theoretical analyses.

Recently, Starodubsky and Hintz [2] extracted new neutron densities of ^{206,207,208}Pb from elastic proton scattering at 650 MeV. While on the whole consistent with the previous ^{208}Pb neutron density of Ray [1], Starodubsky and Hintz [2] quote much lower uncertainties even in the interior of the ^{208}Pb nucleus. This was achieved mainly by eliminating uncertainties associated with the t matrix. The resulting neutron density and its associate uncertainty makes the comparison with model calculations meaningful and allows to treat the proton (charge) and neutron distributions on the same footing.

Theoretically, the neutron density can be calculated using both, the nonrelativistic and relativistic mean–field approaches. It is known [3] that while both

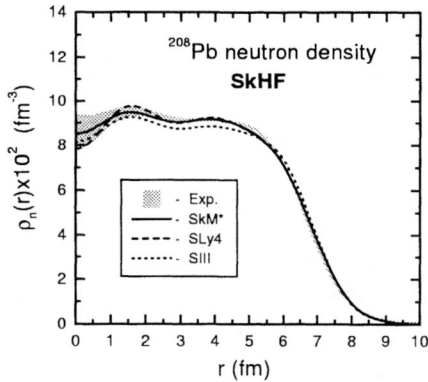 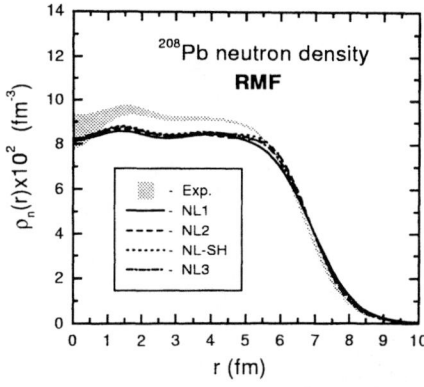

FIGURE 1. Comparison between the experimental neutron density distribution (Ref. 2) and the Skyrme–Hartree–Fock (left panel) and the relativistic mean–field (right panel) calculations using various interactions.

approaches (with common parameterizations) reproduce the ground state binding energies, charge densities and rms charge radii, and energies of single–particle levels almost equally well, they differ in prediction of the neutron densities and radii. The relativistic mean field theory overpredicts the neutron skin thickness of asymmetric nuclei over that of the Skyrme interactions and the empirical data. This is illustrated in Fig. 1 for the case of ^{208}Pb nucleus. The left panel displays the ^{208}Pb neutron density as calculated by the nonrelativistic Skyrme–Hartree–Fock (SkHF) model using various Skyrme forces [4–6]. The SkHF calculations give the neutron rms radius in the range of 5.62–5.65 fm in a close agreement with an empirical value of 5.655(42) fm [2], and the predicted neutron densities for all SkHF parameters used (with a possible exception of an older SIII interaction) follows closely the empirical neutron density [2].

The right panel of Fig. 1 shows the results of the relativistic mean–field (RMF) calculations with some commonly used parameterizations [7–10]. All parametrizations underestimates the neutron density in the interior of the nucleus and overpredict it for large radii. All predictions lie in a narrow band close each to other, and give the rms neutron radii in the interval 5.72–5.74 fm ([8–10]), with a possible exception of the NL1 parameter set [7] that gives 5.79 fm.

The present work deals with a question whether an inclusion of the new empirical neutron density of the ^{208}Pb nucleus into the RMF parameter optimization procedure may resolve this problem and enables us to find new RMF parameters that bring the calculated ^{208}Pb neutron density to closer agreement with the empirical one. We have performed, therefore, extensive multiparameter fits to observables specific to this nucleus.

Our starting point is the standard RMF Lagrangian density

$$\mathcal{L} = \bar{\psi}(i\gamma_\mu \partial^\mu - M)\psi + \frac{1}{2}\partial_\mu\sigma\partial^\mu\sigma - \frac{1}{2}m_\sigma^2\sigma^2 - \frac{1}{3}b_\sigma M(g_\sigma\sigma)^3 - \frac{1}{4}c_\sigma(g_\sigma\sigma)^4 - g_\sigma\bar{\psi}\psi\sigma$$

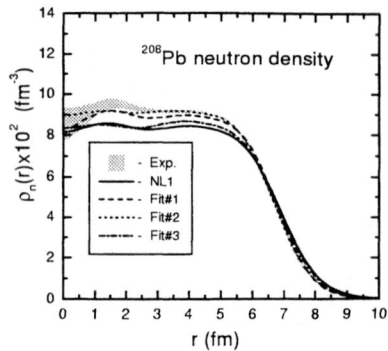

FIGURE 2. Comparison of experimental ^{208}Pb neutron density to some typical results of RMF fits (see text for details).

$$-\frac{1}{4}\omega_{\mu\nu}\omega^{\mu\nu} + \frac{1}{2}m_\omega^2\omega_\mu\omega^\mu + \frac{1}{4}c_\omega(g_\omega^2\omega_\mu\omega^\mu)^2 - g_\omega\overline{\psi}\gamma_\mu\psi\omega^\mu \qquad (1)$$
$$-\frac{1}{4}\boldsymbol{\rho}_{\mu\nu}\boldsymbol{\rho}^{\mu\nu} + \frac{1}{2}m_\rho^2\boldsymbol{\rho}_\mu\boldsymbol{\rho}^\mu - g_\rho\overline{\psi}\gamma_\mu\boldsymbol{\tau}\psi\boldsymbol{\rho}^\mu - \frac{1}{4}F_{\mu\nu}F^{\mu\nu} - e\overline{\psi}\gamma_\mu\frac{(1-\tau_3)}{2}\psi A^\mu,$$

where the symbols used have their usual meaning [11].

The Lagrangian parameters are usually obtained by a fitting procedure to some bulk properties of spherical nuclei. The nuclear properties fitted consist mostly of the binding energies and charge radii. In our case, the set of observables to which the fit is performed consists of the following ground–state properties of the ^{208}Pb nucleus: binding energy, energies of proton and neutron s.p. states of the last valence shells, spin–orbit splittings of the valence s.p. levels, charge density distribution and its rms radius, neutron density distribution [2] and its rms radius.

The typical results obtained are shown in Fig. 2.

The label Fit#1 marks the results obtained with the standard RMFA model with the scalar and vector selfcouplings as given by Lagrangian (1). A good simultaneous fit to both, the charge and neutron densities for the ^{208}Pb nucleus was achieved, however, at the price of destruction of single–particle energies of levels near Fermi energy; the proton levels were shifted up by ∼3 MeV and the neutron ones were pushed down by approx. the same amount.

Fit#2 denotes the results obtained with the isoscalar–scalar $\boldsymbol{\delta}$ included into the Lagrangian. The δ–field is usually not expected to play an important role for nuclei, whose isospin asymmetry is not large. However, for isospin–asymmetric matter the contribution of the δ–field should be considered. We used the standard δ–field Lagrangian with the Yukawa coupling to nucleons. It reads

$$\mathcal{L}_\delta = \frac{1}{2}\partial_\mu\boldsymbol{\delta}\partial^\mu\boldsymbol{\delta} - \frac{1}{2}m_\delta^2\boldsymbol{\delta}^2 + g_\delta\overline{\psi}\boldsymbol{\tau}\psi\boldsymbol{\delta}. \qquad (2)$$

The relevant component for the δ–meson field is the isospin component $\delta^{(3)}$; remaining components vanish, in particular $\delta^{(1)} = \delta^{(2)} = 0$. Its main effect is on the neutron and proton effective masses, respectively, which thus behave differently in the nuclear medium.

By inclusion of the δ–field into the RMFA Lagrangian, the significant improvements in the description of observables were achieved. In particular, the charge and neutron densities and their rms radii follow closely the experimental data, and the energies of single–particle states and spin–orbit splittings are also correctly reproduced. However, the resulting couplings of the isovector meson fields ρ and δ, respectively, are rather strong and imaginary. This changes the character of the corresponding scalar and vector fields (g_ρ^2 and g_δ^2 are negative). It should be noted that the small isovector field in the RMF model without δ–meson now appears as a difference of substantially stronger isovector–vector (ρ) and isovector–scalar (δ) fields, respectively.

The results marked Fit#3 are the same as Fit#1 (standard RMF model), with neutron observables withdrawn from the data set. In this case the results obtained are very close to the ones predicted by the commonly used RMF parameter sets. (The results obtained with NL1 parameters are given for comparison.)

We may conclude, that the RMF theory has serious difficulties in reproducing the neutron density of the ^{208}Pb nucleus simultaneously with the charge density. This is in contrast with the nonrelativistic SkHF model. The approach results either in a corruption of the structure of single–particle levels near Fermi energy (standard RMF model), or in physically unacceptable values of coupling constants of isovector meson–fields (RMF model with the δ–meson included). This behaviour of the RMF theory needs further work to explain.

This work was supported in part by the SGA agency under grant no. 95/5305/645.

REFERENCES

1. Ray, L. and Hodgson, P., *Phys. Rev. C* **20**, 2403 (1979)
2. Starodubsky, V.E. and Hintz, N.M., *Phys. Rev. C* **49**, 2118 (1994)
3. Sharma, M.M. and Ring, P., *Phys. Rev. C* **45**, 2514 (1992)
4. Beiner, M., Flocard, H., Van Giai, N. and Quentin, P., *Nucl. Phys. A* **422**, 103 (1975)
5. Bartel, J., Quentin, P., Brack, M., Guet, C. and Hakanson, H.B., Nucl. Phys. A **386**, 79 (1982)
6. E. Chabanat, E. et al., *Proc. Int. Workshop on Research with Fission Fragments*, eds.: T. von Egidy et al., World Scientific, Singapore, 1997, p. 155
7. Reinhard, P.-G., Rufa, M., Maruhn, J., Greiner, W. and Friedrich, J., *Z. Phys. A* **323**, 13 (1986)
8. Suk-Joon Lee, Fink, J., Balantekin, A.B., Strayer, M.R., Umar, A.S., Reinhard, P.-G., Maruhn, J.A. and Greiner, W., *Phys. Rev. Lett.* **57**, 2916 (1986)
9. Sharma, M.M., Nagarajan, M.A. and Ring, P., *Phys. Lett. B* **312**, 377 (1993)
10. Lalazissis, G.A., König, J. and Ring, P., *Phys. Rev. C* **55**, 540 (1997)
11. Serot, B.D. and Walecka, J.D., *Adv. Nucl. Phys.* **16**, 1 (1986)

Measurement of the Magnetic Moments of ^{7}Be and of the Halo Nucleus ^{11}Be

S. Kappertz[1], W. Geithner[1], G. Katko[1], M. Keim[2], G. Kotrotsios[1],
P. Lievens[3], R. Neugart[1], L. Vermeeren[4], S. Wilbert[1]
and the ISOLDE Collaboration[2]

[1] *Institut für Physik, Universität Mainz, D-55099 Mainz, Germany*
[2] *EP Division, CERN, CH-1211 Geneva 23, Switzerland*
[3] *Laboratorium voor Vaste-Stoffysica en Magnetisme, K.U. Leuven, B-3001 Leuven*
[4] *Instituut voor Kern- en Stralingsfysica, K.U. Leuven, B-3001 Leuven*

Abstract. Recently the CERN/ISOLDE laser ion source was optimized to provide intense beryllium ion beams. This allowed us to perform optical spectroscopy on ^{7,9,10}Be yielding the unknown magnetic moment of ^{7}Be and the isotope shifts in the lithium-like spectrum of Be$^+$. Furthermore we measured the magnetic moment of the neutron halo nucleus ^{11}Be via β-NMR on optically polarized nuclei.

Introduction

Nuclear moment investigations have always contributed considerably to the understanding of nuclear structure. The moments of odd-A radioactive beryllium isotopes are of particular interest. ^{7}Be is a crucial isotope in the solar neutrino problem and the estimates of the electron and proton capture rates would benefit from the determination of its ground state moments, in particular of the quadrupole moment. From the ^{7}Be magnetic moment, together with the known moment of stable ^{7}Li, also the isoscalar and isovector magnetic moments of the lightest T=1/2 mirror pair of the 0p-shell could be derived. On the neutron-rich side, ^{11}Be is the only known one-neutron halo nucleus and it has an anomalous ground-state spin-parity. Several theoretical calculations of the magnetic moment exist, and a measurement would yield information on the relative weights of the $|^{10}Be(0^+) \times 1s_{1/2}>$ and $|^{10}Be(2^+) \times 0d_{5/2}>$ components of the ^{11}Be ground state wave function (1). We measured both magnetic moments using collinear laser spectroscopy with optical detection for ^{7}Be and with β-NMR on optically pumped ions for ^{11}Be.

The Magnetic Moment of ^{7}Be and the Isotope Shifts of ^{7,9,10}Be

The experiments were performed at the ISOLDE mass separator facility at CERN. ^{7,9,10}Be were produced by the impact of the pulsed 1 GeV proton beam from the PS-Booster synchrotron on a ^{12}C target. The reaction products diffused into a tungsten cavity in which resonant laser ionization took place (2). A very efficient two-step ionization scheme was used, including excitation from the $2s^2$ $^1S_0$ atomic ground state to the $2s2p$ $^1P_1$ state and subsequent excitation to an auto-ionizing state. The required laser beams for the two-step ionization process were obtained from copper-vapour-laser pumped dye lasers combined with respectively frequency tripling and frequency doubling. The ions were electrostatically extracted out of the cavity, accelerated to an energy of 40-60 keV, mass separated and guided to the experiment.

The ion beam was merged with a cw laser beam whose frequency was tuned to the (Doppler-shifted) optical resonance line $2s\ ^2S_{1/2} \rightarrow 2p\ ^2P_{1/2}$ at 313 nm. The uv laser light was obtained by intra-cavity frequency doubling in the dye laser running on DCM dye which was pumped by the beam of an Ar^+-laser. For the isotopes mentioned the yield of the order 10^9 atoms per second was sufficiently high to monitor the resonant excitation by observing the fluorescence with a photomultiplier. Figure 1 shows the recorded hyperfine structures of ^7Be and ^9Be. Also the isotope shifts of $^{7,9,10}Be$ were measured by comparing the beam energies required for tuning the Doppler shifts of the different isotopes to resonance. The isotope shifts reveal the magnitude of the specific mass shift (3), facilitating considerably the search for the ^{11}Be resonance frequencies (see next section). This result is also valuable for testing atomic structure calculations using many-body perturbation theory (4). The specific mass shift turns out to be roughly twice as large as the normal mass shift, which is in good qualitative agreement with the theoretical predictions.

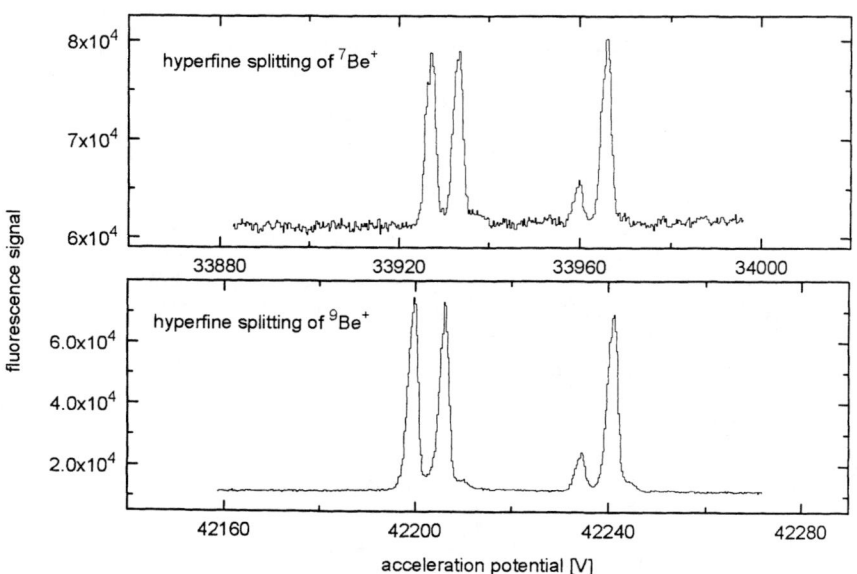

FIGURE 1. Optical hyperfine structures of ^7Be and ^9Be.

From the observed hyperfine structures the previously unknown magnetic moment of ^{7}Be can be deduced (3). A preliminary value of $\mu_I(^7Be) = -1.398(15)\ \mu_N$ has been obtained from a comparison with the hyperfine structure of stable ^{9}Be.

The Magnetic Dipole Moment of ^{11}Be

^{11}Be was produced by fragmentation of uranium in a UC$_2$ target exposed again to the 1 GeV PS-Booster proton beam, and it was ionized in the laser ion source as discussed before. Here the yield was more than two orders of magnitude lower than for the long-lived isotopes directly produced from carbon and not decaying during diffusion out of the target. For ^{11}Be we used the much more sensitive β-NMR method (5,6), not only for reasons of sensitivity, but also to introduce a normalization of the signal fluctuating with the beam intensity. In addition, the accuracy of a g-factor measurement by NMR is by far superior to an optical measurement of hyperfine structure. The ion beam was again superimposed on a cw laser beam, but in this case the laser light was circularly polarized and a weak longitudinal magnetic field was applied to the interaction region. Thus an atomic polarization was created after several cycles of excitation and decay, which was transferred by hyperfine interaction to a nuclear polarization. When entering the strong transverse field of the β-NMR set-up, the angular momenta of the ions were rotated adiabatically by 90^0. After decoupling of the nuclear and electronic spins the ions were implanted into a metallic beryllium single crystal exposed to a homogeneous magnetic field of about 0.3 T. To compensate for the magnetic force on the ions, electrostatic deflectors were adjusted carefully so as to implant the ions centrally into the crystal.

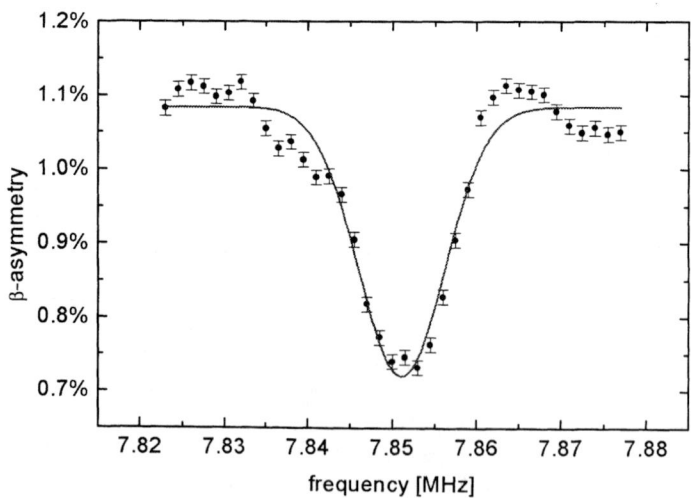

FIGURE 2. β-NMR signal of ^{11}Be in a beryllium single crystal

The nuclear polarization leads to an asymmetry in the angular distribution of the emitted β-decay electrons, which is detected by two scintillator telescopes placed at 0° and 180° relative to the magnetic field. In order to slow down the relaxation of the ^{11}Be nuclear polarization to the order of the half-life ($T_{1/2}$= 13.8 s), the beryllium crystal was cooled to about 50 K. The asymmetry then reached about 1 % for pumping on the strongest hyperfine component, mainly limited by the available low power density of the uv laser beam. Also the other two hyperfine structure components were observed, but only with moderate resolution. In order to obtain an accurate and reliable value for the magnetic moment we observed the destruction of the asymmetry caused by the coupling of the nuclear Zeeman sublevels through rf irradiation at the Larmor frequency. The well-saturated NMR signals were finally narrowed to a few kHz by reducing the rf power (see figure 2). A calibration of the magnetic field at the position of the crystal was obtained by performing a β-NMR experiment on ^{8}Li under the same conditions and using its well-known magnetic moment.

The data are currently being analyzed and the preliminary result of $\mu_I(^{11}Be)$ = −1.6814(13) μ_N shows a quenching of the $s_{1/2}$ Schmidt value by only about 12 %. The absolute value of the magnetic moment is larger than most theoretical calculations predict. In particular, it remains to be investigated to what extent this quenching is due to the admixture of the $d_{5/2}$ configuration and to higher-order configuration mixing. The final results and a discussion on its implications for the structure of this nucleus will be published soon (7).

† Supported by BMBF, Germany (contract no. 06 MZ 866 I)

REFERENCES

1. Suzuki, T., Otsuka, T., and Muta, A., *Phys. Lett. B* **364**, 69 (1995).
2. Lettry, J., et al., *Rev. Sci. Instr.* **69**, 761 (1998); see also Köster, U., et al., Sebastian, V., et al., contributions to this conference.
3. Kappertz, S., et al, to be published.
4. Mårtensson-Pendrill, A.-M., and Salomonson, S., *J. Phys. B* **15**, 2115 (1982) and private communication (1997); Chung, K.T., *Phys. Rev. A* **44**, 5421 (1991); Chung, K.T., and Zhu, X.-W., *Phys. Scr.* **48**, 292 (1993).
5. Arnold, E., et al., *Phys. Lett. B* **197**, 311 (1987).
6. Keim, M., et al., *Hyp. Int.* **97/98**, 543 (1995).
7. Geithner, W., et al., to be published.

Nuclear Mass Formula with the Shell Energies Obtained by a New Method

H. Koura*, M. Uno*†, T. Tachibana* and M. Yamada*

*Advanced Research Institute for Science and Engineering, Waseda University
3-4-1 Okubo, Shinjuku-ku, Tokyo 169-8555, JAPAN
†Ministry of Education, Science, Sports and Culture, 3-2-2 Kasumigaseki, Chiyoda-ku, Tokyo 100-0013, JAPAN

Abstract. Nuclear shapes and masses are estimated by a new method. The main feature of this method lies in estimating shell energies of deformed nuclei from spherical shell energies by mixing them with appropriate weights. The spherical shell energies are calculated from single-particle potentials, and, till now, two mass formulas have been constructed from two different sets of potential parameters. The standard deviation of the calculated masses from all the experimental masses of the 1995 Mass Evaluation is about 760 keV. Contrary to the mass formula by Tachibana, Uno, Yamada and Yamada in the 1987-1988 Atomic Mass Predictions, the present formulas can give nuclear shapes and predict on super-heavy elements.

NEW METHOD

We have been studying a method of determining nuclear shell energies and incorporating them into a mass formula [1–3]. In this method the shell energy of a deformed nucleus is calculated from shell energies of neighboring spherical nuclei by mixing them with appropriate weights. The spherical proton and neutron shell energies are obtained with use of single-proton and single-neutron potentials, respectively, which depend smoothly on proton number Z and neutron number N.

We assume an extended *spherical* Woods-Saxon potential of neutron (or proton), of which the central part is

$$V_{\text{cen}}(r) = V_0 \frac{1}{\{1 + \exp\left[(r - R_v)/a_v\right]\}^{a_v/\kappa}} \left\{ 1 + V_{dp} \frac{1}{1 + \exp\left[(r - R_v)/a_v\right]} \right\}. \quad (1)$$

Here, κ is a parameter governing the behavior at large distances, and V_{dp} is introduced to form a dip in the surface region. When $\kappa = a_v$ and $V_{dp} = 0$, this potential reduces to the ordinary Woods-Saxon potential. The spin-orbit term is

$$V_{\text{ls}}(r) = V_{ls0}\frac{\hbar^2}{2m}\frac{1}{r}\frac{d}{dr}\left\{\frac{1}{1+\exp\left[(r-R_{ls})/a_{ls}\right]}\right\}\boldsymbol{l}\cdot\boldsymbol{s}, \qquad (2)$$

where m is the nucleon mass, and R_{ls} and a_{ls} are not the same as R_v and a_v. In Eqs. (1) and (2), V_0, V_{dp}, R_v, a_v, κ, V_{ls0}, R_{ls} and a_{ls} are functions of Z and N, and include some parameters. The charge symmetry is imposed on these potentials. The potential parameters are determined by comparison with experimental single-particle levels of 15 doubly-magic and magic-submagic nuclides. Among them there are ^{16}O and ^{12}C, and it is difficult to fit well to both of them. Till now, we have chosen two sets of parameter values: one fitted better to ^{16}O (type-I), and the other fitted better to ^{12}C (type-II).

In order to describe a deformed nucleus, we superpose spherical nuclei with the fractions determined from the assumed deformation. The energy which is obtained as a weighted sum of the spherical shell energies is minimized with respect to deformation parameters to get the ground-state shell energy and equilibrium shape. For the nuclear shape we assume axial and reflection symmetry. In this report, we take into account up to α_6, and take the same deformation for protons and neutrons.

RESULTS

Figure 1 (a) and (b) show the differences between experimental and calculated masses for two types of mass formula. These two figures show rather similar tendency at the heavy region. However, the type-I formula gives too low masses in the region $40 < A < 48$, $N - Z < 0$, while the type-II formula gives too high masses around $Z \approx N \approx 18$. The standard deviation of the calculated masses from all the experimental masses of the 1995 Mass Evaluation [4] is 806 keV for type-I and 761 keV for type-II. They might be improved by further adjustment of parameter values. For comparison with other mass formulas, we give the standard deviations of the type-II formula from experimental masses in somewhat smaller nuclidic regions. The Extended Thomas-Fermi plus Strutinsky Integral method (ETFSI) [5] gives about 730 keV for nuclei with $A > 35$, which is somewhat larger than our value of about 710 keV for these nuclei. The Finite-Range Droplet Model (FRDM) [6] gives about 680 keV for nuclei with $N \geq 8$ and $Z \geq 8$, which is somewhat smaller than our value of about 720 keV.

Contrary to TUYY formula [7], our present formula can be applied to superheavy elements. As for the mass of $^{298}_{114}$X, for example, it gives about 194 MeV, which is to be compared with 193.29 MeV of ETFSI and 192.78 MeV of FRDM. The shell energy of this nucleus, which is predicted to be -7.59 MeV by FRDM, is obtained to be about -7.3 MeV by our method. Meanwhile, the dip of our shell energy at $Z = 114$ is not so remarkable as that of FRDM. The location of the neutron-drip line is slightly inner than that of TUYY, and is somewhat outer than those of ETFSI and FRDM.

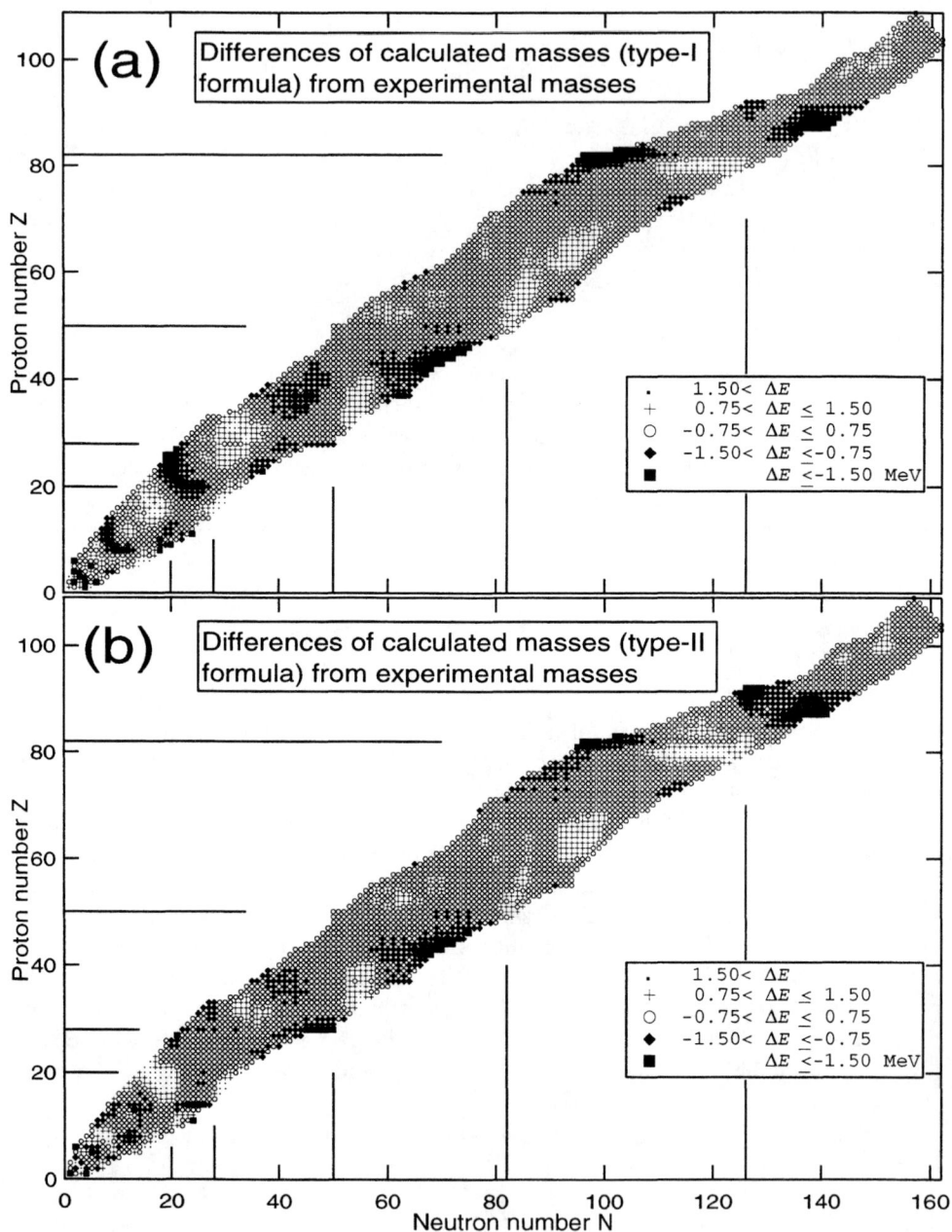

FIGURE 1. Errors of calculated masses $\Delta E = E_{\text{Calc.}} - E_{\text{Exp.}}$ of (a) type-I formula and (b) type-II formula.

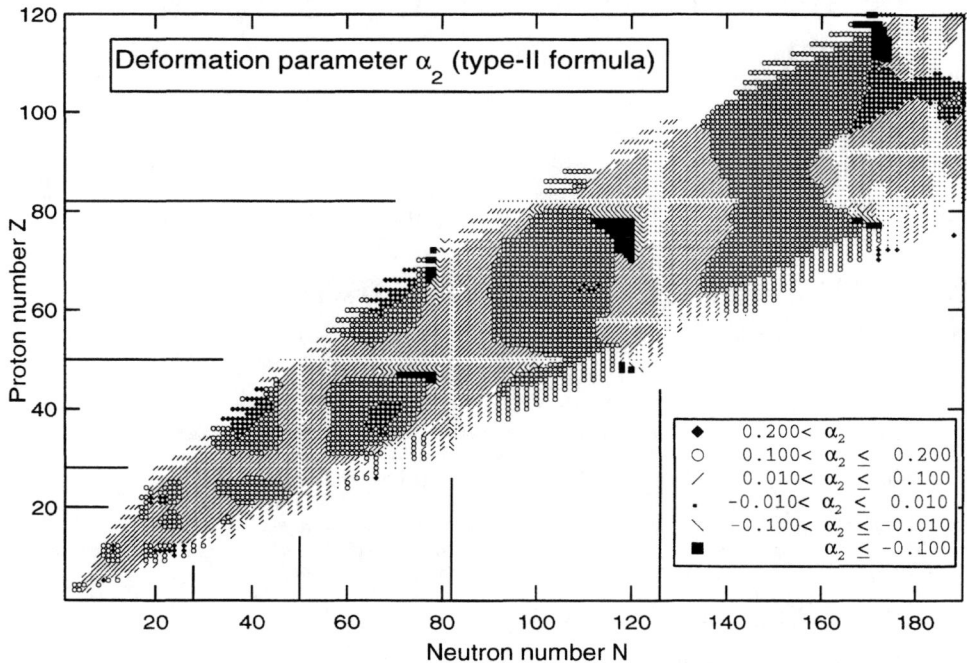

FIGURE 2. Deformation parameter α_2 by type-II formula.

Figure 2 shows the deformation parameter α_2 for the type-II formula. The general trend of the deformation is well represented though, in some nuclei, the absolute magnitudes of the deformation are somewhat too small.

REFERENCES

1. H. Koura, M. Uno, T. Tachibana and M. Yamada, *Technical Report of Advanced Research Center for Science and Engineering, Waseda University*, **No.95-25**, 1 (1995).
2. M. Uno, H. Koura, T. Tachibana and M. Yamada, *International Conference on Exotic Nuclei and Atomic Masses (ENAM 95)*, Arles, France (1995), p. 159.
3. H. Koura, T. Tachibana, M. Uno and M. Yamada, *JAERI-Conf* **96-008**, 284 (1996).
4. G. Audi and A. H. Wapstra, *Nuclear Physics* **A595**, 409 (1995).
5. Y. Aboussir, J. M. Pearson, A. K. Dutta and F. Tondeur, *Atomic Data and Nuclear Data Tables* **61**, 127 (1995).
6. P. Möller, J. R. Nix, W. D. Myers and W. J. Swiatecki, *Atomic Data and Nuclear Data Tables* **59**, 185 (1995).
7. T. Tachibana, M. Uno, M. Yamada and S. Yamada, *Atomic Data and Nuclear Data Tables* **39**, 251 (1988).

Interaction radii of proton-rich radioactive nuclei at A=60-80

G. F. Lima[1], A. Lépine-Szily[1], A. C. C. Villari[1,2], W. Mittig[2],
M. Chartier[2,10], R. Lichtenthaler[1], D. Hirata[3], J. C. Angelique[4],
G. Audi[5], J.M.Casandjian[2], A. Cunsolo[6], C. Donzeaud[7], A. Foti[6],
A. Gillibert[8], M. Lewitowicz[2], S. Lukyanov[9], M. MacCormick[7],
D.J.Morrissey[10], N.A. Orr[4], A.N.Ostrowski[2,11], B.M.Sherrill[10],
C. Stephan[7], T. Suomijarvi[7], L. Tassan-Got[7], D. J. Vieira[12],
J. M. Wouters[12]

1.IFUSP-Universidade de São Paulo, C.P.66318, 05315-970 São Paulo, Brasil. 2.GANIL, Bld Henri Becquerel, BP 5027,14021 Caen, Cedex, France, 3.SPring-8, Kamigori cho, Ako gun, Hyogo, 678-12, Japan, 4.LPC-ISMRA,Bld du Marechal Juin, 14050 Caen, Cedex, France, 5.CSNSM,Batiment 108,91406 Orsay Cedex, France, 6.INFN, Corso Itália 57, 95129 Catania, Italy, 7.IPN Orsay, BP1, 91406 Orsay Cedex,France, 8. CEA/DSM/DAPNIA/ SPhN, CEN Saclay, 91191 Gif-sur-Yvette, France, 9.LNR,JINR, Dubna,P.O.Box 79,101000 Moscow, Russia, 10.NSCL,Michigan State University, East Lansing MI,48824-1321,USA 11.Department of Physiscs & Astronomy, University of Edinburgh, Edinburgh, EH9 3JZ UK 12.Los Alamos National Laboratory, Los Alamos NM, 87545,USA

Abstract. The interaction radii of proton-rich, radioactive $_{31}$Ga, $_{32}$Ge, $_{33}$As, $_{34}$Se, $_{35}$Br isotopes were measured using the direct method. The secondary beams were produced using a ^{78}Kr primary beam of 73 MeV/nucleon in conjunction with SISSI and the SPEG spectrometers at GANIL. Most elements show reduced radii which vary with N, with a minimum around N=36-38. The experimental reduced radii are compared to theoretical values obtained from Glauber reaction cross-section calculations based on Relativistic Mean Field (RMF) nuclear densities.

The region of A $\sim$ 60-80 near the N=Z line presents interesting properties such as shape-transitions, shape-coexistence and the reinforcing or switching of shell-gaps [1,2]. Very near to the spherical magic number N=40 strong ground-state deformations ($\beta_2 \sim 0.44$) were found in the ^{76}Sr (Z=N=38) and ^{74,76}Kr (Z=36, N=38,40) nuclei, while neighbouring nuclides with lower Z values, such as ^{70,72}Ge (Z=32,N=38,40) and ^{72,74}Se (Z=34,N=38,40) present spherical behaviour around N=38 [1]. The deformations parameters measured in this region show rapid variations with N and Z. The even-even $_{30}$Zn, $_{32}$Ge and $_{34}$Se nuclei with N≤38 have spherical ground-states, presenting moderate prolate deformations for higher N values. This region also presents several isomers, which are not resolved in our method (e.g., ^{66}Ga, ^{67,69,71}Ge, ^{69,71,73}Se, ^{72,74}Br). A systematic measurement of nuclear radii can provide an important guide to understanding such effects and the comparison with theoretical calculations is a stringent test of the predictive power of existing nuclear models far from stability.

In our experiment we used a ^{78}Kr primary beam accelerated to 73 MeV/nucleon by the GANIL cyclotrons and a 90 mg/cm^2 ^{nat}Ni target to produce proton-rich radioactive nuclei with masses between A $\approx$ 60-80. The superconducting solenoid device, SISSI, was used to focalize the primary beam onto the production target and the secondary beam into the α spectrometer. An original method [3] based on electron stripping in a thin mylar foil positioned between the dipoles of the α-spectrometer enabled light secondary beams to be eliminated while transmitting the nuclei of interest. A telescope of several cooled Si detectors was positioned in the focal plane of the magnetic spectrometer SPEG. It consisted of three thin Si ΔE detectors (50 μm, 150 μm and 163 μm, respectively), one of these being position-sensitive, and a thick Si-Li detector (4.5 mm), where the secondary beam particles were stopped. An additional thick Si detector was placed behind the telescope to detect the light particles produced by reactions of the beam in the previous detectors. The time-of-flight (*TOF*) of the secondary particles was measured over a 80m long flight path between a fast microchannel-plate "start" detector located at the exit of the α spectrometer and a silicon "stop" detector located in the SPEG focal plane. The m/q identification is achieved using the *TOF* measurement, while the Z identification is obtained from the Si detector telescope by using the energy loss in the first ΔE detector.

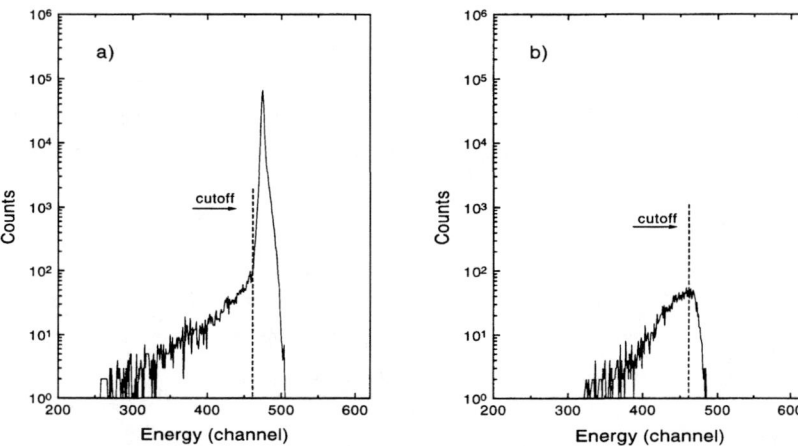

Figure 1: a) Typical energy spectrum obtained in the Si detector telescope. The vertical line indicates the separation between the elastic peak and the tail of reactions that take place in the detector telescope. b) Energy spectrum of secondary particles detected in coincidence with the light particle detector.

The nuclear radius can be obtained from the total reaction cross-section of the nuclides. The so called "Direct Method", already employed in other works [4-6], uses the Si detectors as the target. The energy spectrum obtained by the Si detector telescope is shown in Fig.1a. The low energy tail in the spectrum is due to reactions between the projectiles and Si nuclei. The cut between elastic and reaction events occurs at 130 MeV, and the contribution of reaction events with Q$\leq$130 MeV can be accounted for by detecting the light charged particles in a thick Si detector placed behind the telescope. In Fig.1b we also show the energy spectrum for secondary particles detected in coincidence with the light particle detector, which is characterized by the absence of the elastic peak. The loss due to events with Q≤130 MeV without emission of charged particles (only neutrons) was estimated with CASCADE calculations to be less than 0.2% of the total reaction cross-section. The reaction probability (P_R) is defined as the ratio of the number of reactions divided by the total number of events and it is connected to the integral of σ_R over the range of the projectiles in the target. As the incident particles may undergo reactions in the detector/target with any energy, from E_0 to zero, the mean energy-integrated reaction cross-section $\bar{\sigma}_R$ is defined by the following equation

$$\bar{\sigma}_R = \frac{\int_{E_0}^{0} \sigma_R(E) \left(\frac{dE}{dx}\right)^{-1} dE}{\int_0^{R(E_0)} dx} = -\frac{m \log(1-P_R)}{N_A R(E_0)} \quad (1)$$

where $R(E_o)$ is the range, for this we used the stopping-power tables of Hubert [7] and dE/dx is the stopping power which is approximated by $dx/dE \propto E^{-0.65}$. $\bar{\sigma}_R$ can be calculated from P_R, the experimental reaction probability and the range. The reaction cross-section σ_R and the reduced interaction radius r_0 are connected by the expression, $\sigma_R(E) = \pi r_o^2 f(E)$, where $f(E)$ is a function which depends on the projectile and target mass and also on the beam energy. For $f(E)$ we use the Kox's formula [8] with the parametrization obtained by Fernandez [6], which has been shown to work well for light neutron-rich radioactive nuclei[4,6]. This formula is given as

$$f(E) = \left(A_p^{1/3} + A_t^{1/3} + a \frac{A_p^{1/3} A_t^{1/3}}{A_p^{1/3} + A_t^{1/3}} - C(E) \right)^2 \times \left(1 - \frac{V_{BC}}{E_{cm}} \right) \quad (2)$$

where A_p is the mass number of the projectile; A_t is the mass number of the target; $C(E) = 0.31 + 0.014 \, E/A_p$, is an energy dependent transparency; a=1.85 is an asymmetry parameter; and V_{BC} is the Coulomb barrier. The reduced radius r_0 is extracted by combining Eqs. (1) and (3).

$$\bar{\sigma}_R = \frac{\pi r_0^2 \int_{E_0}^{0} f(E) \left(\frac{dE}{dx}\right)^{-1} dE}{R(E_0)} \quad (3)$$

The results of r_o as a function of the neutron number N are presented in Fig.2 for Ga (Z=31, N=32-36), Ge(Z=32,N=33-39), As(Z=33,N=34-39), Se(Z=34,N=35-40) and Br(Z=35,N=36-41) nuclides. The experimental reduced radii are not constant as a function of N. The decrease in the reduced radius of the spherical Ga, Ge, As and Se isotopes from N=32 to 38 is not related to deformations since all these nuclei present spherical ground-states. However, the possibility of populating an unknown deformed isomeric state in any of these isotopes can not be ruled out.

Spherical relativistic mean field (RMF) calculations were performed for these nuclei with center-of-mass corrections, pairing correlation corrections included [9].

The proton and neutron densities obtained in this RMF description, were the inputs of our Glauber calculations performed for these nuclei on Si at the actual energies to obtain the reaction cross sections and the values of the reduced radius r_o. The spherical RMF calculations are in reasonable agreement with the data.

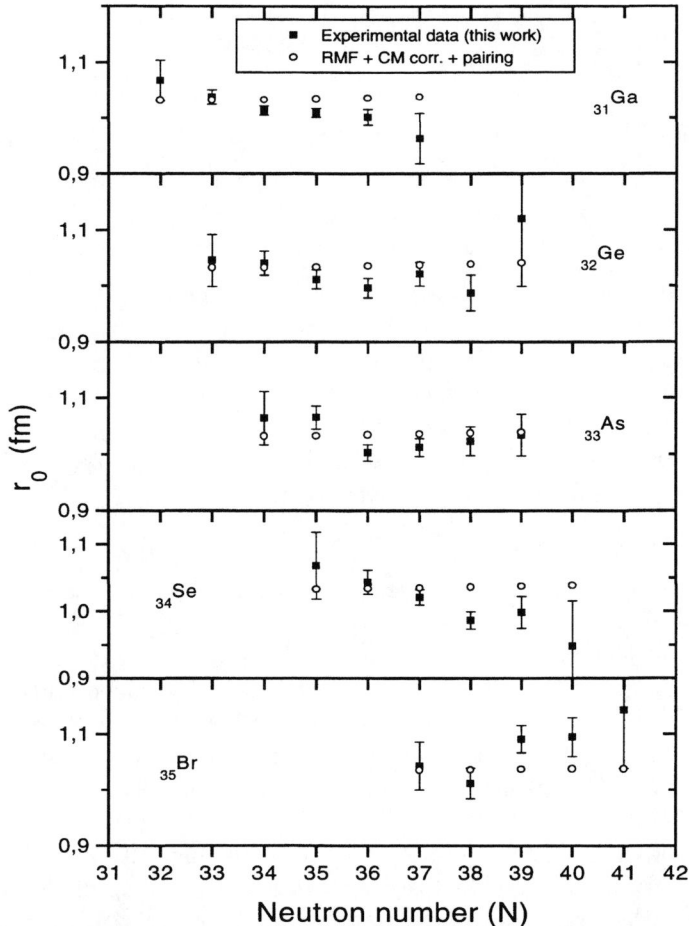

Figure 2: Experimental reduced radii measured in this work compared to theoretical calculations described in the text.

REFERENCES.
1. J.H. Hamilton, *Treatise on Heavy-Ion Science* vol.8. Plenum Press. Ed. D. A. Bromley, 1988.
2. C. Lister et al.,*Phys. Rev.* **C42**(1990)1191.
3. W. Mittig et al., *Nucl. Phys.* **A616**(1997)329c.
4. W. Mittig et al., *Phys. Rev. Lett.* **59**(1987)1889.
5. R.E. Warner et al., *Phys. Rev.* **C 37**(1988)1884 and *Phys. Rev.* **C 40**(1989)2473.
6. A. C. C. Villari et al., *Phys. Lett.* **B268**(1991)345.
7. F. Hubert et al., *At. Data Tables* **46**(1990)1.
8. S. Kox et al., *Phys. Rev.* **C35**(1987)1678.
9. I. Tanihata et al.,*Phys. Lett.* **B289**(1992) and D. Hirata et al., *Phys. Lett.* **B314**(1993)168.

Recent Mass Predictions in the Infinite Nuclear Matter Model

L. Satpathy* and R. C. Nayak[†]

* Institute of Physics, Sachivalaya Marg, Bhubaneswar 751 005, India
[†] Physics Department, G. M. College, Sambalpur - 768 004, India

Abstract: We obtain ground-state masses of ~ 7208 nuclei in the ranges $4 \leq Z \leq 120$ and $8 \leq A \leq 273$ in the infinite nuclear matter model with several improvements incorporated. Here we report two important results emerging from this new mass table. Several new islands of stability in the exotic regions and shell quenching in $N = 82$ and $N = 126$ shells are predicted. The shell quenching points out to a solution of the puzzle existing in r-process nucleosynthesis.

I. INTRODUCTION

The finite nuclear matter [INM] model (1) has been built in terms of the properties of nuclear many-body system by the use of extended Hugenholtz-Van Hove theorem (2). Applying this model in 1988, we had predicted (3) the masses of 3481 nuclei only in the range $18 \leq A \leq 267$. This prediction did not include the far off drip-line regions and should be considered as an interim mass table.

We have now improved (4) the INM model by using better definition of Fermi energies wihich has resulted in the accurate decoupling of the infinite part from the finite part of the ground state energy, together with the cancellation of the exchange Coulomb, finite-size proton form facter and Nollen-Schiffer anomaly terms. In addition, we have developed a new scheme of an interactive network covering the entire nuclear chart to obtain the local energy (comprising shell, deformations etc.) of a nucleus in a consistent manner, using the technique of ensemble averaging. We have now been able to predict masses of 7208 nuclei in the ranges $4 \leq Z \leq 120$ and $8 \leq A \leq 273$ covering the drip-line regions. There are only five parameters, representing the properties of infinite nuclear matter and finite size coefficients which are determined once for all in least square-fit to known nuclear masses. The root-mean-square deviation of the fit to 1844 known masses is 401 keV while the mean deviation is remarkably 9 keV. The details can be seen in Ref. (5).

Here we report two important results emerging from the new mass table, namely new islands of stability in the exotic regions and shell quenching in $N = 82$ and $N = 126$ shells.

II. NEW ISLANDS OF STABILITY

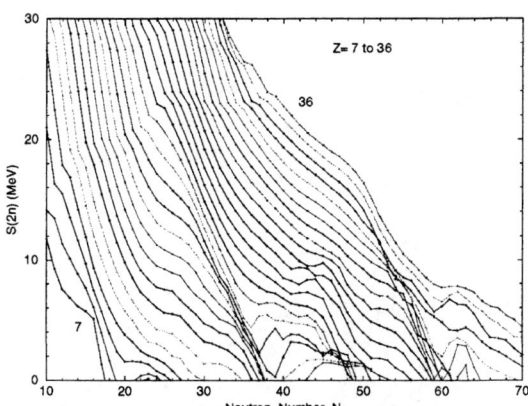

FIGURE 1. S(2n) systematics for $Z = 7$ to 36

The two-neutron seperation energies S(2n) obtained in our prediction are presented in Fig. 1 - 3 for the entire region of nuclear chart extending upto drip-lines. They are smooth and regular. The feature of monotonic decrease with neutron number, getting arrested abruptly for $Z \simeq 10 - 12$ near N = 20, known as "island of inversion" is quite well reproduced in our calculation.

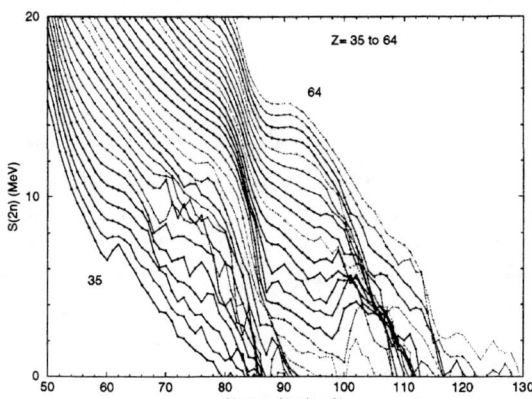

FIGURE 2. S(2n) systematics for $Z = 35$ to 64

This enhances our confidence in the predictive potential of this mass table.Any systematic deviations showing some structure in the above diagrams indicate the possible existence of new islands of stability. This may arise due to new shell closure and/or large deformation. Fig. 1 - 3 show such islands around the following combinations of neutron and proton numbers (N, Z): (40, 19), (46, 32) (104, 50), (120, 64), (126, 66), (132, 76), (144, 76) and (164, 82). It is interesting to find that many of the neutron numbers 64, 126, 164 and proton numbers 94, 66, 86 have been theoretically predicted (6) to show extra stability before.

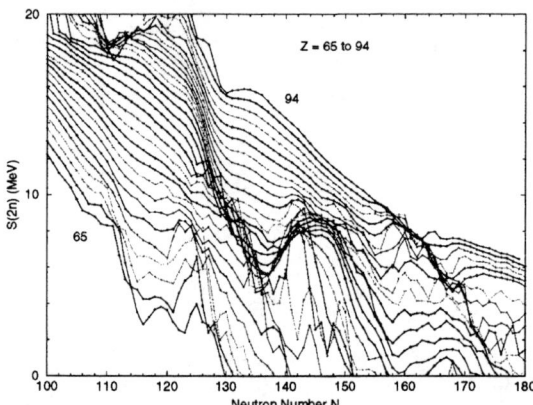

FIGURE 3. S(2n) systematics for Z = 65 to 94

III. SHELL QUENCHING IN N = 82 AND 126 SHELLS

Whether the usual features of shell structure seen in the valley of stability presists in the drip-line region is a fundamental question in the nuclear stucture physics today. It has now been recognised that N = 20 shell gap vanishes naturally towards nuclei with $Z \leq 12$ which are in the drip-line regions. Such shell quenching has also been seen for N = 50 shell gap. The shell quenching in the high shells N = 82 and 126 has not been experimentally observed. Analysis of abundances of various elements in the r-process nucleosynthesis has revealed (7) that shell quenching for N = 82 is essential to explain the peak around A = 120 and 140 seen in such studies. In (Fig. 4), we compare the shell quenching with the aid of neutron separation energies S(n) for N = 81 and N = 83 isotone below $^{122}_{50}Sn_{82}$ for different mass formulas and microscopic calculations. It can be seen that the mass formula based on finite range droplet model [FRDM, (8)] and extended Theorem Fermi model [ETFSI, (9)] don't show shell quenching. The HF + BCS model with Skyrme interaction SIII also does not show shell quenching. However, the Skyrme SkP designed for this purpose shows shell quenching (9). It is interesting that INM model clearly shows quenching. This points out a solutions to the puzzle existing in r-process nucleosynthesis. We have shown (5) that INM model, also predicts shell quenching for $N = 126$ shells while FRDM does not. Thus INM model seems to possess the unique ability to predict this important property of shell quenching in large N shell.

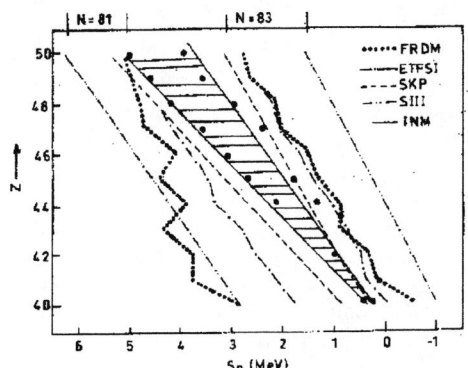

FIGURE 4. Shell quenching in N=82 shell

References
1. Satpathy, L., J. Phys. G: Nucl. Phys. **13**, 761-781 (1987).
2. Satpathy, L., and Nayak, R.C., Phys. Rev. Lett. **51**, 1243-1246 (1981).
3. Satpathy, L., and Nayak, R.C., At. Data and Nucl. Data Tables **39**, 241-249 (1988).
4. Nayak, R., Uma Maheswari, V.S., and Satpathy, L., Phys. Rev. **C52**, 711-717 (1995).
5. Satpathy, L., and Nayak, R.C., J. Phys. G: Nucl. Phys. (in press); Nayak, R.C. and Satpathy, L., *preprint no. IP/BBSR/97-45*; At. Data and Nucl. Data Tables (submitted).
6. Ragnarsson, I., Nilson, S.G., and Sheline, R.K., Physics Reports **45**, 1-87 (1978).
7. Chen, B. et al., Phys. Lett. **B355**, 37-44 (1995).
8. Möler, P. et. al., At. Data and Nucl. Data Tables **59**, 185-381 (1995).
9. Aboussir, Y. et. al., At. Data and Nucl. Data Tables **61**, 127-176 (1995).

Spectroscopic applications of the ISOLDE laser ion source

V. Sebastian[1], R. Catherall[2], V.N. Fedoseyev[3], U. Georg[2], G. Huber[1]
Y. Jading[2], O. Jonsson[2], U. Köster[4], M. Koizumi[5], K.-L. Kratz[6], E. Kugler[2]
J. Lettry[2], V.I. Mishin[3], H.L. Ravn[2], C. Tamburella[2], A. Wöhr[7*]
and the ISOLDE Collaboration[2]

[1]*Johannes-Gutenberg Universität, Institut für Physik, 55099 Mainz, Germany*
[2]*CERN, 1211 Geneva, Switzerland*
[3]*Institute of Spectroscopy, Russian Academy of Sciences, 142092 Troitsk, Russia*
[4]*Technische Universität München, Physik-Department, 85748 Garching, Bavaria*
[5]*Japan Atomic Energy Research Institute, 1233 Watanuki, Takasaki, 370-12 Japan*
[6]*Johannes-Gutenberg Universität, Institut für Kernchemie, 55099 Mainz, Germany*
[7]*KU Leuven, Instituut voor Kern- en Stralingfysika, 3001 Leuven, Belgium*

Abstract. At the ISOLDE facility radioactive ion beams are produced via proton induced reactions in a target which is connected to a laser ion source. For beryllium a two step excitation scheme with laser light at wavelengths of $\lambda=235$ nm and $\lambda=297$ nm has been developed. Efficient laser ionization of beryllium was achieved with a new optical set-up using frequency tripling with two non-linear BBO crystals to generate laser light in the ultraviolet for the first excitation step. The second step was optimized to reach the $2p^2\ ^1S_0$ autoionizing state for high ionization efficiency. The isotope shift of ^{7,9,10,11,12,14}Be could be measured by tuning the wavelength of the first step.
The laser ion source has also been used for the preparation of neutron-rich silver ion beams. Tuning the laser frequency of the first step it was possible to ionize selectively low- and high spin isomers of silver isotopes via the hyperfine structure.
In both cases it was demonstrated that laser spectroscopy of exotic isotopes can be performed directly with the laser ion source. An outlook on other possible elements for laser spectroscopy using the laser ion source will be given.

Production of beryllium with a laser ion source

At ISOLDE, radioactive ions are produced via proton induced reactions in a thick target. The radioactive isotopes diffuse out of the target material and effuse to a hot tungsten cavity, where the resonant laser ionization takes place [1]. The **R**esonance **I**onization **L**aser **I**on **S**ource (RILIS) has been used to ionize the elements silver, manganese and nickel. Recently ionization was extended to beryllium, zinc, cadmium, copper [2], magnesium and tin. These pulsed ion beams are almost free from isobaric

---

* *present address: Oxford University, Department of Physics, Oxford OX1 3PU, United Kingdom*

contamination, apart from the continuous surface ionization of alkalies and other elements with a very low ionization potential.

A pulsed dye laser system is used for the resonant laser ionization. The tunable dye lasers are pumped with a copper vapor laser system at a repetition rate of 10 kHz to achieve a high ionization efficiency. The laser frequency is measured with a commercial lambdameter via the fringe patterns from Fizeau interferometers. For the ionization of beryllium a two step excitation scheme is used and requires laser light in the UV region (λ=235 nm) for the first step to reach the 2s2p $^1P_1$ state. This light is generated via frequency tripling with two non-linear BBO crystals and saturation of the transition was observed. Since the tungsten cavity was kept at a high temperature (above T=2300 °C) for a short release time of beryllium a Doppler broadening of 15 GHz at the first transition was observed for light elements like beryllium. This Doppler width of the resonance line limits the accuracy of the laser ion source for applications in optical spectroscopy. The second step was tuned to reach the $2p^2$ $^1S_0$ autoionizing state which has a linewidth of 74.9 cm^{-1} [3].

The efficiency of the ion source was determined in off-line measurements to at least 3.4(7)% [2]. The yield and release of ^{7,9,10,11,12,14}Be obtained from a 52 g/cm^2 uranium-carbide/graphite target is discussed in [4]. Collinear laser spectroscopy experiments were performed to measure the isotope shift of ^{7,9,10,11}Be$^+$ and the magnetic moment of ^{11}Be was measured with β-NMR [5,6]

Measurement of isotope shift for beryllium

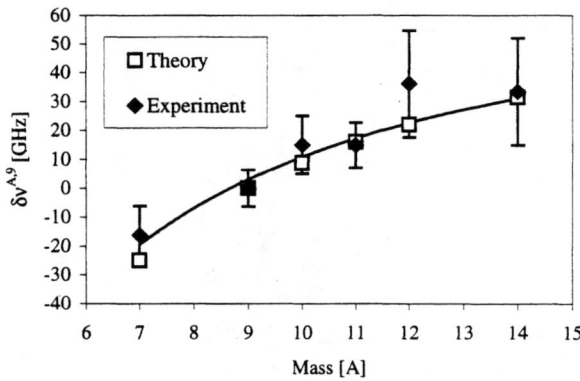

FIGURE 1: Isotope shift of ^{7,9,10,11,12,14}Be. The very low production rate and the limited collection time for ^{12}Be and ^{14}Be give additional statistical errors.

The isotope shift of ^{7,9,10,11,12,14}Be in the first excitation step is measured relative to the stable ^{9}Be (Figure 1). The isotopes ^{7,9,10}Be could be measured directly with a Faraday cup. For ^{11}Be and ^{12}Be the intensity of the β-counts were measured and ^{14}Be was identified via its β-delayed neutron emission. The theoretical value of the mass effect in the isotope shift has been calculated for ^{10}Be with a full-core plus correlation method to $\Delta\nu^{10,9}$ = 8.776 GHz [7]. Since the isotope shift for a light element like beryllium is nearly a pure mass polarization effect, the isotope shift is extrapolated with the mass factor for the other isotopes. We have demonstrated that laser spectroscopy of exotic isotopes can be performed directly with the laser ion source.

Isomer separation in silver ion beams

The laser ion source has been used for the selective ionization of neutron-rich silver isotopes [8]. Due to differences in the hyperfine structure of the resonance line an isomer selective separation could be obtained for the first time at ISOLDE by tuning the laser wavelength of the first excitation step. The hyperfine structure and isotope shift of neutron deficient silver isotopes are known from collinear laser spectroscopy [9].

The principle of isomer separation is shown in Figure 2 at the example of ^{107}Ag. The stable ^{107}Ag has a magnetic moment of $\mu_I=-0.11\mu_N$ and the hyperfine structure of the $5s^2S_{1/2} - 5p^2P°_{3/2}$ resonance line (λ=328.1 nm) cannot be resolved. For the ^{107m}Ag isomer the magnetic moment is large (μ_I=4.4μ_N) and the hyperfine structure of the $5s^2S_{1/2}$ atomic state is resolved with a laser line width of 0.2 cm^{-1}. Tuning the laser to the outer hyperfine components, ^{107m}Ag was selectively ionized.

The principle of isomer separation in a laser ion source can be applied for most of the spin isomers with different magnetic moments. For ^{107}Ag the isomer separation is equivalent to a mass resolution of m/Δm=10^6.

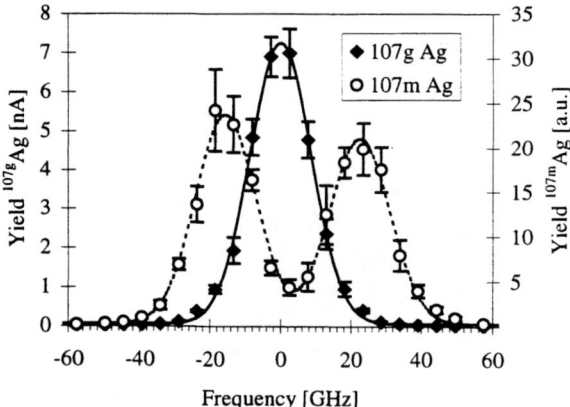

FIGURE 2: Isomer separation of ^{107g,m}Ag. The isomer ^{107m}Ag is selectively ionized at a frequency offset of 20 GHz.

For heavy silver isotopes the delayed neutron was measured as a function of laser frequency and the hyperfine pattern was recorded. A first successful application of isomer separation by laser ionization is the study of gamma ray decay of ^{122}Ag [10]. Detuning the laser frequency of 20 GHz away from the center, ionization of the low-spin isomer was suppressed, and only gamma rays of the high-spin component were observed.

Tuning the laser frequency to the expected hyperfine components of ^{129}Ag allowed an identification of the r-process "waiting-point" nuclide ^{129}Ag [11].

Outlook

The neutron rich tin isotopes are interesting candidates for laser spectroscopy. For tin a 3-step resonant ionization scheme was tested with an efficiency of about 10%. It is of fundamental interest to measure the charge radii of the tin isotopes around the doubly magic ^{132}Sn via the optical isotope shift. Various theoretical mean-field and Skyrme-Hartree-Fock calculations show a large spread in the prediction of the charge radii of the isotopes around ^{132}Sn [12]. The hyperfine structure and isotope shift in tin have been measured with high resolution laser spectroscopy for $^{108-111}$Sn [13] and $^{110-125}$Sn [14]. The laser ion source with its high efficiency and its Doppler limited spectral resolution could allow to measure the isotope shift for the even isotopes very far from stability.

References:

[1] V.I. Mishin et al., *Nucl. Instrum. Methods Phys Res. B 73, 550 (1993)*
[2] J. Lettry et al., *Rev. Sci. Instrum. 69, 761 (1998)*
[3] C.W. Clark et al., *J. Opt. Soc. Am. B 2, 891 (1985)*
[4] U. Köster et al., *contribution to these proceedings*
[5] Stefan Kappertz et al., *contribution to these proceedings*
[6] Matthias Keim, *contribution to these proceedings*
[7] K.T. Kwong et al., *Phys. Rev. A 48, 1945 (1993)*
[8] Y. Jading et al., *Nucl. Instrum. Methods Phys Res. B 126, 76 (1997)*
[9] U. Dinger et al., *Nucl. Phys., A 503, 331 (1989)*
[10] K.-L. Kratz et al., *Proc. Int. Conf. on Fission and Properties of Neutron -Rich Nuclei, December 1997, Sanibel Island, Florida, USA, ed. by Joseph H. Hamilton, World Scientific, Singapore (1998)*
[11] K.-L. Kratz, *contribution to these proceedings*
[12] P.-G. Reinhard et al., *Proc. Int. Workshop on Gross Properties of Nuclei and Nuclear Excitations, Hirschegg, Austria, January 1998*
[13] J. Ebertz et al., *Z. Phys. A 326, 121 (1987)*
[14] M. Anselment et al., *Phys. Rev. C 34, 1052 (1986)*

Status of the Canadian Penning Trap Mass Spectrometer at the Argonne National Laboratories

K.S. Sharma[1], R.C. Barber[1], F. Buchinger[2], J.E. Crawford[2], X. Feng[2], H. Fukutani[1], S. Gulick[2], G. Hackman[3], J.C. Hardy[4], D. Hofman[3], J.K.P. Lee[2], P. Martinez[2], R.B. Moore[2], G. Savard[3], D. Seweryniak[3] and J. Uusitalo[3]

[1] *Department of Physics, University of Manitoba, Winnipeg, MB, Canada R3T 2N2*
[2] *Department of Physics, McGill University, Montreal, PQ, Canada H3A 2T8*
[3] *Physics Division, Argonne National Laboratory, Argonne, IL 60439*
[4] *Cyclotron Institute, Texas A&M University, College Station, TX 77843-3366*

The Canadian Penning Trap (CPT) mass spectrometer has been modified and reassembled at the ATLAS facility of the Argonne National Laboratory. It is currently being commissioned there. With this apparatus we expect to be able to measure the masses of a wide variety of nuclides having half-lives greater that 0.1 s, to an accuracy of better than 10 ppb of the mass. The planned program of measurements is aimed at improving our knowledge of the nuclear mass surface in general and to provide precise mass data of relevance to tests of the Standard Model and fundamental symmetries. This report describes the current status of the instrument.

The Canadian Penning Trap (CPT) mass spectrometer was originally constructed at the TASCC facility of the Chalk River Laboratories (Atomic Energy of Canada Limited). A description of the spectrometer in its original form has been previously published (1,3). After the closure of the TASCC facility, the instrument was modified and reassembled at the ATLAS facility of the Argonne National Laboratory. It is currently being commissioned there. With this apparatus we expect to be able to carry out measurements on a wide variety of nuclides (both stable and unstable) with half-lives greater than 0.1 s. The planned program of measurements is aimed at improving our knowledge of the nuclear mass surface in general and to provide precise mass data of relevance to tests of the Standard Model and fundamental symmetries. The first off-line mass measurements with the CPT are expected to commence in the summer of 1998. Measurements on radioactive species should begin later in the year.

THE INSTRUMENT

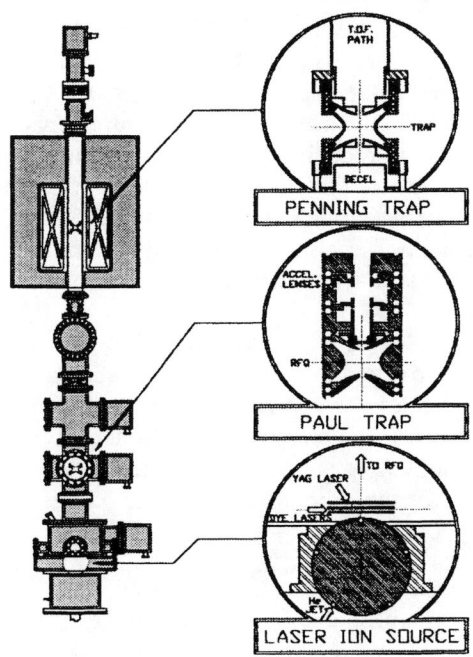

FIGURE 1. The schematic layout of the CPT mass spectrometer.

A schematic view of the CPT spectrometer in its off-line configuration is shown in Fig. 1. In this mode, a pulse of light from a Nd-YAG laser is used to create a burst of ions from a sample mounted on the target wheel. For radioactive species more elaborate procedures are necessary. Originally, we proposed the use of a resonant-ionization ion-source coupled to a He-jet transport system to produce ions of short-lived species. We have replaced this technique with one that makes better use of the facilities available at our new location. Unstable nuclides are produced as recoils from heavy-ion reactions using beams from the ATLAS facility. The products from these reactions are first analyzed in a gas-filled, split-pole, magnetic separator. Once the ions of the nuclide selected for study have been separated from the primary beam and other unwanted species, they pass through a thin window and are thermalized in a cell containing helium gas. The helium gas is continuously pumped away and the ions collected in a linear radio-frequency quadrupole (RFQ) trap. When sufficient numbers of ions have been accumulated they are ejected in a pulse.

Each pulse of ions is guided to the entrance of a Paul trap (or radio-frequency quadrupole trap) by a set of electrostatic lenses and steerers. The time of flight of the ions is used to select the species injected into this trap. Once inside the trap, the ions lose their energy in collisions with the He buffer gas and are trapped. Subsequent bunches may then be trapped and accumulated. When sufficient numbers of ions have been accumulated they are transferred to the precision Penning trap for mass analysis. This initial process of cooling and accumulation offers several advantages that reduce the likelihood of systematic effects in our measurements. It allows the 20 Hz repetition rate of the system producing the ions to be matched to the time required for a typical measurement cycle of the Penning trap (1 s). The rate of production of the ions is averaged over several ion-pulses ensuring a relatively constant number of ions injected into the Penning trap each time. More importantly, the initial conditions of the ions at the point of transfer to the Penning trap are well defined and reproducible.

The determination of the mass of a nuclide will be carried out in the precision Penning trap located in the bore of the, 5.9 T superconducting solenoid. The design of the trap is similar to that used by the ISOLTRAP group at CERN (3) with some exceptions as noted in (2). In a Penning trap the ions are confined by the combined effects of an axial magnetic field and an axially symmetric, quadrupole, electric field. The motion of the ions in the trap can be described in terms of three fundamental frequencies: the reduced cyclotron frequency, ω_+, the magnetron frequency, ω_- and the axial oscillation frequency, ω_z. These frequencies are all affected by the electrostatic fields and are unsuitable for the accurate determination of the mass of the ions. Only the true cyclotron frequency, $\omega_c = \dfrac{qB}{m} = \omega_+ + \omega_-$, may be related to the mass of the ions, independently of the electric fields. The cyclotron frequency is determined by exciting the ions at the cyclotron frequency and detecting the resonance with a time-of-flight technique (4).

CURRENT STATUS

The commissioning of the CPT mass spectrometer at its new location is well underway. Off-line tests, using gold ions, have successfully demonstrated the accumulation and cooling of ions in the Paul trap. The cooled ions have been transferred to the Penning trap and resonances at ω_+ and ω_c have been observed for gold ions. A Fourier limited resolving power of approximately 500,000 (m/δm, FWHM) is routinely obtained. Detailed studies of systematic effects are underway in preparation for our first off-line mass measurements.

On-line tests of the injection system for radioactive species have also begun. Recently, in a series of on-line runs, reaction products have been separated by the split-pole spectrometer, stopped in the gas cell and detected at the exit of the linear RFQ structure. Off-line experiments, with a ^{252}Cf fission-source, indicate efficiencies on the order of 10% for the transfer of ions stopped in the gas to the end of the linear RFQ structure. The beam line to transfer these ions to the CPT is currently being assembled.

CONCLUSION

The CPT mass spectrometer should allow high-accuracy mass measurements to be performed on a wide variety of stable and unstable nuclides. Our new injection system should allow us to inject nuclides with half-lives as low as 100 ms into the spectrometer. Initially, we plan to undertake the measurement of the masses of stable nuclei in the region of A=200 to a precision of better than 10^{-8}. Such measurements will allow us to fully evaluate the performance and capabilities of our new instrument and provide precise masses in a region where longstanding inconsistencies exist. In addition to survey-type measurements among unstable nuclei aimed at improving our general knowledge of the mass surface, we also plan measurements among the nuclides that participate in superallowed beta-decay. We hope to improve the precision to which the Q-values for such decays are known and to extend the dataset to previously inaccessible candidates. A precision of better than 10^{-8} is required. When combined with precise data on half-lives and branching ratios, this should make possible stringent tests of the CVC hypothesis in beta-decay and provide tests of the unitarity of the CKM matrix (5).

REFERENCES

1. Sharma, K. S., Barber, R. C., Buchinger, F., Crawford, J.E., Hagberg, E., Hardy, J.C., Koslowsky, V. T., Lee, J. K. P., Moore, R. B., Savard, G. and Watson, M., in *Proceedings of the International Conference on Exotic Nuclei and Atomic Masses*, Arles, France, ed. by M. de Saint Simon and O. Sorlin (Editions Frontieres) 1995, pp.811-816 .

2. Savard, G., Beeching, D., Hagberg, E., Hardy, J. C., Koslowsky, V. T., Watson, M., Barber, R.C., Feng, X., Sharma, K. S., Buchinger, F., Crawford, J. E., Gulick, S., Lee, J. K. P. and Moore, R. B., *Nuclear Physics* A626 (1997) 353c-356c.

3. Stolzenberg, H. et al, Physical Review Letters 65, 1990, p.3104.

4. Bollen, G., Moore, R.B., Savard, G., Stolzenberg, H., *Journal of Applied Physics* 68, 1990, p4355.

5. Hardy, J.C., these proceedings.

CORRELATION PARAMETERS IN NUCLEAR BINDING ENERGIES

S.I.Sukhoruchkin, D.S.Sukhoruchkin

PNPI, Gatchina, 188350, Russia: e-mail: sergeis@hep486.pnpi.spb.ru

Recently introduced Quark Model with Goldstone boson exchange (GBE) [1] for correct description of baryon properties uses as parameters Δ-excitation of nucleon ΔM_Δ=147 MeV= $(m_\Delta - m_N)/2$ and the initial nucleon mass m_N^{GBE}=1360 MeV close to $9\Delta M_\Delta$. Existence in particle masses (including ΔM_Δ and pion mass) empirical correlations [2,3] common with those observed in energies of nuclear states permits to recall suggestion [4] about the influence of nucleon structure on fine nuclear effects. The Mass Difference File (MDF) was produced from AME-95 [5] for study of these effects: correlations in nuclear binding energies (E_B) and in their differences (ΔE_B) [6-8]. Systematic character of observed effects is seen in distribution of ΔE_B of all N-even nuclei (Fig.1-2) [6,7] where the strongest maximum at 46.0 MeV corresponds to pairs of nuclei with N≤82 differing by ΔZ=2,ΔN=4 ($\Delta E_B(^6He)$). In Table 1 combinations ΔZ,ΔN in nuclei forming maxima in Fig.1-2 are given (with numbers of $\Delta E_B(n\times^6He)$ boxed).

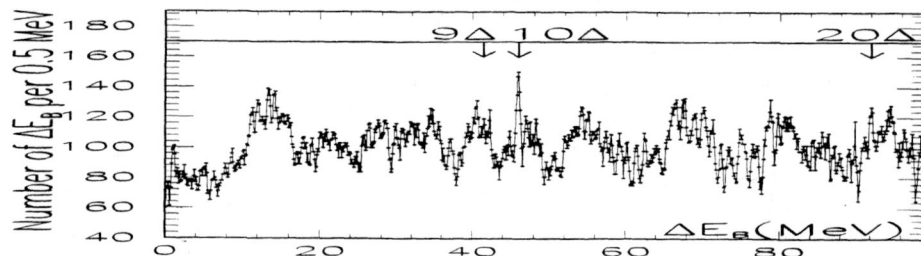

Fig.1. Sum distribution of ΔE_B of N-even nuclei (arrows mark mentioned in text maxima).

It was marked in [6-8] that positions of maxima in ΔE_B-distributions could be expressed as integers (n=9,10,20,41) of the period Δ=4.6 MeV which accounts (9/2) of the period ε_o=1022 keV explicitly seen in the discreteness of two-proton separation energies ($S_{2p} = \Delta E_B$ by ΔZ=2). This period derived from several relation in S_{2p} [6] is seen also in the distribution of four-proton separation energies ($S_{4p} = \Delta E_B$ by ΔZ=4): in Fig.3 arrows indicate period $2\varepsilon_o$ (data for N-even nuclei with N=52-82). The faint maximum in Fig.1 at ΔE_B=41.5 MeV=9Δ could be seen also in nuclei with Z=78-82 (Fig 4). Maxima at ΔE_B=92.1 MeV (2×46.0MeV) and at 188.5 MeV (shifted by 4.5 MeV from 4×46.0 MeV=184.0 MeV) were analyzed.

Table 1. Numbers of pairs of N-even nuclei (with their ΔZ, ΔN) where the difference of binding energies ΔE_B are turned out to be within maxima in sum ΔE_B distributions (Fig.1).

ΔE_B=	46.0 MeV=10Δ			41.5 MeV=9Δ				92.1 MeV=20Δ				188.5 MeV=41Δ		
ΔN	number of ΔE_B			number of ΔE_B			ΔN	number of ΔE_B			ΔN	number of ΔE_B		
6	10	10	0	11	0	0	10	6	3	9	18*	3	2	16*
4	4	48	11	4	31	8	8	8	21	7	16	3	11	2
2	0	0	14	0	0	14	6	0	3	6	14	2	7	4
ΔZ=	0	2	4	0	2	4		2	4	6		6	8	10*

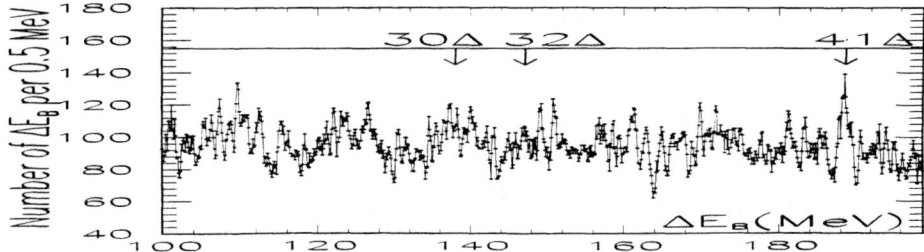

Fig.2. Sum distribution of ΔE_B of N-even nuclei (arrows mark mentioned in text maxima).

By analysis of ΔE_B-distribution in N-even light nuclei (N≤50) differing by $2\Delta=\Delta N=4$ (by ^{6}He) the maximum at 50.7 MeV (n=11 in units Δ) was found [5,7] (see Fig.5). For all N-even nuclei differing by two, three and four ^{6}He-clusters ΔE_B-distributions (shown in [7]) were produced and the observed positions of groupings in ΔE_B were compared with integer numbers of 9Δ and 10Δ. For example in nuclei differing by three ^{6}He-cluster ΔE_B-distribution has structure (with period about 1 MeV) which starts from $\Delta E_B=138.0$ MeV (3×46.0 MeV=$135\varepsilon_o$) and includes ΔE_B close to integers n= 136-137-138 in units of period ε_o. The shift of ΔE_B means additional increasing of binding energies connected with the presence of additional clusters (results of the performed analysis are presented in [7,8]).

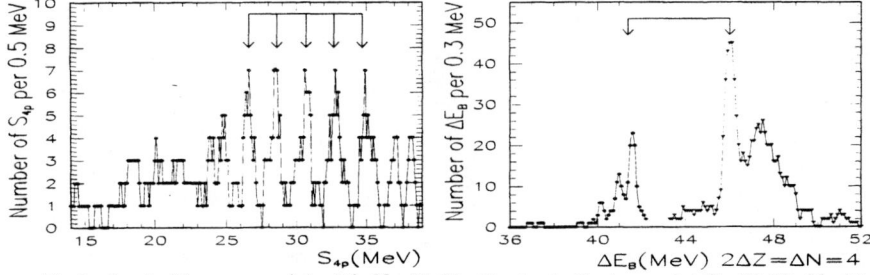

Fig.3. S_{4p} in N-even nuclei with N=52-82. Fig.4. ΔE_B in nuclei Z=78-82, N=51-126.

The next step in data analysis consisted in the division of all nuclei along several large groups (with Z=64-82, Z=40-64 or N=40-82) where again ΔE_B (differences of E_B known with accuracy better then 0.1 MeV) were plotted as histograms (with averaging interval 0.5 MeV) and the relation between ΔE_B and respective combinations $\Delta Z, \Delta N$ were studied. Fixation in each of the group the prominent maxima in ΔE_B-distributions (say 188.5 eV in the first group or 46.0 MeV in the second) means that we mark pairs of nuclei with unusually stable mass differences. We proceed further calculations of ΔE_B between these fixed nuclei and all other nuclei and by such a way the search for additional stable intervals in E_B was done.

This method can be illustrated by one example. Fixing above discussed $\Delta E_B=188.5$ MeV= $=41\Delta$ (see Fig.2) and plotting distribution of all ΔE_B between E_B of the heavier nuclei (forming this interval) and E_B of all other nuclei one gets distribution with the single strong maximum at 147.1 MeV. This value corresponds to n=32 in units of Δ and is marked by arrow in ΔE_B-distribution for all nuclei differing by $\Delta Z=8, \Delta N=14$ (Fig.6). Differences 188.5 MeV-147.1 MeV=41.4 MeV and $\Delta Z(=10-8=2), \Delta N(=18-14=4)$ are in agreement with the observed stability (in nuclei with N≤126) of $\Delta E_B=9\Delta=41.5$ MeV connected with $2\Delta Z=\Delta N=4$ (Fig.4).

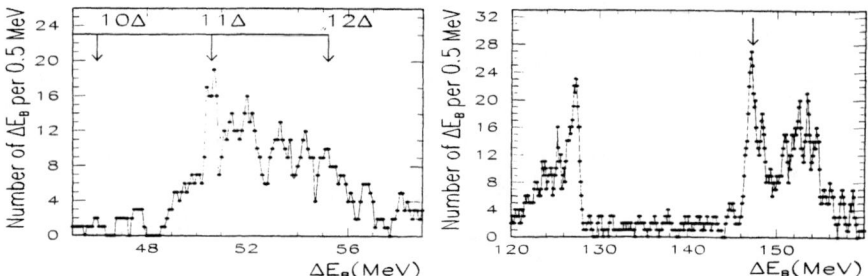

Fig.5. ΔE_B in nuclei N≤50, $2\Delta Z=\Delta N=4$. Fig.6. ΔE_B in nuclei $\Delta Z=8$, $\Delta N=14$.

Some of the noticed long-range correlations in ΔE_B are shown in Fig.7 (maximum at 409 MeV in N-even nuclei) and in Fig.8 (maximum in odd-odd nuclei at 441 MeV=3×147 MeV). The test of the systematic character of these groupings (marked with arrows) is underway [7].

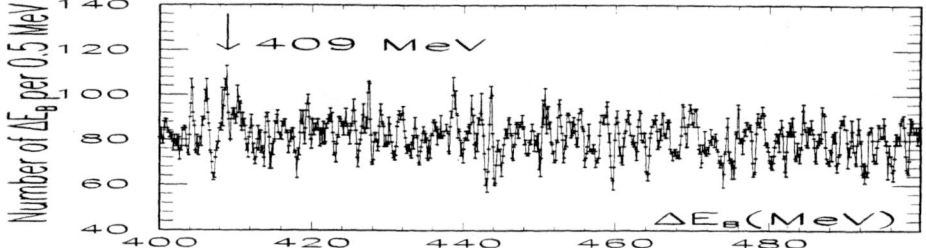

Fig.7. Distribution of ΔE_B in all N-even nuclei in region 400-500 MeV.

The original observation about proximity of a stable interval ΔE_B in light nuclei (147.2 MeV) to the parameter of Δ-excitation of nucleon $\Delta M_\Delta=294$ MeV/2=147 MeV [2,6] agrees with the appearance of the same interval in heavy near-magic nuclei. We should note that in the former case it is connected with four-^{4}He-configuration but in the latter – with quite another ($\Delta Z=8, \Delta N=14$). Introduced earlier parameter of discreteness in particle masses - period $\delta=16m_e$ (from m_π-$m_e=17\delta$ and $m_\mu+m_e=13\delta$) permitted represent ΔM_Δ as 18δ [2,9,10].

In paper [11] the stability of parameters of valence nucleon interaction in nuclei after lead was noticed ($\hat{\varepsilon}_{np}=340$ keV). Later the same nearly constant value (341 keV=$\varepsilon_o/3$) was found in nuclei with other shells [7,12,13]. The possibility to find out this tuning effect (with the same parameter or integer to it) in nuclear excitations [9,12-14] is based on shown in [15] connection of residual nucleon interaction with excitations of near-magic nuclei. The principal possibility to find out common features (or common parameters of the general origin) in nuclear excitations and binding energies was expressed in [16,17] and these results are in line with it.

Nuclear parameters ε_o and Δ are very close to $2m_e$ ($m_e=0.511$ keV) and mass difference of pions $m_{\pi^\pm}-m_{\pi^0}$ [2,9,10]. The latter deviates from $9m_e$ by relative value SFERC (Scaling Factor equal to Electrodynamics Radiative Correction $\alpha/2\pi$, see [9]) introduced [2] for description of the observed ratio $m_e/3\Delta M_\Delta=1/(32\times 27)$ and ratios between stable nuclear intervals (for example $\varepsilon'/\varepsilon_o$=SFERC with fine-structure interval $\varepsilon'=1.2$ keV in highly excited states).

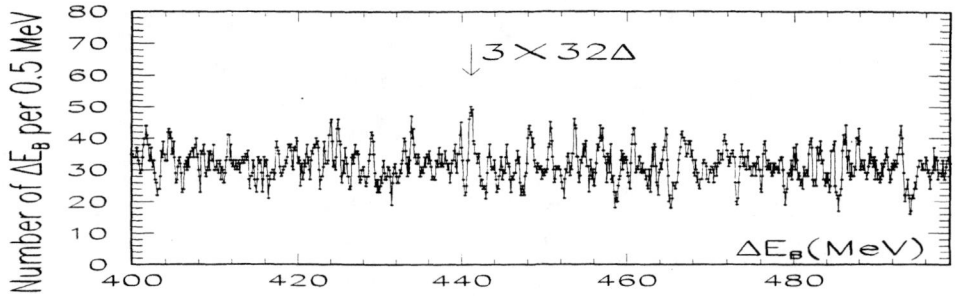

Fig.8 Distribution of ΔE_B in all odd-odd nuclei in region 400-500 MeV.

We should mention SFERC-ratio between masses of muon and vector Z-boson [9,14] which combined with ratio $m_e/3\Delta M_\Delta$ leads to the lepton ratio $L=m_\mu/m_e=M_Z/3\Delta M_\Delta$.

Masses of light quarks (forming nucleon and mesons) estimated in [18] as m_d=9.3 MeV and m_u=5.1 MeV agree with observation [19] that m_d-m_u accounts 4075 keV (near to $8m_e$ [10]).

Above mentioned closeness of the initial nucleon mass to $9\Delta M_\Delta$ (and hence the initial constituent quark mass M_q to $3\Delta M_\Delta$) agrees with some estimations of M_q discussed in literature (see [10]). Independent observation of the closeness of ΔE_B to ΔM_Δ=147 MeV should deserve attention as a signal about the influence of nucleon structure on nuclear data.

REFERENCES

1. L.Ya.Glozman, hep-ph/9711490 (to be publ. Nucl.Phys.A), Phys.Rev.C56,5,2779(1997).
2. S.I.Sukhoruchkin, Proc.Conf.Stat.Prop.Nuclei,Albany,1971,Ed.J.Garg,Pl.Pr.p.215.
3. R.Frosch, Nuovo Cimento A, 104,6,p.913 (1991), and ref. therein.
4. S.Devons, Proc.Rutherford Jubilee Int.Conf.,Manchester,Ed.J.Birks.Heywood,p.611.
5. G.Audi, A.H.Wapstra, AME-95, Nucl.Phys.A595,p.409 (1995).
6. S.I.Sukhoruchkin, Sov. J. Nucl. Phys. 53, 1266 (1991).
7. S.I.Sukhoruchkin, ISINN-6,Int.Sem.Inter.Neutr.Nucl.,Dubna,1998,JINR E3-98-70.Abstr.
8. S.I.Sukhoruchkin, Intern.Conf.Nucl.Struct.at Extremes, Lewes, 1998, Abstracts.
9. S.I.Sukhoruchkin, Symmetry Methods in Phys.Dubna,1993,JINR E2-94-347,p.528,536.
10. S.I.Sukhoruchkin, Symmetry Methods in Phys.Dubna,1995,JINR E2-96-224,p.549.
11. G.Mouze and C.Ythier, Nuovo Cimento, 103A,105 (1990).
12. S.I.Sukhoruchkin, Proc.9-th Int.Sym.Capt.Gamm.Ray Spec.Ed.G.Molnar, Budapest,1996.
13. S.I.Sukhoruchkin, Low Energy Nucl.Dynamics (World Scientif.,1995),p.647.
14. S.I.Sukhoruchkin, Proc.Int.Conf.Nucl.Data,Trieste,1997,p.311,SIF.v.59,Bologna.
15. I.Talmi, Proc.8-th Int.Symp.Capt.Gamma-Ray Spec.Frib.,1993,Ed.J.Kern,p.67.
16. A.Arima, Nucl. Phys. A507 (1990), p.305c.
17. A.Bohr, Nucl.Struct.Dubna Symp.,1968, IAEA,Vienna,.p.179.
18. H.Leutwyler, The ratios of the light quarks masses. CERN-TH/96-44.
19. R.Lebed, Phys.Rev. D 47, 3, 1134 (1993).

The Magnetic and Quadrupole Moment of Oriented Nuclei Measured with the β-LMR-NMR

S.Teughels[1,*], G.Neyens[1,*], N.Coulier[1,+], G.Georgiev[1], S.Ternier[1,+],
K.Vyvey[1], D.L.Balabanski[1,a], R.Coussement[1], W.F.Rogers[2],
D.Cortina-Gil[3], F.De Oliveira[3], M.Lewitowicz[3], W.Mittig[3],
P.Roussel-Chomaz[3], A.Lépine-Szily[4]

[1] *Instituut voor Kern- en Stralingsfysica, University of Leuven,
Celestijnenlaan 200-D, B-3001 Leuven, Belgium.*
[a] *On leave from Faculty of Physics, St.Kliment Ohridsky University of Sofia,
BG-1164 Sofia, Bulgaria.*
[2] *Westmont College, 955 La Paz Road, Santa Barbara, CA 93108 USA.*
[3] *Grand Accélérateur National d'Ions Lourds, B.P. 5027, F-14021 Caen Cedex, France.*
[4] *Institute of Physics, University of São Paulo, C.P. 66318, 05389-970, São Paulo, Brazil.*

Abstract. By combining two well established methods, the β-level mixing resonance (β-LMR) and the β-nuclear magnetic resonance (β-NMR), it is possible to determine the magnetic moment and the quadrupole moment of nuclei with low spin. The combined method can be applied on spin oriented ground state β-decaying nuclei. It was tested on the known ^{12}B and successfully applied on the neutron rich ^{18}N. We found for the magnetic moment μ(^{18}N)=0.157(7)n.m. and for the quadrupole moment Q(^{18}N)=30(3)mbarn.

INTRODUCTION

To determine the quadrupole Q and the magnetic moment μ of nuclei with low spin we implant the oriented nuclei in a non cubic single crystal, the principal axis of the axially symmetric electric field gradient V_{ZZ} (// c-axis) is turned over a small angle β with respect to a holding field B_0. This magnetic field is perpendicular to the incoming beam. A linearly polarized radio-frequency signal is applied in the beam direction and can be split up into two circular polarized fields with a frequency ν_{RF} and amplitude B_1. The Hamiltonian of an ensemble of nuclei with spin I submitted to this three interactions, can be written in the LAB system with $\omega_0 = -\dfrac{g\mu_N B_0}{\hbar}$, $\omega_1 = -\dfrac{g\mu_N B_1}{\hbar}$, $\omega_Q = \dfrac{eQV_{ZZ}}{4I(2I-1)\hbar}$, $\omega_{RF} = 2\pi\nu_{RF}$ as:

$$H = \omega_0 \cdot \hat{I}_z + \omega_Q \cdot \{(1/2)(3\cos^2\beta - 1)(3\hat{I}_z^2 - \hat{I}^2)\}$$
$$+ \omega_1 \cdot \{\hat{I}_x \cos(\omega_{RF}t + \Delta) + \hat{I}_y \sin(\omega_{RF}t + \Delta)\}$$

For small angles β (<5°), only the component of V_{ZZ} in the direction of B_0 is taken into account. The Hamiltonian is time dependent. To make it time independent we transform it to a rotating frame x'y'z' with frequency v_{RF} by the unitary transformation $U(t) = \exp(-i\hat{I}_z(\omega_{RF}t + \Delta)/\hbar)$, which gives us:

$$H' = (\omega_0 - \omega_{RF})\hat{I}_{z'} + \omega_1 \hat{I}_{x'} + \frac{\omega_Q}{\hbar}\{(1/2)(3\cos^2\beta - 1)(3\hat{I}_{z'}^2 - \hat{I}^2)\}$$

It is possible to solve the time-evolution of the Hamiltonian in this system [1]. When we look at the Rabi diagram for a nucleus with spin I=1 (Fig. 1), one can see that the electric field gradient causes the energy splitting in the m-states proportional to m^2, while the static magnetic field causes the regular Zeeman splitting. At two different strengths of the external field B_{NMR} the energy put in by the RF-signal matches the splitting of the levels. In the middle of these two fields the levels cross at B_{LMR} [2].

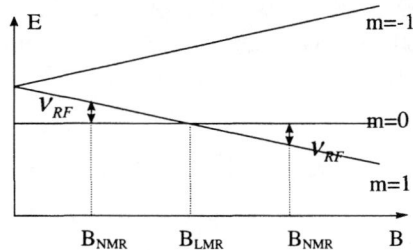

FIGURE 1. Rabi diagram for a nucleus with spin I=1.

The nuclei are oriented by the nuclear reaction. In high-energy reactions, it is favorable to select the reaction products emitted in the forward direction to have the maximum production-rate. This means we start with initial alignment. The axial symmetry axis of the alignment is parallel to the fragment momentum, and the +m and -m quantum states are equally populated. If one selects the nuclei under a certain angle with respect to the primary beam, a polarized ensemble is obtained. In this situation the symmetry axis is perpendicular to the reaction plane and the +m and -m states are unequally populated. In both cases, when the RF-frequency matches the level separation or when the levels cross, the populations of the involved m-states are mixed, which means the orientation will change. Polarization is induced when starting with initial alignment, or polarization is partially destroyed in the latter case [2]. During the experiment we measure this three resonant changes of the β-asymmetry as a function of the static magnetic field. The RF-frequency ν_{RF} and a modulation on this frequency δ_{RF} are kept constant. The ratio of the quadrupole moment to the magnetic moment can be directly derived from the level crossing resonance field B_{LMR}, and the magnetic moment from the nuclear magnetic resonance field B_{NMR}. For a nucleus with spin I=1, the conditions for resonance are

$$\frac{\nu_Q[kHz]}{\mu[n.m.]} = \frac{B_{LMR}[G]}{0.984 \cos \beta} \qquad \mu[n.m.] = \frac{2.624 \nu_{RF}}{\left| B_{NMR}^1 - B_{NMR}^2 \right|}.$$

SPIN POLARIZED ^{12}B: A TEST CASE

In order to demonstrate the feasability of the combined LMR-NMR method, we selected the β-decaying nucleus ^{12}B ($I^\pi=1^+$, $T_{1/2}$=20.20ms, Q_β=13.4MeV, μ=1.003n.m., Q=13.4(1.4)mbarn). The experiments were conducted at the Laboratoire d'Analyses par Réactions Nucléaire in Namur, Belgium. A deuteron beam of 1.5MeV was impinged upon a ^{11}B target (90mg/cm^2 on a gold backing of 25μm). The nuclei were selected under an angle of 45^0 with respect to the beam direction to have the maximum polarization possible, and they were implanted in a Mg single crystal (hcp), ν_Q(B<u>Mg</u>)=46.5(5)kHz at room temperature. The principal axis of the crystal was oriented at an angle β=3(1)^0 with respect to the external magnetic field axis. One can see the three resonances (Fig. 2) for an applied RF-signal with ν_{RF}=5000(10)Hz, δ_{RF}=200(1)Hz and B_1=3.0(5)G. The data are fitted with a numerical calculation using the eigenstates of the interacting Hamiltonian. The only fit-parameter is the initial orientation, determining the amplitude of the resonance.

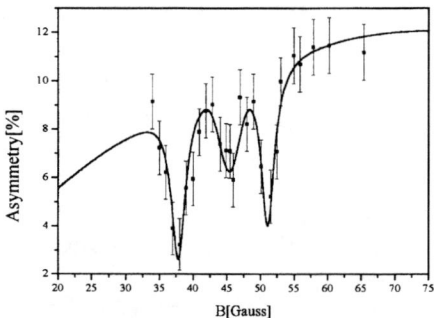

FIGURE 2. The LMR-NMR resonances for spin polarized ^{12}B in a Mg single crystal.

EXPERIMENT ON SPIN ALIGNED ^{18}N

The combined method was successfully applied on ^{18}N ($I^\pi=1^-$, $T_{1/2}$=624ms) implanted in a Mg single crystal at 40K. The nuclei were produced by a fragmentation reaction of ^{22}Ne (60MeV/u) and separated with the LISE III spectrometer at GANIL, Caen, France. Out of the position of the Level Mixing Resonance we derive

$\nu_Q/\mu(^{18}N\underline{Mg})=1311(7)$kHz/n.m. (Fig. 3a) which is in good agreement with an earlier measurement using the LMR-method $\nu_Q/\mu(^{18}N\underline{Mg})=1328(33)$kHz/n.m. [3]. In a next experiment we put an RF-signal, $\nu_{RF}=20(10^{-3})$kHz, $\delta_{RF}=3(10^{-3})$kHz and $B_1=40(1)$G. Due to this big modulation and amplitude, the NMR resonances as function of B_0 are very broad. From the position of the resonances (Fig. 3b) we derived the magnetic moment $\mu(^{18}N)=0.157(7)$n.m. and we can deduce the quadrupole moment $Q(^{18}N)=30(3)$mbarn, knowing $V_{ZZ}(N\underline{Mg})=2.80(20).10^6$V/cm^2 at 40K. Because the resonances were so broad and the magnetic moment is small we had to subtract the LMR-data from the combined LMR-NMR-data to see the two NMR resonances.

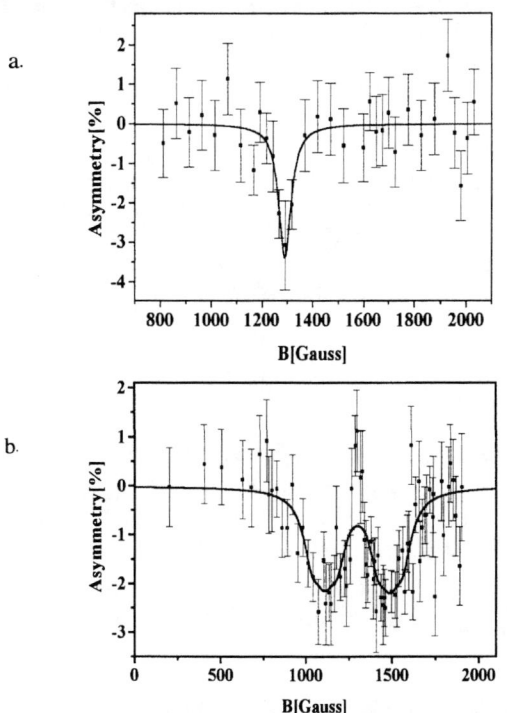

FIGURE 3a. The Level Mixing Resonance and **3b.** the two Nuclear Magnetic Resonances of ^{18}N in a Mg single crystal at 40K.

* G.Neyens is a post-doctoral researcher of the FWO-Vlaanderen (Fund for Scientific Research). + N.Coulier and S.Ternier have a Ph.D.-fellowship from the IWT.

REFERENCES

1. Matthias E., Olsen B., Shirley D.A., Templeton J.E. and Steffen R.M., *Phys. Rev. A* **4**, 1626 (1971)
2. Neyens G., Nouwen R., Coussement R., *Nucl. Instr. and Meth. in Phys. Res. A* **340**, 555 (1994)
3. Neyens G., Coulier N., Ternier S., Vyvey K., Michiels S., Coussement R., Balabanski D.L., Casandjian J.M., Chartier M., Cortina-Gil D., Lewitowicz M., Mittig W., Ostrowski A.N., Roussel-Chomaz P., Alamanos N., Lépine-Szily A., *Phys. Lett. B* **393** 36-41 (1997)

β-NMR Measurement of the Nuclear Quadrupole Moments of [20, 26-31]Na

S. Wilbert[1], B.A. Brown[2], W. Geithner[1], U. Georg[1], S. Kappertz[1], M. Keim[3], P. Lievens[4], R. Neugart[1], M. Neuroth[1], L. Vermeeren[5] and the ISOLDE Collaboration[3]

[1] *Institut für Physik, Universität Mainz, D-55099 Mainz, Germany*
[2] *NSCL, Department of Physics, Michigan State University, East Lansing, MI 48824, USA*
[3] *EP Division, CERN, CH-1211 Geneva 23, Switzerland*
[4] *Laboratorium voor Vaste-Stoffysica en Magnetisme, K.U. Leuven, B-3001 Leuven*
[5] *Instituut voor Kern- en Stralingsfysica, K.U. Leuven, B-3001 Leuven*

Abstract. The quadrupole moments of the neutron-rich isotopes [26-31]Na and of the neutron-deficient [20]Na were measured by β-NMR after implantation of the short-lived nuclei into the host crystal lattices of Mg, LiNbO$_3$ and NaNO$_3$. The ion beams from ISOLDE were neutralized and polarized by in-beam optical pumping. Of particular interest are the isotopes close to N = 20 for which various experimental observations suggest rather strong deformation.

Introduction

In general, the shell model describes quite well the properties of sd-shell nuclei, but for extreme N/Z ratios deviations from simple shell-model predictions are observed. For example, neutron-rich nuclei at the N = 20 neutron shell closure with Z ≈ 11 exhibit unexpectedly large deformations. These are ascribed to the occupation of intruder states from the fp shell. Employing in-beam optical polarization and β-NMR spectroscopy (1) we measured the nuclear quadrupole moments of a number of sodium isotopes.

In-Beam Optical Polarization and Quadrupole Interaction in β-NMR

The experiments were performed at the ISOLDE on-line isotope separator. Radioactive sodium isotopes were produced by fragmentation of uranium, extracted as a beam of singly-charged ions and neutralized in a sodium vapour cell. Optical pumping with circularly polarized laser light was performed along a beam section of 1.5 m which was exposed to a small longitudinal magnetic field. In the transitional field at the entrance of the NMR magnet the polarization was rotated adiabatically to the transverse direction, before the electronic and nuclear spins were decoupled. Finally the atoms were implanted into a suitable single crystal, and the polarization of the nuclei and the NMR signals were detected by measuring the asymmetry of the β-decay with respect to the magnetic field direction.

The quadrupole moments, interacting with the electric field gradient of a non-cubic host lattice at the site of implanted sodium nuclei, give rise to a splitting of the NMR signals. Due to quadrupole interaction the rf transitions between adjacent nuclear Zeeman levels $|m_I\rangle$ in the static magnetic field are separated into 2I equidistant components. For a given isotope one has to find a host with lattice sites providing a suitable field gradient and a relaxation time of the polarization exceeding the half-life. Such properties are usually found for ion crystals or metallic single crystals.

Recent Measurements on $^{29\text{-}31}$Na

At the ENAM95 Conference (2) we reported first successful experiments on $^{26\text{-}28}$Na. These isotopes are produced rather abundantly (10^6 - 10^7 atoms/s), their β-decay asymmetry parameters are large (0.75 - 0.9), and their half-lives span the range from 1s to 30 ms. Thus they were very favourably used to study and optimize the experimental conditions and to develop ideas for an extension of the measurements towards the N = 20 shell closure. These early measurements were performed using a LiNbO$_3$ crystal which has a sufficiently large electric field gradient to resolve the NMR spectra even for the very small quadrupole moments of ^{26}Na and ^{27}Na. For ^{28}Na the splitting already became inconveniently large (1.5 MHz) and the resonances were inhomogeneously broadened because of lattice defects leading to a distribution of field gradients at the sites of Na$^+$ ions.

From shell-model predictions and from the ideas of deformation it was expected that the quadrupole moments would further increase towards N $\approx$ 20. With the production yields decreasing from 6×10^5 to 4×10^2 atoms/s between ^{28}Na and ^{31}Na it was clear that statistics required much narrower NMR structures. These were found in a metallic Mg single crystal. Here the quadrupole splitting of ^{28}Na was about 25 kHz and the line width was less than 10 kHz.

For the low-yield isotopes ^{20,30,31}Na, a further gain in sensitivity had to be achieved by combining the effect of several frequencies. As for all resonances the distance ratios from the Larmor frequency are known, it is possible to scan a pattern of frequencies symmetrically from the Larmor frequency over a point where all transitions with $\Delta m_I = \pm 1$ are induced simultaneously, thus equalizing the population of all m_I sublevels of the nuclear spin system. This established method to enhance the NMR signal in the observed asymmetry has to be applied very carefully: (i) The Larmor frequency has to be known very precisely, because a partial overlap of the resonances would reduce and asymmetrically broaden the signals in an uncontrolled way. (ii) The rf amplitude should nearly saturate the resonances, but should avoid the regime where two- (ore more) quantum transitions become dominant. (iii) The analysis of the spectra requires a realistic model of the resonance line shape which is produced by the interaction of the nucear spin system with several frequencies stepped over different ranges. In particular, it is not expected that the line shape is a trivial superposition of one-resonance signals.

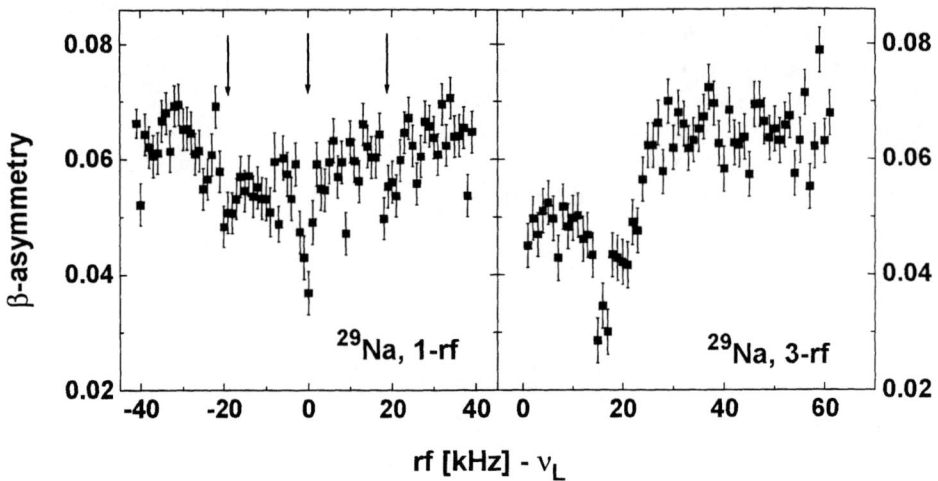

Figure 1. Comparison of the one-frequency and the three-frequency signals for ^{29}Na.

As ^{29}Na and ^{31}Na both have the spin I = 3/2, as their half-lives of 45 and 17 ms are short and not very different, and as even the quadrupole moments of both isotopes turn out to be similar, nature provides us with an ideal test case to study the conditions (ii) and (iii). For ^{29}Na (Fig. 1) it is still possible to take a well resolved spectrum of all three individual resonances. The three-frequency spectrum measured with otherwise unchanged conditions gives a model spectrum for ^{31}Na for which all essential experimental parameters, in particular the quadrupole moment, are known independently. The Larmor frequency forming the center of the structure can be determined by measuring the NMR signal in a cubic crystal lattice. For this purpose an Au foil was most convenient, because it gave narrow resonances on the maximum asymmetry signals.

The final analysis of the three-frequency spectrum of ^{31}Na (Fig. 2) is in progress. ^{30}Na and the neutron-deficient ^{20}Na, both with I = 2, were investigated using a two-frequency scheme. This compromise between the maximum signal achieved with four frequencies and a single-frequency scan still provides information about the sign of the quadrupole moment, which is lost when all resonances coincide.

How can we obtain the sign of the quadrupole moments? From the optical pumping scheme we know the population distribution over the atomic |F, m_F> levels, and we can calculate the distribution over the nuclear spin states |m_I> achieved after decoupling of I and J. Thus from the relative intensities of the individual resonances we know the assignment to (positive or negative) m_I quantum numbers. Then the signature for a positive or negative sign of the quadrupole interaction is whether the intensities increase or decrease from lower to higher frequencies. As the sign of the internal electric field gradient is not known a priori, we obtain relative signs of the quadrupole moments, and the absolute signs depend on the rather safe assumption that model

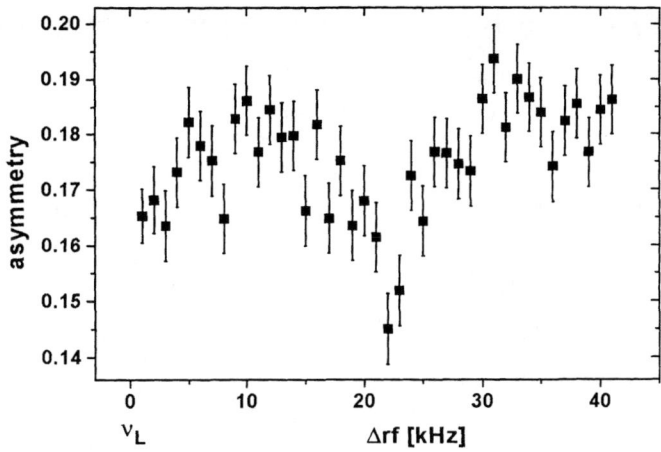

Figure 2. Three-frequency signal of ^{31}Na.

predictions and systematics give the signs correctly in general or alternatively for one selected reference isotope.

The measurements yield the quadrupole interaction constants (i.e. the products $Q \times V_{zz}$) for different isotopes whose quadrupole moments Q interact with the same electric field gradient V_{zz}. An absolute calibration of the quadrupole moments is obtained from measurements of the NMR spectrum of ^{28}Na implanted in a NaNO$_3$ crystal. For this particular crystal the field gradient at the Na lattice site is known from classical NMR experiments (3).

The results in comparison with shell-model predictions are presented graphically in the contribution by M. Keim (4). This comparison shows that intruder states from the fp shell have to be taken into account for the isotopes close to N = 20. The fp-shell admixture has little effect for ^{29}Na, but improves dramatically the theoretical values for ^{30}Na. Even the sign of the ^{30}Na moment from a pure sd-shell calculation (5) is opposite to the measured one, thus stressing the importance of the experimental sign information. For ^{31}Na both predictions happen to be rather similar and somewhat smaller than the preliminary experimental value.

† Supported by BMBF, Germany (contract No. 06 MZ 566) and by N.F.W.O., Belgium

REFERENCES

1. Arnold, E., et al., *Z. Phys. A* **331**, 295 (1988) and *Phys. Lett. B* **281**, 16 (1992).
2. Keim, M., et al., *Proc. Int. Conf. on Exotic Nuclei and Atomic Masses (ENAM 95)*, M. de Saint Simon and O. Sorlin, eds., Editions Frontières (Gif-sur-Yvette, 1995), p.65.
3. Pound, R.V., *Phys. Rev.* **79**, 685 (1950).
4. Keim, M., contribution to these proceedings.
5. Carchidi, M., Wildenthal, B.H., and Brown, B.A., *Phys. Rev. C* **34**, 2280 (1986).

The Wigner Term in Heavy Nuclei

Nissan Zeldes

The Racah Institute of Physics, The Hebrew University of Jerusalem, Jerusalem 91904, Israel

Abstract. We note the origin of the Wigner term in $T = 0$ and $J = 0$ nucleon pair correlations. Empirical evidence indicates that its energy is continuous through a shell region, and it persists through the region 82-126. $Z = 126$ might well be the next spherical proton magic number after $Z = 82$.

INTRODUCTION

The Wigner term $\varepsilon|T_z|$ in nuclear masses is presently directly observable up to $A = 60$. There is recent renewed interest in it in heavier p-rich nuclei. We first point out its relation to ground state (g.s.) $T = 0$ and $J = 0$ pair correlations. Then, combining empirical evidence from masses and charge symmetry, we show its persistence through mixed major shells in heavier nuclei.

TWO-NUCLEON CORRELATIONS AND GROUND STATE MASSES

In the lowest-seniority model with isospin the nuclear g.s. energy of even-even (e-e) and odd-a j^a configurations, and the center of mass (c.m.) energy of g.s. J-multiplets of odd-odd (o-o) configurations with given seniority v and reduced isospin t, can be intuitively written [1, 2] as

$$W^{nucl}(j^a) = N_0(V_0 - \overline{V}_2) + N_1(\overline{V}_1 - \overline{V}_2) + N_t \overline{V}_2 \qquad (1)$$

where $\overline{V}_1$ and $\overline{V}_2$ are the respective $(2J + 1)$-averaged two-body matrix elements $V_{j^2 J}$ for $T = 0$ and for $J_{even} \neq 0$, V_0 denotes the matrix element $V_{j^2 J=0}$, N_t is the total pairs number $\frac{1}{2}a(a-1)$, N_0 the number of $J = 0$ pairs and N_1 the number of $T = 0$ pairs:

$$N_0(j^a, vtT) = \frac{1}{2j+1}\left[\frac{1}{4}(a-v)(4j+8-a-v) - T(T+1) + t(t+1)\right] \qquad (2)$$

$$N_1(j^a, T) = \frac{1}{4}\frac{a(a-1)}{2} - \frac{1}{2}\left[T(T+1) - \frac{3}{4}a\right]. \qquad (3)$$

CP455, *ENAM98: Exotic Nuclei and Atomic Masses*
edited by B. M. Sherrill, D. J. Morrissey, and Cary N. Davids
© 1998 The American Institute of Physics   1-56396-804-5/98/$15.00

The empirical effective shell model interaction [3] satisfies the relations

$$V_0 < 0, \quad \overline{V}_1 < 0, \quad \overline{V}_2 > 0. \tag{4}$$

Consequently eq. (1) is minimized by maximizing the pair numbers N_0 and N_1. This corresponds to lowest T and v and highest compatible t, which for e-e and odd-a nuclei[1] are $T = |T_z|$ and (v, t) equal respectively (0, 0) and (1, 1/2). Eq. (1) then gives for the g.s. energy of a nucleus with $Z_0 = \frac{1}{2}A_0$ protons and $\frac{1}{2}A_0$ neutrons forming a doubly-magic core and a valence nucleons in a j-subshell the well known result [1, 4]:

$$\begin{aligned} E_{g.s.}(A_0 + a, T = |T_z|) = E^{Coulomb}(Z_0 + p) + \\ E_0 + a\left(c + \tfrac{1}{2}\pi\right) + \tfrac{1}{2}a(a-1)b + \left[T(T+1) - \tfrac{3}{4}a\right]\varepsilon + \\ \frac{1-(-1)^a}{2}\left(-\tfrac{1}{2}\pi\right) + \frac{1-(-1)^{np}}{2}\left[(1-\delta_{I,0})\kappa + \delta_{I,0}\lambda\right], \end{aligned} \tag{5}$$

where E_0 and c are the respective nuclear energies of the core and of a single valence nucleon, π, b and ε are given linear combinations of V_0, $\overline{V}_1$, and $\overline{V}_2$, and κ and λ are pairing terms in o-o nuclei. n and p denote the respective numbers of valence neutrons and protons, and $I = n - p$ is the neutron excess. For a multi-subshell valence configuration eq. (5) is a useful approximation near shell region boundaries [4], with a, n and p denoting the total respective numbers of valence nucleons, and the parameters denoting average major shell values.

The linear term $\varepsilon T = \varepsilon|T_z|$ having a cusp line at N = Z is the Wigner term, resulting from maximizing the numbers N_1 and N_0 of strongly interacting T = 0 and J = 0 nucleon pairs by minimizing T.

EMPIRICAL EVIDENCE ON THE WIGNER TERM IN HEAVY NUCLEI

Empirical evidence indicates that the Wigner term energy is continuous through the shell rather than being describable by a cusp limited to the close vicinity of N = Z. The question is best approached by studying long complete isobaric chains extending on both sides of N = Z in a diagonal shell region, obtainable from known $I \geq 0$ isobars by subtraction of the neutron-proton mass differences and the Coulomb energy, and using charge symmetry [2].

Fig. 1 shows plots of such masses and corresponding $Q_{\beta^-\beta^-}$ values for A = 72 and A = 73 isobars. The mass plots are highly charge-symmetric, with a cusp at N = Z. There is a discontinuous decrease of $Q_{\beta^-\beta^-}$ associated with crossing the cusp towards higher Z, and the slopes of the lines on the two sides of N = Z are about equal. This is all in qualitative agreement with eq. (5).

---

[1] In o-o nuclei there is competition between v = 0 and $T = |T_z|$ coupling [4].

Obviously, by eliminating a few masses around the cusp the remaining $Q_{\beta^-\beta^-}$ plot would still comprise two parallel straight lines left and right of N = Z, with a vertical off-set, rather than a single straight line like for a quadratic mass parabola. Thus the effects of the Wigner term are noticeable throughout the shell.

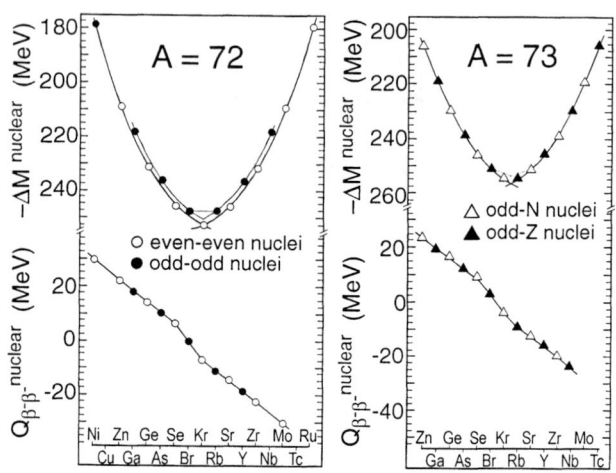

FIGURE 1. Isobaric mass parabolas corrected for Coulomb effects and corresponding $Q_{\beta^-\beta^-}$ values for A = 72 and A = 73 isobars. Reprinted from Phys. Lett. B429, N. Zeldes, The physical origin of the Wigner term and its persistence in heavy nuclei, pages 20-26, copyright 1998, with permission from Elsevier Science.

We use the same method of extrapolation based on charge symmetry to address the possible existence of the Wigner term in the higher diagonal shell regions 50-82 and 82-126. Here available isobaric chains are incomplete, comprising experimentally known n-rich isobars, and their charge-symmetrically extrapolated p-rich mirror nuclei. Intermediate masses around N = Z are lacking.

Fig. 2 shows plots of masses and $Q_{\beta^-\beta^-}$ values corrected for Coulomb effects for A = 207 and A = 208 isobars in the 82-126 shell region. The plots look like those of fig. 1 with the central parts missing. They indicate a cusp intersection of the two quadratic parabolic arcs through the masses when they are extended towards each other all the way to I = 0. Thus, to the extent that extrapolation to I = 0 is justifiable, fig. 2 indicates the persistence of the Wigner term in the highest reachable diagonal shell region 82-126.

This finding has a bearing on the location of the next spherical proton magic number beyond Z = 82. The Wigner term results from T = |T$_z$| isospin coupling of neutrons and protons in the same (mixed) valence subshells. Its occurrence in fig. 2 indicates the occurrence of the same valence subshells for neutrons and protons, with Z = 126 a proton magic number similarly to the neutrons. This agrees with recent self-consistent mean field calculations [5] and with a recent suggestion based on BE(2) systematics [6].

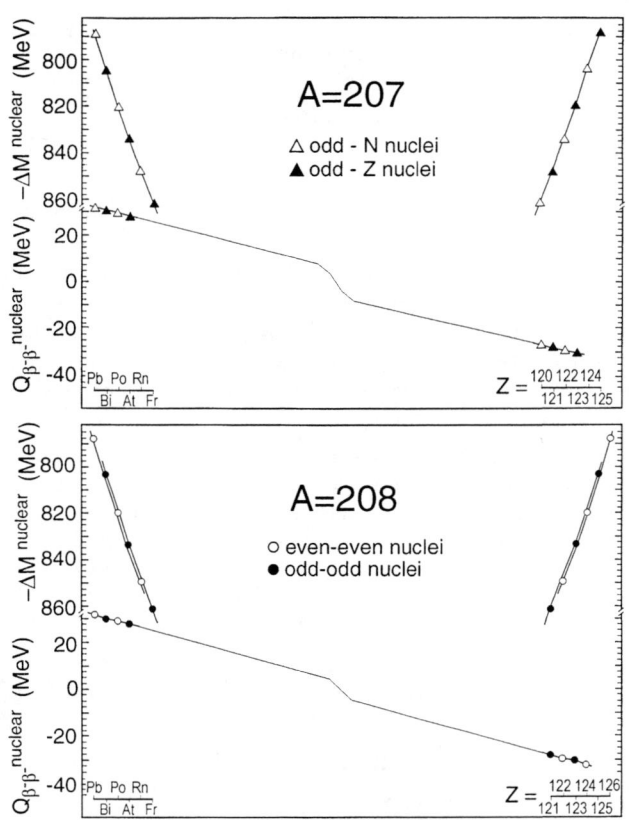

FIGURE 2. The same as fig. 1 for A = 207 and A = 208 isobars in the 82-126 shell region. The thin $Q_{\beta^-\beta^-}$ lines are adjusted to the data. Reprinted from Phys. Lett. B429, N. Zeldes, The physical origin of the Wigner term and its persistence in heavy nuclei, pages 20-26, copyright 1998, with permission from Elsevier Science.

REFERENCES

1. A. de-Shalit and I. Talmi, Nuclear Shell Theory (Academic Press, New York, 1963).
2. N. Zeldes, Phys. Lett. B (in press).
3. W.W. Daehnick, Phys. Rep. **96** (1983) 317.
4. N. Zeldes in Handbook of Nuclear Properties, edited by D.N. Poenaru and W. Greiner (Clarendon Press, Oxford, 1996), 12.
5. S. Ćwiok, J. Dobaczewski, P.-H. Heenen, P. Magierski and W. Nazarewicz, Nucl. Phys. **A611** (1996) 211.
6. N.V. Zamfir, G. Hering, R.F. Casten and P. Paul, Phys. Lett. **B357** (1995) 515.

ISOTOPE SHIFTS OF OPTICAL TRANSITIONS OF Kr I USING COLLINEAR FAST BEAM LASER SPECTROSCOPY

Jens Lassen and Hans A. Schuessler
Dept. of Physics, Texas A&M University, College Station, TX 77843-4242

Fast beam laser spectroscopy of stable and radioactive isotopes has been one of the most productive methods to obtain groundstate properties of nuclei far from stability. The isotope shifts (IS) as measured from optical transitions allow to determine changes in the mean square nuclear charge radii .Such measurements are particularly useful when extended over long chains of isotopes, since then the influence of the number of neutrons on the observed nuclear structure can be studied systematically [1]. For odd isotopes the hyperfine structure (HFS) has to be known. This is particularly important for resonance ionization schemes in which isotope ratios have to be measured.

We have determined the isotope-shifts and hyperfine structure splittings systematically in the various stable isotopes of Kr for the transitions from the metastable Kr states $5s[3/2]_2°$ and $5s'[1/2]_0°$. These metastable states are separated from the ground state by 9.9eV. The levels are accessible by optical spectroscopy in the near infrared spectral region.

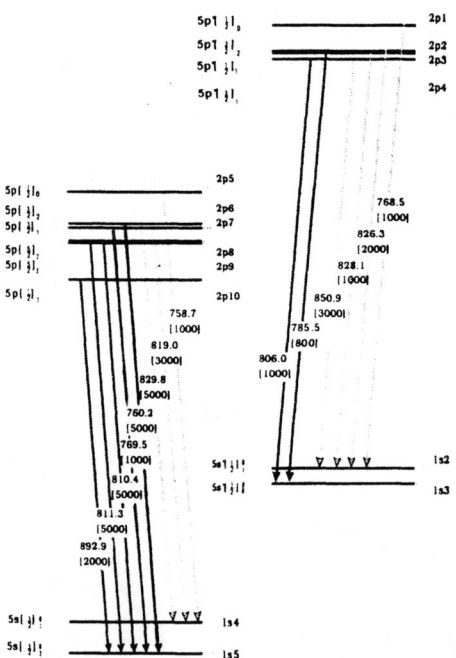

Fig. 1. Energy level diagram for Kr showing the seven measured optical transitions from the metastable $5s[3/2]_2°$ and $5s'[1/2]_0°$ to the 5p and 5p' energy levels.

In addition to laser spectroscopy we also study field ionization for trace detection of the Kr isotopes. The complete fast beam apparatus is shown schematically in Fig. 2. The laser spectroscopy measurements were performed by collinear fast beam laser spectroscopy. A discharge ion-source produces a Kr ion beam with beam energies ranging from 3 to 11keV. The ion beam is overlapped collinearly with a laser beam from a Coherent CR899 Ti-sapphire laser and post accelerated into a charge exchange cell operated with Rb as the charge exchange medium. Rb allows for a near resonant charge exchange into the metastable Kr states with high efficiency [2]. The optical detection (PMT I) for collinear fast beam laser spectroscopy is located directly following the charge exchange cell to minimize optical pumping.

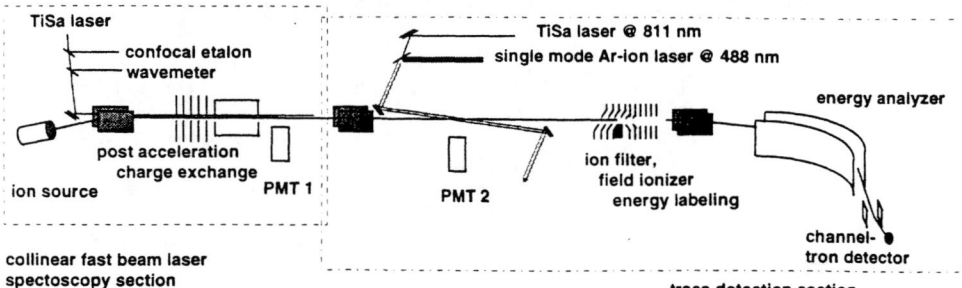

Fig. 2. Collinear fast beam apparatus used for detection of rare Kr isotopes. The first section of the apparatus was used to measure IS and HFS in the 5s to 5p transitions accessible from the metastable Kr states and the fluorescence signal is collected on PMT 1. Field ionized atoms are detected at the channeltron located at the end of the apparatus.

The results of these investigations allow us to directly find the transitions needed for the quasi-collinear excitation needed in the resonance ionization scheme to measure rare Kr isotopes by stepwise excitation and field ionization [3]. PMT 2 is then used for monitoring the excitation into the intermediate level. A comprehensive compilation of IS and HFS in the stable isotopes of Kr is given in Table 1 and Table 2 and for some lines are compared with the data available in the literature. The data will be useful in evaluating the proposed ultra trace analysis schemes for krypton. Trace detection techniques are of interest for extending the present systematic study of IS and HFS to the extremely short lived [71]Kr and [73]Kr isotopes, where beam intensities of only a few ions/sec are expected and where particle detection should yield an increase in the signal to noise ratio by about two orders of magnitude as compared to fluorescence detection. This data can also be used in testing electronic structure calculations of Kr. In the field ionization set-up the excitation is performed under an angle of 2 degrees to avoid light field induced shifts in the field ionization region. Our extended collinear fast beam apparatus with stepwise excitation and field ionization is shown in the right segment of Fig. 2. The IS and HFS measurements were performed with the apparatus shown in the left segment of Fig. 1. The results were then scaled for the quasi collinear excitation. In this way systematic errors due to the determination of the overlap angle of laser beam and atom beam could be avoided. The field ionization arrangement is also used to measure trace amounts of the rare isotopes [81]Kr($T_{1/2} = 10.7$a) and [85]Kr ($T_{1/2} = 2.3*10^5$ a) and of other trace elements for which suitable excitation schemes can be found.

TABLE 1. Compilation of the isotope shifts

transition	wavelength [nm]	ref.	86-84 [MHz]	86-82 [MHz]	86-80 [MHz]	86-78 [MHz]	86-83 [MHz]
5s'[1/2]$_0^o$	785.5		71.7 (0.5)	136.2 (1.1)	214.0 (1.2)	294.8 (1.6)	94.5
5p'[1/2]$_1$		4	69.3 (1.6)	136.3 (1.8)	214.5 (1.9)	296.2 (2.0)	93.3 (1.6)
		5	66.0 (1.2)	132.3 (2.4)	206.7 (2.8)	288.9 (3.4)	83.0 (1.8)
5s'[1/2]$_0^o$	806.0		74.5 (1.0)	139.1 (2.2)	215.9 (2.7)	293.8 (3.0)	87.1
5p'[3/2]$_1$		4	68.6 (2.3)	132.8 (2.0)	209.7 (3.0)	288.1 (3.1)	84.9 (1.9)
		5	69.3 (1.5)	138.0 (2.0)	217.8 (3.0)	300.9 (3.5)	85.1 (2.1)
5s[3/2]$_2^o$	760.2	6	73.7 (0.4)	141.6 (4.0)	221.6 (4.0)	304.3 (0.5)	90.2 (0.6)
5p[3/2]$_2$		4	73.4 (1.1)				94.1 (1.4)
		5	69.0 (1.2)				
		7	72.4 (.05)	139 (2.0)	216 (4.0)	305 (2.0)	89.5 (1.8)
5s[3/2]$_2^o$	769.5		71.9 (1.0)	141.1 (2.0)	220.0 (3.0)	302.7 (3.0)	88.8
5p[3/2]$_1$		4	74.2 (1.2)	143.5 (1.0)	224.9 (2.4)	307.5 (2.6)	90.4
		5	68.1 (1.5)	137.4 (3.0)	215.4 (3.6)	303.0 (4.7)	
5s[3/2]$_2^o$	810.4		70.8 (0.6)	138.4 (0.8)	216.2 (0.9)	297.7 (1.6)	87.6
5p[5/2]$_2$		4	70.0 (1.3)	136.3 (1.)	213.7 (1.9)	292.8 (2.2)	87.2 (1.3)
		5	71.1 (2.4)	132.6 (3.0)	210.0 (3.5)	294.0 (3.8)	
5s[3/2]$_2^o$	811.3		68.2 (0.7)	132.6 (0.5)	207.1 (0.5)	287.5 (0.7)	83.8
5p[5/2]$_3$		4	65.9 (1.3)	129.8 (1.6)	203.8 (1.8)	282.6 (2.0)	80.7 (1.4)
		5	65.7 (1.8)	129.3 (3.5)	205.5 (3.8)	278.4 (4.2)	
5s[3/2]$_2^o$	892.9		63.6 (2.5)	123.7 (2.7)	195.6 (3.5)	246.3 (4.3)	77.6 (3.5)
5p[1/2]$_1$							

TABLE II. Compilation of the hyperfine structure constants A and B in MHz

state		this work	theory[8]	ref.[4]		ref.[9]		ref.[10, 11]		ref.[8]		ref.[12]		ref.[4]		ref.[7]	
5s[3/2]$_2^o$	A			-243.93	(0.04)	-244.0	(0.1)									-243.8	(0.3)
	B			-452.93	(0.60)	-452.0	(0.5)									-454	(4)
5p'[1/2]$_1$	A	226.35 (0.14)	228.4	226.47	(0.16)	226.0	(0.3)	226.8	(0.4)	226.8	(0.2)	228.7	(1.5)	226.6	(0.6)		
	B	27.2 (1.2)	21.0	26.5	(1.2)	25.4	(2.0)	22.0	(2.0)	-----		-----		27.0	(6.0)	21.0	(3.0)
5p'[3/2]$_1$	A	-575.88 (0.24)	-571.11	-576.68	(0.16)			-576.9	(1.2)								
	B	21.7 (1.9)	24.28	18.6	(1.2)			-----		----							
5p[3/2]$_2$	A															-108.3	(0.3)
	B															-86	(5)
5p[5/2]$_2$	A	-156.35 (0.90)	-158.99	-156.49	(0.08)			-156.0	(1.0)								
	B	-409.5 (1.2)	-407.72	-407.7	(1.3)			-410	(20)								
5p[5/2]$_3$	A	-103.74 (0.80)	-103.13	-104.02	(0.06)			-103.0	(1.0)								
	B	-434.4 (2.0)	-431.70	-436.9	(1.7)			-430	(30)								
5p[1/2]$_1$	A	-143.2 (0.7)	-137.6					-143.9	(0.3)	142.6	(0.6)						
	B	-21.3 (4.4)	-20					-21	(1)								

This work was supported in part by DOE and the Energy Resources Program of Texas A&M University.

1. W. H. King, *Isotope shifts in atomic spectra* (Plenum Press, New York, 1984).
2. J. Pan, Dissertation, *Department of Physics* (University of Colorado, Boulder, 1997).
3. K. Stratmann, R. Hohmann, H. J. Kluge, *et al.*, Rev. Sci. Instrum. **65**, 1847 (1994).
4. B. D. Cannon and R. G. Janik, Phys. Rev. A **42**, 397 (1990) and references therein.
5. D. A. Jackson, J. Opt. Soc. Am. **69**, 503 (1979).
6. H. A. Schuessler, A. Alousi, R. M. Evans, *et al.*, Phys. Rev. Lett. **65**, 1332 (1990).
7. M. Keim, E. Arnold, w. Borchers, *et al.*, Nucl. Phys. A **586** (1995).
8. X. Husson, J.-P. Grandin, and H. Kucal, J. Phys. B **12**, 38833889 (1979).
9. H. Gerhardt, F. Jeschoneck, W. Makat, *et al.*, Hyperf. Interact. **9**, 175 (1981).
10. J. R. Brandenberger and S. C. Parker, Phys. Rev. A **44**, 3354 (1991).
11. J. R. Brandenbrger, Phys. Rev. A **39**, 64 (1989).
12. W. L. Faust and L. Y. Chiu, Phys. Rev. **129**, 1214 (1963).

NUCLEI AT THE DRIP LINES

Exploration Beyond the Proton Drip-Line

Philip J. Woods

Department of Physics and Astronomy, Edinburgh University, EH9 3JZ UK

Abstract. This paper reviews the recent developments in the study of nuclei lying beyond the proton drip-line. Particlar emphasis is made of the phenomenon of proton radioactivity for which a large and wide ranging data base now exists. Theoretical models of proton emission from spherical nuclei can predict the detailed trends of proton decay spectroscopic factors very well. The first examples of proton decay from highly deformed nuclei have been discovered and are well reproduced by a theoretical approach based on Nilsson states. Such measurements provide an insight into the fragmentation of single particle strength in deformed nuclei. The technique of Recoil Decay Tagging and its particular application to the study of the structure of deformed proton radioactive nuclei is discussed.

INTRODUCTION

The proton drip-line represents the dividing line between isotopes that are either bound or unbound to the emission of a proton from their ground-states. For light elements, nuclei lying beyond the proton drip-line only exist in the form of resonances, ^{39}Sc being the heaviest system to be studied to date [1]. The Coulomb barrier experienced by an unbound proton increases progressively with element number,Z, until it becomes likely that systems that are proton unbound can be directly detected. For example the isotope ^{77}Y has recently been identified [2] using the LISE3 separator at GANIL although mass formulae suggest the odd proton is energetically unbound. Progressing beyond Z=50 it becomes more probable than not that odd Z nuclei have at least one proton-radioactive isotope. This is due to a combination of the large height of the Coulomb barrier and the decrease in the rate of change of proton decay Q-value with neutron number $\triangle Q_p/\triangle N$ which varies with an approximate inverse dependence on the mass number, A [3]. For proton radioactivity to be observed experimentally it must have a significant decay branch which means that in practise the drip-line must be crossed by several isotopes. For example, ^{171}Ir is the heaviest proton unbound Ir isotope but proton emission is first observed for the isotope ^{167}Ir [4]. A continuous chain of odd-Z proton emitting elements has been identified from Z=67-83 [5], these results along with other

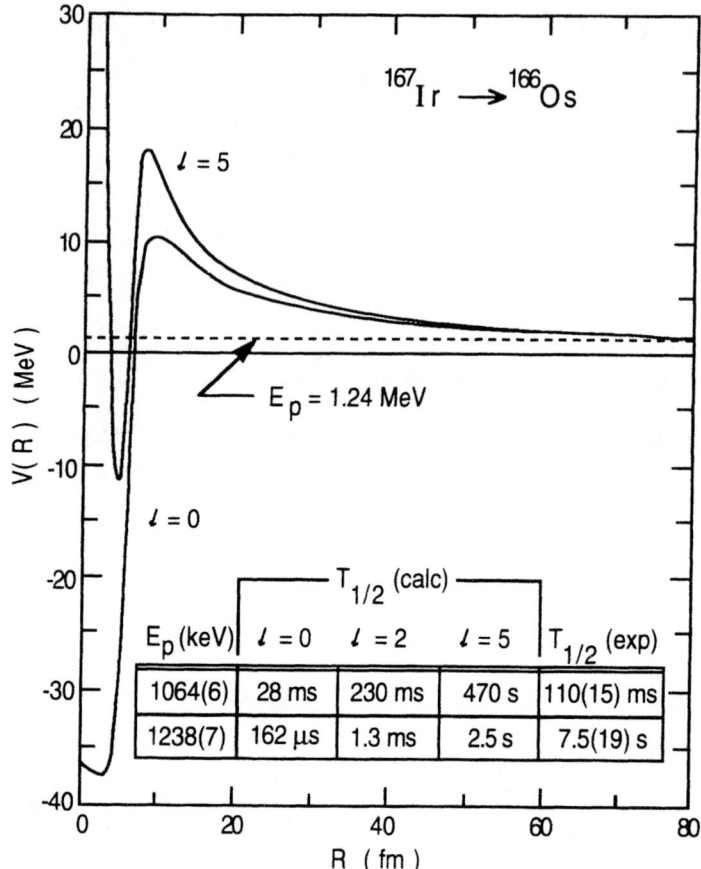

Figure 1. Proton-nucleus potentials calculated for the proton emitter ^{167}Ir. The inset shows proton decay partial half-lives calculated using the WKB approximation for three values of the orbital angular momentum l, compared to the experimental values for the ground and isomeric state configurations.

examples now constitute a large and wide ranging data base of proton transitions from which to explore the phenomenon of proton radioactivity.

SPHERICAL PROTON EMITTERS

In its simplest form the proton decay transition probability can be calculated using a semi-classical WKB approach [6]. A recent theoretical review of proton emission from spherical nuclei by Aberg et al. [7] has shown that such calculations agree surprising well with more exact DWBA and two-potential treatments when the same proton potential parameter set is used. The choice of a realistic potential introduces an uncertainty ~2 into the calculation [8]. A typical potential set is the global scattering potential of Becchetti and Greenlees [9] used in Figure 1 for the cases of ground and isomeric proton emission from ^{167}Ir which occur from $s_{1/2}$ and $h_{11/2}$ orbitals, respectively. As can be seen from the table inset into Figure 1 the proton decay rates are extremely sensitive to the orbital angular momentum of the unbound proton, l, and shell model assignments can be confidently made despite the uncertainty in the choice of potential.

A very attractive aspect of proton decay is that the proton can be considered to be preformed inside the nucleus thereby avoiding some of the uncertainty associated with alpha-decay transition rate calculations. Nonetheless, in order to calculate proton decay rates correctly a spectroscopic factor must be introduced which is defined theoretically as [10]

$$S = \frac{1}{2J_i+1} | <J_i||a^\dagger_{nlj}||J_f> |^2$$

where (nlj) represents a single particle state, J_i and J_f represent the total angular momentum of the nucleus in the parent and daughter systems and j represents the angular momentum of the emitted proton.

Davids et al. [4] used a low seniority shell model calculation to predict proton radioactivity spectroscopic factors for spherical nuclei with 64<Z<82. The calculation assumed degenerate $s_{\frac{1}{2}}$, $d_{\frac{3}{2}}$ and $h_{\frac{11}{2}}$ protons and spectator neutrons and predicted

$$S = \frac{P}{9}$$

where P represents the number of pairs of proton holes in the daughter nucleus wrt the Z=82 closed shell. In Figure 2 this calculation is compared with experimental values for spectroscopic factors defined as the ratio of the theoretical and experimental proton partial half-lives

$$S^{exp} = \frac{t_{\frac{1}{2}}^{wkb}}{t_{\frac{1}{2}}^{exp}}$$

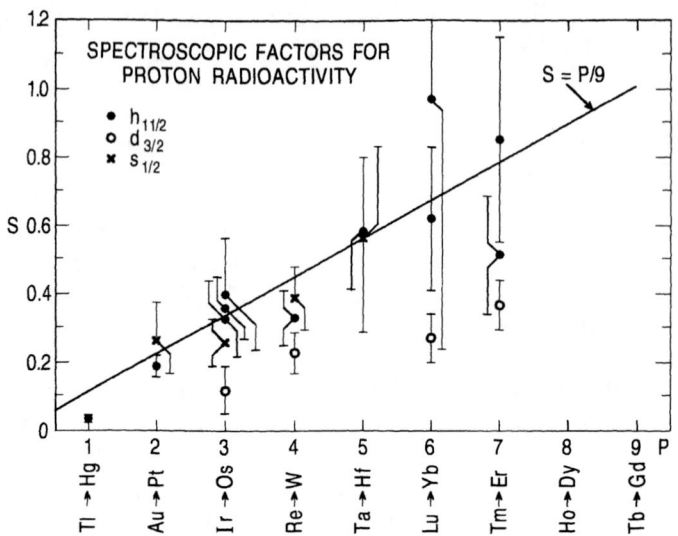

Figure 2. Experimental proton decay spectroscopic factors compared with low seniority shell model calculation predictions for the region 64<Z<82.

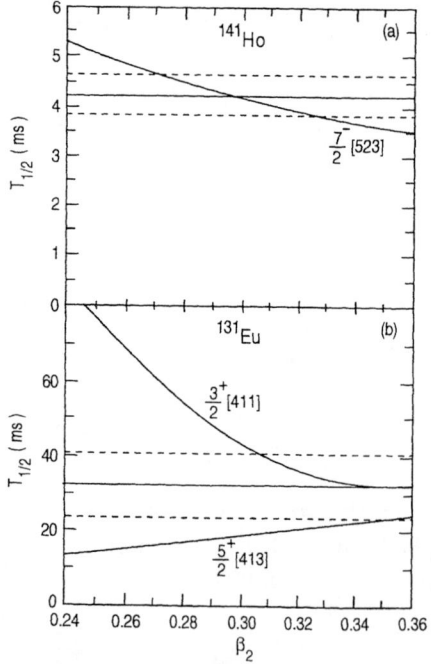

Figure 3. Calculated proton decay half-lives for Nilsson configurations of ^{141}Ho and ^{131}Eu as a function of deformation. Experimental values are shown as a horizontal solid line with a dashed line representing the uncertainty.

In general the spectroscopic factors are reproduced remarkably well although the experimental values for the $d_{3/2}$ transitions are consistently lower than the model predictions. Recent calculations of spectroscopic factors using a more sophisticated BCS approach [7] are also able to reproduce the trends of spectroscopic factors well although the discrepancies for $d_{3/2}$ transitions remain present.

DEFORMED PROTON EMITTERS

It has been known for some time that the decay rates of the proton emitters ^{109}I and ^{113}Cs cannot be reproduced using calculations of the type described above [11]. One possible explanation for this behaviour was the onset of modest prolate deformations in the region of the proton drip-line above Z=50. Bugrov and Kadmensky [12] developed a model for proton emission from deformed nuclei using Nilsson wavefunctions with quadrupole deformation treated as a free parameter. They were able to reproduce the anomalous half-lives of ^{109}I and ^{113}Cs using relatively modest deformations $\beta \sim 0.1$ consistent with values expected for this transitional region. The macroscopic-microscopic mass model of Moller et al. [13] predicted the onset of much higher prolate deformations ($\beta \sim 0.3$) immediately below Z=69 along the region of the proton drip-line. Experiments using the Fragment Mass Analyzer (FMA) [14] at Argonne have recently identified ground-state proton radioactivity from ^{141}Ho (Z=67) lying just inside this region, and ^{131}Eu (Z= 63) [15] lying at the heart of the region, with predicted quadrupole deformation parameters of 0.29 and 0.33 [13], respectively. Spherical proton decay calculations signally fail to reproduce the half-lives. Davids et al. [13] demonstrated that the decay rates could be well reproduced using the calculational approach of Bugrov and Kadmensky obtaining Nilsson configurations and quadrupole deformations consistent with the predictions of Moller et al. [13] (see Figure 3). This successful extension of the theory to highly deformed nuclei represents a significant deepening of our understanding of the proton decay phenomenon. These transitions provide a unique insight into the fragmentation of single particle strength in deformed nuclei.

GAMMA-RAY SPECTROSCOPY OF PROTON EMITTERS USING RECOIL DECAY TAGGING

Although theories of spherical and deformed proton emitters are now being tested over a wide range of nuclei it is desirable to have independent information on the structure of these nuclei. In particular high resolution in-beam gamma-ray studies can provide insights into the nuclear deformation. The technique of Recoil Decay Tagging (RDT) (see Figure 4) [16,17] is an ideal tool that was developed [16] with this particular goal in mind and is now used extensively in the study of neutron-deficient and heavy nuclei exhibiting charged particle decay modes.

In-beam gamma-rays were first identified from the proton emitter ^{109}I [16] at Daresbury, however a direct cascade feeding the ground-state was not established

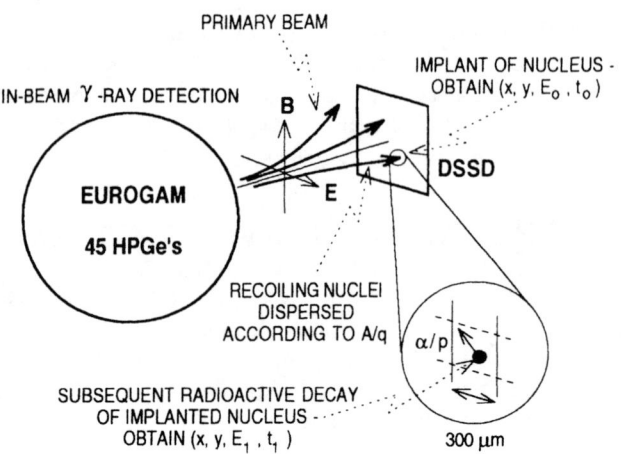

Figure 4. Schematic diagram of the Recoil Decay Tagging (RDT) technique applied using the Daresbury RMS to identify in-beam γ-rays from the ground-state proton emitter ^{109}I

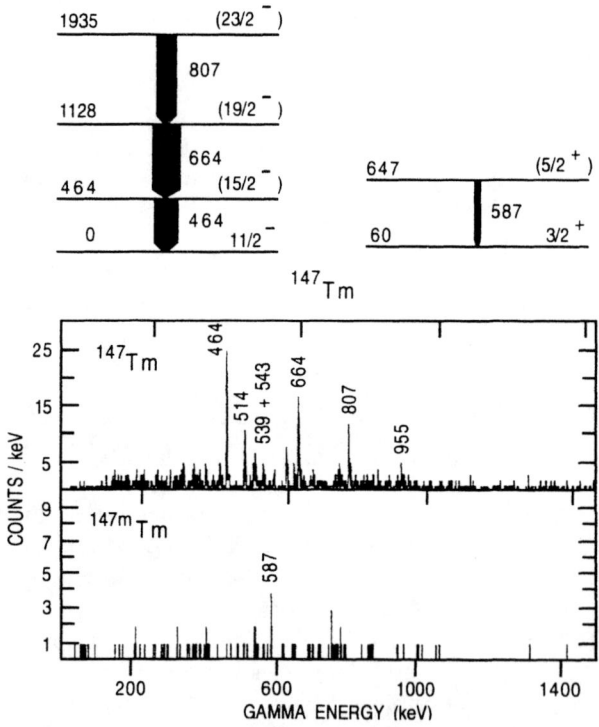

Figure 5. Energy level scheme and gamma spectra for ^{147}Tm obtained using the RDT technique on the Argonne FMA. The upper and lower spectra show gamma-rays preceding the proton-decays of the $h_{11/2}$ ground-state and $d_{3/2}$ isomeric state, respectively.

limiting the conclusions that could be drawn. Following on from this the RDT technique was successfully applied to ground and isomeric proton emission from ^{147}Tm (see figure 5) in an experiment on the Argonne FMA [19]. The gamma-ray band built on the ground-state is consistent with β=0.13. Interestingly, Kadmensky and Bugrov have applied their model for deformed proton emission to this case [20] and the results do not disagree significantly from spherical calculations, unlike the effect of such a deformation on proton decay rates in the region above Z = 50. It appears that Tm and Ho proton emitters lie right at the interface of the region of rapid shape change to high prolate deformations. Clearly it is desirable to identify in-beam gamma-rays from the highly deformed proton emitters. In a very recent RDT experiment using the Argonne FMA coupled to Gammasphere, gamma-rays were successfully identified from the ground and isomeric states in ^{141}Ho, the latter having a cross-section~40nb [21], these data are currently being analysed.

TWO PROTON EMISSION

Two proton radioactivity has yet to be observed, it is expected to occur in highly neutron - deficient even Z nuclei when the proton pairing energy suppresses single proton emission. The most promising candidate isotope discovered to date is ^{45}Fe identified using high energy heavy ion fragmentation on the FRS at GSI by Blank et al. [22], with the yet to be discovered isotope ^{48}Ni being another possible candidate if it indeed exists. As with one proton radioactivity these relatively low Z nuclei will have decay rates very sensitive to the decay Q-value and an alternative approach would be too look in heavier regions using fusion-evaporation reactions, although present techniques will probably have to improve in sensitivity by approximately two orders of magnitude to make such searches feasible. One should also not rule out two proton radioactivity from an isomeric state in a nucleus not lying beyond the proton drip-line, this after all was how the first example of one proton radioactivity was serendipitously discovered [23].

The two proton decay mechanism can be studied from resonant states. Pioneering studies of the beta-delayed two proton decay mechanism showed a predominance of sequential emission [24]. More recent experiments using radioactive beams to populate two proton-unbound resonances [25,26] have also yielded similar results. In order to suppress the sequential process an experiment at MSU has used inelastic scattering of a radioactive ^{17}Ne beam to populate an excited state that is bound to one proton emission. The initial results reported here at this conference by Chromik et al. [?] indicate the existence of a decay branch corresponding to simultaneous two proton emission. A further experiment is planned to study in more detail the decay of this state.

CONCLUDING REMARKS

The recent years have produced an explosion of information on the phenomenon of proton radioactivity with 35 transitions now being known. Theoretical models are able to reproduce in detail the systematic variation of proton decay spectroscopic factors for a wide range of spherical nuclei [4,7] thereby sensitively testing the nuclear shell model at the extreme edge of stability. The first examples of proton emission from highly deformed nuclei have been discovered [15]. These transitions have provided a direct insight into the fragmentation of single particle strength within highly deformed nuclei. Decay rates from these nuclei are found to agree well with theoretical calculations assuming Nilsson states [15,12]. In-beam studies of the gamma-rays using the RDT technique are providing independent complementary nuclear structure information on proton-radioactive nuclei [16,18] that will assist in constraining theoretical calculations of proton decay rates. Furthermore, such studies will provide new insights into the behviour of proton unbound nuclei at high spin and excitation energy. With all these exciting developments occurring it is correct to say that we have now entered the Age of Spectroscopy for nuclei beyond the proton drip-line.

ACKNOWLEDGEMENTS

I would like to use this opportunity to thank all the colleagues I have worked with on experiments at Argonne, Oak Ridge, Louvain-la-Neuve, MSU and Daresbury that are referred to a greater or lesser extent as part of this Review. In particular I wish to register my thanks to Cary Davids for ensuring that the programme of proton radioactivity and RDT work using the Double-sided Silicon Strip Detector system developed in Edinburgh did not terminate prematurely, but thrived in our collaboration at Argonne.

REFERENCES

1. M.F. Mohar et al., *Phys. Rev.* **C38**, 747 (1988).
2. P.H. Regan et al., *Acta Polonika* in Press.
3. P.J. Woods, Proceedings of the 4th International School of Heavy Ion Physics, Erice, 315 (1997).
4. C. N. Davids, P. J. Woods, J. C. Batchelder, C. R. Bingham, D. J. Blumenthal, L. T. Brown, B. C. Busse, L. F. Conticchio, T. Davinson, S. J. Freeman, D. J. Henderson, R. J. Irvine, R. D. Page, H. T. Pentilla, D. Seweryniak, K. S. Toth, W. B. Walters and B. E. Zimmerman, *Phys. Rev.* **C55**, 2255 (1997).
5. P.J. Woods and C.N. Davids, *Ann. Rev. Nucl. Part. Sci.* **47**, 541 (1997).
6. S. Hofmann, W. Reisdorf, G. Munzenberg, F.P. Hessberger, J.R.H. Schneider, P. Armbruster, *Z. Phys.* **A305**, 111 (1982).
7. S. Aberg, P.B. Semmes, W. Nazarewicz, *Phys. Rev.* **C56**, 1762 (1997).

8. P.J. Sellin, P.J. Woods, T. Davinson, N.J. Davies, A.N. James, K. Livingston, R.D. Page, A.C. Shotter, *Phys. Rev.* **C47**, 1933 (1993).
9. F.D. Becchetti and G.W. Greenlees, *Phys. Rev.* **182**, 1190 (1969).
10. A. Bohr and B.R. Mottelson, *Nuclear Structure 1* (W.A. Benjamin, New York 1969).
11. A. Gillitzer, T. Faestermann, K. Hartel, P. Kienle and E. Nolte, *Z. Phys.* **A326**, 107 (1987).
12. V.P. Bugrov and S.G. Kadmensky, *Sov. J. Nucl. Phys.* **49**, 967 (1989).
13. P. Moller, J.R. Nix, K.-L. Kratz, *At. Data Nucl. Data Tables* **66**, 131 (1997)
14. C. N. Davids et al., *Nucl. Instrum. Methods* **B70**, 358 (1992).
15. C.N. Davids, P.J. Woods, D. Seweryniak, A.A. Sonzogni, J.C. Batchelder, C.R. Bingham, T. Davinson, D.J. Henderson, R.J. Irvine, G.L. Poli, J.Uusitalo, W.B. Walters, *Phys. Rev. Lett.* **80**, 1849 (1998).
16. E.S. Paul, P. J. Woods, T. Davinson, R. D. Page, P. J. Sellin, C. W. Beausang, R. M. Clark, R. A. Cunningham, S. A. Forbes, D. B. Fossan, A. Gizon, J. Gizon, K. Hauschild, I. M. Hibbert, A. N. James, D. R. LaFosse, I. Lazarus, H. Schnare, J. Simpson, R. Wadsorth and M. D. Waring, *Phys. Rev.* **C51**, 78 (1995).
17. R. S. Simon, K-H. Schmidt, F. P. Hessberger, S. Hlavae, M. Honusek, G. Munzenberg, H. G. Clerc, U. Gollerthan, W. Schwab, *Z. Phys.* **A325**, 197 (1986).
18. D. Seweryniak, C. N.Davids, W. B. Walters, P. J. Woods, I. Ahmad, H. Amro, D. J. Blumental, L.T. Brown, M. P. Carpenter, T. Davinson, S.M. Fischer, D.J. Henderson, R.V.F. Janssens, T. L. Khoo, I. Hibbert, R.J. Irvine, R. J. Irvine, C.J. Lister, J. A. Mckenzie, D. Nisius, C. Parry, and R. Wadsworth *Phys. Rev.* **C55**, R2137 (1997).
19. S.G. Kadmensky and V.P. Bugrov, *Phys. At. Nucl.* **59**, 399 (1996).
20. P.J. Woods, private communication (1998).
21. B. Blank et al., *Phys. Rev. Lett.* **77**, 2893 (1996).
22. K.P. Jackson et al., *Phys. Lett.* **B33**, 281 (1970).
23. R. Jahn et al., *Phys. Rev.* **C31**, 1576 (1985).
24. R. A. Kryger, A. Azhari, M. Hellstrom, J. H. Kelley, T. Kubo, R. P. Pfaff, E. Ramakrishnan, B. M. Sherril, M. Thoenesson, S. Yokoyama, R. J. Charity, J. Dempsey, A. Kirov, N. Robertson, D. G. Sarantites, L. G. Sobotka, J. A. Winger, *Phys. Rev. Lett.* **74**, 861 (1995).
25. C. R. Bain, P. J. Woods, R. Coszach, T. Davinson, P. Decrock, M. Gaelens, W. Galster, M. Huyse, R. J. Irvine, P. Leleux, M. Loiselet, C. Michotte, R. Neal, A. Ninane, G. Ryckewaert, A. C. Shotter, G. Vancraeynest, J. Vervier and J. Wauters, *Phys. Lett.* **B373**, 35 (1996).
26. M. Chromik, P.G. Thirolf, M. Thoenesson, M. Fauerbach, T. Glasmacher, R. Ibbotson, R.A. Kryger, H. Scheit, P.J. Woods, Conference Proceedings.

Status of neutron drip-line nuclei

B. Jonson

Department of Physics
Chalmers University of Technology
S-412 96 Göteborg, Sweden

Abstract. A brief overview of the present status of neutron drip-line nuclei is given. The emphasis is put on recent achievements obtained in reaction and beta-decay studies of these nuclei.

INTRODUCTION

The neutron drip-line region has been subject for a considerable interest ever since one became aware of a novel structural feature – *the neutron halo* – that occurred for some of the most neutron-rich light nuclides. The experimental investigations of drip-line nuclei, as well as the theories, have become more and more sophisticated and a large body of high-statistics data are now available. This is mainly due to the development of sophisticated methods for the production of radioactive nuclear beams [1].

To study the halo structure experimentally it is necessary to turn either to the static properties of the halo or, more often, to processes in which it is created or destroyed. An early expectation was that one could let the halo free for detailed investigations in the laboratory by some simple dissociation mechanism. It turns out, however, that reaction mechanisms and final-state interactions have very important influences on the experimental results and it is not easy to get a clear halo signal from the data. The weak interaction provides an alternative probe via beta decays from or into halo states and it is especially interesting if the halo is an excited state.

There have been several reviews of the physics of drip-line nuclei and in particular about halo states [2–5], where the main experimental and theoretical developments over the past decade have been treated. In this contribution some examples of experimental results obtained since the latest meeting in this series [6] will be given.

HALO BREAKUP REACTIONS

The main information about the halo wave function has until now been gained from studies of momentum distributions of neutrons and charged fragments from breakup of halo states in different targets. Much effort has gone into the determination and the theoretical interpretation of fragments from break-up reactions. When interpreting the distributions it is important first to understand how to pass the bridge between experiment and theory – the experimental filter. This has to be known in detail in order to make a meaningful comparison. It has also turned out that most distributions are distorted in some way and thus do not reflect the original momentum distribution in the halo. For three-body halos one must in general expect [7] that the two-body subsystems have low-lying continuum structure, i.e. final state interactions are important. There may also be a bias such that reactions leaving the core intact will occur preferentially at large impact parameters. This implies that one does not probe the complete original wavefunction, and this leads to narrower distributions [8–10]. A similar effect is also present in core breakup reactions [11,12] where the halo neutron may be "shadowed" by its own core or by the target. A recent experiment performed at GANIL, where core-breakup reactions of ^{11}Be in a Be target were studied, gave a neutron multiplicity of $M_n = 0.37 \pm 0.10$ [12] which directly can be related to the shadowing effect [8].

Experiments on well-established halo states

^6He. The first drip-line nucleus to be observed was ^{6}He which was identified already in 1936 [13]. This nucleus is well known and the essential features of its halo structure are believed to be understood [2]. There have still been some interesting new results concerning this isotope recently at the ALADIN/LAND setup at GSI where reactions of 240 MeV/u ^{6}He in different targets were studied. The reaction fragments were studied in "complete kinematics" experiments. Such experiments give in general the possibility to obtain excitation energy spectra from the complete system or any subsystem and to begin to search for correlations in a more systematic way. It was found that the dominant reaction mechanism in a carbon target is a two-step process: knock-out of one halo neutron followed by the decay of the ^{5}He resonance. From the shape of the (α-n) two-body invariant mass spectrum [17] it was concluded that it mainly reflected a strong α-n interaction leading to the $3/2^-$ ground state of ^{5}He. This observation seems to offer new possibilities for studying the structure of the binary subsystems of two-neutron halo nuclei. A large spin alignment of ^{5}He in a plane perpendicular to the ^{5}He momentum vector was also observed [14]. From the correlation function a small mixing from the $1/2^-$ excited state in ^{5}He [16,15] could be determined.

In the sudden approximation the momentum distribution of ^{5}He in the projectile system should reflect the internal momentum distribution of the removed neutron in ^{6}He, which in turn is determined by the ^{6}He ground-state wavefunction. The measured distribution [17] was found to be narrower than the Fourier transform of the ^{6}He wave function obtained in an (α+n+n) three-body model [18]. This is again an example of the narrowing of the momentum width when only part of the wavefunction is sampled [8,10].

In experiments [19] at Dubna the differential cross section for elastic scattering of ^{6}He from a He-gas target was studied. The results reveal very large cross sections at backwards angles in the c.m. system. The authors interpreted this as the exchange of two neutrons between ^{6}He and ^{4}He via a direct one-step mechanism. Then the data could be fitted assuming that the the main contribution to the two-neutron exchange comes from the "di-neutron" configuration of the ^{6}He wave function with a spectroscopic factor close to one.

Continuum excitations in weakly bound nuclei in the drip-line region can provide information about their single particle and collective structure. In three-body breakup of 240 MeV/u ^{6}He into (α+n+n) in carbon and lead targets [20] the differential cross sections for inelastic nuclear and electromagnetic excitations into the ^{6}He continuum were measured. A low lying E1 strength was observed at about 1 MeV above the threshold. The strength up to 10 MeV gives about 90 % of the non-energy weighted cluster sumrule [22].

^8He. The most neutron-rich bound He isotope is ^{8}He. Its ground-state wavefunction has been described [21] in a five-body (α+4n) cluster orbital shell-model approximation. In an experiment at GSI with 240 MeV/u ^{8}He on a carbon target halo-neutron knockout reactions were studied. In the ^{6}He channel the intermediate ^{7}He fragment was observed and the angular distribution of the $\mathbf{p}_{^6He-n}$ vector in a coordinate system with the z-axis parallel to the $\mathbf{p}_{^7He}$ direction showed a similar spin alignment as observed for ^{6}He [14]. It was found that the decay proceeds via the $3/2^-$ ground state of ^{7}He [23].

^{11}Li. The best studied nucleus with a halo structure is ^{11}Li but even if one has collected data in many different experiments there are still many unsolved problems. Here I shall give some examples on the different studies that have given some information about the weights of the $(1s_{1/2})^2$ and $(0p_{1/2})^2$ components in the ^{11}Li ground-state wave function.

In an experiment at RIKEN [24] the isobaric analogue state (IAS) of ^{11}Li was studied using a charge-exchange reaction, ^{11}Li(p,n)$^{11}Be^*$, in inverse kinematics. The position and the width of the IAS was measured (21.16 MeV and 0.49 MeV, respectively). The result gives a relatively small Coulomb displacement energy, which is interpreted as due to an extended proton distribution in the IAS. The particle-decay width is in support of a strong mixing of the $(1s_{1/2})^2$ configuration in the halo wave function.

At GSI the ALADIN/LAND setup was used for a detailed study of ^{11}Li. The transverse momentum of the ^{10}Li fragment, reconstructed from the momenta of ^{9}Li and neutrons, was measured after knockout reactions. The width of this distribution, reflecting the neutron momentum in the ground state, was found to be reproduced by assuming about equal amounts of $1s_{1/2}$ and $0p_{1/2}$ single particle momentum distributions, including core shadowing [8] and experimental filter [25].

The transverse momentum distribution is shown in Figure 1.

From the same data the angular correlation between the $\mathbf{p}_{^{10}Li}$ direction and the $\mathbf{p}_{^9Li-n}$ vector was constructed. The differential cross section as a function of $cos(\theta_{^9Li-n})$ has a skew shape signalling a linear term, characteristic of a state with components of different parity. An analysis [25] shows that one from the data may conclude that the $(1s_{1/2})^2$ component contributes somewhere between 25 and 60 %.

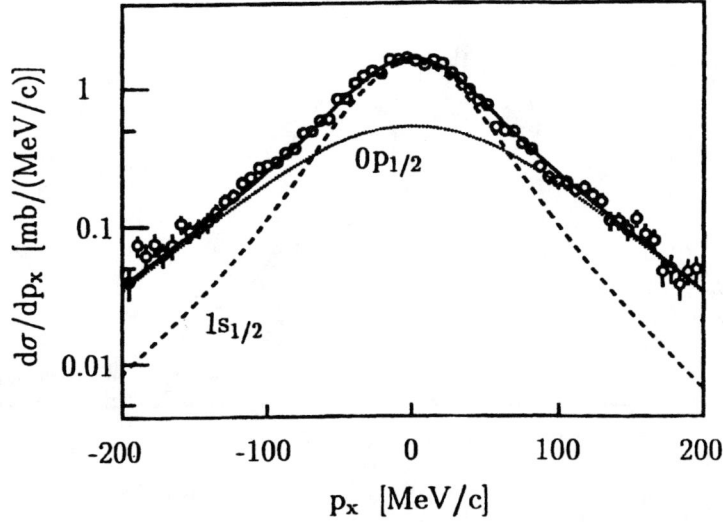

FIGURE 1. Transverse momentum (p_x) distribution of ^{10}Li reconstructed from the momenta of ^{9}Li and neutrons measured in coincidence after neutron knockout from ^{11}Li in a carbon target. The solid line represent a superposition of equal amounts of s- and p-states.

In the beta-decay studies, discussed below, the feeding to the first excited state in ^{11}Be at 320 keV is sensitive to the p part of the ^{11}Li ground state wave function. An analysis [38] indicate about equal mixture of s- and p-components.

There is at present no experiment that gives a precise determination of the $(1s_{1/2})^2$ component in the ^{11}Li wavefunction but all results indicate a contribution somewhere between 40 and 50 %.

HEAVIER HALO NUCLEI

^{14}Be. Fragmentation reactions with a beam of 287 MeV/u ^{14}Be have been studied at the ALADIN/LAND setup at GSI [25]. The heaviest known Borromean nucleus is ^{14}Be and for the understanding of its structure it is important to know the structure of its unbound neighbor ^{13}Be. The data shown in Figure 2 have clear similarities to the invariant mass spectrum of ^{10}Li [27] and may be interpreted as evidence for a low-lying state as its ground state close to the ^{12}Be+n threshold.

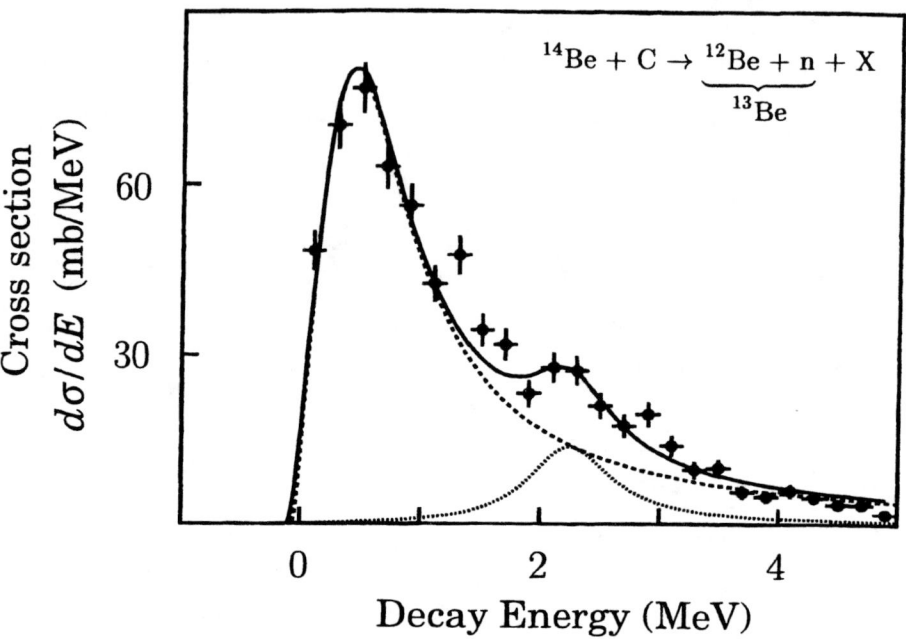

FIGURE 2. Decay energy spectrum of ^{13}Be obtained from an invariant mass analysis of ^{12}Be+n after breakup of ^{14}Be in a C target.

^{19}C. This is the heaviest one-neutron halo nucleus studied until now. It is not clear how well developed the ground-state halo is. Even with a rather uncertain one-neutron separation energy (245±90 keV) the small breakup cross section in a Ta target, σ=0.8±0.3 b measured at GANIL [28], indicates a much less pronounced halo than the one found in ^{11}Be. In recent experiments at MSU and with the FRS at GSI the longitudinal momentum widths of ^{18}C fragments after breakup of ^{19}C were measured at 77 MeV/u [29,30] and 910 MeV/u [31]. In both cases a narrow width was observed for the ^{18}C fragments, 42±4 MeV/c [29] at 77 MeV/u and 69±3 MeV/c at 910 MeV/u. The theoretical interpretations of the data have resulted in several proposals for the structure of the ground state wavefunction for ^{19}C (see [31]). The main consensus seems to be that the $1s_{1/2}$ and the $0d_{5/2}$ states are almost degenerate and that the first excited 2^+-state of the ^{18}C core plays an important role. An experimental challenge is thus to determine the contribution of the excited core configuration.

BETA-DECAY OF HALO STATES

The nuclear beta-decay is well understood and can add valuable information also to halo states. For neutron-rich nuclei the decay is dominated by Gamow-Teller transitions and strong such transitions will also give information on halos. During the last few years several experiments on the beta-decay of ^{11}Li have been performed [32–34] with emphasis on possible signatures due to its neutron-halo structure.

The allowed beta decay of ^{11}Li ($I^\pi=3/2^-$) can only populate one excited particle-bound state in ^{11}Be, namely the $1/2^-$ state at 320 keV. The beta-decay branching ratio to this state is used to give a direct estimate [38] of the $(1s_{1/2})^2/(0p_{1/2})^2$ ratio in the ^{11}Li ground-state wave function as mentioned above.

A strong beta transition to a broad state at about 18 MeV excitation energy was identified by studying beta-delayed charged-particle emission from ^{11}Li [38]. The neutron decay from this state, observed via recoiling ^{10}Be fragments, shows that the main feeding populates the 2^+ state in ^{10}Be. This finding can be explained by assuming a strong admixture of a $(0p_{1/2})^2$ component in the ^{11}Li ground state.

The beta-delayed deuteron emission process is a decay mode that is especially favorable for neutron-rich Borromean nuclei since the process is only energetically possible if the two-neutron separation energy is very low. Beta-delayed deuterons have been observed for both ^{6}He [35,36] and ^{11}Li [37]. An interesting feature of this type of decays is that they might proceed directly to the continuum and are therefore expected to yield much more information about the mother state than decays proceeding through excited intermediate states.

A combined analysis based on the singles neutron spectra, charged-particle and γ-ray data [34] as well as their correlations has shown that a quantitative description of the data requires more than 30 decay branches of ^{11}Li. The components overlap and produce continuous spectra with relatively broad peaks. The low-energy part of the observed spectra of beta-delayed neutrons following the ^{11}Li decay can be described with transitions to known daughter states in ^{11}Be. The data corresponding to higher energies give also indications of new excited states in the energy region above 11 MeV excitation in ^{11}Be. The beta-strength function from [34] is shown in Figure 3.

The new transitions to the high lying excited states in ^{11}Be constitute most of the observed B_{GT}-strength giving $B_{GT} \sim 4$. Shell-model calculations predict a larger GT-strength around 16–18 MeV with total $B_{GT} \sim 10$ [39].

CONCLUSION

In this contribution I have tried to give a flair of the present status of experiments on neutron-rich nuclei. The main experiments that I have discussed have been connected to halo states and I have not discussed unbound nuclei very much. It is, however, clear that the interest in unbound nuclei, just outside the drip-line,

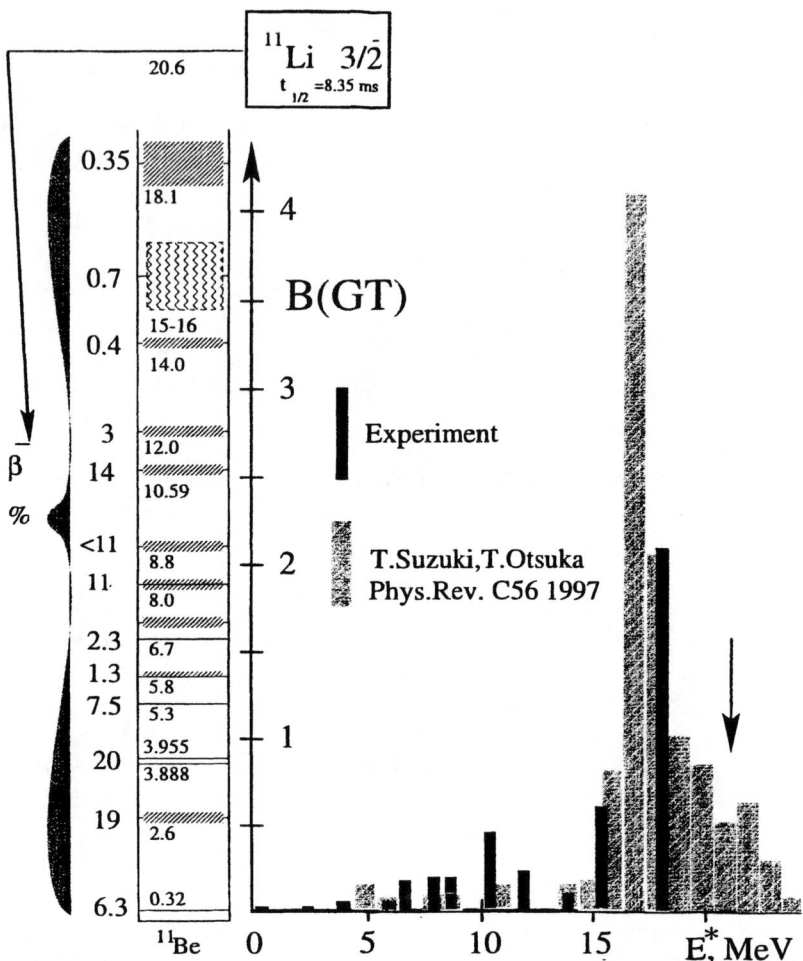

FIGURE 3. Decay scheme and beta-strength function for ^{11}Li obtained from a combined analysis of beta-delayed neutron and charged particle data. The histograms show the beta strength functions from [39] and from the experimental data.

is strongly increasing and the inverse kinematics methods seem here to give the necessary technical tools for such studies [40].

REFERENCES

1. Geissel, H., Münzenberg, G., Riisager, K., *Annu. Rev. Nucl. Part. Sci.* **45**, 113 (1995).
2. Zhukov, M.V., et al., *Phys. Rep.* **231**, 151 (1993).
3. Hansen, P.G., Jensen, A.S., Jonson, B., *Annu. Rev. Nucl. Part. Sci.* **45**, 591 (1995).
4. Tanihata, I., *J. Phys.* **G**, 157 (1996).
5. Jonson, B., Riisager, K., *The Royal Society, Theme Issue "Sci. with Radioactive Beams"*
6. Proc. ENAM95, Arles, *Editions Frontiers* Ed. de Saint Simon, M., Sorlin, O.,
7. Garrido, E., Federov, D.V., Jensen, A.S., *Phys. Rev.* **C55**, 1327 (1997).
8. Hansen, P.G., *Phys. Rev. Lett.* **77**, 1016 (1996).
9. Hencken, K., Bertsch, G., Esbensen, H., *Phys. Rev.* **C54**, 3043 (1996).
10. Esbensen, H., *Phys. Rev. Lett.* **C53**, 2007 (1996).
11. Nilsson, T., et al., Europhys. Lett., **30**, 19 (1995).
12. Grévy, S., Axelsson, L., et al., to be published.
13. Bjerge, T., Borgström, K.J., *Nature* **138**, 400 (1936).
14. Chulkov, L.V., et al., *Phys. Rev. Lett.* **79**, 201 (1997).
15. Chulkov, L.V., these proceedings.
16. Chulkov, L.V., Schrieder, G., *Z. Phys.* **A359**, 231 (1997)
17. Aleksandrov, D. et al., *Nucl. Phys.* **A633**, 234 (1998).
18. Zhukov, M.V., et al., *Nucl. Phys.* **A533**, 428 (1991).
19. Ter-Akopian, G.M., et al., Phys. Lett. **B** in press
20. Aumann, T., et al., to be published
21. Zhukov, M.V., Korsheninnikov, A.A., Smedberg, M., *Phys. Rev.* **C47**, 2937 (1993).
22. Danilin, B.V., et al., *Nucl. Phys.* **A632**, 383 (1998).
23. Eberlein, B., PhD thesis, Mainz 1998, GSI DISS. 98-03
24. Teranishi, T., etal., *Phys. Lett.* **B407**, 110 (1997).
25. Simon, H., PhD thesis, TU Darmstadt, 1998
26. Borge, M.J.G., et al., *Phys. Rev.* **C55**, R8 (1997).
27. Zinser, M., et al., *Nucl. Phys.* **A619**, 151 (1997)
28. Liegard, ., Thesis 1998, LPCC T 97-02, LPC, Caen.
29. Bazin, D., et al., *Phys. Rev.* **C** to be published
30. Bazin, D., these prooceedings.
31. Baumann, T., et al., to be published in Phys Lett.B
32. Aoi, N., et al., *Nucl. Phys.* **A616**, 181c (1997).
33. Morrissey, D.J., et al., *Nucl. Phys.* **A627**, 22 (1997).
34. Mukha, I., et al. these proceedings and to be published.
35. Riisager, K., at al., *Phys. Lett.* **B235**, 30 (1990)
36. Borge, M.J.G., et al., *Nucl. Phys.* **A560**, 664 (1993)
37. Mukha, I., et al., *Phys. Lett.* **B367**, 65 (1996).
38. Borge, M.J.G., et al., *Nucl. Phys.* **A613**, 199 (1997).
39. Suzuki, T., Otsuka, T., *Phys. Rev.* **C56**, 847 (1997).
40. Axelsson, L., et al., *Phys. Rev.* **C54**, R1511 (1996).

Probing Halo Structure with Breakup Reactions

I. J. Thompson

Department of Physics, University of Surrey, Guildford GU2 5XH, UK

Abstract. Halo nuclei very often have only one bound state, the ground state, so breakup reactions are an essential spectroscopic tool. The progressive refinement of reaction models for breakup often leads to significant changes in the structural information that is extracted from such reactions.

Halo nuclei very often have only one bound state, the ground state, and then a continuum with a very limited number of identifiable levels. Consideration of fragmentation reactions is therefore necessary, since conventional nuclear models which deal with energies, spins and parities of excited states have limited applicability. We must therefore measure and model processes which excite the continuum, and the study of breakup reactions is an essential tool for these investigations.

The earliest probes of halo structure examined the total reaction cross sections [1,2], and found that these were considerably larger in comparison with neighbouring nuclei nearer the valley of stability. The large cross section became attributed to the large radial separations of the most weakly bound neutrons, and this interpretation appeared more definite when narrow widths were found in the momentum distributions [3] following fragmentation reactions at high beam velocities. Since the large reaction cross section and the narrow momentum widths are measurements of breakup cross sections, the theory of fragmentation reactions entered into our subject at the beginning, although in a very simplified manner. Similarly, the measurement of large Coulomb breakup cross sections [4] was first interpreted with a breakup theory modelled on excitations to discrete levels. This suggested the possibility of a low-lying 'soft dipole mode' in which the final-state interactions after breakup were presumed to give a resonance state in which the halo nucleon(s) oscillated against the remaining core nucleons. Later, the same semiclassical theory for inelastic excitations was used to connect the $B(E\lambda)$ distributions to the plane-wave continuum, in ^{11}Be for example [5], and although a very narrow peak in the excitation energy spectrum is predicted and found, no evidence was found for any low-lying resonance state.

These first models of breakup, namely the Serber model for momentum distributions and semiclassical theory for Coulomb excitations, are now being supplemented by more accurate theories which do not make (so) many of the initial simplifying assumptions. This is necessary if detailed spectroscopic structure information is to be obtained for halo nuclei. In addition we need to quantify and (if possible) correct uncertainties in the analyses introduced due to the approximations in the reaction theory, in particular in our reaction theory for breakup.

In order to have a (relatively) complete account of breakup reactions of a halo nucleus, considered as a core plus one or two nucleons, we need to take into account the following 'complications' to the simple Serber or semiclassical models:

1. effects of *core excitation* in the halo ground state and/or induced during the breakup reaction;

2. *interference effects* on certain observables from different final-state components;

3. effects of both the *nuclear and Coulomb* components of the actual target-projectile interactions,

4. those *final state interactions* between the broken-up fragments (both with the target, and with each other);

5. *multiple excitation/deexcitation steps*, e.g. from one breakup configuration to another;

6. effects of *absorption and shadowing* in *inelastic* breakup, when not all projectile fragments necessarily survive to the final state.

It is not a trivial theoretical task to develop breakup theories which take into account some (or all) of these complications, but we shall see from the examples below that some attempt is necessary. In the following sections I examine a series of instances where we are discovering that important structural information can be learned when reaction data are analysed more correctly.

1 Total reaction cross sections

The earliest analyses of reaction cross sections [1,2] fitted them with the results of optical-limit Glauber models to determine the matter rms radii of single-particle densities, following very successful applications to stable nuclei [6,7]. This is to average over the many-body ground-state density of a composite projectile such as a halo nucleus, and scatter this average on the target. The correct method for scattering of a halo nucleus in the Glauber model is, however, to scatter the individual fragments (core, halo nucleons) from the target, and to average these scattering amplitudes over the ground state densities of the composite system. The first systematic examination of this effect [8] showed that it is necessary to revise the halo radii fitted to reaction cross sections. The fitted radii for weakly-bound halos are now considerably larger: ^{11}Li, for example, has a radius of 3.55 ± 0.10 fm [8] not 3.10 ± 0.17 fm [2].

2 Momentum widths for nonspherical halos

The measurements of momentum distributions following high-energy fragmentation showed approximately constant widths for a wide range of target nuclei. This lent support to the validity of the Serber model, in which the distributions are related directly to the Fourier transforms of the ground-state wave functions. We know, however, that a wide range of Coulomb and nuclear mechanisms must be responsible for the different targets. Preliminary reaction models [9] show that the widths do depend moderately on target size and on beam velocity.

More dramatically, however, the widths of momentum distributions from the breakup of ^{8}B were found to be much narrower [10,11], by a factor of up to 2, than the predictions of the Serber model. This was initially attributed to an unusually long tail to the wave function of the last proton of ^{8}B [10], but now we realise [12,13] that for all *non-spherical* halo wave functions the momentum widths from nuclear breakup can be much reduced because of selective shadowing of the different m-state components of the wave function. The narrowing will be greater for $\ell = 2$ single-particle wave functions, as in the $A = 17$ O, F and Ne nuclei, and much reduced for s-wave halos [14]. The reaction model calculations of shadowing are more precise than using simply a lower radial cutoff in the Serber model, and are necessary, for instance, to predict the anisotropic breakup distributions.

3 Coulomb E1–E2 interference

The breakup of a halo nucleus, according to the Serber model, gives rise to fragment momentum distributions that are isotropic in the projectile rest frame. More detailed examination [11] of the Coulomb mechanism shows, however, that anisotropies arise from interference of different final-state partial waves. Measurements [11,15] of ^{7}Be longitudinal momenta from ^{8}B breakup verify this effect. The determination of the ratio of even- and odd-parity final-state cross sections gives, to first order, the ratio of E1 to E2 transition strengths [11].

4 Coulomb-nuclear interference effects

The Coulomb breakup of ^{8}B was first measured [18] with the aim of determining the strength of the $E1$ transition strength at low relative p–^{7}Be energies. It soon became clear from considerations of reaction mechanism [19,20] that the extraction of the astrophysical S-factor $S_{17}(0)$ depended on the correction for the $E2$ contribution to the breakup. This lead to further experiments [17,16] with the express aim of determining the $E2$ strengths. These two experiments were analysed first with semi-classical theory, and both came to the conclusion that $E2$ effects are very small.

Better calculations would use the single-particle cluster wave functions of [24] or [25] for ^{8}B, and calculate also the nuclear breakup using either partial-wave [26]

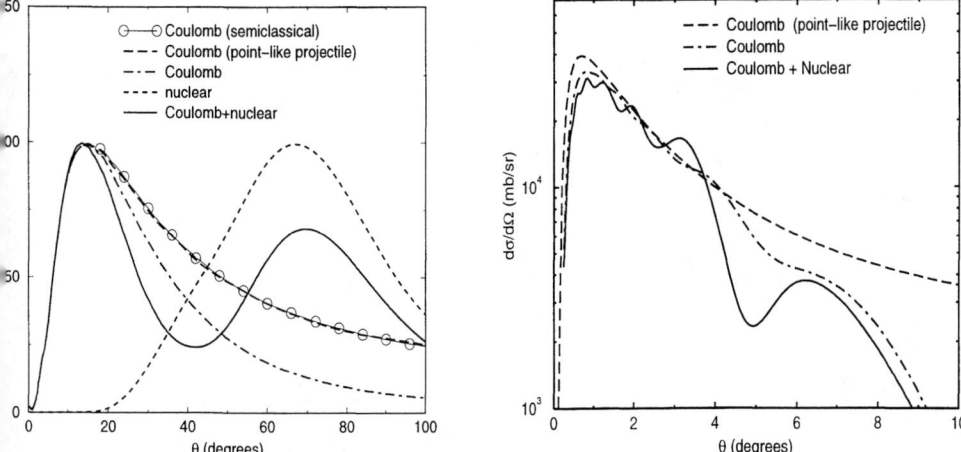

IGURE 1. Comparison of the Coulomb and the nuclear contributions to the differential cross ction for the breakup of ^{8}B: (left) on ^{58}Ni at 25.8 MeV [16] and (right) on ^{208}Pb at 419 MeV 7].

semiclassical [23] theory. Results [21–23] show that nuclear transition potentials om the ground state to the continuum are of much larger range than the size of 3. In fact, it was found that a nuclear-free zone only exists for impact parameters eyond 25 or 30 fm. At smaller radii there begins to be destructive Coulomb-iclear interference, and the breakup cross sections are reduced (Fig. 1). Even the oulomb transition potentials are different from pure $r^{-\lambda-1}$ forms at these radii, ecause of the large magnitude of the tail of the ground-state proton wave function. e conclude that the halo-size and nuclear interference effects are important over e same angular range as populated by the $E2$ transitions, so they need to be etermined together by a more adequate reaction theory.

5 Adiabatic models for Coulomb breakup

The breakup of a halo nucleus comes about because the forces of the target t differentially on the core and valence fragments. If these forces are weak, erturbation theory can be used, while if they are stronger, or if second-order fects such as on elastic scattering are required, then higher-order treatments are cessary. One tractable method of treating higher-order interactions in a system several fragments is to use an *adiabatic* or *sudden* assumption, that the projectile citation energies are much smaller than the beam energy. At high beam velocity, is assumption allows the halo scatterings to be calculated separately for all the re-valence separations, and then averaged over the product of the ground-state d final-state wave functions.

For the case of purely Coulomb breakup of one-neutron halo nuclei, the adiabatic

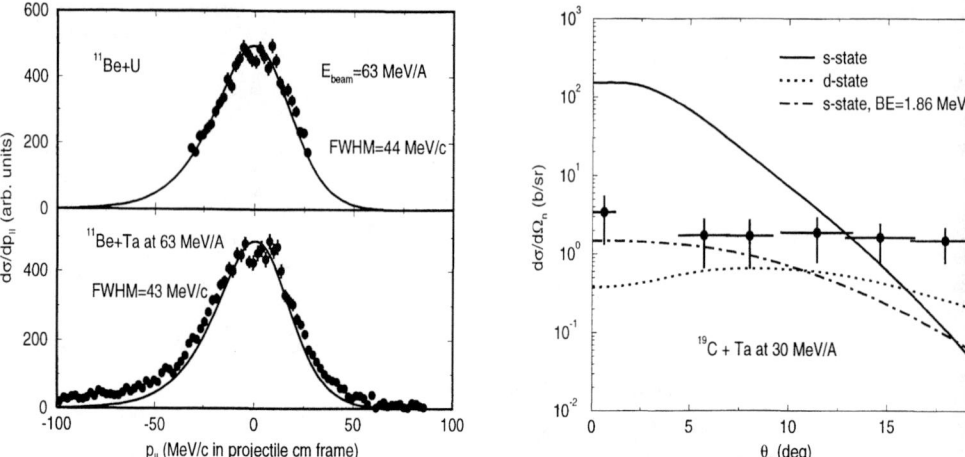

FIGURE 2. Adiabatic-bremstrahlung calculations of the Coulomb contributions to (left) the parallel momenta of ^{10}Be from ^{11}Be, and (right) the angular distributions of neutrons from ^{19}C (both from [29]). The data are from [27] and [28], respectively.

wavefunction may be inserted in a post-form T-matrix integral, and the resulting bremstrahlung integral can be evaluated analytically. This has proved to give a simple but accurate account of the breakup of deuterons [33,34], and of ^{11}Be (see Fig. 2, left) [29]. Application to ^{19}C breakup [29] gives the neutron angular distributions shown in Fig. 2 (right), for s- and d-waves of three plausible structure hypotheses. Comparison with the recent experimental data shown [28] indicates that the ^{19}C g.s. seems most similar to a ^{18}C in a 2^+ state with an s-wave neutron bound state.

The adiabatic-bremstrahlung method can also be used for two-neutron halo nuclei such as ^{6}He. First calculations of selected partial angular distributions are shown in Fig. 3, for two ^{6}He models of different sizes (solid line: 2.50 fm and dashed line 2.35 fm rms matter radius). These calculations automatically include all orders of the Coulomb interaction with the target nucleus. We see from Fig. 3a that point-dineutron model of ^{6}He would greatly overestimate the Coulomb breakup. This is because the more correlated neutrons would give much too large a mean distance of the charged α-particle from the centre of mass of ^{6}He.

Another application of the adiabatic approximation is a direct estimation of the effect of breakup channels on the elastic scattering angular distribution [35], as the higher-order breakup effects amount to a multiplicative factor for the elastic cross section.

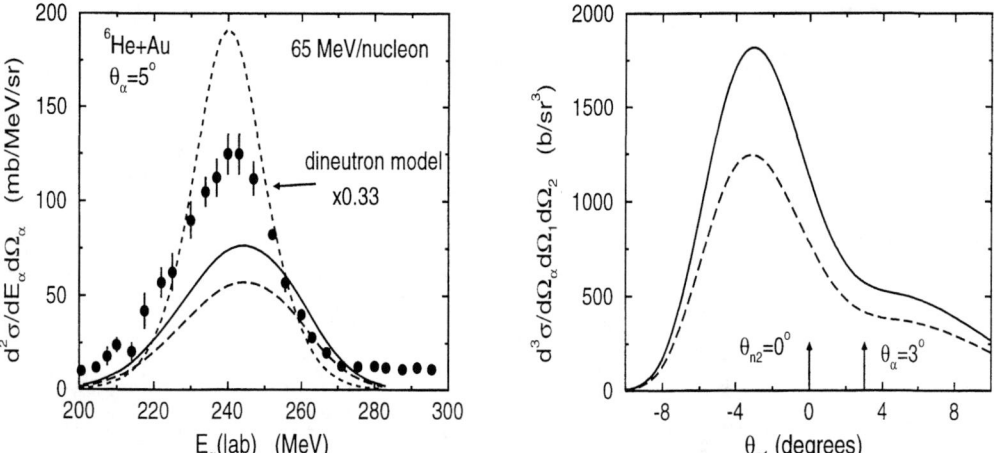

FIGURE 3. Adiabatic-bremsstrahlung calculations of the Coulomb contributions to (left) the energy spectrum of ^{4}He from ^{6}He, and (right) the angular distributions of the second neutron from ^{6}He for the indicated ^{4}He and first-neutron angles (data from [30,31], theory from [32]).

6 Final-state interactions between halo fragments

When a halo nucleus breaks up, any low-lying resonances between the fragments will typically enhance the observed cross sections around the resonance energy. This often gives important information about the spectrum of intermediate nuclei such as ^{10}Li or ^{13}Be. This effect is well-established for the decay of ^{6,8}He, because of low-lying resonances in ^{5,7}He respectively [36].

Of particular interest here are the final-state interactions between *all three bodies* of a two-neutron halo nucleus at an excited continuum energy. The B(E1) distribution is an appropriate observable here, and its calculation requires the solution of the three-body scattering problem for positive energies. This has been done for ^{6}He [37,38] and ^{11}Li [39,40], but theoretical and experimental uncertainties still remain. These works examine the non-resonant continuum, and search for any resonance states at positive energies. Calculations are most progressed for ^{6}He, and there we find that final-state interactions strongly enhance the continuum wave functions of the $n+n+\alpha$ system in the final state, and give rise (Fig. 4b, from [38]) to enhanced $B(E1)$ breakup strengths. The continuum phase shifts are also made more positive (Fig. 4a), but apparently not sufficiently positive to constitute a dipole resonance state. Other calculations [37] find a similar dipole response, but do explain it in terms of S-matrix poles which are (apparently) to be taken as dipole excited states. The resolution of this difference is still underway [40], but may depend on understanding the different treatments of Pauli blocking in the different three-body continuum calculations.

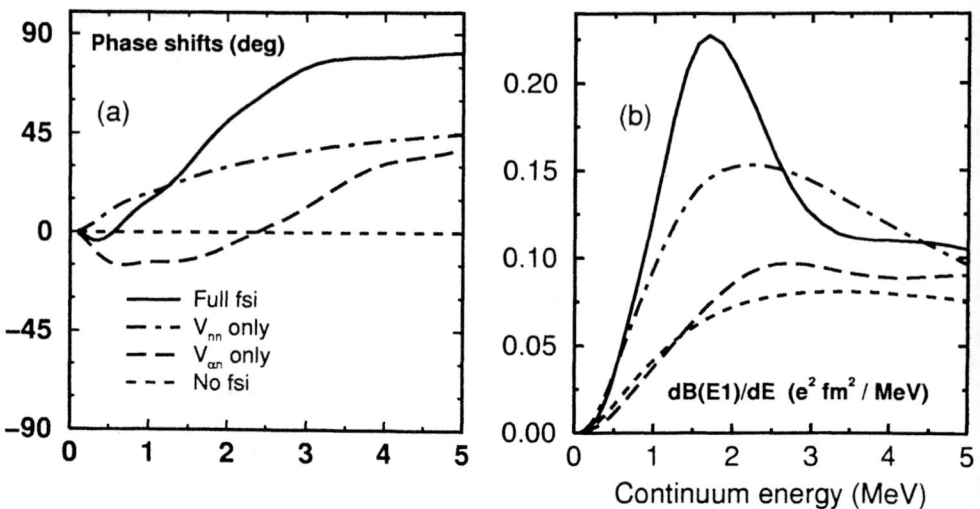

FIGURE 4. Features of the 1^- continuum in ^{6}He, for different final state interactions in the continuum, using the Pauli-projection method with $K_{max} = 10$ (modified from [38]): (a) diagonal $K=1$ phase shifts, and (b) $dB(E1)/dE$ distributions.

Conclusions

The progressive refinement of reaction models for breakup leads often to significant changes in the spectroscopic structural information that may be extracted from breakup reactions. It is feasible for modern applications of breakup theory to take into account the few-body structure of the halo, and make quantitive predictions for the spectroscopy of the ground and continuum states of halo nuclei. The next step is to develop more comprehensive models which take multistep processes into account simultaneously with many of the 'complications' above.

Support from the EPSRC grant GR/J/95867 is acknowledged.

REFERENCES

1. I. Tanihata, H. Hamagaki, O. Hashimoto, S. Nagamiya, Y. Shida, N. Yoshikawa, O. Yamakawa, K. Sugimoto, T. Kobayashi, D.E. Greiner, N. Takahashi and Y. Nojiri, Phys. Lett. **B160** (1985) 380
2. I. Tanihata, T. Kobayashi, O. Yamakawa, S. Shimoura, K. Ekuni, K. Sugimoto, N. Takahashi, T. Shimoda and H. Sato, Phys. Lett. B **206** (1988) 592.
3. T. Kobayashi, O. Yamakawa, K. Omata, K. Sugimoto, T. Shimoda, N. Takahashi and I. Tanihata, Phys. Rev. Lett. **60** (1988) 2599
4. I. Tanihata, Nucl. Phys. **A522** (1991) 275c
5. T. Nakamura et al, Phys. Lett. **B331** (1994) 296
6. S. Kox et al, Phys. Rev. **C35** (1987) 1678
7. S.K. Charagi and S.K. Gupta, Phys. Rev. **C41** (1990) 1610

8. J.S. Al-Khalili and J.A. Tostevin, Phys. Rev. Letts **76** (1996) 3903
9. P.G. Hansen, Proc. of Int. Conf. on Exotic Nuclei and Atomic Masses, June 19-23, 1995, Arles, France, M. de Saint Simon and O. Sorlin (eds), Editions Frontières (1995), p.175.
10. W. Schwab et al., Z. Phys. **350** (1995) 283
11. J. H. Kelley, Sam M. Austin, A. Azhari, D. Bazin, J. A. Brown, H. Esbensen, M. Fauerbach, M. Hellstrm, S. E. Hirzebruch, R. A. Kryger, D. J. Morrissey, R. Pfaff, C. F. Powell, E. Ramakrishnan, B. M. Sherrill, M. Steiner, T. Suomijrvi, and M. Thoennessen, Phys. Rev. Lett. **77** (1996) 5020
12. Y. Ogawa and I. Tanihata, Nucl. Phys. **A616** (1997) 239c
13. H. Esbensen, Phys. Rev. **C53** (1996) 2007
14. Y. Ogawa, Y. Suzuki, and K. Yabana, Nucl. Phys. **A571** (1994) 784
15. B. Davids, this conference.
16. Johannes von Schwarzenberg et al., Phys. Rev. **C53** (1996) R2598
17. T. Kikuchi et al., Phys. Letts. **B391** (1997) 261
18. T. Motobayashi et al., Phys. Rev. Letts. **73** (1994) 2680
19. K.-H. Langanke and T.D. Shoppa, Phys. Rev. **C49** (1994) R1771
20. R. Shyam and I.J. Thompson, Phys. Letts. **B415** (1998) 315
21. F.M. Nunes, R. Shyam and I.J. Thompson, J.Phys. G (Aug 1998), in press.
22. F.M. Nunes and I.J. Thompson, Phys. Rev. **C57** (1998) R2818
23. C.H. Dasso, S.M. Lenzi and A. Vitturi, Nucl. Phys. A, in press.
24. H. Esbensen and G. Bertsch, Nucl. Phys. **A600** (1996) 37
25. L.V. Grigorenko, B.V. Danilin, V.D. Efros, N.B. Shulgina and M.V. Zhukov, Phys. Rev. **C57** (1998) R2099
26. I.J. Thompson, Computer Physics Reports, **7** (1988) 167
27. J. H. Kelley *et al.*, Phys. Rev. Lett. **74** (1995) 30.
28. E. Liegard, N. A. Orr *et al.*, LPC-Caen report LPCC 98-03, to be published; E. Liegard, Thèse, Université de Caen, France (1998), unpublished.
29. P. Banerjee, J.A. Tostevin and I.J. Thompson, Phys. Rev. C (Aug 1998), in press.
30. D. P. Balamuth, K. A. Griffioen, J. E. Bush, K. R. Pohl, D. O. Handzy, A. Aguirre, B. M. Sherrill, J. S. Winfield, D. J. Morrissey, and M. Thoennessen, Phys. Rev. Lett. **72** (1994) 2355.
31. J. E. Bush (private communication)
32. P. Banerjee, I.J. Thompson and J.A. Tostevin, Phys. Rev. C (Aug 1998), in press.
33. J.A. Tostevin, S. Rugmai, R.C. Johnson, H. Okamura, et al, Phys. Lett. **B424** (1998) 219
34. J.A. Tostevin, S. Rugmai, R.C. Johnson, Phys. Rev. **C57** (1998) 3225
35. R.C. Johnson, J.S. Al-Khalili and J.A. Tostevin, Phys. Rev. Lett. **79** (1997) 2771
36. A. Korsheninnikov and T. Kobayashi, Nucl. Phys. **A567** (1994) 97
37. A. Cobis, D.V. Fedorov and A.S. Jensen, Phys. Rev. Lett. **79** (1997) 2411
38. B.V. Danilin, I.J. Thompson, M.V. Zhukov and J.S. Vaagen, Nucl. Phys. **A632** (1998) 383
39. A. Cobis, D.V. Fedorov and A.S. Jensen, Phys. Lett. **424** (1998) 1
40. I.J. Thompson, B.V. Danilin, V.D. Efros, M.V. Zhukov and J.S. Vaagen, J.Phys. G (Aug 1998), in press

Studies of excited state in ^{11}Li. Spectroscopy of ^{7}He.

A. A. Korsheninnikov[1*], M. S. Golovkov[1*],
A. Ozawa[1], E. A. Kuzmin[2], E. Yu. Nikolskii[2], K. Yoshida[1],
B. G. Novatskii[2], A. A. Ogloblin[2], I. Tanihata[1],
Z. Fulop[1†], K. Kusaka[1], K. Morimoto[1], H. Otsu[1], H. Petrascu[1], F. Tokanai[1]

[1] *RIKEN, Hirosawa 2-1, Wako, Saitama 351-0198, Japan*
[2] *The Kurchatov Institute, Kurchatov sq. 1, 123 182 Moscow, Russia*
* *On leave from the Kurchatov Institute, Kurchatov sq. 1, Moscow 123182, Russia*
† *On leave from ATOMKI, H-4001, Debrecen, PoB 51, Hungary*

Abstract. (i) The excited state of ^{11}Li at $E^* \sim 1.3$ MeV observed in the proton inelastic scattering was recently reinterpreted in terms of shakeoff mechanism with proton scattering on the ^{9}Li-core. However, the observed peak has constant excitation energy along its kinematical locus. This fact represents the most important criterion of excited state. This constant energy contradicts to the shakeoff mechanism. (ii) Preliminary results of recent spectroscopic studies of ^{7}He and ^{7}H are reported. The excited state of ^{7}He was observed ($E^* = 2.9 \pm 0.3$ MeV, $\Gamma = 2.2 \pm 0.3$ MeV, $\Gamma_\alpha/\Gamma_{tot} = 0.85 \pm 0.15$). This state has unexpected structure resulting in decay into the channel α+3n, but not into ^{6}He+n. Hint on a possible existence of resonance ^{7}H was obtained.

STUDIES OF EXCITED STATE IN ^{11}Li

The current interest to nuclei far from stability originated in the discovery of giant neutron halo in ^{11}Li (1,2). Hereupon ^{11}Li has become the most intensely studied exotic nucleus. Nevertheless the structure of ^{11}Li is not yet well established and modern understanding suffers first of all from a lack of precise knowledge of low-lying states in ^{10}Li. Another burning matter directly related to the ^{11}Li structure is a question on its excitation. Experimental search for excited state in ^{11}Li was performed in Refs. (3-7). Four measurements showed the state at $E^* \sim 1.2 - 1.3$ MeV (3,5-7).

After experimental studies of inelastic scattering of p+^{11}Li (5,6) theoretical paper (8) was published, where the peak observed in Ref. (5,6) was reinterpreted in terms of shakeoff mechanism. Before to discuss this topic, we would like to comment what was measured in Ref. (5,6).

CP455, *ENAM98: Exotic Nuclei and Atomic Masses*
edited by B. M. Sherrill, D. J. Morrissey, and Cary N. Davids
© 1998 The American Institute of Physics 1-56396-804-5/98/$15.00

The beam of ^{11}Li scattered on the proton target and emitted protons were detected using two telescopes of solid state detectors (Fig. 1). The detectors were position-sensitive and coordinates of each detected proton were measured. At the same time total area of telescopes was large. The image of ^{11}Li-beam on the target was not a point, but had large size (the beam parameters were measured for each collision of ^{11}Li with the target). Thus, the used detection system was not similar to a classical experiment with one small detector and point-like beam-spot. The measuring system of Ref. (5,6) was more powerful and due to this it was possible to study reactions with secondary beams, which have tremendously lower intensity than stable beams. The detection system from Ref. (5,6) is equivalent to a simultaneous usage of numerous amount of small detectors with point-like beam.

The measured distribution represented number of detected protons depending on the proton energy and emission angles. Then one axis was converted into the excitation energy in the residual system ^{11}Li and all detected events were projected on this axis of E^* to catch the whole accumulated statistics. One of the obtained spectra is shown in Fig. 2. Thus it would be incorrect to interpret the obtained spectra in the same way as a spectrum measured by one small detector. To perform comparison of theory with the experimental spectra, one needs to integrate theoretical distribution in the same way, how it was done in the experiment. In Ref. (5,6) some calculations of physical continuum were shown, in particular, phase volumes. They have shapes, which are unusual for spectrum measured by one small detector reflecting importance of the discussed integration.

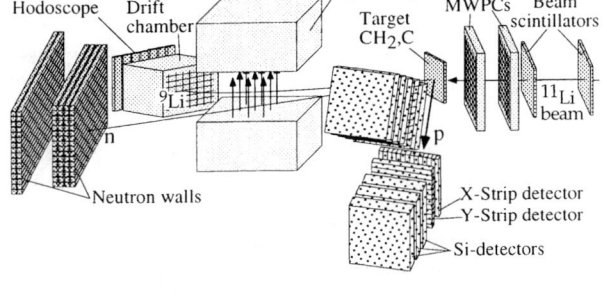

FIGURE 1. Experimental setup for studies of ^{11}Li+p-scattering.

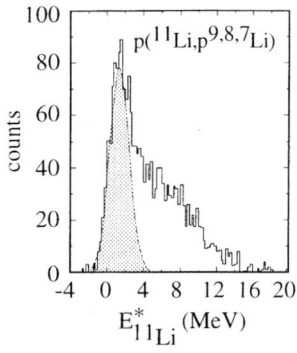

FIGURE 2. Proton spectrum from the reaction p(^{11}Li,p$^{9-7}$Li).

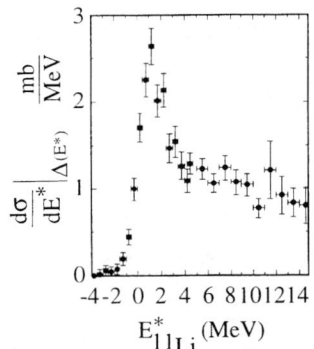

FIGURE 3. Cross section for the reaction p(^{11}Li,p$^{9-7}$Li).

Apart from the spectrum shown in Fig. 2 as a number of counts we extracted cross section in mb/MeV presented in Fig. 3. Again, a comparison of theory with these experimental data requires the integration. Note, that spectra in Fig. 2 and 3 are not equivalent to each other, reflecting that fact, that initial two-dimensional distribution contains more information that any its projection. The spectrum in Fig. 2 presents straight projection of events from two-dimensional plot, while the spectrum in Fig. 3 is the result of projection of these events weighted by solid angles. If some theory describes one of this spectra, it does not automatically mean, that it will describe another spectrum. Summarizing, even in qualitative theoretical treatment of the experimental data one has to be very careful.

Let us return to the paper (8), in which the shakeoff mechanism was suggested to be an explanation of the ^{11}Li-peak observed in Ref. (5,6). The shakeoff mechanism means the following. The incoming proton scatters on the ^{9}Li-core and the system ^{9}Li+2n can transfer into continuum. Then the sudden approximation is used, which means that the observed picture supposed to be just a photograph of the internal motion in ^{11}Li. Since this nucleus is very broad in space, there can be narrow components over momenta and energy. Thus, in sudden approximation there can be a peak, let say, at the energy about 1 MeV. This is the shakeoff considered in Ref. (8). It includes quasielastic scattering of proton on the core plus sudden approximation.

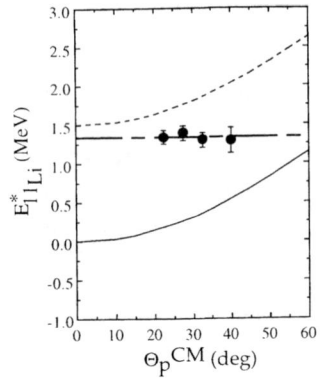

FIGURE 4. Experimental dependence of the peak position on the angle of scattered proton and that for the shakeoff.

The shakeoff mechanism differs from a population of excited state. The most important difference is the following. Excited state must have the same excitation energy along its kinematical locus, the position of the peak must not depend on the angle of scattered proton. In any other mechanism position of some peak must change with the angle. As it is seen in Fig. 4, experimental points show the constant energy. This is the most important criterion for excited state. If the energy of the peak would change along the locus, observation of the excited state of ^{11}Li would not be reported in Ref. (5,6).

Position of bump in the shakeoff model depends on the angle. Shakeoff includes proton scattering on the ^{9}Li-core and the larger scattering angle, the large recoil of ^{9}Li. The resulting dependence in terms of $E^*_{11_{Li}}$ shown in Fig. 4 by solid curve is not a constant. This recoil effect is the most solid thing in the shakeoff model. It means energy-momentum conservation and it is more solid than any approximation used in Ref. (8) for the shakeoff. Thus, the shakeoff contradicts to the experimental observation.

The authors of Ref. (8) declare that in the lowest approximation they obtain constant energy also. After that they write that more accurate calculation gives nonconstant excitation energy. This more accurate dependence in their model should be like the dashed curve in Fig. 4. It is very pity, that the authors of Ref. (8) did not show in their paper an explicit form of this dependence, which would allow to avoid many discussions. Finally note, that also in Ref. (9) scattering of incoming particle on the core of halo nucleus was investigated and in strict quantum mechanical approach the same non-constant energy was obtained as our solid curve in Fig. 4.

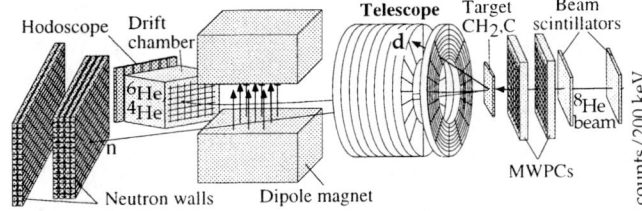

FIGURE 5. Experimental setup for study of the transfer reaction p(^{8}He,d)^{7}He.

SPECTROSCOPY OF ^{7}He

Physics of exotic nuclear beams follows the course of development of the whole nuclear physics. Studies of general properties of nuclei far from stability was followed by a stage of more delicate studies. We report on such a new step on this way, as an experimental study of transfer reactions with exotic nuclear beams.

We investigated the reaction p(^{8}He,d)^{7}He at $E_{^8He}^{lab}$ = 50 A MeV using targets CH$_2$ and C. To study transfer reactions with exotic beams, we designed new detection system. It represents a stack of solid state strip-detectors with large area (Fig. 5). In the middle of the telescope there is a hole and the beam goes through this hole. We detected deutrons at small angles in the laboratory system (~ 5°-25°) corresponding to high cross section. Particles from the ^{7}He decay (^{6}He, ^{4}He and neutrons) were detected also.

The obtained deuteron spectra are shown in Fig. 6 as a function of energy in the center of mass of ^{7}He counted from the threshold n+^{6}He. Apart from the inclusive deuteron spectrum, spectra of deutrons are shown, which were detected in coincidence with both ^{6}He and ^{4}He as well as separately. A peak in Fig. 6A, 6B, 6C represents the ground state of ^{7}He, which is a resonance decaying into n+^{6}He. In Fig. 6D this peak is absent, because ^{7}He$_{g.s.}$ is lower than the ^{4}He-threshold. Another peak is seen in Fig. 6D corresponding to the excited state in ^{7}He. This excited state is not seen in Fig. 6C (compare with Fig. 6B, where in the region of ^{7}He* there are ~ 200 counts per bin). Deutron spectra

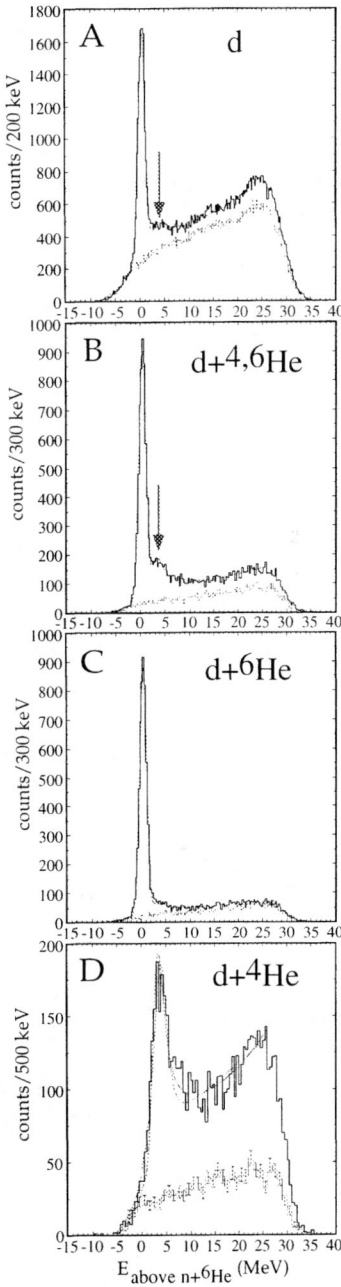

FIGURE 6. Spectra of deuterons from the reactions p(^{8}He,d) and p(^{8}He,d^{k}He). Solid histogram - CH$_2$-target, dotted - C-background.

185

detected in coincidence with neutron in addition are shown in Fig. 7. Again, the ^{7}He excited state observed in the spectra in Figs. 7A, 7B and 7D is not seen in coincidence with ^{6}He. Thus, this excited state decays into the ^{4}He-channel, but not into n+^{6}He. The obtained parameters of the excited state are E_{obs} = 3.3±0.3 MeV (energy above the n+^{6}He-threshold), E^* = 2.9±0.3 MeV (energy above ^{7}He$_{g.s.}$), Γ_{obs} = 2.2±0.3 MeV and the decay branching is $\Gamma_\alpha/\Gamma_{tot}$ = 0.85±0.15.

The ^{7}He decay scheme is shown in Fig. 8. The excited state of ^{7}He decays finally into ^{4}He+3n, but not into ^{6}He+n in spite of larger decay energy in the latter case. This corresponds to a structure of ^{7}He* with excitation of two neutrons (if only one neutron would be excited, the subsystem ^{6}He survives and the decay into ^{6}He+n should be predominant). In the result spin-parity of this excited state should be the same as that of the ground state of ^{7}He.

From the measured decay branching we estimated using R-matrix theory ratio of reduced widths for decays of ^{7}He* into ^{6}He+n and ^{4}He+3n. For the latter channel the decay through the intermiduate state of ^{6}He*(2$^+$) was considered. This estimation has showed that an admixture of the ^{6}He-configuration in ^{7}He* is small, of the order of persent.

That fact, that we don't see a state of ^{7}He having one excited neutron, can be understood if for such a state width for decay into ^{6}He is very large. For example, if the next after the 0P$_{3/2}$-orbital is the orbital 1S$_{1/2}$,

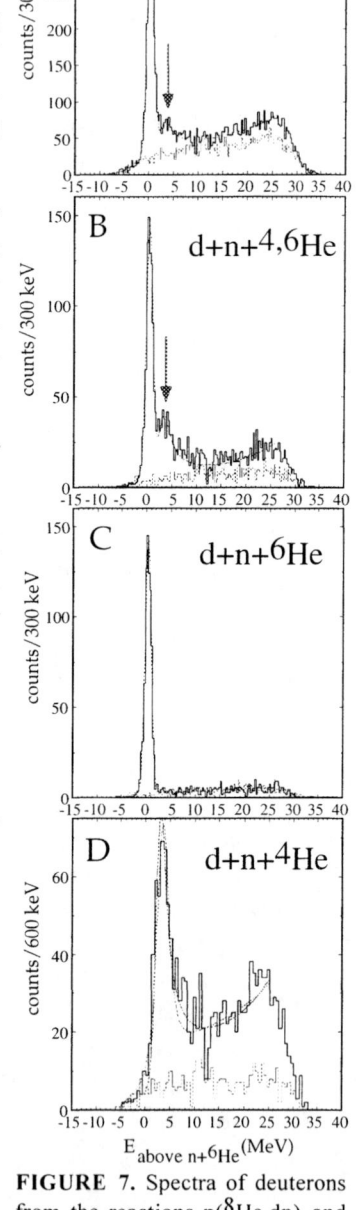

FIGURE 7. Spectra of deuterons from the reactions p(^{8}He,dn) and p(^{8}He,dn^{k}He). Solid histogram - CH$_2$-target, dotted - C-background.

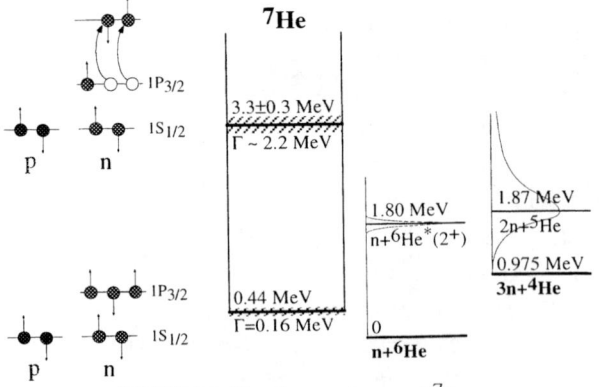

FIGURE 8. The decay scheme of ^{7}He.

single-neutron excited state can not exist, because nothing holds the system against the decay into ^{6}He+n. Finally, note that another spectroscopy of ^{7}He was performed using stable beam (10).

In these measurements we investigated the reaction p(^{8}He,^{2}He)^{7}H also. This is proton transfer from ^{8}He to the proton-target. The virtual state of two protons, ^{2}He, is not bound, but kinematical aspects of this reaction are similar to that for p(^{8}He,d)^{7}He, and these two reactions were investigated simultaneously. We detected both protons from decay of ^{2}He and, by measuring their energies and angles, we determined the energy in the center of mass of the residual system ^{3}H+4n. Thus, this method is like missing mass method, but with detection of unstable particle.

We observed a peculiarity in the spectrum, which could correspond to the resonance ^{7}H. However, it is not reliable statistically. This hint on a resonance ^{7}H provides the following "sea-light" for further experimental search for ^{7}H: energy above threshold ^{3}H+4n ~ 1.7 MeV, $\Gamma \leq 2$ MeV, $\sigma^{CM} \leq 150 - 100$ μb/sr at $\Theta^{CM}_{^2He} \sim 10^o - 25^o$.

SUMMARY

(i) The peak observed in the inelastic scattering of p+^{11}Li has constant excitation energy along its kinematical locus. This fact represents the most important criterion of observation of excited state. This constant energy contradicts to the shakeoff mechanism.

(ii) Using exotic nucleus beam, the nucleon transfer reaction was investigated. The excited state of ^{7}He was observed (E^* = 2.9±0.3 MeV, Γ = 2.2±0.3 MeV, $\Gamma_\alpha/\Gamma_{tot}$ = 0.85±0.15). This state has unusual structure resulting in decay into the channel α+3n, but not into ^{6}He+n. Hint on a possible existence of resonance ^{7}H was obtained.

REFERENCES

1. Tanihata, I., Hamagaki, H., Hashimoto, O. et al., Phys. Rev. Lett. **55**, 2676 (1985).
2. Hansen, P.G. and Jonson, B., Europhys. Lett. **4**, 409 (1987).
3. Kobayashi, T., Nucl. Phys. **A538**, 343c (1992).
4. Bohlen, H.G., Kalpakchieva, R., Aleksandrov, D.V. et al., Z. Phys. **A351**, 7 (1995).
5. Korsheninnikov, A.A., Nikolskii, E.Yu., Kobayashi, T. et al., Phys. Rev. **C53**, R537-540 (1996).
6. Korsheninnikov, A.A., Kuzmin, E.A., Nikolskii, E.Yu. et al., Phys. Rev. Lett. **78**, 2317 - 2320 (1997).
7. Zinser, M., Humbert, F., Nilsson, T. et al., Nucl. Phys. A619, **151** (1997).
8. Karataglidis, S., Hansen, P.G., Brown, B.A., Amos, K., Dortmans, P.J., Phys. Rev. Lett. **79**, 1447 - 1450 (1997).
9. Johnson, R.C., "Elastic scattering and elastic break-up of halo nuclei in a special model", paper presented at the European Conference on Advances in Nuclear Physics and Related Areas, Thessaloniki, Greece, July 8-12 1997
10. Bohlen, H.G., Blazevic., A., Gebaur, B. et al., submitted to Z. Phys.

Study of Exotic Nuclei by Proton Scattering in Inverse Kinematics

G.D.Alkhazov

Petersburg Nuclear Physics Institute, 188350, Gatchina, Russia

Abstract. A short overview of investigations of exotic nuclei by proton scattering at low energy is presented. Proton scattering at intermediate energy as a method to study the structure of exotic nuclei is discussed. The experimental study of nuclear matter distributions in ^{6}He and ^{8}He nuclei by 0.7 GeV proton elastic scattering at small angles is described.

The properties of exotic nuclei were investigated previously mainly by reaction cross section measurements and fragmentation experiments, nuclear targets being used. At the same time, it is well known that proton scattering was used successfully in the past to investigate nuclear structure of stable nuclei. In inverse kinematics, the method of proton scattering may be applied also for studying exotic nuclei. A number of experiments (Table) with proton scattering on exotic nuclei has been completed by now (1-16), and valuable information on the structure of exotic nuclei has been obtained. Theoretical aspects of proton scattering on exotic nuclei have been considered in several papers as well (13,17-31).

Protons as probing particles have an advantage over nuclei since the mechanism of proton-nucleus scattering is relatively simple, effects of distortions of the observed spectra due to absorption and rescattering are relatively small, and as a consequence of

TABLE .

Research center	Measured quantity	Energy	Reference
BEVALAC (USA)	Reaction cross section	0.8 GeV	(1)
RIKEN (Japan)	Spectroscopic data	32-83 MeV	(3,6,8)
	Differential elastic cross section		(2,3,6,8,13)
	Differential inelastic cross section		(3,15)
	Quasielastic cross section		(12)
GSI (Germany)	Differential inelastic cross section	101 MeV	(4)
	Differential elastic cross section	0.7 GeV	(14)
GANIL (France)	Differential elastic cross section	42-72 MeV	(7,16)
	Charge exchange cross section		(7)
MSU (USA)	Charge exchange cross section	93 MeV	(9)

CP455, *ENAM98: Exotic Nuclei and Atomic Masses*
edited by B. M. Sherrill, D. J. Morrissey, and Cary N. Davids
© 1998 The American Institute of Physics   1-56396-804-5/98/$15.00

this, the interpretation of the data is less ambiguous. Another advantage of using hydrogen for a target is that the proton target may not be excited or destroyed at low energy. The recoiled target protons can be easily detected, and their energy and scattering angle may be measured, which facilitates reconstruction of the kinematics of the scattering events. For this reason, the proton scattering in inverse kinematics is an efficient means of spectroscopic studies of exotic nuclei (3,5,6,8,10,12,15). Using this method, one can determine the excitation energies and widths of the resonant unbound states. As an example, an excited state of ^{11}Li at E_{ex}=1.3 MeV has been recently observed (8).

From the absolute values of the inelastic cross sections one can determine the strength or the deformation parameter of the relevant transition (4,23). The shape of this cross section tells us the orbital number of the transition. Thus, the data on excitation of E_{ex}=1.3 MeV state in the case of ^{11}Li correspond to L=1, that is to a dipole type of excitation (15).

Recently a quasielastic proton scattering experiment with exotic nuclei has been performed, and the halo neutron momentum distribution has been determined (12). Unfortunately, the projectile energy of 83 MeV/u in this experiment seems to be at the lowest limit for a quasielastic process. Therefore it would be useful to repeat a similar experiment at a higher energy (29).

Charge exchange reactions (7,9,11) may be also of use in studying exotic nuclei by proton scattering. As an example, from the data on ^{6}He(p,n)^{6}Li scattering one can make a qualitative conclusion (7) that the ^{6}He ground state and its ^{6}Li isobaric analog state have similar structures.

From proton elastic scattering at low energy it is possible in principle to get quantitative information on the radial shape of exotic nuclei. However, it is better to use protons at intermediate energy for this purpose, since at intermediate energy there exist theoretical approaches, such as the Glauber multiple scattering theory, that allow us to describe the proton-nucleus scattering with high accuracy, and to connect the investigated nuclear distributions with the measured cross sections in a quite straightforward way.

To calculate the differential cross section for elastic scattering at intermediate energy using the Glauber theory one needs the many-body ground-state density distribution of the investigated nucleus and the elementary proton-nucleon interaction amplitude, the parameters of the last being taken from the data on free proton-nucleon scattering. In agreement with the Glauber theory, the amplitude of proton elastic scattering on an exotic nucleus that consists of a compact core and an extended nucleon halo may be represented in the form:

$$F(\mathbf{q}) = F_h(\mathbf{q}) + F_c(\mathbf{q}) + F_{hc}(\mathbf{q}) \qquad (1)$$

where $F_h(\mathbf{q})$ and $F_c(\mathbf{q})$ are the terms responsible for scattering on the halo and the core, correspondingly, $F_{hc}(\mathbf{q})$ is a rescattering term, and $\mathbf{q}$ is the momentum transfer. When the halo size is significantly larger than that of the core, the contribution of the last term is relatively small. The amplitudes $F_h(\mathbf{q})$ and $F_c(\mathbf{q})$ may be written (20,31) as:

$$F_h(\mathbf{q}) \approx A_h \, S_h(\mathbf{q}) \, f_{pN}(\mathbf{q}), \tag{2}$$

$$F_c(\mathbf{q}) = S_c(\mathbf{q}) \, F'_c(\mathbf{q}). \tag{3}$$

Here A_h is the number of the halo nucleons, $S_h(\mathbf{q})$ is the halo form factor, $f_{pN}(\mathbf{q})$ is the elementary amplitude of proton scattering on the halo nucleons, $S_c(\mathbf{q})$ is the form factor of the core motion (motion of its center of mass in the center of mass system of the whole nucleus), $F'_c(\mathbf{q})$ is the amplitude of proton elastic scattering on a free core. Since the halo size is significantly larger than that of the core, the amplitude $F_h(\mathbf{q})$ decreases with $\mathbf{q}$ increasing much faster than the amplitude $F_c(\mathbf{q})$ (Fig.1). Therefore, analyzing the differential cross section for elastic scattering at small momentum transfers where the contribution of scattering from the halo is significant, and at bigger momentum transfers where the contribution of scattering from the core dominates, one can extract information on the halo and the core sizes (19). Note that the speed with which the contribution of scattering from the core decreases with $\mathbf{q}$ increasing depends on the effective core size which includes the size of the core center of mass motion (see the form factor $S_c(\mathbf{q})$ in Eq.(3)). The latter depends on the halo size and on the halo nucleon correlation. On the other hand, since $S_c(\mathbf{q})$ is a rather smooth function of $\mathbf{q}$, the position of diffraction minimum in the cross section depends mainly on the core size in its center of mass system (the minimum in $|F_c(\mathbf{q})|^2$ is close to that of $|F'_c(\mathbf{q})|^2$).

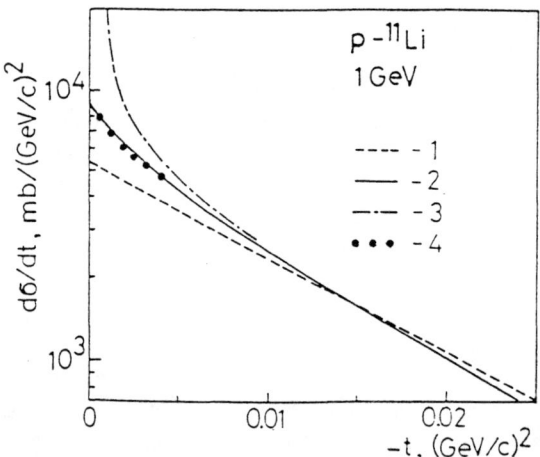

FIGURE 1. Calculated differential cross sections $d\sigma/dt$ for p-^{11}Li elastic scattering at 1 GeV at small momentum transfers ($-t=\mathbf{q}^2$). *1*- the contribution of the scattering due to strong interaction of the scattered protons with the core; *2*- the sum of the contributions of the scattering due to strong interaction of the protons with the core and halo, the core and halo density distributions being parametrized by Gaussian functions with the r.m.s. radii of $R_c = 2.5$ fm and $R_h = 6.8$ fm; *3*- the same as *2*, including the contribution of the Coulomb interaction; *4* - the same as *2*, including a 5% long-tailed halo-density component parametrized by a Gaussian function with the r.m.s. radius of 20 fm, the total ^{11}Li r.m.s. radius in this case being bigger as compared with *2* by 0.4 fm. (A difference between curves *4* and *2* is seen only at small values of $|t|$).

It seems that another useful source of information on spatial structure of halo nuclei may be the differential cross section $d\sigma/d^2q$ for nuclear fragmentation on a hydrogen target at small values of the momentum transfer **q**. If we take into account only the knock-out of halo nucleons, which is the dominant process of nuclear fragmentation, then for one-nucleon halo nuclei it is easy to obtain (31) that

$$d\sigma/d^2q = k^{-2}|f_{pN}(\mathbf{q})|^2 \{ 1 - S_h^2(\mathbf{q}) \}. \tag{4}$$

Here k is the magnitude of the projectile wave vector. The differential cross sections calculated by Eq.(4) for different halo sizes in case of 1GeV/u proton scattering on one-nucleon-halo nuclei are presented in Fig.2a. It is seen that at small values of the momentum transfer **q** there should exist a dip, the width of which may serve as a measure of the halo size. A similar result is obtained for nuclei with 2 (or more) halo nucleons. However, now the shape and absolute values of $d\sigma/d^2q$ depend also on the halo nucleon correlation (32) (Fig.2b).

To study the ground state spatial structure of exotic nuclei, the data on reaction cross section σ_r (which is defined as a difference between the total cross section and the integral elastic cross section) may be also helpful. Note that all the three quantities discussed above, the differential elastic cross section, the differential cross section $d\sigma/d^2q$ for fragmentation, and the reaction cross section σ_r, are determined completely through the ground state wave function of the exotic nucleus under study.

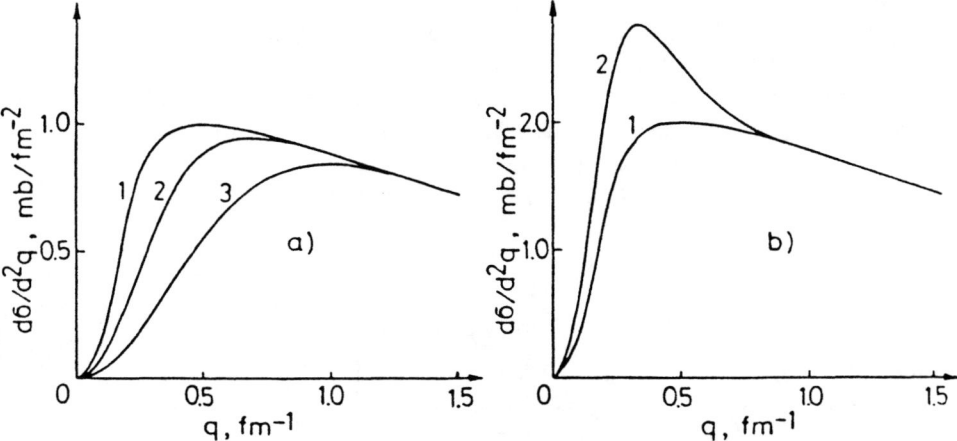

FIGURE 2. The fragmentation cross section $d\sigma/d^2q$ as a function of the value of the transverse momentum **q** for a one-nucleon-halo nucleus (Fig.2a) and for a two-nucleon-halo nucleus (Fig.2b). Curves *1,2,3* in Fig.2a correspond to the halo radius of 8 fm, 5 fm, and 3 fm, respectively. Curve *1* in Fig.2b corresponds to the halo radius of 8 fm for the non-correlated motion of the halo neutrons. Curve *2* (Fig.2b) corresponds to the halo radius of 8 fm, the two neutrons being bound in a di-neutron cluster of 3 fm radius. Note that only the process of knock-out of the halo nucleons is taken into account. The contribution of the Coulomb dissociation is not included either.

The reaction cross section depends mainly on the r.m.s. nuclear matter radius. It depends also on the radial shape of the nuclear density distribution. This dependence is essentially different for a proton target and for nuclear targets of different size (26). It is this dependence that has allowed the authors of Ref.(1) to determine the radial shape of ^{11}Li nuclei using the data on reaction cross sections.

Most of experimental studies of proton scattering on exotic nuclei were performed at RIKEN at low energy of about E_p= 60 MeV/u. Only 2 experiments have been done by now at intermediate energy. The reaction cross section for proton scattering on ^{11}Li at E_p= 0.8 GeV has been measured at BEVALAC (1), and the differential cross sections for proton elastic scattering on heavy helium isotopes have been measured at E_p=0.7 GeV at GSI (14).

The measurements of the elastic cross sections were performed (14) using the secondary He beams from the fragment separator FRS. The basic element of the experimental setup was the ionization chamber IKAR (Fig.3), developed in PNPI (Gatchina, Russia) for small angle hadron scattering investigations. IKAR was filled with hydrogen which served as the target and the working gas of the chamber. The information on the momentum transfer was obtained from the proton recoil energies measured by IKAR, and from the projectile scattering angles determined with the help of multiwire proportional chambers placed in front of and behind IKAR. The measured cross sections for p^6He, p^8He, as well as for p^4He, are shown in Fig.4.

The cross sections were analyzed (14) applying the Glauber multiple scattering theory. As is known, at small scattering angles the calculated cross sections are determined mainly by the one-body densities, the effects from nucleon correlations being relatively small. Only center-of-mass correlations were included in the analysis. It was checked that possible clusterization correlations had little influence on the calculated cross sections and on matter distribution parameters deduced. Four different phenomenological parametrizations, each having 2 free parameters, were used to describe the one-body density distributions. The model parameters were varied to fit the calculated cross sections to the experimental ones.

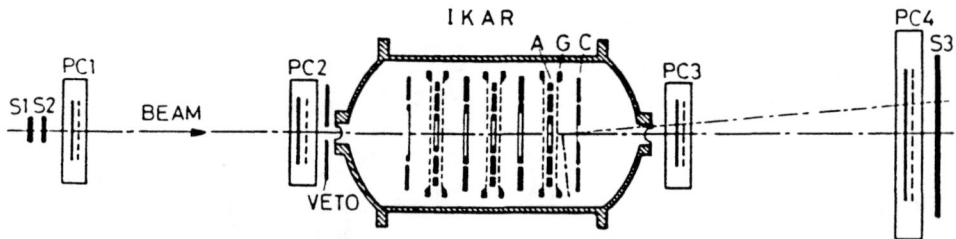

FIGURE 3. A schematic view of the experimental setup. IKAR: ionization chamber filled with hydrogen at 10 bars; A-anode, C-cathode, G-grid (indicated for one of the six identical modules); PC1-PC4: multiwire proportional chambers; S1-S3, VETO: scintillation counters. Tracks for a typical scattering event are shown by dot-dashed lines.

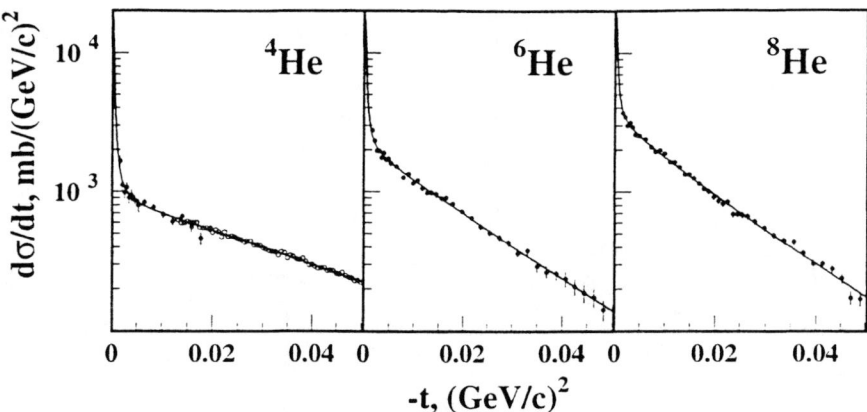

FIGURE 4. The differential cross sections $d\sigma/dt$ versus the momentum transfer squared for 0.7 GeV proton elastic scattering from ^{4,6,8}He isotopes. Full dots show the data of Ref.(14). Open dots in the case of ^{4}He show the data of Ref.(33). Full lines are the theoretical cross sections fitted to the data.

Note that it is for the first time that from the cross sections at small momentum transfers more than 1 free parameter of the density distributions have been determined. This has become possible due to a rather high precision of the cross section measurements. The nuclear matter distributions and nuclear matter radii deduced in the analysis with the four parametrizations mutually agree within small errors. This work (14) shows that both neutron rich helium isotopes ^{6}He and ^{8}He have extended matter distributions, significantly more extended ones than in the case of ^{4}He. The nuclear r.m.s. matter radii deduced are $R_m=2.30(7)$ fm for ^{6}He, and $R_m=2.45(7)$ fm for ^{8}He, to compare with $R_m=1.49(1)$ fm for ^{4}He. The determined values of the core and halo radii are $R_c=1.88(12)$ fm, $R_h=2.97(26)$ fm for ^{6}He, and $R_c=1.55(15)$ fm, $R_h=3.08(10)$ fm for ^{8}He (14). Fig.5 demonstrates the obtained density distributions. The shaded areas reflect the model uncertainties in the density determination and the statistical errors of the model parameters. In the case of ^{6}He, a theoretical density distribution (34,35) is also shown. It is seen that up to $r=5.5$ fm the density deduced from the experiment without any theoretical input on its radial shape is in very good agreement with the theory. Some disagreement between these distributions is observed only at rather big radii. However it may be shown that the differential elastic cross sections have poor sensitivity to a possible density tail at these big radii (see Fig.1), and this tail may not be determined reliably from the data. If one performs an analysis of the data using phenomenological densities with the tails taken from the theory, then the previously determined r.m.s. radii of $R_m=2.30(7)$ fm for ^{6}He and $R_m=2.45(7)$ fm for ^{8}He should be corrected by $+(0.1-0.2)$ fm (30).[1]

---

[1] The influence of the density asymptotics on the R_m values deduced will be discussed also in Ref.(36).

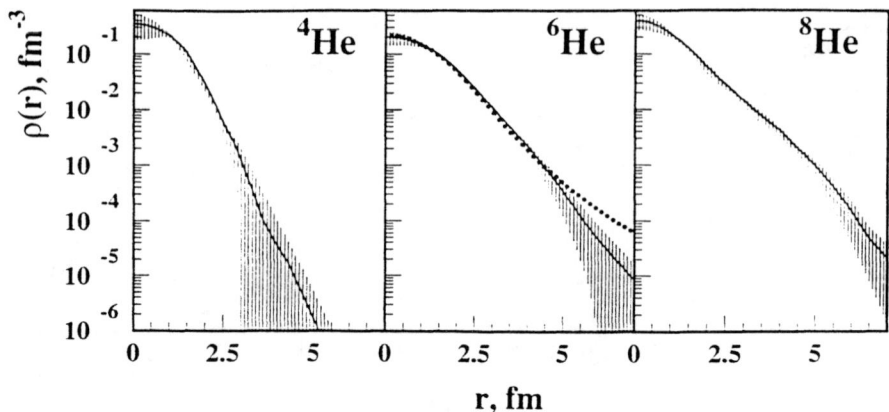

FIGURE 5. Nuclear matter distributions for ^{4,6,8}He isotopes. Solid lines represent the deduced matter distributions (14), averaged over the four models used. The shaded areas reflect the model uncertainties and the statistical errors of the model parameters. For the case of ^{6}He, a theoretical density distribution (34) corresponding to R_m=2.52 fm is also shown (dotted line).

As it has been discussed, to get more precise information on the halo and core sizes, and possibly on the correlation of the halo nucleons, it would be useful to extend measurements of the elastic scattering cross sections to higher momentum transfers, and to measure the differential cross sections for nuclear fragmentation (on a proton target, as a function of the momentum transfer **q**). Experiments on quasielastic proton scattering at intermediate energy for determination of the nucleon momentum distributions are also of interest.

To conclude, the proton scattering has proven to be an effective means of studying exotic nuclei. Proton scattering at low energy has given valuable spectroscopic information. The proton small-angle elastic scattering at intermediate energy has been shown to provide information on spatial structure of halo nuclei. New highly informative experiments with proton scattering at intermediate energy may be performed.

ACKNOWLEDGEMENTS

The author thanks A.A.Vorobyov, A.A.Lobodenko, V.V.Anisovitch, P.Egelhof, M.Mutterer, J.Al-Khalili, J.Tostevin for valuable discussions, and I.Thompson for providing a theoretical density.

REFERENCES

1. I.Tanihata, T.Kobayashi, T.Suzuki et al., Phys. Lett. **B287**, 307-311 (1992).
2. C.B.Moon, M.Fujimaki, S.Hirenzaki, et.al., Phys.Lett. **B297**, 39-43 (1992).
3. A.A.Korsheninnikov, K.Yoshida, D.V.Aleksandrov, et al., Phys.Lett. **B316**, 38-44 (1993).
4. G.Kraus, P.Egelhof, C.Fisher et al., Phys.Rev.Lett. **73**, 1773-1776 (1994).
5. A.A.Korsheninnikov, D.V.Aleksandrov, N.Aoi et al., Nucl.Phys. **A588**, 23c-28c (1995).
6. A.A.Korsheninnikov, E.Yu.Nikolskii, T.Kobayashi et al., Phys.Lett. **B343**, 53-58 (1995).
7. M.D.Kortina-Gil, P.Roussel-Chomaz, N.Alamanos et al., Phys.Lett. **B371**, 14-18 (1996).
8. A.A.Korsheninnikov, E.Yu.Nikolskii, T.Kobayashi et al., Phys.Rev. **C53**, R537-R540 (1996).
9. J.A.Brown, D.Bazin, W.Beneson et al., Phys.Rev. **C54**, R2105-R2108 (1996).
10. A.A.Korsheninnikov, E.A.Kuzmin, E.Yu.Nikolskii et al., Nucl.Phys. **A616**, 189c-200c (1997).
11. M.D.Cortina-Gil, P.Roussel-Chomaz, N.Alamanos et al., Nucl.Phys.. **A616**, 215c-222c (1997).
12. T.Kobayashi, K.Yoshida, A.Ozawa et al., Nucl.Phys. **A616**, 223c-230c (1997).
13. A.A.Korsheninnikov, E.Yu.Nikolskii, C.A.Bertulani et al., Nucl.Phys. **A617**, 45-56 (1997).
14. G.D.Alkhazov, M.N.Andronenko, A.V.Dobrovolsky et al., Phys.Rev.Lett. **78**, 2313-2316 (1997).
15. A.A.Korsheninnikov, E.A.Kuzmin, E.Yu.Nikolskii et al., Phys.Rev.Lett. **78**, 2317-2320 (1997).
16. M.D.Kortina-Gil, P.Roussel-Chomaz, N.Alamanos et al., Phys.Lett. **B401**, 9-14 (1997).
17. A.N.F.Aleixo, C.A.Bertulani, M.S.Hussein. Phys.Rev. **C43**,2722-2727 (1991).
18. S.A.Fayans, S.N.Ershov, E.F.Svinareva. Phys.Lett. **B292**, 239-241 (1992).
19. G.D.Alkhazov, A.A.Lobodenko. Pis'ma v ZhETF **55**, 377-379 (1992).
20. G.D.Alkhazov, A.A.Lobodenko. Yad.Fiz. **56**, 89-99 (1993).
21. S.Hirenzaki, H.Toki, I.Tanihata et al., Nucl.Phys. **A552**, 57-65 (1993).
22. A.K.Chaudhuri. Phys.Rev. **C49**, 1603-1608 (1994).
23. L.V.Chulkov, C.A.Bertulani, A.A.Korsheninnikov. Nucl.Phys. **A587**, 291-300 (1995).
24. R.Crespo, J.A.Tostevin, R.C.Jonson. Phys.Rev. **C51**, 3283-3289 (1995).
25. R.Crespo, J.A.Tostevin, I.J.Thompson. Physs.Rev. **C54**, 1867-1876 (1996).
26. M.P.Bush, J.S.Al-Khalili, J.A.Tostevin, R.C.Johnson. Phys.Rev. **C53**, 3009-3013 (1996).
27. Rituparna Kanungo, C.Samanta. Nucl.Phys. **A617**, 265-281 (1997).
28. S.N.Ershov, T.Rodge, B.V.Danilin et al., Phys.Rv. **C56**, 1483-1499 (1997).
29. G.D.Alkhazov. Pis'ma v ZhETF **66**, 75-80 (1997).
30. J.S.Al-Khalili, J.A.Tostevin. Phys.Rev. **C57**, 1846-1852 (1998).
31. G.D.Alkhazov Pis'ma v ZhETF **67**, 296-301 (1998).
32. G.D.Alkhazov, to be published.
33. O.G.Grebenjuk, A.V.Khanzadeev, G.A.Korolev et al., Nucl.Phys. **A500**, 637-652 (1989).
34. J.S.Al-Khalili, J.A.Tostevin, I.J.Thompson. Phys.Rev. **C54**, 1843-1852 (1996).
35. M.V.Zhukov, D.V.Fedorov, B.V.Danilin et al., Nucl.Phys. **A552**, 353-362 (1993).
36. G.D.Alkhazov, A.V.Dobrovolsky, P.Egelhof et al., to be published.

Energy and Angular Correlations in Peripheral Fragmentation of Halo Nuclei

T. Aumann[1,2,*] and L.V. Chulkov[3]

[1] *Gesellschaft für Schwerionenforschung (GSI), Planckstr. 1, D-64291 Darmstadt, Germany*
[2] *Institut für Kernchemie, Johannes Gutenberg Universität, D-55099 Mainz, Germany*
[3] *Kurchatov Institute, 123182 Moscow, Russia*

Abstract. Results from an experiment studying the fragmentation of halo nuclei, performed recently at GSI, are summarized. Special attention is paid to halo-breakup reactions of ^{6}He on carbon and lead targets, serving as a touchstone in extracting the nuclear structure information from high-energy reactions. We show, that for light targets the dominant break-up mechanism is a two-step process: knock-out of one halo neutron and subsequent decay of the unbound ^{5}He resonance. The momentum distribution of the $\alpha+n$ center-of-mass motion contains the information on the wavefunction of the halo neutrons. The ^{5}He ground state resonance shows up in the relative-energy distribution of the $\alpha + n$ system. The angular correlations observed in the decay of this resonance show a characteristic anisotropy demonstrating the sensitivity to the angular momenta involved. Differential cross sections $d\sigma/dE^*$ for inelastic nuclear and electromagnetic excitations into the ^{6}He continuum are discussed.

I INTRODUCTION

An intense scientific activity, both experimental and theoretical, has over the last decade been performed in order to understand the basic structure of the halo states [1–3]. Special attention has been focused on nuclei showing a halo state consisting of two neutrons. These two-neutron halos are Borromean [3], i.e., three-body systems with no bound binary sub-systems. The lightest Borromean nucleus is ^{6}He. With its two-neutron separation energy of 975 keV, it shows evidence for a halo structure [4,5]. A major source in gaining information on the structure of halo nuclei are fragmentation reactions [1,2]. However, the breakup reaction mechanisms have to be considered in extracting the nuclear structure information. There are essential advantages of a study involving ^{6}He: the basic $\alpha - n$ and $n - n$ interactions are well known, good wave functions built on different microscopic three-body models are available [6–8], and spectroscopic information exists. In this context, we consider the fragmentation of ^{6}He as a touchstone to study the interplay between nuclear structure and reaction mechanisms.

---

*) present address: Michigan State University, East Lansing, MI 48824-1321, USA

The experimental results presented in the present article were obtained from an experiment performed at GSI, Darmstadt, using secondary beams at energies around 250 MeV/u [9–12]. We shall restrict mainly to the break up of ^{6}He into $\alpha + xn$, reactions where the core remains intact. The reaction mechanisms considered, and the experimental technique to disentangle them, are discussed in section 3. Different information on the structure can be deduced from the different breakup mechanisms. The one-neutron knockout reaction, yielding the unbound $\alpha + n$ system, contains information on the momentum content of the halo. In addition, angular correlations and invariant mass analysis can be used for spectroscopy of intermediate unbound nuclei, in this case the ^{5}He resonances. The inelastic excitation by the electromagnetic or nuclear multipole fields leading to the breakup into $\alpha + n + n$ contains information on the continuum structure, for instance the B(E1) strength distribution of the halo nucleus. Correlations between e.g. the $\alpha - n$ or $n - n$ subsystems in the 3-body decay can be studied as well.

II EXPERIMENTAL METHOD

The experiments were performed at GSI, Darmstadt, utilizing a 340 MeV/u ^{18}O primary beam. The ^{6,8}He, ^{11}Li and ^{14}Be beams are produced in a thick (8 g/cm^2) Be target, and subsequently separated using the fragment separator FRS [13]. The ions hitting the C and Pb secondary targets were identified uniquely event-by-event by means of energy-loss and time-of-flight measurements, as well as position measurements in order to reconstruct the trajectories. Behind the target, the fragments were deflected by a large-gap dipole magnet. Position, time-of-flight, and energy-loss measurements were used to identify the fragments and deduce their momenta. The neutrons were detected in the large-area neutron detector LAND [14], allowing for position and time-of-flight determination, as well as multiple hit recognition. The large solid angle acceptance allow for a kinematically complete measurement in the projectile-rapidity domain. Experimental details can be found in [10,11,15].

III REACTION MECHANISMS

The reactions discussed here are 'halo-breakup' reactions, reactions where the core remains intact, e.g. ^{6}He $\rightarrow \alpha + xn$. Two mechanisms are considered to induce the halo breakup: knockout of one or two halo neutrons caused by projectile-target nucleon-nucleon collisions, and the excitation into the continuum induced by the electromagnetic or nuclear multipole fields. In the former case one or both neutrons are scattered to large angles, respectively, and thus are not within acceptance of LAND. For the latter reaction type, both neutrons will have essentially beam velocity and are detected in LAND. The apparent neutron multiplicity can thus be used to disentangle the different contributions [15]. In Fig. 1 the cross sections for the different processes are shown for the halo breakup of ^{6}He on C and Pb targets

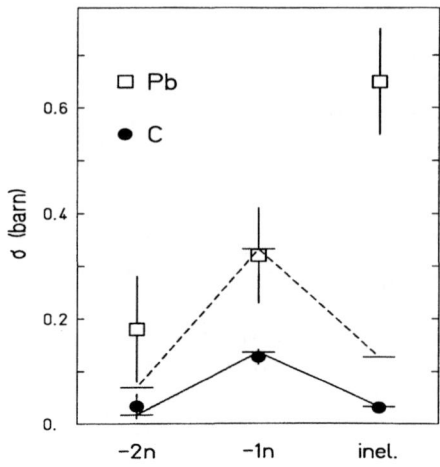

FIGURE 1. Measured integrated cross sections for single- ($-1n$) and two-neutron ($-2n$) knockout, and for inelastic excitation (inel.) in ^{6}He (240 MeV/u) on a C target (solid symbols) and a Pb target (open symbols), leading to break-up into α and neutrons. The solid and dashed lines represent calculations in an eikonal model for the C target [16] and for the Pb target [11], respectively. Electromagnetic excitations are not included in the model calculation. Adopted from [11].

[11]. It can be seen that for light targets, where the electromagnetic interaction is negligible, the dominant process is one-neutron knockout. The cross sections are compared to Glauber calculations (eikonal approximation) [16], resulting in a perfect agreement between experiment and theory for the C target[†]. In contrast, a large excess in the inelastic scattering cross section is found for the Pb target, which can be attributed to electromagnetic induced dissociation, which is not included in the calculation.

IV ONE-NEUTRON KNOCKOUT

A Momentum Distributions

At high beam energies the internal degrees of freedom can be considered as frozen during the reaction. In this 'sudden approximation', the momentum distributions of fragments and neutrons correspond to the initial momentum distribution of the halo neutrons in the projectile. It has been shown, however, that in the case of two neutron halo nuclei, these distributions are strongly effected by final state interactions between the neutron and the fragment [17–19]. By taking such final state interactions into account, the measured distributions could be reproduced. The sensitivity on the original wave function seems to be partially lost, since different wavefunctions yield finally good agreement with the experimental distributions. The center-of-mass motion of the (fragment+neutron) system, however, is not under influence of final state interactions, and thus can directly be compared to the model calculations. This is done in Fig. 2, where the transverse momentum distribution of the $\alpha + n$ system is shown after one-neutron knockout from ^{6}He on a C

[†] In Ref. [16] knock-out reactions and inelastic excitations are referred to as stripping processes and diffractive scattering, respectively.

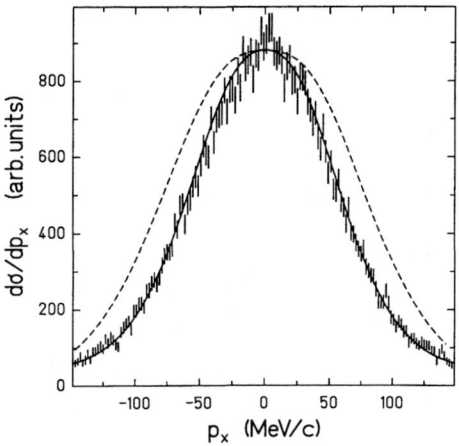

FIGURE 2. Distribution of the $(\alpha + n)$ center-of-mass along one transverse momentum component. The dashed curve represents calculations based on the Fourier transform of the ^{6}He wave function obtained [6] in the three-body model $(\alpha + n + n)$. The data [10] shown by the error bars are reconstructed from the measured momenta of α-particles and neutrons. The solid line is a calculation based on the wavefunction of [6] and taking into account the peripheral nature of the process by introducing a cut-off [23].

target [10]. The dashed curve results from a 3-body model [6] and has obviously a much larger width than the experimental distribution. After introducing a radial cut-off in order to account for the peripheral nature of the reaction process ('shadowing effect') [20–22], very good agreement with the experimental data was achieved [23] (solid curve). The method as outlined by Hansen [21] was used, but applied for a two-neutron halo nucleus. Therefore the asymptotic of the 3-body wave function [6] was approximated by a Hankel function. A neutron separation energy of 1.75 MeV was needed to reproduce the 3-body solution, which corresponds roughly to the binding energy of ^{6}He relative to the separation into ^{5}He and n. The same method was then applied to reproduce the measured ^{10}Li momentum distribution. Agreement could be achieved only if both, contributions from s and p waves are considered with about equal weight [24,12].

B Spectroscopy of Unbound Nuclei

In Fig. 3 the relative energy distribution of the $\alpha - n$ system is shown. A pronounced peak structure coinciding with the energy of the known $p_{3/2}$ ^{5}He resonance is visible. A Monte Carlo calculation starting from a Breit Wigner parameterization of this resonance with parameters from the literature (dotted curve) can reproduce the data very well for energies larger than 0.5 MeV (solid histogram). Additional evidence for the dominance of this resonance is obtained by inspecting the correlation function deduced by dividing the experimental spectrum with a randomized one (see Ref. [10] for details). Moreover, the momentum distributions of the neutron and the α particles can be reproduced by assuming a two-step process: One neutron knockout leading to the unbound ^{5}He resonance (with a comparatively long lifetime of about 300 fm/c), which subsequently decays into $\alpha + n$ (far away from the reaction zone) [10].

The presence of the intermediate ^{5}He resonance can also be seen in the angular correlation observed between the direction of the ^{5}He momentum and the $\alpha - n$

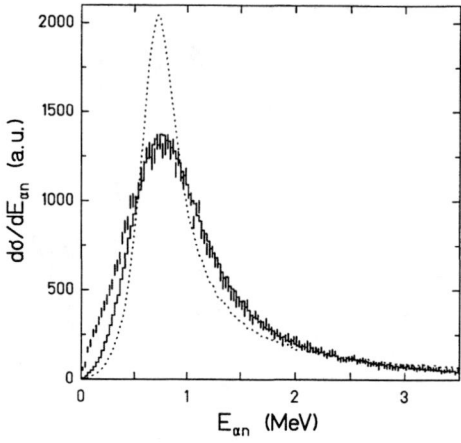
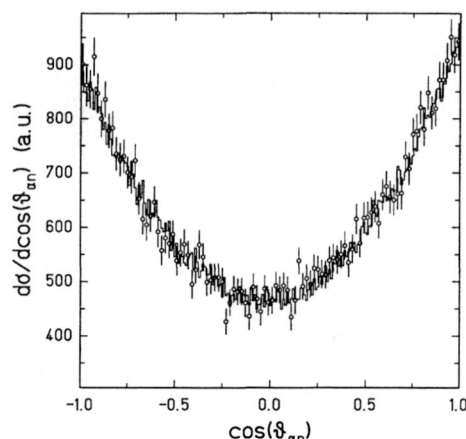

FIGURE 3. Relative kinetic energy distribution of the $\alpha - n$ system (invariant mass spectrum). The experimental data are shown as error bars. The dotted curve displays a Breit-Wigner parameterization with the known parameters for the $p_{3/2}$ resonance. The solid histogram shows the result of a Monte Carlo simulation including experimental resolutions and acceptance. Adopted from [10].

FIGURE 4. Distribution of the angle ϑ_{an} between the direction of the relative momentum $\mathbf{p}_{an}$ and the direction of the $(\alpha + n)$ cm motion $\mathbf{p}_{^5\mathrm{He}}$. The experimental data are shown by error bars. The solid histogram is the result of a Monte Carlo calculation with a correlation function $W \sim 1 + 1.5\cos^2(\vartheta_{an})$ and taking into account experimental resolutions and acceptance. Adopted from [9].

relative momentum $\mathbf{p}_{\alpha n} = \frac{m_n m_\alpha}{m_n + m_\alpha}(\frac{\mathbf{p}_\alpha}{m_\alpha} - \frac{\mathbf{p}_n}{m_n})$. The distribution on this angle $\vartheta_{\alpha n}$ is shown in Fig. 4 exhibiting an anisotropy characteristic for a relative angular momentum $l_n = 1$. The solid histogram, which describes the data very well, results from a Monte Carlo calculation assuming an angular correlation function $W \sim 1 + 1.5\cos^2(\vartheta_{\alpha n})$. Experimental effects are taken into account in the Monte Carlo procedure. A two step process involving only the $p_{3/2}$ resonance, however, would yield a correlation function $W \sim 1 + 3\cos^2(\vartheta_{\alpha n})$, an anisotropy twice as the observed one. Chulkov and Schrieder [25] have shown that the experimental result can quantitatively be understood by assuming a 7 % admixture only of the higher lying $p_{1/2}$ resonance in ^{5}He.

In turn, the presence of the angular correlation between the ^{5}He momentum and the decay direction (or momentum of the second neutron) shows that the $(\alpha + n)$ center-of-mass motion indeed is correlated to the initial momentum of the halo neutron prior to the knock-out reaction, and thus carries the information on the projectile wave function.

For ^{11}Li a similar analysis has been carried out [24,12], and an asymmetric distribution was observed, which has its origin in the interference of states with different parity [25]. The observed distribution again call for s and p wave contributions. Similar the ^{9}Li + n relative energy spectrum [15,16]. The three observables, ^{10}Li invariant mass spectrum, ^{10}Li momentum distribution, as well as the angular cor-

relations observed in the decay of ^{10}Li, can consistently reproduced by considering
s and p wave contributions to the one-neutron knockout cross section with about
equal weight. This corresponds to an occupation probability of about 40 % for the
s state in ^{11}Li [16].

V CONTINUUM EXCITATIONS

In contrast to properties known for stable nuclei, a considerable low-lying multipole strength has been predicted for halo nuclei (see e.g. [29,30,26] and refs. therein) but also for heavier neutron-rich nuclei (see e.g. [31–33] and refs. therein). Experimentally, low-lying dipole components were observed in the neutron-halo nuclei ^{11}Li [34,35,15] and ^{11}Be [36], and in neutron-rich Oxygen isotopes [37]. While the electromagnetic break-up allows to deduce the E1 continuum strength, a mixture of multipolarities is present in the nuclear inelastic scattering.

The measured cross sections $d\sigma/dE^*$ are shown in Fig. 5 for ^{6}He, deduced from the invariant mass of the $\alpha + n + n$ decay channel, obtained with Pb (upper frame) and C (lower frame) targets at 240 MeV/u. A dominant peak is visible at 1.8 MeV in the spectrum obtained with the C target, coinciding with the energy of the known 2^+ resonance in ^{6}He with a width of 110 keV. The observed width of the peak is dominated by the experimental resolution and agrees very well with our simulations. A second 2^+ resonance was predicted by 3-body models at 4.3 MeV [28], but no evidence is found in the experimental spectrum. The higher lying continuum strength can be a mixture of various multipolarities, e.g. soft monopole and dipole strength is expected in addition to quadrupole contributions [28]. An analysis of the differential cross section with respect to scattering angle of ^{6}He (reconstructed from the momenta of fragment and neutrons) is in progress. In addition, angular correlations in the decay can be helpful to get access to the multipolarity of the observed strength. Low-lying strength other than dipole and quadrupole is indicated.

For the Pb target the dominant contribution comes from electromagnetic excitation (see section 3 and Fig. 1). The experimental data are compared to semiclassical calculations based on the strength distributions as predicted by different 3-body models [26] (solid curve), [27] (dashed curve). The experimental response is taken into account. The calculation using the B(E1) distribution of [26] is shown in addition without experimental filter (dotted curve) for comparison. Both calculations reproduce the absolute cross section and the peak position fairly well, the agreement with the model predictions of [26] is somewhat better. According to the eikonal calculation [16] we expect about 20 % nuclear contribution to the cross section. At 1.8 MeV a small peak is visible, which is more pronounced in the correlation function obtained by dividing the experimental spectrum by a randomized one (see Ref. [10,11]). The ratio of the extracted cross section for this peak to the cross section of the 2^+ resonance obtained with the C target agrees well with the scaling for the diffraction cross section as predicted by the calculation [16,11].

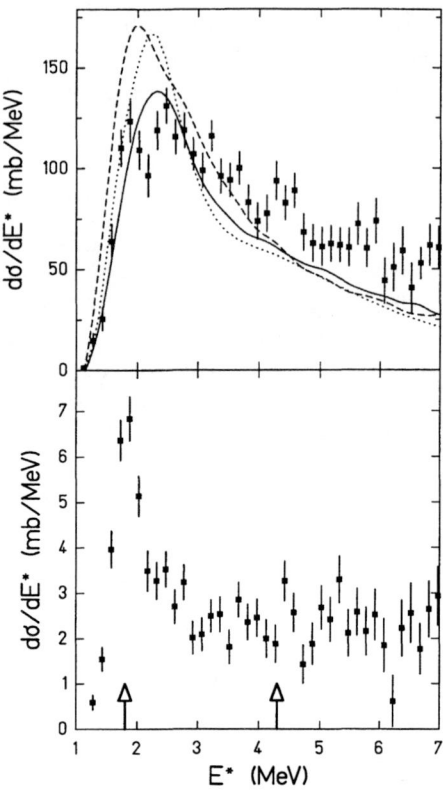

FIGURE 5. Excitation energy (E^*) spectra of ^{6}He deduced from the invariant mass of the α+n+n decay channel, obtained with the Pb target (upper frame) and the C target (lower frame) at 240 MeV/u bombarding energy. The spectra are corrected for detection efficiency and solid angle acceptance, but are not deconvoluted with respect to resolution in E^*. In case of the Pb target, the dotted curve represents the calculated electromagnetic cross section using the $dB(E1)/dE^*$ distribution from the three-body model of [26] and a semi-classical perturbative calculation. The solid curve is obtained by convoluting the dotted curve with the instrumental response. The dashed curve represents calculated cross sections using the $dB(E1)/dE^*$ distribution from the three-body model of [27], including convolution with the instrumental response. The excitation energies of a known (E^* = 1.80 MeV) and a predicted (E^* = 4.3 MeV, [28]) $I^\pi = 2^+$ resonance are indicated by arrows. Adopted from [11].

The electromagnetic contribution to the excitation of the 2^+ resonance can be neglected, it's less than 1 mb if B(E2) values are taken from the 3-body models, or as derived from the cross section with the C target. After subtracting the small nuclear contribution to the cross section, the B(E1) distribution can be deduced. 5% of the TRK sumrule is exhausted for excitation energies below 5 MeV. This corresponds to 40 % of a cluster sumrule, where the strength is splitted into dipole strength associated to the core, and E1 strength associated to the relative motion of valence neutrons and the core. Integrating up to 10 MeV excitation energy, the cluster sumrule is fully exhausted.

Acknowledgement. The authors are very much indebted to their colleagues of the S135 collaboration: D. Aleksandrov, T. Aumann, L. Axelsson, T. Baumann, M.J.G. Borge, L.V. Chulkov, J. Cub, W. Dostal, B. Eberlein, Th.W. Elze, H. Emling, H. Geissel, V.Z. Goldberg, M. Golovkov, A. Grünschloß, M. Hellström, K. Hencken, J. Holeczek, R. Holzmann, B. Jonson, A.A. Korshenninikov, J.V. Kratz, G. Kraus, R. Kulessa, Y. Leifels, A. Leistenschneider, T. Leth, I. Mukha, G. Münzenberg, F. Nickel, T. Nilsson, G. Nyman, B. Petersen, M. Pfützner, A. Richter, K. Riisager, C. Scheidenberger, G. Schrieder, W. Schwab, H. Simon, M.H. Smedberg, M. Steiner, J. Stroth, A. Surowiec, T. Suzuki, O. Tengblad and M.V. Zhukov.

Note. Some of the experimental data shown (Fig. 1, Fig. 5) are not published and have to be considered as preliminary.

REFERENCES

1. P.G. Hansen, A.S. Jensen and B. Jonson, Ann. Rev. Nucl. Part. Sci. **45** (1995) 591.
2. I. Tanihata, J. Phys. (London) **G22** (1996) 157.
3. M.V. Zhukov *et al.*, Phys. Rep. **211** (1993) 151.
4. M.J.G. Borge *et al.*, Z. Phys. **A340** (1991) 255.
5. I. Tanihata *et al.*, Phys. Lett. **B289** (1992) 261.
6. B.V. Danilin *et al.*, Yad. Fiz. **53** (1991) 71; Sov. J. Nucl. Phys. **53** (1991) 45.
7. E. Garrido, D.V. Fedorov and A.S. Jensen, Nucl. Phys. **A617** (1997) 153.
8. J. Wurzer and H.M. Hofmann, submitted to Phys. Rev. C.
9. L.V. Chulkov *et al.*, Phys. Rev. Lett. **79** (1997) 201.
10. D. Aleksandrov *et al.*, Nucl. Phys. **A633** (1998) 234.
11. T. Aumann *et al.*, to be published.
12. GSI Scientific Report 1997, GSI-98-1 (1998) 16.
13. H. Geissel *et al.*, Nucl. Instr. and Meth. **B70** (1992) 286.
14. T. Blaich *et al.*, Nucl. Instr. and Meth. **A314** (1992) 136.
15. M. Zinser *et al.*, Nucl. Phys. **A619** (1997) 151.
16. G. Bertsch, K. Hencken and H. Esbensen, Phys. Rev. **C57** (1998) 1366.
17. A.A. Korsheninnikov and T. Kobayashi, Nucl. Phys. **A567** (1994) 97.
18. F. Barranco, E. Vigezzi and R.A. Broglia, Phys. Lett. **B319** (1993) 387; Z. Phys. **A356** (1996) 45.
19. M. Zinser *et al.*, Phys. Rev. Lett. **75** (1995) 1719.
20. H. Esbensen, Phys. Rev. C **53** (1996) 2007.
21. P.G. Hansen, Phys. Rev. Lett. **77** (1996) 1016.
22. K. Hencken, G. Bertsch and H. Esbensen, Phys. Rev. C **54** (1996) 3043.
23. T. Aumann, L.V. Chulkov, V.N. Pribora and M.H. Smedberg, GSI-Preprint-98-17 (1998), Nucl. Phys. A. (in press).
24. H. Simon, PhD thesis, TU Darmstadt (1998); S135 Collaboration, to be published.
25. L.V. Chulkov and G. Schrieder, Z. Phys. **A359** (1997) 231.
26. B. Danilin, I. Thompson, J. Vaagen and M. Zhukov, Nucl. Phys. **A632** (1998) 383.
27. A. Cobis, D. Fedorov and A. Jensen, Phys. Rev. Lett. **79** (1997) 2411.
28. B.V. Danilin *et al.*, Phys. Rev. **C55** (1997) R577.
29. P.G. Hansen and B. Jonson, Europhys. Lett. **4** (1987) 409.
30. G. Bertsch and J. Foxwell, Phys. Rev. **C41** (1990) 1300.
31. F. Catara, C.H. Dasso and A. Vitturi, Nucl. Phys. **A602** (1997) 181.
32. F. Ghielmetti *et al*, Phys. Rev. **C54** (1996) R2143.
33. I. Hamamoto and H. Sagawa, Phys. Rev. **C53** (1996) R1492.
34. D. Sackett *et al.*, Phys. Rev. **C48** (1993) 118.
35. F. Shimoura *et al.*, Phys. Lett. **B348** (1995) 29.
36. T. Nakamura *et al.*, Phys. Lett. **B331** (1994) 296.
37. T. Aumann *et al.*, these proceedings.

Coulomb Excitation of Exotic Nuclei and Segmented Germanium Detectors

Thomas Glasmacher

*Department of Physics and Astronomy and National Superconducting Cyclotron Laboratory
Michigan State University, East Lansing, Michigan, 48824, USA*

Abstract. The experimental technique of intermediate-energy Coulomb excitation allows for in-beam γ-ray spectroscopy of β-unstable nuclei far from stability. The large velocity ($v/c \approx 0.25$-0.6) of radioactive beams prepared by in-flight separation techniques allows the use of thick secondary targets (100-1000 mg/cm^2) and measurements can be performed in a few days with beam rates as low as 10 particles/s. In a single experiment one can simultaneously measure the energy of excited bound states with respect to the ground state in exotic nuclei as well as the Coulomb excitation cross section to excite these states. This cross section is a direct function of the electromagnetic matrix elements B($E\lambda$) and B($M\lambda$) and can be used – together with the measured excitation energies – to study experimentally the evolution of nuclear shells far from stability.

Since inelastic scattering to a bound state is indicated through the presence of a photon, which is emitted from the moving projectile, the large Doppler shift of the detected photon energy requires the use of position-sensitive photon detectors. Until now, photon detectors used in intermediate-energy Coulomb excitation experiments were primarily based on scintillation detectors. This limited the achievable energy resolution to about $\Delta E/E = 6\%$. With the advent of highly-segmented large-volume germanium detectors it will be possible within the next year to perform in-beam γ-ray spectroscopy experiments with fast exotic beams and to achieve energy resolutions comparable to the intrinsic energy resolution of germanium detectors ($\Delta E/E = 0.2\%$). Segmented germanium detector arrays, including the one under construction at the NSCL, are described and their possible applications at present and future radioactive ion-beam facilities will be discussed.

INTRODUCTION

The electromagnetic excitation of nuclei is a well-established experimental method in nuclear physics. Traditionally, stable targets of interest were bombarded with heavy-ion beams well below the Coulomb barrier. With the advent of radioactive ion-beam facilities in the last decade it has become possible to extend Coulomb excitation studies to particle-unstable nuclei. Here, particle beams of interest are prepared and strike a heavy, stable target and leave the projectile and target in excited states. By measuring either the number of inelastically scattered particles directly or by measuring the γ-rays emitted at the de-excitation of a bound state one can measure the

Coulomb-excitation cross section to individual states. Through the well-established theory of Coulomb excitation (1), this cross section is directly related to the electromagnetic matrix elements $B(E\lambda)$ and $B(M\lambda)$. Coulomb excitation in inverse kinematics has been applied both at beam energies below and above the Coulomb barrier between projectile and target.

INTERMEDIATE ENERGY COULOMB EXCITATION

At intermediate beam energies (30-300 MeV/nucleon) – well above the Coulomb barrier – the technique of intermediate-energy Coulomb excitation has been applied at projectile fragmentation radioactive ion-beam facilities. In this particular technique scattering to an excited bound state is indicated through the presence of a de-excitation γ ray which is detected in a photon detector. The large beam energy allows the use of thick secondary targets (100-1000 mg/cm^2) which are still easily traversed by γ-rays (the possible absorption of low energy γ-rays needs to be accounted for). The use of thick targets offsets the limited efficiency of the γ-ray and also compensates for possibly low primary beam intensities. In order to ensure that electromagnetic interactions are dominant the scattered projectile is identified and detected at very forward angles. The maximum projectile scattering angle θ_{max} is related to the minimum impact parameter b_{min} of the projectile. A more detailed discussion of the experimental technique can be found in (2).

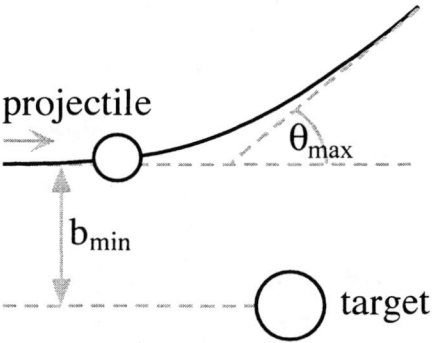

FIGURE 1. Schematic illustration of a projectile being scattering the Coulomb field of an infinitely heavy target nucleus. The minimum impact parameter b_{min} depends on the maximum scattering angle θ_{max}.

The technique of intermediate-energy Coulomb excitation has in the past few years been employed to study the structure of light neutron-rich nuclei. The only excited bound state in ^{11}Be has been extensively studied (3-5) at GANIL, RIKEN and MSU. Other studies have focused on even-even nuclei, among them a measurement of the large deformation in ^{32}Mg (6) and measurements of the quadrupole collectivities in argon, sulfur and silicon isotopes (7-9). The choice of elements lighter than calcium was motivated by the practical consideration that fully stripped secondary beams of sufficient intensity are available in this region and that the π(sd)-shell allows the study of isotopic chains halfway to the neutron dripline at current radioactive ion-beam facilities. Experimental studies performed in the π(sd)-shell are illustrated in figure 2.

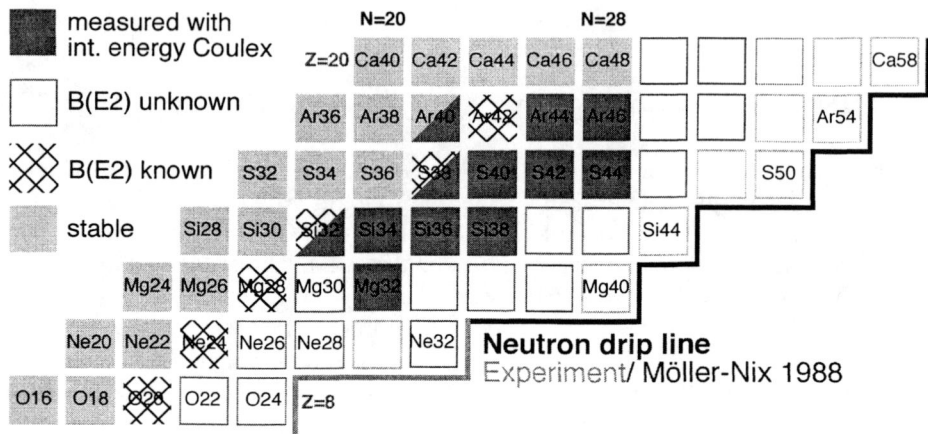

FIGURE 2. Even-even nuclei in the π(sd) shell studied with the intermediate-energy Coulomb excitation technique at RIKEN (6) and at the NSCL (7-9). For calibrations some known nuclei have also been measured.

The recent measurement of ^{56}Ni by Yanagisawa et al. presented at this conference shows that an extension of the technique to heavier nuclei is possible. In addition, the contribution by Ibbotson et al. illustrates that one is not restricted to even-even nuclei and that measurements of odd nuclei in the π(sd)-shell have sufficient energy resolution, as long as the level density in the nuclei studied is low enough.

To obtain a final γ-ray energy spectrum in the frame of the projectile, the photons measured in the laboratory have to be transformed on an event-by-event basis into the projectile frame. The Lorentz transformation depends on the beam velocity β and the scattering angle θ of the photon with respect to the scattered projectile. Thus the achievable γ-ray energy resolution does not only depend on the intrinsic energy resolution of the photon detector used, but also on the uncertainty in the beam velocity (as the projectile slows down in the target) and in the uncertainty in the scattering angle (due to the finite opening angle of the photon detector). In order to achieve large photopeak efficiencies most experiments have chosen NaI(Tl) detectors to measure the γ-rays. The intrinsic energy resolution of such scintillation detectors is about 6% at 662 keV. Detailed contributions to the achievable γ-ray energy resolution of a perfect are shown in figure 3 for secondary beam energies of 50 MeV/nucleon and 150 MeV/nucleon. At the lower beam energy the contributions to the achievable energy resolution are of about the same size as the intrinsic energy resolution for the NaI(Tl) detectors. However, if it is possible to significantly reduce the uncertainty in the scattering angle (by about a factor of 10), then the intrinsic detector resolution dominates the achievable energy resolution.

Large-volume high-purity germanium detectors (which can have intrinsic energy resolutions of 0.2% at 1332 keV) have not been used in intermediate energy Coulomb excitation experiments since their large opening angle (at "reasonable" distances from the target) would have limited the final energy resolution to values similar to the ones achievable with the more economical NaI(Tl) arrays.

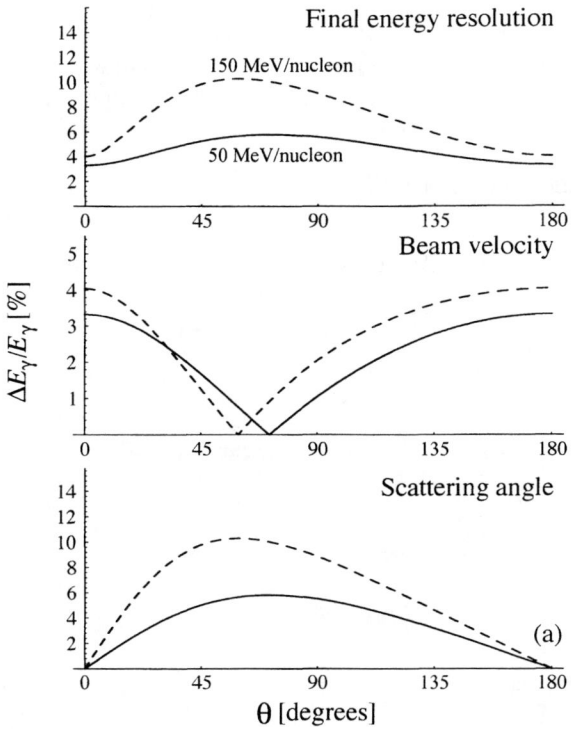

FIGURE 3. Achievable energy resolutions for a detector with perfect intrinsic energy resolution as a function of the angle between detector and scattered beam (top panel). The center panel shows the contribution due to the uncertainty of the velocity of the projectile, assuming that the velocity changes by $\Delta\beta=0.03c$ in the target. The bottom panel illustrates the uncertainty due to the finite opening angle of the detector (assumed to be $\Delta\theta=10°$). The two contributions scale linearly with $\Delta\beta$ and $\Delta\theta$, respectively.

SEGMENTED GERMANIUM DETECTORS

Advances in Germanium detector technology have made it possible to produces large-volume coaxial high-purity Germanium crystals with segmented outer contacts (10). These detectors provide a high-resolution signal in the center and information about the interaction point of the γ-rays on the segmented outer contacts. Just as in unsegmented detectors, it will be difficult to resolve double-hits as they cannot be easily distinguished from Compton scattering events. Arrays build on this technology are being built for experiments at REX-ISOLDE (Miniball) (11) and SPIRAL (Exogam). The beam velocities at these two ISOL facilities are smaller than at projectile fragmentation facilities such as the NSCL. At the NSCL we are assembling an array of eighteen 32-fold segmented detectors for experiments with fast ($\beta \approx 0.4$) exotic beams. A sketch of the detector is shown in figure 4. The large degree of segmentation of the crystals together with the large volume of the crystals will allow for in-beam γ-ray spectroscopy with fast beams and provide an energy resolution of 0.2-0.5% at 1332 keV and a photopeak efficiency of about $\varepsilon_{ph} \approx 1\%$. The array can be set up in different configurations, such that the user can optimize for each experiment the desired energy resolution versus photopeak efficiency.

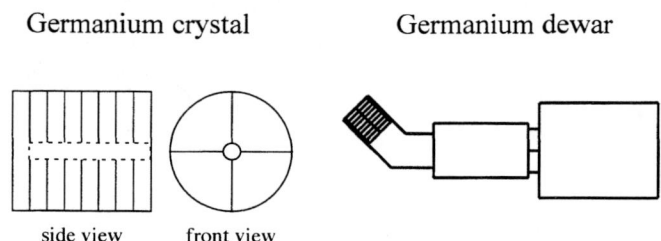

FIGURE 4. Sketch of the 32-fold segmented germanium crystal for the NSCL 18-detector HPGe array. The crystals are 8 cm long and 7 cm in diameter. They are segmented in four quarters and eight longitudinal slices, each of which is 1 cm thick. The crystals will be mounted in a dewar which has a 45 degree bend, such that γ-rays can enter from the side in experiments with fast beams and from the front in stopped-beam experiments.

ACKNOWLEDGEMENTS

This work was supported by the National Science Foundation under grants PHY-9528844 and PHY-9724299.

REFERENCES

1. Alder, K. and Winther, A., *Coulomb excitation* (Academic, New York, 1966).
2. Glasmacher, T., Ann. Rev. Nucl. Part. Phys., in print (1998).
3. Anne, R., Bazin, D., Bimbot, R. *et al.*, Zeitschrift für Physik A **352**, 397 (1995).
4. Nakamura, T., Motobayashi, T., Ando, Y. *et al.*, Physics Letters B , in print (1997).
5. Fauerbach, M., Chromik, M. J., Glasmacher, T. *et al.*, Physical Review C **56**, R1 (1997).
6. Motobayashi, T., Ikeda, Y., Ando, Y. *et al.*, Physics Letters B **346**, 9 (1995).
7. Scheit, H., Glasmacher, T., Brown, B. A. *et al.*, Physical Review Letters **77**, 3967 (1996).
8. Glasmacher, T., Brown, B. A., Chromik, M. J. *et al.*, Physics Letters B **395**, 163 (1997).
9. Ibbotson, R. W., Glasmacher, T., Brown, B. A. *et al.*, Physical Review Letters **80**, 2081 (1998).
10. Eberth, J., Thomas, H. G., Weisshaar, D. *et al.*, "Development of segmented Ge detectors for future γ-ray arrays," in *Progress in Particle and Nuclear Physics* (1997), Vol. 38, pp. 29.
11. Habs, D., Rudolph, D., Thirolf, P. *et al.*, "Physics with Ge-miniball-arrays," in *Progress in Particle and Nuclear Physics* (1997), Vol. 38, pp. 111.

Spectroscopy of ^{25}Al and ^{26,27,28}P using high-energy stripping reactions

Navin Alahari[a1], D. Bazin[a], B.A. Brown[a], B. Davids[a], G. Gervais[a]
T. Glasmacher[a], K. Govaert[a], P.G. Hansen[a], M. Hellström[b]
R.W. Ibbotson[a], V. Maddalena[a], B. Pritychenko[a], H. Scheit[a]
B.M. Sherrill[a], M. Steiner[a], J.A. Tostevin[c] and J. Yurkon[a]

[a] *N.S.C.L, Michigan State University, East Lansing 48824, U.S.A.*
[b] *Department of Physics, Univ. of Lund, Lund, P.O. Box 118 S-22100, Sweden.*
[c] *Department of Physics, Univ. of Surrey, Guildford, GU2 5XH, U.K.*

Abstract. We report here results of an application of single-nucleon stripping reactions at high energies (65 MeV/u) in inverse kinematics to obtain spectroscopic factors. From measurements of the partial cross-sections for ground and excited states in residual nuclei formed in one-proton stripping reactions, single particle orbits and occupancies of light nuclei have been studied in the s-d shell. Single proton stripping cross-sections of ^{25}Al and ^{26,27,28}P on a Be target have been measured using the S800 spectrograph and the NaI(Tl) array at the NSCL. These results indicate that this technique may provide a general tool for the intermediate energy range analogous to transfer (pick-up) reactions at low-energy.

In the past, single-nucleon transfer reactions coupled with a theoretical analysis based on the DWBA approach (for the reaction mechanism) and the shell model (for the nuclear structure) have served as an important tool for obtaining spectroscopic factors for nuclear structure studies [1]. This technique is somewhat less suitable for exotic nuclei produced in fragmentation reactions at energies in the 100 MeV/u range. We present here a novel method of obtaining spectroscopic factors from one-nucleon stripping (often referred to as knock-out reactions) at these energies. The essence of this method is measuring partial cross-sections of the various states in the residual nucleus(A - 1) populated in a one-nucleon stripping reaction from particle - γ coincidences. These are compared with theoretical calculations combining spectroscopic factors from the shell model and single particle stripping cross-sections for various j states obtained in an extension of the eikonal approximation (Glauber model). The next sections describe details of the measurements

---

[1] On leave of absence from Nuclear Physics Division, Bhabha Atomic Research Centre, Bombay 400085, India.

and the analysis. We expect that the new technique of using high-energy stripping reactions[2] will become a valuable supplement to other methods, for studying nuclei far from the valley of stability, such as measurements of interaction cross-sections, momentum distributions (of the valence nucleon or core) and Coulomb excitation.

Radioactive beams of ^{25}Al and ^{26,27,28}P with an energy of approximately 65 MeV/u and a momentum spread of 0.5% were produced in fragmentation reactions using a 100 MeV/u ^{36}Ar beam reacting on a 470mg/cm^2 Be target and were purified using the A1200 fragment separator at the National Superconducting Cyclotron Laboratory. The large acceptance S800 superconducting spectrograph, operated in a dispersion matched mode [2] in conjunction with the focal plane detector setup [3] were used to identify and measure the momentum distributions of the core fragments produced in one-proton breakup reactions on a 14mg/cm^2 Be target. Time of flight information over a distance of 70 m and energy measurements using a segmented ion chamber and a thick (5cm) plastic scintillator were used to identify and measure the yields of the core fragments in the stripping reaction. The two x/y position sensitive cathode readout drift chambers recorded the momentum and angle information of the core fragments at the focal plane of the spectrograph. The momentum and scattering angle of the fragments after the reaction were then reconstructed from the known magnetic field, position and slopes at the focal plane using the ion optics code COSY [4]. The measured parallel momentum distributions for the ^{26,27,28}P isotopes are shown in Fig. 1. Only statistical errors are

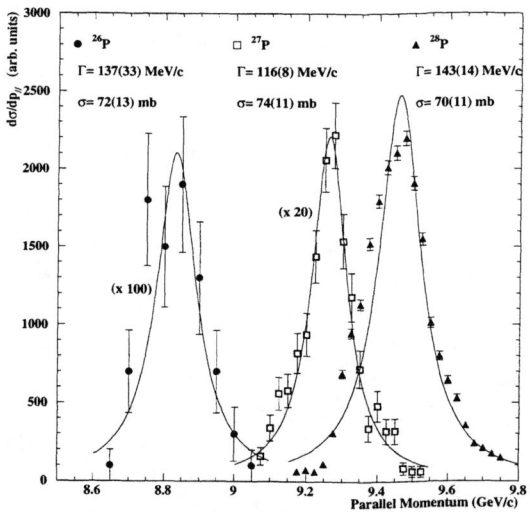

FIGURE 1. Parallel momentum distributions of the ^{25,26,27}Si cores. The continuous line are Lorentzian fits. The extracted widths (Γ) and the one-proton stripping cross-sections are also given.

---

[2]) The usage of the word stripping is historical; its first use was by Serber (1947) to describe the breakup of 90 MeV/u deuterons. See R. Serber, *Ann. Rev. Nucl. Part. Sc.* **44**, 1 (1994).

shown. The measured total cross-sections for one-proton breakup are also indicated. Gamma-rays in coincidence with the breakup events were measured using the NSCL position-sensitive NaI(Tl) array consisting of 38 detectors [5] which were placed around the target chamber. In the ^{26}P case the statistics were insufficient to obtain γ-ray information. A typical γ-ray spectrum obtained in coincidence with ^{24}Mg cores is shown in Fig. 2.

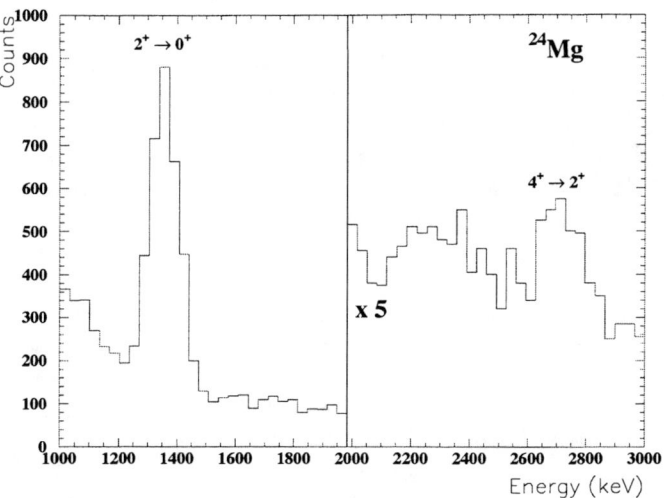

FIGURE 2. A γ-ray spectrum obtained in coincidence with ^{24}Mg cores formed from a single proton stripping from ^{25}Al. The γ-rays arising from the $2^+ \rightarrow 0^+$ and $4^+ \rightarrow 2^+$ transitions are indicated.

The shape (width) of the parallel momentum distributions of the core has been shown to be related to the spatial extent of the halo particle [6]. Shown in Fig. 1 are fits assuming a single Lorentzian form for the measured parallel momentum distribution (valid if the valence nucleon is an s state) and the extracted widths. The measured widths are seen to be around 100 MeV/c for the three isotopes. The proton separation energies S_p for ^{26,27,28}P are known to be [0.14(20)], 0.897(35) and 2.066(4) MeV respectively [7]. The phosphorus isotopes have been predicted to be candidates for proton "halos" [8] due to their expected structure of a valence $s_{1/2}$ proton coupled to the core. The momentum distributions are not expected to be as narrow as in the corresponding neutron case due to the effect of the Coulomb barrier which will create an additional effective binding. The measured widths of these distributions can be qualitatively understood in a simple manner as follows. The tail of the calculated wavefunctions [8] for the valence proton can be approximated as a Yukawa form (the corresponding momentum distribution is a Lorentzian); from its slope ($\alpha = [(2\mu E_b^{eff})]^{0.5}/\hbar$), we have extracted an "effective binding energy", E_b^{eff}, of $\simeq 1.4$ MeV + S_p. Using the extracted values of the E_b^{eff} for the valence proton, the corresponding calculated widths ($\Gamma = 2\hbar c\alpha$) of the Lorentzian are found to be similar to the measured widths. This simple exercise illustrates the important role played by the Coulomb barrier in restricting the radial extension of the valence proton. Detailed calculations using the eikonal approach to understand the shape of the momentum distribution are in progress.

The cross-sections for populating the various excited states in the residual nucleus (core) with mass (A-1) are a measure of their overlap with the initial (ground) state in the nucleus with mass A. To be able to extract spectroscopic factors for the reaction, we need to compute the single particle stripping cross-sections (σ_{sp}). These have been obtained in an eikonal approach and details of these calculations will be discussed elsewhere [9]. The measured partial cross-sections are then obtained using $\sigma(nI^\pi) = \sum_j C^2S(j,nI^\pi)\sigma_{sp}(j)$ where $\sigma(nI^\pi)$ is the stripping cross-section to a final state of spin I and parity π and C^2S is the spectroscopic factor, obtained from shell model calculations [10]. The experimental partial cross-sections have been obtained from the measured intensities of the γ-rays in coincidence with the core fragments combined with the total one-proton stripping cross-section. Shown in Fig. 3 is the simplified level scheme and the experimental and calculated feeding for the states in ^{27}Si which is the core formed in one proton stripping of ^{28}P. The calculation of the indirect feedings require the additional knowledge of γ-ray branching ratios. These and the level energies have been taken from [11]. The measured total cross-sections shown agree well with the calculated cross-sections. Listed in Table 1 are the measured and calculated partial cross-sections for the production of γ-rays from the ^{24}Mg and ^{26}Si cores.

TABLE 1. Partial cross-sections for one proton stripping from a)^{25}Al at $\simeq$ 60 MeV/u and b) ^{27}P on a Be target at 65 MeV/u. The cross-sections (for the various levels in the core) are obtained using $\sigma(n,I^\pi) = \sum_j C^2S(j,n,I^\pi)\sigma_{sp}(j)$ (see text).

γ_{j-i} Transition	Cross-section Calculated	[mb] Exp
Total (^{24}Mg)	47	84(13)
$4^+ \rightarrow 2^+$	9	8(4)
$2^+ \rightarrow 0^+$	36	42(10)
Total (^{26}Si)	82	74(11)
$2^+_2 \rightarrow 2^+_1$	19	27(14)
$2^+_1 \rightarrow 0^+_1$	42	46(18)

The calculated single particle cross-sections, $\sigma_{sp}(j)$, depend on the binding energy of the stripped particle and hence have the behaviour of the valence nucleon ("halo" or normal) built in naturally. The shape of the momentum distribution and stripping cross-sections are both very sensitive probes of the nuclear structure.

The measured momentum distribution for the ^{24}Mg core is very broad and charac-

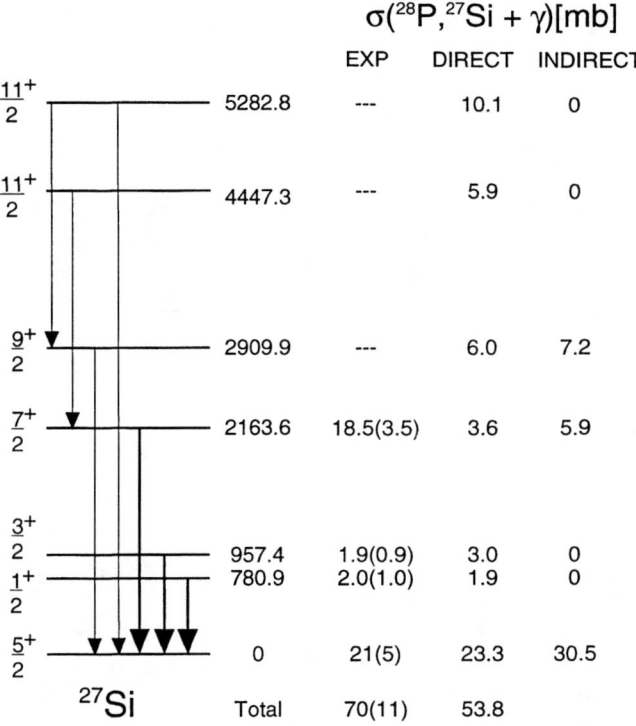

FIGURE 3. Simplified level scheme for the ^{27}Si core. The measured γ-rays are shown as thickened arrows. The partial cross-sections (mb) for the various states are given alongside. The theoretical feedings are obtained from the product of the spectroscopic factors and σ_{sp}. The indirect feedings have been obtained from the direct feeding of the higher lying states and the known γ-ray branchings. The experimental partial cross-sections represent (assuming no feeding from states at higher energy) predominantly the direct population of the relevant state. The measured cross section for the direct population of the ground state is derived from the measured momentum distributions with and without γ-rays in coincidence with the core [13].

teristic of a d wave proton in contrast to the narrower distributions for the s-wave protons in the phosphorus isotopes. But on the other hand the total one-proton removal cross-sections are, it would seem, rather surprisingly, about the same in the two cases. Qualitatively this comes about because of the strong interplay between σ_{sp} and the spectroscopic factors. Therefore one has to be very careful and quantitative in classifying proton halos in these medium mass nuclei.

The experimental results agree well with both the theoretical total stripping cross sections (summed over all bound states) and with the partial cross sections. This suggests that the new approach presented above is capable of severely constraining theoretical models of the structure of nuclei far from the valley of stability. Our method as presented, has two drawbacks. The first, inherent in the γ-ray technique, is that the partial cross sections emerge from an in-out intensity balance, which is very sensitive to possible unobserved transitions. The analysis presented in Table 1, in part, circumvents this problem by using modern theory with its high predictive power to construct the intensity pattern of the most prominent γ-rays. Thus for ^{24}Mg, with the present data, important contributions arising from the unobserved $T=1$ states near 10 MeV excitation energy have been considered. The reason for the apparent excess cross-section to the ground state is still unclear. The second drawback is a temporary one. Our experiments are based on the limited resolution of the NaI(Tl) scintillation detectors which restricts the precise extraction of the measured intensities. The NSCL is now constructing [12] an array of 18 segmented HPGe detectors, which will allow precise corrections for the Doppler broadening and give an excellent energy resolution. This array combined with the high resolution and large acceptance of the S800 spectrometer will be a powerful tool to explore nuclear structure at intermediate energies.

REFERENCES

1. G.R. Satchler, *Direct Nuclear Reactions*, Oxford Univ. Press, 1991.
2. B.L. Cohen, *Rev. of Sci. Inst.*, **30**, 415 (1959).
3. J. Yurkon, D Bazin, W. Benenson, D.J. Morrissey, B.M. Sherrill, D. Swan and R. Swanson, *Nucl. Instr. and Meth.* **A** (1998), to be published.
4. M. Berz *et al*, *Phys. Rev.* **C47**, 537 (1993).
5. H. Scheit, T. Glasmacher, R.W. Ibbotson and P. Thirolf, *Nucl. Instr and Meth. A* (1998), to be published.
6. J.H. Kelley *et al.*, *Phys. Rev. Lett.*, **77**, 5020 (1996); P.G. Hansen and A.S. Jensen, *Ann. Rev. Nucl. Part. Sci.*, **45**, 591 (1995).
7. G. Audi and A. H. Wapstra, *Nucl. Phys.* **A595**, 409 (1995).
8. B.A. Brown and P.G. Hansen, Phys. Lett. **B 381**, 391 (1996).
9. J.A. Tostevin, presented at "Nuclear structure at the Extremes", Lewes, U.K, June 17-19, 1998 and to be published.
10. B.A. Brown and Wildenthal, *Ann. Rev. of Nucl. Part. Sci.* **38**, 29 (1988).
11. P.M. Endt, *Nucl. Phys.* **A 521**, 1 (1990).
12. T. Glasmacher *et al*, to be published.
13. A. Navin *et al*, to be published.

Coulomb Dissociation of ^{19}C

T. Nakamura[1], N. Fukuda[1], T. Kobayashi[2], N. Aoi[5], H. Iwasaki[1],
T. Kubo[3], A. Mengoni[4], M. Notani[3], H. Otsu[2], H. Sakurai[3],
S. Shimoura[5], T. Teranishi[3], Y. Watanabe[1], K. Yoneda[1]
and M. Ishihara[1,3]

1 Department of Physics, University of Tokyo, 7-3-1 Hongo, Bunkyo, Tokyo 113-0033, Japan
2 Department of Physics, Tohoku University, 2-1 Aoba, Aramaki, Aoba, Sendai 980-8578, Japan
3 Institute of Physical and Chemical Research (RIKEN),
2-1 Hirosawa, Wako, Saitama, 351-0198, Japan
4 ENEA, Applied Physics Division, Via Don Fiammelli, 40129 Bologna, Italy
5 Rikkyo University, 3-34-1 Nishi-Ikebukuro, Toshima, Tokyo 171-8501, Japan

Abstract. We present the results on the Coulomb dissociation of the neutron-halo-nucleus candidate ^{19}C, which was studied by a kinematically complete measurement of the breakup of ^{19}C on a Pb target into ^{18}C and neutron at 67·AMeV. A large E1 strength has been observed at the low excitation energy of around 1 MeV to follow the distribution characteristic of the direct Coulomb breakup for the $2s_{1/2}$-dominant configuration of ^{19}C halo wave function. The angular distribution of ^{18}C+n center of mass system has led to the indirect determination of ^{19}C one-neutron separation energy, with which the excitation energy spectrum is well reproduced.

The neutron rich carbon isotope ^{19}C has drawn much attention recently due to its small one-neutron separation energy ($S_n = 240\pm100$ keV [1–4]), and hence for the possibility of a neutron halo formation. The quoted separation energy value is even smaller than the established one-neutron halo nucleus ^{11}Be that has $S_n = 504\pm6$ keV. The small separation energy is, however, not the only factor to facilitate the halo formation. The spin-parity and the configuration of the single particle state also play major roles in the formation of the halo. Shell model calculations [5–7] show that the ground state of ^{19}C has either one of the following three configurations: 1) Spin parity $J^\pi=1/2^+$ with the dominant single particle configuration ^{18}C(0^+)⊗$2s_{1/2}$. In this case, the neutron halo is developed due to the absence of the centrifugal barrier. 2) $J^\pi=5/2^+$ with dominant single particle configuration ^{18}C(2^+)⊗$2s_{1/2}$. In this case the binding energy of the valence neutron is effectively increased by the excitation energy of ^{18}C(2^+) state of 1.62 MeV, thereby causing the halo formation to be hindered. 3) $J^\pi=5/2^+$ with dominant single particle configuration ^{18}C(0^+)⊗$1d_{5/2}$. In this case the large centrifugal barrier hinders the halo formation.

^{19}C has been studied by inclusive measurements of either ^{18}C or of the neutron momentum component following the ^{19}C breakup. The longitudinal momentum distribution of ^{18}C after ^{19}C breakup has been measured at incident energy of 88 MeV/u at MSU [5,6], where a very narrow momentum width of 42MeV/c FWHM was observed. A similar experiment made at very high energy, 914 MeV/u at GSI, shows a rather broad width of 69±3 MeV/c [8]. The neutron momentum distribution with the condition of core breakup was measured at 30MeV/u at GANIL, and a narrow component of FWHM 42±17 MeV/c has been observed [9,10]. At present discrepancies between these experimental values are not well understood, and no definitive conclusion has been drawn for the structure of ^{19}C. The difficult situation may be partly due to the ambiguity of the separation energy of ^{19}C. The previously measured four values 700±240 keV [1], 50±420 keV [2], 230±120 keV [3], and -70±240 keV [4] range from 700 keV to near 0 keV, while the averaged value is 240±100 keV. With such ambiguity, the halo configuration can not be determined with sufficient reliability since the halo property is very sensitive to the separation energy. In this situation an independent experimental study of ^{19}C with a different method has been called for.

The Coulomb dissociation is a process in which the incident particle is excited by the strong Coulomb field of a high-Z target. The excited intermediate state is then broken up into a few particles. In the present experiment the ^{19}C nucleus incident on Pb is broken up into ^{18}C and a neutron. One advantageous feature of the Coulomb dissociation reaction is the simple reaction mechanism. The Coulomb dissociation process is well described by the equivalent photon method [11,12], where Coulomb excitation occurs as being induced by the absorption of a virtual photon. The Coulomb dissociation cross section, $d\sigma_{CD}/dE_x$, is described as a product of the virtual photon numbers $N_{E1}(E_x)$ and reduced transition matrix element $dB(E1)/dE_x$ as follows:

$$\frac{d\sigma_{CD}}{dE_x} = \frac{16\pi^3}{9\hbar c} N_{E1}(E_x) \frac{dB(E1)}{dE_x}. \tag{1}$$

A kinematically complete experiment, where both of momentum vectors for ^{18}C and neutron are measured on a event-by-event basis, allows us to determine the invariant mass $M(^{19}C^*)$ of the intermediate excited state. The relative energy E_{rel} is deduced as $E_{rel}=M(^{19}C^*) - M(n) - M(^{18}C)$. The excitation energy is simply related to the relative energy as $E_x=E_{rel} + S_n$.

The Coulomb dissociation method is suitable for probing the halo structure. The characteristic feature of the Coulomb dissociation of a halo nucleus is its substantial cross section of the order of 1 barn as seen in the case of ^{11}Be [13] and ^{11}Li [14–16]. This feature is attributed to the fact that the B(E1) distribution has a huge peak at very low excitation energy, say about 1 MeV. The $dB(E1)/dE_x$ distribution (and thus the excitation energy spectrum) is well described by a direct breakup mechanism as shown clearly in the case of the one-neutron halo nucleus ^{11}Be [13].

In the direct breakup mechanism, the B(E1) distribution is described by the matrix element as follows:

$$\frac{dB(\mathrm{E1})}{dE_\mathrm{x}} = |<\mathbf{q}\,|\,\frac{Ze}{A}rY_m^1\,|\,\Phi_\nu(r)>|^2 . \tag{2}$$

The initial state $\Phi_\nu(r)$ represents the halo wave function, the E1 operator involves r, the distance between the core and halo neutron, and the final state $\mathbf{q}$ represents the continuum state which is well approximated by a plane wave. An important point to be noted here is that the matrix element shown above has the form of a Fourier transform of the product of r and the halo wave function. The lower-energy part of the B(E1) distribution is in fact the reflection of the tail part wave function, so that the huge B(E1) distribution is a natural feature of the halo nuclei. The sensitiveness of the B(E1) distribution to the tail part of the wave function is useful as a spectroscopic tool of the valence neutron. For example, in the case of $J^\pi=1/2^+$, the B(E1) distribution is dominated by the $^{18}\mathrm{C}(0^+)\otimes 2s_{1/2}$ configuration. Other contributions such as $^{18}\mathrm{C}(2^+)\otimes 1d_{5/2}$ make only negligible contribution since the amplitude of the wave function at large r is small due to the large centrifugal barrier and for larger separation energy. The amplitude of the B(E1) itself thus determines the spectroscopic factor of the $2s_{1/2}$ configuration. In the case of $J^\pi=5/2^+$, its low-dense tail wave function results in a broad and small amplitude in the B(E1) distribution [17]. Such different behavior in the B(E1) distribution enables the determination of the spin parity and single particle configuration.

The Coulomb dissociation of $^{19}\mathrm{C}$ has been measured at the RIKEN projectile-fragment separator RIPS. The $^{19}\mathrm{C}$ secondary beam was produced by the $^9\mathrm{Be}(^{22}\mathrm{Ne},^{19}\mathrm{C})\mathrm{X}$ reaction at 110 MeV/u. The typical intensity of $^{19}\mathrm{C}$ was 300 cps. The $^{19}\mathrm{C}$ ions bombarded the 320 mg/cm^2 thick Pb target at 67 MeV/u. The detector setup is practically the same as the previous $^{11}\mathrm{Be}$ Coulomb dissociation experiment [13]. The outgoing $^{18}\mathrm{C}$ was bent by the dipole magnet, tracked by the drift chamber, and hit the hodoscope. The TOF and tracking information is used for a particle identification and momentum determination of $^{18}\mathrm{C}$. The outgoing neutron was measured by the 64 bars of plastic scintillator hodoscope, where the time of flight information and the position information in the detector was used to deduce the momentum vectors of the neutron. To estimate the nuclear breakup contribution in the Pb target, a carbon target run was separately made.

Figure 1 shows the excitation energy spectrum for the Pb target and C target as a function of the relative energy E_rel. As clearly seen in the spectrum the cross section for the Pb target is substantially larger than that for the C target. The breakup cross section for the Pb and C target was 1.34±0.12 b and 82±14 mb, respectively. These values are consistent with the previous inclusive measurements at MSU of 1.1±0.4 b and 105± 17 mb, respectively. In this comparison it should be noted that the present exclusive measurements favors the Coulomb breakup where both the neutron and $^{18}\mathrm{C}$ particles are emitted in a kinematically small cone, while the nuclear breakup on the C target has rather broad neutron angular distribution so that the part of the cross section for the C target with large neutron scattering

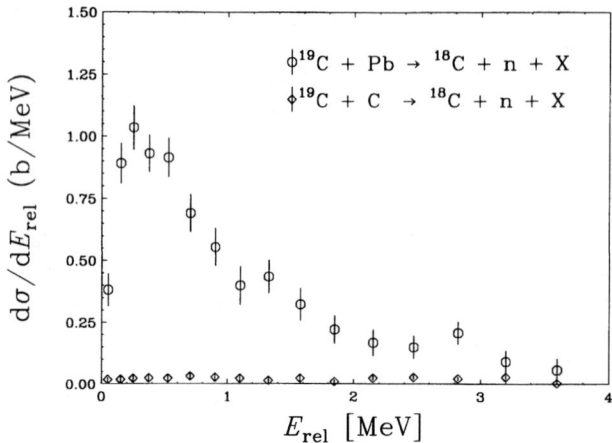

FIGURE 1. Dissociation cross section as a function of relative energy $E_{\rm rel}$ for Pb target and C target.

is not included in the present value. The smaller value of the cross section for the present case in the C target compared to the MSU value is thus reasonably understood.

The Coulomb dissociation part of the breakup cross section in Pb target is deduced tentatively assuming that the nuclear contribution in Pb target is larger by the ratio of the (projectile+target) radii between Pb and C target. The resulting spectrum is shown in fig. 2. The obtained cross section has been found to be very large (1.20±0.11 barn) with a peak at $E_{\rm rel}$ = 300 keV.

The Coulomb dissociation spectrum is analyzed in the frame of a direct breakup mechanism. As a description of the ground state to be inserted in Eq. (2), we apply three different wave functions, i.e., 1)J^{π}=1/2$^+$ with ^{18}C(0$^+$)⊗2$s_{1/2}$ configuration, 2)J^{π}=5/2$^+$ with ^{18}C(2$^+$)⊗2$s_{1/2}$ configuration, and 3)J^{π}=5/2$^+$ with ^{18}C(0$^+$)⊗1$d_{5/2}$ configuration, where each wave function is derived from a Woods-Saxon potential, adjusted to fit the separation energy. The dotted curve in fig. 2 shows the case 1) configuration with S_n=240 keV, dot-dashed curve shows the case 2) and dashed curve shows the case 3) with the same S_n. For each case, the experimental resolution $\Delta E_{\rm rel} = 0.24\sqrt{E_{\rm rel}}$ MeV is folded. The spectroscopic factor S was adjusted to fit the experimental spectrum within the condition $S \leq 1$. This condition is only met for the case 1), while the other two cases are substantially smaller than the experimental spectrum even with S=1. The first conclusion from the comparison with the data is thus that the case 2) and case 3) corresponding to the J^{π}=5/2$^+$ are excluded from the candidate of the ground state wave function of ^{19}C. Even if we modify the S_n from 100 keV to 700 keV, any mixture of 2) and 3) configurations can not reproduce the data. As for the case 1) the spectral shape can not be reproduced with S_n=240 keV. However, when we employ the higher

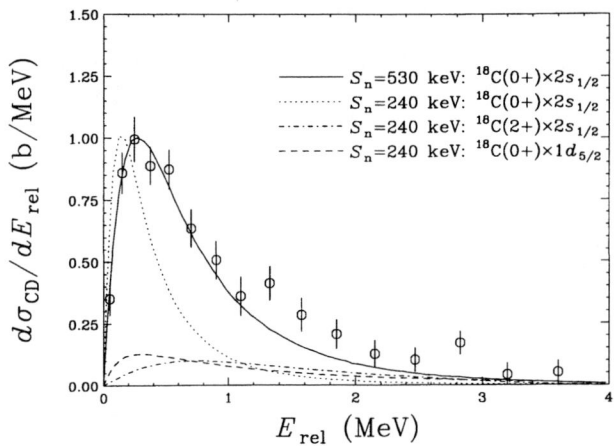

FIGURE 2. Coulomb dissociation cross section as a function of E_{rel}. The spectrum is compared with the single-particle configurations described in the text.

S_n value of 530 keV, the spectrum shape is well reproduced as shown by the solid curve. The S_n value of 530 keV is derived from the independent analysis of the scattering angle distribution of the ^{18}C+n system to be described later. Note that other configurations in $J^\pi=1/2^+$ state make a negligible small contribution to the matrix element as discussed above. The amplitude of the spectrum is thus used to determine the spectroscopic factor for $2s_{1/2}$ configuration. The deduced spectroscopic factor for the solid curve is $S=0.67$. The B(E1) strength up to 4 MeV amounts to $0.71\pm0.07 e^2\text{fm}^2$, corresponding to 1.5 W.u, which are typical values for the formation of the halo in ^{19}C.

Figure 3 shows the angular distribution of the ^{18}C+n system for the data $0 \leq E_{rel} \leq 0.5$ MeV. In the kinematically complete experiment where we measured the momentum vectors of the incident ^{19}C, outgoing ^{18}C and the neutron, the scattering angle of the ^{19}C center of mass system can be determined on a event by event basis. In the Coulomb excitation, the angular distribution is proportional to the angular distribution of the virtual photon spectrum, which is a function of excitation energy E_x, thus a function of $(E_{rel} + S_n)$. When we specify a E_{rel}, the angular distribution has a shape determined by the single parameter S_n. The preliminary analysis with the equivalent photon method shows that the angular distribution is best fitted with $S_n=530\pm130$ keV as indicated by the solid curve in fig. 3, while with $S_n=240$ keV (dot-dashed curve) averaged from the previous four mass measurements, the experimental data cannot be reproduced.

In conclusion, we have made a kinematically complete measurement of the ^{19}C Coulomb dissociation. The large cross section with a peak $E_{rel} = 300$ keV has been observed. The ground state configuration of $J^\pi=5/2^+$ is excluded. The $J^\pi=1/2^+$ with ^{18}C$(0^+)\otimes 2s_{1/2}$ configuration reproduces the data quite well with the sepa-

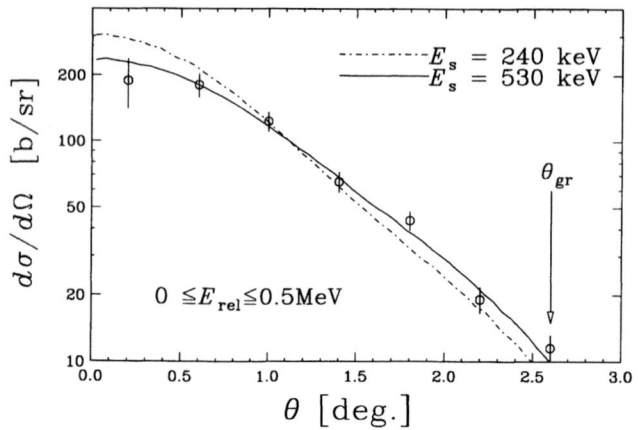

FIGURE 3. Scattering angle distribution of ^{18}C+n system on a Pb target. The scattering angle is obtained as a function of c.m angle of the ^{19}C + Pb target. θ_{gr} represents the grazing angle (2.6 deg.). The angular resolution 8.4 mrad is folded in the spectrum.

ration energy 530±130 keV derived independently by the analysis of the angular distribution of ^{19}C center of mass system. The dominance of $2s_{1/2}$ demonstrates that the ^{19}C is a one-neutron halo nucleus.

REFERENCES

1. D.J. Vieira, et al., *Phys. Rev. Lett.* **57**, 3253 (1986).
2. A. Gillibert et al., *Phys. Lett. B* **192**, 39 (1987).
3. J.M. Wouters et al., *Z. Phys. A* **331**, 229 (1988).
4. N.A. Orr et al., *Phys. Lett. B* **258**, 29 (1991).
5. D. Bazin et al., *Phys. Rev. Lett.* **74**, 3569 (1995).
6. D. Bazin et al., *Phys. Rev. C* **57**, 2156 (1998).
7. T. Tagami, Doctor Thesis, University of Tokyo (1998) (unpublished).
8. T. Baumann et al., submitted to *Phys. Lett. B*, (1998).
9. F.M. Marqués et al., *Phys. Lett. B* **381** 407 (1996).
10. N.A. Orr, *Nucl. Phys. A* **616**, 155c (1997).
11. C. Bertulani, G. Baur, *Phys. Rep.* **163**, 299 (1988).
12. J.D. Jackson, *Classical Electrodynamics, 2nd Edition* (Wiley, New York 1975).
13. T. Nakamura et al., *Phys. Lett. B* **331**, 296 (1994).
14. K. Ieki et al., *Phys. Rev. Lett.* **70**, 730 (1993); D. Sackett et al., *Phys. Rev. C* **48** 118 (1993).
15. S. Shimoura et al., *Phys. Lett. B* **348**, 29 (1995).
16. M. Zinser et al., *Nucl. Phys. A* **619**, 151 (1997).
17. D. Ridikas, M.H. Smedberg, J.S. Vaagen, and M.V. Zhukov, *Nucl. Phys. A* **628**, 363 (1998).

New evidence for parity inversion in ^{10}Li from ^{9}Li and γ-ray coincidences

M. Chartier[1,2], J. R. Beene[2], B. Blank[1,3], L. Chen[1], A. Galonsky[1],
N. Gan[2], K. Govaert[1], P. G. Hansen[1], J. Kruse[1], V. Maddalena[1],
M. Thoennessen[1], R. L. Varner[2]

[1] *National Superconducting Cyclotron Laboratory, Michigan State University,
East Lansing, MI 48824-1321, USA*
[2] *Oak Ridge National Laboratory, P.O. Box 2008, Oak Ridge, TN 37831-6368, USA*
[3] *CENBG, BP 120, Le Haut Vigneau, 33175 Gradignan Cedex, FRANCE*

Abstract. An experiment was performed at the National Superconducting Cyclotron Laboratory of Michigan State University to study the structure of neutron-unbound ^{10}Li importantly associated to the understanding of the structure of two-neutron halo ^{11}Li. The analysis of ^{9}Li-γ coincidences from the single-proton stripping of ^{11}Be demonstrates that the previously observed low decay energy virtual s-state in ^{10}Li is a low-lying state which decays to the ground state of ^{9}Li and therefore shows evidence for parity inversion in ^{10}Li.

I INTRODUCTION

The structure of neutron-unbound ^{10}Li, and especially the nature of its ground state, remains an open question and thus the complete understanding of the two-neutron halo nucleus ^{11}Li. Indeed the knowledge of the (n+^{9}Li) interaction is of essential importance to theoretical calculations of the three-body Borromean system (n+n+^{9}Li) for which the two sub-systems ^{2}n and ^{10}Li are unbound.

The first case of an intruder state in the shell structure of nuclei was the appearance of the $\frac{1}{2}^+$ state from the sd-shell as the ground state in the light p-shell nuclei instead of the $\frac{1}{2}^-$. This parity inversion has long ago been clearly observed experimentally in the case of ^{11}Be where the abnormal parity state ($\frac{1}{2}^+$) is the ground state and the lowest normal parity state is the first excited state at 320 keV. Systematics and theoretical predictions [1] presented in Figure 1 show that the energy of the lowest non-normal parity state with respect to the lowest normal parity state increases systematically with Z for isotones with N = 7 and Z > 4. It is expected that this trend should continue in the lighter p-shell N = 7 isotones, and therefore the ground state of ^{10}Li, as well as the ground state of ^{9}He, is predicted to be a neutron s-state, i.e. a non-normal parity state. Several calculations [2–4] agree on the 2^- ground state for ^{10}Li when the last $s_{\frac{1}{2}}$ neutron is coupled to the $p_{\frac{3}{2}}$ proton.

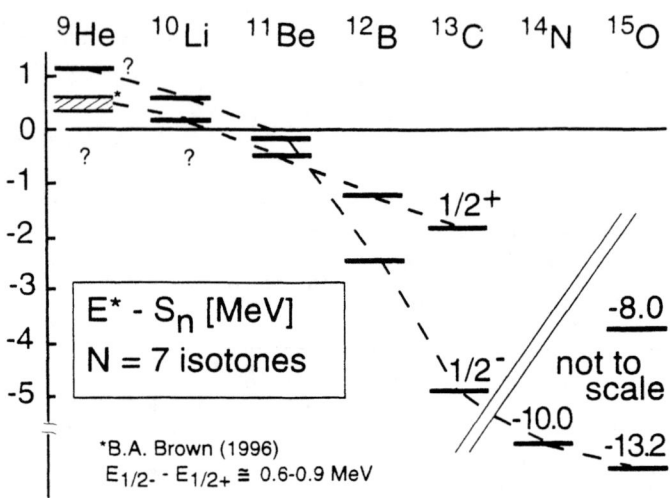

FIGURE 1. N = 7 isotone shell-model predictions [1] for the lowest normal and abnormal parity states. Parity inversion is expected in neutron-unbound ^{10}Li and ^{9}He.

The experimental landscape has so far been much more unclear. Several experiments using transfer reactions have shown evidence for a p-state around 240-540 keV, see Bohlen [5], Young [6], Caggiano [7] et al.. The first experimental evidence for a neutron virtual s-state at very low energy has been provided by the work of Kryger [8] and Yokoyama [9] et al. using the method of sequential neutron decay spectroscopy at $0°$. But the possibility that this virtual s-state could represent a decay of an excited state in ^{10}Li onto the first excited state in ^{9}Li could not be unambiguously refuted. The observation by Zinser et al. [10] of a narrow momentum distribution of the neutrons in coincidence with ^{9}Li nuclei from the single neutron stripping of ^{11}Li has been interpreted as final state interactions [11] and strongly suggested that the s-state couples to the ground state of ^{9}Li (the removal of a halo neutron is not expected to excite the ^{9}Li core). These authors extracted an s-state just above threshold at ~ 50 keV corresponding to a scattering length of ~ -20 fm.

II EXPERIMENTAL DETAILS

An experiment was recently performed at the National Superconducting Cyclotron Laboratory of Michigan State University to investigate the possibility that the low decay energy s-state is not the ground state of ^{10}Li but feeds into the first excited state of ^{9}Li with the subsequent emission of a 2.7 MeV γ-ray.

A secondary beam of ^{11}Be (46 MeV/nucleon, $\approx$ 50000 ions/s) produced from a ^{13}C primary beam and separated in the A1200 spectrometer bombarded a 300 mg/cm^2 ^{9}Be target (see Figure 2).

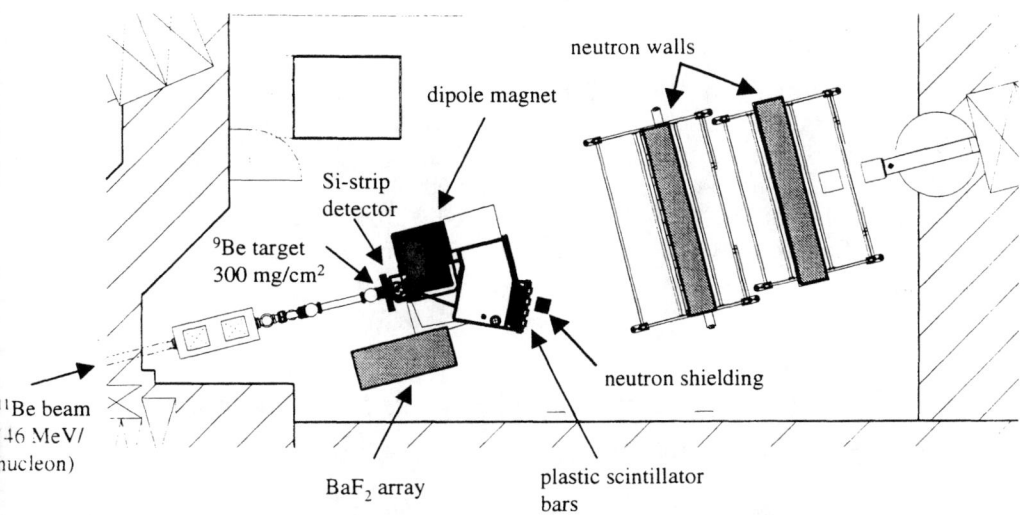

FIGURE 2. Schematic diagram of the experimental set-up.

A double-sided silicon strip detector located just after the target measured the energy loss (ΔE) and position of the charged fragments produced in the break-up of ^{11}Be. Their residual energy (E) was measured in 8 vertical plastic scintillator bars located at the exit of a deflecting magnet. Particle identification was achieved from the ΔE-E matrix, showing a clear Z separation. The mass number identification required a Monte Carlo simulation to help determine the shape of the identification gates to apply to the ΔE-E matrix. This simulation accounts for the energy loss in the target and the silicon detector as well as the angular, energy and reaction straggling.

Neutron and γ-ray coincidences from the breakup of ^{11}Be into the channels of ^{10}Li and ^{10}Be were measured using the MSU neutron walls at 0° and the BaF$_2$ array of ORNL-NSCL-TAMU at 90°. The two liquid scintillator filled neutron walls covered an area of 2 × 2 m^2 with a total efficiency of about 20 % when used in singles, and provided both position and energy measurements. The 144 BaF$_2$ crystals covered a solid angle of approximately 1.5 sr with a total efficiency of about 5 % for γ-rays of 2.615 MeV (^{228}Th) and a resolution of about 12 %.

III DATA ANALYSIS

The γ-ray and charged fragment coincidence analysis required an event-by-event reconstruction of the γ showers to add back the pair production escape peaks located in the nearest neighbouring crystals. Moreover since the γ-rays are emitted by a relatively high energy moving source ($\beta \approx 0.3$) a correction for the Doppler shift was necessary:

$$E'_\gamma = \gamma(1 - \beta cos\theta)E_\gamma \quad (1)$$

The analysis of these coincidences was optimized with the ^{10}Be-γ data which could be easily interpreted using the known level scheme of ^{10}Be. After the reconstruction and corrections mentioned above, clear evidence (see Figure 3) for a 3.37 MeV (γ_1) γ-ray from the 2^+ state and for a 2.90 MeV (γ_2) γ-ray from the decay of the 2^- state at 6.26 MeV onto the first excited state (2^+) was obtained. A weaker indication of a $\sim$ 6 MeV (γ_3) γ-ray non-observed up to now has been attributed to the decay of the 1^- state at 5.96 MeV onto the ground state (0^+).

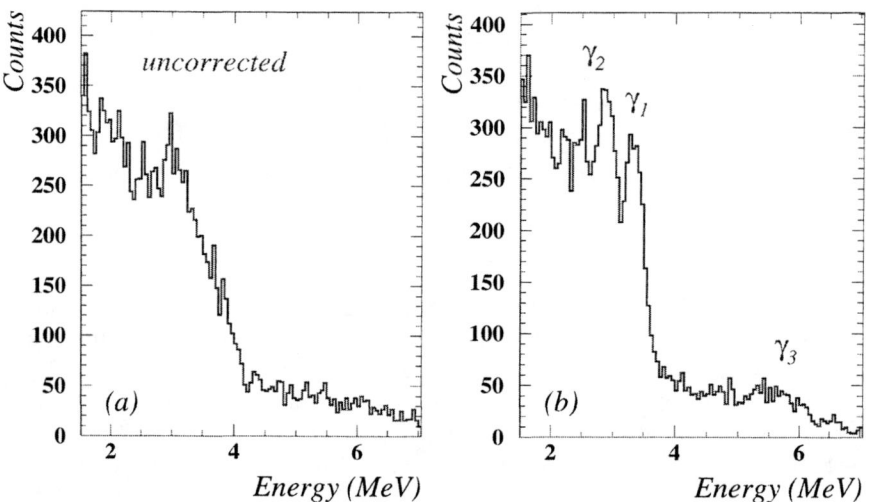

FIGURE 3. (a) Uncorrected ^{10}Be-γ coincidence spectrum. (b) ^{10}Be-γ coincidence spectrum after reconstruction of the showers and Doppler shift correction, showing evidence for 3.33 ± 0.29 (γ_1), 2.88 ± 0.29 (γ_2) and 5.64 ± 0.44 MeV (γ_3) γ-rays.

The study of the ^{10}Be-γ coincidences from the single-neutron stripping of ^{11}Be should shed light on the wave function of the ground state of ^{11}Be, the dominant part of which is known to be a $1s_{\frac{1}{2}}$. Sagawa *et al.* [1] have suggested that the parity inversion in ^{11}Be can only be explained if more complex contributions, such as the coupling of $d_{\frac{5}{2}}$ to the 2^+ of ^{10}Be, are considered. Our data, as well as those from M. Fauerbach et al. [12], show clearly that not only the core excitation to its first excited state should contribute, but also higher states such as the 2^- and the 1^-. Relative intensities for these γ-rays have been preliminary determined ($I_{2.90/3.37} \approx 0.9$, $I_{5.96/3.37} \approx 0.4$) and their final interpretation for the structure of ^{11}Be is under way.

IV RESULTS

Figure 4 shows the efficiency corrected γ-ray spectrum for ^{9}Li-γ coincidences and ^{8}Li-γ coincidences. An excess of 80 ± 10 γ-rays at 2.7 MeV are present in the ^{9}Li spectrum corresponding to population and decay of the first excited state of ^{9}Li. From this yield, a feeding of 4 ± 1 % can be extracted in the single-proton stripping reaction of ^{11}Be. The dominant error comes from the uncertainty of the fragment mass identification (a contamination of about 25 % from ^{8}Li in the number of ^{9}Li fragments used for normalization was determined by the Monte Carlo simulation mentioned above).

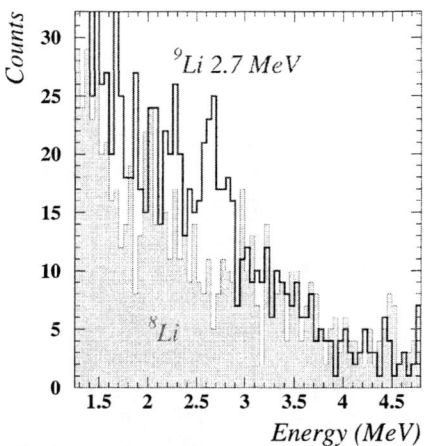

FIGURE 4. ^{9}Li-γ coincidence spectrum (black line) showing an excess of 80 ± 10 γ-rays at 2.7 MeV when the background events of ^{8}Li-γ coincidences (shaded area) are subtracted.

This branching ratio compares very well with a recent p-shell shell-model calculation [13] which leads to the formation of ^{9}Li* in the single-proton stripping of ^{10}Be in 5.2 % of the cases.

As expected, the number of 2.7 MeV γ-rays is small, therefore the triple γ-neutron-fragment coincidences were insufficient to be analyzed. The angular neutron distributions showed, similarly to the GSI and GANIL data, a dependance of the width with respect to the selected fragment (see Table 1). A narrower distribution in the case of ^{9}Li fragments results from final state interactions when an intermediate state is formed in ^{10}Li whereas in the case of ^{10}Be a wider distribution corresponds to the diffraction dissociation of the neutron. The obtained widths also support the mass identification of the fragments.

TABLE 1. Neutron-fragment coincidences: widths of the neutron angular distributions.

Fragment	Γ (MeV/c)	Fragment	Γ (MeV/c)
^{10}Be	64	^{8}Li	42
^{6}He	50	^{9}Li	36

Both, the weak population of the first excited state in ^{9}Li and the narrow momentum distribution of the neutrons in coincidence with the ^{9}Li fragments, indicating the formation of ^{10}Li, lead to the conclusion that the previously observed virtual s-state in ^{10}Li is a low-lying state which decays to the ground state of ^{9}Li. Thus we conclude that it is the ground state of ^{10}Li and that we observe, like in the case of ^{11}Be, a parity inversion in neutron-unbound ^{10}Li.

REFERENCES

1. Sagawa, H. et al., *Phys. Lett.* **B309**, 1 (1993).
2. Brown, B. A. Int. Symp. on Frontiers of Nucl. Struc. Phys., RIKEN, Japan (1993).
3. Kitagawa, H. et al., *Nucl. Phys.* **A551**, 16 (1993).
4. Popelier, N. A. F. M. et al., *Z. Phys.* **A346**, 11 (1993).
5. Bohlen, H. G. et al., *Nucl. Phys.* **A616**, 254c (1997).
6. Young, B. M. et al., *Phys. Rev.* **C49**, 279 (1994).
7. Caggiano, J. A. et al., submitted to *Phys. Rev. Lett.*, Preprint MSUCL-1103 (1998).
8. Kryger, R. A. et al., *Phys. Rev.* **C47**, R2439 (1993).
9. Yokoyama, S. K., *PhD Thesis*, Michigan State University (1996)
10. Zinser, M. et al., *Phys. Rev. Lett.* **75**, 1719 (1995).
11. Barranco, F. et al., *Phys. Lett.* **B319**, 387 (1993).
12. Fauerbach, M. et al., in preparation.
13. Timofeyuk, N., and Thompson, I., private communication (1997).

Sizes of the He isotopes deduced from proton elastic scattering measurements

J.A. Tostevin and J.S. Al-Khalili

Department of Physics, School of Physical Sciences
University of Surrey, Guildford GU2 5XH, United Kingdom[1]

Abstract. Glauber theory provides a microscopic formulation of reactions of composite nuclei at high energies. Two approaches, recently used for the treatment of the proton–He systems, are discussed and contrasted. The observed sensitivity of few-body calculations to the nuclear size and structure inputs used is discussed.

It has been demonstrated that reaction calculations which include an explicit treatment of the few-body nature of halo nuclei result in an increased transparency in their high energy collisions with *massive* targets [1]. The resulting reductions in the calculated cross sections then suggest that larger halo extensions are required to reproduce the already enhanced cross section data. Such an analysis for ^{6}He+^{12}C is consistent with a ^{6}He rms matter radius of order 2.5 fm [2]. This is encouraging since three-body models, with physically sensible inputs in each two-body channel, can produce ^{6}He nuclei which differ appreciably from this size only by over- or under-binding the two halo neutrons.

Stimulated by recent data on elastic ^{6}He and ^{8}He scattering from protons at 700 MeV/nucleon [3] we consider such few-body calculations of observables in the case of a *nucleon* target. An understanding of the sensitivity of elastic scattering and reaction cross sections to the assumed projectile structure for a nucleon target is of interest in assessing the spectroscopic value of such data.

In Glauber's multiple scattering theory the proton+A elastic amplitude, for incident proton wave number k and momentum transfer $\boldsymbol{q}$, is the integral over proton–target center of mass (c.m.) impact parameters

$$f(\boldsymbol{q}) = \frac{ik}{2\pi} \int d^2\boldsymbol{b}\, e^{i\boldsymbol{q}\cdot\boldsymbol{b}} \left[1 - \mathcal{S}_A(b)\right], \qquad \mathcal{S}_A(b) = \langle \Phi_A | \prod_{j=1}^{A} S_j(b_j) | \Phi_A \rangle, \qquad (1)$$

where b_j is the incident proton impact parameter on target nucleon j. The profile function $\mathcal{S}_A$ (the eikonal elastic S-matrix) is thus an A-body matrix element

---

[1] This work is supported by the United Kingdom Engineering and Physical Sciences Research Council through grant GR/J95867

of the (translationally invariant) target ground state wave function Φ_A. $\mathcal{S}_A$ also determines the reaction cross section observable. The $S_j(b_j) = 1 - \Gamma_{pj}(b_j)$ specify completely the nucleon-nucleon scattering/dynamics input, have been determined [4,5] from fits to small angle pp and pn scattering data, and are assumed here to be spin-independent.

The required nuclear structure input is the target *many-body* density integrated over spin coordinates $\rho_A(\boldsymbol{x}_1, \ldots, \boldsymbol{x}_A) \equiv \langle \Phi_A | \Phi_A \rangle_{\text{spins}}$, a function of $A - 1$ target nucleon position coordinates $\boldsymbol{x}_j$ referred to the target center of mass. Dependent on the structure model used, ρ_A will contain cluster, dynamical, antisymmetrisation, and/or short range correlations in addition to the (already assumed) c.m. correlations.

I APPROXIMATION SCHEMES

The p+^{6,8}He scattering data of [3] have been analysed using a minimally (c.m.) correlated density description [3] and a few-body description [6] of the nuclear structure. To be definite we write equations in the case of ^{6}He, a Borromean two-neutron halo nucleus with a well developed α+n+n three-body structure. We think it useful to present clearly the approximations used in the two cases.

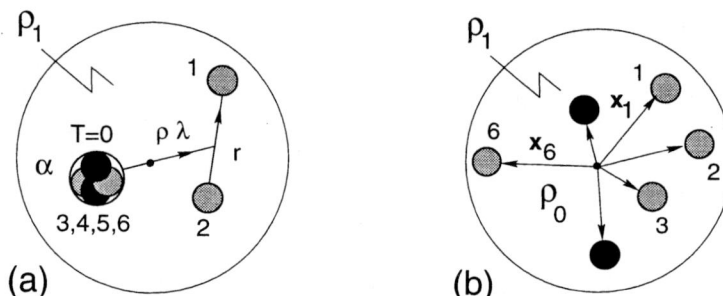

FIGURE 1. Schematic representation of the treatments of ^{6}He (a) in the few-body model, and (b) when including only c.m. correlations (see text). It is understood that the overall ^{6}He one-body density $\rho_1(x)$ is the same in the two cases and that $\sum_i \boldsymbol{x}_i = 0$.

In the few-body model, ^{6}He is treated as shown schematically in Fig. 1(a). The α+n+n relative motion wave function $\psi^{(3)}_{\text{rel}}$ is obtained by solution of a three-body Schrödinger equation. Denoting the orbital angular momenta in coordinates $\boldsymbol{r}$ ($\boldsymbol{\rho}$) by l (λ), with total L, the dominant components ϕ_L in the ^{6}He ground state have $L(=\lambda=l)=0$ and 1. The ^{6}He many-body density is then [7]

$$\rho_6(\boldsymbol{x}_1, \ldots, \boldsymbol{x}_6) = \langle |\psi^{(3)}_{\text{rel}}(\boldsymbol{\rho}, \boldsymbol{r})|^2 \rangle_{\text{spins}} |\Phi_4|^2$$

$$\langle |\psi^{(3)}_{\text{rel}}(\boldsymbol{\rho}, \boldsymbol{r})|^2 \rangle_{\text{spins}} = \frac{1}{(4\pi)^2} \left[\phi_0^2(\rho, r) + \phi_1^2(\rho, r) - \phi_1^2(\rho, r) P_2(\hat{\boldsymbol{\rho}} \cdot \hat{\boldsymbol{r}}) \right] \quad (2)$$

where Φ_4 is a (translationally invariant) α particle wave function and the assumed clustering in the projectile is clear. The corresponding one-body density $\rho_1(x)$ (normalised to unity) and hence the rms radius of ^{6}He can be computed as detailed in [1]. The profile function is

$$\mathcal{S}_6(b) = \int d\boldsymbol{\rho} \int d\boldsymbol{r} \, \langle |\psi_{\text{rel}}^{(3)}(\boldsymbol{\rho},\boldsymbol{r})|^2 \rangle_{\text{spins}} \, \mathcal{S}_4(b_\alpha) \mathcal{S}_n(b_1) \mathcal{S}_n(b_2) \, , \qquad (3)$$

where $\mathcal{S}_4(b_\alpha)$, given by Eq. (1), is the free p+α elastic S-matrix at the same incident energy per nucleon and should be consistent with such data [6]. Alpha particle core polarisation effects are assumed to be negligible [8]. These formulae summarise the physical basis of the presented few-body calculations for ^{6}He. For ^{8}He, Eq. (2) is revised to use instead a $\langle |\psi_{\text{rel}}^{(5)}|^2 \rangle_{\text{spins}}$ in $\rho_8(\boldsymbol{x}_1,\ldots,\boldsymbol{x}_8)$ [6,9].

At the other extreme, if one neglects *all* correlations in the ^{6}He ground state, the uncorrelated 6-body density consistent with a given one-body density $\rho_1(x)$ is $\tilde{\rho}_6(\boldsymbol{r}_1,\ldots,\boldsymbol{r}_6) = \prod_{j=1}^{6} \rho_1(r_j)$, where all $\boldsymbol{r}_j$ are independent and refer to a fixed center. This is certainly inadequate for light systems, e.g. [5]. Including c.m. correlations (only) requires the use of the minimally correlated many-body density

$$\rho_6(\boldsymbol{x}_1,\ldots,\boldsymbol{x}_6) = \mathcal{N} \prod_{j=1}^{6} \rho_0(x_j) \, \delta(\sum_{i=1}^{6} \boldsymbol{x}_i) \qquad (4)$$

where the imposed c.m. constraint is explicit and where $\mathcal{N}$ is a normalisation. Clearly $\rho_0(x)$, see Fig. 1(b), is the assumed position probability density of each nucleon about the ^{6}He c.m. such that the resulting ρ_6 derives a given one-body density $\rho_1(x)$. For a proper assessment of the importance of the cluster and other correlations explicit in Eq. (2) it would be helpful to compare calculations of different observables when using Eqs. (2) and (4). The relationship of ρ_0 to ρ_1 however is non-trivial and calculations using Eq. (4) have not been performed. Only when ρ_1 has a Gaussian form are the c.m. effects easily treated [5].

In the analysis presented in [3] the above mentioned c.m. correlations are included, but only approximately. In that analysis a number of model one-body densities ρ_1 are assumed for the ^{6}He and ^{8}He systems. In all cases the c.m. correlations are nevertheless treated as if ρ_1 *is* a Gaussian distribution with the rms matter radius of ρ_1. The accuracy of this procedure is untested, particularly for halo nuclei like ^{6}He, where an essential feature of realistic wave functions with Borromean three-body asymptotics will be an extended component in the density.

II CALCULATIONS OF OBSERVABLES

In [3], on the of basis fits to the experimental data obtained using the approximate (minimally correlated) theoretical model above, it is concluded that the p+^{6,8}He elastic scattering data determine "essentially model independent" values for the rms radii for the He isotopes. Values are quoted with small errors. We show that this is manifestly not the case and that reaction calculations are highly sensitive to details of the structure (wave function) inputs beyond simply their rms radii.

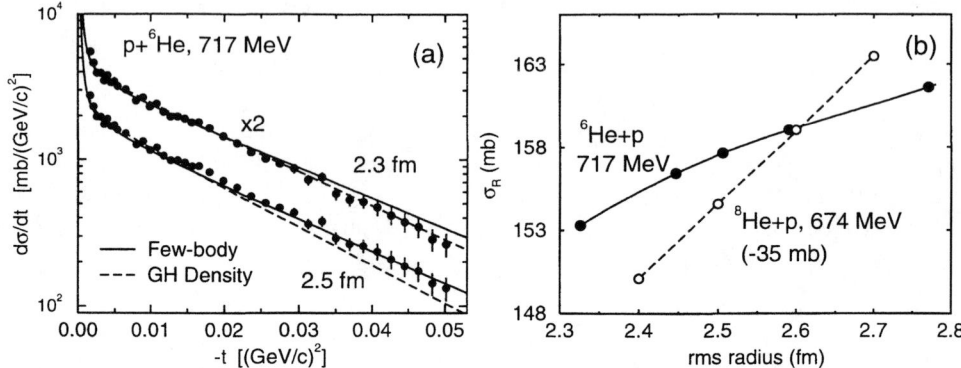

FIGURE 2. (a) Calculated and experimental elastic differential cross sections versus q^2 ($=-t$) for p+^{6}He at 717 MeV. (b) Calculated reaction cross sections for p+^{6}He, ^{8}He as a function of the nuclear rms radius.

Fig. 2(a) contrasts the results of the few-body approach and those of [3] for p+^{6}He. The solid curves are the few-body results for ^{6}He structures with rms radii of 2.33 fm (upper) and 2.5 fm (lower). These include cluster correlations and realistic 2n-halo asymptotics. The dashed curves show the results of the approach of [3] using model (GH) ^{6}He densities with radii of 2.3 fm (upper) and 2.5 fm (lower). The results from the two models are quite different. The GH density-based calculations suggest the radius of 2.5 fm is too large (manifest as too steep a cross section with q^2). On the other hand the ^{6}He wave function with this rms radius reproduces the measured ^{6}He+^{12}C reaction cross section [2] and is consistent with the elastic scattering data within the few-body analysis.

The calculated reaction cross sections from the few-body model are shown in Fig. 2(b) for both ^{6}He and ^{8}He systems. Calculations of this observable when using the approximate treatment of c.m. correlations [3] have not been presented. Based on results for ^{12}C targets [1,2], it is expected that this observable will show considerable sensitivity to the cluster correlations included in the few-body approach. A comparison of such calculations would therefore be interesting. The sensitivity of the reaction cross section to rms radius is significant and a measurement of this observable could be very valuable.

The differences noted in Fig. 2(a) may arise from many sources since, within the few-body model, c.m. correlations are treated exactly, we include the granular nature of the nucleus and so use wave functions with realistic asymptotics. However, since these wave functions are exact solutions of three-body calculations, specific features of the wave function are not easily controlled to assess different sensitivities.

For instance, the wave function with rms radius 2.33 fm used in Fig. 2(a) has 2n-separation energy ≈ 1.2 MeV. While it has realistic three-body asymptotics the spatial fall off will be incorrect in detail.

For ^{8}He we use the COSMA wave function for $\psi_{rel}^{(5)}$ which gives a simple expression for $\rho_8(x_1,\ldots,x_8)$ [9] while including exactly c.m., cluster, and those correlations associated with the antisymmetrisation of the four valence neutrons, amongst themselves. Each is assumed in a $p_{3/2}$ orbital with respect to the alpha core. In the original COSMA model these have oscillator radial wave functions. Here we also match these functions appropriately to (p-wave) Hankel function tails for an assumed single particle separation energy. This simple wave function is now flexible enough to allow construction of families of ^{8}He wave functions with the same asymptotic forms and different rms radii, or with the same rms radius and different asymptotic forms. The results of calculations using such wave functions are

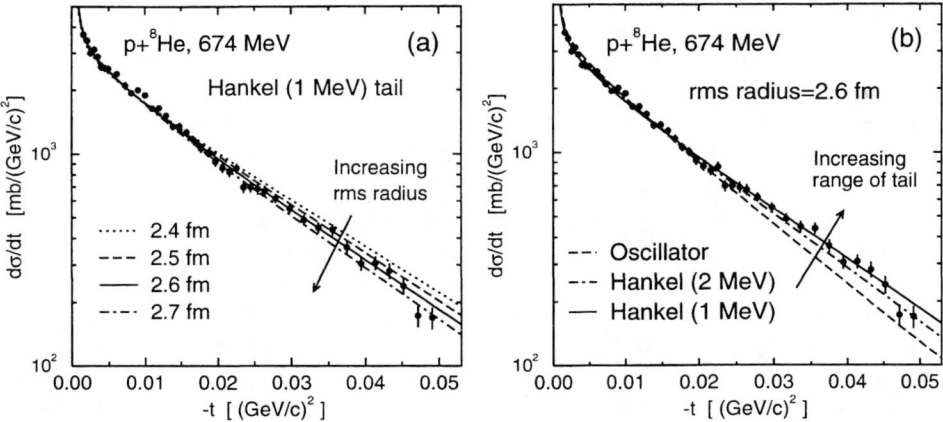

FIGURE 3. Calculated and experimental elastic differential cross sections versus q^2 ($=-t$) for p+^{8}He at 674 MeV for wave functions with (a) different rms radii but a fixed asymptotic form, and (b) a fixed rms radius but different asymptotic forms, for the p-wave valence neutrons.

shown in Fig. 3. Fig. 3(a) shows the calculated differential cross sections for fixed valence nucleon asymptotics, a Hankel function of 1 MeV separation energy, and the rms radii indicated. An increase in the slope of the cross section with rms radius is obtained, suggesting that high quality data might accurately determine an rms size. Evident from Fig. 3(b) however is that calculations for wave functions with this same rms radius, but different functional asymptotic forms for the valence n wave functions, show greater variation. Moreover, extending the range of the asymptotics, from Gaussian toward less weakly bound Hankel forms, the calculated slopes of the differential cross section move in the opposite direction to those in Fig. 3(a). It follows that a suitably chosen wave function with a small rms radius and Gaussian asymptotics can produce a similar result to a wave function with larger rms radius and Hankel function asymptotics.

This clarifies a very basic model dependence in the differential cross section calculations. While the experimental data can be used to assess the consistency of the data and a given structure model, one requires a confidence in the nuclear structure model used, in its asymptotics, and in the full and accurate treatment of this structure in the reaction calculation, to go further. We believe our few-body model treatment of both the structure and the scattering of ^{6}He is the most accurate attempt yet to do so. It would be interesting to engineer three-body wave functions for ^{6}He with the same (physical) 2n-separation energy but different rms sizes to delineate more carefully the sensitivity to the rms size in this case. The practicality of performing 6-body ^{6}He calculations using Eq. (4), to clarify the role of correlations beyond the trivial c.m. effects should also be considered.

III SUMMARY

We have discussed the 'few-body' and 'minimal correlations' models as applied in analyses of high energy proton elastic scattering from the helium isotopes. We show that the few-body calculations reveal very significant dependence on the structure model assumed and that the available elastic scattering data do not, by themselves, determine the rms radii of these isotopes in any model independent sense. Few-body calculations show clear sensitivity to the rms radius but also to the wave function asymptotics assumed in calculating the target many-body density. The use of simple model descriptions of the structures will thus lead to significant ambiguities in extracted spectroscopic information. Within the few-body model description for both the structure and the reaction, the elastic scattering data are consistent with a ^{6}He rms radius of 2.5 fm, and so with few-body calculations of the ^{6}He+^{12}C reaction cross section. To assess the specific role played by the clustering in the system it will be necessary to formulate differential and reaction cross section calculations which are able to treat exactly the c.m. correlations only. The observed sensitivity to the asymptotic behaviour of the wave functions suggests a more sophisticated wave function is needed in the case of the ^{8}He system.

REFERENCES

1. Al-Khalili J.S. and Tostevin J.A. *Phys. Rev. Lett.* **76**, 3903 (1996); Al-Khalili J.S., Tostevin J.A. and Thompson I.J. *Phys. Rev.* C **54**, 1843 (1996).
2. Tostevin J.A. and Al-Khalili J.S. *Nucl. Phys.* A **616**, 418c (1997).
3. Alkhazov G.D. et al., *Phys. Rev. Lett.* **78**, 2313 (1997).
4. Ray L. *Phys. Rev.* C **20**, 1857 (1979).
5. Alkhazov G.D., Belostotsky S.L. and Vorobyov A.A., *Phys. Rep.* **42C**, 89 (1978).
6. Al-Khalili J.S. and Tostevin J.A. *Phys. Rev.* C **57**, 1846 (1998).
7. Al-Khalili J.S., Thompson I.J. and Tostevin J.A. *Nucl. Phys.* A **581**, 331 (1995).
8. Kuo T.T.S., Krmpotić F., and Tzeng Y. *Phys. Rev. Lett.* **78** (1997) 2708.
9. Zhukov M.V., Korsheninnikov A.A. and Smedberg M.H. *Phys. Rev.* C **50**, R1 (1994).

New neutron-rich isotope ^{31}F and particle instability of ^{25}N and ^{28}O

H. Sakurai[1], S.M. Lukyanov[1,2], M. Notani[1], N. Aoi[3],
D. Beaumel[1,5], N. Fukuda[3], M. Hirai[3], E. Ideguchi[1], N. Imai[3],
M. Ishihara[1,3], H. Iwasaki[3], T. Kubo[1], K. Kusaka[1], H. Kumagai[1],
T. Nakamura[3], H. Ogawa[4], Yu.E. Penionzhkevich[2], T. Teranishi[1],
Y.X. Watanabe[3], K. Yoneda[3], A. Yoshida[1]

[1] *The Institute of Physical and Chemical Research (RIKEN), 2-1 Hirosawa, Wako, Saitama 351-01, Japan*
[2] *Flerov Laboratory of Nuclear Reactions, Joint Institute for Nuclear Research, 141980 Dubna, Moscow region, Russia*
[3] *Department of Physics, University of Tokyo, 7-3-1 Hongo, Bunkyo, Tokyo 113, Japan*
[4] *Department of Applied Physics, Tokyo Institute of Technology, Oh-Okayama 2-12-1, Meguro-ku, Tokyo 152, Japan*
[5] *Institute de Physique Nucléaire, Orsay Cedex, France*

Abstract. The neutron drip line determination up to fluorine has been performed by projectile fragmentation of a 94.1AMeV ^{40}Ar beam at the fragment separator RIPS at RIKEN. A new neutron-rich isotope ^{31}F has been observed for the first time while clear evidence for the particle instability of ^{25}N and ^{28}O has been obtained. The sudden change of stability from oxygen to fluorine may demonstrate the onset of a deformation for the neutron-rich fluorine isotopes.

The experimental determination of the neutron drip line is essential for understanding the nuclear stability in extreme isospin asymmetry. It has been believed that the neutron drip line has been mapped to fluorine ($Z=9$) [1] or neon isotopes ($Z=10$) [2], based on the experimentally observed heaviest isotopes in combination with mass predictions [3]. However, the definite experimental proof for the stability beyond these nuclei, in particular for nitrogen, oxygen, and fluorine isotopes, is rather difficult due to the low yields of the extremely neutron-rich nuclei.

At RIKEN, we searched for new neutron-rich isotopes in Ne-Al region, and found new isotopes, ^{31}Ne, ^{37,38}Mg, ^{40,41}Al [4,5]. One of important results is the stability of ^{31}Ne. In a previous work [6], ^{31}Ne had been reported to be particle instable in accordance with many mass formulae [3]. ^{31}Ne is located at the deformation region around $Z{\sim}11$ and $N{\sim}20$, so called "island of inversion" region, where a tendency towards prolate deformation in spite of the effect of spherical stability due to magicity of the neutron number 20 has been emerged [7–10]. A highlight of recent such investigation is the observation of a large deformation of the $N=20$ nucleus ^{32}Mg [11]. Some of theoretical treatments [9,12] incorporating deformation effect tend to predict particle stability of ^{31}Ne.

Of interest is how the deformation region is extended to lower Z at $N \sim 20$. The present adopted heaviest fluorine isotope ^{29}F [13] with $N=20$ is in harmony with various mass formulas [3] even without deformation effects included. It should be noted, however, that the particle stability of ^{31}F with $N=22$ is predicted by some theoretical works including deformation effects [9,12] which also predicts the particle stability of ^{31}Ne.

The stability of a double magic nucleus ^{28}O has recently drawn attentions, even though the particle instability of ^{25,26}O was clearly shown in two experiments [6,14]. Due to the double magicity of ^{28}O, the stability has been studied in several theoretical papers [3,8,12,15–17], but the stability predictions disagree with each other. Recently, an attempt to search for ^{28}O has been made at GANIL by using an enriched ^{36}S beam [18]. The conclusion about particle instability of ^{28}O has been made. However, the expected yield of ^{28}O was estimated in an extrapolation of yields of bound neighboring nuclei by a fitting procedure. More precise arguments on the expected yield may require more statistics.

This report presents the results of a recent experiment to synthesize and identify extremely new neutron-rich nuclei in the region of nitrogen-fluorine with $N \leq 22$ [19]. Such productions and observations have been made possible by combination of a high-energy beam 94.1A MeV ^{40}Ar and the RIKEN Projectile Fragment Separator (RIPS) [20] at RIKEN. The ^{40}Ar projectile is inferior to ^{36}S in terms of production cross sections for very neutron-rich fragments. However, compared with ^{36}S, ^{40}Ar has more dynamic range in production of neutron-rich nuclei up to $N=22$, as well as higher beam intensity and energy. The use of RIPS is powerful in collecting reaction products owing to its large momentum acceptance (6%) and solid angle (5msr), as well as sizable maximum magnetic rigidity of 5.76 Tm. With all of these effects combined, yield rates of about one order of magnitude higher than in the similar experiment [18] were achieved.

The ^{40}Ar beam after acceleration up to 94.1A MeV at the AVF and RIKEN Ring Cyclotron reacted with a 690 mg/cm^2 thick ^{nat}Ta target. The target thickness was chosen according to a semiempirical yield estimation by the INTENSITY code [21] that predicts the secondary beam intensities. The primary beam current was monitored by an array of plastic detectors located near the production target.

The reaction fragments were collected and analyzed by the RIPS operating at an achromatic mode and at the maximum values of momentum acceptance and solid angle. The magnet rigidity of the RIPS spectrometer was optimized for ^{28}O isotope according to the INTENSITY calculation. To reduce the rates of light isotopes, such as proton-lithium isotopes, an aluminium wedge with thickness of 226.2 mg/cm^2 was used at the momentum dispersive focal plane (F1).

Particle identification was performed by a standard method on the basis of the energy loss (ΔE), total kinetic energy (TKE), time-of-flight (TOF) and magnetic rigidity ($B\rho$) measurements for each fragment [4,5].

The positions of the fragments at the momentum dispersive focal plane (F1) were measured using a parallel plate avalanche counter (PPAC) in order to determine the $B\rho$ value. The sensitive area of this PPAC was 15cm(H)×10cm(V), which covered a full rigidity acceptance of 6%. The charge division method was applied for particle position measurement. To minimize δ-ray effects causing worse position resolution the PPAC was equipped with four independent cathodes stripped horizontally. It allows us to reject spurious position information due to δ-rays and to reach a high level of detection efficiency by imposing a condition of more than two hits out of the four cathodes. A typical efficiency for a $Z=4$ particle was over 95%.

The fragmentation products reached the final focal point (F3) of the RIPS spectrometer, where a plastic scintillation counter (F3-Pl) with thickness of 1.0 mm and a telescope consisting of 6 lithium-drift silicon detectors (SSD) were installed. A plastic veto counter with thickness of 1mm was located after the SSD telescope. The thicknesses of the silicon detectors were 1.1, 1.1 , 3.1, 2.9, 4.0 and 4.3 mm, respectively. The thicknesses of the detectors were chosen so as to stop the fragments of interest in the fifth detector. The active area of the first four detectors was 48mm×48mm and last two had sensitive diameter of about 35mm. The time-of-flight (TOF) of each fragment was determined from the F3-Pl timing and the RF signal of the cyclotron. The typical value of TOF was about 250 ns. Each SSD detector provided independent energy-loss values ΔE, while all telescope provided a total kinetic energy (TKE). The veto counter rejected pile-up events due to light fragments.

The atomic number Z of a fragment was determined from ΔE at the first four detectors and from the TOF information. The accuracy in Z determination was about 0.8% for nitrogen isotopes. The mass-to-charge ratio (A/Q) was obtained with an accuracy of 0.5% for the same isotopes. The fragment charge (Q) was obtained from the combination of TKE, $B\rho$ and TOF. In the analysis, we employed fully stripped fragments by imposing a condition ($Z - Q$) with a width of $|Z - Q|/Z \leq 6.5\%$.

Figure 1 shows a two-dimensional scatter plot, A/Z versus Z, obtained from the data accumulated for four days with a mean primary beam intensity of 45 pnA. Numbers of events accumulated for the isotopes ^{19}B, ^{23}N and ^{22}C were about 4400, 7100 and 900, respectively. A new isotope ^{31}F (8 events) was observed for the first time. The absence of events corresponding to ^{26,27,28}O isotopes as well as ^{24,25}N and ^{30}F is clearly visible. For instance, in the case of particle stability of ^{28}O or ^{25}N, the associate events are expected to appear inside the ellipses in Fig.1.

The non-observation of an isotope cannot carry the absolute proof of its unbound character. To achieve a definite evidence, we plotted the observed yields versus Z value for the nuclei with $N=2Z+4$, as shown in Fig.2. A smooth monotonous decrease of the experimental yields is observed. The solid line in Fig.2 represents the yields expected by the INTENSITY that includes the EPAX parameterization for production cross sections [22]. We modify a parameter U in the EPAX from the standard value of 1.5 to 1.6 for this reaction, as required to reproduce yields for very neutron-rich fragments. Such modification has been also made in Refs. [18,23].

The interpolation between the yields of the observed nuclei with $N = 2Z + 4$ enables us to estimate the ^{25}N and ^{28}O yields to be about 240 and 40 events, respectively. The distinct deviation of the estimated event rates as well as the fact of no events in the regions of ^{28}O and ^{25}N provide strong evidence to conclude that ^{28}O and ^{25}N are particle instable. In the recent work [18], the expected yield of ^{28}O was deduced from an extrapolation procedure on the basis of measured yields for $N=20$ isotones. However, this type of extrapolation requires good enough predictions for production cross sections of fragments. In contrast, our evidence obtained from the interpolation is essentially free of any assumptions on fragment production mechanism.

We found that the heaviest nitrogen and oxygen isotopes are ^{23}N and ^{24}O with the same neutron number $N=16$, while the heaviest isotope of fluorine has been

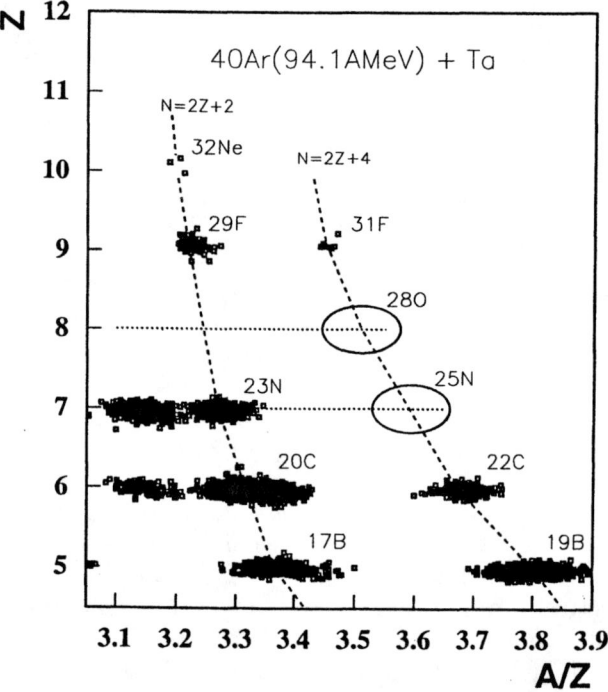

FIGURE 1. Two-dimensional A/Z versus Z plot, which was obtained in the reaction of a 94.1AMeV ^{40}Ar beam on a 690 mg/cm^2 tantalum target during a 4-day run. A new isotope ^{31}F is clearly visible (8 events). No events associated with ^{25}N and ^{28}O as well as ^{24}N, $^{25-27}$O and ^{30}F were obtained. In the case of particle stability of ^{28}O or ^{25}N, the associate events are expected to appear inside the ellipses. The dashed lines are drawn guide the eye for the isotopes with the same neutron numbers $N = 2Z + 2$ and $N = 2Z + 4$.

extended up to ^{31}F with $N=22$. Concerning the instability of ^{24,25}N, mass formula predictions [3] are in agreement with our results. On the other hand, the locations of oxygen and fluorine drip line are not predicted well by mass formulas, most of which predict that the heaviest oxygen isotope is ^{26}O and that of fluorine is ^{29}F.

The large enhancement of nuclear stability from oxygen to fluorine may imply that a shape transition from spherical shape to deformed shape appears between oxygen and fluorine. Due to the double magicity of ^{28}O, ^{28}O can be assumed to have a spherical shape. A few mean field calculations [16,17] predict the binding energy in a spherical coordinate. One of them, the Complex Scale Hartree-Fock model [17], predict that both ^{26}O and ^{28}O are unbound. In addition, ^{28}O are predicted to be unbound by two shell models; a shell model with the limited valence space up to sd shell [15] and another shell model which has wide valence spaces including sd and fp spaces but rather weak interactions [8]. These predictions suggest that ^{28}O instability can be understood in a spherical shape or weak mixing of sd and fp orbits.

On the other hand, ^{31}F stability requires a large deformation in theoretical works. Among the mass formulas, only the finite-range droplet model (FRDM) [12] predicts the stability of ^{31}F. The FRDM includes the nuclear deformation effects for both the macroscopic and microscopic parts. The FRDM also predicts the particle stability of ^{31}Ne, which has been experimentally confirmed in Ref. [4]. The stability of ^{31}F is also predicted by a shell model including fp valence spaces with rather strong interactions [9]. This model requires a large configuration mixing in sd and fp orbitals to explain the known ground-state spin of ^{29,31}Na, and hence predicts

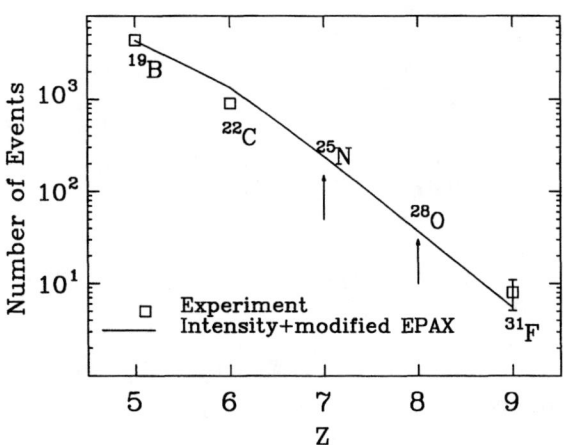

FIGURE 2. Isotopic production rates for nuclei with the neutron number $N = 2Z + 4$. The solid curve presents expected rates according to the INTENSITY using the modified EPAX parameterization. The expected yields for ^{28}O and ^{25}N are indicated by the arrows.

the particle stability of ^{31}F as well as ^{31}Ne. Consequently, both the FRDM formula and the PR shell model demonstrate that the stability of ^{31}F can be related to an enhanced binding energy due to the deformation effects.

In summary, the new neutron-rich isotope ^{31}F was observed for the first time using the RIPS spectrometer at the reaction of 94.1AMeV ^{40}Ar+^{nat}Ta. The clear evidence of the particle-unbound character of ^{25}N and ^{28}O was obtained on the basis of the interpolation of experimental yields between ^{19}B, ^{22}C and ^{31}F. In addition, the particle instability for ^{24}N, ^{25,26,27}O and ^{30}F are clearly shown.

Our sincere gratitude is extended to the staffs of the RIKEN ring cyclotron for their operation of the ECR and accelerator. One of the authors (S.L.) thanks the support from RIKEN. The authors (S.L. and Y.P.) acknowledge the partial support by the Russian Foundation for Fundamental Research (RFFI) under grand No. 96-02-17381a.

REFERENCES

1. I. Tahihata, D. Hirata and H. Toki, Nucl.Phys. A**583** 769 (1995).
2. C. Detraz and D. J. Vieira, Annu. Rev. Nucl. Part. Sci. **39** 407 (1989); A.C Mueller and B.M.Sherrill, Annu. Rev. Nucl.Part. Sci. **43**, 529(1993).
3. *1986–1987 Atomic Mass Predictions*, edited by P.E. Haustein [At. Data Nucl. Data Tables **39** 185 (1988)].
4. H. Sakurai *et al.*, Phys. Rev. C**54** R2802 (1996).
5. H. Sakurai *et al.*, Nucl. Phys. A**616**, 311c (1997).
6. D. Guillemaud-Mueller *et al.*, Phys.Rev. C**41** 937 (1990).
7. X. Campi *et al.*, Nucl.Phys. A**251** 193 (1975).
8. E.K. Warburton, J.A Becker and B.A. Brown, Phys. Rev. C**41** 1147 (1990).
9. A. Poves and J. Retamosa, Nucl.Phys. A**571** 221 (1994).
10. M. Fukunishi, T. Otsuka and T. Sebe, Phys. Lett. B**292** 279 (1992).
11. T.Motobayashi *et al.*, Phys. Lett. B**346** 9 (1996).
12. P. Möller *et al.*, At.Data Nucl. Data Tables **39** 185 (1995).
13. D. Guillemaud-Mueller *et al.*, Z.Phys. A**332** 189 (1989).
14. M. Fauerbach *et al.*, Phys. Rev. C **53** 647 (1996).
15. A. Poves *et al.*, Z.Phys. A **347** 227 (1994).
16. Z. Ren *et al.*, Phys.Rev. C**52** R20 (1995).
17. A. T. Kruppa *et al.*, Phys.Rev.Lett. **79** 2217 (1997).
18. O. Tarasov *et al.*, Phys.Lett. B**409** 64 (1997).
19. H. Sakurai *et al.*, to be submitted.
20. T. Kubo *et al.*, Nucl.Instr. and Methods, B**70** 309 (1992).
21. J.A. Winger, B.M. Sherrill and D.J. Morrissey, Nucl. Instrum. Meth. B**70** 380 (1992).
22. K. Sümmerer *et al.*, Phys.Rev. C**42** 2546 (1990).
23. R. Pfaff *et al.*, Phys.Rev. C**53** 1753 (1996).

Study of ^{11}Be structure via the p(^{11}Be,^{10}Be)d reaction

S. Fortier[a], J.S.Winfield[a] [1], S.Pita[a], W.N.Catford[b], N.A.Orr[c],
Y.Blumenfeld[a], R.Chapman[d], S.P.G.Chappell[e], N.M.Clarke[f],
N.Curtis[b], M.Freer[f], S.Galès[a], K.L.Jones[b], H. Langevin-Joliot[a],
H. Laurent[a], I.Lhenry[a], J. M. Maison[a], P.Roussel-Chomaz[g],
M.Shawcross[b], M.Smith[d], K.Spohr[d], T.Suomijarvi[a], A.de Vismes[g]

a)Institut de Physique Nucléaire, IN2P3-CNRS, 91406 Orsay Cedex, France
b)Department of Physics,University of Surrey,Guildford, Surrey GU2 5XH,UK
c)Laboratoire de Physique Corpusculaire, IN2P3-CNRS, 14050Caen Cedex, France
d)Department of Electronic Engineering and Physics,University of Paisley, Paisley PA1 2BE,UK
e)Department of Nuclear Physics, University of Oxford, Oxford OX1 3RH, UK
f)University of Birmingham, Edgbaston,Birmingham, B15 2TT,UK
g)GANIL(CEA/DSM-CNRS/IN2P3), BP 5027,14076 Caen Cedex 5, France

Abstract. The reaction ^{11}Be(p,d)^{10}Be has been studied for the first time, using a secondary ^{11}Be beam of 35.3 MeV/nucleon. Angular distributions up to about 15°_{cm} were measured by detecting ^{10}Be in a spectrometer and coincident deuterons in a position sensitive silicon detector array. Preliminary analysis provides evidence for a large core excitation component in the structure of ^{11}Be$_{GS}$.

I INTRODUCTION

With the development of intense radioactive beams, one-nucleon transfer reactions will certainly represent one of the major tools for investigating the microscopic structure of nuclei far from stability. The feasibility of such experiments in different mass regions, relying on the use of large position-sensitive light particle detectors, has been examined in ref. [1]. The first such transfer experiment is reported here, namely an investigation of the ^{1}H(^{11}Be,^{10}Be)^{2}H reaction performed with the goal of providing a test of the various competing nuclear models for the structure of the halo nucleus ^{11}Be. Theoretical interest has been focussed on this nucleus for several decades, due to the observation of a "parity inversion" -the $J^\pi=1/2^+$ ground state being in contradiction with the naive shell model prediction of $1/2^-$. Several

---

[1]) Visiting fellow at University of Surrey

calculations of the structure of ^{11}Be have recently been performed, using different theoretical approaches, such as the shell model [2], the variational shell model [3], the Generator Coordinate model [4], and coupling of the neutron with a vibrational core [5–7] or a rotational core [8,9]. Most of them correctly reproduce the parity inversion but make very different predictions about the degree of coupling of the ^{11}Be ground state with the first 2^+ excited state of ^{10}Be at 3.37 MeV, the ratio of spectroscopic factors in the different models $S(2^+)/S(0^+)$ varying from 0.07 to 0.73. In this context, a stringent test of these models can be provided by the experimental determination of this ratio of spectroscopic factors by means of a neutron pick-up reaction.

II EXPERIMENT AND RESULTS

The ^{11}Be(p,d)^{10}Be reaction was studied in inverse kinematics, using a ^{11}Be secondary beam of 35.3 MeV/nucleon produced by fragmentation of ^{15}N in the SISSI device at GANIL. The SPEG spectrometer placed at 0° was used to analyze the ^{10}Be nuclei from the interaction of ^{11}Be with a $(CH_2)_n$ target of thickness 50 μm. Particle identification and momentum and angle measurements were provided by the standard focal plane detection system (two XY drift chambers, an ionization chamber and a plastic scintillator giving energy loss and time-of-flight information). Information on the incident angle at the target position was provided event-by-event by two XY drift chambers, placed in the beam line before the analysing magnet. Deuterons in coincidence with ^{10}Be were detected using CHARISSA. This array of ten position sensitive sheet-resistive silicon detectors (5 x 5 cm^2 and thickness 500 μm), was located in the target chamber at angles between $\sim 5^o$ and 35°. The ^{11}Be intensity was $\sim 3.10^4$ pps, and was monitored throughout the experiment in a small plastic detector placed at the high momentum end of the focal plane.

The ^{10}Be spectra were accumulated at two different magnetic field settings differing by 0.9%, for a total exposure of 3.7×10^9 ^{11}Be nuclei. Focal plane spectra are displayed in figure 1. Three peaks are observed in fig.1(bottom), corresponding to the population of known bound states in the ^{10}Be nucleus by the (p,d) reaction. The 2^+ state at 3.368 MeV is well separated from the 0^+ ground state and the unresolved group of states at 5.958 MeV (2^+), 5.960 MeV (1^-), 6.179 MeV (0^+) and 6.263 MeV (2^-). A large Doppler broadening is observed for excited states in ^{10}Be (v/c=0.28), due to the in-flight emission of γ-rays of several MeV.

The background observed in the singles ^{10}Be spectrum (fig.1-top) is fully removed by the coincidence with CHARISSA detectors. This background primarily originates from the reaction ^{12}C(^{11}Be,^{10}Be)^{13}C on the carbon atoms of the polypropylene target and contributions from breakup of deuterons and ^{11}Be nuclei. The energies of contaminant peaks labelled "C*" in the vicinity of the 0^+ and 2^+ peaks correspond to either excitation of ^{13}C states around 10 MeV or mutual excitation of ^{10}Be and ^{13}C. On the other hand, the (^{11}Be,^{10}Be$_{GS}$) transfer reaction to the ground and first excited states in ^{13}C has a much lower cross section, as the corresponding

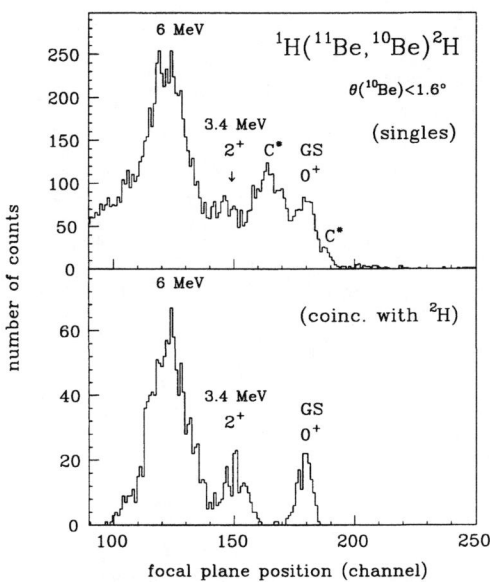

FIGURE 1. Focal plane spectra measured for singles ^{10}Be (top) and ^{10}Be in coincidence with deuterons(bottom), at laboratory angles less than $1.6°$.

peaks are not observed. It is to be noted that the inverse kinematics suppresses the (^{12}C,^{13}C) laboratory cross sections by a factor of 22 compared with (p,d).

The strong peak at 6 MeV is most probably dominated by the pickup of one $p_{3/2}$ neutron from the ^{10}Be core, feeding the $\nu(s_{1/2}\text{-}p_{3/2}^{-1})$ 1^- and 2^- states at 5.960 and 6.263 MeV, with contributions from the second 2^+ and 0^+ states at 5.958 and 6.179 MeV. The peak at 6 MeV is superimposed on a background, corresponding to the high energy tails of the ^{11}Be→^{10}Be+n and d→p+n breakup reactions near threshold. The contribution of the (p,pn) reaction to the ^{10}Be spectra has been removed by energy conditions applied to CHARISSA detectors which select only coincident recoil deuterons. This (p,pn) contribution at threshold was found to be about 25% of the total ^{1}H(^{11}Be,^{10}Be) cross section in the vicinity of the 2^+ peak.

Angular distributions for the 0^+ and 2^+ states and the 6 MeV multiplets within the central region ($\theta_{LAB} < 1.6°$) of the angular acceptance of the SPEG spectrometer are shown in fig.2. Cross sections were obtained by using combined results from the coincident and singles data. Data in coincidence with CHARISSA detectors were corrected for the variation of geometrical detection efficiency as function of angle. This efficiency was calculated using a simulation program, accounting for the large emittance and beam spot size of the secondary beam, and for the uncertainties of scattering angle determination. Error bars in fig.2 are only statistical. Systematic errors coming from uncertainties on the scattering angle determination and the efficiency of the detection system are expected to be of comparable size.

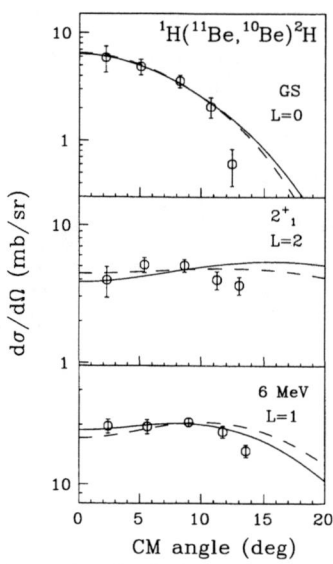

FIGURE 2. *Experimental and calculated angular distributions (see text)*

III ANALYSIS OF ANGULAR DISTRIBUTIONS

Theoretical differential cross sections were calculated using the zero-range DWBA code DWUCK4 [11], with standard corrections for finite-range and non-locality effects. The separation energy method has been used to calculate n+^{10}Be form factors in a Woods-Saxon well with standard geometry r_0=1.25 fm, a=0.65 fm. The halo character of the wave function of the $2s_{1/2}$ neutron coupled to the ^{10}Be ground state comes out naturally from this calculation, due to the low binding energy (0.504 MeV) of the ^{11}Be ground state. One has assumed a pure $d_{5/2}$ transfer to the 2^+ state. Calculations for the 6 MeV peak have assumed a p3/2 neutron pick-up in the ^{10}Be core, exciting the doublet of 1^- and 2^- states. Different combinations of optical potentials for the proton and deuteron channels were used in the calculations in order to test the sensitivity of extracted spectroscopic factors to the input parameters.

Recent proton elastic scattering experiments involving ^{10}Be and ^{11}Be radioactive beams have shown that standard global nucleon-nucleus potentials adjusted for stable nuclei do not reproduce the data unless one reduces the depth of the real Saxon-Woods well [13]. Calculations with potential P1 (table 1) were performed using p+^{11}Be optical parameters derived from the global nucleon-nucleus CH89 [12], with a real well depth reduced by 0.7 [13]. Calculations were also performed for comparison with the proton-nucleus potential from ref. [14], which reproduces elastic scattering data in stable nuclei of the 1p shell (P2), and the global nucleon-nucleus potential from ref [15] (P3).

Standard DWBA calculations were done using the deuteron potential from ref. [16], deduced from elastic scattering experiments (D3). However, it is well known that (p,d) and (d,p) reactions are generally not correctly described by standard DWBA calculations, which do not account for the effects arising from the breakup of the deuteron in the field of the nucleus. The adiabatic deuteron breakup approximation (ADBA) proposed by Johnson and Soper [10] has provided a simple and successful approach to take these effects into account. This approach consists of using a folding potential derived from proton and neutron optical parameters at half the deuteron energy to generate the distorted wave functions in the deuteron channel, in place of the optical potential deduced from elastic scattering data. Calculations within the ADBA approach were performed using the adiabatic potentials D1 and D2, derived from the global nucleon-nucleus CH89 [12] (with a factor of 0.9 applied to the real well depth [13]) and from the optical potential parameters from ref. [14], respectively.

All these various calculations nicely reproduce the slope of the L=0 angular distribution observed experimentally for the ground state transition, and the nearly isotropic angular distributions observed for the 2^+ and 6 MeV states below 10^o_{cm}. The full and dashed lines in fig.2 correspond to ADBA calculations with the combinations of proton and deuteron potentials P1-D1 and P2-D2, respectively. Spectroscopic factors S were extracted using the relation $\sigma_{exp} = NS\sigma_{calc}/(2J_{tr}+1)$ with a normalization factor taken equal to the usual value of 2.29 [11]. The values obtained for the transitions to the 0^+ ground state and 2^+ first excited state of ^{10}Be using different combinations of proton and deuteron optical potentials are displayed in table 1.

TABLE 1. Results of the analysis of the p(^{11}Be,^{10}Be) reaction (see text)

	S(0^+)	S(2^+)	S(2^+)/S(0^+)
P1-D1	0.52	0.46	0.88
P2-D2	0.42	0.32	0.76
P2-D3	0.57	0.57	1.00
P3-D3	0.72	0.71	0.99

Spectroscopic factors for the ground state transition vary from 0.42 to 0.72 and can be compared with the spectroscopic factors extracted for the inverse transition ^{10}Be$_{GS}$ → ^{10}Be$_{GS}$, studied by means of the (d,p) reaction on a radioactive ^{10}Be target, namely 0.73 [17] and 0.77 [18]. This dispersion of spectroscopic factors is outside the commonly accepted value of about 20% on absolute spectroscopic factors extracted from DWBA analyses. On the other hand, ratios of spectroscopic factors do not depend on possible uncertainties on absolute cross sections values, and are typically less dependent on the ingredients of reaction calculations. The ratio $R=S(2^+)/S(0^+)$ deduced from the present analysis ranges between 0.76 and 1.00. Further analysis is in progress in order to estimate the global uncertainty to be assigned to this preliminary R-value of about 0.9.

Such a large value of $S(2^+)/S(0^+)$ greatly exceeds the predictions of refs. [5,8,9], which range from 0.07 to 0.18, and the value of $R = 0.26$ from recent shell model calculations [2]. On the other hand, variational shell model calculations from ref. [3] predict a strong coupling of the ^{11}Be ground state with the 2+ excited core of ^{10}Be, giving a value of R of 0.73, in rather good agreement with the present results.

IV CONCLUSION

The p(^{11}Be,^{10}Be)d reaction has been studied in order to determine the ratio of spectroscopic factors $R=S(2^+)/S(0^+)$, which provides a sensitive test of different models for the structure of ^{11}Be. The preliminary results are consistent with a description of the ^{11}Be ground state being very strongly coupled to the first 2^+ state in ^{10}Be, as given by the predictions of ref. [3].

We acknowledge the contribution of the SPEG crew and cyclotron operation team to this experiment. Our thanks are also due to Dr N.Timofeyuk and Dr J.Van de Wiele for fruitful discussions.

REFERENCES

1. J.S.Winfield, W.N.Catford and N.A.Orr,Nucl.Instr.Meth.A396, 147(1997)
2. E.K.Warburton and Brown B.A., Phys.Rev.C46,923 1992
   B.A.Brown, private communication
3. T.Otsuka, N.Fukunishi and H.Sagawa, Phys.Rev.Lett70,1385(1993)
4. P.Descouvemont,Nucl.Phys.A615,261(1997)
5. N.Vinh Mau, Nucl.Phys.A592,33(1995)
6. T.Bhattacharya and K.Krishan,Phys.Rev.C56,212(1997)
7. H.Lenske,private communication
8. H.Esbensen,B.A.Brown and H.Sagawa,Phys.Rev.C31,1274(1995)
9. F.M.Nunes, I.J.Thompson and R.C.Johnson, Nucl.Phys.A592,33(1995)
10. R.C.Johnson and P.J.R.Soper, Phys.Rev.C1, 976,(1970)
11. P.D.Kunz,private communication
12. R.L.Varner et al.,Phys.Rep.201,57(1991)
13. M.D.Cortina-Gil et al., Phys.Lett.B401,9(1997)
14. B.A.Watson,P.P.Singh and R.E.Segel, Phys.Rev.182,977(1969)
15. F.D.Becchetti Jr. and G.W.Greenless, Phys.Rev.182,1190 (1969)
16. C.M.Perey and F.G.Perey,Phys.Rev.132,755(1963)
17. D.L.Auton, Nucl.Phys.A157,305(1970)
18. B.Zwieglinski et al., Nucl.Phys.A315,124(1979)

Measurement of the E1 strength function of ^{11}Be

R.L. Varner, N. Gan, J.R. Beene, M.L. Halbert, D.W. Stracener, A. Azhari, E. Ramakrishnan, P. Thirolf, M.R. Thoennessen, and S. Yokoyama

Physics Division, Oak Ridge National Laboratory, Oak Ridge, Tennessee, USA
Department of Physics and Astronomy and National Superconducting Cyclotron Laboratory, Michigan State University, East Lansing, Michigan

Abstract. We have investigated the E1 strength function of ^{11}Be by Coulomb excitation and measurement of the subsequent projectile photon decay. The photons were measured in a wall of BaF$_2$ detectors. We have used the virtual photon method to extract the photoabsorption cross section and hence the dipole strength function. We compare our findings with sum rule predictions. This is a first example of techniques we will extend to heavier mass nuclei.

INTRODUCTION

An interesting question in the study of radioactive nuclei is how collective excitations, such as the giant dipole resonance (GDR), evolve as one moves away from β-stability. Various models have been used to explore this question. It has been suggested, for example, that the E1 strength of very neutron-rich nuclei could spread over a very wide energy region, and appear at energies lower than expected from the systematics of stable nuclei [1] [2].

The nucleus ^{11}Be is a single-neutron halo nucleus and has been widely studied, particularly its low-lying states and particle decay modes. The E1 strength at $E_x = 0.6 - 4$ MeV has been investigated through kinematic reconstruction following Coulomb dissociation [7]. However, this strength accounts only about 5% of the total TRK sum rule. Most of the E1 strength is expected to be located at higher excitation energies ($E_x > 8$ MeV). We extend the studies of the E1 strength distribution to the higher excitation energies.

Our experiment measures the photabsorption cross section of ^{11}Be by Coulomb excitation followed by photon-decay to the ground state. This process can be thought of as virtual photon scattering [8]. To select Coulomb excitation events, we detect particles only at very forward scattering angles, which selects large impact parameters. To identify the ground-state GDR γ-ray decays, we usually deduce the total excitation energy from the kinetic energy of the scattered particle, and match it with the energy of the emitted photon. In this case, simply detecting and identifying the scattered ^{11}Be projectile in coincidence with γ-rays is sufficient. This is because ^{11}Be has a neutron separation energy of only 504 keV, and only two bound states, the $1/2^+$ ground state and the $1/2^-$ 320 keV first excited state. Consequently, detection of scattered ^{11}Be implies that one of these two states was populated directly by photon decay following Coulomb excitation. The contribution of decays to the $1/2^-$ 320 keV state is not expected to be significant since E1 excitation and E1 decay dominate. Contributions from higher multipolarity excitations and decays are much smaller. Photon decays to unbound states will be followed by neutron decays, which will change the identity of the ^{11}Be projectile. Therefore, photons in coincidence with the ^{11}Be projectiles are predominately from ground-state GDR γ-ray decays, and photons from decays to the excited states, whether these states are bound or unbound, are strongly suppressed.

MEASUREMENT

The measurement was performed at the National Superconducting Cyclotron Laboratory of Michigan State University. A 100 MeV/nucleon ^{13}C beam from the K1200 cyclotron was used as a primary beam to produce ^{11}Be in the A1200 fragment separator with an energy of 77 MeV/nucleon, an average intensity of 10^6 pps, and a momentum spread of $\approx 3\%$.

Projectile-like particles, mostly ^{11}Be and ^{10}Be, were detected with the zero-degree detector from the MSU 4π array [10]. This detector consists of 8 E-ΔE plastic scintillators, and was mounted 1.35 meter away from target, subtending polar angles from $1.10°$ to $3.24°$. Photons were measured using the 142 element ORNL–TAMU–MSU joint BaF$_2$ array, the elements of which were mostly 6.5 cm $\times$ 20 cm hexagonal BaF$_2$ scintillators. In order to take advantage of the forward-focusing of the photons emitted from the projectile rest frame, these were arranged as a wall covering the polar angular range of $12°$ to $45°$. An estimate of the ^{208}Pb photon yield was made by detecting photons with a small BaF$_2$ array at backward angles, where target photons dominate. Target-out yields were measured and subtracted from the data. The photon detectors were calibrated with discrete γ-rays up to 15.11 MeV.

In our analysis, only events with one photon, and no neutron in the BaF$_2$ array, and one particle detected were accepted. This criterion was chosen to enhance the ground state γ-ray decays from the ^{11}Be nucleus. Random coincidences were corrected by subtracting the data in which particles and γ-rays arose from different beam bursts.

In order to eliminate the background from the ^{10}Be component and continuum we used particle spectra gated by γ-rays with different energies. For the excitation energies below the neutron binding energy of ^{11}Be (504 keV), the separation between ^{10}Be and ^{11}Be is very good, where the 320 keV $1/2^- \rightarrow 1/2^+$ transition in ^{11}Be dominates. The Coulomb excitation of the $1/2^-$ state has been studied at a range of bombarding energies [5] [6], and is in agreement with the B(E1) from the lifetime measurements [3]. We used this transition to determine the absolute normalization for our data.

Unfortunately, at excitation energies between the thresholds of (γ,n) (504 keV) and $(\gamma,2n)$ (8.73MeV), photon spectra are dominated by the γ-rays from the daughter nucleus ^{10}Be. After being excited, almost all the ^{11}Be projectiles emit a neutron, and populate the bound states in ^{10}Be. The yield of photons from these states is much stronger than the yield of single photon decays to the ground state of ^{11}Be. It is very difficult to resolve ^{11}Be from ^{10}Be with the resolution of the present plastic scintillators.

At excitation energies well above the $(\gamma,2n)$ threshold (8.73MeV), the γ-ray branching ratio of ^{11}Be and ^{10}Be will be comparable. Therefore, compared to the lower excitation energy, the photon yields of ^{11}Be will increase relative to that of ^{10}Be.

The photon yields were unfolded from the data, using the simulated response of the BaF$_2$ detector array. The acceptance of the particle detector array was taken into account in the response calculation.

The absolute differential cross section is shown in Fig. 1a. The uncertainties shown are statistical.

We have estimated the ground-state decay contributions of the isoscalar GQR. It is at least one order of magnitude smaller than the contribution from the GDR. The results are shown in Fig. 1a. We have also studied contributions from hadronic excitation by comparing the cross sections for inelastic excitation with and without the nuclear interaction. Within the acceptance of our detector, the differences between the integrated cross sections of these calculations from 8 to 25 MeV are less than 1%.

PHOTO-ABSORPTION CROSS SECTIONS

Bertulani and Nathan [11] have related ground state γ-ray decays from the GDR following the Coulomb excitation to the elastic photon scattering with

$$\frac{d^2\sigma_\gamma}{d\Omega dE_\gamma}(E_\gamma) = \frac{1}{E_\gamma}\frac{dN_{E1}}{d\Omega}(E_\gamma)\sigma_{\gamma\gamma}(E_\gamma) \qquad (1)$$

where N_{E1} is the E1 virtual photon number, and $\sigma_{\gamma\gamma}$ is the cross section of elastic photon scattering, which can be expressed in terms of the photo-absorption cross section $\sigma_{abs}(E_\gamma)$ as [12]

$$\sigma_{\gamma\gamma}(E_\gamma) = \frac{8\pi}{3}\left[\left|\frac{E_\gamma\sigma_{abs}(E_\gamma)}{4\pi\hbar c}\right|^2 + \left|\frac{E_\gamma^2}{2\pi^2\hbar c}P\int_0^\infty \frac{\sigma_{abs}(E'_\gamma)dE'_\gamma}{E'^2_\gamma - E_\gamma^2}\right|^2\right]. \qquad (2)$$

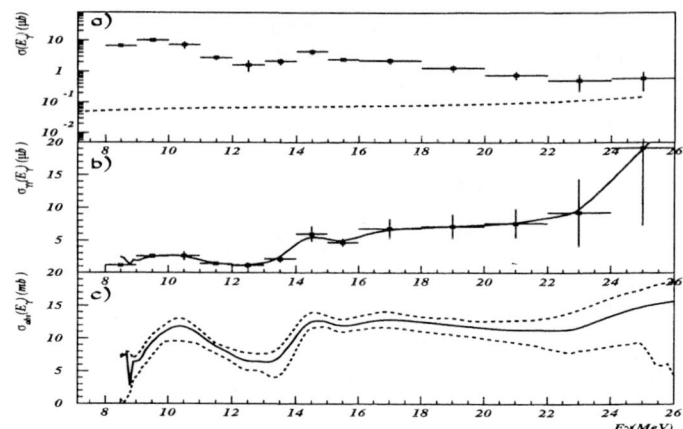

FIGURE 1. Photon cross-sections of ^{11}Be. a) Cross section for ground state γ-decay following Coulomb excitation. The line is a calculation of the GQR contribution. b) The deduced cross section of photon elastic scattering. The line is the fitted result using Eq. 2. c) The unfolded photoabsorption cross section. The dashed lines are uncertainties.

The cross section $\sigma_{\gamma\gamma}$ of the elastic photon scattering is deduced from the measured yields of ground-state γ-ray decay according to the Eq. (1). Fig. 1b presents results for $\sigma_{\gamma\gamma}$.

Extracting the photo-absorption cross sections σ_{abs} from the $\sigma_{\gamma\gamma}$ requires us to invert Eq. (2). This is difficult, because of the infinite range of the integral in Eq. (2), and the finite energy range and discrete nature of the experimental data. We have developed a numerical procedure to solve this problem, in which we approximate $\sigma_{\gamma\gamma}$, the estimate σ_{abs} using on the first term of Eq. (2). Then the integral was evaluated, and used to make correction for the new guess. The procedure was repeated until the differences between the calculated and measured $\sigma_{\gamma\gamma}$ were small.

The solid line in Fig. 1c is the extracted photoabsorption cross section σ_{abs}. The dashed lines in Fig. 1c are an estimate of the uncertainties generated by a Monte Carlo technique.

We estimated that the effect of the low energy dipole strength [7] and the quasi-deuteron [13] above 30 MeV is small ($< 5\%$).

The energy resolution of these data prevent us from studying detailed structures in the GDR region, but the results show that the E1 strength distribution is spread from 8 MeV to 26 MeV, and shows few prominent features except a broad peak near the energy of the 10.59 MeV state in ^{11}Be and a rise near 14.5 MeV. This is qualitatively consistent with the Hartree-Fock calculations for neutron rich nuclei [1] [2]. Unfortunately, we know of no theoretical calculations for the GDR region of ^{11}Be.

Oscillator sum rules allow us to connect the observed strength distribution of collective excitations with bulk and surface parameters of the nuclear medium(nucleus radius, symmetry energy, incompressibility etc.) [14]. Energy weighted moments of the photonuclear cross section defined by

$$\sigma_k = \int_0^\infty \sigma_{abs} E^k dE. \tag{3}$$

can be related to these sum rules [14]. We consider the moments σ_0, σ_{-1}, σ_{-2}. The Thomas-Reiche-Kuhn (TRK) sum rule limits the integrated total cross section σ_0, while the "bremsstrahlung weighted" cross section integral σ_{-1} can be related to a sum rule expression proportioned to the mean square charge radius of the nucleus [14]. The sum rule for the σ_{-2} moment can be shown to be proportional to the dipole polarizability of the nucleus [14]. Table 1 shows experimental values for the cross section moments integrated from 8 to 25 MeV compared with the sum rule limits evaluated [14] using experimental values [15] charge radius (assumed to be the same as ^{9}Be), and experimental constraints on the matter density distribution [16].

Our results shows that the total strength between 8 and 25 MeV exhausts 120% of the TRK sum rule. The total strength at 1 – 4 MeV in ref. [7] shown in Table 1 is only about 5% of the TRK sum rule. The E1 strength of halo nuclei at low excitation is presumably related to the extended distribution of the valence neutron. The observed low energy strength exhausts about 80% of the "cluster sum rule" [17] expected for a neutron weakly coupled to a ^{10}Be) core.

The experimental value of the σ_{-1} is consistent with the corresponding sum rule [14] using the experimental [15] RMS charge radius $\sqrt{\langle r^2 \rangle} = 2.52$ fm. This suggests that the charge distribution for the ^{11}Be nucleus is similar to the one for the ^{10}Be core, and is consistent with the picture that ^{11}Be consists of a core plus a valence neutron.

The σ_{-2} moment should be particularly interesting in this case, since the corresponding sum rule is proportional to the nuclear dipole polarizability, which is extremely sensitive to nuclear surface properties. The Migdal estimate of the σ_{-2} sum rule, which ignores surface effects is [14] 0.11 mb MeV^{-1}. The sum rule limit given in Table 1 improves on the Migdal value by including surface effects in a leptodermous approximation using droplet model expressions [14]. This can be seen to increase the Migdal sum rule value by almost a factor of 5. It should not be expected that the leptodermous approximation would treat surface effects in a halo nucleus adequately. In fact, even the data integrated from 8-25 MeV exceeds the

TABLE 1. Comparison between the sum rules and the experimental values.

Moment	Calculated Sum Rule Limit	Experimental value	E=1-4MeV
σ_0 (MeV-mb)	152	181 ± (+23 -35)	6.7
σ_{-1} (mb)	15.49	11.4 ± (+1.4 -2.2)	4.7
σ_{-2} (MeV^{-1}-mb)	0.484	0.784 ± (+0.095 -0.162)	3.8

droplet model sum rule limit by ∼60%. If the low energy contribution is added in, the data is almost an order of magnitude larger than the sum rule value!

SUMMARY

We have measured the E1 strength of ^{11}Be from 8.5 MeV to 25 MeV. The energy distribution is relatively flat, which is consistent with the theoretical expectation from the Hartree Fock calculations. The total cross section has exhausted $\approx$ 1.2 TRK sum rule. The experimental value of the σ_{-1} moment of the the distribution is about equal to the bremsstrahlung weighted sum rule when the low energy contributions are included. The measured σ_{-2} is much higher than corresponding sum rule estimates.

This measurement illustrates the potential usefulness of ground state γ-ray decay following projectile Coulomb excitation as a tool for study of the GDR of radioactive nuclei. More precise measurements could be made with more intense radioactive beams, using a high resolution spectrograph to identify and detect the scattered projectile. We believe it will be possible to apply this technique to systematically study the isospin dependence of the GDR strength distribution in unstable nuclei.

ACKNOWLEDGEMENTS

Research at the Oak Ridge National Laboratory is supported by the U.S. Department of Energy under contract number DE-AC05-96OR22464 with Lockheed Martin Energy Research Corp This research was supported in part by an appointment to the Oak Ridge National Laboratory Postdoctoral Research Associates Program administrated jointly by the Oak Ridge National Laboratory and the Oak Ridge Institute for Science and Education.

REFERENCES

1. T. Hoshino, H. Sagawa and A. Arima, Nucl. Phys. **A253**, 228 (1991).
2. I. Hamamoto, H. Sagawa, Phys. Rev. **C53**, R1492 (1996).
3. D.J. Millener, J. W. Olness, E. K. Warburton, and S. S. Hanna, Phys. Rev. **C28**, 497 (1983).
4. R. Anne,*et al.* , Z. Phys. **A352**, 391 (1995).
5. T. Nakamura, T. Motobayashi, Y. Ando, A. Mengoni, T. Nishio, H. Sakurai, S. Shimoura, T. Teranishi, Y. Yanagisawa, M. Ishihara, Phys. Lett. **B394**, 11 (1997).
6. M. Fauerbach, M. J. Chromik, T. Glassmacher, P. G. Hansen, R. W. Ibbotson, D. J. Morrissey, H. Scheit, P. Thirolf, and M. R. Thoennessen, Phys. Rev. **C56**, R1 (1997).
7. T. Nakamura, *et al.*, Phys. Lett. **B331**, 296 (1994).
8. C.A. Bertulani and G. Baur, Phys. Rep. **163**, 1 (1988) and references therein.
9. J. Beene, Nucl. Phys. **A583**, 73c (1995).

10. N.T.B. Stone, et al., NSCL Annual Report **93**, 123(1993), and NSCL Annual Report **94**, 181 (1994).
11. C.A. Bertulani and A.M. Nathan, Nucl. Phys. **A554**, 158 (1993).
12. A.M. Nathan, Phys. Rev. **43**, R2479 (1991).
13. J.S. Levinger, Phys. Lett. **82B**, 181 (1979).
14. E. Lipparini and S. Stringari, Phys. Rep. **175**, 103 (1989) and references therein.
15. J.A. Jansen, R.Th. Peerdeman and C. DeVries, Nucl. Phys. **A188**, 337 (1972).
16. I. Tanihata, T. Kobayashi, O. Yamakawa, S. Shimoura, K. Ekuni, K. Sugimoto, N. Takahashi, T. Shimoda and H. Sato, Phys. Lett. **206**, 592 (1988). C.A. Bertulani and H. Sagawa, Nucl. Phys. **A588**, 667c (1995).
17. Y. Alhassid, M. Gai, and G.F. Bertsch, Phys. Rev. Lett. **49**, 1482 (1982)

Do the excited states in the system of two neutrons exist?

D.V. Aleksandrov[1], E.Yu. Nikolskii[1], B.G. Novatskii[1], D.N. Stepanov[1] and R. Wolski[2]

[1] RRC "Kurchatov Institute", Moscow 123182, Russia
[2] Institute of Nuclear Physics, Krakow, Poland

Abstract. The experimental study of 2n-system was carried out in the reaction $T(d,^3He)$. The large cross section unbound state of ^2n was observed with decay width $\Gamma=(1.1\pm0.2)$ MeV. In addition, the assumption of the excitation of two wide resonances in 2n-system was claimed with energies of 3.6 and ~11.8 MeV

From experimental studies of pn and pp scattering there was found out the final state interaction in this systems with izospin T=1 and it was showed that singlet D and 2p-systems have the virtual levels near the threshold of decay. In 60th and 70th the hypotheses about charge independence and the symmetry of the nucleon-nucleon interaction in the measurements of the scattering length was checked up experimentally The small difference in n-n and p-n interaction (a_{nn}=-16.6 fm and a_{nn}=-23.7 fm) was found(review (1), for example). Due to the absence of the neutron target the main way of investigation n-n interaction the D(n,p), T(n,d), $T(d,^3He)$ and $T(t,^4He)$ reactions were used. We should make a point out on the fact that the main aim of this researches was set on proving of the hypotheses of charge independence and symmetry of nucleon-nucleon forces, but not on searching of the excited states of unstable ^2n. As a rule the spectra were received in a narrow energetic region of the zero binding energy of dineutron (2). In the high energy physics were searching for «narrow» dibarion resonance (^2p) with energies tenth or even hundreds MeV, which are traditionally far from the excitation energies, typical for the low energy physics (3).

The aim of this investigations is searching for the excited states of the dineutron in the $T(d,^3He)$ reaction by usual methods of the nuclear spectroscopy in the interval from the ground state ^2n to the maximum possible energy, admitted by the conditions of the experiment ($E^* \sim 15$ MeV). From three variants of searching of the dibarion resonance (2n, 2p, np) we have chosen the first one, because in this case we can see the nuclear interaction between two identical particles.

The investigation was carried out on the RRC " Kurchatov Institute" cyclotron, with a maximum beam energy of deuterons E=31 MeV. The measurements of energy spectrums of ^{3}He from the T(d,^{3}He) reaction were made in the angle interval from 6 to 13 degrees. This limited the intensity of the beam, because of the huge background from the elastic scattering of the deuterons. The mean beam current of the deuterons was ~0.01μA.

Two self-supporting titan foils with the equal width (5.1 mg/cm^2) served as the targets. One of them was saturated with tritium. The amount of tritium in Ti-t target and presence of admixtures in it were find by a special measurements on the cyclotron by elastic scattering of deuterons with E = 13 MeV. The value of the elastic crossection on tritium was taken from (4). The maintenance of tritium were found to be equal 38.4% in Ti-T foil (by the number of atoms in regards to Ti).

The products of the reaction were measured with a telescope of ΔE-E semiconductor silicon detectors with a width 30μm and 1.2 mm corresponding. The solid angle of the detecting system was $1.3*10^{-4}$ sr.

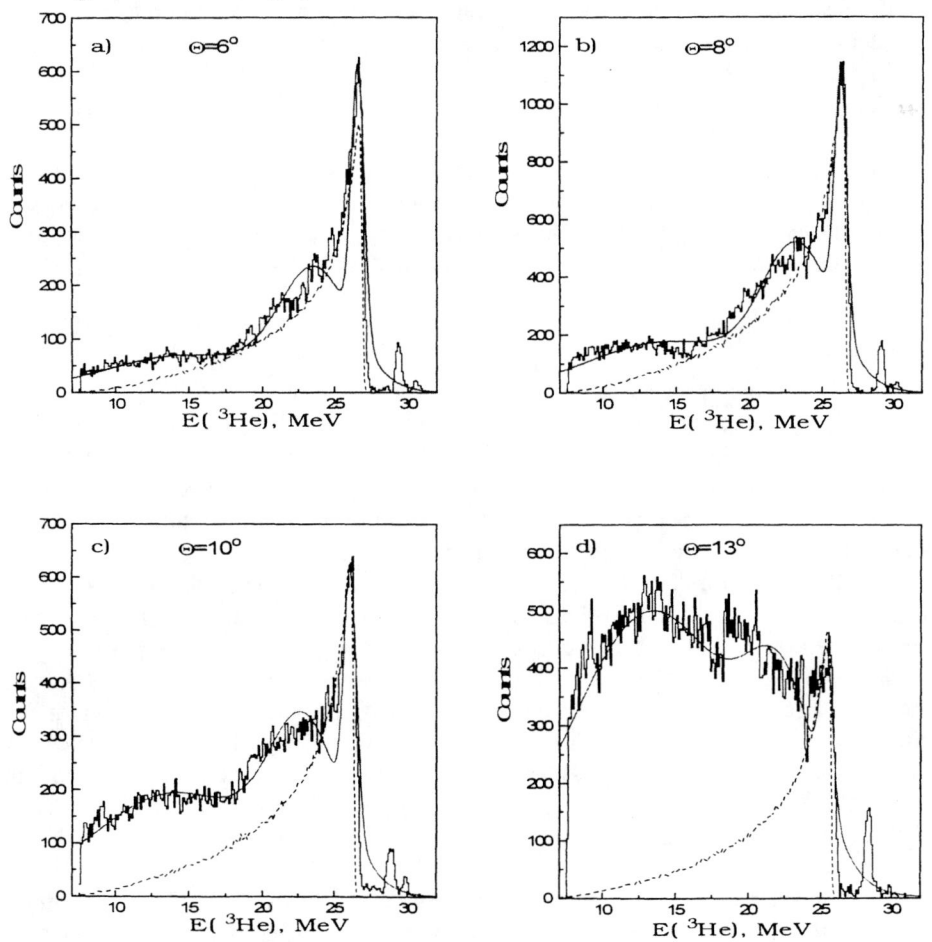

FIGURE 1. The energy spectra of ^{3}He from T(d,^{3}He) at the angles 6-13°.

The spectra of ^{3}He were measured in energetic region from 5 to 35 MeV and at the angles 6,8,10 and 13 degrees with a good statistic on both Ti and Ti-T targets. The energetic calibration was made by using ^{48}Ti(d,^{3}He) reaction and the recoil nuclei from ^{3}He(d,^{3}He)D on the small admixture of ^{3}He(4- 5 %) in Ti-T targets from β-decay of tritium(the peak in discreet part of spectrum at 29 MeV). The energetic spectra of ^{3}He from T(d,^{3}He) reaction after subtraction of admixtures from (d,^{3}He) on Ti target are shown in Fig.1. In the high energy part of the spectrums the intensive peak are clearly seen. It corresponds to the final state interaction of two neutrons. There are two wide distributions in the spectra with maximum lies at the 23 and 14 MeV besides this peak. The contribution of this bumps to the spectra are increase with angle increasing and becomes dominant at 13 degree. The same picture was observed in T(d,^{3}He) reaction at the deuteron energy equaling 11 MeV (5).

We tried to explain this complicated structure by a known interaction mechanism:

1. The resonance-like structures can appear in the spectrum of the products of nuclear reactions from the different two step processes. The only one process, which goes via formation ^{4}He* with corresponding sequential break-up ^{4}He$^* \rightarrow ^3$He+n, should be taken into account in T(d,^{3}He) reaction. The Monte-Carlo simulation of this process made it possible to exclude this channel, because the energies and the widths of the computed peaks did not agree with the observed values.
2. Using Migdal-Watson approximation, the calculation of n-n interaction was done for the all measured angles using the scattering length a_{nn}= -16 fm (dashed line). As it seen, the reliable agreement between experimental and theoretical curves was found only for high energy region.
3. The shape of measured spectra are not reproduced by three particle phase space distribution D+T$\rightarrow^3$He+n+n

So none of described process can explain the measured spectra.

We proposed the hypothesis that in the reaction T(d,^{3}He) besides the population of ground state of dineutron with the decay width Γ=1.1±0.2 MeV, two wide resonances are excited in n-n system. Using χ^2 criteria the measured energy spectra were fitted by three gaussians, where positions, widths and intensities were free parameters. The result of the fitting is shown in the figure by solid line.

After transformation to the c.m. system, energies of the resonances were found to be E*=3.6 ±0.3 MeV and E*~11.8 MeV with respect to ground state of ^{2}n. The same values are obtained at all other angles within experimental errors. The position of the resonances is in a good agreement with known interval rule $E_4:E_2$ ~11.8:3.6=3.3, showing on possible existing of "rotating band" with spin-parities 2$^+$ and 4$^+$ in ^{2}n. The spin and parity of the resonance E*=3.6 MeV not contradicts with the systematic of all even-even nuclei, where the first states, besides magic, have characteristics 2$^+$. From the formula $\Delta E=\hbar^2 l(l+1)/2\mu R^2$, the estimation of dineutron "radius" was obtained to be ~ 8 fm. This value is closed to the radii of such loosely bound neutron-rich nuclei with large neutron halo like ^{11}Li, ^{11}Be.

At the same time the authors would like to mention that to confirm the results of this experiment it is necessary to carry out some extra experiment with higher beam energy (50-60 MeV).

This work was supported by a RFFI grant N^0 96-02-17298a.

REFERENCES.

1. B.Kuhn, Sov.J.Particles Nucl. **6**, 139 (1976)
2. E.Baumgartner, H.E.Conzett, E.Shield, R.J.Slobodrian, Phys. Rev. Lett., **16**, 105(1966)
3. Yu.A.Trojan, A.V.Nikitin, V.N.Pechenov at all, Preprint JINR, P1-90-78, Dubna
4. M.Ivanovich, P.G.Joung, G.G.Ohlsen, Nucl. Phys., **A110**, 441(1968)
5. H.T.Larson, A.D.Bacher, K.Nagatini and T.A.Tombrello, Nucl.Phys., **A149**, 161(1970)
6. D.R.Tilley, H.R.Weller, G.M.Hale, Nucl.Phys., **A541**, 19(1992)
7. P.G.Hansen, A.S.Jensen, B.Jonson, Annual Rev. of Nucl. and Part. Science, **45**, 591(1995)
8. I.Tanihata, J. Phys. G., Nucl. Part. Phys., **22**, 157(1996)
9. F.Ajzenberg-Selove, Nucl.Phys., **A413**,1(1984)

Continuum Excitations in Neutron-Rich Oxygen Isotopes

T. Aumann[1,2,*], A. Leistenschneider[3], K. Boretzky[1,2], D. Cortina[1],
J. Cub[1,4], W. Dostal[2], B. Eberlein[2], Th.W. Elze[3], H. Emling[1],
H. Geissel[1], A. Grünschloß[3], M. Hellström[1], J. Holeczek[5],
R. Holzmann[1], S. Ilievski[1], N. Iwasa[1], M. Kaspar[1], A. Kleinböhl[1],
J.V. Kratz[2], R. Kulessa[6], Y. Leifels[1], E. Lubkiewicz[6],
G. Münzenberg[1], P. Reiter[7], M. Rejmund[1], C. Scheidenberger[1],
Ch. Schlegel[1], H. Simon[4], J. Stroth[3], K. Sümmerer[1], E. Wajda[6],
W. Walus[6], S. Wan[1]

[1] *Gesellschaft für Schwerionenforschung (GSI), Planckstr. 1, D-64291 Darmstadt, Germany*
[2] *Institut für Kernchemie, Johannes Gutenberg Universität, D-55099 Mainz, Germany*
[3] *Institut für Kernphysik, Johann Wolfgang Goethe Universität, D-60486 Frankfurt, Germany*
[4] *Institut für Kernphysik, Technische Universität, D-64289 Darmstadt, Germany*
[5] *Instytut Fizyki, Uniwersytet Śląski, PL-40-007 Katowice, Poland*
[6] *Instytut Fizyki, Uniwersytet Jagelloński, PL-30-059 Kraków, Poland*
[7] *Argonne National Laboratory, Argonne, IL 60439, USA*

Abstract. Electromagnetic and nuclear excitations of the neutron-rich Oxygen isotopes ranging from $A = 17$ to $A = 22$ are studied experimentally in reactions at energies around 600 MeV/u. By measuring the four-momenta of all decay products the excitation energy is determined. From the differential cross sections for electromagnetic excitation, the E1-strength distributions can be deduced. For ^{18,20,22}O, low-lying dipole strength is observed, exhausting about 5% of the energy weighted TRK sumrule for energies up to 5 MeV above the continuum threshold.

I INTRODUCTION

The multipole continuum response of nuclei near the driplines is expected to change considerably in comparison to what is known from stable nuclei. One effect which shows up in all calculations for neutron-rich nuclei is the additional strength below the normal giant resonance region, predicted for different multipolarities [1–4] (and refs. therein). This low-lying strength reflects the multipole response of

---

*) present address: Michigan State University, East Lansing, MI 48824-1321, USA

the loosely bound valence neutrons. An experimental example is the low-lying E1-strength just above the threshold as observed in the halo-nuclei ^{6}He [5], ^{11}Li [6–8], and ^{11}Be [9]. For ^{6}He and ^{11}Li, for instance, about 10% of the Thomas-Reiche-Kuhn dipole sumrule was found below an excitation energy of 5 MeV [5,8].

In the present experiment, the Oxygen isotope chain in the mass range $A = 17$ to $A = 22$ is investigated. Utilizing the electromagnetic excitation process at high energies ($\approx$ 600 MeV/u) the E1 strength distribution can be studied up to the energy region of the giant dipole resonance (GDR). For light nuclei, however, the GDR is located at relatively high excitation energies ($E^* \approx 25$ MeV), which are suppressed by the electromagnetic excitation process due to the adiabatic cut-off. Other multipolarities than E1 can be studied by nuclear excitation processes using light targets. In addition, by measuring the γ-rays originating from excited fragments after fragmentation processes, γ-spectroscopy of unstable nuclei can be investigated.

II EXPERIMENTAL METHOD AND REACTION MECHANISMS

The radioactive beams were produced in a fragmentation reaction of a primary ^{40}Ar beam, delivered by the synchrotron SIS at GSI, Darmstadt, impinging on a beryllium target. The fragments were separated using the Fragment Separator FRS [10]. In two settings, a degrader was inserted in the midplane of the FRS. In these cases only ^{17,18}O and ^{19,20}O were transported to the experimental area, respectively. In a third setting without degrader and optimized for ^{22}O, the beam contained various isotopes with similar A/Z ratio ranging from Be up to O, which were identified uniquely by means of energy loss and time-of-flight measurements. The trajectory of the incident ions was measured by a multi-wire proportional chamber and a position sensitive Si pin-diode. Behind the target, the fragments were deflected by a large-gap dipole magnet. By means of energy-loss and time-of-flight measurements, as well as position measurements before and behind the dipole magnet, the nuclear charge, velocity, scattering angle, and the mass of the fragments are determined. The neutrons stemming from the excited projectile or excited projectile-like fragments are focussed to forward directions and detected with high efficiency in the LAND neutron detector [11], placed at zero degree about 11 m downstream from the target and covering an angular range of about 90 mrad. To detect the γ-rays, the target was surrounded by the 4π Crystal Ball spectrometer, consisting out of 160 NaI detectors.

In Fig. 1, the mass distribution for fragments with nuclear charge $Z_f = 8$ (upper frame) and $Z_f = 7$ (lower frame) in coincidence with at least one neutron is shown for an ^{20}O beam after reacting in C and Pb targets. The distributions obtained with the Pb target are scaled to that obtained with the C target for $Z_f = 7$ (lower frame in Fig. 1). For $Z_f = 7$, the two histograms coincide, thus showing no contributions from electromagnetic excitations in case of the Pb target. Contrary, for the Oxygen

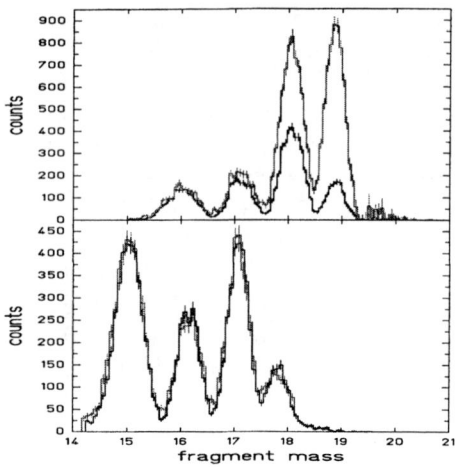

FIGURE 1. Fragment mass distribution for fragment nuclear charges $Z_f = 8$ (upper frame) and $Z_f = 7$ (lower frame) after fragmentation of ^{20}O on C and Pb targets (overlayed histograms) in coincidence with the LAND neutron detector. The histograms obtained with the Pb target are scaled to that obtained with the C target for $Z_f = 7$ (lower part).

fragments (upper panel) with $A_f = 19$ and $A_f = 18$, corresponding to the 1n and 2n removal channels, an enhancement is observed for the Pb target compared to the C target. The fact that, under these conditions, the 1n removal channel is suppressed in case of nuclear reactions can be understood by inspecting the different reaction mechanisms. The dominant process in case of nuclear fragmentation is knock-out of at least one nucleon and subsequent evaporation from the excited pre-fragments. Since only neutrons in the projectile-rapidity domain are within the acceptance of LAND, neutrons originating from the first step (knock-out or abrasion), which are scattered to large angles, are not detected (see [8] for a detailed discussion in the context of fragmentation of halo nuclei). Thus, even the cross sections for few-neutron removal by nuclear and electromagnetic interactions are of similar magnitude, the electromagnetic part can be separated experimentally. Note that also for the 2n channel a similar reduction as for the 1n channel can be obtained by requiring exactly two neutrons detected in LAND. In Fig. 1 not all tracking information available was used, thus in the final analysis the mass resolution will be improved.

III RESULTS AND OUTLOOK

As discussed in the previous section, the nuclear contribution to the cross section measured with the lead target can be reduced to a 10-20% level under certain experimental conditions. Then, the measured differential cross section is directly connected to the B(E1) distribution. For ^{18}O, we obtain about 5% of the TRK energy weighted sumrule by integrating the deduced strength up to 5 MeV above the continuum threshold, consistent with photo-absorption measurements [12]. For ^{20}O and ^{22}O we find a similar exhaustion of the sumrule in the respective energy region, while the maximum of the cross section shifts towards lower excitation energies going from ^{18}O to ^{22}O. The odd Oxygen isotopes are presently analyzed. In a future analysis also the nuclear excitation will be investigated. The reaction mechanism of nuclear excitation can be experimentally disentangled from knock-out processes in a similar manner as the electromagnetic excitation for the Pb target.

The experiment allows also for γ-spectroscopy of unstable nuclei. An example is given in Fig. 2, where the γ sum-energy spectra are shown for fragments ^{18}O (upper

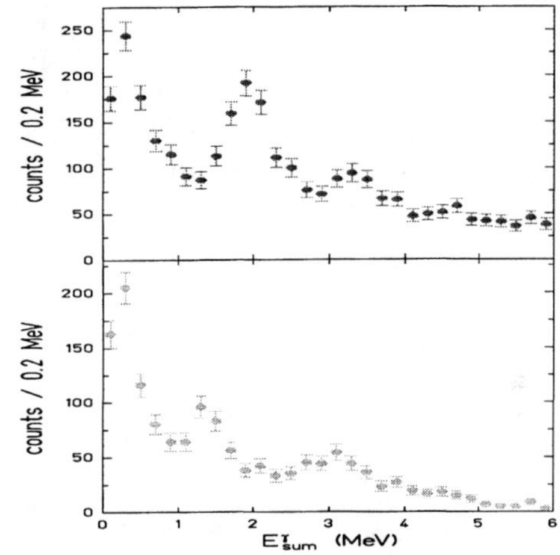

FIGURE 2. γ sum-energy spectra as measured in the fragmentation of 600 MeV/u ^{20}O in a Pb target, yielding the excited fragments ^{19}O (lower frame) and ^{18}O (upper frame). The peaks correspond to the known energy levels at 1.47 MeV and 3.16 MeV in ^{19}O, and 1.98 MeV and 3.56 MeV in ^{18}O.

frame) and ^{19}O (lower frame) produced in reactions of ^{20}O in a Pb target. The first two excited states in these two nuclei are clearly visible. In this case these states are known, but it demonstrates that the fragmentation reactions at high energies are suitable to perform γ-spectroscopy of exotic nuclei. By measuring the partial cross sections for one-nucleon knock-out reactions feeding the ground- and excited states, spectroscopic factors may deduced [13] as well.

REFERENCES

1. F. Catara, C.H. Dasso, A. Vitturi, Nucl. Phys. **A602** (1997) 181.
2. F. Ghielmetti, G. Colò, P.F. Bortignon, R.A. Broglia, Phys. Rev. **C54** (1996) R2143.
3. I. Hamamoto and H. Sagawa, Phys. Rev. **C53** (1996) R1492.
4. B.V. Danilin *et al.*, Phys. Rev. **C55** (1997) R577.
5. S135 Collaboration, to be published.
6. D. Sackett *et al.*, Phys. Rev. **C48** (1993) 118.
7. F. Shimoura *et al.*, Phys. Lett. **B348** (1995) 29.
8. M. Zinser *et al.*, Nucl. Phys. **A619** (1997) 151.
9. T. Nakamura *et al.*, Phys. Lett. **B331** (1994) 296.
10. H. Geissel *et al.*, Nucl. Instr. and Meth. **B70** (1992) 286.
11. T. Blaich *et al.*, Nucl. Instr. and Meth. **A314** (1992) 136.
12. J.G. Woodworth *et al.*, Phys. Rev. **C19** (1979) 1667.
13. N. Alahari *et al.*, these proceedings.

Visualizing ^{11}N by resonance reactions

L. Axelsson[1], K. Markenroth[1]
for the collaboration: M.J.G. Borge[2], S. Fayans[3], V.Z. Goldberg[3],
S. Grévy[4], D. Guillemaud-Mueller[4], B. Jonson[1], K.-M. Källman[5],
T. Lönnroth[5], M. Lewitowicz[6], P. Manngård[5], I. Martel[2],
A.C. Mueller[4], I. Mukha[7], T. Nilsson[1], G. Nyman[1], N.A. Orr[8],
K. Riisager[7], G.V. Rogatchev[3], M.-G. Saint-Laurent[6], I.N. Serikov[3],
O. Sorlin[4], O. Tengblad[2], F. Wenander[1], J.S. Winfield[8], R. Wolski[9]

[1] *Chalmers Göteborg*, [2] *CSIC Madrid*, [3] *RSC Moscow*, [4] *IPN-Orsay*, [5] *Aabo Akademi Turku*,
[6] *GANIL Caen*, [7] *Aarhus University*, [8] *LPC-Caen*, [9] *JINR Dubna*

Abstract. The proton-rich, unbound nucleus ^{11}N has been studied using the technique of elastic resonance scattering in inverse geometry. The spins and parities of the three lowest resonance levels have been determined as $1/2^+$, $1/2^-$ and $5/2^+$, resulting from the orbitals $1s_{1/2}$, $0p_{1/2}$ and $0d_{5/2}$, respectively. This level ordering is the same as the one found in the mirror nucleus of ^{11}N, namely the one-neutron halo nucleus ^{11}Be. Especially, the in ^{11}Be well known level inversion, where the ground state is a $1/2^+$ intruder state below the from a naive shell model expected ground state $1/2^-$, is also found in ^{11}N.

INTRODUCTION

The neutron-rich members of the A=11 isobaric chain has proven to give rich information on the structure of dripline nuclei, and specifically about the halo phenomenon [1] through ^{11}Be and ^{11}Li. On the proton rich side, it is known that the mirror nucleus of ^{11}Be, ^{11}N, is unbound, and the aim of the work which this contribution refers to [2] is to identify the ground state and excited states of ^{11}N. An interesting question in this context is whether ^{11}N shows the same parity inversion that is observed in ^{11}Be in which the ground state is an $1s_{1/2}$ intruder state and the only bound excited state at 320 keV is a $0p_{1/2}$ state [3].

THE EXPERIMENT

The first experimental data for ^{11}N were obtained by Benenson et al. [4] in a study of the reaction ^{14}N(^{3}He,^{6}He). The same reaction was later used by Guimarães et

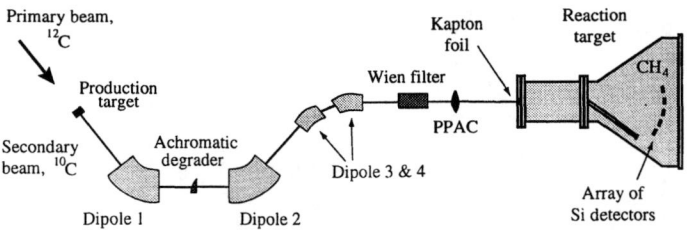

FIGURE 1. The experimental arrangement used in the resonance scattering experiment.

al. [5]. In the last year, two more experimental results have been published [6,7]. Several theoretical publications also indicate the interest in this isotope [11–14]. The necessity to use complicated reactions hampers all spectroscopic investigations of nuclei far from stability, and is one reason that contradictory results are not uncommon, as this specific case also shows. We chose a more direct method to study the resonances in ^{11}N, namely the resonance scattering reaction ^{10}C+p in inverse geometry. This technique uses a radioactive beam (in this case ^{10}C) which is scattered in a thick gas target (here a proton-rich gas). The thickness of the gas target is adjustable through the pressure and is modified as to stop the heavy ions in the gas. The incoming ions are slowed down, and when their energy correspond to a resonance in the compound system, the cross section for elastic scattering increases dramatically. Since the ions are slowed down continuously, the whole energy region is scanned by each ion. The scattered protons are ejected in the forward direction (lab. system) and, having smaller energy loss than the ions, they reach the detectors in the back of the chamber. An important note is that 0° in the lab. system transforms to 180° in c.m.

This simple reaction provided the opportunity to give indisputable spin-parity determinations as well as energies and widths of the lowest three states in ^{11}N. The basic ideas of the method used as well as its application to radioactive beams can be found in [8–10]. The spins and parities can be deduced since the differential cross section is dependent upon these parameters, and in this case the projectile (^{10}C) is spinless. The experiment was carried out at the LISE3 spectrometer at GANIL, using a ^{10}C beam of ≈ 9 MeV/u which was produced from a 75 Mev/u ^{12}C primary beam. The beam intensity was about $7 \cdot 10^3$ pps with a high purity since both degrader and Wien filter were used. A scattering chamber filled with methane (CH_4) gas was used as proton target. The scattered protons were detected in 2-2.5 mm thick Si detectors located at 0° and various angles. These thicknesses allow for measurement of protons to 15-20 MeV. The background arising from the carbon in the CH_4 molecules was determined by measurements with CO_2 gas in the chamber. From this, we estimate that the carbon contributed less than 10 percent to the total intensity and had a constant, featureless energy distribution. To get a calibration, the energy resolution and to test the method we investigated the well known resonances in ^{13}N (by ^{12}C+p). The overlapping levels at 3.50 and 3.55 MeV were clearly seen, and this gave an energy resolution of 30 keV. An additional

TABLE 1. Resonance states in ^{11}N.

Level I^π	Energy[a] MeV	$\Gamma_{c.m}$ keV	SF
$1/2^+$	1.35± 0.1	$1.2^{+0.2}_{-0.1}$	0.9
$1/2^-$	2.04± 0.1	0.7± 0.1	0.9
$5/2^+$	3.77± 0.06	0.6± 0.1	0.73
$3/2^{-[b]}$	4.3	70	–
$3/2^{+[b]}$	5.1	1100	–
$5/2^{+[b]}$	5.5	1500	–

[a] Relative to the ^{10}C+p threshold.
[b] Tentative assignment.

calibration was given by an alpha source.

ANALYSIS AND RESULTS

A potential-well model was used in the analysis of the data. The justification for using this model is that the three lowest levels in ^{11}Be are of predominantly single particle nature [3], which validates the assumption that the same is true for ^{11}N. A central Woods-Saxon potential together with a spin-orbit term, a Coulomb term and a centrifugal term, using conventional values of the radii and diffuseness parameters, was utilized. To get the best fit to the data, the potential depths, radii and diffusenesses were varied for each ℓ-value. The best fit is shown in figure 2 (left part). Note that it is not possible to fit the spectrum at all below 3.5 MeV if one assumes a $1/2^-$ ground state, and we can therefore conclude that ^{11}N exhibit the same intruder state as ^{11}Be. The energies and widths of the resonances are given in Table 1. The three lowest states are found to have the same quantum characteristics in ^{11}Be and ^{11}N. The partial waves are pictured in Figure 2 (right part). For a more detailed discussion, se [2]. The position of a resonance is defined

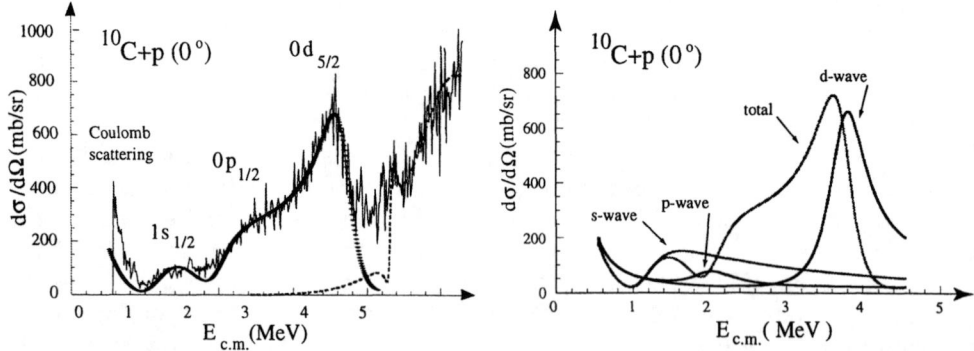

FIGURE 2. Left: The experimental data and the theoretical fit of the levels. Right: The partial waves of the theoretical fit and their sum.

as the energy where the amplitude of the wavefunction has its maximum at 1 fm, and the width as where this amplitude has decreased to $1/\sqrt{2}$ of its maximum value. The only disagreement between the data and the theoretical fit is the dip in the theoretical curve around 1.5 MeV, where there is none in the experimental data. At higher energies we have used a coherent sum of Breit-Wigner resonances to fit the experimental excitation function (dashed line in Figure 3), since there is no reason to assume that these states are of single-particle nature. Also, the solid angle corrections are large here, and the thickness of the Si-detectors cut the spectra around 5 MeV. No interference between the higher and lower part of the spectrum has been considered.

Scattered protons were also registered in detectors placed at different angles (see Figure 1). When using the same potential parameters as for the central detector for the data from the 12° detector, the results agree well. However, one should be aware that the data for $\theta \neq 0°$ are more difficult to interpret since these energy distributions are sensitive to the width of the beam as well as beam straggling which makes protons with the same enery see different detector angles. The final analysis of these data is in progress.

CONCLUSIONS

In conclusion, we have been able to identify the low-energy states in the unbound nucleus ^{11}N and assigned spins and parities to the three lowest resonances. These levels have the same ordering as the corresponding levels in ^{11}Be. In particular, the level inversion of the ground state is reproduced.

REFERENCES

1. P.G. Hansen, A.S. Jensen, B. Jonson, Ann. Rev. Nucl. Part. Sci. **45**, 591 (1995)
2. L. Axelsson et al. Phys. Rev. C **54**, R1511 (1996)
3. F. Ajzenberg-Selove, Nucl. Phys. **A506**, 2 (1990)
4. W. Benenson et al. Phys. Rev. C **9**, 2130 (1974)
5. V. Guimarães et al., Nucl. Phys **A588**, 161C (1995)
6. A. Azhari et al., Phys. Rev. C **57**, 628 (1998)
7. A. Lépine-Szily et al., Phys. Rev. Lett.**80** 1601 (1998)
8. K.P. Artemov et al., Sov. J. Nucl. Phys. **52**, 1460 (1990)
9. V.Z. Goldberg, A.E. Pakhomov, Phys. Atomic Nuclei **56**, 1167 (1993)
10. V.Z. Goldberg et al., Phys. At. Nucl. **60**, 1061 (1997)
11. H.T. Fortune, D. Koltenuk and C.K. Lau, Phys. Rev. C **51**, 3023 (1995)
12. F.C. Barker, Phys. Rev. C **53**, 1449 (1996)
13. P. Descouvemont, Nucl. Phys. **A615**, 261 (1997)
14. S. Aoyama et al., Phys. Rev. C **57**, 957 (1998)

Proton Decay Studies at HRIBF*

J. C. Batchelder[1], C. R. Bingham[2,3], K. Rykaczewski[2,4], K. S. Toth[2],
T. Davinson[5], T. N. Ginter[6], C. J. Gross[7], R. Grzywacz[3], Z. Janas[4],
M. Karny[4], S. H. Kim[3], B. D. MacDonald[8], J. F. Mas[2], J. W. McConnell[2],
A. Piechaczek[9], J. J. Ressler[10], R. C. Slinger[5], J. Szerypo[11], W. B.
Walters[10], W. Weintraub[3], P. J. Woods[5], C.-H. Yu[2], E. F. Zganjar[9]

1 Oak Ridge Associated Universities, Oak Ridge TN, 37831 USA
2 Oak Ridge National Laboratory, Oak Ridge, 37831 TN USA
3 University of Tennessee, Knoxville TN 37996 USA
4 IEP, Warsaw University, 00681 Warsaw, Hoza 69, Poland
5 University of Edinburgh, Edinburgh, EH9 3JZ, United Kingdom
6 Vanderbilt University, Nashville TN 37235 USA
7 Oak Ridge Institute for Science and Education, Oak Ridge, TN 37831 USA
8 Georgia Institute of Technology, Atlanta GA 30332 USA
9 Louisiana State University, Baton Rouge, LA 70803 USA
10 University of Maryland, College Park, MD 20742 USA
11 Joint Institute for Heavy Ion Research, Oak Ridge TN, 37831 USA

A double-sided Si-strip detector system has been installed and commissioned at the focal plane of the Recoil Mass Spectrometer at the Holifield Radioactive Ion Beam Facility. The system can be used for heavy charged particle emission studies with half-lives as low as a few μsec. In this paper we present identification and study of the decay properties of the five new proton emitters: ^{140}Ho, ^{141m}Ho, ^{145}Tm, ^{150m}Lu and ^{151m}Lu.

INTRODUCTION

Nuclei which are energetically unbound to the emission of a proton are located beyond the proton drip line. Observation of protons emitted from these isotopes allows us not only to establish the limits of stability for a given element, but also gives information on the structure and mass of the parent nucleus.

A simple spherical WKB calculation of the expected rate of proton tunneling through the barrier agrees with experimental data for those nuclei that have $Z \geq 69$ (see recent result for ^{145}Tm [1]). However, for those nuclei that lie in the deformed region below $Z = 69$, any calculation would have to take into effect the predicted [2] strong prolate deformation of the nucleus. Recent work by Davids et al. [3] has observed proton emission from the ground state of ^{141}Ho, with a proton energy of 1169(8) keV and half-life of 4.2(4) ms. They showed that the data do not agree with spherical WKB predictions but rather with calculations that take into account deformation..

In the present study, a double-sided Si-strip detector (DSSD) system was used to study proton emission with the Recoil Mass Spectrometer at the Oak Ridge Holifield Radioactive Ion Beam Facility (HRIBF). Technical details of this system and its capacities are given in Refs. [1,4].

New Proton Emitters: ^{145}Tm, ^{151m}Lu, ^{150m}Lu, ^{140}Ho, and ^{141m}Ho

One of the early experiments with this system resulted in the discovery of proton emission from ^{145}Tm [1], with E_p and $T_{1/2}$ of 1.728(10) MeV and 3.5(10) μs. Thulium-145 was produced via the ^{92}Mo(^{58}Ni,p4n) reaction (see Fig. 1a). A 0.91-mg/cm^2 thick target of ^{92}Mo (97 % enrichment) was bombarded with 315-MeV ^{58}Ni ions (307 MeV at the target mid-point) extracted from the HRIBF Tandem Accelerator, with an average beam current on target of ~ 15 pnA over a period of 50 hours. The production cross section for ^{145}Tm is estimated to be 500 nb. The calculated (WKB) half-life from the $0h_{11/2}$ orbital is 1.8(3) μs, which when compared with the experimental half-life yields a spectroscopic factor of 0.51(16). This value is consistent with an overall spherical description for this nucleus [1].

The lutetium isotopes were among the first nuclei where proton emission was observed. However, for both the cases of ^{151}Lu and ^{150}Lu, only one high-spin proton-emitting isomer has been observed. Neighboring nuclei however, have been observed to emit a proton from both a high- and low-spin isomer. With this in mind, we reinvestigated the proton decays of ^{151}Lu and ^{150}Lu. Lutetium-151 and ^{150}Lu were produced via the ^{96}Ru(^{58}Ni,p2n) and ^{96}Ru(^{58}Ni,p3n) reactions with beam energies of 266 and 292 MeV. Figure 1(b) shows the spectrum obtained from the above reaction with a constraint on the time between recoil and decay of ≤ 500 ms. A further constraint on the time of ≤ 50 μs is shown in Fig. 1(c). It clearly shows a peak at 1.311(10) keV which we attribute to the proton decay of an isomer located at an energy of 78(11) keV [5] above the previously known $h_{11/2}$ ground state. The half-life associated with this new radioactivity is 15.6(11) μs. A WKB approximation calculation for protons emitted from the $0h_{11/2}$ ($\Delta \ell = 5$), $1d_{3/2}$ ($\Delta \ell = 2$), and $2s_{1/2}$ ($\Delta \ell = 0$) orbitals of ^{151}Lu was performed with the experimental E_p as input (see table 1). As can be seen from the table, the calculated value that matches most closely is $d_{3/2}$ for the isomer, with resulting spectroscopic factor of 0.27(7). At the time of this writing, the ^{150}Lu data analysis is just beginning, but on-line data indicate values of E_p = ~1.35 MeV and $T_{1/2}$ of the order of 20μs [6].

Additionally, our investigations have led to the discovery of proton decay from a short-lived isomer of ^{141}Ho and from a new isotope ^{140}Ho [7]. Holmium-140 and ^{141}Ho were produced via the ^{92}Mo(^{54}Fe,p5n) and ^{92}Mo(^{54}Fe,p4n) reactions, respectively, using 315-MeV ^{54}Fe ions extracted from the HRIBF Tandem Accelerator. In this experiment, ions of mass 140 were deposited on one side (strips # 20-40) of the DSSD, and mass 141 were deposited on the other (# 1-19). Figures 2(a) and 2(b) show the decay spectrum obtained from A=141, with time constraints of (a) T ≤ 50 μs, and (b) 200 μs ≤ T ≤ 16 ms. This clearly shows that there are two proton peaks from ^{141}Ho. Their preliminary energies and half-lives are [E_p = 1177(10) keV, $T_{1/2}$ = 3.9(5) ms] for the previously known proton arising from the ground state of ^{141}Ho, (which compare well the literature values [3]), and [E_p = 1242(15) keV, $T_{1/2}$= $8.4^{+3.1}_{-1.8}$ μs] for the new isomer of ^{141}Ho. Figure 2(c) shows the decay spectrum

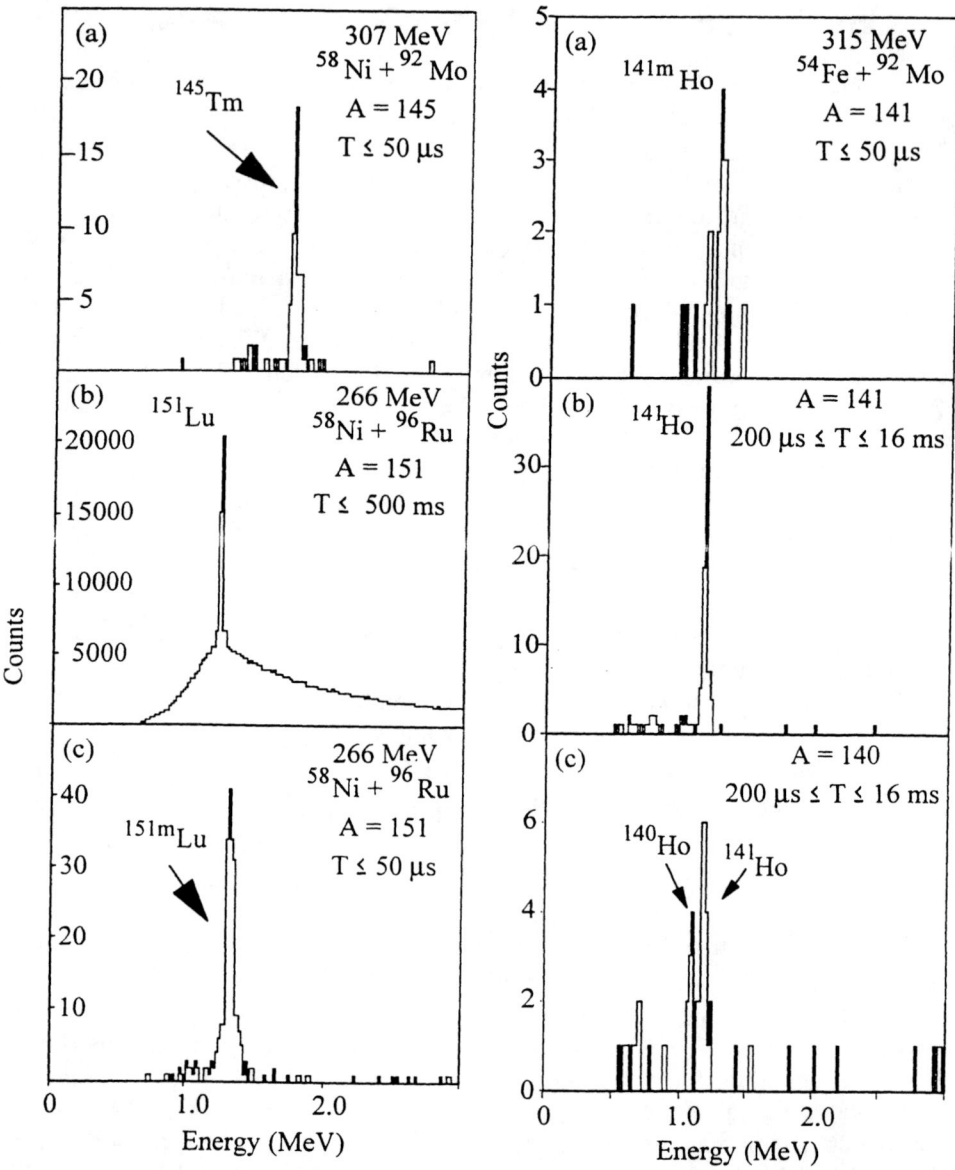

Figure 1. (a) Decay spectrum observed during 307-MeV ^{58}Ni bombardments of ^{92}Mo, with a time between decay and recoil implantation of < 50 μs. (b) Decay spectrum observed during 266-MeV ^{58}Ni bombardments of ^{96}Ru, with a time between decay and recoil implantation of < 500 ms. (c) The decay spectrum from (b) with a further time constraint of < 50 μs.

Figure 2. (a) Decay spectrum from 315-MeV ^{54}Fe bombardments of ^{92}Mo, taken from strips # 1-19 (A = 141) and a time between decay and recoil implantation of ≤ 50 μs. (b) The same spectrum with a time constraint of 200 μs ≤ T ≤ 16 ms. (c) Decay spectrum taken from strips # 20-39 (A =140) with a time constraint of 200 μs ≤ T ≤ 16 ms.

TABLE 1. Observed proton half-lives compared with spherical WKB calculations.

species	E_p (keV)	Exp $T_{1/2}$ (p)	calculated $T_{1/2}$ (WKB)		
			$0h_{11/2}$ ($\Delta l = 5$)	$1d_{3/2}$ ($\Delta l = 2$)	$2s_{1/2}$ ($\Delta l = 0$)
^{145}Tm	1.728(10)	3.5(10) μs	$1.8^{+0.3}_{-0.2}$ μs	0.7(1) ns	80(12) ps
^{151m}Lu	1.311(10)	15.6(11) μs	11(3) ms	4(1) μs	0.5(1) μs
^{141m}Ho	1.242(15)	$8.4^{+3.1}_{-1.8}$ μs	5(2) ms	1.5(6) μs	0.15(7) μs
^{141}Ho	1.177(7)	3.9(5) ms	26(5) ms	7.0(14) μs	0.73(14) μs
^{140}Ho	1.084(10)	$6.1^{+3.1}_{-1.5}$ ms	330(110) ms	86(20) μs	9(2) μs

from mass 140 with the same time constraint of 200 μs ≤ T ≤ 16 ms. Here, two lines are observed; one from the ^{141}Ho ground state (due to overlap of the tail of mass 141), and one from the new isotope ^{140}Ho, with preliminary values of E_p = 1084(15) keV, and $T_{1/2} = 6.1^{+3.1}_{-1.5}$ ms.

As can be seen from Table 1, the spherical WKB approach gives values which can not easily explain the experimental half-lives of the Ho isotopes. In Ref. 3 Davids, et al., demonstrate that the ^{141}Ho ground-state decay can be explained using the formalism developed by Bugrov and Kadmenskii [8], resulting in a good fit for $\beta_2 \sim 0.3$ for proton emission from the $\pi 7/2^-$ [523] state originating from the $h_{11/2}$ orbital. Our calculations, which minimize the energies of the ^{141}Ho states in deformation space indicates that the $\pi 1/2^+$[411] (originating from the $d_{3/2}$ orbital) and the $\pi 7/2^-$ [523] have nearly the same energy and deformation ($\beta_2 \sim 0.3$ and $\beta_4 \sim -0.07$, $\beta_6 \sim 0.1$). The next excited state, $\pi 5/2^-$ [532] is predicted to be 250 keV higher. Thus we assign the 65(15) keV isomer to the $\pi 1/2^+$[411] state. Proton emission from the odd-odd nucleus ^{140}Ho is probably arises from either the $\pi 7/2^-$[523]$\otimes\nu 9/2^-$[514] or $\pi 7/2^-$[523]$\otimes\nu 5/2^+$[402] orbital as predictions indicate that these two levels should be nearly degenerate.

ACKNOWLEDGEMENTS

* UNIRIB is a consortium of universities, Oak Ridge Associated Universities, and is supported by them and the U.S. Department of Energy under contract No. DE-AC05-76OR00033 with the Oak Ridge Associated Universities. ORNL is managed by Lockheed Martin Energy Research under contract number DE-AC05-96OR22464 with the U.S. Department of Energy.

REFERENCES

[1] J. C. Batchelder et al., Phys Rev C **57**, R1042 (1998).
[2] P. Möller, et al., At. Data Nucl. Data Tables **59**, 185 (1995).
[3] C. N. Davids et al., Phys Rev Lett. **80**, 1849 (1998).
[4] C. J. Gross et al., Application of Accelerators in Research and Industry, AIP conference proceedings # 392, Woodbury, NY, Vol 1, (1997) p 401.
[5] C. R. Bingham et al., in preparation
[6] T. N. Ginter et al., in preparation
[7] K. Rykaczewski et al., in preparation
[8] S. G. Kadmenskii and V. P. Bugrov, Phys. of Atomic Nuclei **59**, 399 (1996).

Probing the nuclear structure of ^{8}B and ^{19}C

T. Baumann[a], H. Geissel[a], H. Lenske[b], K. Markenroth[c],
T. Aumann[a], L. Axelsson[c], U. Bergmann[d], D. Cortina-Gil[a],
L. Fraile[e], M. Hellström[a], M. Ivanov[f], N. Iwasa[g], R. Janik[f],
B. Jonson[c], G. Münzenberg[a], F. Nickel[a], T. Nilsson[c], A. Ozawa[g],
A. Richter[h], K. Riisager[d], C. Scheidenberger[a], G. Schrieder[h],
W. Schwab[a], H. Simon[h], B. Sitar[f], M. H. Smedberg[c], P. Strmen[f],
K. Sümmerer[a], T. Suzuki[g], M. Winkler[a], M. V. Zhukov[c]

[a] *Gesellschaft für Schwerionenforschung, 64291 Darmstadt, Germany*
[b] *Institut für Theoretische Physik I, 35392 Giessen, Germany*
[c] *Fysiska Institutionen, Chalmers Tekniska Högskola, 412 96 Göteborg, Sweden*
[d] *Institut for Fysik og Astronomi, Aarhus Universitet, 8000 Aarhus C, Denmark*
[e] *Instituto Estructura de la Materia, CSIC, 28006 Madrid, Spain*
[f] *Comenius University, 84215 Bratislava, Slovakia*
[g] *RIKEN, 2-1 Hirosawa, Wako, Saitama 351-01, Japan*
[h] *Institut für Kernphysik, Technische Universität, 64289 Darmstadt, Germany*

Abstract. The nuclei ^{8}B and ^{19}C were investigated in breakup reactions at relativistic energies. The fragment separator FRS at GSI was used as an energy-loss spectrometer to measure the longitudinal momentum distributions of the breakup fragments after one-nucleon removal. For the case of ^{8}B, the measured momentum distribution is in agreement with our previous experiment, but in this measurement the momentum range extends up to 175 MeV/c in the projectile frame. For ^{18}C fragments from the breakup of ^{19}C, we extract a much larger momentum width than the value obtained in a measurement at lower projectile energies [1].

Nuclear halos are a well-established phenomenon for a few light nuclei close to the neutron drip-line [2], with ^{11}Be and ^{11}Li being the classical cases. These nuclei are well-described as two or three body systems, reflecting the halo structure in a rather simple way. But there are cases where the label "halo" is not so easily applied, among these are ^{8}B, a nucleus at the proton drip-line with the small binding energy of 137 keV, and ^{19}C, the last bound carbon isotope with odd neutron number. The question about the existence of a proton halo in ^{8}B and the difficulties to describe the structure of ^{19}C make these nuclei interesting cases to study in detail.

CP455, *ENAM98: Exotic Nuclei and Atomic Masses*
edited by B. M. Sherrill, D. J. Morrissey, and Cary N. Davids
© 1998 The American Institute of Physics   1-56396-804-5/98/$15.00

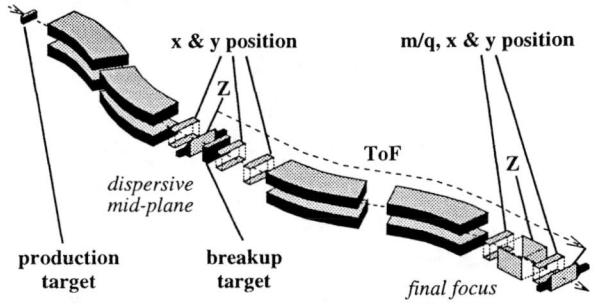

FIGURE 1. Schematic setup of the experiment at the FRS.

In our study, we measured the longitudinal momentum distributions of ^{7}Be and ^{18}C fragments after breakup of ^{8}B and ^{19}C, respectively. By measuring the momentum distribution of the core fragment of a one-nucleon halo-system, one can obtain information about the wave function of the removed valence nucleon. If the valence nucleon is removed in an ideal way, the remaining core carries the momentum of this nucleon, whose longitudinal component can be measured in a magnetic spectrometer. In this case, the observed momentum distribution should be directly related to the Fourier transformed wave function of the valence-nucleon. A narrow longitudinal momentum distribution would correspond to a spatially extended wave function according to the Heisenberg uncertainty principle.

However, in reality the measured momentum distribution includes contributions from various final state interactions and reaction dynamics. Also, parts of the halo wave function are shielded by the core and not accessible experimentally. Though these effects tend to reduce the width of the momentum distribution, the relation to the valence wave function persists, and conclusions on the spatial shape of the halo wave function, especially at high incident energies, still can be drawn.

The experiment was carried out at the relativistic radioactive beam facility of GSI. For the production of ^{8}B and ^{19}C, we used primary beams of ^{12}C at 1.5 GeV/u and ^{40}Ar at 1.0 GeV/u, respectively. Production targets of ^{9}Be were placed at the entrance of the magnetic spectrometer FRS, which was operated as an energy-loss spectrometer. A carbon breakup target, placed at the dispersive mid-plane, was used in both measurements. We employed time projection chambers for high-resolution position measurements at the mid plane and the final focus of the FRS. A Z-identification before the breakup target and a full particle identification at the final focus were achieved using plastic scintillators and an ionization chamber (see Fig. 1). With the particle identification and the m/q selection of the spectrometer, the reaction channels were well-defined.

The momentum distribution of the core fragments is directly deduced from the position measurements at the final focus, using the experimentally determined dispersion. In order to extend the range of momenta, the part of the spectrometer following the breakup target was scaled to higher magnetic rigidities. The recorded

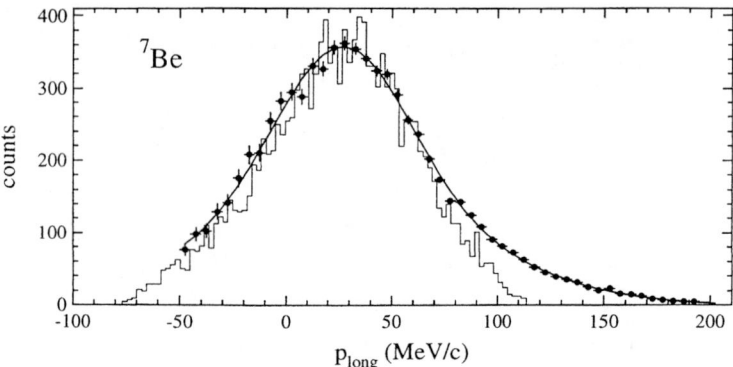

FIGURE 2. The measured ^{7}Be longitudinal momentum distributions from a breakup reaction of ^{8}B in a carbon target. The filled datapoints mark the recent measurement, also plotted is a fit (bold line); the histogram shows the distribution from a previous measurement [3] (absolute values on the axes refer to the filled datapoints).

position distributions at each setting were combined to one momentum distribution.

Figure 2 shows the recently measured longitudinal momentum distribution of ^{7}Be from the ^{8}B breakup at 1440 MeV/u in a 4.45 g/cm^2 carbon target. The comparison with the distribution obtained in our previous experiment [3] shows the improvement—in statistics and in the range of momenta—represented by this new measurement, while both distributions are in good agreement in the central region. The recently measured distribution, extending to 175 MeV/c from the center, determines not only the FWHM of 91 ± 5 MeV/c (projectile frame), but also the shape of the tails of the momentum distribution.

The longitudinal momentum distribution for ^{18}C fragments arising from the breakup of ^{19}C at 914 MeV/u in a 4.45 g/cm^2 carbon target is displayed in the left

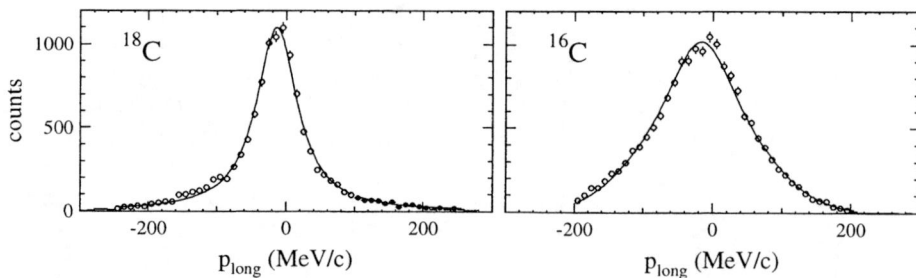

FIGURE 3. The measured ^{18,16}C longitudinal momentum distribution after one-neutron removal from ^{19,17}C in a carbon target. The filled symbols in the left panel represent datapoints from a second spectrometer setting to extend the measurement to higher momenta. For ^{18}C, a Lorentzian fits the data, in the case of ^{16}C two Gaussians were added.

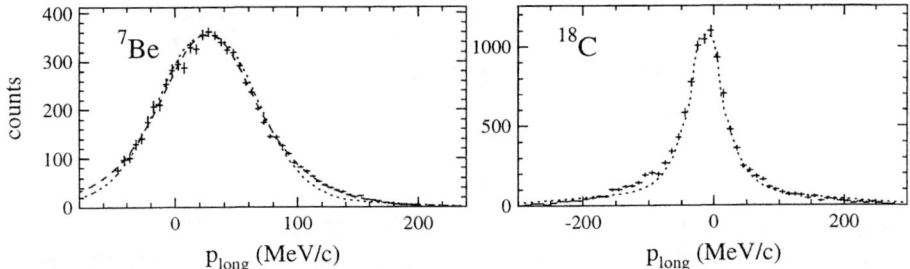

FIGURE 4. The measured longitudinal momentum distributions in comparison to theoretical calculations. Left panel: the dashed curve results from a calculation in the three-body cluster model [4], whereas the dotted curve represents a mean-field plus RPA calculation [5]. Right panel, dotted curve: RPA calculation including dynamical core polarization [7].

panel of Fig. 3. This distribution has a Lorentzian shape, and the FWHM amounts to 69 ± 3 MeV/c. In the right panel of Fig. 3, we show the momentum distribution of ^{16}C from the breakup of ^{17}C that was measured in the same experiment. A comparison with the widths parameters obtained in experiments at lower projectile energies [1] shows an interesting fact: While there is a perfect agreement for the case of ^{17}C (143 ± 6 MeV/c compared to 145 ± 5 MeV/c from [1]), the width for the ^{19}C breakup, 42 ± 5 MeV/c, is considerably narrower than our value. An energy dependence of the breakup reaction for ^{19}C could be one possible explanation for the observed difference.

Theoretical calculations, as shown in Fig. 4, are able to reproduce the observed momentum widths for ^{8}B and ^{19}C. Although completely different in their approaches, a calculation in the three-body cluster model as well as a mean-field plus RPA calculation result in an extended proton wave function in ^{8}B. The theoretical description of ^{19}C is still a challenge, especially if one considers a possible energy dependence of the momentum width. But the recent investigations seem to indicate that ^{19}C has a less pronounced halo than, e.g., ^{11}Be.

The data presented here give a precise information about the shape of the distributions, and this, rather than just a width parameter, should be the object of comparison between measured datasets and theoretical calculations.

REFERENCES

1. Bazin, D., et al., *Phys. Rev. C* **57**, 2156 (1998).
2. Hansen, P. G., et al., *Annu. Rev. Nucl. Part. Sci.* **45**, 591 (1995).
3. Schwab, W., et al., *Z. Phys.* **A350**, 283 (1995).
4. Grigorenko, L., *Phys. Rev. C* **57** (1998), and Smedberg, M. H., private communication.
5. Lenske, H., private communication.
6. Bazin, D., et al., *Phys. Rev. Lett.* **74**, 3569 (1995).
7. Baumann, T., et al., submitted to *Phys. Lett. B*, (1998).

Measurement of Parentage in Stripping Reactions of Halo Nuclei

Daniel Bazin[a], N. Alahari[a], B. Blank[b], J.E. Bush[c], J.A. Caggiano[a], L. Chen[a], B. Davids[a], T. Glasmacher[a], K. Govaert[a], V. Guimares[d], P.G. Hansen[a], R.W. Ibbotson[a], D. Karnes[a], J.J. Kolata[d], V. Maddalena[a], B. Pritychenko[a], H. Scheit[a], B.M. Sherrill[a]

[a] N.S.C.L., Michigan State University, East Lansing, MI 48824-1321, USA
[b] C.E.N. Bordeaux-Gradignan, Le Haut Vigneau, F-33175 Gradignan Cedex, France
[c] Dept. of Physics, Univ. of Pennsylvania, Philadelphia, PA 19104-6396, USA
[d] Dept. of Physics, Univ. of Notre Dame, Notre Dame, IN 46556, USA

Abstract. Recent experiments performed using the S800 spectrograph at the National Superconducting Cyclotron Laboratory have shed a new light on techniques to measure the valence wave function of halo nuclei. The momentum of the heavy fragment resulting from the stripping of the halo particle is measured by the spectrograph, and its final excited state is measured by means of a NaI γ-detector array located around the target. The large angular and momentum acceptances of the S800 provide a measurement of the full momentum vector. The analysis of the shape of the momentum distributions corresponding to different core fragment final states distinguishes between core excitations coming from reactions where a particle from the core - rather than the halo - is removed in the reaction, and components of the halo wave function where the halo particle is coupled to an excited state of the core.

INTRODUCTION

Stripping reactions of valence particles in halo nuclei may be a way to probe the spacial extent of their wave functions. Typically, the momentum distribution of the remaining core is measured at forward angles, and interpreted as the Fourier transform of the spacial wave function, with the appropriate core absorption effects taken into account. However, a careful comparison with calculations shows that discrepancies occur in the tails of the distributions, which suggests that the core is not merely a spectator of the reaction, but might also play an active role in the reaction (1). The detection of γ-rays in coincidence with the core fragments provides an identification of the reaction process and of the different components of the valence wave function. The momentum distributions corresponding to nucleons removed from different orbits and different core fragment final states will exhibit different shapes, and from the absolute calibrations of both the cross sections and the γ-rays array efficiency, as well as an appropriate reaction model, spectroscopic factors can be deduced. For example in the case of ^{11}Be, the ground state wave function of the valence neutron is known to be predominantly s-wave but may have some d-wave coupled to the 2^+ state of ^{10}Be. The ground state wave function also has 4 neutrons in the $0p_{3/2}$ shell and removal of these neutrons produces a hole state in ^{10}Be, either a 1^- or 2^- at around 6 MeV. In this contribution, we show that the measured core recoil momenta in coincidence with different final states of the core is consistent with this picture.

TWO TEST CASES: ^{11}Be AND ^{15}C

Both ^{11}Be and ^{15}C are interesting to study because they are known to have a dominant s-wave component in their ground state wave functions (2,3), but rather different binding energies (0.5 MeV and 1.2 MeV respectively). Fig. 1 shows the momentum distributions of the ^{10}Be core fragments after the breakup of ^{11}Be on a CH$_2$ target. The losange distribution is gated on the Doppler corrected 3.37 MeV

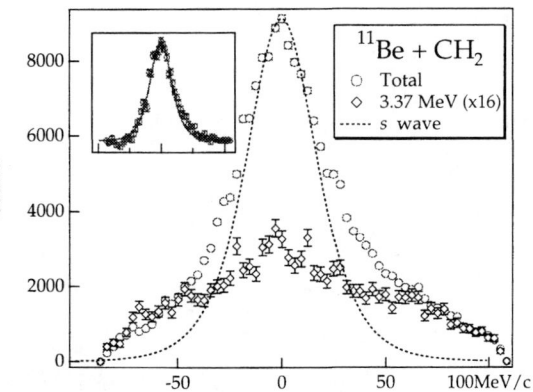

FIGURE 1. Momentum distributions of ^{10}Be after the breakup of ^{11}Be on a CH$_2$ target. See text for details.

$2^+\rightarrow 0^+$ transition in ^{10}Be. The dotted line is an $l=0$ Hankel wave function calculation (4) and clearly doesn't account for the total distribution, especially in the tails. In the inset the difference between the total distribution and the one gated on the γ-ray (multiplied by 16 to account for the efficiency) is in much better aggreement with the calculation. One possible interpretation of this result is that a neutron from the ^{10}Be core rather than the halo is removed in the reaction. In that case the neutron is removed from the $0p_{3/2}$ shell, which means that the shape of the momentum distribution should that of a p-wave. On the other hand, the excited ^{10}Be core could result from a parentage to its 2^+ in the wave function of ^{11}Be. In this case however, because the ground state of ^{11}Be is $1/2^+$, the parentage would be $^{10}\text{Be}^{2+} \otimes 0d_{5/2}$ and one would expect to observe a d-wave for the gated momentum distribution. Fig. 2 shows the data compared to the two calculations, which indicates that these events come from reactions where a neutron is removed from the ^{10}Be core rather than the halo.

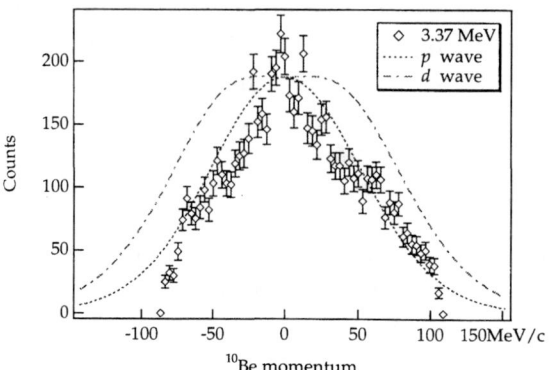

FIGURE 2. Momentum distribution of ^{10}Be gated on the 3.37 MeV γ-ray. The calculations correspond to two possible interpretations (see text).

A similar situation occurs in ^{15}C, where the observed γ-ray at 6.09 MeV corresponds to the $1^- \to 0^+$ transition in ^{14}C. In the case of a neutron removal from the ^{14}C core, this neutron would have been in the $0p_{1/2}$ shell, whereas a contribution from a parentage to the 1^- state in ^{14}C in the wave function of ^{15}C would lead to $^{14}\text{C}^{1-} \otimes 1s_{1/2}$ and/or $^{14}\text{C}^{1-} \otimes 0d_{5/2}$ configurations. Fig. 3 shows the data with the $l=0$ Hankel wave function calculation, as well as the p-wave corresponding to the removal of a neutron from the ^{14}C core. The results obtained on these two nuclei (^{11}Be and ^{15}C) indicate that knockout reactions where a neutron is removed from the core are not uncommon, and have to be taken into account for a precise analysis of the momentum distributions.

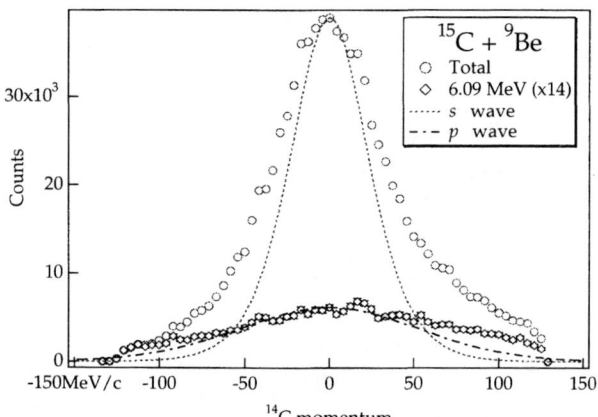

FIGURE 3. Momentum distributions of ^{14}C fragments after the breakup of ^{15}C on a Be target. Total (circle) and gated on the 6.09 MeV γ-ray (losanges). The dotted line corresponds to a pure s-wave calculation, and the dashed-dotted line to a p-wave if a neutron is removed from the core rather than the halo.

A MORE COMPLEX CASE: ^{17}C

Shell model calculations suggest that the ground state of ^{17}C is more complex than that of ^{11}Be or ^{15}C, as the $0d_{5/2}$ and $1s_{1/2}$ orbits are close to each other, and that both s and d single-particle components could be present in the wave function. Parentages of 1.58 and 0.16 for the $0d_{5/2}$ and $1s_{1/2}$ orbits are obtained with the WBP interaction (5), assuming a coupling to the 2^+ state of ^{16}C and a $3/2^+$ ground state for ^{17}C. The momentum distribution calculated for this configuration is shown in fig. 4 together with the data (circles), as well as an $l=2$ Hankel wave function calculation. The shell model calculation agrees remarkably well with the data and futhermore, the momentum distribution gated on the $2^+ \to 0^+$ transition in ^{16}C (losanges) has basically the same shape as the total distribution (see figure). This observation is radically different from the behavior observed in ^{11}Be and ^{15}C. It points towards the shell model description of this nucleus, where the valence (halo) neutron is coupled to an excited ^{16}C core. Moreover, the intensity of the 1.77 MeV peak accounts for about 80% (preliminary number) of the total cross section, whereas the branching observed for ^{11}Be and ^{15}C is only about 5-10%.

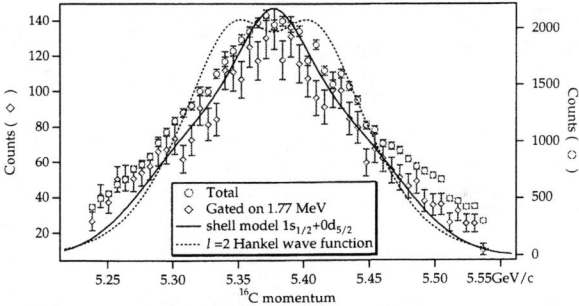

FIGURE 4. Momentum distributions of ^{16}C after the breakup of ^{17}C on a Be target. See text for details.

The few results presented here were obtained on the large acceptance and high resolution spectrograph S800. The qualities of this instrument, used in combination with a γ-ray detection array, allow a very detailed study of the wave functions of halo nuclei, as well as the discrimination of the various processes taking place in knockout reactions.

REFERENCES

1. D. Bazin *et al.*, Phys. Rev. C **57**, 2156 (1998).
2. J.H. Kelley *et al.*, Phys. Rev. Lett. **74**, 30 (1995).
3. J.D. Goss *et al.*, Phys. Rev. C **8**, 514 (1973).
4. P.G. Hansen, Phys. Rev. Lett. **77**, 1016 (1996).
5. E.K. Warburton and B.A. Brown, Phys. Rev. C **46**, 923 (1992).

On the Road to Doubly-magic ^{48}Ni

B. Blank[a], J. Benlliure[b], F. Boué[b], R. Collatz[b], S. Czajkowski[a],
F. Davi[a], R. Del Moral[a], J.P. Dufour[a], A. Fleury[a], A. Heinz[b],
M. Lewitowicz[d], C. Marchand[a], M. Hellström[b], Z. Hu[b], Z. Janas[c],
M. Karny[c], M. Pfützner[c], E. Roeckl[b], M.S. Pravikoff[a], M. Shibata[b],
K. Sümmerer[b]

[a] *Centre d'Etudes Nucléaires de Bordeaux-Gradignan, F-33175 Gradignan Cedex, France*
[b] *Gesellschaft für Schwerionenforschung, Planckstr. 1, D-64291 Darmstadt, Germany*
[c] *Inst. of Exp. Physics, University of Warsaw, PL-00-681 Warsaw, Hoża 69, Poland*
[d] *Grand Accélérateur National des Ions Lourds, B.P. 5027, F-14021 Caen Cedex, France*

Abstract. A relativistic primary beam of ^{58}Ni from the SIS synchrotron at GSI was used to produce proton-rich isotopes in the titanium-to-nickel region by projectile fragmention at the FRS. We report here on the first observation of the $T_z = -7/2$ nuclei ^{45}Fe and ^{49}Ni. In addition, the new isotope ^{42}Cr ($T_z = -3$) was identified. This opens the route to the yet unobserved doubly-magic nucleus ^{48}Ni.

I INTRODUCTION

Doubly-magic nuclei have a magic attraction for nuclear physicists. These nuclei exhibit a remarkable stability compared to neighboring isotopes and are, at least if not too far from stability, spherical. Two doubly-magic nuclei have been observed for the first time recently: ^{100}Sn [1,2] and ^{78}Ni [3]. There is probably only one doubly-magic nucleus, ^{48}Ni, yet unobserved, which is within experimental reach in the near future. The particular interest of this nucleus is that it is the only case over the entire chart of nuclei where the mirror nucleus, ^{48}Ca, is also bound. In addition, the element nickel is the only one where three doubly-magic nuclei should exist, i.e. ^{48}Ni, ^{56}Ni, and ^{78}Ni.

II EXPERIMENTAL TECHNIQUE AND RESULTS

In experiments [4,5] performed at the FRS of GSI, we used relativistic primary beams of ^{58}Ni to produce and separate proton-rich exotic fragments. The isotopes selected by the FRS were identified by the standard detection set-up of the FRS.

In a first experiment, production cross sections for proton-rich ^{58}Ni fragments were systematically measured. A comparison with model prediction of the EPAX parametrisation shows that the experimentally observed production rates far from stability are much higher than predicted. This discrepancy reaches a factor of 750 for ^{46}Fe. These experimental results allowed us to extrapolate to the production cross sections of ^{45}Fe of 5 pb and of ^{48}Ni of 0.1 pb.

In a second experiment, we searched for these two nuclei. In a three-days setting optimized for ^{45}Fe, we obtained the reults presented in Fig. 1. The nuclear charge Z and the mass-to-charge ratio A/Z are determined by means of an energy-loss measurement, the TOF, and the $B\rho$ as calculated from the positions in the intermediate and final focal planes as well as from the respective magnetic fields.

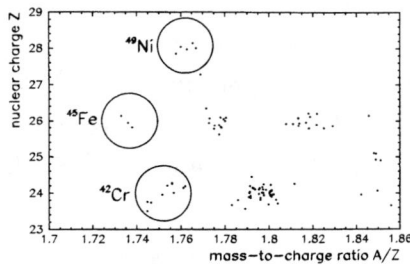

FIGURE 1. Two-dimensional plot of the nuclear charge Z versus the mass-to-charge ratio A/Z for a setting optimized for ^{45}Fe.

We observed 10 events of ^{42}Cr, 3 events of ^{45}Fe, and 5 events of ^{49}Ni. These three isotopes have been identified for the first time in the present experiment. As the production cross sections decrease by about a factor of 20 per mass unit [4], we do not expect any count for the doubly-magic nucleus ^{48}Ni.

^{45}Fe and ^{49}Ni are the first $T_z = -7/2$ nuclei experimentally identified. It is very likely that they are the only nuclei of the $T_z = -7/2$ series which are experimentally accessible. The observation of ^{45}Fe is of prime interest for the search for 2p radioactivity. According to recent mass predictions, it is the best candidate for the 2p radioactivity because it is within experimental reach and its $T_{1/2}^{2p}$ value is predicted to be short as compared to its β-decay half-life. Similarly to ^{45}Fe, ^{49}Ni and ^{42}Cr are predicted to be 2p unbound by all commonly used mass models, however, most mass predictions give them unbound only by a few hundred keV, so that its decay should be dominated by β decay.

III SUMMARY AND OUTLOOK

We reported on the first observation of the two $T_z = -7/2$ nuclei ^{45}Fe and ^{49}Ni, the most proton-rich nuclei ever identified with an excess of seven protons. These nuclei, together with the new $T_z = -3$ nucleus ^{42}Cr, are all predicted to be two-proton unbound, ^{45}Fe being probably the best candidate for 2p radioactivity.

Another important motivation for pursuing these studies is the conclusion that the doubly-magic nucleus ^{48}Ni is reachable in fragmentation experiments. In deed, with the installation of the SIS cooler, primary-beam intensity increases of a factor of 10 are expected for ^{58}Ni. This should yield prodcution rates for ^{48}Ni of about 1-3 counts per day.

REFERENCES

1. R. Schneider *et al.*, Z. Phys. A **384**, 241 (1994).
2. M. Lewitowicz *et al.*, Phys. Lett. B **332**, 20 (1994).
3. C. Engelmann *et al.*, Z. Phys. A **352**, 351 (1995).
4. B. Blank *et al.*, Phys. Rev. C **50**, 2398 (1994).
5. B. Blank *et al.*, Phys. Rev. Lett. **77**, 2893 (1996).

Toward a Complete Picture of ^{11}Be Interaction with Silicon

C. Borcea[1], F. Carstoiu[1], F. Negoita[1], M. Lewitowicz[2],
M.G. Saint-Laurent[2], R. Anne[2], D. Guillemaud-Mueller[3],
A.C. Mueller[3], F. Pougheon[3], O. Sorlin[3], A. Fomitchev[4],
S. Lukyanov[4], Yu. Penionzhkevich[4], N. Skobelev[4], Z. Dlouhy[5]

[1]) *IFIN-HH, P.O. Box MG-6, 76900 Bucharest-Magurele, Romania*
[2]) *GANIL(IN2P3/CNRS,DSM/CEA) BP 5027, 14021 Caen Cedex, France*
[3]) *IPN, CNRS-IN2P3, 91406 Orsay Cedex, France*
[4]) *FLNR, JINR, 141980 Dubna, Moscow Region, Russia*
[5]) *NPI, 250 68 Rez, Czech Republic*

Abstract. For the interaction of ^{11}Be with silicon, the dissociation and stripping cross sections and momentum distributions have been studied in a wide energy range. Cross sections for other reaction channels and angular distributions of coincident neutrons completed the picture of this interaction.

Various reaction channels in the interaction of the halo nucleus ^{11}Be with silicon have been studied by means of a multiple telescope coupled to an array of neutron detectors. The 43 MeV/nucleon ^{11}Be secondary beam has been obtained at GANIL using the LISE3 spectrometer; it was stopped inside this telescope and reactions occurred at various depths which means various energies of ^{11}Be. The reaction products have been identified in successive detectors. Therefore the telescope served as energy degrader, target and detector/identifier. The set-up permitted to extract simultaneously many different experimental information [1,2], as cross sections for different reaction mechanism or momentum distributions.

An extended Glauber model that uses realistic wave functions (w.f.) for ^{11}Be [3] has been developed with the intention to describe all the data without free parameters. A more detailed presentation will be given in a forth comming publication. This model shows that while parallel momentum distribution of the core is sensitive mainly to the asymptotic *slope* of the w.f., the breakup cross section is sensitive to the *normalization* of the w.f. Therefore, measuring *both* characteristics in the same experiment one could provide a rather complete information on the w.f. In the present experiment, stripping and diffraction/coulomb dissociation, the two main mechanisms that contribute to the one neutron breakup of ^{11}Be were separately

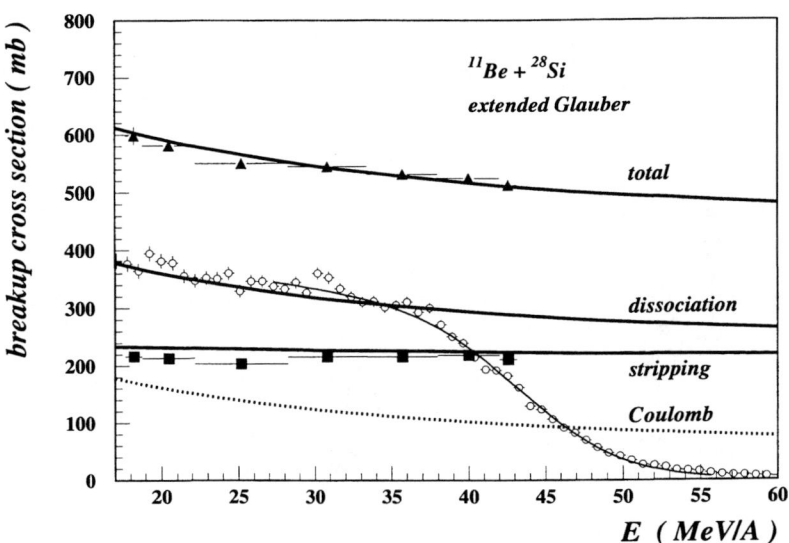

FIGURE 1. Dissociation, stripping and total breakup cross sections. The full lines are the result of an extended Glauber model calculations. The doted line indicates the Coulomb part of the dissociation cross section. The thin line through the high energy part of the dissociation breakup data shows the fit that accounts for the parallel momentum distribution of the ^{10}Be.

measured [2] in the energy range 20–40 MeV/nucleon. Both the magnitude and the trend of the data are satisfactorily reproduced by the model, as can be seen in Fig. 1. One observes that the energy dependence is essentially due to the decrease of the Coulomb contribution when the energy increases.

A third mechanism which we call friction breakup has been observed. It involves a strong core-target interaction that however preserves the core integrity. The energy damping of the core may amount to 70 MeV or more in this case. Most of these events are associated with the emission of a neutron. For this mechanism no significant energy dependence has been observed. Its contribution to the breakup cross section is rather weak: 40±10 mb.

Longitudinal momentum distribution of the ^{10}Be core has been determined for stripping by measuring its energy distribution at the exit from the reaction detector. A tail extending towards lower energies and which corresponds to the mechanism described above has to be subtracted before comparing to model calculations that accounted for the reaction mechanism (see Fig. 2). The enlarging effects of the detector thickness, straggling and finite energy resolution have been properly taken into account. The FWHM found in this case is 44±3 MeV/c while the parameter free calculation gives 40 MeV/c, slightly under the experimental value. Though the subtraction of the "tail" in Fig. 2 has been guided by the coincident neutron data (one remind that coincident halo neutrons should be absent for stripping), the low statistics may imply some uncertainties of this procedure that will reflect themselves in the width of the distribution (they have been included in the above error).

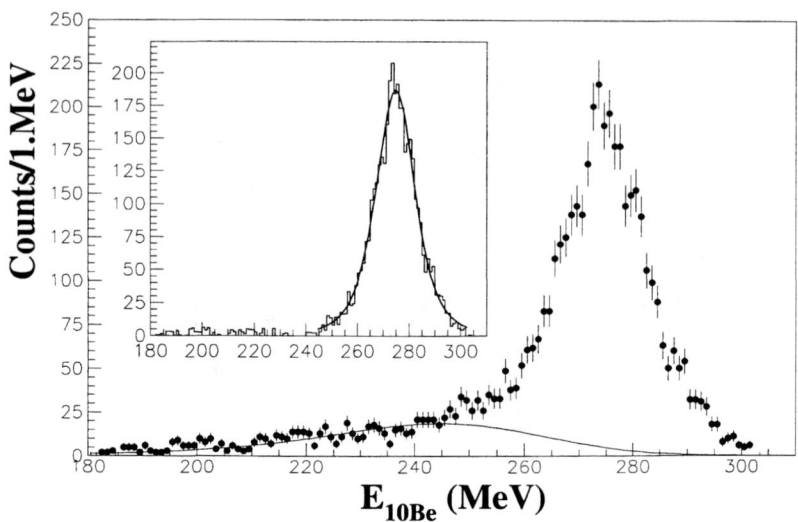

FIGURE 2. Energy distribution of ^{10}Be after the detector in which it was produced in an stripping breakup event. The "tail" towards the low energies corresponds to friction breakup as described in text. The inset shows the distribution after the "tail" subtraction and the fit (continuous line) for determining the parallel momentum distribution of ^{10}Be.

It should be mentioned that the functional used for describing the momentum distribution accounts for the reaction mechanism and is able to reproduce the whole experimental distribution including the tails, which was not the case with lorentzian or gaussian shapes. The procedure has been applied to four successive detectors, i.e. to four different energies, finding similar values.

For dissociation, the longitudinal momentum distributions of the emerging ^{10}Be has been determined from the shape of the fall off at the highest energies in Fig. 1, as described in [2]. Indeed, in the absence of such a distribution, the fall off should be much sharper and essentially due to instrumental effects (finite resolution, straggling). The momentum distribution of ^{10}Be makes the descent much slower, adding a long tail toward higher energies. The distribution that satisfies the present data has a FWHM of 41±2 MeV/c. Our model calculations for this distribution include contributions due to nuclear diffraction and to Coulomb dissociation and yield a value of 40 MeV/c, in good agreement with the measured data.

The experimental device permitted to measure some other interesting quantities. One of them is the two neutron removal cross section for ^{11}Be (in this case ^{9}Be has been identified in the exit channel) which accounts for about 50 mb. Due to the halo character of ^{11}Be, one expects this value to be close to the one neutron removal cross section of ^{10}Be core. Another quantity is the charge changing cross section of ^{11}Be as a function of energy; at 40 MeV/nucleon it amounts to 1450±50 mb. Lower limits have been determined for the reaction cross section of channels in which Li or He isotopes have been identified as final products: they are in both cases equal to 250 mb.

^{10}Be has been present as a small impurity in the incident beam; that allowed to measure its charge changing cross section whose value was found to be very close to that of ^{11}Be.

The neutron detection array provided informations concerning the angular distributions of neutrons coincident with various kinds of reactions in the telescope. In principle, these data can provide informations about the transverse momentum distribution of the neutron. However, the influence of the reaction mechanism on the observed distributions is important and not simply to evaluate. Coincident measurements have been performed for the following types of processes identified by the telescope: diffraction/coulomb dissociation, friction breakup, removal of two neutrons (i.e. ^{9}Be in the exit channel), other reactions in the telescope and in particular those in which Li and He isotopes were present in the exit channel. To all these data, in order to render them comparable with each other, a standard fitting procedure have been applied: two lorentzian distributions of different widths have been considered for the momentum distribution of the neutron in the ^{11}Be frame. After transforming to the lab system and converting into angular distributions they have been compared to the data. The results concerning the width (in MeV/c) of the narrow component which is of interest for the present discussion are the following: 49 for difraction/Coulomb dissociation, 35 for friction breakup, 36 for two neutron removal, 34 for reactions, 33 for reaction having a Li in the exit channel and 36 for reactions having a He in the exit channel. With the exception of dissociation, all other data are grouped around the value of 35 MeV/c. For them, the core has suffered a rather violent interaction with the target which is not the case for dissociation. Therefore, the domain of impact parameters in the two cases is rather different which could possibly explain the observed difference.

In conclusion, for the two main breakup mechanisms (diffraction/dissociation and stripping), the cross sections and longitudinal momentum distribution of the core have been measured in a wide energy range (20–40 MeV/nucleon). Taken together they provide informations on both normalization and asymptotic slope of the wave function. Model calculations that use realistic wave functions and account for the reaction mechanism are in agreement with the ensemble of these data. The angular distributions of neutrons coincident with different reactions in the telescope are very much influenced by the specific reaction mechanism. Though calculations are not available for these distributions yet, they may stimulate further theoretical work.

REFERENCES

1. Corre, J. M., et al., Nucl. Instrum. Meth. **A359**, 511 (1995).
2. Negoita F., et al., Phys. Rev. **C54**, 1787 (1996).
3. Nunes, F. M., Thompson, I. J., and Johnson, R. C., Nucl. Phys. **A596**, 171 (1996).

Deformation dependence of proton decay rates and angular distributions in a time-dependent approach

N. Carjan*, P. Talou*, D. Strottman[†]

* Centre d'Etudes Nucléaires de Bordeaux-Gradignan, 33175 Gradignan cedex, France
[†] Theoretical Division, Los Alamos National Laboratory, Los Alamos, NM 87545, USA

Abstract. A new, time-dependent, approach to proton decay from axially symmetric deformed nuclei is presented. The two-dimensional time-dependent Schrödinger equation for the interaction between the emitted proton and the rest of the nucleus is solved numerically for well defined initial quasi-stationary proton states. Applied to the hypothetical proton emission from excited states in deformed nuclei of ^{208}Pb, this approach shows that the problem cannot be reduced to one dimension. There are in general more than one directions of emission with wide distributions around them, determined mainly by the quantum numbers of the initial wave function rather than by the potential landscape. The distribution of the "residual" angular momentum and its variation in time play a major role in the determination of the decay rate. In a couple of cases, no exponential decay was found during the calculated time evolution (2×10^{-21}sec) although more than half of the wave function escaped during that time.

I FORMALISM

The hamiltonian $\mathcal{H}_{pA}$ describing the interaction between the emitted proton and the axially symmetric core-daughter nucleus can be written in cylindrical coordinates as

$$\mathcal{H}_{pA}(z,\rho) = -\frac{\hbar^2}{2\mu}\left[\frac{1}{\rho}\frac{\partial}{\partial \rho} + \frac{\partial^2}{\partial \rho^2} + \frac{\partial^2}{\partial z^2} - \frac{\Lambda^2}{\rho^2}\right] + \mathcal{V}_{pA}(z,\rho) \qquad (1)$$

with $\mathcal{V}_{pA}$ the (stationary) potential interaction. Λ represents the projection of the proton total angular momentum on the axis of symmetry (z axis) and is a good quantum number. The total proton wave function is $\psi(z,\rho,\phi,t) = f(z,\rho,t)exp(i\Lambda\phi)$ and the two-dimensional Time-Dependent Schrödinger Equation (TDSE) can then be written as

$$i\hbar\frac{\partial}{\partial t}f(z,\rho,t) = \mathcal{H}_{pA}(z,\rho)f(z,\rho,t) \qquad (2)$$

To obtain the time evolution of an initial wavefunction $\psi(\vec{r},0)$, this equation has been integrated numerically within a finite spatial grid using a multistep predictor method called MSD2 [1]. Knowing $\psi(\vec{r},t)$ at any time t, the following physical quantities can be estimated :

• **Total tunneling probablity** given by the fraction of the wave function located outside the deformed nuclear surface at time t

$$P_{tun}(t) = \int_{V_{out}} |\psi(\vec{r},t)|^2 d^3r \qquad (3)$$

• **Tunneling angular distribution** (or differential cross-section of tunneling) estimated in spherical coordinates (r,θ,ϕ) and defined as

$$P_{tun}(t,\theta) = \frac{dP_{tun}}{d\Omega} = \int_{r_{out}(\theta)}^{\infty} |f(r,\theta,t)|^2 r^2 dr \qquad (4)$$

where $r_{out}(\theta)$ is the radial position of the potential ridge in the direction θ.

• **Total decay rate** related to the total tunneling probability by

$$\lambda(t) = \frac{1}{1 - P_{tun}(t)} \frac{dP_{tun}}{dt} \qquad (5)$$

The potential of interaction V_{pA} has been chosen as a Woods-Saxon nuclear potential correlated to the shape of the daughter nucleus, plus a Coulomb part assuming a uniformly charged nuclear volume (for more details, see [3]). A convenient parametrization of the nuclear shapes has been used such that only one parameter ε is needed to describes shapes from the spherical case ($\varepsilon = 0$) to the point of two tangent fragments ($\varepsilon = 1$). Once the shape of the nucleus is defined, quasi-stationary states are obtained in a Nilsson-like model, and injected as initial wave functions in the TDSE.

II NUMERICAL RESULTS

Although we have applied the above formalism to the hypothetical case of proton emission from excited states of ^{208}Pb, the qualitative results found are independent of this choice. No comparison with experiment was attempted.

Figs. 1 and 2 represent the quasi-stationary states chosen as initial wavefunctions in TDSE for a spherical potential as well as for a slightly deformed one.

In the deformed case, the degeneracy is well removed but the wavefunction changed noticeably only for $\Lambda = 0$ and 1. For $\varepsilon = 0$, in spite of very different spatial distributions for different Λ values of the single-proton wavefunctions and of very different effective potentials in Eq. (1), the calculated decay rates are identical (cf. Fig. 3). For $\varepsilon = 0.1$, the different states don't evolve in the same way. After the

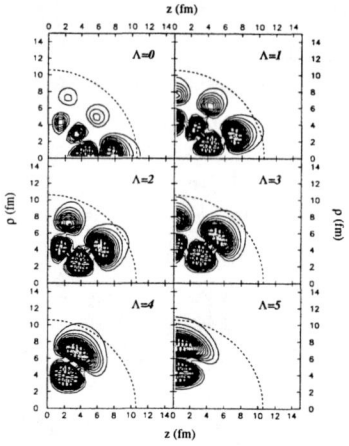

FIGURE 1. $|\psi|^2$ of the $2h$ proton states ($l = 5$, $E = 12.48$ MeV) in a ^{208}Pb spherical nucleus. The dashed lines represent the spherical potential ridge.

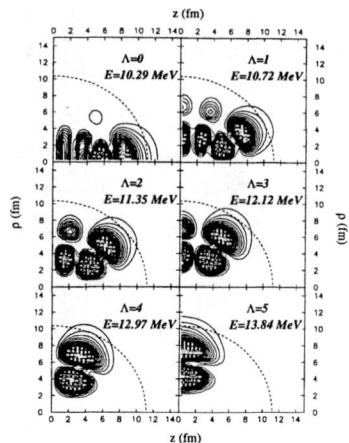

FIGURE 2. Same as in Fig. 1 but for a slightly deformed potential ($\varepsilon = 0.1$).

initial transient period observed in the spherical case too [2], some decay rates reach a stationary period (exponential decay) while others still evolve. This behaviour seems to be related to the number of branches in the wavefunction, and to the width of the initial distribution of angular momenta (cf. Fig. 4).

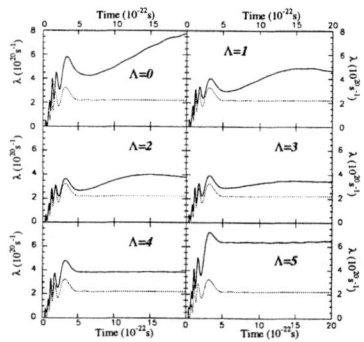

FIGURE 3. Time-dependent decay rates λ_{TD} and tunneling probabilities P_{tun} obtained with the $2h$ initial states in a slightly deformed potential. The same quantities for the states of Fig. 1 evolving in a spherical potential are plotted in dotted lines.

In Fig. 5 are represented the tunneling angular distributions calculated at different times for the "$2h$" states evolving in the deformed potential. The most probable directions of emission are dictated by the orientation of the main branches of the quasi-stationary proton states and they do not always coincide with the direction of lowest barrier. There is a broad distribution of emission angles around these directions that unavoidably interfere. Obviously, the problem cannot be reduced to one dimension.

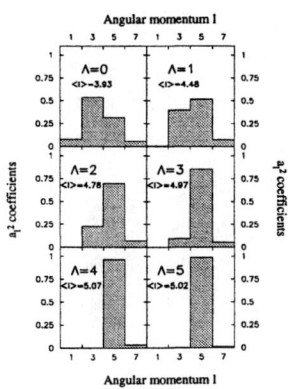
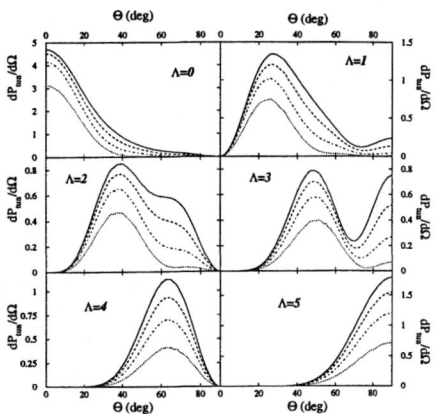

FIGURE 4. Angular momentum distributions for the "2h" states at $\varepsilon = 0.1$ (calculations performed at $t = 0$). Only odd l-values are permitted because of parity conservation.

FIGURE 5. Time evolution of the tunneling angular distributions for the "2h" states at $\varepsilon = 0.1$. Curves are equally spaced in time from $t = 2.5 \times 10^{-22}s$ to $t = 20 \times 10^{-22}s$.

III CONCLUSION

The numerical scheme developed in this work, i.e., the numerical resolution of the time-dependent Schrödinger equation for well defined initial metastable states, has been proved to be very fruitful. Informations on the very dynamics of the multidimensional tunneling process have been pointed out. Applied to the proton emission from deformed nuclei, it has been shown that a simple one-dimensional semi-classical approach is not enough to take into account the different degrees of freedom of the process, and that a more elaborate scheme is needed. This approach seems very promising and could be applied to many related problems : proton emission from highly deformed nuclei [4], $2p$ emission [5], chaos-assisted tunneling [6], to name but a few.

REFERENCES

1. T. Iitaka, Phys. Rev. E (May 1994)
2. O. Serot, N. Carjan, D. Strottman, Nucl. Phys. A569 (1994) 562
3. P. Talou, N. Carjan, D. Strottman. Submitted to Nuclear Physics A.
4. C.N. Davids et al., Phys. Rev. Lett. 80,#9 (1998) 1849
5. C. Detraz et al., Nucl. Phys. A519 (1990) 529
6. S. Tomsovic and D. Ullmo, Phys. Rev. E 50,#1 (1994) 145

Evidence for 2p-Radioactivity in ^{17}Ne

M.J. Chromik[1,2], P.G. Thirolf[1,2], M. Thoennessen[1],
M. Fauerbach[1], T. Glasmacher[1], R. Ibbotson[1], R.A. Kryger[1],
H. Scheit[1], and P.J. Woods[3]

National Superconducting Cyclotron Laboratory and Department of Physics & Astronomy [1]
Ludwig-Maximilian-Universität München, Germany [2]
University of Edinburgh, Scotland [3]

Abstract. The first excited state of ^{17}Ne has been populated via intermediate energy Coulomb excitation with a radioactive beam of ^{17}Ne on a ^{197}Au target to search for the 2p-decay of this state, which is in competition with the γ-decay back to the ground state in ^{17}Ne. The reconstructed invariant mass spectrum of the outgoing decay products ^{15}O in coincidence with 2 protons shows evidence for a 2p-transition from the first excited state in ^{17}Ne as well as for transitions from higher excited states in ^{17}Ne, which will decay via sequential 2p-emission.

So far all experimental attempts to identify 2p-radioactivity at or near the proton dripline have been unsuccessful (e.g. [1]). A promising candidate is ^{17}Ne, where the first excited state ($J^\pi = 3/2^-$, $E^* = 1.288$ MeV) is bound by 168 keV with respect to one-proton emission but unbound with respect to two-proton emission by 344 keV (for details see [2,3]). Therefore this state can decay via a simultaneous two-proton decay to ^{15}O, since the widths of the low lying states in ^{16}F are too small ($\simeq$ 40keV) for a sequential decay via their tails. The 2p-decay is in competition with the γ-decay to the ground state in ^{17}Ne. In a recent intermediate energy Coulomb excitation experiment the γ-decay from the first excited state to the ground state ($J^\pi = 1/2^-$) has been measured and the experimental yield has been compared to the theoretically expected cross section. The measured γ-ray yield accounts for only 43% of the predicted yield from an excitation cross section of 18.7 mbarn, thus encouraging the investigation of a potential two-proton decay branch.

The experiment to search for the 2p-decay of ^{17}Ne was performed at the National Superconducting Cyclotron Laboratory at Michigan State University. The radioactive ^{17}Ne beam with 60MeV/u was produced using the A1200 fragment separator and a Wien filter to further purify the secondary beam. We achieved with a beam intensity of $\simeq$ 5000 ^{17}Ne part./s a purity of 90%. In order to identify the 2p-decay from the first excited state in ^{17}Ne a complete reconstruction of the decay kinematics in the CM-System is necessary. Thus the interaction point on the target as well as the energies and directions of all outgoing decay particles had to be measured.

Fig. 1 shows the experimental setup. In front of the target two PPAC's served to determine the interaction point of the ^{17}Ne in the target and a thin scintillator 40m upstream was used for TOF-measurements to identify the incoming particle. The target was surrounded by the MSU-NaI-array [4]. The decay fragments were then analysed in a multiple stage particle telescope, which was positioned at 0° relative to the beam axis. The position of the heavy fragment was measured in a 500μm thick 40x40 double-sided silicon strip detector and the ΔE in a Si-Pin Diode (500μm), which also delivered a time signal. The heavy fragments were then stopped in a 5mm thick Si(Li)-Detector. The light fragments penetrated these detectors and were then detected in a 300 μm thick 16x16 double-sided silicon strip detector, which served as a ΔE-detector as well as for position measurements. Finally the protons were stopped in a 4x4 CsI array, (consisting of 16 crystals 1.7cm x 1.7cm x 5cm, read out by photodiodes). These detectors were packed in a close geometry and placed 16.3cm behind the target, thus covering an opening angle from 0° to 7°. A 6.5cm thick lead collimator (with a central opening adapted to the size of the 40x40 strip detector) was placed in front of the detectors in order to shield the surrounding NaI-array from γ-rays originating from interactions with the detector material.

The incoming particles were identified by the TOF (see fig. 2(a)), between the plastic scintillator (40m upstream) and the Si-PIN. Since the ^{17}Ne was spatially separated from the beam contaminant ^{15}O after a bending magnet at the end of the Wien-filter we could also apply a geometrical gate on the the 40x40 strip detector which complemented the TOF-gate. Fig. 2(b) shows the ΔE(PIN)-E(Si5) spectrum. One can clearly see the peaks from the primary ^{17}Ne and the small contaminant of primary ^{15}O and also channeling events in the 5mm thick Si-detector (Si5) and in the PIN-detector. Below the ^{17}Ne peak there is a small enhancement corresponding to ^{15}O fragments from the (^{17}Ne,^{15}Opp) reaction. Applying a gate on incoming ^{17}Ne-particles confirms this identification.

In fig. 2(c) the sum of the ΔE in all the strips of the 16x16 strip detector is plotted against the sum of the deposited energy in all CsI detectors. In a spectrum gated on incoming ^{17}Ne, one can clearly see the p- and the d-band, and barely

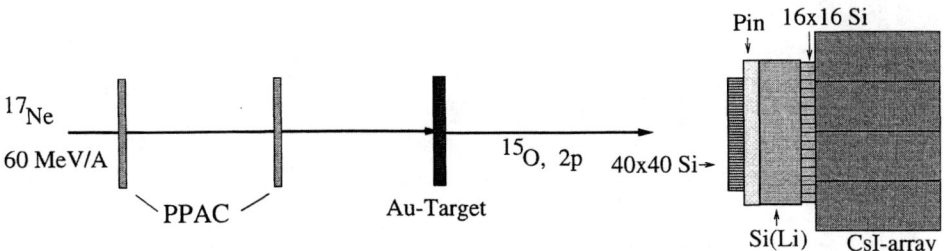

FIGURE 1. Experimental Setup

the t-band. An intensive band, consisting of events with twice the ΔE and E-values of the proton band can be identified with the 2p-band. At even larger energy losses and energies one can see also the ^{3}He and ^{4}He bands. The band which is marked with "X" corresponds to events where more than 2 particles hit the 16x16 strip detector. Using the geometric correlation between the 16x16 strip detector and the CsI-array one can then extract the energies and directions for each proton. With the information of the energies of the outgoing fragments and their trajectories one can perform the transformation into the CM-System in order to obtain the invariant mass spectrum. This spectrum is shown in fig. 3(a). The gaussian fit curves indicate the preliminary assigned peaks, the lowest one at an energy of $295 \pm 40^{stat} \pm 50^{syst}$ keV corresponding to the simultaneous 2p-decay of the first excited state in ^{17}Ne. The measured decay energies agree with the

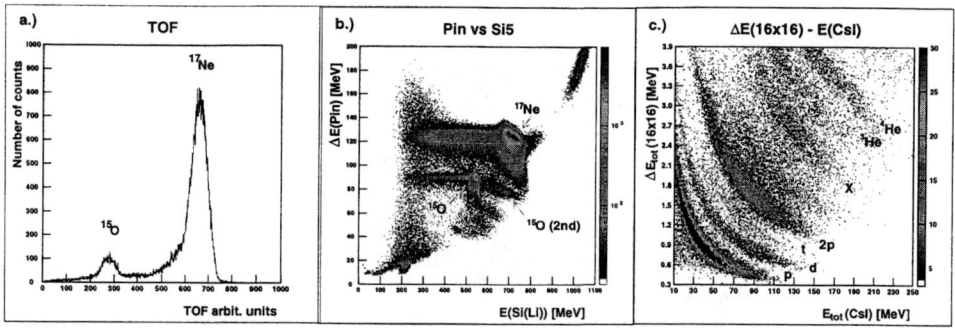

FIGURE 2. Particle Identificat6ion of the decay channels

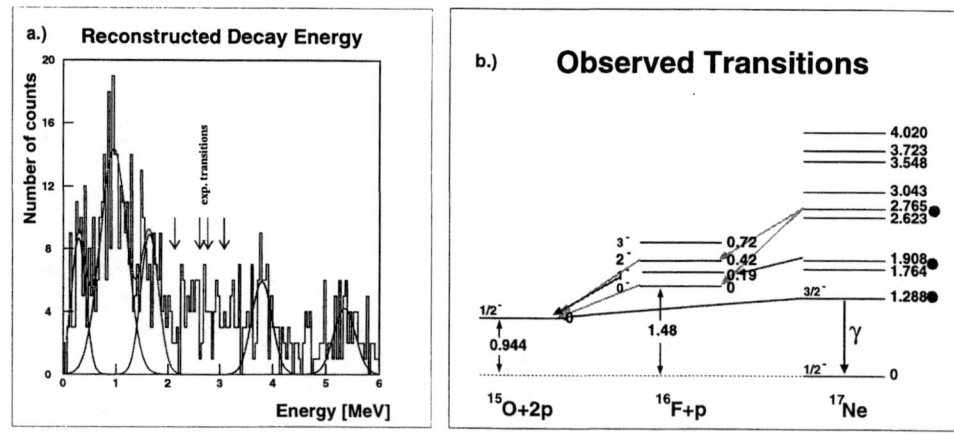

FIGURE 3. Reconstructed decay energy (a) and preliminary decay scheme of ^{17}Ne (b). The dots and their arrows indicate observed states in ^{17}Ne and preliminary dentified decay steps, respectively.

known values [5]. The arrows between 2 and 3 MeV indicate positions of known transitions, which could not be resolved. The energy resolution is on the order of 60-100 keV, mainly influenced by the error in the determination of the interaction point on the target. The decays of the higher lying states proceed predominantly sequential via intermediate states in ^{16}F. This can be extracted from the center-of-mass spectra of the individual protons. The preliminary decay scheme with the identified transitions is shown in fig. 3(b).

The decay of the first excited state can only be explained by a simultaneous 2p-decay. The preliminary analysis of the opening angle of the protons in the CM-system indicate an isotropic distribution, which would be consistent with a simultaneous uncorrelated 2p-emission. A preliminary analysis of the lifetime indicates a 2p-lifetime in the order of picoseconds, compared to a lifetime of $\simeq 0.1$ ps for the γ-decay from the first excited state in ^{17}Ne.

In conclusion, we observed several sequential 2p-transitions at their expected energies as well as evidence for 2p-radioactivity of the first excited state in ^{17}Ne. The preliminary lifetime of picoseconds is too fast for the emission of two protons from the $1d_{5/2}$ shell, which is calculated to be a factor of $\simeq 300$ slower [2]. Using an improved experimental setup with optimized efficiency and energy resolution will help to clarify the remaining uncertainties in the context of the reported first evidence for two proton radioactivity.

One of us (MJC) acknowledges the support and hospitality of the NSCL and the support of the "Studienstiftung des Deutschen Volkes".
We acknowledge the help of A. Azhari and S. Yokoyama during the experiment and thank J. Brown, D.J. Morrissey and M. Steiner for producing the radioactive ^{17}Ne beam. This work was supported by the National Science Foundation under grant PHY95-28844.

REFERENCES

1. R.A.Kryger *et al.*, Phys. Rev. Lett. **74**, 860 (1995)
2. M.J.Chromik *et al.*, Phys. Rev. C **55**, 1676 (1997) and references therein.
3. P.Woods *et al.* Proposal for CYCLONE, Louvain-la-Neuve 1995, unpublished.
4. H.Scheit *et al.*, submitted to Nucl. Instr. Meth. A (1998)
5. V.Güimaraes *et al.*, Z. Phys. A **353**, 117 (1995)

Cross-section studies of light exotic nuclei

D. Cortina-Gil, K. Sümmerer[1], T. Baumann, H. Geissel

Gesellschaft für Schwerionenforschung, 64291 Darmstadt, Germany

Abstract. Production cross-sections for some exotic fragments produced in the reaction ^{40}Ar + Be at 1000 MeV/nucleon have been studied. These data, together with new available data at higher projectile Z, and with new data for the same reaction at lower energies, allowed a revision of the semi-empirical EPAX formula. A discussion of some of the improvements introduced in this semi-empirical parameterization will be presented in this contribution.

The production and study of nuclei far from stability has become one of the most important topics in nuclear physics in the last 20 years. This subject is still of high actuality when very exotic nuclei close to the drip-line are involved. In this paper we will only speak about the projectile fragmentation, which is the main technique used at the fragment separator (FRS) at GSI for the production of secondary beams. In general, the main characteristics describing the fragmentation process are quite well understood. These "basic features" have been deduced from a large amount of experimental data. Relying on these systematics of fragmentation cross-sections and under certain conditions of "limiting fragmentation" (independence of the cross-section from the energy above a certain threshold) and "factorization", a semi-empirical parameterization of the fragmentation yields was deduced in the 80's [1]. Recently, the existence of new high-quality results for production cross-sections at high [4,5] and intermediate [6] energies, motivated an attempt to improve this formula [2].

Several experiments involving light exotic nuclei produced by fragmentation have been performed at FRS last year. Some interesting production cross-sections for exotic nuclei could be extracted. These data were also included in the revision of the EPAX formula [2].

The experimental set-up at the FRS that was used for the measurement described in this contribution is shown in Fig. 1.

---

[1] e-mail address : K.Suemmerer@gsi.de

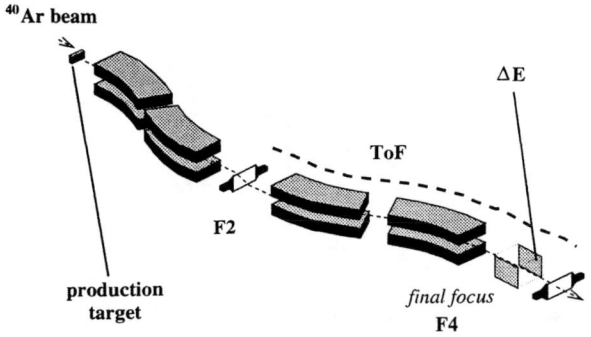

FIGURE 1. Schematic setup of the experiment at the FRS.

The primary beam was ^{40}Ar at 1000 MeV/nucleon. This beam hit a thick Be target producing a composite secondary beam via projectile fragmentation. The fragments are emitted in the forward direction and those inside the acceptance of the fragment separator are transmitted to the final focal plane. Both sides of the chart of nuclides, proton-rich and neutron-rich, have been explored. The results of this investigation are presented on the following pages. The spectrometer ensures a complete identification of the fragments arriving at its final focal plane (F4), by the combined measurements of ToF between two plastic scintillators placed at F2 and F4 and ΔE measured in an ionisation chamber placed at F4.

The number of incident particles was measured with a secondary electron transmission monitor. The production cross-sections are calculated from the ratio between the fragments detected at the final focal plane F4, corrected for the aquisition dead time, and transmission [3] between the target and the final focal plane, and the incident primary projectiles.

In Fig. 2, the results obtained in the reaction ^{40}Ar+Be at 1000 MeV/nucleon for proton and neutron-rich fragments are shown. In both cases, the production cross section in mb are plotted as a function of the mass number for each isotope. The experimental data are represented by black circles and they are compared with the results of the new EPAX parameterization. Some of these experimental points have been used to fix a new EPAX parameterization, in particular the very exotic cases of ^{19}C and ^{22}N.

The EPAX formula proposes a parameterization depending on the A and Z of both, projectile and target. The main features of this formula are described in Ref. [1]. One of the weak points of this formula was that it could not reproduce the distribution for fragments with A close to the projectile. In this context recent data [4,5] are very important, because they provide new input data allowing to obtain a better description of this effect. In addition slightly modified parameters were used describing the absolute normalization, width, and asymmetry of these

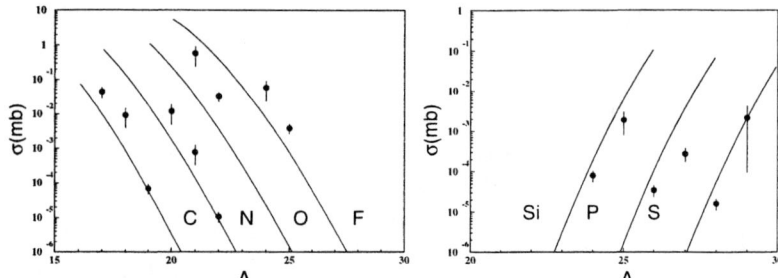

FIGURE 2. a) Production cross-sections for the ^{40}Ar + Be reaction at 1000 MeV/nucleon, as a function of the mass number for different neutron-rich fragments. The circles represent the experimental points and the lines correspond to the predictions of the revised EPAX formula [2]. b) Same representation for some proton-rich fragments.

distributions. The functional form of the EPAX parameterization is preserved.

The best example of the effect introduced by this new description can be observed in Fig. 3, in which we compare data from Ref. [4] with both versions of the EPAX formula. Another advantage of the revised formula concerns the energy range. The previous version was constructed mostly with data at relativistic energies, and in consequence was assumed to provide a good description only for fragmentation reactions at high incident energies. The publication of new experimental data for the reaction ^{40}Ar on Al target at 99 MeV/nucleon from Ref. [6] opened new perspectives. The predictions of the revised EPAX remain still good for the primary beam at intermediate energies as is shown in Fig. 4, in which data from Ref. [6] and those presented in this contribution are presented together.

In conclusion, we can say that this work is still in progress, but the expectations

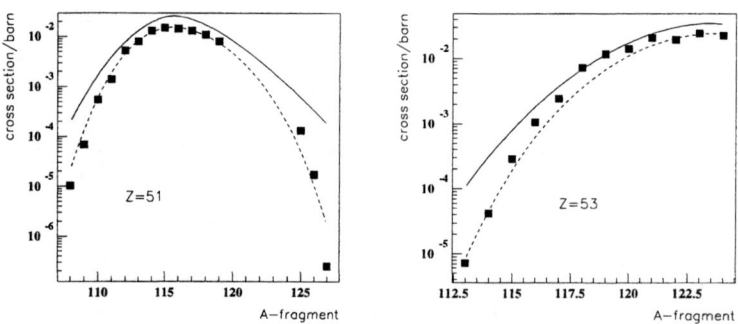

FIGURE 3. Production cross-sections for the reaction ^{129}Xe+Al at 790 MeV/nucleon for fragments with Z=51 and Z=53. The black squares correspond to the experimental data from Ref. [4], the full curve corresponds to the old EPAX formula [1], and the dashed curve corresponds to the revised EPAX formula [2]

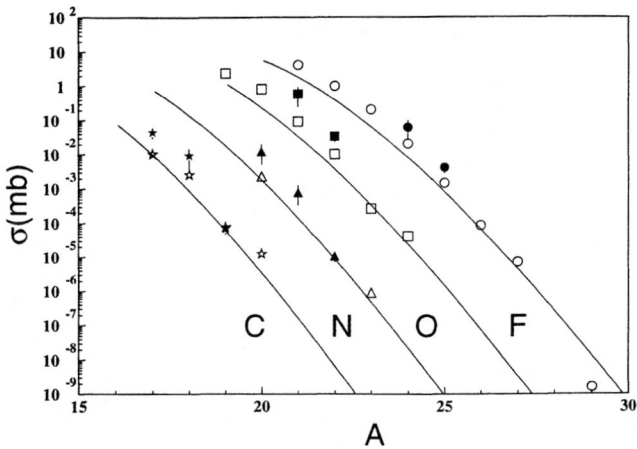

FIGURE 4. Comparison between the production cross-sections for the reaction ^{40}Ar+Al at 99 MeV/nucleon from Ref. [6] (open symbols) and those presented in this contribution for the reaction ^{40}Ar+Be at 1000 MeV/nucleon (full symbols). All the data are compared to the new EPAX formula (full curves) [2].

are rather optimistic. A better description of the projectile fragmentation process has been obtained by fitting the new experimental data. The most obvious improvement is for fragments with A close to the projectile, the second feature is the extension of the energy range covered by this formula to intermediate energies. The new EPAX parameterization gives a good description for a wide range of projectiles A=40-208 and energies of 90-1500 MeV/nucleon.

REFERENCES

1. K. Sümmerer et al., Phys. Rev C42 (1990) 2546
2. K. Sümmerer et al., in preparation
3. N. Iwasa et al., Nucl. Instr. Meth. B 126 (1997)284
4. J. Reinhold et al., Phys. Rev. C (July98), in print
5. M. De Jong et al., Nucl. Phys. A628(1998)479-492
6. H. Iwasaki et al., RIKEN Accel. Prog. Rep 31 (1998)

New insight in halo fragmention

B.V.Danilin*, S.N.Ershov[1], T.Rogde, J.S.Vaagen[2]

*RRC The Kurchatov Institute, Moscow 123182, Russia
[1] JINR, Dubna, Russia
[2] SENTEF, Department of Physics, University of Bergen, Norway

Abstract. A microscopic approach to fragment momentum distributions in Borromean two-neutron halo nuclei has been developed, which takes the correlated structure of the halo continuum into account. The crucial role of the low-lying three-body resonances and soft modes is demonstrated.

An unusual type of nuclear structure with large extension and spatial granularity manifested in a halo, has been discovered in weakly bound light dripline nuclei. The peculiarities of the halo structure are directly reflected in continuum spectra, where the strength concentration for transitions with different multipolarities, so called soft modes, are expected at low excitation energies near the breakup threshold. The study of fragment correlations in breakup reactions may shed light on the nature of soft modes; are these collective excitations or direct transitions to continuum, and clarify the role of correlations in the ground state. For such purposes one needs a reaction theory where characteristic features of halo structure in bound and continuum states intertwine with the reaction mechanism.

A simple and extensively used model for energetic fragmentation reactions is the Serber model [1], based on the sudden approximation. The momentum distributions of the fragments are squares of Fourier transforms of the ground state wave function and a measured fragment momentum distribution gives a direct information on constituents in the ground state of the projectile. At least two points should be mentioned. First, the interaction with target should be included at a dynamical level. The second point concerns distortions from final state interactions (FSI) between decaying fragments. Microscopic four-body theories of elastic scattering and direct inclusive reactions with two neutron halo nuclei has only been published recently [2,3].

In the course of breakup the projectile nucleus is excited above the three-body threshold by the interaction of the projectile constituents with target and then it decays. We will consider only the processes where the one-step reaction mechanism dominates. Taking into account the weak binding of halo nuclei this reaction scenario is physically reasonable at collision energy higher then a few tens MeV per

CP455, ENAM98: Exotic Nuclei and Atomic Masses
edited by B. M. Sherrill, D. J. Morrissey, and Cary N. Davids
© 1998 The American Institute of Physics   1-56396-804-5/98/$15.00

nucleon and can be described in the framework of distorted wave (DW) theory. In the center of mass frame the exclusive cross section (when energies and momenta of all particles are observed) can be written as [3]

$$\frac{d^8\sigma}{d\hat{k}_f d\vec{k}_x d\vec{k}_y dE_\kappa} = (2\pi)^4 \frac{\mu_i \mu_f}{\hbar^4} \frac{k_f}{k_i} \sqrt{2} \left(\frac{\mu_x}{\hbar^2}\right)^{\frac{3}{2}} \sqrt{E_\kappa - \varepsilon_y} \frac{1}{2(2J_i+1)} \sum |T_{fi}|^2. \quad (1)$$

The $\vec{k}_{i,f}$ and $\mu_{i,f}$ are relative momenta and reduced masses of colliding nuclei in initial and exit channels, $\vec{k}_{x,y}$ and $\mu_{x,y}$ are Jacobi momenta and reduced masses of projectile constituents, $E_\kappa = \varepsilon_x + \varepsilon_y = \frac{\hbar^2 k_x^2}{2\mu_x} + \frac{\hbar^2 k_y^2}{2\mu_y}$ is the total projectile excitation energy measured from the breakup threshold. The factor $\sqrt{E_\kappa - \varepsilon_y}$ gives the phase-space accessible for breakup. The transition amplitude T_{fi} includes all the interaction dynamics, and is given in [3].

To get the momentum distributions the exclusive cross section (1) has to be integrated over unobserved coordinates of breakup fragments. For Borromean two-neutron halo nuclei an understanding of the essential halo structure has been obtained in the framework of the three-body model [4]. The method of hyperspherical harmonics (HH) [5,6] has been used to treat the three-body interaction dynamics for both bound and scattering states and has given [7,8,3] a comprehensive description of data for A=6 systems. It is important to underline that in calculating continuum wave functions the final state interaction (the pair interactions between all projectile constituents) were fully taken into account.

Usually the experimental data include contributions for all excitation energies allowed by the experimental setup. Hence, we need additionally to integrate momentum distributions over projectile excitation energy from the breakup threshold up to some maximal energy defined by experimental conditions.

The cross section is an incoherent sum over all quantum numbers which characterize unobserved fragments and there are the interference terms between excitations with different total momenta J_f reached by the different transferred momenta j. The interference is expressed in different ways for the longitudinal and transverse distributions. Hence, the interference between nuclear states with different parity excited at the same energy with comparable intensity may cause deviations from symmetrical shape in longitudinal distributions. Studying such distortions may give valuable informations on the structure of halo nuclei.

As example, we discuss the ^{6}He breakup reactions on proton target at collision energies 50 and 200 MeV per nucleon. The decay of ^{6}He continuum structures (J_f^π = 0^+, 1^-, 2^+, 3^- and 1^+ from the breakup threshold up to 10 MeV excitation energy) which dominate [3] the low energy part of the continuum were taken into account.

To clarify the nature of momentum distributions the contributions from three different intervals of the ^{6}He continuum have been considered. The well known

2^+_1 resonance at $E_\kappa = 0.83$ MeV which dominates in ^{6}He energy spectra (see, for example, the experimental data on ^{6}He inelastic scattering on proton at 73 MeV/A [12]) defines the contribution from the first interval. In the second interval the excitation of different soft modes is expected [8,3]. The third contains contributions from the highest excitation energies taken into account in our model. The fragment momentum for a fixed excitation energy is restricted by allowed phase space. At 10 MeV the core (neutron) fragment cannot have momenta larger then ~ 160 (130) MeV/c. The most remarkable feature in the distributions is the significant ($\sim \frac{1}{3}$) contribution from excitation of the 2^+_1 resonance which defines the most narrow part. The contribution from the second interval has the same magnitude but is more broad. The highest excitation energies define the distribution wings.

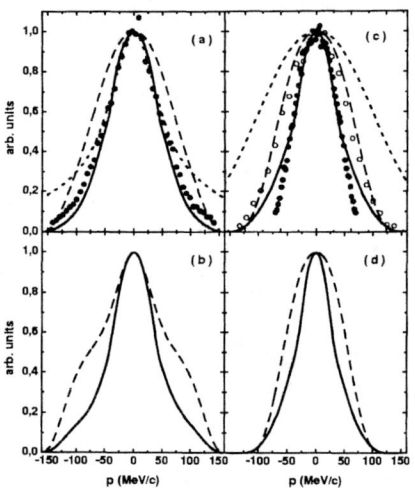

The Figs. 1a (1b) and 1c (1d) show the core and neutron transverse (longitudinal) distributions, respectively, normalized to unity at zero momentum for 200 MeV/A. The calculated shape of distribution change a little from 50 to 200 MeV/A. The solid (dashed) lines correspond to calculations with (without) final state interactions, the dotted one show the transverse distributions from Serber model [13]. The experimental data for transverse distributions in the fragmentation of ^{6}He on a carbon target at energy 240 MeV/A [14] are shown by solid circle, and for p(^{6}He,pn) at 83 MeV/A [15] by open cycles. We see that the Serber model does not describe simultaneously the core and neutron momentum distributions. For the core it overestimate the distribution wings what demands cutting the contribution for high momentum. For the neutron it gives significantly broader distribution. Calculations (dashed lines) which take into account the reaction mechanism and the correlation in ground state but neglect FSI are also unsufficient and overestimate the distribution width. Due to the final state interaction a redistribution of transition strength over continuum excitation energy occurs, the low energy part is enhanced, the well pronounced resonances like the 2_1^+ may be created. As a result, the widths of the momentum distributions are substantially reduced from ~ 150 (110) to ~ 85 (65) (MeV/c) for the core (neutron) distribution and good agreement with experimental data is obtained.

In conclusion, we have developed a microscopic four-body distorted wave theory for breakup reactions of two-neutron Borromean halo nuclei which is appropriate for analysis of processes where the low energy part of the continuum spectrum of the halo nucleus gives significant contributions to the fragment momentum distributions. It was shown that the decays of the 2_1^+ resonance at 0.83 MeV above the three-body threshold and soft modes define the width of the momentum distributions. The final state interaction plays a crucial role in formation of resonance and the transition strength accumulation near the breakup threshold.

REFERENCES

1. R.Serber, Phys. Rev. **72** (1947) 1008.
2. J.S. Al-Khalili, I.J. ThompsonJ.A. and Tostevin, Nucl. Phys. **A581** (1995) 331.
3. S.N. Ershov, T. Rogde, B.V. Danilin, J.S. Vaagen, I.J. Thompson and F.A. Gareev, Phys. Rev. **C56** (1997) 1483.
4. M.V. Zhukov, B.V. Danilin, D.V. Fedorov, J.M. Bang, I.J. Thompson, J.S. Vaagen, Phys. Rep. **231** (1993) 151.
5. B.V. Danilin, M.V. Zhukov, A.A. Korsheninnikov, L.V. Chulkov, V.D. Efros, Sov. Jour. Nucl. Phys. **49** (1989) 351, 359; ibid **53** (1991) 71.
6. B.V. Danilin, M.V. Zhukov. Yad. Fiz. **56** (1993) 67 [Sov. J. Nucl. Phys. **56** (1993) 460].
7. B.V. Danilin, M.V. Zhukov, S.N. Ershov, F.A. Gareev, R.S. Kurmanov, J.S. Vaagen and J.M. Bang, Phys. Rev. **C43** (1991) 2835.
8. B.V. Danilin, T. Rogde, S.N. Ershov, H. Heiberg-Andersen, J.S. Vaagen, I.J. Thompson and M.V. Zhukov, Phys. Rev. **C55** (1997) R577.
9. M.A. Franey and W.G. Love, Phys. Rev. **C31**, (1985) 488.
10. K.H. Bray et al., Nucl. Phys. **A189** (1972) 35.
11. C.W. Glover et al., Phys. Rev. **C41** (1990) 2487.
12. A.A. Korsheninnikov et al., Nucl. Phys. **A616** (1997) 189c.
13. M.V. Zhukov, L.V. Chulkov, B.V. Danilin and A.A. Korsheninnikov, Nucl. Phys. **A553** (1991) 428.
14. D. Aleksandrov et al., Nucl. Phys. **A** in press.
15. T. Kobayashi, K, Yoshida, A. Ozawa, I. Tanihata, A. Korsheninnikov, E. Nikolski and T. Nakamura, Nucl. Phys. **A616** (1997) 223c.

Spacial Probability Distribution of Nucleons and Nuclear Binding-Energy Losses in ^{6}He and ^{8}He Associated with Halo Formation

Friedrich Everling

Ringheide 24 f, 21149 Hamburg, Germany

Abstract. Coulomb-energy differences of mirror nuclides are evaluated in order to investigate the spacial nucleon distribution for the $0d_{3/2}$ subshell. The data support the hypothesis that the ^{32}S core and the spatial extension of the subshell remain constant during the build-up process and that the nucleons are distributed over that area. Assuming this also for the $0p_{3/2}$ subshell, the binding energy losses for the halo nuclides ^{6}He and ^{8}He are determined. A diagram for the average binding energies of the evaluated mirror nuclides up to $T_z = \pm 2$ is given.

TEST OF A HYPOTHESIS FOR THE $0d_{3/2}$ SUBSHELL, A=32-40

The spacial nucleon distribution is investigated by using the Coulomb repulsion of the protons. Binding-energy differences of mirror nuclides plotted up to $T_z = \pm 2$ in **figure 1b** for the $0d_{3/2}$ subshell represent the Coulomb energies between 1, 2, 3, and 4 excess protons and all others, together with their mutual interaction energies. The mirror pairs

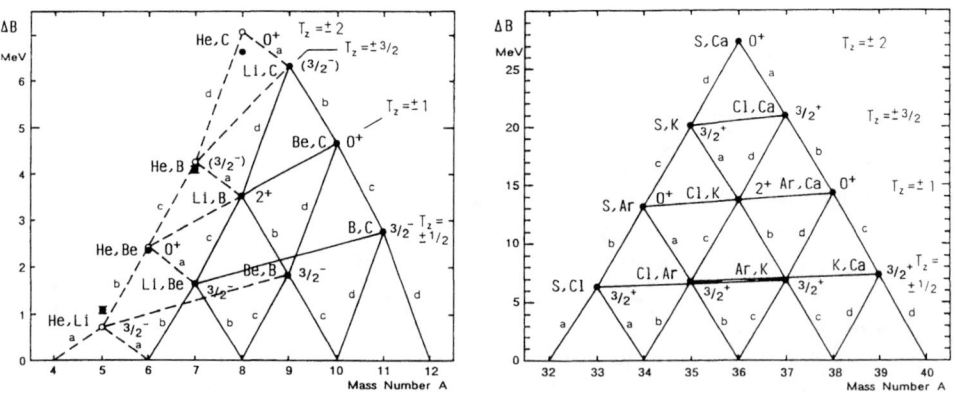

FIGURES 1a (left) AND 1b (right). Coulomb-energy differences of mirror nuclides belonging to the $0p_{3/2}$ and $0d_{3/2}$ subshells.

CP455, *ENAM98: Exotic Nuclei and Atomic Masses*
edited by B. M. Sherrill, D. J. Morrissey, and Cary N. Davids
© 1998 The American Institute of Physics   1-56396-804-5/98/$15.00

are chosen so that the excess protons (and neutrons) belong to the same subshell. For ^{36}Cl,^{36}K the 2$^+$ ground states are used in recognition of known coupling rules.

The hypothesis is set up that
(1) the ^{32}S core does not change its spacial distribution during the build-up process,
(2) the spacial extension of the $0d_{3/2}$ subshell does not change, no matter how many protons (or neutrons) populate it, and
(3) each proton is distributed over the same region of space allocated to the subshell.

Together with the known fact that proton and neutron pairs are formed, it follows that every value of the diagram consists of 3 types of Coulomb interaction between the protons: the number of interactions with the ^{32}S core, the number of pair interactions, and the number of non-pair interactions. They are combined to be components of a triple as shown in **figure 2**. The three types of interaction are schematically visualized as straight lines, bold half-circles, and thin half-circles. The nuclides shown are the proton-rich members of the mirror pairs. The proton excess is found on the left side of the dotted line. The 3 kinds of lines mentioned show the Coulomb interactions created when the neutron excess is replaced by the proton excess. In contrast to what the schematic drawing suggests, the straight lines represent interactions of the same size, just as the thin lines of different curvature do.

In case of A=9 and 37 at T_z =1/2, the small curved line is not a pair interaction, because the two protons on the right side form a pair already.

Differences of values ΔB of figure 1 lead to equalities shown by lowercase letters there. They can be verified by comparing differences of the triples in figure 2. There are 10 values of ΔB and 3 unknowns u, v, and w in the order given above. The system of

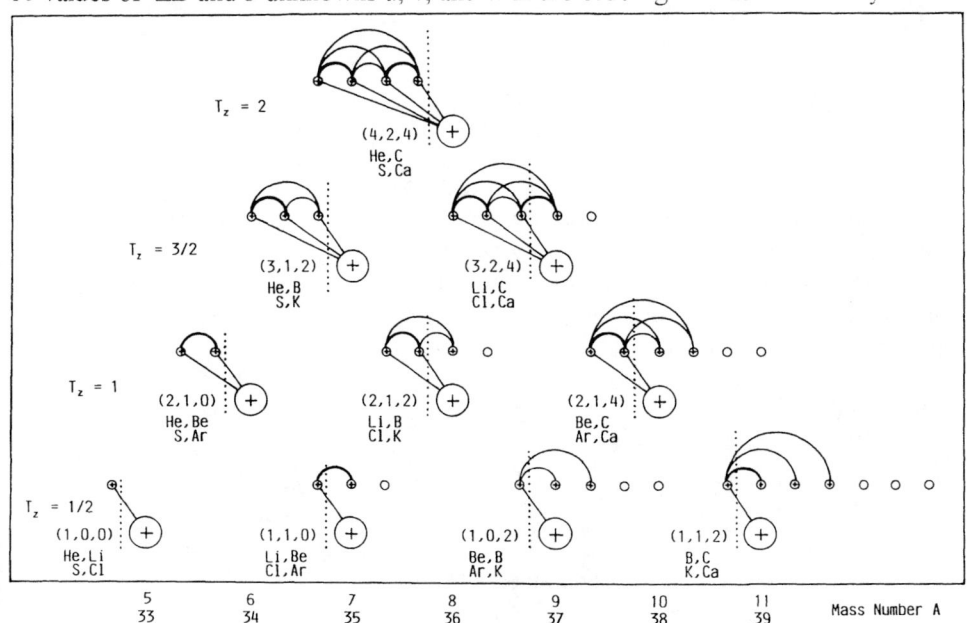

FIGURE 2. Schematic representation of 3 types of interaction.

linear equations is thus 7-fold overdetermined. The 7 independent equations could also be created by combining a special case of Garvey-Kelson mass relations if written for binding energies.

With the definition $\Delta B(A,T) = B(A,T_z=T) - B(A, T_z=-T)$ the equations are
1) $\Delta B(33, 1/2) = \Delta B(34, 1) - \Delta B(35, 1/2)$;   2) $\Delta B(33, 1/2) = \Delta B(35, 3/2) - \Delta B(36,1)$
3) $\Delta B(33, 1/2) = \Delta B(36, 2) - \Delta B(37, 3/2)$;   4) $\Delta B(35, 1/2) = \Delta B(36, 1) - \Delta B(37, 1/2)$
5) $\Delta B(35, 1/2) = \Delta B(37, 3/2) - \Delta B(38, 1)$;   6) $\Delta B(37, 1/2) = \Delta B(38, 1) - \Delta B(39, 1/2)$
7) $\Delta B(36, 2)\ \ = 2 \cdot \Delta B(36, 1)$

In addition, the values for the three interactions in the order given above can be obtained by the most precise values of ΔB, making a least-squares adjustment unnecessary. The number in italics is the standard deviation in the last digit.
1) $u = \Delta B(33, 1/2)$ $\qquad\qquad\qquad\qquad$ $u = 6.365,1\ \ 5$ MeV
2) $v = \Delta B(35, 1/2) - \Delta B(33, 1/2)$ $\qquad$ $v = 0.382,6\ \ 9$ MeV
3) $w = \frac{1}{2} \cdot [\Delta B(37, 1/2) - \Delta B(33, 1/2)]$ $\qquad$ $w = 0.283,1\ \ 3$ MeV

The differences of the right and left sides of the 7 equations are in the order of 20 keV, four of them within the standard deviation, which is very little because the hypothesis and correspondingly the Garvey-Kelson relations are not expected to hold exactly because of small superimposed effects of nuclear structure.

By using equations 3 to 6, the 7th equation can be transformed into $\Delta B(37, 1/2) - \Delta B(33, 1/2) = \Delta B(39, 1/2) - \Delta B(35, 1/2)$ which describes the parallelity of the two overlapping lines at $T_z = \pm 1/2$ in figure 1b. The left side of the equation exceeds the right side by only **0.000,8   2,1   MeV**, which means they **agree within the standard deviation of 2.1 keV**.

APPLICATION TO THE $0p_{3/2}$ SUBSHELL, A = 4 TO 12

The diagram for the $0p_{3/2}$ subshell region was shown above in **figure 1a**. For ^{8}Li, ^{8}B the 2^+ ground states are used. Figure 1a is distorted much more than 1b because the interaction with only 2 instead of 16 core protons no longer dominates. If all values containing He isotopes and the value of $B(^9Be) - B(^9B)$, lowered 41 keV possibly by the dumpbell structure of ^{8}Be, are omitted, only 2 relations for checking remain, which deviate from zero by only **−0.003,5   1,5** and **−0.005,9   3,0   MeV**. The 3 unknowns are determined to be u' = 0.774,6 *1,3*, v' = 0.869,6 *1,3*, and w' = 0.559,5 *4* MeV.

The difference between the computed value (open circle) and the empirical one (dot) for $B(^8He) - B(^8C)$ is **0.451   24 MeV**. It is interpreted as binding energy loss of ^{8}He associated with the halo formation. The 4 neutrons do not only occupy the region of the $0p_{3/2}$ subshell as the 4 protons supposedly do. For $B(^7He) - B(^7B)$ the difference of 0.21 *8* MeV is not precise enough for any corresponding conclusion. For $B(^6He) - B(^6Be)$ a loss of **0.074 4 MeV** is obtained. Like ^{8}He, ^{6}He is known to have a double-neutron halo. The deviation at $B(^5He) - B(^5Li)$ in the other direction is interpreted as caused by the unrearranged ^{4}He core as explained below.

AVERAGE BINDING ENERGIES OF THE MIRROR NUCLIDES USED

Figure 3b shows the negative values of the average ground-state binding energies of the mirror nuclides used, together with the ground states of ^{32}S, ^{36}Ar, and ^{40}Ca. A term linear in A was added to compensate the steep decrease.

In earlier diagrams of the mass excess including excited states, a rearrangement at and near A=16 by 6.05 MeV, the energy of the first excited 0^+ level of ^{16}O, was observed (1), after which the $0d_{5/2}$ subshell is accomodated regularly. **Figure 3a** shows a rearrangement of 7.46 MeV in ^{4}He although there is no exited state. The ground state of ^{4}He is not directly connected with the one of ^{8}Be because of the almost linear trends for subshells in the light mass region (1, 2). If the two lines at T = 1/2 are drawn parallel, the rearrangement energy at A=5 happens to be 6.03 4 MeV as in ^{16}O.

As indicated by a circle in figure 1a, ΔB at A=5 is expected to be 0.774,6 $1,3$ MeV. The empirical value is 0.30 7 MeV higher, apparently because the rearrangement of the ^{4}He core has not yet taken place. The comparison of figures 3a and b shows that it was necessary to exclude the value $\Delta B(5, 1/2)$ from the calculations.

REFERENCES

1. Everling, F., Nucl. Phys. **40**, 670-689 (1963).
2. Everling, F., submitted for publication.

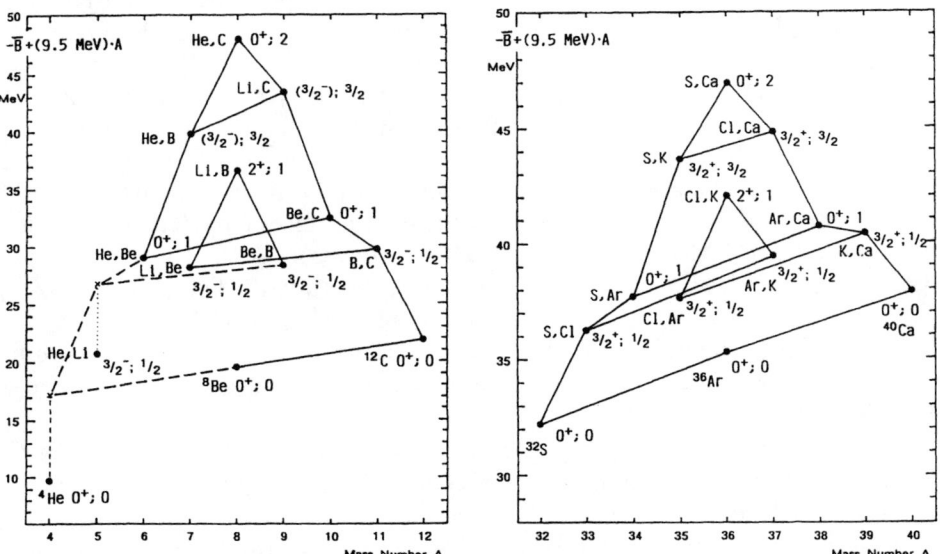

FIGURE 3a (left) **AND 3b**(right). Averages of the binding energies of which the differences were shown in figure 1. The 0^+, T=1 levels are connected directly, not via the 2^+ ground state average.

Density-Dependent Pairing in Nuclei far from Stability

S. A. Fayans*, S.V. Tolokonnikov*, E. L. Trykov[†] and D. Zawischa[‡]

*Russian Research Centre - "Kurchatov Institute" - 123182 Moscow, Russia
[†] Institute of Physics and Power Engineering - 249020 Obninsk, Russia
[‡] Institut für Theoretische Physik, Universität Hannover - D-30060 Hannover, Germany

Abstract. An approach based on the local energy-density functional method with the Green's function formalism for describing the ground-state properties of superfluid nuclei is presented. The Gor'kov equations are treated exactly in the coordinate-space representation. The method is used to calculate the odd-even mass differences and odd-even effects in charge radii which occur to be very sensitive to the density dependence of the effective pairing force. Knowledge of this density dependence allows us to make predictions for the pairing gap at the Fermi surface as a function of nuclear matter density.

In spite of much effort, the effective nucleon-nucleon interaction in the pairing channel suitable for nuclear structure calculations and for obtaining an accurate value of the pairing energy gap in infinite matter is not yet well established. The major difficulties are connected with the consistent account for the in-medium renormalizations [1,2] and for finite-range and nonlocal effects [3]. The knowledge of such effects is crucial to elucidate the proper dependence of the effective force on the density ρ and its gradients. The empirical information gained from the studies of laboratory nuclei seems to be indispensable in this respect. The presently most successful *simultaneous* description of the *bulk* nuclear properties, such as binding energies and radii, is achieved with a ρ-dependent effective force (e.g. [4,5]). Thus, the ρ-dependence of the particle-hole effective interaction, or the Hartree-Fock part of the energy-density functional (EDF), is more or less known. On the same ground, one may expect that *simultaneous* description of the *differential* observables, such as odd-even mass differences and the odd-even effects in radii, would shed light on the ρ-dependence of the effective interaction in the particle-particle channel and on the pairing part of the EDF.

To reveal the form of Δ as a function of ρ through the ρ-dependence of the underlying effective pairing force, the observed changes of geometrical characterstics of nuclei, first of all in *charge radii*, should be analysed [6]. The density of nuclear matter is sensitive to the derivative $d\Delta/d\rho$ near the saturation point

$\rho \approx 0.16$ fm^{-3} [7]. A negative slope in Δ causes a decrease of ρ, i.e. an expansion of the system. In finite systems this effect leads to the increase of radii as was confirmed by the self-consistent EDF calculations [7-9].

Here, to calculate the differential observables, we use an approach [10] based on the general variational principle for the EDF with a fixed energy cutoff $\epsilon_c > \epsilon_F$ (ϵ_F is the Fermi energy). It involves the integration of the Green's functions, obtained from the coordinate-space Gor'kov equations, in the complex energy plane [11]. This technique is appropriate for the correct treatment of the coupling with particle continuum, especially in weakly bound nuclei. Compared to refs. [7,8,12] where the calculations were done mostly in the HF+BCS framework, the present approach corresponds to a full treatment of the HFB problem and permits direct comparison of the pairing gap extracted from the analysis of nuclear data with nuclear matter calculations. It is described in detail in our forthcoming paper [10]. Here we would like to mention only the well known fact that the major pairing effects are developed near the Fermi surface and the pairing energy is defined by a sum concentrated near the Fermi surface. In infinite matter the pairing energy per particle is $E_{pair}/N = -3\Delta^2(k_F)/8\epsilon_F$. It occurs that, with cutoff EDF, this leading pairing contribution to the energy of the system is exactly accounted for. The formulated approach is in line with the Hohenberg-Kohn theorem which specify that the EDF may be chosen to be of a local form, i.e. dependent on the normal and anomalous local real densities $\rho(\vec{r})$ and $\nu(\vec{r})$. Having found the pairing field $\Delta(\vec{r})$, the mean-field potential $U(\vec{r})$, and the chemical potential μ, the Gor'kov equations can be solved exactly by using the coordinate-space technique [11]. The generalized Green's function is integrated along a suitably chosen contour in the energy plane up to ϵ_c to yield both the normal and anomalous densities ρ_c and ν_c which are used then to compute the energy of the system. Our approach does not imply a cutoff of the basis since the general variational principle is formulated with a "cutoff" local-density functional from which the ground state characteristics of a superfluid system can be calculated by using the generalized Green's function expressed through the four linear-independent solutions of the Bogoliubov equations at the stationary point [11], but only those solutions from the whole set are needed which correspond to the eigenenergies E_α of the HFB hamiltonian up to the cutoff $\epsilon_c > \epsilon_F$ (the HFB hamiltonian is the matrix of the first variational derivatives of the EDF).

The calculations were performed with the density functional DF3 [9]. In the anomalous part of the EDF the "gradient" pairing force was used [8]:

$$\mathcal{F}^\xi(x) = C_0 f^\xi(x)\delta(\vec{r}-\vec{r}'), \quad f^\xi(x) = f^\xi_{ex} + h^\xi x^q + f^\xi_{grad} r_0^2 \left(\vec{\nabla} x\right)^2, \tag{1}$$

where $C_0 \approx 305$ MeV·fm^{-3} is the inverse density of states at the Fermi surface, $q = 2/3$, $x = (\rho_n + \rho_p)/\rho_0$ is the dimensionless isoscalar density. The superscript ξ indicates that the interaction corresponds to an energy cutoff ϵ_c (in our case $\epsilon_c = 40$ MeV while the Fermi energy $\epsilon_{0F} = 36.6$ MeV).

Typical results obtained for the lead isotope chain are shown in fig. 1. It is seen that the neutron separation energies S_n are described reasonably well with any

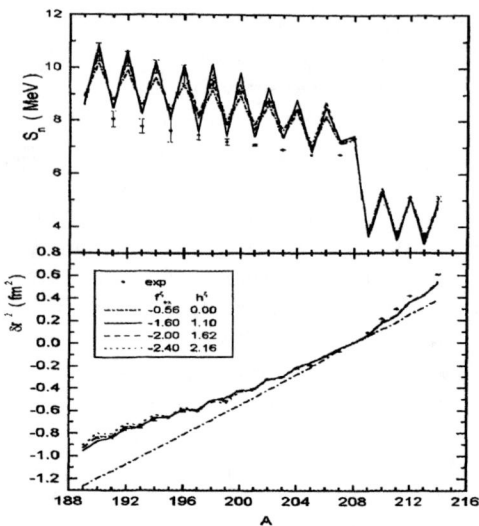

FIGURE 1. Neutron separation energies (top) and differences of mean squared charge radii (bottom) for lead isotopes with respect to ^{208}Pb as reference nucleus in comparison with experimental data. All calculations are done self-consistently using the EDF method with different parameter sets of the pairing force of eq. (1). The gradient strength $f^\xi_{\text{grad}} = 2$ for all sets except the case of "constant" pairing with $f^\xi_{\text{ex}} = -0.56$.

set of parameters just sufficient to produce an "experimental" average gap at the Fermi surface but the $\delta\langle r^2\rangle_{\text{ch}}$ values can be explained only if the pairing force (1) contains density dependence and if its parameters are taken in a certain ratio. The physics behind this is discussed in [8]. The success in the simultaneous description of both observables, S_n and $\delta\langle r^2\rangle_{\text{ch}}$, in finite nuclei is due to the gradient term $\propto f^\xi_{\text{grad}} \approx 1$. This term vanishes in infinite uniform matter. Now the sets of the other two deduced parameters (f^ξ_{ex}, h^ξ) can be used to predict the pairing gap in nuclear matter by solving the gap equation $\Delta = C_0 f^\xi \nu_c$. The results are shown in fig. 2. The curve for "constant" pairing $f^\xi_{\text{ex}} = -0.56$ stands by itself with a positive derivative everywhere. In this case no acceptable description of $\delta\langle r^2\rangle_{\text{ch}}$ is obtained (see fig. 1). An interesting observation is that all sets of the deduced parameters which give a satisfactory description of S_n and $\delta\langle r^2\rangle_{\text{ch}}$ yield about the same value of $\Delta \approx 3.3$ MeV at $k_F \approx 1.16$ fm^{-3} (at ≈ 0.66 of the equilbrium density). These sets produce for $\Delta(k_F)$ a characteristic bell form known from the calculations for infinite matter. Our preferred set which gives a slightly better fit compared to others is ($f^\xi_{\text{ex}} = -1.6$, $h^\xi = 1.10$). The corresponding pairing gap, shown in fig. 2 by solid line, occurs to be in a qualitative agreement with the most recent calculations [2].

The formulated approach based on the local energy density functional method with coordinate-space technique is shown to be quite successful in describing the ground state properties of superfluid finite nuclear systems. The combined analysis

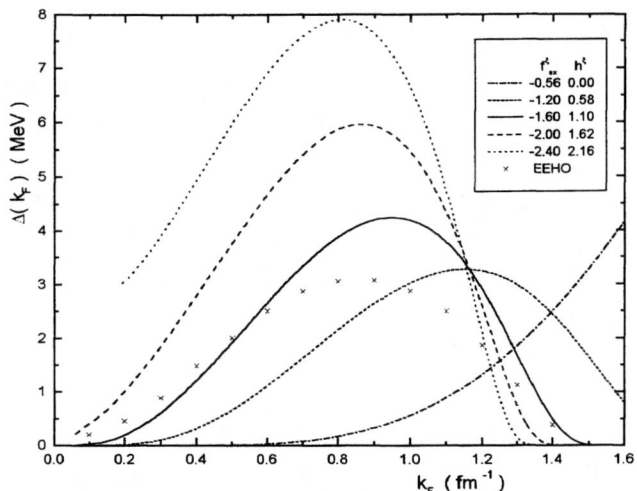

FIGURE 2. Pairing gap at the Fermi surface in infinite nuclear matter as a function of the Fermi momentum for different parameter sets deduced from the EDF calculations for the lead isotopes. Crosses represent nuclear matter calculation from ref. [2].

of the differential observables such as neutron separation energies and isotopic shifts in charge radii with this approach gives hope to construct a universal density-dependent effective interaction which would allow one to predict pairing properties both for exotic nuclei very far from stability and for nuclear matter.

This work was supported in part by the DFG (Deutsche Forschungsgemeinschaft) and by the RFBR (Russian Foundation for Basic Research) through the Grant 98-02-16979.

REFERENCES

1. Wambach J., Ainsworth T. L., and Pines D., *Nucl. Phys. A* **555** 128 (1993).
2. Elgarøy Ø. et al., *Nucl. Phys. A* **604** 466 (1996).
3. Baldo M. et al., *Phys. Lett. B* **350** 135 (1995).
4. Vautherin D., and Brink D. M., *Phys. Rev. C* **5** 626 (1972).
5. Dechargé J., and Gogny D., *Phys. Rev. C* **21** 1568 (1980).
6. Regge U., and Zawischa D., *Comments At. Mol. Phys.* **23** 257 (1989).
7. Fayans S. A. et al., *Phys. Lett. B* **338** 1 (1994).
8. Fayans S. A., and Zawischa D., *Phys. Lett. B* **383** 19 (1996).
9. Krömer E. et al., *Phys. Lett. B* **363** 12 (1995).
10. Fayans S. A., Krömer E., Tolokonnikov S. V., Trykov E. L., and Zawischa D., to be submitted to *Nucl. Phys. A*.
11. Belyaev S. T. et al., *Sov. J. Nucl. Phys.* **45** 783 (1987).
12. Smirnov A. V. et al., *Sov. J. Nucl. Phys.* **48** 995 (1988).

Semimicroscopic calculations of total reaction cross sections for light exotic nuclei

Sergej A.Fayans[*], Dmitrij V.Bolotov[**], Oleg M.Knyazkov[**] and Inna N.Kuchtina[***]

[*] *Kurchatov Institute, Moscow 123182, Russia*
[**] *St.Petersburg State University, St.Petersburg 198904, Russia*
[***] *Joint Institute for Nuclear Research, Dubna 141980, Russia*

Abstract. Total reaction cross sections and elastic scattering angular distributions for light exotic nuclei are calculated using both the double-folding model at low energies and the Glauber approximation at intermediate energies.

INTRODUCTION

Investigation of the total reaction cross sections and elastic scattering is an important tool to study properties of light exotic nuclei (see e.g. [1,2,3] and references therein). From the corresponding data valuable information concerning optical potential (OP) and nuclear density distributions in the domain far from stability can be obtained. A semimicroscopic approach (SMA) based on the double-folding model for the real part of OP and on the energy-density functional method for neutron and proton densities has been used in [4] to analyse quasielastic scattering data for a few systems including exotic nuclei such as ^{8}B and ^{11}Li.

Here we present SMA calculations of the total reaction cross sections for light exotic nuclei using the double-folding model at lower energies and the Glauber approximation at higher energies. The SMA is applied also to analyse the elastic scattering angular distributions of ^{6}He on ^{12}C at E=41.6 MeV/n [5] and of ^{11}Be on ^{12}C at E=49.3 MeV/n [6].

The real part of the OP is represented by the sum:

$$U(R) = U^D(R) + U^{EX}(R), \qquad (1)$$

where $U^D(R)$ is the direct double-folding potential [7] and $U^{EX}(R)$ is the exchange potential [8]. The total optical potential is:

$$U_{tot}(R) = U(R) + i[N_w \tilde{U}(R) - \alpha R \frac{dU(R)}{dR}], \qquad (2)$$

where U(R) is given by (1) and N_w and α are the volume and surface absorption factors.

Table 1: Total reaction cross sections (in mbn)

Target	E/A (MeV/n)	N_w	α	σ_R	Nucleus	σ_R, theor.	σ_I, exp.
^{12}C	40	0.3	0.03	1200	^{20}N	1227	1121(17)
^{12}C	60	0.3	0.03	1080	^{20}O	1181	1078(10)
^{9}Be	40	0.5	0.02	1173	^{20}F	1162	1113(11)
^{9}Be	60	0.5	0.02	1058	^{20}Ne	1163	1144(10)
^{27}Al	40	0.3	0.045	1730	^{20}Na	1172	1094(11)
^{27}Al	60	0.3	0.03	1530	^{20}Mg	1207	1150(12)

Table 2: Total reaction cross sections (in mbn) calculated in the Glauber approximation

System	E/A (MeV/n)	σ_R, theor.	σ_I, exp.
^{8}B+^{12}C	800	838	798±6 [14]
^{11}Li+^{12}C	400	959	989±21 [15]
^{11}Li+^{12}C	800	1050	1056±14 [15]

The double–folding potential U(R) is calculated using the M3Y effective nucleon–nucleon interaction [9] and nuclear densities obtained within the energy-density functional approach with a fixed parameter set for all colliding nuclei. Total reaction cross sections and angular distributions are calculated with the coupled-channels code ECIS–88 [10].

TOTAL REACTION CROSS SECTIONS

The SMA calculations of total reaction cross sections for the ^{8}B projectile are presented in Table 1 (fifth column). The values of N_w and α parameters are close, for the ^{12}C target to those obtained in the SMA analysis of quasielastic scattering angular distributions. Unfortunately, there are no measurements for the ^{9}Be and ^{27}Al targets. A few values of the N_w and α were chosen to calculate reaction cross sections for these targets. Our results for the ^{12}C target are in close agreement with the preliminary experimental data obtained at E=38.5 MeV/n and 59.7 MeV/n in [11].

To check the neutron and proton densities of light exotic nuclei obtained within the energy–density functional approach we calculated also the total reaction cross sections at intermediate energies in the optical limit of the Glauber theory [12]. The free NN cross sections at appropriate relative energies were taken from [13]. Such calculations do not contain additional free parameters.

This approach was used to calculate the total reaction cross sections for the A=20 isobars on the ^{12}C target. The results together with experimental data obtained at energy E=950 MeV/n [14] are listed also in Table 1 (correspondingly, seventh and eighth columns).

The theoretical cross sections are somewhat greater than experimental values, with mean deviation within 6 per cent. Other results for intermediate energies are given in Table 2. It can be seen that theoretical results agree well with experimental data [15,16].

ELASTIC SCATTERING ANGULAR DISTRIBUTIONS

Angular distributions of (quasi)elastic scattering are very sensitive to the OP used in the analysis. To check our semimicroscopic OP we have analysed the elastic scattering angular distributions of ^{6}He on ^{12}C at E=41.6 MeV/n and of ^{11}Be on ^{12}C at E=49.3 MeV/n. The calculated angular distributions in comparison with experimental data [5,6] are shown in Fig.1. Here the absorption parameters are N_w=0.5, α_w=0.03 for the ^{6}He+^{12}C case and N_w=0.4, α_w=0.03 for the ^{11}Be+^{12}C system. To obtain a acceptable fit to experimental data, one needs a surface term in the real part of the semimicroscopic OP. It has the same form as the surface term in the imaginary part (see eq.(2)). The dashed curves in Fig.1 were obtained without such term in the real part of the OP, while for the solid lines it was included with parameters α_v=0.15 for the ^{6}He projectile and α_v=0.07 for the ^{11}Be nucleus. An essential role of this additional term in the OP to get good agreement with experiment is obvious. We also found a larger real surface term for the ^{6}He projectile than for the ^{11}Be. At the present stage of our calculations, the surface term in the real part of the OP is introduced phenomenologically as a substitute for a dynamical polarization potential (DPP).

CONCLUSIONS

The total reaction cross sections σ_R and elastic scattering angular distributions for light exotic nuclei are analysed within a semimicroscopic double-folding approach. The σ_R values are also calculated at higher energies within the optical limit of the Glauber theory. The calculations with nucleon densities obtained by the energy–density functional method for the ^{8}B and ^{11}Li projectiles, and also for

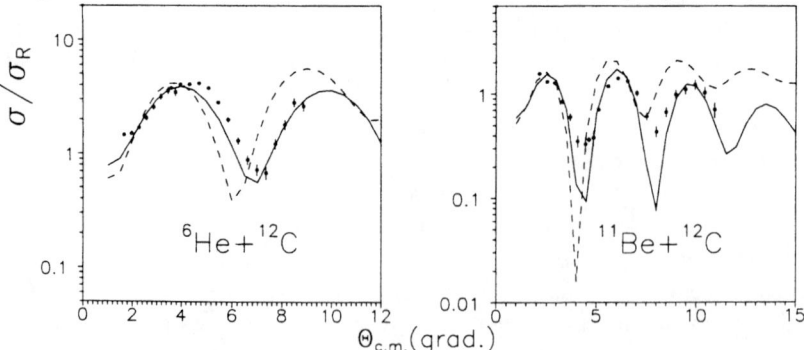

Figure 1. Theoretical and experimental angular ditributions of elastic scattering.

Notations are in the text.

the A=20 isobars, is shown to be in quite reasonable agreement with the experimental cross sections both at low and intermediate energies.

It is demonstrated that a good agreement of the SMA angular distributions with the experimental data for the elastic scattering of ^{6}He on ^{12}C at E=41.6 MeV/n and of ^{11}Be on ^{12}C at E=49.3 MeV/n can be achieved by adding a rather large surface real term to the double–folding OP. The presence such a dynamical polarization potential may indicate a cluster structure of the projectiles.

ACKNOWLEDGEMENTS

This work was supported in part by the RFBR (Russian Foundation for Basic Research) through the grant 98-02-16979.

REFERENCES

1. Bertulani C.A., Canto L.F. and Hussein M.S., Phys. Rep. **226**, 281–376 (1993)
2. Tanihata I., J.Phys. **G22**, 157–198 (1996)
3. Knyazkov O.M., Kuchtina I.N., Fayans S.A., Phys. Part. Nucl. **28**, 418–439 (1997)
4. Fayans S.A., Knyazkov O.M., Kuchtina I.N. et al., Phys. Lett. **B357**, 509–514 (1995)
5. Al-Khalili J.S., Cortina-Gil M.D., Roussel-Chomaz P. et al., Phys. Lett. **B378**, 45–49 (1996)
6. Roussel-Chomaz P. (private communication)
7. Satchler G.R. and Love W.G., Phys. Rep. **55**, 183–253 (1979)
8. Chaudhuri A.K. and Sinha B., Nucl. Phys. **A455**, 169–178 (1986)
9. Bertsch G.E., Borysowicz J., McManus H. et al., Nucl. Phys. **A284**, 399–419 (1977)
10. Raynal J., Phys. Lett., **B196**, 7–21 (1987)
11. Fukuda M. et al. The Fourth Int. Conf. on Radioactive Nuclear Beams, June 4-7, 1996, Omiya, Japan, Abstract Book, p.141
12. Glauber R.J., in Lectures in Theoretical Physics, ed.by W.E.Brittin (Interscience, New York, 1959), V.1, p.315
13. Charagi S.K. and Gupta S.K., Phys. Rev. **C41**, 1610–1618 (1990)
14. Chulkov L., Kraus G., Bochkarev O. et al., Nucl. Phys. **A603**, 219–237 (1996)
15. Obuti M.M. et al., Nucl. Phys. **A609**, 74–90 (1996)
16. Tanihata I. et al., Phys. Lett. **B287**, 307–311 (1992)

Nolen-Schiffer Anomaly and Atomic Masses

S. A. Fayans

Russian Research Centre - "Kurchatov Institute" - 123182 Moscow, Russia

Abstract. A new form of the nuclear energy-density functional for describing the ground state properties of finite nuclei up to the drip lines and beyond is proposed. The surface energy-density term has a fractional form containing $(\nabla\rho)^2$ both in the numerator and in the denominator. An effective ρ-dependent Coulomb-nuclear correlation term is added. A fit to the nuclear masses and radii shows that the latter term gives contribution of the same order of magnitude as the Nolen-Schiffer anomaly in Coulomb displacement energy. The self-consistent run with the suggested functional, performed for about 100 spherical nuclei, has given the rms deviations from the experiment of ≈ 1.2 Mev in masses and ≈ 0.01 fm in radii. The extrapolation to the drip lines goes in between the ETFSI and the macroscopic-microscopic model predictions.

The most successful microscopic approaches for describing the nuclear ground state properties are the self-consistent mean-field models based on the effective energy-density functionals (EDF) incorporating forces of Skyrme type with zero range or of Gogny type with finite range, and also the relativistic mean field (RMF) model. These models can reproduce the masses and radii of measured nuclei with the respective rms deviations of about 2 MeV and 0.02 fm from experiment [1]. But their predictions, already for nuclei not too far from stability [2], are in striking deviation from those of the macroscopic-microscopic (MM) models [3] or of the extented Tomas-Fermi model with Strutinsky integral (ETFSI) [4]. These latter models are able to reproduce nuclear masses and charge radii with the rms error down to ≈ 0.6 MeV and ≈ 0.02 fm, respectively, and their predictions are currently considered to be the most reliable. Such a large disagreement between the two approaches may indicate that some important physical ingredients are missing in the EDF construction, and perhaps the form of the EDF used so far in the microscopic calculations is not flexible enough to effectively incorporate them. Here a new EDF is proposed. The total energy density of a nuclear system is represented as

$$\varepsilon = \varepsilon_{\text{kin}} + \varepsilon_{\text{v}} + \varepsilon_{\text{s}} + \varepsilon_{\text{Coul}} + \varepsilon_{sl} + \varepsilon_{\text{anomal}} , \qquad (1)$$

where ε_{kin} is the kinetic energy term taken with the free operator $t = p^2/2m$, i.e. the effective mass $m^* = m$; all the other terms are discussed below.

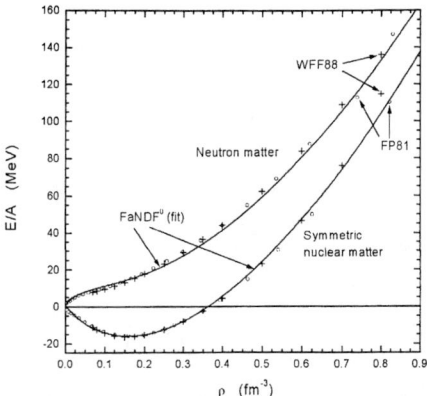

FIGURE 1. Binding energy per nucleon in symmetric nuclear matter and in neutron matter. Open circles and crosses are the calculations of ref. [6] and ref. [7], respectively. The solid lines show the fit by fractional expressions of eq. (2).

The volume term in (1) is chosen to be in the form

$$\varepsilon_v = \frac{2}{3}\epsilon_F^0 \rho_0 \left[a_+^v \frac{1 - h_{1+}^v x_+^\sigma}{1 + h_{2+}^v x_+^\sigma} x_+^2 + a_-^v \frac{1 - h_{1-}^v x_+}{1 + h_{2-}^v x_+} x_-^2 \right]. \qquad (2)$$

Here and in the following $x_\pm = (\rho_n \pm \rho_p)/2\rho_0$, $\rho_{n(p)}$ is the neutron(proton) density, $2\rho_0$ is the equilibrium density of symmetric nuclear matter with $\epsilon_F^0 = (9\pi/8)^{2/3}\hbar^2/2mr_0^2$, the Fermi energy and $r_0 = (3\pi\rho_0/8)^{1/3}$, the radius parameter. The fractional expressions of the type of eq. (2) were introduced in [5]. Such expressions allow an extrapolation of nuclear equation of state (EOS) to very high densities while preserving causal behavior. This might be of advantage since the available microscopic nuclear matter EOS often violate causality at $\rho > 1$ fm^{-3}. Thus, in deriving the parameters of eq. (2), we shall use the EOS of refs. [6,7] only in the region of up to about six times the saturation density. The four parameters in the isoscalar part $\propto a_+^v$ are fixed by fitting to the EOS of symmetric nuclear matter [6,7] for the UV14 plus TNI model. The result shown in fig. 1 is obtained with $\sigma = 1/3$, the compression modulus $K_0 = 220$ MeV, the equilibrium density $2\rho_0 = 0.16$ fm^{-3} ($r_0 = 1.143$ fm) and the chemical potential $\mu_0 = -16.0$ MeV. The dimensionless parametres are $a_+^v = -9.559$, $h_{1+}^v = 0.633$, $h_{2+}^v = 0.131$. Keeping them fixed, a fit to the neutron matter EOS is performed to determine the three parameters of the isovector part $\propto a_-^v$ in eq.(1). Good description is obtained with $a_-^v = 4.428$, $h_{1-}^v = 0.250$, $h_{2-}^v = 0.130$ (the asymmetry energy coefficient is $\beta_0 = 30.0$ MeV).

The surface term in eq. (1) is meant to describe the finite-range and nonlocal in-medium effects. It is taken as follows:

$$\varepsilon_s = \frac{2}{3}\epsilon_F^0 \rho_0 \frac{a_+^s r_0^2 (\vec{\nabla} x_+)^2}{1 + h_+^s x_+^\sigma + h_\nabla^s r_0^2 (\vec{\nabla} x_+)^2}, \qquad (3)$$

with $h^s_+ = h^v_{2+}$, a^s_+ and h^s_∇ the two free parameters. Such a form is obtained by adding the terms $\propto (\vec{\nabla} x_+)^2$ *both* in the numerator *and* in the denominator of the isoscalar volume energy density of eq. (2). Alternatively, this peculiar surface term may be regarded as the Padé approximant for the (unknown) expansion in $(\vec{\nabla}\rho)^2/(1 + h^s_+ x^\sigma_+)$ where the form factor $1/(1 + h^s_+ x^\sigma_+)$ imitates a transformation to the Migdal's quasiparticles (cf. ref. [8]). In fact, h^s_+ is also a free parameter but here we prefer to keep it fixed by the above condition.

The Coulomb part in eq. (1) is approximated by

$$\varepsilon_{\text{Coul}} = 2\pi e^2 \rho_{\text{ch}} \left(\frac{1}{r} \int_0^r \rho_{\text{ch}} r^2 \, dr + \int_r^\infty \rho_{\text{ch}} r \, dr \right) - \frac{3}{4}\left(\frac{3}{\pi}\right)^{1/3} e^2 \rho_p^{4/3}(1 - h_{\text{Coul}} x^\sigma_+), \quad (4)$$

where the first term is the direct Coulomb contribution (expressed through charge density ρ_{ch} and written, for simplicity, for the case of the spherical symmetry), while the second term is the exchange part taken in the Slater approximation and combined with the Coulomb-nuclear correlation term $\propto h_{\text{Coul}}$. The latter is believed to account for the correlated motion of protons in nuclei beyond the direct (Hartree) and exchange (Fock) Coulomb interaction [9].

The spin–orbit term ε_{sl} in eq. (1) comes from the two-body isoscalar spin-orbit interaction with the strength derived from an average description of the splitting of the single-particle states in ^{208}Pb. The last term in eq. (1), the anomalous energy density, is chosen as in ref. [10].

The three parameters a^s_+, h^s_∇ and h_{Coul} remain to be defined. This was done through a χ^2 fit to the masses and radii of about 100 spherical nuclei from ^{38}Ca to ^{220}Th with the result $a^s_+ = 0.600$, $h^s_\nabla = 0.440$ and $h_{\text{Coul}} = 0.941$, the rms deviations being 1.2 MeV and 0.01 fm for masses and radii, respectively. We shall call the EDF in the suggested form, with the just extracted parameters, the Nuclear Density Functional FaNDF^0. Typical results of the spherical HFB calculations with FaNDF^0 are shown in fig. 2 for even Pb isotopes, from the proton drip line to the neutron drip line (47 nuclides), in comparison with experimental data and other model predictions. The ETFSI model is chosen as a reference. The nuclei in the $A \approx 222$ to 248 region might have a static deformation [3,4], so one expects that, with deformed code, the results for FaNDF^0 in this region will be shifted down closer to the MM or ETFSI results. It is seen that the predictions obtained with the Gogny force just outside the measured regions are in strong disagreement with other models. Approaching the neutron drip line, the masses obtained with FaNDF^0 fall in between the MM and ETFSI predictions.

Finally, the mass differences for the mirror nuclei ^{17}F-^{17}O and ^{41}Sc-^{41}Ca calculated with FaNDF^0 are 3.546 MeV and 7.174 MeV, respectively, whereas the corresponding experimental values are 3.543 MeV and 7.278 MeV. Omitting the Coulomb-nuclear correlation term by setting $h_{\text{Coul}} = 0$, the calculated mass differences for these mirror pairs would be respectively 3.300 MeV and 6.872 MeV leading to a 6-7% discrepancy; this kind of discrepancy is known as the Nolen-Schiffer anomaly. It follows that Coulomb-nuclear correlations play an important

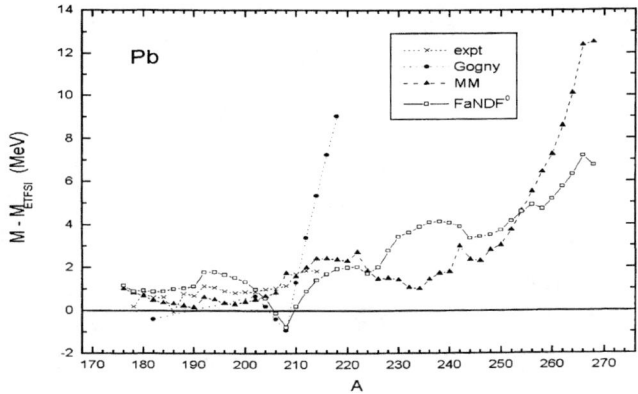

FIGURE 2. Deviations of various theoretical masses from the ETFSI mass [4] for a long chain of lead isotopes. Black triangles correspond to the MM model [3], solid dots to the Gogny force (from [2]). The results obtained with FaNDF^0 are shown by open squares. The experimentally known masses including those derived from systematics are presented by crosses.

role in finite nuclei. Incorporating the corresponding term in the EDF improves the description of nuclear ground state properties and greatly reduces the severity of the Nolen-Schiffer anomaly.

The agreement between microscopic self-consistent theory and experiment is significantly improved with the proposed EDF in which the volume part fits the microscopic EOS for infinite uniform matter, the surface term has a peculiar form as given by eq. (3) and, for finite systems, the Coulomb part contains an additional Coulomb-nuclear correlation term. The first resuts obtained with the FaNDF^0 parametrization are encouraging. The proposed construction of the EDF seems to be an important step towards a universal nuclear density functional.

This work is supported by the RFBR (Russian Foundation for Basic Research) through the Grant 98-02-16979.

REFERENCES

1. Pomorski et al., *Nucl. Phys. A* **624** 349 (1997).
2. Patyk Z. et al., GSI-Preprint-97-40, August 1997.
3. Möller P. et al., *At. Data Nucl. Data Tables* **59** 185 (1995).
4. Aboussir Y. et al., *At. Data Nucl. Data Tables* **61** 127 (1995).
5. Smirnov A. V., et al., *Sov. J. Nucl. Phys.* **48** 995 (1988).
6. Friedman B., and Pandharipande V. R., *Nucl. Phys. A* **361** 502 (1981).
7. Wiringa R. B., Fiks V., and Fabrocini A., *Phys. Rev.* **38** 1010 (1988).
8. Khodel V. A., and Saperstein E. E., *Phys. Reports.* **92** 183 (1982).
9. Shaginyan V.R., *Sov. J. Nucl. Phys.* **40** 728 (1984).
10. Fayans S. A., and Zawischa D., *Phys. Lett. B* **383** 19 (1996).

Exotic Molecular and Halo States in ^{12,14}Be

M. Freer[a,b], N.A. Orr[b], M. Labiche[b], F.M. Marqués[b], J.C. Angélique[b], L. Axelsson[c], B. Benoit[d], U. Bergmann[e], M.J.G. Borge[j], W.N. Catford[f], S.P.G. Chappell[g], N.M. Clarke[a], G. Costa[h], N. Curtis[f], A. D'Arrigo[d], F. D'Oliviera[m], E. de Goes Brennard[d], O. Dorvaux[h], B.R. Fulton[a], G. Gardina[i], C. Gregori[j], S. Grévy[b,k], D. Guillemaud-Mueller[k], F. Hanappe[d], B. Heusch[h], B. Jonson[c], G. Kelly[l], C. Le Brun[b], S. Leenhardt[k], M. Lewitowicz[m], K. Markenroth[c], M. Motta[i], A.C. Mueller[k], J.T. Murgatroyd[a], T. Nilsson[c], A. Ninane[n,b], G. Nyman[c], I. Piqueras[j], K. Riisager[e] M.G. Saint Laurent[m], F. Sarazin[m,b], S. Singer[a], O. Sorlin[k], L. Stuttgé[h], D.L. Watson[o]

[a] School of Physics and Astronomy, University of Birmingham, Birmingham B15 2TT, U. K.
[b] LPC-ISMRA, Bd Maréchal Juin, 14050 Caen Cedex, France.
[c] Fysiska Institutionen, Chalmers Tekniska Högskola, S-412 96 Göteborg, Sweden.
[d] Université Libre de Bruxelles, CP 226, B-1050 Bruxelles, Belgium.
[e] Det Fysiske Institut, Aarhus Universitet, DK 8000 Aarhus C, Denmark.
[f] Department of Physics, University of Surrey, Guildford, Surrey, GU2 5XH, U. K.
[g] Department of Nuclear Physics, University of Oxford, Keble Road, Oxford OX1 3RH, U. K.
[h] IReS, B.P.28, F-67037 Strasbourg Cedex, France.
[i] Departimento di Fisica, Università di Messina, Salita Sperone 31, I-98166 Messina, Italy.
[j] Instituto Estructura de la Materia, CSIC, E-28006 Madrid, Spain.
[k] Institut de Physique Nucléaire 91406 Orsay Cedex, France.
[l] School of Sciences, Staffordshire Univeristy, College Road, Stoke-on-Trent, ST4 2DE, U. K.
[m] GANIL, BP 5027, 14076 Caen Cedex, France.
[n] Institut de Physique, Université Catholique de Louvain, B-1328 Louvain-la-Neuve, Belgium.
[o] Department of Physics, University of York, York, YO1 5DD, U. K.

Abstract. The two nuclei ^{12}Be and ^{14}Be have been studied using breakup reactions on p, ^{12}C and ^{208}Pb targets. The decay of ^{12}Be into two helium clusters (^{6}He+^{6}He and ^{4}He+^{8}He) was observed from a series of excited states between 10 and 25 MeV, with spins in the range 4^+ to 8^+. The single neutron angular distributions for ^{14}Be exhibit the narrow forward peak characteristic of a halo. The widths of these distributions in coincidence with ^{12}Be fragments are $\Gamma_L = 78 \pm 6$ and 80 ± 1 MeV/c for breakup on carbon and lead.

I INTRODUCTION

The stability of the α-cluster has a strong influence on the structure of light nuclei and has spawned an industry devoted to the understanding of the role of clustering in s-d shell nuclei dating from the early 1960's when heavy-ion beams were first exploited. The advent of radioactive beams has also revealed clustering in light, neutron-rich nuclei, in the guise of the halo, where the nucleus is composed of a core and valence halo neutrons, and the molecular type cluster states composed of α-particles bound by the valence neutrons.

In the 1960's, Ikeda [1] developed a classification of clustering in light A=4n, α-conjugate nuclei, in which cluster structures appeared at, or close to, the decay threshold for the particular cluster partition. This scheme, which is on the whole verified by experimental observations, is based on the premise that to create the internal cluster structures an energy equivalent to the binding energy of the constituents is required. Thus systems that include weakly bound neutrons, e.g. ^{11}Be (S_n=0.5 MeV) and ^{11}Li (S_{2n} ≃0.3 MeV), should show strong clustering characteristics. Clustering is amplified in these systems by reduced centrifugal barriers for the weakly bound valence neutrons allowing a greater decoupling of the core and halo neutrons. The heaviest particle stable Be isotope, ^{14}Be (S_{2n} = 1.34 ± 0.11 MeV), is Borromean and is known to exhibit a two-neutron halo. In contrast to other halo systems, the configuration of the valence neutrons is expected to contain a significant d-wave admixture. Additionally, ^{14}Be is the heaviest two-neutron halo nucleus currently known. The investigation of ^{14}Be, presented here, thus provides an opportunity to study the evolution of halo systems with binding energy, angular momentum and mass.

Another extreme manifestation of clustering is the formation of chain and ring structures composed of individual cluster units, such as α-particles. Indeed there is some evidence for such structures in light nuclei. Similar arrangements have also been observed in atomic systems, for example carbon clusters [2] which are believed to be related to particular spectral lines in astronomical observations of stellar dust clouds. Another nuclear analogue of such phenomena are α-ring and chain structures covalantly bonded by valence neutrons, predicted by Wilkinson [3], much as the binding of atomic molecules through the exchange of electrons. Von Oertzen [4] has characterized the structure of the sequence of Be isotopes ^{9}Be to ^{11}Be in terms of dinuclear 2α-Xn structures, and certain carbon isotopes as trinuclear molecules. The present contribution reports a measurement of the 2α-4n cluster system ^{12}Be.

II EXPERIMENTAL MEASUREMENTS

The secondary beams of ^{12,14}Be ($\overline{E}_{^{12}Be}$ = 31.5 MeV/nucleon, $i \approx 2.10^4$ pps, $\overline{E}_{^{14}Be}$ = 41.3 MeV/nucleon, $i \approx 10^2$ pps) were prepared from the fragmentation of an ^{18}O beam (63 MeV/nucleon) on a thick Be target using the LISE3 spectrometer. In

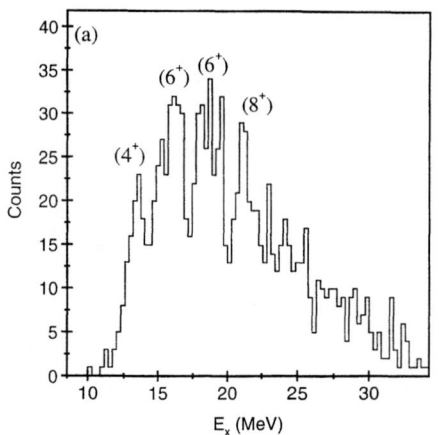 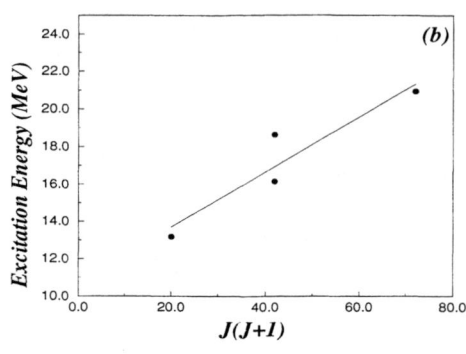

FIGURE 1. (a) ^{12}Be excitation energy spectrum for ^{6}He+^{6}He coincidences, (b) the energy-spin systematics of the breakup states.

the case of the ^{14}Be studies, measurements were carried out on secondary reaction targets of carbon and lead ($\overline{E}_{^{14}Be} = 35$ MeV/nucleon at target mid-point) in an attempt to disentangle the effects of nuclear and Coulomb induced breakup. For the ^{12}Be measurements, ^{12}C and $(CH_2)_n$ target foils were used, with beam tracking and time-of-flight measurement provided by two parallel plate avalanche counters. The breakup of the ^{12}Be nucleus into two helium fragments was detected in an array of ten Si-CsI telescopes placed around the beam axis. The silicon elements were two dimensional position sensitive detectors (2DPSDs), proving a measurement of the angle of the incident particles with a resolution of better than 0.5°. The telescopes provided a clean identification of the ^{4}He, ^{6}He and ^{8}He nuclei of interest, and a measurement of the energy with a resolution of 1.5%. For the ^{14}Be experiment the charged reaction products were again detected using the 10 element position sensitive Si-CsI array with one of the elements placed at zero degrees. Neutrons were detected using the 99 modules of the DEMON array. A staggered arrangement

Reaction	Target	σ (mb)
p(^{12}Be,^{6}He^{6}He)p	$(CH_2)_n$	0.76 (0.05)
^{12}C(^{12}Be,^{6}He^{6}He)^{12}C	$(CH_2)_n$	0.26 (0.06)
^{12}C(^{12}Be,^{6}He^{6}He)^{12}C	^{12}C	0.28 (0.04)
p(^{12}Be,^{4}He^{8}He)p	$(CH_2)_n$	2.91 (0.06)
^{12}C(^{12}Be,^{4}He^{8}He)^{12}C	$(CH_2)_n$	1.62 (0.11)
^{12}C(^{12}Be,^{4}He^{8}He)^{12}C	^{12}C	0.79 (0.07)

TABLE 1. ^{12}Be breakup reaction cross sections.

Channel	$\sigma_{telescope}$ [mb]		Γ_n [MeV/c]	
	C	Pb	C	Pb
^{12}Be	460±40	2300±400	—	—
^{12}Be+n	—	—	78±6	80±1
^{11}Be	85±15	—	—	—
^{10}Be	145±20	—	—	—
^{10}Be+n	—	—	119±44	—

TABLE 2. Preliminary results for the breakup of ^{14}Be on C and Pb targets. The single neutron angular distributions have been characterized in terms of a Lorentzian distribution.

for the neutron detectors was chosen so as to maximize the coverage at forward angles whilst minimizing the effects of cross talk (both geometrically and in the off-line analysis).

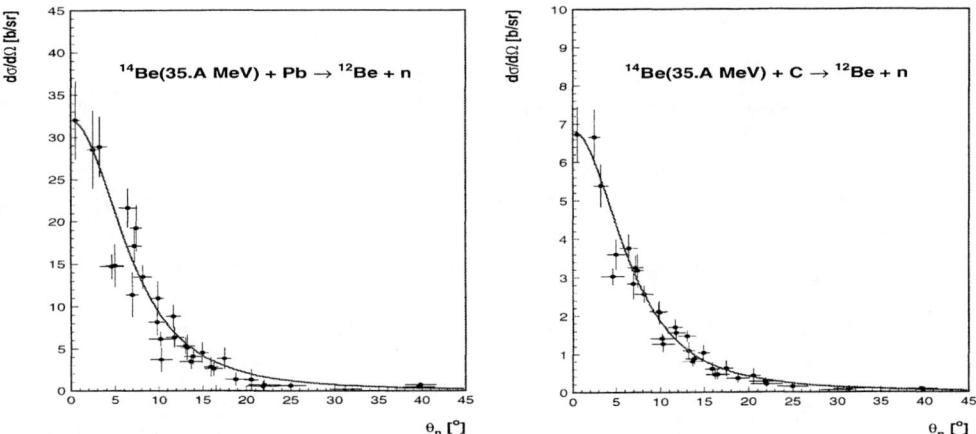

FIGURE 2. Single neutron angular distributions for the breakup of ^{14}Be (E_n=10 to 100 MeV).

A ^{12}Be breakup

The Q-values for the reactions ^{12}C(^{12}Be,^{x}He^{y}He)^{12}C and p(^{12}Be,^{x}He^{y}He)p were calculated from the energies of the two detected helium nuclei and the energy of the undetected recoil-like particle inferred from the measured momenta of the beam and detected reaction products. In the case of the (CH$_2$)$_n$ target this technique was able to resolve the reactions taking place on the carbon and hydrogen. The cross sections for the various reactions are given in Table 1. Selecting events identified with peaks in the Q-value spectra for the above reactions, it is possible to calculate the excitation energy, or invariant mass, of the ^{12}Be nucleus prior to decay using the relationship

$$E_x = \frac{1}{2}\mu v_{rel}^2 + Q_{bu} \qquad (1)$$

where v_{rel} is the relative velocity of the two breakup fragments, μ the reduced mass and Q_{bu} the breakup Q-value, which for the ^{6}He+^{6}He and ^{8}He+^{4}He channels is 10.11 and 8.95 MeV respectively. Figure 1a shows the ^{12}Be excitation energy spectrum for decays into two ^{6}He nuclei. These states, which span the excitation energy region 10 to 25 MeV, are the first definitive evidence for the 2α-4n, molecular cluster structure in this nucleus [5]. Angular correlation measurements of the breakup products indicate that the states may be associated with spins from 4$^+$ to 8$^+$, and the inferred excitation energy-spin sequence appears to be consistent with a rotational band with a large moment of inertia (Figure 1b).

B ^{14}Be breakup

The ^{14}Be reaction cross sections derived from the telescope data confirm that the two-neutron removal reaction channel is dominant, as seen in an earlier experiment [6]. The single neutron angular distributions (shown in Figure 2) exhibit the narrow, forward peaked form characteristic of a halo. Interestingly the characteristic widths of the distributions for the two targets are very similar and thus may suggest, as in the case of ^{11}Li [7], the existence of a very low lying state in ^{13}Be. The analysis of the ^{12}Be+n invariant mass spectrum, presently underway, should shed further light on this conjecture. The ^{12}Be+n+n invariant mass spectrum is being investigated in parallel with the objective of extracting the low-lying dipole strength function, $dB(E1)/dE_x$. The two-neutron correlations — relative momenta and correlation function, $C(q)$ — are also under analysis, with the present effort concentrating on inclusion of the effects of neutron-neutron final state interactions. The results of these analyses when compared with realistic three-body models are expected to provide important insights into the halo structure of ^{14}Be.

[1] K. Ikeda, Suppl. Prog. Physics (Japan) Extra Numbers, 464 (1968).
[2] S. Yang, et al., Chem. Phys. Letts. **144**, 431 (1988).
[3] D.H. Wilkinson, Nucl. Phys. **A 452**, 296 (1986).
[4] W. von Oertzen, Z. Phys. **A 354**, 249 (1996), Z. Phys. **A 357**, 355 (1997).
[5] A.A. Korshinnikov, et al., Phys. Letts. **B 343**, 53 (1995).
[6] K. Riisager et al., Nucl. Phys. **A540**, 365 (1992).
[7] F. Barranco et al., Phys. Lett. **B319**, 387 (1993).

Resonance Scattering to Study Nuclei at the Borders of Nuclear Stability

Vladilen Z. Goldberg*

*Russian Research Centre "Kurchatov Institute",
Institute of General and Nuclear Physics, Moscow, Russia, 123182

Abstract. Application of a new method of elastic resonance scattering to study drip-line nuclei are considered.

During the last few years it has become evident that the main interest in studies of nuclear structure has shifted to drip-line nuclei [1]. The most direct way to reach border nuclei is given by different facilities producing r/a beams. Mainly these beams leave much to be desired from the point of view of nuclear spectroscopy: the intensity is low, the beam energy spread is large, and there is no possibility to change the the energy of the beams by spectroscopically reasonable steps. On the other hand there is a great need of spectroscopical information because spin-parities even of the ground states of drip-line nuclei as a rule are unknown. Surprisingly some problems can be solved by means of elastic resonance scattering at available r/a beams due to a new method developed recently. The method is based on the dominance of the scattering through resonances over various sources of background. A simplified outline of the measurements looks like the following. A beam of heavy ions enters the scattering chamber through a thin window. The chamber is filled by a gas (hydrogen or methan, helium). The ions slow down and stop in the gas, and the light recoil nuclei (protons or alpha particles) are detected by Si detectors positioned around direction of the beam in the gas. The reverse kinematic of the reactions is crucial to provide for the possibility of observation of recoil particles at zero degree and good resolution. The basic ideas and details can be found in [2,3]. The most evident source of background is the inelastic resonance scattering. It is also the most dangerous effect, because it can have rather large probability and produces the same recoils. However to produce the same energy of recoils (at the same angle) in elastic and inelastic scattering (E and I S) the energy of incident heavy ions should be higher for the latter and hence this process should take place closer to the entance to the chamber. As a result of this effect a different time delay is appeared for elastically and inelastically scattered recoils relative to the incident heavy ions.

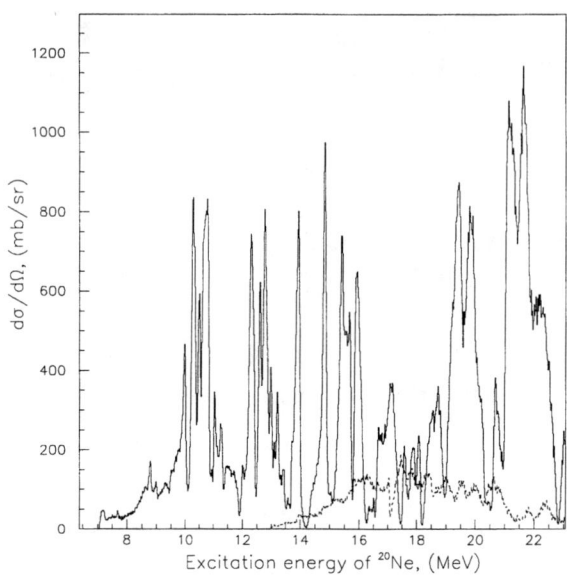

FIGURE 1. Elastic and inelastic scattering of ^{16}O on ^{4}He at 180°.

Figure 1 [4] demonstrates the selection of E and I S (low cross section curve) for case of ^{16}O (110 MeV) interaction with helium with the delay difference about 2 ns. The very deep minima in Figure 1 are the evidence of low background contribution. The effective energy resolution in the spectrum is about 45 KeV. Generally the energy resulution depends upon various parameters of the experiment, but mainly upon the scattering angle and upon the energy losses of heavy ions in the gas before scattering. For He recoils the resolution of about 30 keV is quite feasible and for protons it could be 15 keV due their small specific energy losses. The check of resolution was made at Ganil with a beam of ^{12}C produced exactly as a r/a beam [6]. The obtained resolution was about 25 keV.

PROTON RICH NUCLEI

The application of the method to study proton rich nuclei is straighforward. The resonance S of proton rich nuclei on hydrogen should be used. Due to various practical reasons methan is used instead of pure hydrogen.

Figure 2 shows the excitation function of E S of ^{7}Be on hydrogen obtained at the Kurchatov Institute cyclotron with 31 MeV ^{7}Be beam [5]. The broad structure seen in Figure 2 after the known 3+ resonance should be interpreted as the manifestation of a broad 2s-state and the weak 1+ state in ^{8}B. Probably the latter is the mirror to 1+ state at 2.3 MeV in ^{8}Li. The 2s-state found in ^{8}B is important to understand the nature of intruder 2s-states in light drip-line nuclei and to evaluate the solar

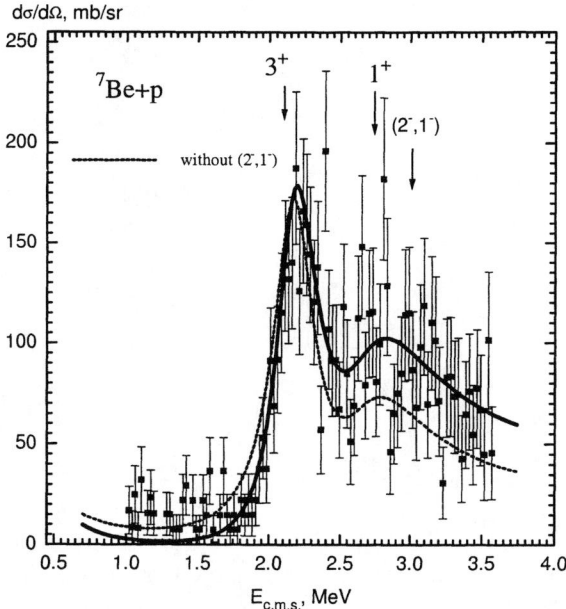

FIGURE 2. Excitation function of elastic scattering of ^{7}Be on protons at 180°

neutrino flux. Evidently it is very difficult to "feel" the presence of this state in the mirror ^{8}Li nucleus due to absence of any barrier for s-neutrons. In a similar way the ^{11}N nucleus was investigated at Ganil [6]. A number of states were identified in this nucleus and spin parities as well as widths, and positions were assigned to the three lowest states (including 2s ground state). The field of investigation of light proton rich nuclei is promising. The nuclei from ^{9}C and up to ^{21}Al can be studied with available r/a beams. Increasing with Z the Coulomb scattering should put the limit of investigation for the low energy part of excitation fuctions for heavier nuclei.

ANALOG STATES OF NEUTRON RICH NUCLEI

There is no mirror reflection of the method for application on the neutron rich side of nuclear stabilty. However the needed information can be obtained through isobar analog states in the neighboring nuclei with the same mass number but greater charge. Due to isotopic spin conservation the energy of the ground state of a neutron rich nucleus relative to the threshold of decay by a neutron, and the energy of the corresponding isobar analog state relative to the threshold of decay by a proton differs by the Coulomb energy. Because the Coulomb energy is positive there are many cases where the neutron rich drip-line nuclei are stable to nuclear decay while the corresponding isobar analod states are unstable to T-allowed proton decay and hence available for resonance scattering study. Fig. 3

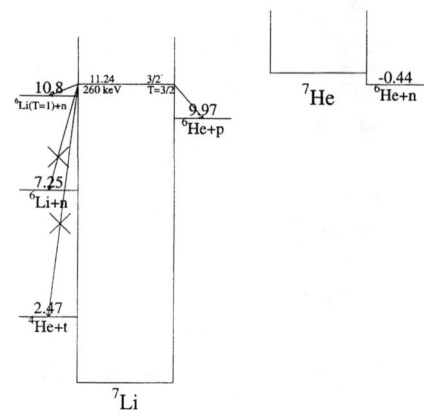

FIGURE 3. T=3/2 level in ^{7}Li

illustrates the situation. The energies of the ground state in ^{7}He and its analog state in ^{7}Li are known. As it is expected the corresponding analog state in ^{7}Li is more unstable with respect to the isospin-allowed proton decay with Q=1.27 MeV. There are many open channels for nuclear decay of the T=3/2 state but only two of them are allowed by T-conservation. These are the proton decay to ^{6}He and the "analog", neutron decay to the excited (T=1) state in ^{6}Li (the Coulomb energy differences lock the latter channel in heavier nuclei). Both allowed channels are characterized by low penetrabilities, and the T=3/2 state appeared to be rather narrow. That is why it should dominate in E S of ^{6}He on hydrogen. In the same way the isobar analog states corresponding to the lowest states of ^{9}He, ^{10}Li, and up to ^{19}C can be investigated by the method in question.

Many groups [6] (besides the Russian) participated in the development and applications of the method, but the contributions of the Göteborg group (B. Jonson) and the group from Finland (W.Trzaska, T. Lönnroth) can not be overevaluated. Part of the work was made under support of the grant 97-02-17113 of RFBR.

REFERENCES

1. Hansen, P.G., Jensen, A.S., Jonson, B., Ann. Rev. Nucl. Part. Sci., **45**, 591, (1995)
2. Artemov, K.P., et al., Sov. J. Nucl. Phys. **52**, 1460 (1990)
3. Goldberg, V.Z. and Pakhomov, A.E., Phys. At. Nucl. **56**, 1167, (1993)
4. Goldberg, V.Z., et al., JYFL Ann. Rep. 1997, Dep. of Physics, Univ. of Jyväskylä
5. Goldberg, V.Z., et al., Lett. to JETP, **67**, 953 (1998)
6. Axelsson, L., et al., Phys. Rev, **C54**, R1511 (1996).

Final-state interactions in the system ^{8}He+n

L. Chen[a], K. Govaert[a], B. Blank[a], M. Chartier[a], A. Galonsky[a],
P.G. Hansen[a], J. Kruse[a], V. Maddalena[a], M. Thoennessen[a], K. Ieki[b],
Y. Iwata[b], Y. Higurashi[b], S. Takeuchi[b], F. Deak[c], A. Horvath[c], A. Kiss[c],
Z. Seres[d]

a. National Superconducting Cyclotron Laboratory, Michigan State University, East Lansing, Michigan, USA
b. Department of Physics, Rikkyo University, Tokyo, Japan
c. Department of Atomic Physics, Eötvös University, Budapest, Hungary
d. Research Institute for Particle and Nuclear Physics, Hungarian Academy of Sciences, Budapest, Hungary

Abstract. Final-state interactions in the systems ^{6}He+n and ^{8}He+n have been studied via the breakup of 30 MeV/u ^{10,11,12}Be beams on a ^{9}Be target. The results for the longitudinal and transverse momentum distributions of the outgoing neutron and fragment are presented and discussed.

The neutron rich isotopes ^{6}He and ^{8}He are known to be halo nuclei with an extended valence neutron distribution. These even helium isotopes are particle stable. The odd isotopes ^{7}He and ^{9}He on the other hand are unstable against neutron emission. While the ground state of ^{7}He is known to be a 3/2⁻ state, the situation for ^{9}He is less clear. Some shell model calculations suggest that the ground state of ^{9}He should be 1/2⁺ instead of 1/2⁻ according to the normal shell order. This parity inversion is well established for ^{11}Be and there is mounting evidence that the same effect occurs for the lighter N=7 isotone ^{10}Li (1). If the trend continues a (1/2⁺,1/2⁻) pair should be separated by ~ 0.5-1 MeV for ^{9}He (2).

The nucleus ^{9}He has already been studied in a pion double charge exchange reaction on ^{9}Be (3). The results were interpreted in terms of a number of sharp resonances of which the lowest was 1.1 MeV above the neutron emission threshold. This picture was confirmed by two-body reactions which were performed at the HMI (4). Since an s state would not give the observed resonance-like structure we can assume that this state is the p state. A lower-lying s state can then be expected close to the particle threshold which should give a clear signal in an experiment measuring the final-state interaction between the neutron and the ^{8}He fragment.

We have studied the breakup of ^{10,11,12}Be on a light ^{9}Be target. The longitudinal and transverse momentum distributions of the outgoing neutron and fragment and the invariant mass spectra provide information on s and p state final-state interactions. To clarify the influence of the initial state, two projectiles ^{11,12}Be with very different sets of valence neutrons were used. The valence neutron in ^{11}Be is a relatively pure $s_{1/2}$ state while ^{12}Be in spite of its magic neutron number has a complex structure involving s,p and d neutron orbits. The projectile ^{10}Be was included to provide information on the relative contributions of the core and valence neutrons.

The experiment was performed at the NSCL using 30 MeV/u beams of ^{10,11,12}Be impinging on a 200 mg/cm^2 ^{9}Be target. The ^{10,11}Be nuclei were produced by fragmentation of a 80 MeV/u ^{13}C beam in a 1900 mg/cm^2 ^{9}Be target. For the case of ^{12}Be a primary beam of 80 MeV/u ^{18}O bombarded a 1455 mg/cm^2 ^{9}Be production target. The fragments were identified and separated in the A1200 fragment separator. The intensity of the secondary beams on target varied between about 20-80·10^3 particles per second.

A kinematically complete experiment requires the measurement of the angle and energy of the incident beam particle hitting the target and the emission angles and energies of the products leaving the target. The trajectory of the projectile was measured by a set of two PPACS in front of the target. The neutrons in coincidence with ^{8}He and other charged fragments were detected by means of the MSU neutron walls (5). The position in the neutron cells provided the angle and the time of flight the energy of the neutron. For the charged fragments a double sided silicon strip detector behind the target provided angle and energy loss information. The particles were then swept by a dipole magnet into a plastic scintillator array yielding the fragment energy while the neutrons leaving the target traveled straight and arrived at the neutron walls which were centered at zero degrees (6). For the identification of the incident beam particle the time of flight between a thin plastic scintillator after the A1200 focal plane and the fragment array was measured. The combination of energy loss and total energy allowed us to select the reaction channel of interest.

The analysis of the data is underway. The ^{6}He and ^{8}He fragments could easily be separated. We have extracted total reaction cross sections and compared them to previous results obtained at about 40 MeV/u at Ganil (7). For the He isotopes a total cross section of about 70 mb was measured for all three projectiles with about 10 mb for the interesting case of ^{8}He. At this moment the analysis of the neutron coincidence data is in progress. Investigation of the momentum distributions of the neutrons revealed a very narrow distribution for the ^{11}Be beam when selecting the ^{8}He+n exit channel. Fig. 1 shows the parallel momentum distribution of neutrons in coincidence with ^{6}He and ^{8}He fragments. The experimental data could be fitted with a Lorentzian distribution with width parameters of $\Gamma = 51$ MeV/c and $\Gamma = 39$ MeV/c respectively. The narrow distribution in the latter case can be interpreted as an indication of a stronger final-state interaction in the ^{8}He+n system. For the ^{12}Be projectile on the other hand, the distribution becomes broader and very similar to the ^{6}He+n case. Similar results were obtained for the transverse momentum distributions. Removing for the ^{11}Be beam the neutrons coming from the core fragmentation of ^{10}Be results in

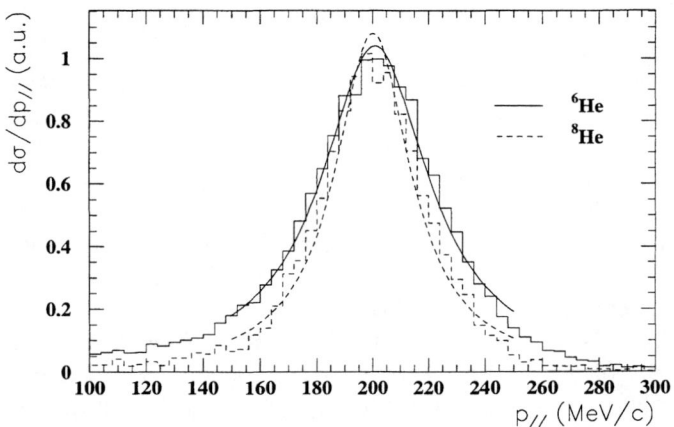

FIGURE 1 Parallel momentum distribution of neutrons in coincidence with ^{6}He and ^{8}He fragments for the 30 MeV/u ^{11}Be beam. The histograms display the experimental data while the curves represent fits using a Lorentzian distribution with width parameter $\Gamma = 51$ MeV/c for ^{6}He and $\Gamma = 39$ MeV/c for ^{8}He.

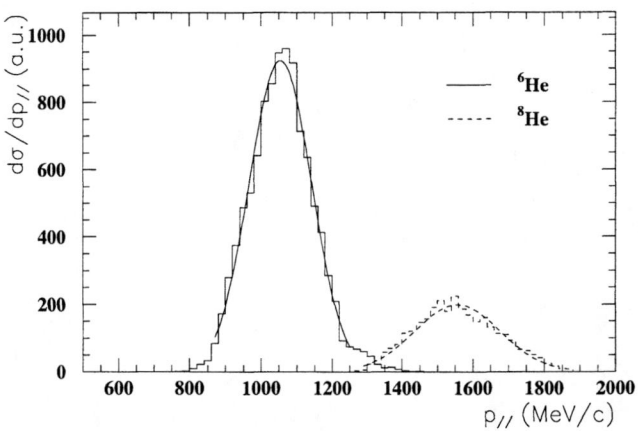

FIGURE 2 Parallel momentum distribution of ^{6}He and ^{8}He fragments for the 30 MeV/u ^{12}Be beam. The histograms display the experimental data while the curves represent fits using a Gaussian distribution with FWHM = 204 MeV/c for ^{6}He and FWHM = 284 MeV/c for ^{8}He.

more narrow momentum distributions for neutrons in coincidence with ^{6}He since we now select the halo neutrons.

The momentum distributions of the fragments can be compared to the predictions of the projectile fragmentation model proposed by Goldhaber (8). Fig. 2 shows the parallel momentum distribution of ^{6}He and ^{8}He fragments for the 30 MeV/u ^{12}Be beam. Fitting the experimental data with a Gaussian distribution yields, after correction for energy loss in the target, a FWHM of respectively 196 MeV/c for ^{6}He and 283 MeV/c for ^{8}He for the case of the ^{10}Be projectile, 174 MeV/c and 269 MeV/c for the case of ^{11}Be and 200 MeV/c and 277 MeV/c for the case of ^{12}Be. In comparison the theoretical model predicts for a typical width parameter σ_o of 85 MeV/c, valid for beam energies above about 50 MeV/u, larger values for the widths of the order of 300-350 MeV/c. Our results are better described by a width parameter σ_o of 65 MeV/c confirming the observed drop of σ_o with decreasing beam energy (9).

Besides the parallel and transverse momentum distributions we plan to extract in the future the invariant mass spectra which correspond to the decay energy of the system in question. These can shed more light on the s and p state final-state interactions.

REFERENCES

1. M. Zinser et al., Phys. Rev. Lett. 75 (1995) 1719.
2. A. Brown, private communication (1996).
3. K. Seth et al., Phys. Rev. Lett. 58 (1987) 1930.
4. W. Von Oertzen et al., Nucl. Phys. A588 (1995) 129.
5. P.D. Zecher et al., Nucl. Instr. Meth. Phys. Res. A401(1997) 329.
6. J. Kruse et al., NSCL Annual Report (1995) 261.
7. S. Grévy, PhD thesis Orsay (1997) and to be published.
8. A.S. Goldhaber, Phys. Lett. B53 (1974) 306.
9. M.J. Murphy et al., Phys. Rev. C28 (1983) 428.

NEUTRON ANGULAR DISTRIBUTIONS FROM THE CORE BREAK-UP REACTIONS OF THE ^{11}Be AND ^{11}Li HALO NUCLEI

S. Grévy[1,3], L. Axelsson[2], J. C. Angélique[3], R. Anne[4],
D. Guillemaud-Mueller[1], P. G. Hansen[5], P. Hornshoj[6],
B. Jonson[2], M. Lewitowicz[4], A. C. Mueller[1], T. Nilsson[2],
G. Nyman[2], N. Orr[3], F. Pougheon[1], K. Riisager[6],
M. G. Saint-Laurent[4], M. Smedberg[2], O. Sorlin[1].

[1]Institut de Physique Nucléaire, IN2P3-CNRS, F-91406 Orsay Cedex, France
[2]Fysiska Institutionen, Chalmers Tekniska Högskola, S-41296 Göteborg, Sweden
[3]LPC, IN2P3-CNRS, ISMRA, F-14050 Caen Cedex, France
[4]GANIL, F-14021 Caen Cedex, France
[5]NSCL, Michigan State University, East Lansing, MI48824-1322, USA
[6]Institut for Fysik og Astronomi, Aarhus Universitet, DK-8000 Aarhus C, Denmark

Abstract

The halo nuclei ^{11}Be and ^{11}Li have been studied through core break-up reactions, where the halo neutrons are detected in anti-coincidence with the core of the halo nucleus. In this particular channel, the halo neutrons are not expected to participate in the reaction and should therefore show the same properties as when situated inside the halo nucleus. The widths of the halo neutron momentum distributions have been extracted in coincidence with He fragments, $\Gamma = 32 \pm 4$ MeV/c, and Li fragments, $\Gamma = 42 \pm 4$ MeV/c for ^{11}Be and with He fragments, $\Gamma = 42 \pm 6$ MeV/c for ^{11}Li. An experimental value of the shadow effect for ^{11}Be when breaking up to Li and He fragments has been obtained to be 0.63. A simple theoretical calculation to reproduce this value is given.

It is now well etablished that, for some light very rich neutron nuclei, the valence neutron(s) form a distribution which extends well beyond that expected from systematics [1]. Since their discovery in the middle of the 80's, ^{11}Be and ^{11}Li halo nuclei have been studied through dissociation reactions in which the neutron coming from the halo was detected in coincidence with the core. It has been shown that, in this particular channel, neutron distributions were modified by reaction mechanisms and final state interactions [1,2,3]. In this new experiment, core break-up reactions, in which the halo neutron is detected in coincidence with charged fragments coming from the core, have been used. Since the impact parameter is small, the halo neutron is supposed not to participate in the reaction and then, to carry out its properties as in the halo nucleus. These channels are more exclusive that the *restricted-inclusive* one used for the study of the ^{19}C [4]. In core break-up reactions, neutrons coming from the core are detected together with the neutrons coming from the halo. Therefore, as it was shown for the ^{19}C, it is essential to separate the two components and remove the one associated with the the fragmentation of the core. For this purpose the experiment was also performed within the same kinematics conditions for the core nuclei ^{10}Be and ^{9}Li.

The experiment was performed at the GANIL facility with the LISE3 spectrometer. The various secondary beams (^{11,10}Be, ^{11,9}Li) were produced by fragmentation of a ^{18}O primary beam with an energy of 76 MeV/u. To purify the beams, a wedge-shaped achromatic degrader of Al (1100 μm) was mounted in the intermediate focal plane between the two dipole magnets. At the end of the spectrometer, the beams were focused onto a reaction target of Be (2000 μm) placed inside a detector telescope. The incoming particles were identified through their energy loss in a Si detector (300 μm) and their time of flight in the spectrometer. The charged reaction products were identified through their energy loss in a Si detector (300 μm) and their residual energy in a CsI detector (4500 μm). The neutrons were detected in an hodoscope of 29 liquid scintillators NE213 of varying sizes in order to get higher granularity at smaller angles and covering a large angular domain (-25° to 49°).

The heavy ions cross-sections from the break-up reactions in the Be target obtained for each secondary beam are presented in table 1, showing good agreement with the previous measurement [2]. As expected, almost identical values were found for the halo nucleus (^{11}Be or ^{11}Li) and its corresponding core nucleus (^{10}Be or ^{9}Li). The neutron angular distributions have been extracted in coincidence with Li and

σ_r (mb)	He fragments	Li fragments
^{11}Be	125±20	245±25
^{10}Be	135±15	260±25
^{11}Li	180±25	-
^{9}Li	190±25	-

Table 1: Measured heavy ions cross sections for ^{11}Be and ^{10}Be in coincidence with Li and He fragments and for ^{11}Li and ^{9}Li in coincidence with He fragments.

He fragments for the ^{11,10}Be, and in coincidence with the He fragments for the ^{11,9}Li. The remaining distributions after the subtraction of the core component, representing the distributions of the halo neutrons, are presented for the ^{11}Be on the figure 1. The full lines represent fits using a Lorentzian, which corresponds to the Fourier transformation of a Yukawa wave function, commonly used to describe the halo nucleus. The widths Γ of the momentum distributions and the halo-neutron cross-sections obtained by integration of the neutron angular distributions are given in table 2 for the ^{11}Be and the ^{11}Li.

The width Γ=42±4 MeV/c for the ^{11}Be in coincidence with Li fragments is in agreement with the width extracted from the core distributions in dissociation reactions (Γ=47±7 MeV/c [5]). This value is also compatible with recent theoretical predictions which converge to a width of approximately 46 MeV/c [6,7]. The small difference can be understood as a *shadow effect* which leads to a decrease of the

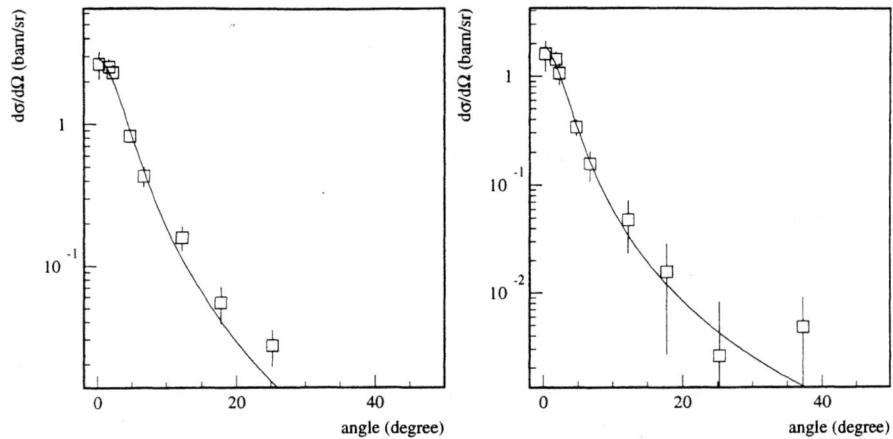

Figure 1: Angular distributions of the halo-neutrons coming from the ^{11}Be in coincidence with Li fragments (left) and with He fragments (right). The full lines represent a fit with a Lorentzian function, giving a width Γ of the distribution of 42±4 MeV/c and 32±4 MeV/c in coincidence with Li and He fragments respectively.

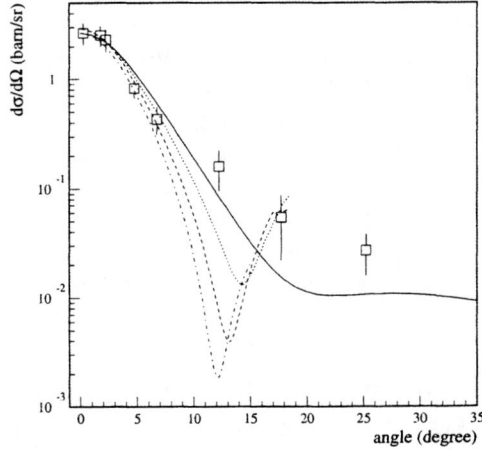

Figure 2: Comparaison of the experimental (open square) angular distribution of the halo neutrons coming from the ^{11}Be in coincidence with Li fragments and the calculated distributions without shadow effect (full line) and with 3 different cutoff radii for the halo wave function (3, 3.86 and 4.5 fm for the dotted, dashed ans dot-dashed lines respectively)

	^{11}Be → Li	^{11}Be → He	^{11}Li → He
Γ(MeV/c)	42±4	32±4	42±6
$\int \frac{d\sigma_n}{d\Omega} d\Omega$ (mb)	100±25	40±15	100±40

Table 2: Extracted widths and integrated neutron cross-section from the neutron halo distribution of ^{11}Be in coincidence with Li or He fragments and from ^{11}Li in coincidence with He fragments.

experimental width of approximately 10 %. This shadow effect corresponds to the removal of the central part of the halo wave function in the core-target collision. There are some cases in which the neutron can not be considered as a spectator of the reaction (for more details, see [5]) and a scattering or an absorption of this neutron can occur [2]. Figure 4 shows calculations [8] for the halo neutron angular distribution without shadow effect (full line) and for several estimations of the shadow effect in the halo wave function (dashed lines). In this case, the neutron is supposed to be located outside of the potential well ($r_{perpandicular} > r_{cutoff}$). The experimental value for the halo neutron multiplicity M_n, given by the ratio of the integrated neutron cross section by the heavy ions cross section, is 0.37 for the ^{11}Be in the core break-up channel (the shadow effect S=1-M_n). This value can be compared to the multiplicity given by our model and a good agreement is found for r_{cutoff}=3.86 fm with a multiplicity of 0.41 whereas the 3 and 4.5 fm cutoff radii give a multiplicity of 0.58 and 0.30 respectively. The only same shadow effect cannot explain the narrower width Γ=32±4 MeV/c in coincidence with He fragments. With the present statistics, the widths are not significantly different and more precise number is needed to discuss the importance of Final State Interactions or an impact parameter dependance in the He channel. The width of 42±6 MeV/c found for the halo of ^{11}Li is in good agreement with both core momentum distributions (Γ=47±6 MeV/c [5]) and previous measurements in the *restricted-inclusive* channel (Γ=43±3 MeV/c [9]). The shadow effect is in this case more difficult to define since we detect only one of the two halo-neutrons. The shadow effect could be expected to be less important since, even if the halo neutron are correlated, one of them could still be situated outside of the shadow region.

[1] P. G. Hansen, A. S. Jensen and B. Jonson, Ann. Rev. Nucl. Part. Sci. **45**(1995)591.
[2] R. Anne et al., Nucl. Phys. **A575**(1994)125.
[3] F. Barranco, E. Vigezzi and R. A. Broglia, Phys. Lett. **B319**(1993)387.
[4] M. Marqués et al., Phys. Lett. **B381**(1996)407.
[5] S. Grévy, Thesis of Paris XI university, Orsay, France(1997),IPNO-97-24.
[6] P. G. Hansen, Phys. Rev. Lett. **77**(1996)1016.
[7] H. Esbensen, Phys. Rev. **C53**(1996)2007.
[8] S. Grévy, L. Axelsson et al., unpublished.
[9] T. Nilsson et al., Europhys. Lett. **30**(1995)19.

THE DESCRIPTION OF EXOTIC NUCLEI ELASTIC SCATTERING IN THE FRAMEWORK OF GLAUBER MODEL WITH NON-EIKONAL CORRECTIONS

K.A. Gridnev, T.V. Taroutina

Saint-Petersburg State University, Saint-Petersburg, Russia

Abstract

Proton elastic scattering from exotic nuclei was calculated in the framework of Glauber model. The inclusion of large-angle scattering was made on the basis of exact impact parameter by the eikonal expansion in terms of the inverse powers of momentum.

The appearance of radioactive nuclei beam accelerators has stimulated the intensive studies of nuclei near the drip line. It has been showed that some nuclei exhibit a halo of valence nucleons and have a large matter radii in comparison with their stable isotopes. Two-neutron halo has been observed in several light neutron rich nuclei, such as ^{11}Li and ^{8}He. One of the methods of investigating the exotic nuclei is Glauber model of elastic scattering (eikonal approximation). However, one casts some doubt upon the use of Glauber approach over the range of large scattering angles. The successful application of eikonal approximation for the analysis of experimental data of elastic and some kinds of inelastic processes makes it actual to specify the eikonal approximation for the large angles. One of the methods of correcting the eikonal approximation was suggested by Andrianov A.A.[2]. In this method the analysis of scattering process for large scattering angles is made in the basis of exact impact parameter by means of the eikonal expansion in terms of the inverse powers of momentum along the fixed direction [2].

In our calculations we ignore spin-orbit effects. Because of the logarithmic divergence of Coulomb phase it is convenient to write the scattering amplitude $f(\theta)$ [4] as,

$$f(\theta) = -ik \int J_0(\mathbf{q}\mathbf{b})(e^{i\chi(b)} - 1)b\,db + f_C(\theta) = f_N(\theta) + f_C(\theta), \quad k^2 = 2\mu E_{c.m.}/\hbar^2, \quad (1)$$

where μ is the reduced mass, J_0 is the ordinary Bessel function of order zero, b is the impact parameter and q is the momentum transfer, $q = 2k\sin(\theta/2)$. The phase $\chi(b)$ contains a nuclear and a Coulomb parts and is given by

$$\chi(b) = \chi_N(b) + \chi_C(b); \quad \chi_N(b) = -\frac{1}{\hbar v}\int_{-\infty}^{\infty} dz V_N[(b^2 + z^2)], \quad (2)$$

$$\chi_C(b) = (2Z_1Z_2e^2/\hbar v)[\ln(kb) + \frac{1}{2}E_1(b^2/r_{m.s.}^2)], \quad E_1(x) = \int_x^{\infty}\frac{e^{-t}}{t}dt, \quad (3)$$

where V_N is the complex nuclear potential, v is the relative velocity, $r_{m.s.}$ is the root mean square radius of the nucleus. The Coulomb amplitude $f_C(\theta)$ is given by the usual expression. Nuclear part of the phase is determined by optical potential [4]

$$V_N = <t_{pn}>\rho_n + <t_{pp}>\rho_p, \quad <t_{pN}> = -\hbar v\overline{\sigma_{pN}}(\alpha_{pN} + i)/2, \quad (4)$$

CP455, *ENAM98: Exotic Nuclei and Atomic Masses*
edited by B. M. Sherrill, D. J. Morrissey, and Cary N. Davids
© 1998 The American Institute of Physics 1-56396-804-5/98/$15.00

where ρ_p and ρ_n are the proton and neutron nuclear densities. The Pauli-corrected total cross-section, $\overline{\sigma_{pN}} = \sigma_{pN}P(\varepsilon_{NF}/\varepsilon)$ depends on the Fermi energy of target nucleon,ε_{NF}, calculated for the proton and neutron nuclear densities [5]. The proton and neutron densities for ^{11}Li were obtained in [3]in three-body model $^{11}Li + n + n$. These densities describe neutron halo in ^{11}Li, provide ^{11}Li $R_{mat}(^{11}Li) = 3.2$ fm in agreement with the experiment and contain the mixture of 0p- and 1s-orbitals for the valence neutrons. The densities for ^8He were obtained in [7] in five-body model $\alpha+4n$. These densities reproduce the experimental radius ^8He $R_{mat}(^8He) = 2.52$ fm and contain spreaded distribution of 4 valence neutrons.

$$\rho_i(r) = N_{ci}\frac{\exp(-r^2/a^2)}{\pi^{3/2}a^3} + N_{vi}\frac{2\exp(-r^2/b^2)}{3\pi^{3/2}b^5}[Ar^2 + B(r^2 - \frac{3}{2}b^2)^2], \quad i = n, p, \quad (5)$$

The parameters of ^{11}Li densities are as follows: $N_{cp} = 3$, $N_{cn} = 6$, $N_{vp} = 0$, $N_{vn} = 2$, $a = 1.89$ fm, $b = 3.68$ fm, $A = 0.81$, $B = 0.19$. The parameters of ^8He densities are $N_{cp} = 2$, $N_{cn} = 2$, $N_{vp} = 0$, $N_{vn} = 4$, $a = 1.38$ fm, $b = 1.99$ fm, $A = 1.0$, $B = 0.0$.

In this work we used Glauber amplitude correction method, suggested in [2] and based on the construction of eikonal perturbation theory along the fixed direction (the direction of the average momentum $\frac{1}{2}(\mathbf{k_i} + \mathbf{k_f})$). In this method the scattering amplitude in the representation of exact impact parameter is expanded in terms of the inverse momentum k^{-1}, in which the first term with k^0 coincides with the phase in the Glauber approach and the terms with k^{-1}, k^{-2}, ... are the corrections:

$$f(\mathbf{b}) = \int e^{-i\mathbf{q}\mathbf{b}} f'(\mathbf{q}) d^2\mathbf{q} = -ik \sum_{n=0} \left(\frac{i}{2k}\right)^n \tau_n(b). \quad (6)$$

In the case of the spherical-symmetric potential the corrections in the frame of method [2] are the same as Wallace corrections [8]. and the correspondent scattering amplitudes with the inclusion of non-eikonal corrections of the 1st,2nd and 3d order are the following:

$$f^I(\theta) = -ik \int J_0(qb) \left(e^{i[\chi_0(b)+\tau_1(b)]} - 1\right) b\,db, \quad (7)$$

$$f^{II}(\theta) = -ik \int J_0(qb) \left(e^{i[\chi_0(b)+\tau_1(b)+\tau_2(b)]}e^{-\omega_2(b)} - 1\right) b\,db, \quad (8)$$

$$f^{III}(\theta) = -ik \int J_0(qb) \left(e^{i[\chi_0(b)+\tau_1(b)+\tau_2(b)+\tau_3(b)+\phi_3(b)]}e^{-[\omega_2(b)+\omega_3(b)]} - 1\right) b\,db, \quad (9)$$

$$\tau_1(b) = -\frac{1}{k}\left(1 + b\frac{d}{db}\right)\int_0^\infty dz \left(\frac{1}{\hbar v}V_N(b,z)\right)^2, \quad (10)$$

$$\tau_2(b) = -\frac{1}{k^2}\left(1 + \frac{5}{3}b\frac{d}{db} + \frac{1}{3}b^2\frac{d^2}{db^2}\right)\int_0^\infty dz(\frac{1}{\hbar v}V_N(b,z))^3 - b[\chi_0'(b)]^3/24k^2, \quad (11)$$

$$\tau_3(b) = -\frac{1}{k^3}\left(\frac{5}{4} + \frac{11}{4}b\frac{d}{db} + b^2\frac{d^2}{db^2} + \frac{1}{12}b^3\frac{d^3}{db^3}\right) \times$$
$$\int_0^\infty dz(\frac{1}{\hbar v}V_N(b,z))^4 - b\tau_1'(b)\frac{[\chi_0'(b)]^2}{8k^2}, \qquad (12)$$

$$\phi_3 = -\frac{1}{k}\left(1 + \frac{5}{3}b\frac{d}{db} + \frac{1}{3}b^2\frac{d^2}{db^2}\right)\int_0^\infty dz\left(\frac{1}{2k}\frac{dV_N(r)}{dr}\right)^2, \qquad (13)$$

$$\omega_2(b) = \frac{\chi_0'(b)[b\chi_0''(b) + \chi_0'(b)]}{8k^2}; \quad \omega_3(b) = \frac{b\chi_0'(b)\tau_1''(b) + b\tau_1'(b)\chi_0''(b)}{8k^2}, \qquad (14)$$

In the work [6] J.S. Al-Khalili et al. analysed the elastic scattering ^{11}Be + ^{12}C in the framework of few-body Glauber model with the modification of each constituent eikonal phase to account for curvature of its trajectory. It was shown that the inclusion of non-eikonal corrections leads to a significantly improved description of the experimental data.

In the present work we calculated the differential cross-sections of proton elastic scattering from the nuclei ^{11}Li and ^{8}He for different energies in the frame of ordinal Glauber model and with the inclusion of 1st, 2nd and 3d order non-eikonal corrections (see Fig.1).

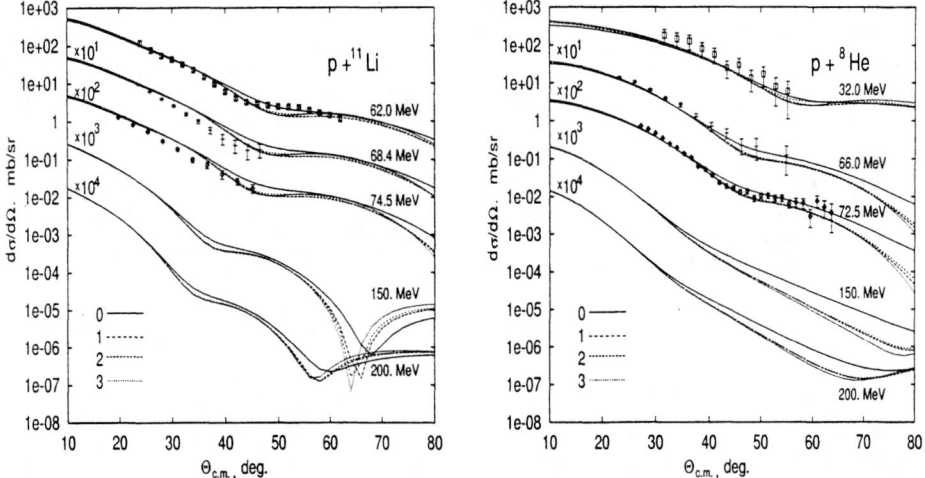

Fig. 1. Differential cross sections for ^{11}Li + p and ^{8}He + p elastic scattering in inverse kinematics. "0" – calculations in the frame of ordinal Glauber model, "1" – with the 1st order non-eikonal correction, "2" – with the 1st and 2nd order non-eikonal corrections, "3" – with the 1st, 2nd and 3d order non-eikonal corrections. The experimental data are taken from the work [1].

The inclusion of non-eikonal corrections strongly affects the differential cross section at large scattering angles. It is seen that the effect of the corrections is decreasing with the increase of the projectile energy. The effective optical potentials of proton interaction with exotic nuclei were reconstructed from the phase-shifts containing non-eikonal corrections.

The calculations showed that imaginary part of optical potential is strongly affected by non-eikonal corrections (see Fig 2). From Fig.2 it is seen that the potentials are converging with the increase of energy.

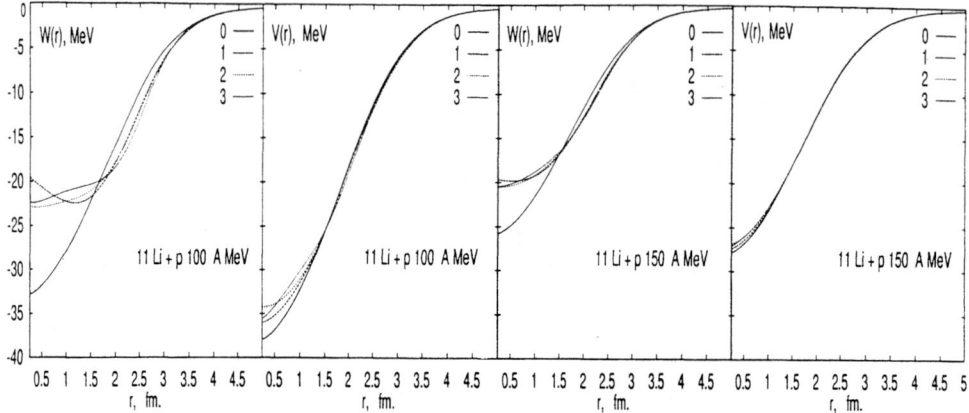

Fig. 2. Reconstructed effective optical potentials from the phases, containing non-eikonal corrections. W(r) – imaginary part of the potential, V(r) – real part of the potential. "0" – potential, reconstructed from the ordinal Glauber phase, "1" – potential, reconstructed from the phase, containing the 1st order non-eikonal correction, "2" – potential, reconstructed from the phase, containing the 1st and 2nd order non-eikonal corrections, "3" – potential, reconstructed from the phase, containing the 1st, 2nd and 3d order non-eikonal corrections.

We have estimated root-mean-square radius of ^{11}Li by calculating total cross section for the energy 74.5 A MeV with the non-eikonal corrections to the scattering phase. The estimated rms radius from ordinal Glauber approximation is 3.12 fm, the rms radius from phases, containing non-eikonal corrections is about 3.20 fm.

The authors are grateful to Korsheninnikov A.A. for the presented experimental data.

References

[1] A.A. Korsheninnikov, E.Yu. Nikolskii, T.Kobayashi et al. Phys. Rev. 1996 **C53** R537
[2] A.A. Andrianov Yadernaja physika 1975 **22** 385, A.A. Andrianov Theor. math. phys. 1977 **33** 337
[3] M.V. Zhukov, B.V. Danilin, D.V. Fedorov, J.M. Bang, I.J. Thompson J.S. Vaagen Phys. Rep. 1993 **231** 151.
[4] C.A. Bertulani, L.F. Canto, M.S. Hussein Phys. Rep. 1993 **226** 281
[5] E. Clementel and C. Villi Nuovo Cimento 1955 **II** 176
[6] J.S. Al-Khalili, J.A. Tostevin, and J.M. Brooke Phys. Rev 1997 **C55** R1018
[7] M.V. Zhukov, A.A. Korsheninnikov and M.H. Smedberg Phys. Rev. 1994 **C50** R1
[8] S.J.Wallace, Ann. Phys. (N.Y.) **78**,190 (1973)

Two-neutron removal reactions for three-body halo nuclei

E. Garrido*, D.V. Fedorov[†] and A.S. Jensen[†]

*Instituto de Estructura de la Materia, CSIC, Serrano 123, E-28006 Madrid, Spain
[†]Institute of Physics and Astronomy, University of Aarhus, DK-8000 Aarhus C, Denmark

Abstract. We formulate a method to compute breakup processes of three-body halo systems reacting with a target nucleus. The final state two-body interaction is also used for the three-body bound state wave function. The interaction between one of the three particles and the target is described by the optical model. Both absorption and diffraction processes are treated. The differential and total two-neutron cross sections for ^{6}He and ^{11}Li are calculated and compared to available experimental data.

INTRODUCTION

Halo nuclei are studied in detail by use of the radioactive beam facilities [1–4]. Lots of data demand analyses and interpretation and few-body descriptions seem to provide an overall basic understanding. Breakup reactions of three-body halo systems involve four particles, i.e. the three halo projectile particles and the target. A four body model is therefore in principle needed. However, the large beam energy allows efficient approximations directly or indirectly based on the high-energy eikonal approximation and the adiabatic assumption of frozen intrinsic degrees of freedom during the collision.

The crucial ingredients in these processes are the initial three-body wave functions, the final state interaction and the two-body reaction between one of the three particles and the target [5–8]. The remaining two particles are essentially spectators and reactions simultaneously involving more than one particle are much less important. The large spatial extension guaranties that this model is a very good approximation. We shall here fully exploit these simplifying, valid assumptions to keep the approach practical enough to compute many observables.

In this report we shall first briefly describe a general, practical and accurate method to compute cross sections for breakup reactions of spatially extended three-body systems and afterwards we shall show results for two-neutron (2n) removal from ^{6}He (n+n+α) and ^{11}Li (n+n+^{9}Li) by fragmentation reactions on a light target. We shall describe the reaction between each particle and the target by the well-studied phenomenological optical model, where elastic scattering is fully treated and all other processes are included as absorption removing probability from the elastic channel. Thus, the corresponding optical potentials are needed. This formulation instantaneously introduces a dependence on the beam energy and the target structure via the optical potentials describing the particle–target interactions. The introduction of the beam direction produces a difference between longitudinal and transverse momentum distributions.

The second indispensable ingredient is the final state interaction which, at least for Borromean systems where low-lying resonances almost inevitably are present, is especially important for the differential cross sections. These two-body interactions are also used to compute the bound state structure of the reacting three-body system. The final state is made of two separable two-body subsystems, the undisturbed two-particle subsystem and the scattering state made by the target and the removed particle, see Fig. 1.

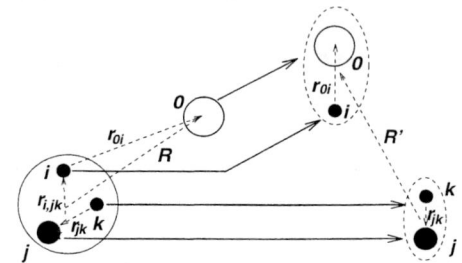

FIGURE 1. Sketch of the reaction and the coordinates used. The target is labelled by 0 and $\{i, j, k\}$ label the particles within the three-body projectile.

THEORY

When a spatially extended three body halo hits a relatively small target at high energy the probability that more than one of the constituents interacts strongly with the target is small. The cross section can then be written as a sum of three terms where each term describes the contribution to the reaction caused by the interaction between the target and the corresponding particle

$$d\sigma = \sum_{i=1}^{3} d\sigma^{(i)} = \sum_{i=1}^{3} \frac{1}{v}\frac{2\pi}{\hbar}|T^{(i)}|^2 d\nu_f^{(i)}, \qquad (1)$$

where v is the relative projectile-target velocity, $T^{(i)}$ is the transition matrix element and $d\nu_f^{(i)}$ is the density of final states. Each of the three terms includes elastic and inelastic scattering of the corresponding particle on the target. Neglecting the Coulomb interaction and assuming that the target has zero spin we obtain after summing over final and averaging over initial states the diffraction cross section for a non-polarized projectile as

$$\frac{d^9\sigma_{el}^{(i)}(\mathbf{P}', \mathbf{p}'_{jk}, \mathbf{p}'_{0i})}{d\mathbf{P}' d\mathbf{p}'_{jk} d\mathbf{p}'_{0i}} = \frac{d^3\sigma_{el}^{(0i)}(\mathbf{p}_{0i} \to \mathbf{p}'_{0i})}{d\mathbf{p}'_{0i}} \frac{1}{2J+1} \sum_{Ms_{jk}\Sigma_{jk}\Sigma_i} |M_{s_{jk}\Sigma_{jk}\Sigma_i}^{JM}|^2 \qquad (2)$$

$$\frac{d^3\sigma_{el}^{(0i)}(\mathbf{p}_{0i} \to \mathbf{p}'_{0i})}{d\mathbf{p}'_{0i}} = \frac{1}{(2\pi\hbar)^3}\frac{1}{v}\frac{2\pi}{\hbar}\frac{1}{2s_i+1}\sum_{\Sigma_i\Sigma'_i}|T_{\Sigma_i\Sigma'_i}^{(0i)}|^2 \delta(E'_{0i} - E_{0i}), \qquad (3)$$

where the coordinates are defined in Fig. 1 and the conjugate momenta are denoted by the corresponding **p**. Primes are used for the final states. The transition matrix elements M and T with spin and spin-projection indices refer to the two undisturbed particles and the interacting particle-target interaction [7].

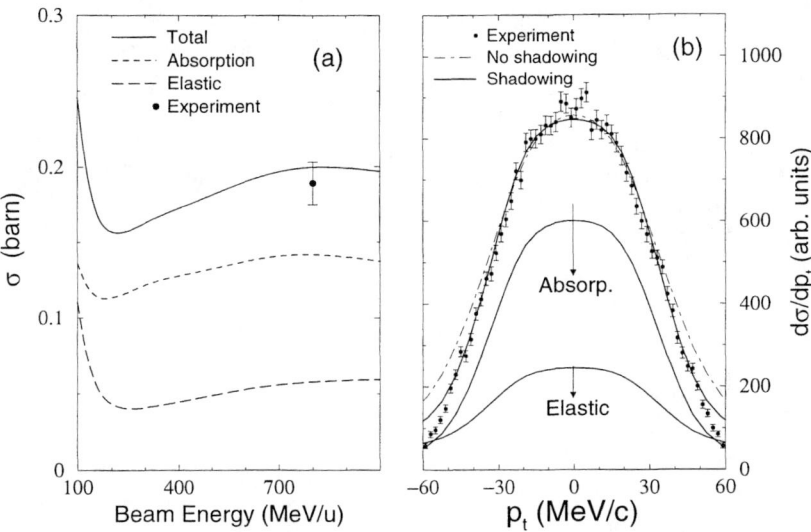

FIGURE 2. Calculated two-neutron removal cross sections for ^{6}He fragmentation on ^{12}C. Spheres of radius 3 fm around the undisturbed particles are removed in the initial wave function.

The absorption cross section where only two projectile particles survive in the final state is simply obtained by replacing the elastic σ_{el} by σ_{abs}, i.e.

$$\frac{d^6\sigma_{abs}^{(i)}(\mathbf{P}',\mathbf{p}'_{jk})}{d\mathbf{P}'d\mathbf{p}'_{jk}} = \sigma_{abs}^{(0i)}(p_{0i})\frac{1}{2J+1}\sum_{Ms_{jk}\Sigma_{jk}\Sigma_i}|M_{s_{jk}\Sigma_{jk}\Sigma_i}^{JM}|^2 \quad (4)$$

The total cross section arising from the i'th particle is the sum of eqs.(2) and (4).

NUMERICAL RESULTS

We investigate the nucleus ^{6}He (n+n+α) and ^{11}Li (n+n+^{9}Li) with the wave functions obtained by solving the Faddeev equations in coordinate space [7]. The interaction parameters reproduce the three-body ground state structure and the measured neutron-target cross sections. The detected two-neutron removal contribution from the interaction between core (α-particle or ^{9}Li) and target is expected to be relatively small. The final state two-body continuum wave functions are calculated with the appropriate boundary conditions and normalization.

Then eqs.(2) and (4) are integrated over the unobserved quantities and different momentum distributions or differential cross sections are obtained. However, the so-called core shadowing problem remains [8,9]. Basically this means that the finite sizes of the target and the core prohibit some of the neutron-target reactions namely those where the core also interacts with the target.

Final state interactions and shadowing are crucial for the neutron momentum distributions and the absolute cross sections, respectively. We compute absolute values of differential and total two-neutron removal cross sections including dependence on beam energy and target. In Figs. 2 and 3 we show the absolute values of the total two-neutron removal cross sections divided into contributions from scattering ($\sim 40\%$) and absorption ($\sim 60\%$). In Fig. 2 we also show the transverse momentum distribution. The longitudinal distribution is similar but distinguishable. The realistic beam-energy dependence of the cross section arises from the phenomenological optical potential. For a ^{11}Li projectile the dominant components in the neutron-core relative wave function are s-waves and the neutron momentum distributions (not shown) are narrower than for ^{6}He where p-waves dominate. Remarkable agreement with the experimental data is found in all cases for absolute as well as for differential cross sections.

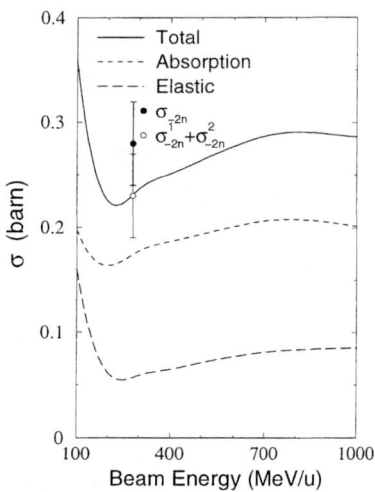

FIGURE 3. Calculated two-neutron removal cross sections for ^{11}Li fragmentation on ^{12}C. Spheres of radius 4 fm around the core and the surving neutron are removed in the initial wave function. The experimental points are from [3].

REFERENCES

1. Hansen P.G., Jensen A.S., and Jonson B., *Ann. Rev. Nucl. Part. Sci.* **45**, 591 (1995).
2. Humbert F. et al., *Phys. Lett.* **B347**, 198 (1995).
3. Zinser M. et al., *Nucl Phys.* **A619**, 151 (1997).
4. Aleksandrov D. et al., *Nucl. Phys.* **A633**, 234 (1998).
5. Korsheninnikov A.A., and Kobayashi T., *Nucl. Phys.* **A567**, 97 (1994).
6. Barranco F., Vigezzi E., and Broglia R.A., *Phys. Lett.*, **B319**, 387 (1993).
7. Garrido E., Fedorov D.V., and Jensen A.S., *Phys. Rev.*, **C53**, 3159 (1996); *ibid.* **C55**, 1327 (1997); *Nucl. Phys.*, **A617**, 153 (1997); *Europhys. Lett.*, **36**, 497 (1996).
8. Bertsch G.F., Hencken K., and Esbensen H., *Phys. Rev.* **C57**, 1366 (1998).
9. Hansen P.G., *Phys. Rev. Lett.*, **77**, 1016 (1996).

Phase equivalent potentials for three-body halos

E. Garrido[a], D. V. Fedorov[b] and A. S. Jensen[b]

[a] *Instituto de Estructura de la Materia, CSIC, Serrano 123, E-28006 Madrid, Spain*
[b] *Institute of Physics and Astronomy, University of Aarhus, DK-8000 Aarhus C, Denmark*

Abstract. We consider a potential with Pauli forbidden states and its phase equivalent partner in a three-body problem using the adiabatic hyperspherical method of [1]. We prove that the adiabatic spectra of these two potentials are asymptotically equivalent provided the adiabatic term corresponding to the forbidden state is removed.

The potentials with Pauli forbidden states are often used to describe interactions between composite clusters containing identical particles, in particular in few-body models of neutron halos [1,2]. Although it is fairly simple to account for a forbidden state in a two-body system, for example by using the phase equivalent potential [2], the three-body system imposes a much harder and more ambiguous problem where several different prescriptions have so far been used. In [1] a new method was suggested within the adiabatic hyperspherical approach basically amounting to a complete removal of the adiabatic term which corresponds to the forbidden state.

Here we check the consistency of the new method considering two phase equivalent potentials, one of them with a forbidden state, within the adiabatic hyperspherical approach of [1]. The forbidden state potential produces the adiabatic spectrum with one additional term resulting from the forbidden bound state. We show that removing this term makes the adiabatic spectra for the two potentials asymptotically identical both at small and large distances. This consequently results in similar properties of the three body systems and therefore lends additional support to the new method.

In the **adiabatic hyperspherical method** the coordinates of the three-body system are divided into "fast" variables (five angles Ω) and a "slow" variable (hyperradius ρ). For a system of particles with masses m_i and coordinates $\mathbf{r}_i$ the hyperradius is defined in terms of an arbitrary mass scale m as $\rho^2 = (\sum_{i=1}^{3} m_i r_i^2)/m$. The eigenfunctions of the fast variables $\Phi_n(\rho, \Omega)$ are used as a basis for expansion of the total wave function $\Psi(\rho, \Omega) = \sum_n f_n(\rho) \Phi_n(\rho, \Omega)$. The corresponding eigenvalues $\lambda_n(\rho)$ of the fast subsystem (divided by ρ^2) serve as adiabatic potentials in the radial equations

$$\left(-\frac{\hbar^2}{2m}\frac{\partial^2}{\partial\rho^2} + \frac{\hbar^2}{2m\rho^2}\left(\lambda_n(\rho) + \frac{15}{4}\right) - Q_{nn}(\rho) - E\right) f_n(\rho) \quad (1)$$
$$= \sum_{n'\neq n}\left(-2P_{nn'}(\rho)\frac{\partial}{\partial\rho} - Q_{nn'}(\rho)\right) f_{n'}(\rho),$$

where P and Q are the nonadiabatic terms [3]. The great advantage of this basis is that it correctly describes all types of the long range asymptotic behavior including the pathological Efimov effect. Clearly the behavior of the angular eigenvalues $\lambda_n(\rho)$ is crucial for the properties of the system.

Suppose the potential $V(r)$ provides an s-wave bound state of a system of two particles with the reduced mass μ, the binding energy $E = \hbar^2\kappa^2/2\mu$ and the bound state wave function $\psi_0(\mathbf{r}) \equiv u_0(r)(\sqrt{4\pi}r)^{-1}$ with the asymptotics $u_0(r \to 0) \propto r$, $u_0(r \to \infty) = \sqrt{2\kappa}\, C_0 e^{-\kappa r}$. The corresponding **phase equivalent potential** without this bound state is then [2]

$$V_{pep}(r) = V(r) + \Delta V(r), \quad \Delta V(r) = -\frac{\hbar^2}{\mu}\frac{d^2}{dr^2}\ln\left(\int_0^r u_0(r')^2 dr'\right). \quad (2)$$

The additional term ΔV diverges as r^{-2} at small distances and vanishes exponentially at large distances

$$\Delta V(r \to 0) \to \frac{\hbar^2}{\mu}\frac{3}{r^2}, \quad \Delta V(r \to \infty) = \frac{\hbar^2}{\mu} 4\kappa^2 C_0^2 e^{-2\kappa r}. \quad (3)$$

Considering the **asymptotic behavior of the eigenvalues** we shall assume for simplicity that the potentials only act on s-waves and that the forbidden state is only in one of the three two-body subsystems. The angular eigenproblem where only s-waves are included is then very simple [3]

$$\left(-\frac{\partial^2}{\partial\alpha^2} + \frac{2m\rho^2}{\hbar^2}\sum_{i>j} V_{ij}\right)\Phi_n = \nu_n^2 \Phi_n, \quad (4)$$

where V_{ij} is the potential between particles i and j, $\nu_n^2 = \lambda_n + 4$ and α is the hyperangle ($\rho\sin\alpha = r\sqrt{\mu/m}$). At **short distances** ($\rho \to 0$) the spectrum is

$$\lambda_n = (2n)^2 - 4, \quad n = 1, 2, \ldots. \quad (5)$$

The eigenproblem with phase equivalent potential has an additional term

$$\frac{2m\rho^2}{\hbar^2}\Delta V \sim \frac{2m\rho^2}{\hbar^2}\frac{6\hbar^2}{\mu r^2} = \frac{6}{\sin^2\alpha} \equiv \frac{2(2+1)}{\sin^2\alpha}. \quad (6)$$

This term leads to a modified spectrum

$$\lambda_n = (2(n+1))^2 - 4, \quad n = 1, 2, \ldots \quad (7)$$

which is, in fact, identical (in the limit $\rho \to 0$) to the spectrum (5) except for the lowest eigenvalue which is missing. Removing the lowest eigenvalue from the basis (5) makes the spectra for the forbidden state potential (5) and its phase equivalent partner (7) equal within two terms in the small ρ expansion

$$\nu_n^2 = (2(n+1))^2 + \frac{2m\rho^2}{\hbar^2}\sum_{i>j} V_{ij}(0) \, , \, \tilde{\nu}_n^2 = (2(n+2))^2 + \frac{2m\rho^2}{\hbar^2}\sum_{i>j} V_{ij}(0) \, . \quad (8)$$

At **large distances** if all the three scattering lengths a_i are finite two types of solutions exist in the limit $\rho \gg \max(|a_{ij}|)$ [3]. The first type corresponds to eigenvalues λ_n which approach constants at $\rho \to \infty$ (compare with (5))

$$\lambda_n = (2n)^2 - 4 \, , \, n = 1, 2, \ldots \, . \quad (9)$$

The second type of eigenvalues solution is only present when there is a bound state in a two-body subsystem. This two-body bound state with the binding energy $E_{binding}$ generates an additional asymptotically diverging eigenvalue $\lambda_0 = -2m\rho^2 E_{binding}/\hbar^2$. Therefore the phase equivalent potential without bound states has at large distances the eigenspectrum (9) while the forbidden state potential has the same spectrum plus the additional eigenvalue λ_0 resulting form the (forbidden) bound state. Again removing this eigenvalue makes the spectra asymptotically identical within two terms in the $1/\rho$ expansion

$$\lambda_n = \nu_n^2 - 4 = (2n)^2 - 4 - \frac{16(n+1)^2}{\pi}\sum_{i>j}\frac{a_{ij}}{\rho}\sqrt{\frac{1}{m}\frac{m_j m_k}{m_j + m_k}} \, , \, n = 1, 2, \ldots \, . \quad (10)$$

As a **numerical illustration** we consider the neutron halo nucleus ^{11}Li within a three-body model (^{9}Li+n+n). We calculate (using Faddeev equations with only s-waves in all Faddeev components) the angular eigenvalues for this system for two phase equivalent potentials in the n-n subsystem

$$V_{deep}(r) = \frac{\hbar^2}{2\mu_{nn}}2\beta\alpha(1 - \coth(\beta r)) \quad (11)$$

$$V_{pep}(r) = V_{deep}(r) + \frac{\hbar^2}{\mu_{nn}}8\alpha\beta^2(\alpha^2 - \beta^2)\sinh(\beta r)$$
$$\times \frac{\alpha(e^{2\alpha r} + 1)\sinh(\beta r) - \beta(e^{2\alpha r} - 1)\cosh(\beta r)}{(\alpha^2(\cosh(2\beta r) - 1) - \beta^2(e^{2\alpha r} - 1) + \alpha\beta\sinh(2\beta r))^2} \quad (12)$$

where μ_{nn} is the reduced n-n mass and where all distances are in fm and all energies in MeV. The parameters $\alpha = 4.151\text{fm}^{-1}$ and $\beta = 1.063\text{fm}^{-1}$ are chosen to reproduce the n-n scattering length of 18.45 fm and effective range 2.83 fm. The neutron-core potential $V_{nc} = -7.8\exp(-(r/2.55)^2)$ is taken from [4]. The resulting spectrum of the angular eigenvalues is shown in Fig. 1. The lowest eigenvalue of the deep potential with the forbidden state is removed. Since the potential V_{deep} behaves as r^{-1} at the origin the eigenvalues start linearly from zero according to (8).

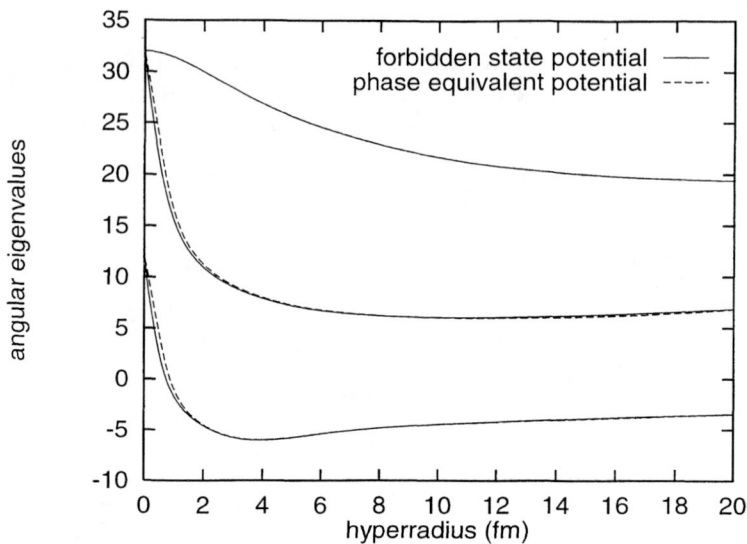

FIGURE 1. The lowest angular eigenvalues for the ^{9}Li+n+n system. The solid lines correspond to the deep n-n potential V_{deep} with a forbidden state. The eigenvalue corresponding to this forbidden state has been removed from the spectrum. The dashed lines correspond to the phase equivalent potential V_{pep}.

This very good agreement between the two spectra will be somewhat violated when more than one two-body subsystem has forbidden states because the simple formula (8) will be modified. The large distance expansion (10) however will still be valid.

In **conclusion**, we consider a three body system where one of the binary potentials has a Pauli forbidden state. Compared to its phase equivalent partner without a forbidden state this potential produces an additional adiabatic term in the hyperspherical adiabatic expansion method. Removing this term from the basis makes the resulting adiabatic spectra asymptotically identical and in general very close. We conclude therefore that this method to account for the forbidden states in a three-body system has firm mathematical foundations and can be considered as a valuable alternative to the other existing methods.

REFERENCES

1. E. Garrido, D. V. Fedorov, A. S. Jensen, Nucl. Phys. A **617** (1997) 153.
2. D. Ridikas, J. S. Vaagen, J. M. Bang, Nucl. Phys. A **609** (1996) 21.
3. D. V. Fedorov, A. S. Jensen and K. Riisager, Phys.Rev.C **50** (1994) 2372
4. L. Johannsen, A. S. Jensen and P. G. Hansen, Phys. Lett. B **244** (1990) 357.

Angular correlation in breakup of three-body halo nuclei

E. Garrido*, D.V. Fedorov† and A.S. Jensen†

*Instituto de Estructura de la Materia, CSIC, Serrano 123, E-28006 Madrid, Spain
†Institute of Physics and Astronomy, University of Aarhus, DK-8000 Aarhus C, Denmark

Abstract. We use a previously formulated three-body model to compute angular correlations in high-energy fragmentation reactions of two-neutron halos on light targets. The neutron absorption process contributes much more than neutron scattering. We compare the model predictions with the available experimental data for ^{6}He and the consequences for ^{11}Li are derived. Various reaction mechanisms are investigated but the obvious choice leads to disagreement with the data. Other physically less intuitive possibilities are also investigated.

MODEL DESCRIPTION

Nuclear halos were discovered about ten years ago [1]. They are spatially extended states with small one or two-nucleon separation energies. The weakly bound nuclear three-body halos have been remarkably well described by three-body models. This includes bound and continuum structure as well as details of neutron removal reactions [2–4]. However the recently measured angular correlations in fragmentation reactions of Borromean two-neutron halos are apparently sensitive to both the reaction mechanism and the halo structure [5]. A satisfactory description is so far not available. Before invoking new reaction mechanisms it is essential to test the existing successful models [2] which can reproduce other measured distributions. In this contribution we shall investigate these angular correlations using the three-body model and the sudden approximation, study the dependence on various physical parameters and reaction mechanisms and in this way explain or predict the data, see how far the model can go and possibly suggest specific reaction mechanisms which might reproduce the available data.

The constituents of the spatially extended three-body halo are assumed to interact individually with a small target. The differential cross section $d\sigma$ is then a sum of three terms $d\sigma^{(i)}$ each describing the independent contribution to the process from the interaction between the target and the halo particle i. We also neglect the binding energy of the initial three-body bound state compared to the high energy of the beam. The process is described as removal of one particle (participant) while the other two particles (spectators) both survive essentially undisturbed. The participant is either absorbed or scattered by the target with respectively two and three halo particles in the final state. We neglect the Coulomb interaction and assume that the target has zero spin. With these assumptions we arrive at manageable expressions for the differential elastic and absorption cross sections [4].

In the experiment the core and a neutron are detected in coincidence with velocities approximately equal to the beam velocity assuming the participant-spectators approximation, namely that only one halo particle reacts at a time with the target without destroying or affecting the other two. The finite extension of projectile and target therefore requires that only configurations where the participant is sufficiently far away from the spectators can contribute. We account for this by omitting those geometric configurations in the initial wave function where the participant n is closer to the two spectators n and c than a cut off distances r_{nn} and r_{nc} which are treated as parameters. This shadowing effect substantially reduces the absolute values of the cross sections [2,6]. The two other halo particles could still react simultaneously with the target, but here we neglect these smaller contributions.

The projectile looses one of the neutrons and a specific angular correlation, namely the angular distribution of the relative momentum of the remaining neutron-core system is measured in a coordinate system with the z-axis along its total center of mass momentum [5].

NUMERICAL RESULTS

We consider the nuclei ^{6}He (n+n+α) and ^{11}Li (n+n+^{9}Li) with the wave function obtained by solving the Faddeev equations in coordinate space [2] by using the potentials from [3]. These two-body interactions between the halo particles in initial and final states reproduce all the known low-energy properties. The binding energies and sizes of the initial states and the experimental neutron and core momentum distributions are essentially reproduced with these parameters. The interaction between each halo particle and the target is described by the phenomenological optical model where both elastic scattering and absorption is treated [7]. Furthermore, shadowing parameters of about 4 fm maintain or even improve the quality of all the previous results and in addition the available absolute two-neutron removal cross sections are reproduced.

We compare in Fig. 1 the rather different measured and computed angular correlations in the projectile center of mass. Both the asymmetry and the variation with angle are much larger for the computed curves. We then went through a number of calculations to test dependence and sensitivity. The final state interaction is essential for the neutron momentum distribution, but here we find a significant contribution although still substantially smaller than the difference to the data.

Shadowing is simply removal of unwanted geometric configurations in the initial wave function. We obtain an essentially unchanged distribution for ^{6}He, since the relative neutron-core s-wave only contributes marginally and the p-waves with different angular momentum projections are all reduced by the same amount. For ^{11}Li these shadowing effects are more visible, since the predominant removal of the dominating s-states now produce a more symmetric p-like structure.

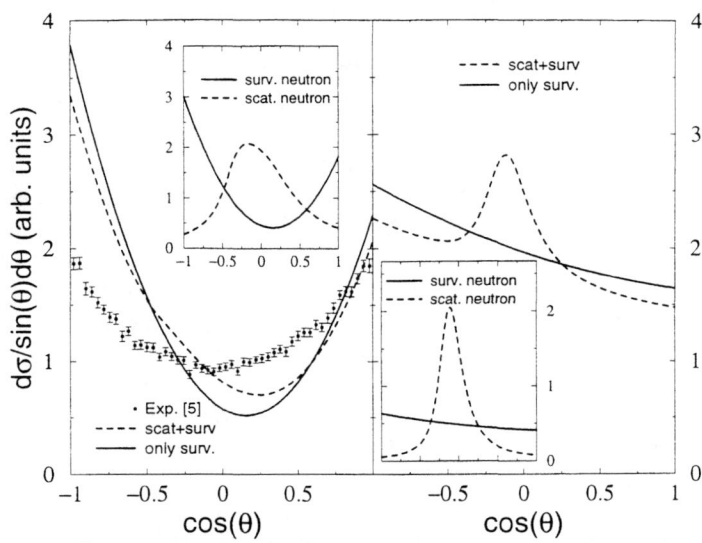

FIGURE 1. The neutron-core angular correlation after fragmentation of ^{6}He (left) and ^{11}Li (right) on ^{12}C at 240 MeV/u with $r_{nn} = r_{nc} = 3$ fm and 4 fm for the two nuclei. The external parts give the total contributions when the elastic part is obtained by neglecting the contribution from the scattered neutron (solid) and when 50% of the elastic part is obtained from the scattered neutron and 50% from the undisturbed neutron (dashed). The insets show the scattering contributions from the undisturbed neutron (solid) and the scattered neutron (dashed).

The s-waves are angle-independent and insignificant for ^{6}He and dominating for ^{11}Li. The p-waves vary symmetrically with angle and the asymmetric sp-interference terms change sign for $\cos\theta = 0$ and would therefore not contribute to the total cross section obtained by integrating over the angle. Thus the angular correlation is for ^{6}He essentially determined by the contributions from the p-wave and is therefore symmetric. For ^{11}Li this correlation is essentially due to the sp-interference contribution and is therefore very asymmetric.

The contributions arising from three different angular momentum projections along the neutron-core center of mass momentum differ substantially. The asymmetry is due to the sp-interference appearing only for $m_l = 0$, which contains part of the p-wave and all the s-wave contributions. The terms with $m_l = \pm 1$ arise entirely from p-waves and is therefore symmetric due to the lack of sp-interference. Furthermore, we obtain relatively small angular variation for the terms with $m_l = \pm 1$ and a strong variation of opposite curvature from the terms with $m_l = 0$.

We varied the energy of the neutron-core relative state from zero to above the lowest resonances and computed the angular correlation when only that energy is populated in the reaction. For ^{6}He all these distributions resemble the energy averaged p-wave distribution in disagreement with the data. For ^{11}Li we find a rather flat and asymmetric distribution at low energy, where the s-waves dominate completely. As the excitation energy increases the p-waves contribute more and more. An increased asymmetry due to the interference term and at higher energies the p-wave contribution is clearly pulling the distribution towards symmetry.

When the participant neutron is scattered by the target the detected neutron can be either spectator or participant. We therefore have to add the two distributions with equal probability provided they both arrive within the forward angle where the detection takes place. The sum of these contributions amounts to about 30% of the total contribution. We find almost identical distributions for the spectator whether the participant neutron is scattered or absorbed. The remaining contribution of $\approx 15\%$ arising from the participant neutron has in our approximation a completely different shape, see the inset of Fig. 1. The distribution from scattering is almost symmetric and peaked at $\pi/2$. This corresponds to a preferred direction perpendicular to the beam direction consistent with the forward scattering of a neutron on a target seen from the neutron rest system. For ^{6}He the scattering only changed the total distribution very little. For ^{11}Li a peak appeared due to the rapid angular variation of the scattering contribution compared to the almost constant background from the dominating s-waves. These estimates are crude and all contributions are included independent of possible additional experimental selection.

In conclusion, the scattering and absorption contributions are qualitatively different. The scattering process is only expected to contribute by less than 15% due to experimental selection within a narrow cone around the beam direction. The correlations for ^{6}He and ^{11}Li are qualitatively different and the computed distribution vary too strongly compared to the data for ^{6}He. Provided the experimental correlation is correct this indicates that a subtle reaction mechanism is at work. From the present investigation this can be either a constant background missing for example due to suppression of p-waves or a preferred selection of s waves or a relative suppression of zero angular momentum projections.

REFERENCES

1. Hansen P.G., Jensen A.S., and Jonson B., *Ann. Rev. Nucl. Part. Sci.* **45**, 591 (1995).
2. Garrido E., Fedorov D.V., and Jensen A.S., *Phys. Rev.*, **C53**, 3159 (1996); *ibid.* **C55**, 1327 (1997); *Nucl. Phys.*, **A617**, 153 (1997); *Europhys. Lett.*, **36**, 497 (1996).
3. Cobis A., Fedorov D.V., and Jensen A.S., *Phys. Rev. Lett.* **79**, 2411 (1997); *Phys. Lett.* **B424**, 1 (1998).
4. Garrido E., Fedorov D.V., and Jensen A.S., *Contribution to these proceedings.*
5. Chulkov L.V. et al., *Phys. Rev. Lett.* **79**, 201 (1997).
6. Bertsch G.F., Hencken K., and Esbensen H., *Phys. Rev.* **C57**, 1366 (1998).
7. Udías J.M., Sarriguren P., Moya de Guerra E., Garrido E. and Caballero J.A., *Phys. Rev.* **C51**, 3246 (1995).

Study of the unbound nucleus ^{11}N by the ^{12}C(^{14}N,^{15}C)^{11}N transfer reaction

A.Lépine-Szily[1], J.M.Oliveira Jr[1,2], A.N.Ostrowski[3], H. G. Bohlen[4], R.Lichtenthaler[1], A.Blazevic[4], C.Borcea[5], V.Guimarães[6], R.Kalpakchieva[7], V.Lapoux[8], M.MacCormick[9], F.Oliveira[9], W.von Oertzen[4], N.A.Orr[10], P.Roussel-Chomaz[9], Th.Stolla[4], J.S.Winfield[10]

1. IFUSP-Universidade de São Paulo, C.P.66318, 05389-970 São Paulo, Brazil 2. Departamento de Ciências e Matemática da Universidade de Sorocaba, Sorocaba, Brazil 3. Department of Physics & Astronomy, University of Edinburgh, Edinburgh, EH9 3JZ UK 4. Hahn-Meitner Institut, Glienicker Strasse 100, D-14109 Berlin, Germany 5. Institute of Atomic Physics, Bucarest, Romania 6. UNIP-Objetivo, R. Dr. Bacelar 1212, 04026-002, São Paulo, Brazil 7. Flerov Laboratory of Nuclear Reactions, JINR, Dubna, 141980 Dubna, Russia 8. CEA/DSM /DAPNIA/SPhN, CEN Saclay, 91191 Gif-sur-Yvette, France 9. GANIL, Bld Henri Becquerel, BP 5027, 14021 Caen Cedex, France 10. LPC-ISMRa, Bld du Maréchal Juin, 14050 Caen Cedex, France

Abstract. A spectroscopic study of the proton-rich, particle unstable nucleus ^{11}N has been performed using the multi-nucleon transfer reaction ^{12}C(^{14}N,^{15}C)^{11}N at 30 AMeV incident energy at GANIL. Levels of ^{11}N are observed as well defined resonances in the spectrum of the ^{15}C-ejectiles. They are localised at 2.18(5), 3.63(5), 4.39(5), 5.12(8) and 5.87(15) MeV above the ^{10}C+p threshold. The comparison of the measured widths with R-matrix calculations allows the estimation of spins and parities for these resonances.

Light proton-rich nuclei were much less studied than their neutron-rich neighbours, in particular the proton-unbound nucleus ^{11}N was almost unknown until recently. Theoretical calculations of Fortune et al. [1], Barker [2] and of Grévy et al. [3] using a simple potential model predict level energies and widths for the first three levels of the ^{11}N, and demand a comparison with reliable and precise experimental data. Until recently only one resonance, with spin $1/2^-$ ($\Gamma = 0.74(10)$ MeV) has been observed in the proton-unbound nucleus ^{11}N, lying at 2.24 Mev above the ^{10}C+p threshold [4] and the unobserved $1/2^+$ ground state in ^{11}N was supposed to lie at 1.9 MeV decay energy. Recent measurements [5-7] on the ^{11}N nucleus claim to observe the the s1/2 ground-state below the p1/2 level and some new excited states are also observed. Recently new experimental efforts shed more light on the spectroscopy of ^{11}N. The multi-nucleon transfer reaction [8] ^{12}C(^{14}N,^{15}C)^{11}N was measured at GANIL to undertake the spectroscopic study of ^{11}N, where the ^{11}N is the recoiling nucleus. The Q-value for the three-particle threshold of ^{15}C$_{g.s.}$+^{10}C$_{g.s.}$+p is Q = −29.997 MeV.

The ^{14}N beam had an energy of 30 AMeV, and the thickness of the ^{12}C target was 0.5 mg/cm^2. The ejectiles were analysed by the high-precision magnetic spectrometer SPEG. The laboratory angles subtended by SPEG were $\theta = 2.5 \pm 1.2°$ and $\phi = 0 \pm 2.0°$ in the horizontal and vertical planes, respectively. The standard SPEG detection system was used; it includes two drift-chambers, an ionisation chamber and a plastic scintillator for the measurements, respectively, of the focal plane position, the energy-loss(ΔE) and the residual energy. The time-of-flight (TOF) was measured using the fast scintillator signal with respect to the cyclotron radio-frequency. The two-dimensional particle identification spectrum (Z vs. A/q), where Z and A/q are calculated from ΔE and TOF, allows a clear separation of all mass groups due to its very good resolution. The reaction products were momentum analysed by the horizontal and vertical position measurement carried out by the two drift chambers. The incident position (x,y) and the incident angles (θ, ϕ) of each particle in the focal plane were reconstructed by two position measurements at a distance of 1.2 m. The scattering angle Θ was calculated from the measured (θ, ϕ) angles.

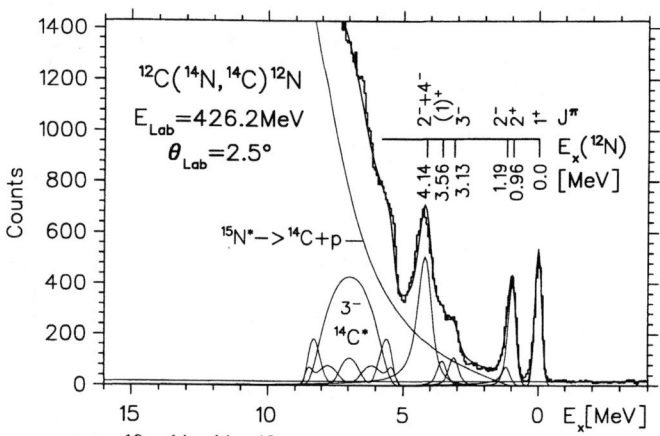

Figure 1: Spectrum of the $^{12}C(^{14}N,^{14}C)^{12}N$ reaction, used for calibration purposes. See text for details.

The projection of the kinematically corrected spectra on the momentum axis yielded the one-dimensional spectra used in the following discussions. The results of the $^{12}C(^{14}N,^{14}C)^{12}N$ reaction has been used for momentum and energy calibration purposes, the spectrum is shown in Fig. 1. The lower lying well defined peaks are the bound ground state and the unbound unresolved doublet states with E^*=0.96 MeV / 1.19 MeV of ^{12}N. The width of the ground state peak represents our energy resolution of 270 keV. Excited unbound resonances are fitted by Breit-Wigner line shapes. The background is mainly originating from the decay of an intermediately formed excited $^{15}N^*$ nucleus which decays into ^{14}C+p. The relative population of the ^{12}N levels illustrates directly the difficulties to observe a $2s_{1/2}$ resonance in the presence of a background. The first negative parity level (2^-, 1.19 MeV) formed by the coupling of the $p_{3/2}$ neutron hole with a $2s_{1/2}$ proton is a tiny peak in the spectrum and its coupling partner, the 1^--state at 1.80 MeV, cannot be observed in the presence of the background. On the other hand, the intensely populated positive parity levels in ^{12}N (1^+ ground state and 2^+ first excited state at 0.96 MeV) can be described by the coupling of a $p_{1/2}$ proton to the $p_{3/2}$ neutron hole.

Fig. 2 shows the results for the $^{12}C(^{14}N,^{15}C)^{11}N$ reaction. The ^{15}C ejectile has two particle-stable states, the $1/2^+$ ground state, and the $5/2^+$ first excited state at 0.74 MeV excitation energy. The transfer to the $5/2^+$ excited state of the ejectile ^{15}C, due to the transfer dynamics and angular momentum transfer- and spin-weighting factors is enhanced by a factor of about 7. The origin of the energy axis is set to the ^{10}C+proton decay threshold, *with the ^{15}C-ejectile in its $5/2^+$ excited state at 0.74 MeV*. The most prominent peaks in the spectrum of Fig. 2 are resonances of ^{11}N situated at ^{11}N decay energies of $E_{decay} = 2.18(5), 3.63(5), 4.39(5), 5.12(8)$ and $5.87(15)$ MeV with experimental widths of $0.44(8), 0.40(8), \leq 0.22(10), \leq 0.22(10)$ and $0.7(2)$ MeV respectively. The statistical significances of these peaks are respectively, 20, 22, 8, 3, 5σ, giving a strong confidence in their existence.

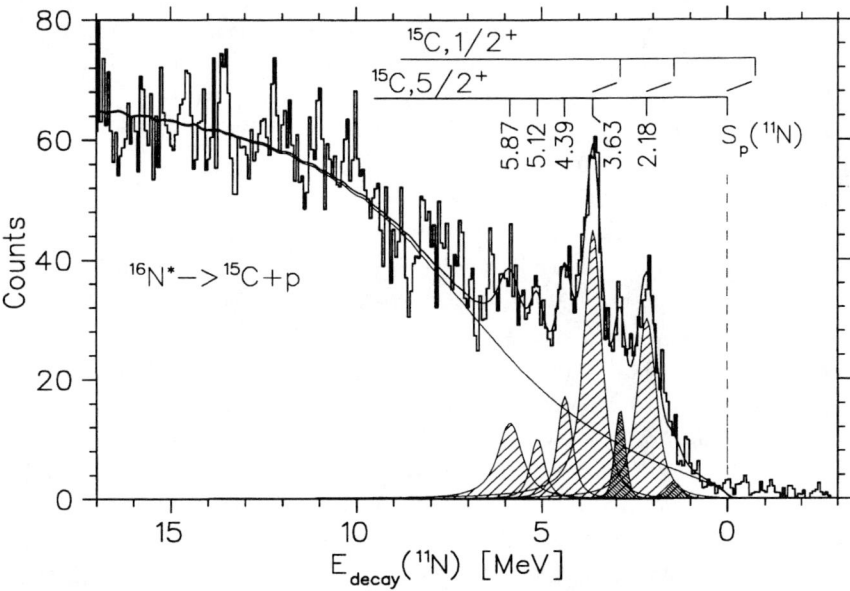

Figure 2: Spectrum of the ^{12}C(^{14}N,^{15}C)^{11}N reaction obtained at 2.5°. The origin of the energy axis is given by the proton decay threshold (^{11}N→^{10}C+p) in combination with the ^{15}C$_{5/2+}$ excited state. The resonances populated in combination with the ^{15}C ground state and with the excited state at 0.74 MeV are hatched respectively by darker and lighter filling. The corresponding scales are shown in the upper part of the figure, the energies indicated are decay energies of the ^{11}N nucleus in MeV.

The 3-body sequential decay background was calculated using a Breit-Wigner shaped resonance in ^{16}N at 17 MeV excitation energy (with a width of 6 MeV), that could be part of E1 or E2 giant resonances. The strength in ^{16}N is populated in a direct 2n pick-up with a relatively large cross-section. The differential cross-sections of the ^{12}C(^{14}N,^{15}C)^{11}N reaction populating the levels at 2.18 MeV, 3.63 MeV and 4.39 MeV are 0.6(3), 0.9(3) and 0.26(10) $\mu b/sr$, respectively.

A small peak between the 2.18 MeV and 3.63 MeV resonances was attributed to the resonance at 3.63 MeV *with ^{15}C in its ground-state*, since the small peak has exactly the distance of 0.74 MeV from the 3.63 MeV resonance and the cross-section ratio follows the spin-weighting factor ratio (7). This peak is more narrow, because it does not involve the Doppler-broadening effect, since ^{15}C is in the ground state. The presence of this peak imposes the observation of a small peak at an energy in the plotted scale at 2.18−0.74 MeV = 1.44 MeV, with an intensity approximately seven times smaller than the 2.18 MeV resonance, which corresponds to the ^{15}C in its ground state. With this interpretation all the observed counting rate around 1.44 MeV is exhausted, leaving no room for a possible $2s_{1/2}$ ground state resonance of ^{11}N. The peaks at 2.18 and 3.63 are respectively the $p_{1/2}$ and $d_{5/2}$ resonances of ^{11}N. R-matrix calculations were performed for the observed ^{11}N resonances, obtaining Γ=0.46 MeV for an l=1 level at 2.18 MeV (1/2$^-$) and 0.60 MeV for an l=2 level at 3.63 MeV (5/2$^+$), confirming our findings.

As discussed above, we have no clear evidence of the population of the $2s_{1/2}$ ground state of ^{11}N, however this does not exclude the existence of the 1/2$^+$ resonance in ^{11}N. Shell-model calculations [6,9] predict a $p_{3/2}$ state (3/2$^-$) around 4.6 MeV decay energy and a $f_{5/2}$ state (5/2$^-$) at 5.7 MeV. The energy separation between $p_{1/2}$ and $p_{3/2}$ states in other nuclei of the A=11 chain are respectively 2.37 MeV in ^{11}Be, 2.125 MeV in ^{11}B, 2.00 MeV in ^{11}C. The separation between the 4.39 MeV state and the $p_{1/2}$ state in ^{11}N is 2.21 MeV, strongly suggesting that this is the $p_{3/2}$ state. However, as a p state its width should be larger ($\Gamma \geq$ 500 keV). Axelsson et al also observed a narrow state (Γ = 70 keV) at 4.32 MeV, and in ^{11}Be there is a narrow (Γ = 15 keV) 3/2$^-$ state at 3.9 MeV. The structure of the 5.12 and 5.87 MeV levels could be the ^{10}C core excited to its 2$^+$-state at E*=3.35 MeV, coupled to a $p_{1/2}$ proton at 2.18 MeV resonance energy with the sum energy of 5.53 MeV. This coupling results in two levels with 3/2$^-$ and 5/2$^-$, which could be respectively the peaks at 5.12 and 5.87 MeV. The width of the 5.87 MeV peak is in agreement with this interpretation, while the the 5.12 MeV peak is somewhat too narrow. An alternative description for this peak is given in the shell-model frame by Millener [9]. More experimental and theoretical efforts are required to establish the nature of the higher resonances in ^{11}N.

ACKNOWLEDGEMENTS: We thank Dr. D.J. Millener for enlightening discussions and informations prior to publication.

REFERENCES.
1. H.T.Fortune, D.Koltenuk, C.K.Lau, *Phys. Rev.* **C51**(1995)3023.
2. F.C.Barker, *Phys. Rev.***C53**(1996)1449.
3. S.Grévy, O.Sorlin, N.Vinh-Mau, *Phys. Rev.***C56** (1997)2885.
4. W.Benenson, E.Kashy, D.H.Kong-A-Siou et al., *Phys. Rev.***C9**(1974)2130.
5. V.Guimarães, S.Kubono, M.Hosaka, S.C.Jeong et al., *Nucl. Phys.***A588**(1995)161c.
6. A.Azhari, T.Baumann, J.A. Brown, M.Hellstron et al., *Phys. Rev.***C57**(1998)628.
7. L. Axelsson, M. J. G. Borge, S. Fayans, V. Z. Goldberg et al., *Phys. Rev.***C54**(1996)R1511.
8. A. Lépine-Szily, J.M.Oliveira, A.N.Ostrowski, H.G.Bohlen et al., *Phys. Rev. Lett.***80**(1998)1601.
9. D. J. Millener, private communication

Survey of the Beta-Strength of the Halo Nucleus ^{11}Li

I. Mukha[1,†], M.J.G. Borge[2], D. Guillemaud-Mueller[3], P. Hornshøj[1], B. Jonson[4], H. Fynbo[1], T. Leth[1], T. Nilsson[4,5], G. Nyman[4], K. Riisager[1], G. Schrieder[6], M.H. Smedberg[4] and O. Tengblad[2]

[1] Institut for Fysik og Astronomi, Aarhus Universitet, Denmark; [2] Insto. Estructura de la Materia, CSIC, Madrid, Spain; [3] Institut de Physique Nucléaire, IN2P3-CNRS, Orsay, France; [4] Fysiska Institutionen, Chalmers Tekniska Högskola, Göteborg, Sweden; [5] PPE division, CERN, Genève, Switzerland; [6] Institut für Kernphysik, Technische Hochschule, Darmstadt, Germany; [†] on leave from Kurchatov Institute, Moscow, Russia

The two-neutron halo nucleus ^{11}Li, which has been actively studied through nuclear reactions in many experiments, still presents a challenge due to unknown contributions of the $(s_{1/2})^2$ and $(p_{1/2})^2$ components in the halo wave-function. A complementary method of probing the ^{11}Li ground state is to measure its β-decay where theoretical calculations [1–4] show a sensitivity to the structure of the ^{11}Li wave-function.

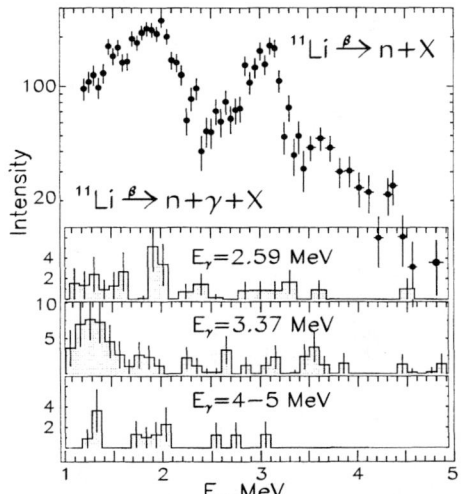

FIGURE 1. Neutron spectra from the β-decay of ^{11}Li (dots). Insets show spectra of neutrons measured in triple coincidences with $\beta + \gamma$ with shown selected γ-energies.

A comprehensive experimental study of β-delayed charged particles (H, He, Be), neutrons and γ-rays that follow the ^{11}Li β-decay has been carried out at the ISOLDE PSB facility at CERN. Singles spectra as well as γ + neutron, γ + charged particle and neutron + charged particle correlations have been measured in coincidences with β-particles [5].

The singles neutron spectra (measured in the 0.01–12 MeV range) have revealed a very complex feeding in the ^{11}Li β-decay. For example, the n-spectrum shown by dots in fig. 1 show peaks which are too broad to be explained by single β-transitions. Thus the correlation measurements became very important for correct assignments. We concentrated on the decay branches which have been observed in several independent measurements especially in the triple coincidences.

CP455, ENAM98: Exotic Nuclei and Atomic Masses
edited by B. M. Sherrill, D. J. Morrissey, and Cary N. Davids
© 1998 The American Institute of Physics   1-56396-804-5/98/$15.00

The spectrum of singles γ-rays following the β-decay of ^{11}Li was measured several times, see for example [3,7]. Figure 2a shows the $(\beta)\gamma$-spectrum measured with the NaI-detector. Four peaks in the right part of the spectrum belong to the known 3368 keV and 2590 keV γ-transitions in ^{10}Be and their sigle and double escape peaks. We have used the known intensities of the 3368 and 2590 keV γ-peaks [7] for the analysis of the triple $\beta+\gamma+$neutron and $\beta+\gamma+$charged particle coincidences.

An illustration of the measured triple coincidences is shown in fig. 1, where one can compare the singles neutron spectrum with projections of the triple $\beta+\gamma+$neutron events on the neutron energy axis. Three neutron spectrum histograms in the inset correspond to the selected 2.59 MeV γ-line (upper panel, the admixture of Compton scattered $\gamma i(3.37)+\beta+n$ was not subtracted) and to the 3.37 MeV γ-line (middle panel). The lower panel corresponds to the Compton scattered γ's of high energy (most likely from the 5959 or 6260 keV states), registered in the 4–5 MeV range, in coincidences with $\beta+$neutrons. Though statistics is low, we may distinguish the main 1–1.5 and $\sim$2 MeV components in the upper panel and the dominating $\sim$1.3 MeV component in the middle panel. This gives us a clue about the presence of transitions ^{11}Be*(8.0 MeV)$\to$ ^{10}Be*(5.958)+n$\to$ ^{10}Be*(3.37)+γ(2.59 MeV)+n, ^{11}Be*(8.82 MeV)$\to$ ^{10}Be*(5.958)+n$\to$ ^{10}Be*(3.37)+γ(2.59 MeV)+n and ^{11}Be*(5.24 MeV)$\to$ ^{10}Be*(3.37)+n$\to$ ^{10}Be+γ(3.37 MeV)+n, respectively. The two neutron peaks in the lower panel in fig. 1 at 1.3 and 2 MeV are assumed to be due to the transitions ^{11}Be*(8.82 MeV)$\to$^{10}Be*(5.959)+n$\to$ ^{10}Be+γ(5.96 MeV)+n and

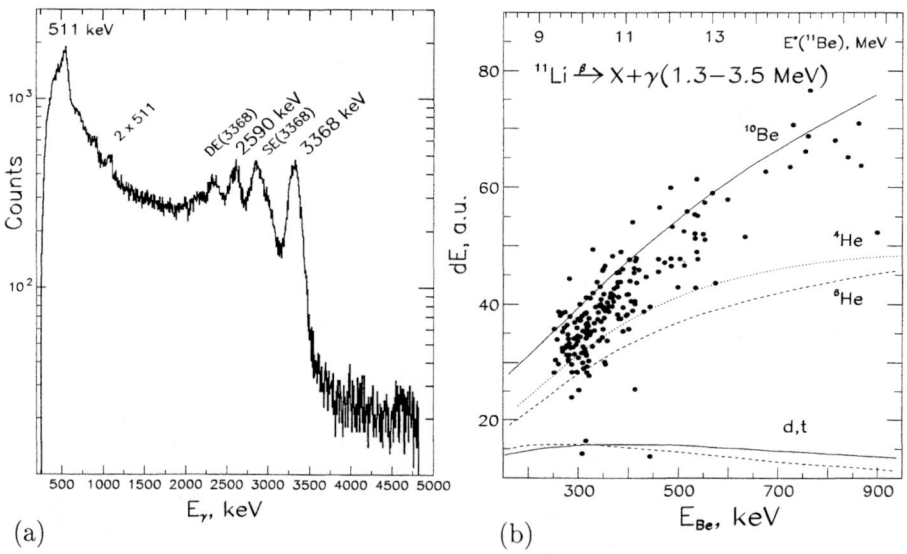

FIGURE 2. (a) – the $\beta\gamma$ spectrum from ^{11}Li. The identified peaks are marked by energies of respective γ-transitions; (b) – the ΔE-E plots of charged particles measured in coincidences with $\beta+\gamma$ corresponding to the selected γ energies 1.3 – 3.5 MeV. Upper axis shows excitation energies in ^{11}Be for the case of the transition to ^{10}Be*(5.96 MeV).

^{11}Be*(8.0 MeV)→ ^{10}Be*(5.959)+n→ ^{10}Be+γ(5.96 MeV)+n which are measured as Compton scattered γ(4–5 MeV) in coincidences with β+n.

Another illustration is the ΔE-E plot of charged particles measured in triple coincidences with $\beta + \gamma$. It is shown in fig. 2b. The γ's were selected in the broad energy range from 1.3 MeV to 3.5 MeV to have larger statistics. Comparing with the calculated energy losses of different charged particles (see details in [7]) one can see that registered ions are mainly ^{9,10}Be. Thus the lab energy (lower x-axis) was calculated assuming the final nucleus is ^{10}Be. The upper x-axis is the calculated excitation energy of the ^{11}Be nucleus if it decays by one neutron emission to the ^{10}Be*(5.95 MeV) state. In fig. 2, on right, one can see two groups of Be-events corresponding to a β-feeding of the states in ^{11}Be above 11 MeV, which have not been seen in the triple β+γ+neutron coincidences. A combined analysis of the singles neutron spectra with the charged particle and γ-ray data [7,8] as well as their correlations has shown that a quantitative description of the data may require more than 30 decay branches of ^{11}Li [6]. A summary of the assumed decay branches is given in the Table 1, where the excitation energies and decay channels of the daughter states in ^{11}Be are listed as well as branching ratios and partial strengths of the corresponding Gamow-Teller transitions. The data are normalized to the ^{11}Li decay probabilities to the 0, 1, 2 and 3 neutron branches: P_{0n}=6.3%, P_{1n}=86.7%, P_{2n}=4.2% and P_{3n}=1.9%, respectively, see [7].

TABLE 1. Assumed branches of the ^{11}Li β- decay; E* is the ^{11}Be excitation energy, b is a branching ratio, B_{GT} – Gamow-Teller factor.

E*, MeV	Decay channel	b, %	B_{GT}
18.15[a]	n,d,t,α,^{6}He	0.35	2.1
15–16 ?	^{10}Be*(5.95)+n	0.7(4)	0.57[b]
14.0 ?	^{10}Be*(3.37)+n	0.4(1)	0.101
12.0 ?	^{10}Be+n	0.27(2)	0.208
	^{10}Be*(5.95)+n	0.7(5)	
	n+^{10}Be*(9.64)→		
	→ α+^{6}He+n	0.9(4)	
	→ 2α+3n	0.3(2)	
	→^{9}Be+2n	0.7(4)	
10.59	^{10}Be+n	0.4(1)	0.475
	^{10}Be*(3.37)+n	0.6(1)	
	^{10}Be*(5.95)+n	0.9(2)	
	^{10}Be*(6.18)+n	2.0(3)	
	n+^{10}Be*(9.64)→		
	→ α+^{6}He+n	4.9(8)	
	→ 2α + 3n	1.6(3)	
	→^{9}Be+2n	3.6(5)	
8.82	^{10}Be+n	0.62(3)	0.134
	^{10}Be*(5.95)+n	3.8(8)	
	^{10}Be*(1$^-$,5.96)+n	3.9(15)	
8.0	^{10}Be+n	1.2(1)	≥0.09
	^{10}Be*(3.37)+n	3.2(3)	
	^{10}Be*(1$^-$,5.96)+n	3.3(15)	
	^{10}Be*(5.95)+n	≤4.	
6.7	^{10}Be+n	1.1(1)	0.162
	^{10}Be*(3.37)+n	1.2(7)	
5.85	^{10}Be+n	0.6(1)	0.009
	^{10}Be*(3.37)+n	0.7(4)	
5.25	^{10}Be+n	1.0(1)	0.031
	^{10}Be*(3.37)+n	6.4(6)	
3.96	^{10}Be+n	9.0(5)[c]	0.032
	^{10}Be*(3.37)+n	1.8(2)	
3.89	^{10}Be+n	see 3.96	
	^{10}Be*(3.37)+n	8.3(10)	0.024
2.64	^{10}Be+n	19(2)	0.039

[a]See details about all decay channels in [7].
[b]Provided the excitation energy is 15.5 MeV.
[c]Unresolved with the 3.89 MeV state decay.

The ^{11}Li GT-strength function has been determined over almost the full excitation spectrum. The accuracy of the assignments depends on our observation limit in branching ratios which is of ~0.3%. More precise measurements may change the absolute values of the data listed in Table 1 due to their normalisation to the P$_n$-values. The ^{11}Li β-decay is described via transitions mainly to the known daughter states in ^{11}Be. Indications of β-feeding of three unknown states in ^{11}Be at excitation energies above 11 MeV (they are marked by the "?" sign in the E* column) are found as well. The suggested transitions to the high lying excited states in ^{11}Be contribute most of the observed GT-strength which in total amounts to ~4. Shell-model calculations predict much more intensive strength distributions peaked at excitation energy 16–18 MeV (the giant Gamow-Teller resonance) with total B$_{GT}$ ~10 [3,4].

According to the $p(sd)$ shell-model calculations [4] the narrowing of the gap between the p and the sd shells can change the structure of the GT-strength distribution. Namely, as (the probability of the $\nu p_{1/2}^2$ component in an admixture of $\nu p_{1/2}^2$ and $\nu 1s_{1/2}^2$ configurations decreases, the main peak becomes wider and more strength emerges at energies above the GT-peak and beyond the β-decay energy threshold. Thus, the measured GT-strength might indicate a probability much lower than 50% for the halo neutrons being in the $\nu p_{1/2}^2$ configuration which could be interpreted as the significant lowering of the sd shell in ^{11}Li. However, the large mismatch between experimental data and calculations could be explained in the alternative way.

The most interesting information related to the ^{11}Li structure (namely, β-transitions with large B$_{GT}$ values which correspond to large overlaps of mother and daughter state wave-functions) is found at excitation energies above 11 MeV where the measured branching ratios are of 0.3–1.% where the measured yields have large statistical errors. In the expected region of the giant GT-resonance the measured strength function has an accuracy which is not fine enough to make a quantitative comparison with the theoretical predictions and more correlation measurements, which should be sensitive to the ^{11}Be excitation energies 10–19 MeV, are needed. In particular, the charged particle ^{6}He+α and $\alpha + \alpha$ correlations are expected to be most adequate in a search of the β-feeding to such ^{11}Be states.

REFERENCES

1. M.V. Zhukov et al., *Phys. Rev.* **C52**, 2461 (1995).
2. Y. Ohbayasi and Y. Suzuki, *Phys. Lett.* **B346**, 223 (1995).
3. D.J. Morrissey et al., *Nucl. Phys.* **A627**, 222 (1997).
4. T. Suzuki and T. Otsuka, *Phys. Rev.* **C56**, 847 (1997).
5. M.J.G. Borge et al., ENAM-95 *Proceedings*, Arles, France, 19-23 June 1995, Edition Frontieres, 1996, pp. 285-289.
6. M.J.G. Borge et al., in preparation.
7. M.J.G. Borge et al., *Nucl. Phys.* **A613**, 199 (1997); *Phys. Rev.* **C55**, R8 (1997).
8. I. Mukha et al., *Phys. Lett.* **B367**, 65 (1996).

Shell-Quenching in the Infinite Nuclear Matter Model

R. C. Nayak

Dept. of Physics, G.M. College, Sambalpur-768004,India.

Abstract. The behavior of shell gaps at the magic neutron numbers for nuclei towards the neutron-drip line is studied in the recently developed Infinite Nuclear Matter model of atomic nuclei. It is shown that this model supports quenching of these gaps in agreement with the reported findings of the recent calculations for the r-process nuclidic abundance pattern of heavier elements.

INTRODUCTION

The basic question that has been under the critical scrutiny in the recent years, is the so-called quenching of shell gaps at the magic number nuclei towards the neutron-drip line. This question has particularly arisen, after the recent observations (1) in the realm of nucleo-astrophysical calculations, for the possible r-process nuclidic abundance pattern. The most important ingredients of this calculation are the nuclear masses lying in the n-drip region. Since these are not known experimentally, one has to use predicted values of some model. In this connection, Chen et al (1) observed deep and narrow troughs in front of each of the $A \simeq 130$ and $A \simeq 195$ peaks calculated using the predictions of both finite-range droplet model (FRDM) (2) and the Extended Thomas-Fermi plus Strutinsky Integral (ETFSI) (3) model. A critical examination (1) of this anomaly was traced down to the occurrence of relatively large shell gaps for the drip-line nuclei, at N=82 and 126 magic shells. It was therefore been suggested (1), that this problem could be resolved by invoking less pronounced shell gaps. This has been confirmed by repeating the above calculations with the masses obtained with SKP force in the frame work of spherical Hartree-Fock-Bogolyubov (HFB) (4) model. However some (5) are sceptical of this inference, as they argue that the very method of nuclear abundance calculation may be too simplistic to warrant such a strong observation on shell quenching. Nevertheless, as the shell-quenching resolves the reported anomaly, it can be considered as one of the possible solutions and hence may not be ruled out as a basic requirement of nuclear structure. Even a model calculation (6) with the Bogolyubov shell-quenching empirically included in the ETFSI model, has succeeded in resolving the reported anomaly. The last but not the least, is the decrease of shell-gaps in the low mass region calculated using the known masses.

In the light of this situation, it would be worth analyzing the shell gaps obtained in the recently improved (7,8) INM model (9,10), which, unlike the macroscopic-microscopic models is on a different theoretical basis. Fundamentally, this model has outgrown from the generalized version (11) of a theorem of many-body theory, well-known in literature as the Hugen Holtz -Van Hove (HVH) theorem (12). Since this theorem has been shown (13) to be independent of the form of the nuclear interaction and remains valid even with multi-body forces, the model would be certainly on a sound conceptual basis. Apart from these aspects, the model does not explicitly take the existence the magic numbers and the consequent shell corrections unlike other models. As a result, this model is more suitable for the present study.

In the following, we discuss the INM model in brief to facilitate our discussion, followed by the results on the shell quenching.

THE INM MODEL

Recalling the essential features we note, that in this model, the ground-state energy $E^F(A,Z)$ of a nucleus with neutron number N, proton number Z, mass number A and asymmetry $\beta=(N-Z)/(N+Z)$, is considered equivalent to the energy of a perfect sphere made up of infinite nuclear matter at its saturation density plus the residual local energy η, which corresponds to the characteristic property of the nucleus and most possibly consists of the so-called shell and deformation. Then taking the appropriate finite-size corrections for the sphere of nuclear matter as

$$f(A,Z) = a_s A^{2/3} + a_C(Z^2 - 5(3/(16\pi))^{2/3} Z^{4/3})A^{-1/3} - \delta(A,Z) + ..., \quad (1)$$

we can write E^F as,

$$E^F(A,Z) = E(A,Z) + f(A,Z) + \eta(A,Z), \quad (2)$$

where E refers to the energy of infinite nuclear matter contained in the sphere and $\delta(A,Z)$ is the usual pairing energy term. The quantity E being the property of nuclear matter at ground state, should satisfy the generalized HVH theorem (11)

$$\frac{E}{A} = \frac{1}{2}[(1+\beta)\epsilon_n + (1-\beta)\epsilon_p], \quad (3)$$

where $\epsilon_n = (\partial E/\partial N)_Z$ and $\epsilon_p = (\partial E/\partial Z)_N$ are the neutron and proton Fermi energies of nuclear matter. These relations are found to lead to the following INM equations,

$$\frac{E^F}{A} = \frac{1}{2}[(1+\beta)\epsilon_n^F + (1-\beta)\epsilon_p^F] + S(A,Z), \quad (4)$$

$$\frac{\eta(A,Z)}{A} = \frac{1}{2}[(1+\beta)(\partial\eta/\partial N)_Z + (1-\beta)(\partial\eta/\partial Z)_N] \quad (5)$$

and

$$-a_v + a_a\beta^2 = \frac{1}{2}\left[(1+\beta)\epsilon_n^F + (1-\beta)\epsilon_p^F\right] - \left[\frac{N}{A}(\frac{\partial f}{\partial N})_Z + \frac{Z}{A}(\frac{\partial f}{\partial Z})_N\right], \quad (6)$$

where, $S(A,Z) = f/A - (N/A)(\partial f/\partial N)_Z - (Z/A)(\partial f/\partial Z)_N$ is a function exclusively of finite size coefficients. a_v and a_a are the other two global parameters, which can be identified as the volume and asymmetry energy coefficients. We have demonstrated elsewhere (7,13), that the S-function referred above becomes an exclusive function of only two parameters, namely a_s and a_c. The other terms such as exchange Coulomb, finite size proton form factor and Nolen-Schifer charge anomaly that could occur in equation (1), are found to cancel and the terms like surface symmetry and curvature also cancel to a major extent in equation (4). Thus all the global parameters can be estimated with the least correlation by fitting equations (4) and (6) to known masses of nuclei. The remaining quantity η, is obtained from the recursion relations derived from equation (5), from the values of neighboring known ones. Recently we (8) have developed a new scheme of an interactive network utilizing these recursion relations. The essential feature of this new treatment is, the ensemble averaging procedure for the η of a given nucleus, from of a set of alternative values generated by repeated application of the recursion relations in all possible directions of the nuclear chart. This method has enabled us to successfully generate masses of nuclei (about 7188 in number) lying in the range $4 \leq Z \leq 128$ and $8 \leq A \leq 270$.

RESULTS AND DISCUSSIONS

We all know, that nuclei having magic neutron and proton numbers are relatively more bound than the neighboring nuclei. This is reflected quantitatively in terms of a quantity termed as the shell gap, which can be defined as

$$\Delta(N_m) = S_n(N_m - 1) - S_n(N_m + 1), \qquad (7)$$

where N_m is the magic neutron number at a particular shell and S_n is the neutron separation energy. Usually this quantity is relatively large at the magic neutron number compared to neighboring nuclei. In the light and medium mass nuclei (N_m up to 50), the shell gaps can be calculated from the known masses and the quenching of the shell-gaps can be evidently seen. However in high mass region, this calculation is not feasible as the masses are not experimentally known.

TABLE 1. Shell gaps Δ at various neutron magic numbers N_m

Z	$\Delta(N_m = 50)$	Z	$\Delta(N_m = 82)$	Z	$\Delta(N_m = 126)$
40	1.6 (2.1)	50	1.1	82	1.3
38	1.4 (2.1)	48	1.3	80	1.3
36	1.3 (1.6)	46	1.1	78	-0.5
34	1.3 (1.2)	44	0.8	76	0.6
32	0.7 (-)	42	0.9	74	0.0
30	0.6 (-)	40	0.2	72	-0.3

For the INM model, we calculate the shell gaps (Eqn. 7) from the already generated mass table, at all the N_m magic numbers over a wide range of proton numbers toward the n-drip line. The results are presented in Table 1 for all even-Z isotope series at $N_m = 50,82,126$. The corresponding experimental quantities where ever available (mostly at $N_m = 50$) are also presented within parenthesis. From the results we can clearly see that the shell gaps go on decreasing with decreasing proton number (equivalently increasing neutron excess). The experimental quantities at $N_m = 50$ supports the decreasing trend. It must be noted that, the results of INM model for the odd-Z isotone series qualitatively remain similar, and hence these are not presented due to lack of sufficient space.

Concluding the present work we can say, that the results of INM model quite convincingly show, the decreasing trend of shell gaps at all the magic numbers with the increase of neutron excess. This is in agreement with the recent findings of the r-process abundance pattern from astrophysical studies. Spherical HFB calculations carried out by others, also support this conclusion. Experimentally, shell quenching at magic numbers up to $N_m = 50$ is well supported. In view of all these evidences, we can conclude that, perhaps this feature is inherently present in nuclear structure for all nuclei lying in the neutron drip-line region.

ACKNOWLEDGMENTS

The author acknowledges the generous help of the Computer Centers of both Institute of Physics, Bhubaneswar and G. M. College, Sambalpur, where different parts of the work were carried out.

REFERENCES

[1] Chen, B. et al ,*Pays. Left. B* **355** , 37 (1995).
[2] Moeller, P. et al, *At. Data NaCl. Data Tables* **59** , 1 (1995).
[3] Aboussir, Y. et al *At. Data Nucl. Data Tables* **61** , 61 (1995).
[4] Dobaczewski, J. et al, *Phys. Scr.* **T 56**, 15 (1995).
[5] Goriely, S. and Arnould, M., *Astrn. Astro.* **262**, 73 (1992)
[6] Pearson, J. M., Nayak, R. C. and Goriely, S., *Phys. Lett. B* **387**, 455 (1996).
[7] Nayak, R.C., V. S., Uma Maheswari and Satpathy, L. , *Phys. Rev. C* **52**, 711 (1995).
[8] Nayak, R.C. and Satpathy, L., *At. Data and Nucl Data Tables* (1988) , In press.
[9] Satpathy, L. *J. Phys. G* **13**,761 (1987). .
[10] Satpathy, L. and Nayak, R.C. , *At. Data and Nucl. Data Tables*
[11] Satpathy, L. and Nayak, R.C. , *Phys. Rev. Lett.* **51**, 1243 (1983).
[12] Hugenholtz, N. H. and Van Hove, W., *Physica (Utrecht)***24**, 363 (1958) .
[13] Satpathy, L., V.S., Uma Maheswari and Nayak, R. C. , *Phys. Rep.* (1997)in press.

Half-life Measurements of ^{31,32}Ne

M. Notani[a,1], N. Aoi[c], N. Fukuda[c], H. Iwasaki[c], K. Yoneda[c],
H. Ogawa[d], T. Teranishi[a], S.M. Lukyanov[a,b], Yu.E. Penionzhkevich[b],
T. Nakamura[c], H. Sakurai[a], E. Ideguchi[a], A. Yoshida[a],
Y. Watanabe[a], T. Kubo[a] and M. Ishihara[a,c]

[a] *The Institute of Physical and Chemical Research (RIKEN)*
2-1 Hirosawa, Wako, Saitama 351-01, Japan
[b] *Flerov Laboratory, Joint Institute for Nuclear Research*
141980 Dubna, Moscow, Russia
[c] *Department of Physics, University of Tokyo*
7-3-1 Hongo, Bunkyo, Tokyo 113, Japan
[d] *Department of Applied Physics, Tokyo Institute of Technology*
2-12-1 Oh-okayama, Meguro-ku, Tokyo 152, Japan

Abstract. The first half-life measurement of two extremely neutron-rich nuclei, ^{31,32}Ne, has been performed by means of the Riken Projectile Fragment Separator (RIPS). The two nuclei are the heaviest stable neon isotopes observed experimentally [1] [2].

INTRODUCTION

Due to the development of heavy-ion accelerators, extremely neutron-rich nuclei in $9 \leq Z \leq 12$ have been identified in the projectile-like fragments, recently. In the new neutron-rich isotopes, we made an attempt to measure the beta-decay half-lives of ^{31,32}Ne for the first time.

EXPERIMENTAL SETUP

The radioactive ion beams were produced by the fragmentation of a $95A$ MeV primary ^{40}Ar beam on a ^{181}Ta target at RIKEN Ring Cyclotron facility. Very neutron-rich nuclei in $9 \leq Z \leq 11$ were separated by the Riken Projectile Fragment Separator (RIPS) [3] with a 221mg/cm^2 aluminium energy-degrader at the momentum dispersive focal plane.

---

[1] E-mail address: notani@rikaxp.riken.go.jp

The fragments were identified in TOF-ΔE measurement by using a 1 mm plastic scintillator(PL) and a set of silicon detectors with thickness of 0.5,0.5,1, 3 and 1 mm. ^{31}Ne and ^{32}Ne fragments were implanted into the 3mm silicon detector. The last three silicon detectors were also used to detect β-rays. When fragments of interest reached the detectors, the primary beam was turned off. Then, we measured time interval from the fragment arrival timing to the β-ray emission timing during a 100 ms period.

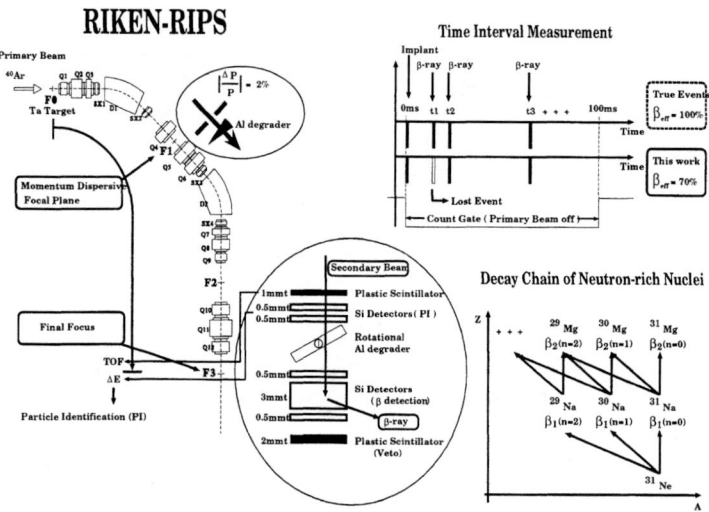

FIGURE 1. Experimental setup for the particle identification of projectile-like fragments and β-ray detection. The β-rays are emitted throught the process of β-decay chain by a extremely neutron-rich nuclei.

DECAY CURVE

In Fig.2, the time-interval histogram indicates the first β events measured by the silicon detectors. Since the real first β events are only caught by 70% due to the β-detection efficiency of the silicon detectors, we must include the contribution from the daughter nuclei for a correct fitting.

$$\frac{dN_1}{dt}(t) = -\lambda_1 N_1(t) \tag{1}$$

$$\frac{dN_2}{dt}(t) = -\lambda_2 N_2(t) + \lambda_1 N_1(t) \tag{2}$$

Where N_1 and N_2 are the number, λ_1 and λ_2 the decay constant of parent and daughter, respectively. These simultaneous equations are solved with an initial condition:

$$N_1(t=0) = A \tag{3}$$
$$N_2(t=0) = 0 \tag{4}$$

The solution of these equations are shown in (5,6).

$$N_1(t) = A exp(-\lambda_1 t) \tag{5}$$
$$N_2(t) = \frac{\lambda_1 A}{\lambda_2 - \lambda_1}\{exp(-\lambda_1 t) - exp(-\lambda_2 t)\} \tag{6}$$

In the time-interval histogram for ^{31,32}Ne, we draw three curves as the result of a likelihood fitting with these functions. The solid line is a fitting with sum of two functions (5,6). The dashed and dotted curves are drawn as parent (5) and daughter (6) components, respectively.

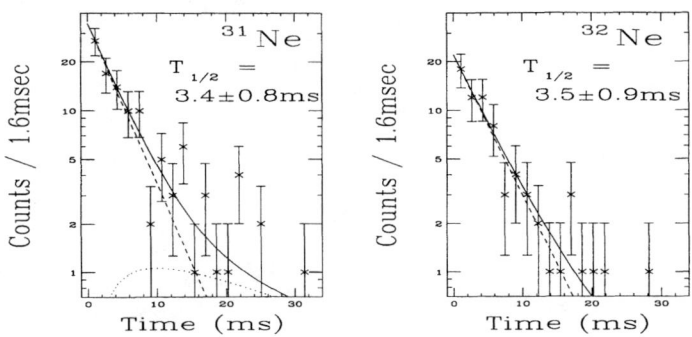

FIGURE 2. Time interval histogram of ^{31,32}Ne

RESULTS AND DISCUSSION

The half-lives measured by this work are listed together with the values measured by other groups in Table 1 for comparison.
1) We succeeded in measuring half-lives of ^{31,32}Ne for the first time.
2) We have determined half-lives of ^{27,29}F , ^{29,30}Ne with smaller measurement error compared with the previous work.
3) Our measured values are consistent with that from Tarasov's group.

Figure 3 shows the investigation of neutron number dependence of β-decay half-life for three sets of isotopes Na,Ne and F.

1) As a main trend, half-lives decrease along with the increasing neutron number.
2) Sudden changes of the trend can be seen around(but not exactly at) N=20, which is probably due to odd-even mass effect rather than the shell effect?

Mass measurement experiment for the isotopes of our interest is expected in order to understand the above point.

Nuclei (A,Z)	T$_{1/2}$ [ms] This work	Tarasov [4]	Table'97 [5]
^{27}F	4.9 ± 0.2	5.3 ± 0.9	-
^{29}F	2.6 ± 0.4	2.4 ± 0.8	-
^{29}Ne	15.6 ± 0.5	15 ± 3	200 ± 100
^{30}Ne	5.8 ± 0.2	7 ± 2	-
^{31}Ne	3.4 ± 0.8	-	-
^{32}Ne	3.5 ± 0.9	-	-
^{31}Na	19 ± 4	18 ± 2	17.0 ± 0.4
^{32}Na	11.5 ± 0.8	-	13.2 ± 0.4
^{33}Na	8.5 ± 0.4	-	8.2 ± 0.4

TABLE 1. Experimental half-lives

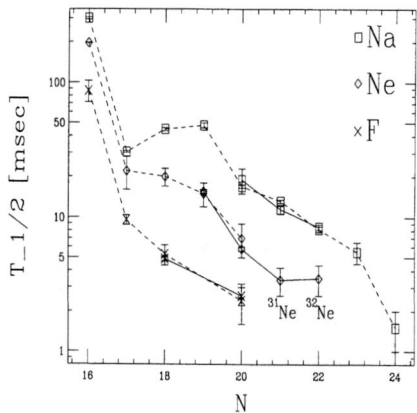

FIGURE 3. Half-lives of neutron-rich nuclei, F, Ne and Na

REFERENCES

1. D. Guillemaud-Mueller et al.: *Phys. Rev. C* **41**, 937 (1990).
2. H. Sakurai et al.: Phys. Lett. C **54**, R2802 (1996).
3. T. Kubo et al.: Nucl. Instrum. Methods Phys. Res. B **70**, 309 (1992).
4. O. Tarasov et al.: *Phys. Lett. B* **409**, 64 (1997).
5. G. Audi et al.: *Nucl. Phys. A* **624**, 1 (1997).

Elastic Transfer in ^{4}He(^{6}He,^{6}He)^{4}He

A. Piechaczek[a*], R. Raabe[a], A. Andreev[a], D. Baye[b], W. Bradfield-Smith[c],
T. Davinson[c], M. Gaelens[a], W. Galster[d], M. Huyse[a], J. McKenzie[c],
A. Ninane[d], A.C. Shotter[c], G. Vancraeynest[a], P. Van Duppen[a], A. Wöhr[a]

[a]) *Instituut voor Kern- en Stralingsfysika, Katholieke Universiteit Leuven, B-3001 Heverlee*
[b]) *Universite Libre de Bruxelles, B-1050 Brussels*
[c]) *University of Edinburgh, Edinburgh EH9 3JZ*
[d]) *Universite Catholique de Louvain, B-1348 Louvain-la-Neuve*
*) present address: Louisiana State University, Baton Rouge, LA 70803-4001

Abstract. We have investigasted the reaction ^{4}He(^{6}He,^{6}He)^{4}He at c.m. energies of 11.6 and 15.9 MeV and determined the differential elastic cross section in the angular range between 50 and 145 degree. The comparison of the measured data with DWBA calculations using a double folding potential for the ^{6}He-^{4}He scattering shows evidence for the elastic transfer of a di-neutron cluster between the two ^{4}He cores.

INTRODUCTION

The neutron halo nucleus ^{6}He ($T_{1/2}$ = 0.807 s) has been investigated repeatredly by reaction studies at intermediate energies of several ten to several hundred MeV/u (see, e.g. [1,2]). Mainly recently, a number of experiments were carried out using radioactive ^{6}He beams at energies between ten and a few ten MeV. Smith et al. [3] and Warner et al. [4] investigated the elastic scattering of a 8-9 MeV and 10.2-MeV ^{6}He beams on targets ranging from ^{9}Be to ^{197}Au. Skobelev et al. [5] and Penionzkhevich et al. [6] studied the elastic scattering of a 55-MeV ^{6}He beam bombarding a ^{208}Pb target. Fomichev et al. and Penionzkhevich et al. [8,6] measured fission cross sections for the bombardment of a ^{209}Bi target by a ^{6}He beam between 30 and 70 MeV (c.m.). We have investigated the elastic scattering ^{4}He(^{6}He,^{6}He)^{4}He at beam energies of 30 and 40 MeV, corresponding to c.m. energies of of 11.6 and 15.9 MeV, respectively. ^{6}He nuclei may be described as an inert ^{4}He core surrounded by two weakly bound valence neutrons, and elastic two neutron transfer is expected to occur between the ^{4}He-cores. Its observation and analysis may allow one to distinguish between different spatial and momentum correlations of the two valence neutrons in ^{6}He.

EXPERIMENT

The experiment was performed at the ARENAS3 Radioactive Ion Beam Facility in Louvain-La-Neuve, Belgium, which supplies radioactive beams of low and intermediate masses for experiments in astrophysics, nuclear and solid state physics. The experimental set-up consisted of a ^{4}He gas target, bombarded by a ^{6}He^{++} beam of 30 {40}[1] MeV total energy in combiation with an array of Segmented Silicon Strip

---

[1] Numbers referring to the 40-MeV experiment are given in brackets {}.

Detectors (SSSD). ^{6}He nuclei were produced in a liquid LiF target via the reaction ^{7}Li(p,2p)^{6}He, induced by a 30 MeV proton beam of 110 μA which was delivered by the K = 30 CYCLONE30 cyclotron. The ^{6}He atoms diffused out of the target, which was heated by the impact of the proton beam and were carried by a ^{4}He gas flow into an ECR ion source. The emerging ions were injected into the K= 110 CYCLONE110 cyclotron which was tuned for the acceleration of ^{6}He^{++} in the third harmonic mode to an energy of 30 {40} MeV. The intensity of the obtained ^{6}He^{++} secondary beam at the target position in the experimental area amounted to 0.35×10^5 {0.7×10^5} pps. Tuning CYCLONE110 for the accelleration of the isobaric contaminant ^{6}Li^{++}, we did not observe any beam related signals in our detectors, which is crucial for the particle identification via the method of kinematic coincidences (see below). The non-observation of ^{6}Li is due to the fact that this reactive element is released only weakly from the production target and that the mass difference $\Delta A/A$ between ^{6}Li and ^{6}He is 6.26×10^{-4}, which is well beyond the mass resolving power of CYCLONE110 of 1×10^{-4}. For the investigation of the ^{4}He(^{6}He,^{6}He)^{4}He reaction a gas target was used which contained a ^{4}He gas layer of 1 cm thickness at 500 mb pressure between an 2.33 mg/cm^2 aluminum entrance window ($\varnothing$=11 mm) and 1.84 mg/cm^2 Kapton® {2.33 mg/cm^2 aluminum} exit windows ($\varnothing$=30 mm). A total of 38 {34} hours of data were taken with the filled gas target and several hours for the determination of background reactions with the empty gas target. The products of reactions betweeen the ^{6}He beam and the target were registered in two SSSDs arrays, LEDA and LAMP. The LEDA array is placed at a distance of 500 mm from the target and comprises 126 silicon detectors. It is assembled from of eight 300-μm thick silicon segments, which cover each an azimuthal angle φ = 45°. Each segment is subdivided into 16 concentric detectors subtending the polar angles θ between 6° and 16°. The LAMP array with 96 silicon detectors is composed of 6 such segments which are inclined with respect to the beam axis by 45° and thus form a lamp-shade shaped structure, covering the polar angles from 20° to 70° degree at a distance from the target of 100 mm. The probability $\varepsilon_2(\Theta_{cm})$ to detect the ejectile as well as the recoil of the reaction ^{4}He(^{6}He,^{6}He)^{4}He was calculated as a function of the center-of-mass scattering angle. As a result, scattering events between 50° and 140° c.m. are detected with an efficiency $\varepsilon_2(\Theta) \cong 0.8$ as LAMP-LAMP coincidences. The unambiguous identification of ^{6}He and ^{4}He, the separation of the elastic scattering on ^{4}He from elastic and inelastic background processes and the determination of the elastic scattering cross section were based on the two-dimensional energy-energy plot of multiplicity-2 events, where both particles were detected in LAMP. An unambiguous assignment E_1, $E_2 \rightarrow$ ^{6}He, ^{4}He was achieved for 95% of the events with full energy deposition by combining the information of the measured scattering angle and energy and comparing them to the calculatrions taking into account the reaction kinematics and energy losses in the gas target. The center-of-mass scattering angle was determined by calculating the deposited energies E_1 and E_2 as a function of Θ_{cm} and sorting the multiplicity-2 events into angular bins. Figure 1 shows the differential cross section for the 30 {40} MeV

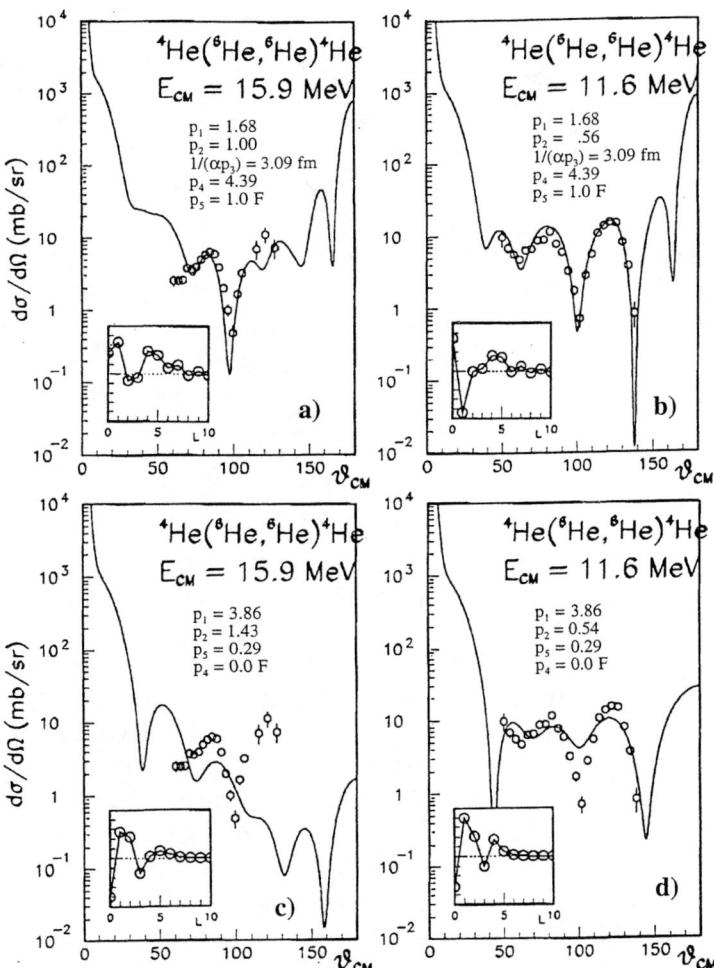

FIGURE 1. Best fits *with* (a,b) and *without* (c,d) parity term. p_1, and p_3-p_5 were forced to be identical for the two energies. F: fixed parameter. Insets: The resulting phase shifts $\delta(L)$ on a $\pm\ 100°$ scale.

data determined by combining the number of events per angular bin, the beam intensity, the target thickness, the measuring time, and dividing by $\varepsilon_2(\Theta_{cm})$.

DISCUSSION

The data are compared with the differential cross sections calculated in the framework of the DWBA. We used the double-folding potential V(R) [9], based on calculated ^{6}He and ^{4}He wave functions, and the Minnesota effective nucleon-nucleon interaction, augmented by a parity term taking into account the exchange of a di-neutron cluster:

$$V(R) = -(p_1 + i\ p_2)\ V_G \exp(-R^2/a_G^2) - p_5\ V_S\ \{1 + \exp[(R-R_S)/a_S]\}^{-1}$$
$$+ (-)^L\ p_4\ V_P\ \{\theta(1-R/R_P) + \theta(R/R_P - 1)\ \exp[-p_3\alpha(R-R_P)]/(R-R_P)\} \quad (1)$$

with $V_G = 41.2$ MeV, $a_G = 1.95$ fm, $V_S = 58.52$ MeV, $a_S = 0.725$ fm, $R_S = 2.647$ fm, $V_P = 4.39$ MeV, $R_P = 2.50$ fm, and $1/\alpha = 3.20$ fm. V_G, a_G, V_S, a_S, and R_S are the parameters of the potential as calculated in the folding approach. V_P is the height of the parity dependent term (at a distance of 2.5 fm), which was assumed to be of Yukawa shape and was regularized for radii < 2.5 fm by using the unit step function θ. $1/\alpha$ was calculated according to [10] to be 2.30 fm taking the 2-n sparation energy of ^{6}He as the effective binding energy of the transferred di-neutron cluster. Fig. 1 shows the best fits to the data, obtained by varying p_1 to p_4 with $p_5 = 1$ (with exchange interaction), and p_1, p_2, and p_5 with $p_4 = 0$ (no exchange interaction). The values for p_1 and $p_3 - p_5$ were forced to be identical for both energies. The description of the data *with exchange interaction* is excellent for the $E_{c.m.} = 11.6$ MeV data and good for the $E_{c.m.} = 15.9$ MeV data. p_1 may be allowed to deviate from 1 in order to take care of antisymmetrization effects not otherwise included in V(R) [14]. p_2, representing the absorption from the elastic channel, is expected to be energy dependent and was allowed to vary independently for the 11.6 and 15.9 MeV data. The value for $1/(\alpha p_3) = 3.09$ fm is very satisfactory, coming close to the estimate of $1/\alpha = 3.20$ fm [10]. Also the amount of 0.4 MeV for the modulation of the interaction barrier at a distance of 7 fm, induced by the alternating sign of the parity term is of reasonable magnitude. The reflection coefficients $|\eta_L|$ and the phase shifts δ_L (insets in Fig. 1) show that partial waves with $L \leq 5$ are strongly absorbed and that the exchange term acts on the higher partial waves by modulating their phase shifts δ_L. Despite the smaller value for p_2 the $|\eta(L)|$ are smaller in the 15.9- than in the 11.6- MeV data for $L \leq 5$. The best fits to the data *without exchange interaction* require strong, unphysical, modifications of both V_G and V_S, $p_1 = 3.86$ and $p_4 = 0.29$, and are of considerably less quality. The observed behaviour of the fits may be interpreted as evidence for the presence of the elastic transfer of a di-neutron cluster in the ^{4}He(^{6}He,^{6}He)^{4}He reaction at center-of-mass energies of 11.6 and 15.9 MeV.

This work was in part supported by the D.O.E. under grant DE-FG02-96ER-40978.

REFERENCES

[1] I. Tanihata et al., Phys. Lett. B **160** (1985) 380, Phys. Lett. B **289** (1992) 261
[2] M. D. Cortina-Gil et al., Phys. Lett B **371** (1996) 14
[3] R. J. Smith et al., Phys. Rev. C **43** (1991) 761
[4] R. E. Warner et al., Phys. Rev. C **51** (1995) 178
[5] N. K. Skobelev et al., Z. Phys. A **341** (1992) 313
[6] Yu. E. Penionzkhevich et al., Nucl. Phys. A **583** (1995) 791c
[7] J. Jänecke et al., Phys. Rev. C 54 (1996) 1070
[8] A. S. Fomichev et al., Z. Phys. A **351** (1995) 129
[9] D. Baye et al., Phys. Rev. C **54** (1996) 2563
[10] W. von Oertzen and H. G. Bohlen, Phys. Rep. **19** (1975) 1

Proton Radioactivity from Highly Deformed Nuclei

A.A. Sonzogni[1], C.N. Davids[1], P.J. Woods[2], D. Seweryniak[1,3], J.C. Batchelder[4], C.R. Bingham[5], T. Davinson[2], D.J. Henderson[1], R.J. Irvine[2], G.L. Poli[6], J. Uusitalo[1] and W.B. Walters[3].

[1] *Argonne National Laboratory, 9700 South Cass Avenue, Illinois 60439*
[2] *University of Edinburgh, Edinburgh EH9 3JZ, United Kingdom*
[3] *Department of Chemistry, University of Maryland, College Park, MD 20742*
[4] *UNIRIB, Oak Ridge Associated Universities, Oak Ridge, TN 37381*
[5] *Department of Physics and Astronomy, University of Tennessee, Knoxville, TN 37996*
[6] *Istituto di Fisica Generale Applicata, University of Milano, I-20133 Milano, Italy*

Abstract

Proton emission half-lives are calculated within the DWBA formalism for ^{131}Eu and ^{141}Ho assuming permanent quadrupole deformation. The decay rates are consistent with a decay from either the $3/2^+[411]$ or $5/2^+[413]$ Nilsson states for ^{131}Eu and the $7/2-[523]$ Nilsson state for ^{141}Ho.

The proton decay of ^{131}Eu and ^{141}Ho has been recently reported [1]. The predicted quadrupole deformations for these nuclei are 0.331, 0.37 for ^{131}Eu and 0.29, 0.35 for ^{141}Ho [2, 3]. Although gamma-ray information for these nuclei has not been reported yet, extrapolation of the first 2^+ energy from nearby nuclei is consistent with the predictions. These values of β_2, larger than those of the other proton emitters [4], makes these nuclei particularly interesting. As was pointed out in ref. [1], the experimental results couldn't be explained in terms of a spherical WKB calculation of proton decay. It can be argued that since proton emission is sensitive to the potential through which the proton must tunnel, which in turn depends on the nuclear shape, it is expected that proton radioactivity will give very valuable information about nuclear shapes.

The proton emission from spherical nuclei has been recently discussed in ref. [5]. Bugrov and Kadmensky developed a formalism to explain the proton decay of ^{109}I and ^{113}Cs nuclei under non-spherical nuclear shapes [6]. The same formalism will be used here to calculate half-lives for ^{131}Eu and ^{141}Ho.

For a proton moving in a non-spherical potential, the wave function is expanded in terms of spherical wave functions:

$$|\Psi(N\Omega)> = \sum_{j\ell} C_{j\ell}(N\Omega)|N\ell j\Omega>. \qquad (1)$$

The coefficients $C_{j\ell}(N\Omega)$ were first obtained by Nilsson using a harmonic oscillator potential. The approach followed here is that of ref. [8], the spherical wave functions are obtained numerically (the Becchetti-Greenlees parametrization [7] is used), the Hamiltonian matrix is calculated and then diagonalized. The $C_{j\ell}(N\Omega)$ coefficients as well as Nilsson diagrams are calculated this way.

In non-spherical DWBA, the decay amplitude is given by:

$$B = (\frac{2(2J_f+1)}{2J_i+1})^{1/2} <J_f j_p 0 K_i | J_i K_i> \sum_{\ell j, m_s} C_{\ell j}(N\Omega) \times$$

$$<\ell_p \frac{1}{2}(K_i-m_s)m_s|j_p K_i><\ell\frac{1}{2}(K_i-m_s)m_s|jK_i> \times$$

$$<Y_{\ell_p}^{K_i-m_s}(\Omega')\frac{F_{\ell_p}(k,\eta)}{r}|V_{DWBA}(\vec{r})|\frac{R_{n\ell j}(r)}{r}Y_{\ell}^{K_i-m_s}(\Omega')>, \quad (2)$$

where $R_{nlj}(r)/r$ is the radial part of the spherical proton wave function, $F_l(r)$ is the regular Coulomb function and $V_{DWBA}(r)$ is the total potential minus the monopole Coulomb term.

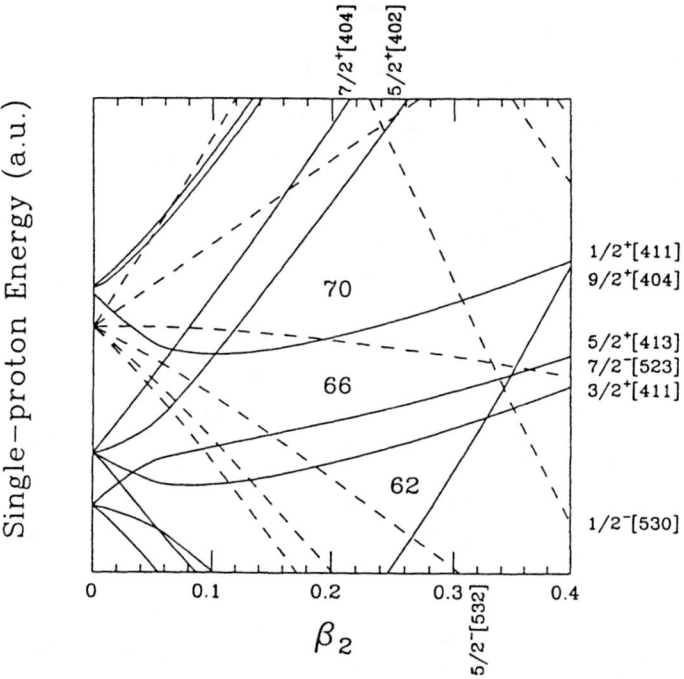

Fig. 1 *Proton orbits for nuclei in the vicinity of* ^{131}Eu *and* ^{141}Ho

The half-life is related to the decay amplitude by:

$$T_{1/2} = \frac{\hbar ln2}{2\pi B^2} \quad (3)$$

As outlined in ref. [6], the observed energy and half-life of the proton decay can be used to determine from which Nilsson orbital the proton originated from. The strategy is to use eq. 2 for a large number of Nilsson orbitals, and presumably, only a few of them will agree with the observations. A Nilsson diagram for protons can be seen in Fig. 1. For quadrupole deformations between 0.2 and 0.4, the odd proton for Eu and Ho nuclei will most likely decay from the $1/2^+[411]$, $3/2^+[411]$, $5/2^+[413]$, $5/2^+[402]$, $7/2^+[404]$ $5/2^-[532]$ and $7/2-[523]$ orbitals. We won't consider any nuclear structure effects, i.e. we will take the spectroscopic factor equal to 1. A value of 0.5 was used in ref. [6]. The use of another set of nuclear potential systematics can produce differences by a factor of 2 between calculations. Therefore, we will accept as final candidates those orbitals that are within a factor of 2 of the measured half-lives.

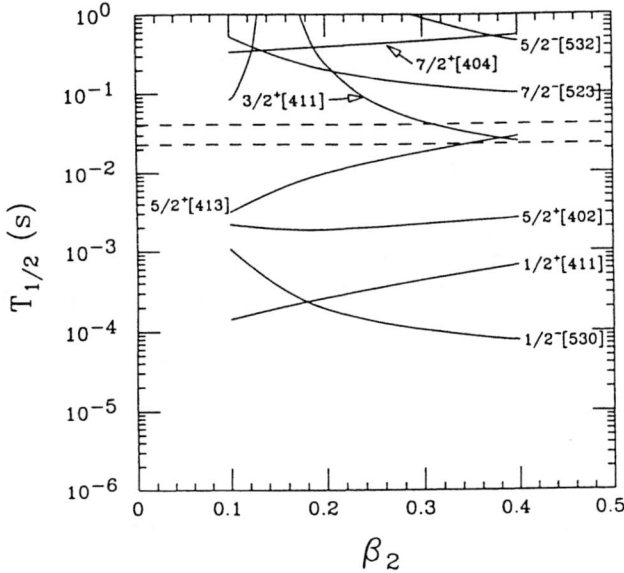

Fig. 2 *Half-life as a function of deformation for different Nilsson orbits for* ^{131}Eu.

The results of the calculations can be seen in figs. 2 and 3 for ^{131}Eu and ^{141}Ho respectively. The measured half-lives are indicated with dashed lines. For ^{131}Eu, only calculations from 2 orbitals, $3/2^+[411]$ and $5/2^+[413]$, match the experimental results at $\beta_2 \sim 0.35$. For ^{141}Ho, the orbitals $7/2-[523]$ at $\beta_2 \sim 0.3$ and $3/2^+[411]$ at $\beta_2 \sim 0.16$ reproduce the measured half-life. The $7/2-[523]$ orbital is preferred over the $3/2^+[411]$ orbital since it is closer to the Fermi level and in fact, it is observed for the deformed ground states of odd-A Ho with $A \geq 157$ [9].

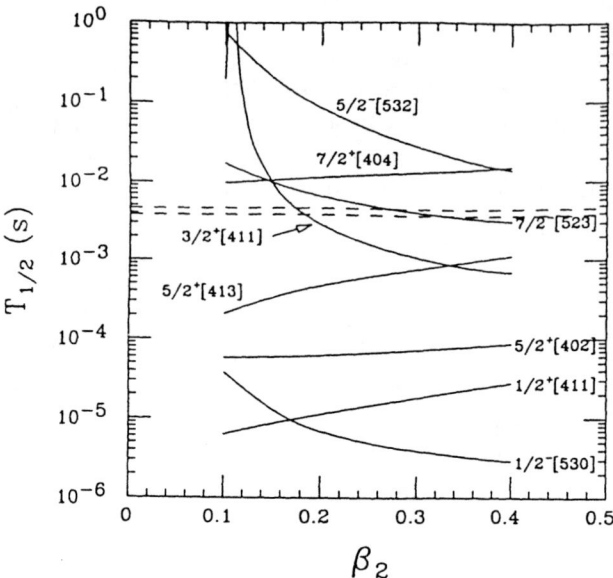

Fig. 3 *Half-life as a function of deformation for different Nilsson orbits for* ^{141}Ho.

In summary, using the measured proton energy and half-life, the Nilsson orbits $3/2^+[411]$ or $5/2^+[413]$ were assigned to the decay of ^{131}Eu while the $7/2-[523]$ was assigned to ^{141}Ho.

References

[1] C.N. Davids *et al*, Phys. Rev. Lett. **80**, 1849 (1998).

[2] P. Moller *et al*, At. Data Nucl. Data Tables **59**, 185 (1995).

[3] Y. Aboussir *et al*, At. Data Nucl. Data Tables **61**, 127 (1995).

[4] P.J. Woods and C.N. Davids, Annu. Rev. Nucl. Part. Sci **47**, 541 (1997).

[5] S. Aberg, P.B. Semmes and W. Nazarewics, Phys. Rev **C56**, 1762 (1997).

[6] V.P. Bugrov and S.G. Kadmensky, Sov. J. Nucl. Phys. **52**, 229, (1990). S.G. Kadmensky and V.P. Bugrov. Phys. of At. Nuclei **59**, 399 (1996).

[7] F.D. Bechetti and G.W. Greenlees, Phys. Rev. **182**, 1190 (1969).

[8] B.L. Andersen, B.B. Back, and J.M. Bang, Nucl. Phys. **A147**, 33 (1970).

[9] R.B. Firestone, *Eight Edition of the Table of Isotopes*, edited by V.S. Shirley (Wiley, New York, 1996).

Quasifree scattering of neutron in ^{11}Li from deuteron

S. Takeuchi[1], S. Shimoura[1], T. Teranishi[2], Y. Ando[1], M. Hirai[3],
N. Iwasa[2], T. Kikuchi[1], S. Moriya[1], T. Motobayashi[1],
H. Murakami[1], T. Nakamura[3], T. Nishio[1], H. Sakurai[2],
T. Uchibori[1], Y. Watanabe[2], Y. Yanagisawa[1] and M. Ishihara[2,3]

[1] *Rikkyo University, Nishi-Ikebukuro, Toshima, Tokyo 171-8501, JAPAN*
[2] *RIKEN (Institute of Chemical and Physical Research),
Wako, Saitama 351-0198, JAPAN*
[3] *University of Tokyo, Hongo, Bunkyo, Tokyo 113-8654,JAPAN*

Abstract. Quasifree scattering (QFS) of a neutron in ^{11}Li by a deuterium target have been observed in the ^{11}Li+d reaction at 64 MeV/u. QFS events have been identified by a velocity spectrum of deuteron. The momentum distribution of scattered neutron and excitation energy spectrum of the ^{10}Li as a spectator of the reaction were obtained.

INTRODUCTION

A neutron drip-line nucleus ^{11}Li has been investigated by using various secondary reactions [1–6]. This nucleus exhibits irregular behaviors in various nuclear properties such as large interaction cross section [1], narrow momentum distributions of projectile fragments [2] and large Coulomb dissociation cross sections at low excitation energies [3–5]. They are attributed to a neutron-halo structure, which arises from a particular configuration consisting of a ^{9}Li core and two valence neutrons with an extremely small binding energy. The charge exchange reaction (p,n) and (d,2n) of ^{11}Li with inverse kinematics has been measured to investigate isospin response of the halo nucleus ^{11}Li [7,8].

In the present experiment with ^{11}Li beams onto deuteron target, deuterons having the velocity of $\frac{2}{3}$ of incident beams were observed. It is understood from the kinematics that these deuterons are derived from QFS of the deuteron and the valence neutron of ^{11}Li with a large differential cross section of ^{2}H(n,n)^{2}H [9] at backward scattering angle (see Fig.1). In the present report, we discuss the momentum distribution of the knocked-out particle (neutron) in the

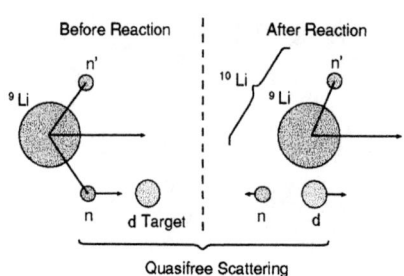

FIGURE 1. n-d Quasifree Scattering

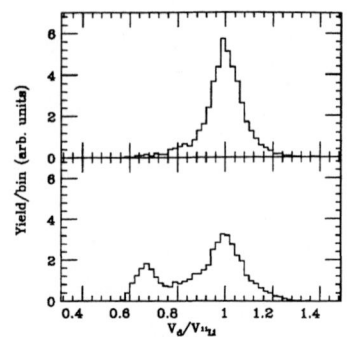

FIGURE 2. Spectra of the ratio between velocities of the deuteron and the beam for ^{1}H (upper) and ^{2}H (lower) targets.

nucleus (^{11}Li) and the energy spectrum of ^{10}Li.

EXPERIMENTAL PROCEDURE

The experiment was performed using the accelerator facility at RIKEN. A primary ^{18}O beam of 100 MeV/u bombarded a production target ^{9}Be of 1.4 g/cm^2 thickness. Reaction products were analyzed by the projectile fragment separator RIPS [10] to obtain a radioactive ^{11}Li beam. A energy of an incident ^{11}Li beam was determined event by event by a measurement of the time of flight (TOF) between RF and 0.5 mm thick F2 plastic scintillator. A beam size and its direction were measured with two sets of position sensitive detectors consisting of two parallel plate avalanche counters (PPAC). After traversing beam detectors at F2 and F3, the ^{11}Li beam was focused on a (CD$_2$)$_n$ and C secondary targets with the thicknesses of 206 mg/cm^2 and 188 mg/cm^2, respectively. The average beam energy was 64 MeV/u at the middle of the target with a spread of 6% (FWHM) and a typical intensity was 2×10^4 cps. The C target was used for subtracting contributions from the carbon nucleus in (CD$_2$)$_n$ target.

Outgoing charged particles were detected by ΔE and E hodoscopes which located at 314 cm downstream of target. The hodoscopes consisted of vertically segmented 5 mm thick plastic scintillators and two layers of horizontally segmented 6 cm thick plastic scintillators, respectively. Charged particles were identified by using the ΔE - E and TOF-E methods. The E hodoscopes were also used for the detection of neutrons, where the ΔE hodoscope worked as charged particles vetoes. Velocities of charged particles and neutrons were determined by the TOF over the 314 cm flight path between the target and the hodoscope. The resolution of the TOF was 0.5 ns (FWHM). The detection efficiency of neutrons for each layer of plastic scintillator was $\sim$ 6.6%. A He

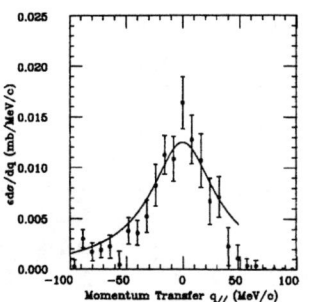
FIGURE 3. Momentum distribution of the scattered neutron in ^{11}Li.

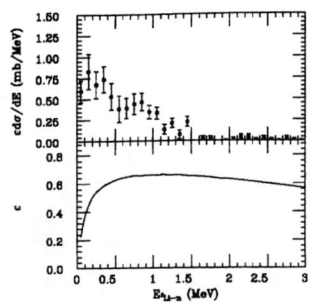
FIGURE 4. Relative energy of ^{9}Li and n and efficiency ϵ.

gas chamber was placed between the target and the particle detectors in order to minimize the energy losses and the multiple scattering.

RESULTS AND DISCUSSION

Figure 2 shows the ratio between velocities of the deuteron and the beam for events coincident with ^{9}Li. The upper spectrum denotes the reaction for the ^{1}H target and the lower one for the ^{2}H target. The peak at the ratio of unity in the spectra is considered to be deuterons through charge exchange reaction, e.g. ^{11}Li $\rightarrow$ ^{11}Be* $\rightarrow$ d+^{9}Li, where all particles detected in this reaction have almost same velocity as the beam. The peak at the ratio of $\frac{2}{3}$ in the spectrum for the ^{2}H target is derived from the QFS between valence neutron of ^{11}Li and deuteron (see Fig.1).

The width of the QFS peak in Fig.2 relates to the momentum distributions of the scattered neutron. Figure 3 shows the momentum distribution $q_{//}$ which is the parallel component of the momentum transfer to direction of deuteron by using the QFS kinematics. This distribution provides us the information of the momentum fluctuation of the valence neutron in ^{11}Li. The present detection system is insensitive for the region $q_{//} >40$ MeV/c, which corresponds to lower energy region of the deuteron.

The relative energy spectrum of ^{9}Li and neutron was obtained from the momentum vectors of them. The deuteron having $\frac{2}{3}$ velocity of a beam was used to distinguish QFS from all of reaction process. The spectator of unbound nucleus ^{10}Li was observed as the scattering state of ^{9}Li and neutron. It is noted that the ^{9}Li-n relative energy spectra in the fragmentation or Coulomb dissociation reaction [11–14] may be affected by the final state interactions (FSI) of ^{9}Li or neutron with other particles having almost same velocity as those of ^{9}Li and/or neutron. On the other hand, since the present QFS reaction involves only the ^{9}Li and n with almost same velocities, the effect of the FSI could be minimized. Figure 4 shows the relative energy spectrum for the ^{2}H target and the lower panel shows the detector efficiency ϵ. The peak of 0.3 MeV and shoulder of around 0.8 MeV are observed in the spectrum. The spectrum are qualitatively consistent with the result of other experiments [7,13].

The knocked-out neutron is considered to be bound by the ^{9}Li+n system (^{10}Li) with a binding energy of $S_{2n}+E_{rel}$. Assuming a Yukawa wave function of neutron, momentum distribution is characterized by a Lorentzian. Solid curve in Fig.3 shows weighted mean of Lorentzians by the yield of the relative energy spectrum. The curve is consistent with the present data.

SUMMARY

We have observed the deuterons having the $\frac{2}{3}$ velocity of the beam in the reaction of ^{11}Li of 64 MeV/u incident to a deuteron target. This is identified as a quasifree scattering of the valence neutron and the deuteron. The momentum distribution of scattered neutron in ^{11}Li was obtained, and we have obtained the peak of 0.3 MeV and shoulder of around 0.8 MeV in the relative energy spectrum , which are qualitatively consistent with recent works.

REFERENCES

1. I. Tanihata et al., Phys. Rev. Lett. **55**(1985)2676
2. T. Kobayashi et al., Phys. Rev. Lett. **60**(1988)2599
3. K. Ieki et al., Phys. Rev. Lett. **70**(1993)730
4. D. Sackett et al., Phys. Rev. **C48**(1993)118
5. S. Shimoura et al., Phys. Lett. **348B**(1995)29
6. T. Kobayashi, Nucl. Phys. **A538**(1992)343c
7. S. Shimoura et al., Nucl. Phys. **A616**(1997)208c
8. T. Teranishi et al., Phys. Lett. **407B**(1997)110
9. J.L. Romero et al., Phys. Rev. **C2**(1970)2134
10. T. Kubo et al., Nucl. Instr. Meth. **B70**(1992)309
11. A.I. Amelin et al., Yad. Fiz. **52**(1990)1231[Sov. J. Nucl. Phys. **52**(1990)782]
12. H.G. Bohlen et al., Z. Phys. **A344**(1993)381
13. K.H. Wilcox et al., Phys. Lett. **59B**(1975)142
14. M. Zinser et al., Phys. Rev. Lett. **75**(1995)1719

Proton decay of the closed neutron shell nucleus ^{155}Ta*

J. Uusitalo[1], C. N. Davids[1], P. J. Woods[2], D. Seweryniak[1,3], A. A. Sonzogni[1],
J. C. Batchelder[4], C. R. Bingham[5], T. Davinson[2], J. DeBoer[6], D. J. Henderson[1],
H. J. Maier[6], J. Ressler[3], R. Slinger[2], and W. B. Walters[3]

[1] *Argonne National Laboratory, Argonne, IL 60439*
[2] *University of Edinburgh, Edinburgh EH9 3JZ, U. K.*
[3] *Dept. of Chemistry, University of Maryland, College Park, MD 20742*
[4] *UNIRIB, Oak Ridge Associated Universities, Oak Ridge, TN 35996*
[5] *Dept. of Physics and Astronomy, University of Tennessee, Knoxville, TN 37996*
[6] *Sektion Physik, Univ. Munich, Am Coulombwall 1, D-85748 Garching, Germany*

Abstract

The new proton radioactivity ^{155}Ta has been observed. It was produced via the $p4n$ fusion evaporation channel using a ^{58}Ni beam on a ^{102}Pd target. The measured decay properties were: $E_p = (1765 \pm 10)$ keV and $t_{1/2} = (12^{+4}_{-3})$ μs. Using the WKB approximation a spin and parity of $J^\pi = 11/2^-$ and a spectroscopic factor of $S_p^{exp} = 0.58^{+0.22}_{-0.17}$ were determined.

Nuclear structure studies of nuclei beyond the proton drip line have been intensively pursued lately. Many new ground-state proton emitters have been found. In addition, in-beam γ-ray studies using the Recoil Decay Tagging (RDT) method have been performed or are planned for a number of these nuclei. The proton decay energy yields the mass difference between the initial and the final states, and combining it with the partial proton decay half-life, one can obtain the total angular momentum change in the decay process. So far direct proton decays have been investigated for nuclei 50<Z<84 beyond the proton drip line. Below Z = 69 the proton drip line nuclei are predicted to be well deformed, while the heavier ones are predicted to be at most slightly deformed [1]. The WKB barrier penetration approximation combined with a low-seniority shell-model calculation of the spectroscopic factors have succesfully reproduced the proton decay rates when applied to spherical or at most slightly deformed nuclei [2], [3]. In these calculations the model space consists of 18 particles in the degenerate $s_{1/2}$, $d_{3/2}$, and $h_{11/2}$ proton orbitals above the magic core 64, with neutrons considered to be spectators. In the case of well-deformed nuclei a more complex method using the Nilsson model has recently been developed and applied [4], [5]. For more details, see also the contribution of Sonzogni at this conference. In the present paper, the discovery of the proton-unbound closed shell nucleus $^{155}_{73}\text{Ta}_{82}$ will be presented and discussed.

The proton rich nucleus $^{155}_{73}\text{Ta}_{82}$ was produced via the $p4n$ fusion evaporation channel using a ^{58}Ni beam on a ^{102}Pd target. The beam was delivered by the Argonne ATLAS accelerator, and bombarding energies of 315 MeV and 320 MeV were used. The target thickness was 1.0 mg/cm², leading to excitation energies of 77 MeV and 80 MeV, respectively, in the middle of the target. The total doses transported to the target were 1.3×10^{15} particles and 1.7×10^{15} particles, respectively. The Fragment Mass Analyzer (FMA) was used to separate the reaction products from the beam, and to disperse them by mass/charge state. The FMA was set to focus mass A=155 and charge state Q=28, with Q = 27 and 29 collected at the same time. After passing through the focal plane the products were implanted into a Double-sided Silicon Strip Detector (DSSD), with thickness 65 μm, area 16×16 mm², and having 48 orthogonal strips on the front and rear. The time of arrival, position and energy of the implants were recorded. This information was then used to correlate with the position of the subsequent decays. At the front of the DSSD were placed four silicon detectors, forming a seven cm-deep box. These detectors served as veto detectors, providing a decrease in the background caused by escaping alphas in the region where the possible discrete proton decay energy lines were expected to be. Behind the DSSD was placed a large (5×5 cm²) 500 μm thick silicon detector, used to reduce the background caused by electrons and β-delayed protons. The decay energy calibration was performed using the known proton decay lines of ^{147}Tm, E_p = 1051 keV [6] and ^{147m}Tm, E_p = 1119 keV [7], produced in a separate reaction ^{58}Ni+^{92}Mo.

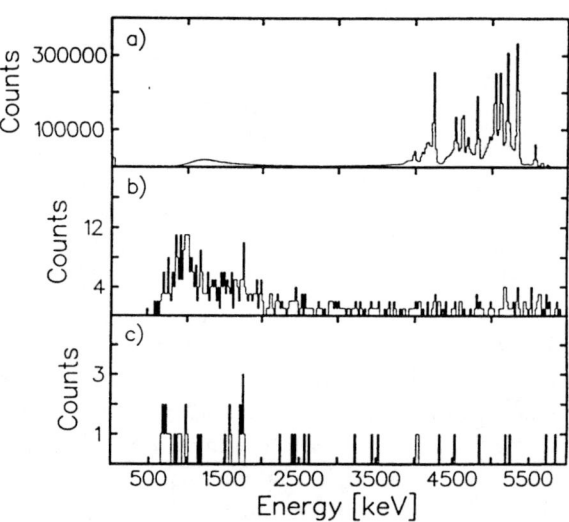

Fig. 1. *a.) The total decay spectrum vetoed with the box detectors and with the back detector is shown. b.) The decay spectrum correlated with implants of mass A = 155 and with maximum time between implant and decay of 50 μs is shown. c.) Same as b) but in addition a second decay was demanded as described in the text. In spectra b) and c) the vetos are still in effect.*

When correlated with mass A=155 and using a maximum time between implant and decay of 50 µs, a peak can be seen as shown in the decay spectrum in fig 1b. The energy of this peak was determined to be $E_p=1765\pm10$ keV. This peak is assumed to originate from the proton decay of ^{155}Ta. The measured half-life for this decay was $12\pm^4_3$ µs. The continuum background is caused by real correlations between ^{155}Lu-implants and escaped alpha particles. The full energy part of the ^{155}Lu decay alphas was used to obtain a correction for the pile-up effect caused by the fast decay energy signal falling on the tail of the implant signal. This effect is significant for decays faster than 20 µs.

The theoretical partial β-decay half-life is of the order of $t^{beta}_{1/2} \sim 1$ s [8] which is too long to be significant. The daughter ^{154}Hf β-decays with a half-life of $t_{1/2} = 2$ s [9] to ^{154}Lu which in turn β-decays with a half-life of $t_{1/2} = 1.12$ s [9]. The grand-grand daughter ^{154}Yb has a 93 % alpha branch, and it decays with a half-life of $t_{1/2} = 0.41$ s [9] (see fig. 1c). A total of seven correlations between ^{154}Yb alphas and candidate ^{155}Ta protons were found. A maximum time between implant and these alphas of 15 seconds was used, while at the same time rejecting those cases where there was an A = 154 implant in a 4 second time window preceding the alpha decay in the same detector position. The expected number of such random correlations was calculated to be 2, based on the method given in ref. [10]. This leads to an error probability of less than 0.5 % to accidentally produce at least seven of the above-mentioned correlations. The relatively high calculated number of random correlations is due to the high counting rates in the detector setup and the required long correlation time of the second decay. The cross section for producing ^{155}Ta was deduced to be 60 nb at a bombarding energy of 320 MeV.

Table 1. *Measured and theoretical decay properties of ^{155}Ta are shown. The predicted ground state to ground state decay Q_p-value is obtained using the mass table from ref. [11].*

Measured Q_p [keV]	Liran-Zeldes Q_p [keV]	Measured $T_{1/2}$ [µs]	WKB-$T_{1/2}$ [µs]	Proton orbital
1776 ± 10	1371	$12\pm^4_3$	4.3×10^{-4}	$s_{1/2}$
			3.5×10^{-3}	$d_{3/2}$
			7.0	$h_{11/2}$

Using the measured proton decay energy and WKB approximation the calculated half-life is $t_{1/2} = 7.0$ µs, leading to a spin and parity of $J^\pi = 11/2^-$ for the decaying state. From the ratio of theoretical and measured half-life a spectroscopic factor of $S^{exp}_p = 0.58^{+0.22}_{-0.17}$ is obtained. This is in good agreement with the theoretical spectroscopic factor of $S^{th}_p = 0.56$ predicted by the low-seniority shell model [3]. The measured excitation energies of isomeric states in heavier Tantalum isotopes are 102 keV (9^+ i.s. - 2^- g.s.) [13], 22 keV ($11/2^-$ i.s. - $1/2^+$ g.s.) [12],

141 keV (9^+ i.s. - 2^- g.s.) [3], and 64 keV ($11/2^-$ i.s. - $1/2^+$ g.s.) [3] for ^{156}Ta, ^{157}Ta, ^{158}Ta, and ^{159}Ta, respectively, suggesting that the same isomeric state lies at most some tens of keV above the ground state in ^{155}Ta. Using this estimate the WKB-approximation gives a partial half-life much less than 1 μs for $\ell = 0$ proton decay (see table 1.), which is too short to be observed.

*This work was supported by the U. S. Department of Energy, Nuclear Physics Division, under Contract W-31-109-ENG-38.

References

[1] P. Möller, J. R. Nix, W. D. Myers, and W. J. Swiatecki, At. Data and Nucl. Data Tables **59**, 185 (1995)

[2] P. J. Woods and C. N. Davids, Annu. Rev. Nucl. Part. Sci. **47**, 541 (1997)

[3] C. N. Davids et al., Phys. Rev. C **55**, 2255 (1997)

[4] V. P. Bugrov and S. G. Kadmenskii, Sov. J. Nucl. Phys. **49**, 967 (1989)

[5] C. N. Davids et al., Phys. Rev. Lett. **80**, 1849 (1998)

[6] O. Klepper et al., Z. Phys. **A305**, 125 (1984)

[7] S. Hoffman et al., Proceedings of the 7th International Conference on Atomic Masses and Fundamental Constants, AMCO-7 Darmstadt, 1984, edited by O. Klepper (THD, Schriffenreiche Wissenschaft und Technik, Darmstadt, 1984), Vol. 26, p.184

[8] K. Takahashi, M. Yamada, and T. Kondoh, At. Data and Nucl. Data Tables, **12**, 101 (1973)

[9] R. B. Firestone, Table of Isotopes, Eighth Edition, Volume 2 (1996)

[10] K. -H. Schmidt, C. -C. Sahm, K. Pielenz, and H. -G. Clerc, Z. Phys. **A316**, 19 (1984)

[11] S. Liran and N. Zeldes, At. Data and Nucl. Data Tables, **17**, 431 (1976)

[12] R. J. Irvine et al., Phys. Rev. C **55**, R1621 (1997)

[13] R. D. Page et al., Phys. Rev. C **53**, 660 (1996)

# NUCLEAR STRUCTURE AND SHAPES

Exotic Nuclei from a Theoretical Perspective

Witold Nazarewicz

Department of Physics, University of Tennessee, Knoxville, Tennessee 37996[1]
Physics Division, Oak Ridge National Laboratory, Oak Ridge, Tennessee 37831[2]
Institute of Theoretical Physics, University of Warsaw, ul. Hoża 69, PL-00-681 Warsaw, Poland

Abstract. One of the main frontiers of nuclear structure today is the physics of radioactive nuclear beams. Experiments with radioactive beams will make it possible to look closely into many aspects of the nuclear many-body problem. What makes this subject both exciting and difficult is: (i) the weak binding and corresponding closeness of the particle continuum, implying a large diffuseness of the nuclear surface and extreme spatial dimensions characterizing the outermost nucleons, and (ii) access to the exotic combinations of proton and neutron numbers which offer prospects for completely new structural phenomena.

INTRODUCTION

The field of radioactive nuclear beams (RNB) is one of the main frontiers of nuclear science today. One of the indications of the potential of this field is the large international interest in the development of facilities with RNB capabilities. At present there are only a few laboratories with radioactive ion beam capabilities. However, the prospects for new experiments and the success of the current programs have led to a number of RNB facilities under development, and a number of further proposals, including the construction of the next-generation facilities in Europe, U.S., and Japan.

Theoretically, exotic nuclei represent a formidable challenge for the nuclear many-body theories and their power to predict nuclear properties in nuclear "terra incognita". It is important to remember that the lesson learned by going to the limits of the nuclear binding is also important for "normal" nuclei from the neighborhood of the beta stability valley. And, of course, radioactive nuclei are crucial astrophysically; they pave the highway along which the nuclear material is transported up in the proton and neutron numbers during the complicated synthesis process in stars.

---

[1] Research supported by the U.S. Department of Energy under Contract DE-FG02-96ER40963.
[2] Research supported by the U.S. Department of Energy under Contract DE-AC05-96OR22464 with Lockheed Martin Energy Research Corp.

Recent research relating to exotic nuclei has already demonstrated the potential for exciting new nuclear physics. Examples include the exotic structure of halo nuclei, which present us with new forms of nuclear matter, the surprising fragility of magic numbers, which hints at some of the marked changes in the underlying foundations of nuclear structure, and the production and study of the long-searched doubly closed shell nuclei ^{78}Ni and ^{100}Sn. In the following, I shall briefly comment on some of the themes related to the RNB physics.

THEORETICAL CHALLENGES FAR FROM STABILITY

In the description of weakly bound systems, the major theoretical challenge is the correct treatment of the particle continuum. For weakly bound nuclei, the Fermi energy lies very close to zero, and the decay channels must be taken into account explicitly. As a result, many cherished approaches of nuclear theory must be modified. (For an extensive discussion of the theoretical perspectives far from stability, see the recent review [1].) But there is also a splendid opportunity: the explicit coupling between bound states and continuum, and the presence of low-lying scattering states invite strong interplay and cross-fertilization between nuclear structure and reaction theory.

How can one extend traditional tools of nuclear theory to account for the scattering of nucleons from bound single-particle orbitals to unbound states? The closeness of the particle continuum reverberates in two aspects of the theoretical description. Firstly, the particles forming a bound nuclear state can virtually scatter back and forth into the particle continuum phase space. This process must conserve the localization of the nuclear wave function which remains bound even with such a *virtual* scattering taken into account. A theoretical description of this kind of effect still remains virgin territory, although some progress has been made in the analysis of the virtual pair scattering [2,3]. Secondly, nucleons can very easily leave the nucleus altogether and enter the particle continuum through the *real* scattering. This is an old problem which, in the context of excited states near or above the particle threshold, has been addressed by the continuum shell model (CSM) [4–6]. In the CSM, the continuum states (decay channels) and bound states are treated on equal footing. Consequently, correlations due to the coupling to resonances, the spatial extension effects in weakly bound states, the structure of resonances, and the structure of particle transfer form factors are properly described by the CSM. Unfortunately, in many shell-model calculations for weakly bound nuclei (including these presented at this meeting!) the continuum aspects are completely disregarded; hence their conclusions should be taken with a grain of salt (see, however, the recent study [7]).

Often, particle continuum is approximated by the *quasibound states*, i.e., the states resulting from the diagonalization of a finite potential in a large basis [8,9] or by enclosing the finite nuclear potential within an infinite well with walls positioned at a large distance from the nuclear surface [10,11]. More sophisticated methods of discretizing continuum include the Sturmian function expansions [12–15] and resonant (Gamow) state expansions.

APPLICATIONS OF GAMOW STATES TO WEAKLY BOUND NUCLEI

The Gamow states are eigenstates of the time-independent Schrödinger equation with complex eigenvalues [16–18]. They are regular at $r=0$ and satisfy a purely outgoing wave type of asymptotics with the complex energy eigenvalue. The real part of the complex energy eigenvalue is the expectation value of the one-body Hamiltonian, while the imaginary part is related to the total decay width of the quasi-stationary state. The Gamow states are the poles of the S-matrix on the complex energy plane lying below the positive real axis. The closer they lie to the real axis, the more they resemble the bound states, and they can be associated with narrow resonances.

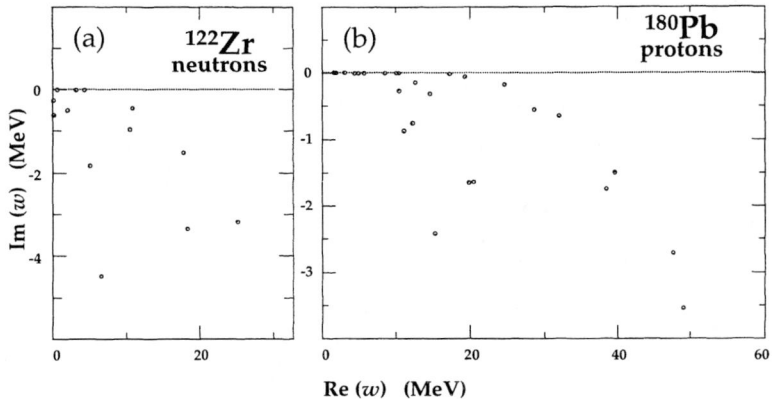

FIGURE 1. The distribution of Gamow energy eigenvalues w_i in the $(\text{Re}(w), \text{Im}(w))$ plane for (a) neutron drip-line nucleus ^{122}Zr (neutron eigenvalues), and (b) proton-rich nucleus ^{180}Pb (proton eigenvalues). (From Ref. [19].)

A generalized completeness relation proposed by Berggren [17] paved the way for using Gamow states as basis states in a similar way as the ordinary bound states are used. Using the generalized completeness relation, one can treat a selected set of resonant states on the same footing as bound states. The remaining part of the continuum is treated by means of the integral along a path in a complex energy plane. Recently, this technique has been applied to calculate the single-particle level density and shell corrections for finite depth potentials [20,19]. Figure 1 illustrates the distribution of spherical Gamow eigenvalues for ^{122}Z and ^{180}Pb (all partial waves are shown). While for the neutrons in ^{122}Zr the large-width Gamow states appear just above the Re(w)=0 threshold, for the protons in ^{180}Pb the particle continuum is shifted effectively by ∼8 MeV due to the presence of the Coulomb barrier. The proton Gamow states that appear at low energies are extremely narrow resonances, usually discussed in the context of proton emitters.

MICROSCOPIC DESCRIPTION OF NUCLEAR MASSES

In the description of weakly bound systems, pair scattering plays a unique role. In standard methods based on bound and quasibound states the virtual scattering of nucleonic pairs from bound states to the positive-energy states leads to the presence of a "particle gas" surrounding the nucleus [2]. This problem is overcome in the HFB method with a realistic pairing interaction in which the coupling of bound states to the particle continuum is correctly taken into account [2,3,21,22]. Consequently, for large exotic nuclei, the self-consistent HFB treatment is not a matter of choice, it is a must. The calculations are not easy, especially if the self-consistent symmetries (e.g., spherical symmetry) are broken.

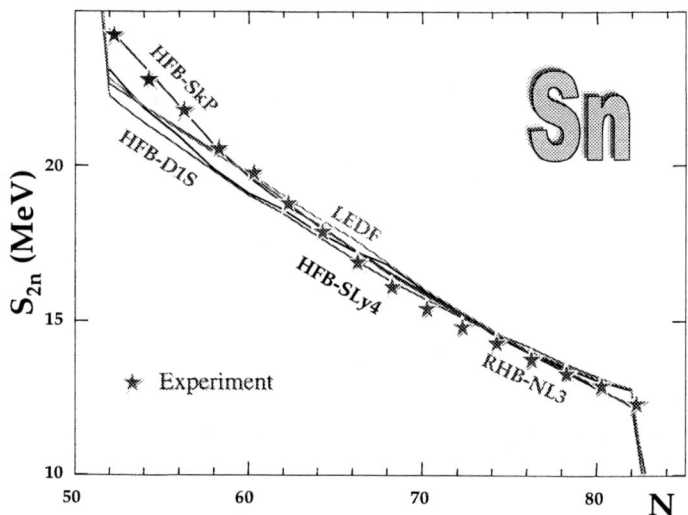

FIGURE 2. Two-neutron separation energies S_{2n} for the Sn isotopes calculated in five microscopic models: HFB-D1 (courtesy of Jacques Dechargé), HFB-SkP and HFB-SLy4 (courtesy of Jacek Dobaczewski), LEDF (courtesy of Sergei Fayans), and RHB-NL3 (courtesy of Georgios Lalazissis). The experimental data are indicated by stars.

For medium-mass and heavy nuclei, no-core microscopic-mass calculations (i.e., based on microscopic effective interactions and employing many-body theory) are usually performed within the mean-field approach. One has to bear in mind, however, that in the mean-field (i.e., independent-quasiparticle) theory, the correlations beyond pairing are not considered, and this gives rise to systematic deviations between experimental and calculated masses [23]. On the other hand, when comparing mass differences such as separation energies, the contributions originating from correlations beyond the mean field tend to cancel out. Figure 2 displays two-neutron separation energies for the Sn isotopes calculated in five state-of-the art microscopic models: HFB-D1S (based on finite-range Gogny interaction D1S [24]), HFB-SkP and HFB-SLy4 (based on zero-range Skyrme parametrizations SkP [2] and SLy4 [25]), LEDF (local energy-density functional model with parametrization FaNDF⁰ [26]), and RHB-NL3 (relativistic Hartree-Bogolyubov model with NL3 parametrization [27]). All models give a very good description of existing experimental data; some interesting deviations are seen when approaching the proton drip line.

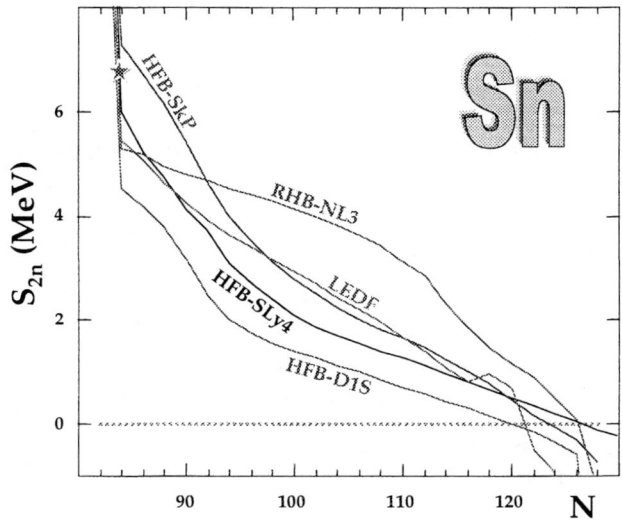

FIGURE 3. Similar to Fig. 2 except for very neutron-rich Sn isotopes.

How do all these microscopic models perform when extrapolated to exotic neutron-rich nuclei? This is presented in Fig. 3. Clearly, the differences between forces are greater in the region of "terra incognita" than in the region of known masses shown in Fig. 2. As seen in Fig. 3, the position of the neutron drip line for the Sn isotopes slightly depends on the effective interaction used; it varies between $N=120$ (HFB-D1S) and $N=126$ (RHB-NL3). Hence, even if the theoretical method used to calculate nuclear masses is reliable near the drip line, the uncertainty due to the largely unknown isospin dependence of the force gives an appreciable theoretical "error bar". A detailed analysis of the force dependence of results may give us valuable information on the relative importance of various force parameters and many-body approximations.

ODD-EVEN STAGGERING OF NUCLEAR MASSES

The presence of the odd-even staggering (OES) in nuclear binding energies [28] is usually attributed to the existence of nucleonic pairing correlations [29,30]. Recently, a similar effect has been observed for ultra-small superconducting metallic grains [31] and it is believed to result from the superconducting correlations [32]. Although the motion of electrons in metals is very different from that of nucleons in nuclei, the mechanism behind electronic and nucleonic superconductivity (the presence of an attractive residual interaction which gives rise to a correlated many-fermion system) is indeed very similar [33,34]. On the other hand, no evidence has been found for superconductivity in metal clusters, and the OES of binding energies in such systems is attributed to the Jahn-Teller effect which, by breaking the spherical symmetry of the mean field, gives rise to deformed single-particle orbitals [35,36].

In a recent study [37], the phenomenon of OES in nuclei has been analyzed using the self-consistent HF method. In the independent quasiparticle picture, the gap parameter is often related to the binding energies of three adjacent systems:

$$\Delta^{(3)}(\mathcal{N}) \equiv \frac{(-1)^{\mathcal{N}}}{2}[B(\mathcal{N}-1) + B(\mathcal{N}+1) - 2B(\mathcal{N})], \quad (1)$$

where $\mathcal{N}$ is the particle number (i.e, N or Z). As shown in Ref. [37], the quantity $\Delta^{(3)}(n)$ calculated in the HF method *without pairing* is nearly zero for the odd values of n and is positive for even values of n, reflecting the presence of the deformed mean field. This result suggests that both pairing and mean-field components of OES can be extracted from binding energies by using the three-point filter, $\Delta^{(3)}$. Namely, the values of $\Delta^{(3)}$ calculated at odd values of n can be, roughly, associated with the pairing effect,

$$\Delta_\nu(\mathcal{N}) \equiv \Delta_\nu^{(3)}(\mathcal{N} = 2n+1), \quad (2)$$

while the differences of $\Delta^{(3)}$ at adjacent even and odd values of n give information about the deformed single-particle spectra [37],

$$e_{n+1} - e_n = 2\left[\Delta_\nu^{(3)}(\mathcal{N} = 2n) - \Delta_\nu^{(3)}(\mathcal{N} = 2n+1)\right]. \quad (3)$$

The neutron staggering parameter, $\Delta^{(3)}$, extracted from the experimental binding energies is shown in Fig. 4. The values of $\Delta_\nu^{(3)}$ at even neutron numbers are almost twice as large as those at odd neutron numbers. Therefore, in light nuclei the mean-field and pairing effects contribute almost equally to the staggering of nuclear masses.

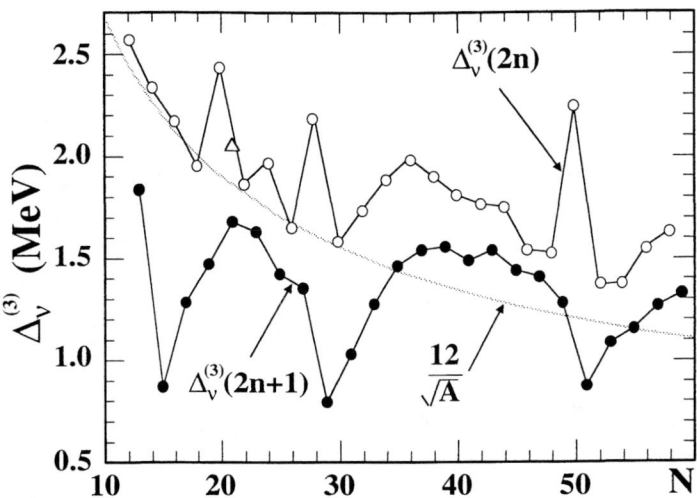

FIGURE 4. Experimental values of $\Delta_\nu^{(3)}(N)$. The thick gray line indicates the average trend, $\tilde{\Delta}=12/\sqrt{A}$. Each point represents the arithmetic mean over several even-Z isotones. (From Ref. [37].)

SPECTROSCOPY OF PROTON EMITTERS

Proton radioactivity is an excellent example of the elementary three-dimensional quantum-mechanical tunneling. Lifetimes of proton emitters directly provide an indication of the angular momentum content of the narrow proton resonance [38]. Experimental and theoretical investigations of proton emitters (or theoretically predicted ground-state di-proton emitters) are just opening up a wealth of exciting physics associated with the residual interaction coupling between bound states and extremely narrow resonances in the region of very low density of single-particle levels.

Recently, a method of calculating deformed proton resonances by means of the coupled-channel technique with Gamow states has been proposed [39,40]. In very deformed nuclei, proton resonances can be treated by means of the strong coupling approach. However, to investigate the influence of the angular momentum dependence of the proton decay width, this formalism has to be extended to account for the Coriolis coupling.

Another exciting avenue is the competition between gamma-radiation and the emission of prompt protons. Here, spectacular examples are proton-emitting intruder bands in ^{58}Cu [41] and ^{56}Ni [42]. In ^{56}Ni, where two intruder bands have been observed, the lower rotational band can be explained by large-scale shell-model calculations in the pf shell. Also the results of cranked mean-field calculations indicate that this band is built upon a 4p-4h excitation within the pf shell (see Fig. 5, 4^04^0 band). The second band, however, is expected to involve particles in the $1g_{9/2}$ orbit, which is supported by its nearly identical behavior to the band in ^{58}Cu. However, the best scenario for this band is based on *one* proton promoted to the $1g_{9/2}$ orbit (4^04^1 band in Fig. 5), while a neutron *and* a proton occupies this orbit in ^{58}Cu. The fact that our best theoretical scenario for deformed identical

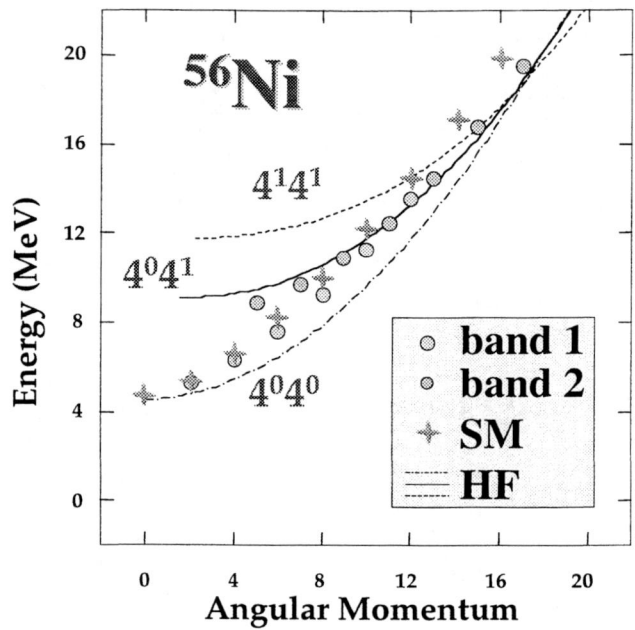

FIGURE 5. Excitation energy versus angular momentum for experimental and calculated intruder bands in ^{56}Ni. Lines indicate the 4^04^0, 4^04^1, and 4^14^1 bands calculated in the cranked HF+SLy4 model, while the KBF shell-model calculations (SM) are indicated by crosses.

bands in ^{56}Ni and ^{58}Cu involves structures with *different* intruder content is very puzzling and requires further investigations.

CONCLUSIONS

Studies of properties of nuclei far from the beta stability line, both the ground-state characteristics (masses, radii, deformations...) and the properties of excited states of exotic nuclei, are crucial for our understanding of the effective nucleon-nucleon interaction and the behavior of the nuclear many-body system [43]. In this context, the study of matter with radioactive beams of nuclei is one of the most exciting challenges of nuclear physics and nuclear astrophysics today. The field is extremely rich and has a truly multidisciplinary character. Experiments with radioactive beams will make it possible to look closely into many exciting aspects of the nuclear many-body problem. Although an experimental excursion into new territories of the chart of the nuclides will offer many excellent opportunities for traditional nuclear structure, there are many unique features of exotic nuclei (associated with weak binding, large diffuseness, and large spatial dimensions) that give very good prospects for entirely new phenomena likely to be different from anything we have observed to date. A broad international community is enthusiastically using existing RNB facilities and hoping and planning for future-generation tools.

REFERENCES

1. J. Dobaczewski and W. Nazarewicz, Phil. Transactions, in press, 1998; nucl-th/9707049.
2. J. Dobaczewski, H. Flocard, and J. Treiner, Nucl. Phys. **A422**, 103 (1984).
3. J. Dobaczewski, W. Nazarewicz, T.R. Werner, J.-F. Berger, C.R. Chinn, and J. Dechargé, Phys. Rev. **C53**, 2809 (1996).
4. U. Fano, Phys. Rev. **124**, 1866 (1961).
5. C. Mahaux and H. Weidenmüller, *Shell Model Approaches to Nuclear Reactions* (North-Holland, Amsterdam, 1969).
6. R.J. Philpott, Fizika **9**, suppl. 3, 21 (1977).
7. K. Bennaceur, F. Nowacki, J. Okołowicz, and M. Płoszajczak, J. Phys. **G24**, 1631 (1998).
8. M. Bolsterli, E.O. Fiset, J.R. Nix, and J.L. Norton, Phys. Rev. **C5**, 1050 (1972).
9. W. Nazarewicz, T.R. Werner, and J. Dobaczewski, Phys. Rev. **C50**, 2860 (1994).
10. J.R. Bennett, J. Engel, and S. Pittel, Phys. Lett. **B368**, 7 (1996).
11. F. Ghielmetti, G. Colo, E. Vigezzi, P.F. Bortignon, and R.A. Broglia, Phys. Rev. **C54**, R2143 (1996).
12. W. Glöckle, J. Hufner, and H.A. Weidenmüller, Nucl. Phys. **A90**, 481 (1967).
13. J.S. Vaagen, B.S. Nilsson, J. Bang, and R.M. Ibarra, Nucl. Phys. **A319**, 143 (1979).
14. G. Rawitscher, Phys. Rev. **C25**, 2196 (1982).
15. M. Buballa, S. Dróżdż, S. Krewald, and J. Speth, Ann. Phys. **208**, 346 (1991).
16. J. Humblet and L. Rosenfeld, Nucl. Phys. **26**, 529 (1961).
17. T. Berggren, Nucl. Phys. **A109**, 265 (1968).
18. T. Vertse, P. Curutchet, and R.J. Liotta, Lecture Notes in Physics **325** (Springer Verlag, Berlin 1987), p. 179.
19. T. Vertse, R.J. Liotta, W. Nazarewicz, N. Sandulescu, and A.T. Kruppa, Phys. Rev. **C57**, 3089 (1998).
20. N. Sandulescu, O. Civitarese, R.J. Liotta, and T. Vertse, Phys. Rev. **C55**, 1250 (1997).

21. S.T. Belyaev, A.V. Smirnov, S.V. Tolokonnikov, and S.A. Fayans, Sov. J. Nucl. Phys. **45**, 783 (1987).
22. W. Poschl, D. Vretenar, and P. Ring, Comp. Phys. Commun. **103**, 217 (1997).
23. Z. Patyk, A. Baran, J.F. Berger, J. Dechargé, J. Dobaczewski, R. Smolańczuk, and A. Sobiczewski, Acta Phys. Pol. **B27**, 457 (1996).
24. J. Dechargé and D. Gogny, Phys. Rev. **C21**, 1568 (1980).
25. E. Chabanat, *Interactions effectives pour des conditions extrêmes d'isospin*, Université Claude Bernard Lyon-1, Thesis 1995, LYCEN T 9501, unpublished.
26. S.A. Fayans, private communication; ENAM98, Abstracts, PB19, PB21.
27. G.A. Lalazissis, J. König, and P. Ring, Phys. Rev. **C55**, 540 (1997).
28. W. Heisenberg, Z. Phys. **78**, 156 (1932).
29. A. Bohr, B.R. Mottelson, and D. Pines, Phys. Rev. **110**, 936, (1958).
30. A. Bohr and B.R. Mottelson, Nuclear Structure, vol. 1 (W.A. Benjamin, New York, 1969).
31. C.T. Black, D.C. Ralph, and M. Tinkham, Phys. Rev. Lett. **76**, 688 (1996).
32. R. Rossignoli, N. Canosa, and P. Ring, Phys. Rev. Lett. **80**, 1853 (1998).
33. A. Bohr and B.R. Mottelson, Nuclear Structure, vol. 2 (W.A. Benjamin, New York, 1975).
34. H.J. Lipkin, APS News, January 1998.
35. K. Clemenger, Phys. Rev. **B32**, 1359 (1985).
36. M. Manninen, J. Mansikka-aho, H. Nishioka, and Y. Takahashi, Z. Phys. **D31**, 259 (1994).
37. W. Satuła, J. Dobaczewski, and W. Nazarewicz, Phys. Rev. Lett. (1998).
38. P.J. Woods and C.N. Davids, Ann. Rev. Nucl. Part. Sci. **47**, 541 (1997).
39. E. Maglione, L.S. Ferreira, and R.J. Liotta, Phys. Rev. Lett. **81**, 538 (1998).
40. A. Kruppa *et al.*, in preparation.
41. D. Rudolph, C. Baktash, J. Dobaczewski, W. Nazarewicz, W. Satuła, M.J. Brinkman, M. Devlin, H.-Q. Jin, D.R. LaFosse, L.L. Riedinger, D.G. Sarantites, and C.-H. Yu, Phys. Rev. Lett. **80**, 3018 (1998).
42. D. Rudolph, C. Baktash, M.J. Brinkman, E. Caurier, D.J. Dean, M. Devlin, J. Dobaczewski, P.-H. Heenen, H.-Q. Jin, D.R. LaFosse, W. Nazarewicz, F. Nowacki, A. Poves, L.L. Riedinger, D.G. Sarantites, W. Satuła, and C.-H. Yu, submitted to Phys. Rev. Lett.
43. *Scientific Opportunities With an Advanced ISOL Facility*, Report, November 1997.

Monte Carlo shell model calculations for exotic nuclei

Takaharu Otsuka[1,2], Takahiro Mizusaki[1] and Yutaka Utsuno[1]

[1] *Department of Physics, University of Tokyo, Hongo, Tokyo 113-0033, Japan*
[2] *RIKEN, 2-1 Hirosawa, Wako, Saitama 351-0198, Japan*

Michio Honma

Center for Mathematical Sciences, University of Aizu, Tsuruga, Ikki-machi Aizu-Wakamatsu, Fukushima 965, Japan

Abstract. The formulation and recent applications of Quantum Monte Carlo diagonalization method (QMCD) are reported. The QMCD has been proposed for solving the quantum many-body interacting systems. The level structure of low-lying states can be studied with realistic interactions, providing a useful tool for studying physics of unstable nuclei. We report that the doubly closed shell probability of a proton-rich unstable nucleus ^{56}Ni is shown to be only 49 % in a full pf shell calculation, in contrast to the corresponding probability of ^{48}Ca which reaches 86 %. The level scheme and E2 transition probabilities of neutron-rich nuclei around ^{32}Mg are discussed.

I INTRODUCTION

The nuclear shell model has been successful in the description of various aspects of nuclear structure, partly because it is based on a minimum number of natural assumptions. Although the direct diagonalization of the Hamiltonian matrix in the full valence-nucleon Hilbert space is desired, the dimension of such a space is too large in many cases, preventing us from performing the full calculations. In order to solve this problem, the Shell Model Monte Carlo (SMMC) method has been proposed successfully [1]. However, the SMMC is basically restricted to ground-state and thermal properties, and is not a proper tool for studying excited states, i.e., level scheme and E2 properties. Moreover, the SMMC suffers from the so-called minus sign problem for realistic interactions. On the other side, the Quantum Monte Carlo Diagonalization (QMCD) method has been proposed [2-5]. The QMCD can describe not only the ground state but also excited states, including their energies and transition matrix elements. The sign problem does not exist in the QMCD.

II FORMULATION OF QMCD

We first sketch the QMCD method briefly [2–5]. In the QMCD method, we create many-body basis states in a stochastic way, and then diagonalize, within the Hilbert space spanned by these bases, the shell model Hamiltonian, H, consisting of single particle energies and a two-body interaction. First, the following states are introduced [2]:

$$|\Phi(\sigma)\rangle \propto \prod_{n=1}^{N_t} e^{-\Delta\beta h(\vec{\sigma}_n)} |\Psi^{(0)}\rangle, \qquad (1)$$

where $h(\vec{\sigma})$ is a one-body operator obtained from H, σ means a set of vectors of random numbers $\{\vec{\sigma}_1, \cdots, \vec{\sigma}_{N_t}\}$, N_t and $\Delta\beta$ are parameters, and $|\Psi^{(0)}\rangle$ is an initial state. We examine the states generated by eq. (1), and retain only those that lower sufficiently the energy of the eigenstate of being obtained. The states thus generated and selected are called QMCD bases. Their number is called the QMCD basis dimension. It is convenient to adopt the QMCD bases in the form of Slater determinants. As one of the advantages of using the Slater determinants, we mention that the form of a Slater determinant is kept in eq.(1) because $h(\vec{\sigma})$ is a one-body operator.

The developments of the formulation of the QMCD method are divided into three phases. In its first version, i.e., phase I, the QMCD bases are generated according directly to eq.(1) [2]. The phase I has been shown to be good for simple systems [2]. It was realized, however, that the phase I is not efficient enough for handling realistic shell model systems [4], even after implementing the projection onto a good quantum number of the z-component of angular momentum, M (i.e., M-projection) [3].

We then introduced several improvements, moving to phase II [4], so as to enhance the efficiency of the QMCD calculations. Since the number of manageable bases is finite in practice, we should first select bases of higher importance. We then rewrite $h(\vec{\sigma})$ so that the sampling of the bases is made around a mean-field solution, for instance, a Hartree-Fock (HF) local minimum.

The other improvement is made on the restoration of symmetries [3,4]. The nucleus is an isolated system and conserves several symmetries. Their restoration is practically impossible in stochastic processes except for extremely simple cases as treated in [2], and one has to enforce the restoration. The rotational symmetry is restored by the J-drive and the M-projection [3]. The J-drive means generation of the QMCD bases with the same intrinsic structure but different orientations. The isospin can be restored too.

The phase II means combination of all the above improvements, and enabled us to perform various full one-major-shell calculations with realistic effective interactions [4]. Although decent solutions have then been obtained for most cases, it turned out that the calculation cannot be achieved with tractable QMCD dimensions in some cases. We therefore improve the method, resulting in phase III [5].

If one generates basis states by a stochastic method, each basis should contain unnecessary fluctuations which are nothing but noise components. This deficiency is inherent to the stochastic process. We therefore revise the basis-generation method so that each basis contains more relevant components lowering the energy and less irrelevant components to be cancelled by other bases. The Hilbert space used for the Hamiltonian diagonalization can then be much compressed. In fact, this *basis compression* enables us to carry out some QMCD calculations which are otherwise practically infeasible. This compression process is one of the characteristic differences of the QMCD method from other quantum Monte Carlo approaches: In the latter, a much larger number of states in the form of eq. (1) are taken so as to evaluate the effects of their proper superposition, whereas one can vary them and select good ones in phase III of the QMCD.

We now move on to the other major improvement, i.e., the restoration of the angular momentum, $\boldsymbol{J}$. In phase III, all QMCD basis vectors are projected onto good J and M when their matrix elements are calculated. The projection is carried out by rotating about the three axes of Euler angles. The K-mixing amplitudes are evaluated, for instance, so as to minimize the energy when the basis is added. Thus, the uncertainty about angular momentum is removed completely in practice.

In phase III calculations, the energy monitored in the basis generation is evaluated after projecting onto good J and M or a good M. For more details, please refer to [5]. We emphasize that, at each QMCD dimension, once the last basis is fixed by the above process, the Hamiltonian matrix elements are projected onto good J and M, and are diagonalized. The isospin is treated exactly by utilizing the method in [1].

By combining all the above improvements, phase III has been constructed, and has been proved to be quite efficient in solving large-scale shell model problems [5].

III TEST OF QMCD IN *PHASE* III FOR ^{48}CR

The validity of phase III has been confirmed by comparing to the result of the exact diagonalization of the same Hamiltonian. Here, the nucleus ^{48}Cr is taken, and the exact result is obtained from Ref. [6] where an excellent agreement with recent experimental data [7] is presented. Figure 1 shows energies of yrast states of ^{48}Cr.

As shown in the top part of Fig. 1, the ground-state energy has been reproduced within 130 keV with the QMCD dimension 40. This result is already ~100 keV below the lower edge of the error bar of the SMMC result with the temperature $T=0.5$ MeV [8]. One finds in the bottom part of Fig. 1 that the back-bending pattern can be well reproduced by this QMCD calculation.

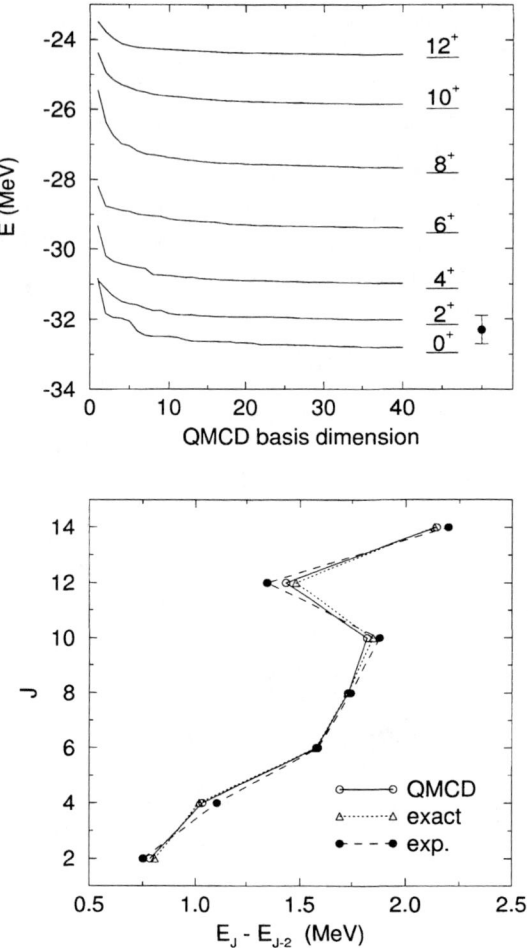

FIGURE 1. Energies of yrast states of ^{48}Cr obtained by QMCD calculation with KB3 Hamiltonian compared with the energies obtained by the exact diagonalization [6]. In the top figure, the energy eigenvalues are shown as functions of the QMCD dimensions. Since the angular momentum projection is made for each basis, the addition of a new basis (i.e., increase of the dimension) implies inclusion of more dynamical degrees of freedom. The point with error bar at far right is the ground state energy of the SMMC calculation with finite temperature T=0.5 MeV [8]. In the bottom figure, a back-bending plot is made for the QMCD (open circle) and exact (triangle) results in comparison with experimental data (closed circle) [7].

IV APPLICATION OF QMCD IN *PHASE* III TO ^{56}NI

We apply the QMCD shell model in phase III to an $N=Z$ unstable nucleus, ^{56}Ni, where the $N=Z=28$ closed-shell structure has been expected due to the spin-orbit splitting [9]. Since this closed shell can be destroyed by mixing within the same major shell, the calculation with the full pf-shell configurations is crucial. Such calculations, however, have been limited to lighter pf-shell nuclei. The QMCD calculation presented below [5] is the first full pf-shell calculation for ^{56}Ni.

The single-particle energies and two-body interaction are those called FPD6 [11] and KB3 [10]. The FPD6 is an empirical two-body interaction given in the form of analytic functions with parameters adjusted for $A=41\sim49$ nuclei [11]. The KB3 is based upon the G-matrix in [12] with empirical improvement for the monopole interaction [10]. In both cases, the single-particle energies are obtained from experimental levels of nuclei around ^{40}Ca. We stress that both the FPD6 and KB3 interactions have been designed for full pf-shell calculations, and should be used with the full pf-shell configurations.

Because of better agreement to experimental data as shown later, we discuss calculations with the FPD6 unless otherwise stated, while comparing to experiment.

Figure 2 shows energy levels of ^{56}Ni. One sees a good agreement between calculated levels and experiment [13]. Recently $B(E2; 0_1^+ \to 2_1^+)=600\pm120$ e^2fm^4 has been measured, which is rather large [14]. The FPD6 effective charges [11] produce a somewhat too large value, and the isoscalar charge is re-adjusted by multiplying by a factor of 0.9, resulting in $e_p=1.23e$ and $e_n=0.54e$. One then obtains $B(E2; 0_1^+ \to 2_1^+)=610$ e^2fm^4. The $B(E2)$'s are calculated also for the high spin states with these charges, as shown in Fig. 2. Figure 2 also includes results obtained with KB3. Note that the $M=0$ Hilbert space has a dimension of about 1.1 billion for ^{56}Ni in the full pf shell, precluding a conventional shell model calculation.

One of the salient advantages of the QMCD method over other quantum Monte Carlo approaches including the SMMC method is its capability of direct analysis of the wave function. This is particularly important in the present case in clarifying the $N=Z=28$ closed shell structure: We compute the probability of the $N=Z=28$ closed shell component in the wave function of ^{56}Ni ground state. The result is only 49 %. This is rather small compared to what would be expected for a closed shell nucleus. The occupation probability of $f_{7/2}$ is 0.91 for the ground state. This means that, in the non-doubly-magic part of the wave function, about 3 nucleons are excited from $f_{7/2}$ on the average. Thus, if a truncated shell-model calculation were attempted, 6p-6h excitations from $f_{7/2}$ should be included at least.

We now discuss the structure of ^{48}Ca for comparison [11]. The wave function of the ^{48}Ca ground state contains the $N=28$ and $Z=20$ closed shell component with 86 % probability. This is much larger than the corresponding value for ^{56}Ni. Thus, a sizable breaking of the $N=Z=28$ doubly magic is seen in ^{56}Ni, especially compared to ^{48}Ca. If the $N=Z=28$ shell of ^{56}Ni were broken by the same mechanism as the $N=28$ shell of ^{48}Ca, the closed-shell probability of ^{56}Ni would be given by the square of the corresponding value of ^{48}Ca: $(0.86)^2=0.74$. Clearly, the actual value,

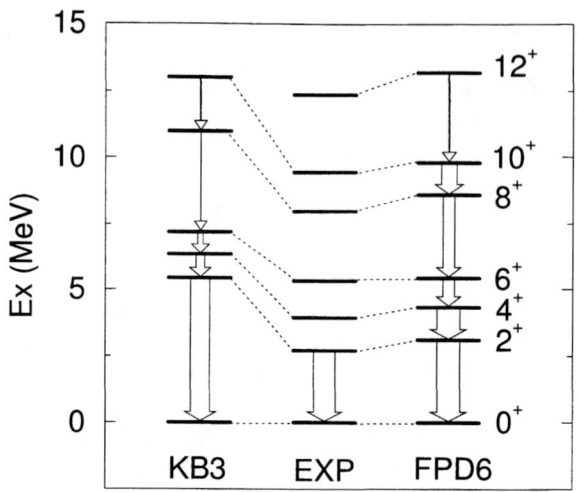

FIGURE 2. Experimental (EXP) yrast levels of ^{56}Ni compared with QMCD results with FPD6 and KB3 Hamiltonians. The $B(E2;(L+2)_1^+ \to L_1^+)$ value is indicated by the width of the arrow, which is so that the experimental $B(E2;2_1^+ \to 0_1^+)$ value takes its mean value, i.e., 120 e^{2}fm^4 [14].

0.49, is much smaller. This is because the $N=Z=28$ shell of ^{56}Ni is broken largely due to interactions between a valence proton and a valence neutron, particularly terms with a quadrupole nature. This seems to be a consequence of strong proton-neutron correlations characterizing $N=Z$ nuclei. On the other hand, the pairing-type interaction should be the major cause of breaking the $N=28$ shell in ^{48}Ca.

V APPLICATION TO THE STRUCTURE OF UNSTABLE NUCLEI AROUND ^{32}MG

The QMCD shell model has been applied to the structure study of extremely neutron-rich unstable nuclei around ^{32}Mg. Since the major issue is the breaking the $N=20$ closed shell [15,16], one has to include both the sd shell and the pf shell. In such calculations, the spurious center-of-mass motion has to be removed. We developed a method suitable for the QMCD calculations, and reduced the spurious component to practically zero. Figure 3 shows recent QMCD result for excitation energies of the first 2^+ and 4^+ states of even-A Mg isotopes, exhibiting a nice agreement to experiment. Figure 4 shows a similar result for $B(E2)$ values. This type of systematic calculations are very important and crucial, but were not feasible even in the largest truncated shell model calculation [17], because of exploding dimension. Note that no particle-hole truncation is imposed in QMCD calculations.

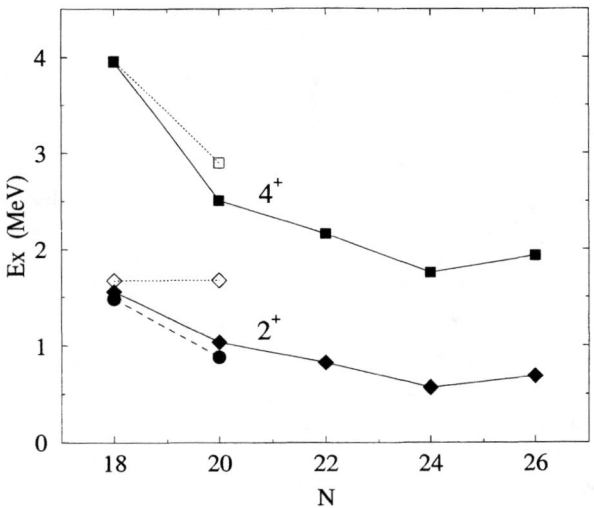

FIGURE 3. Calculated energy levels of the first 2^+ (closed diamond) and 4^+ (closed square) states of even-A Mg isotopes as functions of the neutron number, N. Experimental 2^+ levels are also shown by closed circles. Open symbols indicate results of the calculation within the sd shell.

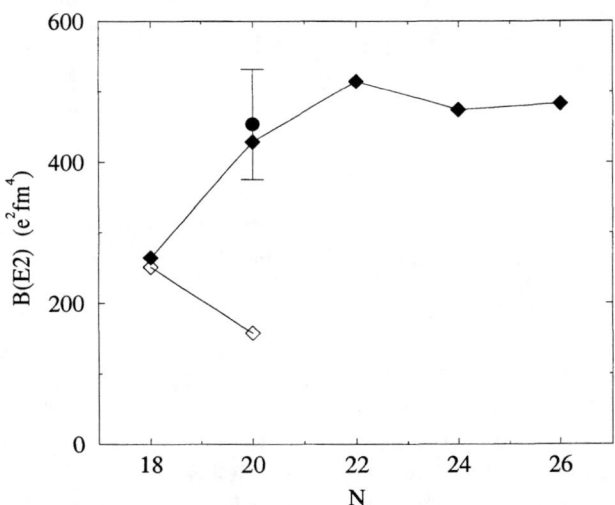

FIGURE 4. $B(E2; 0_1^+ \to 2_1^+)$ values Mg isotopes. QMCD results are shown by closed diamonds, while open diamonds indicate the results obtained within the sd shell. The experimental value is shown by closed circle [16].

VI SUMMARY

In summary, we have presented the basic points of the QMCD formulation, up to its latest revision, phase III characterized by the compression of the basis space and the precise treatment of the angular momentum. Thus, in the QMCD calculation, favorable bases are generated based upon their contribution to the energy eigenvalue of interest, and quite naturally some of such bases or their seeds can be taken from mean-field solutions. The lowest levels of ^{56}Ni are then well described with the FPD6 interaction. It has been shown that the doubly closed shell structure is substantially broken in ^{56}Ni, in contrast to ^{48}Ca. The irregular level structure of higher-spin yrast states of ^{56}Ni is also reproduced, thus ensuring the validity of the present conclusion. In view of this study, the doubly closed shell structure of ^{100}Sn can be questioned and is becoming a more intriguing issue. The structure of neutron-rich unstable nuclei around ^{32}Mg is discussed. A large set of the QMCD results for unstable O, Ne, Mg and Si isotopes has been obtained, and is being prepared for publication with many stimulating features.

We thank Professors B.A. Brown and A. Poves for providing FPD6 and KB3 two-body matrix elements, respectively. This work was supported in part by Grant-in-Aid for Scientific Research (B) (No. 08454058) from the Ministry of Education, Science and Culture.

REFERENCES

1. S. E. Koonin, D. J. Dean, and K. Langanke, Phys. Repts. **278**, 1 (1997); references therein.
2. M. Honma, T. Mizusaki, T. Otsuka (1995): Phys. Rev. Lett. **75**, 1284
3. T. Mizusaki, M. Honma, T. Otsuka (1996): Phys. Rev. **C53**, 2786
4. M. Honma, T. Mizusaki and T. Otsuka (1996): Phys. Rev. Lett. **77**, 3315
5. T. Otsuka, M. Honma and T. Mizusaki (1998): Phys. Rev. Lett. **81**, 1588
6. E. Caurier, *et al.*, (1994): Phys. Rev. **C50**, 225
7. S.M. Lenzi, *et al.*, (1996): Z. Phys. **A354**, 117
8. K. Langanke, *et al.*, (1995): Phys. Rev. **C52**, 718
9. For instance, A. Bohr and B. R. Mottelson, *Nuclear Structure* Vol.1, (Benjamin, New York, 1969).
10. A. Poves, A. Zuker (1981): Phys. Rep. **70**, 235
11. W.A. Richter, *et al.*, (1991): Nucl. Phys. **A523**, 325
12. T. T. S. Kuo and G. E. Brown, Nucl. Phys. **A114**, 241 (1968).
13. R. B. Firestone, *et al.*, (ed.), Table of Isotopes, (Wiley, New York, 1996).
14. G. Kraus, *et al.*, Phys. Rev. Lett. **73**, 1773 (1994).
15. D. Guillemaud-Mueller, *et al.*, (1984): Nucl. Phys. **A426**, 37
16. T. Motobayashi, *et al.*, (1995): Phys. Lett. **B346**, 9
17. N. Fukunishi, T. Otsuka and T. Sebe, (1992): Phys. Lett. **B296**, 279

Quantum Monte Carlo Calculations of Light Nuclei

Steven C. Pieper

Physics Division, Argonne National Laboratory, Argonne IL 60439

Abstract.
Quantum Monte Carlo calculations using realistic two- and three-nucleon interactions are presented for nuclei with up to eight nucleons. We have computed the ground and a few excited states of all such nuclei with Greens function Monte Carlo (GFMC) and all of the experimentally known excited states using variational Monte Carlo (VMC). The GFMC calculations show that for a given Hamiltonian, the VMC calculations of excitation spectra are reliable, but the VMC ground-state energies are significantly above the exact values. We find that the Hamiltonian we are using (which was developed based on ^{3}H, ^{4}He, and nuclear matter calculations) underpredicts the binding energy of p-shell nuclei. However our results for excitation spectra are very good and one can see both shell-model and collective spectra resulting from fundamental many-nucleon calculations. Possible improvements in the three-nucleon potential are also be discussed.

INTRODUCTION

A major goal in nuclear physics is to understand how nuclear binding, stability, and structure arise from the underlying interactions between individual nucleons. To achieve this goal, we must both determine the Hamiltonian to be used, and devise reliable methods for many-body calculations with it. In principle quantum chromodynamics can prescribe the nuclear Hamiltonian, but it will be a long time before this will be done with useful accuracy. Thus the nuclear Hamiltonian is determined phenomenologically, and our knowledge of it is refined, in part, by the many-body calculations we make with it. A large amount of empirical information about the nucleon-nucleon scattering problem has been accumulated over time, resulting in ever more sophisticated NN potential models. These models have strong spin and isospin dependence, and spin and orbital angular momentum are mixed by a strong tensor interaction. In addition the three-nucleon interaction must be considered in realistic calculations, however there is very little experimental knowledge of it.

Exact (defined here to mean an error of less than 1% in the computed binding energy) many-body calculations with such a Hamiltonian are very complicated, and it is only in the last three years that results have been obtained for A≥6; these are briefly discussed in this contribution. The nuclear Hamiltonian is presented in the next section, variational Monte Carlo (VMC) for nuclei is presented in Sec. 3 and Green's function Monte Carlo (GFMC) in Sec. 4. Finally a very few recent results are given in the last section. The specific methods and results presented here are from the work of the Argonne, Los Alamos, and Urbana groups; a complete description of our VMC and GFMC calculations may be found in [1].

I THE NUCLEAR HAMILTONIAN

Our Hamiltonian includes a nonrelativistic one-body kinetic energy, the Argonne v_{18} two-nucleon potential [2] and the Urbana IX three-nucleon potential [3],

$$H = \sum_i (-\frac{\hbar^2}{2m}\nabla_i^2) + \sum_{i<j} v_{ij} + \sum_{i<j<k} V_{ijk} \ . \tag{1}$$

The difference between proton and neutron masses is included in our calculations, but ignored above. The Argonne v_{18} potential is one of a number of new, highly accurate NN potential models developed since 1990. It can be written as a sum of electromagnetic and one-pion-exchange terms and a shorter-range phenomenological part,

$$v_{ij} = v_{ij}^\gamma + v_{ij}^\pi + v_{ij}^R \ . \tag{2}$$

The electromagnetic terms include one- and two-photon-exchange Coulomb interactions, vacuum polarization, Darwin-Foldy, and magnetic moment terms, with appropriate proton and neutron form factors.

The one-pion-exchange part contains the normal Yukawa and tensor functions with a short-range cutoff. This and the remaining phenomenological part of the potential can be written as a sum of eighteen operators, which is where the name v_{18} comes from:

$$v_{ij}^\pi + v_{ij}^R = \sum_{p=1,18} v_p(r_{ij}) O_{ij}^p \ . \tag{3}$$

The first fourteen are charge-independent,

$$\begin{aligned} O_{ij}^{p=1,14} &= [\ 1, (\sigma_i \cdot \sigma_j), S_{ij}, (\mathbf{L} \cdot \mathbf{S}), \mathbf{L}^2, \mathbf{L}^2 (\sigma_i \cdot \sigma_j), (\mathbf{L} \cdot \mathbf{S})^2] \\ &\otimes [1, (\tau_i \cdot \tau_j)] \ , \end{aligned} \tag{4}$$

and the last four break charge independence. The $v_p(r)$ are determined by fitting NN scattering data.

The potential was fit directly to the Nijmegen NN scattering data base [4,5], which contains 1787 pp and 2514 np data in the range $0 - 350$ MeV, with a χ^2 per datum of 1.09. It was also fit to the nn scattering length measured in $d(\pi^-,\gamma)nn$ experiments and the deuteron binding energy.

The Urbana series of three-nucleon potentials is written as a sum of two-pion-exchange and shorter-range phenomenological terms,

$$V_{ijk} = V_{ijk}^{2\pi} + V_{ijk}^{R} \, . \tag{5}$$

The two-pion-exchange term is that of the original Fujita-Miyazawa model [6] and contains both spin (tensor) and isospin dependence. The shorter-range phenomenological term is purely central. The parameters for model IX have been determined by fitting the density of nuclear matter and the binding energy of ^{3}H in conjunction with the Argonne v_{18} interaction. The V_{ijk}^R certainly should have non-central terms [7], and our recent work has been concerned with developing an improved model.

In light nuclei we find

$$\langle V^{NNN} \rangle \sim (.02 - .05) \langle v^{NN} \rangle \sim (.15 - .3) \langle H \rangle \, . \tag{6}$$

where the large fraction of $\langle H \rangle$ is due to the large cancellation of kinetic and potential energy. We expect a similar ratio for the four-body potential:

$$\langle V^{4N} \rangle \sim (.02 - .05) \langle V^{NNN} \rangle \sim .01 \langle H \rangle \, . \tag{7}$$

Calculations are just approaching this accuracy.

II VARIATIONAL MONTE CARLO

The variational method can be used to obtain approximate solutions to the many-body Schrödinger equation, $H\Psi = E\Psi$, for a wide range of nuclear systems, including few-body nuclei, light closed-shell nuclei, and nuclear and neutron matter [8]. A suitably parameterized trial wave function, Ψ_T, is used to calculate an upper bound to the exact ground-state energy,

$$E_T = \frac{\langle \Psi_T | H | \Psi_T \rangle}{\langle \Psi_T | \Psi_T \rangle} \geq E_0 \, . \tag{8}$$

The parameters in Ψ_T are varied to minimize E_T, and the lowest value is taken as the approximate ground-state energy.

Here I discuss a simplified form of Ψ_T; see [1] for the complete form used in current calculations,

$$|\Psi_T\rangle = \mathcal{S} \prod_{i<j}^{A} F_{ij} |\Phi\rangle \, ; \tag{9}$$

The complete Ψ_T also includes three-body correlations induced by V_{ijk}.

The Φ is the one-body part; it determines the J, M_J, T, T_z of the nucleus, and is completely antisymmetric. For A≤4, it does not have any spatial dependence in our formulation, but more than 4 nucleons cannot be fully antisymmetrized in just spin-isospin space. Our Φ for p-shell nuclei is the product of the $\Phi(^4\text{He})$ for four nucleons and p-wave solutions of a one-body well for the remaining nucleons; the result is then antisymmetrized. These p-wave solutions are coupled in a LS representation, and (with some exceptions for $A = 8$ levels) all spatial symmetries allowed in a 1–$\hbar\omega$ p-shell basis are used. We compute both diagonal and off-diagonal expectations in this basis and then make a diagonalization to obtain the best variational wave functions. Such a procedure allows us to obtain variational energies for several levels with the same quantum numbers. We are also investigating Φ that have the correct asymptotic decomposition into clusters (e.g. d+^{4}He for ^{6}Li).

The F_{ij} are the two-body correlations induced by v_{ij} and are written in terms of a subset of the operators in v_{ij}:

$$F_{ij} = \sum_p f_p(r_{ij}) O_{ij}^p. \tag{10}$$

The $f_p(r)$ are the solutions of Euler-Lagrange equations; see [9] for an introduction to these and [10] for details on the complete $f_p(r)$ used in modern calculations.

The number of spin-isospin components in Ψ_T grows rapidly with the number of nucleons. Calculations of the sort being described here are currently feasible up to only $A = 8$. Cluster methods have been used for VMC calculations of larger nuclei [11], and we are now developing these for the GFMC.

III GREEN'S FUNCTION MONTE CARLO

GFMC projects out the lowest-energy ground state using

$$\Psi_0 = \lim_{\tau \to \infty} \exp[-(H - E_0)\tau]\Psi_T. \tag{11}$$

The eigenvalue E_0 is calculated exactly while other expectation values are generally calculated neglecting terms of order $|\Psi_0 - \Psi_T|^2$ and higher. In contrast, the error in the variational energy, E_T, is of order $|\Psi_0 - \Psi_T|^2$, and other expectation values calculated with Ψ_T have errors of order $|\Psi_0 - \Psi_T|$. Here I present a simplified overview of nuclear GFMC; a rather complete discussion may be found in [1].

We use the Ψ_T of Eq. (9) as our initial trial function and define the propagated wave function $\Psi(\tau)$ as

$$\Psi(\tau) = e^{-(H-E_0)\tau}\Psi_T; \tag{12}$$

obviously $\Psi(\tau = 0) = \Psi_T$ and $\Psi(\tau \to \infty) = \Psi_0$. Introducing a small time step, $\Delta\tau$, $\tau = n\Delta\tau$, gives

$$\Psi(\tau) = \left[e^{-(H-E_0)\Delta\tau}\right]^n \Psi_T = G^n \Psi_T. \tag{13}$$

where G is the short-time Green's function. The $\Psi(\tau)$ is represented by a vector function of $\mathbf{R}$, and the Green's function, $G_{\alpha\beta}(\mathbf{R}', \mathbf{R})$ is a matrix function of $\mathbf{R}'$ and $\mathbf{R}$ in spin-isospin space (labeled by the subscripts α, β), defined as

$$G_{\alpha\beta}(\mathbf{R}', \mathbf{R}) = \langle \mathbf{R}', \alpha | e^{-(H-E_0)\Delta\tau} | \mathbf{R}, \beta \rangle. \tag{14}$$

It is calculated with leading errors of order $(\Delta\tau)^3$ as discussed below. Omitting spin-isospin indices for brevity, $\Psi(\mathbf{R}_n, \tau)$ is given by

$$\Psi(\mathbf{R}_n, \tau) = \int G(\mathbf{R}_n, \mathbf{R}_{n-1}) \cdots G(\mathbf{R}_1, \mathbf{R}_0) \Psi_T(\mathbf{R}_0) d\mathbf{R}_{n-1} \cdots d\mathbf{R}_1 d\mathbf{R}_0, \tag{15}$$

and the integration is done by Monte Carlo.

The short-time propagator is approximated as

$$G_{\alpha\beta}(\mathbf{R}', \mathbf{R}) = G_0(\mathbf{R}', \mathbf{R}) \langle \alpha | \left[\mathcal{S} \prod_{i<j} \frac{g_{ij}(\mathbf{r}'_{ij}, \mathbf{r}_{ij})}{g_{0,ij}(\mathbf{r}'_{ij}, \mathbf{r}_{ij})} \right] | \beta \rangle, \tag{16}$$

where g_{ij} is the exact two-body propagator,

$$g_{ij}(\mathbf{r}'_{ij}, \mathbf{r}_{ij}) = \langle \mathbf{r}'_{ij} | e^{-H_{ij}\Delta\tau} | \mathbf{r}_{ij} \rangle, \tag{17}$$

$$H_{ij} = -\frac{\hbar^2}{m} \nabla^2_{ij} + v_{ij}. \tag{18}$$

Here $g_{0,ij}$ is the free two-body propagator,

$$g_{0,ij}(\mathbf{r}'_{ij}, \mathbf{r}_{ij}) = \left[\sqrt{\frac{\mu}{2\pi\hbar^2\Delta\tau}} \right]^3 \exp\left[-\frac{(\mathbf{r}'_{ij} - \mathbf{r}_{ij})^2}{2\hbar^2\Delta\tau/\mu} \right], \tag{19}$$

where $\mu = m/2$ is the reduced mass, and G_0 is the free A-nucleon propagator. All terms containing any number of the same v_{ij} and ∇^2_{ij} are treated exactly in this propagator, as we have included the imaginary-time equivalent of the full two-body scattering amplitude. Eq. (16) has errors of order $(\Delta\tau)^3$, however they are from commutators of terms involving potentials for two different pairs of a triple of nucleons; these commutators become large only when all three nucleons are close. Since this is a rare occurrence, the $\Delta\tau$ of Eq. (17) can be five or more times larger than that used with simpler approximations of G. The free Green's function, $G_0(\mathbf{R}', \mathbf{R})$, is sampled to obtain the Monte Carlo configuration R' from R.

The exact propagator of Eq. (17) can be computed for the full v_{18} potential, however the $\mathbf{L}^2$ and $(\mathbf{L} \cdot \mathbf{S})^2$ terms in the potential correspond to state-dependent changes of the mass appearing in the free Green's function. Since we do not know how to sample such a free Green's function, we cannot use the exact g_{ij} for the full potential, but rather must use one constructed for an approximately equivalent potential that does not contain quadratic $\mathbf{L}$ terms. The difference between the desired and approximate potentials is computed perturbatively; comparisons with

Faddeev solutions for ^{3}H suggest this results in errors of less than 0.5% for the total energy.

For more than four nucleons, GFMC calculations suffer significantly from the well-known Fermion sign problem. The resulting exponential growth of the statistical errors as one propagates to larger τ limits our calculations to $\tau \leq 0.1$ MeV^{-1}, and for most cases we do not go beyond $\tau = 0.06$ MeV^{-1}. This means that any errors in Ψ_T corresponding to excitations of less than ~ 10 MeV will be damped out by less than $1/e$. For this reason the diagonalization described in the previous section is very important; it removes excitations of only a few MeV. In practice we find that the dependence of the GFMC energy on τ corresponds to the removal of small admixtures of states with very high energies (~ 0.1 to 2 GeV).

IV SOME RESULTS

Only a few of our results can be presented here; many other results may be found in [1,12]. Figure 1 compares many of our computed GFMC and VMC energies with experiment. Only statistical errors are shown for the Monte Carlo calculations; we believe that the GFMC calculations are converged to ~ 0.3 MeV for $A = 6$ and ~ 0.6 MeV for $A = 8$. The GFMC ground state results show that the Hamiltonian being used underbinds the p-shell nuclei more and more as A increases. Also the underbinding becomes worse as one moves away from $Z = N$, indicating an

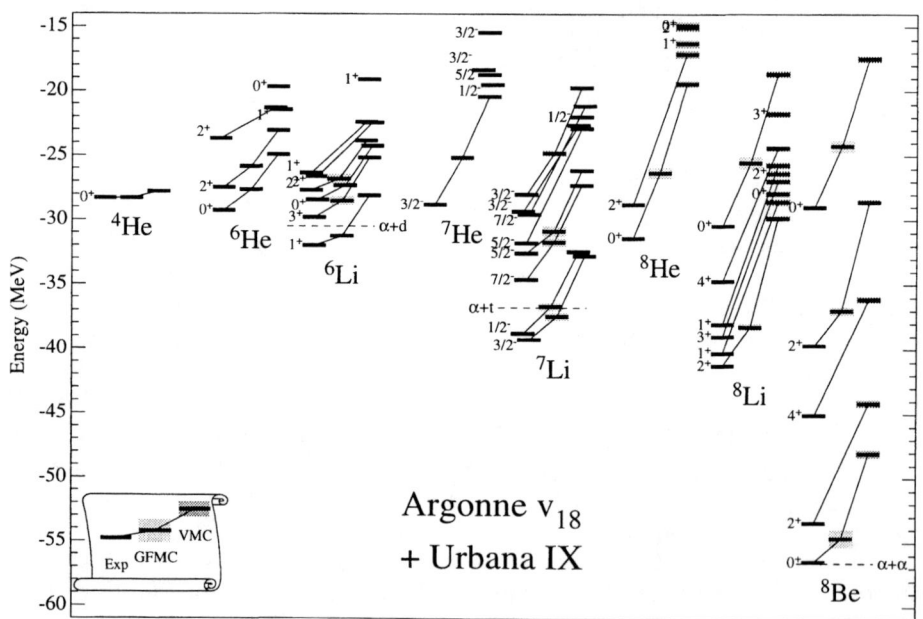

FIGURE 1. Experimental and computed levels of light nuclei.

isospin problem. A comparison of the VMC and GFMC results shows that our variational wave functions are much poorer for the p-shell than for the 3- and 4-body nuclei. However, wherever we have done GFMC calculations for excited states we find that the VMC excitation energies (relative to the ground-state energy) are reliable. Thus the failure of the variational wave function appears to be a bulk property with little state dependence. The computed excitation energies are in quite good agreement with experiment, although spin-orbit splitting appears to be systematically underpredicted. Thus the failure of the Hamiltonian is also principally a bulk and isospin feature.

We are presently attempting to improve our Hamiltonian by adding a few additional terms to the NNN potential. We are examining terms that are suggested by meson-exchange and related arguments. Examples of contributions that we are considering include several Z-diagrams with π-, σ- and ω-meson exchanges. The σ,ω diagram provides a three-nucleon spin-orbit force that improves the details of the predicted excitation spectra. We are also computing the dominant terms of a three-pion, three-nucleon force; this is relatively more attractive in $T = \frac{3}{2}$ than in $T = \frac{1}{2}$ triples; thus it improves the bulk and isospin energies.

There has been some interest in this conference in the reliability of models that treat nuclei such as ^{6}He as an inert ^{4}He core and two neutrons. Figure 2 shows our computed proton-proton pair distributions in ^{4,6,7}He. If the ^{4}He core were truly inert, these distributions would all be the same. Instead it is more peaked in ^{4}He than in the p-shell nuclei, showing the effects of either polarization of the ^{4}He core by the valence neutrons or charge-exchange interactions between the protons and valence neutrons; presumably both are significant.

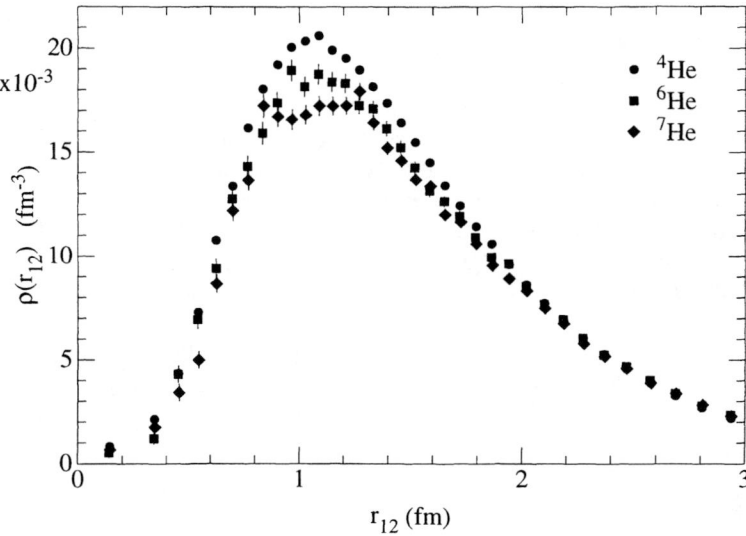

FIGURE 2. Computed pp-pair distributions in ^{4,6,7}He.

The last few years have seen much progress in nuclear QMC calculations. The energy predictions of a given realistic Hamiltonian can be found with 1% accuracy for up to eight nucleons. The resulting wave functions can be used to compute both elastic and inelastic electron-scattering form factors; our recent results [12] for ^{6}Li agree very well with the data without the need for an effective charge. We are also computing low-energy reaction rates for processes such as ^{4}He$(d,\gamma)^6$Li for primordial (big bang) nucleosynthesis studies. We are also improving the Hamiltonian used in these calculations and extending the GFMC method to larger nuclei by developing a cluster-model GFMC.

ACKNOWLEDGMENTS

This work was done in collaboration with J. Carlson, J. L. Forest, B. S. Pudliner, V. R. Pandharipande, R. Schiavilla, and R. B. Wiringa. It was supported by the U. S. Department of Energy, Nuclear Physics Division, under contract No. W-31-109-ENG-38.

REFERENCES

1. B. S. Pudliner, V. R. Pandharipande, J. Carlson, S. C. Pieper, and R. B. Wiringa, Phys. Rev. C **56**, 1720 (1997).
2. R. B. Wiringa, V. G. J. Stoks, and R. Schiavilla, Phys. Rev. C **51**, 38 (1995).
3. B. S. Pudliner, V. R. Pandharipande, J. Carlson, and R. B. Wiringa, Phys. Rev. Lett. **74**, 4396 (1995).
4. J. R. Bergervoet, P. C. van Campen, R. A. M. Klomp, J.-L. de Kok, T. A. Rijken, V. G. J. Stoks, and J. J. de Swart, Phys. Rev. C **41**, 1435 (1990).
5. V. G. J. Stoks, R. A. M. Klomp, M. C. M. Rentmeester, and J. J. de Swart, Phys. Rev. C **48**, 792 (1993).
6. J. Fujita and H. Miyazawa, Prog. Theor. Phys. **17**, 360 (1957).
7. S. A. Coon, M. T. Peña, and D. O. Riska, Phys. Rev. C **52**, 2925 (1995).
8. R. B. Wiringa, Rev. Mod. Phys. **65**, 231 (1993).
9. J. A. Carlson and R. B. Wiringa, in *Computational Nuclear Physics 1*, edited by K. Langanke, J. A. Maruhn, and S. E. Koonin (Springer Verlag, Berlin, 1991).
10. R. B. Wiringa, Phys. Rev. C **43**, 1585 (1991).
11. S. C. Pieper, R. B. Wiringa, and V. R. Pandharipande, Phys. Rev. C **46**, 1741 (1992).
12. R. B. Wiringa and R. Schiavilla, submitted, available at http://xxx.lanl.gov/ps/nucl-th/9807037.

Superdeformation and Smooth Band Termination in $A \sim 60$ Nuclei

C. E. Svensson

*Department of Physics and Astronomy, McMaster University,
Hamilton, Ontario, Canada L8S 4M1*

Abstract. High-spin states in the proton-rich nuclei ^{62}Zn and ^{60}Zn have been studied with Gammasphere and the Microball charged-particle detector array. Two sets of strongly coupled rotational bands in ^{62}Zn have been observed up to the terminating states of their respective configurations and lifetime measurements for these bands confirm the predicted loss of collectivity associated with the smooth termination of rotational bands. Superdeformed (SD) bands have been observed in both ^{62}Zn and ^{60}Zn, establishing a new region of superdeformation for nuclei with particle numbers $N, Z \approx 30$. Linking transitions connecting the doubly-magic SD band in ^{60}Zn to the yrast line have been identified. These linking transitions not only provide the first spin, parity, and excitation energy measurements for SD states in the $A \sim 60$ mass region, but their stretched $E2$ character and relatively large $B(E2)$ values and intensities suggest that the decay-out process for $A \sim 60$ SD bands differs substantially from that observed in heavier systems.

INTRODUCTION

Recent high-spin studies of proton-rich nuclei in the $A \sim 60$ mass region have revealed a rich pattern of shape transformations and noncollective-collective transitions with increasing angular momentum. With a limited number of valence particles outside of doubly-magic ^{56}Ni, the low-spin decay schemes of these nuclei are dominated by spherical shell model states. At intermediate spins, the presence of high-j orbitals both below ($f_{7/2}$) and above ($g_{9/2}$) the spherical shell gaps at $N, Z = 28$ lead to well-deformed ($\beta_2 \sim 0.3$) collective rotational bands in these nuclei built on particle-hole excitations across these shell gaps. However, with only a small number of particles and holes occupying these high-j orbitals, the maximum spin available in these rotational bands in limited to ~ 20–30 $\hbar$. These bands thus terminate in fully-aligned non-collective states at experimentally observable spins. At still higher spins, the large superdeformed (SD) shell gaps in the single-particle orbitals for particle numbers $N, Z = 30$ lead to the prediction [1,2] of highly collective superdeformed bands in this mass region.

This presentation focuses on recent high-spin studies of two proton-rich Zn isotopes. Lifetime measurements for two sets of strongly coupled rotational bands observed in ^{62}Zn [3] have confirmed the predicted [4] loss of collectivity associated with the phenomenon of smooth band termination, superdeformed bands have been identified in both ^{62}Zn [5] and the $N = Z$ nucleus ^{60}Zn [6], and linking transitions connecting the ^{60}Zn SD band to the yrast line suggest that the decay-out process in this nucleus is non-statistical.

EXPERIMENTS

The results presented here were obtained in two experiments with the Gammasphere [7] HPGe multi-detector array and the Microball [8], a 4π charged-particle detector consisting of 95 CsI(Tl) scintillators. In the first experiment, high-spin states in ^{62}Zn and ^{60}Zn were populated via the ^{40}Ca(^{28}Si,$\alpha 2p$)^{62}Zn and ^{40}Ca(^{28}Si,2α)^{60}Zn reactions with a 125-MeV ^{28}Si beam provided by the 88-Inch Cyclotron at Lawrence Berkeley National Laboratory. In this experiment Gammasphere comprised 83 Ge detectors and 2.5×10^9 particle–γ^3 and higher fold coincidence events were recorded. Based on measured γ-ray yields, the $\alpha 2p$ and 2α evaporation channels leading to ^{62}Zn and ^{60}Zn were estimated to represent $\sim 10\%$ and $\sim 0.1\%$ of the total fusion cross section, respectively. In the second experiment, ^{60}Zn was populated (again at the $\sim 0.1\%$ level) via the ^{40}Ca(^{32}S,3α)^{60}Zn reaction with a 134-MeV ^{32}S beam provided by the ATLAS facility at Argonne National Laboratory. Gammasphere comprised 101 Ge detectors and 1.7×10^9 particle–γ^4 and higher fold coincidence events were recorded in this experiment.

In both experiments the Hevimet collimators were removed from the BGO Compton suppression shields of the Gammasphere detectors to provide γ-ray multiplicity and sum-energy measurements for each event [9]. Combined with energy measurements for the charged particles provided by the Microball, the γ-ray sum-energy measurements enabled highly efficient channel selection based on total energy conservation requirements [10]. The very clean channel selection provided by this method was particularly important in isolating the γ rays belonging to the weakly populated $N = Z$ nucleus ^{60}Zn.

SMOOTH BAND TERMINATION IN ^{62}ZN

A partial decay scheme for ^{62}Zn is shown in Fig. 1. In addition to previously observed [11,12] shell-model states, two sets of strongly coupled rotational bands (labeled bands 1 and 2 in Fig. 1) were identified in this nucleus. To assign configurations to these bands, theoretical calculations for ^{62}Zn were carried out employing the configuration-dependent shell-correction approach [4] with the cranked Nilsson potential [13]. Based on the excellent agreement between these calculations and the experimental observations [3], configurations involving one $f_{7/2}$ proton hole, one $g_{9/2}$ proton, and one (two) $g_{9/2}$ neutrons were assigned to bands 1 (2) (the [11,01]

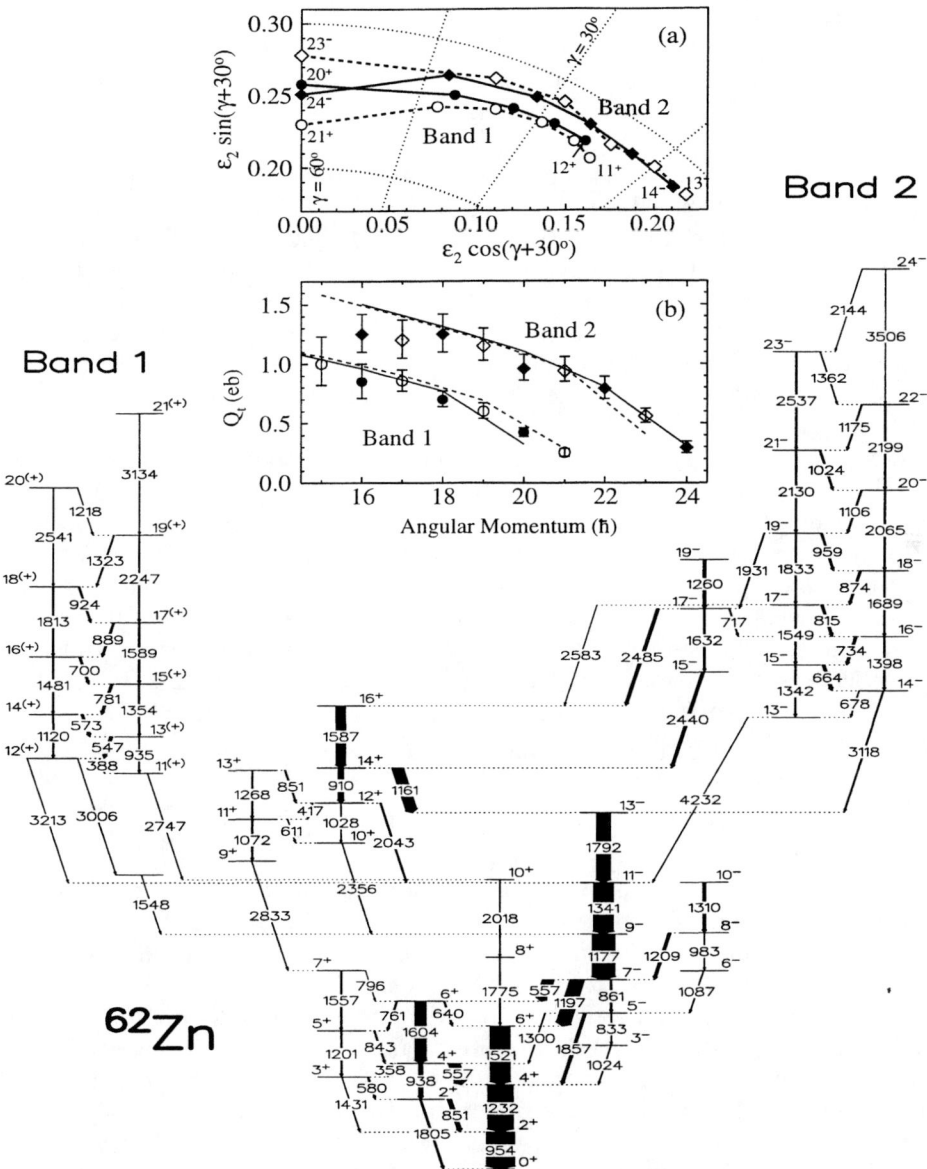

FIGURE 1. Partial decay scheme for ^{62}Zn. Transition energies are given to the nearest keV. The unlinked superdeformed band observed in ^{62}Zn is not shown in this figure. Inset (a) shows calculated shape trajectories in the (ε_2, γ) plane for the smoothly terminating bands, and inset (b) shows Q_t values for these bands from experiment (symbols) and theory (lines). In both insets, circles (diamonds) are used for band 1 (2), and filled (open) symbols and solid (dashed) lines are used for the $\alpha = 0$ ($\alpha = +1$) signatures of these bands.

and [11,02] configurations in shorthand notation[1]). These assignments indicate that bands 1 and 2 have been observed up to their respective terminating states in which all of the angular momentum available from their single-particle configurations has been exhausted. This represents the first observation of the terminating states of rotational bands in the $A \sim 60$ mass region.

Bands 1 and 2 in ^{62}Zn are examples of smoothly terminating bands [14] which are observed over an extended spin range prior to termination. As shown in Fig. 1(a), these bands are calculated to be triaxial ($\gamma \sim 15$–$45°$) at intermediate spins and to change shape gradually, terminating in noncollective oblate ($\gamma = 60°$) states. This predicted transition to an oblate terminating state implies a loss of collectivity [4], and hence transition quadrupole moment Q_t values for these bands which decrease as termination is approached. Although smoothly terminating bands have been identified in many nuclei in the $A \sim 110$ mass region [15], Q_t measurements up to the terminating states of these bands have not been possible. The loss of collectivity inherent in the present theoretical interpretation of smoothly terminating bands had thus not been tested. The relatively low spins at which bands 1 and 2 in ^{62}Zn terminate, however, enabled Q_t values for these bands to be measured all the way to termination. These Q_t values, extracted from thin-target Doppler shift attenuation measurements [3], are shown in Fig. 1(b). They do indeed decrease with increasing angular momentum, in excellent agreement with theoretical calculations for ^{62}Zn (lines in Fig. 1(b)), and reach values corresponding to noncollective transitions strengths of ~ 1 Weisskopf unit (W.u.) ($Q_t \sim 0.2$ eb) at the highest spins. These results thus confirm the predicted loss of collectivity associated with the smooth termination of rotational bands[2].

SUPERDEFORMATION IN ^{62}ZN

In addition to the smoothly terminating normal deformed (ND) bands discussed above, a superdeformed band was observed in ^{62}Zn. The γ-ray spectrum obtained by summing coincidence gates set on the members of this band, which was populated with an intensity of $\sim 1\%$ of the ^{62}Zn channel, is shown in Fig. 2. Although an additional γ ray at 3986 keV was observed in coincidence with the band and is a candidate for a decay-out transition, this band could not be linked conclusively into the remainder of the ^{62}Zn decay scheme. The parity, spins, and excitation energies of the SD states are thus uncertain. However, coincidences between the members of the SD band and high-spin ND transitions in the ^{62}Zn level scheme suggest that the observed SD transitions cover the spin range from approximately

---

[1] A useful shorthand configuration notation in this mass region is to label the high-j particles and holes by $[p1p2, n1n2]$ where $p1$ ($n1$) is the number of proton (neutron) $f_{7/2}$ holes and $p2$ ($n2$) is the number of proton (neutron) $g_{9/2}$ particles. The remaining (unlabeled) particles are distributed over the (mixed) $f_{5/2}$, $p_{3/2}$, and $p_{1/2}$ orbitals.

[2] Recent transition quadrupole moment measurements approaching the terminating states of rotational bands in ^{108}Sn and ^{109}Sb [16] have also confirmed this predicted loss of collectivity.

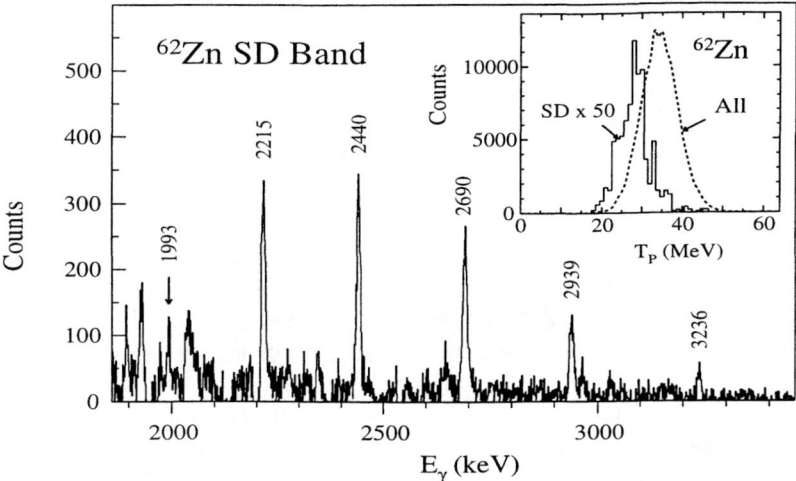

FIGURE 2. γ-ray spectrum of the ^{62}Zn superdeformed band obtained by summing coincidence gates set on all of the band members. The inset shows spectra of total center of mass charged-particle energy T_p in coincidence with triple gates set on members of the SD band (solid line) and low-spin transitions in ^{62}Zn (dashed line).

$18\hbar$ to $30\hbar$. In addition, the excitation energy of the entry region which feeds this band could be accurately determined by measuring the total center of mass energy T_p of the evaporated charged particles in coincidence with γ rays in the SD band. As shown in the inset of Fig. 2, the mean T_p of 28.8 MeV for events feeding the SD band was 6.2 MeV below the average for the $\alpha 2p$ channel. The corresponding mean excitation energy of 32.0 MeV for events feeding the SD band was thus 6.2 MeV higher than the ^{62}Zn average entry excitation energy of 25.8 MeV. These measurements clearly demonstrate that the SD band was populated by only the highest excitation energy, and hence highest spin, components of the ^{62}Zn entry distribution. The large deformation and high collectivity of this band were established by a thin-target Doppler shift attenuation measurement of its transition quadrupole moment. Assuming an axially symmetric shape, the measured Q_t of $2.7^{+0.7}_{-0.5}$ eb corresponds to a quadrupole deformation $\beta_2 = 0.45^{+0.10}_{-0.07}$.

The properties of this band are in excellent agreement [5] with theoretical calculations which indicate that SD bands with deformations $\beta_2 = 0.41$–0.49 become yrast in ^{62}Zn for spins $I \geq 24$ $\hbar$. Although the lack of exact spin, parity, and excitation energy measurements for this band prevented a definite configuration assignment, comparisons with these calculations favor the assignment of this band to a configuration with two $f_{7/2}$ holes in both the proton and neutron subsystems and two proton and three neutron $g_{9/2}$ particles (the [22,23] configuration). However, the [22,22] and [22,24] configurations cannot be ruled out. Despite the uncertainty in the exact configuration of this band, these results do establish a new region of superdeformation for nuclei with proton and neutron numbers $N, Z \approx 30$.

SUPERDEFORMATION IN THE $N = Z$ NUCLEUS ^{60}ZN

The doubly-magic superdeformed band in the $A \sim 60$ mass region, corresponding to filling the single-particle energy levels up to the large SD shell gaps at $N, Z = 30$, occurs in the $N = Z$ nucleus ^{60}Zn. The study of high-spin states in $N = Z$ nuclei in this mass region is, however, hindered by the very small cross sections for populating these nuclei in fusion-evaporation reactions with stable beams and targets. As noted above, ^{60}Zn was populated with only $\sim 0.1\%$ of the total fusion cross section in these experiments. However, theoretical calculations for ^{60}Zn [6], indicate that the doubly-magic SD band in this nucleus becomes yrast at relatively low spin ($I \sim 16\hbar$) and by $I \sim 24\hbar$ is separated from excited states by a large (~ 2 MeV) energy gap. If ^{60}Zn is populated at high-spin a substantial fraction of the channel intensity is thus expected to feed this band and, in fact, this doubly-magic SD band was populated in these experiments with a larger fraction of the channel intensity (60(4)% and 34(3)% in the first and second experiments) than any previously observed SD band.

Figure 3 shows the γ-ray spectrum obtained by summing coincidence gates set on the members of the ^{60}Zn SD band and the proposed decay scheme for ^{60}Zn [6]. In addition to observing the doubly-magic SD band, linking transitions connecting this band to the yrast line were identified. Angular distribution measurements for the 3184 and 3656 keV transitions shown in Fig. 3 indicate that both of these γ rays have stretched $E2$ character. These linking transitions thus establish the spins, parity, and excitation energies of the yrast SD states in ^{60}Zn, the first such measurements in the $A \sim 60$ mass region.

A particularly interesting feature of these results is that the decay out of this SD band differs substantially from that observed in the decay of $A \sim 190$ SD bands. Unlike the $A \sim 190$ region, where the decay out is dominated by the $E1$ transitions expected in a statistical decay process [17], all of the linking transitions in ^{60}Zn are assigned stretched $E2$ character. Furthermore, the $B(E2)$ values for these decay-out transitions, obtained from branching ratios and a measurement of the in-band transition quadrupole moment [6], range from ~ 0.01 to 1 W.u., ie. 2–4 orders of magnitude larger than the *upper limit* of $\sim 10^{-4}$ W.u. set on the decay-out $B(E2)$'s in ^{194}Pb [18]. These "large" decay-out $B(E2)$ values in ^{60}Zn are difficult to reconcile with the weak-mixing statistical model [19] which has proved so successful in describing the decay out of SD bands in the $A \sim 190$ region, and suggest that the decay out in ^{60}Zn proceeds by a non-statistical process (eg. Ref. [20]). Detailed theoretical calculations tailored to the much lower level density in the $A \sim 60$ mass region and, in the case of the $N = Z$ nucleus ^{60}Zn, including possible effects due to isospin suppression [21] of $\Delta T = 0$ dipole transitions are clearly required to obtain a more complete understanding of the decay-out process in this mass region. These first results on the decay out of the doubly-magic SD band in ^{60}Zn do, however, suggest that the decay-out process in these light nuclei may differ considerably from that observed in heavier systems.

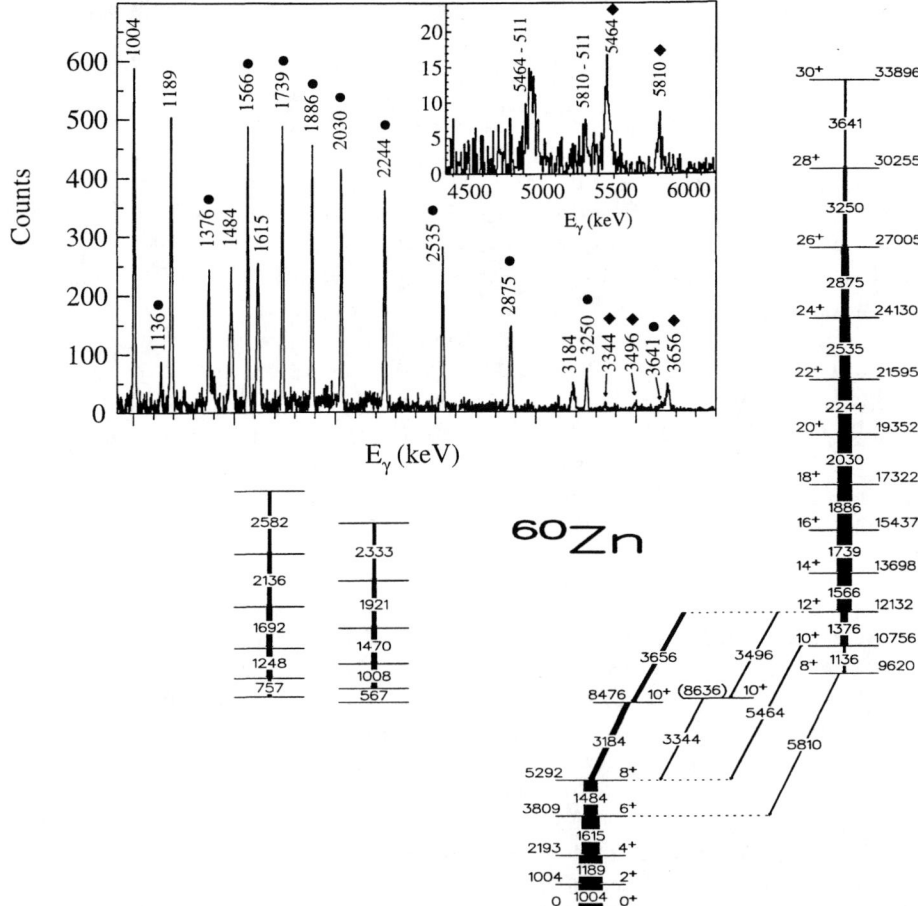

FIGURE 3. Partial decay scheme for ^{60}Zn. The order of the 3496 and 3344 keV linking transitions is uncertain, as are the spins and excitation energies of the unlinked normal deformed bands. The inset shows the γ-ray spectrum in coincidence with a sum of gates on the members of the SD band (circles). Diamonds label the linking transitions connecting the SD band to the yrast line.

SUMMARY

High-spin states in the proton-rich nuclei ^{62}Zn and ^{60}Zn have been studied in two experiments with the Gammasphere array and the Microball charged-particle detector. Some of the results presented here include (i) transition quadrupole moment measurements for terminating rotational bands in ^{62}Zn which confirm the predicted loss of collectivity associated with the phenomenon of smooth band termination, (ii) the observation of superdeformed bands in both ^{62}Zn and ^{60}Zn, and (iii) the identification of linking transitions in ^{60}Zn which establish the spins, parity, and excitation energies of the yrast SD states and suggest that the decay-out process in this nucleus is non-statistical.

ACKNOWLEDGEMENTS

The results presented here have involved the collaboration of a large number of researchers from several institutions. Special thanks for a successful collaboration go to all of my colleagues at Argonne National Laboratory, Lawrence Berkeley National Laboratory, Lund Institute of Technology, McMaster University, Oak Ridge National Laboratory, Staffordshire University, Universität zu Köln, Universität München, University of Tennessee, and Washington University. This work has been partially funded by NSERC (Canada), the DOE under Contracts Nos. DE-AC05-96OR22464, DE-AC03-76SF00098, W-31-109-ENG-38, DE-FG05-88ER40406, and DE-FG05-93ER40770, EPSRC (U.K.), BMBF (Germany) under Contract Nos. 06-OK-668 and 06-LM-868, the Swedish Natural Science Research Council, and the Royal Swedish Academy of Sciences.

REFERENCES

1. I. Ragnarsson, in *Proceedings of the Workshop on the Science of Intense Radioactive Ion Beams*, Los Alamos National Laboratory Report LA-11964-C, p. 199 (1990).
2. R. K. Sheline, P.C. Sood, and I. Ragnarsson, Int. J. Mod. Phys. **A6**, 5057 (1991).
3. C. E. Svensson *et al.*, Phys. Rev. Lett. **80**, 2558 (1998).
4. A. V. Afanasjev and I. Ragnarsson, Nucl. Phys. **A591**, 387 (1995).
5. C. E. Svensson *et al.*, Phys. Rev. Lett. **79**, 1233 (1997).
6. C. E. Svensson *et al.*, (to be published).
7. I.-Y. Lee, Nucl. Phys. **A520**, 641c (1990).
8. D. G. Sarantites *et al.*, Nucl. Instrum. Methods Phys. Res., Sect. A **381**, 418 (1996).
9. M. Devlin *et al.*, Nucl. Instrum. Methods Phys. Res., Sect. A **383**, 506 (1996).
10. C. E. Svensson *et al.*, Nucl. Instrum. Methods Phys. Res., Sect. A **396**, 228 (1997).
11. M. M. King, Nucl. Data Sheets **60**, 337 (1990).
12. K. Furutaka *et al.*, Z. Phys. A **358**, 279 (1997).
13. T. Bengtsson and I. Ragnarsson, Nucl. Phys. **A436**, 14 (1985).
14. I. Ragnarsson *et al.*, Phys. Rev. Lett. **74**, 3935 (1995).
15. V. P. Janzen *et al.*, in *Proceedings of the Conference on Nuclear Structure at the Limits*, Argonne National Laboratory Report ANL/PHY-97/1, p. 171 (1997).
16. R. Wadsworth *et al.*, Phys. Rev. Lett. **80**, 1174 (1998).
17. R. Krücken *et al.*, Phys. Rev. C **54**, 1182 (1996).
18. R. Krücken *et al.*, Phys. Rev. C **55**, R1625 (1997).
19. E. Vigezzi, R. A. Broglia, and T. Døssing, Phys. Lett. B **249**, 163 (1990).
20. P. Bonche *et al.*, Nucl. Phys. **A519**, 509 (1990).
21. E. K. Warburton and J. Weneser, in *Isospin in Nuclear Physics*, North-Holland, Amsterdam (1969), ch. 5.

In-beam γ-ray spectroscopy in the vicinity of ^{100}Sn

Dariusz Seweryniak

Argonne National Laboratory, Argonne, IL 60439 USA
and
University of Maryland, College Park, MD 20742 USA [1]

Abstract.
In recent years, in-beam γ-ray experiments supplied a vast amount of data on high-spin states in nuclei in the vicinity of ^{100}Sn. The present contribution reviews spectroscopic information obtained recently for N≥50 nuclei around ^{100}Sn, with emphasis on isomer studies, and discusses selected results in the frame of the shell model.

Introduction

The heaviest existing doubly-magic self-conjugated nucleus, ^{100}Sn, is situated in the nuclidic chart at the point where the N=Z line crosses the proton drip-line. It is a unique ground for studying single-particle energies and residual interactions far from the line of stability. In addition, neutron-proton correlations are expected to play a significant role in ^{100}Sn and its properties might possibly be affected by low lying continuum states.

In-beam γ-ray spectroscopy has been a very important tool in studies of nuclear structure in the ^{100}Sn region. It provides information on energies, spins and parities, and half-lives of yrast, high-spin states, which is very often complementary to β-decay studies. However, the interpretation of the data is complicated by the fact that so far very little is known about one- and two-quasi-particle neighbors of ^{100}Sn. Multi-particle configurations observed in nuclei located further from ^{100}Sn are used to place constraints on the single-particle energies and residual interactions with respect to the ^{100}Sn core.

The status of experimental studies around ^{100}Sn is shown in Fig. 1. A summary of the results obtained recently using in-beam spectroscopic methods, both on prompt γ-ray transitions and isomers will be presented in the next section and will

---

[1] supported by the U.S. Department of Energy, Nuclear Physics Division, under contracts No. W-31-109-ENG-38 and No. DE-0202-94-ER49834.

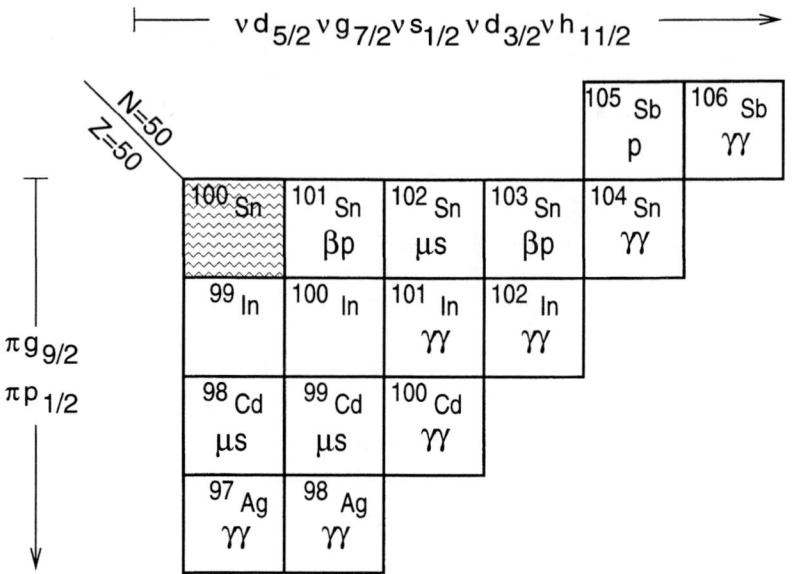

FIGURE 1. The status of experimental studies of N≥50 nuclei situated in the vicinity of ^{100}Sn. Nuclei for which prompt γ rays were observed are marked with "γγ", whereas nuclei where only isomers are known are labeled by "μs". "β p" and "p" denote beta-delayed proton precursors and a ground-state proton emitter, respectively.

be followed by a shell-model discussion of selected nuclei. The last section contains an outlook for in-beam spectroscopy in the ^{100}Sn region.

Experimental techniques and results

Heavy-ion beams of ^{50}Cr and ^{58}Ni combined with neutron-deficient targets are used to access nuclei around ^{100}Sn. There are two experimental difficulties one has to face when trying to approach ^{100}Sn using heavy-ion induced fusion-evaporation reactions. First, mainly protons are evaporated, bringing the system back to the line of stability, and the most interesting channels, which require evaporation of neutrons, are produced with very low cross sections. Second, the total reaction cross section is highly fragmented (typically about 20 reaction channels are open) resulting in a huge prompt γ-ray background associated with deexcitation of strongly populated nuclei. To overcome these obstacles, large, efficient arrays of Ge detectors are used to detect in-beam γ rays. In addition, charged-particle detectors and neutron detectors are employed to count light evaporated particles, i.e. protons, α particles and neutrons, and thus assign observed γ rays to individual reaction channels. A typical experimental set-up is schematically shown in Fig. 2(a). Alternatively, one could use a recoil mass separator for the reaction channel selection,

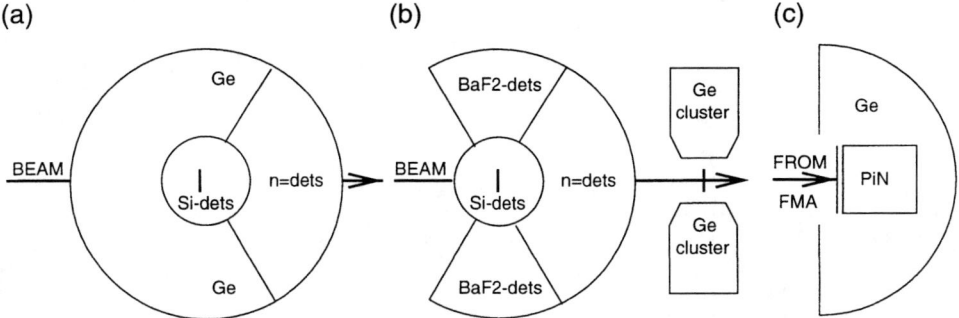

FIGURE 2. (a) Typical in-beam detection system used for studies in the ^{100}Sn region. (b) The experimental setup used at the NBI to study isomers around ^{100}Sn. (c) The detector system used at the ANL to study delayed conversion electrons in ^{102}Sn.

but the full potential of this approach has not yet been explored.

The first experiments in the ^{100}Sn region using this technique were carried out at the Hahn-Meitner Institute in Berlin with the OSIRIS array of Ge detectors and were continued at the Tandem Accelerator Laboratory of the Niels Bohr Institute (NBI) in Riso, Denmark, using NORDBALL. The summary of the results obtained by the two collaborations can be found in Ref. [1]. Among $T_z=3/2$ nuclei, transitions in ^{97}Ag [2] and ^{101}In [3] were found. Excited states in ^{103}Sn have not been found yet despite relatively large calculated cross section. There exist extensive data on $T_z=2$ nuclei ^{96}Pd [4], ^{98}Ag [5], ^{100}Cd [6], ^{102}In [7], and ^{104}Sn [8]. Above the Z=50 gap ^{106}Sb and ^{107}Sb are the lightest Sb isotopes studied in-beam.

Following the success of the early experiments, attempts were made to extend spectroscopic information to nuclei located even closer to ^{100}Sn using more efficient Ge arrays such as PEX at the NBI or GASP at the Laboratori Nazionali de Legnaro. Because of the rapidly decreasing cross sections with the departure from the line of stability these experiments did not provide data on new nuclei, but the level schemes for already known nuclei were significantly extended. For example, a regular band of M1 transitions was found in ^{105}Sn [9] and was proposed to originate from magnetic rotation.

The search for isomers in the ^{100}Sn region turned out to be more successful. Already standard in-beam experiments indicated much better sensitivity for delayed γ rays. For example, the $17/2^+$ isomer was observed in ^{99}Cd [10]. In an experiment performed at the Niels Bohr Institute, in order to optimize the detection of delayed γ rays, a catcher foil was placed downstream from the target and was surrounded with Ge cluster detectors. Gamma rays were assigned to individual reaction channels using again charged-particle and neutron detectors. The experimental set-up is shown schematically in Fig. 2(b). As a result, isomers were found in the two $T_z=1$ nuclei ^{98}Cd [11] and ^{102}Sn [12]. Four delayed transitions were observed in ^{98}Cd and were proposed to form the $8^+ \rightarrow 6^+ \rightarrow 4^+ \rightarrow 2^+ \rightarrow 0^+$ cascade. In ^{102}Sn, two delayed γ

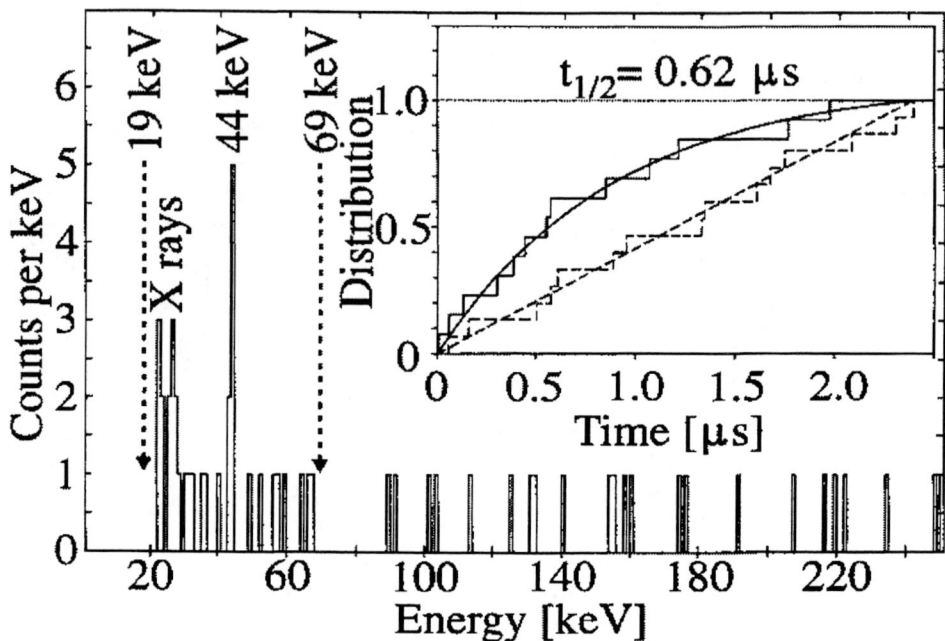

FIGURE 3. Spectrum of electrons detected in coincidence with the two delayed $4^+ \to 2^+$ and $2^+ \to 0^+$ transitions in ^{102}Sn. The Kolgomorow plot of the decay times for events associated with the 6^+ isomer (solid line) is compared with background (dashed line) in the inset.

rays were assigned as the $4^+ \to 2^+$ and $2^+ \to 0^+$ transitions and were suggested to be situated below the 6^+ isomeric state, but the $6^+ \to 4^+$ isomeric transitions was not found. In a follow-up experiment carried out at the Argonne National Laboratory (ANL) [13], a box consisting of 5 PiN diodes surrounded by 4 Ge detectors was placed behind the catcher foil at the focal of the Fragment Mass Analyzer to search for conversion electrons corresponding to the $6^+ \to 4^+$ isomeric transitions in ^{102}Sn. The experimental set-up is shown schematically in Fig. 2(c). Fig.3 shows electron spectrum detected in coincidence with the two known delayed γ-ray transitions in ^{102}Sn. The line at 44 keV was interpreted as the L-conversion line and the energy of 48 keV was proposed for the $6^+ \to 4^+$ isomeric transition. In a very recent work [14] half-lives of several high-spin states in ^{104}Sn were measured using GASP equipped with a plunger. In addition, the level scheme for ^{104}Sn was considerably extended and another case of a regular M1 band was found.

It is worth noting that the β-decay of ^{101}Sn [15] and ^{103}Sn [16] is followed by proton emission, and that protons emitted from the ground-state of ^{105}Sb [17] were observed, reflecting close the proximity of the proton drip-line.

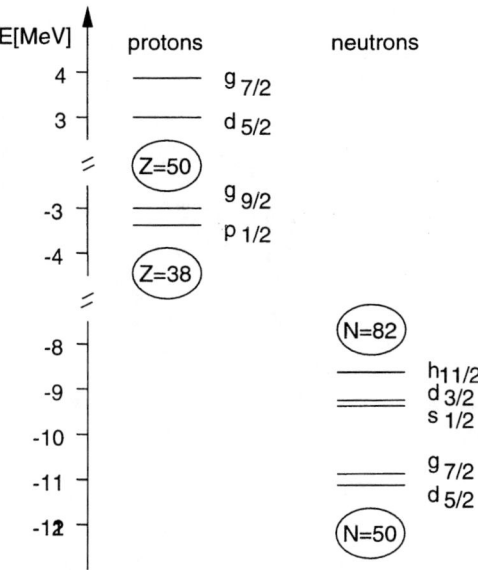

FIGURE 4. Calculated and extrapolated single-particle energies with respect to the ^{100}Sn core.

Shell-Model calculations

Excited states of nuclei in the vicinity of ^{100}Sn can be described using the shell model. However, single-particle energies and residual interactions between valence nucleons are not known. In the shell-model calculations presented in Ref. [1] the ^{88}Sr core was used to describe neutron-deficient N≥50, Z≤50 nuclei. The configuration space consisted of the $\pi p_{1/2}$ and $\pi g_{9/2}$ proton orbitals, which span the Z=38-50 shell, and the $\nu g_{7/2}$, $\nu d_{5/2}$, $\nu s_{1/2}$, $\nu d_{3/2}$ and $\nu h_{11/2}$ neutron orbitals, which fill the N=50-82 shell, with some restrictions on the occupation numbers of the higher lying $\nu s_{1/2}$, $\nu d_{3/2}$ and $\nu h_{11/2}$ orbitals. A mixture of measured, adopted from the ^{208}Pb region, and calculated two-body matrix elements was used in these calculations.

Fig. 4 shows calculated or extrapolated single-particle energies for orbitals which play an important role in the ^{100}Sn region. Energies of the $\pi p_{1/2}$ and $\pi g_{9/2}$ proton orbitals are known well from the fit to the data on N=50 isotones. The energies of the $\pi d_{5/2}$ and $\pi g_{7/2}$ proton orbitals, which are important above the Z=50 gap, were extrapolated from the light odd Sb isotopes and are not very well known. For neutrons, the two lowest lying orbitals $\nu d_{5/2}$ and $\nu g_{7/2}$ are almost degenerate. The two low-spin orbitals $\nu s_{1/2}$ and $\nu d_{3/2}$ are located about 2 MeV higher together with the high-spin intruder orbital $\nu h_{11/2}$. The yrast, high-spin states in the neutron-deficient nuclei with N>50 can be interpreted as due to the $(\pi g_{9/2}^{-1})^m \otimes \nu(d_{5/2}, g_{7/2})^n$ configuration at lower spins with the $\nu h_{11/2}$ orbital playing an increasingly important role for higher spins.

Usually, good agreement is found between the measured and calculated level energies, spins and parities for nuclei around ^{100}Sn, and in most cases it is possible to assign specific configurations to the observed states, especially to the ones with low excitation energies, where the density of the states is still low.

Comparison between calculated and measured transition strength constitutes a more stringent test. The 8^+ isomer proposed in ^{98}Cd was interpreted as the $(\pi g_{9/2}^{-1})^2{}_{8^+}$ configuration decaying to the 6^+ member of the same multiplet. A B(E2,$8^+\to 6^+$) transition strength of $0.44(^{+20}_{-10})$ W.u. was extracted from the data. A small proton effective charge of $0.85(^{+20}_{-10})$e was deduced based on a comparison with the shell-model calculations. It implied a very small proton polarization charge suggesting a stiff underlying core. In ^{102}Sn, the proposed 6^+ isomer is dominated by the $(\nu d_{5/2}, \nu g_{7/2})_{6^+}$ configuration, but the final state has more complicated structure. In contrast to ^{98}Cd, an unusually large neutron effective charge of between 1.6 and 2.3 e was obtained for ^{102}Sn from B(E2,$6^+\to 4^+$)=$4.0^{+2.4}_{-1.1}$ W.u. implying an E2-soft core. The latter result is supported by recent half-life measurements of high-spin states in ^{104}Sn [14] which also indicated a large neutron effective charge. The high neutron effective charge can be partially ascribed to deficiencies of shell-model calculations, in particular to the positions of the $d_{3/2}$ and $s_{1/2}$ neutron orbitals which are not known very well. The remaining discrepancy, however, is not presently understood and further effort is needed to improve available experimental information. In Ref. [18] the polarization charges for the orbits near N=Z=50 are calculated using the Hartree-Fock method with the Skyrme interaction. Using the values tabulated in Ref. [18] for the SkM* interaction, the neutron polarization charge corresponding to the $6^+\to 4^+$ transition in ^{102}Sn comes to 1.5 e. By scaling the calculated energy of the 2^+_1 state from 4.5 MeV to 2.7 MeV, and assuming that the isovector part of the polarization charge and the coupling to the isoscalar giant quadrupole vibrations remain unchanged, the calculated polarization charges of Ref. [18] will scale such that 2.0 e [19] is obtained for the transition in ^{102}Sn discussed above, i.e. closer to the experimentally deduced value. It is worth noting that first evidence for E3 collective excitations of the ^{100}Sn core was also found in Ref. [14].

Outlook

A new generation of large Ge arrays, such as EUROBALL and GAMMASPHERE, equipped with state-of-the-art ancillary detectors, offer one more chance to study in-beam immediate neighbors of ^{100}Sn. In fact, several experiments are planned in the ^{100}Sn region in the near future. Among the most interesting topics are: the study of excited states in ^{103}Sn which escaped detection so far; a search for states above the isomers in ^{98}Cd and ^{102}Sn, which involve core excitations; the study of neutron-proton multiplets in ^{100}In. The next logic step on the way to ^{100}Sn would be ^{99}In and ^{101}Sn, which offer direct measurement of some of the single-particle energies with respect to ^{100}Sn, but this might be very difficult because of the very small cross sections involved (of the order of 10 nb). The studies of ^{100}Sn itself are a matter of a more distant future. More detailed spectroscopy of already known nuclei might supply information about core excitations and about more exotic modes of excitations which might appear at high spin, such as rotation or magnetic rotation. The fact that the decay of some of the nuclei in the ^{100}Sn region is accompanied by proton emission offers the possibility of using the Recoil-Decay Tagging method, with its unprecedented selectivity, to study their excited states.

REFERENCES

1. H. Grawe, R. Schubart, K.H. Maier, and D. Seweryniak, *Phys. Scr.* **T56**, 71 (1995).
2. D. Alber et al., *Z.Phys.* **A335**, 265 (1990).
3. J. Cederkall et al., *Phys. Rev.* **C53**, 1955 (1996).
4. D. Alber et al., *Z. Phys.* **A332**, 129 (1989).
5. R. Schubart et al., *Z. Phys.* **A352**, 373 (1995).
6. D. Alber et al., *Z. Phys.* **A327**, 127 (1987).
7. D. Seweryniak et al., *Nucl. Phys.* **A589**, 175 (1995).
8. R. Schubart et al., *Z. Phys.* **A352**, 373 (1995).
9. A. Gadea et al., *Phys. Rev.* **C55**, R1 (1997).
10. M. Lipoglavsek et al., *Phys. Rev. Lett.* **76**, 888 (1996).
11. M. Gorska, M. Lipoglavsek et al., *Phys. Rev. Lett.* **79**, 2415 (1997).
12. M. Lipoglavsek, M. Gorska et al., *Z. Phys.* **A356**, 239 (1996).
13. M. Lipoglavsek, D. Seweryniak et al., submitted to *Phys. Lett.* **B**.
14. M. Gorska, H. Grawe et al., accepted for publication in *Phys. Rev.* **C**.
15. Z. Janas et al., *Phys. Scr.* **T56**, 262 (1995).
16. P. Tidemand-Petersson et al., *Z. Phys.* **A302**, 343 (1981).
17. R.J. Tighe et al., *Phys. Rev.* **C49**, R2871 (1994).
18. I. Hamamoto, H. Sagawa, *Phys. Lett.* **B394**, 1 (1997).
19. I. Hamamoto, private communication.

Nuclear structure information from Recoil Decay Tagging experiments

M. Leino, R. Julin, J.F.C. Cocks, P.A. Butler[1], O. Dorvaux, K. Eskola[2],
P.T. Greenlees[1], P. Jones, S. Juutinen, K. Helariutta, H. Kankaanpää,
H. Kettunen, P. Kuusiniemi, M. Muikku, R.D. Page[1], P. Rahkila,
A. Savelius, W.H. Trzaska, J. Uusitalo[3]

Department of Physics, University of Jyväskylä, P.O. Box 35, FIN-40351 Jyväskylä, Finland
[1] *Oliver Lodge Laboratory, Department of Physics, University of Liverpool, Liverpool, L69 3BX, U.K.*
[2] *Department of Physics, University of Helsinki, FIN-00014 University of Helsinki, Finland*
[3] *Present address: Argonne National Laboratory, Argonne, Illinois 60439, USA*

and the JUROSPHERE collaboration.

Abstract. Nuclear structure studies of very neutron-deficient nuclei produced in heavy-ion induced fusion-evaporation reactions have been performed. The γ-rays have been observed using the JUROSPHERE array of escape-suppressed γ spectrometers and assigned to specific nuclei using the recoil decay tagging (RDT) technique.

INTRODUCTION

In-beam experiments are a powerful method of obtaining nuclear structure information. Radiation from the target under irradiation is observed using suitable detector systems, often large arrays of efficient escape-suppressed Ge γ-ray spectrometers (1). Recently, multi-detector in-beam conversion electron measurements have also become possible (2). Since the detector arrays consist of separate detectors, which may even be segmented, coincidence measurements with high folds are possible. Consequently, one obtains information on band structures, and the development of collective phenomena in nuclei is one of the most important fields of study using in-beam methods.

If one is interested about the structure of a nucleus situated far from stability, the radiation from the nucleus will be swamped by stronger reaction channels. In extreme cases the yield of the wanted nuclide is only 10^{-7} of the total reaction yield. In the region of heavy elements ($Z \geq 82$) the background in the Ge detectors is dominated by fission products.

The use of recoil separators in connection with heavy ion fusion evaporation reactions has provided an efficient method of suppressing the γ-ray background. A thin target, with surface density typically of 100-1000 $\mu g/cm^2$, is used. The fusion products, recoiling out of the target foil with a velocity on the order of a few percent of the speed of light, are separated in-flight from unwanted reaction products and from the primary beam using electromagnetic fields. The flight time through such devices is about 1 μs, so very short-lived nuclei can be studied. Furthermore, the short flight time provides the possibility to demand a coincidence between radiation detected around the target and a separated nucleus observed at the focal plane of the separator. This recoil gating method efficiently cleans the γ-ray spectrum from radiation originating from fission products and from coulomb excitation.

Recoil separators often have large acceptance, which provides high transmission of the wanted reaction products from target to focal plane but leads to poor mass resolution. This can be improved, but with a loss in transmission. Examples of such devices are the Daresbury recoil separator (3) and the Argonne Fragment Mass Analyzer or FMA (4). The problem of poor mass resolution can be circumvented by the new method of recoil decay tagging or RDT. The method was first used at GSI in the study of nuclear shape coexistence in ^{180}Hg (5) where the velocity filter SHIP (6) served as recoil separator. The basis of the method lies in the observation of position and time correlated decay chains in the focal plane semiconductor detector where the separated nuclei are stopped (7). Events observed in the detector are recorded together with energy and position information and stamped with the time of occurrence. The first event in a decay chain is the arrival of an evaporation residue (EVR) and the subsequent chain signals are due to its radioactive decay and possible further decays.

Good position resolution in the stop detector is essential for the efficient use of this method. The average time difference between events classified as EVR or decay events in any pixel must be significantly longer than the mean life of the state in question. The division of the Si detector into pixels has been achieved in two ways. The double-sided silicon strip detectors, or DSSD (8), make use of separate segments in the xy-plane of the detector surface. Typical values are 48x48 strips in a detector with surface area of 16mmx16mm. Charge division in the resistive layer of a PIPS detector provides another method. The PIPS detectors used at the gas-filled recoil separator RITU (9) in Jyväskylä have an area of 80mm (hor.) by 35 mm (vert.), are divided horizontally into 16 strips, and have a vertical position resolution of typically 300 μm.

RDT measurements combine the observation of correlated chains at the focal plane with the observation of prompt radiation from the target: All EVR events are recorded, along with any coincident γ-ray events of fold 1 or higher. All decay events are also recorded. Both EVR and decay events are time stamped. Only such γ-ray events are then accepted which are coincident with the wanted recoil nuclei, identified on the basis of their characteristic decay. Most often, the identification of the nuclide is based on α decay but proton decay has also been made use of (10,11).

THE JUROSPHERE COLLABORATION

During the year 1997, a series of in-beam γ-ray measurements were performed at the Physics Department, University of Jyväskylä (JYFL). Some 3500 h of beam time were devoted to 25 experiments, most of them RDT-measurements. The γ-ray array called JUROSPHERE was provided by the French/UK Loan Pool.

The JUROSPHERE spectrometer array consisted of 15 Eurogam phase I detectors (12) and 10 TESSA detectors (13). The Eurogam detectors have a relative efficiency of 70 % (compared with a 76mmx76mm NaI(Tl) detector at 1.3 MeV) and were positioned at 134° and 158° relative to the beam direction. The TESSA detectors have a relative efficiency of 25 % and were placed at 79° and 101°. The spectrometer array had a total photopeak efficiency of ≈ 1.5 % at 1.3 MeV. In all experiments, beams provided by the JYFL ECR ion source were accelerated with the K = 130 MeV cyclotron. Fusion products were separated from unwanted background using the gas-filled recoil separator RITU (9). Typical beam intensities used were 5-15 pnA.

During the seven-month campaign, excited states were observed for the first time in several very neutron-deficient isotopes in the range Z = 74-92. Highlights from the project include structure studies of the following nuclides:

- ^{226}U from the reaction ^{208}Pb(^{22}Ne,4n)   (14)

- ^{206}Ra from the reaction ^{170}Yb(^{40}Ar,4n)   (15)

- ^{198}Rn from the reaction ^{166}Er(^{36}Ar,4n)   (16)

- ^{168,170}Pt from the reaction ^{112}Sn(^{58,60}Ni,2n)   (17)

Typical cross sections for the production of the most neutron-deficient nuclides were on the order of a few μb with the exception of ^{198}Rn for which the production cross section was about 200 nb. Examples of RDT work performed using JUROSPHERE will be discussed in the following.

SHAPE COEXISTENCE IN VERY NEUTRON-DEFICIENT MERCURY ISOTOPES

The occurrence of different shapes in one nucleus at low excitation energy and angular momentum is a well-known feature of neutron-deficient nuclides around Z = 82 (18). Mercury isotopes provide one of the best examples of the phenomenon of shape coexistence. The known even-even Hg isotopes are weakly oblate ($\beta_2 \approx -0.15$) in their ground state, and the yrast structure displays the corresponding oblate rotational bands down to ^{190}Hg. In lighter isotopes, the prolate intruder band with $\beta_2 \approx 0.25$ and with a larger moment of inertia becomes yrast at moderate spin. In accordance with expectations, the prolate excited minimum reaches its lowest energy close to mid-shell at N = 102 (19). In very light e-e Hg isotopes, according to recent Nilsson-Strutinsky calculations (20), the ground state is expected to be spherical while the prolate structure might disappear or give way to a superdeformed structure with $\beta_2 \approx 0.5$.

In a recent RDT experiment, the increase in the excitation energy of the prolate band, as one proceeds towards the proton drip line, was verified for ^{178}Hg (21). Furthermore, three γ-ray transitions were tentatively assigned to an E2 cascade de-exciting the lowest 2^+, 4^+, and 6^+ states in ^{176}Hg. The question of the disappearance of the prolate minimum remained unanswered, however. In the present work (22), we have performed an extended study of the level structure of ^{176}Hg, produced in the reaction ^{144}Sm(^{36}Ar,4n)^{176}Hg at a bombarding energy of 190 MeV. The estimated cross section was about 7 μb, and altogether 90000 ^{176}Hg α decay events with full α particle energy were observed. The α particle energy spectrum is shown Fig. 1 and the recoil gated singles γ-ray spectrum is shown in Fig. 2a. Both spectra are dominated by the 2p2n evaporation product ^{176}Pt. The power of the RDT method can be seen from Fig. 2b which displays the γ-ray spectrum gated on those evaporation residues which are correlated within a maximum time interval of 100 ms with 6.75 MeV ^{176}Hg α particles.

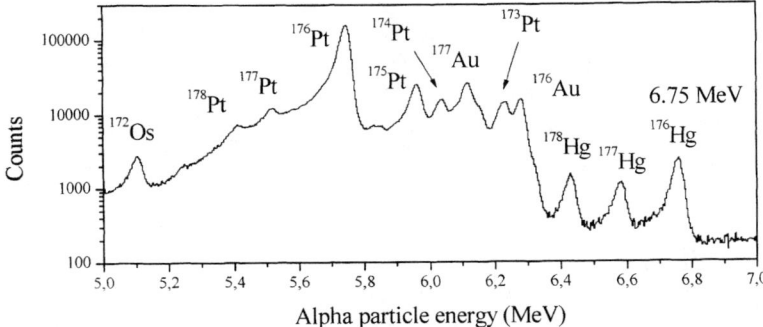

FIGURE 1. The α particle energy spectrum from the reaction 190 MeV ^{36}Ar + ^{144}Sm. Only those α particle events were accepted which could be correlated with evaporation residues within 100 ms.

The tentative level scheme of ^{176}Hg, based on preliminary results from the present work, including RDT tagged γ-γ coincidence data, confirms the assignments of Carpenter et al. (21) for the lowest three levels. The energy level systematics of a range of Hg isotopes is shown in Fig. 3. The rise of the lowest 2^+ and 4^+ level energies in ^{176}Hg is an indication of a transition towards a spherical ground state in accordance with the predictions (20). The compression of the energy differences between the positive parity yrast levels above the 6^+ and 8^+ state is interpreted to be the result of the crossing of the prolate band. The two-band mixing model (23) was used to extract the undisturbed energy difference between the prolate and oblate band heads. Variable moment of inertia (VMI) parameters and the prolate-oblate interaction strength (≈ 100 keV) which reproduced the ^{178}Hg level scheme (21) were used. A value of about 1300 keV was deduced for the energy difference. This is approximately 600 keV higher than in ^{178}Hg, revealing a rapid increase of the excitation energy of the prolate structure with decreasing neutron number.

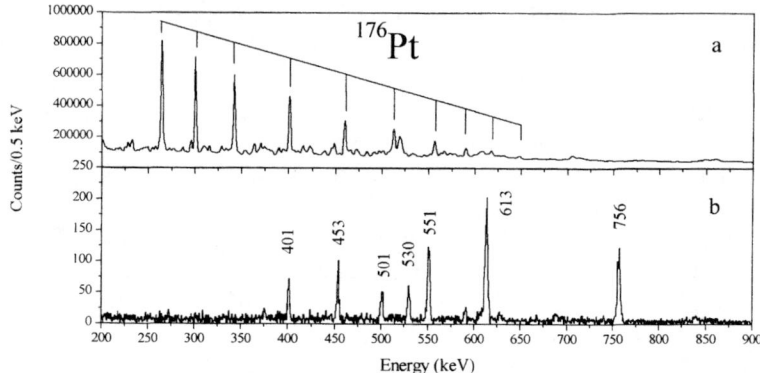

FIGURE 2. a) Singles γ-ray spectrum gated by evaporation residues from the reaction ^{36}Ar + ^{144}Sm detected at the focal plane. b) Singles γ-ray spectrum of ^{176}Hg. The transitions were identified using the RDT method.

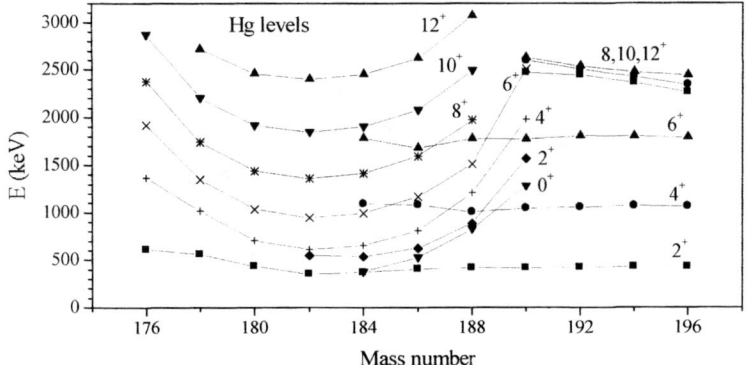

FIGURE 3. Energy level systematics of neutron-deficient even isotopes of Hg.

EXCITED STATES IN ^{226}U; EVIDENCE FOR OCTUPOLE STRUCTURE

Recent experiments have provided information on octupole structures in the actinide region (24). Nuclei with N ≈ 134 and Z = 88-90 are known to have the largest octupole correlations in their ground state. On the basis of Strutinsky-type potential energy calculations, Nazarewicz et al. (25) predicted that ^{224,226}U should have a deep minimum in the potential energy surface for non-zero β_3 with a ≈ 0.2 MeV gain in energy relative to the reflection-symmetric shape.

To produce ^{226}U nuclei, the reaction 112 MeV ^{22}Ne + ^{208}Pb was used. The efficiency of RITU is estimated to be 10 % for this reaction. A run with a duration of 320 h produced some 60000 full energy α particles from the 7.57 MeV ^{226}U → ^{222}Th ground state to ground state decay. To search for ^{226}U γ-rays, the RDT method was used with a maximum EVR-α time difference of 800 ms, i.e. about three ^{226}U half-lives. The energy spectrum of γ-rays which are in prompt coincidence with ^{226}U nuclei is shown in Fig. 4a. The partial level scheme deduced from the observed transitions in ^{226}U is shown in Fig. 4b. The arguments used in setting up the level scheme were based on energy sum, intensity balance, and γ-γ coincidence relations. The spin-parity values are tentative and are mainly based on systematics of the N = 134 isotones. For further information, see Ref. (14). The tentative assignment of the 80.5 keV $2^+ \rightarrow 0^+$ transition is consistent with preliminary data from a recent experiment on fine structure in the α decay of ^{230}Pu (26). As a conclusion, the angular momentum alignment properties (24) as a function of increasing rotational frequency suggest that ^{226}U can be regarded as a rotating reflection-asymmetric shape.

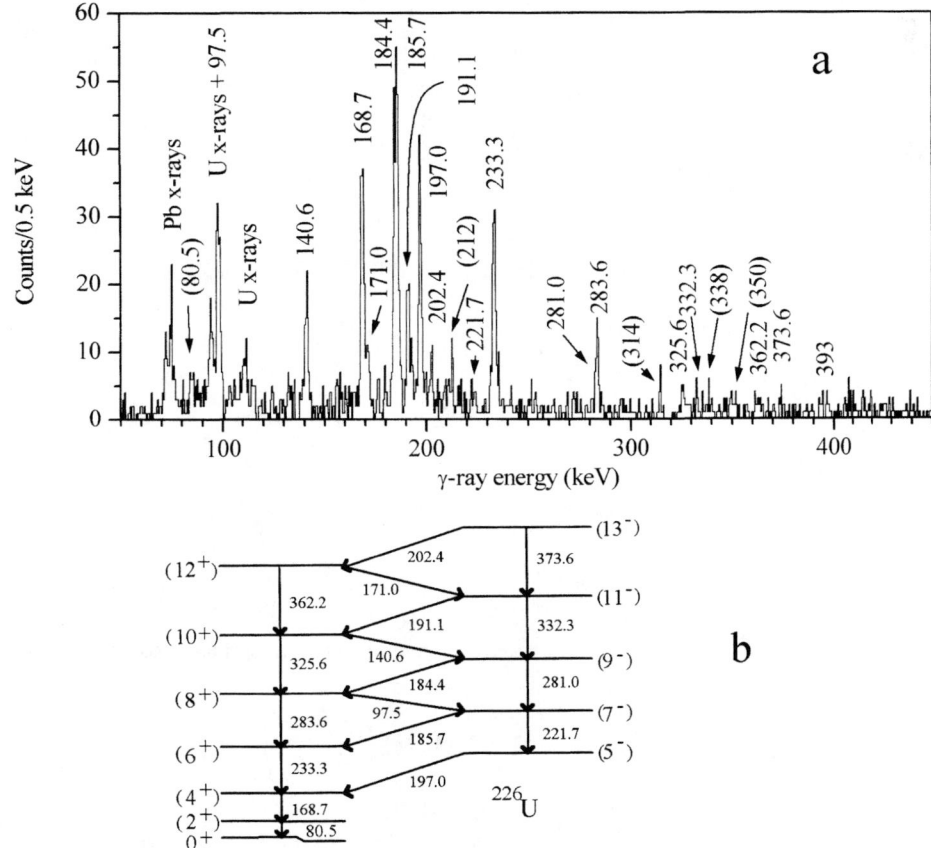

FIGURE 4. a) RDT gated singles γ-ray spectrum of ^{226}U. b) The partial decay scheme of ^{226}U.

STRUCTURE STUDIES OF NEUTRON-DEFICIENT RADIUM ISOTOPES

Neutron-deficient Ra nuclei are characterised by six valence protons and several valence neutron holes outside the closed Z = 82 and N = 126 shells. Prior to the present work (27), the lightest Ra isotope studied using in-beam γ-ray spectroscopy was ^{212}Ra (28). More neutron-deficient Ra isotopes are of interest due to the predicted region of enhanced deformation with a rather sharp border line close to ^{200}Rn and ^{204}Ra (29).

The neutron-deficient Ra isotopes ^{206,208,210}Ra were produced in 4n neutron-evaporation exit channels using the reactions 183 MeV ^{40}Ar + ^{170,172,174}Yb, respectively. The RITU transmission was estimated to be 25 % in each case, and the cross section for the production of ^{206}Ra was estimated to be 5 μb.

In addition to prompt information acquired with JUROSPHERE, delayed singles and γ-γ coincidence data were measured at the focal plane and made use of to construct the ^{208}Ra level scheme. For ^{210}Ra, delayed γ-γ coincidences were used. Fig. 5 displays energy levels in ^{206,208,210}Ra, observed for the first time in the present work, along with those of ^{212}Ra (28) and ^{214}Ra (30) studied earlier. The excitation energy of the lowest 2^+ state decreases smoothly with decreasing neutron number showing no evidence for a sudden increase in ground state deformation in the nuclei studied.

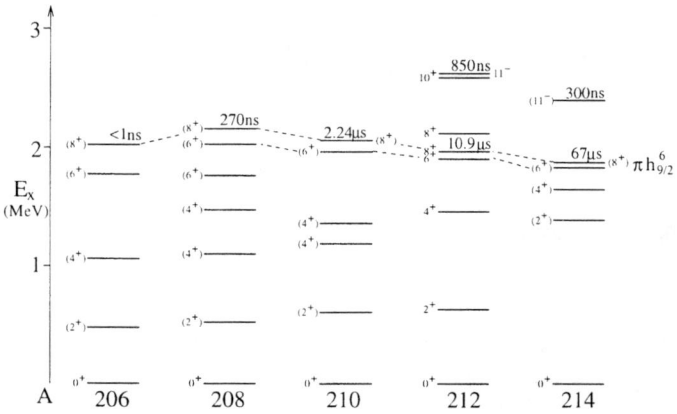

FIGURE 5. Energy level systematics for neutron-deficient even Ra isotopes

OUTLOOK

Presently, a series of RDT measurements are being performed at JYFL using segmented clover detectors both around the target and at the RITU focal plane. One of the main studies concerns structure of ^{254}No produced in the reaction ^{48}Ca + ^{208}Pb. RDT experiments will continue to be a major project at JYFL. Experimental developments include a detector system that allows the use of beta decay for tagging of fusion products. Due to wide interest in nuclides produced in symmetric reactions, design of an improved separator with better background conditions but high transmission will be undertaken. Recoil gated or RDT tagged in-beam electron spectroscopy measurements will also be performed.

ACKNOWLEDGEMENTS

This work was supported by the Academy of Finland and by the Access to Large Scale Facility program under the Training and Mobility of Researchers program of the European Union. PTG and RDP acknowledge funding by the Engineering and Physical Sciences Research Council of the UK.

REFERENCES

1. Beausang C.W., Simpson J., J. Phys. G **22**, 527-558 (1996).
2. Butler P.A. et al., Nucl. Instr. Meth. **A381**, 433-442 (1996).
3. James A.N. et al., Nucl. Instr. Methods **A267**, 144-152 (1988).
4. Davids C.N. and Larson J.D., Nucl. Instr. Meth. **B40/41**, 1224-1228 (1989).
5. Simon R.S. et al., Phys. A **325**, 197-202 (1986).
6. Münzenberg G. et al., Nucl. Instr. Meth. **186**, 423-433 (1981).
7. Schmidt K.-H. et al., Nucl. Phys. **A318**, 253-268 (1979).
8. Sellin P.J. et al, Nucl. Instr. Meth. **A311**, 217-224 (1992).
9. Leino M. et al., Nucl. Instr. Meth. **B99**, 653-656 (1995).
10. Paul E.S. et al., Phys. Rev. C **51**, 78-87 (1995).
11. Seweryniak D. et al., Phys. Rev. C **55**, R2137-2141 (1997).
12. Beausang C.W. et al., Nucl. Instr. Meth. **A313**, 37-49 (1992).
13. Nolan P.J., Gifford D.W., Twin P.J., Nucl. Instr. Meth. **A236**, 95-99 (1985).
14. Greenlees P.T. et al., submitted to Phys. Lett. B.
15. Cocks J.F.C. et al., to be published.
16. Taylor R.B.E. et al., to be published.
17. King S.L. et al., submitted to Phys. Lett. B.
18. Wood J.L., Heyde K., Nazarewicz W., Huyse M., Van Duppen P., Phys. Rep. **215**, 101-201 (1992).
19. Dracoulis G.D. et al., Phys. Lett. B **208**, 365-368 (1988).
20. Nazarewicz W., Phys. Lett. B **305**, 195-201 (1993).
21. Carpenter M.P. et al., Phys. Rev. Lett. **78**, 3650-3653 (1997).
22. Muikku M. et al., to be published.
23. Van Duppen P., Huyse M., Wood J.L., J. Phys. G **16**, 441-450 (1990).
24. Butler P.A., Nazarewicz W., Rev. Mod. Phys. **68**, 349-421 (1996).
25. Nazarewicz W. et al., Nucl. Phys. **A429**, 269-295 (1984).
26. Greenlees P.T. et al., to be published.
27. Cocks J.F.C. et al., to be published.
28. Kohno T. et al., Phys. Rev. C **33**, 392-395 (1986).
29. Möller P. et al., Atomic Data and Nuclear Data Tables **59**, 185-381 (1995).
30. Horn D. et al., Nucl. Phys. **A317**, 520-534 (1979).

Microsecond isomers studies

Robert Grzywacz

University of Tennessee, Knoxville, Knoxville, TN 37996, US

for a Collaboration of:
University of Warsaw, GANIL Caen, University of Surrey, CEN Bordeaux-Gradignan, IPN Orsay, FLNR-JINR Dubna, LPC Caen, GSI Darmstadt, CEN Bruyéres-le-Chatel, IPNL Lyon, IAP Bucharest, KUL Leuven, NORDITA Copenhagen, ORNL Oak Ridge, CSNSM, Orsay, Daresbury Laboratory, University of Göttingen, University of Tennesse, University of Brighton, University of Liverpool, University of York

INTRODUCTION

Isomer spectroscopy following fragmentation or fission of heavy-ion projectiles with full in-flight fragment identification provides access to excited states of nuclei unreachable with any other methods. Wide mass and charge distribution of the fragmentation products together with very high isomer detection selectivity allows the systematic search for isomers over large parts of the nuclidic chart. Their main decay properties can be studied even at the extremely low production rates, which are even below 1 atom/s.

Several experiments [1–7] have been performed in the last four years which were aimed at discovery of new isomers. About seventy isomers have been detected, out of which more then twenty were new. The main effort was concentrated at GANIL laboratory, with experiments using the intermediate energy (60-70 MeV/u) neutron deficient beams of ^{78}Kr, ^{92}Mo, ^{106}Cd, and ^{112}Sn and using the LISE3 fragment separator. Several new isomers close to the proton drip line between Z=50 (^{102m}Sn) and Z=33 (^{66m1,2}As) were observed, see fig. 1. The observation of the new isomer ^{32m}Al [2] demonstrated that microsecond isomer spectroscopy can be an effective method to study the structure of the very neutron-rich nuclear systems. The follow - up discovery of a new island of isomers located near the ^{68}Ni, from fragmentation of ^{86}Kr beam, was an important step forward towards the study of the doubly magic ^{78}Ni. The spectroscopic investigations of neutron rich nuclei has been reinforced by the recent results obtained at GSI using the FRS separator with fragmentation and fission of the high energy (1 GeV/u) ^{238}U beam [7]. Evidence for new isomers close to the ^{208}Pb has been obtained. Viewed in a broad context of experimental studies of the neutron-rich nuclei, our method should be seen as complementary

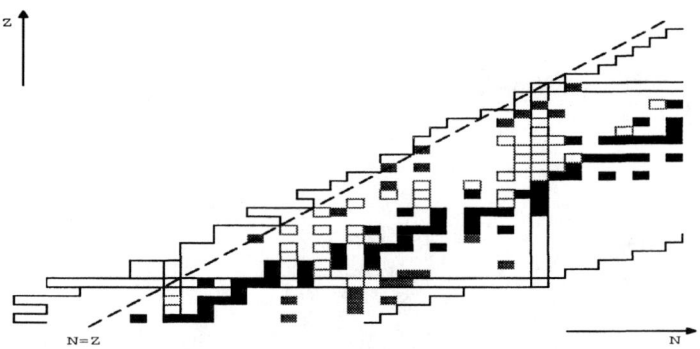

FIGURE 1. The microsecond isomers observed in the fragmentation of the intermediate energy heavy-ion beams at GANIL. Grey filled squares symbolize the nuclei for which the first evidence for the existence of isomerism has been found, open squares - known isomers detected.

to beta decay studies [8], virtually taking the role of in-beam spectroscopy on the neutron rich side of nuclide chart.

EXPERIMENTAL METHOD

The experimental method which allows a global search for isomers requires use of a fragment separator with a gamma detection set-up installed at the final focus, around the ion-stopping detector. It is based on the time correlation between electronic signals generated during ion flight and stopping - and those from the gamma detector when the isomer decay is registered. The first ones give the unambigous information on the mass (A), atomic number (Z) and charge (Q) of the ion, the second - unique gamma decay spectrum of the isomer. The microsecond correlation allows for one-to-one correspondance between these two groups of identification signals. Thus, it is possible to investigate the existence of isomers for all produced and transmitted fragments in the same setting of the spectrometer, still preserving the background-free isomeric spectra. The typical lifetime of investigated isomers, which is of the order of one microsecond, is comparable to the ion's time-of-flight (TOF) through the spectrometer. In some cases one could observe isomers with a much shorter lifetime, due to an in-flight decay blocking mechanism. Since the ions emitted from the target, are usually fully stripped of electrons, the lifetime of the isomeric state can be longer compared to that of nuclei with all electrons present, due to the cancelation of the internal conversion (IC) decay channel. This effect is particularly strong for highly converted transitions and can increase the lifetime of a state as much as 10 times, see the case of the ^{92}Tc [1]. The most sensitive to the blocking are the 0^+ isomers with a dominating E0 decay branch, which for E^* <1022 keV can decay only by the IC channel. Such a case has been found in

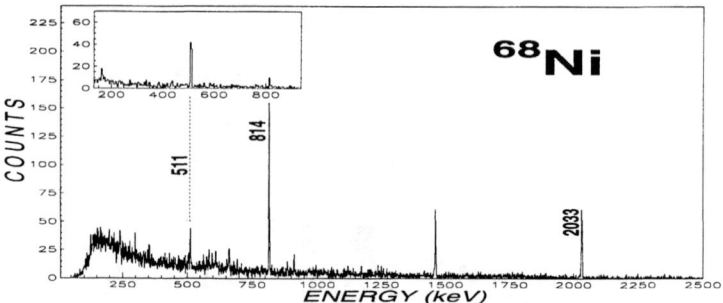

FIGURE 2. The gamma lines observed in coincidence with ^{68}Ni ions. The total projection for the times between 0 and 70 µs after implantation is compared with the short (1 µs) time gate (shown in the inset). The 814 and 2033 belong to the decay of the 5$^-$ isomer and 511 keV line to the decay of the short lived 0$^+$ isomer.

^{74}Kr [3], where the lifetime of the isomeric 0$^+$ level is extended at least 50 times for the nucleus fully stripped of electrons. The blocking mechanism is not as important for a class of 0$^+$ isomers with a strong pair-production decay branching ratio (E* >>1022 keV). Such an isomer (with E*=1.77 MeV in ^{68}Ni [10,11]) has been observed in the experiment with ^{86}Kr beam [5]. The 511 keV line from positron anihilation (see inset in fig. 2) with a characteristic isomeric half-life of 340(30) ns is a clear experimental evidence for the decay of this metastable state. The decay of the 5$^-$ isomer at 2033 MeV [9] has also been observed, see fig. 2. Its half-life of 860(50) µs was the longest isomeric half-life directly observed within the present method .

The unique conditions of fragmentation type experiments allow for the direct measurement of the "ionic" half-life. Thus, it makes possible the independent determination of conversion coefficient. This method would be particulary important for the 0$^+$ isomers where the conversion may be the only decay channel. The method requires the measurement of the isomer to total ratios at implantation points with two different time-of-flights (TOF). A unique combination of spectrometers existing at GANIL allows for such a measurement. The production target can be placed either in front of the ALPHA spectrometer, with the TOF base of 118 m, or in front of LISE3, with a TOF-base of 43.9 m. We have performed a test measurement for the ^{72m}Cu isomer which decays via cascade of three transitions of 51, 82 and 138 keV with a measured half-life after implantation of 1.76(3) µs [5]. The half-life for the isomer with the nucleus fully stripped of electrons was measured to be greater than 7.2 µs, which suggests the quadrupole character for the isomeric 51 keV line. The independently measured conversion coefficient for this line determines the transition type to be E2.

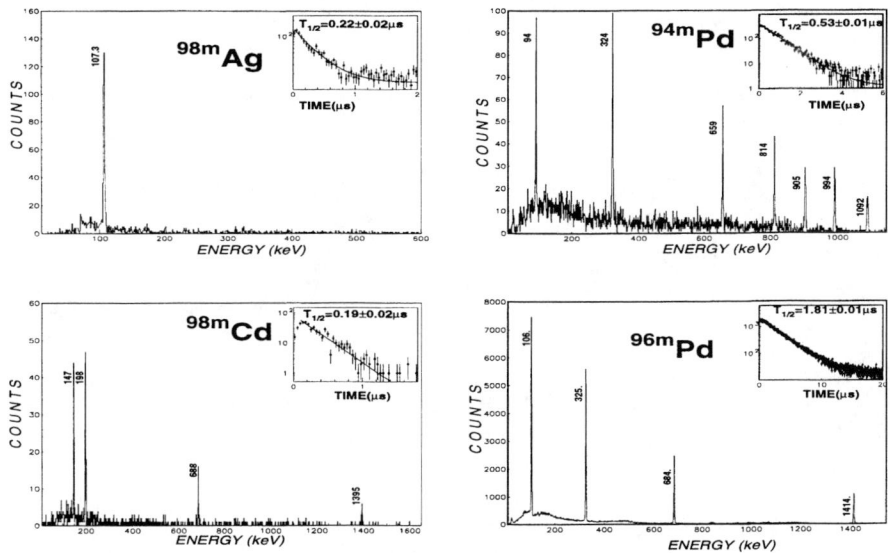

FIGURE 3. Gamma spectra and decay patterns (in the insets) showing the decays of ^{98m}Cd, ^{98m}Ag, ^{94m}Pd, and ^{96m}Pd observed among the fragmentation products of ^{106}Cd.

ISOMERS IN $T_Z=1$ NUCLEI CLOSE TO ^{100}SN

The isomers in the ^{100}Sn region have been studied in several experiments using ^{112}Sn and ^{106}Cd beams. The number of known isomers detected in the first experiment [1] gave a very rich dataset on the population of isomeric state in the fragmentation reaction. The follow-up experiment [4] reported the first evidence for the existence of $T_Z=1$ isomers like ^{102}Sn, ^{98}Cd, ^{96}Ag, and ^{94}Pd. All of them, except for ^{96}Ag, have been identified also in the fusion-evaporation reactions [12–14]. The experiment performed with ^{106}Cd beam confirmed the previous results, however reporting systematically shorter half-lives than those obtained in fusion (see fig. 3). Compare 2.2(3) μs [15] with 1.81(1) μs for ^{96}Pd, 0.8(2) μs [12] with 0.53(1) μs for ^{94}Pd and 0.48(8) μs [14] with 0.19(2) μs for ^{98}Cd. Note that the systematically shorter lifetimes have been reported already in [4]. The divergence may point to systematic experimental errors in one of the methods. Therefore, a new and independent measurement is necessary. The consequences of the difference observed for the ^{98m}Cd decay are of utmost importance for the discussion of the systematics of the proton effective charge [14] in this region of nuclide chart.

As a by-product of this experiment, the isomer in the ^{98}Ag was identified. We measured the two characteristic gamma rays deexciting the 167 keV level, 107.3 and 60.6 keV. The half-life of this state, 220(20) ns is compatible with the Weisskopf estimate for an E2 type for the 107.3 keV transition. The measurements in the ^{100}Sn

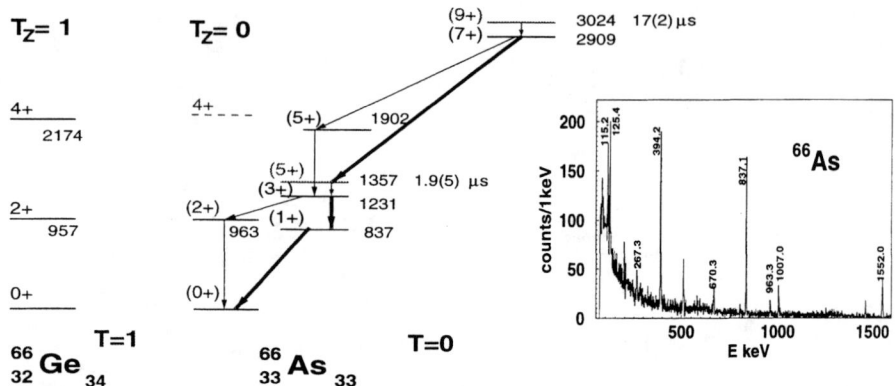

FIGURE 4. The level scheme and tentative spin assignments of the excited states observed in the ^{66}As. The dashed line shows the expected isobaric analogue state of the 4$^+$ level observed in ^{66}Ge.

region are still to be continued with the fragmentation reaction. In particular the still unresolved decay scheme of ^{96}Ag should be studied. With improved intensities of ^{112}Sn or ^{106}Cd beams, one may attempt in the future to observe experimentally the 8$^+$ isomer in ^{98}Sn, a mirror nucleus of the ^{98}Cd.

ISOMERS CLOSE TO THE N=Z LINE: T=1 VS T=0 PAIRING

The discovery of isomers in ^{66}As [6] was another step in studying self-conjugate nuclei. The structure of the isomer reveals the particular role played by the proton-neutron correlation in odd-odd nuclei [16–18]. The decay scheme shown in the fig. 4 was obtained from the data collected in the fragmentation of ^{78}Kr beam. The estimated cross section for the production of ^{66}As is about 1 μb. With this cross section we were able to measure the gamma decay spectrum and half-lives, and also perform $\gamma - \gamma$ coincidence analysis which allowed us to construct the decay scheme.

From the proposed level sequence, spin and parity values (fig. 4), we can speculate about the possible configurations of the states of the odd-odd N = Z nucleus ^{66}As. The detailed argumentation for the chosen spin assignment can be found in [6]. In a first attempt, one can consider the two particle excitations in the spherical $f_{5/2}$, $p_{3/2}$, $p_{1/2}$ and $g_{9/2}$ subshells. The observed states can be classified according to their isospin. The states with J^π=1$^+$,3$^+$,5$^+$ are the T=0 two-particle states in the fp subshell. The J^π=9$^+$ state also has isospin T=0 but with protons and neutrons occupying the $g_{9/2}$ orbital in aligned configuration. The T=1 are the 0$^+$ (ground) and 2$^+$ states which are the isobaric analogue states to the ground and first excited states of the even-even T_Z=1 nucleus ^{66}Ge. As shown by Vincent et al. [19] the

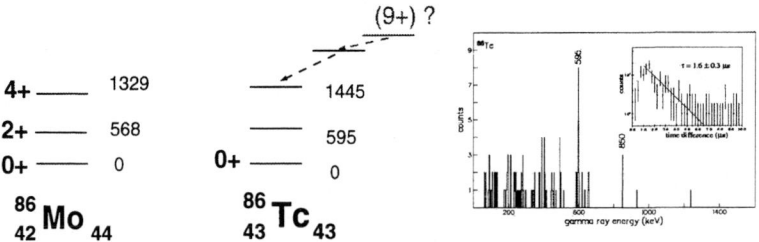

FIGURE 5. The gamma decay of an isomer in ^{86}Tc [20]. Only two strong gamma lines are unambigously identified. Gamma transitions in ^{86}Tc are compared to those deexciting 2^+ and 4^+ states in ^{86}Mo.

present experimental data on ^{62}Ga, the next odd-odd N=Z nucleus reveals only the T=0 cascade, while the non-yrast T=1 levels are not populated in the in-beam experiment. Little is known about the structure of the excited states in ^{70}Br. In view of the ^{62}Ga results, the observation of the non-yrast 4^+ state in an in-beam type experiment is desired to confirm our speculative assignment but not very likely in the immediate future. The other evidence of such of a collective T=1 cascade is suggested in the decay of ^{86m}Tc, which has been observed in the experiment with ^{92}Mo fragmentation [20]. The beta-decay of the ground state of this nucleus has been recently obsreved [21] strongly suggesting the superallowed Fermi type transition. Thus, the ground state of ^{86m}Tc has $J^\pi=0^+$ - and it is an IAS to the ground state of ^{86}Mo. The observed two gamma lines of 850 and 595 keV may belong to the T=1 cascade conecting 4^+, 2^+ and 0^+ states – IAS of those in ^{86}Mo (see fig. 5). This finding has to be confirmed in a more complete experiment.

ISOMERS IN NEUTRON-RICH NUCLEI

The experimental difficulties of nuclear structure exploration on the exotic neutron rich side with other production methods makes information obtained from the decay of microsecond isomers very valuable. We selected the region of nuclei around ^{68}Ni for several reasons. The closed shell Z=28 should allow for reliable theoretical interpretation and a direct comparison with various existing shell-model calculations. The presence of the high-spin $g_{9/2}$ orbital in the neighborhood of $p_{1/2}$ one generates isomerism, similarly to the valence mirror nuclei with N=50. In a pioneering experiment [1], the known proton-rich "island of isomers" with N≈50, $40 < Z < 50$ was investigated. The isomerism in this region is due to the occupation of the proton $\pi g_{9/2}$ orbital, forming *seniority isomers* in stretched $\pi g_{9/2}^n$ configurations, and *single particle isomers* due to the large spin difference of the neighboring $\pi p_{1/2}$ orbital. This gives rise to the 8^+ E2 isomers in the N=50 isotones ^{92}Mo, ^{94}Ru, ^{96}Pd and ^{98}Cd, and the well known $\pi p_{1/2} - \pi g_{9/2}$ M4 isomerism. The neutron-rich nuclei studied in [5] – with Z≈28, $40 < N < 50$, are the "valence

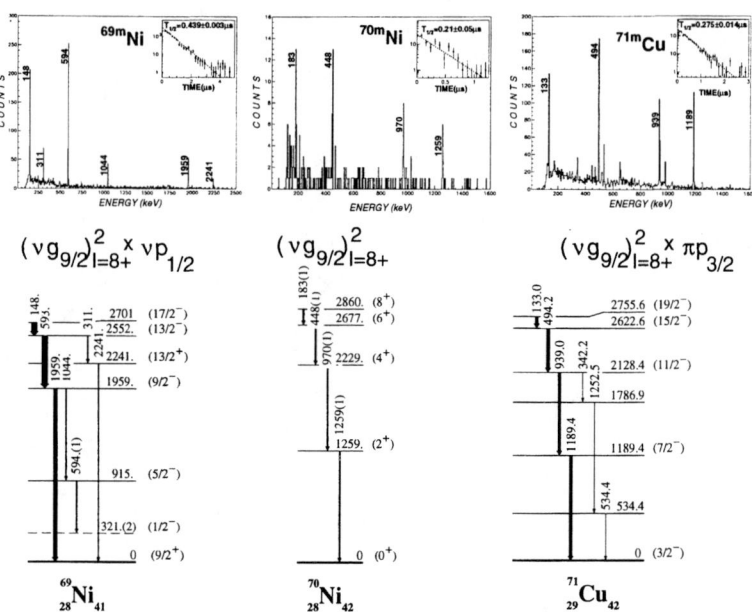

FIGURE 6. The observed gamma spectra and deduced decay schemes of the isomers ^{69}Ni, ^{70}Ni and ^{71}Ni. The dominant configuration of the isomer wave function is shown.

mirrors" (to the N=50, 40 < Z < 50 region) with the role of protons and neutrons outside the closed shell interchanged. To date information about excited states in the region of the neutron rich isotopes in the vicinity of the magic Z=28 number is very scarce. Previous to the present experiment data about excited states for the nickel isotopes were obtained using multinucleon transfer reactions [9,22,23]. The 17/2$^-$ and 1/2$^-$ isomers ^{69m1m2}Ni and 8$^+$ ^{70m}Ni (fig. 6) have been observed showing the apparent resemblance of the order of the excicted states to their valence mirror partners: ^{91}Nb and ^{92}Mo. The dominating role of $(\nu g_{9/2})^2$ coupling in the structure of the excited states in ^{69}Ni and ^{70}Ni and ^{71}Cu can be clearly observed (see fig. 6). The comparison of the experimental data obtained for the nickel isotopes with the shell-model calculations [25] shows that the good agreement between theory and experiment obtained for the less exotic Ni isotopes [22] breaks down beyond ^{68}Ni.

PERSPECTIVES

Future measurements on the neutron-rich nuclei are foreseen, as their results can be particularly important for a complete understanding nuclear structure. We plan to continue our search for more neutron rich isomers beyond ^{70}Ni and in the ^{208}Pb region. For nuclei with very large neutron excess dramatic changes in the shell structure such as disappearance of the shell gap [24] and appearance of new shell closures are predicted, which are intimately related to the small neutron binding energy. The isomers with their additional excitation energy can simulate phenomena occurring in loosely bound yet unaccessible nuclei. Beyond the aspects of nuclear structure studies in neutron-rich nuclear matter and/or at low nuclear density, experimental data in this region are of key importance for understanding the astrophysical r-process [26].

The interesting cases of the isomers along the N=Z line need to be further studied as they may allow the direct insight into the proton-neutron interaction in heavy nuclei. The fragmentation reaction can certainly reach beyond the N=Z line below Z=50 presenting the fascinating prospect of studying the isomers in mirror nuclei beyond N=Z line.

This work has been supported by IN2P3 and Polish Commitee for Scientific Research (KBN).

REFERENCES

1. R. Grzywacz et al., Phys. Lett. **B355** (95) 439
2. M. Robinson et al., Phys. Rev. **C53** (96) R1465
3. Ch. Chandler et al., Phys. Rev. **C56** (97) R2924
4. R. Grzywacz et al., Phys. Rev. **C55** (97) 1126
5. R. Grzywacz et al., Phys. Rev. Lett. in print
6. R. Grzywacz et al., Phys. Lett. B in print
7. M. Pfützner et al., Proc. of Int. Conf. on Fission and Properties of Neutron-Rich Nuclei, Eds. J. H. Hamilton and A. V. Ramayya, World Scientific (Singapore) 1998.
8. S. Franchoo et al., submitted to Phys. Rev. Lett.
9. R. Broda et al., Phys. Rev. Lett. **74** (95) 868
10. M. Bernas et al., Phys. Rev. **C24** (81) 756
11. M. Bernas et al., Phys. Lett **B113** (82) 279
12. M. Górska et al., Z. Phys. **A353** (1995) 233
13. M. Lipoglavsek et al., Z. Phys. **A356** (1996) 239
14. M. Górska et al., Phys. Rev. Lett. **79** (1997) 2415
15. H. Grawe and H. Haas, Phys. Lett. **B120** (1983) 63
16. D. Rudolph et al., Phys. Rev. Lett. **76** (1996) 2415
17. D. Dean et al., Phys. Lett. **B 399** (1997) 1
18. P. Van Isacker and D.D. Warner et al., Phys. Rev. Lett. **78** (1997) 3266
19. S. Vincent et al., this conference
20. P. Regan et al., Acta Phys. Pol. **28** (97) 431
21. Ch. Longour et al., submitted to Phys. Rev. Lett.
22. T. Pawłat et al., Nucl. Phys. **A574** (94) 623
23. R. Broda et al., in [7]
24. J. Dobaczewski et al., Phys. Rev. Lett. **72** (94) 981
25. H. Grawe et al., Prog. Part. Nucl. Phys. **38** (97) 15
26. F.K. Thielemann et al., Nucl. Phys. **A570** (1994) 329c

Shell-Model Monte Carlo Studies of Nuclei Far From Stability

David J. Dean

*Physics Division, Oak Ridge National Laboratory, Oak Ridge, TN
and Physics Department, University of Tennessee, Knoxville, TN* [1]

Abstract. The shell-model Monte Carlo (SMMC) technique transforms the traditional nuclear shell-model problem into a path-integral over auxiliary fields. I describe below the method and its applications to two physics issues: calculations of electron-capture rates, and exploration of pairing correlations in unstable nuclei.

Studies of nuclei far from stability have long been a goal of nuclear science. Nuclei on either side of the stability region, either neutron-rich or deficient, are being produced at new radioactive beam facilities across the world. At these facilities, and with the help of advances in nuclear many-body theory, the community will address many of the key physics issues, including: mapping of the neutron and proton drip lines, thus exploring the limits of stability; understanding effects of the continuum on weakly bound nuclear systems; understanding the nature of shell gap melting in very neutron-rich systems; determining nuclear properties needed for astrophysics; investigating deformation, spin, and pairing properties of systems far from stability; and analyzing microscopically unusual shapes in unstable nuclei.

The range and diversity of nuclear behavior, as indicated in the above list of ongoing and planned experimental investigations, have naturally engendered a host of theoretical models. Short of a complete solution to the many-nucleon problem, the interacting shell model is widely regarded as the most broadly capable description of low-energy nuclear structure, and the one most directly traceable to the fundamental many-body problem. Difficult though it may be, solving the shell-model problem is of fundamental importance to our understanding of the correlations found in nuclei.

My research over the past few years has been in the area of the nuclear shell model solved not by diagonalization, but by integration. In what follows, I will describe the shell-model Monte Carlo (SMMC) method, and discuss two recent and interesting

---

[1] Oak Ridge National Laboratory (ORNL) is managed by Lockheed Martin Energy Research Corp. for the U.S. Department of Energy under contract number DE-AC05-96OR22464. This work was supported in part through grant DE-FG02-96ER40963 from the U.S. Department of Energy.

CP455, *ENAM98: Exotic Nuclei and Atomic Masses*
edited by B. M. Sherrill, D. J. Morrissey, and Cary N. Davids
© 1998 The American Institute of Physics   1-56396-804-5/98/$15.00

results obtained from theory. These include calculations of electron-capture rates in fp-shell nuclei, and pairing correlations in medium-mass nuclei near N=Z.

Investigations of both ground state and thermal properties of nuclei have been described using the SMMC technique [1]. This method offers an alternative way to calculate nuclear structure properties, and is complementary to direct diagonalization. SMMC cannot find, nor is it designed to find, every energy eigenvalue of the Hamiltonian. It is designed to give thermal or ground-state expectation values for various one- and two-body operators. Indeed, for larger nuclei, SMMC may be the only way to obtain information on the thermal properties of the system from a shell-model perspective. The partition function of the imaginary-time many-body propagator, $U = \exp(-\beta \hat{H})$ where $\beta = 1/T$ and T is the temperature of the system in MeV, is used to calculate the expectation values of any observable $\hat{\Omega}$ with

$$\langle \hat{\Omega} \rangle = \frac{\text{Tr}\hat{U}\hat{\Omega}}{\text{Tr}\hat{U}} . \tag{1}$$

Since $\hat{H}$ contains many terms that do not commute, one must discretize $\beta = N_t \Delta \beta$. Finally, two-body terms in $\hat{H}$ are linearized through the Hubbard-Stratonovich transformation, which introduces auxiliary fields over which one must integrate to obtain physical answers. The method can be summarized as

$$Z = \text{Tr}\hat{U} = \text{Tr}\exp(-\beta \hat{H}) \rightarrow \text{Tr}\left[\exp(-\Delta\beta\hat{H})\right]^{N_t}$$
$$\rightarrow \int \mathcal{D}[\sigma] G(\sigma) \text{Tr} \prod_{n=1}^{N_t} \exp\left[\Delta\beta \hat{h}(\sigma_n)\right] , \tag{2}$$

where σ_n are the auxiliary fields (there is one σ-field for each two-body matrix-element in $\hat{H}$ when the two-body terms are recast in quadratic form), $\mathcal{D}[\sigma]$ is the measure of the integrand, $G(\sigma)$ is a Gaussian in σ, and $\hat{h}$ is a one-body Hamiltonian. Thus, the shell-model problem is transformed from the diagonalization of a large matrix to one of large dimensional quadrature. Dimensions of the integral can reach up to 10^5 for rare-earth systems, and it is thus natural to use Metropolis random walk methods to sample the integrands. Such integration can most efficiently be performed on massively parallel computers.

The SMMC method is also used to calculate the response function $R_\mathcal{A}(\tau)$ of an operator $\mathcal{A}$ at an imaginary-time τ. The operator $\mathcal{A}$ may be of the one-body form, $a^\dagger a$, as is the case with Gamow-Teller distributions, or of the pair creation (or annihilation) form, $a^\dagger a^\dagger$, as is the case for pair transfer calculations. The response describes the dynamical behavior of the nucleus under the influence of the operator, and is given by

$$R_\mathcal{A}(\tau) \equiv \langle \hat{\mathcal{A}}^\dagger(\tau)\hat{\mathcal{A}}(0) \rangle = \frac{\text{Tr}[e^{-(\beta-\tau)\hat{H}} \hat{\mathcal{A}}^\dagger e^{-\tau \hat{H}} \hat{\mathcal{A}}]}{\text{Tr}[e^{-\beta \hat{H}}]} . \tag{3}$$

The strength distribution, $S_\mathcal{A}(E)$, is related to $R_\mathcal{A}(\tau)$ by a Laplace Transform,

$$R_{\mathcal{A}}(\tau) = \int_{-\infty}^{\infty} S_{\mathcal{A}}(E) e^{-\tau E} dE. \tag{4}$$

Note that E is the energy transfer within the parent nucleus, and that the strength distribution $S_{\mathcal{A}}(E)$ has units of MeV^{-1}.

The impact of nuclear structure on astrophysics has become increasingly important, particularly in the fascinating and presently unsolved problem of type-II supernovae explosions. One key ingredient of the precollapse scenario is the electron capture cross section on nuclei [2,3]. An important contribution to electron capture cross sections in supernovae environments is the Gamow-Teller (GT) strength distribution. This strength distribution, calculated in SMMC using Eqs. (3,4) above, is used to find the energy-dependent cross section for electron capture. In order to obtain the electron capture rates, the cross section is then folded with the flux of a degenerate relativistic electron gas [4]. Note that the Gamow-Teller distribution is calculated at the finite nuclear temperature which, in principle, is the same as the one of the electron gas.

It is important to calculate the GT strength distributions reasonably accurately for both the total strength, and the position of the main GT peak in order to have a quantitative estimate for the electron capture rates. For astrophysical purposes, calculating the rates to within a factor of two is required. I concentrate here on mid-fp shell results for the electron capture cross sections [4]. The Kuo-Brown interaction [5], modified in the monopole terms by Zuker and Poves [6], was used throughout these pf-shell calculations. This interaction reproduces quite nicely the ground- and excited-state properties of mid-fp shell nuclei [7,8], including the total Gamow-Teller strengths and distributions, where the general overall agreement between theory and experiment [9] is quite reasonable. The SMMC technique allows one to probe the complete $0\hbar\omega$ fp-shell region, without any parameter adjustments to the Hamiltonian, although the Gamow-Teller operator has been renormalized by the standard factor of 0.8.

Do the electron capture rates presented here indicate potential implications for the pre-collapse evolution of a type II supernova? To make a judgement on this important question, one should compare in Table I the SMMC rates for selected nuclei with those currently used in collapse calculations [3]. For the comparison, I choose the same physical conditions as assumed in Tables 4–6 in [3]. Table I also lists the partial electron capture rate which has been attributed to Gamow-Teller transitions in Ref. [3]. Note that for even parent nuclei, the present rate approximately agrees with the currently recommended *total* rate. A closer inspection, however, shows significant differences between the present rate and the one attributed to the Gamow-Teller transition in [3]. The origin of this discrepancy is due to the fact that Ref. [10] places the Gamow-Teller resonance for even-even nuclei systematically at too high an excitation energy. This shortcoming has been corrected for in Refs. [10,3] by adding an experimentally known low-lying strength in addition to the one attributed to Gamow-Teller transitions. However, the overall good agreement between the SMMC results for even-even nuclei and the recommended rates indicates

TABLE 1. Comparisons of the SMMC electron capture rates with the total (λ_{ec}) and partial Gamow-Teller (λ_{ec}^{GT}) rates as given in Ref. [3]. Physical conditions at which the comparisons were made are given in the last column.

Nucleus	λ_{ec} (sec^{-1}) (SMMC)	λ_{ec} (sec^{-1}) (Ref. [3])	λ_{ec}^{GT} (sec^{-1}) (Ref. [3])	Conditions
^{55}Co	3.89E-04	1.41E-01	1.23E-01	$\rho_7 = 5.86, T_9 = 3.40, Y_e = 0.47$
^{57}Co	3.34E-06	3.50E-03	1.31E-04	$\rho_7 = 5.86, T_9 = 3.40, Y_e = 0.47$
^{54}Fe	7.83E-05	3.11E-04	9.54E-07	$\rho_7 = 5.86, T_9 = 3.40, Y_e = 0.47$
^{55}Fe	1.20E-08	1.61E-03	1.16E-07	$\rho_7 = 5.86, T_9 = 3.40, Y_e = 0.47$
^{56}Ni	3.47E-02	1.60E-02	6.34E-03	$\rho_7 = 5.86, T_9 = 3.40, Y_e = 0.47$
^{58}Ni	1.01E-03	6.36E-04	4.04E-06	$\rho_7 = 5.86, T_9 = 3.40, Y_e = 0.47$
^{60}Ni	7.39E-05	1.49E-06	4.86E-07	$\rho_7 = 5.86, T_9 = 3.40, Y_e = 0.47$
^{59}Co	3.44E-07	2.09E-04	6.37E-05	$\rho_7 = 10.7, T_9 = 3.65, Y_e = 0.455$
^{57}Co	2.06E-05	7.65E-03	3.69E-04	$\rho_7 = 10..7, T_9 = 3.65, Y_e = 0.455$
^{55}Fe	1.07E-07	3.80E-03	5.51E-07	$\rho_7 = 10.7, T_9 = 3.65, Y_e = 0.455$
^{56}Fe	9.80E-06	4.68E-07	6.60E-10	$\rho_7 = 10.7, T_9 = 3.65, Y_e = 0.455$
^{54}Fe	3.84E-04	9.50E-04	3.85E-06	$\rho_7 = 10.7, T_9 = 3.65, Y_e = 0.455$
^{51}V	1.06E-06	1.24E-05	9.46E-09	$\rho_7 = 10.7, T_9 = 3.65, Y_e = 0.455$
^{52}Cr	1.32E-04	2.01E-07	1.59E-10	$\rho_7 = 10.7, T_9 = 3.65, Y_e = 0.455$
^{60}Ni	3.61E-04	7.64E-06	2.12E-06	$\rho_7 = 10.7, T_9 = 3.65, Y_e = 0.455$

that the SMMC approach also accounts correctly for this low-lying strength. This has already been deduced from the good agreement between SMMC Gamow-Teller distributions and data including the low-energy regime [9]. Thus, for even-even nuclei, the SMMC approach is able to predict the *total* electron capture rate rather reliably, even if no experimental data are available. Note that the SMMC rate is somewhat larger than the recommended rate for ^{56}Fe and ^{60}Ni. In both cases the experimental Gamow-Teller distribution is known and agrees well with the SMMC results [9]. While the proposed increase of the rate for ^{60}Ni is not expected to have noticeable influence on the pre-collapse evolution, the increased rate for ^{56}Fe makes this nucleus an important contributor in the change of Y_e during the collapse (see Table 15 of [3]).

For electron capture on odd-A nuclei, observe that the SMMC rates, derived from the Gamow-Teller distributions, are significantly smaller than the recommended total rate. This is due to the fact that for odd-A nuclei the Gamow-Teller transition peaks at rather high excitation energies in the daughter nucleus. The electron capture rate on odd-A nuclei is therefore carried by weak transitions at low excitation energies. Comparing the SMMC rates to those attributed to Gamow-Teller transitions in Refs. [10,3] reveals that the latter have been, in general, significantly overestimated which is caused by the fact that the position of the Gamow-Teller resonance is usually put at too low excitation energies in the daughter. The SMMC calculation implies that the Gamow-Teller transitions should not contribute noticeably to the electron capture rates on odd-A nuclei at the low temperatures studied in Tables 14–16 in [3]. Thus, the rates for odd-A nuclei given in these tables should generally be replaced by the non-Gamow-Teller fraction.

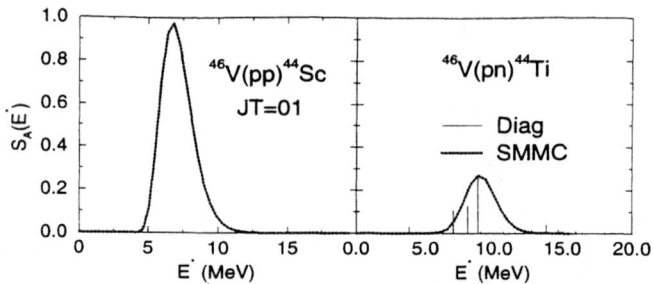

FIGURE 1. Left: proton pair strength distribution for ^{46}V. Right: proton-neutron (T=1) pair strength distribution. SMMC: thick line; direct diagonalization: impulses.

I would now like to turn to the subject of pair correlations in nuclei, and calculations aimed at their understanding. Nuclei near N=Z offer a unique place to study proton-neutron pairing, particularly in the isospin T=1 channel. In fact, most heavy odd-odd, N=Z nuclei beyond ^{40}Ca have total spin J=0, T=1 ground states. Theoretical studies have shown that many of these nuclei have enhanced T=1 proton-neutron correlations when compared to their even-even counterparts. These correlations are to a lesser extent present in even-even systems, but tend to decrease as one moves away from N=Z. In at least one nucleus in the mass 70 region, ^{74}Rb, there is experimental evidence for a ground state $T = 1$ band [11].

Experimentally, pair correlations can best be measured by pair transfer on nuclei. Although total cross sections are typically underpredicted when one employs spectroscopic factors computed from the shell model, relative two-nucleon spectroscopic factors within one nucleus are more reliable. Therefore, it is necessary for one to calculate and measure pair transfer from both the ground and excited states in a nucleus.

The SMMC method may be used to calculate the strength distribution of the pair annihilation operator A_{JTT_z}, as defined in [1]. The total strength of these pairing operators, i.e. the expectation $\langle A^\dagger_{JT} A_{JT} \rangle$, has been studied previously as a function of mass, temperature [12,13], and rotation [14]. I would like to briefly present here for the first time the strength distributions of the pair operators as calculated in SMMC. The strength distribution for the pair transfer spectroscopic factors is proportional to $\langle A - 2 \mid A_{JT} \mid A \rangle$, and is calculated by the inversion of Eq. (4).

In future work, I will discuss the strength distributions in detail. Here I would like to briefly conclude by demonstrating that the SMMC results and the direct diagonalization results agree very nicely for the proton pair strength distributions in the ground state of ^{46}V. This is demonstrated in the left panel of the figure. Shown in the right panel is the isovector proton-neutron pairing strength distribution with respect to the daughter nucleus. Notice that the overall strength is much larger in the proton-neutron channel, as discussed previously in [13], and that the peak is several MeV lower in excitation relative to the like-particle channel. In both cases the strength distribution in ^{46}V differs significantly from that found in ^{48}Cr, where one finds that the dominant component is a ground-state to ground-state transition involving mainly particles in the $0f_{7/2}$ single-particle state. In both odd-odd N=Z channels, the distribution is fairly highly fragmented.

In this proceedings, I have used two specific examples (there are several others) for which the SMMC calculations have proven very useful in understanding the properties of nuclei in systems where the number of valence particles prohibits the use of more traditional approaches. The method has proven to be a valuable tool to further our understanding of nuclear structure and astrophysics. While I have concentrated mainly on stable or proton-deficient nuclei in this contribution, continuing developments in the areas of multi-major shell calculations will, in the near future, provide much information on very neutron-rich nuclei.

REFERENCES

1. Koonin S.E., Dean D.J., and Langanke, K., *Phys. Repts.* **278**, 1 (1997), and references therein.
2. Bethe H.A., *Rev. Mod. Phys.*, **62**, 801 (1990).
3. Aufderheide M.B., Fushiki I., Woosley S.E., Hartman D.H., *Astrophys. J. Supp.* **91**, 389 (1994).
4. Dean D.J., Langanke K., Chatterjee L., Radha P.B., and Strayer M.R., *Phys. Rev.* **C58**, 536 (1998).
5. Kuo T. T. S. and Brown G. E., *Nucl. Phys.* **A114**, 241 (1968).
6. Poves A. and Zuker A., *Phys. Repts.* **70**, 235 (1981).
7. Caurier E., Zuker A., Poves A., and Martinez-Pinedo G., *Phys. Rev.* **C50**, 225 (1994).
8. Langanke K., Dean D.J., Radha P.B., Alhassid Y., and Koonin S.E., *Phys. Rev.* **C52**, 718 (1995).
9. Radha P.B., Dean D.J., Koonin S.E., Langanke K., and Vogel P., *Phys. Rev.* **C56**, 3079 (1997).
10. Fuller G.M., Fowler W.A., and Newman M.J., *Astrophys. J.* **48**, 279 (1982).
11. Rudolph D., et al, *Phys. Rev. Lett.* **76**, 376 (1996).
12. Langanke K., Dean D.J., Radha P.B., and Koonin S.E., *Nucl. Phys.* **A602**, 244 (1996).
13. Langanke K., Dean D.J., Radha P.B., and Koonin S.E., *Nucl. Phys.* **A613**, 253 (1997).
14. Dean D.J., Koonin S.E., Langanke K., and Radha P.B., *Phys. Lett.* **B399**, 1 (1997)

In-beam γ-ray spectroscopy in the ground-state proton emitter ^{113}Cs

C. J. Gross[1,2], Y. A. Akovali[2], C. Baktash[2], J. C. Batchelder[1],
C. R. Bingham[3], M. P. Carpenter[4], C. N. Davids[4], T. Davinson[5],
D. Ellis[3], A. Galindo-Uribarri[2], T. N. Ginter[6], R. Grzywacz[3],
R. V. F. Janssens[4], J. W. Johnson[1], J. F. Liang[2], C. J. Lister[4],
J. Mas[7], B.D. MacDonald[8], S. D. Paul[2], A. Piechaczek[9],
D. C. Radford[2], W. Reviol[3], K. Rykaczewski[2], W. Satuła[7],
D. Seweryniak[4], D. Shapira[2], K. S. Toth[2], W. Weintraub[3],
P. J. Woods[5], C.-H. Yu[2], E. F. Zganjar[9], J. Uusitalo[4]

[1] Oak Ridge Institute for Science and Education, Oak Ridge, TN 37831, USA
[2] Physics Division, Oak Ridge National Laboratory, Oak Ridge, TN 37831, USA
[3] Department of Physics and Astronomy, University of Tennessee, Knoxville, TN 37996, USA
[4] Physics Division, Argonne National Laboratory, Argonne, IL 60439 USA
[5] Department of Physics, University of Edinburgh, Edinburgh EH9 3JZ, UK
[6] Department of Physics and Astronomy, Vanderbilt University, Nashville, TN 37235, USA
[7] Joint Institute for Heavy Ion Research, Oak Ridge, TN 37831
[8] School of Physics, Georgia Institute of Technology, Atlanta, GA 30332 USA
[9] Department of Physics and Astronomy, Louisiana State University, Baton Rouge, LA 70803, USA

Abstract. Gamma-ray transitions in the ground-state proton emitter ^{113}Cs have been identified using the reaction ^{58}Ni(^{58}Ni,p2n) at a beam energy of 230 MeV and the recoil decay tagging technique. The first experiment was done using the Recoil Mass Spectrometer at the Holifield Radioactive Ion Beam Facility where γ-ray transitions were detected with 6 Clover and 5 Duet Ge detectors. A follow-up experiment using the GAMMASPHERE-FMA combination at Argonne National Laboratory was performed. Ninety-six Ge and 4 LEPs detectors were used to record recoil-$\gamma\gamma$ coincidences. Both experiments employed standard recoil mass separation techniques which resulted in the implantation of A=113 reaction products into a double-sided silicon-strip detector. By gating on the energy of the emitted proton and decay time, the correlation between γ rays and the implanted ^{113}Cs could be observed. Initial analysis and comparison with energy level systematics of the Cs isotopes reveal a decay sequence based on the $h_{\frac{11}{2}}$ bandhead. Further analysis is required to determine the decay of this sequence to the expected positive parity ground state.

INTRODUCTION

Ground-state proton emitters are undergoing great scrutiny experimentally and theoretically [1-5]. Located beyond the proton drip line, these weakly bound systems are held together by the Coulomb and angular momentum barriers. Through the measurement of the proton decay energy and half-life, the valence configuration for spherical nuclei can be determined using such simple approaches as the WKB method [5].

There are, however, a few proton emitters such as ^{113}Cs which have anomalous half-lives and cannot be described using the spherical WKB approach. It has been suggested that deformation is the cause for these anomalies and the theoretical approach of Bugrov and Kadmenskii [4] which includes configuration mixing associated with deformation can be applied to these nuclei. Situated just above the $Z = 50$ closed shell, candidate valence configurations for the odd proton are $d_{\frac{5}{2}}$, $g_{\frac{7}{2}}$, $g_{\frac{9}{2}}$ (extruder), and $h_{\frac{11}{2}}$ (intruder). The 18-μs half-life [3,6] of the 0.96 MeV proton radioactivity is not adequately described by the calculated half-life of a spherical $d_{\frac{5}{2}}$ state shown in column 1 of table 1 taken from ref. [5]. The half-lives in columns 2 and 3 which include the possibility of deformation given in column 4 are from ref. [4]. Thus, the ground state of ^{113}Cs is interpreted to be built upon the $[422]\frac{3}{2}^{+}$ state with a deformation of $\beta_2 \approx 0.2$.

However, other experimental evidence of deformation that would support this conjecture is lacking. One way to determine the extent to which these nuclei are deformed is to study the excited states of these nuclei and compare the observed decay properties to simple, well established criteria for measuring collectivity. The technique known as recoil decay tagging (RDT) [7] is ideal for the detection of γ rays correlated with the proton radioactivity. To this end, we have performed two experiments to identify γ-ray transitions and their coincidence relationships in ^{113}Cs.

TABLE 1. Calculated half-lives for the $d_{\frac{5}{2}}$ orbitals ^{113}Cs from ref. [4,5]

$T_{\frac{1}{2}}$ WKB (μs)	$T_{\frac{1}{2}}[420]\frac{1}{2}^{+}$ (μs)	$T_{\frac{1}{2}}[422]\frac{3}{2}^{+}$ (μs)	Deformation β_2
0.51			0.0
	0.4-6.8	29-41	0.10
	0.3-1.1	16-17	0.20
		14-15	0.24
		12-14	0.27

EXPERIMENT DETAILS

The experiments used 230-MeV ^{58}Ni beams on ^{58}Ni targets. Identification of 10 γ rays was performed using the Recoil Mass Spectrometer (RMS) [8] and the new Clover Ge array presently being installed at the Holifield Radioactive Ion Beam Facility (HRIBF) and the cascade relationship between these transitions was achieved using the GAMMASPHERE [9] and Fragment Mass Analyzer (FMA) [10] combination at the Argonne Tandem Linear Accelerator System (ATLAS) facility. Cesium-113 is produced via the p2n channel and its production cross-section is on the order of 25 μb. Both experiments used standard recoil-γ coincidence techniques and thin, highly segmented, double-sided silicon strip detectors [11]. The time correlations between the γ rays detected at the target, the implantation of the recoiling nucleus, and its subsequent proton decay, permit a clean assignment of the γ rays to ^{113}Cs. The short half-life of this nucleus minimizes the chance of random correlations affecting our assignments.

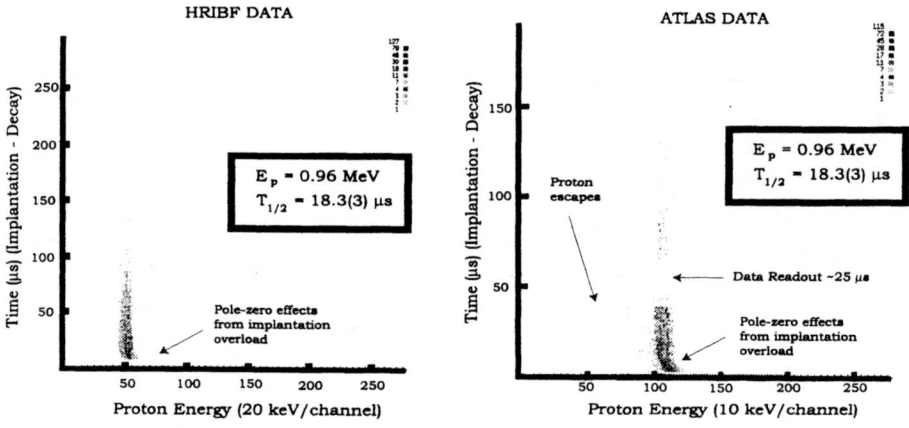

FIGURE 1. Total decay time versus proton energy spectra of ^{113}Cs from the HRIBF (left) and ATLAS (right). Note the "hole" in the ATLAS data arising from the limited delay range of the GAMMASPHERE data acquisition.

One critical aspect central to the success of these experiments is the the relationship between the readout of the data and the detection of the decay event. In both experiments, the data acquisition systems have a programmable time period to delay the readout of the implantation events. This is important since it takes several tens of microseconds for the data to be read out before new events may be accepted. During this time, decay events may not be recorded properly and are, therefore, suppressed at the end of the event. At HRIBF, the time delay was set to 80 μs before the first data were read. At GAMMASPHERE, the data acquisition was set to 50 μs, and this is the reason for the "hole" in the proton decay time vs.

energy spectrum shown on the right side of fig. 1. At the end of this experiment, the time was extended to 100 μs by replacing an integrated circuit delay chip in the data acquisition system.

DISCUSSION

From the 5500 proton decay events recorded in the HRIBF experiment, a half-life of 18.3 ± 0.3 μs was measured and is in agreement with previous measurements. The total γ-ray spectra gated by ^{113}Cs decay events are shown in fig. 2. Despite the lack of anti-Compton shields on the Clover Ge detectors and the lower overall detection efficiency, the HRIBF data contains almost all of the transitions detected by GAMMASPHERE. The largest discrepancy occurs below 200 keV where identification of the 92 and 166 keV transitions is hindered by the high background. However, these transitions can be observed in the data if one knows they are there. The 72 keV transition is not observed in the HRIBF data probably due to intensity and absorption associated with the temporary, thick-walled target chamber. A charge-state reset foil was not used at the HRIBF while one was used during the ATLAS experiment. Since all the strong γ-ray transitions are observed in both experiments we can conclude that these transitions have half-lives less than a few nanoseconds and have small electron conversion branches. Thus, we can confine ourselves to consider that all observed transitions are E1, M1, or E2. We note that under the above criteria, M2 radiation is possible above 250 keV and E3 radiation above 1 MeV [12].

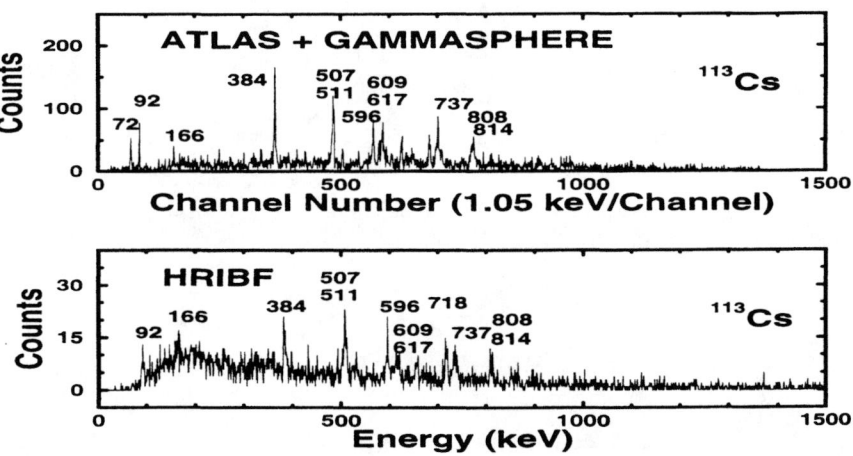

FIGURE 2. Total projection of γ-ray spectra of ^{113}Cs from both experiments.

An examination of the γ-ray coincidences recorded in the ATLAS experiment revealed a cascade relationship between the 384-596-737-814 keV transitions. A comparison of the energy level systematics [13,14] for the $h_{\frac{11}{2}}$ bands in the cesium isotopes is shown in fig. 3; it reveals a similar decay pattern and we tentatively assign this cascade to the $h_{\frac{11}{2}}$ bandhead. Further evidence supporting this assignment is that this is the most intensely populated cascade which is typical in this mass region for the $h_{\frac{11}{2}}$, high-j, intruder orbital. The decay of this band to the expected positive parity ground state is, at this time, unclear. It depends strongly on the location of the lowest energy $\frac{9}{2}^+$ state. Should this state lie above the $\frac{11}{2}^-$ state, then the decay would proceed through M2 or E3 transitions, and depending upon the energy and half-life, decay in flight outside the focus of the Ge array. Such a scenario does not necessarily lead to a large reduction in recoil-γ efficiency if the conversion electron branch of this decay is small.

The cesium isotope which most closely resembles ^{113}Cs is ^{125}Cs [14] which is only 4 neutrons above mid-shell (^{121}Cs$_{66}$). The similarity between these two isotopes can be explained by the deeper penetration into the single-particle energy spectrum of low-K intruder orbitals. In ^{125}Cs, the decay of the $h_{\frac{11}{2}}$ band is isomeric since the $\frac{9}{2}^+$ state lies above the $\frac{11}{2}^-$ state.

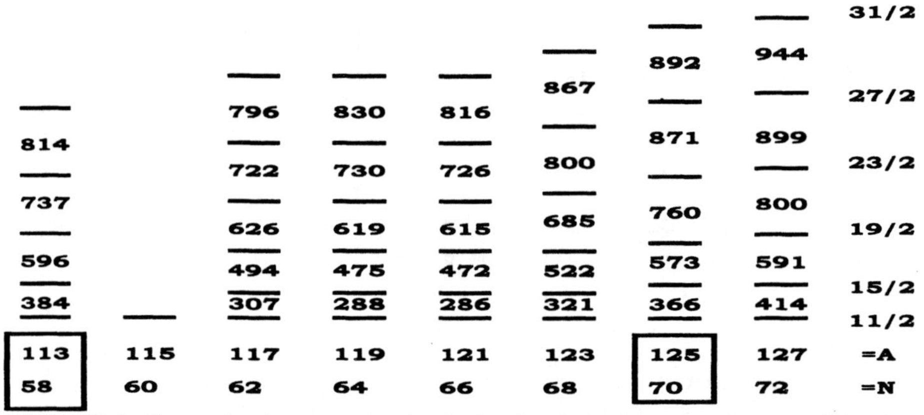

							944	31/2
					867	892		
		796	830	816				27/2
814						871	899	
		722	730	726	800			23/2
737						760	800	
		626	619	615	685			19/2
596		494	475	472	522	573	591	
384		307	288	286	321	366	414	15/2
								11/2
113	115	117	119	121	123	125	127	=A
58	60	62	64	66	68	70	72	=N

FIGURE 3. Energy level systematics for the $h_{\frac{11}{2}}$ bands in the cesium isotopes. The data were taken from Sun, et al. [12] and the present experiment.

SUMMARY

The method known as recoil decay tagging has been used successfully at the HRIBF. A subset of the new Clover Ge array currently being installed at the RMS target position has been used to identify up to 10 γ-ray transitions in the ground state proton emitter ^{113}Cs. In a follow-up experiment using the GAMMASPHERE-FMA combination, one cascade has been tentatively identified as the $h_{\frac{11}{2}}$ band using arguments based on the energy level systematics of the Cs isotopes. Further analysis is required to identify the decay out of this band and to characterize the other low spin states in the suspected deformed nucleus.

ACKNOWLEDGEMENTS

This work was supported by the U. S. Department of Energy under contracts DE-AC05-76OR00033 (ORISE), DE-AC05-96OR22464 (ORNL), W-31-109-ENG-38 (ANL), DE-FG02-96ER40983(UT), DE-FG02-96ER40978 (LSU), DE-FG05-88ER40407 (VU), DE-FG05-87ER40361(JIHIR), and DE-FG02-96ER40958 (GIT). ORNL is managed by Lockheed Martin Energy Research. ORISE is managed by Oak Ridge Associated Universities.

REFERENCES

1. Davids C.N., et al., *Phys. Rev C* **55**, 2255 (1997).
2. Woods P.J., and Davids C.N., *Annu. Rev. Nucl. Part. Sci.*, **47**, 541 (1997) and references therein.
3. Batchelder J.C., et al., *Phys. Rev. C* **57**, R1042 (1998).
4. Bugrov V.P., and Kadmenskii S.G., *Sov. J. Phys.* **49**, 967 (1989).
5. Aberg S., Semmes P. B., and Nazarewicz W., *Phys. Rev. C* **56**, 1762 (1997) and references therein.
6. Page R.D., et al., *Phys. Rev. Lett.* **72**, 1798 (1994).
7. Paul E.S., et al., *Phys. Rev. C* **51**, 78 (1995).
8. Gross C.J., et al., *Application of Accelerators in Research and Industry*, AIP Conf. Proc. No 392 (AIP, Woodbury, NY, 1997), Vol. 1, p. 401.
9. Lee I.Y., *Nucl. Phys.* **A520**, 361 (1990).
10. Davids C.N., et al., *Nucl. Instrum. Methods Phys. Res.* **70**, 358 (1992).
11. Sellin P.J., et al., *Nucl. Instrum. Methods A* **311**, 217 (1992).
12. *Table of Isotopes, 7th Edition* edited by Lederer C. M., and Shirley V. S., (Wiley, New York, 1978).
13. Sun X., et al., *Phys. Rev. C* **51**, 2803 (1995).
14. Hughes J.R., et al., *Phys. Rev. C* **44**, 2390 (1991).

Probing the Structure of the N=Z=31 Nucleus $^{62}_{31}$Ga

S.M. Vincent[1], P.H. Regan[1], D.D. Warner[2], W. Gelletly[1],
J. Simpson[2], R. Bark[3], D. Blumenthal[4], M.P. Carpenter[4],
C.N. Davids[4], D.J. Henderson[4], R.V.F. Janssens[4], C.J. Lister[4],
D. Nisius[4], C.D. O'Leary[5], C.J. Pearson[1], T. Saitoh[3], J. Schwartz[4,6],
D. Seweryniak[4], and S. Törmänen[3,7]

[1] *Department of Physics, University of Surrey, Guildford GU2 5XH, UK*
[2] *CCLRC Daresbury Laboratory, Warrington WA4 4AD, UK*
[3] *Niels Bohr Institute Tandem Accelerator Laboratory, Risø, DK-4000, Roskilde, Denmark*
[4] *Argonne National Laboratory, 9700 South Cass Ave, Il., 60439*
[5] *School of Sciences, Staffordshire University, Stoke-on-Trent, ST4 2DE, UK*
[6] *A.W. Wright Nuclear Structure Laboratory, Yale University, New Haven, Connecticut, 06511*
[7] *Department of Physics, University of Jyväkylä, Jyväkylä, Finland*

Abstract. The decay scheme of the $N = Z$ odd–odd nucleus, ^{62}Ga has been deduced for the first time using the reactions ^{40}Ca$(^{24}$Mg,pn$)^{62}$Ga and ^{40}Ca$(^{28}$Si,αpn$)^{62}$Ga at beam energies of 65 MeV and 88 MeV respectively. Transitions from excited states in ^{62}Ga were identified using the PEX apparatus at the Niels Bohr Institute Tandem Accelerator Laboratory. Gamma ray angular correlation data collected using the AYEBALL array at the Argonne National Laboratory are consistent with a cascade of stretched E2 transitions which decay from a spin 1^+ bandhead via a pure magnetic dipole decay, directly linking a proposed $T = 0$ structure with the $T = 1$, $J^\pi = 0^+$ ground state.

INTRODUCTION

The role of proton–neutron pairing correlations is of fundamental interest in nuclear structure, and it is in the medium mass, $N = Z$ odd–odd nuclei where this phenomenon will manifest itself most evidently [1]. With this in mind, the decay scheme of the $N = Z$ odd–odd nucleus, ^{62}Ga has been deduced for the first time. This type of study can provide an important insight into the evolution of the $T = 0$ and $T = 1$ modes with increasing mass along the $N = Z$ line. In particular, it has recently been demonstrated [2] that the energy separation of the two sets of states is linked to the relative importance and purity of the $T = 0$ and $T = 1$ pairing modes.

EXPERIMENTAL TECHNIQUE

Excited states in $^{62}_{31}$Ga were populated following two separate fusion-evaporation reactions. Identification of transitions in ^{62}Ga was achieved using the reaction ^{40}Ca(^{28}Si,αpn)^{62}Ga, performed at the Niels Bohr Institute using an 88 MeV beam bombarding a 1 mg/cm^2 self supporting enriched (99.96%) target of ^{40}Ca. The gamma-ray detection was afforded by the PEX spectrometer array [3], consisting of four, seven-element germanium cluster detectors [4] each with a BGO suppression shield. Two of the cluster detectors had their central crystals at an angle of 105° to the beam direction, with the other two being at 146°. The production cross section of this nucleus represented less than 5% of the total, and so information on evaporated charged particles for channel selection was obtained using a 31-element silicon inner ball [5] surrounding the target position, in conjunction with an array of 15 BC501 liquid scintillator neutron detectors [6] positioned at forward angles.

High resolution γ-γ data was also collected at the Argonne National Laboratory using the reaction ^{40}Ca(^{24}Mg,pn)^{62}Ga at a beam energy of 65 MeV. A target of natural Ca was used with a thickness of 500 μg/cm^2 with 300 μg/cm^2 gold coating (to reduce oxidization) and a thick (20 mg/cm^2) gold backing which stopped the residual nuclei at the target position. Gamma rays from excited states were detected using the AYEBALL array [7] which consisted of 18 high purity, Compton suppressed germanium detectors, mounted in four annular rings, at angles 79°, 101°, 134° and 158° to the beam direction.

RESULTS

Figure 1-a shows a total gamma ray spectrum from the PEX experiment and figure 1-b shows the projection of the gamma-gamma coincidence matrix formed with 1α, 0 or 1p and 1n detection criteria. The spectrum clearly identifies transitions at 246, 376, 571 and 1179 keV, which we assign to the αpn channel, ^{62}Ga. Contaminant lines from the α2p channel (^{62}Zn), and the 2αp channel (^{59}Cu), which appear from a combination of the finite detection efficiency of the silicon ball and the misidentification of gamma-ray events in the neutron detectors are also clearly marked. Figure 1-c is a sum of gamma-gamma coincidence spectra gated by the 246, 376 and 571 keV transitions identified in figure 1-b, and projected from the 1α, 0 or 1p and 1n matrix. The spectrum identifies other transitions in coincidence, which are thus also assigned to ^{62}Ga. A tentative γ-ray spectrum for ^{62}Ga has recently been reported by de Angelis et al. in a conference contribution [8], and although no decay scheme was presented, the spectrum is consistent with figure 1-c.

An isomeric state with a mean-lifetime of 4.6±1.6 ns was identified feeding the 246 and 571 keV transitions using the recoil distance decay technique with the PEX data. Angular gated spectra, shown in figure 2, were sorted with gamma rays detected in the pairs of clusters centred at either (a) 105° or (b) 146°. These spectra show the 571 keV peak centred at 146° to have both a Doppler shifted and

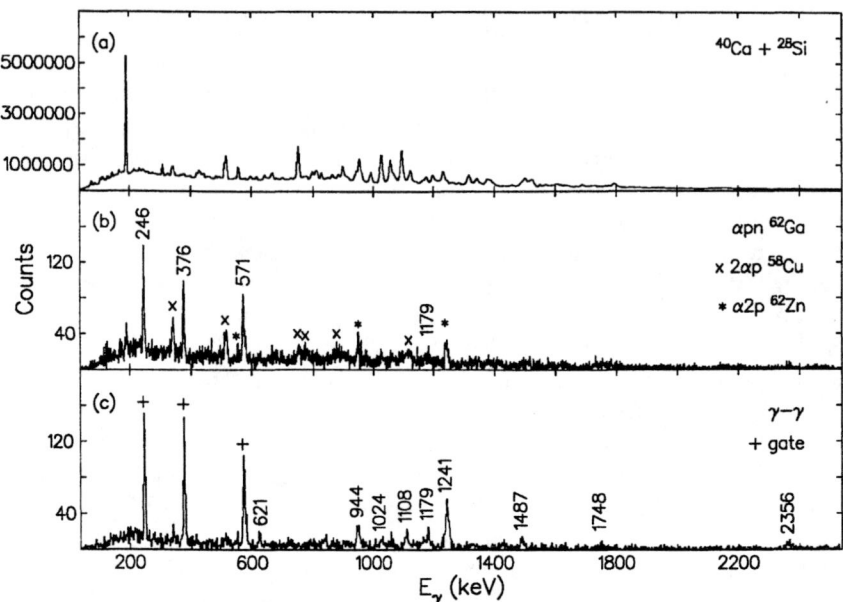

FIGURE 1. (a) Total projection from the PEX experiment; (b) Projection from 1α, 0 or 1p and 1n gated γ-γ matrix, identifying strongest transitions in ^{62}Ga, lines from contaminant masses are marked; (c) Sum of 246, 376 and 571 keV gates (labelled +) on 1α, 0 or 1p and 1n gated γ-γ matrix, showing all the transitions assigned to ^{62}Ga in this work.

an unshifted component in its lineshape. Analysis of the peak shape of the 246 keV transition shows that this is also of a two-component nature. The unshifted line comes from ^{62}Ga recoils which have been implanted in the stopper foils placed in front of the silicon detectors in the charged particle detector. The measured ratio of counts in the moving and stopped peaks of the 571 keV transition, corrected for loss of recoils through the exit hole in the silicon ball, combined with the target-stopper distance and the recoil velocity, yield the lifetime of the isomeric state.

Multipolarity assignments were made using a combination of gamma-ray anisotropies from the PEX data and a DCO ratio analysis from the AYEBALL data. The angle gated spectra from the PEX experiment (shown in figure 2), were used to obtain gamma-ray anisotropies using the method described in reference [9]. A gamma-ray anisotropy A was defined such that $A = 2\left(\frac{W(146°)-W(105°)}{W(146°)+W(105°)}\right)$, where $W(146°)$ and $W(105°)$ are the efficiency corrected intensities measured in the cluster detectors centred at 146° and 105°, respectively. A clear separation was shown between previously identified gamma rays of a stretched E2 and pure dipole nature. Weighted averages of the anisotropies measured for known stretched E2 and $\Delta I = 1$ dipole transitions in ^{62}Zn [10] and ^{65}Ga [11] are plotted with anisotropy values obtained for the 246, 376 and 571 keV transitions is ^{62}Ga in figure 3-a. The anisotropy value obtained for the 376 keV line was found to be consistent with a stretched E2 decay, while the 246 and 571 keV transitions have ratios which are

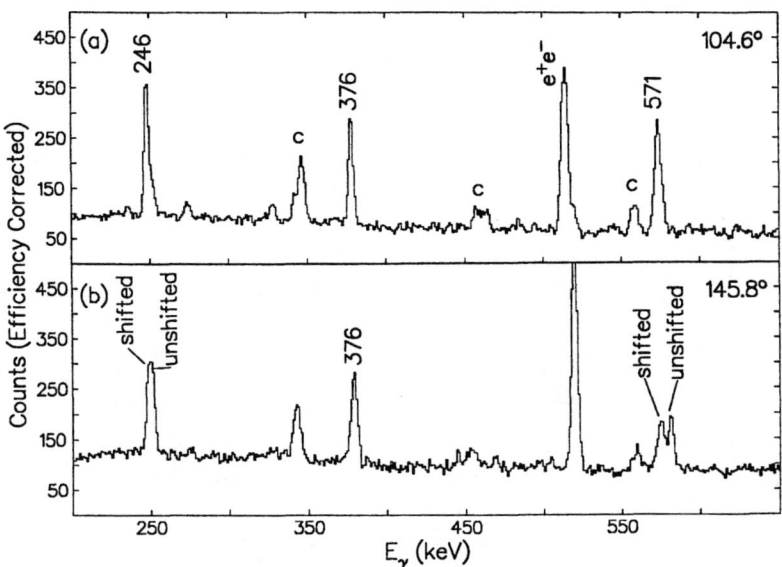

FIGURE 2. Expanded spectra of the angular projections from the PEX data gated by 1 alpha, 1 neutron and either 0 or 1 protons, showing lines in ^{62}Ga, (a) from the clusters centred at 105° and (b) from the clusters centred at 146°. The lines marked 'c' are contaminants from other channels, the line marked 'e^+e^-' is due to pair production in the target. The 246 keV and the 571 keV peaks have a shifted and an unshifted component due to an isomer.

consistent with an isotropic decay.

Multipolarities for the other transitions in ^{62}Ga were then derived from DCO ratios [12] using the AYEBALL data. In this analysis, a γ-γ coincidence matrix was constructed with events measured in detectors at 158° (θ_1) versus events in any of the 79°, 101° or 134° (θ_2) detectors. By gating on a transition of known multipolarity on each axis of this matrix in turn and measuring the efficiency corrected relative intensity of the projected γ rays on the other axis, a DCO intensity ratio, $R_{dco} = \frac{I_{\gamma_1}(\theta_1) \text{ gated by } \gamma_2(\theta_2)}{I_{\gamma_1}(\theta_2) \text{ gated by } \gamma_2(\theta_1)}$, could be obtained. Again, coincidences between previously assigned stretched E2, $\Delta I = 1$ E1 and $\Delta I = 1$ M1 transitions from the strongly populated channel of ^{61}Cu [13] were used to verify and calibrate this method. Weighted averages obtained for known E2,E2 and E2,E1 coincidences are plotted in figure 3-b, along with DCO ratios measured for the 246, 571 and 1241 keV transitions in ^{62}Ga, gated by the 376 keV transition. Assuming a stretched E2 multipolarity for the gating 376 keV transition (from its measured anisotropy in the PEX data), the 571 keV transition has a DCO ratio consistent with $\Delta I = 1$. By comparison, the ratios obtained for the 246 keV and the 1241 keV transitions are consistent with stretched quadrupole decays.

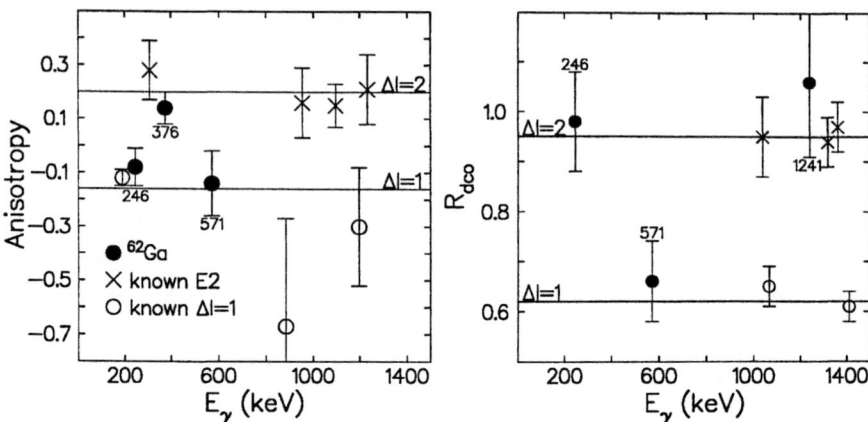

FIGURE 3. (a) Anisotropy values measured from PEX data for 246, 376 and 571 keV transitions in ^{62}Ga. Also plotted are known transitions in ^{62}Zn and ^{65}Ga, with weighted averages indicated for $\Delta I=1$ and $\Delta I=2$ transitions marked. (b) DCO ratios for 246, 571 and 1241 keV transitions in ^{62}Ga from the AYEBALL data, gated by the 376 keV E2 identified in (a). Ratios for known $\Delta I=1$ and $\Delta I=2$ transitions in ^{61}Cu gated by the 1310 keV E2 transition, along with their weighted averages are also plotted.

DECAY SCHEME

The decay scheme for ^{62}Ga obtained in the current work is shown in figure 4-a. The intensity of the 946 keV and 2356 keV transitions are the same within experimental uncertainties and hence their ordering is tentative. The spin parity assignments are made assuming the 571 keV transition feeds the $T = 1$, $I^\pi = 0^+$ ground state identified in previous beta decay studies [14] and that near yrast states are preferentially populated in fusion evaporation reactions. Given the difficulty of generating low-lying negative parity states from the available single particle orbitals, the 571 keV dipole transition would correspond to a pure magnetic dipole $1^+ \to 0^+$ decay with no E2 admixture. The experimentally derived DCO ratio for this transition when gated by an E2 transition is 0.66 ± 0.08, which is consistent with this picture.

Figure 4-b shows for comparison the low-lying states in the $T_z = 1$ isobar, ^{62}Zn [10]. In contrast to the recently reported ^{74}Rb/^{74}Kr pair [15], there is no observable population of the isobaric analogue states in the odd-odd nucleus. However, due to its lower deformation, the first excited state in ^{62}Zn lies at 954 keV so that if the assumption of the direct feeding of a 0^+ ground state for ^{62}Ga is correct, then the first 2^+, $T = 1$ state is likely to be non-yrast. The excited states are therefore interpreted as being a $T = 0$ band, built on an $I^\pi = 1^+$ bandhead, which decays directly into the $T = 1$, $I^\pi = 0^+$ ground state. This nucleus has been interpreted in the context of the shell model and the IBM-4 [16], an interacting boson model with $T = 0$ and $T = 1$ bosons.

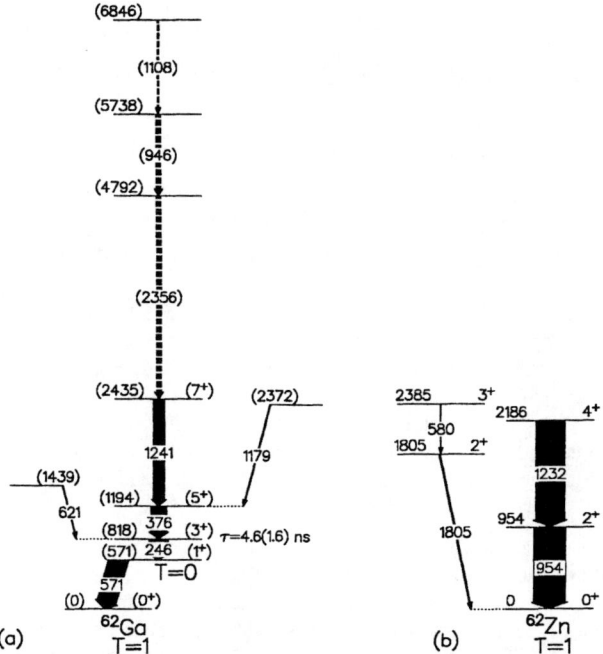

FIGURE 4. (a) Partial decay scheme for ^{62}Ga deduced from this work. Tentatively ordered transitions are dashed, spins and parities are made from anisotropy and DCO arguments. (b) Low lying yrast states in the $T_z=1$ isobaric anologue nucleus ^{62}Zn.

REFERENCES

1. D.D. Warner, *Perspectives for the Interacting Boson Model*, edited by R.F. Casten et al. World Scientific, Singapore, 373 (1988).
2. Van Isacker P. and Warner D.D., *Phys. Rev. Lett.* **78**, 3266 (1997).
3. Grawe H. et al., *Z. Phys.* **A358**, 185 (1997).
4. Beausang C.W. and Simpson J., *J. Phys.* **G22**, 527 (1996).
5. Kuroyanagi T. et al., *Nucl. Instr. and Meth.* **A316**, 289 (1992).
6. Arnell S.E. et al., *Nucl. Instr. Meth.* **A300**, 303 (1991).
7. Carpenter M.P., *Z. Phys.* **A358**, 261 (1997)
8. de Angelis G. at al., *Nucl. Phys.* **A630**, 426c (1998).
9. Pohl K.R. et al., *Phys. Rev.* **C53**, 2682 (1996).
10. Svensson C.E. et al., *Phys. Rev. Lett.* **80**, 2558 (1998).
11. Shengjiang Z., *Chin. Jour. Nucl. Phys.* **13**, 331 (1991).
12. Krane K.S., Steffen R.M. and Wheeler R.M., *Nucl. Data Tables* **11**, 352 (1973).
13. Hatsukawa Y., *Z. Phys.* **A359**, 3 (1997).
14. Davids C.N. et al. *Phys. Rev* **C19**, 1463 (1979).
15. Rudolph D. et al., *Phys. Rev. Lett.* **76**, 376 (1996).
16. Vincent S. et al., *Submitted to Phys. Lett.* **B**

Decay of Very Neutron-Rich Mn Nuclides and Vanishing of the N=40 Subshell Closure in ^{66}Fe

A. Wöhr[1,2,*], M. Hannawald[1], W.B. Walters[3], T. Kautzsch[1], B. Pfeiffer[1], K.-L. Kratz[1], V.N Fedoseyev[4], V.I. Mishin[4], D. Forkel-Wirth[5], V. Sebastian[6], M. Koizumi[5], U. Köster[7], J. Lettry[5], H.L. Ravn[5] and the ISOLDE-Collaboration[5]

[1] Institut für Kernchemie, Universität Mainz, D-55099 Mainz, Germany
[2] Instituut voor Kern- en Strahlingsfysica, University of Leuven, B-3001 Leuven, Belgium
[3] Department of Chemistry, University of Maryland, College Park, MD 20742, USA
[4] Institute of Spectroscopy, Russian Academy of Sciences, RUS-142092 Troitzk, Russia
[5] CERN, CH-1211 Geneva 23, Switzerland
[6] Institut für Physik, Universität Mainz, D-55099 Mainz, Germany
[7] TU München, Physik-Department, D-85748 Garching, Bavaria, Germany

Abstract. The use of chemically selective laser ionization combined with beta-delayed neutron counting at ISOLDE has permitted identification and half-life measurements for ^{61}Mn up through ^{69}Mn. The 14(4)-ms half life for ^{69}Mn is one of the shortest determined for any nuclide beyond the sd shell. Gamma-ray singles and coincidence spectra have been determined for the decays of ^{64,66}Mn to levels of ^{64,66}Fe, revealing strong deformation and vanishing of the N = 40 subshell in the Fe isotopes.

INTRODUCTION

Until recently, the principal data available for neutron-rich nuclides between ^{48}Ca and ^{78}Ni were the γ-spectroscopic data obtained in the 1980's at GSI (1,2) and the nuclear masses reported by Seifert et al. (3). During the past three years, a number of new experimental studies on level structures and decay properties on Fe group nuclei have been performed (4-9). On the theoretical side, nuclear masses, deformation values and β-decay properties were calculated by Möller et al. (10) on the basis of the macroscopic-microscopic Finite Range Droplet Model (FRDM) and the Quasi-Particle Random Phase Approximation (QRPA). Similarly, Aboussir et al. (11) also calculated masses and ground-state (g.s.) deformation values on the basis of the Extended Thomas-Fermi plus Strutinsky Integral (ETFSI-1) approach that are not always in agreement with those of the FRDM for the Fe group nuclei considered here. Earlier, Richter et al. (12) had performed fully microscopic shell-model calculations in this mass region and significantly underpredicted the binding energies for the Cr to Fe isotopes with N>36. In addition to the clear nuclear-structure interest in the neutron-rich Fe group nuclei may also play an important role as possible seed nuclei in the astrophysical r-process (13). In the present paper, we report new measurements for the half-lives of heavy Mn nuclides up to ^{69}Mn and for the level structure of even-even ^{64,66}Fe populated in the decays of ^{64,66}Mn.

CP455, *ENAM98: Exotic Nuclei and Atomic Masses*
edited by B. M. Sherrill, D. J. Morrissey, and Cary N. Davids
© 1998 The American Institute of Physics   1-56396-804-5/98/$15.00

EXPERIMENTAL PROCEDURES AND RESULTS

The neutron-rich isotopes of Mn were produced at the CERN-ISOLDE facility by 1-GeV proton-induced spallation of unarium in a thick UC_2 target. Ionization was accomplished using a chemically selective laser ionization scheme described previously (14). Beams of mass-separated Mn nuclides with masses differing by $\Delta A \geq 4$ were transported separately to two different beam lines equipped with moving tape systems where delayed-neutron counting and γ-ray singles and coincidence measurements could be performed independently. In both cases, counting took place at the point of deposit owing to the short half-lives of the nuclides under study, and the tape systems were used to remove the daughter nuclides and unavoidable surface-ionized isobaric Ga activities. Because the Mn half-lives being sought are in the millisecond range, counting in both systems was initiated by the proton pulses, separated by a multiple of 1.2 seconds, and continued for 1.0 s for each cycle.

Beta-delayed neutron data of high statistical quality were collected by multiscaling measurements using the Mainz ^{3}He neutron counter. The decay curves were fitted with a constant small background component up through A=65. The fits for the heavier isotopes took into account predicted β-delayed neutron branching (10b) together with the known half-lives (5c,7,15) for the Fe daughter and Co grand-daughter isobars. For A=66 to 68, the contributions from delayed-neutron branching of the Fe and Co isobars are quite small and actually do not affect the Mn half-life fits. For A=69, however, a three-component fit was necessary to account for the significant Fe and Co neutron branches. The resulting data are summarized in Table 1, and are compared to literature values and QRPA predictions using experimental masses as far as they are available (15) and g.s. deformations from the FRDM calculations (10).

Table 1. Experimental and theoretical half-lives for the neutron-rich Mn nuclides.

Mass	Half-life[ms]				Mass	Half-life[ms]			
	This work	Lit.	Ref.	QRPA		This work	Lit.	Ref.	QRPA
61	623(10)	710(10)	(2)	234	65	88(4)	160(30)	(7a)	39
62	671(5)	880(150)	(15)	267			110(20)	(7b)	
63	275(4)	282(18)	(15)	76			88(7)	(5c)	
		321(22)	(5c)		66	66(4)	220(40)	(7a)	21
64	89(4)	240(30)	(7a)	67			90(20)	(7b)	
		140(30)	(7b)				62(13)	(5c)	
		91(7)	(5c)		67	42(4)			25
					68	26(4)			18
					69	14(4)			16

For ^{61}Mn, the 623(10) ms half-life that we have observed may be compared with the spread of half-life values that range from 540 ms to 850 ms reported in (2) for γ-rays in the β-decay daughter ^{61}Fe. For the strongest lines, however, a half-life of 710(10) ms was obtained which is listed in NUBASE (15). The 671(5) ms half-life that we have determined for ^{62}Mn is at the lower edge of the 880(150) ms value reported in (2). Our half-life of 275(4) ms for ^{63}Mn is in good agreement with the value of 282(18) ms reported in (1), but is somewhat lower than the recent value obtained in (5c).

Our new half-lives for ^{64,65,66}Mn are in good agreement with the recent values reported by Sorlin *et al.* (5c), but are in disagreement with the initial values reported by Ameil *et al.* (7a) and cited in NUBASE (15). In a subsequent publication the same authors report somewhat shorter half-lives for these nuclides (7b), however still exceeding our and Sorlin's values. Similarly, Franchoo *et al.* found systematic differences between their Ni half-lives (8) and those of Ameil *et al.* (7). These discrepancies do serve to point out possible difficulties of assessing the reliability of results from techniques where the nuclidic identification is not unambiguous. We would like to note again in this context, that in our work we take advantage of element-selective laser ionization as well as of isobar-selective delayed-neutron detection. By means of γ-spectroscopy it has been verified that the only isobar-contamination comes from surface-ionized neutron-deficient Ga isotopes.

For ^{67,68,69}Mn, the new half-life values are quite short. For nuclides beyond the sd shell, ^{69}Mn$_{44}$ has one of the largest relative isospin values (N-Z)/Z=1.76 and one of the shortest half-lives of any nuclide for which β-decay properties have been measured.

The γ-ray data were written in event-by-event mode for β-gated γ singles as well as γγ(t) coincidences with time recorded relative to each proton pulse. In this way, 8 consecutive spectra of 150 ms time intervals were collected. As considerable data exist for the structure of even-even Fe nuclides up to A=62, in the present study we focused on the γ-spectra for decay of 89-ms ^{64}Mn and 66-ms ^{66}Mn.

At A=64, γ-ray peaks up to 4.2 MeV could be assigned to the decay of ^{64}Mn. More than 20 lines have been incorporated into a decay scheme of at least 8 excited levels. The most intense line in the early ^{64}Mn spectra is at 746 keV and is taken to be the 2^+ to 0^+ transition in the even-even daughter ^{64}Fe. As five of the eight levels appear to depopulate to both the g.s. and the first 2^+ level, a low spin for the g.s. of ^{64}Mn is indicated. No candidate has been identified so far for the 4^+ to 2^+ transition. More than 20 γ-ray lines with energies up to at least 4.2 MeV have also been observed in the decay of 66-ms ^{66}Mn. The strongest line by far is at 573 keV which is – as in the above case of ^{64}Fe – taken as the g.s.-transition from the first 2^+ level in even-even ^{66}Fe. A relatively weak γ-line of 840 keV depopulates a level at 1414 keV and feeds the first 2^+

state. On the basis of the observed coincidences, the γ-branching and their intensities, we take this level to be the first 4+ state in ^{66}Fe. A number of high-energy g.s.-transitions (E>2.7 MeV) indicates strong Gamow-Teller feeding to that energy region, which is in perfect agreement with the predominant $\nu f_{5/2} - \pi f_{7/2}$ decay predicted by the QRPA, nearly independent of the assumed deformation. More detailed information will be given in a forthcoming paper (16).

The 2_1^+ energies and the E_{4+}/E_{2+} ratios for even-even Cr, Fe, Ni, Zn, and Ge isotopes are shown in Figure 1. As is well established, the 2^+ energy in the Z=28 Ni nuclides rises sharply at N=40 and exhibits clear evidence for a partial double shell closure, similar to that for Z=40, N=50 in ^{90}Zr. Recent studies of the structure of ^{69}Ni and ^{69}Cu are consistent with the closed shell character of ^{68}Ni (8,9).

The sudden drop in the energy of the 2_1^+ level for ^{64,66}Fe suggests an increase in deformation(β_2) at N=40 as indicated by the calculations of ETFSI-1 (11) rather than the near zero value for β_2 predicted by FRDM (10). Indeed, the energies of the yrast 2^+ and 4^+ levels and the E_{4+}/E_{2+} ratio for ^{66}Fe$_{40}$ are almost identical to those for ^{76}Ge$_{44}$. Extensive studies of the structure of the even-even Ge nuclides by Lecomte et al. (17) indicated deformation values for β_2 of 0.27. Such a value appears to be a good estimate for the g.s.- deformation in ^{66}Fe and can be contrasted with the more modest deformation of the lighter Fe nuclides where values of β_2 are ~0.18.

We attribute this deformation to the strong proton-neutron (pn) interaction between the two $f_{7/2}$ proton holes and the $g_{9/2}$ neutrons which results in a dramatic lowering of the energy of the $\nu g_{9/2}$ orbital. As a consequence, the downsloping 1/2+[440] and 3/2+[431] orbitals lie at a lower energy and can be occupied by the 38th and 40th neutrons in Fe rather than by the 42nd and 44th neutrons in the Ge nuclides. The observation of an M2 isomeric transition in ^{61}Fe$_{35}$ attributed to a 9/2$^+$ isomer at 860 keV is consistent with such an idea (9). For comparison, in ^{65}Zn$_{35}$ the 9/2$^+$ state lies at 1065 keV. As collectivity usually increases with the distance from closed shells, as demonstrated by the deformation of ^{48}Cr$_{24}$, it is likely that ^{64}Cr$_{40}$ with four proton holes in the Z=28 closed shell will be even more deformed than ^{66}Fe.

In summary, we have extended accurate half-life measurements up to ^{69}Mn whose relative isospin of [(N-Z)/Z]=1.76 is among the highest so far investigated. Gamma-ray studies of the structure of the heavy Fe nuclides reveal a sharp downturn in the energies of the first 2^+ states of ^{64}Fe and ^{66}Fe which we attribute to a vanishing of the spherical N=40 subshell and the onset of a new region of significant deformation, probably being most pronounced around ^{64}Cr. Together, these new results which complement other recent studies (4-9), and the previously published spectroscopic (1,2) and mass measurements (3), provide a solid basis on which to evaluate the various theoretical models for nuclear structure and decay in this mass region.

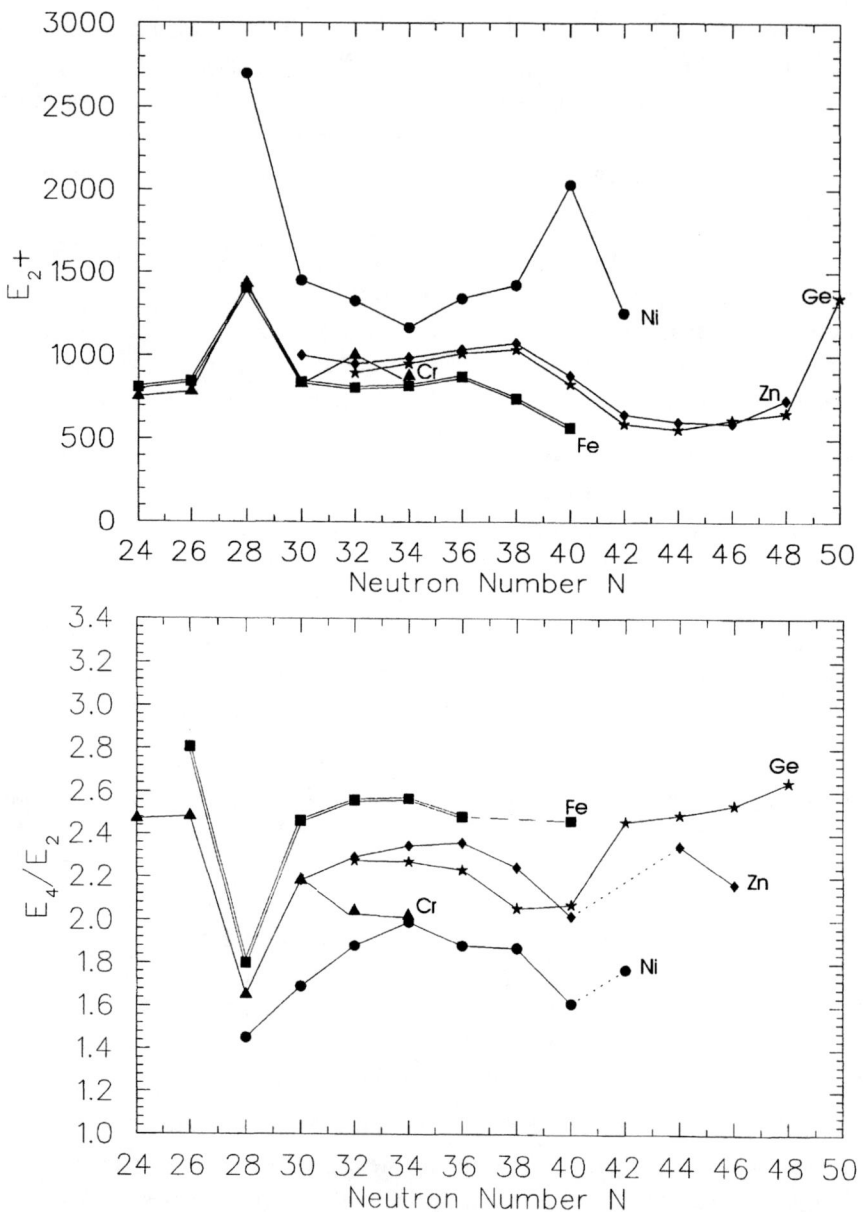

Figure 1. Systematics of the 2$^+$ energies (upper part) and E_{4+}/E_{2+} ratios (lower part) for even-even Cr, Fe, Ni, Zn, and Ge

ACKNOWLEDEMENTS

This work was supported by the German BMBF (06MZ864) and DFG (436RUS/17/40/97 and Kr806/3), the Russian Foundation for Basic Research (96-02-18331), the Belgian FWO, and the U.S. Department of Energy.

* Present address: Physics Dept., Oxford University, Oxford OX1 3PU, UK

REFERENCES

1. U. Bosch et al., Phys. Lett. **B164**, 22 (1985); and Nucl. Phys. **A477**, 89 (1988).
2. E. Runte et al., Nucl. Phys. **A399**, 163 (1983).
3. H.L. Seifert et al., Z. Phys. **A 349**, 25 (1994).
4. T. Pawlat et al., LNL 1995 Progress Report, LNL-INFN 105/96, 7 (1996); and R. Broda et al., Il Nuovo Cimento, in press (1998).
5. O. Sorlin et al., Proc. ENAM'95, 603 (1996); Nucl. Phys. **A632**, 205 (1998); and Il Nuovo Cimento, in press (1998).
6. T. Dörfler et al., Phys. Rev. **C54**, 2894 (1996).
7. F. Ameil et al., Proc. ENAM'95, 537 (1995); and Eur. Phys. J. **A1**, 275 (1998).
8. S. Franchoo et al., contrib. to this conf.; and subm. to Phys. Rev. Lett. (1998).
9. R. Grzywacz, et al., contrib. to this conf., and subm. to Phys. Lett. B.
10. P. Möller et al., At. Data Nucl. Data Tables **59**, 183 (1995), and **66**, 131 (1997).
11. Y. Aboussir et al., At. Data Nucl. Data Tables **61**, 127 (1995).
12. W.A. Richter et al., Nucl. Phys. **A586**, 445 (1995).
13. K. Takahashi et al., Astron. Astrophys. **286**, 857 (1994).
14. V.I. Mishin et al., Nucl. Instr. and Meth. **B73**, 550 (1993), and **B126**, 88 (1997).
15. G. Audi et al., Nucl. Phys. **A634**, 1 (1997).
16. M. Hannawald et al., to be published.
17. R. Lecomte et al., Phys. Rev. C **22**, 1530, and 2420 (1980).

Fine structure in the alpha-decay of the neutron-deficient ^{191}Po isotope

A.N.Andreyev*, N.Bijnens*, J.F. Cocks†, K.Eskola‡, K. Helariutta†, M.Huyse*, H. Kettunen†, P.Kuusiniemi†, M. Leino†, W.H. Trzaska†, P. Van Duppen* and R. Wyss∥

*Instituut voor Kern- en Stralingsfysica, University of Leuven, B-3001 Leuven, Belgium
†Department of Physics, University of Jyväskylä, FIN-40351 Jyväskylä, Finland
‡Department of Physics, University of Helsinki, Helsinki, Finland
∥Royal Institute of Technology, Department of Physics Frescati, Kungl. Frescativ. 24, 104 05 Stockholm, Sweden

Abstract.
The α-decay properties of the neutron-deficient ^{191}Po nucleus have been investigated at the RITU gas-filled separator using the ^{36}Ar+^{160}Dy→^{191}Po+5n reaction channel. In contrast to previous studies two α-decaying isomeric states were observed in ^{191}Po. By analysing α-γ coincidences for both isomers evidence was found for fine-structure decay to excited levels in the ^{187}Pb daughter nucleus. The tentative interpretation of these levels could be a weak coupling of the odd neutron to an intruder 2p-2h deformed 0^+ state in the even-even core. The difference in the reduced α widths of the α transitions towards the ground state and excited state points to a deformed $13/2^+$ isomeric state in ^{191}Po.

Recent in-beam studies of light odd-mass Pb and Po isotopes identified the structure of the lowest excited states in the ^{187,189,191}Pb [1,2] and in the ^{193,195,197}Po isotopes [3] and gave further evidence that the structure of the first excited states in the odd-mass Pb and Po nuclei can be interpreted as a weak coupling of the odd $i_{13/2}$ neutron hole to the low-lying near-spherical states of the even-even core. A similar coupling of the odd neutron to the members of the deformed band resulting from a proton-pair excitation across the $Z = 82$ closed shell has been observed in ^{191,195,197}Pb [2,4,5]. Such deformed π(2p-2h) states have earlier been identified in even Pb isotopes from $A = 208$ down to $A = 188$ [6,7]. The α decay of ^{192}Po revealed a strong feeding to this 0_2^+ state in ^{188}Pb which indicates that in ^{192}Po the ground state has a dominant π(4p-2h) intruder configuration. In this study the α-decay properties of odd-mass ^{191}Po will be investigated. In previous experiments on ^{191}Po only 'one' α-decaying state was observed (E_α = 7314(20) keV and $T_{1/2} = (15.5^{+6.0}_{-2.5})$ ms [8], E_α = 7330(15) keV and $T_{1/2} = (27^{+22}_{-8})$ ms [9]) in contrast to all heavier odd-mass Po isotopes with masses $193 \leq A \leq 201$ (see e.g. [10]).

The experiments have been performed at the gas-filled recoil separator RITU [11]. A pulsed beam of ^{36}Ar was used with an initial energy of 196 MeV on a ^{160}Dy target (thickness = $500\mu g/cm^2$, 67.1% enriched). Nickel degrader foils were used to change the beam energy. The evaporation residues (EvRs) were implanted into a position-sensitive silicon detector (PSSD). This detector is divided into sixteen independent strips. The identification of the nuclides has mainly been done by the observation of genetically correlated α-decay chains and by the behavior of the excitation functions. Other technical details about the separator and the electronics can be found elsewhere [11,12]. A high-pure germanium detector was installed 2 cm behind the PSSD, outside the vacuum chamber, giving the possibility to take α-γ and α-X ray coincidences.

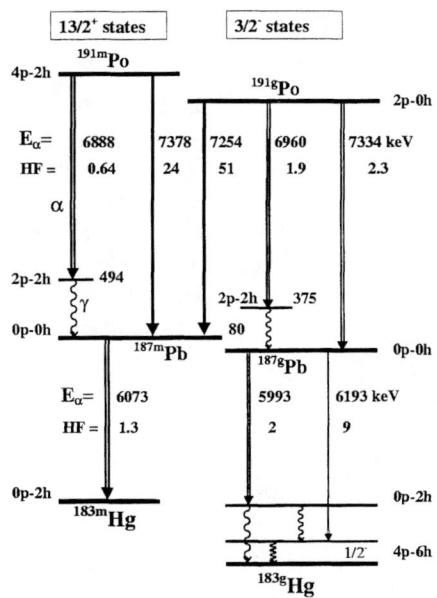

FIGURE 1. Proposed alpha-decay scheme of ^{191m,g}Po.

In the energy region where ^{191}Po was expected, two α-lines were observed with a similar intensity of ≈ 1500 counts. The α line at 7334(10) keV showed a half-life of 22(1) ms while the α line at 7378(10) keV showed a half-life of 98(8) ms. Their excitation functions are in agreement with a 5n reaction channel. Due to their difference in half-life values, these two α-transitions are assigned to the α decay of two isomers in ^{191}Po. Further proof for this assignment and establishment of the decay pattern is provided by the correlations of these alpha-decays with known alpha-decays of the daughter ^{187m,g}Pb isotope. The decay scheme of ^{191}Po, based on our data and on the data from [13,14] for daughter activities is shown in Fig. 1. It should be noted that not only three-fold (recoil-α_1-α_2) but also four-fold

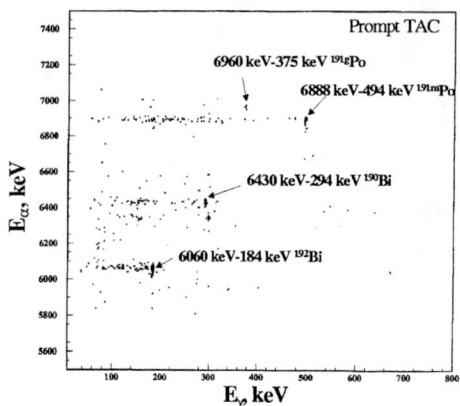

FIGURE 2. Prompt α-γ coincidence matrix.

correlations were observed where the fourth registered events were the gamma-decays of the excited levels, populated by the alpha-decay of ^{187g}Pb (see [13] and Fig. 1). Based on the systematics for heavier odd-mass Hg, Pb and Po isotopes and on the hindrance factors, a low spin value of $3/2^-$ and a high spin value of $13/2^+$ have been tentatively assigned to the ground and to the isomeric states in ^{191}Po, respectively. A weak α-line with the energy of 7254(15) keV and a half-life value of 22(6) ms has been assigned as a crossover transition from the low spin $3/2^-$ ground state in ^{191}Po to the high spin $13/2^+$ isomeric state in ^{187}Pb (see Fig. 1).

Fine structure α-lines in the decay of ^{191m,g}Po were identified from the analysis of the α-γ coincidence matrix, shown in Fig. 2. Except for strong α-γ transitions corresponding to known cases (lower groups) one sees α-γ coincident pairs with E_α = 6888(15) keV and E_γ = 494(1) keV, and E_α = 6960(15) keV and E_γ = 375(1) keV. From the time difference for pairs of EvR-[α-γ] events the half-life value of 116(30) ms has been deduced for the 6888-keV α line and a half-life value of 15^{+8}_{-4} ms for the 6960-keV α line. On the basis of the sum-energy balance, half-life values, and the behavior of the excitation functions the α-line with E_α = 6888(15) keV has been assigned to ^{191m}Po. An admixture of an E0-component was found in the 494-keV transition based on α-Pb X-ray coincidences. This results in a spin value of $13/2^+$ for the excited state at 494 keV and a hindrance factor (HF) of 0.64(15) for the 6888-keV α line. The hindrance factor is defined as the ratio of the reduced α width, using the method of Rasmussen [15], of the transition relative to the reduced α width of the ground-state to ground-state transition in the even-even neighbours.

The α line at 6960 keV has tentatively been assigned to the decay of the ^{191}Po ground state on the basis of its half-life behavior and the sum-energy (6960 + 375 = 7335 keV). By using similar arguments, this results in an $3/2^-$ excited state at 375 keV and a HF of 1.9 for the 6960-keV fine-structure α line.

Although the α-decay energy of the $13/2^+$ isomeric state in ^{191}Po is slightly higher than the decay energy of the $3/2^-$ ground state, its half life is 4 times longer. This is due to the strong hindrance (HF = 24) of the transition from the $13/2^+$ isomer in Po to the $13/2^+$ isomer in Pb. On the contrary, the decay of the $13/2^+$ isomer to the second $13/2^+$ state in ^{187}Pb is slightly faster than the ground-state to ground-state decay of the even-even neighbours. Total potential-energy calculations indicate that for all odd Po nuclei ($189 \leq A \leq 199$) a slightly deformed ($\beta_2 \sim 0.1$) configuration is lowest in energy as well for the positive as for the negative parity states. But there are two exceptions: in ^{191}Po the positive parity states have their minimum at $\beta_2 = 0.22$ and $\gamma = -60$, and in ^{189}Po the negative parity states have a similar minimum. Such a minimum can be explained by the specific interaction of the proton 4p-2h configurations with the valence neutrons. This 4p-2h configuration can decay to a $13/2^+$ state originating from the coupling of an odd neutron to the 2p-2h excited state in the even-even Pb neighbours, but not to the 0p-0h $13/2^+$ isomeric state. The ^{191}Po $3/2^-$ ground state is expected to have a 2p-0h configuration and can thus decay to the 0p-0h ground state in Pb and to the $3/2^-$ excited state originating from the coupling of the $3/2^-$ neutron to the 2p-2h 0^+ intruder state in the even-even Pb neighbour.

The observation of two α-decaying isomeric states in ^{191}Po with similar energy but different half life together with the evidence for fine structure in both decays leads to the identification of proton 2p-2h based states in ^{187}Pb. Due to the observation of strongly retarded $l = 0$ α decay between the two $13/2^+$ isomers in the ^{191}Po-^{187}Pb α-decay chain, and of normal $l = 0$ α decay between the $13/2^+$ isomer and the $13/2^+$ excited state and between the $3/2^-$ ground states, evidence is found for shape staggering in ^{191}Po.

REFERENCES

1. Baxter, A. et al., *ANU Annual Report*, 33 (1997).
2. Fotiades, N. et al., *Phys. Rev. C.* **57**, 1624–1633 (1998).
3. Fotiades, N. et al., *Phys. Rev. C.* **56**, 723–729 (1997).
4. Griffin, J.G. et al., *Nucl.Phys., A* **530**, 401 (1991).
5. Vanhorenbeeck, J. et al., *Nucl.Phys., A* **531**, 63 (1991).
6. Wood, J.L. et al., *Phys.Rep.* **215**, 101 (1992).
7. Bijnens, N. et al., *Z. Phys.* **A356**, 3 (1996).
8. Quint, A.B. et al., *GSI Report* **88-1**, 16 (1988), and *Z. Phys.* **A346**, 119 (1993).
9. Batchelder, J.C. et al., *Phys. Rev. C* **55**, R2142 (1997).
10. Wauters, J. et al., *Phys. Rev. C* **47**, 1447 (1993).
11. Leino, M. et al., *Nucl. Instr. & Meth.* **B99**, 653 (1995).
12. Andreyev, A. et al., *Z. Phys.* **A358**, 63 (1997).
13. Misaelidis, P. et al., *Z.Phys.* **A301**, 199–208 (1981).
14. Lane, G.J. et al., *Nucl. Phys.* **A589**, 129 (1995).
15. Rasmussen, J.O., *Phys.Rev.* **111**, 1593 (1959).

Total Coulomb Excitation Cross Section Measurements of Radioactive Nuclear Beams in Low Energy Inverse Kinematics

C.J. Barton,[1] D.S. Brenner,[1] R.F. Casten,[2] N.V. Zamfir,[1,2,3,4] R.L. Gill,[3] D. Shapira[5]

[1]*Clark University, Worcester, MA 01610 USA*
[2]*WNSL Yale University, New Haven, CT 06520 USA*
[3]*Brookhaven National Laboratory, Upton, NY 11973 USA*
[4]*Institute of Atomic Physics, Bucharest Magurele, Romania*
[5]*Physics Division, Oak Ridge National Laboratory, Oak Ridge, TN 37831 USA*

Abstract. Coulomb excitation is a powerful, well-understood, and selective technique to excite levels in nuclei primarily through E2 transitions with relatively large cross sections (typically 10s of mbarns). A description of an apparatus to measure total Coulomb excitation cross sections of radioactive nuclear beams will be presented. The apparatus uses inverse kinematics in order to overcome the difficulties presented by limited beam intensities and high background rates. A high efficiency NaI(Tl) through-well detector is used to observe the deexcitation γ-rays. The discussion will address the challenges of doing low energy Coulomb excitation experiments of radioactive beam nuclei and a recent experiment with radioactive ^{69}As nuclei at the Holifield Radioactive Ion Beam Facility (HRIBF).

Coulomb excitation is a powerful, well-understood, and selective technique for exciting a nucleus. The excitations occur primarily through E2 transitions with relatively large cross sections, around 10s of mbarns, for the low lying levels in a nucleus. Radioactive Nuclear Beam (RNB) facilities, such as the proposed NISOL facility, will produce beams of exotic nuclei for study making it advantageous to use the well-understood technique of Coulomb excitation to extend our knowledge of nuclear structure out toward the drip lines.

Experiments with RNBs will present at least two challenges. Beam intensities will be low compared with what is now achievable at stable beam facilities and, since the beam will be comprised of radioactive nuclei, there will be significant residual background. These limitations require a combination of the use of highly efficient detectors, modified experimental techniques, and the measurement of observables that will be very sensitive to structural features of nuclei.

CP455, *ENAM98: Exotic Nuclei and Atomic Masses*
edited by B. M. Sherrill, D. J. Morrissey, and Cary N. Davids
© 1998 The American Institute of Physics   1-56396-804-5/98/$15.00

A modification of the traditional Coulomb excitation experiment is now needed in order to overcome the limitation posed by the radioactive background. In order to remove the activity from the decay of the primary beam from the vicinity of the gamma detectors a thin, low Z excitation target is used. A target of about 1 mg/cm^2 of ^{12}C is chosen for ease of use. This target is too thin to stop the beam nuclei, so there will be no build-up of activity at the target. The kinematics of the reaction will also ensure that all scattered nuclei, even those that are Coulomb excited and scatter at back angles in the center of mass reference frame, will be forward scattered in the laboratory frame. An example showing the kinematics of ^{69}As(^{12}C,^{12}C*)^{69}As* and ^{12}C(^{69}As,^{69}As*)^{12}C* and, for comparison, examples where a target is infinitely massive and where a target and beam have the same mass is shown in Fig. 1. These examples were calculated for elastic scattering events. Coulomb excitation is sufficiently elastic to satisfy this condition. It is important to note that only in cases where the beam nuclei are more massive than the target nuclei is there a maximum angle for scattering in the laboratory. In experiments with RNBs, it is important to exploit inverse kinematics to ensure that the scattered activity will leave the target area in a narrow, forward focused cone.

An apparatus for measuring Coulomb excitation cross sections is shown in Fig. 2. A through-well NaI(Tl) detector with a 90% geometric efficiency is used to detect γ-rays from the Coulomb excited beam nuclei. The ^{12}C target is placed near the center of the NaI(Tl) detector. To measure the total beam flux, a microchannel plate- aluminized mylar foil assembly is placed in the beam line. There is also a plastic scintillator beam stop that is positioned downstream of the target. Both particle detectors are used for coincident measurements with the γ detector.

We have performed an experiment at the HRIBF at Oak Ridge with a ^{69}As beam. One level below 1 MeV in the known level scheme seemed a likely candidate for Coulomb excitation. This level would be characterized with the decay of this level primarily to the ground state via the emission of a 863.9 keV γ-ray. Data was collected and analyzed for evidence of Coulomb excitation in this odd-A nucleus. The analysis is for a 10.5 hour live-time run of 160 MeV ^{69}As with an intensity of 1.5*10^5 nuclei/sec. The raw NaI(Tl) energy spectra and the projected NaI(Tl) energy spectra, gated with the NaI(Tl) – Plastic Scintillator TAC and background subtracted, are show in Fig. 3.

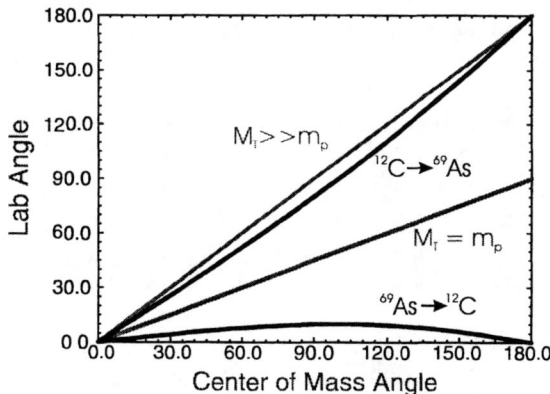

FIGURE 1. Graph of various elastic scattering events. The kinematics for each event is labeled.

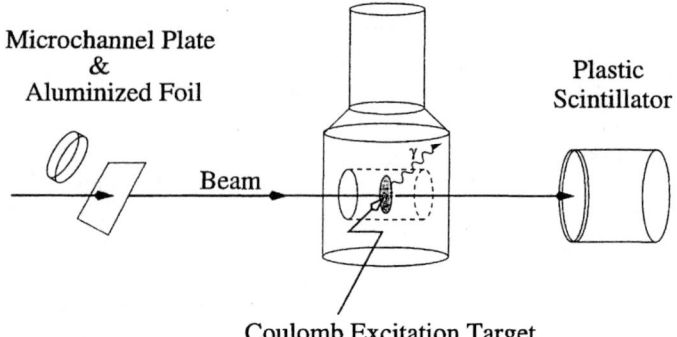

FIGURE 2. Drawing of experimental apparatus for measuring Coulomb excitation cross sections of RNBs in inverse kinematics.

In the NaI(Tl) energy spectrum little evidence was seen for Coulomb excitation of the 863.9 keV level. A peak suggestive of Coulomb excitation is seen in the gated and background subtracted spectrum. This data gives an upper limit of 75 s.p. units for the B(E2) for the transition, assuming a pure E2 transition.

As can be seen in the raw NaI(Tl) spectrum, there is a large amount of γ-ray background from the daughters of ^{69}As that complicate the analysis. The reason for this complication is that the Plastic Scintillator was placed about 5 ns in flight time downstream of the NaI(Tl) detector in order to accept the full angular range of the scattered nuclei. This small time separation makes the time separation of Coulomb excitation events and decay events difficult. Further complications arose from the difficulties of steering the low intensity radioactive beam correctly onto the center of the target. This deposited activity on the target holder and increased background. The plastic scintillator was also radiation damaged by the primary beam shortening its useful lifetime.

For future measurements of B(E2) values, modifications to the previously mentioned experimental apparatus will be made. Principally, a new scattering chamber and scintillator beam stop needs to be made in order to increase the distance between the two detectors. This change in position removes the deposited activity from the immediate vicinity of the NaI(Tl) detector thus providing the time resolution necessary to separate Coulomb excitation events from β^+ decay events. It also reduces the solid angle subtended by the NaI(Tl) detector at the scintillator and allows for additional shielding between the two. Also needed are low intensity beam diagnostics to ensure that the two. Also needed are low intensity beam diagnostics to ensure that the primary beam is centered on target.

The new plastic scintillator detector is positioned to be about 30 cm downstream of the target in this geometry ensuring about 20 ns flight time for scattered nuclei. The

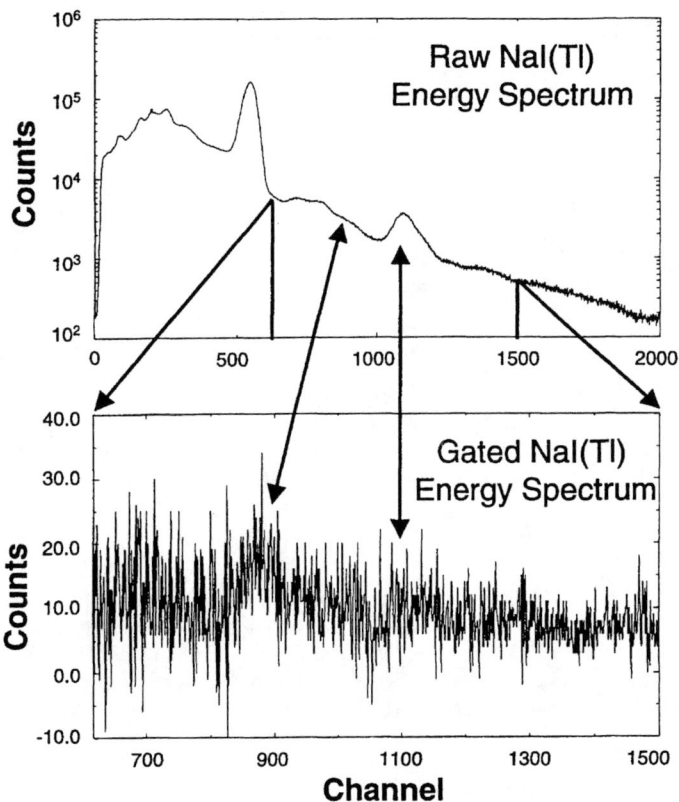

FIGURE 3. NaI(Tl) spectra obtained at the HRIBF. There is a large background present in the raw NaI(Tl) spectrum. The gated and background subtracted spectrum is shown in the expanded region below. Note that the background peak is greatly diminished while the suspected peak from Coulomb excitation begins to rise out of the background where it was not apparent before.

scintillator is also segmented into four quadrants allowing low intensity beam diagnostics to be performed from the count rate in each segment. A small graphite plug is placed along the beam axis over the scintillator in order to stop the primary beam from damaging the scintillator at beam intensities of 10^6 nuclei/sec or greater. This does not interfere with the measurement of the Coulomb excitation cross section since, due to inverse kinematics, most of the Coulomb excited nuclei will exit the target near that maximum kinematically allowed scattering angle. These modifications should ensure that future work will be completed without the background difficulties encountered during the run at the HRIBF.

This work was supported by the Department of Energy under grants and contracts: DE-GF02-91ER40609, DE-FG02-88ER40417, DE-AC02-76CH00016 and DE-AC-96OR22464.

Spectroscopy of N=82,83 ^{136,137}Xe Isotopes From ^{248}Cm Fission

P. Bhattacharyya[1], C.T. Zhang[1], P.J. Daly[1], Z.W. Grabowski[1], R. Broda[1*], B. Fornal[1*], I. Ahmad[2], T. Lauritsen[2], L.R. Morss[2], W.R. Phillips[3], J.L. Durell[3], M.J. Leddy[3], A.G. Smith[3], W. Urban[3†], B.J. Varley[3], N. Schulz[4], E. Lubkiewicz[4¶], M. Bentaleb[4], J. Blomqvist[5]

[1] *Purdue Univeristy, West Lafayette, IN 47907, USA.*
[2] *Argonne National Laboratory, Argonne, IL 60439, USA*
[3] *University of Manchester, M13 9PL Manchester, UK*
[4] *Institut de Recherches Subatomiques, Universite Louis Pasteur, F-67037 Strasbourg, France*
[5] *Royal Institute of Technology, S-10405 Stockholm, Sweden*

Abstract. Prompt γ-ray cascades in neutron-rich nuclei around doubly magic ^{132}Sn have been studied at Eurogam II using a ^{248}Cm fission source. Here we report results for the four-valence-proton N=82 nucleus ^{136}Xe and for its N=83 neighbor ^{137}Xe. For both nuclei, the yrast level spectra have been considerably extended, and empirical nucleon-nucleon interactions have been used to assign probable shell model configurations for most of the observed levels.

The advent of large multidetector γ-ray arrays has made it possible to study prompt and delayed γ-ray cascades in fission product from actinide sources [1]. We have been investigating the yrast excitations of few-valence-particle nuclei around doubly magic ^{132}Sn by analyzing fission product γ-ray data recorded at Eurogam II using a ^{248}Cm source. First results for the two- and three-proton N=82 nuclei ^{134}Te and ^{135}I [2], for the two-neutron nucleus ^{134}Sn [3], and for the N=83 nuclei ^{134}Sb, ^{135}Te, and ^{136}I [4] have already been published.

We have now extended these studies to the four-valence-proton N=82 nucleus ^{136}Xe, and to its N=83 neighbor ^{137}Xe. The investigation of ^{136}Xe faced initial difficulties since its predicted yield in ^{248}Cm fission is only 0.4%, and the 3 μs half-life of the ^{136}Xe yrast 6$^+$ state ruled out the possibility of identifying higher-lying γ-rays from the $\gamma\gamma\gamma$ data. A second 6$^+$ state in ^{136}Xe, known from β-decay [5,6], de-excites to the 6$^+$ isomer by a 370 keV M1/E2 transition but no states with I > 6 have been placed in ^{136}Xe scheme up to now. In the present work, double gating on 381 and 1313 keV γ-rays de-exciting the lowest 4$^+$ state in ^{136}Xe showed in coincidence (Fig. 1(a)) a few strong ^{136}Xe γ-rays, which feed the 4$^+$ state directly, as well as cross coincident $^{106-109}$Mo γ-rays from complementary fission fragments. Indeed the strongest peaks in Fig. 1(a) are the 193, 371 and 527 keV γ-rays of ^{108}Mo, the 4n fission partner of ^{136}Xe. Double gating on ^{108}Mo γ-rays showed in coincidence (Fig. 1(b)) known γ-rays of ^{108}Mo, ^{136}Xe, ^{137}Xe and ^{138}Xe, and some other lines. Of these, the 370 keV γ-ray in Fig. 1(b) was taken to be the ^{136}Xe $6_2^+ \rightarrow 6_1^+$ transition previously mentioned, while the 968 and 975 keV γ-rays also appeared attractive candidates for placement in ^{136}Xe above the 3 μs isomer; the 370 and 968 keV γ-rays were subsequently found to be in prompt coincidence. Double gating on the 370 and 968 keV γ-rays gave the Fig. 1(c) coincidence spectrum, where the relative intensities of cross-coincidence $^{106-109}$Mo γ-rays are seen to be closely similar to those in Fig. 1(a), thus providing vital support for placement of the 370 and 968 keV γ-ray pair in the ^{136}Xe scheme. Additional γ-rays now assigned to ^{136}Xe are also labelled in Fig. 1(c). These and other coincidence spectra showed

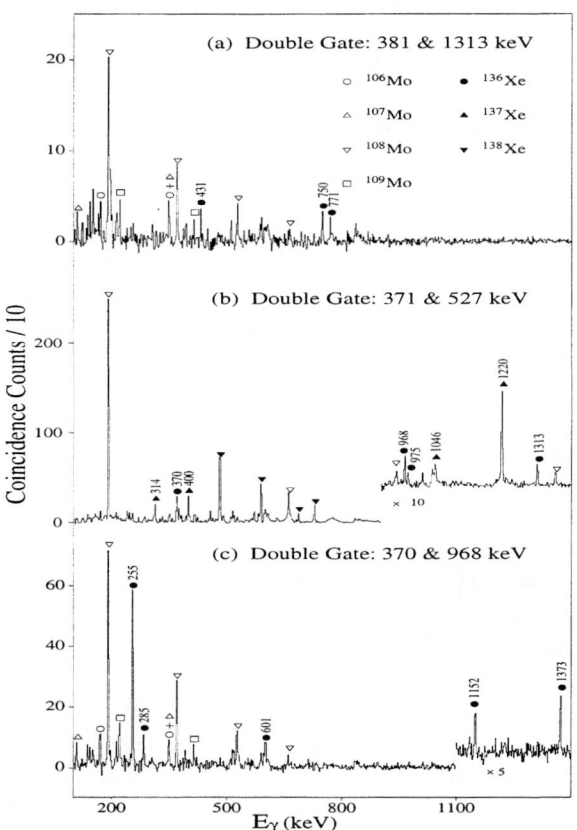

FIGURE 1. Key γ-γ coincidence spectra for the isotopic identification of ^{136}Xe transitions and construction of the ^{136}Xe level scheme.

clearly that two main cascades feed the ^{136}Xe 6^+ isomer, one consisting of 255, 968 and 370 keV γ-rays, the other in parallel of 618 and 975 keV transitions as shown in the extended ^{136}Xe level scheme (Fig. 2). Tentative spin-parity assignments and probable four-proton configurations are also indicated in Fig. 2. The highest levels located at 5953 and 6173 keV may be core-excited $\pi g_{7/2}^4 \nu f_{7/2} h_{11/2}^{-1}$ states similar to the yrast 12^+ and 13^+ states at ~ 6 MeV in ^{134}Te.

Many years ago a study of delayed γ-rays following ^{252}Cf fission assigned to the N=83 nucleus ^{137}Xe a cascade of 314, 400, and 1221 keV γ-rays de-exciting an isomer with $t_{1/2} \sim 8$ ns [7]. The same γ-rays were previously reported in Ref. [8], where they were not assigned. The levels of ^{137}Xe populated in β^- decay of ^{137}I includes a 1220 keV level that de-excites to the $7/2^-$ ground state [9]; this level was later identified as the $11/2^-$ level [10]. First inspection of the data from the present fission product measurements revealed strong 314.0, 400.2 and 1220.0 keV γ-rays in mutual coincidence, which provided an excellent starting point for further investigation of the ^{137}Xe yrast levels. Detailed analyses of the $\gamma\gamma\gamma$ data then led to the ^{137}Xe scheme shown in Fig. 2. In this case, the multipole order of the strongest transitions could be determined from DCO measurements and these results are also included in Fig. 2. The broad similarity between the ^{136}Xe and ^{137}Xe yrast level structures is evident as most of the levels in ^{137}Xe can be explained by the coupling of the $f_{7/2}$ neutron with the ^{136}Xe levels.

One of the main aims in studying the spectroscopy of few-valence-particle nuclei around ^{132}Sn is to characterize the nucleon-nucleon interactions in this region. To perform shell model calculations for the N=82 isotones, the simplest method is

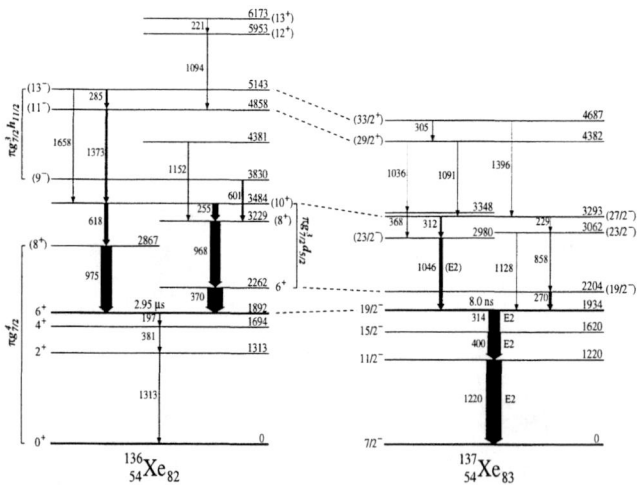

FIGURE 2. Proposed level schemes of ^{136}Xe and ^{137}Xe. The corresponding states have been connected by dashed lines.

to adopt two-body interactions from the experimental level spectrum of the two-proton nucleus ^{134}Te. This approach, which takes account of diagonal matrix elements only, and thereby neglects configuration mixing, provided valuable guidance in the interpretation of the observed ^{136}Xe levels. A few years ago, Wildenthal [11] performed comprehensive shell model calculations for all N=82 nuclei then known, and used an iterative procedure to obtain a best fit set of 160 two-body matrix elements, both diagonal and off-diagonal. When the ^{136}Xe level energies were calculated using Wildenthal's parameters rather than those directly from ^{134}Te, the mean deviation between theory and experiment decreased from 92 to 28 keV. Since the results for ^{135}I and ^{136}Xe attained in our work are significant addidtions to the data base and first results for the five-proton N=82 nucleus ^{137}Cs are forthcoming [12], it will now be possible to update the nucleon-nucleon interaction matrix elements in this region.

This work was supported by the U.S. Department of Energy under Contracts No. DE-FG02-87ER40346 and No. W-31-109-ENG-38, by the Science and Engineering Council of the U.K. under Grant No. GRH 71161, and by the Polish Scientific Committee Grant No. 1044-PO3-96-10. The authors are indebted for the use of ^{248}Cm to the Office of Basic Energy Sciences, U.S. Department of Energy through the transplutonium production facilities at the Oak Ridge National Laboratory.

* On leave from Institute of Nuclear Physics, PL-31342 Cracow, Poland

† Present address: Institute of Experimental Physics, Warsaw University, PL-00681 Warsaw, Poland

¶ On leave from Institute of Physics, Jagellonian University, Reymonta 4, PL-30-059 Cracow, Poland

REFERENCES

1. Ahmad, I., and Phillips W.R., *Rep. Prog. Phys.* **58**, 1375 (1995).
2. Zhang, C.T. et al., *Phys. Rev. Lett.* **77**, 3743 (1996).
3. Zhang, C.T. et al., *Z. Phys.* **A358**, 9 (1997).
4. Bhattacharyya, P., et al., *Phys. Rev.* **C56**, R2363 (1997).
5. Mantica, Jr. P.F. et al., *Phys. Rev.* **C43**, 1696 (1991).
6. Keyser, U. et al., *Proc. Intern. Conf. Atomic Masses and Fundamental Constants* 6th, East Lansing (1979), J.A. Nolen Jr., W. Benenson Eds., Plenum Press, New York, 495 (1980).
7. Clark, R.G., Glendenin, L.E. and Talbert, Jr. W.L., *Proc. Symp. Phys. Chem. Fission*, Rochester (1973), IAEA, Vinenna, **2** 221 (1974).
8. John, W., Guy, F.W. and Weselowski, J.J., *Phys. Rev.* **C2**, 1451 (1970).
9. Fogelberg, B., and Tovedal, H., *Nucl. Phys.* **A345**, 13 (1980).
10. Hoff, P., Ekstrom, B. and Fogelberg, B., *Z. Phys.* **A332**, 407 (1989).
11. Wildenthal, B.H., *Understanding the Variety of Nuclear Excitations*, ed. A. Cavello, 1991 World Scientific Publishing Company (1991).
12. Broda, R., et al., (to be published).

Alpha Decay of the $h_{9/2}$ Ground and $s_{1/2}$ Intruder States in Light Bi and At Isotopes[1]

C. R. Bingham[1,2], J. C. Batchelder[3], J. A. Cizewski[4,5], C. N. Davids[4], R. J. Irvine[6], W. Reviol[1], D. Seweryniak[4], K. S. Toth[2], W. B. Walters[7], J. Wauters[1], J. L. Wood[8], X. J. Xu[1], J. Uusitalo[4], E. F. Zganjar[9]

[1] *University of Tennessee, Knoxville, Tennessee 37996*
[2] *Oak Ridge National Laboratory, Oak Ridge, Tennessee 37831*
[3] *Oak Ridge Associated Universities, Oak Ridge, Tennessee 37831*
[4] *Argonne National Laboratory, Chicago, Illinois 60439*
[5] *Rutgers University, Piscataway, New Jersey 08854*
[6] *University of Edinburgh, Edinburgh, EH9 3JZ United Kingdom*
[7] *University of Maryland, College Park, Maryland 20742*
[8] *Georgia Institute of Technology, Atlanta, Georgia 30332*
[9] *Louisiana State University, Baton Rouge, Louisiana 70803*

Abstract.
The $1/2^+$ state in the odd-A Bi isotopes is interpreted as an intruder shell model state with one hole in the $s_{1/2}$ level just below the Z = 82 closed shell. Likewise, the $9/2^-$ state in the odd-A Tl isotopes is the intruder level corresponding to one proton in the otherwise empty shell above Z = 82. These intruder states come quite low near the middle of the neutron shell and are observed as isomeric states. The positions of the intruding isomeric states as a function of the neutron number are important in understanding the shell model interactions producing the relative stability of the intruding states. In this region, the middle of the neutron shell is also quite far from stability, approaching the proton drip line, and hence, α emission is a dominant decay mode. In a series of α-decay fine structure measurements of Bi isotopes, the location of the $s_{1/2}$ intruder level has been determined down to mass 187. The ground states of the odd-A At isotopes from ^{197}At upward have been assigned a spin of $9/2^-$, but the expected $1/2^+$ intruder state is still relatively unknown in these nuclei. In this paper we report our most recent results from the α decay of ^{187}Bi, ^{189}Bi, and ^{191}Bi and show the systematic trend of these intruder states in Bi and Tl.

---

[1] Research sponsored by the U. S. Department of Energy under contracts DE-FG02-96ER40983 (Tennessee), DE-AC05-96OR22464 with Lockheed Martin Research Corporation (ORNL), DE-AC05-76OR-00033 (ORAU), W-31-109-ENG-38 (ANL), DE-FG05-88ER00033 (Maryland), DE-FG02-96ER40958 (Georgia Tech.) and DE-FG02-96ER40978 (LSU).

I INTRODUCTION

The presence of $h_{9/2}$ intruder states in the odd-A Tl isotopes and $s_{1/2}$ intruders in the odd-A Bi isotopes has been known for some time [1], though the excitation energies have been more difficult to determine. In a series of experiments, we have measured α decay fine structure in an effort to determine the locations of these important intruding levels. The α transitions between states of the same spin and parity are generally unhindered, while highly hindered transitions occur for transitions between intruder states and ground states in these nuclei. The use of mother-daughter correlations can be used to isolate which α-decay peaks belong to the unhindered $1/2^+ \rightarrow 1/2^+$ and $9/2^- \rightarrow 9/2^-$ transitions. The fine structure peaks corresponding to decay between an intruder state and the ground state provide a means of determining the excitation energy of the intruder. In a systematic study we have made measurements on the α decay of ^{187,189,191}Bi and ^{195,197}At and established the location of the $1/2^+$ isomer down to ^{187}Bi [2].

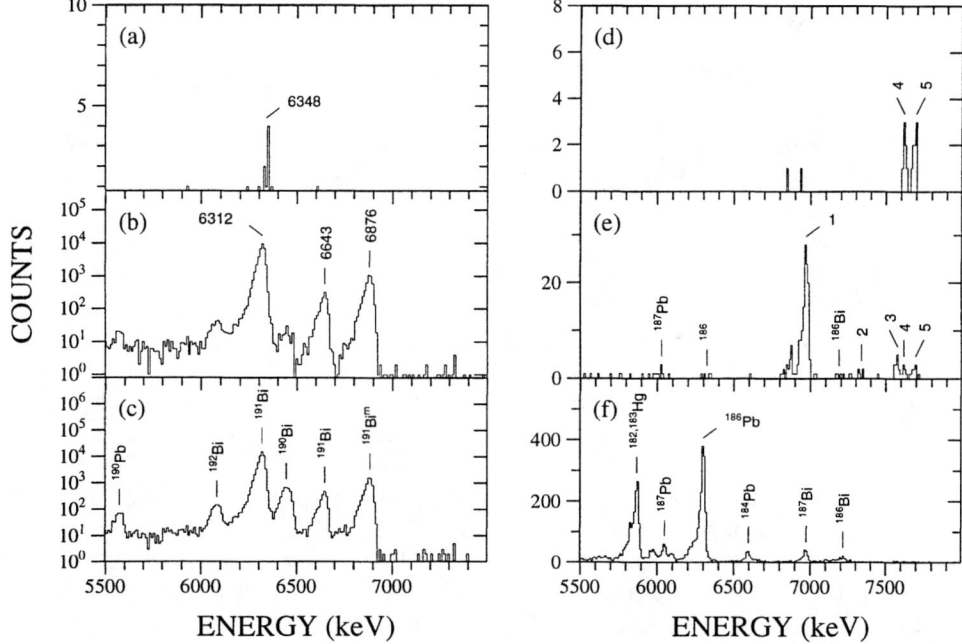

FIGURE 1. Alpha spectra obtained from the bombardment of ^{97}Mo with 420-MeV ^{92}Mo(right side) and ^{159}Tb with 175-MeV ^{36}Ar (left) under the given experimental conditions: (c) & (f) totals, (b) gated with mass 191, (a) in coincidence with 299.5-keV γs, (e) A = 187, t < 250 ms, and (d) A = 187, t < 1.0 ms.

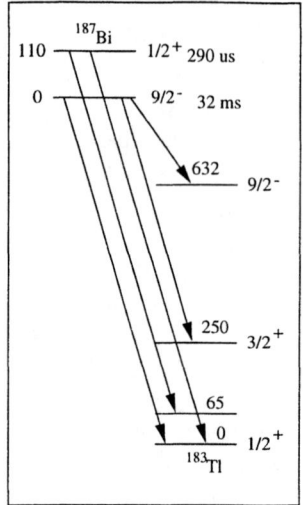 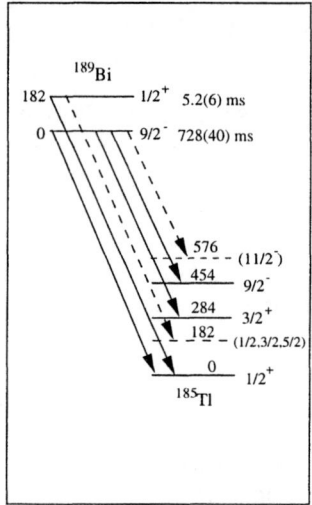 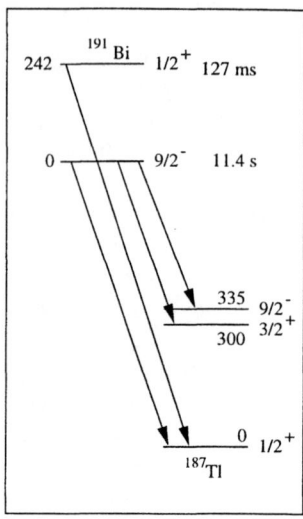

FIGURE 2. Alpha decay schemes for ^{187}Bi, ^{189}Bi, and ^{191}Bi.

II EXPERIMENTAL METHOD AND RESULTS

The activities were produced by bombarding $\sim$ 1 mg/cm^2 thick targets of ^{97}Mo, ^{96}Mo, ^{159}Tb, and ^{165}Ho with ^{92}Mo, ^{95}Mo, and ^{36}Ar ions from the Argonne Tandem Linear Accelerator System (ATLAS). The fusion evaporation residues were separated from the beam and mass-analyzed with the Fragment Mass Analyzer and deposited in a double-sided silicon strip detector which was used to measure the α spectra. Individual decay events were correlated with specific implants by recording the time and mass for the implant events and subsequently the time and energy of the decay events. Thus, one can deduce the parent identity, the energy, and the half-life of each α line observed. The identity of the specific isomer can often be deduced from correlation of the first α decay after an implant with a subsequent α decay of the daughter.

Spectra from ^{187}Bi are shown in the righthand panels of Fig. 1. Panels (d) and (e) reveal 3 α-particle peaks (7006, 7379, 7624 keV) associated with a half-life of 32(3) ms and two peaks (7670, 7734 keV) decaying with a $t_{1/2}$ of $290^{+90}_{-50}\mu$s. The longer half-life is that of the 9/2$^-$ ground state of ^{187}Bi and the other is that of the 1/2$^+$ isomer. The placement of these 5 α transitions in a level scheme is shown in Fig. 2. The ground state decays to both the 1/2$^+$ ground state and 9/2$^-$ isomer of ^{183}Tl, thus confirming the excitation energy of the isomer. The 1/2$^+$ isomer in ^{187}Bi also decays to the ground state, thus establishing its excitation energy of 110(30) keV. The unhindered transition from the 1/2$^+$ isomer to the level at 65(25) keV in ^{183}Tl is not completely understood, but perhaps is associated with a 3p-4h configuration in Tl. Panels (a) and (b) of Fig. 1 reveal at least 3 peaks (6312, 6348, 6643 keV) associated with the decay of the 9/2$^-$ ground state of ^{191}Bi

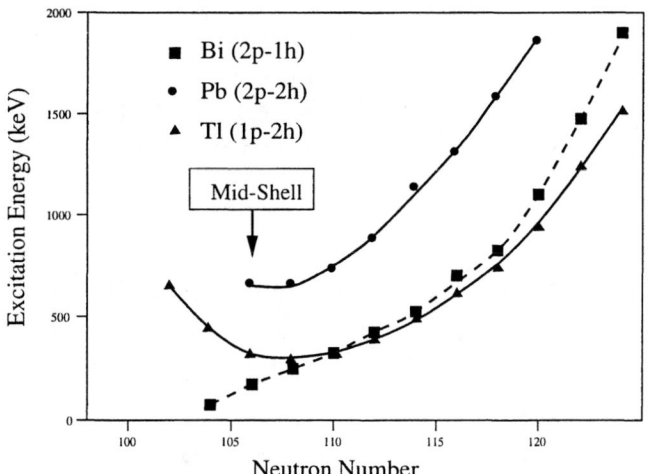

FIGURE 3. Systematics of the $9/2^-$ and $1/2^+$ levels in Tl and Bi in comparison with those of the 0_2 levels in Pb

and one (6876 keV) from the $1/2^+$ isomer (see Fig. 2). Because of its relative weakness, the $9/2^- \rightarrow 3/2^+$ α transition at 6348 keV could only be separated from the strong 6312-keV transition by requiring a coincidence with the 299.5-keV γ ray which depopulates the $3/2^+$ level (Fig. 1(a)). Similar data for ^{189}Bi was recently published [3]; the decay scheme, shown in Fig. 2, is similar to the one for ^{187}Bi.

The systematics of the intruder levels in Bi and Tl are shown in Fig. 3 along with the position of the 2p-2h 0^+ states in Pb. It is seen that the minimum in excitation energies for Tl and Pb intruders is near midshell, while the $1/2^+$ intruder in Bi seemingly continues a downward trend at N = 104. This tendency is probably related to the onset of a low-lying prolate minimum in the Pb nuclei near mass 186 [4].

REFERENCES

1. Wood, J. L., Heyde, K., Nazarewicz, W., Huyse, M., and Van Duppen, P., Phys. Rep. **215**, 101 (1992).
2. Batchelder, J. C., *et al.*, submitted for publication in Phys. Rev. C
3. Wauters, J., *et al.*, Phys. Rev. C **55**, 1192 (1997).
4. Nazarewicz, W., Phys. Lett. **B305**, 195 (1993).

ELASTIC AND INELASTIC PROTON SCATTERING ON THE UNSTABLE ^{20}O NUCLEUS MEASURED WITH THE "MUST" DETECTOR ARRAY

E. Khan[a], Y. Blumenfeld[a], T. Suomijärvi[a], N. Alamanos[b], F. Auger[b], N. Frascaria[a], A. Gillibert[b], T. Glasmacher[c], M. Godwin[a], V. Lapoux[b], I. Lhenry[a], F. Maréchal[a], D.J. Morissey[c], A. Musumara[b], N. Orr[d], S. Ottini[b], P. Piattelli[e], E.C. Pollacco[b], P. Roussel-Chomaz[f], J.C. Roynette[a], D. Santonocito[a], J.E. Sauvestre[g], and J.A. Scarpaci[a]

a) Institut de Physique Nucléaire, IN$_2$P$_3$-CNRS, 91406 Orsay, France
b) SPhN, DAPNIA, CEA Saclay, 91191 Gif sur Yvette Cedex, France
c) NSCL, Michigan State University, East Lansing, Mi 48824, USA
d) LPC, IN$_2$P$_3$-CNRS, Bd. Maréchal Juin, 14050 Caen Cedex, France
e) INFN-Laboratorio Nazionale del Sud, Via S. Sofia 44, Catania, Italy
f) GANIL, BP 5027, 14021 Caen Cedex, France
g) DPTA/SPN-CEA Bruyères, 91680 Bruyères-le-Châtel Cedex 12, France

Abstract.
A modular array named MUST, based on Silicon strip technology has been built and will be devoted to the measurement of recoiling light particles in direct reactions induced by radioactive beams. The detector is described and the performances illustrated with preliminary results from a (p,p') scattering experiment recently performed with a secondary ^{20}O beam at GANIL.

INTRODUCTION

The availability of radioactive nuclear beams with reasonable intensity and optical quality furnishes the tantalizing opportunity to study direct nuclear reactions, such as elastic and inelastic scattering and transfer reactions, induced by unstable nuclei on light particles. Such reactions are performed in inverse kinematics, where the radioactive nucleus of interest bombards a target containing the light particles. An efficient method to gain access to the excitation energy and scattering angle characterizing the reaction is to measure the energy and angle of the recoiling particle. Experiments of this type, concerning elastic and inelastic proton scattering, have already been performed at RIKEN [1] and the NSCL/MSU [2,3]. The aim of the MUST detector is to utilize this method under optimal conditions in a wide range of experimental configurations. In this contribution, the detector will be described and its performances illustrated with preliminary results from an ^{20}O scattering experiment performed at the GANIL facility.

THE MUST ARRAY

In inverse kinematics reactions, the energy and angular ranges of the recoiling light partner depend on the type of reaction studied. In order to fulfill the experimental requirements for elastic, inelastic and nucleon transfer reactions, the detection system must be modular to cover various angular ranges and present a dynamic range for light particles between 500 keV and 50 MeV with an angular resolution better than $1°$ and an energy resolution of the order of 100 keV. Identification of the light particles is also crucial in order to suppress background from different reaction mechanisms and target contaminants.

We have chosen to implement an array of eight silicon-strip detectors backed by Si(Li) diodes and CsI crystals read out by photodiodes. The double sided silicon strip detectors made by CANBERRA semiconductor are 6X6 cm^2 in size, 300 μm thick with 60 strips 1 mm wide on each side allowing a localization in X and Y. The 6X6 cm^2, 3mm thick Si(Li) detectors were built at the IPN Orsay, while the

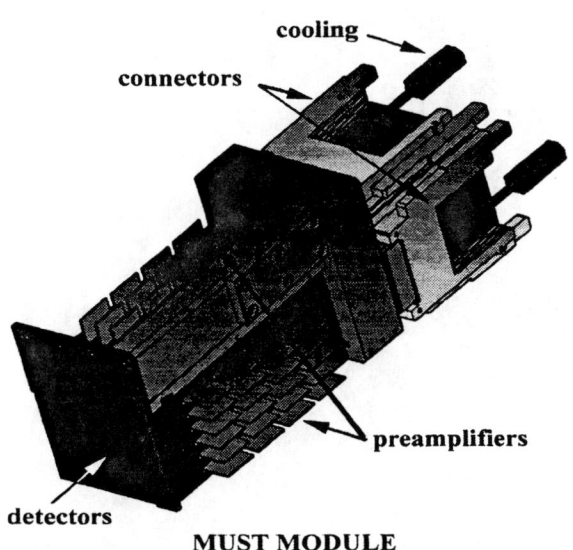

MUST MODULE

FIGURE 1. One module of the MUST detector

6X6 cm^2, 15 mm thick tapered CsI crystals were furnished by SCIONIX. Each strip is equipped with a preamplifier located under vacuum, mounted immediately adjacent to the detectors. A schematic representation of one module is displayed on fig.1.

Integrated VXIbus electronics have been constructed, dedicated to the array. Each strip detector is associated with one VXI module which houses all the necessary amplifiers and converters. A Digital Signal Processor in each module furnishes calibrated data to the acquisition system, based on a SUN SPARC5 workstation. All the electronics for the array are contained in one VXI size D crate located adjacent to the reaction chamber.

Low energy particles which stop in the strip detector are identified by an energy and a time of flight measurement between the strip detector and a start detector which can be a beam tracking detector. The time resolution of the strip detector has been measured to be better than 1ns. Higher energy particles are identified by the ΔE-E method.

THE ^{20}O(P,P') EXPERIMENT

A subject of current interest is the evolution of density and transition density distributions for nuclei far from stability. The measurement of the angular dis-

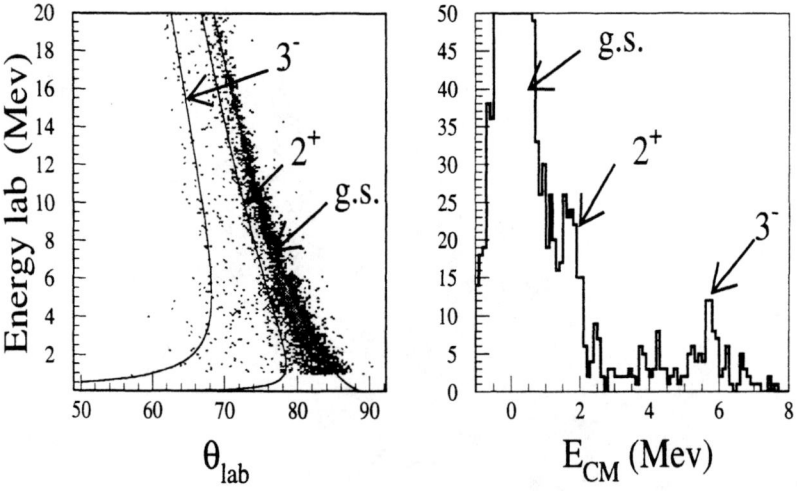

FIGURE 2. Left panel : Angle vs. energy plot for recoiling protons from the ^{20}O + CH$_2$ reaction at 45 MeV/nucleon. Solid lines are the expected kinematic loci for scattering to the states indicated. Right panel : Corresponding experimental excitation energy spectrum.

tributions of elastic and inelastic proton scattering allows us to test theoretical predictions of these quantities through microscopic folding model analyses. As a first experiment with the MUST detector, we have measured at GANIL ^{20}O(p,p') angular distributions of elastic scattering and inelastic scattering towards the first collective 2^+ and 3^- states. ^{18}O scattering was also measured for comparison. The secondary beams, produced by fragmentation of a 77 MeV/u ^{40}Ar beam and refocused with the SISSI solenoid, impinged on a 1mg/cm^2 CH$_2$ target. The ^{20}O beam had an intensity of approximately 5×10^3 pps and was 98% pure. Two low pressure multi-wire proportional chambers [4] were used for beam tracking and one of them furnished the start signal for the time of flight measurement of the recoiling light particles which were detected with MUST in coincidence with the scattered projectiles in the SPEG spectrometer. Elastic and inelastic angular distributions beween $\Theta_{c.m.} = 15°$ and $65°$ have been measured.

Fig.2 shows some preliminary data from this experiment. The left hand panel displays a scattering angle vs. proton energy plot measured in coincidence with forward going ^{20}O nuclei. Corrections for the trajectories of the incident beam particles have been performed. Elastic and inelastic scattering are clearly observed in agreement with the predicted kinematical loci, shown as solid lines. This is confirmed on the right hand side of the figure which shows the excitation energy spectrum after transformation into the center of mass frame. The total excitation energy resolution measured on the elastic peak is approximately 600 keV. The 2^+ and 3^- collective excitations of ^{20}O are observed at their known excitation energies.

OUTLOOK

The newly built silicon strip array MUST is a powerful instrument for the study of direct reactions induced by radioactive beams. Apart from the ^{20}O(p,p') experiment described above, measurements of proton scattering on ^{6}He, ^{30}S and ^{34}Ar have recently been performed at GANIL. The versatility of the MUST array will allow studies of transfer reactions and 2-proton decay when the SPIRAL facility comes on-line in the near future.

REFERENCES

1. Korsheninnikov, A.A., et al. *Phys. Rev. Lett.* **78**, 2317 (1997).
2. Maréchal, F. et al., *these proceedings*.
3. Cottle, P.D. et al., *these proceedings*.
4. Ottini, S. *PhD thesis, CEA Saclay, internal report Dapnia/SPhN 98-01-T* (1998).

THE ISOMERIC CROSS SECTION RATIOS (ICSR) METHOD AT THE INVESTIGATION OF THE ROLE OF ISOSPIN AT THE POPULATION COMPOUND STATES CLOUSED YRAST BAND IN THE NUCLEAR REACTIONS WITH NEUTRON HEAVY IONS

Chuvilskay T.V.[*], Seleznev Yu.G.[*], Shirokova A.A.[*],
and Herman M.[**]

[*]Institute of Nuclear Physics Moscow State University
[**]IAEA Nuclear Data Section, Vienna, Austria

Abstract. The calculations of the isomeric ratios σ_m/σ_g for the reactions pairs $^{128,130}Te(^{6,4}He,n)^{133mg}Xe$ and $^{128,130}Te(^{8,6}He,3n)^{133mg}Xe$ were performed. The recently developed statistical model code EMPIRE-2.9 was used. It was included the built-in parameter libraries and systematics. For the reactions induced by alpha-particles we used the Gilbert-Cameron level densities and Multi-Step Compound approach. The fusion cross sections of the reactions induced by $^{6,8}He$ projectiles were calculated in terms of the coupled channels method. It was shown by us the important role of the reduction of the Coulomb barrier, so ions $^{6,8}He$ seems more profitable than ^4He for population of the residual nucleus high spin states. It is based on the hypotheses that the study of the high spins of the nucleus is most important to perform in case when the compound nucleus arising in the processes of the collision have got minimum excitation energy E^* which is possible with given momentum J. It decreases the length of the statistical n-and/or gamma-cascade. For the simulation of the processes of population the further development of the ICSR method is the analysis of the isospin role in the reactions with the radioactive beams. Those investigations are important problem of the nuclear reactions as $^{6,8}He$ ions insert the large isospin.

INTRODUCTION

The isomeric cross section ratios (ICSR) for a pair of isomeric states is known to depend strongly on the spins of the isomers and of the mechanism of those reactions. Experimental and theoretical studies on ICSR[1], especially as a function of the incident particle energy, provide useful information on the level density $\rho(E^*,J)$, the spin-cutoff parameter, the moment of the inertia of the excited nucleus, the spectra of the reaction products. The calculated angular momentum population of the compound nucleus by various light particles and heavy ions versus is presented. For instance, it is maximum angular momentum square induced by the particles or ions into nucleus as a function of excitation energy for the ^{120}Sn compound nucleus. It is shown that the alpha-particle carried in momentum is the largest one in the energy range from 30 to 60 MeV. We have measured ICSR produced by the $^{130}Te(^4He,n)^{133mg}Xe$ reaction up to energy $E_\alpha=31$ MeV[2]. The resolution of the gamma ray spectrometer is 3 keV on the gamma-lines of ^{60}Co isotopes. A good resolution is necessary because we need to divide the products of the more than 40 reactions sometimes. We also used the enriched targets in our experiments. In table 1. the results is demonstrated. It is seen the good agreement is achieved with factor 0.4 to $J_{r.b.}$. There is a satisfactory accordance for this mass region. Low energy results is not difficult explain in terms of Fermi-gas model with a residual nucleon-nucleon interaction. Because of the calculated ICSR is sensitive to the low-energy level density $\rho(J_{fk})$, where is known, this interaction plays an important role. The influence of this reduce calculated values of ICSR because of the finished distribution of the residual nucleus on spins is :

$$W(J_{fk}) = \rho(J_{fk}) \sum_{Jc, Jf \ldots Jfk1} \frac{\sigma(Jc,E)}{\rho(Ec,Jc)} \tau(J_c, J_{f1} \ldots J_{fk}),$$

Where $\rho(J_{fk})$ is level density of the finished state, $\rho(E_c, J_c)$ – level density of the compound nucleus characterized by spin J_c and excitation energy E_c, τ is a product of the barrier transmission coefficient, $\sigma(J_c, E)$ – the distribution of the angular momentum J_c of the compound nucleus. The cutoff parameter $\sigma \sim \sqrt{aE}$ is much more for $\rho(J_c)$ than for $\rho(J_{fk})$ i.e. $\rho(J_{fk})$ is more sensitive on spin.

The capability experimental data and the looking for the discrepancies between statistical theory and experiments will be useful for the development of the nuclear reaction models.

TABLE 1. The new results of the calculations ICSR(σ_m / σ_g) on the statistical model code EMPIRE-2.9[3]

MeV	EXPER	1[2] J=0.4J r.b.		1		3		4	
E ion	σ_m / σ_g	E^*	σ_m / σ_g	E^*	σ_m / σ_g	E^*	σ_m / σ_g	E^*	σ_m / σ_g
15	0.5± 0.1	17.8	0.8						
17	1± 0.2	19.7	0.9						
20	1.8± 0.4	22.6	1.0	22.6	9.8				
23	1.8± 0.4	25.5	1.2		2				
24	2± 0.4	26.5	2.0	22.1	98				
25	2.4± 0.5	28.2	2.4					23.0	cr.sect.0
27	1.8± 0.4	30.2	2.6			22.8	cr.sect.0	24.9	1.4
28	2.4± 0.4	31.2	2.7			23.8	2.3	25.8	5.2
29	2.0± 0.4	32.2	2.8	26.8	∞	24.7	1.76	26.8	4.1
31	2.0± 0.4	34.2	2.8			26.6	5.3	28.7	7.0

1. ^{130}Te (^{4}He,n) ^{133mg}Xe
2. ^{128}Te (^{6}He,n) ^{133mg}Xe
3. ^{128}Te (^{8}He,3n) ^{133mg}Xe
4. ^{130}Te (^{6}He,3n) ^{133mg}Xe

SHORT DESCRIPTION OF EXPERIMENTAL SET-UP

The investigation of the reactions:
^{128,130}Te(^{6,4}He,n)^{133mg}Xe ($T_{1/2}$=5.25d; 2.19d) and ^{128,130}Te(^{8,6}He,3n)^{133mg}Xe by means of activation technique is planned. $n\sigma\Phi/\lambda = N(t); t > 2T_{1/2}$
where n -number of the stabile nuclei of the target,

σ - cross-section of the activation,
Φ - the flux of the particles,
N(t) - number of the activation nuclei after irradiation.

As an example, for ^{128}Te(^{8}He,3n)^{133mg}Xe reaction estimated cross-section in the most interesting energy region E(He)=20-25 MeV
σ = 3*10^{-1} mb[4] and the beam, presented in the project of SPIRAL (GANIL,FRANCE) 2.2*10^6 1/s and typical target thickness 10^{-6} m we have max obtained activity about 10^{-2} decay/s. Such activity of the isomeric pairs can be indicated if the optimal back-ground conditions are provided.

These experiments are simple and suitable for the beginning stage of SPIRAL beams by means of gamma-ray Ge(Li) spectrometer.

TABLE 2. Characteristics of the nuclei in alpha-n reaction

Inreach, %	98.1	
J^π target	0$^+$	
Q$_{reaction}$, MeV	-5.3	
Isomer	^{133m}Xe	^{133g}Xe
J^π isomer	11/2 -	3/2 +
T$_{1/2}$	2.19 d	5.25 d
E$_γ$, keV	233(IT) *	81
I$_γ$, % decays	10.0	37.0
*IT – isomeric transition		

CONCLUSION

It was indicated by us the interest at the investigation with the radioactive beams of the ions of ^{6,8}He the reactions producing the isomeric pairs. It was shown by the calculations[5] the important role of the reduction of the Coulomb barrier at the population high spin low energy states. The further development of the ICSR method it seems to be important to analyse of the isospin role in the reaction with the production of isomers by compound nucleus stage[6]. The calculation of the ICSR includes the

three stage: the producing of the compound nucleus, the evaporation of the particles and the gamma-cascade emitted by the residual nucleus. Earlier the isospin was not included at the codes. The idea is to take into account it because at the present time it is admitted that the isospin is a good quantum number as for a heavy nuclei as up to excitation energy about tens MeV. As the ions ^{6,8}He insert the larger isospin than alpha-particle the purity of the isospin is kept[7].

A << 1/(T$_0$ +1), where T$_0$ =(N-Z)/2.

So it is hope to reveal such breach of the purity of the isospin compared the experimental and the calculated values of ICSR because of the peculiarities of the distribution of the level density on the isospin and the nuclear reactions mechanisms with radioactive ions. Such investigations are the important problem of the nuclear physics - the breach of the isobaric symmetry and the role of the isovector members of the Coulomb potential. On the other hand the competition to the process to populate the high spin states is the weak bound of the nucleons in the ^{6,8}He than that in alpha-particle.

REFERENCE

1. Vandenbosch R., Huizenga J.R. Phys.Rev.,v.120,no 4,1960, p.1313
2. Glebov N.K., Tulinov A.F., Khodirev V.A., Chuvilskaya T.V., Shavtvalov L.Ya., Shirokova A.A. Izv.Ac.Nauk.Ser.Fiz. Russia, v.55,1,1991,p.141
3. Herman M., to be published
4. Sitenko A.G. The theory of the nuclear reactions, Moscow, Energoatomizdat,1983, p.148
5. Chuvilskaya T.V. Abstract of International Conference «Nuclear Structure at the Limits», Argonne, July 22-26, 1996, ANL, ILLINOIS, p.172
6. Ignatjuk A.V. Statistical properties of the exciting nuclei, Moscow, energoatomizdat, 1983, p.143
7. Gopich P.M. and Zaljubovsky I.I. Nuclear spectroscopy,Kharkov, visha skola, 1980, p.26

Evolution of Collective Motion in Light Polonium Nuclei

J. A. Cizewski,[1,2] K. Y. Ding,[1] N. Fotiades,[1] D. P. McNabb,[1] W. Younes,[1,3]
R. Julin,[4] M. Leino,[4] J. Cocks,[4] P. Greenlees,[4] K. Helariutta,[4] P. Jones,[4]
S. Juutinen,[4] A. Kankaanpää,[4] H. Kettunen,[4] P. Kuusiniemi,[4] M. Muikku,[4]
P. Rahkila,[4] A. Savelius,[4] C. N. Davids,[2] R. V. F. Janssens,[2] D. Seweryniak,[2]
M. P. Carpenter,[2] H. Amro,[2] P. Decrock,[2] P. Reiter,[2] D. Nisius,[2] L. T. Brown,[2]
S. Fischer,[2] T. Lauritsen,[2] J. Wauters,[5] C. R. Bingham,[5] M. Huyse,[6] A. Andreyev[6]

[1] Department of Physics and Astronomy, Rutgers University, New Brunswick, NJ 08903
[2] Physics Division, Argonne National Laboratory, Argonne, IL 60439
[3] N-Division, Lawrence Livermore National Laboratory, Livermore, CA 94550
[4] Department of Physics, University of Jyväskylä, Jyväskylä, Finland
[5] Department of Physics and Astronomy, University of Tennessee, Knoxville, TN 37996
[6] Instituut voor Kern- en Stralingsfysica, K.U. Leuven, Leuven, Belgium

Abstract. The γ-ray spectroscopy of even- and odd-mass isotopes of polonium have been studied using arrays of Ge detectors coupled to recoil-mass analyzers, including recoil-decay tagging techniques. The level energies and B(E2) branching ratios can be reproduced by theoretical frameworks which do not explicitly include proton particle-hole excitations across the Z=82 shell, conclusions in contrast to those deduced from alpha-decay measurements.

The nuclei in the lead region exhibit a wide variety of shapes as a function of neutron and proton number, as well as excitation energy and angular momenta. As the numbers of valence protons and neutrons increase, the structure evolves from shell model behavior near N=126 to moderate collective oblate shapes. As the middle of the N=82-126 neutron shell is approached, proton particle-hole excitations across the Z=82 shell gap give rise to strongly collective oblate excitations, which coexist with the more moderate ground-state shapes [1].

In recent years, the structure of the lightest polonium isotopes has been probed using recoil separators coupled to arrays of Compton-suppressed Ge detectors. The use of recoil separators was critical to identify gamma rays associated with the evaporation residues, which are populated with cross sections considerably less than the dominant fission channel. For the lightest isotopes, the recoil decay tagging technique was used to identify prompt transitions in specific alpha-decaying isomers of the evaporation residues. Such studies have been completed on $^{192-197}$Po [2-5], and in the case of the odd-A isotopes, transitions associated with both $13/2^+$ and $3/2^-$ isomers were identified.

More recently, high-spin excitations were probed in ^{196}Po using the ^{172}Yb(^{28}Si,4n) reaction at 143 MeV using the Jurosphere array of Compton-suppressed Ge detectors coupled to the gas-filled recoil separator RITU at the University of Jyväskylä.

TABLE 1. B(E2) branching ratios for non-yrast states in ^{196}Po.

Transition	Delayed	Prompt	B(E2) ratios Vibrator	Rotor	$4p - 2h$
$4_2^+ \to 2_2^+ / 4_2^+ \to 4_1^+$	1.41(27)	1.42(23)	1.10	> 20	> 20
$2_2^+ \to 2_1^+ / 2_2^+ \to 0_1^+$	20.9(46)	21.5(58)	∞	2.04 ($K = 0$)	Large $E0$

In addition to recording prompt γ rays at the target position in coincidence with recoils at the RITU focal plane, transitions associated with the decay of long-lived isomers, $t_{1/2} \approx 1\mu s$, were measured with a single Ge detector behind the focal plane. The level spectrum deduced for the decay of the 856(17)ns 11⁻ isomer in ^{196}Po is displayed in Fig. 1. These measurements essentially confirm the previous results [6] for the decay of this high-spin isomer, although additional non-yrast states have been observed following weak population in the isomer decay. The placements of the transitions, except those which directly depopulate the isomer, are confirmed by prompt $\gamma - \gamma$ coincidence measurements at the target position.

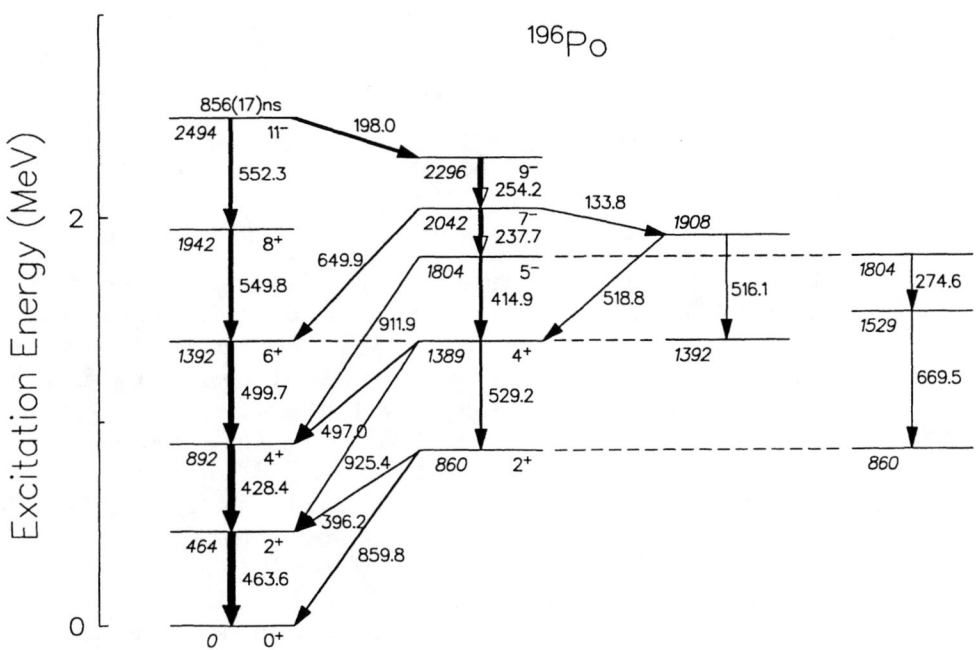

Fig. 1: Decay of 11⁻ isomer in ^{196}Po.

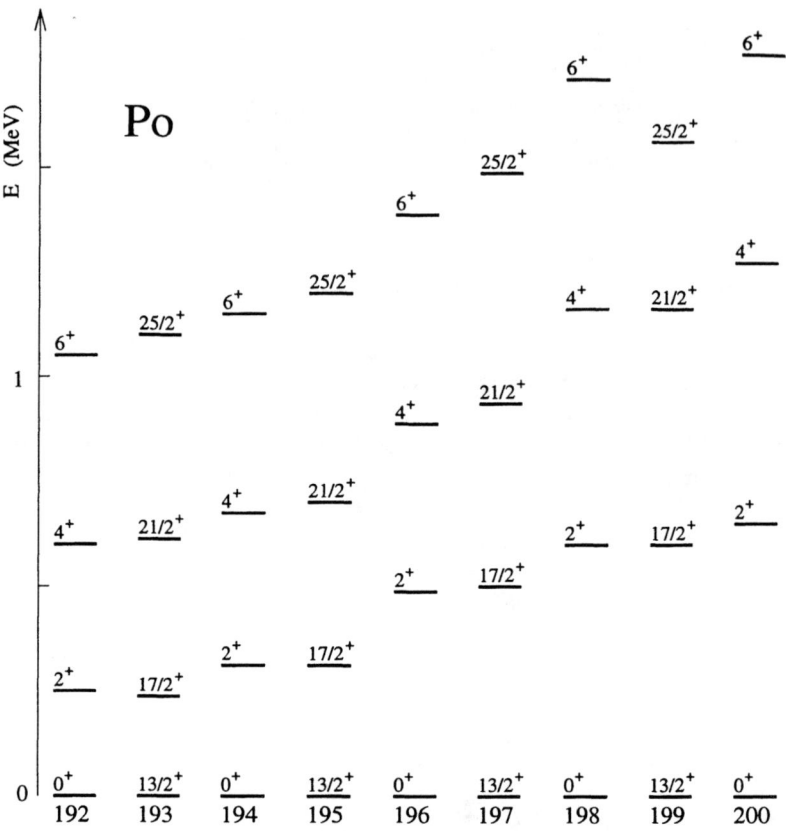

Fig. 2: Level systematics for $^{192-200}$Po.

In Table 1 the relative B(E2) branching ratios associated with the decay of the non-yrast states in ^{196}Po are summarized. The data from both delayed measurements and prompt spectroscopy gated on the 414-keV transition are tabulated; pure E2 transitions are assumed. These experimental results are compared with predictions for vibrational and rotational models, as well as expectations for 4-particle, 2-hole (4p-2h) proton excitations across the Z=82 shell gap. For "intruder" 4p-2h excitations, the non-yrast positive-parity states are candidates for the quasiband built on the "intruder" configuration, and allowed, intra-band transitions should be a factor of 20-50 times larger than forbidden, inter-band transitions. The experimental results are only consistent with the expectations of the vibrational model. In addition, the imbalance in γ-ray intensity required for the E0 transitions expected to characterize the $\Delta J=0$ transitions between "normal" 2-particle, and "intruder" 4p-2h configurations is not observed. The present results confirm those of ref. 7, but are in marked disagreement with those of ref. 6, which reported intensity imbalances. The level spacings, with nearly degenerate members of the 2- and 3-phonon multiplets, and the relative B(E2) branching ratios support a vibrational structure for ^{196}Po. Detailed spectroscopy of the lighter isotopes, in particular ^{194}Po, would further probe the structure of the Po nuclei and the role of 4p-2h intruder configurations in the yrast and non-yrast states. A new analysis of ^{194}Po is in progress [8].

The low-spin levels of the $^{200-192}$Po isotopes, including the levels above the $13/2^+$ isomers in the odd-A nuclei, are summarized in Fig. 2. The gamma-ray data suggest that the lightest Po isotopes evolve from a harmonic vibrational structure in ^{196}Po to a more collective, anharmonic vibrator in ^{192}Po. The structure of the excitations built on the $13/2^+$ isomers in the ^{193,195,197}Po isotopes, which follow a simple weak-coupling pattern, support this interpretation. If the yrast states in the core were complicated admixtures of "normal" and 4p-2h "intruder" configurations, which changed as a function of neutron number, the level spacings above the $13/2^+$ isomers would not be expected to follow those of the core.

The evolution of low-lying structure from ^{208}Po to ^{192}Po has been successfully modeled in the particle-core coupling framework and the microscopic composition of the 2^+ states determined from quasiparticle RPA calculations [9]. The theoretical analyses suggest that it is the opening of the $\nu i_{13/2}$ shell in the lighter isotopes, and the attractive interaction between $i_{13/2}$ neutrons and the valence $h_{9/2}$ protons, which result in increased collectivity, which remains essentially of anharmonic vibrational character, in the lighter isotopes.

These conclusions are in contrast to results from the alpha decay of these Po nuclei, where the fine structure suggests that proton particle-hole excitations are a dominant component in the ^{192}Po ground state [10].

This work is supported in part by the U.S. National Science Foundation and Department of Energy.

REFERENCES

[1] J. L. Wood et al., Phys. Rep. **215**, 101 (1992).
[2] W. Younes et al., Phys. Rev. C **52**, R1723 (1995)
[3] K. Helariutta et al., Phys. Rev. C **54**, R2799 (1996).
[4] N. Fotiades et al., Phys. Rev. C **55**, 1724 (1997).
[5] N. Fotiades et al., Phys. Rev. C **56**, 723 (1997).
[6] D. Alber et al., Z. Phys. A **339**, 225 (1991).
[7] L. A. Bernstein et al., Phys. Rev. C **52**, 621 (1995).
[8] K. Helariutta et al., to be published.
[9] W. Younes and J. A. Cizewski, Phys. Rev. C **55**, 1218 (1997).
[10] N. Bijnens et al., Z. Phys. A **356**, 3 (1996).

Inelastic proton scattering on the radioactive nuclei ^{18}Ne and ^{20}O in inverse kinematics

P.D. Cottle[1], L.A. Riley[1], J.K. Jewell[1], T. Glasmacher[2,3], K.W. Kemper[1], Y. Blumenfeld[4], M. Chromik[2], S.E. Hirzebruch[2,4], R.W. Ibbotson[2], F. Maréchal[4], D.J. Morrissey[2,5], H. Scheit[2], T. Suomijärvi[4]

[1] *Department of Physics, Florida State University, Tallahassee, Florida 32306-4350*
[2] *National Superconducting Cyclotron Laboratory, Michigan State University, East Lansing, Michigan 48824*
[3] *Department of Physics and Astronomy, Michigan State University, East Lansing, Michigan 48824*
[4] *Institut de Physique Nucléaire, IN$_2$P$_3$-CNRS, 91406 Orsay, France*
[5] *Department of Chemistry, Michigan State University, East Lansing, Michigan 48824*

Abstract.
Elastic and inelastic scattering to the 2_1^+ state of the single closed shell radioactive nuclei ^{18}Ne and ^{20}O have been measured in inverse kinematics with a beam energies of 30 MeV/u. The matrix element determined for the $0_{gs}^+ \to 2_1^+$ transitions in these reactions are compared with the corresponding electromagnetic matrix elements to determine M_n/M_p, the ratio of the neutron and proton multipole matrix elements. A comparison between the M_n/M_p values in ^{20}O and ^{18}O suggests that the matrix element ratio is increasing as the neutron number increases. The M_n/M_p result for ^{18}Ne is consistent with one obtained from a comparison of electromagnetic data from the mirror nuclei ^{18}O and ^{18}Ne.

The determination of proton and neutron multipole matrix elements for transitions between nuclear states provides one of the most important tools for understanding the relative importance of valence and core contributions to these transitions. The competition between valence and core contributions is of particular interest in single-closed-shell nuclei, where the low-lying excitations would be composed exclusively of the valence neutrons or protons if the closed core were truly inert. Methods for determining proton and neutron matrix elements generally involve the comparison of measurements of a transition using two experimental probes with different sensitivities to proton and neutron contributions. Studies of $0_{gs}^+ \to 2_1^+$ transitions and others in stable nuclei have been performed using a variety of combinations of experimental probes (for example, see [1]). However, until recently it has been impossible to examine neutron and proton contributions in this way in short-lived radioactive nuclei, where such studies would be of particular interest because of the relatively small binding energies of some of the valence nucleons. Data on the electromagnetic matrix elements for $0^+_{gs} \to 2^+_1$ transitions has been available for some short-lived even-even nuclei for some time [2], and recent advances in intermediate energy Coulomb excitation (for example, see [3]) have made even more information of this type available. These electromagnetic data provide information on the proton contributions to the $0^+_{gs} \to 2^+_1$ matrix elements. With recent advances in techniques for providing intense beams of radioactive nuclei, inverse kinematics proton scattering provides a way to determine the neutron contributions to these matrix elements. At center-of-mass energies less than 50 MeV - corresponding to radioactive beam energies of less than 50 MeV/nucleon - inelastic proton scattering is much more sensitive to the neutron contributions in transitions than those of the protons, and therefore can be used together with electromagnetic data to understand the relationship between proton and neutron contributions [4,5].

Here we report the results of inverse kinematics proton scattering study of the $0^+_{gs} \to 2^+_1$ transition in the single-closed-shell radioactive nuclei ^{18}Ne and ^{20}O. ^{18}Ne is the heaviest proton-rich radioactive nucleus in which proton scattering has been measured, while ^{20}O is two neutrons from stability but only four neutrons from the heaviest bound oxygen isotope (^{24}O) [6]. Electromagnetic data are already available for these transitions (see [2] for a compilation), so the proton scattering data provide determinations of M_n/M_p, the ratio of the neutron and proton multipole matrix elements [1].

We measured angular distributions of protons scattered from the ground states and 2^+_1 states in ^{18}Ne and ^{20}O in inverse kinematics. The 30 MeV/nucleon beams were produced in the A1200 fragment separator [7] at the National Superconducting Cyclotron Laboratory. The secondary beam intensities were greater approximately 30,000 particles/sec. Scattered protons were detected using the FSU-MSU array of 8 Si strip-Si PIN-CsI particle telescopes. We detected protons in the laboratory scattering angle range 65° - 80°, corresponding to a center-of-mass angular range of approximately 20° - 45°. The angular distributions for ^{20}O are shown in Figure 1.

To extract the strength of the $0^+_{gs} \to 2^+_1$ transitions in the (p,p') reactions measured here, we used the coupled channels code CHUCK [8] and a standard vibrational form factor. The elastic scattering data taken here are not sufficient to reliably extract parameters for optical model potentials; consequently, the potential parameters for both nuclei were taken from a study of the ^{20}Ne(p,p') reaction at 30 MeV [9]. We obtained $\beta_2 = 0.50(4)$ for ^{20}O and $\beta_2 = 0.46(4)$ for ^{18}Ne.

Inelastic proton scattering at low energies (≤ 50 MeV) is much more sensitive to the neutron contribution than the proton contribution to a transition. If $b^{(p,p')}_{n(p)}$ is the external-field neutron (proton) interaction strength for low-energy proton scattering, then the ratio $b_n^{(p,p')}/b_p^{(p,p')}$ is approximately 3 [1]. In contrast, the electromagnetic matrix element $B(E2; 0_{gs}^+ \rightarrow 2_1^+)$ measures the proton matrix element M_p. We used equation (7) from [1] to obtain M_n/M_p from the (p,p') and electromagnetic measurements. Our results, $M_n/M_p=2.9(4)$ for ^{20}O and $M_n/M_p=0.36(11)$ for ^{18}Ne, clearly deviate from the values of $M_n/M_p = N/Z$ (1.5 for ^{20}O, 0.80 for ^{18}Ne) that would indicate purely isoscalar transitions. In the case of ^{20}O, we would expect the neutrons to dominate the $0_{gs} \rightarrow 2_1^+$ transition (giving $M_n/M_p > N/Z$) because the proton shell is closed. For the closed-neutron shell nucleus ^{18}Ne, we would expect $M_n/M_p < N/Z$, indicating the dominance of the proton contribution in the transition. Both expectations are borne out by the data.

Given the proximity of the neutron drip line, it is interesting to compare the values of M_n/M_p for the $0_{gs}^+ \rightarrow 2_1^+$ transitions in ^{18}O and ^{20}O to look for suggestions of a trend. Since ^{18}O is stable, there are many more probes available to examine proton and neutron contributions in this nucleus. In Figure 2, we illustrate M_n/M_p results for ^{18}O obtained using a variety of methods [10–14], including a comparison of electromagnetic data and low energy proton scattering [15] identical to that used here for ^{20}O and ^{18}Ne. The comparison of ^{20}O to various results for ^{18}O suggests that the relative role of the valence neutrons in the $0_{gs}^+ \rightarrow 2_1^+$ transition is growing with increasing neutron number. However, confirmation of this will require inverse kinematics proton scattering measurements of the corresponding transitions in ^{22,24}O. These measurements will become possible as new radioactive beam facilities come on line. In addition, Figure 2 shows values for M_n/M_p in ^{18}Ne obtained here and using the mirror method of [11] in which electromagnetic matrix elements from the mirror nuclei ^{18}O and ^{18}Ne are compared. The two ^{18}Ne results are consistent with each other, providing some confidence in our method of analysis.

This work was supported by the National Science Foundation and the State of Florida.

REFERENCES

1. A.M. Bernstein, V.R. Brown, V.A. Madsen, Comments Nucl. Part. Phys. **11**, 203 (1983).
2. S. Raman, C.H. Malarkey, W.T. Milner, C.W. Nestor, Jr., P.H. Stelson, At. Data and Nucl. Data Tables **36**, 1 (1987).
3. R.W. Ibbotson, T. Glasmacher, B.A. Brown, L. Chen, M.J. Chromik, P.D. Cottle, M. Fauerbach, K.W. Kemper, H. Scheit, and M. Thoennessen, Phys. Rev. Lett. **80**, 2081 (1998).
4. V.A. Madsen, V.R. Brown, and J.D. Anderson, Phys. Rev. **C12**, 1205 (1975).
5. A.M. Bernstein, V.R. Brown, V.A. Madsen, Phys. Lett. **103B**, 255 (1981).
6. M Fauerbach, D.J. Morrissey, W. Benenson, B.A. Brown, M. Hellström, J.H. Kelley, R.A. Kryger, R. Pfaff, C.F. Powell, and B.M. Sherrill, Phys. Rev. **C53**, 647 (1996).

7. B.M. Sherrill, D.J. Morrissey, J.A. Nolen Jr., N. Orr, J.A. Winger, Nucl. Inst. Meth. **B56/57**, 1106 (1991).
8. P.D. Kunz, University of Colorado report (unpublished).
9. R. de Swiniarski, A. Genoux-Lubain, G. Bagieu, J.F. Cavaignac, Can. J. Phys. **52**, 2422 (1974).
10. P. Grabmayr, J. Rapaport, and R.W. Finlay, Nucl. Phys. **A350**, 167 (1980).
11. A.M. Bernstein, V.R. Brown ad V.A. Madsen, Phys. Rev. Lett. **42**, 425 (1979).
12. J. Kelly, W. Bertozzi, T.N. Buti, J.M. Finn, F.W. Hersman, M.V. Hynes, C. Hyde-Wright, B.E. Norum, A.D. Bacher, G.T. Emery, C.C. Foster, W.P. Jones, D.W. Miller, B.L. Berman, J.A. Carr and F. Petrovich, Phys. Lett. **169B**, 157 (1986).
13. S. Iversen, H. Nann, A. Obst, K.K. Seth, N. Tanaka, C.L. Morris, H.A. Thiessen, K. Boyer, W. Cottingame, C.F. Moore, R.L. Boudrie, and D. Dehnhard, Phys. Lett. **82B**, 51 (1979).
14. S.J. Seestrom-Morris, D. Dehnhard, M.A. Franey, D.B. Holtkamp, C.L. Blilie, C.L. Morris, J.D. Zumbro, and H.T. Fortune, Phys. Rev. **C37**, 2057 (1988).
15. J.L. Escudié, R. Lombard, M. Pignanelli, F. Resmini, and A. Tarrats, Phys. Rev. **C10**, 1645 (1974).

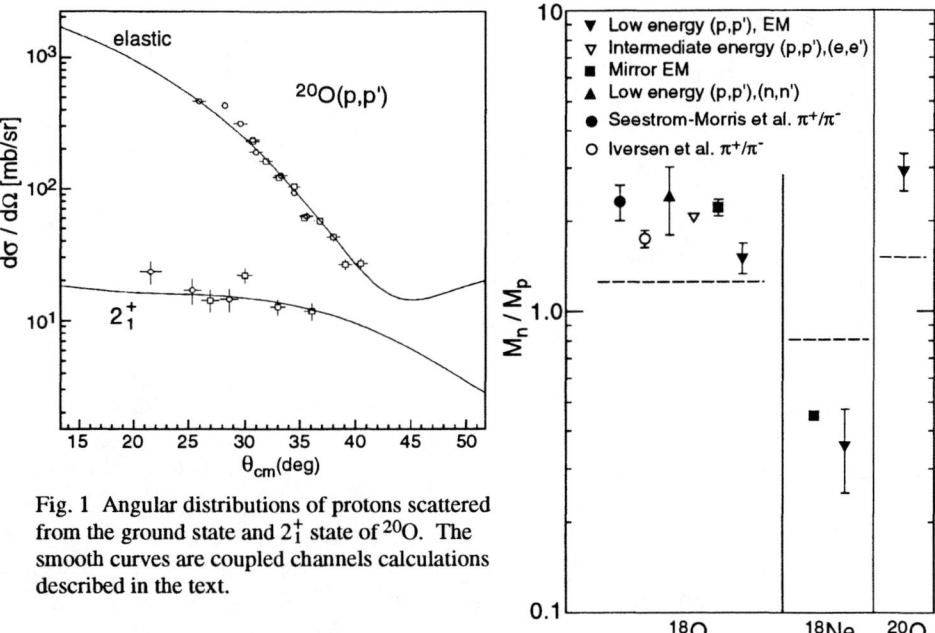

Fig. 1 Angular distributions of protons scattered from the ground state and 2_1^+ state of ^{20}O. The smooth curves are coupled channels calculations described in the text.

Fig. 2 M_n/M_p values for the $0_{gs}^+ \to 2_1^+$ transitions in ^{18}O, ^{18}Ne and ^{20}O. The dashed lines correspond to $M_n/M_p = N/Z$.

Interplay between Nuclear Structure and Reaction Mechanism in the Production of Projectile-like Short-lived Isomers

J.M. Daugas[a], M. Lewitowicz[a], R. Anne[a], J.C. Angélique[b],
L. Axelsson[c], R. Béraud[d], C. Borcea[e], E. Chabannat[d],
Th. Ethvignot[e], S. Franchoo[f], M. Glogowski[g], R. Grzywacz[g],
H. Grawe[h], D. Guillemaud-Mueller[i], M. Huyse[f], Z. Janas[g],
M. Karny[g], C. Longour[j], M.J. Lopez-Jimenez[a], A.C. Mueller[i],
A. Nowak[g], F. de Oliveira-Santos[a], N.A. Orr[b], A. Płochocki[g],
M. Pfützner[g], K. Rykaczewski[g], M.G. Saint-Laurent[a],
J.E. Sauvestre[k], O. Sorlin[i], P. Van Duppen[f], J.S. Winfield[a]

[a] GANIL BP 5027,14076 Caen Cedex 5, France, [b] LPC Caen, [c] Dept. of Physics Göteborg, [d] IPN Lyon, [e] IAP Bucharest, [f] IKS KU,Leuven, [g] IFD Warsaw University, [h] GSI Darmstadt, [i] IPN Orsay, [j] IReS Strasbourg, [k] CEA Bruyères le Châtel

Abstract. Isomeric ratios and the momentum distributions of nuclei produced in the fragmentation of a 60AMeV ^{92}Mo beam on a thin ^{27}Al target have been studied. A strong dependence of the isomeric ratio on the structure of the isomer and on the reaction mechanism has been observed for the first time at intermediate energies.

The isomeric ratio F (number of nuclei produced in the isomeric state divided by the total number of detected nuclei for a given A and Z) observed in fragmentation-like reactions varies from several to almost 100% [1–3]. In principle, the value of F should depend strongly on the structure of nucleus. Indeed, the population of the Yrast and non-Yrast isomeric states has been observed to be significantly different [2]. The ratio F should also depend on the reaction mechanism, especially for projectile - like fragments. Up to now no quantitative explanation of the variation of F has proven possible within the framework of existing models [4,5]. From the experimental point of view, the roles of the reaction mechanism, deexcitation of the fragments and the structure of the isomeric and ground states have not been clearly separated. At relativistic energies, it has been found that the isomeric ratio F is also related to the alignment and polarisation of outgoing fragments [3]. However, this effect had not been observed at intermediate energies. A better understanding of the fragmentation-like reaction mechanism and, in particular, of the transfer of angular momentum, can be achieved by measuring the isomeric ratio F as a function of the velocity of outgoing fragment, which to some extent should be related to the impact parameter.

To address these questions a dedicated experiment has been performed at GANIL using the LISE3 spectrometer and a 60AMeV ^{92}Mo beam. In order to measure precisely the velocity distributions a thin (6μm) ^{27}Al target was employed and the magnetic rigidity of the spectrometer was changed by steps of about 0.2 %. In addition the momentum acceptance of LISE3 was limited to about 0.1 %. The detection set-up at the final focus consisted a three-element Si-detector telescope. The unambiguous identification of fragments in mass, atomic number and atomic charge was achieved by means of energy-loss, total-kinetic-energy and time-of-flight measurements. The typical time-of-flight of the fragments through the spectrometer was about 400 ns. The silicon telescope was surrounded by four high efficiency Ge detectors and one clover LEPS detector (see fig.1).The total photopeak efficiency of the γ-detectors was 4.3 % at 220 keV and 2 % at 1 MeV.

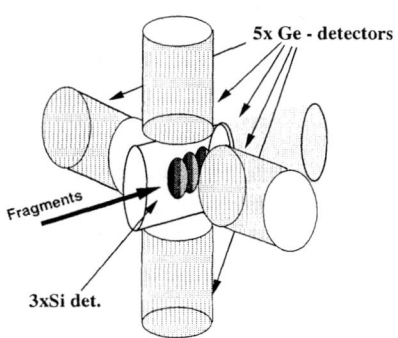

FIGURE 1. Experimental set-up placed at the final focus of the LISE3 spectrometer.

A wide range of fragments and short-lived isomers from Germanium to Technetium were studied. Almost all of the short-lived isomers were produced in two or three ionic charge states. Some selected results are presented in figures 2 and 3. The production yields and the isomeric ratios F are plotted versus the ratio of the fragment velocity to the beam velocity v/v_0. Here F represents the isomeric ratio (in %) at the production target, ie. taking into account the in-flight decay losses. In figure 2 the results are shown for (a) ^{92}Tc, (b) ^{90}Mo and (c) ^{69}Ge where the isomeric and ground states are respectively (a) E=270.2 keV, J^π=4$^+$, $T_{1/2}$=1.03μs and J^π=8$^+$; (b) E=2874.73keV, J^π=8$^+$, $T_{1/2}$= 1.12μs and J^π=0$^+$; and (c) E=397.95 keV, J^π=9/2$^+$, $T_{1/2}$=2.81μs and J^π= 5/2$^-$.The data clearly show a strong dependence of the isomeric ratio F on the fragment velocity v with a pronounced minimum or maximum value close to the beam velocity v_0. The F(v) dependence becomes almost flat for fragments far from projectile. Indeed, it may be noted in figure 2 that F is about 15 times greater at $v=v_0$ than in the low-velocity tail of the fragment distribution for ^{92}Tc, while remaining almost constant for ^{69}Ge. This effect suggests that the reaction mechanism strongly influences the observed isomeric ratio. The high selectivity of the few - nucleon transfer reactions with respect to the angular momentum transfer is shown in figure 3, where the isomeric ratio F is plotted for two different isomeric states of the same nucleus, namely ^{90m1}Nb ($J^\pi=11^-$, E=1880.21keV), figure 3a and ^{90m2}Nb ($J^\pi=6^+$, E=122.37keV), figure 3b. The ground state of ^{90}Nb is $J^\pi=8^+$. The F distribution displays a completely different behavior for these isomers which may be qualitatively explained by the important spin differences for the two isomers with respect to the spin of the ground state. If the spin of the isomeric state is greater than the spin of the ground state, the minimum value of the isomeric ratio F is found at $v=v_0$, while the opposite is the case if the spin of the isomeric state is lower than the ground state one,ie. F has a maximum close to $v=v_0$. In both cases the angular momentum transfer is expected to have a minimum at $v=v_0$. A similar effect but for two different nuclei may be noticed in comparing figures 2(a) and 2(b). The high sensitivity of the F

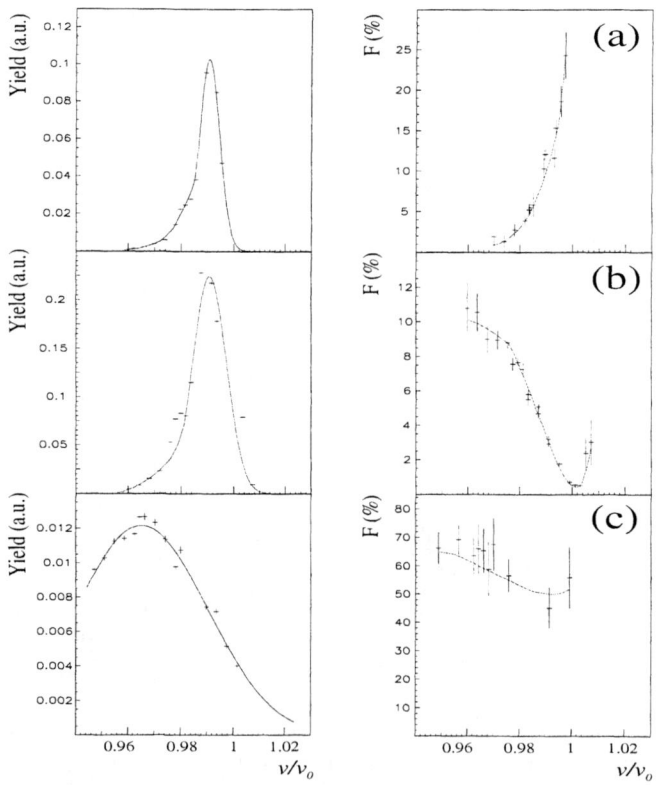

FIGURE 2. Yields (left column) and isomeric ratios (right column) for ^{92}Tc (a), ^{90}Mo (b) and ^{69}Ge (c) measured as a function of the ratio of the fragment velocity v to the beam velocity v_0.

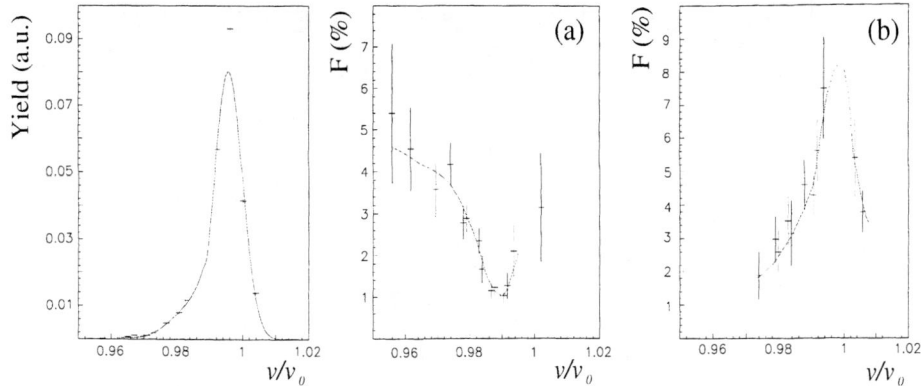

FIGURE 3. Production yield (left panel) and isomeric ratio F of ^{90}Nb. Two different isomeric states are represented, where (a) corresponds to E=1880.21keV, $J^\pi=11^-$, $T_{1/2}$=472ns and (b) to E=122.37keV, $J^\pi=6^+$, $T_{1/2}$=65μs.

distributions with respect to the spin of isomer might thus prove to be a means to determine the spins of nuclei far from stability produced in fragmentation-like reactions.

In summary, the results obtained in the present work have demonstrated the strong influence of the initial angular momentum transfer on the formation probability of isomeric states produced at intermediate energies. Both nuclear structure and reaction mechanism effects have been observed. A quantitative description of the isomeric ratios demands further development of models of nuclear reactions at intermediate energies. Within the framework of a relatively simple model of fragmentation at relativistic energies proposed recently by de Jong et al. [4], it is possible to calculate an average isomeric ratio. However, the validity of the above model as well as its capability to describe the observed distributions of the isomeric ratio at lower incident energies remains to be proven.

The observed behavior of the isomeric ratios suggests the existence of a strong alignment and polarisation of the quasi-projectile. This in turn, could allow for future measurements of the spin, quadrupole and magnetic moments of isomeric states far from stability.

REFERENCES

1. Young B.M., et al., *Phys. Lett.* **B311**, 22 (1993).
2. Grzywacz R., et al., *Phys. Lett.* **B355**, 439 (1995).
3. Schmidt-Ott W.D., et al., *Z. Phys.* **A350**, 215 (1994).
4. de Jong M., et al., *Nucl. Phys.* **A613**, 435 (1997).
5. Asahi K., et al., *Phys. Rev.*, **C43**, 456 (1991).

Mass dependence of the effective isovector charge in the sd-shell

M. Fauerbach[†], P.D. Cottle[†], T. Glasmacher[*,††], R.W. Ibbotson[*], K.W. Kemper[†], B. Pritychenko[*,††], H. Scheit[*,††], M. Steiner[*]

[†] *Department of Physics, Florida State University
Tallahassee, FL 32306, USA*
[*] *National Superconducting Cyclotron Laboratory, Michigan State University,
East Lansing, Michigan 48824, USA*
[††] *Department of Physics and Astronomy, Michigan State University,
East Lansing, Michigan 48824, USA*

Abstract. In order to map out a possible mass and isospin dependence of the effective charges in the sd-shell, we performed an intermediate energy Coulomb excitation experiment with the semi-magic, proton-rich nucleus ^{38}Ca. The transition between the first excited state (2_1^+, $E^* = 2206$ keV) and the ground state (0_{gs}^+) was studied, and the reduced transition strength and the proton transition matrix element were deduced. The nucleus ^{38}Ca was chosen for this study, as it is at the upper mass end of the sd-shell, and will together with already existing high precision data for the lower mass end of the shell most clearly show a systematic trend in the effective charges. Our study also helps to fill the void of available high quality data for proton-rich nuclei in this mass region. We compare our data to theoretical shell model predictions with mass dependent, and mass independent effective charges, and to experimental values of the neutron-rich mirror nucleus ^{38}Ar.

INTRODUCTION

Shell model calculations have been very successful in calculating transition matrix elements for a variety of nuclei in the sd-shell (see eg [1–3]). Whereas the measured proton (neutron) transition matrix elements can contain contributions from excitation of all protons (neutrons) in the nucleus over indefinitely many shell-model orbits, this is –due to computational limitations– not the case for the 'model space' transition matrix elements. Due to this truncated model space, one has to renormalize the calculated transition matrix elements, in order to be able to reproduce the 'total', measured matrix elements. Therefore, the concept of effective charges was successfully introduced in order to reproduce core polarization effects generally associated with electric transitions.In principle one would expect the effective charges to be state and mass dependent. Brown *et al.* [1] found no evidence for a state or mass dependence when comparing their shell-model calculations to experimental data in the *sd*-shell. However, the authors also point out the lack of available data from proton-rich nuclei, which makes a firm conclusion difficult.

EXPERIMENTAL METHOD AND RESULTS

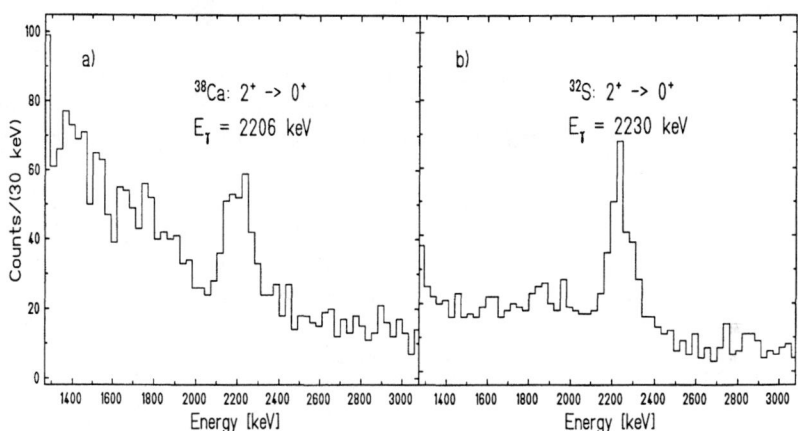

FIGURE 1. Doppler-shifted γ-ray spectrum gated on ^{38}Ca (a) and ^{32}S fragments (b).

A 80 MeV/nucleon ^{40}Ca beam from the K1200 cyclotron at the NSCL irradiated a 202 mg/cm^2 target of ^{9}Be located at the mid-acceptance target position of the A1200 fragment separator [4]. The energy spread of the resulting ^{38}Ca fragments was limited to $\pm 1\%$ with an aperture and separation was obtained by placing a thin, achromatic wedge (^{27}Al, 64 mg/cm^2) at the second dispersive image of the A1200. A 'cocktail' beam containing several fragment species was used to perform the experiment in order to study other nuclei in the vicinity simultanously. This could be done, as the counting rate was not a limiting factor, and the fragment identification was unambiguous. About 20% of the mixed beam was ^{38}Ca ($\approx$ 12kps), and the average energy of the ^{38}Ca particles was 56.1 MeV/A. The beam was transported to the experimental station were it interacted with a 184 mg/cm^2 gold target. Scattered beam particles were detected in the zero-degree detector, which consisted of a fast-slow plastic phoswich detector. This detector defines a half-cone opening angle of $\theta_{\text{lab}} < 4°$. The secondary target was surrounded by an array of 38 position sensitive NaI(Tl) detectors arranged in three concentric rings around the target and shielded from background photons by 16.5 cm thick walls of low-background lead. A more detailed description about the experimental procedure can be found in [3,6–8]. One of the beam contaminants was ^{32}S, which has a ($2_1^+ \rightarrow 0_{gs}^+$) transition with an energy of E$_\gamma$ = 2230.5 keV. This is very close to the ($2_1^+ \rightarrow 0_{gs}^+$) transition energy in ^{38}Ca (E$_\gamma$ = 2206 keV). As the reduced transition strength in ^{32}S is known to high precision [9] we can use our measured γ-ray intensities combined with the number of incoming particles to deduce the transition strength in ^{38}Ca without having to rely on efficiency measurements. The Doppler-corrected γ-ray energy spectra, recorded under the condition that a ^{38}Ca (^{32}S) fragment was detected in the zero degree-detector, are shown in Fig. 1. The photopeaks centered around γ-ray energies of 2206 keV and 2230 keV in the projectile frame ($\beta \approx 0.32(0.30)$) – corresponding to the $2_1^+ \to 0_{gs}^+$ transitions are clearly visible for both beams.

We measure a reduced transition strength B(E2, $2_1^+ \to 0_{gs}^+$) = (12.7 ± 2.96) $e^2 fm^4$ for ^{38}Ca. From this reduced transition strength we extract the proton transition matrix element M_p. Our result of M_p = (7.93 ± .97) fm^2 has to be compared to the results of shell-model calculations with mass independent and mass dependent effective charges.

TABLE 1. Extracted effective charges from Ref. [2], and the corresponding transition matrix elements. The first two lines show the values extracted from fits to nuclei with positive (negative) T_z. The third line shows the values without a T_z separation, and the fourth line shows the results for a mass dependent fit. Here, n denotes the number of nucleons outside the ^{16}O core (n = 22 in our case). In the last row the values predicted in Ref. [1] are shown.

			^{38}Ca		^{38}Ar	
	e_n [e]	e_p [e]	M_p(th.) [fm^2]	M_p(exp.) [fm^2]	M_p(th.) [fm^2]	M_p(exp.) [fm^2]
$T_z > 0$	0.403	1.169	3.3	7.97(93)	10.2	10.9(4)
$T_z < 0$	0.752	0.945	6.2		8.2	
all	0.561	1.030	4.6		9.0	
A dep.	0.317 + 0.012n	1.188 + 0.0035n	4.8		11.0	
independent	0.45	1.15	3.1		8.3	

Brown et al. [1] use mass independent effective charges and a finite-well, 'local Woods-Saxon' potential (see [1] for details), and derive a proton transition matrix element of M_p = 3.1 fm^2, which is more than a factor of two smaller than the measured value. A possible explanation for this discrepancy might be the use of the (simpler) finite well potential for this calculation. This seems to affect the nuclei at the upper end of the sd-shell more than the lighter sd-shell nuclei, which can be seen clearly for the mirror nucleus ^{38}Ar. For ^{38}Ar, Brown et al. performed both a 'full blown' calculation, using a harmonic oscillator potential, and one with a 'local Woods-Saxon' potential (quoted above). Whereas, they get excellent agreement with the experimental value for M_p when using the 'full' calculation, the value calculated using the 'local Woods-Saxon' potential, underestimates the experimental M_p by 25%. For ^{38}Ca this effect might even be enhanced due to the relatively small proton binding energy in this nucleus.

Alexander, Castel, and Towner [2] analyzed a similar set of data, however, they derived not only mass dependent effective charges, but also found evidence for a isospin (T_z) dependence of the effective charges. The different derived effective charges and the corresponding transition matrix elements can be found in Table I. A comparison to our experimental value shows very good agreement with the theoretical calculations using the isospin dependent effective charges.

CONCLUSIONS

Our measurement of the reduced transition strength in the proton-rich nucleus ^{38}Ca provides essential information about the mass and isospin dependence of the effective charges in the sd-shell. Our deduced proton transition matrix element for ^{38}Ca, is more than a factor of two larger than the value deduced by Brown et al. [1], using mass and isospin independent, constant effective charges. We find good agreement with the theoretical values derived by Alexander et al. [2], when using their isospin dependent description for the effective charges. Furthermore, our data indicates that a combination of the isospin dependence with a mass dependence of the effective charges would lead to an even better agreement, between the theoretical and experimental data. Finally, our result also seems to confirm the previously suggested decrease of the isovector effective charge with mass. The lack of high quality data of proton-rich nuclei in this mass region made a conclusion about the mass dependence of the isovector effective charge prior to this measurement impossible. We were able to contribute substantially in filling this void, however, further investigation of other proton-rich nuclei would be helpful in order to adequately describe this mass dependence.

Due to the conservation of isospin in light nuclei, our result on ^{38}Ca can be combined with the known transition strength of the mirror transition in ^{38}Ar to provide the ratio of M_n/M_p using only electromagnetic probes. A more detailed discussion about this subject can be found in [10].

REFERENCES

1. B.A. Brown, B.H. Wildenthal, W. Chung, S.E. Massen, M. Bernas, A.M. Bernstein, R. Miskimen, V.R. Brown, V.A. Madsen, Phys. Rev. **C26** (1982) 2247.
2. T.K. Alexander, B. Castel, L.S. Tower, Nucl. Phys. **A445** (1985) 189.
3. R.W. Ibbotson et al., Phys. Rev. Lett. **80** (1998) 2081.
4. B.M. Sherrill, D.J. Morrissey, J.A. Nolen Jr., and J.A. Winger, Nucl. Instr. Methods **B56** (1991) 1106.
5. T. Glasmacher, P. Thirolf, and H. Scheit, to be published.
6. H. Scheit et al., Phys. Rev. Lett. **77**, 3967 (1996).
7. T. Glasmacher et al., Phys. Lett. B **395**, 163 (1997).
8. P.M. Endt, Nucl. Phys. **A521**, 1 (1990).
9. M. Fauerbach et al., to be published.

On the Q_β-puzzle near ^{132}Sn

B. Fogelberg [1], K.A. Mezilev [2], H. Mach [1] and V.I. Isakov [2]

[1] *Department of Neutron Research, Uppsala University, S-61182 Nyköping, Sweden*
[2] *St. Petersburg Nuclear physics Institute, 188350 Gatchina, Russia*

Abstract: The Q_β values for 14 nuclides in the vicinity of ^{132}Sn have been measured using high resolution $\beta\gamma$-coincidence spectroscopic methods. The new results for the atomic masses in this region are significantly more precise than the previously accepted values, and differ significantly from them in a few cases. The new data is used to resolve the recently observed Q_β puzzle near ^{132}Sn.

INTRODUCTION

It is believed that the simple excitations near the doubly magic nuclei can be precisely interpreted in the shell model, provided that crucial experimental information is known. Generally, the new experimental facts can test the basic shell model parameters, but in the rare but most interesting cases, it is the shell model calculations that can challenge the experimental results. Such has been the case for the Q_β puzzle near ^{132}Sn.

The accepted masses of the neutron rich N=82 isotones in the vicinity of ^{132}Sn have recently been questioned (1) using predictions based on shell model systematics. These authors define a mass "window" W, comprising a specific combination of N=82 ground state masses (of ^{132}Sn, ^{133}Sb, ^{134}Te, and ^{135}I), that can be related to the experimental excitation energies of the 4^+ and 6^+ states in ^{134}Te and the $15/2^+$ state in ^{135}I:

$$W = M(^{132}Sn) - 3M(^{133}Sb) + 3M(^{134}Te) - M(^{135}I)$$

$$= E(15/2^+) - 3(c.f.p.)^2 E(4^+, 6^+)$$

A significant difference of almost 500 keV has been noted between the W = −3570 keV from spectroscopy and W = −3080 (150) keV from N=82 masses. Since such a comparison gives agreement to within 5 keV for the N=126 isotones at ^{208}Pb, the authors conclude (1) that the N=82 mass values could be inaccurate by considerably more than the estimated errors. However, this particular conclusion has later been challenged by the first theoretical work (2) employing a realistic effective interaction for shell model calculations of the N=82 isotones. There is an obvious need for an experimental clarification of the mass data in a close vicinity of ^{132}Sn for the dual purpose of understanding the inconsistency between experimental data and a seemingly well founded empirical systematics, and also to give a firm basis for tests of the increasingly more accurate full scale shell model calculations in this neutron rich region.

EXPERIMENTAL PROCEDURES

We have therefore undertaken a major re-investigation of the total β-decay energies, Q_β, in the isobaric chains involving ^{132}Sn and the nearest N=82 isotones. The atomic masses of far-from-stability nuclides are obtained by adding the Q_β energies to the known mass values of nuclides placed closer to the stability line. This standard procedure was used at our laboratory in the most recent study (3) of the mass data near ^{132}Sn. We have to note, however, that the values derived from this procedure depend critically on the accuracy of the previously accepted mass data, as well as on the correctness of the existing information on the decay schemes. A substantial part of the present experimental work thus consisted of $\gamma\gamma$-coincidence measurements performed to verify the level structure of importance for the Q_β determinations in a number of nuclides. The $\gamma\gamma$-coincidence results also allowed for a critical examination of the possible γ-ray impurities in the spectra projected from gates employed in the Q_β measurements, and proved to be of vital importance in one of the cases studied. Another important ingredient in the current investigation has been the application of a high resolution Si(Li) detector for very accurate measurements of the low energy β-transitions.

The method employed for the Q_β determinations is $\beta\gamma$-coincidence spectroscopy, where the end point energies of selected β-transitions are measured using solid state spectrometers. The key ingredient in the analysis of such measurements is a transformation of the observed pulse-height distribution to a β-spectrum. A precise knowledge of the spectrometer response to mono-energetic electrons of different energies is vital for high accuracy determinations of the end point energies. The response of the HPGe spectrometer employed in most of our far-from-stability end point measurements (3) near ^{132}Sn had been carefully studied at the BILL electron spectrometer (4) (previously in operation at ILL in Grenoble) using electrons in the range 1 – 8 MeV. The HPGe spectrometer is particularly useful for relatively high β-energies, well exceeding 1 MeV, and was employed in most of our previous Q_β determinations as well as in the present control measurements of all energetic β-transitions of interest here. The control measurements of β-transitions having energies well below 2 MeV (which included a few cases with high total decay energies, but having low energy β-branches) were all performed using the high resolution Si(Li) diode. The response of this spectrometer was determined using conversion electrons from thin sources of ^{137}Cs and ^{207}Bi. The thickness of the Si(Li) diode was 2 mm. Consequently, the useful range of β end point energies for this spectrometer was restricted to less than about 1.5 MeV. As in the previous Q_β determinations (3), the spectrometers were operated at modest counting rates in order to reduce pulse pile-up effects, and the electronic system included active pile-up-rejection circuitry. Such measures are of substantial importance for preventing distortion of the critical high-energy part of the β-spectra.

The current re-investigation of the Q_β and atomic mass data included the decays of isotopes of Sn, Sb, Te and I in the range A=131–135. All nuclides were obtained as mass separated fission products at the OSIRIS ISOL facility (5) at Studsvik, Sweden. The low energy radioactive ion beam was collected on a thin movable tape used for removal of long-lived daughter products when necessary. The Si(Li) detector was placed inside the vacuum system to view the beam deposition spot from an angle of about 45 degrees. The HPGe β-spectrometer, on the other hand, was separated from the vacuum by windows of 0.08 mm Al and 0.25 mm Be. The electron energy loss in these windows has been measured as function of the electron energy and is known with an uncertainty of less than about 5–10 keV depending on the electron energy. This uncertainty is actually one of the main contributions to the total uncertainty of β end point energies up to about 4–6 MeV. At higher energies, the total uncertainty increases due to an incomplete knowledge of the response function, to become about 30 keV in the vicinity of 10 MeV electron energy. This level of precision is thus comparable to the best values expected from direct mass measurements using Penning traps or RF spectrometers. The energy scale of the HPGe detector was calibrated by γ-rays from standard sources including the 6.129 MeV line of ^{16}O. The Si(Li) detector was calibrated on line using known conversion electron lines from fission product nuclei, and also off line by placing a ^{207}Bi source at the location of the beam spot. The possible uncertainty from a distortion of the Si(Li) spectra due to coincidence summing of a β-particle event with events from conversion electrons or X-rays was investigated by a computer simulation. The maximum influence on the end point energies was found to be less than 3 keV in all cases of interest here.

The data analysis was performed by gating individual γ-rays and projecting (background-subtracted) β-distributions from the $\beta\gamma$-coincidence matrix. These were subsequently transformed to β-spectra using the empirical response functions, and converted to Fermi-Kurie distributions. Finally, the end point energy determinations were performed by a least squares fit. Whenever possible, several independent γ-ray gates were used for each Q_β-determination.

The concurrent analysis of the new $\gamma\gamma$-coincidence data was used to examine the γ-rays selected as gates for β-spectra with respect to their positions in the relevant decay schemes. Severe problems were found in the case of the ^{134}I decay, where for example, the 1136.2-keV γ-ray (which was used as a gate in the previous Q_β analysis (3)) was found to be a doublet, while other spectra have shown strong true coincidence summing effects from a cascade of lower lying transitions. A set of new measurements was therefore performed using the 2 mm Si(Li), placed at varying distances from the ^{134}I source. The data obtained at solid angles exceeding a few percent has shown a conclusive evidence that for some choices of gating transitions, the true coincidence summing of β-particle and Compton events distorted the β-spectra towards higher energies. Consequently, the final analysis of the decay energy of ^{134}I was performed on the data taken at a solid angle of less than about 3%, where the summing contribution was found negligible.

DISCUSSION OF THE RESULTS

A selection of the new data on decay energies and mass excess values in the ^{132}Sn region is given in Table 1. The full set of results includes data on eight nuclides studied previously by us, and on six nuclides placed closer to the stability line, as well as data on ^{131}In, ^{132}In, ^{134}Sb, and ^{134}Sn, not studied presently but subject to significant shifts in the mass excess values. The precision in the mass excess values measured here has been strongly improved, and is now of the order of $\delta m/m \sim 3 \times 10^{-7}$. The glaring discrepancy with the previous value for ^{134}I can be understood from the preceding discussion. Another significant discrepancy was found for the ^{132}Te decay energy, which had been earlier determined (6) at one occasion only. The full account of this set of measurements will be given elsewhere (8).

TABLE 1. Selected beta decay energies and mass excess values.

Nuclide	Q_β (MeV) this work	ME (MeV) this work	ME (MeV) Audi-Wapstra'95	DME (keV)
^{135}I	2.627 (6)	-83.793 (8)	-83.788 (23)	5
^{134}Te	1.513 (7)	-82.559 (11)	-82.399 (34)	160
^{134}I	4.052 (8)	-84.072 (8)	-83.949 (15)	123
^{133}Sb	4.002 (7)	-78.951 (28)	-78.957 (76)	-6
^{132}Sn	3.115 (10)	-76.577 (24)	-76.621 (26)	-44

The improved mass data give an excellent basis for a precise derivation of binding and interaction energies in the ^{132}Sn region. In the particular case of the N=82 isotones, we can now use the new mass excesses of ^{132}Sn, ^{133}Sb, ^{134}Te, and ^{135}I to deduce the mass "window" defined by Zhang et al. (1) as W = −3608 (94) keV. This gives an agreement, within the uncertainty limit of 1σ, with the value of W = −3570 keV obtained from the spectroscopy of excited states.

REFERENCES

1. Zhang, C.T., et al., Phys. Rev. Lett. **77**, 3743 (1996).
2. Andreozzi, F., et al., Phys. Rev. C **56**, R16 (1997).
3. Mezilev, K.A., et al., Phys. Scripta T **56**, 272 (1995).
4. Mampe, W., et al., Nucl. Instr. and Methods **154**, 127 (1978).
5. Fogelberg, B., et al., Nucl. Instr. and Methods **B70**, 137 (1992).
6. Ivanov, Y.F., et al., Izv. Akad. Nauk SSSR, ser. Fiz. **29**, 157 (1965).
7. Audi, G., and Wapstra, A.H., Nucl. Phys. **A595**, 409 (1995).
8. Fogelberg, B., et al., to be published.

High-spin states in ^{71}As, ^{72}Se, and ^{72}Br

N. Fotiades,[1] J. A. Cizewski,[1,2] C. J. Lister,[2] C. N. Davids,[2] R. V. F. Janssens,[2]
D. Seweryniak,[2] M. P. Carpenter,[2] T. L. Khoo,[2] T. Lauritsen,[2] D. Nisius,[2]
P. Reiter,[2] J. Uusitalo,[2] I. Wiedenhover,[2] A. O. Macchiavelli,[3] and R. W. McLeod[4]

[1] *Department of Physics and Astronomy, Rutgers University, New Brunswick, NJ 08903*
[2] *Physics Division, Argonne National Laboratory, Argonne, IL 60439*
[3] *Nuclear Science Division, Lawrence Berkeley National Laboratory, Berkeley, CA 94720*
[4] *Thomas Jefferson National Laboratory, Newport News, VA*

Abstract. The ^{16}O + ^{58}Ni reaction was used to study yrast and non-yrast excitations in ^{71}As, ^{72}Se, and ^{72}Br. High-spin yrast and negative-parity non-yrast bands were observed in ^{72}Se. The $f_{7/2}$ proton extruder orbital was identified in ^{71}As. The odd-even staggering in the $\pi g_{9/2} \nu g_{9/2}$ decoupled band in ^{72}Br is compared with similar structures in heavier Br isotopes.

The nuclei in the A$\approx$70 mass region exhibit a complicated interplay of single-particle and collective degrees of freedom, with spherical structures coexisting with more deformed shapes associated with the proton intruder $g_{9/2}$ orbital. Prolate, oblate, and triaxial shapes, both of collective and non-collective character, are predicted. Which shape is favored changes rapidly as a function of neutron and proton number and angular momentum, often with several shapes coexisting in the same nucleus. A highly sensitive γ-ray spectrometer and a relatively light, heavy-ion beam could extend the knowledge of both yrast and non-yrast excitations in this region of rapid shape changes and shape coexistence.

Excited states in ^{71}As, ^{72}Se, and ^{72}Br have been investigated using the ^{16}O + ^{58}Ni reaction at 59.5 MeV at ATLAS with the Gammasphere array at the target position coupled to the Fragment Mass Analyzer. This was the first experiment with this particular configuration. Gamma-gamma coincidences, as well as gamma-recoil mass coincidences, have been established. Measurements of the directional correlations (DCO) of the transitions have allowed angular momentum assignments.

The yrast band in ^{72}Se has long been understood as evidence for shape coexistence [1,2]. The yrast states at low angular momentum are weakly collective oblate states which evolve to a collective, prolate band for I>6$\hbar$. The predicted [2] positive-parity non-yrast states have to date not been observed. In the present work the negative-parity bands have been extended to 17$\hbar$. The kinematic moments of inertia, $\mathcal{J}^{(1)}$, as function of rotational frequency, $\hbar\omega$ are displayed in Fig. 1 for the positive-parity yrast band and two negative-parity bands. The negative-parity band in ^{74}Se [3] is shown for comparison. The negative-parity band-1 in ^{72}Se is built on the 3$^-$ state which has been identified as a collective excitation in heavier Se isotopes [3]. The negative-parity band-2 is probably a two-quasineutron excitation involving $g_{9/2}$ and fp-shell neutrons [3].

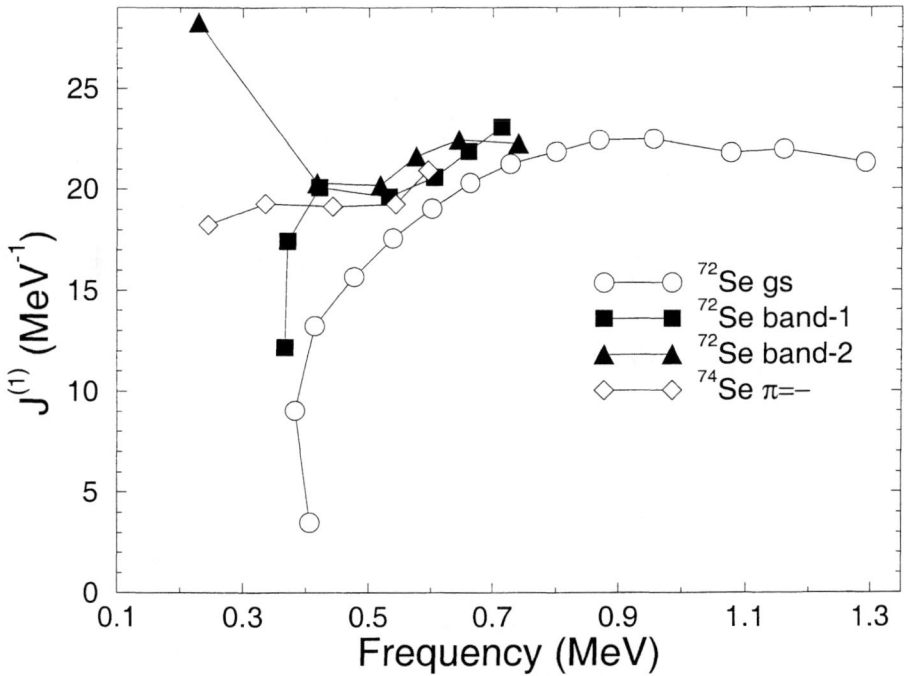

Fig. 1: $\mathcal{J}^{(1)}$ vs. $\hbar\omega$ for yrast and negative-parity bands in ^{72}Se and the negative-parity band in ^{74}Se.

The positive-parity yrast structure in ^{71}As is based on the $\pi g_{9/2}$ configuration which is low-lying for large prolate deformations. Total Routhian Surface (TRS) calculations [4] predict this band to be γ-soft at low rotational frequencies, becoming more rigid, but highly triaxial, at higher frequencies. In the present work the negative-parity yrast band has been extended to $(37/2)^-$ and a new negative-parity $\Delta I=1$ band identified, as shown in Fig. 2. This new negative-parity band has the characteristics expected for the $7/2^-[303]$ extruder orbital from the $\pi f_{7/2}$ shell, which is expected to be near the Fermi surface for the large, $\beta >0.3$, deformations which are known to characterize ^{71}As. The experimental B(M1)/B(E2) ratios for this band, which vary between 4 and 6 $\mu_N^2/(eb)^2$, are in good agreement with theoretical values of this ratio expected for the proton $f_{7/2}$ extruder orbital. This is the first identification of this orbital in this mass region and further supports arguments for large prolate deformations in ^{71}As.

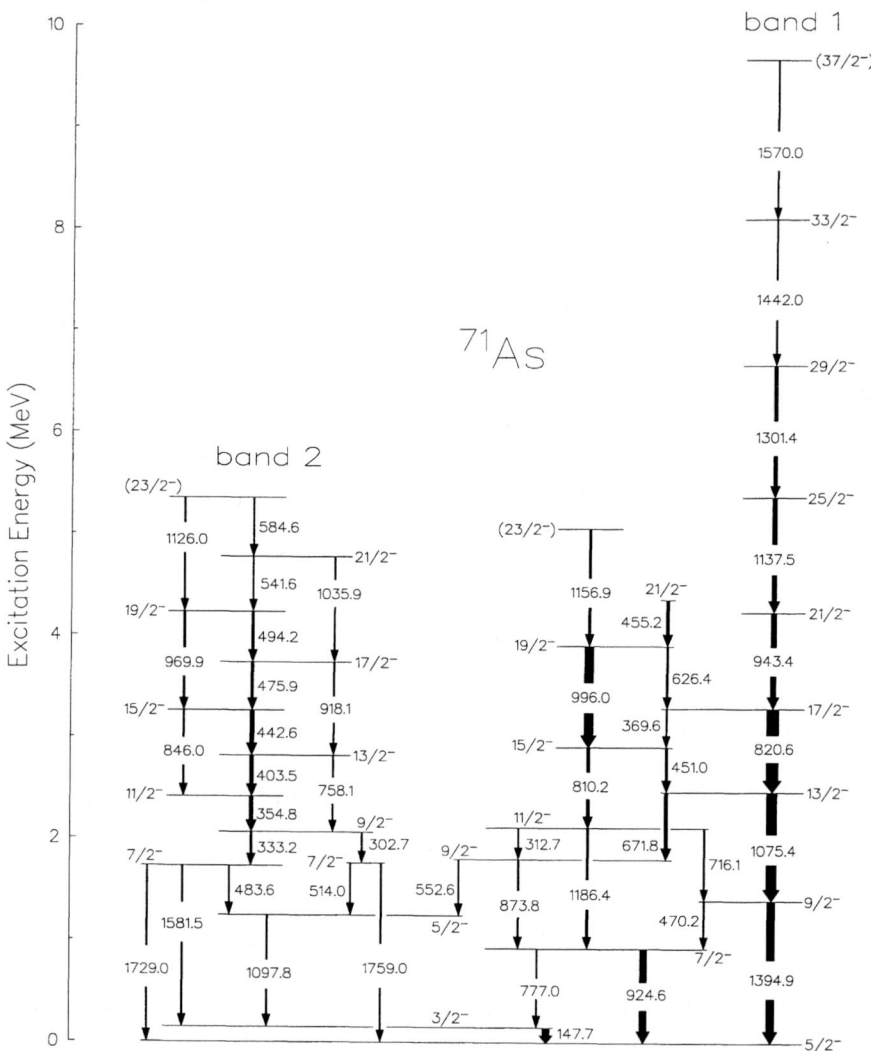

Fig. 2: Partial level scheme of ^{71}As highlighting negative-parity excitations.

The low-spin members of the positive-parity yrast band in ^{72}Br had been previously established [5,6]. The ground-state 3^+ assignment can be understood as the coupling of both the proton and neutron in the low-Ω $g_{9/2}$ orbitals, $3/2^+[431]$. In the present work both signature partners are observed and extended to (18^+), for $\alpha=0$, and (21^+), for $\alpha=+1$.

Signature inversion at low spin for the positive-parity bands in the heavier Br isotopes has been observed and interpreted as a change in the sign of the triaxial deformation as a function of rotational frequency [7-9]. In Fig. 3 the signature splittings, the energy differences of the states with I and I-1 divided by twice the spin, for the positive-parity bands in ^{72,74,76}Br are compared. In ^{72}Br no signature inversion at low rotational frequency is observed. This suggests that the $\gamma < 0$, collective deformation persists to low frequency in ^{72}Br, with low-Ω $g_{9/2}$ neutrons *and* protons. This is in contrast to the heavier isotopes, with higher-Ω $g_{9/2}$ neutrons, where non-collective shapes are important at the lowest frequencies [8,9].

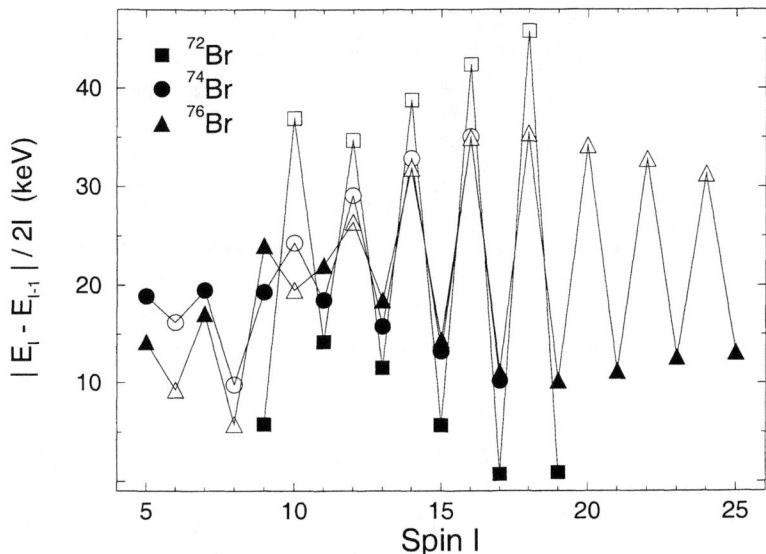

Fig. 3: Signature splitting for positive-parity bands in ^{72,74,76}Br. Open symbols are $\alpha=0$; closed symbols are $\alpha=+1$.

In summary, both yrast and non-yrast structures have been observed in several A≈72 nuclei up to high angular momenta. Large deformations, $\beta > 0.3$, are necessary to reproduce the observed band structures, as well as non-axial shapes, which are more rigid in ^{72}Br with N=37.

This work supported in part by U.S. National Science Foundation and Department of Energy.

REFERENCES

[1] T. Mylaeus *et al.*, J. Phys. G **15**, L135 (1989).
[2] L. Chaturvedi *et al.*, Phys. Rev. C **43**, 2541 (1991), and references therein.
[3] T. Matsuzaki and H. Taketani, Nucl. Phys. A **390**, 413 (1982).
[4] R. S. Zighelboim *et al.*, Phys. Rev. C **50**, 716 (1994).
[5] G. Garcia Bermudez *et al.*, Phys. Rev. C **25**, 1396 (1982).
[6] S. Ulbig *et al.*, Z. Phys. A **329**, 51 (1988).
[7] J. Döring *et al.*, Phys. Rev. C **47**, 2560 (1993).
[8] Q. Pan *et al.*, Nucl. Phys. A **627**, 334 (1997).
[9] D. F. Winchell *et al.*, Phys. Rev. C **55**, 111 (1997).

Analysis of the Elastic $^6Li + ^{12}C$ Scattering: Energy Dependence, "Abnormal Dispersion" and Dynamic Polarization Potential

S.A.Goncharov[*], A.S.Dem'yanova[†], A.A.Ogloblin[†]

[*] *Skobeltsyn Institute of Nuclear Physics, Moscow State University,119899 Moscow,Russia* and [†] *RSC Kurchatov Institute,123182 Moscow,Russia*

Abstract

Abstract. The analysis of the elastic scattering of ^6Li by ^{12}C at the lab energies 30.6, 59.8, 90, 99, 156, 210 and 318 MeV and by $^{13,14}C$ at 93 MeV have been carried out in the frame of the semimicroscopic approach accounting for dispersion relations. It indicates the abnormal dispersion behaviour at energies near 15 MeV/nucl.

In recent paper (1) the analysis of the data (2-8) on elastic scattering of 6Li by 12C at the energies $E_{^6Li}$ = 30.6, 59.8, 90, 99, 156, 210 and 318 MeV as well as by $^{13,14}C$ at 93 MeV have been carried out in the frame of the phenomenological approach. The each data set was individually analyzed using Woods-Saxons potentials and then the dispersive analysis of the obtained volume integrals was carried out. This analysis shows a clear change of the Airy–minima positions and nonmonotonous behaviour of the obtained integral characteristics in the lab energy range 70–100 MeV. This "abnormal dispersion" differs from well–known threshold anomaly.

In the present work we fulfilled the semimicroscopic optical model analysis of the same data sets. The optical model potential developed in (9) was used (see the next section) and mutual analysis of all data sets was carried out which allowed to get an unambiguous parameter set.

The optical model potential have been constructed (9) by combination of the microscopic mean field components (real central V_F and Coulomb V_C) and the phenomenological dynamic polarization potential (DPP). DPP contained both volume ($W_S \cdot f(r, r_S, a_S)$) and surface ($W_D \cdot g(r, r_D, a_D)$) Woods–Saxons forms for the absorption, and the dispersion correction (V_P) to real part.

$$U_{opt}(r) = V_F(r) + V_C(r) + V_P(r) + iW_S(r) + iW_D(r) \quad (1)$$

We did not include the spin-orbital interaction because did not treat the spin observables.

The mean field components were calculated in the frame of the folding model explicitly including the single nucleon knock-on exchange (SNKE) effects:

$$V_F = \sum_{T=0,1} (V_T^D + V_T^E). \quad (2)$$

$$V_T^D(r) = \int\int v_T^D(\mathbf{s})\rho_T^a(\mathbf{r_1})\rho_T^A(\mathbf{r_2})d\mathbf{r_1}d\mathbf{r_2}, \quad \mathbf{s} = \mathbf{r_2} - \mathbf{r_1} + \mathbf{r} \quad (3)$$

$$V_T^E(r) = \int\int v_T^E(\mathbf{s})\rho_T^a(\mathbf{r_1}, \mathbf{r_1} + \mathbf{s})\rho_T^A(\mathbf{r_2}, \mathbf{r_2} - \mathbf{s})exp\left(\frac{i\mathbf{K}(\mathbf{r})\mathbf{s}}{\zeta}\right)d\mathbf{r_1}d\mathbf{r_2}, \quad (4)$$

$$K^2(\mathbf{r}) = \frac{2m\zeta}{\hbar}\Big(E - V(\mathbf{r}) - V_C(\mathbf{r})\Big), \quad \zeta = \frac{aA}{a+A}. \quad (5)$$

It should be noted that we have treated only isoscalar ($T = 0$) components assuming small isovector contributions.

We used $M3Y$ (Reid–Elliott) parametrization (10) of the effective nucleon–nucleon interactions $v_T^i(s)$ and some realistic models of the density distribution ρ_T^i for target nuclei and 3-particle model by Danilin et al. (11) for ^6Li.

The Coulomb potential was determined like the central one by folding the two particle Coulomb interaction and empirical charge densities.

For construction of the dispersion correction we assumed very week energy dependence of the Woods-Saxon forms geometrical parameters r_i and a_i, (i=S,D). Then accounting for corresponding dispersion relations one can present

$$V_P(r, E) = \alpha(E) \cdot W_S(E) \cdot f(r, r_S, a_S) + \beta(E) \cdot W_D(E) \cdot g(r, r_D, a_D), \quad (6)$$

where α, β, W_S, W_D are free parameters for each energy, and r_i, a_i are mutually adjusted for all energies.

Both the previous phenomenological analysis (1) and the present semimicroscopic one gave consistent results with practically the same angular distributions for corresponding energies and rainbow (Airy) extremum positions, and also the similar energy dependence of the volume integrals.

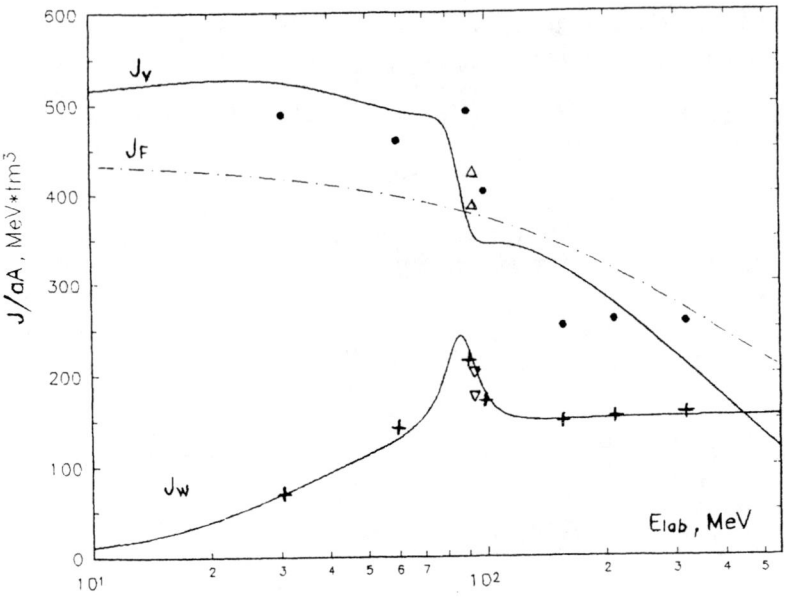

FIGURE 1.

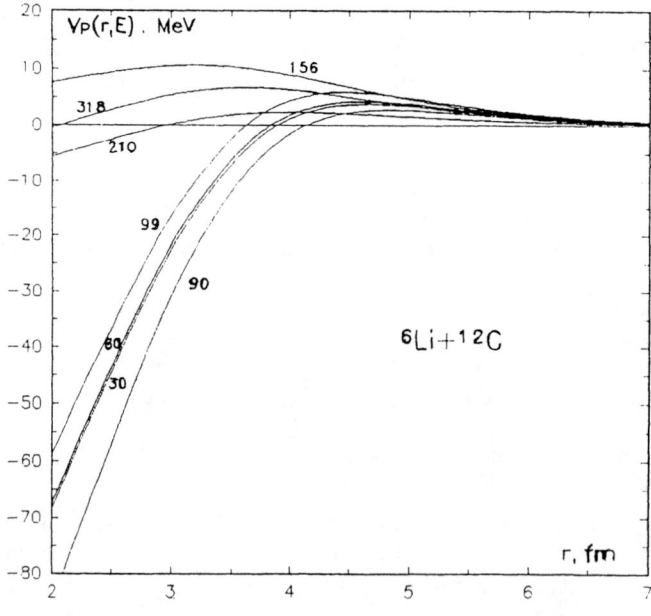

FIGURE 2.

In **Figure 1.** the real (J_V) and imaginary (J_W) volume integrals per nucleon pair as a function of the incident energy are shown. The black circles (J_V) and crosses (J_W) are the results of our semimicroscopic analyses for $^6Li+^{12}C$. The results for $^6Li+^{13}C$ and $^6Li+^{14}C$ scattering at 93 MeV are shown by triangles (up–J_V and down–J_W). We also did simple parametrization of $J_W(E)$ for the points obtained from analysis similar to (12) (solid curve in the lower part). The solid curve in the upper part shows the results of the dispersion integral prediction based on this $J_W(E)$-parametrization. The real volume integrals of the "bare" potential calculated in frame of the folding model are shown by the dot–dashed curve. One can see that our model potential volume integrals J_V are well consistent with dispersive curve in the interesting energy range.

The correlated disturbances in the smooth energy dependencies for J_V and J_W in the lab energy range 70–100 MeV are observed as well as in phenomenological analysis (1): the sharp increase of the $J_V(E)$ toward smaller energies corresponds to the maximum in $J_W(E)$. The physical reason of this effect is not completely clear. Possibly it is connected with $^6Li->^3He+^3H$ cluster structure.

The DPP radial form and energy dependence were also obtained. In **Figure 2** the real parts of DPP (dispersion corrections) obtained in our semimicroscopic calculations are shown for $^6Li+^{12}C$ at examined energies. It is seen that DPP behaviour is naturally connected with observed anomaly.

For all considered energies DPP has a repulsive part in the surface region (> 4 fm), but at the lower energies a remarkable attractive part at smaller distances appears. It becomes particularly large in the "abnormal dispersion" region. It was also found that the smallest contribution of dispersion correction to the real potential is in the range near 35 MeV/nucl.

REFERENCES

1. Dem'yanova,A.S., Goncharov,S.A., Ogloblin,A.A., Contribut. Abstr. VI Int.Conf. on Nucl.–Nucl. Collis., Gatlinburg, USA, June 2–6, 1997. p.P-029, and Contribut. Int.Conf. Nucl.Struct. and Related Topics", Dubna,Russia, Sept.9–14,1997. p.28.
2. Chuev,V.I., et al., J. de Phys. **32**,C6-161(1971).
3. Bingham,H.G., et al., Phys. Rev. **C11**,1913(1975).
4. Glukhov,Yu.A., et al., Yadernaya Fizika(Sov.J Nucl.Phys) **34**,312(1981).
5. Schwandt,P., et al., Phys. Rev. **C24**,1522(1981).
6. Cook,J., et al., Nucl. Phys. **A388**,173(1982).
7. Nadasen,A., et al., Phys. Rev. **C37**,132(1988).
8. Nadasen,A., et al., Phys. Rev. **C47**,674(1993).
9. Goncharov,S.A., Knyaz'kov, O.M., Kolozhvari, A.A., Yadernaya Fizika (J. of

Nucl. Phys.) **59**,666(1996).
10. Bertsch,G., et al., Nucl. Phys. **A284**,399(1977).
11. Danilin,B.V., et al., Nucl. Phys. **A533**428(1991).
12. Carlson,B.V., et al., Phys. Rev. **C41**742(1990).

Symmetry Structure In Neutron Deficient Xenon Nuclei

I. M. Govil
Department of Physics, Panjab university, Chandigarh-160014.

Abstract. The paper describes the measurements of the lifetimes of the excited states in the ground state band of the Neutron deficient Xe nuclei (^{122,124}Xe) by recoil Distance Method (RDM). The lifetimes of the 2^+ state in ^{122}Xe agrees with the RDM measurements but for ^{124}Xe it does not agree the RDM measurements but agrees with the earlier Coulomb-excitation experiment. The experimental results are compared with the existing theories to understand the changes in the symmetry structure of the Xe-nuclei as the Neutron number decreases from N= 76 (^{130}Xe) to N=64 (^{118}Xe)

INTRODUCTION

The lifetimes of the excited states in the ground state band of the neutron deficient ^{122,124}Xe isotopes were measured because of the large discrepancies reported in the literature (1-9). The B(E2) values for the even-even nuclei are important in studying the nuclear structure because these have the direct signature of the intrinsic quadrupole moment and hence the intrinsic deformation of the nucleus. Most of the Xe isotopes are in the 50<N<82 region. A small but gradual increase in the Xenon B(E2) values is expected as the neutron number is decreased from N=82 to the midshell value of N=66, a flattening at Midshell and finally a gradual decrease below N=66. Instead, the recent values for the three Xenon isotopes (N=64,66,68) constitute a small peak which can not be explained by the standard models (5-9). We (10) have, therefore, compared the lifetimes of the ground state band of these two nuclei to confirm the validity of the existing Algebraic (5,8-9) and the Multishell Models (11-12) to understand the symmetry structure in these nuclei.

EXPERIMENTAL DETAILS

The experiment was performed at the University of Notre Dame Nuclear Structure Laborartory using the ^{110}Pd(^{16,18}O,4n)^{122,124}Xe reaction at a beam energy of 66 MeV. The target was a stretched self supporting ^{110}Pd foil of 1.2 mg/cm² thickness. Data was collected for eighteen target to stopper distances ranging from 14 μm to a maximum distance of 10,000 μm. The lifetimes of 2^+, 4^+, 6^+ and 8^+ states of the ground state band were extracted using the computer code 'LIFETIME'. The corrections for the effects of the cascade and the side-feeding, both observed and unobserved from higher states and nuclear deorientation effects were made.

RESULT AND DISCUSSION

A summary of the lifetimes and B(E2) values alongwith the deformation parameters for ^{122,124}Xe are presented in the Tables 1(a) and 1(b). As can be seen from the Table 1(a), results obtained in the present work for ^{122}Xe are in good agreement with that of Petkov et al. (1) but are in disagreement with the earlier results of Kutschera (2) and others (3,4). However, for the 2^+ state of ^{124}Xe our result of 82 ps is in good agreement with the result from Coulomb excitation experiment by Gordon et al. (13) rather than the RDM measurement of 60 ps reported by Dewald et al. (14). The B(E2) values for the 2^+ - 0^+ transition are of the order of 46 W.u. (β_2=0.20) and 80 W.u. (β_2=0.27) for ^{124}Xe and ^{122}Xe respectively. This shows the increased deformation as one moves towards the midshell. In the Algebraic models, Otsuka et al. (15) have carried out IBM-2 calculations in an attempt to describe the systematic behaviour of the B(E2) values for the even-even Xe, Ba and Ce nuclei. They were particularly interested with the apparent saturation of the B(E2) strength as the midshell is approached. They made a strong case for the inclusion of the Pauli blocking

TABLE 1(a). Lifetimes and B(E2) values of ^{122}Xe

Transitions	B(E2) (e^2b^2)	Lifetimes τ (ps)				
		Present work	Ref. (1)	Ref. (2)	Ref. (3)	Ref. (4)
2^+ - 0^+	0.29	72(4)	70(2)	89(8)	51.0(+10.0,-6)	108(19)
4^+ - 2^+	0.33	8.0(4)	6.5(3)	8(1.2)	5.3(+3.0,-0.6)	9(2)
6^+ - 4^+	0.26	2.9(2)	1.5(2)	3.9(7)	9(3)	3.6(5)
8^+ - 6^+	0.19	1.8(3)	0.7(2)	<3.5	-	-
10^+ - 8^+	0.36	<0.6	<0.5(2)	-	-	

TABLE 1(b). Lifetimes and B(E2) values of ^{124}X

Transitions	B(E2) (e^2b^2)	Lifetimes τ (ps)		
		Present work	Ref. (14)	Ref. (13)
2^+ - 0^+	0.17	82(4)	60(5)	81(7)
4^+ - 2^+	0.67	3.0(2)	5.6(5)	
6^+ - 4^+	0.60	1.0(1)	1.1(4)	-
8^+ - 6^+	0.39	0.7(2)	0.5(1)	-
10^+ - 8^+	0.32	<0.6	<1.4	

effect while calculating the spectral properties and the agreement with the experiment was excellent for N>70 Xenon isotopes for which the dominant symmetry is SU(5). However, for more deformed midshell Xenon isotopes, where dominant symmetry is O(6), there is a marked underprediction because the Hamiltonian with SU(5) symmetry may not be the true representative for the entire region. Our B(E2) value as shown in Fig. 1a , for ^{124}Xe is in agreement with calculations of Otsuka et al. including the Pauli blocking effect but for ^{122}Xe it is found to be higher because of the possible change of symmetry from SU(5) to O(6) for ^{122}Xe. In the fermion dynamic symmetry model (FDSM) calculations (16-17), the group structure has been changed from O(5) to the O(6) when going from the heavier (>^{126}Xe) to the lighter Xenon nuclei (<^{124}Xe). The experimental values for ^{122}Xe obtained by us and for ^{120}Xe obtained by J. C. Walpe et al. (16) earlier are in good agreement with these predictions. However, for ^{124}Xe our B(E2) value is lower than the predictions of the FDSM with O(6) symmetry. Therefore, it confirms that the symmetry structure changes as one goes from the heavier (> ^{124}Xe) to the lighter Xenon isotopes (<^{122}Xe) from O(5) to O(6) at ^{122}Xe rather than at ^{124}Xe as assumed in the FDSM calculations earlier (17).

In the geometrical nuclear models, we have compared our results with the single shell Nilsson asymptotic model (SSNAM) (8), the finite range droplet model (FRDM) (11) and the Hartree-Fock calculations (12). The experimental B(E2) values are compared with the above theoretical models in Fig. 1b. It is clear that the FRDM and the Hartree-Fock calculations give reasonably good agreement for both the nuclei as compared to the SSNAM.

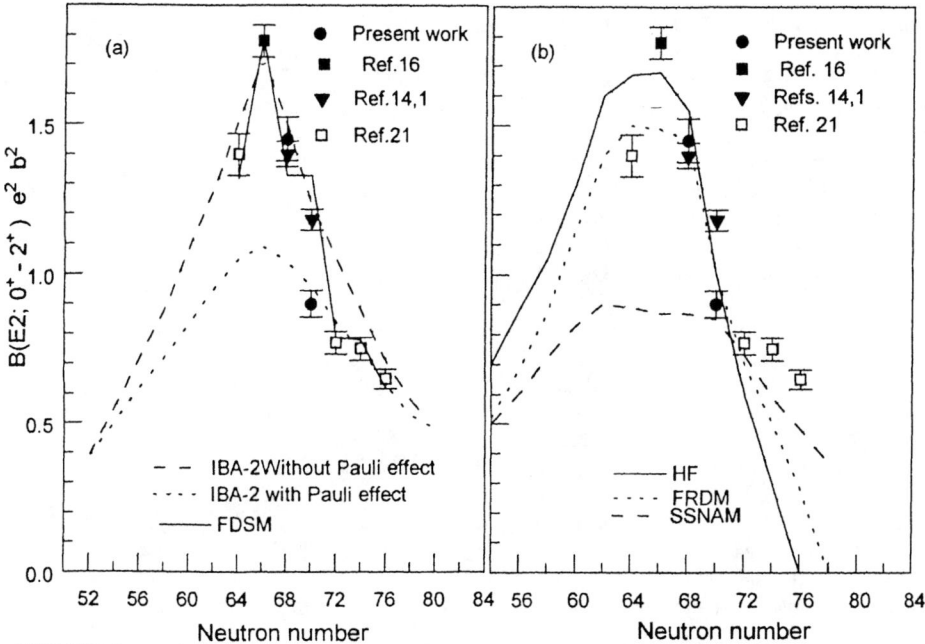

FIGURE 1. A comparison of the experimental B(E2) values with theoretical predictions.

SUMMARY

Lifetimes of the ground state band upto 10^+ spin in ^{122,124}Xe nuclei have been measured with the recoil-distance technique. Our results for ^{122}Xe show good agreement with the more recent experimental measurements of RDM rather than with the old values reported in the literature. For ^{124}Xe the lifetime of the 2^+ state is in agreement with the Coulomb excitation measurement rather than the RDM value. The present B(E2) value for the 2^+- 0^+ transition in ^{124}Xe agrees very well with the prediction of IBM-2 calculations with Pauli blocking effect assuming SU(5) symmetry. On the other hand, the B(E2) value for the 2^+ -0^+ transition in ^{122}Xe agrees with the FDSM calculations with O(6) symmetry. This indicates that the symmetry structure changes from O(5) to O(6) at ^{122}Xe as one goes from heavier (N>70) to lighter (N< 68) Xenon isotopes. In the geometrical models, the multi-shell FRDM and the Hartree-Fock calculations give a better agreement for both ^{122,124}Xe nuclei as compared to the single shell Nilsson asymptotic model (SSNAM).

ACKNOWLEDGEMENTS

The Authors acknowledge with thanks the Notre Dame Accelerator crew for providing the excellent oxygen beam . This work has been supported by the National Science Foundation grant No. INT-9309296 and INT-9215295.

REFERENCES

1. Petkove, P., et al. Nucl. Phys. A **589**, 341 (1995).
2. Kutschera W.,et al. Phys. Rev. C **5**, 1658 (1972).
3. Droste, Ch. , et al. J. Phys. G. Nuclear Part. Phys. **18**, 1763 (1992).
4. Choudhury, A., Nucl. Structure Lab., Uni. Of Notre Dame Biennal Report No. 1983/85, p.17
5. Feng, D. H.,et al. Phys. Lett. B **205**, 156 (1998)
6. Raman, S., et al.Phys. Rev. C **43**,556 (1991)
7. Bhatt,K.H., et al. Phys. Rev. C **49** ,808 (1994)
8. Raman, S., et al.Phys. Rev. C **52** , 1380 (1995).
9. Mantica, P.F., Phys. Rev. C **45** , 1586 (1992).
10. I. M. Govil et al. Phys. Rev. C **57**, 632 (1998)
11. Moller, P., et al. Nucl. Phys. A **536**, 20 (1992).
12. Tajima, N., et al. (private communications).
13. Gordon, D. M., et al. Phys. Rev. C **12**,628 (1975)
14. Dewald, A., et al. , Proc. of Zakopane School on Phys. Poland Vol. 2., 28th April -12th May, (1990)
15. Otsuka,T.,et al. Phys. Lett. B **247**, 191 (1990).
16. Walpe, J. C.,et al. Phys. Rev. C **52** , 1792 (1995)
17. Wu, C. L., et al. Advances in Nuclear Physics, edited by J. W. Negel and E. Vogt (Plenum, New York, Vol. 21, p.227 (1994)
18. Moller. P., et al. At. Data Nucl. Data Tables **39**, 213 (1988).
19. Bonche, P., et al. Nucl. Phys. A443, 39 (1985)
20. Quentin,P., et al. Annu. Rev. Nucl. Sci. **238**, 523 (1985).
21. Raman, S., Atomic Data and Nuclear Data Tables, **36**, 1 (1987).

FRAGMENTATION OF ALPHA-CLUSTER STATES IN ^{32} S. BOZONIZATION

K.A.Gridnev, M.Brenner*, A.E.Antropov, S.E.Belov, B.Z.Taibin, K.N.Ershov, D.K.Gridnev,
M.P.Kartamishev, I.V. Krouglov, T.V.Taroutina

Saint-Petersburg State University, Saint-Petersburg, Russia
*Department of Physics, Åbo Akademi University, Turku, Finland

There are bosons under the study in different fields of physics. In superconductivity - these are Cooper pairs, in semiconductors - these are excitons and biexcitons, in superfluidity these are atoms of ^{4}He, in nuclear physics - these are interacting bosons, in the theory of vacuum this is the chiral condensate and Higgs boson. Genuine consideration is an effect of exchange coupling. The last experiment on anomalous large angular scattering (ALAS) of alpha-particles on ^{28}S, ^{32}S gives a signal of bosonization in light nuclei [1].

Iachello [2] introduced a new phenomenological model based on a group theory. To the s-boson he added p-boson to describe the dipole degrees of freedom from the dipole character of the relative motion (relative vector) of the di-nuclear molecule. In the frame of $O(4)$ symmetry he derived the following expression for energy spectrum:

$$E(n,L) = -D + a(n+1/2) - b(n+1/2)^2 + cL(L+1) \tag{1}$$

Very close spectrum was obtained by K. Baktibaev [3] for the system of p-bosons in the reduction of $U(4) \supset U(3)$. This model gives a possibility to include d-bosons. In the frame of the group theory approach J. Cseh et all [4] got the same formula for the spectrum.

There is a question: how such spectrum is related to the potentials? The estimate of such potential is given in the work [5]:

$$\sim \frac{m\omega}{2\hbar}(r-r_0)^2, \text{ where } r_0 = \sqrt{\frac{\hbar}{m\omega}} n_0, \tag{2}$$

$\sqrt{\hbar/m\omega}$ - is the oscillator length defined by the united system and n_0 - is the minimal number of p-boson required by the Wildermuth condition.

Greiner and Cindro [6] used the Morse plus centrifugal potentials to generate the considered spectrum:

$$V_\mu(r) = A + B(e^{-2\beta x} - e^{-\beta x}), \quad \frac{\hbar^2 L(L+1)}{2\mu r^2} \approx \frac{\hbar^2 L(L+1)}{2\mu}(C_0 + C_1 e^{-\beta x} + C_2 e^{-2\beta x}). \tag{3}$$

This approach was criticised by K. Kato and J. Abe [7]. B. Sahu et. all [8] suggested semiempirical formula for the description of the resonances observed in ^{12}C+^{12}C and ^{12}C+^{16}O systems:

$$E(n,l) = V_B^{(l)} + a + bl + dl^2 - c(n+3/4)^2, \text{ where } V_B^{(l)} = V_B + \frac{\hbar^2 l(l+1)}{2\mu R_B^2(l)}, \tag{4}$$

V_B is generated by the potential $V(r) = u_0/\cosh^2\alpha(r-R)$.

Upper presented spectrum has the same character as for the collective excitations of a trapped Bose-condensed gas [9]:

$$\omega = \omega_0(2n^2 + 2nl + 3n + l)^{1/2}. \tag{5}$$

CP455, *ENAM98: Exotic Nuclei and Atomic Masses*
edited by B. M. Sherrill, D. J. Morrissey, and Cary N. Davids
© 1998 The American Institute of Physics   1-56396-804-5/98/$15.00

For the non-interacting harmonic oscillator model spectrum will be:

$$\omega = \omega_0(2n+l), \qquad (6)$$

where ω_0 - frequency of harmonic trap, interacting with repulsive forces. It was obtained as a solution of Gross-Pitaevski (GP) equation:

$$\left[-\frac{\hbar^2}{2m}\nabla^2 + V_{ext} + 4\pi\frac{\hbar^2 a}{m}|\psi(r)|^2\right]\psi(r) = \mu\psi(r), \qquad (7)$$

V_{ext} - is the external confining potential, μ - is the chemical potential, a - is the s-wave scattering length. The condensate wave function $\psi(r)$ is related to the atomic density through

$$\rho(r) = |\psi(r)|^2. \qquad (8)$$

For the isotopic harmonic trap [10]:

$$V_{ext}(r) = \frac{1}{2}m\omega_{HO}^2 r^2. \qquad (9)$$

The coincidence between spectrum of the vibrations of the Bose condensate in the frame of the GP equation and spectrum of bosons in the frame of algebraic approach is not accidental. The dynamics of the quantum state, populated by α-particles, can be described within the Hartree-Fock approximation by the Gross-Pitaevskii equation:

$$\psi_t + \Delta\psi - 2k\psi|\psi|^2 - V(x,t)\psi = 0, \text{ where } V(x) = \frac{\omega^2 x^2}{4}. \qquad (10)$$

If we put $\omega=0$ we shall have the Nonlinear Schrodinger Equation (NOSE). We investigated the stationary solution of the NOSE in the work [11] for the interaction of the heavy ions with nuclei. A. Rybin et all [12] showed that the GP equation can be mapped into the NOSE. This result gives a possibility to describe the $2\pi/\omega$ periodic oscillations of the soliton-like wave function of the Bose condensate in the parabolic potential basing on the analysis of all approaches such as algebraic and the GP equation for the description of α-cluster spectrum ^{32}S. We suggested such equation for the α-cluster states in ^{32}S:

$$V = A + BN + CN^2 + DL(L+1) + Fn(n+1), \qquad (11)$$

where the F-term is appearing due to the soliton vibration of the Bose condesate.

We revised Fourier analysis of the experimental spectrum, which was done in the work [1] (Fig.1). On the Fig.2 one can see the Fourier spectrum in the range 13.5 - 16.3 MeV. In this case we have used the discrete form of Fourier amplitude:

$$F_k = \dot{F}(\omega_k) = \sum_i^N f(x_i)e^{-ix_i\omega_k}\Delta x_i;$$

$$\text{Re} f(x_i) = \frac{2}{\hbar}\sum_k^N \text{Re}\, F(x_i)\cos(x_i\omega_k)\Delta\omega_k; \qquad (12)$$

$$\text{Im} f(x_i) = -\frac{2}{\hbar}\sum_k^N \text{Im}\, F(x_i)\sin(x_i\omega_k)\Delta\omega_k,$$

where $\omega_k = k\Delta\omega$ ($\Delta\omega$ is a constant value).

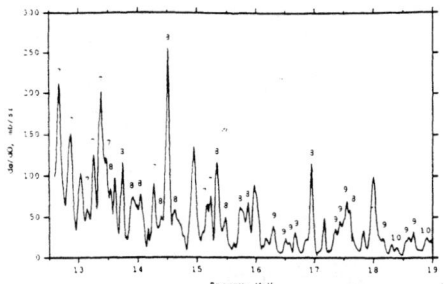

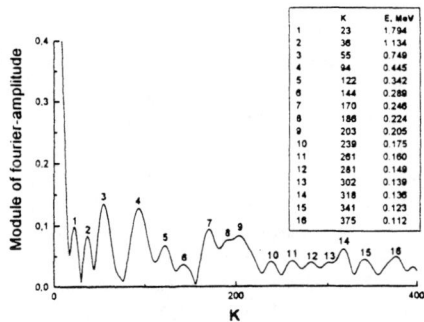

Fig. 1. The excitation function of a particles scattered by ^{28}Si in $\Theta=173°$. The number denote the spins of the resonance states as obtained from angular distributions fitted by squared Legendre polynomials at backward angles.

Fig.2. Module of Fouier-amplitude versus K for energies range 13.5 -16.3 MeV

The prominent picks on the Fig. 2 correspondent to the periods in the energy spectrum to the following values: 749, 445, 246, 205 KeV respectively. These values are more close to the AVR (anharmonic vibrator+rotator) model [13] and to the approach of Sahu et. all [8].

On the Fig. 3 one can see the fitting of the experimental cluster spectrum obtained with the help of formula (4). The fragmentation of cluster states can be obtained only with the adding of the solitonic term $F(N+1/2)$. The value of the constant F is in the accordance with the Rybin and et all theory [12].

Cluster states in ^{32}S have a qvasimolecular behaviour due to a tidal interaction having the form [14]:

$$\Phi(r,\theta,\varphi) = - \sum_{m=-2,0,2} \phi^{(m)}(r) Y_2^{(m)}(\theta,\varphi). \tag{13}$$

In the astronomy if the star interacting with other star the pulsation can be described by the superposition of two harmonic [15] On the Fig. 4, taken from the work [16] one can see that spectrum of ^{32}S has a tidal character.

Our investigation of the spectrum of ^{32}S showed that the fragmentation of cluster states in ^{32}S is due to the vibrations of the α-particles (like a condensate) located in the atmosphere of the nucleus.

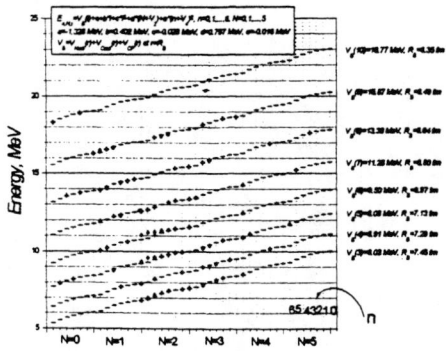

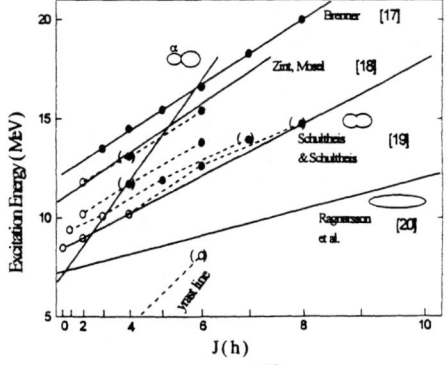

Fig. 3. Energy levels of ^{28}Si+α system

Fig. 4. Levels of ^{32}S

References

[1] A.E. Antropov, M. Brenner, V.Z. Goldberg, W. Greiner, K.-M. Källman, T. Lonroth, A. Ludu, P. Manngard, A.E. Pakhomov, V.V. Pankratov, Proc. of the 7-th Inter. Conf. on Nuclear Reaction Mechanisms, Varenna, June 6-11, 1994, p.430
[2] F. Iachello, *Phys. Rev.* C **23** (1981) 2778
[3] K.B. Baktibaev, A.K. Kabulov, G.S. Kabulov, K.E. Ramankulov, Notes of Academy of Sciences of the USSR, ser. phys., V.60, №5, (1996) p.118
[4] Cseh, G. Levai and A. Aalgora Revista Mexicana de Fisica, 43, Supplemento, 1, (1997), 69–77
[5] P.O. Hess, G. Levai, J. Cseh, *Phys. Rev.* C **54** (1996) 2345
[6] N. Cindro, W. Greiner, *J. of Phys.* **69** (1983) 175
[7] K. Kato, J.Abe, *Progr. Theor. Phys.* v.80 (1988) 119
[8] B. Sahu, B.M. Jyrwa, P. Susan, C.S. Shastry, *Phys. Rev.* C **44** (1991) 2729
[9] S. Stringari, *Phys. Rev. Lett.* **77** (1996) 2360
[10] F. Dalfovo, L. Pitaevskii, S. Stringari, *Phys. Rev.* A **54** (1996) 4213
[11] K.A. Gridnev, E.F. Hefter, K. Mikulas, V.M. Semjenov, V.B. Subbotin, Nuovo cimento, 93 (1986) 135
[12] A. Rybin, M. Lindbery, J. Tumonen, Biennial report of Åbo Akademi, Dpt. of Physics, 1995 - 1997 p. 16
[13] N. Cindro, W. Greiner, *J. Phys. G.: Nucl. Part. Phys.* **9** (1983) 175
[14] H. Lamb, Hydrodynamics, 6-th. ed.-Cambridge University Press, 1932
[15] P. Ledoux, *Astrophysical Journal*, 114 (1951) 373
[16] S. Kubono, N. Ikeda, T. Nomura, M. Oyaizi, J. Tanaka, M.H. Tanaka, J. Funatsu, H. Miyataka, S. Kato, T. Yamaya, S. Ohami, Proc., Fifth, Int., Conf., Clustering Aspects in Nucl. and Subnucl. Systems, Kyoto, 1988, p.583
[17] M. Brenner, E. Indola, K.-M. Källman, T. Lönnroth, P. Mangård, M. Halldorsdottir, Th. Karlsson, Z. Mate, L. Zolnai, V.Z. Goldberg, G.V. Rogatchev, M.V. Rojkov, L.N. Sterikov, W. Trzaska, R. Wolski, Heavy Ion Physics 7, 1998, 1
[18] P.G. Zint, U. Mosel, *Phys. Rev.* C **14** (1974) 1488
[19] H. Schultheis, R. Schultheis, *Phys. Rev.* C **25** (1982) 387
[20] I. Rognarsson, S. Aberg, *Phys. Lett.* B **114** (1982) 387

Identification of ^{162}Gd and a New Type of Identical Bands

E.F. Jones[1], P.M. Gore[1], J.H. Hamilton[1], A.V. Ramayya[1], R.S. Dodder[1], C.J. Beyer[1], J.K. Hwang[1], X.Q. Zhang[1], S.J. Zhu[1,2,3], A.P. de Lima[1,4], J. Kormicki[1], J.D. Cole[5], R. Aryaeinejad[5], W.C. Ma[6], G.M. Ter-Akopian[1,3,7], Yu. Ts. Oganessian[7], A.V. Daniel[1,3,7], J.O. Rasmussen[8], S.J. Asztalos[8], I.Y. Lee[8], A.O. Macchiavelli[8], M.A. Stoyer[9], R.W. Lougheed[9], S.G. Prussin[10], R. Donangelo[11]

[1] *Physics Department, Vanderbilt University, Nashville, TN 37235, USA*
[2] *Physics Department, Tsinghua University, Beijing, P.R. China*
[3] *Joint Institute for Heavy Ion Research, Oak Ridge, TN 37831, USA*
[4] *Physics Department, University of Coimbra, 3000 Coimbra, Portugal*
[5] *Idaho National Engineering Laboratory, Idaho Falls, ID 83415-2114, USA*
[6] *Physics Dept., Mississippi State University, Mississippi State, MS 39762, USA*
[7] *Joint Institute for Nuclear Research, Dubna 141980, Russia*
[8] *Lawrence Berkeley National Laboratory, Berkeley, CA 94720, USA*
[9] *Lawrence Livermore National Laboratory, Livermore, CA 94550, USA*
[10] *Nuclear Engineering Dept., University of California, Berkeley, CA 94720, USA*
[11] *Physics Department, University of Rio de Janeiro, Rio de Janeiro, Brazil*

Abstract. From γ-γ-γ coincidence measurements in spontaneous fission of ^{252}Cf, level energies were established to spins 14^+ to 20^+ in many neutron-rich nuclei. New isotope ^{162}Gd was identified. Yrast bands of ^{152,154,156}Nd, ^{156,158,160}Sm, and ^{160,162}Gd exhibit near-identical transition energies and moments of inertia (MOI) shifted by the same constant amounts for every spin state from 2^+ to 12^+ or 14^+ for nuclei differing by 2n, 4n, 2p, 4p, α, α+2p, and α+2n. These shifted identical bands (SIB) are a new phenomenon. Analysis of all known even-even proton- to neutron-rich nuclei from Ba(Z=56) to Os(Z=76) reveal no SIB for proton-rich nuclei and few cases of SIB for the most neutron-rich pairs around N = 98-102 separated by 2n and 2p.

IDENTIFICATION OF ^{162}Gd

Spontaneous fission of ^{252}Cf was studied with 72 large volume Compton suppressed Ge detectors and two X-ray detectors in Gammasphere at Lawrence Berkeley National Laboratory. The total projection of the γ-γ matrix gated on the Gd

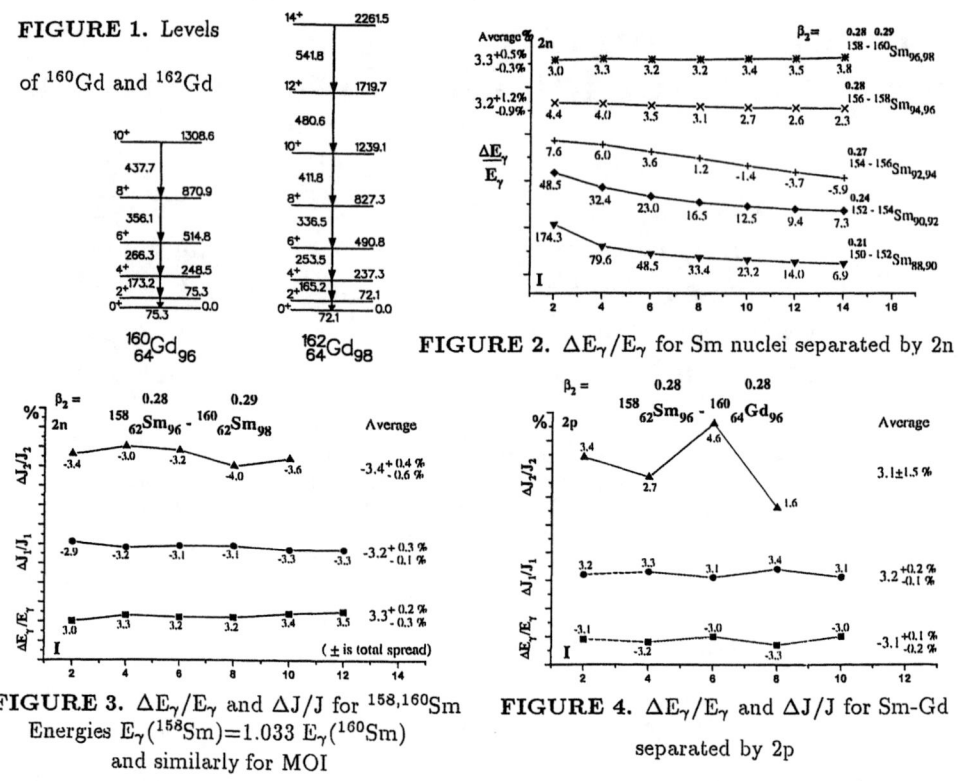

FIGURE 1. Levels of ^{160}Gd and ^{162}Gd

FIGURE 2. $\Delta E_\gamma / E_\gamma$ for Sm nuclei separated by 2n

FIGURE 3. $\Delta E_\gamma / E_\gamma$ and $\Delta J/J$ for ^{158,160}Sm Energies $E_\gamma(^{158}\text{Sm})=1.033\, E_\gamma(^{160}\text{Sm})$ and similarly for MOI

FIGURE 4. $\Delta E_\gamma / E_\gamma$ and $\Delta J/J$ for Sm-Gd separated by 2p

X-rays was used to verify known transitions in ^{160}Gd and to identify for the first time transitions in ^{162}Gd. The levels of ^{160,162}Gd are shown in Fig. 1 [1].

SHIFTED IDENTICAL BANDS

Identical bands (IB) are an important new discovery in proton-rich [2] and neutron-rich [1] nuclei. Comparing our data on transition energies and MOI's in n-rich Nd, Sm, and Gd yrast bands [1,3], we found many examples of a new type of identical bands with constant differences (shifts) of a few percent in transition energies and MOI between two nuclei separated by 2n, 2p, 4n, 4p, α, α+2n, and α+2p for all transitions from 2^+- 0^+ to 12^+- 10^+ or 14^+- 12^+. Examples of SIB are given in Figs. 2-8 (where $\pm$ indicates total data spread, not uncertainties). No SIB occur for p-rich to n-rich Ba, Ce, and Nd nuclei separated by 2n nor for Sm [Fig. 2] and Gd nuclei for N $\leq$92. However, there is remarkably constant shift for the first six yrast transitions between ^{158}Sm and ^{160}Sm [Fig. 3]. Identical MOI shifts in Sm and Gd 2n-separation cases are consistently larger than expected from mass differences. The 4n Sm and Gd cases likewise have some SIBs, e.g. $^{156-160}$Sm have percentage differences in E_γ, J_1, and J_2 of $6.6^{+0.9\%}_{-0.5\%}$, $-6.3^{+0.5\%}_{-0.7\%}$, and $-5.6^{+0.6\%}_{-1.2\%}$,

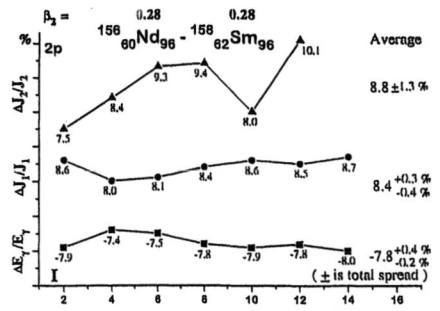

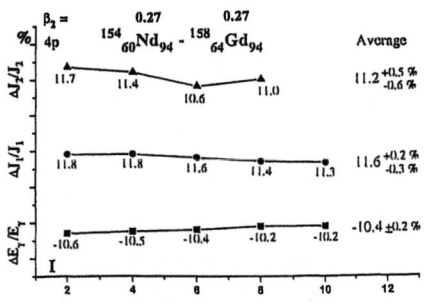

FIGURE 5. $\Delta E_\gamma/E_\gamma$ and $\Delta J/J$ for Nd-Sm separated by 2p

FIGURE 6. $\Delta E_\gamma/E_\gamma$ and $\Delta J/J$ for Nd-Gd separated by 4p

respectively. It is surprising that $^{158-162}$Gd form SIBs since $^{158-160}$Gd do not form SIBs (spreads greater than $\pm 1\%$ there).

Fig. 4 shows SIBs for 2p-separated nuclei $^{158}_{62}$Sm$_{96}$- $^{160}_{64}$Gd$_{96}$ where $\Delta E_\gamma/E_\gamma$, $\Delta J_1/J_1$, and $\Delta J_2/J_2$ are remarkably constant to 10^+. Nd nuclei which have no SIB for 2n separation do exhibit SIB for 2p and 4p. For $^{156}_{60}$Nd$_{96}$- $^{158}_{62}$Sm$_{96}$, $\Delta E_\gamma/E_\gamma$ and $\Delta J/J$ are remarkably constant to 14^+ [Fig. 5].

For 2p separation, lighter mass Nd nuclei have larger J_1's than Sm and similarly Sm than Gd, contrary to mass expectations. For the N=96 $^{156}_{60}$Nd-$^{158}_{62}$Sm pair, lighter mass and lighter Z ^{156}Nd has 8.4% larger MOI and presumably larger deformation than ^{158}Sm. Likewise for the N = 96,98 2p-separated ^{158,160}Sm-^{160,162}Gd pair, lighter mass ^{158,160}Sm have 3.1 and 1.2% larger J_1's and presumably larger β_2's than the ^{160,162}Gd. For 2n SIBs, the heavier mass nuclei have larger J_1's. This switch for 2p is undoubtably related to the Z=64 spherical subshell gap [see ref. 4] persisting to N=98, keeping the same N Gd nuclei more spherical than Sm nuclei and likewise for Nd and Sm. The large percentage difference Nd-Sm (8.4%) than Sm-Gd (3%) is consistent with the picture [4] that the influence of the Z=64 spherical subshell on the onset of deformation is felt most strongly for nuclei with Z=64$\pm$2 and essentially vanishes for heavier and lighter Z. This is the first demonstration of the continued importance of the Z=64 spherical subshell gap to N=98. Some 4p cases like ^{154}Nd- ^{158}Gd [Fig. 6] have remarkably constant SIB from 2^+ to 10^+, surprisingly with smaller spreads in their shifts than the two 2p cases.

We searched all other known p- and n-rich Ba, Ce, Dy, Er, Yb, Hf, W, and Os [5] separated by 2n, 2p, and α for SIB. No SIBs were found for p-rich to n-rich Ba or Ce. For nine 2n Dy pairs only the most n-rich N = 96-98 pair is SIB. For 10 Er pairs only the N = 96-98 to 12^+ and 100-102 (most n-rich known) are SIB [Fig. 7]. For 12 Yb pairs only the 102-104 and 104-106 pairs are SIB to 14^+ and 10^+, respectively, and for 12 Hf pairs only the 104-106 pair is barely SIB to 12^+. For 11 W pairs only the 110-112(most n-rich known) pair is nearly SIB, $\Delta E_\gamma/E_\gamma$ = $-7.4^{+0.7\%}_{-1.9\%}$ to 12^+. The 11 Os pairs have no SIB with less than 5% total spread. Except for Hf and one case in Er, it is always the most n-rich one or two pairs with N = 96-108 that are SIB, with the smallest spreads for N = 94-98.

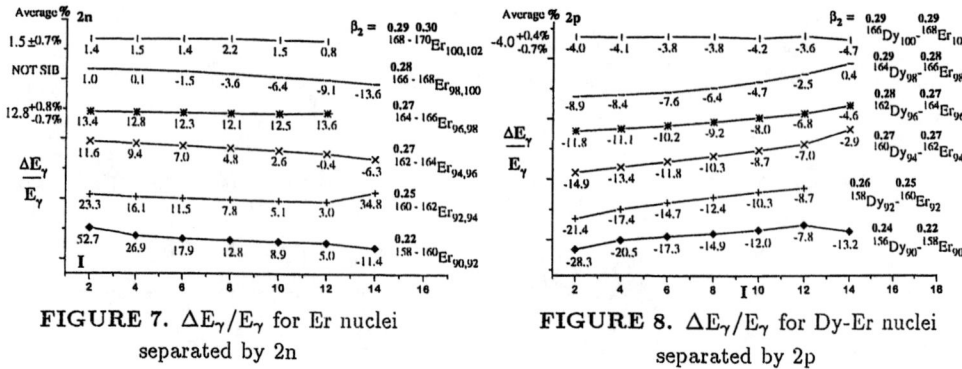

FIGURE 7. $\Delta E_\gamma/E_\gamma$ for Er nuclei separated by 2n

FIGURE 8. $\Delta E_\gamma/E_\gamma$ for Dy-Er nuclei separated by 2p

For 2p separations in 8 Gd-Dy pairs only the most n-rich N = 98 pair form a SIB $\Delta E_\gamma/E_\gamma$ = $-1.9^{+0.5\%}_{-0.3\%}$ to 10^+, in 10 Dy-Er pairs, only the N = 100 pair form a SIB to 14^+ [Fig. 8], for 10 Er-Yb pairs no SIB but one IB for N = 102 (E_γ shift 0.3±0.5%), for 11 Yb-Hf pairs only the last N = 108 nearly meets our SIB criteria to 10^+ (E_γ shift $-9.3^{+1.2\%}_{-0.7\%}$ to 10^+). For 12 Hf-W cases the most n-rich case N = 112 is barely SIB but only four states are known and the trend suggests no SIB to 10^+. There are no SIBs in W-Os. There are no 4n or 4p cases in any Dy to Os nuclei.

The origin of these SIB is not known. The largest number of different types of SIB (e.g. 2n, 4n, 2p, 4p, α) are for the most n-rich Nd, Sm, Gd known with a few cases in the most n-rich Dy, Er, Yb, Hf nuclei. The appearance of SIBs in only the most n-rich, well-deformed nuclei examined from Ba (Z = 56) to Os (Z = 76) and not in more weakly (β <0.2) deformed nuclei as well as the fact that no SIBs appear in any p-rich nuclei out of the over 100 cases we analyzed would argue against their being accidental. Clustering of SIBs occurs in the most n-rich Nd, Sm, Gd, and Dy nuclei with N = 94-98 where shift spreads are smallest. Influence of the deformed shell gap at N=98 may be important for these SIB.

VU, INEL, LBNL and LLNL work supported by U.S. DOE, by contracts DE-FG05-88ER40407, DE-AC07-76ID01570, DE-AC03-76SF00098 and W-7405-ENG48; Tsinghua by Nat. Natural Science Found. of China and Science Found. for Nucl. Industry. Joint Inst. for Heavy Ion Res. is supported by U.TN, VU, ORNL, and DOE by contract DE-FG05-87ER40361 with U.TN. VU acknowledges World Scientific Pub. Co. and Societa Italiana di Fisica for use of some figures.

1. Hamilton, J.H., et al., *Prog. in Particle and Nucl. Phys.*, **38** (1997) 263.
2. Backtash, C., et al., *Ann. Rev. Nucl. Part. Sci.*, **45** (1995) 485.
3. Hamilton, J.H., et al., *Prog. in Particle and Nucl. Phys.*, **35** (1995) 635.
4. Hamilton, J.H., *Structures of Nuclei Far From Stability*, edited by Allan Bromley (New York: Plenum Press) 1989, pp.2.
5. Table of Isotopes, 8th ed., Firestone, R.B., et al., eds. (NY: Wiley & Sons) 1996.

A Fully Relativistic Hartree-Bogoliubov Approach for Deformed Nuclei

D. Hirata[1] and B.V. Carlson[2]

[1] Spring-8, 323-Mihara, Mikazuki-cho, Sayo-gun, Hyogo, Japan 679-5198.
[2] ITA - CTA, Departamento de Física, São José dos Campos, São Paulo, Brazil

Abstract. We present a formalism of the relativistic Hartree-Bogoliubov (RHB) approximation to study the ground state properties of deformed nuclei. An application of such formalism for Sr isotopes is shown.

We present a fully relativistic Hartree-Bogoliubov (RHB) approach for nuclei using an extension of a treatment applied recently to symmetric nuclear matter [1,2]. Our starting point is similar to that of reference [3] but the pairing interaction is treated in completely different manner. The pairing interaction possesses a Dirac structure and axial deformation of the nuclei is allowed.

In recent years relativistic many body theories have been applied to nuclei with success. The relativistic mean field model (RMF), has successfully described various ground-state properties of nuclei [4]. The RMF approach has also been applied to describe the structure of very exotic nuclei. However, the pairing interaction has been neglected or treated by a non-relativistic BCS type of approximation [5]. For nuclei along or nearby the stability line, the BCS approach provides a reasonably description of the pairing properties. However, for drip-line nuclei, the Fermi level is close to the continuum and the coupling between bound and continuum states must be taken into account explicitly. The pairing correlations and the mean fields must be calculated simultaneously in order to obtain the correct description of the ground state properties.

Recently, a spherical RHB approximation in coordinate space was presented [3]. There, however, the pairing interaction was approximated by a nonrelativistic two-body finite-range Gogny interaction and spherical symmetry was employed.

1. The Mean Field Equations for the Self-energy (Σ) and Pairing (Δ) Fields.

We start with a covariant Lagrangian density, $\mathcal{L} = \mathcal{L}_o + \mathcal{L}_{int}$ which contains contributions of the mesons σ, ω, ρ and photon γ with field operators $\sigma(x)$, $\omega^\mu(x)$, $\vec{\rho}^\mu(x)$ and A^μ, respectively. The free Lagrangian density, $\mathcal{L}_o$, is defined as,

$$\mathcal{L}_o(x) = \overline{\psi}(x)[i\partial - M]\psi(x) + \tfrac{1}{2}\left[\partial_\mu \sigma(x)\partial^\mu \sigma(x) - m_\sigma^2 \sigma^2(x)\right] + \tfrac{1}{2}m_\omega^2 \omega_\mu(x)\omega^\mu(x) \\ -\tfrac{1}{4}\Omega_{\mu\nu}\Omega^{\mu\nu} + \tfrac{1}{2}m_\rho^2 \vec{\rho}_\mu(x)\cdot\vec{\rho}^\mu(x) - \tfrac{1}{4}\vec{G}_{\mu\nu}\vec{G}^{\mu\nu} - \tfrac{1}{4}F_{\mu\nu}F^{\mu\nu} \qquad (1)$$

with vector field tensors $\Omega_{\mu\nu} = \partial_\mu \omega_\nu - \partial_\nu \omega_\mu$, $G_{\mu\nu} = \partial_\mu \vec{\rho}_\nu - \partial_\nu \vec{\rho}_\mu$ and $F_{\mu\nu} = \partial_\mu A_\nu - \partial_\nu A_\mu$. The interacting terms are given by $\mathcal{L}_{int}$ as,

$$\mathcal{L}_{int}(x) = g_\sigma \overline{\psi}(x)\sigma(x)\psi(x) - g_\omega \overline{\psi}(x)\gamma_\mu \omega^\mu(x)\psi(x) - g_\rho \overline{\psi}(x)\gamma_\mu \vec{\tau}\cdot\vec{\rho}^\mu(x)\psi(x) \\ - e\overline{\psi}(x)\tfrac{(1+\tau_3)}{2}\gamma_\mu \omega^\mu(x)\psi(x) \qquad (2)$$

Here, g_σ, g_ω and g_ρ are the effective meson-nucleon coupling constants and m_σ, m_ω and m_ρ are their respective bare masses and M is a nucleon bare mass. The meson fields are assumed to be the free Feynman propagators $(-i\mathcal{D}_j^{ab}(x-y)$, $j=\sigma,\omega$ and $\rho)$ of their sources. The effective action of Eq. (2) is written as,

$$S_{int}=\int d^4x \mathcal{L}_{int}=-i\sum_j \int d^4x d^4y \overline{\psi}(x)\Gamma_{j\alpha}(x)\psi(x)D_j^{\alpha\beta}(x-y)\overline{\psi}(y)\Gamma_{j\beta}(y)\psi(y). \quad (3)$$

where $\Gamma_j(y)$ is the meson-nucleon coupling of the meson j and α,β designate all indices necessary for a correct description of the meson propagation and coupling.

Now, we make the Gorkov[6] mean-field approximation by replacing all possible pair of fermion field with their vacuum expectation value

$$(S_{int})_{eff}=\tfrac{i}{2}\sum_j d^4x d^4y D_j^{\alpha\beta}(x-y)\{2\overline{\psi}(x)\Gamma_{j\alpha}(x)\psi(x)\langle\overline{\psi}(y)\Gamma_{j\beta}(y)\psi(y)\rangle$$
$$+2\overline{\psi}(x)\Gamma_{j\alpha}(x)\langle\psi(x)\overline{\psi}(y)\rangle\Gamma_{j\beta}(y)\psi(y)$$
$$-\overline{\psi}(x)\Gamma_{j\alpha}(x)\langle\psi(x)\psi^T(y)\rangle\Gamma_{j\beta}^T(y)\overline{\psi}^T(y) \quad (4)$$
$$-\psi^T(x)\Gamma_{j\alpha}^T(x)\langle\overline{\psi}^T(x)\overline{\psi}(y)\rangle\Gamma_{j\beta}(y)\psi(y)\}$$

where <...> is the time-ordered expectation value in the interacting nuclear matter ground state.

At this point we want to introduce the notion of time-reversed states, $\psi_T(x)$ as, $\psi_T(x)=\mathcal{A}\overline{\psi}^T(x)=i\tau_2\otimes\gamma_5\gamma^2\gamma^0\overline{\psi}^T(x)$ and the generalized baryon (quasi-particle) propagator or full Hartree-Fock-Bogoliubov propagator, $S(x,y)$ as,

$$S(x,y)=\begin{pmatrix} G(x,y) & F(x,y) \\ \tilde{F}(x,y) & \tilde{G}(x,y) \end{pmatrix}=-i\left\langle \begin{pmatrix} \psi(x) \\ \psi_T(x) \end{pmatrix}(\overline{\psi}(y)\ \overline{\psi}_T(y)) \right\rangle. \quad (5)$$

Comparing Eq. (4) with Eq. (5), we obtain the Hartree, Fock and pairing (Bogoliubov) terms as,

$$(S_{int})_{eff}^{Hartree}=i\sum_j \int d^4x d^4y \overline{\psi}(x)\left[\delta(x-y)\Gamma_{j\alpha}(x)\int d^4z \mathcal{D}_j^{\alpha\beta}(x-z)\text{Tr}\{\Gamma_{j\beta}(z)G(x,x^+)\}\right]\psi(y)$$
$$(S_{int})_{eff}^{Fock}=-i\sum_j \int d^4x d^4y \overline{\psi}(x)\left[\Gamma_{j\alpha}(x)\mathcal{D}_j^{\alpha\beta}(x-y)G(x,y)\Gamma_{j\beta}(y)\right]\psi(y) \quad . \quad (6)$$
$$(S_{int})_{eff}^{pair1}=\tfrac{i}{2}\sum_j \int d^4x d^4y \overline{\psi}_T(x)\left[\mathcal{A}\Gamma_{j\alpha}^T(x)\mathcal{A}^+ \tilde{F}(x,y)\mathcal{D}_j^{\alpha\beta}(x-y)\Gamma_{j\beta}(y)\right]\psi(y)$$
$$(S_{int})_{eff}^{pair2}=\tfrac{i}{2}\sum_j \int d^4x d^4y \overline{\psi}(x)\left[\Gamma_{j\alpha}(x)F(x,y)\mathcal{D}_j^{\alpha\beta}(x-y)\mathcal{A}\Gamma_{j\beta}(y)\mathcal{A}^+\right]\psi(y)$$

Now, we write the total effective action (kinetic+ interacting) in a simplified way, given in terms of two fields, Σ and Δ. Σ would describe the average interaction of a nucleon with the surrounding matter. Δ and $\overline{\Delta}$ would describe the formation and destruction of pairs during the propagation.

$$S_{eff}^{tot}=\tfrac{1}{2}\int d^4x d^4y (\overline{\psi}(x)\ \overline{\psi}_T(x))\mathcal{M}(x,y)\begin{pmatrix} \psi(y) \\ \psi_T(y) \end{pmatrix} \quad (7)$$

where

$$\mathcal{M}(x,y)=\begin{pmatrix} [i\partial-M+\gamma_0\mu+\gamma_0\tau_3\delta\mu]\delta(x-y)-\Sigma(x,y) & \Delta(x,y) \\ \overline{\Delta}(x,y) & [i\partial+M-\gamma_0\mu+\gamma_0\tau_3\delta\mu]\delta(x-y)+\Sigma_T \end{pmatrix}.$$

The term $\mu+\tau_3\delta\mu$ is the isospin matrix of chemical potentials which will be used as Lagrange multipliers to fix the average proton and neutron numbers. The Eqs. (6) and

(7) lead to a set of equations of motion for the fields ψ and ψ_T, called the Dirac-Gorkov equation,

$$\int d^4y \begin{pmatrix} [i\partial - M + \gamma_0\mu + \gamma_0\tau_3\delta\mu]\delta(x-y) - \Sigma & \Delta \\ \overline{\Delta} & [i\partial + M - \gamma_0\mu + \gamma_0\tau_3\delta\mu]\delta(x-y) + \Sigma_T \end{pmatrix} \begin{pmatrix} \psi(y) \\ \psi_T(y) \end{pmatrix} = 0,$$

with a self-energy of $\Sigma(x,y) = -(S_{int})_{eff}^{Hartree} - (S_{int})_{eff}^{Fock}$ and a pairing field of $\Delta(x,y) = -2(S_{int})_{eff}^{pair}$.

The symmetries of the effective mean-field Lagrangian yield to the following properties of the mean fields: $\overline{\Delta}(x,y) = \gamma_0 \Delta^\dagger(x,y)\gamma_0$ and $\Sigma_T(x,y) = A\Sigma^T(x,y)A^\dagger$.

2. The static solutions and the self-consistency equations for axially symmetric nuclei

We are looking for a static, ground state solution to the self-consistency equations. Hence, we write the temporal Fourier transform of the full Hartree-Fock-Bogoliubov propagator in terms of two Dirac spinors $U_{\alpha\beta}$ and $V_{\alpha\beta}$. They are the normal and time-reversed components of the positive (ε_α) and negative (ε_β) frequency solutions of the Dirac-Gorkov equation,

$$\int d^3y \begin{pmatrix} (\varepsilon + \mu + \tau_3\delta\mu)\delta(\bar{x}-\bar{y}) - h(\bar{x},\bar{y}) & \overline{\Delta}^\dagger(\bar{x},\bar{y}) \\ \overline{\Delta}(\bar{x},\bar{y}) & (\varepsilon - \mu + \tau_3\delta\mu)\delta(\bar{x}-\bar{y}) + h_T(\bar{x},\bar{y}) \end{pmatrix} \begin{pmatrix} U(\bar{y}) \\ \gamma_0 V(\bar{y}) \end{pmatrix} = 0, \quad (8)$$

where the single particle Hamiltonian is $h(\bar{x},\bar{y}) = (-i\vec{\alpha}\cdot\nabla + \gamma_0 M)\delta(\bar{x}-\bar{y}) + \gamma_0\Sigma(\bar{x},\bar{y})$.

We evaluated Eq. (8) in the frequency representation in the static limit of the meson propagators. The self-consistency equations contain contributions from both the Dirac and Fermi seas. To avoid the complications of renormalization, we simply discard the contribution of the states in the Dirac sea. It is important to notice that the wavevector (U_γ, V_γ) is not the unique solution Eq. (8). The wavevector (BU_γ, BV_γ) is also a solution of the same equation with the same energy eigenvalue. Hence, either both or neither of them will enter in the contributions of the propagator to the self-consistency equations and these contributions will indeed be invariant under time-inversion of their Dirac structure.

We assume pure proton-proton and neutron-neutron pairing. The isospin dependent Dirac-Gorkov equation decouples into independent equations for neutrons and protons. To obtain a set of coupled equations with an axial deformation degree of freedom, we take the z-axis as the symmetry axis and use cylindrical coordinates (r,z,φ). The projection of j along the symmetry axis Ω, the parity π, and the isospin projection t are good quantum numbers. The 4-components Dirac spinors U_γ and V_γ are

$$U_{t\gamma}(\bar{x}) = \begin{pmatrix} uf_{t\gamma}(\bar{x}) \\ iug_{t\gamma}(\bar{x}) \end{pmatrix} = \frac{1}{\sqrt{2\pi}} \begin{pmatrix} uf_{t\gamma}^+(r,z)e^{i\Omega_{t\gamma}^-\varphi} \\ uf_{t\gamma}^-(r,z)e^{i\Omega_{t\gamma}^+\varphi} \\ iug_{t\gamma}^+(r,z)e^{i\Omega_{t\gamma}^-\varphi} \\ iug_{t\gamma}^-(r,z)e^{i\Omega_{t\gamma}^+\varphi} \end{pmatrix}, \gamma_0 V_{t\gamma}(\bar{x}) = \begin{pmatrix} vf_{t\gamma}(\bar{x}) \\ ivg_{t\gamma}(\bar{x}) \end{pmatrix} = \frac{1}{\sqrt{2\pi}} \begin{pmatrix} vf_{t\gamma}^+(r,z)e^{i\Omega_{t\gamma}^-\varphi} \\ vf_{t\gamma}^-(r,z)e^{i\Omega_{t\gamma}^+\varphi} \\ ivg_{t\gamma}^+(r,z)e^{i\Omega_{t\gamma}^-\varphi} \\ ivg_{t\gamma}^-(r,z)e^{i\Omega_{t\gamma}^+\varphi} \end{pmatrix},$$

with t = neutron and proton and $\Omega_{t\gamma}^\pm = \Omega_{t\gamma} \pm \frac{1}{2}$. For each solution with positive $\Omega_{t\gamma}$, there is a time reversed solution with same energy but a negative $\Omega_{t\gamma}$.

The densities that enter in the Hartree contribution to the self-energy are,

$$\rho_s(r,z)=2\sum_{\substack{\varepsilon_{r_p}<0 \\ \varepsilon_{r_p}>0'}}\left(\sum_{j=\pm}|uf_{rY}^j|^2-\sum_{j=\pm}|ug_{rY}^j|^2\right),\ \rho_c(r,z)=2\sum_{\substack{\varepsilon_{r_p}<0 \\ \varepsilon_{r_p}>0'}}(t_y+\tfrac{1}{2})\left(\sum_{j=\pm}|uf_{rY}^j|^2+\sum_{j=\pm}|ug_{rY}^j|^2\right)$$

$$\rho_B(r,z)=2\sum_{\substack{\varepsilon_{r_p}<0 \\ \varepsilon_{r_p}>0'}}\left(\sum_{j=\pm}|uf_{rY}^j|^2+\sum_{j=\pm}|ug_{rY}^j|^2\right)\ \text{and}\ \rho_3(r,z)=2\sum_{\substack{\varepsilon_{r_p}<0 \\ \varepsilon_{r_p}>0'}}2t_y\left(\sum_{j=\pm}|uf_{rY}^j|^2+\sum_{j=\pm}|ug_{rY}^j|^2\right). \quad (9)$$

and the local Hartree contribution to the Hamiltonian is,

$$\Sigma_H(\bar{x},\bar{y})=\delta(\bar{x}-\bar{y})\int d^3z\left[-g_\sigma^2 d_\sigma(\bar{x}-\bar{z})\rho_s(\bar{z})+g_\omega^2 d_\omega(\bar{x}-\bar{z})\rho_B(\bar{z})+g_\rho^2\tau_3 d_\rho(\bar{x}-\bar{z})\rho_3(\bar{z})\right]$$
$$+\delta(\bar{x}-\bar{y})\int d^3z\left[e^2\tfrac{(1+\tau_3)}{2}d_\gamma(\bar{x}-\bar{z})\rho_c(\bar{z})\right], \quad (10)$$

where the meson propagators have been reduced to the form $d_j(\bar{x}-\bar{z})=\frac{1}{4\pi}\frac{\exp(-m_j|\bar{x}-\bar{z}|)}{m_j|\bar{x}-\bar{z}|}$.

The Fock exchange term will not be included in the present calculation. For the pairing field, we take the zero-range limit for the meson propagators and obtain,

$$\bar{\Delta}_t^+(\bar{x},\bar{y})=\delta(\bar{x}-\bar{y})\left[\frac{g_\sigma^2}{m_\sigma^2}\gamma_0 f_t(\bar{x},\bar{y})\gamma_0-\left(\frac{g_\omega^2}{m_\omega^2}+\frac{g_\rho^2}{m_\rho^2}\right)\gamma_0\gamma^\mu f_t(\bar{x},\bar{y})\gamma_\mu\gamma_0\right], \quad (11)$$

where $f_t(\bar{x},\bar{y})=\sum_{\substack{\varepsilon_{r_Y}<0 \\ \varepsilon_{r_Y}>0}}\left[U_{rY}(\bar{x})V_{rY}^*(\bar{y})+\mathcal{B}U_{rY}^*(\bar{x})V_{rY}^T(\bar{y})\mathcal{B}^\dagger\right]\gamma_0$ with $\mathcal{B}=\gamma_5\gamma^2\gamma^0$.

The method used to solve the autoconsistency equations is similar to that used by Ghambir et al. [2].

4. An Application

Nuclei around $Z=40$ sub-shell closure exhibit variations of the deformation when they deviate from the neutron number $N=50$ and have been extensively studied theoretically in non-relativistic[7] and relativistic frameworks[8]. The ground state properties of these nuclei are very sensitive on changes in the proton and neutron numbers. Experimental data showed an abrupt change of deformation at $N\sim60$ for all elements around $Z=40$ except Krypton ($Z=36$) isotopes [10]. The RMF+BCS approach has been applied to these nuclei and gave a very good description of the ground state properties[9]. We want to study these nuclei using the Hartree-Bogoliubov trying to improve the description of the binding energy. Here, we show the results for Krypton isotopes.

The mean-field equations contain the meson masses and meson-nucleon couplings as free parameters. We use the parameter the same parameter as in ref. [9] and take up to 16 shells for the wavefunctions and meson fields expansions, which guarantee the convergence of the binding energy and deformation in that region.

Figure 1 shows the differences between the theory and the experiment al data[11] for the total binding energy as a function of the neutron number, N. The differences for Hartree-Bogoliubov (HB-exp) are shown by dots and for Hartree+BCS (H+BCS-exp) by diamonds ones. The inclusion of the Bogoliubov improves the results at the neutron rich side but differences as large as 4 MeV are still found.

We compare in Fig. 2 the results for the isotope shifts, which are the differences of the charge radii from that of ^{86}Kr. The open circles are the experimental data and are connected by a dashed line. The Hartree-Bogoliubov results appear as closed circles and the Hartree+BCS ones as diamonds. The smooth increase of the isotope shifts from $N=50$ to $N=60$ is well described by the theory. Our results are slighter larger than the ones obtained in ref. [9]. The smooth increase of the isotope shift with the decrease of the neutron number from $N=50$ until $N=42$ and then the decrease of the isotope shift from $N=40$ until $N=36$ observed experimentally is well described by the calculations. The present calculation gives better results in the latter region in comparison with H+BCS of ref. [9].

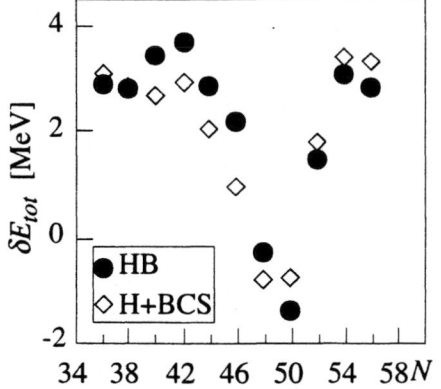

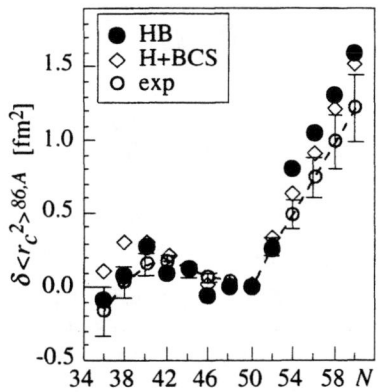

FIGURE 1. Total binding energy differences in MeV as a function of the neutron number.

FIGURE 2. The isotope shifts of Kr as a function of the neutron number.

In general, the Hartree-Bogoliubov approach with NL-SH provides the general feature of the structure of the Kr isotopes as seen in the isotope shift measurements. The description of the experimental binding energies improved with the inclusion of the Bogoliubov in the neutron rich side but large differences still remain. The search for a new parameter set that includes the Bogoliubov would certainly improve the approach.

Acknowledgments

B.V.C. acknowledges partial support from the CNPq-Brazil. All calculations have been done using the supercomputer Fujitsu VPP500 at RIKEN.

References

[1] F.B. Guimarães, B.V. Carlson and T. Frederico, Phys. Rev. **C 54**, 2385 (1996).
[2] F. Matera, G. Fabbri and A. Dellafiore, Phys. Rev. **C 56**, 228 (1997).
[3] J. Meng and P. Ring, Phys. Rev. Let. **77**, 3963 (1996).
[4] Y. K. Gambhir et al., Ann. of Phys. **198**, 132 (1990).
[5] D. Hirata et al. Nucl. Phys. **A616**, 438c (1997).
G. A. Lalazissis, A. R. Farhan, M. M. Sharma, Nucl. Phys. (1998).
[6] L.P. Gorkov, Sov. Phys. JETP **34(7)**, 505 (1958).
[7] P. Bonche et al., Nucl. Phys. **A443**, 39 (1985).
P. Bonche et. al., Nucl. Phys. **A530**, 149 (1991).
[8] D. Hirata, H. Toki, I. Tanihata and P. Ring. Phys. Lett. **B314**, 168 (1993).
[9] G.A. Lalazissis and M.M. Sharma, Nucl. Phys. **A586**, 201 (1995).
[10] M. Keim et al., Nucl. Phys. **A586**, 236 (1995).
[11] G. Audi and A.H. Wapstra, Nucl. Phys. **A595**, 409 (1995).

Decay studies of neutron-rich $A \approx 70$ nuclei produced by fragmentation of 70 MeV/A ^{76}Ge projectiles

J.I. Prisciandaro[1,2], P.F. Mantica[1,2], D.W. Anthony[1,2], M. Huhta[1], P.A. Lofy[1,2], R.M. Ronningen[1], M. Steiner[1], and W.B. Walters[3]

[1] *NSCL, Michigan State University, East Lansing, MI 48824 USA*
[2] *Dept. of Chemistry, Michigan State University, East Lansing, MI 48824 USA*
[3] *University of Maryland, College Park, MD 20742 USA*

Abstract. Neutron-rich nuclides in the vicinity of the semi-magic nucleus ^{68}Ni have been produced by fragmentation of a 70 MeV/nucleon ^{76}Ge beam in a Be target. Beta-delayed γ ray studies were performed for these nuclides using a collection wheel apparatus, two thin plastic scintillators, and two large volume Ge detectors. A 195-keV isomeric transition was identified in ^{73}Zn having $T_{1/2} = 13.0(2)$ ms. A 1296-keV γ-ray transition with $T_{1/2} = 3.4(7)$ s has also been identified. The half-life of this transition is not consistent with the known beta-decay half-lives of nuclides in this region and may be evidence for a new β-decaying isomeric state in ^{69}Ni or ^{71}Cu.

INTRODUCTION

The experimental study of β unstable nuclei in the region $Z \approx 28$, $N > 40$ is important for the testing and further development of theoretical models to better describe the properties of exotic neutron-rich nuclides towards doubly-magic ^{78}Ni. The decay properties of nuclides in this region are also relevant to the refinement of network calculations describing the astrophysical r-process. These nuclides, however, have proven difficult to study due to their low production rates. The recent improvements in the intensity of metal primary beams at the National Superconducting Cyclotron Laboratory at Michigan State University has allowed access to this region of the chart of the nuclides for nuclear structure measurements.

The nuclides in the range ^{67}Co to ^{75}Zn were produced by the fragmentation of a ^{76}Ge primary beam at 70 MeV/nucleon in a Be target. The A1200 fragment separator, with a 70 mg/cm^2 Al wedge at its second dispersive image, was used to separate these fragments from other reaction products. Further M/q separation was achieved using the Recoil Product Mass Separator (RPMS). The resulting secondary beam was implanted into one of nine aluminum foils equally spaced around the circumference of a collection wheel. A stepper motor interfaced with the data acquisition system controlled movement of the wheel. The wheel was stepped at the completion of each implantation + decay cycle. Data acquisition was enabled during both the implantation and decay periods. Beta- and γ-ray emissions were monitored using two 3 mm thin plastic scintillators and two large volume Ge detectors. The β and γ ray detectors were arranged as scintillator-Ge pairs, with the plastic β detectors placed immediately in front of the Ge detectors. The total β efficiency in the above geometry was 40(2)%.

195-keV ISOMER IN ^{73}Zn

Collecting γ-ray singles data during both the implantation and decay periods made it possible to detect short-lived (millisecond) isomers. With the A1200 tuned to the peak production of ^{73}Zn, a short-lived 195-keV transition was identified. The growth and decay curve for this transition is shown in Fig. 1. The measured half-life of 13.0(2) ms is several orders of magnitude faster than the $T_{1/2} = 5.8(8)$ s 195-keV isomeric transition previously assigned to ^{73}Zn [1]. Our value for the half-life restricts the multipolarity of this transition to $M2$ using the Weisskopf estimates, which establishes the spin and parity of the 195-keV state in ^{73}Zn to be $J^{\pi} = 5/2^{+}$.

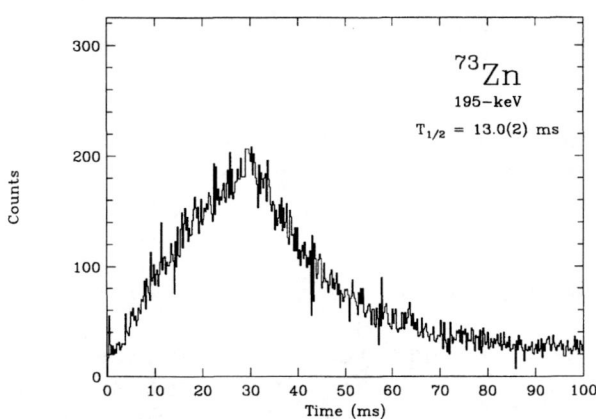

FIGURE 1. Growth and decay curve for the 195-keV isomeric transition in ^{73}Zn

We have identified additional excited states in ^{73}Zn by measuring β-delayed γ rays following the decay of ^{73}Cu. Transitions were identified at 307, 450, 502, 674, and 1559 keV. Both the 674- and 1559-keV transitions were observed to be coincident with the 450-keV ground state transition in ^{73}Zn. The proposed low-energy level structure of ^{73}Zn is given in Fig. 2 and agrees well with the levels obtained using two-proton-one-neutron pickup reactions on a ^{76}Ge target [2]. The low-energy stucture of ^{73}Zn$_{43}$ is similar to ^{75}Ge$_{43}$, which is proposed to be prolate deformed ($\beta_2 \approx 0.2$) in its ground state. We have performed particle-triaxial rotor model calculations [3] in an attempt to reproduce the low-energy level structure of ^{73}Zn and ^{75}Ge. The low energy features of these nuclides were best reproduced assuming an axially symmetric shape with $\beta_2 = 0.20$ and 0.22 for ^{73}Zn and ^{75}Ge, respectively. The above deformation parameters are close to the values predicted by the global Extended Thomas-Fermi Strutinsky Integral (ETFSI) calculations [4] for these nuclides. The moderate quadrupole deformation in the ground state of ^{73}Zn suggests a weakening of the $N = 40$ subshell in close proximity to the $Z = 28$ proton shell closure.

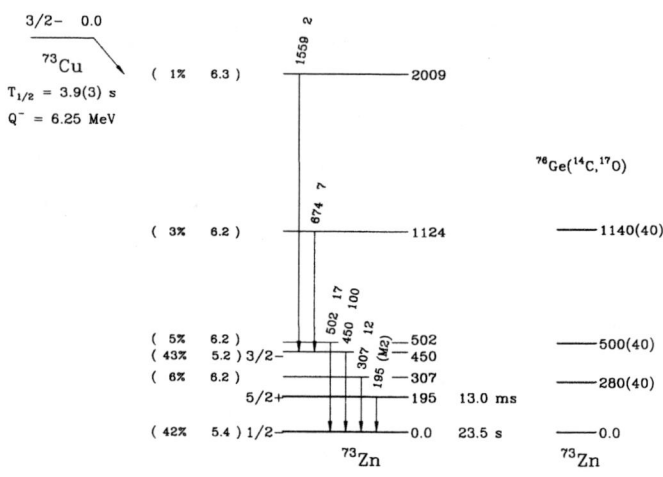

FIGURE 2. Proposed low-energy level structure of ^{73}Zn. The levels identified in two-proton-one-neutron pickup reactions on ^{76}Ge by Bernas et al. [2] are also shown.

1296-keV ISOMERIC TRANSITION

For one series of decay measurements, the A1200 was tuned for the implantation of ^{67}Co, ^{68,69}Ni, ^{70,71}Cu and ^{72}Zn. Known γ-ray transitions were observed for each of these decays, confirming our on-line identification of these isotopes using beam time-of-flight and energy loss measurements. A 1296-keV γ ray transition was identified, and the measured decay curve for this transition was fit considering both a long ($T_{1/2} = 19.5$ s) and short ($T_{1/2} = 3.4(7)$ s) component in the decay (see Fig. 3). The long component can be attributed to a 1298-keV transition recently identified by our group following ^{71}Cu β decay. This transition was found to be coincident with the 489-keV ground state transition in ^{71}Zn. Several other inconsistencies with the known [5] low-energy level scheme of ^{71}Zn have been observed, and will be detailed elsewhere [6]. The short decay component observed for the 1296-keV transition cannot be attributed to the ground state decay of any of the species implanted during these measurements. It is $> 1\sigma$ shorter than the $T_{1/2} = 4.5$ s measured for ^{70g}Cu. The ^{70g}Cu β decay is known [5] to feed only the ground and first excited (885 keV) states of ^{70}Zn. There was no evidence of a 1296-885 coincidence in our γ-γ data, and the relative peak intensities of these transitions would imply a direct β feeding of $> 10\%$ if the 1296-keV transition directly populated the ground state of ^{70}Zn.

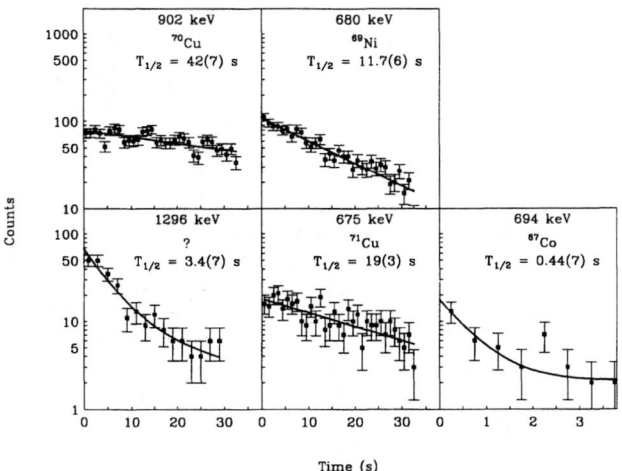

FIGURE 3. Decay curves for selected γ-ray transitions identified during implantation of ^{67}Co, ^{68,69}Ni, ^{70,71}Cu, and ^{72}Zn.

Since the 1296-keV transition was observed in the β-delayed γ ray spectrum, it may arise from a previously unidentified β-decaying isomer in either ^{67}Co, ^{68,69}Ni, ^{70,71}Cu or ^{72}Zn. This transition was also identified using a different A1200 tune where the nuclides ^{68}Co,^{69,70}Ni,^{71,72}Cu, and ^{73}Zn were simultaneously implanted. Therefore, we can narrow the source of this new β-decaying isomer to ^{69}Ni or ^{71}Cu. Further experiments are required to identify the source of the $T_{1/2} = 3.4(7)$ s activity.

REFERENCES

1. E. Runte et al., Nucl. Phys. **A441**, 237-260 (1985).
2. M. Bernas et al., Nucl. Phys. **A413**, 363-374 (1984).
3. I. Ragnarsson and P.B. Semmes, Hyperfine Int. **43**, 425-440 (1988).
4. Y. Aboussir et al. At. Data Nucl. Data Tables **61**, 127-176 (1995).
5. *Table of Isotopes*, 8th edition, ed. R.B. Firestone (Wiley and Sons, New York, 1996).
6. J.I. Prisciandaro et al., in preparation.

In-Beam Coulomb-Excitation Studies of Odd-A $\nu(fp)$– and $\pi(sd)$–Shell Nuclei

R.W. Ibbotson[1], T. Glasmacher[2], P.F. Mantica[3], H. Scheit[2]

1) National Superconducting Cyclotron Laboratory,
2) Department of Physics and Astronomy, and
3) Department of Chemistry,
Michigan State University, East Lansing MI 48824, USA

Abstract. A group of odd-mass neutron-rich nuclei with neutron excesses of 9–11 in the $13 \leq Z \leq 17$ region has been produced and studied by in-beam intermediate-energy Coulomb excitation using an array of NaI(Tl) detectors. The excitation energies of the observed states and B(E2) values connecting these states to the ground-states have been measured. For the ^{41}S and ^{43}S cases the measurements have been compared to particle-rotor and particle-vibrator calculations. The measurements for ^{41}S are consistent with an interpretation of the low-energy behavior of this nucleus as rotations of a deformed core, whereas for ^{43}S no distinction can be made between the deformed (rotational) and spherical (vibrational) calculations.

The experimental study of the low-lying structure of nuclei has been extended in recent years to regions of very neutron-rich and proton-rich nuclei by the availability of beams of β-unstable nuclei with relatively short half-lives. Several studies in the neutron-rich $10 < Z < 20$ region have been performed, revealing several regions of large quadrupole collectivity. A region of large deformation has been discovered for $N = 20$, $Z < 14$, due to the intrusion of the $\nu(f_{7/2})$ orbital into the (sd) shell [1–3]. Another region of large quadrupole collectivity has been discovered for $N \approx 28$, $Z \approx 16$ nuclei [4] although this collectivity has been predicted to be vibrational (based on a spherical ground state) in some models and rotational (based on a deformed ground state) in others [5,6].

Very little other information is known about the nuclei in the $N \approx 28$, $Z \approx 16$ region, however. These nuclei are at the edge of the region of measured mass excess ($N - Z \leq 11$), and well past the last odd-mass neutron-rich nuclei for which J^π assignments have been made ($N - Z \leq 7$ for $12 \leq Z \leq 18$). A measurement of the lowest excitations in the odd-mass nuclei in the $N \approx 28$, $Z \approx 16$ region may prove useful in interpreting the collectivity found in the even-even nuclei. A group of odd-mass nuclei with neutron excesses of 9–11 in the $13 \leq Z \leq 17$ region was therefore produced and studied by Coulomb excitation in order to determine the energies and excitation cross-sections for the lowest excited states.

The nuclei studied in the present work were produced simultaneously by fragmentation of a 70 MeV/A ^{48}Ca beam provided by the K1200 cyclotron at the NSCL at Michigan State University in a 285 mg/cm^2 ^{9}Be target and separated in the A1200. A set of 18 nuclei in the range $12 \leq Z \leq 17$ with $7 \leq N - Z \leq 12$ reached the focal plane at rates of 30 particles per second or greater, where they were identified by energy-loss in a 300μm Si PIN detector and by time of flight with respect to the cyclotron radiofrequency.

This mixed-particle beam was transported to the experimental station where it impinged on a 532 mg/cm^2 ^{197}Au target. Scattered beam particles were detected in a fast/slow plastic phoswich detector. The maximum scattering angle from the target was restricted to $\theta_{lab} < 3.8°$, which corresponds to a distance of closest approach larger the sum of the two nuclear radii by more than 3.1 fm in all cases. An array of 38 cylindrical NaI(Tl) detectors [7] centered about the target position was used for detection of γ rays in coincidence with scattered projectile ions.

For each of the 18 nuclear species in the beam, the spectrum of Doppler-corrected γ rays in coincidence with the scattered nucleus of interest was collected. Since the excitation probability of the observed states is $\leq 10^{-3}$ at all scattering angles, no multiple excitation is expected. This implies that the γ-rays observed correspond to excitations of states at excitation energies equal to the γ-ray energies. Assuming excitation of these states by E2 mode only, the B(E2) values corresponding to these excitations have been extracted using the method of Winther and Alder [8] which involves calculation of the excitation in first-order perturbation theory.

The γ-ray data taken for five of the even-even nuclei in this group have been analyzed separately [5,9]. Of the remaining 13 nuclei for which > 5 million fragments were produced, projectile γ-rays were observed in 7 cases. The γ-ray peak observed in coincidence with scattered ^{43}S fragments was considerably larger than the detector resolution, indicating at least two states close in energy. A summed E2 strength for these γ-rays was measured. Since the energy resolution of the detectors is $\approx 8\%$, the other transitions observed in the present work may of course consist of more than one γ-ray separated by $< 8\%$.

The measured level energies and B(E2;$0^+ \rightarrow 2^+$) values for ^{41,43}S and ^{45}Cl are shown in Table 1 along with the previous measurements for the even-mass sulfur isotopes [4,5,10]. The E(2^+) and B(E2;$0^+ \rightarrow 2^+$) values for $^{44}_{16}$S$_{28}$ are clearly more suggestive of a collective system than would be expected for a closed-shell system such as $^{36}_{16}$S$_{20}$. The data obtained for ^{41}S, ^{43}S and ^{45}Cl exhibit several interesting features; the level energies clearly do not behave as expected if $N = 28$ is a closed shell for $Z = 16$. The presence of an excited state at very similar excitation energies (≈ 900 keV) in ^{40,41,42}S and ^{43}S is unexpected, since this would seem to indicate a region with little change in the collectivity. It is interesting that for ^{41}S, 110% of the B(E2) strength in the even-even neighbors is observed in two clearly resolved states at 449 keV and 904 keV. In ^{43}S and ^{45}Cl (both neighbors of the $N = 28$ nucleus ^{44}S), however, less than half of the B(E2) strength in the neighboring nuclei was observed in this experiment.

TABLE 1. Odd-A and Odd-Z fragments produced at rates greater than 30/sec, listed with the total number observed and average energy per nucleon.

Nucleus	# observed (10^6)	E/A (MeV)	E_γ (keV)	B(E2↑) (e^2fm^4)	Nucleus	E_γ (keV)	B(E2↑) (e^2fm^4)
	This Work				Previous Measurements		
^{41}S	75.1	47.4	449 (3)	167 (65)	^{36}S	3290.9 (3)	96 (26)
			904 (4)	232 (56)	^{38}S	1286 (19)	235 (30)
^{43}S	22.2	42.0	≈940	175 (69)	^{40}S	891 (13)	334 (36)
^{45}Cl	170.8	43.0	929 (6)	87 (24)	^{42}S	890 (15)	397 (63)
					^{44}S	1297 (18)	314 (88)

In order to investigate the nature of the E2 collectivity for these nuclei, the measured level energies and E2 strengths for ^{41}S and ^{43}S have been compared to predictions assuming a statically-deformed core (the particle+rotor model, see e.g. [11]) and (separately) assuming a spherical, vibrating core (the particle+vibrator model). For the particle+rotor calculations, the single-particle energies and eigenfunctions have been calculated using a deformed Woods-Saxon potential assuming a quadrupole deformation parameter which was extracted from the measured $B(E2; 0^+ \to 2^+)$ value in the neighboring even-even nuclei. The moment of inertia has been taken from the known energies of the 2^+ states in the neighboring even-even nuclei. For the particle+vibrator calculations, the one-phonon energy has been taken from the energies of the 2^+ states in the even-even neighbors.

The low-lying negative-parity states calculated by these methods for ^{41}S and ^{43}S are shown in Figure 1. Since the present experiment selectively populated states connected to the ground-state by a large E2 transition strength, only states with a $B(E2; gs \to I^\pi)$ value greater than 30 e^2fm^4 have been shown in this figure. States which are predicted to lie below 300 keV have been considered as possible ground-states, and are also shown.

The particle-rotor calculations for ^{41}S assuming a prolate deformation show most of the E2 strength concentrated in 3 states, one at ≈800 keV, and two at ≈1200 keV. Although these energies are 300 keV higher than observed, the pattern of E2 strength reproduces the observed strengths assuming that the transitions from the ≈1200 keV states are unresolved in our measurements. With the assumption of an oblate deformation for this nucleus, the E2 strength is predicted to be concentrated in 3 states at roughly the same excitation energy (≈1200 keV). The E2 strength involving the lower-energy $5/2^-$ state is much smaller than the measured strength involving the 449 keV state. The particle-vibrator calculations also predict a concentration of E2 strength in a few states at ≈800 keV, which does not adequately reproduce the observations in ^{41}S. Although the calculations for ^{41}S suggest an interpretation of this nucleus as prolate deformed, it appears to be impossible to draw any conclusions for ^{43}S, since adequate agreement can be obtained between the observed E2 strength and any of the calculations performed.

These simple calculations are, of course, merely an attempt to understand the

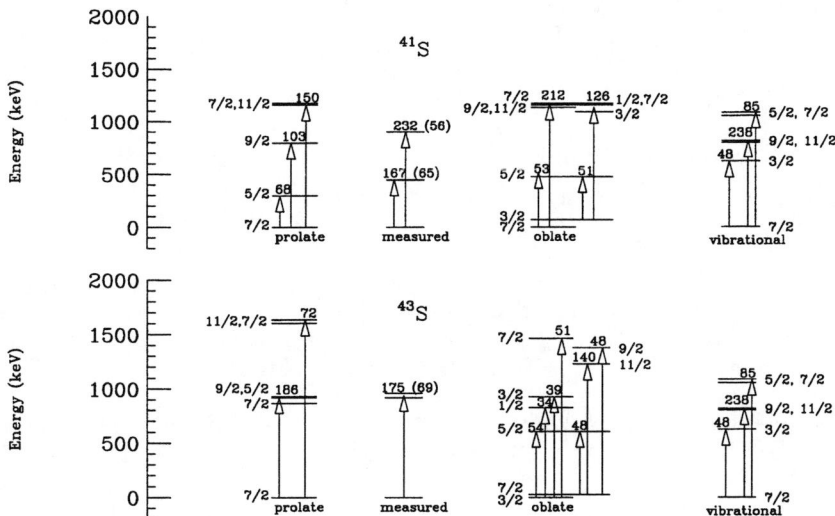

FIGURE 1. Levels and $B(E2; g.s. \rightarrow J^\pi)$ values calculated for ^{41}S and ^{43}S using the particle-rotor model and particle-vibrator model (see text).

gross features of these nuclei. In order to gain a deeper understanding of the nature of the collectivity in these nuclei and the validity of the N=28 shell closure in this region, it will be necessary to determine more information concerning the ground-states and low-energy excited states in these nuclei. A determination of the spins and parities of the ground states and several excited states, possibly through β-decay γ-ray measurements of ^{41}P and ^{43}P, would be helpful in understanding the properties of these nuclei.

This work was supported by the National Science Foundation under grant PHY-9528844.

REFERENCES

1. Thibault, C. et al., *Phys. Rev.* **C12**, 644 (1975).
2. Motobayashi, T. et al., *Phys. Lett.* **B346**, 9 (1995).
3. Guillemaud-Mueller, D. et al., *Nucl. Phys.* **A426**, 37 (1984).
4. Scheit, H. et al., *Phys. Rev. Lett.* **77**, 3967 (1996).
5. Glasmacher, T. et al., *Phys. Lett.* **395B**, 163 (1997).
6. Werner, T.R. et al., *Phys. Lett.* **B335**, 259 (1994).
7. Scheit, H. et al., submitted to *Nucl. Inst. Meth. A*.
8. Winther, A. and Alder K. *Nucl. Phys.* **A319**, 518 (1979).
9. Ibbotson, R.W. et al., *Phys. Rev. Lett.* **80**, 2081 (1998).
10. Endt, P.M., *Nucl. Phys.* **A521** 1 (1990).
11. Ragnarsson, I. and Semmes, P.B., *Hyperfine Int.* **43**, 425 (1988).

β-decay half-lives of new neutron-rich lanthanide isotopes

S. Ichikawa[1], K. Tsukada[1], M. Asai[1], A. Osa[2], M. Sakama[3], Y. Kojima[4], M. Shibata[4], I. Nishinaka[1], Y. Nagame[1], Y. Oura[3], and K. Kawade[4]

[1] *Japan Atomic Energy Research Institute, Tokai-mura, Ibaraki 319-1195, Japan*
[2] *Japan Atomic Energy Research Institute, Takasaki 370-1292, Japan*
[3] *Department of Chemistry, Tokyo Metropolitan University, Hachioji, Tokyo 192-0364, Japan*
[4] *Department of Energy Engineering and Science, Nagoya University, Nagoya 464-8603, Japan*

Abstract. New neutron-rich lanthanide isotopes produced in the proton-induced fission of ^{238}U have been identified using the JAERI on-line isotope separator (JAERI-ISOL) coupled to a gas-jet transport system. The observed K x-rays following the β^--decay of products in the mass separated fraction provided direct isotopic identifications. New isotopes observed, with values of their half-lives given in parentheses, are ^{159}Pm (2 ± 1 s), ^{161}Sm (4.8 ± 0.8 s), ^{165}Gd (10.3 ± 1.6 s), ^{166}Tb (21 ± 6 s), ^{167}Tb (19.4 ± 2.7 s) and ^{168}Tb (8.2 ± 1.3 s). The half-life values are compared to the results of theoretical predictions.

INTRODUCTION

Much progress has been made in exploring the frontier of neutron-rich lanthanide isotopes owing to the improvement of both separation techniques and production methods. By developing the separation technique, about ten neutron-rich lanthanide isotopes produced in fission of relativistic ^{238}U projectiles were discovered by the Fragment separator at GSI (1). The multinucleon-transfer reaction with heavy ions were greatly contributed to produce the isotopes in the mass region around A ≈ 170 (2,3). There still remain many isotopes having a half-life of a few seconds to be discovered in the region with A = 160 - 170 and Z = 61 - 65. To identify these unknown isotopes produced in the proton-induced fission of ^{238}U, we developed the gas-jet coupled on-line isotope separator (JAERI-ISOL) system (4), and the new isotopes of ^{161}Sm, ^{165}Gd and ^{166}Tb were successfully identified (5,6). The produced lanthanide isotopes were mass-separated as the monoxide ions (^{A}M^{16}O$^+$), being well separated from other molecular and metallic species. Thus the K x-rays following the β^--decay of the isotope were clearly observed. By taking advantage of this technique, further investigation is being performed to understand the half-life systematics of the neutron-rich lanthanide isotopes. In this paper, we report the recent results of the new isotope search in the mass region A= 160 - 170. The comparison between the experimental half-lives and those of theoretical predictions is also presented.

EXPERIMENT

Experiments were performed at the JAERI (Japan Atomic Energy Research Institute) tandem accelerator facility. A stack of eight ^{238}U targets (4 mg/cm^2 thick each) set in a multiple-target chamber was bombarded with 20 MeV proton beams with the intensity of about 1 μA. Fission products emitted isotropically from the targets were thermalized in argon gas loaded with an aerosols (KCl or PbI$_2$), then transported into the ISOL through a capillary. The transported nuclides were ionized in the thermal ion source and mass-separated electromagnetically. A mass resolution of about 900 (M/ΔM) around A $\approx$ 100 was achieved in the present system. The detailed performance of the ISOL and the gas-jet coupled thermal ion source is described elsewhere (4).

The mass-separated products were collected on an aluminized Mylar tape and transported to the detection port at prescribed time intervals. The detection port was equipped with two plastic scintillators, 30 mm x 30 mm x 1 mm, for the measurement of β rays through the Mylar window and two Ge detectors set behind the scintillators; one was a planar detector with 5 cm in diameter and 1 cm thick having an energy resolution of 0.61 keV FWHM at 122 keV and the other was a coaxial detector with 28% in relative efficiency and 1.8 keV FWHM at 1.33 MeV. β-γ and γ-γ coincidence data were taken with these four detectors and recorded event by event together with time information.

RESULTS AND DISCUSSION

Recently, the three new neutron-rich lanthanide isotopes of ^{159}Pm (2 $\pm$ 1 s), ^{167}Tb (19.4 $\pm$ 2.7 s) and ^{168}Tb (8.2 $\pm$ 1.3 s) have been identified from the analysis of β-gated x/γ-ray spectra obtained at the mass-A+16 (^{A}M^{16}O$^+$) fraction. In Table I the present results together with previous ones are compared with the theoretical predictions by the gross theory (GT) with new one-particle strength function (7) and the proton-neutron quasiparticle random-phase approximation (pn-QRPA) using three different mass formulae (8). The ratios of the calculation with GT and pn-QRPA using the mass formula of Hilf to the measured half-lives, denoted by $y_i = T_{1/2}^{cal}/T_{1/2}^{exp}$, are shown in Fig.1.

To examine the predictive power of both theories, we calculate the average deviation $\bar{x}$ with following the equation.

Table 1: Comparison between the measured and the predicted half-lives.

Isotope	Experiments	Calculations			
		Gross theory[7]	pn-QRPA[8]		
			Hilf	Groote	Möller
^{159}Pm	2 ± 1^a	3.08	2.93	2.54	2.80
^{161}Sm	4.8 ± 0.8	6.72	10.7	13.0	12.6
^{165}Gd	10.3 ± 1.6	16.0	20.6	27.4	18.4
^{166}Tb	21 ± 6	33.6	83.7	166	82.8
^{167}Tb	19.4 ± 2.7	18.2	67.3	130	63.0
^{168}Tb	8.2 ± 1.3	7.25	37.1	68.4	28.6

^{a}Tentative result.

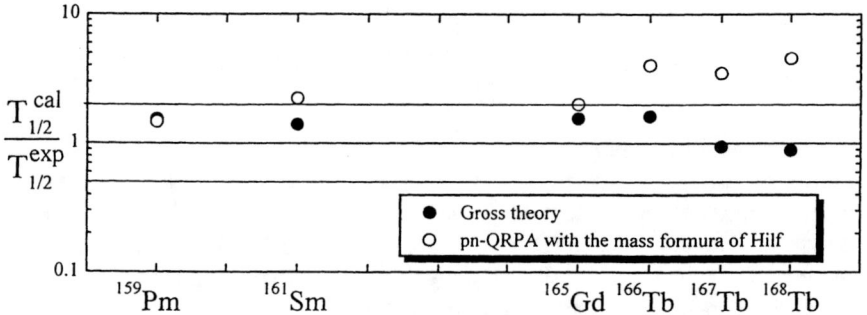

Figure 1: Ratios of the theoretical calculations to the experimental half-lives.

$$\bar{x} = \sum_{i=1}^{n} x_i,$$

where

$$x_i = \begin{cases} y_i, & (y_i \geq 1) \\ 1/y_i, & (y_i < 1) \end{cases}$$

and

$$y_i = T_{1/2}^{cal}/T_{1/2}^{exp}.$$

The average deviation for these new isotopes are given by $\bar{x}(\text{GT}) = 1.38$ and $\bar{x}(\text{pn-QRPA}) = 2.95$, respectively. Although the calculated half-life values depend considerably on the input data of the Q_β values as well as a number of factors associated with the β decay strength function, the predictions with GT agree very well with the experimentally measured half-lives, while the pn-QRPA slightly overestimates.

The large deviation $\bar{x}(\text{pn-QRPA}) = 4.0$ is observed for the half-lives of terbium isotopes. According to Staudt et al. (8), such a large deviation was not obtained in their calculation for the 27 new neutron-rich unknown isotopes: the mean deviation is given as $\bar{x}(\text{Hilf}) = 1.56 \pm 0.54$.

In order to check the predictive power of both theories in the region around A ≈ 60, we calculate the average deviation $\bar{x}$ using the measured half-lives of very neutron-rich isotopes of elements from Ti to Ni (9). The half-lives calculated with GT for these isotopes are good agreement with the experimental results with the average deviation of $\bar{x}(\text{GT}) = 1.90$. The average deviation of $\bar{x}(\text{Hilf}) = 2.39$ with the pn-QRPA is observed except for vanadium isotopes. The very large deviation $\bar{x}(\text{Hilf}) = 20.7$ is found for the half-lives of vanadium isotopes. To further the understanding of the β decay half-life systematics of nuclides far from the β stability, experimental determination of half-life and Q_β values and the improvement of theoretical predictions are needed.

ACKNOWLEDGMENTS

We would like to thank the crew of the JAERI tandem accelerator for supplying intense and stable beams for the experiments. This work was carried out along a line of the JAERI-University Collaboration Research Project.

REFERENCES

1. S. Czajkowski et al., in Proceedings of International Conference on Exotic Nuclei and Atomic Masses, edited by M. de Saint Simon and O. Sorlin (Editions Frontieres 1995), p.553.
2. R. M. Chasteler et al., Nucl. Phys. **A522**, 557 (1991).
3. K. Becker et al., Nucl. Phys. **A522**, 557 (1991).
4. S. Ichikawa et al., Nucl. Instrum. Methods Phys. Res. A **374**, 330 (1996).
5. M. Asai et al., J. Phys. Soc. Jpn. **65**, 1135 (1996).
6. S. Ichikawa et al., Phys. Rev. C in press.
7. T. Tachibana and M. Yamada, in Proceedings of International Conference on Exotic Nuclei and Atomic Masses, edited by M. de Saint Simon and O. Sorlin (Editions Frontieres 1995), p.763.
8. A. Staudt et al., At. Data and Nucl. Data Tables **44**, 79 (1990).
9. F. Ameil et al., Eur. Phys. J. A **1**. 275 (1998).

Conversion Electron Measurements in ^{125,127}Ba

H. Iimura*, S. Ichikawa*, T. Sekine*, M. Oshima* and M. Miyaji†

Japan Atomic Energy Research Institute, Tokai, Ibaraki 319-1195, Japan
†Toshiba Corporation, Kawasaki, Kanagawa 210, Japan

Abstract. Parities of low-lying levels in ^{125,127}Ba have been investigated by measuring conversion electrons following the ^{125,127}La decay. From the measurements, 16 conversion coefficients have been determined for ^{127}Ba and 12 for ^{125}Ba. The results support previous band assignments. No evidence for alternating parity structure, which is indicator of static octupole deformation, has been observed.

INTRODUCTION

Neutron-deficient nuclei in the A=120-130 mass region exhibit a transition from triaxial to prolate shapes, providing a good chance of testing various nuclear models. It has also been suggested that a number of nuclei in the neighborhood of ^{128}Ba possess stable octupole deformations (1). In ^{125,127}Ba, high-spin level structure was studied by in-beam γ-ray spectroscopic method, as summarized in ref. 2 and 3, and several rotational bands were proposed. However, multipolarities of transitions in these nuclei have not been established yet. In this work we measured conversion electrons in ^{125,127}Ba following the ß-decays of ^{125}La ($T_{1/2}$=1.3 m) and ^{127}La ($T_{1/2}$=5.1 m) to obtain information on low-lying levels. The ß-decay of ^{127}La was studied by Gizon also (4). Preliminary results of this work for ^{125}Ba have already appeared (5).

Experiment

The experiment was performed by using the on-line mass separator at tandem accelerator facility in JAERI. The ^{125,127}La nuclei were produced by the reaction ^{nat}Mo(^{32}S,pxn)^{125,127}La with 160-MeV ^{32}S beam. The thickness of molybdenum targets was about 4 mg/cm^2. Reaction products were ionized in a surface-ionization ion source, and mass-separated electromagnetically. In order to obtain ^{125,127}La activities free from the Cs and Ba isobars, the monoxide ions ^{125,127}Ce^{16}O$^+$ were separated by setting the magnetic field at the mass number 141 or 143. The activities were collected on aluminized Mylar tape, which was moved every 2.5 min or 10 min to take away long-lived daughter activities. Electrons and γ-rays were measured simultaneously with a Si(Li) detector (500 mm^2 x 3mm^t, 2.5 keV FWHM at 976 keV) and a 20% HPGe detector, respectively. Source-to-detector distances were 2.5 cm, and 180° geometry was used. The Si(Li) detector was calibrated for relative intensity with ^{133}Ba and ^{207}Bi sources. Electron intensities were normalized to γ-ray intensities using the $2_1^+ \rightarrow 0_1^+$ E2 transition in ^{124}Ba. From the measurements, 16 conversion coefficients were determined for ^{127}Ba and 12 for ^{125}Ba. The K conversion coefficients for ^{127}Ba and ^{125}Ba are plotted with theoretical values in Fig. 1 (a) and (b), respectively.

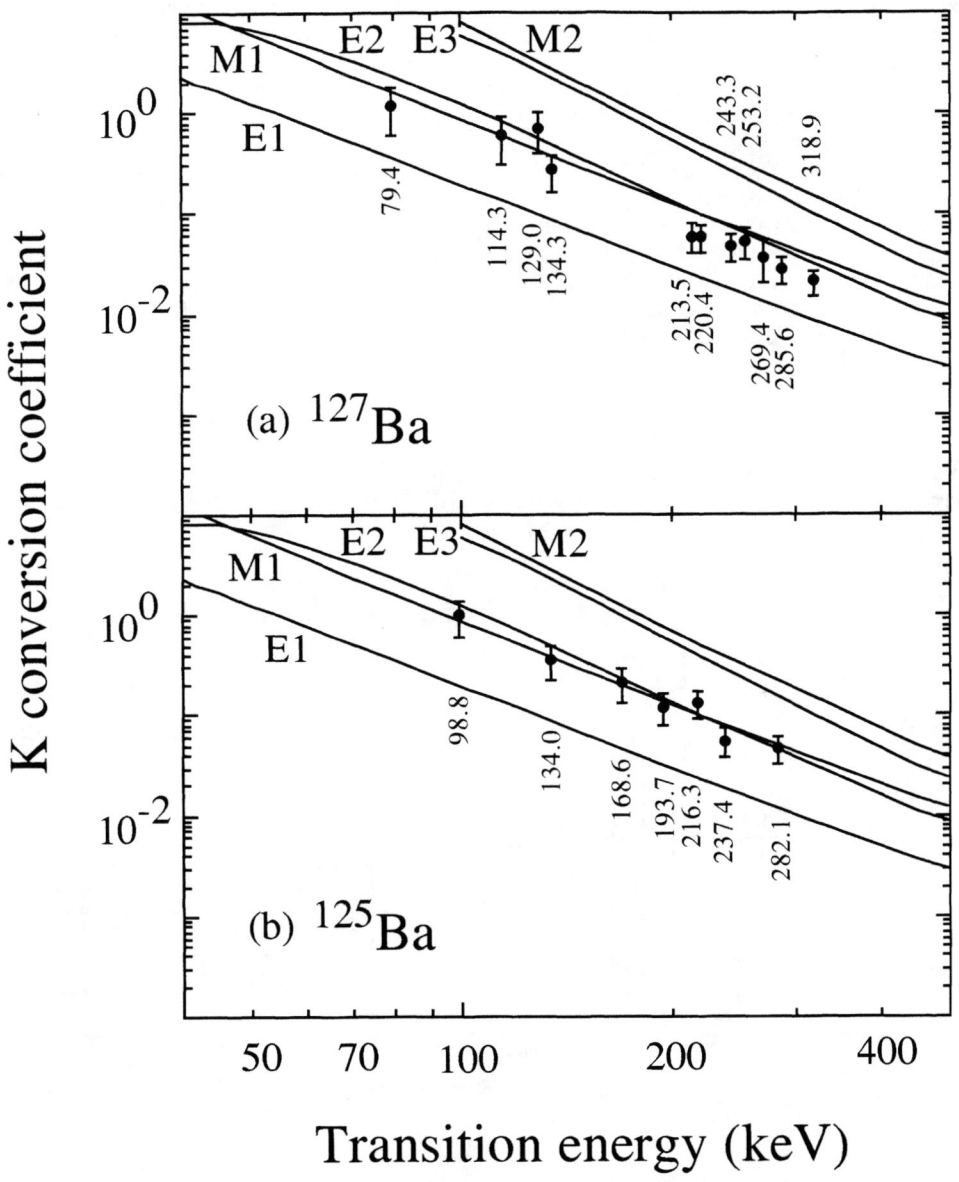

FIGURE 1. Internal conversion coefficients of (a) ^{127}Ba and (b) ^{125}Ba transitions.

Discussion

As seen in Fig. 1(a), all of the K conversion coefficients measured for ^{127}Ba were consistent with M1 and/or E2 multipolarities. Our results confirm the multipolarities of 134.3-, 220.4- and 253.2-keV transitions reported by Cottle *et al.* (6), while others have been determined for the first time. From the multipolarities of transitions, the 195.4-, 324.2-, 415.7- and 669.0-keV levels have been determined to have the same parities as the 81.1-keV level. In the same way, the parities of the 269.5- and 375.1-keV levels have been assigned as positive, and those of 159.7-, 293.9- and 579.5-keV levels as negative. Those parities are shown in Fig. 2, in which the parities of the 0-, 56.2-, 80.3- and 81.1-keV levels and the spins of all levels are from ref. 3.

In the same way, all of the K conversion coefficients measured for ^{125}Ba have M1 and/or E2 character as seen in Fig. 1(b). Those multipolarities have been determined for the first time. From the multipolarities, the 43.7-, 237.4- and 325.9-keV levels have been determined to have the same parities as the 0 keV level. It has also been determined that the 168.6+x keV and 384.9+x keV levels have the same parities as the 0+x keV level, and that the 166.3+x keV and 300.3+x keV have the same parities as the 67.6+x keV level. Those parities are shown in Fig. 3, in which the parities of the 0 keV, 0+x keV and 67.6+x keV levels and the spins of all levels are from ref. 2. As we reported in our previous publication (5), the multipolarity of the 67.6 keV transition was determined to be E1 from the intensity ratio of the K X-ray to the γ-ray in the coincidence spectra. Thus, the parity of the 67.6+x keV level is different from that of the 0+x keV level.

Present results are consistent with the previous assignments of 5/2[402], 1/2[411] and 7/2[523] Nilsson orbitals to band 1, 2 and 3, respectively (3). It has been confirmed that the low-spin states within each rotational band have the same parity. Thus, alternating

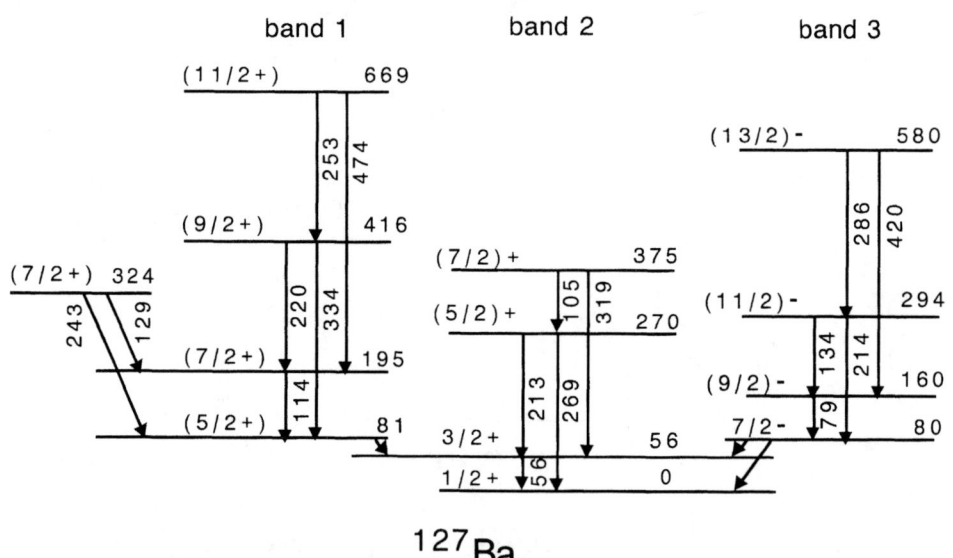

FIGURE 2. Low-lying levels in ^{127}Ba

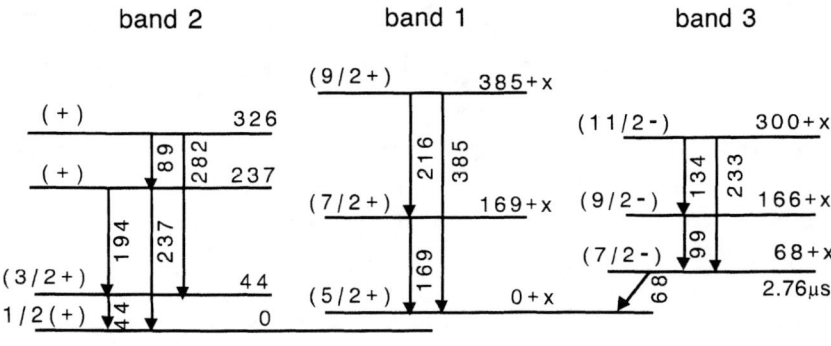

FIGURE 3. Low-lying levels in ^{125}Ba

parity structure, which is indicator of static octupole deformation, has not been observed. No evidence for such structure has either been found in recent studies of spins and parities in ^{128,129,130}La (6,7).

Octupole deformed nuclei in the A=145 and 225 regions often have strong E1 transitions, Weisskopf hindrance factor of which is about $F_W \sim 10^2$. As shown in Fig. 3, our β-γ delayed coincidence measurement yielded $t_{1/2}$= 2.76±0.14 μs as the half-life of the 67.6 keV level in ^{125}Ba (5). The measured half-life value gives a hindrance factor F_W = 5.2×10^6 to this E1 transition, which is much higher value than some of those observed in the A=145 and 225 regions.

In conclusion, it becomes more unlikely that static octupole deformation occurs in the A=130 mass region.

REFERENCES

1. Cottle, P. D., *Z. Phys.* A**338**, 281-283 (1991).
2. Katakura, J., Oshima, M., Kitao K. and Iimura, H., *Nucl. Data Sheets* **70**, 217-314 (1993).
3. Kitao, K. and Oshima, M., *Nucl. Data Sheets* **77**, 1-124 (1996).
4. Gizon, A., *Proc. of 25th Zakopane School on Physics*, World Scientific, 1990, Vol.2, p. 93-111.
5. Iimura, H. *et al.*, *JAERI Tandem Ann. Rept. 1988*, p.136-138.
6. Cottle, P.D., Glamacher, T., Johnson, J. L. and Kemper K. W., *Phys. Rev.* C**48**, 136-139 (1993).
7. Cottle, P.D., Glamacher and Kemper K. W., *Phys. Rev.* C**45**, 2733-2739 (1992).

Fusion Reactions of Deformed Nuclei near Coulomb Barriers

H. Ikezoe[1], T. Ikuta[1], S. Mitsuoka[1], T. Kuzumaki[2], J. Lu[1,3], Y. Nagame[1], I. Nishinaka[1], K. Tsukada[1] and T. Ohtsuki[2]

[1] *Japan Atomic Energy Research Institute, Tokai, Naka, Ibaraki 319-11 Japan*
[2] *Laboratory of Nuclear Science, Tohoku University, Mikamine, Sendai 982, Japan*
[3] *Institute of Modern Physics, Chinese Academy of Sciences, 73000 Lanzhou, China*

Abstract. The fusion evaporation residues were measured in the ^{232}Th + ^{30}Si reaction near the Coulomb barrier. The measured xn cross sections were compared with statistical model calculations where the fusion cross section was calculated by taking into account the target deformation. It was found that the calculated $3n$ and $4n$ cross sections were considerably overestimated and an extra-extra push energy was needed for the complete fusion. The xn cross sections of other fusion reactions with ^{232}Th and ^{238}U targets and deformed rare earth targets were also analyzed. It was found that the obtained extra-extra push energies were well parameterized by a quantity R_L/R_S, the ratio of the center-to-center distance R_L in the collision with the tip of a prolate target to the saddle point radius R_S of the compound nucleus.

INTRODUCTION

Recently, anomalously large anisotropies of the fission fragment angular distributions have been observed in the reaction systems of light projectiles ^{12}C, ^{16}O and ^{19}F on deformed targets ^{232}Th and ^{238}U (1). These large anisotropies, which are 1.5 - 2 times larger than the predictions of the transient state model, become evident near and below Coulomb barriers. Hinde et al. (2) have pointed out that the quasi-fission is the main reaction process below Coulomb barriers and the nuclear orientation of the deformed target with respect to the projectile plays important role in the formation of the compound nucleus, that is, the collisions with the tips of the deformed nuclei lead to quasi-fission, while the collisions with the sides lead to the complete fusion process. In order to investigate this phenomena, we measured the fusion-evaporation residues produced in the ^{232}Th + ^{30}Si reactions at the bombarding energy near the Coulomb barrier. Since the $3n$ and $4n$ cross sections in the fusion reactions have their maximum values near the Coulomb barrier, these are very sensitive quantities to the sub-barrier fusion enhancement due to the target deformation.

EXPERIMENTAL RESULTS

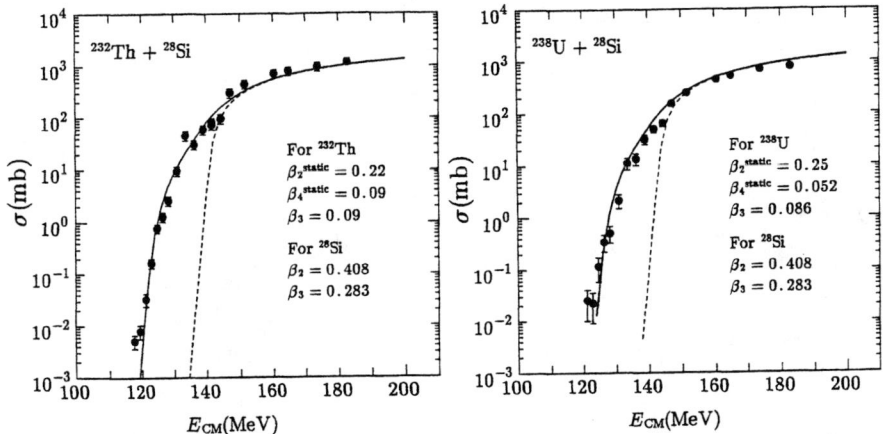

FIGURE 1. Measured fission cross sections for the reactions ^{232}Th + ^{28}Si and ^{238}U + ^{28}Si. The solid lines represent the CCDEF calculations taking into account the couplings of the inelastic excitations of 2^+ and 3^- for the targets in addition to the target deformations. The dashed lines represent the calculated results without taking into account the inelastic coupling and also the target deformations.

A target of ^{232}Th (the thickness of 400 μg/cm^2) was bombarded by ^{30}Si beams of 150.2, 152.2, 153.8, 155.4 and 157.4 MeV from the tandem accelerator of the Japan Atomic Energy Research Institute. The evaporation residues emitted from the target to the beam direction were separated in flight from the primary beams by the recoil mass separator (JEARI-RMS) (3) and implanted in a double-sided position sensitive strip detector. Their energies and subsequent α-decays were measured. The measured xn cross sections were compared with the calculated results of the statistical model using the code HIVAP (4). The fusion cross section which was the input quantity to the HIVAP code was calculated using the code CCDEF (5) by taking into account the static deformation of the target and the coupling of inelastic excitations of the projectile and the target to the fusion process. The calculated fusion cross sections were compared with the measured fission cross sections for the reactions ^{232}Th + ^{28}Si and ^{238}U + ^{28}Si as shown in Fig. 1. Since the calculated results agreed with the data very well, the fusion cross sections calculated by the CCDEF code were used for all the reactions considered here.

In the statistical model calculation, the level density parameters a_n at the ground state deformation and a_f at the saddle point deformation were varied with the excitation energy according to the prescription of (6). The fission barrier heights were adjusted to reproduce the $6n$ channel cross sections measured in the fusion reactions with ^{232}Th and ^{238}U targets. The calculated fusion and evaporation residues cross sections are shown in Fig. 2a for the reaction ^{232}Th + ^{30}Si together with the observed data. Since the calculated fusion barrier is lowered by the amount of more than 10 MeV from the spherical Coulomb barrier (Bass barrier) due to the collision of the projectile with the tip of the deformed target, the statistical model predicts the sizable $3n$ and $4n$ cross sections below the Coulomb barrier. In the present experiment, the evaporation residues originated from the $3n$ or $4n$ channels were not observed in the energy region of $E_{cm} <$

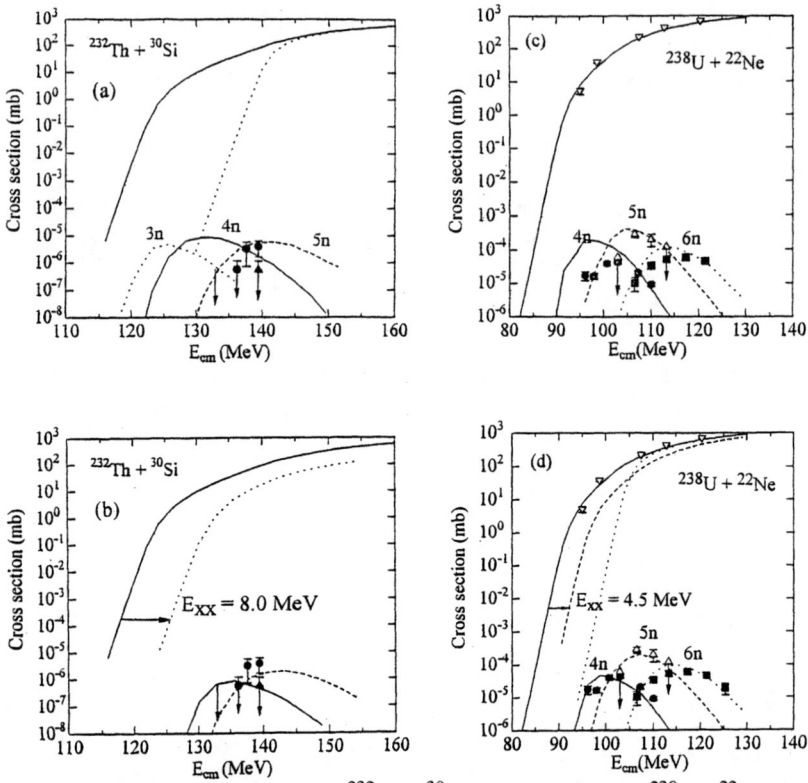

FIGURE 2. xn cross sections for the reactions ^{232}Th + ^{30}Si (present work) and ^{238}U + ^{22}Ne (7). (a) The measured 4n and 5n are shown as the solid triangle and dot, respectively. The calculated 3n, 4n and 5n cross sections are shown as the dotted, solid and dashed lines, respectively. (b) The calculated results with E_{XX} = 8 MeV are shown. (c) The same as (a) in the reaction ^{238}U + ^{22}Ne together with the fission cross sections (open triangle) of (10). (d) The calculated results with E_{XX} = 4.5 MeV.

138 MeV. This fact means that the projectile and the deformed target can not completely fuse together as predicted by the calculation in this energy region. It was also found that an additional energy of 8 MeV above the calculated fusion barrier was needed to reproduce the observed 4n and 5n cross sections as shown in Fig. 2b.

We also compared the xn cross section data for the various reactions with deformed targets with the present calculations. The typical example is shown in Fig. 2c in the fusion reaction ^{238}U + ^{22}Ne (7). The calculated 4n and 5n cross sections are overestimated by an order of magnitude at the low energy compared with the measured data. If the fusion barrier is shifted up to the high energy by the amount of 4.5 MeV, the agreement between the calculated result and the data becomes excellent as shown in Fig. 2d. We can determine the energy shift to reproduce the xn cross sections for the other reaction systems (7-8). Here we define this energy shift as the extra-extra push energy E_{XX}. The obtained E_{XX} are shown in Fig. 3a as a function of the mean fissility parameter x_m. As for the comparison, the extra-extra push energies obtained for asymmetric reaction systems with spherical targets (9) are also shown as open circles. The E_{XX}

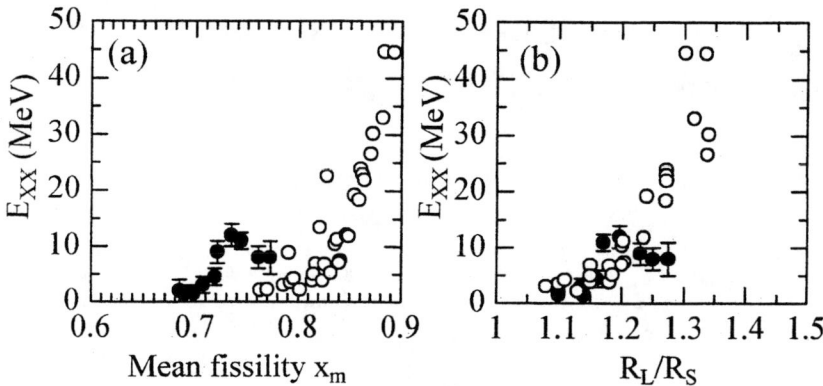

FIGURE 3. (a) Obtained extra-extra push energies (solid points) are plotted as a function of the mean fissility. The open points are the extra-extra push energies for asymmetric reaction system with spherical targets (10). (b) Extra-extra push energies are plotted as a function of R_L/R_S (see text).

values obtained in the present analysis for the fusion reactions with deformed targets shows a threshold value x_m = 0.70. This threshold value is smaller than 0.78 obtained for the asymmetric reaction systems with spherical targets.

In order to understand the difference of the reaction mechanism between the reaction systems with spherical targets and deformed targets, we plotted E_{XX} as a function of a parameter R_L/R_S, where R_L is the center-to-center distance in the collision of a projectile with the tip of a prolate target and R_S the saddle point radius of the compound nucleus. The R_S was calculated by the rotating liquid drop model. As shown in Fig. 3b, the extra-extra push energy is well parameterized by the quantities R_L/R_S with the common threshold value 1.1. This means that the relative value of R_L to R_S is important for the fusion of heavy reaction system and the compact di-nuclear shape at the contact point has a tendency to evolve more easily into the formation of the compound nucleus.

REFERENCES

1. Samant, A. M. and Kailas, S., *Z. Phys.* **A356** (1996) 309 and see references therin.
2. Hinde, D. J., et al., *Phys. Rev. Lett.* **74** (1995) 1295.
3. Ikezoe, H., et al., *Nucl. Instr. and Meth.* **A376** (1996) 420.
4. Reisdorf, W. and Schadel, M., *Z. Phys.* **A343** (1992) 47.
5. Fernandez, J., Dasso, C. H. and Landowne, S., *Comp. Phys. Comm.* **54** (1989) 409.
6. Ignatyuk, A. V., Smirenkin, G. N. and Tishin, A. S., *Yad. Phys.* **21** (1975) 485.
7. Donets, E. D., Shchegolev, V. A. and Ermakov, V. A., *Sov. Jour. Nucl. Phys.* **2** (1966) 723.
8. Amdreyev, A. N. et al., *Z. Phys.* **A345** (1993) 389, Akap'ev, G. N. et al., *Atomnaya Energiya* **21** (1966) 243, Reisdorf, W. R. et al., *Nucl. Phys.* **A438** (1985) 212, Sikkeland, T. et al., *Phys. Rev.* **C1** (1970) 1564, Vermeulen, D. et al., *Z. Phys.* **A318** (1984) 157, Sahm, C.-C et al., *Nucl. Phys.* **A441** (1985) 316, Donets, E. D. et al., *Sov. Phys. JETP* **16** (1962) 7, Sikkeland, T. et al., *Phys. Rev.* **169** (1968) 1000, Mikheev, V. L. et al., *Atomnaya Energiya* **22** (1964) 90, Lazarev, Yu. A. et al., *Phys. Rev. Lett.* **73** (1994) 624.
9. Froblich, P., *Phys. Lett.* **B215** (1988) 36.
10. Viola, V. E. and Sikkeland, T., *Phys. Rev.* **128** (1962) 767.

Properties of two neutron-hole ^{130}Sn and two neutron-particle ^{134}Sn

V. I. Isakov [1], K. I. Erokhina [2], B. Fogelberg [3] and H. Mach [3]

[1] *St. Petersburg Nuclear Physics Institute, Russian Academy of Sciences, 188350 Gatchina, Russia*
[2] *Ioffe Physicotechnical Institute, Russian Academy of Sciences, 194021 St. Petersburg, Russia*
[3] *Department of Neutron Research, Uppsala University, S-61182 Nyköping, Sweden*

Abstract: The excitation energies and E2 transition rates in ^{130}Sn and ^{134}Sn – nuclei having either two neutron holes or two neutron particles above the doubly magic core of ^{132}Sn, are calculated in the framework of the particle-particle RPA method. The calculations are strongly grounded on the most recent experimental data on the structure of single neutron states in the odd-neutron nuclei close to ^{132}Sn. The discussion of the results is focused on the inter-comparison of the effective charges in the *^{132}Sn core ± 2 neutrons* systems.

INTRODUCTION

The present work is part of systematical theoretical studies (1–5) of simple nuclear systems close to ^{132}Sn. Recent experimental investigations dramatically improved the knowledge and characteristics of the isotopes at the doubly magic ^{132}Sn. In particular, study of the excited states in ^{133}Sn (6) has established the neutron single particle energies of states above the shell gaps at Z=50 and N=82. The presence of new experimental data has motivated us to undertake a new theoretical study of the nuclei close to ^{132}Sn that have more than one neutron above the ^{132}Sn core. Here we investigate ^{130}Sn and ^{134}Sn – nuclei having either two neutron holes or two neutron particles above the doubly magic core of ^{132}Sn.

ENERGY SPECTRA

The energy spectra of ^{130}Sn and ^{134}Sn have been calculated using the RPA particle-particle method well proven in our previous studies (7–10). The neutron single-particle basis included 8 levels below and 8 states above the Fermi level (including also some quasi-stationary states). We took the experimental single particle energies, when available, and otherwise the energies generated by the phenomenological potential of the Woods-Saxon type. The effective interaction (Interaction I) had been previously derived in calculations on nuclei close to ^{208}Pb (7,8). In order to investigate the sensitivity of our results on the choice of the effective interaction, we have repeated the calculations using effective forces (Interaction II) defined in (11,12) for the shell-model description of the multiproton systems with Z>50. A comparison of the calculated and experimental excitation energies shows that the theoretical spectra for the Interactions I and II qualitatively lead to similar results. However, Interaction I seems to provide a better description of the experimental energies, particularly for the ground states, which in the case of Interaction II occur to be strongly overbound. The level energies calculated with Interaction I are in a good agreement with our previous calculations for ^{130}Sn (13) but not for the levels with excitation energies above 1.5 MeV in ^{134}Sn. This reflects the fact that our present calculations incorporate the experimental neutron single particle energies above the N=82 shell, and thus represent a more precise approach in comparison to (13), where theoretical energies and more limited basis were used.

E2 TRANSITION PROBABILITIES

We have calculated the E2 electromagnetic transition probabilities in ^{130}Sn and ^{134}Sn. The RPA method for the "magic core ± 2n" nuclei includes the correlations between the two particles, as well as correlations of 4 particles – 2 holes, 6 particles – 4 holes, and so on, but does not take into account the 1 particle – 1 hole admixtures to the two-particle state. This leads to the inevitable introduction of an effective charge, which (in similarity to the simplest systems with either one particle or hole above the core) accounts for the particle-hole core polarization. It is known from our previous work (5) that $e_p(E3) \sim 3.1$ |e|, $e_p(E2) \sim 1.7$ |e| in the region of ^{208}Pb and $e_p(E3) \sim e_p(E2) \approx 1.9$ |e| near ^{132}Sn.

EFFECTIVE CHARGES

The present investigation provides an opportunity to define the effective charges for neutrons in the neutron-rich doubly magic region of ^{132}Sn. By now we have rather extensive experimental information on the spectra and transition probabilities in ^{130}Sn (see Refs.(14,15)). Combined with our most recent experimental results, there are several experimentally established E2 transition rates in ^{130}Sn. The B(E2) rates for the $10_1^+ \rightarrow 8_1^+$ and $5_1^- \rightarrow 7_1^-$ transitions (equal to 15 e^2fm^4 and 52 e^2fm^4, respectively) do not only have very small experimental uncertainties but also (especially for the first transition) refer to states with small configuration mixing where the theoretical description is very reliable. These rates yield the values of effective charge equal to $e_n(E2) = 0.87$ |e| and 0.90 |e|, respectively, which are almost identical to the value of $e_n(E2) \sim 0.9$ |e| for nuclei close to ^{208}Pb (16).

Recently first experimental information became available on the "magic core + 2n" nucleus of ^{134}Sn (17). Our calculations, which use the value of effective charge taken from the ^{130}Sn data discussed above, $e_n(E2) = 0.87$ |e|, give the value of B(E2) ≈ 60 e^2fm^4 for the $6_1^+ \rightarrow 4_1^+$ transition in ^{134}Sn, which is much higher than the experimental quantity of 36(7) e^2fm^4 (17). This disagreement means in reality that the value of the effective charge defined for ^{134}Sn from the intercomparison of the present calculations and the experimental data (17), should be much lower and equal to $e_n(E2) = 0.67$ |e|.

One would like to comment on the fact that the effective charge of $e_n(E2) = 0.67\,|e|$ obtained here for ^{134}Sn, differs significantly from the value of $e_n(E2) = 1.01(10)\,|e|$ defined in (17) for the same transition. There are two contributing factors: the calculations (17) were held in the diagonal approximation and also using the value of $\langle v2f7/2\,|r^2|\,v2f7/2\rangle = 32$ fm^2. In our computations we took into account correlations, i.e. configuration mixing, which rather strongly enhance the $6_1^+ \to 4_1^+$ transition in ^{134}Sn. In ^{134}Sn the situation is different from that in ^{130}Sn, where the 10_1^+, 8_1^+ and partly 6_1^+ states (but not the 4_1^+, 2_1^+ and 0_1^+ ones) are practically pure components of the $\{v1h11/2\}^2$ multiplet. Secondly, the parameters of the single particle field used in our calculations give $\langle v2f\,7/2\,|r^2|\,v2f7/2\rangle = 36.09$ fm^2.

On the other hand, the disagreement between the values of $e_n(E2)$ equal to about $0.87\,|e|$ and $0.67\,|e|$ obtained in the present study from the $10_1^+ \to 8_1^+$ and $5_1^- \to 7_1^-$ transitions in ^{130}Sn and the $6_1^+ \to 4_1^+$ transition in ^{134}Sn, has a simple explanation within the idea of an effective charge, if one notes that the effective charge in a finite system remains only approximately constant. Using an example of the particle-surface vibrations model, one can easily obtain an expression defining the effective charge for the $n_1l_1j_1 \to n_2l_2j_2$ $E\lambda$-transition in the $\omega_\gamma \to 0$ limit as:

$$\left[e_{eff} - |e|\left(\frac{1}{2} - t_3\right)\right] \langle n_2l_2j_2|r^\lambda|n_1l_1j_1\rangle = -\frac{2V_0}{\omega_\lambda}\left[\beta_\lambda^0\right]^2 \frac{3|e|ZR^\lambda}{4\pi}$$
$$\times \langle n_2l_2j_2|R\frac{df}{dr}|n_1l_1j_1\rangle; \qquad f = \frac{1}{1+\exp\left(\frac{r-R}{a}\right)}. \qquad (1)$$

Here ω_γ is the phonon energy of the core and β_λ^0 is the dimensionless amplitude of the core zero-point vibrations of multipolarity λ, connected with the probability of the $E\lambda$ transition in the core by the relation:

$$B(E\lambda; \omega_\lambda \to 0_1^+) = \left(\frac{3eZR^\lambda}{4\pi}\right)^2 \left[\beta_\lambda^0\right]^2. \qquad (2)$$

Using the value of $e_n(E2) = 0.87\,|e|$ for the $1h11/2 \to 1h11/2$, $2d3/2 \to 2d3/2$, $3s1/2 \to 2d3/2$ and $2d5/2 \to 2d3/2$ single particle transition matrix elements involved in the $10_1^+ \to 8_1^+$ and $5_1^- \to 7_1^-$ transitions in ^{130}Sn we obtain from eqn.(1) the $[\beta_\lambda^0]^2$ values equal to $\sim 1.5 \times 10^{-3}$, 1.7×10^{-3}, 1.7×10^{-3} and 1.7×10^{-3}. This corresponds to the transition rates from the 2^+ phonon state to the ground state in ^{132}Sn equal to 9–10 Wu. From the side of ^{134}Sn, using the value of $e_n(E2) = 0.67\,|e|$ for the $2f7/2 \to 2f7/2$ matrix element involved in the $6_1^+ \to 4_1^+$ transition, we obtain $[\beta_\lambda^0]^2 \sim 1.8 \times 10^{-3}$, which corresponds to the B(E2) value in ^{132}Sn equal to ~ 11 Wu. Thus we see that although the effective charges in ^{130}Sn and ^{134}Sn are different, they correspond to the same E2 transition rate in ^{132}Sn. Inversely, taking the value of the E2 transition rate in the core nucleus equal to ~ 9 Wu we obtain from eqn.(1) the magnitudes of $e_n(E2)$ in

^{130}Sn for single particle orbital ν1h11/2 equal to about 0.9 |e| and e_n(E2) in ^{134}Sn for the ν2f7/2 orbital about 0.6 |e|.

One should note, that a smaller value of an effective neutron charge for the outer orbitals than for the inner ones, is quite natural in nuclei close to the neutron drip line. So, for the ν3p3/2 orbital having a very small binding energy the mean square radii is very large, ⟨ν3p3/2 |r²| ν3p3/2⟩ = 58.57 fm², while the radial integral entering the right-hand side of eq.(1) is relatively small (~0.7), thus by using the E2 transition rate in ^{132}Sn equal to 9 Wu., we obtain e_n(E2) ~ 0.3 |e| !

The present estimate of B(E2; $2_1^+ \rightarrow 0_1^+$) in ^{132}Sn of ~9 Wu. is close to our previous estimate (5) derived in the study of ^{134}Te, as well as to the calculations (19) where the continuum was included into consideration.

SUMMARY

Our new experimental results on the level scheme and transition probabilities in ^{130}Sn, as well as the experimental data on ^{134}Sn obtained in the recent experimental works of other authors, are well described in the framework of theoretical approach based on the particle-particle RPA approximation. Our results manifest the tendency of a decreasing value of the neutron effective charge in the neutron rich nuclei when approaching the neutron drip line. A full presentation of our results is given in (18).

REFERENCES

1. Mach, H., *et al.,* Phys. Rev. C **51**, 500 (1995).
2. Erokhina, K. I., and Isakov, V. I., Yad. Fiz. **57**, 212 (1994).
3. Erokhina, K. I., and Isakov, V. I., Yad. Fiz. **59**, 621 (1996).
4. Erokhina, K. I., *et al.,* PNPI Report **No. 2136**, 36 (1996).
5. Omtvedt, J. P., *et al.,* Phys. Rev. Lett. **75**, 3090 (1995).
6. Hoff, P., *et al.,* Phys. Rev. Lett. **77**, 1020 (1996).
7. Isakov, V. I., *et al.,* Izv. AN SSSR, ser. fiz. **41**, 2074 (1977).
8. Artamonov, S. A., *et al.,* Izv. AN SSSR, ser. fiz. **48**, 71 (1984).
9. Artamonov, S. A., *et al.,* Yad. Fiz. **36**, 829 (1982).
10. Artamonov, S. A., *et al.,* Yad. Fiz. **45**, 33 (1987).
11. Waroquier, M., and Heyde, K., Nucl. Phys. A **164**, 113 (1971).
12. Heyde, K., and Waroquier, M., Nucl. Phys. A **167**, 545 (1971).
13. Erokhina, K. I., and Isakov, V. I., Izv. RAN, ser. fiz. **56**, 78 (1992).
14. Sergeenkov, Yu. V., Nucl. Data Sheets **58**, 765 (1989).
15. Jerrestam, D., *et al.,* Prog.Rep. from Studsvik Neutron Res.Lab. (1990-1991) p. 90
16. Decman, D. J., *et al.,* Phys. Rev. C **28**, 1060 (1983).
17. Zhang, C. T., *et al.,* Z. Phys. A **358**, 9 (1997).
18. Erokhina, K. I., *et al.,* PNPI Report **No. 2225** (1998).
19. Saperstein, E. E., *et al.,* Elem. Part. and Nucl. **9**, 221 (1978)

Study of the First Excited $K^\pi=0^+$ Band in ^{162}Dy

Benyuan Liu,[1] N.V. Zamfir,[1,2,3] D.S.Brenner,[2] R.F. Casten,[1] G. Cata-Danil,[1,3] C.W. Beausang,[1] R. Krücken,[1] J.R. Cooper,[1] J.R. Novak,[1] C.J. Barton,[2] R.L. Gill[4]

[1] *WNSL, Yale University, New Haven, CT 06520*
[2] *Clark University, Worcester, MA 01610*
[3] *Institute of Atomic Physics, Bucharest Magurele, Romania*
[4] *Brookhaven National Laboratory, Upton, NY 11973*

Abstract. There has been recently much discussion about the nature of the first excited $K^\pi=0^+$ band in deformed nuclei. An experiment ^{159}Tb$(\alpha,n)^{162}$Ho was conducted at Yale to study the $K^\pi=0_2^+$ band in ^{162}Dy. Preliminary data analysis results of the experiment and future plans with the moving tape collector (MTC) will be discussed.

The nature of the excited $K^\pi=0^+$ bands in deformed nuclei remains enigmatic, despite much study, both experimental and theoretical. Traditionally, this band was interpreted as a β-vibrational band, but there has recently been much discussion about its nature (1). It could also be a γ-γ double phonon state or a two-quasi-particle state or some mixture. Each of these possibilities has different decay properties.

To investigate the character of the $K^\pi=0_2^+$ band, the transitions to the other low-lying positive parity bands, such as $K^\pi=0_1^+$ (g.s. band) and $K^\pi=2^+$ (γ band) are of great importance.

If the $K^\pi=0^+_2$ band is a pure β-vibration, the phonon selection rules for E2 transitions allow decay to the ground state band by destroying one β phonon. The transition to the γ-band is forbidden since the process requires destroying one β phonon and creating one γ phonon simultaneously. If the $K^\pi=0^+_2$ band has a pure double-γ character, the transition to the γ-band should have a collective E2 matrix element but the decay to the ground state band is forbidden due to a similar argument. Therefore, the branching ratio $R^0_{\gamma g}$

$$R^0_{\gamma g} = \frac{B(E2; 0_2^+ \to 2_\gamma^+)}{B(E2; 0_2^+ \to 2_1^+)}$$

is a key signature of the character of the 0_2^+ level. If the ratio is greater than 1, it may indicate a double-γ character, whereas several kinds of intrinsic excitation could produce branching ratios less than 1. Similar ratios could be defined also for the other members of the $K^\pi = 0^+_2$ band.

TABLE 1. Results from different experiments for transitions from the $K^\pi = 0^+_2$ band to the γ band and ground state band in ^{162}Dy showing the discrepancy for the $2^+(K=0_2) \rightarrow 4^+_\gamma$ transition.

Transitios	γ-ray energies (keV)	Relative intensities		
		(n n' γ) (ref. 3)	β decay (ref. 4)	β decay (ref. 5)
$2^+_{K=0} \rightarrow 4^+_\gamma$	392.8	--------	--------	54.6(79)
$2^+_{K=0} \rightarrow 4^+_g$	1187.5	69.4(70)	67.1(28)	71.2(67)
$2^+_{K=0} \rightarrow 2^+_g$	1372.5	100(7)	100(4)	100(7)
$2^+_{K=0} \rightarrow 0^+_g$	1453.2	5.8(35)	3.6(3)	4.3(24)

To measure the branching ratios, it is necessary to measure the γ rays connecting the $K^\pi = 0^+$ band and the γ-band. Usually these two bands lie close in energy so that the E_γ^5 dependence of the E2 transition rate attenuates those γ rays relative to other higher energy γ-decay transitions even if the intrinsic matrix element is large, thus making such studies difficult. In many cases, the $K^\pi=0^+$ bands can be populated by β decay. However, many existing β decay experiments were conducted one or two decades ago with first generation γ-ray detectors and at a time when not much attention was paid to the study of the decay of $K^\pi = 0^+_2$ bands to the γ-band.

Today, modern large high efficiency Ge detectors (2) and data analysis techniques for γ-γ coincidence spectra make it easier to establish decay schemes and extract branching ratios. Hopefully, such long unsolved questions can be addressed anew and resolved.

For ^{162}Dy, several different types of experiments (3,4,5) have been conducted to study the level scheme and structure of the nucleus. However, the data are not consistent among these experiments (Table 1). For example, the γ-ray of 392.8 keV decaying from $2^+(K=0_2)$ to 4^+_γ was reported in one β decay experiment (5), but it was not seen in the other β decay experiment (4) or in the (n, n' γ) experiment (3). Since this γ-ray connects the $K^\pi = 0^+_2$ band to the γ-band, the discrepancy leads to a different branching ratio $R^0_{\gamma g}$, and hence a different interpretation of the nature of the $K^\pi = 0^+_2$ band.

It is the purpose of this contribution to study the characteristics of the $K^\pi = 0^+_2$ band and clarify the discrepancies among previous experiments on ^{162}Dy. In the experiment performed at Yale, ^{162}Dy was populated via the electron capture decay of ^{162}Ho, which was produced at the WNSL tandem accelerator by the ^{159}Tb(α,n)^{162}Ho reaction at a beam energy of 20 MeV. One clover detector and one 72% Ge detector from the YRAST Ball array (6) were used to measure the decay γ-rays.

A partial level scheme of ^{162}Dy established from our study based on γ-γ coincidence spectra is shown in Figure1. We identified several new γ-rays and levels which were not reported in previous β-decay experiments (4,5) and corrected some previous placements.

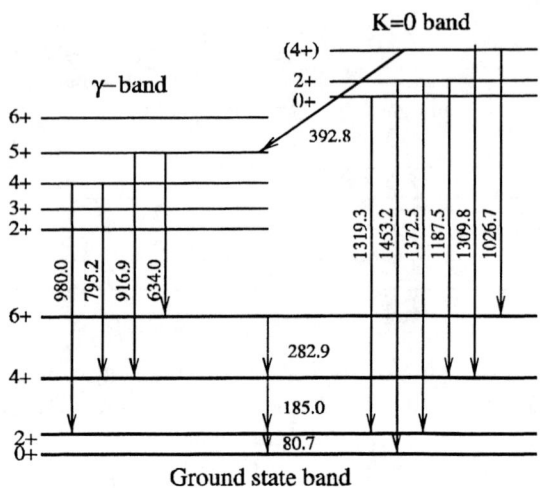

FIGURE 1. Partial decay scheme of ^{162}Dy.

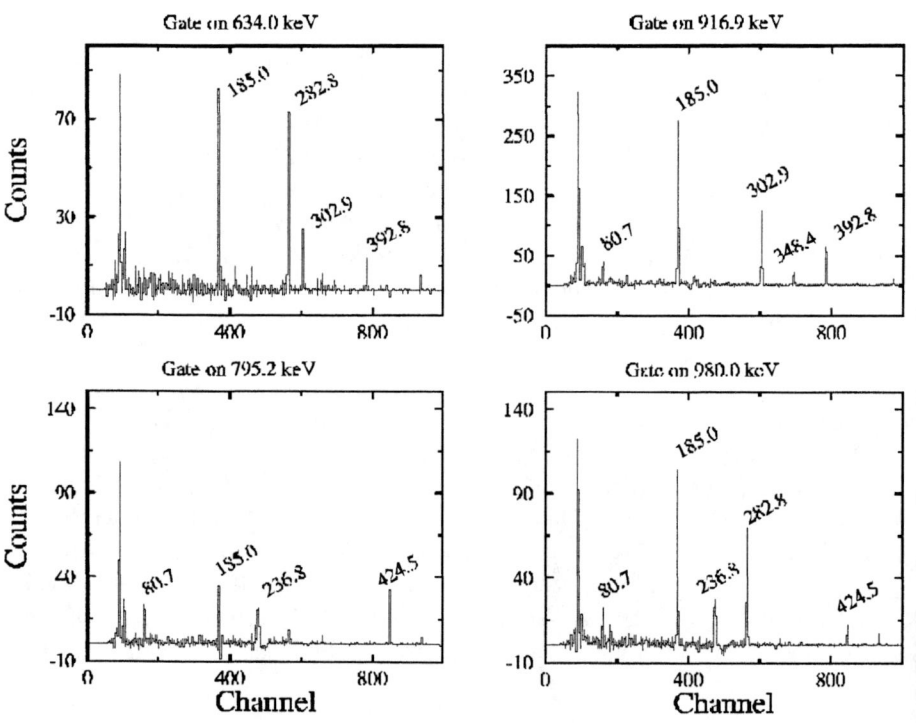

FIGURE 2. Coincident γ-ray spectra supporting the new placement of the 392.8 keV γ-ray.

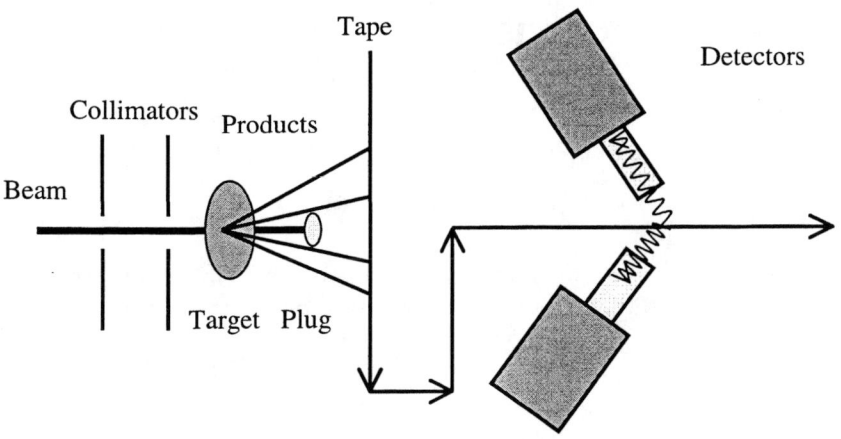

FIGURE 3. Schematic setup of the moving tape collector.

One of the main differences from the previous results is the placement of the 392.8 keV transition as a transition between the $4^+(K=0_2^+)$ and 5^+_γ levels. The gated spectra supporting this placement and ruling out the earlier $2^+(K=0_2^+) \rightarrow 4^+_\gamma$ placement are shown in Figure 2. Due to the existence of a 6^- level 1.3keV above the 4^+ level and the multiple doublets of γ-rays deexciting these two states (3), the spin assignment is tentative, as is the association of this level with the $K=0^+_2$ band. Assuming that the 392.8 keV transition deexcites the $4^+(K=0^+_2)$ level, we calculated the B(E2) ratio from the $4^+(K=0_2)$ level to the 5^+_γ and 4^+(g.s) levels. This gives a matrix elements ratio of 16, which suggests that there may be a sizeable γ-γ double phonon state character in the $K^\pi = 0^+_2$ band of ^{162}Dy.

A moving tape collector (MTC) system was recently built at Yale (Figure 3). The recoil products of fusion-evaporation reactions are deposited in the aluminum layer of the moving tape which transfers the activity to a low-background counting station. To prevent the primary beam hitting and burning the tape, a plug is inserted downstream of the target . Most of the reaction products have a component of transverse momentum and will bypass the plug which is slightly bigger than the dimensions of the beam. This device will be used to make a systematic study of the first excited $K^\pi = 0^+$ bands in deformed rare-earth nuclei.

Work supported by USDOE contract numbers DE-FG02-91ER40609, DE-FG02-88ER40417, and DE-AC02-76CH00016.

References

1. R.F. Casten and P.von Brentano, Phys. Rev. C **50**, 1280(1994)
2. C.W. Beausang and J Simpson, J. Phys. G: Nucl. Part. Phys. **22,** 527(1996)
3. J. Berzins et al, Nucl. Phys. A**584**, 413(1994)
4. L.O. Edvardson et al, Nucl. Phys. A**252**, 103(1975)
5. J.L. Wood and D.S. Brenner, Nucl. Phys. A**174**, 353(1971)
6. C.W. Beausang et al, to be published.

Shape coexistence in the light Po isotopes

A. M. Oros[a,b], K. Heyde[a], C. De Coster[a1], B. Decroix[a]
R. Wyss[c], B. Barrett[a,d] and P. Navratil[d]

[a] *Institute for Theoretical Physics, Proeftuinstr. 86, B-9000 Gent, Belgium*
[b] *Institute for Nuclear Physics and Engineering, Bucharest-Magurele, Romania*
[c] *KTH Stockholm, Sweden*
[d] *Department of Physics, University of Arizona, Tucson, AZ, 85721 USA*

Abstract. We discuss the experimental data on the neutron-deficient Po isotopes in the light of the shape coexistence phenomenon. The existing level systematics can be viewed as resulting from the interplay of two structures with different deformation and their mixing. Combined results of several nuclear models support this interpretation.

Shape coexistence is well documented in the Pb region [1] and supported by a large body of experimental data. Nevertheless, the situation is significantly clearer for the nuclei with $Z \leq 82$ than in the region above the closed proton shell, partly due to the experimental difficulties met in the study of very neutron-deficient nuclei.

Recent experimental data [2-10] have shed new light on the nuclear structure of the neutron deficient Po isotopes. The evolution with the neutron number of the different aspects which come into play in this region can now be followed from the closed shell nucleus ^{210}Po (N=126) – a typical example for a "two-proton" system – down to ^{192}Po (N=108). A clear change in the trend followed by the excitation energy of the yrast and yrare levels can be noticed below mass number A=200, as illustrated in Fig. 1.

As to the cause for this change, two interpretations have been offered until now: (i) one based on the appearance of shape coexisting configurations [3,4,7-9], as substantiated by the variation in the excitation energy of the 0_2^+ state; and (ii) the other based on Particle-Core Model [11,12] calculations [13,14], expecting an increase in collectivity with increasing number of valence particles and the subsequent evolution towards stable deformation in the ground-state [2,5,10].

A careful inspection of the Po isotopes in the Particle-Core Model (PCM) [11,12] performed in the present study leads to quite different conclusions than the ones reached in Refs. [14,10,13]. The model describes two interacting particles (protons

---

[1]) postdoctoral fellow of the Fund for Scientific Research - Flanders (Belgium)

in this case) coupled to vibrational quadrupole and octupole states of a closed-shell (Pb in this case) core. The different conclusions reached are most probably due to the larger configuration space used in the present calculations, as well as to the different approach used in the determination of the model parameters (single-particle energies, coupling strengths, phonon energies) for the calculations on the Po nuclei.

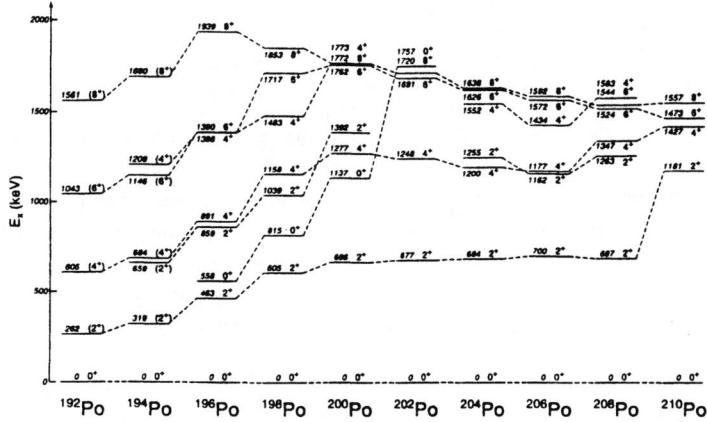

FIGURE 1. Systematics of the yrast and yrare levels of the light Po isotopes. Figure reproduced from Ref. [14].

In Refs. [13,14], *zero* coupling of the two protons to the vibrational core is obtained for all the isotopes above A=200. By comparing the present calculations with the experimental data it can be concluded that the effects of the Particle-Core Coupling are clearly present in the Po isotopes with mass number A=208–200 [15]. These effects are even more clearly seen in the odd-mass Bi isotopes involving the same Pb cores, which were studied simultaneously [15] in the framework of the Particle-Core Model. Below mass A=200, the experimental data on the Po as well as on the Bi isotopes become too scarce to allow for a clear-cut determination of the model parameters. The evolution of the yrast line as a function of the different model parameters is studied with the aim to determine if the experimental yrast states of ^{192}Po can be, by any means, described by this model. The results are summarized in Figure 2. The only parameter combination which gives a rather good description of the yrast line of ^{192}Po involves an extremely low energy ($\hbar\omega_2 \simeq 350$ keV) of the quadrupole phonon in the Pb core. Since $\hbar\omega_2$ should be approximately equal to the experimental energy of the spherical 2^+ state in the core ^{190}Pb, which lies above 1 MeV [16], this solution is clearly unacceptable. The Particle-Core Model, which describes very satisfactorily the heavier Po isotopes (A=200–210), as well as the adjacent Bi isotopes (A=201–209), is most probably not able to account for the mechanism which produces the noticed important perturbation in the experimental systematics (see Fig. 1).

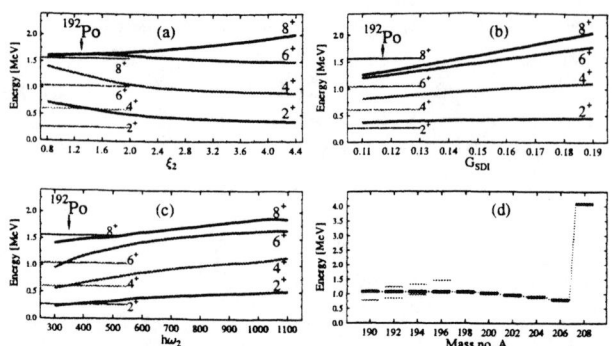

FIGURE 2. Evolution of the energies of the yrast 0^+–8^+ states as a function of the different model parameters: quadrupole coupling strength ξ_2 in part (a), strength of the Surface Delta Interaction G_{SDI} in part (b) and energy of the quadrupole phonon $\hbar\omega_2$ in part (c). The experimental energies of the yrast states in ^{192}Po are shown for comparison on the left side of each figure, plotted with dashed lines. In part (d) we show the evolution with mass number of the energy of the spherical 2^+ state in the Pb isotopes. The energy was corrected for the effects of the mixing between the spherical and intruder 2^+ states, as described in Ref. [16].

In an alternative approach, we have performed Potential Energy Surface (PES) calculations using the method outlined in Ref. [17] and Refs. therein. The predictions concerning energy minima are summarized in Fig. 3. A nearly spherical minimum is present in all the isotopes studied, but it is replaced by the oblate minimum in ^{192}Po as the ground-state configuration. Prolate minima appear as well, in the very neutron-deficient isotopes. The energies predicted for the oblate minima, relative to the nearly spherical ones, correspond very well to the experimental energies of the 0_2^+ states which have been observed in $^{202-196}$Po [7,8]. The PES calculations are in good agreement with the experimental data for the Pb and Hg isotopes as well [17] and seem to have remarkable quantitative predictive power in this mass region.

Following these predictions, we performed a mixing calculation for the region which covers the transition from the "Particle-Core" regime, observed above mass A=200, to the more collective one observed in ^{192}Po. Besides the experimental energies of the yrast and yrare states, hindrance factors measured in α-decay experiments were used together with a method [4,8] which allows to estimate the mixing between the 0^+ configurations in a Po parent, once the mixing of the analogous configurations in the Pb daughter is known. The unperturbed energies of the spherical and intruder quasi-bands and the mixing matrix elements can be determined. The 'reconstructed' spherical states show good agreement with the PCM predictions. The above shape coexistence can also be described by the proton-neutron Interacting Boson Model, using the configuration mixing approach of Ref. [18] for a spherical ground state band and an oblate intruder band.

In conclusion, the present study strongly enforces the interpretation of the various

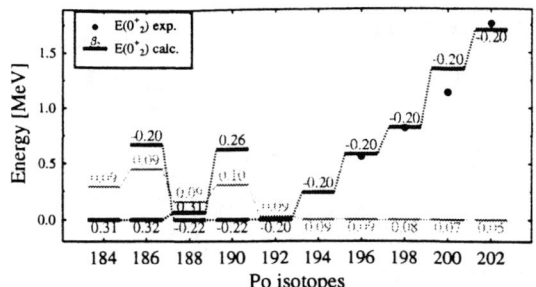

FIGURE 3. Summary of the predictions of the Potential Energy Surface calculations for $^{202-184}$Po. The oblate configuration is predicted to replace the nearly spherical configuration as the ground state in ^{192}Po, and be afterwards replaced by the strongly deformed prolate configuration in ^{186}Po. No oblate minimum is found in the PES for the nuclei lighter than ^{186}Po. The experimental 0_2^+ states are plotted with dots.

aspects met in the light Po isotopes in terms of shape coexistence. Oblate states are strongly lowered in energy with decreasing mass number and their gradually increasing mixing with the spherical configurations seems to be able to account for the situation observed in experiment.

REFERENCES

1. J. Wood, K. Heyde, W. Nazarewicz, M. Huyse and P. Van Duppen, Phys. Rep. 215, 101(1992).
2. A. Maj et al., Nucl. Phys. **A509**, 413(1990).
3. D. Alber et al., Z. Phys. **A 339**, 225(1991).
4. J.Wauters et al., Phys.Rev. **C50**, 2768 (1994).
5. L.A. Bernstein et al., Phys.Rev. **C52**, 621 (1995).
6. W. Younes et al., Phys.Rev. **C52**, R1723 (1995).
7. N. Bijnens et al., Phys. Rev. Lett. **75**, 4571(1995).
8. N.Bijnens et al., Phys.Scr. T56, 110 (1995).
9. K. Helariutta et al., Phys. Rev. **C54**, R2799(1996).
10. N. Fotiades et al., Phys. Rev. **C55**, 1724(1997).
11. A. Bohr and B. Mottelson, *Nuclear Structure*, Vol. 2, W. A. Benjamin Inc. , London, Amsterdam, 1975.
12. K. Heyde and P.J. Brussaard, Nucl. Phys. **A104**, 81(1967).
13. J.A.Cizewski and W.Younes , Z.Phys. A **358**, 133 (1997).
14. W.Younes and J.A.Cizewski, Phys.Rev. **C55**, 1218 (1997).
15. A.M. Oros et al., to be published.
16. P.Van Duppen, M. Huyse and J.L. Wood, J. Phys. G **16**, 441 (1990).
17. W. Nazarewicz, Phys. Lett. B **305**, 195(1993).
18. P.D. Duval and B.R. Barrett, Nucl. Phys. **A376**, 213(1982).

Fine Structure in ^{192}Po α-decay and Shape Coexistence in ^{188}Pb

R.D. Page[1], R.G. Allatt[1], T. Enqvist[2,*], K. Eskola[3], P.T. Greenlees[1], P. Jones[2], R. Julin[2], P. Kuusiniemi[2], M. Leino[2], J. Uusitalo[2,+]

[1] *Department of Physics, Oliver Lodge Laboratory, University of Liverpool, Liverpool L69 7ZE, UK.*

[2] *Department of Physics, University of Jyväskylä, P.O. Box 35, FIN-40351 Jyväskylä, Finland.*

[3] *Department of Physics, University of Helsinki, FIN-00014 Helsinki, Finland.*

Abstract. Excited $J^\pi=0^+$ states in ^{188}Pb populated in the α-decay of ^{192}Po have been identified through α-particle – conversion electron coincidences. The level scheme has been interpreted using a configuration mixing calculation, providing estimates of the mixing matrix elements, mixing amplitudes and the energies of unperturbed and unobserved levels.

INTRODUCTION

^{188}Pb is a pivotal nucleus for the shape coexistence of lead isotopes. In all cases investigated to date, the interactions have occurred between two coexisting shapes, but for ^{188}Pb a range of calculations predict *three* different shape configurations at relatively low excitation energies: a spherical ground state associated with the Z=82 shell closure, an oblate two particle- two hole (2p2h) proton configuration and a more strongly deformed prolate configuration predicted for N≤106, which can be associated with both 4p4h and 6p6h configurations [1]. ^{188}Pb is therefore potentially an exceptionally rich source of information on both shape coexistence and particle-hole excitations across the Z=82 shell gap.

α and β decay studies have mapped the systematics of first excited 0^+ states in lead isotopes, which are attributed to a deformed oblate minimum in the potential energy surface [2]. The minimum excitation energy in this sequence is expected to occur around ^{186}Pb (N=104) but at the time of the present experiment the measurements only extended as far as ^{190}Pb [3]. Extrapolating the systematics suggests an excitation energy of ∼580keV for this oblate state in the next even-even lead isotope, ^{188}Pb, placing it below the lowest 2^+ level at 724keV [4]. This

0^+ level would therefore represent the lowest-lying excited state in ^{188}Pb and could only decay by E0 conversion electron emission.

The in-beam γ-ray spectroscopy of ^{188}Pb has been studied by Heese et al., who observed a low-lying collective band and interpreted it as a deformed prolate configuration similar to that observed in the mercury isotone ^{186}Hg [4]. However, γ-decays to/from the 0^+ bandhead were not identified. Extrapolating the energy level sequence for the collective band in ^{188}Pb, one would expect the prolate 0^+ bandhead to lie at an excitation energy of $\sim$710keV. This state would also be expected to decay mainly by E0 conversion electron emission.

Although the non-yrast 0^+ levels were not strongly populated in the in-beam measurement, α-decay provides a suitable method for populating these low-lying low spin levels. The aim of the present work was therefore to identify α-decays of ^{192}Po to the prolate and oblate excited 0^+ levels in the daughter nuclide ^{188}Pb through α-particle − conversion electron coincidences.

EXPERIMENTAL DETAILS

The experiment was performed at Jyväskylä to study the α-decay fine structure of ^{192}Po nuclei produced as 4n evaporation residues in the reaction ^{36}Ar + ^{160}Dy $\longrightarrow$ ^{196}Po*. The average beam intensity was 38pnA, with the beam energy varied between 172MeV and 184MeV in 4MeV steps. The ^{36}Ar beam was pulsed (10ms on/10ms off) to allow clean decay energy spectra to be obtained during the beam off periods. The isotopically enriched 500μgcm^{-2} thick target comprised 67% ^{160}Dy.

The evaporation residues were separated from unreacted beam using the gas-filled separator RITU [5]. Evaporation residues passing through the separator undergo many charge changing collisions so they can be characterised by an average charge state, following the same average trajectory through the dipole magnet, rather than being dispersed according to their ionic charge state. Consequently RITU has a high transmission efficiency, which in the case of the present experiment was $\sim$25%.

The separated ions were implanted into a 35mm$\times$80mm position sensitive silicon strip detector, which was used to measure the position of their implantation and the position and energy of subsequent α-decays. All events were time stamped, allowing time gated α-particle energy spectra to be generated. The strip detector was surrounded by an array of six 500μm thick, 5cm$\times$5cm silicon detectors mounted perpendicular to the strip detector and used to detect conversion electrons in coincidence with α-particles emitted by the implanted nuclei.

RESULTS

The α-particle energy spectrum recorded in the strip detector during the beam off periods is shown in figure 1. The ^{192}Po α-decay line at 7.17MeV to the spherical ^{188}Pb ground state contains $\sim$13000 events. The large yields of heavier α-emitting polonium isotopes arise from reactions on isotopic contaminants in the target.

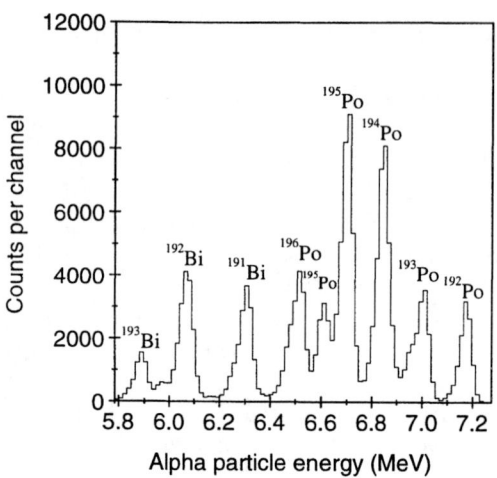

FIGURE 1. Energy spectrum of α-particles occurring within 200 ms of the implantation of an ion at the same location in the strip detector for beam off periods only.

Figure 2 shows the energies of α-particles measured in the strip detector in coincidence with a conversion electron in selected electron detector elements. The clear line at ~6.6MeV in the coincident spectrum has a half-life consistent with the value of 34ms for the ground state ^{192}Po line and fits in well with the systematics of *oblate* intruder states observed in radioactivity studies. There is also a line at ~6.4MeV, which agrees with the energy expected for the deformed *prolate* bandhead of the collective band [4].

The level scheme obtained by combining the 0^+ levels observed in the present work and the positive parity levels deduced from the in-beam spectroscopy study has been interpreted in terms of a three level mixing calculation following the procedure of Dracoulis [6]. These calculations provide estimates of the mixing matrix elements, as well as the energies of unobserved levels and of the levels before mixing. The spherical states mix with the other shapes with a mixing strength of ~50keV, while the oblate − prolate mixing is stronger at ~80keV. These values are compatible with values deduced for neighbouring nuclides and indicate that the ground state of ^{188}Pb is essentially pure, with admixtures of <1% of the other configurations, whereas the oblate 0^+ state comprises an admixture of ~20% of the prolate structure and vice versa.

These mixing admixtures for the 0^+ states can be combined with measured hindrance factors in order to deduce the mixing in ^{192}Po and ^{184}Hg [3,7]. For ^{184}Hg, the results are consistent with previous estimates, while for ^{192}Po the ground state appears to comprise a majority component of the 4p2h *intruder* configuration.

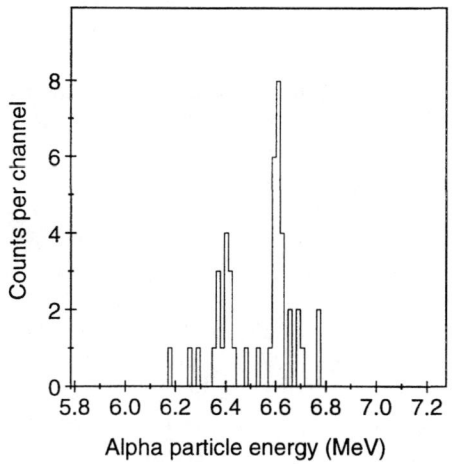

FIGURE 2. Energy spectrum of α-particles detected in coincidence with a low energy signal (≤1 MeV) in selected electron detector elements. This spectrum is for both beam phases since the coincidence requirement removes the beam related background.

ACKNOWLEDGMENTS

This work was supported by the Academy of Finland and by the Access to Large Scale Facility programme under the TMR programme of the European Union. RDP and PTG would like to acknowledge financial support from the U.K. Engineering and Physical Sciences Research Council and PJ acknowledges the receipt of a Marie Curie Research Training Grant. We are grateful to the staff at JYFL for providing excellent beams and technical support.

REFERENCES

* Present address: GSI, Planckstraße 1, D-64291 Darmstadt, Germany.
+ Present address: Argonne National Laboratory, Argonne, Illinois 60439, USA.
1. W. Nazarewicz, Physics Letters B**305** 195 (1993).
2. P. Van Duppen et al., Physics Letters B**154** 354 (1985).
3. N. Bijnens et al., Physica Scripta T**56** 110 (1995).
4. J. Heese et al., Physics Letters B**302** 390 (1993).
5. M. Leino et al., Nuclear Instruments and Methods in Physics Research B**99** 653 (1995).
6. G.D. Dracoulis, Physical Review C**49** 3324 (1994).
7. D.S. Delion et al., Physical Review C**54** 1169 (1996).

Some regularities in the production of isotopes in ^{32,34,36}S - induced reactions in the energy range 6-75 A MeV

O.B. Tarasov[a], Yu.E. Penionzhkevich[a], R. Anne[b], D.S. Baiborodin[a], A.V. Belozyorov[a],
C. Borcea[c], Z. Dlouhy[d], D. Guillemaud-Mueller[e], R. Kalpakchieva[a], M. Lewitowicz[b],
S.M. Lukyanov[a], V.Z. Maidikov[a], A.C.Mueller[e], Yu.Ts.Oganessian[a],
M.G.Saint-Laurent[b], N.K. Skobelev[a], O. Sorlin[e], V.D. Toneev[f], W. Trinder[b]

[a] *Flerov Laboratory of Nuclear Reactions, Joint Institute for Nuclear Research, 141980 Dubna, Moscow region, Russia*
[b] *Grand Accelerateur National d'Ions Lourds, BP 5027, 14076 Caen Cedex 5, France*
[c] *Institute of Atomic Physics, Bucharest-Magurele, P.O.Box MG6, Rumania*
[d] *Nuclear Physics Institute, 250 68 Rez, Czech Republic*
[e] *Institut de Physique Nucleaire, CNRS-IN2P3, 91406 Orsay Cedex, France*
[f] *Bogoliubov Laboratory of Theoretical physics, Joint Institute for Nuclear Research, 141980 Dubna, Moscow region, Russia*

The investigation of the mechanism of nuclear reactions is closely connected with the projects for radioactive nuclear beam facilities, which will open new possibilities for the study of exotic nuclei and which will generate radioactive secondary beams by using primary beams of very different energies. Questions arise concerning to the extent of coexistence of different reaction mechanisms (e.g. multi-nucleon transfer reactions and fragmentation) at various energies, the dependence of the production rates on the isospins, of the projectile and target etc. Some regularities in the production of the isotopes with $6 \leq Z \leq 14$ are investigated in the reactions induced by ^{32,34,36}S beams. The results, discussed in the present work have been obtained in a very broad range of the beam energy (6 < E < 75 A MeV) with various targets: ^{12}C, ^{181}Ta and ^{197}Au. The isotope yields and the most probable fragment mass are studied in relation to the mass and energy of the target and projectile. The experiments with ^{32,34}S beam energy were E<20 A MeV carried out at the U-400 cyclotron of the Flerov Laboratory of Nuclear Reactions (JINR). The yields of the various isotopes were measured using the MSP-144 magnetic spectrometer [1]. The ^{36}S (75 A MeV) beam was provided by the GANIL accelerator facility (France); the isotope yields were measured by the LISE fragment-separator [2]. The distributions of the carbon, oxygen, neon, magnesium and silicon isotopes produced in ^{32}S and ^{34}S (6.3; 9.1 and 16 A MeV) and ^{36}S (75 A MeV) induced reactions on three targets (^{12}C, ^{181}Ta, ^{197}Au) were obtained. The experimental data at low energy were compared with the calculation within the framework of the dynamical model of deep inelastic collisions [3]. The yields of the isotopes in the intermediate energy region were calculated with the LISE-code. The isotopic distributions of the final (experimentally observed) nuclei were calculated within the framework of the statistical theory of decay of excited primary fragments. The comparison of the experimental data and the calculations shows (Fig. 1a) that the contribution of deep inelastic reactions to the production cross section of both neutron-rich and neutron-deficient isotopes is dominant at low energies (the dashed lines), while at intermediate energies the main contribution is difined by fragmentation reactions.

On the basis of the data obtained the following conclusions can be drawn:
- In the energy range 7-10 MeV/A quite an abrupt decrease in the cross section is observed in the case of a light target as the number of transferred protons increases. At high energies this difference decreases and is negligible at intermediate energies.

CP455, *ENAM98: Exotic Nuclei and Atomic Masses*
edited by B. M. Sherrill, D. J. Morrissey, and Cary N. Davids
© 1998 The American Institute of Physics   1-56396-804-5/98/$15.00

- At high energies the isotopic content of the projectile plays a dominant role in the production of nuclei close to the projectile ($Z \geq 12$), while at energies $E < 20$ MeV/A the production cross sections for ^{32}S and ^{34}S beams are comparable.

A difference in the cross sections is only observed in the region of nuclei heavier than the projectile, where pick-up reactions prevail.

The isotope production cross sections are seen to rise for energies up to about 15-20 MeV/A, after which they either flatten or pass through the maximum and drop in the energy regions where fragmentation is expected to prevail.

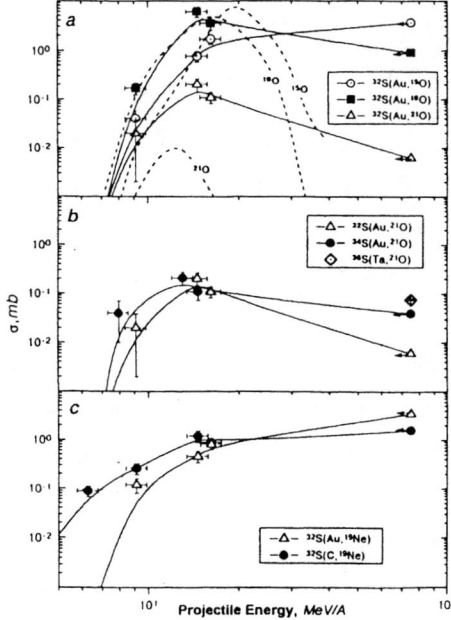

Fig. 1. Total production cross section for different isotopes as a function of energy. The solid curves are the results of the calculations using a modification of the empirical parametrization. The dashed lines present the calculations of the transfer reaction products.

References

1. V.Z.Maidikov et al. Pribori i Tech, Expt. 4 (1979) 68.
2. R.Anne et al. Nucl. Instr. and Meth. A257 (1987) 215.
3. R.Schmidt et al. Nucl. Phys. A311 (1978) 247.

Study of neutron-rich nuclei near the N=20 neutron closed shell

R.Allatt[b], J.C.Angelique[c], R.Anne[d], C.Borcea[e], Z.Dlouhy[f], C.Donzaud[g], S.Grevy[g], D.Guillemaud-Mueller[g], M.Lewitowicz[d], S.Lukyanov[a], A.C.Mueller[g], F.Nowacki[d], N.A.Orr[c], Yu.E.Penionzhkevich[a], R.D.Page[b], F.Pougheon[g], A.Reed[b], M.G.Saint-Laurent[d], W.Schwab[g], E.Sokol[a], O.Tarasov[a], W.Trinder[d], J.S.Winfield[d]

[a] *Flerov Laboratory of Nuclear Reactions, Joint Institute for Nuclear Research, 141980 Dubna, Moscow region, Russia*
[b] *Dept. of Physics, University of Liverpool, Liverpool, L69 7ZE, UK*
[c] *Laboratoire de Physique Gorpusculaire, CNRS-IN2P3, ISMRA ef Universite de Caen, Boulevard du Marechal Juin, 14050 Caen Cedex, France*
[d] *Grand Accelerateur National d'Ions Lourds, BP 5027, 14076 Caen Cedex 5, France*
[e] *Institute of Atomic Physics, Bucharest-Magurele P.O. Box MG6, Rumania*
[f] *Nuclear Physics Institute, 250 68 Rez, Czech Republic*
[g] *Institut de Physique Nucleaire, CNRS-IN2P3, 91406 Orsay Cedex, France*

An interesting aspect of the region of N=20 nuclei is the transition from spherical to deformed shapes in the so-called "island of inversion". The deformations in this region can also result in the appearance of the isomeric states of extremely neutron-rich isotopes. Such effects may influence the decay properties of these nuclei, such as half-life, neutron emission probability. The lack of experimental information on the very neutron-rich isotopes in the C-Al region is mainly due to the very low production cross section. Therefore a very exotic primary beam of ^{36}S (78 AMeV) ions, which give an opportunity to study the β-delayed neutron emission from neutron-rich nuclei with the magic neutron number N=20, such as ^{29}F, ^{30}Ne and ^{31}Na, was used in the experiment. The experiment was carried out at GANIL using the Si(Li) detectors telescope at the focal point. The implantation detectors were surrounded by 3He filled neutron detectors. This detector system was served as both the fragment identification and β-delayed neutron decay measurement. For the first time the β-decay half-lives and neutron emission probability were measured for ^{30}Ne, ^{26,27,29}F. Additionally, the cases of ^{22}N, ^{24}O, $^{24-29}$Ne, ^{25}F, ^{30,32}Na were re-examined (see Table) [1].

The measured half-lives for ^{28}Ne and ^{30,31}Na agree within the error bars with the previous experiments. The only important discrepancy is observed for ^{29}Ne. The experimental half-lives obtained here are in good agreement (within a factor of two) with the sd shell-model calculations of Wildenthal et al. [2] including the values for ^{27,29}F and ^{29,30}Ne. The last suggest that the deformation phenomena, predicted and observed in the Mg - Na region, disappears below Z=11. Thus the standard shell-model space seems to be sufficient to predict half-lives of fluorine and neon isotopes in vicinity of N=20.

The attempt to synthesize ^{28}O was carried out at GANIL using the LISE spectrometer, which collects projectile-like fragments. The fragmentation of a ^{36}S^{16+} (78.1 AMeV) beam with a mean intensity 800 enA was expected to increase the production rate of the neutron-rich isotopes near N=20. Measurements of the momentum distributions of all fragments with N=20 and an optimization of the target material (Be, C, Ni, Ta) and thickness were undertaken to determine the best setting of the LISE spectrometer for ^{28}O. It was found that the Ta target produced the highest rates of the neutron-rich nuclei.

During 53-hours measurement with this average beam intensity no events corresponding to ^{26}O and ^{28}O have been obtained. In the addition to A/q and Z identification, a horizontal coordinate in the intermediate dispensive plane of the LISE spectrometer was in agreement with the computer simulation of horizontal images in the focal point. According to the estimation given by the modified formula of Summerer et al.c [2] one could expect about 11 events corresponding to ^{28}O.

The results of the present experiment point to the particle instability of the ^{28}O isotope as well as for ^{26}O. An upper limit for the cross section of the formation of the oxygen isotopes extracted from the data is estimated to be 0.7 pb and 0.2 pb for ^{26}O and ^{28}O, respectively.

Table. Experimental values of the β-decay half-lives and neutron emission probability of neutron-rich nuclei close to N=20

Isotope	Experimental results			
	This work		Table of Isotopes 1996	
	$T_{1/2}$ ms	P_n %	$T_{1/2}$ ms	P_n %
^{22}N	31 (5)	37 (14)	24 (7)	35 (5)
^{24}O	67 (10)	12 (8)	61 (26)	58 (12)
^{25}F	70 (10)	14 (5)	59 (4)	15 (10)
^{27}F	9.6 (0.8)	11 (4)		
^{29}F	-2.4 (0.8)	100 (80)		
^{27}Ne	22 (6)	0 (3)	32 (2)	2 (0.5)
^{28}Ne	20 (3)	11 (3)	17 (4)	22 (3)
^{29}Ne	15 (3)	27 (9)	200 (10)	
^{30}Ne	7 (2)	9 (17)		
^{30}Na	50 (4)			
^{31}Na	18 (2)		48 (2)	30 (4)

References
1. O.Tarasov et al., Phys. Lett. B409 (1997) p.64.
2. B.H.Windehthal et al. Phys. Rev. C28 (1983), p.1343.
3. K.Summerer et al. Phys. Rev. C42 (1990) p.2546.

Spectroscopy of Neutron Rich Nuclei in the Vicinity of N=20

A.T. Reed, R.D. Page, R.G. Allatt, P.J. Nolan
Dept. of Physics, University of Liverpool, Liverpool, L69 7ZE, UK.

D. Guillemaud-Mueller, C. Donzaud, S. Grévy, A.C. Mueller, F. Pougheon, O. Sorlin
Institut de Physique Nucléaire, CNRS-IN2P3, F-91406 Orsay Cedex, France.

Yu. Penionzhkevich, S. Lukyanov, E. Sokol, O. Tarasov
Flerov Laboratory of Nuclear Reactions, JINR, 141980 Dubna, Russia.

J.C. Angélique, F.M. Marques, N.A. Orr
LPC, Bld. du Marechal Juin, 14050 Caen Cedex, France.

R. Anne, M. Lewitowicz, G. Martinez, M.G. Saint-Laurent, W. Trinder
GANIL, BP 5027, F-14021 Caen Cedex, France.

C. Borcea
Institute of Atomic Physics, P.O. Box MG-6, 76900 Bucharest-Magurele, Romania.

V. Burjan, Z. Dlouhý, J. Novák
Nuclear Physics Institute, CZ-25068 Řež, Czech Republic.

W.N. Catford, P.H. Regan, S.M. Vincent
Dept. of Physics, University of Surrey, Guildford, GU2 5XH, UK.

Abstract The results of a fragmentation reaction studying neutron-rich nuclei in the region N = 20 will be summarised. The various β-delayed decay modes have been analysed and the measurement of half-lives of some of the nuclei have been performed for the first time.

Nuclei in the vicinity of the neutron drip line have revealed a rich variety of novel phenomena which have revolutionised our understanding of nuclear physics. However, the inaccessibility of these short-lived nuclides mean that very little is known about their spectroscopy[1]. Spectroscopic measurements can reveal details of the underlying microscopic structures and have proved essential for understanding these unexpected properties as well as the precise location of the neutron drip line itself. Furthermore, they potentially provide a stringent test of the modern large scale shell model calculations applied to elucidate these problems.

In the present experiment peformed at GANIL using the LISE3 spectrometer[2], extremely neutron rich nuclides were produced by the quasi-fragmentation of a ^{36}S beam and individual ions were identified unambiguously through time of flight and energy loss measurements. By correlating the implantation of the selected ions with their subsequent β-delayed γ-decays and neutron emission, this highly selective experiment has succeeded in identifying γ-rays emitted by previously unstudied nuclides as well as measuring neutron emission probabilities and half-lives.

The half-life and β-delayed γ-decay and neutron emission probabilities of ^{31}Na have been measured previously[3,4]. This nuclide was chosen as a reference nucleus and a total of 100000 ^{31}Na nuclei were produced during the experiment. The corresponding background subtracted β-delayed γ-ray spectrum is shown in Fig. 1. The energies labelled correspond to the γ-rays already known from the previous work. A half-life analysis of this nucleus has yielded a value of 17.9 ± 1.1 ms which is consistent with previously deduced values.

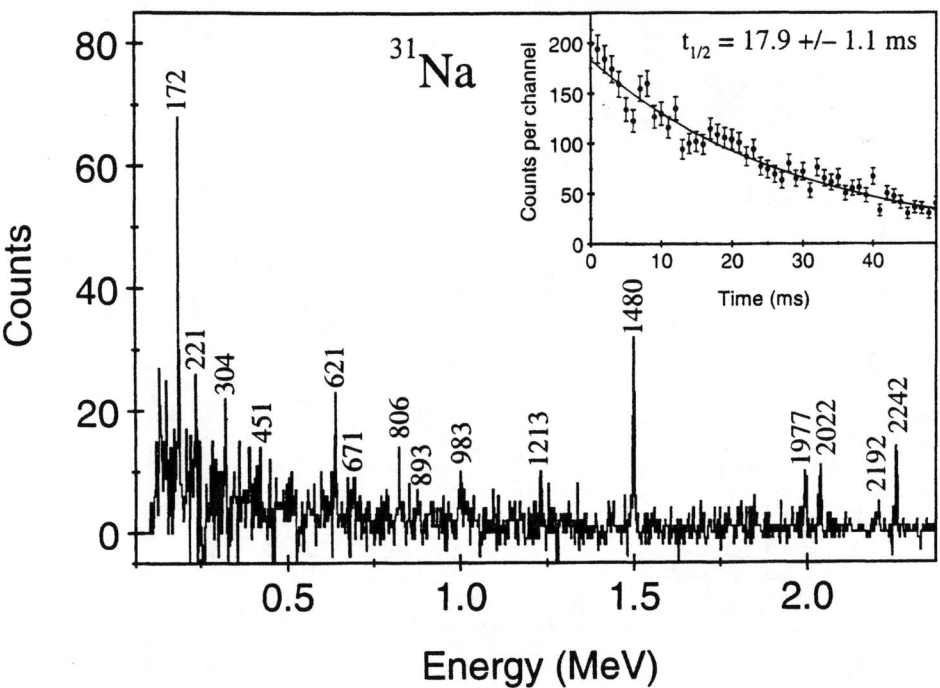

Fig. 1. The β-delayed γ-rays observed in coincidence with the β-particles occuring within 50 ms of the detection of a ^{31}Na ion. The insert shows the corresponding half-life curve obtained for the β-decay of ^{31}Na.

The new measurements obtained in the present experiment are summarised in Fig. 2, which shows the nuclei for which it has been possible, for the first time, to obtain half-life and β-delayed γ-decay and neutron emission measurements. The region studied extends from the putative halo nucleus ^{17}B to with N = 20.

573

We are grateful for the excellent beam and technical support provided to us by staff at the GANIL facility. Support for this work was provided by the Access to Large Scale Facility programme under the TMR programme of the EU. RDP acknowledges receipt of an EPSRC Advanced Fellowship. We also thank the French/UK (IN2P3/EPSRC) Loan Pool for the Eurogam detectors.

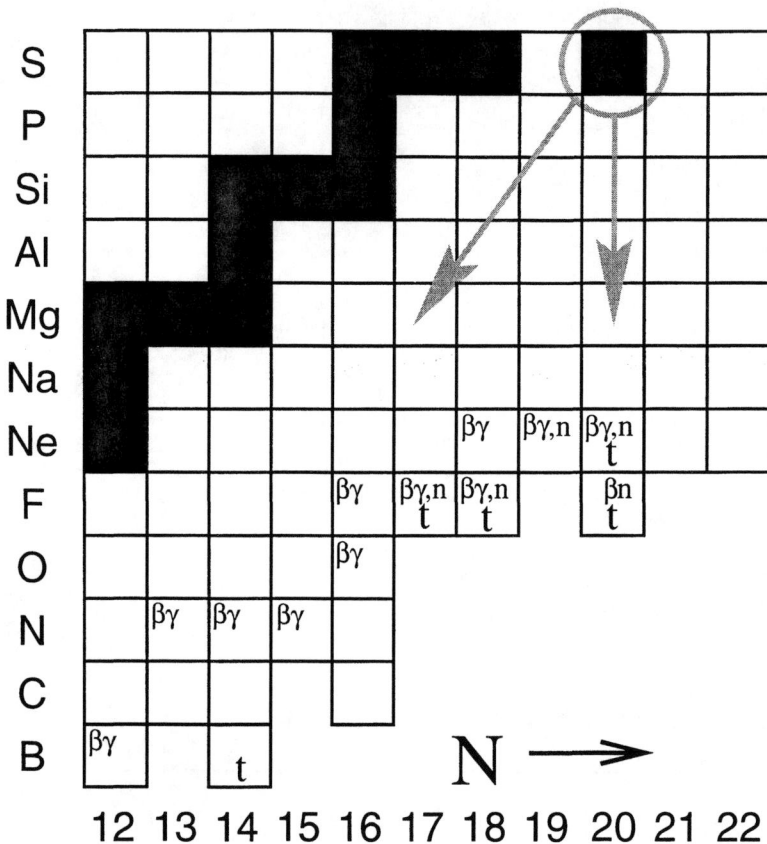

Fig. 2. A section of the chart of the nuclei showing the nuclei for which measurements have been performed for the first time. The black squares correspond to stable nuclei and the circle indicates the beam nucleus fragmented to produce the nuclei studied in this work. The labels $\beta\gamma$, βn and t indicate nuclei for which first measurements of β-delayed γ-rays, β-delayed neutron emission and half-lives have been obtained respectively.

1. P.M. Endt, Nuclear Physics A521 (1990) 1.
2. R. Anne et al., Nucl. Inst. and Meth. A257 (1987) 215.
3. D. Guillemaud-Mueller et al., Nucl. Phys. A426 (1984) 37.
4. G. Klotz et al., Phys. Rev C47 (1993) 2502.

Single-Particle and Cluster Levels in ^{107,108,109}Sn Populated in the Decay of ^{107,108,109}Sb

Jennifer Jo Ressler, Darek Seweryniak, Louis Conticchio, Diana Ciurczak, Joseph Swider, Heikki Penttilä, William B. Walters

UNIVERSITY of MARYLAND, COLLEGE PARK, MARYLAND, 20742 USA

Jan Wauters, Carrol Bingham

UNIVERSITY of TENNESSEE, KNOXVILLE, TN 37999 USA

Robert de Haan

UNIVERSITY of NOTRE DAME, SOUTH BEND, IN 46556 USA

Brian Foy

CLARK UNIVERSITY, WORCESTER, MA 01610 USA

Cary N. Davids

ARGONNE NATIONAL LABORATORY, ARGONNE, IL 60439 USA

Abstract. New features of the level structures of ^{107,108,109}Sn that have been observed following the beta decay of ^{107,108,109}Sb are reported. The recoiling Sb nuclides were isolated using the Fragment Mass Analyzer at the ATLAS accelerator facility. Low-energy levels with strong configuration admixtures of the $(\nu g_{7/2})^3{}_{5/2}$ are identified in both ^{107}Sn and ^{109}Sn.

INTRODUCTION

In this paper, new data will be presented for the structure of the neutron-deficient Sn nuclides, ^{107,108,109}Sn. Of interest are the structures with the three-particle cluster structure described by Kisslinger (1) and discussed in detail by Paar (2).

EXPERIMENTAL PROCEDURES

The neutron deficient ^{107,108,109}Sb were produced at Argonne National Laboratory by bombarding and ^{54}Fe target with beams of ^{58}Ni from the ATLAS accelerator. The recoiling reaction fragments were separated by their A/Q values and implanted onto the tape of a Moving Tape Collector that was used to move the reaction products to a position surrounded by Ge detectors. The results are presented in Table 1.

CP455, ENAM98: Exotic Nuclei and Atomic Masses
edited by B. M. Sherrill, D. J. Morrissey, and Cary N. Davids
© 1998 The American Institute of Physics   1-56396-804-5/98/$15.00

Table 1. Gamma rays observed following decay of Sb-107, 108, 109.

Energy (keV)	rel. int.	from	to	coinc.	Energy (keV)	rel. int.	from	to	coinc.
Gamma rays following decay of 4-s Sb-107									
151.1	42	151	0	667	969.5	67	970	0	none
666.9	25	818	151	151	1280.1	100	1280	0	none
703.9	53	704	0	none	1454.3	68	1454	0	none
818.4	92	818	0	none					
Gamma rays following decay of 7-s Sb-108									
253.4	3.8	2364	2111	260, 491, 618, 905, 1206	1272.8	18	2479	1206	1206, 1372
260.3	1.4	2625	2364	253, 905, 1206	1373	1	3852	2479	1206, 1273
490.6	1.9	2855	2364	253, 905, 1206	1395.1	1			1206
529.7	1.9	2640	2111	905, 1206	1413.3	1.4			1206
617.5	1.7	2982	2364	253, 905, 1206	1434.5	3.3	2640	1206	1206
693.7	2.3	2804	2111	905, 1206	1537.0	0.7	4177	2640	1206
743.9	1.4	2855	2111	905, 1206	1598.8	18	2804	1206	1206
820.7	5.5	2976	2155	949, 1206	1648.7	8.2	2855	1206	1206
826		2982	2155		1770.1	3.9	2976	1206	1206
864.9	6.4	2976	2111	905, 1206	1869.0	4.6	3075	1206	
871		2982	2111		1970.8	5.0			
904.8	26	2111	1206	253, 491, 618, 694, 744, 865, 1206	2155.2	4.7	2155	0	
					2327.8	3.3			
949.1	6.3	2155	1206	821, 1206					
1205.7	100	1206	0	253, 491, 618, 694, 744, 821, 865, 905, 949, 1273, 1434, 1599, 1649, 1770					
Gamma rays following decay of 17-s Sb-109									
246.8	1.8	925	679	665, 679	1101	1.3			
261.2	4.4	925	664	402, 650, 664	1106		1650	545	
402.4	4.9	1328	925	261, 664	1206	9.8			
544.9	11	545	0	951, 1106	1228.9	9.1			
650	≤1	664	14		1273.5	1.8			
663.5	60	664	0		1328	6.7	1343	14	
663.6	4.4	1343	679		1343.2	11	1343	0	
664.5	3.3	679	14		1436	11.0			
678.7	3.3	679	0		1481.8	2.4	1496	14	
752.5	3.9				1495.6	33	1496	0	
831.8	11	1496	664		1600.9	8.2			
925.2	100.0	925	0		1636	0.8	1650	14	
950.5	2.3	1496	545		1650.1	4.0	1650	0	
1047.5	7.9	1061	14		1760	4.2			
1061.4	69	1061	0		1969.9				
1077.8	2.3	1078	0						

Uncertainties in the energy values are about 1 unit in the last digit shown. Uncertainties in the intensity values are ~10% on the stronger lines and 20% for the weaker lines.

RESULTS

The singles spectrum that we observe for decay of 4-s ^{107}Sb to levels of ^{107}Sn is consistent with that reported by Shibata et al., who proposed one new level in ^{107}Sn at 1280 keV (3). The coincidence spectrum obtained by a gate on the 151-keV gamma ray is shown in Figure 1. This gate reveals one peak at 667 keV that supports placement of a level at 818 keV. No other peaks are observed at the positions of other gamma rays found in the singles spectrum nor at positions 151-keV below lines found in singles. No peaks were observed in gates on the other reported gamma rays. Hence, we have assigned the gamma rays at 704, 969, and 1454 keV as ground state transitions.

The results for ^{108}Sn are complementary to new results from in-beam studies (4) and radioactive decay studies (3). Several additional gamma lines not reported by Shibata et al are observed and additional coincidence evidence is noted in Table 1. Additional low-spin levels are proposed at 2982 and 3852 keV.

In previous studies of the decay of ^{109}Sb to levels of ^{109}Sn, the presence of the level at 14 keV recently identified through in-beam studies was not recognized. (5,6) Portions of the gamma-ray spectra in coincidence with the 247- and 261-keV transitions are shown in Figure 2. These spectra and the gate on the 665- and 679-keV peaks indicate that the gamma-ray peak at 665 keV is a triplet and show that the level at 679 keV decays primarily to the newly identified 14-keV level, whereas the level at 664 keV decays primarily to the ground state. A new low-spin level is identified at 1650 keV as shown in Figure 3.

SUMMARY

The low-energy structures of the light Sn nuclides are summarized in Figure 3. In ^{109}Sn, the 545-keV level is the most likely choice for the $1/2^+$ level as it appears to be the only level that does not have a branch to the $7/2^+$ level at 14 keV. In that context, we also propose a $1/2^+$ assignment for the 704-keV level in ^{107}Sn, but we cannot completely rule out the 970-keV level. The principal new feature of these results is to establish the presence of three new energy levels in ^{107}Sn below 1 MeV. These new levels appear to be comparable to the three levels below 980 keV in ^{109}Sn. We suggest that the levels at 665 and 679 keV in ^{109}Sn arise from the $(g_{7/2})^3{}_{5/2,3/2}$ configurations and similar structures for the levels at 818 and 970 keV in ^{107}Sn. These states do not appear to be well reproduced in large-scale shell-model calculations. (7) Of note is that the separation between the $g_{7/2}$ level and the $(g_{7/2})^3{}_{5/2}$ cluster state is almost constant for these light Sn nuclides at a value of just over half of the 2^+ energy in the adjacent even-even Sn nuclides. This behavior may be compared to the separation between the $h_{11/2}$ single-neutron states and the $(h_{11/2})^3{}_{9/2}$ cluster states in ^{125,127,129}Sn. The low energy of the $s_{1/2}$ level in these nuclides is of importance as the $d_{5/2} - s_{1/2}$ E2 component of the transition in ^{102}Sn has been found to be enhanced in recent studies of the structure of ^{102}Sn. (8)

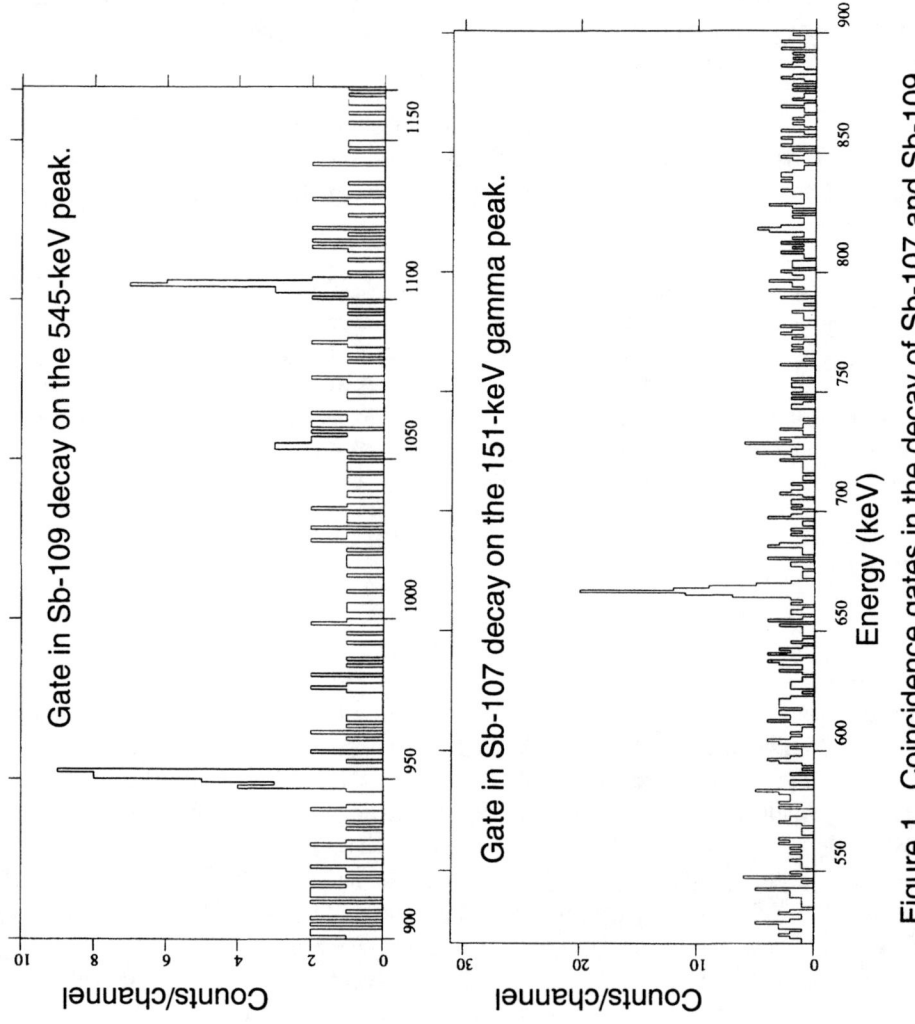

Figure 1. Coincidence gates in the decay of Sb-107 and Sb-109.

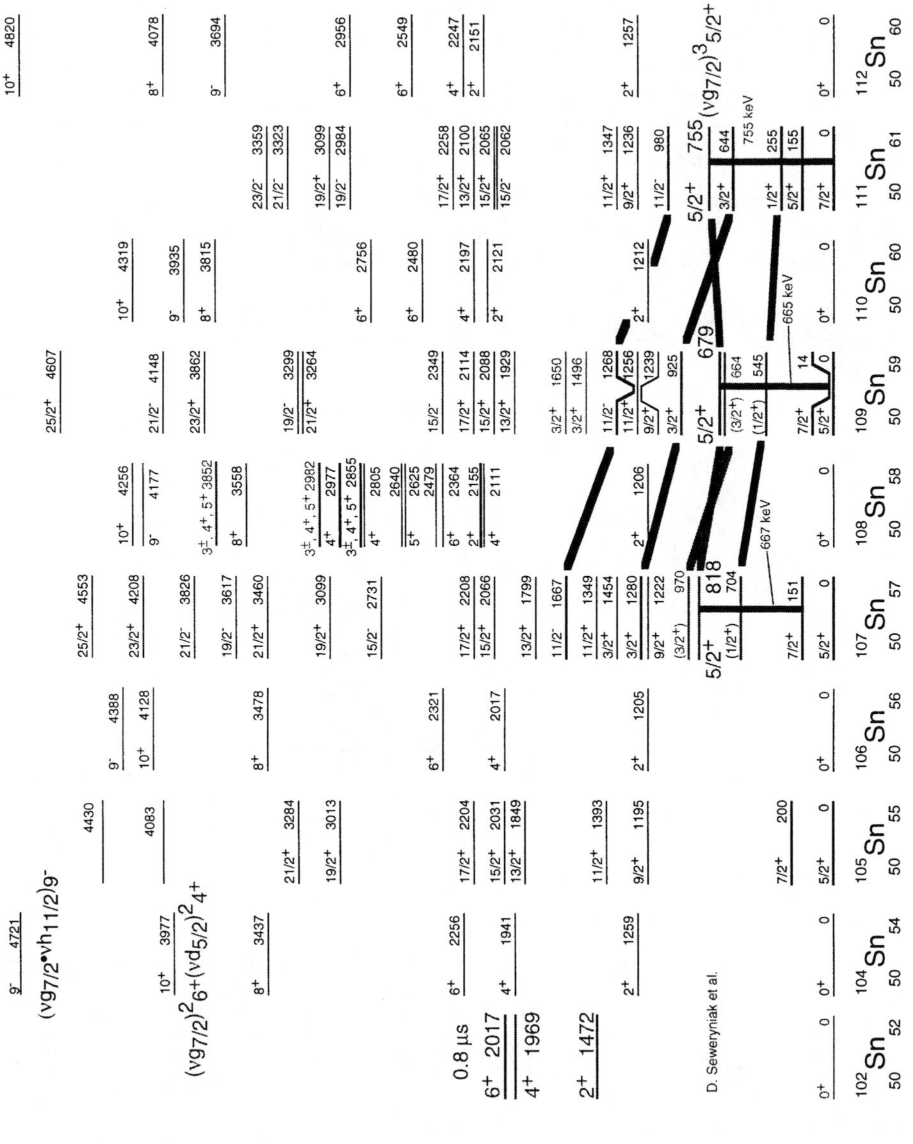

Figure 3. Evolution of single-particle and cluster structures in the light Sn nuclides

ACKNOWLEDGMENTS

The authors are appreciative of the support of the United States Department of Energy for this work and for helpful discussions with Profs. V. Paar, K. Heyde, and J. Rikovska.

REFERENCES

1. L. S. Kisslinger, Nuclear Physics **78**, 341 (1966)
2. V. Paar, Nuclear Physics **A211**, 29 (1973)
3. M. Shibata *et al*, Physical Review C **55**, 1715 (1997).
4. S. Juutinen *et al* Nuclear Physics **A617**, 74 (1997).
5. J.Blachot Nuclear Data Sheets **64**, 913 (1991).
6. L.Kaubler. *et al* Zeitschrift für Physik. A**351**, 123 (1995)
7. T. Engeland, M. Hjorth-Jensen, A. Holt, E. Osnes, Phys. Scripta **T56**, 58 (1995).
8. D. Seweryniak *et al.*, Proc. Gatlinberg Conf. Nucl. Stucture, 1998.

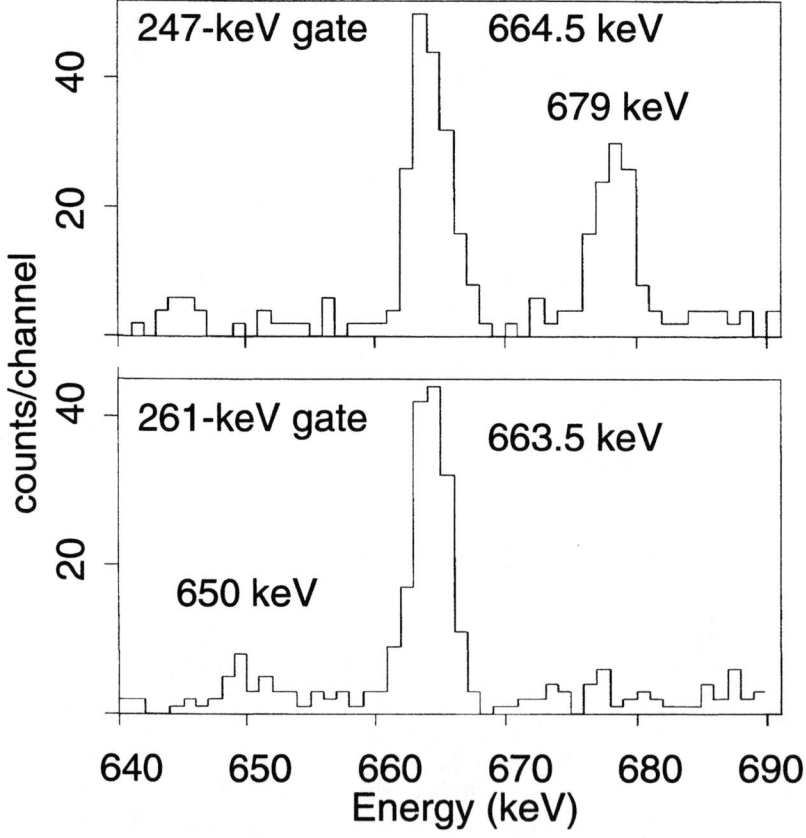

Figure 2. Gates in Sb-109 decay

Production and identification of new, neutron–rich nuclei in the ^{208}Pb region

K. Rykaczewski[a,b,c], J. Kurpeta[c], A. Płochocki[c], M. Karny[c,d],
J. Szerypo[b,c,d], A.-H. Evensen[b], E. Kugler[b], J. Lettry[b], H. Ravn[b],
P. VanDuppen[e], A. Andreyev[e], M. Huyse[e], A. Wöhr[e], A. Jokinen[f],
J. Äystö[f], A. Nieminen[f], M. Huhta[f], M. Ramdhane[g], G. Walter[g],
P. Hoff[h], and ISOLDE Collaboration

[a] ORNL, Physics Division, Oak Ridge, TN 37831, USA
[b] ISOLDE-CERN, CH-1211 Geneva, Switzerland
[c] IEP, University of Warsaw, PL-00681 Warsaw, Poland
[d] JIHIR, Oak Ridge, TN 37831 USA
[e] IKS, University of Leuven, B-3001 Leuven, Belgium
[f] Dept. of Physics, University of Jyväskylä, 40351 Jyväskylä, Finland
[g] IPN, CNRS-IN2P3, Université Louis Pasteur, F-67037 Strasbourg, France
[h] Dept. of Chemistry, University of Oslo, N-0315 Oslo, Norway

Abstract. The recently developed methods allowing the experimental studies on new neutron-rich nuclei beyond doubly-magic ^{208}Pb are briefly described. An identification of new neutron–rich isotopes ^{215}Pb and ^{217}Bi, and new decay properties of ^{216}Bi studied by means of a pulsed release element selective technique at PS Booster-ISOLDE are reported.

Introduction

The level properties of nuclei near the doubly–magic systems are expected to be well understood within a shell-model approach. However, with a departure from the beta–stability line, in particular towards very neutron-rich nuclei, the established shell structure might be modified or even vanish near the neutron-drip line [1,2]. An extension of experimental nuclear structure studies towards the nuclei characterized by exotic isospin values is crucial for verification of such predictions as well as for the r-process nucleosynthesis scenario [3]. Heavy neutron-rich nuclei, 'south-east' of doubly-magic ^{208}Pb, are very difficult to produce and investigate. As an example, one can refer to the experiments employing the on-line mass separator technique and reactions on heavy targets like ^{232}Th and ^{238}U. Such studies often suffered from an isobaric contamination of much more strongly produced and efficiently released elements like francium or radon and their decay products. A new method, based on the pulsed release element selective method recently developed at the PS Booster-ISOLDE at CERN [4] greatly reduces the contamination of these very short–lived α–emitters ($Z \geq 84$) for the isobaric mass chains A=215 to A=218. The production of radioactive nuclei at PS Booster-ISOLDE target occurs during the very short (2.4 μs), intense ($\approx 10^{13}$ protons per pulse) 1 GeV protons pulse. The reduction of isobaric contamination of short-lived activities is due to the waiting time of about 200 milliseconds, applied between the proton beam impact and the release of the radioactive beams from the target-ion source system. A large fraction of the longer-lived nuclei, like beta–decaying neutron–rich bismuth, lead and thallium isotopes, can still be released efficiently [4]. For ^{215}Pb, an isotope having a 36 seconds halflife, over 90% of the produced nuclei will be released. The production rate at the 55 g/cm^2 ^{232}ThC$_2$ target irradiated with a 1 GeV pulsed proton beam (3*10^{13} pps) and combined with a hot plasma ion source (see [5] and earlier refs therein), was as high as 10^8 atoms at 1 μC for ^{213}Bi and still at the level of a few hundred atoms per μC for the new isotope ^{217}Bi. For lead isotopes, the rates were about 5×10^4 atoms per μC for ^{214}Pb and about a few thousands per μC for the new isotope ^{215}Pb, see Fig.1. These production rates are much higher than the yields obtained in a recent experiment employing a fragmentation of relativistic 1 GeV/u ^{238}U beam at FRS-GSI [6]. For example, the rate of ^{215}Pb was about eight orders of magnitude lower ! However, this experiment

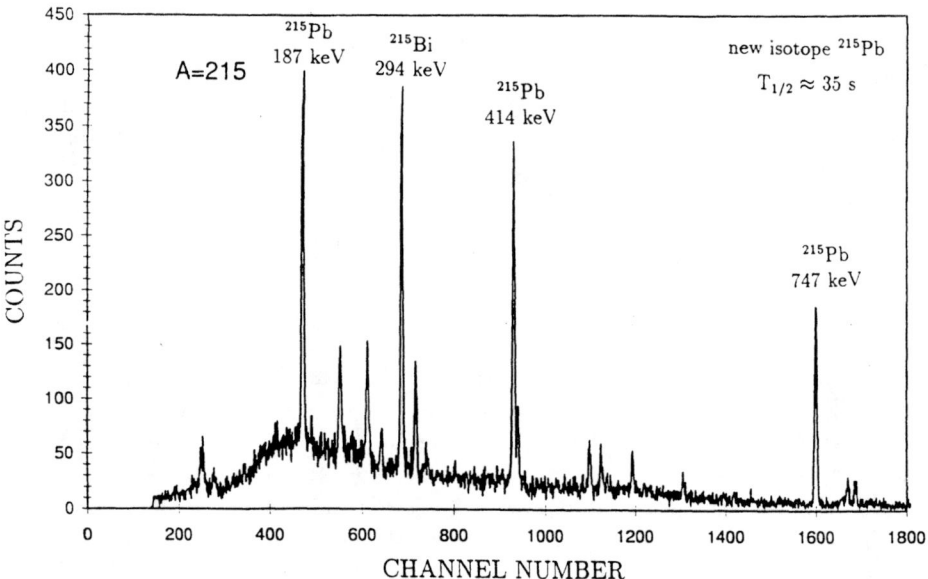

Fig.1. The β–gated γ–spectrum, showing the transitions assigned to the decay of a new neutron–rich isotope ^{215}Pb, recorded at PS Booster-ISOLDE facility at CERN. The collection time of this A=215 mass–separated sample was twelve (!) seconds and the measurement lasted 29 seconds.

was performed at FRS with only 5 × 10^6 ^{238}U ions per second. An increase of the primary beam intensity by a few orders of magnitude seems to be possible for such fragmentation experiments. The rates of exotic neutron–rich nuclei might go up accordingly. Already with relatively low production rates, the GSI study resulted in the identification of seven new isotopes, ^{209}Hg, ^{210}Hg, ^{211}Tl, ^{212}Hg, ^{218}Bi, ^{219}Po and ^{220}Po, and four new isomers ^{203m}Tl, ^{204m}Tl, ^{212m}Pb and ^{211m}Bi [6]. These isomers have a halflife in the μs range, not accessible with the pulsed release method at PS Booster - ISOLDE. Both techniques, which allow the approach to new neutron-rich nuclei beyond doubly–magic ^{208}Pb, a chemically selective pulsed release method at PS Booster-ISOLDE and fragmentation of relativistic ^{238}U beam studied by means of FRS, are complementary at the present time.

Results of PS Booster-ISOLDE experiments

The decay properties of A=215, A=216 and A=217 samples were measured by α-, β-, γ- and X-ray spectroscopy methods. For ^{215}Pb, the halflife of 36.5(3) s was determined from the decay pattern of the strongest γ-transitions at 187, 414 and 747 keV. Bismuth KX-rays were observed in coincidence with these transitions as well as the γ–β concidences recorded in an earlier experiment, see Fig.1. However, to confirm unambigously the identification of the new neutron–rich isotope, an interpretation of this activity as a new isomer in ^{215}Bi has to be ruled out. This requires a measurement with an absorber in front of a beta counter to stop the hypothetical conversion electrons from isomeric decay.

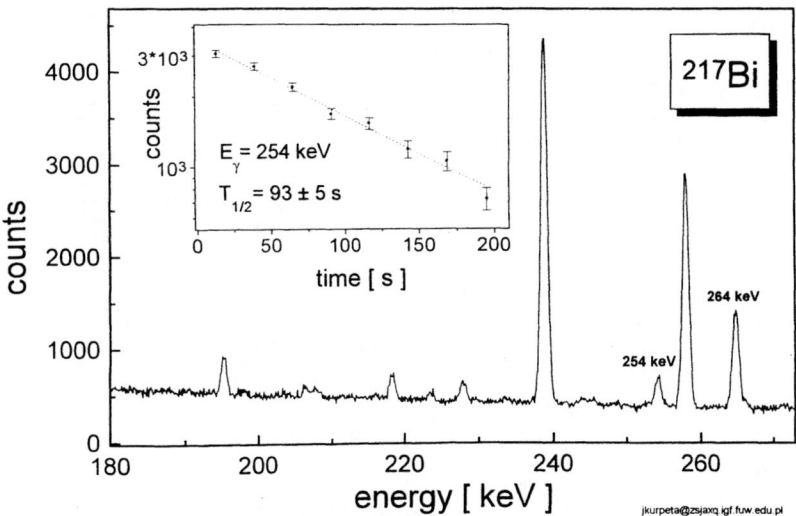

Fig.2. A first 26–second subgroup of γ-multispectrum and a decay pattern for 254 keV line measured for A=217 samples with 3850 mm^2 LOAX detector.

During the studies of A=217 samples of ^{217}Bi, two γ–transitions known from earlier α–decay study of ^{221}At as de–exciting the 254 and 264 keV levels in ^{217}Po, were observed in a β–coincident spectrum. Polonium KX–rays coincident with these γ–transitions confirmed the identification of a new isotope. The halflife of 93(3) s is an average resulting from the halflives of these γ–transitions, see Fig.2. In the same run, $T_{1/2}$=1.46(5) s of ^{217}Po was determined for the first time from the α-decay time pattern.

A reinvestigation of the decay of ^{216}Bi gives a halflife value of 132(3) s, shorter than the previously reported 6.6(2.1) min [7] and 3.6(0.4) min [8]. The four strongest transitions following the β–decay, namely the new lines at 223 and 360 keV, and those previously reported [8] at 419 and 550 keV, are interpreted as a cascade de–exciting the 8^+ state in ^{216}Po. Two weak γ–transitions at 42 and 149 keV, having $T_{1/2} \approx 16$ s, found at A=216 samples, might represent an indication for a decay of new isotope ^{216}Pb. The further analysis and intepretation of spectroscopic data is in progress [9].

In summary, the previously experimentally unaccesible region of neutron–rich nuclei 'south–east' of the doubly–magic ^{208}Pb has been approached via two complementary methods. The extension of FRS based studies is foreseen in 1999. At ISOLDE, the next experiment based on thermal ion source should allow the enhancement of thallium against bismuth and lead isobars, in addition to already achieved chemical selectivity. Both techniques contribute to the tracking of nuclear structure evolution along closed shell Z=82. The rates and release time profiles measured at ISOLDE during the described studies are also important for the design of post–accelerated radioactive ion beam facilities.

Acknowledgements

This work was partially funded by the US DOE under DE-AC05-96OR22464. ORNL is managed by LMER Corporation. Warsaw team was supported by KBN.

References

[1] J.Dobaczewski et al., Phys. Rev. Lett. **72** (94) 981
[2] J.Dobaczewski et al., Phys. Rev. **C53** (96) 2809
[3] B.Chen et al., Phys. Lett. **B355** (95) 37
[4] P. Van Duppen et al., Nucl. Instr. Meth. **B 134** (98) 267
[5] J. Lettry et al., Nucl. Instr. Methods **B 126** (97) 130
[6] M. Pfützner et al., submitted to Phys. Lett. **B**
[7] D.G. Burke et al., Zeit. Phys. **A333** (89) 131
[8] E. Ruchowska et al., J.Phys. **G16** (90) 255
[9] J. Kurpeta, A. Płochocki et al., to be published

Deformation Change between isomeric and ground states in the ^{184}Au and ^{183}Pt isotones

J. Sauvage[1], N. Boos[2], L. Cabaret[3], J. Crawford[4], H.T. Duong[3], J. Genevey[5], M. Girod[6], G. Huber[2], F. Ibrahim[5], M. Krieg[2], F. Le Blanc[1], J.K.P. Lee[4], J. Libert[7], J. Obert[1], J. Oms[1], J. Pinard[3], J.C. Putaux[1], B. Roussière[1], V. Sebastian[2], S. Zemlyanoi[8] and the ISOLDE collaboration[9]. †

[1] *Institut de Physique Nucléaire, 91406 Orsay cedex, France*
[2] *Institut für Physik der Universität Mainz, 55099 Mainz, Germany*
[3] *Laboratoire Aimé Cotton, 91405 Orsay cedex, France*
[4] *Foster Radiation Laboratory, McGill University, H3A2T8 Montréal, Canada*
[5] *Institut des Sciences Nucléaires, 38026 Grenoble cedex, France*
[6] *Service de Physique Nucléaire, CEA, BP 12, 91680 Bruyères-le-Châtel, France*
[7] *Centre d'Etudes Nucléaires de Bordeaux Gradignan, 33175 Gradignan cedex, France*
[8] *Flerov Laboratory of Nuclear Reaction, JINR, Dubna 141980, Moscow Region, Russia*
[9] *CERN, 1211 Geneva 23, Switzerland*

Abstract. Deformation changes $\delta\beta$ have been found out between isomeric and ground states of the ^{183}Pt and ^{184}Au isotones. Atomic spectroscopy measurements by laser were carried out using the COMPLIS setup. Hyperfine structure (HFS) spectra and isotope shift (IS) were obtained for the $5d^96s\ ^3D_3 \rightarrow 5d^96p\ ^3P_2$ and $5d^{10}6s\ ^2S_{1/2} \rightarrow 5d^{10}6p\ ^2P_{3/2}$ optical transitions in Pt and Au atoms respectively, providing deformation parameters β and nuclear moments μ and Q_s. The influence of the proton-neutron coupling on the $\delta\beta$ value in ^{184}Au relatively to its isotone ^{183}Pt has been determined. Besides, the $h9/2$ proton state that is decoupled from the core in ^{183,185}Au, becomes the $3/2[532]$ state ($h9/2$ parentage) strongly coupled in the doubly-odd ^{184}Au nucleus.

INTRODUCTION

The neutron-deficient Au and Pt nuclei are located in the middle of a mass region where nuclear shape instabilities exist. Nuclear spectroscopy measurements performed on the N=105 isotones [1-5] added to theoretical calculations suggested that the abnormal relative location in energy of the isomeric and ground states of ^{184}Au could be due to a deformation change either between ^{184}Au ^{m} and ^{184}Au^g or between ^{184}Au and ^{183}Pt. To explain this anomaly it was actually important to confirm the structure proposed for ^{184}Au^{g,m} and ^{183}Pt^{g,m} and to precisely determine their deformation parameters β. Atomic spectroscopy is a very powerful method to study the deformation changes and structure of isomeric and ground states. But Au and Pt atoms are only available from β^+/EC decays of Hg ion beam. Therefore, we used the COMPLIS setup installed on a beam line of the ISOLDE-BOOSTER facility.

EXPERIMENTAL PROCEDURE AND RESULTS

The COMPLIS setup has been designed to perform Resonance Ionization Spectroscopy (RIS) studies on a pulsed secondary atomic beam produced by laser desorption. The Hg ions are collected and after their decay, Au or Pt are desorbed as atoms and then ionized in two (Au) or three (Pt) atomic steps. The resulting ions are detected and mass-identified by their time-of-flight. The frequency scan is performed on the first step of the RIS process that corresponds to the $5d^{10}6s\ ^2S_{1/2} \to 5d^{10}6p\ ^2P_{3/2}$ optical transition at 243 nm for Au and $5d^96s\ ^3D_3 \to 5d^96p\ ^3P_2$ transition at 306 nm for Pt.

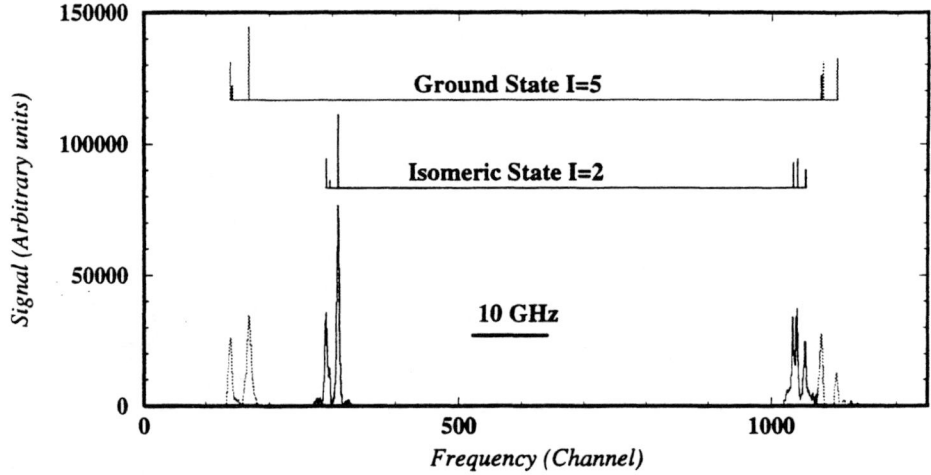

FIGURE 1. HFS spectrum of ^{184}Au^{g+m}. The two spectra above the experimental one have been calculated using the A, B and IS values extracted from the HFS spectrum.

The HFS spectrum of ^{184}Au^{g+m} is shown in figure 1. The magnetic A, A' and electrostatic B, B' hyperfine constants are extracted from the HFS spectra. They are used to determine the magnetic and spectroscopic quadrupole moments μ and Q_s. Assuming a strong coupling scheme and axial symmetry we can calculate the intrinsic quadrupole moment Q_0 and its corresponding β value. From IS the mean square charge radius change $\delta <r_c^2>$ is deduced as well as the nuclear deformation

$<\beta^2>^{1/2}$. All results are given in Table 1. More details about experimental procedure and result extraction can be found in References [6,7].

TABLE 1. Nuclear moments and deformation parameters.

Nucleus	I^π	$\mu_{exp}(\mu_N)$	$\mu_{cal}(\mu_N)$ $[g_{sfree}(0.6g_{sfree})]$	Q_s [b]	$Q_0{}^a$ [b]	β	$<\beta^2>^{1/2b}$
^{183}Pt^g	$1/2^-$	+0.502(5)	+0.63(+0.37)				0.227 (3)
^{183}Pt^m	$7/2^-$	+0.782(14)	+1.46(+0.99)	+3.71(13)	+7.22(58)	+0.242(18)	0.246(3)
^{184}Au^g	5^+	+2.07(2)	+2.3(2.2)	+4.65(26)	+8.06(45)	+0.264(14)	0.255(3)
^{184}Au^m	2^+	+1.44(2)	+1.2(1.4)	+1.90(16)	+6.65(56)	+0.221(17)	0.249(3)

[a] Q_0 is calculated with I=K, and a Q_s value corrected for core polarization effects in Pt.
[b] β values calculated using $\delta <r_c^2>$ values deduced from IS, ^{194}Pt and ^{197}Au have been used as references.

DISCUSSION

In Table 1, one can see that the β and $<\beta^2>^{1/2}$ values deduced from Q_0 values and IS results are very close for ^{183}Pt^m and ^{184}Au^g, which shows that the axial symmetry assumed to calculate Q_0 is fully justified. Since the states in Table 1 have almost pure $K=I$ wave functions they are confirmed as the 1/2[521] and 7/2[514] states in Pt and correspond very likely to the π 3/2[532] $\otimes \nu$1/2[521] and π 3/2[532] $\otimes \nu$7/2[514] configurations for the isomeric and ground states of Au. Thus, the $h9/2$ proton, decoupled from the core by the Coriolis force in ^{183,185}Au [5,8], becomes the strongly coupled 3/2[532] state, arising from the $h9/2$ subshell, in ^{184}Au. Furthermore, the magnetic moments μ_{cal} of these states calculated using the rotor-quasiparticle coupling model [9] added to the method given in reference [10] for a doubly-odd nucleus, are in very good agreement with the μ_{exp} values (see Table 1), which strongly supports the structure given above.

In order to know if the state inversion observed in ^{184}Au (see Figure 2a) is due or not to deformation changes between ^{183}Pt^{g,m} and ^{184}Au^{g,m}, the energy evolution of the 1/2[521] and 7/2[514] neutron states has been calculated as a function of the β deformation (see Figure 2b). We can see in Figure 2b that the β values measured for ^{183}Pt and ^{184}Au allow us to qualitatively explain the state inversion observed in ^{184}Au since the 1/2[521] state lies below the 7/2[514] state for the ^{183}Pt deformations and above it for the ^{184}Au ones. It is worth noting that the deformation change between isomeric and ground states in ^{184}Au is clearly smaller than that in ^{183}Pt. This illustrates the coupling effect of the 3/2[532] proton state to the 1/2[521] and 7/2[514] neutron states, on the nuclear deformation.

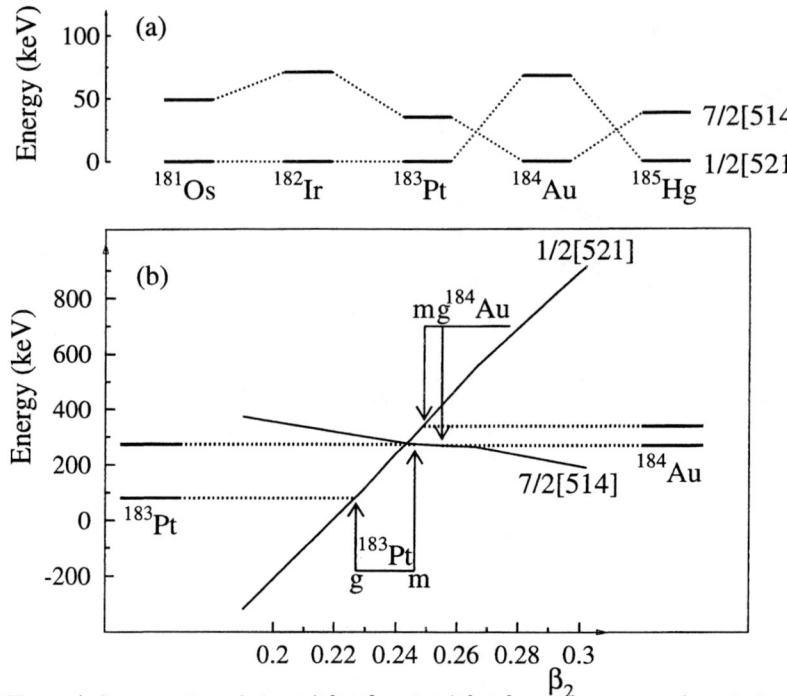

FIGURE 2. a) Systematics of the 1/2[521] and 7/2[514] neutron states (coupled to the proton state from the h9/2 subshell in the case of the doubly-odd nuclei) for the $N = 105$ isotones. b) Energy evolution of the 1/2[521] and 7/2[514] neutron states calculated, using the axial-rotor-one-quasiparticle coupling model (2) with the ^{184}Pt core, as a function of β values.

REFERENCES

1. Roussière B. *et al.*, Z. Phys. A **351**, 127 (1995).
2. Sauvage J. *et al.*, Nucl. Phys. A **592**, 221 (1995).
3. Roussière B.*et al.*, Nucl. Phys. A **504**, 511 (1989).
4. Ibrahim F. *et al.*, Z. Phys. Lett. A **350**, 9 (1994).
5. Bourgeois C. *et al.*, Nucl. Phys. A **386**, 308 (1982).
6. Le Blanc F. *et al.*, Contribution to this conference.
7. Le Blanc F. *et al.*, Phys. Rev. Lett. **79**, 2213 (1997).
8. Macias-Marques M.I. *et al.*, Nucl. Phys. A **247**, 205 (1984).
9. Libert J. *et al.*, Phys. Rev. C **25**, 586 (1982).
10. Ekström C. *et al.*, Phys. Scr. **14**, 199 (1976).

Coulomb Excitation of a ^{78}Rb Radioactive Beam

J. Schwartz[1],[2], C. J. Lister[1], D. H. Henderson[1],
S.M. Fischer[1],[3], P. Reiter[1], A. Aprahamian[4], J. A. Cizewski[5],
C. N. Davids[1], R. deHaan[4], R. V. F. Janssens[1], D. Nisius[1],
D. Seweryniak[1], S.M. Vincent[6]

[1] *Argonne National Laboratory, 9700 South Cass Ave, IL 60439.,* [2] *WNSL, Yale University, 272 Whitney Avenue, CT 06511.,* [3] *Physics Department, University of Pennsylvania, 3400 Walnut St, PA 19104,* [4] *Physics Department, University of Notre Dame, IN 46556 ,* [5] *Physics Department, Rutgers University, NJ 08855,* [6] *Department of Physics, University of Surrey, Guildford GU2 5XH, England*

Abstract. We have produced a secondary radioactive beam of ^{78}Rb and Coulomb re-excited it. The beam was produced in the reaction ^{24}Mg(^{58}Ni,3pn)^{78}Rb at a beam energy of 260 MeV, using the ANL ATLAS accelerator. The residues of interest were separated from other reaction products and non-interacting beam using the Fragment Mass Analyzer (FMA). The beam leaving the FMA was ^{78}Kr and ^{78}Rb^{gs,m1,m2}, which was refocused onto a ^{58}Ni secondary target. We have extracted a spectrum of γ-rays associated with re-excitation of A=78 isobars. The re-excitation of stable ^{78}Kr was observed, which serves as a reference. γ-rays associated with excitation of ^{78}Rb^{gs,m1,m2} were also seen. The measured yields indicate that all the ^{78}Rb states are highly deformed.

INTRODUCTION

Nuclei which lie far from the valley of beta stability are important for reaching a more complete understanding of nuclear structure physics and nucleosynthesis. In the intermediate mass region, studies of proton-rich nuclei with N≈Z can improve our understanding of the evolution of shape coexistence and shell stabilized deformation. These nuclei also lie along the rp-nucleosynthesis path, which is sensitive to shapes and binding energies. They can also shed light on issues of isospin purity and mixing of T=0 and T=1 pairing modes.

N≈Z nuclei can be produced and studied in fusion-evaporation and fragmentation reactions. However, these techniques mainly populate yrast configurations. In contrast, the production of radioactive beams of these nuclei and their subsequent Coulomb re-excitation can be a powerful probe into non-yrast collectivity.

CP455, *ENAM98: Exotic Nuclei and Atomic Masses*
edited by B. M. Sherrill, D. J. Morrissey, and Cary N. Davids
© 1998 The American Institute of Physics   1-56396-804-5/98/$15.00

At present, radioactive beam accelerators are not widely available, so it is necessary to prepare the radioactive beam as part of any experiment. One method of production is to use the kinematic properties of a primary reaction to produce a secondary radioactive beam. Beams prepared in this manner generally have low intensity, poor emittance, low beam purity, and high backgrounds due to the natural radioactivity of the beam. There is also little control over the secondary beam energy. Nevertheless, it is an experimental challenge to investigate whether new data can be obtained on non-yrast collectivity from experiments done with these modest secondary radioactive beams. To date, there have been rather few attempts at re-excitation following fusion-evaporation production of the radioactive beam [1-3]. The experiment reported here represents a considerable development of the technique.

A test experiment should involve nuclear states which can be strongly populated and easily re-excited. Several candidates can be found in the A=80 region. The Rubidium isotopes are known to be deformed, with $0.3 < \beta_2 < 0.4$. We chose to study the nucleus ^{78}Rb which has been studied in atomic beam and fusion-evaporation experiments, through which the spin and parity of the ground state and excited states have been assigned [4,5]. The ground state of ^{78}Rb has $J^\pi=0^+$ and $T_{1/2}$=17.66min. ^{78}Rb also has two isomers: ^{78}Rb^{m1} with $J^\pi=4^-$, E*=111keV, and $T_{1/2}$=5.74min; and ^{78}Rb^{m2} $J^\pi=4^+$, E*=115keV, $T_{1/2} \geq 7$ μsec. The spectroscopic quadrupole moment of ^{78}Rb^{m1} has been measured to be Q_s=-0.814eb [4]. Transition rates between yrast states have also been measured [5]. The goal of this experiment was to investigate the relationship between the shapes of these states.

A beam of $\sim$ 20pnA ^{58}Ni at 260 MeV was used to produce ^{78}Rb, through the reaction ^{24}Mg(^{58}Ni,3pn)^{78}Rb. The ^{24}Mg target had a thickness of 0.82 mg/cm^2. This inverse reaction was particularly well suited to our requirements. The residues were produced [7] with a cross-section of 239mb (29mb for ^{78}Rb^{gs}, 83mb for ^{78}Rb^{m1}, and 127mb for ^{78}Rb^{m2}), in a recoil cone with $\theta_{1/2}^{max}$ $\sim 3^o$ with a mean energy of 150 MeV. The reaction had the added advantage that it provides a stable beam of ^{78}Kr (98mb) for reference.

After the production target, the recoiling reaction products passed through a charge-resetting foil of 0.02mg/cm^2 ^{12}C to equilibrate their charge-state distribution. The residues of interest were separated from the direct beam and other reaction products using the Argonne Fragment Mass Analyzer (FMA) [6]. The FMA separated recoils by A/q in a mode that was non-dispersive in energy. The central A/q setting was chosen to be A=78, q=25, with E_R=150 MeV. The efficiency of the FMA was $e_{FMA}^{3pn} \approx 77\%$ for ^{78}Rb, $e_{FMA}^{4p} \approx 52\%$ for ^{78}Kr. The fraction of recoils with q=25 was $\sim$ 24% for ^{78}Rb and $\sim$ 21% for ^{78}Kr. The setting of the FMA was adjusted to provide a beam which was parallel beyond the focal plane. Suppression of the primary beam was $\geq 10^7$. A 2.5cm square aperture allowed only particles with A/q=78/25 to enter the secondary beam-line. Particles were detected in two 9cmx8cm multi channel plate (MCP) detectors which were position sensitive in two dimensions with a resolution of 2.2mm, and which were placed 20cm apart.

The two-MCP combination was used to characterize the secondary beam at the re-excitation target, placed 92cm behind MCP1. The MCP system could handle event rates $\geq 10^5$/sec. The position of each ion impinging on the target was determined by applying a beam tracking method. The beam spot size at the secondary target was determined to have a FWHM of 2.24cm (Fig. 1). In the data analysis, ions were required to pass through a circle of diameter 3.7cm, centered on the beam-line.

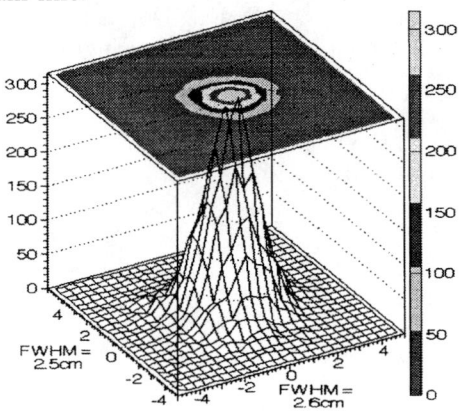

Figure 1. *The reconstructed beam profile at the re-excitation foil. The profile had a FWHM of 2.5cm, of which 1.1 cm was due to the tracking procedure.*

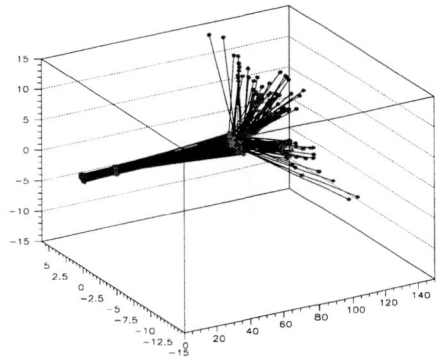

Figure 2. *Event-by-event ray-tracing of A=78 ions along the secondary beam-line. For clarity, only ions scattering into the beam-right and upper MWPC's are shown.*

The re-excitation target was 1.1mg/cm^2 ^{58}Ni, 4.3cm in diameter. Inverse kinematics were again selected in the re-excitation reaction to insure a high probability of detecting the scattered particles downstream. ^{58}Ni was a mechanically convenient re-excitation foil, which was not easily Coulomb excited, and from which A=78 ions were scattered in a recoil cone with $\theta_{lab}^{max} = 48^o$. The scattered ions were detected in a 4-quadrant, position sensitive multi-wire proportional counter (MWPC), placed 19cm beyond the target and subtending laboratory angles $6^o \leq \theta_{det} \leq 45^o$. Each quadrant of the MWPC was a separate detector, with all four quadrants sharing a common gas volume of isobutane at 3 Torr. The scattered-ion rate was ~3/sec. The detection of a scattered ion provided the main experimental trigger. Ions scattering at angles less than 3° in the laboratory frame were not intercepted and passed into a beam dump 20cm behind the MWPC. Photons associated with Coulomb excitation were detected in two 70% HPGe detectors, which were a distance of ~ 7cm from the target, and at an angle $\theta_{Ge} \sim 112^o$ with respect to the beam axis. From these data, complete event reconstruction was possible, allowing Doppler correction (Fig. 2). A γ-ray singles spectrum from the experiment is shown in figure 3. The spectrum is dominated by beta decay of ^{78}Rb ions which stopped in the target chamber. The suppression of these background events, and the extraction of the γ-rays associated with Coulomb excitation of the beam came from analysis of space and time correlations between scattered ions and γ-rays.

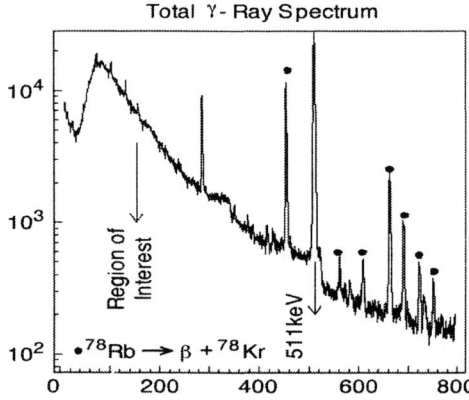

Figure 3. *A scaled-down singles spectrum from both of the 70% HpGe detectors. It is dominated by radioactive decay of the stopped ^{78}Rb beam particles and room-background radioactivity, which are marked with a filled circle.*

Figure 4 shows a time-correlation spectrum in which the real coincidences are evident above a flat background of random events. After time-random subtraction, a clean γ-ray spectrum was obtained, which is shown in figure 5. This spectrum is devoid of characteristic 511keV γ-rays associated with the β^+ decay of the beam. Coulomb excitation of ^{78}Kr and ^{78}Rb^{gs,m1,m2} are evident. The very intense peak at ~154keV corresponds to the excitation and decay of states built on both ^{78}Rb^{m1} and ^{78}Rb^{m2} isomers. Preliminary Coulomb excitation calculations indicate that states built on the ^{78}Rb^{gs} have a similar collectivity to that measured for ^{78}Rb^{m1}. The collectivity of ^{78}Rb^{m1} and ^{78}Rb^{m2} also appear to be similar, in agreement with recent in-beam measurements [5]. A quantitative analysis is in progress.

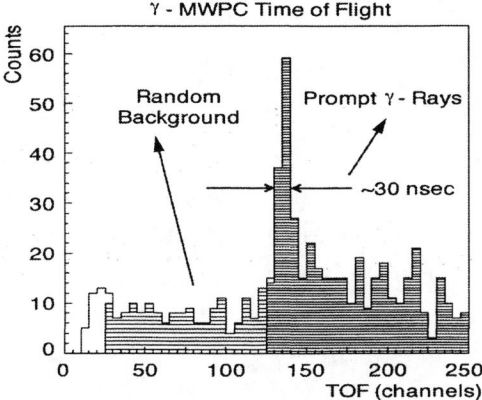

Figure 4. *Time correlations between scattered heavy-ions and gamma-rays. The events to the right of the prompt peak correspond to slow rise-time (low energy) pulses in the large Ge detectors.*

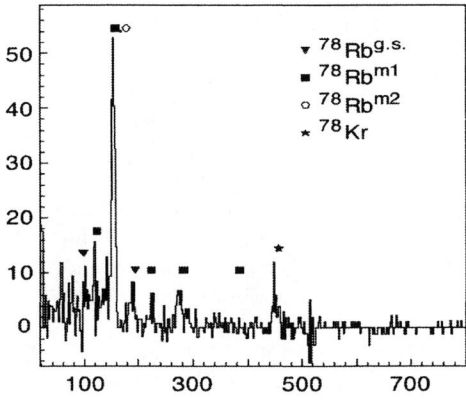

Figure 5. *The time-random subtracted spectrum of γ-rays associated with re-excitation of $A=78$ ions. Peaks associated with re-excitation of ^{78}Kr and $^{78}Rb^{gs,m1,m2}$ are all evident.*

This work is supported by the US Department of Energy, contract nos. W-31-109-ENG-38 and DE-FG-91-40609, and grants from the NSF and U.K. EPSRC.

REFERENCES

1. B . J. Varley *et al* Private communication (1988).
2. M. Oshima *et al* Nuc.Ins.Met.Phy.Res. A312, 425 (1992).
3. Y. Gono *et al* Nuc. Phys. A588, 241c-246c (1995).
4. C. Thibault *et al*, Phys.Rev. C23, 2720 (1981) .
5. R. A. Kaye *et al* Phys. Rev. C54, 1038 (1996), and to be published (1998).
6. C. N. Davids and J. D. Larson, Nucl. Instrum. and Meth. B40/41, 1224 (1989).
7. C. J. Lister *et al.* Private communication (1981). Cross-sections are accurate to $\leq 10\%$.

Magnetic Dipole Bands in ^{82}Rb, ^{83}Rb and ^{84}Rb[1]

R. Schwengner[1], H. Schnare[1], S. Frauendorf[1], F. Dönau[1],
L. Käubler[1], H. Prade[1], E. Grosse[1], A. Jungclaus[2], K. P. Lieb[2],
C. Lingk[2], S. Skoda[3], J. Eberth[3], G. de Angelis[4], A. Gadea[4],
E. Farnea[4], D.R. Napoli[4], C. A. Ur[4], G. Lo Bianco[5]

[1] *Institut für Kern- und Hadronenphysik, FZ Rossendorf, 01314 Dresden, Germany*
[2] *II. Physikalisches Institut, Universität Göttingen, 37073 Göttingen, Germany*
[3] *Institut für Kernphysik, Universität zu Köln, 50937 Köln, Germany*
[4] *INFN, Laboratori Nazionali di Legnaro, 35020 Legnaro, Italy*
[5] *INFN, Sezione di Milano, 20133 Milano, Italy*

Abstract. We have studied the isotopes ^{82}Rb$_{45}$, ^{83}Rb$_{46}$ and ^{84}Rb$_{47}$ to search for magnetic rotation which is predicted in the tilted-axis cranking model for a certain mass region around $A = 80$. Excited states in these nuclei were populated via the reaction ^{11}B + ^{76}Ge with $E = 50$ MeV at the XTU tandem accelerator of the LNL Legnaro. Based on a γ-coincidence experiment using the spectrometer GASP we have found magnetic dipole bands in each studied nuclide. The regular $M1$ bands observed in the odd-odd nuclei ^{82}Rb and ^{84}Rb include $B(M1)/B(E2)$ ratios decreasing smoothly with increasing spin in a range of $13^- \leq J^\pi \leq 16^-$. These bands are interpreted in the tilted-axis cranking model on the basis of four-quasiparticle configurations of the type $\pi(fp) \, \pi g_{9/2}^2 \, \nu g_{9/2}$. This is the first evidence of magnetic rotation in the $A \approx 80$ region. In contrast, the M1 sequences in the odd-even nucleus ^{83}Rb are not regular, and the $B(M1)/B(E2)$ ratios show a pronounced staggering.

INTRODUCTION

In the tilted-axis cranking (TAC) model [1], which considers the rotation of the nucleus about axes tilted with respect to the principal axes, a new rotational mode referred to as magnetic rotation has been established. This mode is expected to appear in nuclei with small deformation, if multi-quasiparticle configurations are formed from high-j proton particles and high-j neutron holes or vice versa. The coupling of these configurations results in a large transverse magnetic moment. The

---

[1] Supported by the German Ministry of Education and Research (BMBF)

rotating magnetic dipole gives rise to the emission of magnetic dipole ($M1$) radiation in contrast to the electric quadrupole ($E2$) radiation induced by the rotating deformed electric charge distribution in the case of conventional rotation. In the case of magnetic rotation the total spin is built up by the gradual alignment of the spins of the high-j nucleons ("shears mechanism"). This concept has been applied for the first time to the $M1$ bands ("shears bands") discovered in nuclei around ^{200}Pb [2]. The predicted decrease of the $M1$ transition strength with increasing spin caused by the gradual alignment of the individual spin vectors (closing of the shears) has recently been experimentally proven for the $M1$ bands in ^{198,199}Pb [3]. Magnetic rotation is also predicted for other mass regions of the nuclear chart [4]. Indeed, it has recently been observed in ^{105}Sn [5], ^{110}Cd [6] and ^{139}Sm [7].

Among the mass regions, where magnetic rotation is predicted to occur, there is also the region around $A = 80$ [4]. There, the particle-like protons fill successively the fp and the high-j intruder $g_{9/2}$ levels while hole-like neutrons occupy the $g_{9/2}$ level. Indeed, sequences of intense $M1$ transitions starting at about $E \approx 3$ MeV have been found in several Br, Rb and Kr isotopes (see, e.g., [8] and Refs. therein) but there is too little experimental information so far to prove the appearance of magnetic rotation. To search for experimental evidence of the predicted magnetic rotation in this region we have investigated the nuclides ^{82}Rb$_{45}$, ^{83}Rb$_{46}$ and ^{84}Rb$_{47}$.

EXPERIMENTAL RESULTS

Excited states in ^{82,83,84}Rb were populated via the reaction ^{11}B + ^{76}Ge at $E = 50$ MeV using the ^{11}B beam of the XTU tandem accelerator of the LNL Legnaro. γ rays were detected with the spectrometer GASP. A total of 1.5×10^8 three-fold coincidence events was recorded in a thin-target experiment. On the basis of this experiment we have found several new band structures with respect to previous work [9,10]. In particular, $M1$ bands have been found for the first time in each studied nuclide. Partial level schemes including these bands found in the present experiment are shown in Fig. 1. These level schemes result from γ-γ and γ-γ-γ coincidence relations and γ-ray intensities. Spin and parity assignments are based on γ-γ directional correlations and deexcitation modes.

INTERPRETATION

The $M1$ bands of negative parity observed in the odd-odd nuclei ^{82}Rb and ^{84}Rb are regular ($E_\gamma \propto J$). The $B(M1)/B(E2)$ ratios deduced from the intensities of transitions deexciting a certain state of the $M1$ band reach values up to 25 $(\mu_N/eb)^2$ and decrease smoothly with increasing spin in a range of $13 \leq J \leq 16$. This is an important characteristic of magnetic rotation. Thus, we have interpreted these bands in the framework of the TAC model [1]. In the calculations, the lowest-lying four-quasiparticle ($4qp$) configuration for $Z = 37$ and $N = 45, 47$ turns out to be $\pi(fp) \pi g_{9/2}^2 \nu g_{9/2}$, which has been adopted. The parameter κ of the QQ

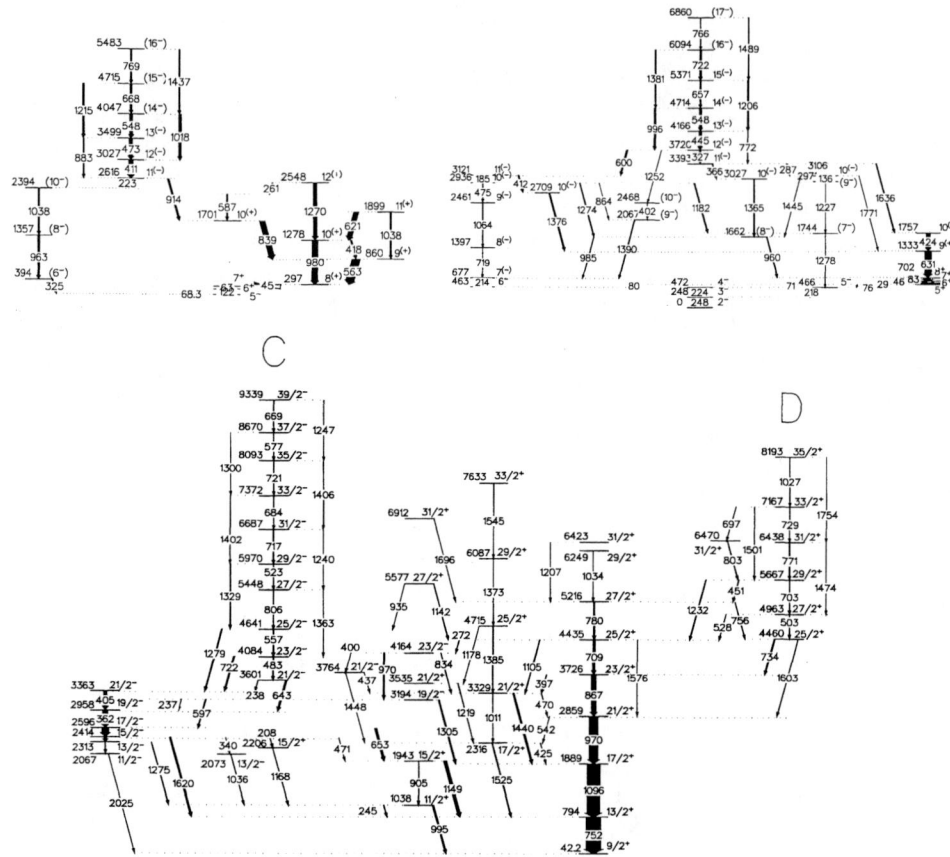

FIGURE 1. Partial level schemes of ^{82}Rb (top left) and ^{84}Rb (top right) and ^{83}Rb (bottom) deduced from this work.

interaction was adjusted such that in a calculation for the even-even neighbor ^{82}Kr the experimental $B(E2, 2^+ \to 0^+)$ [11] value is reproduced and in the case of ^{84}Rb scaled according to $\kappa \propto A^{-5/3}$. An equilibrium deformation of $\epsilon_2 = 0.16$ was obtained for the adopted 4qp configuration in both ^{82,84}Rb. The nuclei turn out to be very soft with respect to γ deformation with a tendency to positive values in ^{82}Rb but negative values in ^{84}Rb. The values of $\gamma = 20°$ and $\gamma = -10°$ are used for ^{82}Rb and ^{84}Rb, respectively. The experimental and calculated $B(M1)/B(E2)$ ratios are compared in Fig. 2. The experimental values in ^{82}Rb are well reproduced in the calculations. This is also the case for ^{84}Rb up to $\hbar\omega \approx 0.7$ MeV. The increase of the experimental values at higher frequency can not be described within the assumed 4qp configuration. It is probably due to a change to a 6qp configuration.

The $M1$ bands C and D in ^{83}Rb are irregular. Moreover, the experimental $B(M1)/B(E2)$ ratios of these bands shown in Fig. 2 display a pronounced stagger-

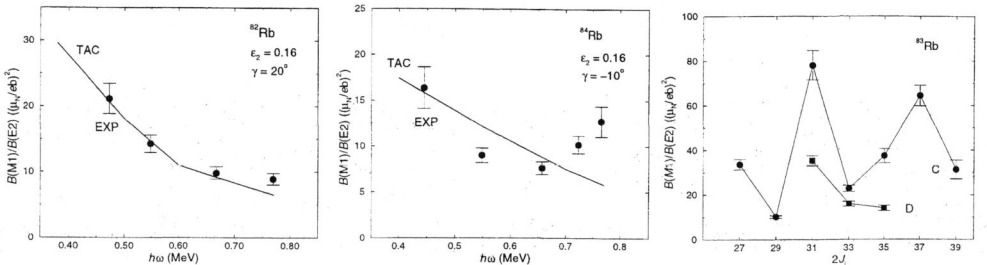

FIGURE 2. Experimental and calculated $B(M1)/B(E2)$ ratios of the negative-parity $M1$ bands in ^{82}Rb (left panel) and ^{84}Rb (middle panel). Experimental $B(M1)/B(E2)$ ratios of the $M1$ bands C and D in ^{83}Rb (right panel).

ing which is not compatible with regular shears bands. In contrast to the odd-odd nuclei, the breakup of a pair of neutrons is necessary in ^{83}Rb to generate $3qp$ or $5qp$ configurations of the shears type. This may drive the nuclear shape to very small quadrupole deformation, which is incapable to establish a stable shears mechanism.

Summarizing, we have observed $M1$ bands in ^{82}Rb, ^{83}Rb and ^{84}Rb for the first time. The $B(M1)/B(E2)$ ratios are of the order 10 - 20 $(\mu_N/eb)^2$ and *decrease* with the angular momentum. This is characteristic for Magnetic Rotation. Thus, first evidence for the predicted existence of this new mode near $A = 80$ has been provided. The $M1$ bands in the doubly odd nuclei ^{82}Rb and ^{84}Rb can be described in the TAC model on the basis of a $4qp$ shears configuration. In contrast, the $M1$ bands in the odd-even nucleus ^{83}Rb are not regular. The difference may be caused by the breakup of a neutron pair driving the nucleus to substantially smaller deformation, which is incapable to sustain the shears mechanism.

REFERENCES

1. Frauendorf, S., *Nucl. Phys.* **A557**, 259c (1993).
2. Baldsiefen, G, et al., *Nucl. Phys.* **A574**, 521 (1994).
3. Clark, R.M., et al., *Phys. Rev. Lett.* **78**, 1868 (1997).
4. Frauendorf, S., *Z. Phys.* **A358**, 163 (1997).
5. Gadea, A., et al., *Phys. Rev.* **C55**, R1 (1997).
6. Juutinen, S., et al., *Nucl. Phys.* **A573**, 306 (1994).
7. Brandolini, F., et al., *Phys. Lett.* **B388**, 468 (1996).
8. Tabor, S.L., and Döring, J., *Phys. Scr.* **T56**, 175 (1995).
9. Döring, J., et al., *Z. Phys.* **A338**, 457 (1991), and *Z. Phys.* **A339**, 425 (1991).
10. Gast, W., et al., *Phys. Rev.* **C22**, 469 (1980).
11. Kemnitz, P., et al., *Phys. Lett.* **B125**, 119 (1983).

Proton Scattering on ^{40}S

F. Maréchal[1,†], T. Suomijärvi[1], Y. Blumenfeld[1], A. Azhari[2,3],
D. Bazin[2], J.A. Brown[2], P.D. Cottle[4], M. Fauerbach[2,3],
T. Glasmacher[2,3], S.E. Hirzebruch[1,2], J.K. Jewell[4],
K.W. Kemper[4], P.F. Mantica[2,5], D.J. Morrissey[2,5], L.A. Riley[4],
J.A. Scarpaci[1] and M. Steiner[2]

[1] *Institut de Physique Nucléaire, IN$_2$P$_3$-CNRS, 91406 Orsay, France*
[2] *NSCL/MSU, East Lansing, MI 48824, USA*
[3] *Depart. of Physics and Astronomy, MSU, East Lansing, MI 48824, USA*
[4] *Depart. of Physics, FSU, Tallahassee, FL 32306, USA*
[5] *Depart. of Chemistry, MSU, East Lansing, MI 48824, USA*

Abstract. We have recently studied the structure of the neutron rich sulfur isotope ^{40}S by using elastic and inelastic proton scattering in inverse kinematics. Optical potential and folding model calculations are compared with the elastic and inelastic angular distributions. Using coupled-channel calculations, the β_2 value for the 2_1^+ excited state is determined to be 0.35±0.05. The extracted value of M_n/M_p ratio indicates a small isovector contribution to the 2_1^+ state of ^{40}S. The microscopic analysis of the data is compatible with the presence of a neutron skin for this nucleus.

INTRODUCTION

Inelastic scattering towards low lying collective states gives access to transition probabilities and nuclear deformations, and is a well suited tool to scan new regions of deformation and the modification of shell strucure far from stability.

We have undertaken a study of neutron rich sulfur isotopes through elastic and inelastic scattering of protons in inverse kinematics. The first experiment was performed on ^{38}S and the results have been reported in ref. [1]. Here we report the very recent results we obtained for the proton scattering on the ^{40}S nucleus.

EXPERIMENTAL SETUP

The secondary ^{40}S beam was produced by fragmenting a 60 MeV/A ^{48}Ca beam, provided by the K1200 cyclotron at the National Superconducting Cyclotron Laboratory, in a ^{9}Be target. The resulting beam was analysed and purified by using the A1200 fragment separator [2]. A final intensity of about 2000 pps for the ^{40}S beam at 30 MeV/A was obtained.

The ^{40}S beam was scattered on a very thin 1.6 mg/cm^2 (CH$_2$)$_n$ target. The recoiling protons were detected using the telescopes of the FSU-MSU silicon strip array [1] positioned 29 cm from the target and covering laboratory angles between 61° and 84°. The particle identification in the recoiling telescopes was performed either by a time-of-flight measurement or by a ΔE-E measurement depending on the particle energy.

The data in the silicon strip telescopes were taken in coincidence with a zero degree ΔE-E plastic detector which identified the outgoing fragments and allowed us to select the reaction channel of interest, thus very effectively reducing the background. Due to the poor emittance of the secondary ^{40}S beam, two Parallel Plate Avalanche Counters (PPAC), were used to measure event by event the incident beam angle and beam position on the target.

RESULTS

Figure 1 displays the excitation energy spectrum for ^{40}S, first integrated over the total measured center of mass angular range $20^o \leq \theta_{cm} \leq 46^o$ (left) and then integrated over the 28°-36° angular bin where the elastic scattering cross-section exhibits a minimum (right). We clearly identify the first 2^+ excited state located at 860±90 keV which is in good agreement with the value obtained in a previous coulomb excitation experiment [3].

Figure 2 shows the angular distributions for the ground state and the first 2^+ excited state of ^{40}S. The elastic angular distribution was obtained by projecting the contents of the corresponding contour in the excitation energy vs. θ_{cm} plane. The two points of the inelastic angular distribution were obtained by fitting the inelastic peak with gaussians for two different angular bins and then integrating the number of counts. The very low statistics do not allow us to clearly identify the inelastic contribution for other angles where the elastic scattering dominates the count rate.

Coupled-channel calculations using the ECIS code [4] are shown in comparison with the data. Note that no arbitrary normalization is involved here. The calculation based on the Becchetti-Greenlees parameterization [5], which was developed for (p,p) scattering on A$\geq$40 nuclei, is shown by the solid lines in fig. 2. The shape of the experimental ground state angular distribution is in full agreement with the calculation. The β_2 value was extracted by normalizing the coupled-channel calculation to the 2_1^+ state cross-section. This yields

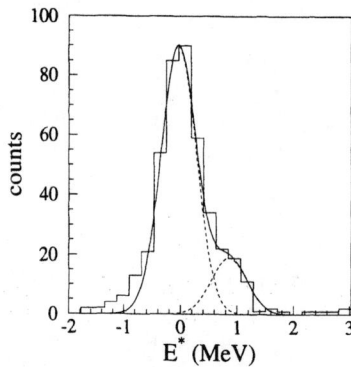 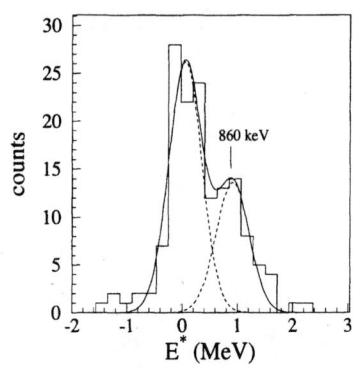

FIGURE 1. Excitation energy spectrum for ^{40}S scattering on protons at 30 MeV/A integrated over the total cm angular range $20° \leq \theta_{cm} \leq 46°$ (left) and over the $28°$-$36°$ angular bin where the elastic contribution is minimum (right). The dashed lines are gaussian fits to the elastic scattering peak and first excited state. The solid line is the sum of the two contributions.

a value of $\beta_2 = 0.35 \pm 0.05$, larger than the value obtained from the COULEX experiment [3].

Combined measurements of inelastic hadron scattering and electromagnetic excitation are intersting since they can provide information to disentangle proton and neutron contributions to the studied transition. For instance, information on the isoscalar or isovector character of the 2^+ excitation is given by the ratio of the neutron to proton multipole transition matrix elements M_n/M_p [6]. This ratio can be calculated from the β_2 values measured by electromagnetic and nuclear excitations using the formula derived in ref. [7]. Using the β_2 value from coulomb excitation measured in ref. [3] this yields $M_n/M_p = (1.50 \pm 0.20)$N/Z for ^{38}S [1] and $M_n/M_p = (1.25 \pm 0.25)$N/Z for ^{40}S. In both cases, M_n/M_p is greater than N/Z which is the value expected for a pure isoscalar excitation, thus indicating an isovector contribution to the 2^+ excitation. However, in the case of ^{40}S, the M_n/M_p value does not increase with respect to ^{38}S as would be expected from the simple picture of a ^{36}S core with valence neutrons driving the excitation. This saturation could be due to the core polarization which becomes more important when the neutron number increases.

A microscopic analysis of the data has also been performed. In that case, the interaction and transition potentials are obtained by folding the ground state and the transition densities with an effective nucleon-nucleon interaction. Hartree-Fock nuclear densities and shell model transition densities were calculated [8] and clearly exhibit the presence of a neutron skin. These densities were folded with the JLM nucleon-nucleon potential. As shown by dashed lines on fig. 2, a good agreement with data is obtained at the expense of a

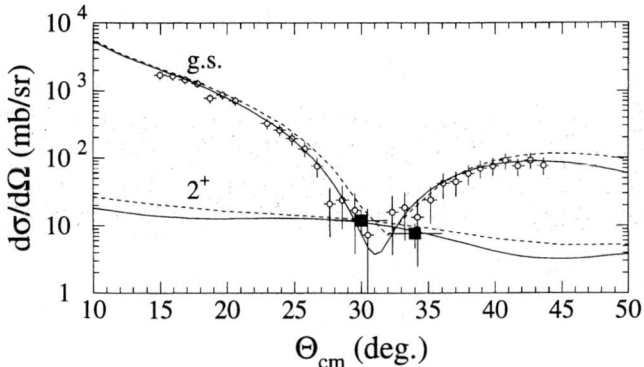

FIGURE 2. Angular distributions for the ground state (circles) and the first excited state (squares) in the ^{40}S(p,p') reaction at 30 MeV/A. The solid lines are coupled-channel calculations with Becchetti-Greenlees optical potential. The dashed lines are folding model calculations (see text for details).

slight renormalization of the neutron densities by a factor of 0.88.

CONCLUSIONS

We have measured angular distributions for elastic scattering and inelastic scattering of protons on the unstable ^{40}S nucleus. The β_2 value has been extracted and compared with the value obtained by coulomb excitation. The deduced M_n/M_p ratio indicates a small isovector contribution to the 2^+ excitation whereas a microscopic analysis is compatible with the presence of a neutron skin in the ^{40}S nucleus.

REFERENCES

† Present address Dept. of Physics, Florida State University, Tallahassee, FL 32306.

1. J.H. Kelley et al., *Phys. Rev.* **C56**, R1206 (1997).
2. B.M. Sherill et al., *Nucl. Instr. and Meth.* **B70**, 298 (1992).
3. H. Scheit et al., *Phys. Rev. Lett.* **77**, 3967 (1996).
4. J. Raynal, *Phys. Rev.* **C23**, 2571 (1981).
5. F.D. Becchetti Jr., and G.W. Greenlees, *Phys. Rev.* **182**, 1190 (1969).
6. M.A. Kennedy et al., *Phys. Rev.* **C46**, 1811 (1992).
7. A.M. Bernstein et al., *Comments Nucl. Part. Phys.* **11**, 203 (1983).
8. E.K. Warburton, J.A. Becker, and B.A. Brown, *Phys. Rev* **C41**, 1147 (1990); B.A. Brown, and W.A. Richter, *Phys. Rev* **C54**, 673 (1996).

Decay properties of ground–state and isomer of ^{103}In

J. Szerypo^{a,b}, R. Grzywacz^a, Z. Janas^a, M. Karny^a, M. Pfützner^a,
A. Płochocki^a, K. Rykaczewski$^{a,b,\,1}$, J. Żylicz^a, M. Huyse^b,
G. Reusen^b, J. Schwarzenberg$^{b,\,2}$, P. Van Duppen^b, A. Woehr^b,
H. Keller^c, R. Kirchner^c, O. Klepper^c, A. Piechaczek^c, E. Roeckl^c,
K. Schmidt^c, L. Batist^d, A. Bykov^d, V. Wittman^d, B. A. Brown^e

^{a} Institute of Experimental Physics, Warsaw University, PL-00681 Warsaw, Poland
^{b} Instituut voor Kern- en Stralingsfysica, University of Leuven, B-3001 Leuven, Belgium
^{c} Gesellschaft für Schwerionenforschung, D-64220 Darmstadt, Germany
^{d} St. Petersburg Nuclear Physics Institute, 188-350 Gatchina, Russia
^{e} NSCL, Department of Physics and Astronomy, MSU, East Lansing, MI 48824-1321, USA

Abstract. The β–decay properties of ground–state and isomer of ^{103}In were investigated by means of γ–ray spectroscopy. Their half-lives were determined to be 60 ± 1 s and 34 ± 2 s, respectively. 149 γ transitions were ascribed to the decay of ^{103g}In. The main part of the resulting distribution of the Gamow–Teller strength B(GT) is associated with the feeding of ^{103}Cd levels at excitation energies around 3 MeV. The sum of the B(GT) values deduced from the present γ–ray data amounts to 0.34, which provides a lower limit to the total Gamow–Teller strength. The β–branching ratio for the ^{103m}In decay is estimated and compared with the relevant values for the neighbouring indium isotopes including ^{101}In whose half-life was determined to be 14.9 ± 1.2 s.

INTRODUCTION

The ground–state of odd–A indium isotopes in the vicinity of doubly–magic ^{100}Sn is expected to have spin and parity $I^\pi = 9/2^+$ due to the coupling of an unpaired g$_{9/2}$ proton to the cadmium even–even core. Within the Elementary Single Particle Shell Model (ESPSM), two GT decay modes are expected for odd–A neutron-deficient indium isotopes, namely: (i) a transformation of an unpaired (odd) g$_{9/2}$ proton into a g$_{7/2}$ neutron, and (ii) a transformation of a g$_{9/2}$ core proton into a

---

[1] Present address: Oak Ridge National Laboratory, Physics Division, PO Box 2008, Oak Ridge, Tennessee 37831-6371, USA
[2] Present address: Nuclear Structure Laboratory, University of Notre Dame, Notre Dame, Indiana 46556, USA

$g_{7/2}$ neutron. The interpretation of the GT–strength distribution B(GT) and, in particular, the ratio between the B(GT) values of the two expected decay modes, should allow for a better understanding of the GT quenching phenomenon.

Another important nuclear structure information can be gained for odd–A indium from M4 isomers, lying at excitation energies of about 600–700 keV. These M4 proton–hole isomers, with spin and parity $I^\pi = 1/2^-$, are formed by promoting one of the core $p_{1/2}$ protons into the $g_{9/2}$ orbital. The excitation energy of such isomers is directly related to the energy gap between $p_{1/2}$ and $g_{9/2}$ proton orbitals, and thus allows to study the gap evolution with decreasing neutron number.

For very neutron–deficient odd–mass indium isotopes, the β–decay of the $1/2^-$ isomer can compete with the M4 deexcitation. Decay studies of both ground–state and isomer would give a direct comparison between the GT transformation of the even–even core plus a single $g_{9/2}$ proton (g.s.) and the core alone (isomer).

For such complex decays the total absorption spectrometer (TAS) [1] is a suitable spectroscopic tool, as the GT–strength is spread over many highly excited daughter states, which results in statistical γ–ray cascades and hence in difficulties of establishing the true β–feeding pattern. However, without the information on the discrete part of the decay scheme, the deconvolution of TAS spectra is unreliable.

Based on the above arguments, we decided to reinvestigate, using the high–resolution γ–ray spectroscopy, the decay properties of the ground and $1/2^-$ isomeric states of ^{103}In and ^{101}In [2].

EXPERIMENTAL PROCEDURE

The experiment was performed at the GSI on–line mass separator. ^{103}In and ^{101}In were produced in ^{50}Cr(^{58}Ni,3p2n) and ^{50}Cr(^{58}Ni,3p4n) reactions, respectively. A 5.9 MeV/u ^{58}Ni beam of about 20 particle nA impinged on an enriched ^{50}Cr target (3.3–4.6 mg/cm^2, enrichment 91% or 97%, on 2 or 2.3 mg/cm^2 ^{nat}Mo backing) that was mounted in front of a FEBIAD–E [3] plasma–discharge ion source.

The mass–separated A=103 activities were implanted in a transport tape. After a collection time of 80 s, the sources were transported to the counting station and measured during 80 s. This station included two Ge detectors, positioned in close geometry to the source, on opposite sides of the tape.

The γ–ray singles spectra were measured in multispectrum mode. Coincidence data were stored event–by–event.

EXPERIMENTAL RESULTS

β–**decay of** ^{103g}**In:** Prior to our work, 13 γ–lines were assigned to the decay of ^{103}In [4]. Our measurements allowed us to identify, from a detailed analysis of γ–singles and $\gamma\gamma$–coincidence data, 149 γ–rays as being due to the β^+/EC–decay of ^{103g}In. The weighted average of the ^{103g}In half–life, obtained from the multispectrum analysis for the 24 most intense lines, is 60 ± 1 s. This result is in agreement with the previously obtained values of 64.8 ± 6.6 s [5] and 70.8 ± 6.0 s [4], with a 1.6σ deviation in the latter case.

Decay properties of the 1/2⁻ isomer in ^{103}In: The presence of a long–lived isomer ^{103m}In has already been reported in the work of Decrock [6] who identified the isomer by its internal transition (IT) and measured the isomeric energy and the half–life to be 630.6 ± 0.5 keV and 35 ± 5 s, respectively. The time analysis of the KX–rays permitted Decrock to estimate the β–decay branch of ^{103m}In to be 67%, leaving 33% for the IT. In our experiment, the the energy of the IT was determined as 631.7 ± 0.1 keV. Its time analysis yielded a half–life of 34 ± 2 s.

β–decay of ^{101}In: For the β–decay of ^{101}In, four γ–transitions were identified and a half–life value of 16 ± 3 s was measured at the Louvain–la–Neuve on–line mass separator [7]. From our data we are able to confirm two of these γ–lines, namely those at 252 and 892 keV, and to show that they are in coincidence with each other. The half–life of ^{101}In was redetermined with improved accuracy to be 14.9 ± 1.2 s. This activity is assigned to ^{101g}In, wheras we have not obtained any evidence for the existence of an isomeric state in ^{101}In.

DISCUSSION

Gamow–Teller strength: On the basis of the ^{103}In decay scheme, a γ–intensity balance was performed for each level, ascribing the difference between the γ–intensity depopulating and populating a given level to its β–feeding. Using the β–feeding results and Q_{EC} = 6.05 ± 0.02 MeV [8], we determined logft values and transformed them into a B(GT) distribution according to [9]

$$B(GT) = \frac{3860 \text{ s}}{ft} \qquad (1)$$

The resulting B(GT) distribution indicates the dominant role of the even–even core decay, which populates mainly the $3qp$ states at the energy of about 3 MeV. The total strength B_Σ(GT), taken as the sum of the partial strengths for different energy bins, is equal to 0.34. Since the numerous weak γ–rays were probably unobserved in our measurements, the B(GT) value should be regarded as lower limit only.

The half–life of ^{101g}In is predicted, within the advanced spherical shell model (ASSM), [10], to be 13 s, in agreement with the experimental result of 14.9 ± 1.2 s. Unfortunately, the ASSM calculations are not yet possible for ^{103}In.

M4 isomerism in odd–A indium isotopes: The properties of $^{101m-109m}$In from measurements or semiempirical estimates are compiled in Table 1. The data on E_γ^{IT} and $T_{1/2}^{exp}$ for $^{105m-109m}$In are taken from the literature. As can be seen, the β–decay becomes an important disintegration mode already for ^{105m}In and dominates the decay of ^{103m}In and ^{101m}In. One should note the good agreement of our I_β estimate

TABLE 1. Data on β-decay branching ratios for M4 isomers in odd-A indium isotopes.

Nuclide	E_γ^{IT} (keV)	$T_{1/2}^{exp}$ (s)	I_β (%)
^{109m}In	650.1	80.4(42)	0
^{107m}In	678.5	50.4(6)	0
^{105m}In	674.1	48(6)	31.9–22.4
^{103m}In	631.7	34(2)	70.6–66.6
^{101m}In	~600	<15	>92–91

for ^{103m}In with the experimental result of about 67% obtained by Decrock et al. [6]. The full discussion of our results is presented in [2].

Acknowledgements: This work was supported: by the Polish Committee of Scientific Research grant KBN 2 P03B 039 13; by the Russian Fund for Basic Research and Deutsche Forschungsgemeinschaft Contract No. 436 RUS 113/201/0(R); by the EC Contracts No. ERBFMGECT950083, ERBCIPD–CT–940091 and ERBCIPD–CT–950083. U. S. NSF grant 9605207; partially by U. S. DOE under DE-AC05-96OR22464 and ORNL managed by LMER Corporation;

REFERENCES

1. Karny, M., Nitschke, J. M., Archambault, L. F., Burkard, K., Cano–Ott, D., Hellström, M., Hüller, W., Kirchner, R., Lewandowski, S., Roeckl, E. and Sulik, A., Nucl. Instr. Meth. **B126**, 411 (1997).
2. Szerypo, J., Grzywacz, R., Janas, Z., Karny, M., Pfützner, M., Płochocki, A., Rykaczewski, K., Żylicz, J., Huyse, M., Reusen, G., Schwarzenberg, J., Van Duppen, P., Woehr, A., Keller, H., Kirchner, R., Klepper, O., Piechaczek, A., Roeckl, E., Schmidt, K., Batist, L., Bykov, A., Wittman, V. and Brown, B. A., Z. Phys. **A359**, 117 (1997).
3. Kirchner, R., Burkard, K., Hüller, W., and Klepper, O., Nucl. Instr. Meth. **186**, 295 (1981).
4. Verplancke, J., Coenen, E., Cornelis, K., Huyse, M., Lhersonneau, G., and Van Duppen, P., Z. Phys. **A315**, 307 (1984).
5. Lhersonneau, G., Dumont, G., Cornelis, K., Huyse, M., and Verplancke, J., Phys. Rev. **C18**, 2688 (1978).
6. Decrock, P., Licentiaats Thesis, Leuven University, 1988, unpublished.
7. Huyse, M., del Marmol, P., Coenen, E., Deneffe, K., Van Duppen, P., and Vanhorenbeeck, J., Z. Phys. **A330**, 121 (1988).
8. Audi, G., and Wapstra, A. H., Nucl. Phys. **A595**, 409 (1995).
9. Towner, I. S., Nucl. Phys. **A444**, 402 (1985).
10. Brown, B. A., and Rykaczewski, K., Phys. Rev. **C50**, R2270 (1994).

Low-Energy Structure of Neutron-Rich S, Cl and Ar Nuclides Through β Decay

J. A. Winger, H. H. Yousif, W. C. Ma, V. Ravikumar, W. Lui,
S. K. Phillips, R. B. Piercey

*Department of Physics and Astronomy,
Mississippi State University, Mississippi State, MS 39762-5167*

P. F. Mantica[a,b], B. Pritychenko[b], R. M. Ronningen[b], M. Steiner[b]

*[a] Department of Chemistry and [b] National Superconducting Cyclotron Laboratory,
Michigan State University, East Lansing, MI 48824*

Abstract.
Detailed nuclear structure studies of $20 \leq N \leq 28$, $14 \leq Z \leq 20$ nuclides have been limited until recently due to the lack of a good production mechanism. With the advent of projectile fragmentation facilities these nuclides can now be produced, separated, and studied in detail using several different techniques. Two recent experiments conducted at the NSCL have provided information on the β decays of ^{39,40,41}P, ^{40,41,42,43}S, and ^{42,43,44,45}Cl, which will be used to establish level schemes for the daughter nuclides. These will provide a better understanding of the systematic change from spherical to deformed shapes within the proton sd and neutron fp shells. Presented here are preliminary results from these experiments with an emphasis placed on the structure of the deformed nucleus ^{40}S.

INTRODUCTION

Nuclei far from stability provide a unique opportunity to increase our understanding of nuclear interactions in extreme conditions and often challenge our theoretical models. An example is the well known deformation near singly magic ^{32}Mg. Theoretical calculations by Werner *et al* have indicated the possibility of deformation in the region near singly magic ^{44}S [1,2]. However, information on low-energy nuclear structure in this region has been quite limited primarily due the inability of most traditional methods to produce these nuclei. Starting about ten years ago a number of experiments to measure half-lives and β-delayed neutron probabilities were performed at GANIL using nuclei produced through projectile fragmentation [3–6]. More recently, experiments have been performed at the NSCL to measure $B(E2, 0^+_{g.s} \to 2^+_1)$ values using intermediate-energy Coulomb excitation [7,8]. This advent of quality radioactive beams from projectile fragmentation facilities has opened this region to further study. Discussed here are preliminary results from two experiments performed at the NSCL to study the properties of neutron-rich S, Cl, and Ar isotopes through the β decay of their parents. This information will be used in a systematic study of the change from spherical to deformed structure in this region.

The nuclides studied in the two experiments were ^{39,40,41}P, ^{40,41,42,43}S, and ^{42,43,44,45}Cl. Of these, only ^{42,43}Cl had been previously studied in detail [9], while four γ rays have been assigned to and the half-life measured for ^{40}S decay [10], and the neutron emission probabilities and half-lives have been measured for ^{39,40,41}P, ^{42,43}S, and ^{44,45}Cl [3-6]. Preliminary level schemes have been established for ^{40}S, ^{40,42}Cl, and ^{42,43}Ar based on information from our first experiment, and have been presented previously [11]. The results from the second experiment are in the early stages of analysis.

EXPERIMENTAL SETUP

The experiments were performed at the NSCL using the A1200 fragment separator to provide separated radioactive beams. The availability of a high quality ^{48}Ca beam, which was fragmented to produce the desired nuclides, at the NSCL [12] was a major reason for the sucess of these experiments. In the first experiment, a 70 MeV/A ^{48}Ca beam was fragmented with a 254 mg/cm^2 Be target with additional separation provided by a degrader wedge. For the second experiment, a slightly higher beam energy of 80 MeV/A and thicker target of 376 mg/cm^2 were used. Typical primary beam intensities of $\sim$5 pnA were available in both experiments.

The basic procedure used to study a group of nuclides was as follows. The A1200 would be set for maximum production of the most exotic nuclide (e.g. ^{40}P). For this setting, additional nuclides would be present in the beam, usually with a higher intensity than the most exotic nuclide. For example, when tuned for maximum ^{40}P production the beam had strong contaminants of ^{42}S and ^{43}Cl. However, since little was know about the other nuclides composing the beam, it was advantageous to commit some portion of the experiment to studying these nuclides. Hence the use of a cocktail beam containing several nuclides of interest made it possible to study the whole group simultaneously. Furthermore, since the constituents of the beam have different masses and the ratio of intensities of the components could be easily varied with small changes in the tuning of the A1200, there would be no ambiguity in the assignment of γ rays to a particular decay. After separation, the beam was transported to a low-background location where the ions were identified using a thin Si detector before being implanted into Al targets. The targets were attached to a rotatable wheel for removal of long-lived daughters, and viewed by two Ge detectors and a thin plastic β detector. The detectors provided information on the identity of each ion implanted, γ-ray energies, and $\gamma\gamma$ and $\beta\gamma\gamma$ coincidences

TABLE 1. β decay properties of nuclides in the current study.

Nuclide	$t_{1/2}$ This Work	$t_{1/2}$ Others	Primary γ rays (keV)
^{39}P	320_{30} ms	160^{+300}_{-100} ms [3]	339.8_7, 398.2_7, 1126.2_7, 1524.6_7
^{40}P	146_{10} ms	260^{+100}_{-60} ms [4]	903.5_3, 1351.0_4, 3233.8_4, 3488.5_4
^{41}P	150_{15} ms	120_{20} ms [4]	329.1_7, 501.9_7, 569.6_7, 903.5_7, 1307.5_7, 1613.7_7
^{40}S		8.8_{22} [10]	211.9_5, 431.8_5, 677.9_7, 889.2_8, 1013.7_7, 1081.6_8
^{41}S	1.99_5 s		130.2_7, 553.6_7, 761.0_7, 1266.2_7, 1404.8_7
^{42}S	1.013_{15} s	560_{60} ms [6]	118.6_5, 470.7_6, 723.5_6, 1281.5_6
^{43}S	260_{15} ms	220^{+80}_{-50} ms [4]	328.7_7, 611.6_7, 878.5_7, 1339.6_7, 2021_1
^{42}Cl		6.8_3 [13]	1207.7_6, 1277.3_5, 2837.3_5, 3208.3_5
^{43}Cl	3.07_7 s	3.3_2 [13]	679.1_7, 761.6_7, 1031.6_7, 1381.7_7
^{44}Cl	640_{40} ms	434_{60} [6]	852.5_7, 965.1_7, 1157.5_7, 2010_1, 2796_2
^{45}Cl	420_{30} ms	400_{43} [6]	541.7_7, 1157.5_7, 1192.6_7, 1228.1_7, 1524.4_7, 2751_2

while changes in the beam-on/beam-off timing cycle and beam-line tuning allowed association of γ rays to specific decays and measurement of the β-decay half-lives.

EXPERIMENTAL RESULTS

Reported in Table 1 are some of the results from analysis of the data. Analysis of the first experiment (^{40}P, ^{40,42}S, ^{42,43}Cl) is nearly complete, while the analysis of the second experiment is in its early stages and the results should be considered very preliminary. The half-lives of these nuclides, except ^{40}S and ^{42,43}Cl, had been measured with low statistics, hence our results are a significant improvement. In addition to the half-lives and level schemes, we will be able to extract β-delayed neutron-emission probabilities for many of these nuclides.

CONCLUSION

The initial goal of this series of experiments was to search for deformation among the S isotopes with the information obtained for the other nuclides being an added bonus. Shown in Fig. 1 is the level scheme for ^{40}S. Notice that the average energy of the second and third excited states lies at ≈ 2.3 time the energy of the 2_1^+ thus indicating primarily vibrational states for the low-energy structure. These two states are probably the 2^+ and 4^+ of the two-phonon triplet. The deviation of the average energy from twice that of the 2_1^+ and the separation between these states of 338 keV indicates some moderate amount of deformation in the nucleus in agreement with the results of Scheit et al [7]. Further analysis will concentrate on theoretical calculations to determine the deformation needed to yield the observed level spacings.

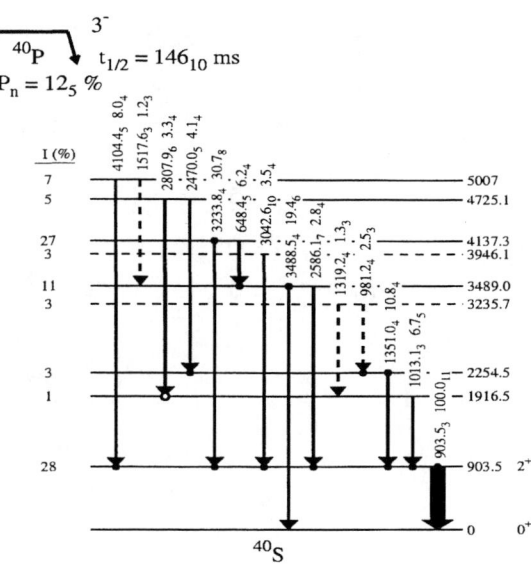

FIGURE 1. Experimentally observed level scheme of ^{40}S from ^{40}P β decay.

ACKNOWLEDGMENTS

This work is supported under DOE Grant Nos. DE-FG02-96ER41006 and DE-FG05-95ER40939, and NSF Grant No. PHY-95-28844.

REFERENCES

1. Werner, T. R. et al, Phys. Lett. **335B**, 259-265 (1994).
2. Werner, T. R. et al, Nucl. Phys. **A597**, 327-340 (1996).
3. Mueller, A. C. et al, Z. Phys. **A330**, 63-68 (1988).
4. Lewitowicz, M. et al, Nucl. Phys. **A496**, 477-484 (1989).
5. Sorlin, O. et al, Phys. Rev. **C47**, 2941-2953 (1993).
6. Sorlin, O. et al, Nucl. Phys. **A583**, 763-768 (1995).
7. Scheit, H. et al, Phys. Rev. Lett. **77**, 3967-3970 (1996).
8. Glasmacher, T. et al, Phys. Lett. **395B**, 163-168 (1997).
9. A. Huck, A. et al, in *Proceedings of the 4th International Conference on Nuclei Far from Stability*, Helsingor, Denmark, Vol. 2, pp. 378-386 (1981).
10. Dufour, J. P. et al, Z. Phys. **A324**, 487-488 (1986).
11. Winger, J. A., Mantica, P. F., and Ronningen, R. M., in *Proceedings of the International Conference on Fission and Properties of Neutron Rich Nuclei*, Sanibel Island, Florida, November 10-15, 1997.
12. Harkewicz, R., Rev. Sci. Instrum. **67**, 2176-2178 (1996).
13. Vosicki, B. et al, Nucl. Instrum. Methods **186**, 307-313 (1981).

Coulomb Excitation of ^{56}Ni

Y. Yanagisawa[1], T. Motobayashi[1], S. Shimoura[1], Y. Ando[1],
H. Fujiwara[1], I. Hisanaga[1], H. Iwasaki[3], Y. Iwata[1],
H. Murakami[1], T. Minemura[1], T. Nakamura[3], T. Nishio[1],
M. Notani[2], H. Sakurai[2], S. Takeuchi[1], T. Teranishi[2],
Y.X. Watanabe[3], and M. Ishihara[2,3]

[1] *Rikkyo University, Nishi-Ikebukuro, Toshima, Tokyo 171-8501, JAPAN*
[2] *RIKEN (Institute of Chemical and Physical Research),
Wako, Saitama 351-0198, JAPAN*
[3] *University of Tokyo, Hongo, Bunkyo, Tokyo 113-8654, JAPAN*

Abstract. We have performed an experiment of Coulomb excitation of a doubly magic nucleus ^{56}Ni to its first excited 2^+ state at 2.7 MeV in inverse kinematics. Radioactive ^{56}Ni particles at an incident energy 70.7MeV/nucleon with 1500 s^{-1} was obtained via fragmentation of ^{58}Ni at RIPS facility of RIKEN. The E2 transition probability $B(E2)$ between the 0^+ ground state to the first 2^+ state was deduced to be 580 ± 70 e^{2}fm^4 by a coupled channel analysis assuming Coulomb excitation mechanism. The present result agrees with the value deduced from the ^{56}Ni + p inelastic scattering experiment at GSI.

INTRODUCTION

The $B(E2)$ value of ^{56}Ni nucleus reported from a lifetime measurement [1] was about 400 e^{2}fm^4, consistent with the predicted value by a shell-model calculation [2]. Recently, a large $B(E2)$ value of 600 ± 120 e^{2}fm^4 was reported from a deformation parameter extracted by a DWBA analysis on the $0^+ \to 2^+$ excitation induced by the p(^{56}Ni, p')^{56}Ni* reaction measurement at GSI [3]. However, the $B(E2)$ extracted from the proton inelastic scattering, in which the interaction is dominated by the nuclear force and not by the electromagnetic force, might have systematic uncertainty. In this report, an experimental study on the Coulomb excitation of ^{56}Ni is discussed. We have measured the γ rays from ^{208}Pb(^{56}Ni,^{56}Niγ)^{208}Pb reaction. Since the reaction is dominated by electromagnetic interaction, a more direct determinant of $B(E2)$ is possible compared with the proton inelastic scattering.

EXPERIMENTAL PROCEDURES

The experiment was performed at the RIKEN Accelerator Research Facility. Primary beams of ^{58}Ni, typically 2pnA at 95 MeV/u, incident on a ^{9}Be target. A fragment separator RIPS [5] analysed the fragments, and provided a beam of the unstable nucleus ^{56}Ni with 70.7 MeV/u. The secondary ^{56}Ni beam was discriminated from contaminants, such as ^{55}Co and ^{54}Fe, by the time-of-flights (TOF) technique. The TOF was measured by the signal from 0.5 mm thick plastic scintillator behind the target and the cyclotron RF signal. A 260 mg/cm^2 ^{208}Pb target was located at the focal point of the RIPS system, and was bombarded ^{56}Ni beam with 1500 s^{-1} intensity. Empty target run was also performed to estimate background which is resulted from reactions due to materials other than the target.

The inelastically scattered ^{56}Ni particles were detected by a ΔE-E silicon detector telescope, placed 35 cm from the target after passing through a 50 μm thick mylar window. It consisted of two silicon detectors of 92mm in diameter, the first one (ΔE) is with 504μm thickness, and the second one (E) is with 1048 μm thickness. The ^{56}Ni particle was identified by the $\Delta E - E$ method. The excitation to the 2$^+$ state in ^{56}Ni was identified by measuring the γ-ray deexcitation in coincidence with the scattered ^{56}Ni particles.

Fifty-six NaI(Tl) scintillators were placed around the target to detect the γ rays. Each scintillator crystal is of rectangular shape with size 6×6×12 cm^3 coupled with 5.1 cm ϕ photomultiplier tube. Its energy resolution was typically 7.5% for the 662 keV γ ray. The high granularity of the setup allows one to measure the angle of the γ-ray emission, which was used to correct for the large Doppler shift of the γ rays from the excited ^{56}Ni in flight.

RESULTS

Figure 1 shows a spectrum for γ-rays associated with the ^{56}Ni + ^{208}Pb inelastic scattering. It is the sum of spectra for all the NaI scintillators where the Doppler shift is corrected. The background yields are subtracted using the empty-target runs. As seen in Fig.1, a full-energy peak at 2.7 MeV corresponding to the 2$^+$ $\rightarrow$ 0$^+$ transition is clearly observed. The peak was fitted with a response curve for 2.7 MeV photons and a continuous background of the form aE$_\gamma$+b. The solid curve in the figure represents the result of the fit. To extract the reaction cross section from the peak yield, the γ-ray efficiency was simulated by a code GEANT [7].

The calculation was able to reproduce the efficiencies for 662 keV and 1275 keV γ rays emitted from ^{137}Cs and ^{22}Na standard sources respectively within uncertainty of 9 %. A theoretical E2 coulomb excitation cross section of the ^{56}Ni + ^{208}Pb inelastic scattering to its 2$^+$ state was calculated by a coupled channel code ECIS88 [8]. The optical potential which describes the ^{40}Ar +

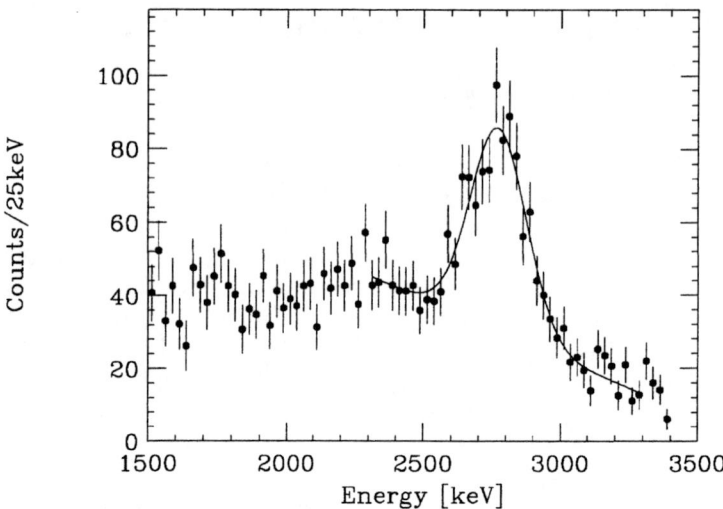

FIGURE 1. Doppler-corrected spectrum for γ-ray energy for the ^{56}Ni + ^{208}Pb inelastic scattering at E_{in} = 70.7MeV/u incident energy.

^{208}Pb elastic scattering data was used [6]. The cross section due to the nuclear excitation was also calculated by employing the collective deformation model. The result yields considerably small contribution (2%) of the nuclear excitation cross section. Both the nuclear deformation parameter β_N and Coulomb one β_C were set to the same magnitude. By comparing the predicted cross section with measured one, a value β_C =0.144 ± 0.034 was obtained. The error includes statistical and systematic uncertainties. The latter is the due to ambiguities in estimating the amount of secondary in the silicon detectors, possible loss of the yield caused by misidentifying the particles, accidental coincident yield, and the γ-ray detection efficiency. Since the parameter β_C is related $B(E2)$, a value of $B(E2: 0^+ \rightarrow 2^+) = 580 \pm 70$ e^{2}fm^4 was deduced for the ^{56}Ni 2^+ excitation.

Among various nuclei with N or Z = 20 or 28, the doubly magic nuclei ^{40}Ca and ^{48}Ca have the excitation energies higher than others by a factor of 2-3 [9] and lower values for $B(E2)$. Various experimental and theoretical $B(E2)$ values for the doubly magic nuclei ^{56}Ni ($N=Z=28$) is shown in Fig 2. The $B(E2)$ value of ^{56}Ni is slightly different between calculations of the shell model. If ^{56}Ni has a closed shell structure as other doubly magic nulcleus, $B(E2)$ is expected to be aroud the 400 e^{2}fm^4 [1]. The present result agrees with the one deduced from the (p,p') data [3], $B(E2) = 600$ e^{2}fm^4. The ^{56}Ni nucleus may not be simply dealt with a doubly magic nucleus in contrast ^{40}Ca, ^{48}Ca.

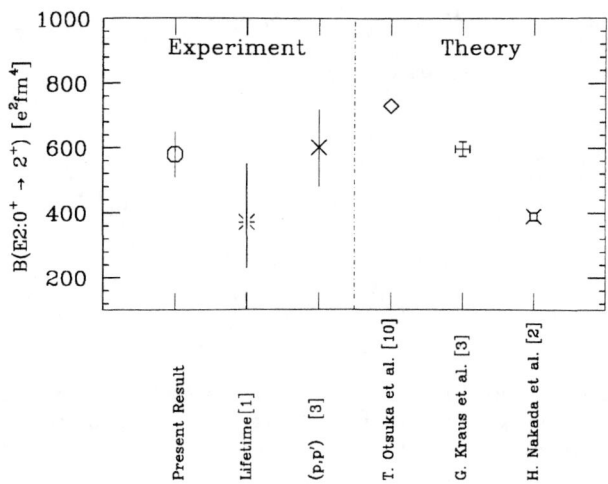

FIGURE 2. Various experimental result and theoretical calculation of $B(E2)$ values for the $0^+ \to 2^+$ transition in ^{56}Ni.

SUMMARY

A Coulomb excitation measurement was performed with radioactive ^{56}Ni beams at incident energy 70.7MeV/u and a ^{208}Pb target. The extracted $B(E2)$ value for the $0^+ \to 2^+$ transition in ^{56}Ni was determined by measuring γ ray from Coulomb-excitated of ^{56}Ni. The $B(E2)$ value, 580 ± 70 e^{2}fm^4, is consistent with the value obtained from a ^{56}Ni + p inelastic scattering experiment. The large $B(E2)$ value indicates that ^{56}Ni may not be simply treated as a doubly closed-shell nucleus.

REFERENCES

1. N. Schulz et al., Pys. Rev. **C8** (1973) 1779.
2. H. Nakada et al., Nucl. Phys. **A571** (1994) 467.
3. G. Kraus et al., Phys. Rev. Lett. **B264** (1994) 1773.
4. T. Motobayashi et al., Phys. Lett. **B346** (1995) 9.
5. T. Kubo et al., Nucl. Instr. Meth. **B70** (1992) 309.
6. N. Alamanos et al., Phys. Lett. **B137** (1984) 37.
7. GEANT, detector description and simulation tool, No.Q123, CERN Program Library.
8. J. Raynal, Coupled channel code ECIS88, unpublished.
9. S. Raman et al., At Data Nucl. Data Tables **36** (1987) 1.
10. T. Otsuka private communication.

New Signatures of phase transitional behavior in nuclei

N.V. Zamfir[1,2,3], R.F. Casten[1], R. Krücken[1], C. W. Beausang[1],
G. Cata-Danil[1,3], J.R. Cooper[1], Benyuan Liu[1], J.R. Novak[1], and C.J. Barton[2]

[1] *WNSL Yale University, New Haven, Connecticut 06520, USA*
[2] *Clark University, Worcester, Massachusetts 01610, USA*
[3] *Institute of Atomic Physics, Bucharest, Romania*

Abstract. The phase/shape transitional behavior of different observables is discussed, with emphasis on the $2^+_\gamma \rightarrow 0^+_2$ transition rate. The existing data on transitional Sm isotopes show that the vibrator-rotor shape transition behaves as a critical phase transition despite the finite-body nature of the nuclear environment.

The development of nuclear structure in shape transitional regions and the study of signatures of this structural evolution are very important, but yet unresolved, issues in nuclear physics. In addition, the study of new transitional regions far off stability, where experimental conditions are difficult, would benefit greatly if shape/phase transition were well understood in existing nuclei.

Correlations of collective observables either with external quantities or with other collective observables have been shown to be a very useful tool in investigating the evolution of nuclear structure (1).

The apparently complex phenomenology of $E(2^+_1)$ and $E(4^+_1)$ across a broad range of nuclei (see, for example, in fig.1a the variation of $E(4^+_1)$ with Z) is seen to comprise a simple bilinear trajectory when the correlation of these two collective observables is displayed (fig.1b). This correlation is characterized by two segments. Each segment has a specific, physically meaningful slope: at the lowest 2^+_1 energies the data are well reproduced by a slope of 3.33 and this segment corresponds to the rotor region. The next set containing all the nuclei from $E(2^+_1)$=0.150 to 0.800 MeV lies on a linear trajectory with a slope of 2.00 which corresponds to an anharmonic vibrator behavior. The two regions are separated by a narrow region [$E(2^+_1)$=0.10- 0.15 MeV] in which the structure varies rapidly (2).

It is the purpose of this contribution to focus on this region of rapid structural change from spherical to deformed shapes and, in particular on the Nd-Sm-Gd nuclei with N>82.

The rapid change in slope in the correlation of $E(4^+_1)$ with $E(2^+_1)$ is matched by a change in the evolution of $E(0^+_2)$ and $E(2^+_\gamma)$ with $E(2^+_1)$, as shown in fig. 2. The sudden change appears at $E(2^+_1)$ = 0.120 MeV which corresponds, in the Nd-Gd region, to N=88-90.

CP455, *ENAM98: Exotic Nuclei and Atomic Masses*
edited by B. M. Sherrill, D. J. Morrissey, and Cary N. Davids
© 1998 The American Institute of Physics   1-56396-804-5/98/$15.00

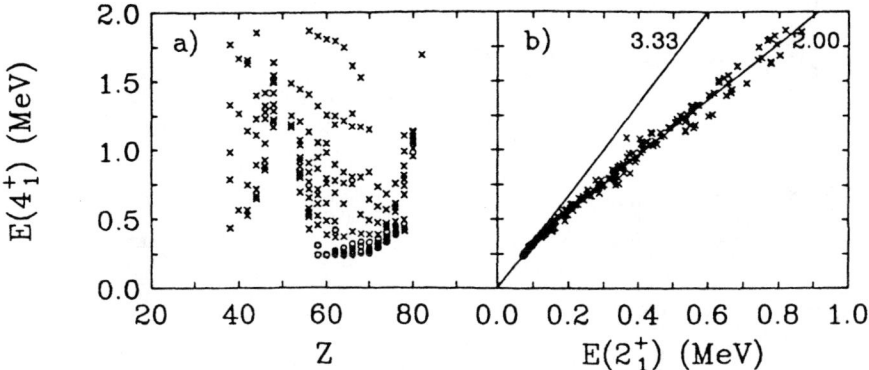

Figure 1. Correlation of $E(4^+_1)$ with a) Z and with b) $E(2^+_1)$ for all collective nuclei with Z=38-82.

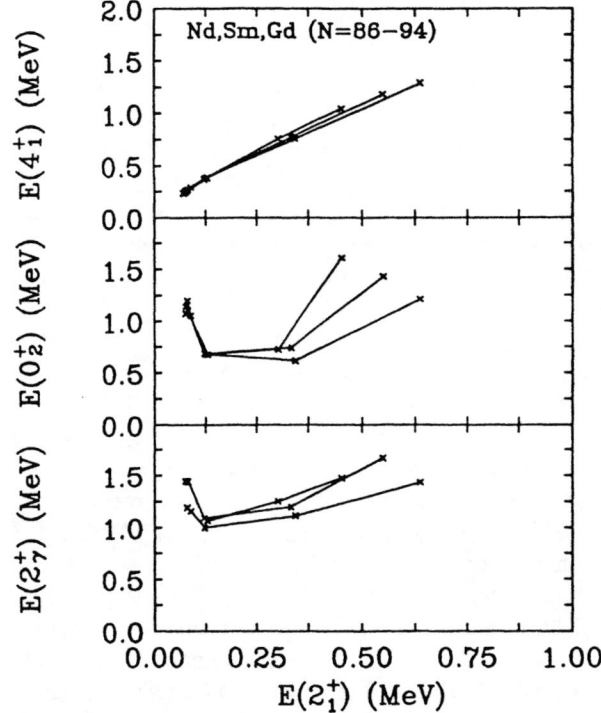

Figure 2. Correlation of $E(4^+_1)$, $E(0^+_2)$ and $E(2^+_\gamma)$ with $E(2^+_1)$ for Nd,Sm,Gd (N=86-94) nuclei.

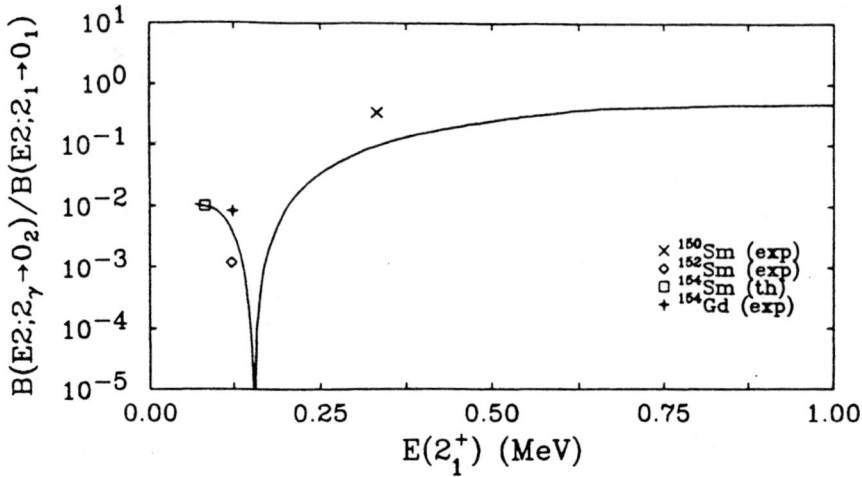

Figure 3. The behavior of the $2^+_\gamma \to 0^+_2$ transition rate (normalized to $B(E2;2^+_1\text{-}0^+_1)$). The continuous line is a schematic calculation in the framework of the IBM (10 bosons, $\chi=-\sqrt{7}/2$) and the symbols show the experimental values for ^{150,152}Sm, ^{154}Gd (3,5) and the predicted value for ^{154}Sm.

A recent study (3) suggests that a very sensitive signature of a phase/shape transition is the matrix element between the 2^+_γ and 0^+_2 states. Calculations in the framework of the Interacting Boson Model (IBA) provide evidence for coexisting deformed and spherical phases at low energy in ^{152}Sm (4). Consequently, rapid changes in the electromagnetic transition rate 2^+_1 to 0^+_1 may point to phase coexistence as well. The behavior of this transition rate is shown in fig. 3. using a schematic IBA calculation. To eliminate the effect of the effective charge on the transition rate, we show the ratio of this B(E2) value to the $B(E2; 2^+_1 \to 0^+_1)$ value: the latter is known to vary smoothly with $E(2^+_1)$. It is evident that the $B(E2;2^+_\gamma \to 0^+_2)$ value varies extremely rapidly near the phase transitional point.

Unfortunately, the γ transition connecting the 2^+_γ and 0^+_2 states is much lower in energy than the transitions from the 2^+_γ level to the ground state band and is therefore hindered by typically 2-4 orders of magnitude. By exploiting the sensitivity of high efficiency γ-ray detector systems, however, such weak transitions can now be sought for with much higher sensitivity. Recently, the transition rate in ^{152}Sm was measured by using the OSIRIS Koln array resulting in a value $B(E2; 2^+_\gamma \to 0^+_2)=0.17(2)$ W.u. (3). The only experimentally known data for this transition in the transitional rare-earth region are for ^{150,152}Sm and ^{154}Gd (3,5). These data are shown in fig. 3. The square symbol in fig. 3 is plotted on the curve at the value of $E(2^+_1)$ for ^{154}Sm: it therefore provides a prediction for the $B(E2; 2^+_\gamma \to 0^+_2)$ in that nucleus based on these schematic calculations. Clearly, an experimental value would help establish the evolution of the phase transitional region. Therefore, two experiment to measure this transition were performed at Yale using the YRAST Ball array (6). The 2^+_γ level was populated by Coulomb excitation with a ^{58}Ni beam at 250 MeV and an ^{16}O beam at 65 MeV. Due to the ^{152}Sm impurity in the target, we could only establish an upper limit of the ratio $B(E2; 2^+_\gamma \to 0^+_2)/ B(E2; 2^+_1 \to 0^+_1) < 0.24$ (ref. 7).

In addition, we measured the lifetimes of the 2^+_γ and 0^+_2 levels. In the experiment with the ^{16}O beam we used a set of eight 1cm^2 solar cells at backward angles to detect backscattered ^{16}O particles. In this way we selected the ^{154}Sm nuclei recoiling forward into a 15mg/cm^2 Au stopper mounted directly on top of the 3 mg/cm^2 ^{154}Sm foil. Using the lineshapes for the 1440 keV ($2^+_\gamma \rightarrow 0^+_1$) and the 1017 keV ($0^+_2 \rightarrow 2^+_1$) transitions we extracted the lifetimes of the 2^+_γ and 0^+_2 levels: 0.61(4)ps and 1.2(2)ps, respectively (7).

The structure of ^{152}Sm and its neighboring nuclei suggest new ideas concerning the nature of phase transitions in finite nuclei and the need to obtain new data on key transition rates and for new theoretical studies.

We are grateful to F. Iachello for the collaboration in ref. 4 and for many discussions fundamental to the topic of this paper. Work supported by the US-DOE under contracts DE-FG02-91ER40609 and DE-FG02-88ER40417.

REFERENCES

1. Casten, R.F. and Zamfir, N.V., *J.Phys.* **G** 22, 1521 (1996)
2. Casten, R.F., Zamfir, N.V., and Brenner, D.S., Phys. Rev. Lett., 71, 227 (1993)
3. Casten, R.F., Wilhelm, M., Radermacher, E., Zamfir, N.V., and von Brentano. P., Phys. Rev. C57, R1553 (1998)
4. Iachello, F., Zamfir, N.V., and Casten, R.F., Phys. Rev. Lett. submitted
5. Nuclear Data Sheets 69, 507 (1993), 75, 827 (1995),
6. Beausang, C.W. et al., to be published
7. Krücken, R. et al., to be published

Extraction of Cluster Spectroscopic Factors from Anomalous Large-Angle Scattering of ^{3}He and α-Particles

N.Burtebaev, B.A.Duisebaev, A.Duisebaev, G.N.Ivanov and S.B.Sakuta*

*National Nuclear Center of Kazakhstan, Almaty 480082 and *Russian Research Center "Kurchatov Institute", Moscow 123182*

Abstract. Anomalous large-angle scattering of ^{3}He and α-particles on strongly clusterized nuclei such as ^{6}Li and ^{7}Li is in the main determined by the clusters exchange mechanism. This gives an opportunity to use the anomalous scattering for investigation of the cluster structure of nuclei in addition to more traditional methods such as transfer reactions (for example, (^{6}Li,d), (^{6}Li,α), (^{7}Li,α)) and quasielastic knockout reactions. The present contribution reports on the results of the investigation of the elastic and inelastic scattering of ^{3}He and α-particles on the ^{6}Li and ^{7}Li nuclei measured at the ^{3}He energies of 50, 60 and 72 MeV and α-particles energy of 50.5 MeV in the full angular range. It was shown that experimental angular distributions can be described only if the exchange mechanism of the elastic and inelastic transfer of the deuteron and triton clusters was taken into account. Cluster spectroscopic factors were extracted for the ground and low-lying excited states of the ^{6}Li and ^{7}Li nuclei. For the ground states the experimental values of the spectroscopic factors agree well with theoretical predictions and data from the quasielastic knockout reactions. However, for the excited states the experimental values of the d- and t-spectroscopic factors differ from theoretical ones by factor of 2-3.

INTRODUCTION

The anomalous increase in cross sections at back-ward angles is often observed in various nuclear collisions. The nature of this phenomenon has been investigated in greatest detail in the elastic channel (1). A large number of mechanisms explaining large-angle scattering have been proposed, but a unified approach has not been developed; most probably, such an approach does not exist. The problem to be solved is therefore the determination of the dominant mechanism in each individual case. For example, cluster-exchange effects are expected to manifest themselves most clearly in backward scattering on light nuclei with strong clustering. Indeed, the analysis of elastic scattering of α-particles and ^{3}He on ^{6}Li nuclei (2-5), the low-lying states of which are well described by overlapping configurations ($\alpha + d$) and (^{3}He + t) has shown that the angular distribution can be described over the full angular range simply by taking into account elastic deuteron or triton transfer.

In inelastic channels the anomalous behavior of the cross sections has been investigated to a considerably lesser extent. The available information, which mainly concerns reactions induced by α particles, was analyzed in the (6). It was shown that, in a number of cases, traditional calculations carried out in the framework of the stripping mechanism are unable to describe the angular distributions over the full angular range without invoking exchange mechanisms.

In this study, we investigate the elastic and inelastic scattering of 50-, 60- and 72 MeV ^{3}He ions on ^{6}Li nuclei with excitation of the $J^\pi=3^+$ state ($E_x=2.185$ MeV) as well as the elastic and inelastic scattering of 50.5 MeV α particles by ^{6}Li and ^{7}Li nuclei with the excitation of the 3^+ state in ^{6}Li and the $1/2^-$ ($E_x=0.478$ MeV) and $7/2^-$ ($E_x=4.63$ MeV) states in ^{7}Li.

EXPERIMENTAL PROCEDURE

Investigation was carried out using 50-, 60- and 72-MeV ^{3}He and 50.5 MeV α particles beams extracted from the isochronous cyclotrons of the National Nuclear Center of Kazakhstan and the Russian Research Center "Kurchatov Institute".

For target preparation, we used metallic lithium enriched in isotopes and evaporated onto very thin (20 μg/cm^2) organic films. After evaporation, the targets were transported to a scattering chamber without violating a vacuum. Their thicknesses ranged from 100 to 500 μg/cm^2 and were determined by weighing immediately after measurements. The enrichment was 90% for ^{6}Li and 99% for ^{7}Li. In the angular range corresponding to the forward hemisphere, scattering nuclei and other charged reaction products were detected by a telescope of silicon counters. The separation of scattering particles from other charged reaction products was accomplished with an electronic system of two-dimensional analysis. Measurements at larger angles involved serious difficulties because of the very low energy of outgoing particles. In this case, single and thin surface barrier counters were used as a detector.

Measurements were carried out in the range of laboratory angles from 10° to 170°. In the case of ^{6}Li we observed two intensive peaks corresponding to elastic and inelastic scattering with the excitation of the 3^+($E_x=2.185$ MeV) state. For ^{7}Li, states at excitation energies $E_x=0.478$ MeV ($1/2^-$) and 4.63 MeV ($7/2^-$) were observed in addition to the elastic peak.

TABLE 1. Woods-Saxon OM Potentials for Scattering of ^{3}He and α Particles by ^{6}Li and ^{7}Li Nuclei

$A+a$	E MeV	$-V$ MeV	r_V fm	a_V fm	$-W$ MeV	r_W fm	a_W fm	r_C fm
^{6}Li+^{3}He	50	105.4	1.15	0.755	26.5	1.70	0.93	1.3
	60	111.6	1.15	0.755	24.4	1.70	0.93	1.3
	72	112.5	1.15	0.782	40.4	1.46	0.83	1.3
^{6}Li+α	50.5	98.7	1.245	0.80	23.17	1.75	0.70	1.3
^{7}Li+α	50.5	86.6	1.245	0.80	18.0	1.70	0.80	1.3

ANALYSIS OF RESULTS AND DISCUSSION

The elastic scattering was analyzed on the basis of the optical model with the central potential in the Woods-Saxon form. The potential parameters were determined by fitting theoretical angular distributions to experimental data with the aid SPI-GENOA code. The sets of values found for the potential parameters for each projectile and target nucleus are summarized in Table 1.

The differential cross sections for inelastic scattering were calculated in the DWBA in the framework of the collective model using the DWUCK4 code.

To estimate the cross sections for d- and t-cluster exchange, we used the DWBA implemented in the DWUCK5 code exactly taking into account finite-range effects. The distorted waves for the entrance and exit channels were calculated with the optical potentials from Table 1. The cluster spectroscopic factors are determined as phenomenological parameters from comparison of the calculated and experimental cross sections at large angles. The resulting values of the cluster spectroscopic factors for the ground and excited states of ^{6}Li and ^{7}Li nuclei are given in Table 2. One can see the more detailed analysis in (7,8).

Analysis shows that the theoretical angular distributions for the elastic and inelastic scattering that are calculated on the basis of the optical and collective models reproduce well the experimental cross sections only in the angular range corresponding to the forward hemisphere. The cross sections behavior in the backward direction can be described only by taking into account the mechanism of the deuteron- and triton-cluster exchange. As the result of analysis, we extracted the values the deuteron and triton spectroscopic factors for the ground and low-lying excited states of ^{6}Li and ^{7}Li nuclei. The experimental spectroscopic factors for the ground states, as it is seen from Table 2, agree with the theoretical values. However, for the excited states, the experimental values differ from the ground state ones by factor of 2-3. This result is unexpected because according to the theoretical predictions, they must be equal. It should be emphasized that we mean the ratio of the spectroscopic factors for the ground and excited states (S_{gr}/S_{exc}) rather than the absolute values. This ratio is less sensitive to the parameters of the optical potentials. Most likely, the reason behind the discrepancy between the theory and experiment is associated with the form factor or with the effects of

TABLE 2. Spectroscopic Factors Obtained from Analysis of the Experimental Data on Elastic and Inelastic Scattering of ^{3}He and α Particles by ^{6}Li and ^{7}Li Nuclei

A	E_x, MeV	J^π	Conf.	S_{exp}	$S_{th}(9)$
^{6}Li	0	1^+	$2S(\alpha+d)$	1.39	0.93-1.07
			$2S(^3\text{He}+t)$	0.5	0.5-0.9
	2.185	3^+	$2D(\alpha+d)$	0.5	1.0
			$1D(^3\text{He}+t)$	1.14	1.0
^{7}Li	0	$3/2^-$	$2P(\alpha+t)$	1.03	1.19
	0.478	$1/2^-$	$2P(\alpha+t)$	0.26	1.0
	4.63	$7/2^-$	$1F(\alpha+t)$	0.28	1.0

channel coupling. Therefore, calculations with form factors using the microscopic wave functions of ^{6}Li and ^{7}Li nuclei and calculations based on the coupled-channel method are of great interest. Ignoring the possible inaccuracy of our analysis, we can also assume that the role of the component corresponding to dissociated deuteron and triton clusters increases for excited states.

ACKNOWLEDGMENTS

We are indebted to V.Z.Goldberg, Yu.M. Tchuvilsky and V.I.Kukulin for discussion of the results.

REFERENCES

1. Gridnev, K.A., and Ogloblin, A.A., *Fiz. Elem. Chastits At. Yadra* **6**, 393(1975).
2. Bragin, V.N., Burtebaev, N., Duisebaev, B.A., Duisebaev, A., Ivanov, G.N., Sakuta, S.B., Chuev, V.I., and Chulkov, L.V., *Yad.Fiz.* **44**, 312(1986).
3. Goldberg, V.Z., Gridnev, K.A., Hefter, E.F., and Novatskii, B.G., *Phys.Lett.* **58**, 405(1975).
4. Bernas, M., DeVries, R., Harvey, H.L., Hendrie, D., Mahoney, J., Sherman, J., Steyart, J., and Zisman, V.S., *Nucl.Phys.A.* **242**, 149(1975).
5. Bechelier, D., Bernas, M., Boyard, J.L., Harvey, H.L., Jourddain, J.S., Radvany, P., and Roy-Stephan, H., *Nucl.Phys.A.* **195**, 361(1972).
6. Zelenskaya, N.S., and Teplov, I.B., *Obmennye Protsessy v Yadernykh Reaktsiakh* (Exchange Processes in Nuclear Reactions), Moscow: Mos.Gos. Univ., 1985.
7. Burtebaev, N., Duisebaev, A.D., Ivanov, G.N., and Sakuta, S.B., *Yad.Fiz.* **58**, 596(1995).
8. Burtebaev, N., Duisebaev, B.A., Duisebaev, A.D., Ivanov, G.N., and Sakuta, S.B., *Yad.Fiz.* **59**, 33(1996).
9. Nemets, O.F., Neudachin, V.G., Rudchik, A.T., Smirnov, Yu.F., and Tchuvilsky. Yu.M., *Nuklonnye Assotsiatsii v A tomnykh Yadrakh i Yadernye Reaktsii Mnogonuclonnykh Peredach* (Nucleon Associations in Atomic Nuclei and Multi-Nucleon Transfer Reactions), Kiev: Naukova Dumka, 1988.

HEAVY ELEMENTS, FISSION, CLUSTER RADIOACTIVITY

New Elements Produced at GSI

Sigurd Hofmann

*Gesellschaft für Schwerionenforschung (GSI),
Planckstraße 1, D-64220 Darmstadt, Germany*

Abstract. In two series of experiments at SHIP, six new elements (Z=107-112) were synthesized via fusion reactions using lead or bismuth targets and 1n-deexcitation channels. The isotopes were unambiguously identified by means of α-α correlations. Not fission, but alpha decay is the dominant decay mode. Cross-sections decrease by two orders of magnitude from bohrium (Z=107) to element 112, for which a cross-section of 1 pb was measured. Based on our results, it is likely that the production of isotopes of element 114 close to the island of spherical *SuperHeavy Elements* (SHE) could be achieved by fusion reactions using ^{208}Pb targets. Systematic studies of the reaction cross-sections indicate that the transfer of nucleons is an important process for the initiation of fusion. The data allow for the fixing of a narrow energy window for the production of SHE using 1n-emission channels. The likelihood of broadening the energy window by investigation of radiative capture reactions, use of neutron deficient projectile isotopes and use of actinide targets is discussed.

THE NEW ELEMENTS 110, 111 AND 112

In recent years the experimental set-up at SHIP was upgraded in order to allow for investigation of new elements on a cross-section level of 1 pb [1]. The preparatory experiments started in June 1994 with a study of excitation functions for production of rutherfordium (Z=104) and hassium (Z=108) in cold fusion reactions. The knowledge of the exact position of the cross-section maximum as function of the beam energy was a necessary prerequisite to search for the new elements beyond meitnerium [2].

Experiments aiming at identification of elements Z=110 and 111 were carried out in November-December 1994. In November-December 1995 the reaction ^{82}Se + ^{208}Pb → ^{290}116* was investigated in the search for a fusion process by radiative capture. Element 112 was identified in February-March 1996 [3, 4]. Beams of ^{40}Ar, ^{50}Ti, ^{51}V, ^{58}Fe, ^{62}Ni, ^{64}Ni, ^{68}Zn, ^{70}Zn and ^{82}Se were used with currents up to $\approx$0.5 pμA. Targets of lead isotopes and ^{209}Bi were irradiated. The isotopes ^{269}110, ^{271}110, ^{272}111 and ^{277}112 were identified by 4, 9, 3 and 2 decay chains, respectively. The mean values of the decay data are shown in the chart of nuclei, Fig. 1.

The most neutron-rich isotopes of the elements bohrium to 110, ^{268}Mt, ^{264}Bh, ^{273}110 and ^{269}Hs could be identified as daughter products of the α decay of ^{272}111 and ^{277}112. Neutron deficient isotopes of elements from mendelevium to seaborgium were investigated in order to explore the stability against fission. Several new isotopes were identified [5].

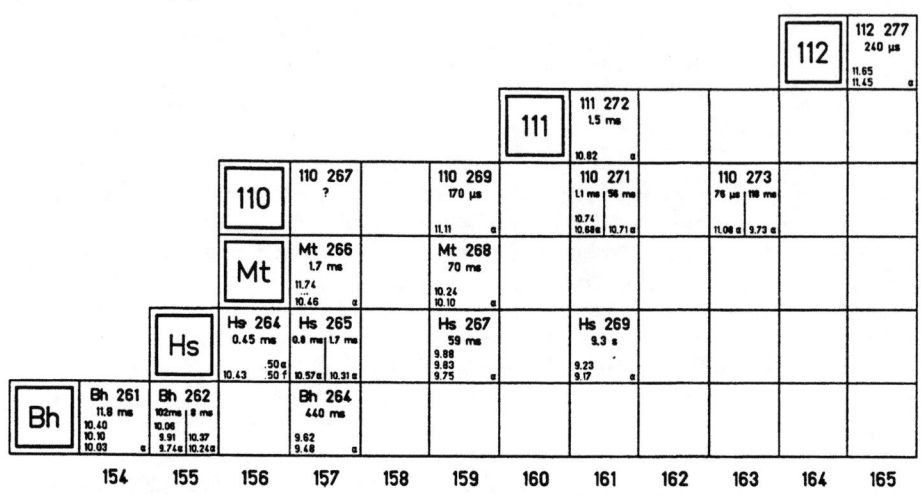

FIGURE 1. The presently known isotopes of the heaviest elements. The numbers given are the total half-life, the α energy in MeV and the branching ratio for electron capture, α decay and fission.

A search experiment for element 113 was started on March 5, 1998. On the basis of the previously measured data a cross-section between 0.3 and 1 pb was expected. During an irradiation period of 46 days a ^{209}Bi target (450 μg/cm^2) was irradiated with ^{70}Zn at two slightly different beam energies: 4.97 MeV/u (25 days, 4.5×10^{18} projectiles) and 5.00 MeV/u (21 days, 3.0×10^{18} projectiles). No event was observed that could be assigned to a decay chain starting at element 113. The obtained cross-section limits are 0.9 and 1.4 pb, respectively, resulting in an averaged value of 0.6 pb at a probability level of 68 %.

Shortly before the search for element 110 was initiated at SHIP, the possible evidence for observation of the α decay of ^{267}110 produced by the reaction ^{59}Co + ^{209}Bi was reported by Ghiorso et al. [6]. In a series of experiments the elements Sg, Hs and 110 were also investigated by a Dubna-Livermore collaboration at the U400 cyclotron at Dubna using a gas-filled separator [7].

GROUND-STATE PROPERTIES OF SHE

Shell-model calculations based on the Strutinsky approach [8, 9, 10, 11] are most successful in reproducing the measured nuclear binding energies. These are obtained by correlation of the α-decay data to decay-chain nuclei of known masses. Recently, also attempts at understanding the shell structure properties of SHE based on the self-consistent theory were made [12, 13], which differ significantly from previous predictions concerning the position of the closed proton shell.

The reason for the uncertainty is the difficulty to locate the three low spin proton subshells $2f_{5/2}$, $3p_{3/2}$ and $3p_{1/2}$, which are filled between Z=114 and 126. Qualitatively, we may expect a wide and less deep minimum of the negative shell-correction energies in the case that the low spin proton levels are equally distributed in energy between Z=114 and 126. Then, also the fission barriers will be flat and narrow, their height and width is mainly determined by the ground-state shell-correction energy. As a result, the fission half-lives will be relatively short. On the other hand, if a wide energy gap will exist beyond one of the proton numbers 114, 120 or 126, then the shell-correction energies will be pronounced for that element. In combination with the neutron shell effect at N=184 a sharp and deep minimum will be formed, similar to that of the double magic ^{208}Pb, resulting in a high fission barrier and relatively long fission half-life. Also the α half-lives will be more strongly modulated by great shell effects resulting in long α half-lives below and short half-lives above the magic number.

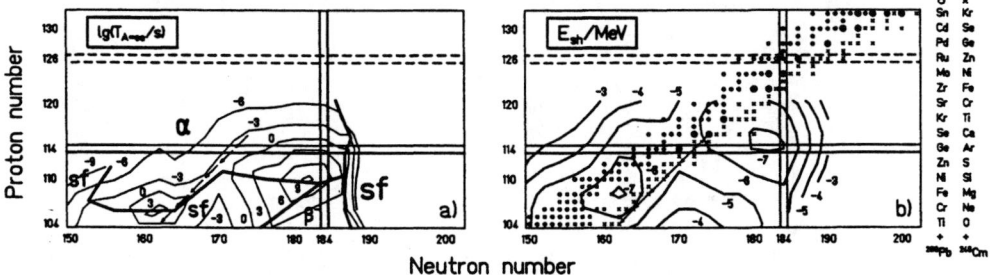

FIGURE 2. Dominating partial α, β or fission half-lives for even-even nuclei (a), as theoretically predicted [9, 10]. The bold lines separate regions of dominantly α decay, β decay and spontaneous fission. Diagram (b) shows the ground-state shell-correction energy and compound nuclei, which can be reached in reactions with targets of ^{208}Pb or ^{248}Cm and stable projectile isotopes. The sequence of arrows indicate the hypothetical decay chain of ^{290}116.

In order to estimate the decay properties of SHE, the predictions set forth by the macroscopic-microscopic models are extremely useful for the curbed extrapolation into this region [9, 10, 11]. The dominating partial half-life is shown in Fig. 2a for even-even nuclei. The two regions of deformed heavy nuclei and spherical SHE merge and form a region of α emitters surrounded by fissioning nuclei. The longest half-lives are 1000 s for deformed heavy nuclei and 30 y for spherical SHE. Fig. 2b shows the compound nuclei, which could be formed by reactions with ^{208}Pb or ^{248}Cm targets and stable projectile isotopes plotted on the contour map of shell-correction energies.

SYNTHESIS OF SHE

A summary of recently measured even-element excitation functions or cross-section values of cold fusion reactions is shown in Fig. 3. In all cases, where excitation functions are known, the cross-section maxima on the right in Fig. 3 are approximately centered between zero and the interaction barrier according to the Bass model [14]. This empirical result seems to present a sound means for the determination of the position of the cross-section maximum in cold fusion reactions.

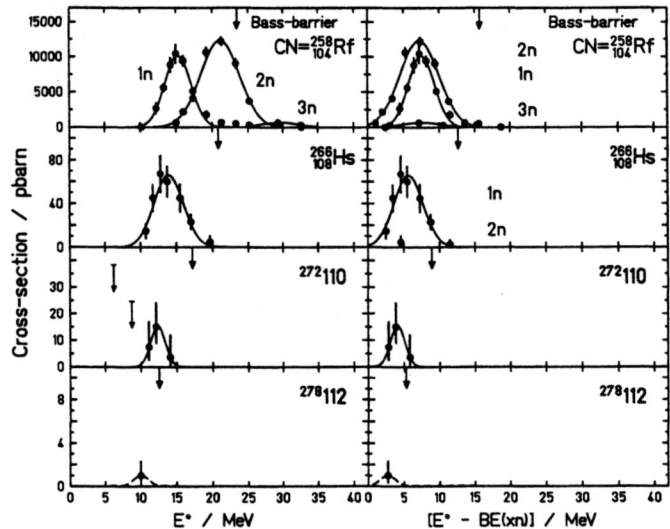

FIGURE 3. Measured even-element excitation functions. On the left part, the cross-sections are plotted as a function of the dissipated energy E*. The mass prediction of Myers and Swiatecki [11] has been used for the compound nucleus. The bigger arrows mark the interaction barriers of the reaction according to the fusion model by Bass [14]. On the right part, the neutron binding energies according to Myers and Swiatecki [11] are subtracted.

A comparison of excitation energies at the barrier for cold and hot fusion reactions over a wide range of SHE is shown in Fig. 4. A remarkable transition is observed from a region of high excitation energies (>40 MeV) for reactions with ^{238}U target resulting in elements up to ≈114 into a region of low excitation energies, down to 6 MeV for element 126. This reflects a change from hot fusion to cold fusion with regard to the excitation energy.

For element 116, the cold fusion reactions with ^{82}Se and ^{80}Se already give rise to excitation energies at the barrier that are smaller than the 1n-binding energy. In this case the free energy can be emitted only by γ rays. Alternatively, the 1n channel may be investigated with neutron deficient projectiles resulting in excitation energies greater than the 1n-binding energy. This choice may become important if the reaction does not allow for an increase in the kinetic energy beyond the value determined by the Bass interaction barrier as discussed in the following.

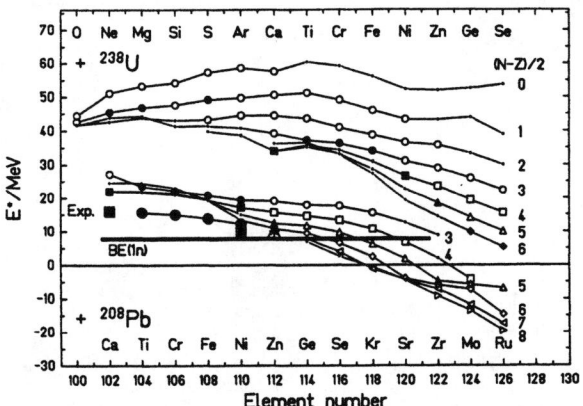

FIGURE 4. Diagrams of excitation energies at the interaction barrier according the model of Bass [14]. The upper cluster shows the trend for hot fusion reactions with ^{238}U targets and projectiles between O and Se, the lower, for cold fusion reactions with ^{208}Pb targets and projectiles from Ca to Ru. The large symbols mark stable projectile isotopes, the little symbols radioactive isotopes. The filled symbols of $T_z=3-6$ nuclei in the upper part are enhanced, because they impressively mark the transition from hot to cold fusion with increasing element number using actinide targets. The other filled symbols up to element 112 mark reactions investigated at SHIP by cold fusion or in Dubna by hot fusion. The star symbols mark the excitation energies, at which the maximum cross-section was observed in cold fusion reactions. The 1n-binding energies are in a range from 7.5 to 8.2 MeV for the investigated cold fusion compound-nuclei and are marked by the horizontal bar.

In all of the investigated cold fusion reactions the largest cross-section was measured 'below the barrier', see Fig. 5. A tunneling process through the barrier based on the model by Bass [14] cannot explain the measured cross-sections. Attempts to improve the heavy element fusion-barrier calculations were recently published by Möller et al. [15].

At and below the barrier, the kinetic energy in the center-of-mass system is converted into potential energy, and the reaction partners come to rest in a central collision in a touching configuration. In the case of ^{64}Ni + ^{208}Pb, the initial kinetic energy of 236.2 MeV, at which the cross-section maximum was measured, is exhausted by the Coulomb potential at a distance of 14.0 fm between the reaction partners. At that distance only nucleons on the outer surface are in contact.

We recall that the kinetic energy at the surface of orbiting nucleons is low. Therefore, at the point of contact of two nuclei in a central collision the probability of nucleons or pairs of nucleons leaving the orbit of one nucleus and move into a free orbit of the reaction partner is high, when the reaction partners are at rest. The process is shown schematically on the right side in Fig. 5. An adequate theoretical description could be obtained by use of the two-center shell model [16]. An approach that reproduces the measured cross-sections within the framework of a dinuclear system (DNS-model) was given by Antonenko et al. [17].

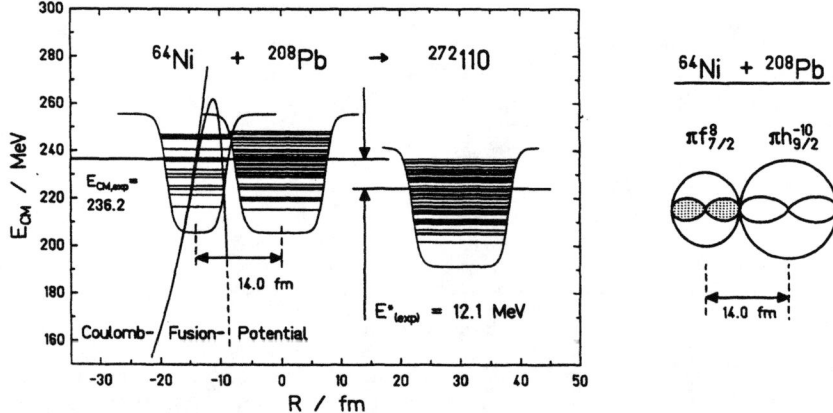

FIGURE 5. Energy against distance diagram for the reaction of an almost spherical ^{64}Ni projectile with a spherical ^{208}Pb target nucleus resulting in the deformed fusion product ^{271}110 after emission of 1 neutron. On the right the outermost proton orbitals are shown at the contact point. For the projectile ^{64}Ni an occupied $1f_{7/2}$ orbit (hatched area) is drawn and for the target ^{208}Pb an empty $1h_{9/2}$ orbit. The protons circulate in a plane perpendicular to the drawing.

Because of pairing energies and high orbital angular momenta involved, the transfer of pairs is more likely than that of single nucleons. The described process is a frictionless pair transfer occurring at the contact point in a central collision at zero longitudinal momenta in the irradiation of ^{208}Pb targets. This capture process seems to be disturbed in the case that the projectile and target nuclei do not come to rest. Already at small kinetic energies left, the wave functions of the outer nucleons do not fit in an optimum way, which results in energy loss by dissipation, so that the system cannot fuse and separates. To compensate the energy loss by increasing the kinetic energy (extra push) leads to increased fission probabilty and reduction of cross-section.

After transfer of 2 protons from ^{64}Ni to ^{208}Pb the repulsive Coulomb force is decreased by 4.9 % allowing for the maintenance of the reaction partners in close contact and for continuation of *fusion initiated by transfer* (FIT). Important factors, which influence the cross-section at the very beginning of the fusion process, are 1. The probability of a head-on collision. and 2. The probability of proton transfer in competition with separation of the reaction partners.

The importance of cold multinucleon transfer for the synthesis of heavy nuclei was determined by von Oertzen [18]. The transfer of massive clusters for production of heavy elements, especially neutron rich species, was considered by Magda and Leyba [19].

In the case of hot fusion reactions using actinide targets, the target nucleus is deformed, and at barrier energies only a fraction of certain orientations will lead to fusion. Nevertheless, the prospect of decreasing excitation energy with increasing element number makes also an investigation of hot fusion reactions necessary (see also M. Itkis, contribution to this conference).

OUTLOOK

The recent technical developments serve as a basis for further experiments in the region of superheavy elements. Important questions are still open and need to be answered. A short list of future experiments is herewith given:

1. Proof of the shell effect at Z=114 to establish the location of SHE.
2. Ground-state to ground-state α decay of even-even nuclei for more accurate evaluation of nuclear binding energies.
3. Search for α transitions of even-even nuclei into rotational levels for determination of the degree of deformation in the region near N=162.
4. Fission branchings of even-even nuclei, for comparison of the extracted partial fission half-lives with the results of nuclear models.
5. Extension of the cross-section data by measurement of complete excitation functions. Comparison of cross-sections of various combinations of odd and even reaction partners may be the best approach to understanding the cold-fusion reaction mechanism on a microscopic level.
6. Fusion with more neutron-rich radioactive projectiles and improved excitation-function systematics for hot-fusion reactions.
7. Search for radiative capture processes (0n channel).
8. Deexcitation of the compound nucleus by in beam γ spectroscopy using the recoil tagging technique.
9. Gamma spectroscopy of separated fusion products after electron capture.
10. Trapping of separated ions for precise mass measurement and investigation of the electron configuration by laser spectroscopy.

11. Chemical properties of elements beyond seaborgium and further studies of chemical properties of seaborgium and the lighter transactinide elements.
12. Microscopic description of the fusion process is needed for an effective explanation of the measured phenomena in the case of low dissipative energies. Then, also relationships between fusion probability and stability of the fusion products may become apparent.

An opportunity for the continuation of experiments in the region of SHE at decreasing cross-sections will be afforded by further accelerator developments. Radioactive beams and high current beams are the options for the future. At increased beam currents, values of a few 10 pμA may become possible, the cross-section level for the performance of experiments can be shifted down into the region of 1 fb. These high currents, in turn, require the development of a new target and improvement of the separator.

REFERENCES

[1] Hofmann S., *Journal of Alloys and Compounds* 213/214, 74-80 (1994).
[2] Münzenberg, G., *Rep. Prog. Phys* 51, 57-104 (1988).
[3] Hofmann S. et al., *Z. Phys. A* 350, 277-280 (1995); 350, 281-282 (1995); 354, 229-230 (1996); 358, 377-378 (1997).
[4] Hofmann S., *Rep. Prog. Phys.* 61, (1998) to be published.
[5] Heßberger F.P. et al., *Z. Phys. A* 359, 415-425 (1997).
[6] Ghiorso A. et al., *Nucl. Phys. A* 583, 861c-866c (1995); *Phys. Rev. C* 51, R2293-R2297 (1995).
[7] Lazarev Yu.A. et al., *Phys. Rev. Lett.* 73, 624-627 (1994); 75, 1903-1906 (1995; *Phys. Rev. C* 54 620-625 (1996).
[8] Strutinsky V.M., *Nucl. Phys. A* 95, 420-442 (1967).
[9] Sobiczewski, A., contribution to this conference.
[10] Möller P. et al., *Atomic Data and Nucl. Data Tables* 59, 185-381 (1995).
[11] Myers W.D. and Swiatecki W.J., *Nucl. Phys. A* 601, 141-167 (1996).
[12] Ćwiok S. et al., *Nucl. Phys. A* 611, 211-246 (1996).
[13] Rutz K. et al., *Phys. Rev. C* 56, 238-243 (1997).
[14] Bass R., *Nucl. Phys. A* 231, 45-63 (1974).
[15] Möller P. et al., *Z. Phys. A* 359, 251-255 (1997) and contrib. to this conference.
[16] Scharnweber D. et al., *Z. Phys. A* 228, 257-278 (1971).
[17] Antonenko N.V. et al., *Phys. Rev. C* 51, 2635-2645 (1995).
[18] von Oertzen, W., *Z. Phys. A* 342, 177-182 (1992).
[19] Magda, M.T. and Leyba, J.D., *Int. J. Mod. Phys E* 1, 221-247 (1992).

The FLNR (JINR) Experiments on Synthesis of Superheavy Nuclei with ^{48}Ca Beam[1]

Yu.Ts.Oganessian, A.V.Yeremin, M.G.Itkis, G.G.Gulbekian, V.B.Kutner

Flerov Laboratory of Nuclear Reactions, JINR, 141980 Dubna, Moscow region, Russia

Abstract. Investigation of the decay properties and formation cross sections of the heaviest isotopes of element 112 with atomic numbers 282 and 283 has been performed at the FLNR (JINR) with the use of the extracted high intensity ^{48}Ca beams. Using a ^{238}U target at the beam energy $E_{targ.}=231\pm3$ MeV ($E^*=31$ MeV) two spontaneous fission events have been detected which corresponds to the cross section of 5.0 ± 2 pb. The result could be explained as spontaneous fission of the even-odd isotope ^{283}112 with the half-life $T_{1/2}=117.2^{+282.9}_{-48.5}$ seconds. The probability of the α decay of ^{283}112 is not excluded and the possible branching ratio could be $b_\alpha \approx b_{sf} \approx 50\%$.

INTRODUCTION

According to the macro-microscopic theory predictions, the stability of the super heavy nuclei has to increase sharply, while approaching the spherical neutron shell N=184. The synthesis of the super heavy spherical nuclides, even if they are distant from the shell N=184, will take place only with a significant neutron excess in colliding nuclei. Thus, the isotopes of the element 112, which have the spherical shape and which are therefore relatively stable, could be produced in the reaction ^{48}Ca+^{238}U (1).

Formed in this reaction compound nucleus ^{286}112 appears to be weakly excited due to the significant mass defect of the double magic ^{48}Ca; the excitation energy at the Coulomb barrier is only $E_x=33$ MeV. At such excitation energy, the shell effects are still presented in the heated nucleus. It may increase the survival probability of evaporation residues (EVR). The high mass asymmetry in the entrance channel ($A_1/A_2 = 0.2$, and $Z_1Z_2=1840$) seems to decrease the dynamic limitations for the fusion of the interacting nuclei, early observed for the more symmetric cold fusion reactions (2).

---

[1] The work has been performed under a partial financial support of the Russian Foundation for Basic Research, contract N 96-02-17209 and the INTAS, contract N 96-662.

The expected decay properties of the isotopes with Z=112 are quite specific. According to the calculations provided by R. Smolanczuk (3), the even-even isotopes ^{282}112 and ^{284}112 have the partial half-life for the α-decay: T_α = 0.05 sec and 1.0 sec, respectively. Their spontaneous fission half-life ($T_{s.f.}$) is slightly higher than T_α.
In the mean time, the daughter nuclei ^{280}110 will mainly decay by spontaneous fission. As for the chain of sequential decays of ^{282}112, all products with Z<106 will have $T_{s.f}$ << T_α For the isotope ^{283}112 the predictions are less clear, because the odd number of neutrons may lead to significant limitations for both α-decay and spontaneous fission. Here, the strong competition between the two modes of decay is also highly expected.

It should be noted, that other calculations, performed by P. Moeller and R. Nix (4), provide the half-life against α-decay for the same nuclei in hundreds and thousands times higher than T_α from (3). However, it seems less significant, since it does not change the main decay modes of the isotopes Z=112, produced in the reaction ^{48}Ca + ^{238}U. Either they will decay by spontaneous fission, or by the chains of their α-decay (one or several) will end by the spontaneous fission anyway. On the other hand, the modern separation technique allows one to synthesize and identify the recoil nuclei, if their half-life is more than one microsecond. It exceeds all uncertainties of theoretical calculations.

^{48}Ca BEAM

The production of the intensive ion beam from rare and extremely expensive ^{48}Ca isotope presents itself as a key problem in our issue on isotope synthesis of the 112 element. Practically all experiments and tests on production and acceleration of the multicharged ions of Ca with both type ion sources PIG and ECR were performed on the cyclotron U-400. The best results were achieved with the ion source ECR-4M (5).

The neutral atoms of calcium were injected into the plasma by the process of controlled heating the metallic sample of ^{48}Ca(50mg) with the enrichment of 70%. The sample was made from the calcium oxide right before being placed in the oven of ECR source. The whole preparation procedure of metallic Ca as well as the recuperation of the material from the ion source chamber was controlled by measuring the yield of γ-rays from the isotope of ^{47}Ca (4.5 d), produced in the ^{48}Ca(γ,n)^{47}Ca reaction.

The new injection system and the modified beam optics of the cyclotron allowed to obtain the internal beam of ^{48}Ca^{5+} ions with intensity up to 1 pμA at the material consumption of about 0.3 mg/h. The beam extraction on the cyclotron U-400 was provided by the stripping method. The variation of the energy of the extracted beam in the range from 200 MeV to 280 MeV was controlled by smooth variation of the magnetic field and with the precise positioning of the stripping foil. The mean beam intensity of ^{48}Ca on the target was $2.2*10^{12}$/sec.

RECOIL SEPARATOR

The recoil nuclei within the angular acceptance $\vartheta_L = \pm\ 2.5^0$ were separated from the beam and other reaction products with the electrostatic separator "VASSILISSA" (6). Enriched isotope of ^{238}U (99.999%) with thickness of 0.3 mg/cm^2 has been used as a target material. It has been deposited uniformly on the aluminum disk with 125mm in diameter and 1.6 mg/cm^2 thickness. The disk was rotating with frequency ω=2000 rpm. The same target design with enriched isotopes of Tb, Yb, and Pb has been used in the several test experiments.

For registration and identification of the recoils the system of time-of-flight detectors and silicon position sensitive strip detector array[2] was installed in the focal plane of the separator. After registration in TOF detectors, the recoils were implanted into the frontal 16-strip silicon detector with an active area 60x60 mm^2. Each of the strips had a longitudinal position sensitivity. The special measurements of the position resolution along the each strip were performed. For sequential α-α decays it was 0.6 mm, for correlated recoil and α-particle - 1.0mm, and for correlated recoil and spontaneous fission events it was 1.5mm. The energy resolution for α-particles within energy range from 6 to 9 MeV was 20 keV. The time accuracy for the recorded events was about 1μsec. The frontal detector was surrounded with four other identical silicon detectors and the entire array had a shape of cube with dimensions 60x60x60mm^3. The geometrical efficiency of the silicon array was 85% of 4π. In the most cases position sensitivity for the backward detectors was not required. In present series of experiments each of the four neighboring strips was connected galvanicly and formed 16 energy sensitive segments.

Scattered low energy projectiles were the main contribution to the background. It was significantly reduced by the additional bending dipole magnet, installed at the very end of the separator. This allowed one to place the detectors at the angle of 8° out of the primary beam direction. After all, the counting rate of the frontal detector was only 25-30Hz.

The separation efficiency of EVRs from the reaction ^{48}Ca+^{238}U has been obtained in the test experiments. The cross sections of xn-evaporation channels were measured in the bombardment of targets ^{159}Tb, ^{174}Yb and ^{206,208}Pb by the ^{48}Ca projectile in a wide energy range. These measurements showed that about 25% of the recoils $^{286-x}$112, produced on the U-target, would be implanted in frontal detector. The TOF detectors were determining an appearance time-mark of recoils. The signals from the time-of-flight detectors were used both for measurements of the velocity of the recoils and for distinguishing decays of the previously implanted nuclei. The high efficiency of TOF detectors ($\varepsilon \cong 99.995\%$) allowed obtaining very clean decay spectra and significant widening (up to few tens of minutes) the time window for measuring decay chains. The latter was particularly important for identification of the EVRs with a long half-life and the persistent time structure of the beam.

---

[2] Canberra Semiconductors

RESULTS

To estimate the yield of EVRs in the reaction ^{238}U (^{48}Ca, xn) $^{286-x}$112 and to chose the optimal beam energy for ^{48}Ca, a set of additional experiments has been provided. On the recoil separator VASSILISSA, the excitation functions for the reaction ^{206,208}Pb(^{48}Ca,xn) $^{254,256-x}$No were measured. At the same time, on the time-of-flight fission fragment spectrometer CORSET (7) there were obtained data of the cross-sections and mass distribution of fission fragments in reactions ^{48}Ca + ^{206,208}Pb and ^{48}Ca + ^{238}U. For all the reactions the compound system was formed at low the excitation energies, close to the Coulomb barrier.

These experiments will be described in full in the separate article; here we will limit ourselves to the main conclusions drawn from the data. As it follows from the analysis of data observed above, the highest cross-section for the reaction ^{48}Ca + ^{238}U is expected for 3n and 4n-evaporation channels. Excitation functions reached their maximums at the E_x=31 and 39 MeV, which corresponds to the beam energy in the middle of the target E_L=231 MeV, and 239MeV, respectively. The absolute value of the cross-section for xn-evaporation channels can be estimated much less definitely. According to different calculations, the cross-section at the maximum of the 3n-evaporation channel varies from 2pb to 20pb; although, it still remains higher in 3-5 times than for the 4n evaporation channel.

On the basis of the above presented data, two long-term irradiations of ^{238}U target with ^{48}Ca at beam energies of 231±3 MeV and 238±3 MeV were performed. The doze of 3.5∗10^{18} projectiles was collected in the first experiment in March 1998. Two spontaneous fission events were detected at that time. For the both events the fission fragments were registered as two coinciding signals with a high energy deposition in both front and backward detectors.

The total kinetic energy (TKE) for the events has also been measured (Fig.1). Detectors were calibrated by fragments from spontaneous fission of the implanted recoils of ^{252}No, produced in the reaction ^{206}Pb (^{48}Ca, 2n). The absolute value of the TKE with its variations for the spontaneous fission ^{252}No was taken from (8) It should be noted, however, that for the nucleus, heavier than No, the TKE must be higher than the value shown in Fig.1.

The analysis of all events collected in the experiment has been provided in order to find genetic decay links of implants. For that purpose, there were used the data of the position and energy resolutions of the correlated events, measured by the frontal detector for the various modes of decay. Thus, for recoil-α-α as well as recoil-α-...-SF for an α-particle in the energy range from 8 to 13 MeV and in the time interval up to 10000 sec no correlation were found. But at the same time, there were two correlated signals from the recoil and spontaneous fission found in the strips #12 and #15 within the position window ±0.8 mm. In the first case, the time interval between the signals "recoil" and "spontaneous fission" was 182.388 sec, while in the second case (strip #15), the time interval was 52.040 sec. As it became evident from the data of the long-term detection, at the position window ±0.8mm the signals, similar to EVRs, were detected in the strip #12 with mean frequency 0.001Hz, while in the strip #15 the frequency is 0.005 Hz.

In the second experiment, the doze of $2.2*10^{18}$ projectiles was collected at the beam energy $E_L=238$ MeV ($E_x=39$MeV). No spontaneous fission events were detected and none α-α correlation were observed in the whole α-particle energy range from 8 to12 MeV in the time interval up to 1000 sec. The upper limit for EVRs cross-sections at this excitation energy is 3.0 ± 1.5 pb.

It is very important to note that the two events observed in these experiments were collected practically with no background conditions. The fact that no SF- isomers as well as any other known SF-nuclei were observed in the two experiments with the total beam doze $5.7*10^{18}$ makes evident an extremely high selectivity of the device for evaporation residues detection. That is why we also have to exclude a possibility to explain the effect in terms of exotic decay mode of known nuclei formed in the reaction ^{48}Ca + ^{238}U. The most reasonable explanation is that we are observing the spontaneous fission of the heaviest unknown even-odd isotope ^{283}112 with the half-life $T_{1/2}=117^{+283}_{-50}$ sec. Despite the fact that the only spontaneous fission was observed in the experiment, we could not also exclude the α-decay with the branching ratio $b_\alpha \leq 50\%$. The cross-section for a new isotope is 5 ± 2 pb (the statistical error is presented here; the absolute cross-section accuracy value is within a factor 2). The absence of any effect at the excitation energy $E_x=39$ MeV is not surprising. The cross-section of the 3n-evaporation channel decreases with the increase of the excitation energy, while the cross-section of 4n-evaporation channel is several times lower than it is for the 3n-channel, as it was already mentioned above.

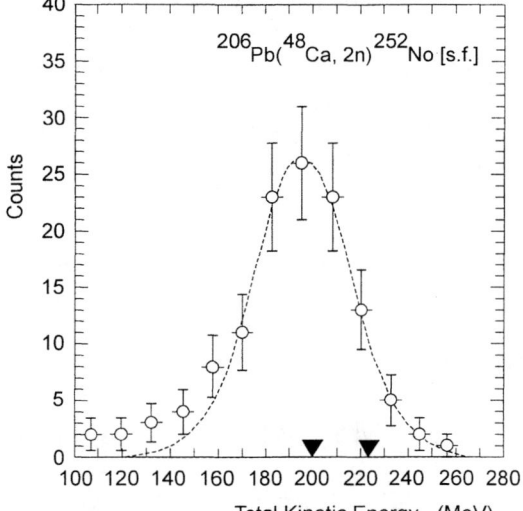

FIGURE 1. The spectra of fission fragment total kinetic energy for spontaneous fission of the isotope ^{252}No. The experimental points- data obtained for the reaction ^{206}Pb(^{48}Ca,2n)^{252}No with recoil separator "VASSILISSA" and normalized on the known distribution (dashed line) taken from literature (7). The data for two events observed in the reaction ^{48}Ca + ^{238}U shown by black arrows.

CONCLUSION

The present article describes a new attempt to synthesize super heavy spherical nuclei in the reaction ^{48}Ca + ^{238}U. Comparing to all previous experiments with ^{48}Ca ions involved, the present one has more than 100 times higher sensitivity. Two spontaneous fission events were detected at the beam energy below the Coulomb barrier. As it follows from the data analysis, the events were most probably triggered by the decay of the even-odd isotope ^{283}112 formed in 3n-evaporation channel. The half-life of the new isotope is about 100 sec. It could be that it also undergoes α-decay with $b_\alpha \leq 50\%$.

The half-life of the new isotopes is more than 5 orders of magnitude longer than the already known isotope ^{277}112 ($T_{1/2} \sim$ 0.4 ms), obtained in the cold fusion reaction (9). The difference could indicate that the stability increases sharply when we enter a domain of the spherical shell corrections. The present experiment is the first one in our long-term of super heavy elements research program with ^{48}Ca beam. We have a strong intention to increase the intensity of the beam, which will allow us to try to synthesize other isotopes with Z=110 and 114 as well as, perhaps, to reproduce the present data with a higher beam doze.

ACKNOWLEDGEMENTS

We would like to thank Dr. R. Smolanczuk and Dr. B.I. Pustylnik for the fruitful discussions, Drs. R.N. Sagaidak, O.N. Malyshev, A.G. Popeko, A.P. Kabachenko, V.I. Chepigin, J. Rohach, V.A. Gorshkov, A.Yu. Lavrentiev, B.N. Gikal, A.V. Tikhomirov, S.L. Bogomolov, S. Hofmann, G. Muenzenberg, M. Veselsky, K. Morita and N. Iwasa for their participation in the experiments. We also thank the U-400 crew for providing stable ^{48}Ca beams of high intensity.

REFERENCES

1. Oganessian, Yu.Ts., "Synthesis and Radioactive properties of the Heaviest Nuclei", Proc. In Conf. "Nuclear Physics at the Turn of the Millenium: Structure of the Vacuum and Elementary Matter", Wilderness, South Africa, (Word Scientific, Singapure), 11, (1996).
2. Blocki, J.P., *Nucl. Phys.* **A 459**, 145 (1985).
3. Smolanczuk, R., *Phys. Rev.* **C 56**, 812 (1997).
4. Moeller, P., et al., *Atomic Data and Nuclear Data Tables* **59**, 185 (1995), Moeller, P., et al., Nuclear Properties for Astrophysics, in http://t2.lanl.gov/publications/astro/normal.html.
5. Kutner, V.B., et al., in Proceedings of the 15th International Conference on Cyclotrons and their Application, 1998, to be published.
6. Yeremin, A.V., et al., Nucl. Instr. And Meth. B 126, 329 (1997).
7. Itkis, M.G., et al., (see talk presented on this Conference).
8. Hulet, E.K., Yad. Fiz. 57, 1165 (1994).

Stability of the heaviest elements

Adam Sobiczewski

Soltan Institute for Nuclear Studies, Hoża 69, PL-00-681 Warsaw, Poland

Abstract. Recent theoretical studies of stability of heaviest elements are shortly reviewed. Alpha-emission and spontaneous-fission decay modes are discussed. Even-even nuclei with proton number $Z = 82 - 120$ and neutron number $N = 126 - 190$ are considered.

INTRODUCTION

The objective of this paper is to give a short review of recent theoretical studies of stability of heviest nuclei. Their structure and such properties as mass, main modes of decay (α decay and spontaneous fission) and respective half-lives are discussed. Even-even nuclei with proton number $Z = 82 - 120$ and neutron number $N = 126 - 190$ are considered.

We concentrate on the studies performed in a macroscopic-microscopic approach (e.g. [1–4], cf. also the reviews [5–7]), which seem to be, at the moment, most complete. Some properties of heaviest nuclei have been, however, also studied by fully microscopic methods, like Hartree–Fock–Bogolubov and Relativistic Mean Field approaches (e.g. [8–11]). We discuss some of the results of these studies and compare them with the results obtained within the macroscopic-microscopic approach.

The theoretical studies are closely connected with, and motivated by, an intensive experimental activity in this field (e.g. [12–20]).

IMPORTANT ROLE OF SHELL EFFECTS

Shell effects are important for all nuclei. Their role for the heaviest nuclei is, however, essential, as many of them would not exist at all without these effects [21].

The analysis of shell effects, performed in [21], has shown that these effects elongate the α-decay half-lives T_α of known even-even heavy nuclei by up to about 5 orders of magnitude, and the spontaneous-fission half-lives T_{sf} by up to about 15 orders of magnitude.

A particular feature of the considered region of nuclei is that some deformed nuclei show shell effects which are similarly strong as the effects observed in spherical magic nuclei, i.e. that we observe deformed shells in these nuclei. Specifically, effects of the deformed neutron shell at the neutron number $N = 152$ are experimentally observed for a long time. There is also an increasing experimental evidence for the existence of even stronger deformed shells at $N = 162$ and $Z = 108$, predicted theoretically. The nucleus $^{270}108$ (^{270}Hs) is expected theoretically [22,1] to be a doubly magic deformed nucleus. To get in a theory, however, these strong shells in a deformed nucleus, one needs to allow the nucleus to deform as it likes, to take the shape comfortable for it. In other words, one needs to consider the properties of a nucleus in a sufficiently large, multidimensional deformation space [23,1,24].

THEORETICAL DESCRIPTION

As mentioned in the Introduction, we concentrate mainly on the macroscopic-microscopic approach. The macroscopic part is described by the Yukawa-plus-exponential model [25]. The microscopc part is the Strutinski shell correction, based on the Woods–Saxon single-particle potential [26].

The α-decay half-lives are described by the phenomenological formula of Viola and Seaborg, but with its free parameters readjusted to account for recent data. Details of the calculations are given in [1,2].

The spontaneous-fission half-lives are analyzed in a dynamical way, with the mass tensor (describing the inertia of a nucleus with respect to its deformation) taken into account. Details of the calculations are given in [3].

The 4-dimensional deformation space $\{\beta_\lambda\}$, $\lambda = 2, 4, 6, 8$, is used, where β_λ are the usual deformation parameters, appearing in the expression for nuclear radius (in the intrinsic frame of reference) in terms of spherical harmonics.

RECENT THEORETICAL RESULTS

Shell correction to energy (mass) of a nucleus

Shell correction to the ground-state mass of a heavy nucleus gives us a first orientation in the stability of this nucleus. Figure 1, taken from [2], shows the shell correction, E_{sh}, calculated for the large region of nuclei under consideration. One can see that E_{sh} has three minima in this region. The first one, which is the deepest (E_{sh}=-14.3 MeV), is obtained for the doubly magic spherical nucleus ^{208}Pb. The second one (E_{sh}=-7.2 MeV) appears at the nucleus $^{270}108_{162}$, which is predicted to be a doubly magic deformed nucleus. The third minimum, with the same depth (E_{sh}=-7.2 MeV) as that of the second minimum, is obtained for the nucleus $^{296}114_{182}$, which is close to the nucleus $^{298}114_{184}$ predicted [27,28] to be a doubly magic spherical nucleus, the next one to the last experimentally known ^{208}Pb.

One can see in Fig. 1 that some of the already synthesized nuclei profit by 6–7 MeV in their binding energy from the shell correction. Without this profit they could not exist, as already mentioned in Sec. 2.

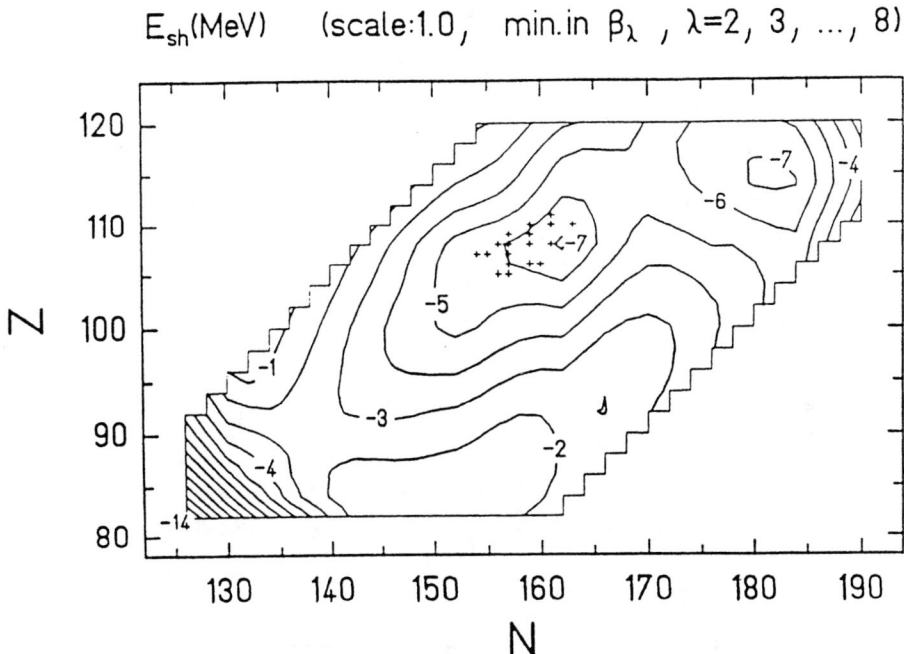

FIGURE 1. Contour map of the shell correction to energy, E_{sh}. Crosses denote the heaviest nuclides synthesized up to now [2].

The appearance of the region of nuclei around the second minimum (deformed superheavy nuclei) constitutes the main change in our view of stability of heaviest nuclei in recent years. Earlier, it was believed for a long time that spherical superheavy nuclei, predicted to be situated around the third minimum, would constitute an island, separated from the usual peninsula of relatively long-lived nuclei by an "ocean" of full instability. After the appearance of deformed superheavy nuclei, however, the peninsula is expected to be extended, to include also the spherical superheavy nuclei.

Single-particle structure

It is instructive to look at the single-particle spectra of the doubly magic nuclei: ^{208}Pb, ^{270}Hs and ^{298}114, for which the three minima of the shell correction E_{sh} have been obtained in Fig. 1. The spectra are shown in Fig. 2, for protons.

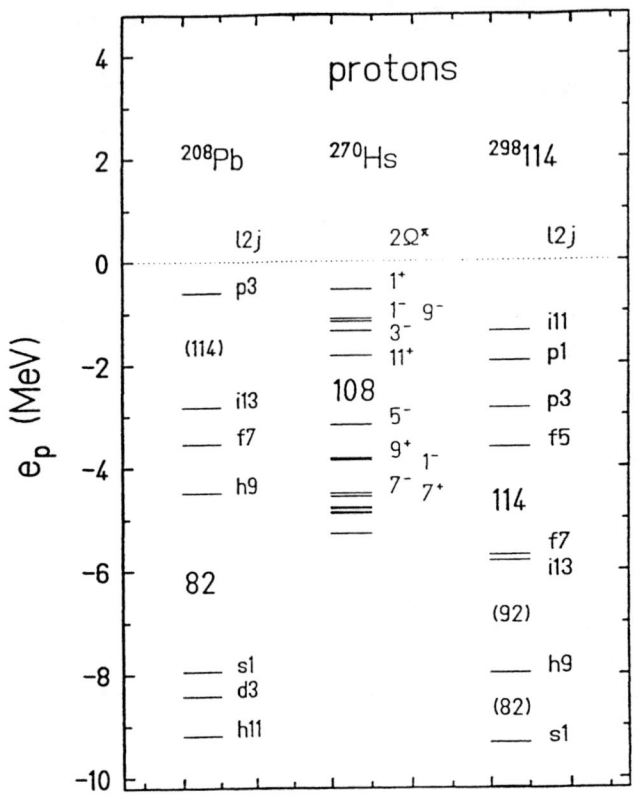

FIGURE 2. Proton single-particle levels calculated for the doubly magic nuclei: ^{208}Pb, ^{270}Hs and ^{298}114. Spectroscopic symbol for the orbital angular momentum l and total spin (multiplied by two) $2j$ are given at each level of the spherical nuclei ^{208}Pb and ^{298}114. Projection of spin (multiplied by two) 2Ω and parity π are shown at each level of the deformed nucleus ^{270}Hs.

One can see a large energy gap (about 3.4 MeV) for the experimentally well known shell closure at $Z=82$ in ^{208}Pb. The gap appearing at $Z=108$ in ^{270}Hs is much smaller (about 1.3 MeV) and corresponds to a deformed shell. One can observe here, however, a low density of levels between $Z=106$ and $Z=110$, which increases the effect of a rather small gap at $Z=108$. A medium-size gap (about 2.1 MeV), obtained at $Z=114$ in the hypothetical spherical nucleus ^{298}114, is created by the spin-orbit splitting of the $2f$ level. Thus, its value strongly depends on the strength of the spin-orbit interaction. For a much smaller strength it would simply disappear.

While the creation of a significant gap in a low-density spectrum of (degenerate) levels of a spherical nucleus seems to be relatively easy, this is rather difficult in a high-density spectrum of (non-degenerate) levels of a deformed nucleus. Probably due to this, large gaps in deformed nuclei are rather unusual. There is only one gap (at $N=152$, for elements around Fm) known experimentally, and only two additional ones (at $N=162$ and $Z=108$, for nuclei around ^{270}Hs) predicted theoretically, in deformed nuclei, while rather many of them are known in spherical nuclei.

In the Hartree–Fock calculations for spherical nuclei [10], using the effective Skyrme interaction, a large gap in the proton spectrum is obtained at $Z=126$. The gap at $Z=114$ remains, however, when one variant of the interaction (SLy7) is used, while it disappears when the other one (SkP) is taken. Still, in both these variants, the α-decay half-lives T_α predicted for nuclei with $Z=114$ and $N\approx 184$ are rather large. They are larger than those obtained in the macroscopic-microscopic calculations [29,30], where the gap at $Z=114$ (as seen in Fig. 2) appears.

Half-lives for deformed superheavy nuclei

Theoretical half-lives for nuclei situated around the doubly magic deformed nucleus ^{270}Hs (deformed superheavy nuclei), i.e. around the second minimum in Fig. 1, have been given and extensively discussed in [1–3]. Here, we only illustrate a comparison of these results with the experimental values obtained for the element 110. This is shown in Fig. 3. The experimental values are taken from [14] for ^{257}110, from [13] for ^{259,261}110, and from [17] for ^{263}110.

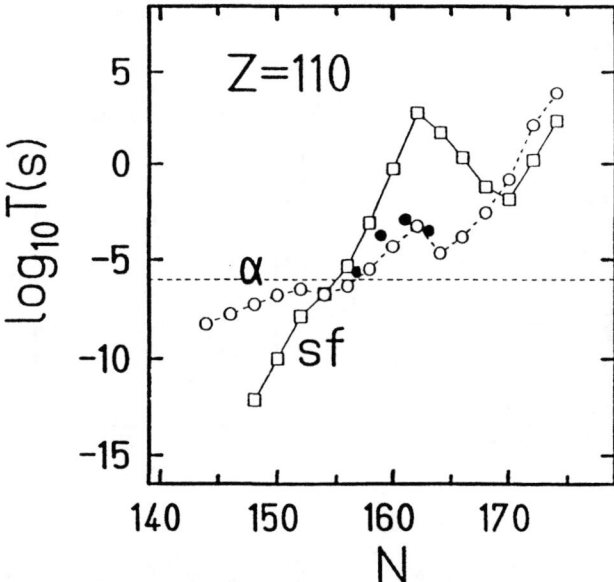

FIGURE 3. Comparison between predicted theoretically (open circles) and measured (full circles) α-decay half-lives, for isotopes of the element 110.

One can see that the measured values are rather close to the predicted ones. In particular, they seem to confirm the existence of the predicted neutron deformed shell at $N = 162$. They also confirm the prediction that T_{sf} is larger than T_α for respective nuclei, as only α-decay has been observed for them.

Half-lives for "spherical" superheavy nuclei

Figure 4 gives half-lives [29,30] for nuclei situated around the third minimum in Fig. 1, i.e. around the hypothetical doubly magic spherical nucleus $^{298}114$. Certainly, only nuclei close to this nucleus are expected to be spherical; this is the reason for which we put the word "spherical" in the title of the subsection into quotation marks. The isotopes with the neutron number N=176–184 of the elements 110–118 are considered in the figure. Only for the element 114, the isotope with N=186 is also shown, to see the behaviour of the half-lives above the shell closure at N=184.

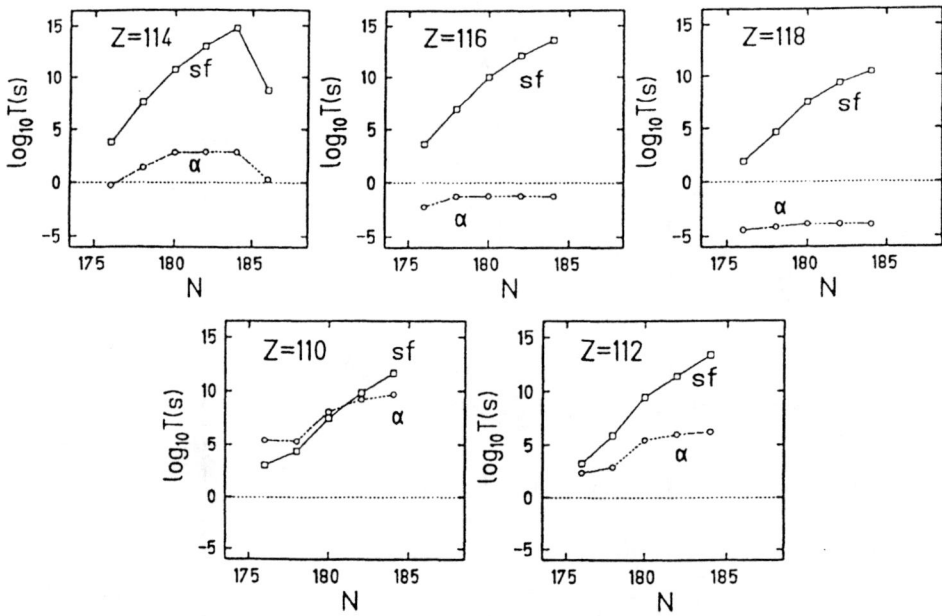

FIGURE 4. Logarithms of the calculated spontaneous-fission (sf) and α-decay (α) half-lives (given in seconds), as functions of the neutron number N, for the elements: 110–118. The horizontal dashed line indicates about the lowest half-life (1 μs) of a nucleus, which can be detected in a present-day set-up, after its synthesis [30].

One can see that the fission half-life is longer than that of α-decay for all considered isotopes of the elements 112–118. The opposite relation is only obtained for the three lightest isotopes of the element 110. As illustrated for the element 114, the half-lives decrease above the shell closure at $N=184$, with the fission half-life decreasing especially fast.

One can also see in Fig. 4 that the longest half-lives, with respect to both decay modes, are expected to appear for nuclei around the nucleus $^{292}110$ and to be of the order of one hundred years. As the nucleus $^{292}110$ is expected to be β-stable (e.g. [31]), these longest half-lives are expected to be the total half-lives. Nuclei with so long half-lives could be accumulated (in distinction to those situated around the doubly magic deformed nucleus $^{270}108$, which have short half-lives). This would give a chance for extensive studies of physical and chemical properties of these exotic nuclei and elements. Certainly, on the condition that cross sections for their synthesis appear sufficiently large.

Fission half-lives of "spherical" superheavy nuclei have been also considered recently in [32] applying the approach used in [3], i.e. the dynamics in two-dimensional deformation space. They have been also studied in [33].

In the Hartree–Fock calculations with a finite-range (Gogny) effective interaction [8], an increased stability of nuclei around the nucleus $^{298}114$ is also obtained. The total half-lives (T_α, because $T_{sf} \gg T_\alpha$) calculated there [8] are by more than one order of magnitude larger than those obtained in the macroscopic-microscopic calculations [29,30].

ACKNOWLEDGEMENTS

The author would like to thank P. Armbruster, S. Hofmann, G. Münzenberg, W. Nörenberg, Yu. Ts. Oganessian and Z. Patyk for helpful discussions. Support by the Polish Committee for Scientific Research (KBN), grant no. 2 P03B 117 15, is gratefully acknowledged.

REFERENCES

1. Z. Patyk and A. Sobiczewski, *Nucl. Phys.* **A533**, 132 (1991).
2. R. Smolańczuk and A. Sobiczewski, Proc. XV Nucl. Phys. Conf.:*Low Energy Nuclear Dynamics*, St. Petersburg (Russia) 1995, eds. Yu.Ts. Oganessian, W. von Oertzen, R. Kalpakchieva (World Scientific, Singapore 1995) p. 313.
3. R. Smolańczuk, J. Skalski and A. Sobiczewski, *Phys. Rev.* **C52**, 1871 (1995).
4. Z. Patyk, J. Skalski, R.A. Gherghescu and A. Sobiczewski, *APH N.S., Heavy Ion Physics* **7**, 13 (1998).
5. A. Sobiczewski, *Fiz. Elem. Chastits At. Yadra* **25**, 295 (1994); *Phys. Part. Nucl.* **25**, 119 (1994).
6. P. Möller and J.R. Nix, *J. Phys.* **G20**, 1681 (1994).
7. A. Sobiczewski, *Usp. Fiz. Nauk* **166**, 943 (1996); *Physics–Uspekhi* **39**, 885 (1996).

8. J.-F. Berger et al., Proc. 24th Intern. Workshop: *Extremes of Nuclear Stucture*, Hirschegg (Austria) 1996, eds. H. Feldmeier, J. Knoll and W. Nörenberg (GSI, Darmstadt 1996) p. 43.
9. G.A. Lalazissis, M.M. Sharma, P. Ring and Y.K. Gambhir, *Nucl. Phys.* **A608**, 202 (1996).
10. S. Ćwiok et al., *Nucl. Phys.* **A611**, 211 (1996).
11. K. Rutz et al., *Phys. Rev.* **C56**, 238 (1997).
12. Yu.A. Lazarev et al., *Phys. Rev. Lett.* **73**, 624 (1994).
13. S. Hofmann et al., *Z. Phys.* **A350**, 277 (1995); **A350**, 281 (1995).
14. A. Ghiorso et al., *Phys. Rev.* **C51**, R2293 (1995).
15. Yu.A. Lazarev et al., *Phys. Rev. Lett.* **75**, 1903 (1995).
16. S. Hofmann et al., *Z. Phys.* **A354**, 229 (1996).
17. Yu.A. Lazarev et al., *Phys. Rev.* **C54**, 620 (1996).
18. M. Schädel et al., *Radiochim. Acta* **77**, 149 (1997).
19. A. Türler et al., *Phys. Rev.* **C57**, 1648 (1998).
20. S. Hofmann, *these Proceedings*.
21. Z. Patyk, A. Sobiczewski, P. Armbruster and K.-H. Schmidt, *Nucl. Phys.* **A491**, 267 (1989).
22. Z. Patyk and A. Sobiczewski, *Phys. Lett.* **B256**, 307 (1991).
23. A. Sobiczewski, Z. Patyk, S. Ćwiok and P. Rozmej, *Nucl. Phys.* **A485**, 16 (1988).
24. R. Smolańczuk, H.V. Klapdor-Kleingrothaus and A. Sobiczewski, Acta Phys. Pol. **B24** (1993) 685
25. H.J. Krappe, J.R. Nix and A.J. Sierk, *Phys. Rev.* **C20**, 992 (1979).
26. S. Ćwiok et al., *Comput. Phys. Commun.* **46**, 379 (1987).
27. A. Sobiczewski, F.A. Gareev and B.N. Kalinkin, *Phys. Lett.* **22**, 500 (1966).
28. H. Meldner, *Ark. Fys.* **36**, 593 (1967).
29. R.A. Gherghescu, Z. Patyk and A. Sobiczewski, *Acta Phys. Pol.* **B28**, 31 (1997).
30. R.A. Gherghescu, Z. Patyk, J. Skalski and A. Sobiczewski, Proc. Intern. Workshop: *Research with Fission Fragments*, Benediktbeuern (Germany) 1996, eds. T. von Egidy, D. Habs, F.J. Hartmann, K.E.G. Löbner and H. Nifenecker (World Scientific, Singapore 1997) p. 116.
31. W.D. Myers and W.J. Swiatecki, *Nucl. Phys.* **A601**, 141 (1996).
32. R. Smolańczuk, *Phys. Rev.* **C56**, 812 (1997).
33. A. Staszczak et al., Proc 3rd Int. Conf.:*Dynamical Aspects of Nuclear Fission*, Casta-Papiernicka (Slovakia) 1996, eds. J. Kliman and B.I. Pustylnik (JINR, Dubna, 1996) p. 22.

Fission Studies of Nuclei far from Stability

K.-H. Schmidt[a], J. Benlliure[a], C. Böckstiegel[b], H.-G. Clerc[b],
A. Grewe[b], A. Heinz[a], M. de Jong[b], A. R. Junghans[a],
J. Müller[b], M. Pfützner[c], S. Steinhäuser[b]

[a]*Gesellschaft für Schwerionenforschung m. b. H., Planckstraße 1, 64291 Darmstadt, Germany*

[b]*Institut für Kernphysik, Technische Universität Darmstadt,
Schloßgartenstraße 9, 64289 Darmstadt, Germany*

[c]*Institute of Experimental Physics, University of Warsaw,
Ul Hoza 69, 00-381 Warszawa, Poland*

Abstract. Nuclear fission from excitation energies around 11 MeV was studied at GSI Darmstadt for 76 neutron-deficient actinides and pre-actinides by use of relativistic secondary beams. The characteristics of multimodal fission of nuclei around ^{226}Th were systematically investigated. The yields of even-Z elements in near-symmetric charge splits were found to be enhanced by about 10% for long isotopic chains of uranium and thorium fissioning nuclei. Conclusions on the dissipated energy from saddle to scission are drawn.

INTRODUCTION

Nuclear fission provides unique information on the reordering of nucleons in a large-scale collective motion. The signatures of shell effects and pairing correlations show up in fission from low excitation energies. Such data are of general importance for the role of shell structure and viscosity in nuclear dynamics.

In a recent experiment, relativistic secondary beams were used to overcome restrictions of conventional experimental techniques which were limited to spontaneously fissioning nuclei and to nuclei in the vicinity of long-lived isotopes, used as target material. The new data systematically cover short-lived neutron-deficient actinides and pre-actinides. In this contribution, part of the large body of data acquired is presented and the resulting progress in the understanding of fission dynamics is sketched.

EXPERIMENT

The secondary-beam facility of GSI, Darmstadt, offers unique possibilities to provide secondary beams of neutron-deficient actinides and pre-actinides produced by fragmentation of relativistic ^{238}U projectiles. Within the limits given by the primary-beam intensity and the fragmentation cross sections (1,2), nuclear charge and mass number of the secondary projectiles can freely be selected by tuning the fragment separator (3). Intensities of typically 100/s were reached for the most abundantly produced isotopes.

Fission from the desired excitation-energy range slightly above the fission barrier was induced in flight by electromagnetic interactions in a heavy target material. In the next chapter we give a detailed description of this excitation mechanism.

The experiment was performed at the fragment separator. As secondary target we used a stack of lead foils with a total thickness of 3 g/cm^2. The average energy of the secondary projectiles in the lead target was about 430 A MeV. The differential energy loss of each fission fragment was measured separately with an horizontally subdivided twin ionisation chamber. In order to correct the energy loss for the velocity dependence, the time-of-flight of the fission fragments was measured by means of a scintillation detector placed in front of the lead target and a (1m × 1m) scintillator wall placed 5.5 m downstream.

Due to the high centre-of-mass velocities, an excellent charge resolution ($Z/\Delta Z \approx 120$) was achieved. Events stemming from reactions at lower impact parameters with nuclear contact were suppressed. For details of the analysis procedure see refs. (4,5).

EXCITATION MECHANISM

The electromagnetic excitation in-flight in the secondary target is one of the most important ingredients of the experiment, ideally adapted to the kinematic properties and to the low intensities of the secondary beams. It populates states in the vicinity of the fission barrier with a large cross section of a few barns. Although the excitation energy acquired is not precisely known for a single event, the excitation-energy distribution can be calculated with rather good precision. The electromagnetic field of a lead target nu-

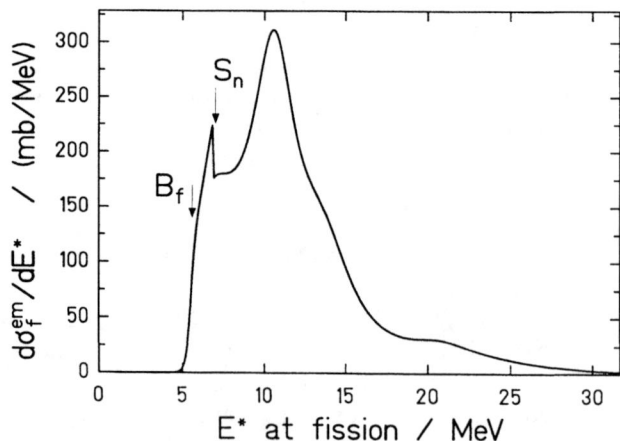

Figure 1. Calculated distribution of excitation energies at fission after electromagnetic excitation of ^{234}U projectiles at 430 A MeV in a lead target.

cleus, seen by the projectile, can be represented by a flux of equivalent photons of different energies and multipolarities according to ref. (6). The projectiles are excited according to the energy-dependent nuclear photo-absorption cross section which is dominated by the giant dipole resonance (GDR) with small contributions of the giant quadrupole resonances. First-chance fission represents the main source of fission, but also fission after evaporation of one or two particles (mostly neutrons) occurs with a probability of about 20%. The excitation-energy distribution at fission after electromagnetic excitation of ^{234}U in the passage of a lead target at 430 A MeV is shown in Figure 1. Although in the deformed ^{234}U nucleus the GDR is split, the distribution shows only one peak at 11 MeV, because the high-energy component of the GDR is strongly suppressed by the decreasing equivalent photon spectrum. For details of the calculation see ref.(4). By considering the small systematic variations of the photo-absorption cross sections as a function of mass number and deformation, the calculated excitation-energy distributions of the other nuclei investigated turned out to be rather similar. In all cases the peak of the excitation-energy distribution is predicted at less than 5 MeV above the fission barrier.

RESULTS AND DISCUSSION

In the present experiment, the elemental yields and the total kinetic energies of long isotopic chains from ^{205}At to ^{234}U have been determined. Part of the elemental yields are shown in Figure 2. The transition from symmetric fission in the lighter systems to asymmetric fission in the heavier systems is systematically covered for the first time. In the transitional region, around ^{226}Th, triple-humped distributions appear, revealing comparable intensities for symmetric and asymmetric fission. In particular for uranium and thorium isotopes strong even-odd effects are observed.

Fission channels

Turkevich and Niday (7) already noticed that different components which they named fission modes appear in the fission-fragment yields and in the kinetic-energy distributions. Later, models were proposed to deduce the fission characteristics from the properties of the scission configuration, e.g. ref. (8). However, these do not explicitly consider the dynamical evolution of the system from saddle to scission which seems to be very important (9,10,11). The concept of independent fission channels has been developed (12,13) according to which the fissioning system follows specific valleys in the potential energy in the direction of elongation. In this model, the yields of the fission channels are related to the properties of the fissioning nucleus near the fission barrier were the fission valleys appear. Several properties (e.g. average mass or charge split, mass or charge width, total kinetic energy) could be related to calculated properties of the highly deformed fissioning system near scission. Still, the mechanisms which determine the yields of the fission channels are not sufficiently well understood to allow for quantitative predictions. Therefore, they are usually deduced from experiment. The

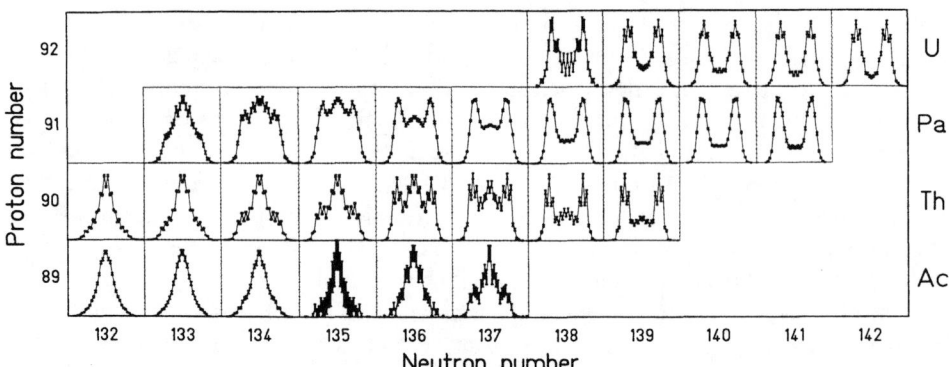

Figure 2. Measured fission-fragment charge distributions from ^{221}Ac to ^{234}U are depicted on a chart of the nuclides. (The figure shows only 28 out of the 76 systems investigated.)

present survey on fissioning systems with strongly varying charge distributions (see Figure 2) provides a systematic view on how the fission-fragment yields vary as a function of the nuclear composition. These data substantially increase the experimental knowledge relevant for a better theoretical understanding of the fission dynamics.

At first glance, two fission components appear in the measured charge distributions, a symmetric and an asymmetric one. The transition is rather smooth, and the weights of the two fission components scale approximately with the mass of the fissioning nucleus. The widths (standard deviation) of the symmetric and the asymmetric components were found to be close to 4.0 charge units and 2.2 charge units, respectively, for all nuclei for which they could be extracted.

In detail, the charge-yield distributions and the total kinetic energies of ^{228}Th, ^{226}Th, and ^{223}Th are shown in Figure 3. In order to consider the trivial variation of the total kinetic energy as a function of mass and charge split, the Coulomb repulsion V_C in the scission-point configuration was parametrized by the following expression, introduced in ref. (8):

$$V_C = \frac{Z_1 \cdot Z_2 \cdot e^2}{r_0 \left(A_1^{1/3} \cdot \left[1 + \frac{2\beta_1}{3} \right] + A_2^{1/3} \cdot \left[1 + \frac{2\beta_2}{3} \right] \right) + d} \quad (1)$$

Z_i, A_i and β_i are nuclear-charge numbers, mass numbers and deformations of the fission fragments, $r_0 = 1.16$ fm is the nuclear-radius constant, e the elementary charge, and $d = 2$ fm the "tip distance" which effectively takes into account the neck formation at scission. The mass numbers were deduced from the charge numbers by the UCD assumption. The deformation parameters were fixed at $\beta_i = 0.6$ for both fragments as predicted by the liquid-drop model (8).

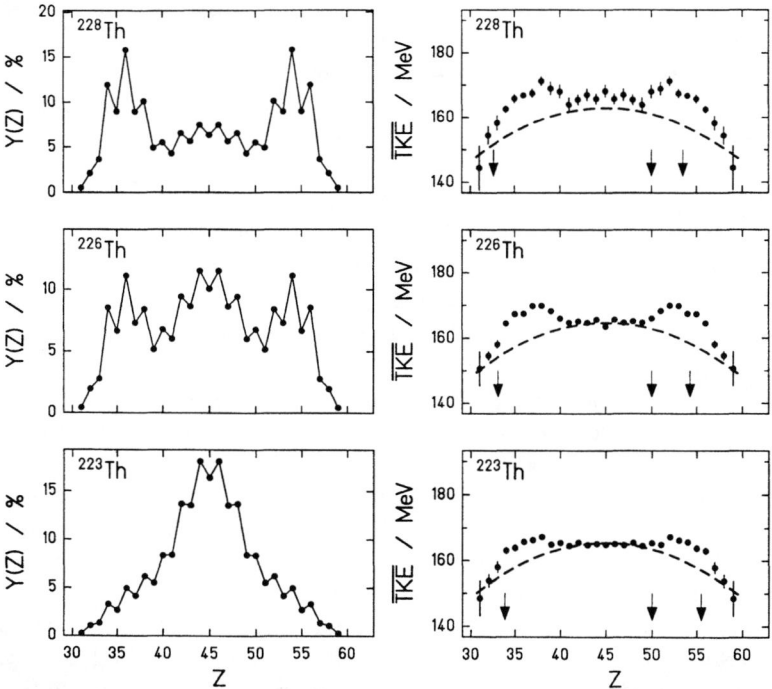

Figure 3. Measured elemental yields (left part) and average total kinetic energies (right part) as a function of the nuclear charge measured for fission fragments of several fissioning nuclei. Only statistical errors are given. The total kinetic energies are subject to an additional systematic uncertainty of 2 %, common to all data (5). Arrows indicate the positions of neutron ($N = 50$, 82) and proton shells ($Z = 50$). The positions of the neutron shells are calculated from the proton numbers by assuming an unchanged charge density (UCD). The predictions of the liquid-drop model for the total kinetic energies are shown as dashed lines.

Figure 3 reveals that the global structural effects observed in the charge yields strongly differ from those showing up in the total kinetic energies. From ^{228}Th to ^{223}Th, the weight of the asymmetric fission component decreases strongly, while the enhancement of the total kinetic energies at asymmetry with respect to the macroscopic model is nearly preserved. Moreover, the position of the enhanced kinetic energy rather seems to follow the position of the $N = 82$ shell in the heavy fragment. The $Z = 50$ shell does not have any noticeable influence neither on the yields nor on the kinetic energies. These findings impose important constraints on an improved dynamical description of the fission process which could provide a quantitative prediction of the characteristics of the fission products.

Dissipation in fission

Data on elemental fission-fragment yields were difficult to obtain in conventional fission experiments. The rather scarce data on total even-odd effects in elemental yields measured previously show a systematic variation with the fissility of the fissioning system (see Figure 4). From this variation, the intrinsic excitation energy acquired by dissipation up to scission was deduced to grow with increasing fissility (14). The excellent nuclear-charge resolution of the present experiment allowed to determine the even-odd structure in the elemental yields for a large number of fissioning systems. The new results of the present work on total even-odd effects for thorium and uranium isotopes which are also shown in Figure 4 seem to be compatible with the previously measured data: The higher excitation energies present in electromagnetic-induced fission (see Figure 1) lead to a reduction by about 40 %, but the dependence on the fissility parameter Z_{cn}^2 / A_{cn} observed previously seems to be confirmed. However, a closer look to the even-odd structure as performed in the following section will reveal that the situation is much more complex.

The present experiment for the first time provides data on even-odd structure in <u>symmetric</u> charge splits for long isotopic chains. Figure 5 shows the local proton even-odd effect as defined by Tracy et al. (15) for thorium isotopes both at symmetry and in the maximum of the asymmetric fission component. In contrast to the total even-odd effect shown in Figure 4, the local even-odd effect at symmetry remains constant over nine isotopes. The local even-odd effect at asymmetry is much larger but it also stays almost constant over the range investigated. We now understand the true reason for the strong variation of the even-odd effect for thorium isotopes (and also for uranium isotopes) investigated here: It is the strong variation of the relative weights of symmetric and asymmetric components in the fission yields, see Figure 2. With increasing neutron number (decreasing fissility), the relative weight of the asymmetric component grows, leading to a larger total even-odd effect.

As already reported (16), the secondary-beam experiment also revealed strong local

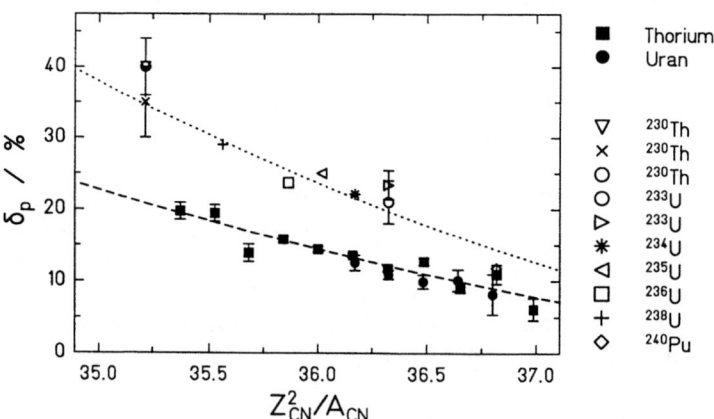

Figure 4. Total even-odd effect in the nuclear-charge yields observed in the fission of different nuclei. Data from thermal-neutron-induced fission (open symbols) from literature are compared to results of the present experiment for thorium and uranium isotopes (full symbols) as a function of the fissility parameter Z_{cn}^2 / A_{cn}.

652

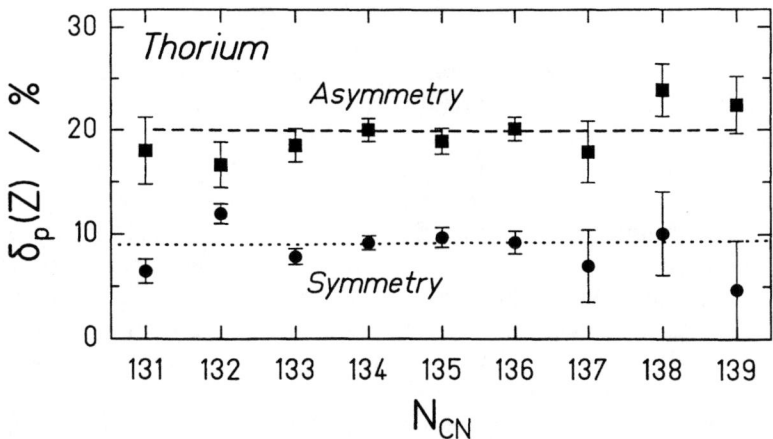

Figure 5. Local even-odd effect for symmetric and the most abundant asymmetric charge splits found in the fission of thorium isotopes after electromagnetic excitation.

even-odd structures for all investigated odd-Z fissioning systems. A similar observation for a few other odd-Z systems was reported by Stumpf et al. (17). We attributed this finding to an enhanced sticking probability of unpaired protons to the heavy fragment due to phase-space arguments in asymmetric charge splits (16). This implies that an enhanced production of fission products with even proton numbers cannot directly be related to the probability of a fully paired proton configuration to survive up to the scission point as was done before (18). Only for symmetric charge splits, the analysis used so far is valid. For asymmetric charge splits, however, part of the measured even-odd structure has to be attributed to the statistical contribution which explains great part of the higher even-odd effect found in asymmetry.

From these findings we conclude that for all neutron-deficient thorium and uranium isotopes investigated in the secondary-beam experiment the dissipated energy from saddle to scission is about the same. In contrast to the first impression that we got from Figure 4, these systems do not follow the general trend deduced previously from the measured even-odd effect of heavier fissioning systems. Unfortunately, for those systems no information on local even-odd structure for near-symmetric charge splits is available due to the low yields at symmetry. Final conclusions on dissipation from the even-odd structure in asymmetric charge splits, the only information available in heavier systems, will require a far more detailed analysis and a full quantitative understanding of the statistical contribution to the even-odd structure in fission along the lines given in ref. (16), including the influence of shell structure.

CONCLUSION

A new experimental technique was applied to measure elemental yields and total kinetic energies after low-energy fission of short-lived radioactive nuclei. In contrast to most secondary-beam experiments of the first generation which are designed to determine one specific property of the secondary projectiles like binding energy or total interaction cross section, the present experiment is a quite elaborate secondary-reaction study where many characteristics of the fission process could be investigated. The beautiful data nicely demonstrate the decisive influence of nuclear structure on the fission process in a particular interesting transitional region around ^{226}Th. In contrast to the total kinetic energies, the element distributions were found to vary strongly, essentially as a function of mass number of the fissioning system. The relative weights of the asymmetric fission channels decrease with decreasing mass of the fissioning nucleus, but the scission configurations remain almost unchanged.

The new findings on even-odd effects in the nuclear-charge distributions after low-energy fission require a revised understanding of pair breaking in fission. Strong variations of the observed even-odd effect in the elemental yields as a function of fissility for different systems and as a function of asymmetry for specific systems are partly attributed to a statistical contribution to this quantity rather than to a variation of the intrinsic excitation energy at scission. Conclusions on the viscosity of cold nuclear matter have to be reconsidered.

The detailed features of the fission fragments measured in the present work for a large number of fissioning systems in a continuously covered region of the chart of the nuclides provide an important test for elaborate models of low-energy nuclear dynamics to be developed.

REFERENCES

1. H.-G. Clerc et al., Nucl. Phys. **A590** (1995) 785
2. A. R. Junghans et al., Nucl. Phys. **A629** (1998) 635
3. H. Geissel et al., Nucl, Instr. Meth. **B70** (1992) 286
4. A. Grewe et al., Nucl. Phys. **A614** (1997) 400
5. C. Böckstiegel et al., Phys. Lett. **B398** (1997) 259
6. B. Baur, C. A. Bertulani, Phys. Rev. **C34** (1986) 1654
7. A. Turkevich, J. B. Niday, Phys. Rev. **84** (1951) 52
8. B. D. Wilkins, E. P. Steinberg, R. R. Chasman, Phys. Rev. **C14** (1976) 1832
9. P. Möller, S. G. Nilsson, Phys. Lett. **31B** (1970) 283
10. G. A. Kudyaev, Yu. B. Ostapenko, G. N. Smirenkin, Sov. J. Nucl. Phys. **45** (1987) 951
11. H. Kudo et al., Phys. Rev. **C25** (1982) 3011
12. V. V. Pashkevich, Nucl. Phys. **A169** (1971) 275
13. U. Brosa, S. Grossmann, A. Müller, Phys. Rep. **197** (1990) 167
14. F. Gönnenwein, "The Nuclear Fission Process", CRC Press, London, 1991, C. Wagemans ed., pp. 409
15. B. L. Tracy et al., Phys. Rev. **C5** (1972) 222
16. S. Steinhäuser et al., Nucl. Phys. **A634** (1998) 89
17. P. Stumpf et al., Proc. "Seminar on Fission", C. Wagemans (ed.), Pont D'Oye, Belgium, 1995, pp. 196
18. H. Nifenecker et al., Z. Phys. **A308** (1982) 39

Structure of Neutron-Rich Nuclei and Rare Processes in Spontaneous Fission of ^{252}Cf

A.V. Ramayya*, J.H. Hamilton*, J.K. Hwang*
and GANDS95 Collaboration

*Department of Physics and Astronomy, Vanderbilt University, Nashville, TN37235, USA.

Abstract. The high selectivity and sensitivity offered by large detector arrays such as Gammasphere and Eurogam enable one to identify gamma rays from individual neutron-rich nuclei produced in the spontaneous fission (SF) of ^{252}Cf and ^{248}Cm. We have identified new γ transitions in ^{101,102,103}Nb, ^{98,100}Y, ^{103,107,109}Mo and ^{109,111}Ru. In Nb nuclei we observed pairing-free rotational bands with $\Im_{exp}/\Im_{rigid} \approx 1$. In ^{103}Mo and ^{109,111}Ru we identified bands with close doublet structures which may be a general consequence of rotation alignment for configurations of half-filled j-shells, which are weakly coupled to the deformed shapes. In addition, rare processes such as cold neutronless binary fission, α-accompanied ternary fission and ^{10}Be-accompanied ternary fission are observed.

Search for $K^\pi=1^+$ Pairing-free Rotational Bands in A$\approx$100 Neutron-rich Nuclei

In the fission of ^{252}Cf about 100 different final fragments are produced. In recent times, many new experimental data concerning the structures of these neutron-rich nuclei have been obtained by studying the prompt γ rays emitted by correlated pairs produced in spontaneous fission (SF). Neutronless cold binary fragmentation and cold neutronless ternary fragmentation also takes place with very low SF yields. These process can also be investigated with these large detector arrays. The primary fragments emit several neutrons until the exitation energy is below the binding energy of a neutron. Then secondary fragments decay to the ground state by the emission of γ rays. The Z and A of the secondary fragments is also identified by comparing the yields with the calculated yields by Wahl [1]. In our experimets a ^{252}Cf source of strength $\approx 28\mu$Ci was sandwiched between two Ni foils of thickness 11.3 mg/cm^2 and then sandwiched between 13.7 mg/cm^2 thick Al foils and placed at the center of the Gammasphere. This experiment was carried out with 72 Compton-suppressed Ge detectors.

It is well known that adding an odd nucleon to a special group of even-even nuclei with unusually weak pairing in their ground states may further reduce the pairing and one may be able to observe pairing-free rotational bands. Deformed odd-odd nuclei with A≈100 are ideal candidates. Peker et al. [2] interpreted the low spin members (1^+, 2^+, and 3^+) in ^{100,102}Y and ^{102,104}Nb as $K^\pi=1^+$ pairing-free rotational bands by comparing their moments of inertia to rigid body values. In these cases $\Im_{exp}/\Im_{rigid} \approx 1$. Pairing collapse for their lowlying states is attributed to the strong n-p coupling of $\pi g_{9/2}$ and $\nu g_{7/2}$ orbitals in the neutron rich region with A≈100. The new partial level schemes of ^{102}Nb and ^{100}Y from our cube are shown in Figure 1 [3]. In ^{102}Nb the ground band is extended to 5^+ with five additional higher energy states and in ^{100}Y to 7^+. Additional transitions in ^{100}Y confirm the $K^\pi=1^+$ band which was proposed by Peker [2]. In ^{102}Nb, even though the moment of inertia approaches the rigid body value, the γ transition energy differences suggest the states above 5^+ may have a change in structure from the ground states. In Table 1, a comparison between the experimental energy levels and theoretical energy levels of the possible pairing-free bands is presented. The theoretical energy levels are obtained from the adiabatical rotational formula with K=1. Table 1 shows the calculated moments of inertia for these two bands This good fit to the energy levels shows that all higher order terms can be neglected at low spin for a deformed nucleus with A≈100 nuclei and N>60. This assumption should be especially true for a pairing-free band since other effects such as Coriolis band mixing cannot increase the already maximal value of moment of inertia. Also, the linear term aI in the rotational formula can be taken as zero since one expects 'a' to be small for pairing-free $K^\pi=1^+$ bands. But a value of 'a' of a few keV would increase the moment of inertia by about 10

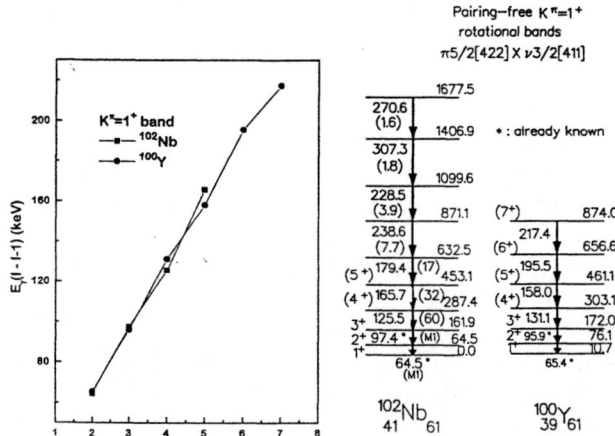

FIGURE 1. $E_\gamma(I \to I-1)$ versus spin in $K^\pi=1^+$ bands in ^{102}Nb and ^{100}Y and partial level scheme of ^{102}Nb and ^{100}Y.

TABLE 1. Comparison between the experimental energy levels and theoretical energy levels. The theoretical energy levels are obtained from the rotational formula of E= $E_0+A[I(I+1)-K^2]+aI=B+AI(I+1)+aI$ (keV) with K=1 and a=0.

I^π	^{102}Nb		^{100}Y	
	$E(I)_{exp}$	$E(I)_{theo}$	$E(I)_{exp}$	$E(I)_{theo}$
1^+	0.0	-0.29	10.7	12.4
2^+	64.5	64.3	76.1	76.5
3^+	161.9	161.1	172.0	172.6
4^+	287.4	290.2	303.1	300.7
5^+	453.1	451.6	461.1	460.9
6^+			656.6	653.1
7^+			874.0	877.3
	A=16.14, B=-32.57		A=16.02, B=-19.61	

Nucleus	E_{1^+}	$2\Im_{exp}/\hbar^2$ (MeV^{-1})	$\Im_{exp}/\Im_{rigid}$
^{100}Y	10.7	62.4	0.95
^{102}Nb	0.0	62.0	0.92

%. From this point of view, the experimental values obtained for the moments of inertia in Table 1 can be regarded to be the lower limits to the actual moment of inertia values. In Table 1, the ratios of $\Im_{exp}/\Im_{rigid}$ are shown. The ratios are 0.95 for ^{100}Y and 0.92 for ^{102}Nb. The present observation of additional membes of the $K^\pi=1^+$ band in ^{100}Y support the pairing-free nature. However, in ^{102}Nb the situation is not clear above the 5^+ level. Usually the pairing strength is quenched as quasiparticles are added. This apparent pairing collapse may be attributed to the strong proton $(g_{9/2})$-neutron$(g_{7/2})$ coupling. Pairing-free states can be obtained by increasing the rotational angular momentum in a deformed nucleus to a large value. At high spin (for example, $> 34\ \hbar$ for ^{168}Hf), the Coriolis antipairing force can overcome the residual pairing force and result in pairing-free states. The extension of the rotational bands in these A≈100 odd-odd nuclei, ^{100}Y and ^{102}Nb, in the present work provide unique support for the existence of pairing-free states at low angular momentum.

Close Doublet Structures in ^{103}Mo, ^{109,111}Ru and Neighbors

The studies of beta decay and prompt fission gamma spectra in spontaneous fission afford unique access to a strongly prolate region of the Z=42-44 nuclei. The large gamma detector arrays are producing high quality data that permit study of odd-A nuclei. Level schemes of ^{103}Mo, ^{109}Ru and ^{111}Ru have been significantly extended recently [4]. These new data provide evidence for a close doublet structure in odd parity bands in these nuclei. In the normal Bohr-Mottelson strong coupling model, one would expect mid-shell bands of odd parity in the N=50-82 shell to show regular structures with I(I+1) spacing. Only ^{101}Zr and ^{105}Mo approach this behavior. In most of the other nuclei shown in Fig. 2, the bands establish a doublet structure between the signature partners as the general pattern. What can be the reason for such a widespread occurrence of doublet structure? Examination of Nilsson level diagram for the region of $61 \leq N \leq 67$ shows that for prolate deformations of β between 0.3 and 0.5, there are usually three orbitals of other types between the $h_{11/2}5/2[413]$ and $h_{11/2}7/2[523]$ orbitals. That is to say the $h_{11/2}$ family will remain half-filled over this region of neutron numbers. A half-filled j-shell will have about the same total energy in oblate as in prolate or spherical shapes. Meyer-ter-Vehn [5] studied the problem of one odd nucleon in a triaxial potential well, but we are not aware of theoretical work on odd systems near the half-filled shell. From his single particle energy diagram for a j=11/2 shell in a triaxial potential [5] we can infer that half-filled shells are rather independent of deformation parameters β and γ. Removing or adding one nucleon from or to the half-filled shell will leave $h_{11/2}$ configurations that are still only weakly coupled to the nuclear shape. Thus, we would expect the yrast bands for odd-parity states to be rotation-aligned bands built on 5/2$^-$ and 7/2$^-$ bandheads, because the parallel alignment of rotational and odd-nucleon angular momentum minimizes the rotational energy for a given I value. That is, the Coriolis coupling dominates. The quasiparticle energies of the 5/2$^-$ (hole) and 7/2$^-$ (particle) states will be similar for the $61 \leq N \leq 67$ region, where the Fermi energy lies between $h_{11/2}5/2[413]$ and $h_{11/2}7/2[523]$ levels. Here, one will get a doublet structure for these nearly half-filled j-shells, regardless of the deformation or shape of the core. In conclusion we have found a number of new γ transitions and new band structure in ^{103,105}Mo and ^{109,111}Ru. These data show that the band structure of ^{103}Mo$_{61}$ is different from that of ^{101}Zr$_{61}$ in that ^{103}Mo shows

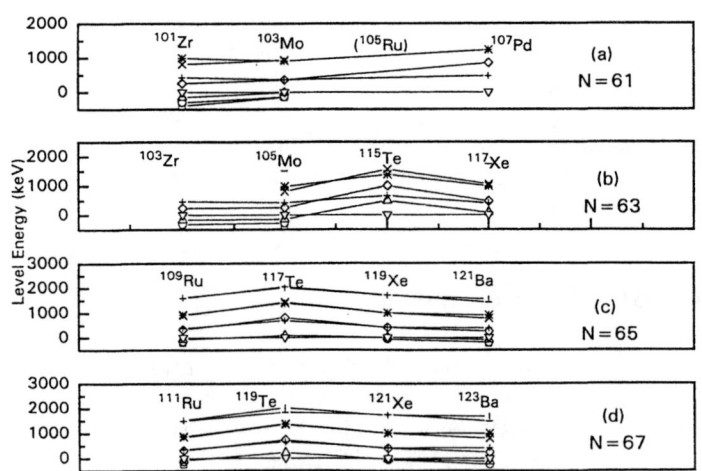

FIGURE 2. Comparisons of odd-parity level energies in nuclei N=61 - 67. □=5/2, ○=7/2, △=9/2, ▽=11/2, ◇=13/2, + =15/2, ×=17/2, *=19/2, - =21/2, and |=23/2.

TABLE 2. Average cold binary fission yields from gates on two light fragment and two heavy fragment transitions. The total theoretical values are renormalized to the total experimental value as shown in column 5.

	A_L / A_H	Y_{exp}	Y_{the}	$Y_{the}^{(ren)}$
Zr / Ce	100/152	0.010(2)	0.38	0.004
	102/150	0.020(4)	2.82	0.033
	103/149	0.030(6)	4.21	0.049
	104/148	0.010(2)	1.03	0.012
Mo / Ba	104/148	0.010(2)	0.47	0.005
	105/147	0.040(8)	5.39	0.063
	106/146	0.040(8)	0.61	0.007
	107/145	0.070(14)	3.07	0.036
	108/144	0.030(6)	7.45	0.087
Tc / Cs	109/143	0.090(18)	11.03	0.128
Ru / Xe	110/142	0.060(12)	3.78	0.044
	111/141	0.10(2)	7.12	0.083
	112/140	0.020(4)	0.59	0.007
	114/138	0.020(4)	1.17	0.014
Pd / Te	116/136	0.050(20)	2.35	0.027

a more pronounced close doublet structure. The odd-parity bands of ^{109,111}Ru exhibit similar intruder band character at higher spins, but ^{111}Ru shows more regular spacing at low spins. Further experimental work and theoretical modeling studies are needed.

Cold binary fission

No direct measurements of yields of correlated pairs in cold binary fission have been made prior to our work. Since the neutronless binary events are much smaller than those with neutrons emitted, double gating techniques have been employed to extract the yields for the cold binary fission. Earlier we reported the first results for the correlated pairs in cold binary fission in ^{252}Cf [6–8] and ^{242}Pu [9].

Subsequently we extracted more detailed yields of cold binary fission As an example for cold binary fission, a double gate was set on the 181.0 and 332.6 keV transitions in ^{146}Ba. One can clearly see the 171.6 keV $2^+ \rightarrow 0^+$ transition in ^{106}Mo, the zero neutron emission partner of ^{146}Ba. By determining the intensities of γ transitions in both fragments and knowing the branching ratios between different transitions, relative binary yields were extracted again where the total yield was normalized to Wahl's table [1]. Presently, many of the spectra of the odd-Z nuclei are not known, so that one could not determine experimentally most odd-Z isotopic yields. The cold binary fission yields are shown in Table 2 along with the theoretical values predicted by Sandulescu et al. [10]. In Table 2, the first report of the cold binary fission of an odd-Z - odd-Z fragmentation is shown for the Tc and Cs pair.

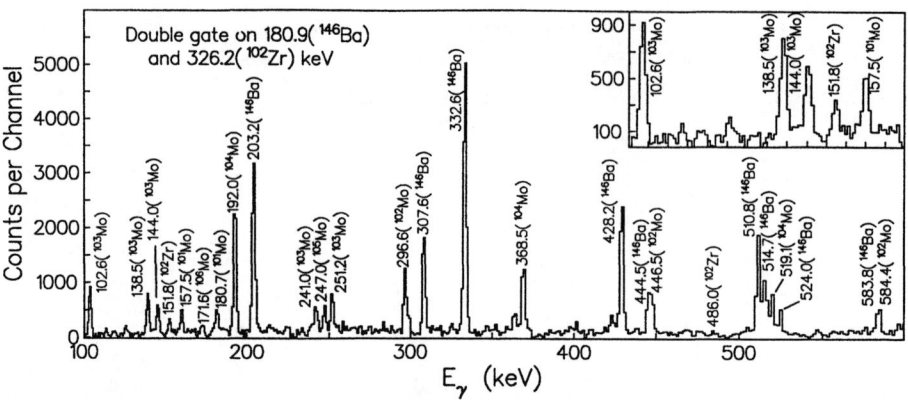

FIGURE 3. The triple γ-coincidence spectrum obtained with a double gate set on the $2^+ \rightarrow 0^+$ transition at 180.9 keV in ^{146}Ba and on the $4^+ \rightarrow 2^+$ transition at 326.2 keV in ^{102}Zr. Note the clear peak at 151.8 MeV which correspond to the $2^+ \rightarrow 0^+$ transition in ^{102}Zr. The peaks at 203.2, 307.6, 332.0, 428.2, 444.5, 510.8, 514.7, 524.1 and 583.8 keV correspond to different transitions in ^{146}Ba because of the 324.2 keV ($7^- \rightarrow 5^-$) transition of ^{146}Ba in the gate.

Cold ternary fission

Fig. 3 shows an example for cold α-ternary fission. By setting double gates on the 326.2 keV (^{102}Zr) and 181 keV (^{146}Ba), one can clearly see the 151.8 keV ($2^+ \rightarrow 0^+$) transition in ^{102}Zr. In Table 3, the cold neutronless alpha ternary fission yields observed in the spontaneous fission of ^{252}Cf are presented [11]. The highest experimental yields were found for the Zr + Ba isotopes. Significant yields also were observed for the Mo + Xe and Sr + Ce ternary fragmentations. The average values of the α ternary fission yields obtained by double-gating on the heavy fragments or light fragments are listed in Table 3. In the case of cold ternary fission, Sandulescu et al. [10] considered that close to the scission configuration a few of the nucleons form a short neck. At a given value of the neck radius a double scission takes place, a third light fragment is formed between the two heavier ones and from this point starts the ternary fission barrier corresponding to a given mass and charge splitting. The width of the potential barrier between the heavier fragments is only 2.5 to 3 fm with an exit point at a tip distance between the fragments of 3.5 to 4 fm, indicating that initially the two heavier fragments are penetrating their relative potential barrier. Later on as the heavier fragments move away, the potential well for the light fragment disappears (for tip distances of about 7 to 8 fm) and it becomes free being repelled mostly along the y axis by the two heavier fragments which have already acquired large kinetic energies. The theoretical values calculated by Sandulescu et al. [10] are shown in columns 3 and 6 of Table 3. Since a different normalization is used for the theoretical calculations, the total theoretical yield is normalized to the total experimantal yield, and the reported values are the renormalized relative theoretical yields. In general, there is good agreement between the relative theoretical and experimental yields.

The emission of an α particle in neutronless ternary fission [11] may be a special case because of its possible preexistence in the nucleus. To definitively establish cold ternary fission or cold multi-fragmentation it is necessary to observe experimentally cold neutronless ternary fission (triple fission) with the third particle being a heavier cluster such as ^{10}Be [12]. Contrary to α-ternary fission whose excited state energies are very high (>20 MeV), ^{10}Be has an excited state (2^+) at an energy of 3.368 MeV. This is not expected for α-emission. Furthermore, ^{10}Be and other heavier ternary fragments are more easily deformable to enhance collective effects. This could lead to hyper-deformed nuclear configurations, of a molecular type, in the exit channel. It will be very interesting to see whether modes of relative vibration and rotation of the three fragments can be seen in future. The excitation of such modes may determine the angular distribution of the emitted fragments.

In Fig. 4a, we see a peak at an energy of ≈3368 keV (marked as 3362 keV) region coincident with the gate transitions. Assuming that the peak at ≈3368 keV is the $2^+ \to 0^+$ transition in ^{10}Be, we set one gate on this transition and the other gate on the $2^+ \to 0^+$ transitions in ^{96}Sr. We observed in our coincidence spectra the 977.5

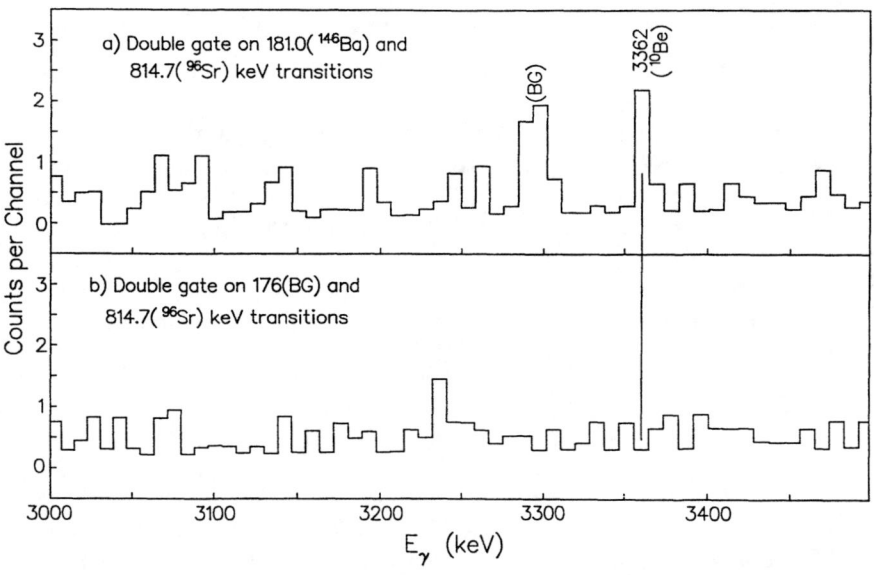

FIGURE 4. a) Coincidence γ-ray spectrum obtained by double gatting on the 814.7, $2^+ \to 0^+$ transition in ^{96}Sr and 181.0 keV, $2^+ \to 0^+$ transition in ^{146}Ba. b) Coincidence γ-ray spectrum obtained by double gatting on the 814.7, $2^+ \to 0^+$ transition in ^{96}Sr and on 176.0 keV background peak, which is shifted from 181.0 keV $2^+ \to 0^+$ transition in ^{146}Ba. BG : Background.

TABLE 3. The alpha ternary isotopic yields Y_{exp} obtained per 100 fission events.

α Partner nuclei	Y_{exp} (%)	Y_{the}^{ren}	α Partner nuclei	Y_{exp} (%)	Y_{the}^{ren}
$^{92}_{36}$Kr–$^{156}_{60}$Nd	0.002(1)	0.002	$^{96}_{38}$Sr–$^{152}_{58}$Ce	0.008(3)	0.010
$^{98}_{38}$Sr–$^{150}_{58}$Ce	0.014(6)	0.017	$^{99}_{38}$Sr–$^{149}_{58}$Ce	0.018(9)	0.016
$^{100}_{38}$Sr–$^{148}_{58}$Ce	0.021(10)	0.027	$^{101}_{38}$Sr–$^{147}_{58}$Ce	0.014(11)	0.010
$^{100}_{40}$Zr–$^{148}_{56}$Ba	0.038(12)	0.017	$^{101}_{40}$Zr–$^{147}_{56}$Ba	0.082(10)	0.058
$^{102}_{40}$Zr–$^{146}_{56}$Ba	0.009(4)	0.017	$^{103}_{40}$Zr–$^{145}_{56}$Ba	0.084(29)	0.050
$^{104}_{40}$Zr–$^{144}_{56}$Ba	0.017(8)	0.016	$^{106}_{42}$Mo–$^{142}_{54}$Xe	0.018(7)	0.031
$^{107}_{42}$Mo–$^{141}_{54}$Xe	0.030(14)	0.017	$^{108}_{42}$Mo–$^{140}_{54}$Xe	0.007(3)	0.014
$^{112}_{44}$Ru–$^{136}_{52}$Te	0.011(6)	0.028	$^{116}_{46}$Pd–$^{132}_{50}$Sn	0.006(3)	0.048

keV, $4^+ \rightarrow 2^+$ transition in ^{96}Sr and the 181.0 and 332.6 keV peaks in ^{146}Ba. Further when we set the double gate on the $\approx$3368(^{10}Be) and 181.0(^{146}Ba) keV peaks, we observed the 814.7 keV ($2^+ \rightarrow 0^+$) transition in ^{96}Sr. Our measured energy for the $\approx$3368 keV peak is 3362(4) keV. The energy values calculated from the level scheme range from 3362.8 to 3367.4 keV [13]. The value of 3362.8 keV is the difference between the transition energies of 5955.4 and 2592.6 keV levels. Hence, we believe that our measured value is within the range of the reported values and it belongs to ^{10}Be. Since we did not observe any Doppler broadening, we propose that ^{10}Be after it is formed is held in the potential well of ^{96}Sr and ^{146}Ba until the 3362 keV γ-ray is emitted.

The yield for the ^{10}Be ternary process is obtained from the double gate on the 181.0 (^{146}Ba) and 814.7 keV (^{96}Sr) transitions. The width of the peak for a stopped recoil is reasonable based on our energy resolution. The number of counts in the 3362 keV peak is corrected for the detector efficiency and gating conditions. Also the efficiency corrected count for the ^{10}Be is normalized to the previously determined binary yields of Ba and Mo pairs. The yield to the first excited state of ^{10}Be is $\approx 4(2) \times 10^{-4}$ per 100 fission fragments. These new results provide significant insight into the cold rearrangement of clusters of nucleons and multifragmentation. They also point the way to exploring a variety of cold ternary spontaneous fission modes including odd and even Z from ^{3}H, ^{6,7}Li to ^{14}C.

ACKNOWLEDGMENTS

The work at Vanderbilt University was supported in part by the US Department of Energy under grant No. DE-FG05-88ER40407. Complete list of authors of GANDS95 collaboration can be found in Phys. Rev. C**56**, 1344 (1997).

REFERENCES

1. A.C. Wahl, *Atom. Data and Nucl. Data Tables* **39**, 1 (1988).
2. L.K. Peker, F.K. Wohn, J.C. Hill and R.F. Petry, *Phys. Lett.* **B169**, 323 (1986).

3. J.K. Hwang et al., To be published (1998).
4. J.K. Hwang et al., *J. Phys.*, **G24**, L9 (1998).
5. J. Meyer-Ter-Vehn, *Nucl. Phys.* **A249**, 111 (1975).
6. G.M. Ter-Akopian et al., *Phys. Rev. Lett.*, **77**, 32 (1996).
7. A. Sandulescu et al., *Phys. Rev.* **C54**, 258 (1996).
8. J.H. Hamilton et al., *J. Phys.* **G20**, L85 (1994).
9. Y.X. Dardenne et al., *Phys. Rev.* **C54**, 206 (1996).
10. A. Sandulescu et al., *Int. J. Mod. Phys.* , **E**, in press (1998).
11. A.V. Ramayya et al., *Phys. Rev.* **C57**, 2370 (1998).
12. A.V. Ramayya et al. and GANDS95 Collaboration, *Phys. Rev. Lett.*, in press (1998).
13. R.B. Firestone and V.S. Shirley, *Table of Isotopes*, 8th ed. (John Wiley and Sons, Inc., New York, 1996).

Fission processes in ^{238}U-collisions on Pb and Be targets at relativistic energies

Monique Bernas

Institut de Physique nucléaire, IN2P3-CNRS, F-91406 Orsay

Abstract.
Independent isotopic yields can be determined for the first time for all U-fission products by a physical method. The neutron rich isotopes of elements between Zn and Te were measured on Pb target and neutron-rich light fragments between Ca and Nb on Be target. Shell closure effects are clearly assigned by enhancements of production yields. The enhancement of very asymmetric mass splittings observed on Pb is found still larger by using a Be target. The known gap between the lightest fragments from binary fission and the heaviest tri-partition fragments has been filled. Our measurements stress the capability of U-collisions on target-nuclei to produce neutron rich species in a large domain of elements.

INTRODUCTION

Fission induced by electromagnetic and nuclear interaction of a 750 A.MeV U-projectile on Pb and Be target-nuclei has been investigated. The analysis of the data has been pursued to probe the ability to produce neutron-rich nuclei outwards from the valley of stability and to characterize the fission processes occuring at intermediate excitation energies,

The main results of the analysis obtained since ENAM-95 are summarized here.

EXPERIMENT

In inverse kinematics, fission fragments from U-projectiles are emitted forwards and are efficiently transmitted to a separator. The resolving power of the fragment separator FRS [1], equipped with time-of-flight scintillation detectors and ionization chambers MUSIC for energy loss measurement, reaches $A/\Delta A = 500$ and $Z/\Delta Z = 130$. For the first time, each of the 2000 fragments produced by fission and fragmentation can be identified unambiguously. Moreover, their velocity is determined by recoil separation techniques.

At those velocities ($\beta = 0.83$ for E=750 A.MeV) ions are totally stripped of their electrons up to Yb. For elements beyond Yb the degrader tecnique [2] allows to insure the identification as well as population rates of ions over the three atomic charge states, q = Z, q = Z-1 and q = Z-2. Therefore fragmentation and fission cross-sections can be accurately measured. These are keys to investigate systematically the production modes involved in the collision. A wide program concerning Pb + p and d, U + p and d and Fe + p and d collisions at 1 A.GeV and lower energies is presently under way at GSI [16]. The cross sections for the spallation and fission measured by inverse kinematics with a high efficiency are needed for developing new methods of incineration and transmutation of nuclear waste.

RESULTS

low energy fission

In thermal neutron induced fission of ^{235}U, at excitation energy of $E_{ex} \approx 6$ MeV after neutron-capture, the double humped structure of the mass (or Z) distribution is strongly marked. It is evaluated by the peak-to-valley ratio, P/V = 400. This shape is governed by spherical (N=82) and deformed (N=88) shell configurations [3]. There is a weak contribution of the symmetric breaking. The three regions are associated respectively to channels SI, SII and SL [4]. The odd-even staggering of the yields due to the resistance to pair breaking between saddle and scission points reaches 24 % [5]. The stiffness of the nucleus towards a change of isotopic spin is expressed by the isobaric width σ_Z, the local width for a given mass A. This width is showing 30% oscillations around a mean value of 0.6. Maxima are related to odd values of Z.

With U-projectiles on a Pb-target, the fission is mainly due to electromagnetic interaction. The ratio P/V = 6 indicates a large increase of the symmetric breaking, i.e. of the SL-channel. The spectrum of excitation energies in U involves a main contribution of the GDR at 10 MeV, a much smaller one from the isovector GQR at 22 MeV and a third contribution from the (GDR)^2 centered at 25 MeV [6]. The mean odd-even effects on the yields of 7% shows that the low excitation energy channel (10 MeV) contributes predominantly. In the detailed analysis of yield distributions as function of mass [7] the three channels SI, SII and SL are clearly seen. The wings of the mass distribution are wider than in thermal fission. The yields are increased for elements lighter than Br. For Ni this enhancement reaches the value of 10^3.

With a Be target the mass distribution is yet more softened. The P/V ratio decreases to ≈ 2.5 and the relative yields on the wings are still larger. The elemental yield of Ni, for example, still increases by a factor of 4 as compared to the Pb target. In the frame of the scission point model, the neutron shell corrections of Wilkins [3] indicate a minimum in the potential at N = 102 (J) for very large deformations. This channel could combine with the N=40 minimum in the region of the light fragment already mentioned in the work of Sida [8] on thermal fission of ^{235}U and could result in a wide asymmetric enhancement of the mass yields. However the enhancement might also come from an overwhelming contribution of the SL channel, which, wider than expected, would cover towards the region of very large asymmetries.

In a recent study of 25 MeV p induced fission of ^{238}U [9], elemental yields for a very asymmetric breaking were found close to our results from U+Pb for isotopes of elements between nickel and germanium. According to the authors, a superasymmetric fission channel at A = 80 related to Z = 28, N = 50 opens when the excitation energy is increased. Until now, such a channel could not be assigned from our results in ^{235}U thermal induced fission [8] or ^{238}U-projectile fission on Pb [10].

Another important result is that with Pb as well as with Be target, the value of σ_Z remains constant at a small value of 0.6 ; it means that extremely large beam luminosities would be required to produce, via fission, isotopes with neutron excess as large as those about which theoreticians do speculate, namely ^{150}Sn or ^{86}Ni. Moreover this small value of σ_Z does not favour a configuration as Z = 28, N = 50, so far apart from the U-projectile values.

Cross sections of 2.1 ± 0.2 b and 0.115 ± 0.04 b were measured on Pb and Be respectively for the low excitation fission regime. Since there is twenty times more atoms in a Be target than in a Pb target of 1g.cm^{-2} which both introduce similar target location straggling effects, Be is choosen to search for new products in regions of large asymmetry. Elemental and isotopic yield distributions obtained on the Be target are discussed below.

high energy fission

When the impact parameter decreases, nuclear collisions in the U+Pb system become more violent. Nucleons are abraded, leaving an excited quasi-projectile which cools down by emitting particles, mostly neutrons. Near the end of the deexcitation process, the residual nucleus may undergo fission. Finally, a few more neutrons are emitted by fission fragments. The total kinetic energy of fission in the center of mass (TKE) remains constant at rising excitation energies. Therefore, laboratory velocities of those fission products are almost the same as in case of low excitation fission. The ratio A/Z is reduced because of the many neutrons released. This regime was unambiguously assigned since the velocity distribution measured by the F.R.S. is characteristic of fission [12].

Indeed a series of intermediate fragments produced by fragmentation at different excitation energies are decaying by fission. The resulting fragment distribution were measured to be centered around mean values $\overline{Z} = 42.9$ and $\overline{N} = 58.1$. The fissioning nuclei are centered around $^{208}_{86}$Rn$_{122}$ which is assumed to undergo fission at an excitation energy of 40 MeV. The mass distribution resulting from the superposition of symmetric mass-splitting of the many parent-nuclei is almost symmetric, odd-even effects on the yields are washed out and $<\sigma_Z> = 1.3$ is twice larger than at lower energies.

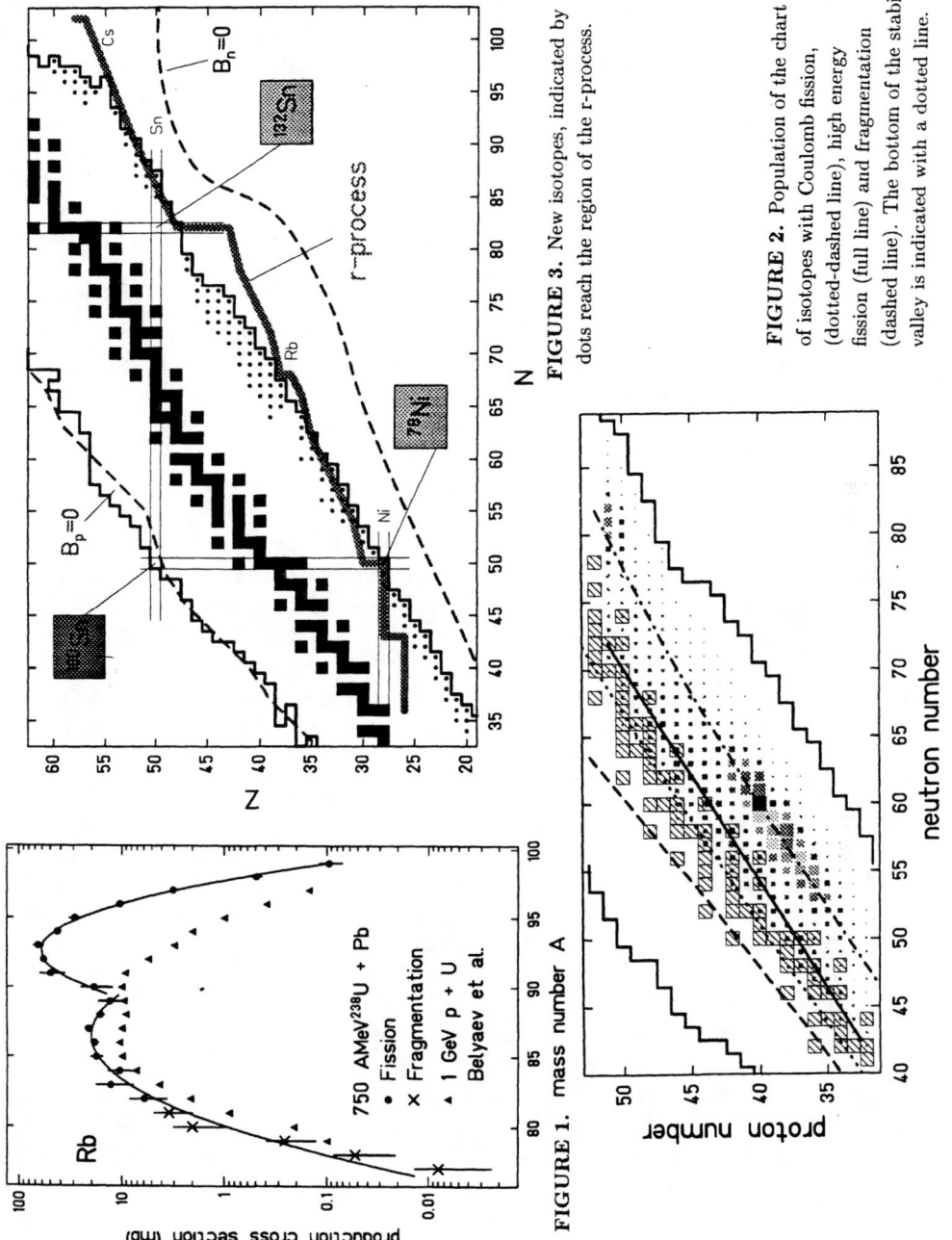

FIGURE 1.

FIGURE 2. Population of the chart of isotopes with Coulomb fission, (dotted-dashed line), high energy fission (full line) and fragmentation (dashed line). The bottom of the stability valley is indicated with a dotted line.

FIGURE 3. New isotopes, indicated by dots reach the region of the r-process.

In Fig. 1 the total isotopic distribution is given for the element Rb. It is compared with previous results from 1 GeV p + U collisions investigated using an on-line mass separator [13]. The yields of very neutron rich isotopes are multiplied by a factor of 20 in the U+Pb system while lighter isotopes are only two times more abundant.

The yields of the three processes for elements between Ge and Te are presented on the chart of nuclei (Fig. 2). Maximum production are indicated as ridges for the three regimes; fission at low excitation [7], fission of excited fragments [12] and fragmentation [14,16].

new isotopes

After the discovery of 56 new n-rich fission-fragments in U+Pb [18], a new measurement was resumed in 1995 to extend the findings by using a Be target and investigating the region of elements around Ni.

Starting from the region of maximum production yields by fission (A/Z= 2.55) a Bρ scanning over increasing magnetic rigidities was performed for fission fragments produced on Pb and Be targets. At the highest field value of the FRS set in this experiment, where isotopes with very large A/Z values are transmitted, data were collected on the Be target over a long period of three days, corresponding to a total dose of 10^{13} U projectiles on target. The FRS was then tuned on the isotope ^{78}Ni, and isotopes of all elements up to Mo were transmitted. 61 new n-rich isotopes were discovered between Ca and Nd [19] and their production yields were measured. The neutron-excess measured by the A/Z ratio reaches presently a value of 2.8 and the new isotopes cover the chart of nuclei up to the r-process path from Ni to Br and in the region of Te, see Fig. 3. Note that some 50 heavy partners of those remain to be identified if a similar dose can be accumulated with the FRS tuned on ^{170}Gd.

The distributions of isotopic cross sections are reported in Fig. 4 for the 22 elements transmitted together with Ni. They are covering seven orders of magnitude, falling down to the lowest values of 2 10^{-10} barn. The variation of the yields seem compatible with a linear dependence on the semilogarithmic plot, cross sections being divided by 8 for each added neutron.

Two questions arise which are relevant for future, when trying to produce more n-rich isotopes;

a) How do cross section distributions compare in fission and fragmentation? Isotopes of Ni are used for the comparison. Up to A = 78 cross-sections are larger in U-fission on Be than in ^{86}Kr fragmentation but the distribution drops more abruptly for fission. However, for more n-rich isotopes, the fragmentation yields are vanishing for N $\geq$ N$_{proj}$ i.e. 50. Then a heavier projectile like ^{96}Zr should be chosen to induce fragmentation and the whole set of isotopic cross-sections would decrease by an order of magnitude as compared with Kr-fragmentation; both fission and Zr-fragmentation cross sections would be the same for ^{80}Ni. From there, yields for production of ^{81,82,83}Ni would be larger using Zr-fragmentation as compared

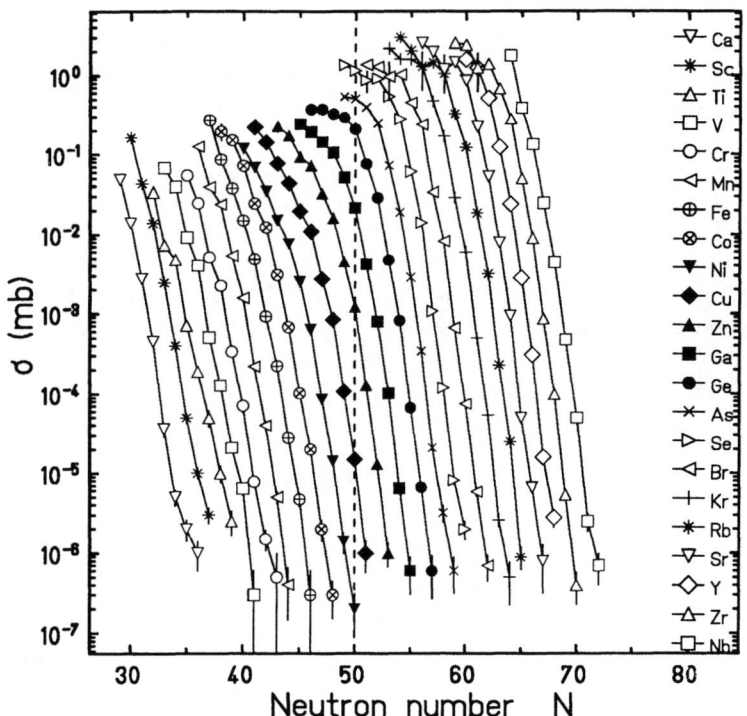

FIGURE 4. Production cross section in Be(^{238}U,f) at 750 A.MeV

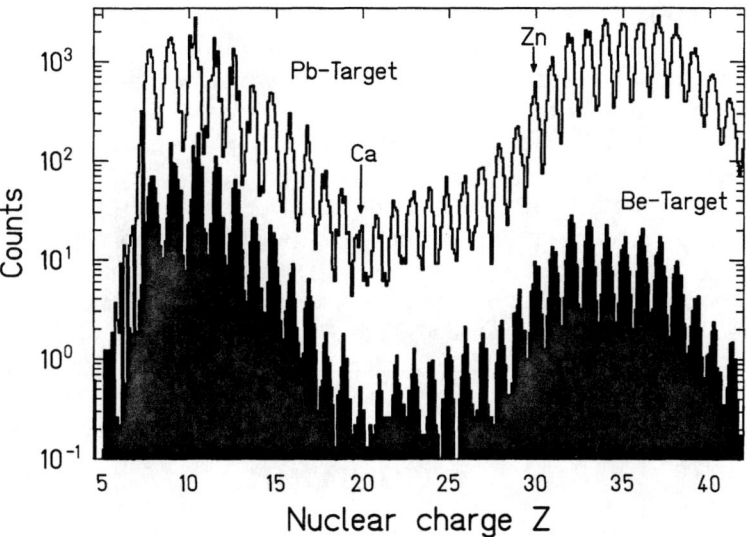

FIGURE 5. Nuclear charge distribution for light isotopes separated with the FRS at large rigidities

with U-projectile fission.

b) Our measurement of fission using inverse kinematics allows for the first time a comprehensive study of the yields for closed shell nuclei at N = 82, 50 and Z = 50. After removing the odd-even effects due to proton-pairs and neutron-pairs, an enhancement of 18 ± 3 % was observed at N=82 and 48 ± 5 % at Z=50 [17]. At N = 50, a few % enhancement was also observed for ^{83}As and ^{82}Ge [11]. On Be target a slight enhancement shows up as well for N = 50 at Z=30 (the region of N=82 Z=50 was not yet covered by the Be-target measurements). Local Q effects contribute to odd-even staggering and closed shell enhancement. Only an accurate measurement at lower excitation could reveal such effects in this region of mass. Promising in this context are the investigations planned to identify and measure the yields for fragments around ^{78}Ni in thermal induced neutron fission at "Lohengrin" (ILL), since the details of structure effects are known to increase at low excitation energy.

small fragments

Nuclear charge distributions measured with Pb and Be-targets around the element Ni at large magnetic rigidities shows all elements down to very light nucleides (Fig. 5). The rates of elements produced by fission decrease with Z from Zr down to Sc. Then counting rates increase by more than two orders of magnitude between Ca and Ne. Elements between Fe and Si are filling the gap between the lightest fragment ever observed in binary fission, ^{68}Fe [20] and ^{34}Si the heaviest fragment produced by ternary fission, [21].

The reaction initiating the small fragments is likely to be a multi-fragmentation process. Given the large magnetic rigidities, the separated isotopes [11] come with a large neutron excess. In this region the A/Z ratio reaches three, a value never obtained for any of the fission products. Such a large n-excess is not compatible with the image of a dense excited piece of nuclear matter which would be formed in head-on collisions. With on-line mass separator techniques, light very neutron-rich clusters were already observed and ^{35}Na with A/Z>3 could be separated in p+Ir collisions [22].

One should notice that production yields for the very light fragments is not as large as it appears on Fig. 4 and 5. The multifragments are probably emitted within a wide range of center of mass velocities centered around 0. The related FRS-transmission is then multiplied by a factor of 10 as compared with the transmission of fission fragments. Energy and target-mass dependences of A/Z and velocity distributions of fragments will be further investigated in the frame of the systematic measurements on hydrogen and deuterium targets presently under way.

OUTLOOK

The structure of very neutron rich nuclei regains considerable interest. For example the quenching of spherical shells, predicted by self-consistent mean field theories deserves experimental tests. An experimental approach to the shell-closure effects is the study of the isomeric- decays -when they do exist-. In a preliminary measurement using the U + Be system we have checked that known fission isomer, ^{98}Y, was populated with rates up to 30 % as in low energy fission. Moreover the 8^+ isomeric decay in ^{70}Ni already observed [23] was clearly identified as well. Similar investigations of isomers in very n-rich isotopes close to Sn or to the N = 60 and 82 regions which look more specific to the FRS are planned.

The recent success of measuring mass excesses using the FRS and ESR in the isochronous mode [24], demonstrate the possibilities to obtain neutron separation energies and β-decay half-lives of the new isotopes. Fission fragments of a defined A/Z region are selected by the FRS with the same momenta. Thus they will be simultaneously transmitted to the ESR. No doubt that the answer to the question of shell quenching effect around ^{78}Ni will be solved in a near future.

The experiment was performed at GSI Darmstadt, with the FRS group in collaboration with IPN-Orsay, IReS-Strasbourg and TU-Darmstadt laboratories.

REFERENCES

1. H. Geissel et al. Nucl. Instr. and Meth. B70, 286 (1992)
2. K.-H. Schmidt et al. Nucl. Phys. A542 (1992) 699
3. B. Wilkins et al. Phys. Rev. C14 (1976) 1832
4. U. Brosa, Phys. Reports 197 (1990) 167
5. J.-P. Bocquet et al. Nucl. Phys. A502 (1989) 213
6. T. Aumann et al. Z. Phys. A352 (1995) 163
7. C. Donzaud et al. Eur. Phys. J A1 (1998) 407
8. J.-L. Sida et al. Nucl. Phys. A502, 233 (1989)
9. M. Huhta et al. Phys. Lett. B405 (1997) 230
10. P. Armbruster et al. Z. Phys. A355, 191 (1996)
11. C. Engelmann Ph. D. Thesis Uni of Tubingen (1998) and to be published.
12. W. Schwab in press in Eur. J. of Phys. (1998)
13. B. N. Belyaev et al.Nucl. Phys. A438 (1980) 479
14. J. Benlliure in press in Eur. J. of Phys. (1998)
15. J. Benlliure et al. poster contribution at this conference.
16. T. Enqvist et al. poster contribution at this conference
17. P. Armbruster, ENAM-95 Conference, Arles France, Edt Frontieres, p. 343
18. M. Bernas et al. Phys. Lett. B331 (1994) 19
19. M. Bernas et al. Phys. Lett. B415 (1997)
20. M. Bernas et al. Phys. Rev. Lett. 67 (1991) 3661
21. M. Hesse, Ph. D. thesis Uni. of Tübingen (1997) and to be published.
22. M. Langevin et al Phys. Lett. B125 (1983) 116
23. R, Grzywacz et al. Phys. Lett 355 B (1995)
24. H. Geissel et al., Invited talk, this conference

PROTON and CLUSTER RADIOACTIVITY and NUCLEUS SHAPES

Stanislav G. Kadmensky

Department of Nuclear Physics, Voronezh State University, Russia, 394000

Abstract. The influence of nucleus shapes and their reconstructions in processes of proton, alpha-particle and cluster decays on decay widths have been investigated.

1. INTRODUCTION

Theories of proton, alpha-particle and cluster radioactivity of atomic nuclei, based on the integral formula for deep subbarrier decay widths [1,2], have the essential advantage over the R-matrix theories of the radioactivity [3] because of the absence of the problem of the channel radius determination and therefore allow consistently to take into account the influence of parent and daughter nucleus shapes and of their reconstruction in decay processes on the decay probabilities.

The width Γ_c of the state i of a parent nucleus A, described by internal wave function $\Psi_{\alpha_i}^{I_i M_i}$ with spin I_i, it's projection M_i and additional quantum numbers α_i, with respect to the transition through the $c \left(c \equiv x I_f j_x L \right)$ channel, in which a daughter nucleus f $\left(A_f = A - A_x \right)$ with internal wave function $\Psi_{\alpha_f}^{I_f M_f}(\zeta_f)$ and the particle x (nucleon, alpha-particle, cluster) with internal wave function $\Psi_{\alpha_x}^{I_x M_x}(\zeta_x)$ (ζ_f and ζ_x being the total sets of intrinsic coordinates of the daughter nucleus and particle) are formed, can be represented as [1-2]

$$\Gamma_c = 2\pi \left| \int \hat{A} \left[U_c^* \frac{F_L(R)}{R} \tilde{V}_{fx} \right] \Psi_{\alpha_i}^{I_i M_i} d\zeta_f d\zeta_x d\vec{R} \right|^2 \qquad (1)$$

where

$$U_c = \left\{ \Psi_{\alpha_f}^{I_f M_f} \left\{ \Psi_{\alpha_x}^{I_x M_x} Y_{LM}(\Omega_{\vec{R}}) \right\}_{j_x m_x} \right\}_{I_i M_i} \qquad (2)$$

is the channel function (braces denote the vector coupling of the momenta), $\hat{A}$ is the antisymmetrization operator, $F_L(R)$ is the regular radial Coulomb function, normalized to the delta function of energy, and $\tilde{V}_{fx} = V_{fx} - Z_x Z_f e^2 / R$, where V_{fx} is the total potential of interaction between the decay fragments. The formfactor $\Psi_c(R)$ and spectroscopic factor W_c for channel c are presented as

$$\Psi_c(R) = R \int \hat{A} \{U_c\}^* \Psi_{\alpha_i}^{I_i M_i} d\zeta_f d\zeta_x d\Omega_{\vec{R}} \quad , \quad W_c = \int_0^{R_1} \left[\Psi_c(R) \right]^2 dR \qquad (3)$$

Following [2], we break down the range of the variable R in the integral in (1) into the shell, cluster, and intermediate domains and introduce shell, intermediate and cluster spectroscopic factors. The shell domain corresponds to the internal region of the parent nucleus ($0 \leq R \leq R_{sh}$)- here, the multiparticle shell model with limited single-particle basis can be applied to the formation of internal wave function $\Psi_{\alpha_i}^{I_i M_i}$, shell formfactor $\Psi_c^{sh}(R)$ and spectroscopic factor W_c^{sh}. In the cluster domain ($R_{cl} \leq R \leq R_1$), corresponding to the external region of the parent nucleus, the decay fragments are completely formed and the cluster formfactor $\Psi_c^{cl}(R)$ obeys the set of the coupled equations. Substituting the amplitudes of experimental widths $\sqrt{\Gamma_c^{exp}}$ in the boundary conditions and integrating these equations, experimental cluster spectroscopic factors $W_c^{cl.exp}$ can be found. Then it is possible to give the classification of different x-transitions on the base of hindrance factors F_c, which are determined as inverse relations of values of $W_c^{cl.exp}$ to $W_{c0}^{cl.exp}$, where $c0$ is the channel index for the favored transition.

The principal problem of decay widths calculations is to determine the regime of the interpolation between the shell and cluster domain, because this regime physically corresponds to reconstruction of the parent nucleus in the decay process.

2. THE PROTON RADIOACTIVITY

In work [4], where the multiparticle theory of the nucleon radioactivity of spherical nuclei was developed, it was shown that because of the presence of nucleons as basic particles in the structure of the parent and daughter nuclei, the shell domain plays the principal role in the function $\Psi_{\alpha_i}^{I_i M_i}$. It means that the reconstruction of the parent nucleus in the proton decay process has a very weak influence on the formation of proton widths. In works [5] this result was generalized on nucleon decays of deformed nuclei.

The high exactitude of the theoretical calculation scheme gives the unique possibility to estimate deformation values and shell model proton configurations of the parent and daughter nuclei far from stability on the base of the experimental proton half-times and energies. In works [5,6] it was shown that the experimental data can be described using the deformation parameter β_2 =0.05-0.1 for ^{109}I, β_2 =0.1-0.1 for ^{113}Cs, β_2 =0.25-0.3 for ^{121}Pr, β_2 =0.1-0.2 for $^{147}Tm, ^{147m}Tm, ^{150}Lu, ^{151}Lu$, β_2 for ^{141}Ho and ^{131}Eu. These values of β_2 are in good accordance with predictions of the macroscopic-microscopic mass model [7].

It is interesting to repeat the measurement of the proton decay of nucleus ^{121}Pr, about which we have only preliminary experimental information [5]. It is important to find further examples of proton decays, especially not only from ground and single particle isomeric states, but from multiparticle isomeric states (type of ^{53m}Co [8])

of spherical and deformed nuclei. In addition, studies of the thin structure of emitted proton spectra, especially connected with transitions to ground rotation band of even-even deformed daughter nuclei, when from the theoretical point of view, the visible yields of protons are expected, are of great interest. The investigations of gamma-rays, following after proton decays, would be extremely valuable to directly determine the deformation parameters of nuclei far from stability.

3. THE ALPHA - RADIOACTIVITY

Alpha-widths are determined completely by the cluster domain of parent nuclei [2] and therefore the central problem of the alpha-decay theory is to find the interpolation regime between shell and cluster domains. At the present time a definite success was achieved on the base of the following interpolation scheme

$$W_c^{cl} = t_{fx} W_c^{sh}, \qquad (4)$$

where the coefficient t_{fx} takes into account in principle the effects of antisymmetrization [9] and approximately does not depend on the channel index c. It means that the reconstruction of the parent nucleus in the alpha-decay process does not lead to loss of memory on the shell spectroscopic factor W_c^{sh} in the cluster domain. This formula allows to make the classification of alpha-transitions on the base of the superfluidity nuclear model and to calculate the absolute and relative alpha-widths of spherical and in principle deformed nuclei [2,10]. On the base of this interpolation scheme (4) in case of α - decay of neutron resonance states for parent nuclei with $50 \leq A \leq 180$ it was shown [11] that only taking into account the high deformation of nuclei ^{156}Gd and ^{172}Yb, the «black-nucleus» α-particle strength-function of neutron resonances can be estimated.

It is important to continue theoretical investigations of different alpha-transitions for spherical and especially for nonspherical nuclei with the aim to test the interpolation scheme (4).

4. THE CLUSTER RADIOACTIVITY

The investigated cluster decays of heavy nuclei with emission of $^{14}C, ^{24}Ne,...$ clusters are connected with a strong changes of nucleus shapes : parent nuclei have visible deformations, the daughter nuclei are spherical. The phenomenon of the cluster radioactivity physically is similar to the fission of heavy nuclei [12], but in contrast to the fission the cluster decay has the deep subbarrier character and is connected only with small numbers of opened channels. Therefore it is possible to develop the microscopic quantum-mechanical theory of the cluster decay. The first step in this direction is done in works [13-15], where the theory of cluster decay is built on the base of the theoretical scheme analogous the α-decay theory, because similar to the α-decay the cluster decay width (1) are determined completely by the cluster domain of the parent nucleus [13]. In this approach the classification of cluster transitions and the scale of absolute cluster width values for even-even parent nuclei and for different clusters x was received. At the same time this theory has no possibility to explain simultaneously the experimental prohibition factors for the cluster decay of odd nuclei $^{223}Ra, ^{225}Ac, ^{233}U$. Actually theoretical prohibition factors F_c^{th} [16] for transitions in ^{223}Ra with the emission of ^{14}C $3/2^+ \rightarrow 9/2^+$ and $3/2^+ \rightarrow 11/2^+$, in ^{225}Ac with the emission of ^{14}C $3/2^- \rightarrow 9/2^-$ and in ^{233}U with the emission of ^{24}Ne $5/2^+ \rightarrow 9/2^+$ are equal correspondingly to 10 ; 3 ; 1.3 and 4, as experimental prohibition factors are equal correspondingly to 500 ; 2.6 ; 1.5 and 50. It means that in the interpolation formula (4) the coefficient t_{fx} essentially depends on the type of channel c and the parent nucleus reconstruction in the cluster decay process is very strong. It is interesting to find configurations of parent nuclei and especially of the odd nucleon for odd parent nuclei in the cluster domain, which correspond to experimental prohibition factors.

5. CONCLUSION

The classification of proton, alpha-particle and cluster decays on the base of the type of parent nucleus reconstruction in decay processes allows to unite physically these phenomena with the phenomenon of fission and gives the possibility to find the new microscopic quantum mechanical characteristics of all these decays.

REFERENCES

1. Kadmensky ,S.G. , Kalechitz , V.E. , Martinov , A.A. , Sov.J. Nucl. .Phys. , **16**,717-724 (1972).
2. Kadmensky ,S.G., Furman, V. I., Alpha Decay and Related Nuclear Reactions ,Moskow: Energoatomizdat,1985.
3. Lein , A. , Thomas, R. The Theory of Nuclear Reactions for Low Energies, Moskow: Izdat. Inostr. Liter., 1960.
4. Bugrov ,V. P. , Kadmensky, S.G. et. al. , Sov.J. Nucl. .Phys. ,**41**, 1123-1133 (1985).
5. Bugrov , V. P. , Kadmensky ,S.G. et. al. , Sov.J. Nucl. .Phys. ,**49**, 967-975 (1989) ; Bogdanov , D. D., Bugrov ,V. P. , Kadmensky ,S.G. et. al. , Sov.J. Nucl. .Phys. , **52**, 358-363 (1990) ; Kadmensky , S.G., Bugrov ,V. P. , Sov.J. Nucl. .Phys. , **59**, 424-427 (1996).
6. Davids ,C.N. et.al. , Phys. Rev. Let. , **80** , 1849-1852 (1998).
7. Moller , P., Nix ,J.R. , Mayers, W. D. and Swiatecki , W.J. , At. Data Nucl. Data Tables, **59**,185 (1995).
8. Bugrov , V. P. et. al. , Sov.J. Nucl. .Phys. , **42** , 57-66 (1985).
9. Fliessbach ,T., Mang , H.J. , Nucl.Phys. , **A263**,75-85 (1982).
10. Kadmensky , S.G. , Zeit. Phys., **A312**, 113-120 (1983).
11. Kadmensky , S.G., Kurgalin ,S. D., Furman , V. I. , Sov.J. Nucl. .Phys. , **35**, 823-832 (1982).
12. Sandulescu ,A., Poenaru , V. N. , Grainer, W. , Particles and Nuclei ,**11**, 1334-1368 (1980).
13. Kadmensky , S.G. , Furman , V. I. , Tchuvilsky , Yu. M. , Izv. Ak. Nauk USSR , ser. Fiz. , **50**, 1786-1795 (1986).
14. Blendovske , R. , Fliessbach , T., Walliser , H., Nucl. Phys., **A 464**, 75-86 (1987).
15. Zamyatnin , Yu. S. et.al., Particles and Nuclei , **21** ,537-594 (1990).
16. Kadmensky ,S.G., Furman ,V. I., Tchuvilsky,Yu. M., Sov.J. Nucl. Phys. , **60**, 16-27 (1997).

Fission of Nuclei with Z=102-112 Produced in Reactions with ^{22}Ne and ^{48}Ca Ions

M.G.Itkis[1], Yu.Ts.Oganessian[1], E.M.Kozulin[1], N.A.Kondratiev[1], L.Krupa[1], I.V.Pokrovsky[1], A.N.Polyakov[1], V.A. Ponomarenko[1], E.V. Prokhorova[1], B.I.Pustylnik[1], A.Ya. Rusanov[2], V.I.Vakatov[1]

[1]*Laboratory f Nuclear Reactions, Joint Institute for Nuclear Research, 141980, Dubna, Moscow region, Russia*
[2]*Institute of Nuclear Physics of the National Center of Kazakhstan, Alma-Ata, Kazakhstan*

Abstract. The talk presents new results obtained in the study of fission of superheavy nuclei ^{256}No, ^{270}Sg and ^{286}112 formed in reactions with ^{22}Ne and ^{48}Ca ions at energies near or considerably lower than the Coulomb barrier. The experiments have been performed at the U-400 accelerator of the Flerov Laboratory of Nuclear Reactions (FLNR) with the use of the time-of-flight spectrometer of fission fragments CORSET.

INTRODUCTION

One of the most interesting aspects of modern nuclear physics is synthesis of heavy and superheavy elements and study of their properties. In this connection experimental investigation of the fusion-fission process of compound nuclei with Z>100 at low excitation energies is of special importance. With this purpose at the U-400 accelerator of FLNR the fission cross sections, mass and energy distributions of fission fragments were measured in the reactions ^{208}Pb+^{48}Ca→^{256}No, ^{248}Cm+^{22}Ne→^{270}Sg and ^{238}U+^{48}Ca→^{286}112 in the compound nucleus excitation energy range from 12 to 60 MeV. The interest in the reaction ^{248}Ca+^{22}Ne is connected first of all with the fact that the number of neutrons in the compound nucleus ^{270}Sg is 164. In this case one can expect that at low excitation energies the fission fragment mass distribution demonstrates properties of bimodal fission determined by structural peculiarities of the deformation potential energy surface, much as it takes place in spontaneous fission of ^{258}Fm and ^{261}Md (Z≥ 50, N≅ 160).

The interest in reactions with ^{48}Ca ions is explained by their importance for the presently realized at the FLNR programme of the superheavy ion synthesis in the region of nuclei with Z=114-116 and N=182-184, for which a considerable increase in their stability to spontaneous fission and α-decay has been predicted. That's why it is very important to measure the compound nucleus cross section for reactions of the U,P,Cm+^{48}Ca type, on the basis of which one is able to predict the fusion cross section and survivability of superheavy nuclei.

EXPERIMENTAL AND RESULTS

The experiments were carried out with the use of the time-of-flight spectrometer CORSET, in which microchannel plates were used as start and position-sensitive stop detectors. Layers of enriched ^{208}Pb (98%) and ^{248}Cm (97%) and ^{238}U (99.9%) isotopes with the thicknesses of 220, 200 and μg/cm^2, respectively, were put on a carbon backing and used as targets. In the reaction camera two semiconductor monitors were installed for registration of elastic ion scattering and determination of their energy.

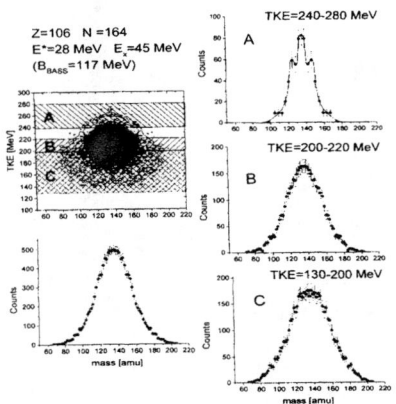

FIGURE 1. Two-dimensional plot of TKE/m and mass yields in different TKE ranges for the reaction ^{248}Cm+^{22}Ne→^{270}Sg (E_{Ne}=102 MeV).

Fig. 1 presents results of measurements of the mass and energy distribution Y(M,TKE) for the ^{270}Sg compound nucleus fission fragments at the ^{22}Ne ion energy of 102 MeV, which corresponds to the excitation energy E*=28 MeV. Note that the excitation energy of the compound nucleus at the Coulomb (Bass) barrier was E*=45 MeV. Thus, we managed to measure the distribution Y(M,TKE) deep below the Coulomb barrier and register $1.2 \cdot 10^4$ events.

Considering separate events of the spectrum at different values of the total kinetic energy TKE, the free energy Q-TKE released during the fission process being changed too, one can see that the fission fragment mass distributions (which on the whole are of symmetric character) undergo considerable changes. When Q-TKE→0 (Fig.1 A) the mass distribution is of the form characteristic of bimodal fission, i.e. it is a narrow mass distribution with $\sigma_M \approx 4.3$. Thus, it has been shown for the first time that bimodal fission is also observed in the case of induced fission of superheavy nuclei, and not only in spontaneous fission of nuclei in the vicinity of Fm.

Let us consider the investigation results of the reactions ^{208}Pb+^{48}Ca and ^{238}U+^{48}Ca leading to formation of the nuclei ^{256}No and ^{286}112 at different excitation energies. Note that in both cases we managed to study the mass and energy distribution of fission fragments and the fission cross section σ_f at energies under the Coulomb barrier and reach very low excitation energies, namely E_x^{min} =12 MeV in the case of the ^{256}102 nucleus and E_x^{min} =21.5 MeV in the case of the ^{286}112 nucleus.

One can note that in the case of ^{256}No, Y(M,TKE) at any excitation energy is of distinctive triangle shape, which is typical for the classic symmetric fission of a compound nucleus. In the case of the ^{286}112 decay, the Y(M,TKE) matrices have a more complicated structure, determined by domination of quasifission reactions with a characteristic grouping of fragments around the fission fragment mass 208 and complementary to it masses, and thus extraction of events of the compound nucleus fission becomes quite a complicated task. Nevertheless, at the excitation energy E_x=26 MeV, between the quasifission peaks one can clearly see the region, which can be assigned to the fission of the compound nucleus ^{286}112, and it is possible to determine the fission cross section. For clarifying the contribution of fission at higher energies by

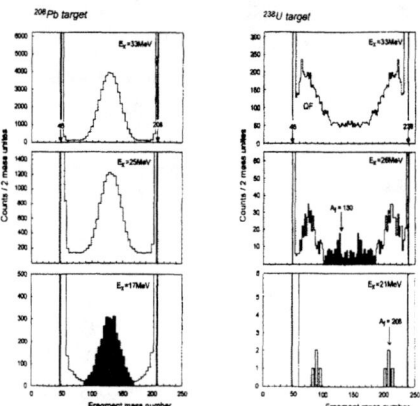

FIGURE 2. Fragment mass distribution in the ^{48}Ca induced reactions.

means of extracting the compound nucleus formation region we used our empiric systematics (1) concerning dependence of the width of the fission fragment mass distribution on the nucleon composition, excitation energy and angular momentum of the fissioning nucleus.

The above-mentioned facts are illustrated by Fig. 2, presenting the yields of the fission fragment masses for both the reactions at several excitation energies. The results of the $\sigma_f(E)$ determination for both reactions are presented in Fig. 3. Note that in the present work we managed to go down in energy much lower than in (2,3).

Another important result is connected with the mass distribution of the ^{286}112 fission fragments at the energy E_x=26 MeV which has been presented above. Despite of the insufficient number of events related, from our point of view, to the compound nucleus fission, one can still see that the mass distribution is asymmetric with masses of the light fragment M_L=130 and of the heavy one M_H=156, which is characteristic of the actinide nuclear fission, whereas symmetric fission prevails in spontaneous fission of nuclei with Z=100-104 (4) and in induced fission of ^{270}Sg. This result allows us to anticipate that nuclei with Z>110 undergo asymmetric fission at low energies, and as opposed to actinide nuclei, the light fragment with M≈130 plays in this case a stabilizing role instead of a heavy one with M≈140 as it takes place in the U-Cf

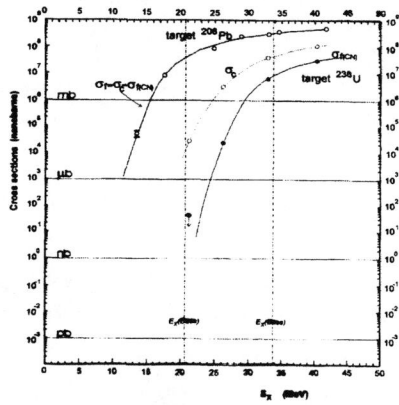

FIGURE 3. The fission cross section σ_f and the evaporation residue cross section σ_{ev} for the reactions ^{48}Ca+^{208}Pb and ^{48}Ca+^{238}U.

region. Of course, such an important conclusion requires that the experiment should be repeated aiming at accumulating better statistics.

CONCLUSION

For the first time measured were the characteristics of induced fission of nuclei with Z>100 at considerably low excitation energies, at which the shell effects play an important role and lead to new quite unexpected properties. It is important to note that obtained results concerning the fission fragment mass and energy distribution and fusion-fission cross sections offer possibilities of predicting properties of superheavy nuclei and planning new experiments.

REFERENCES

1. Itkis, M.G., and Rusanov, A.Ya., *Physics of elementary particles and atomic nucleus* **29(2)**, 389 (1998).
2. Sheh, W., et al., *Phys.Rev. C* **36**, 115 (1987).
3. Bock, R., et al., *Nucl.Phys. A* **388**, 344 (1982).
4. Hulet, E., et al., *Phys.Rev. C* **40**, 770 (1989).

Study of the High-*j* States in ^{249}Cm

I. Ahmad,* B. B. Back,* A. Bacher, G. P. A. Berg,† R. R. Chasman,*
C. C. Foster,† J. P. Greene,* T. Ishii,*,‡ W. R. Lozowski,† L. R. Morss,*
W. Schmitt,†,§ E. J. Stephenson† and T. Yamanaka†,¶

Argonne National Laboratory, Argonne, IL 60439
†Indiana University Cyclotron Facility, Bloomington, IN 47405

Abstract: We have performed the reaction ^{248}Cm(^{4}He,^{3}He) using 98.5-MeV alpha particles from the IUCF cyclotron to populate high-*j* states in ^{249}Cm. A tentative assignment of the $k_{17/2}$ component of the 1/2$^+$[880] Nilsson state has been made.

INTRODUCTION

The position of the $k_{17/2}$ ($\ell=8$) orbital plays an important role in determining the stability of superheavy elements. Because of its large degeneracy, it has a significant influence on the magnitude of shell corrections in this region. The lowest component of the $k_{17/2}$ orbital, in a deformed prolate potential, is the 1/2$^+$[880] Nilsson state [1] which is expected to lie below 2 MeV in nuclei with N≥152. As yet, this orbital has not been observed. In an earlier high resolution (d,p) study [2] of ^{251}Cf, we identified all of the neutron single-particle states between the deformed gaps at N=152 and N=164. Using a momentum-dependent single particle potential, which reproduces the observed level energies in ^{251}Cf, we calculate the energy of the 1/2$^+$[880] orbital at ~1400 keV in ^{251}Cf. An angular momentum decomposition of this orbital indicates that it is 74% $k_{17/2}$. Such orbitals are expected to be strongly populated in (^{4}He,^{3}He) reactions [3,4] due to the angular momentum mismatch. Because of the intense radioactivity associated with ^{250}Cf, it is not possible to study ^{251}Cf levels in the reaction ^{250}Cf(^{4}He,^{3}He). For this reason, we have carried out a study of the (^{4}He,^{3}He) reaction with the longer-lived isotope ^{248}Cm. The nuclide ^{249}Cm is an isotone of ^{251}Cf and hence the ^{249}Cm level ordering is expected to be similar to that of ^{251}Cf. A detailed article has been submitted for publication [5].

EXPERIMENTAL RESULTS

The ^{248}Cm(^{4}He,^{3}He) measurement was performed at the Indiana University Cyclotron Facility. A beam of 98.5-MeV α-particles with currents up to 25 pnA was incident on a 110-μg/cm^2 ^{248}Cm target on a 75-μg/cm^2 carbon backing. The emerging ^{3}He ions were momentum analyzed by the K600 high-resolution magnetic spectrometer. For this experiment, the spectrometer was equipped with a magnetic septum at the entrance to allow operation at small scattering angles close to the beam. A brass collimator at the front of the septum set the solid angle to 0.47 msr. Measurements were made at θ_{cm}=4.1°, 6.1°, 10.2°, 12.2°, and 16.2°. ^{3}He ions that passed through the spectrometer were detected by a series of wire chambers and scintillators located parallel to the spectrometer focal plane. Tuning of the beam line and spectrometer magnetic elements for dispersion matching and focus yielded a minimum resolution of about 45 keV (FWHM). Momentum, as measured by position along the K600 focal plane, was calibrated by observing the excited state spectrum from the ^{208}Pb(^{4}He,^{3}He)^{209}Pb reaction without changing the spectrometer magnetic field settings. Transitions to the ground and first two excited states at 779 and 1423 keV in ^{209}Pb were used to produce a quadratic relationship between focal plane position and ^{3}He momentum that was then applied to the ^{249}Cm spectra.

DISCUSSION

The assignments of levels in ^{249}Cm, which are strongly populated in the ^{248}Cm(^{4}He,^{3}He) reaction, are based on a comparison of the cross sections measured for this reaction with those measured for known high-j states in ^{209}Pb and ^{233}Th. The cross sections were also calculated with the DWUCK4 code [6]. The magnitude of the cross section and the angular distribution are tracked by the calculation but the agreement is only qualitative. Thus it is not possible to firmly distinguish ℓ-transfer or parity from a comparison of measured and calculated cross sections alone.

The ^{232}Th(^{4}He,^{3}He) reaction (Fig. 1) shows only two strong peaks, which are associated with the $j_{15/2}$ orbital. Because of the large gap at N=152 [1], only states above N=152 gap are populated with substantial intensities in ^{249}Cm. In the ^{248}Cm(^{4}He,^{3}He) spectrum, shown in Fig. 2, a strong peak at 593 keV was observed at all angles. This peak is identified as the 15/2 member of the 11/2⁻[725] band. The same level has been identified at 570 keV in the isotone ^{251}Cf [3]. The peak at 664 keV has been assigned to the 11/2 member of the 9/2⁺[615] orbital because this state is expected to have a large cross section and it has been identified in ^{251}Cf at 680 keV. For the 1898 keV level, the observed cross section is about one half the calculated value as is the case for the 988 keV level in ^{233}Th. The measured cross section for the 1560 keV level agrees well with the value calculated for the j=17/2 component of the 1/2⁺[880] band. It is possible that for the higher $j_{15/2}$ states the cross section is reduced because the state is mixed with many other states. On this basis it is plausible that the 1898 keV state would be the 15/2 member of the 13/2⁻[716] band and the 1560 keV level would be the 17/2 member of the 1/2⁺[880] band.

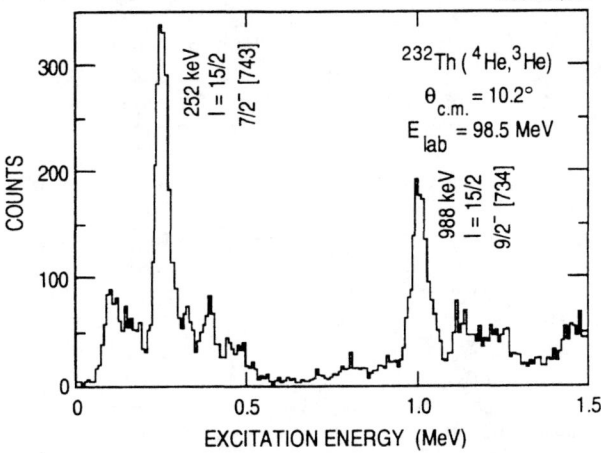

Figure 1. A portion of the ^{3}He excitation energy spectrum from the reaction ^{232}Th(^{4}He,^{3}He) measured with the K600 magnetic spectrometer.

To get some idea of the uncertainties involved in theoretical assignments, we have made calculations of level orderings and spacings using a momentum-independent (m.i.) Woods-Saxon potential, in addition to the calculations carried out with the momentum-dependent (m.d.) potential discussed in [2]. Using the m.i. potential, we are able to carry out Strutinsky method calculations to determine the deformation. We have done such calculations in a three dimensional deformation space and find that the

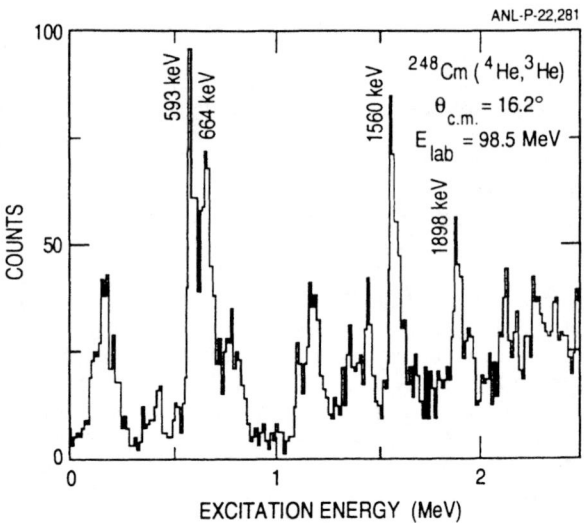

Figure 2. A portion of the ^{3}He spectrum from the reaction ^{248}Cm(^{4}He,^{3}He) measured with the K600 magnetic spectrometer.

minimum in the energy surface is at $v_2=0.24$, $v_4=-.01$ and $v_6=0.02$; in rather good agreement with the deformation inferred [2] from the observed low-lying levels ($v_2=0.25$, $v_4=-.01$ and $v_6=+.02$). In Fig. 3, we show the single-particle spectra calculated with the two potentials and compare them with the experimentally known levels in ^{251}Cf [2] and ^{249}Cm [7,8]. The level orderings and spacings obtained from the momentum independent and momentum dependent potentials are in fairly good agreement with each other, as well as with the experimental assignments below 800 keV. Both potentials give a large gap at N=164. Above the gap, the levels calculated with the m.i. potential are typically 100-200 keV higher than those obtained with the m.d. potential. The notable exception to the agreement in level positions obtained from the two potentials is the position of the 1/2$^+$[880] orbital, which is about 600 keV higher in the m.i. calculation. We note that the 1/2$^+$[880] configuration is at a slightly larger deformation than the other configurations because of its very strong deformation driving character. With these differences in the theoretical estimates, we feel that either of the large peaks at 1560 and 1898 keV could be associated with the 1/2$^+$[880] orbital. Assuming that the level assignments are consistent with the m.i. potential, we feel that we can calculate [9] the properties of the superheavy elements with a considerable degree of confidence using the Strutinsky prescription.

ACKNOWLEDGMENTS

This work was supported by the U.S. Department of Energy, Nuclear Physics Division, contract No. W-31-109-ENG-38 and by the National Science Foundation Grant PHY 48-308-44. The calculations reported here were carried out on the SP computer of the MCS Division of the Argonne National Laboratory and the NERSC facility at Livermore and Berkeley. The authors are also indebted for the use of ^{248}Cm to the Office of Basic Energy Sciences, U.S. Department of Energy, through the transplutonium element production facilities at Oak Ridge National Laboratory.

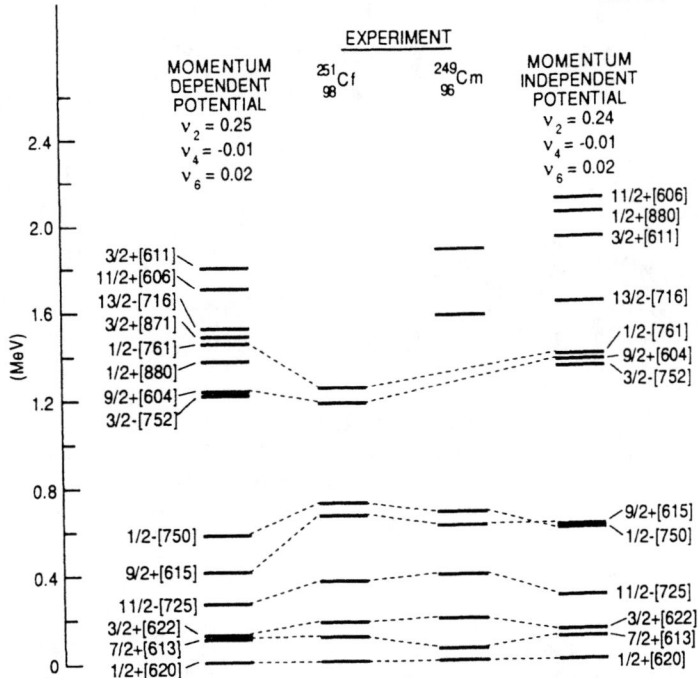

Figure 3. Experimental bandhead energies of neutron orbitals in ^{249}Cm [7,8] and ^{251}Cf [2] along with theoretical values.

REFERENCES

‡Permanent address: JAERI, Tokai, Ibaraki, Japan
§Present address: Massachusetts Institute of Technology, Cambridge, MA 02139
¶Exchange student from the Research Center for Nuclear Physics, Mihogaoka 10-1, Ibaraki, Osaka 567, Japan

1. Chasman, R. R., Ahmad, I., Friedman, A. M., and Erskine, J. R., Rev. Mod. Phys. **49**, 833-891 (1977).
2. Ahmad, I., Chasman, R. R., Friedman, A. M., and Yates, S. W., Phys. Lett. **251**, 338-342 (1990).
3. Massolo, C. P. et al., Phys. Rev. **C34**, 1256-1261 (1986).
4. Tickle, R., and Gray, W. S., Nucl. Phys. **A247**, 187-194 (1975).
5. Ahmad, I., submitted to Nucl. Phys. A.
6. Kunz, P. D. and Rost, E., University of Colorado, code DWUCK4.
7. Braid, T. H., Chasman, R. R., Erskine, J. R., and Friedman, A. M., Phys. Rev. **C4**, 247-262 (1971).
8. Hoff, R. W. et al., Phys. Rev. **C25**, 2232-2253 (1982).
9. Chasman, R. R., and Ahmad, I., Phys. Lett. **B392**, 255-261 (1997).

Experiments on projectile fission and fragmentation relevant for accelerator-driven systems

T. Enqvist[1], J. Benlliure[1], F. Farget[1], J. Taieb[1], K.-H. Schmidt[1],
P. Armbruster[1], C. Böckstiegel[2], M. de Jong[2], M. Bernas[3],
B. Mustapha[3], C. Stéphan[3], L. Tassan-Got[3], A. Boudard[4],
S. Leray[4], R. Legrain[4], C. Volant[4], W. Wlazlo[4], S. Czajkowski[5],
M. Pravikoff[5], and J.P. Dufour[5]

[1] *GSI, Planckstraße 1, D-64291 Darmstadt, Germany*
[2] *TH Darmstadt, Schloßgartenstraße 9, D-64289 Darmstadt, Germany*
[3] *IPN Orsay, IN2P3, F-91406 Orsay, France*
[4] *Dapnia/SPhN, CEA/Saclay, F-91191 Gif sur Yvette, France*
[5] *CENBG, IN2P3, F-33175 Gradignan, France*

Abstract. Spallation experiments in inverse kinematics have been performed at GSI. The momentum distributions and the isotopic production cross sections have been measured. The data are relevant for the design of accelerator-driven subcritical reactors.

I INTRODUCTION

Nuclear collisions of heavy ions at relativistic energies have revealed new aspects of the fragmentation process [1–6] as well as of nuclear fission from low [7–11] and high excitation energies [12–14].

Nuclear collisions at relativistic energies also play an important role in a recently discussed option for an efficient and clean energy production [15,16] by coupling of a high-intensity and high-energy proton accelerator with a subcritical nuclear reactor. This device is also considered to be suited for the incineration of nuclear waste. In these accelerator-driven systems (ADS) proton-induced spallation reactions in lead or lead-bismuth targets would be used to produce additional neutrons needed for the operation of the reactor.

In order to improve the experimental knowledge of spallation reactions relevant for ADS, a program was started at GSI to measure reaction kinematics and the residue cross sections down to 0.1 mbarn with an accuracy of 10% for a series of different projectile-target combinations.

II EXPERIMENT

The reaction products were identified in-flight by the Fragment Separator (FRS) at GSI [17]. The full (Z,N) map of residues was scanned down to a production cross section of the order of 0.1 mbarn using 2% steps in magnetic rigidity. The FRS was equiped with two plastic scintillation detectors [18] at the center and at the exit of the spectrometer which measured the time-of-flight as well as the horizontal positions to provide the magnetic rigidity. An Ionisation Chamber (MUSIC) [19] was mounted at the exit in order to determine the energy loss of the reaction products. For the heaviest elements, close to the projectile, the energy loss in a thick degrader was used to determine the nuclear charge. The primary-beam intensity was monitored using a secondary-electron beam monitor (SEETRAM) [20]. A precision of better than 5% in beam intensity was obtained.

The primary beam of ^{238}U with an energy of 1 A GeV impinged on a 50.5 mg/cm^2 lead target. The liquid hydrogen target of thickness of 85 mg/cm^2 was used with ^{197}Au (0.8 AGeV) and ^{208}Pb, ^{238}U (1 AGeV) beams. The reaction rates in the titanium windows of the target system were also investigated using a dummy target, consisting only of a 36 mg/cm^2 titanium foil.

III DATA ANALYSIS

Once the nuclear charge Z is determined, the mass number A of any reaction product is obtained from the magnetic rigidity Bρ and the velocity v = $\beta \cdot$ c by using the relation

$$B\rho = \beta\gamma \cdot \frac{A}{Z} \cdot c \cdot m_0/e_0, \qquad (1)$$

where γ is the Lorentz parameter, c the velocity of light, m_0 the nuclear mass unit, and e_0 the charge unit.

For an unambiguous isotopic identification of the reaction products, the analysis was restricted to the fully stripped ions. The losses in counting rate due to the fraction of incompletely stripped ions were corrected for. Another correction to be applied considered the losses due to secondary reactions in the scintillation detector and in the degrader material. Finally, the isotopic productions in the titanium vacuum window of the SIS accelerator and in the aluminium foils of the beam monitor were corrected for. The angular transmission of the fragment separator was estimated using the method explained in refs. [10,14].

IV RESULTS

An important improvement brought by the present experiments in inverse kinematics is the knowledge of the reaction kinematics. Figure 1 demonstrates the different kinematical properties of fission and fragmentation products. While the velocities of the fragmentation products, corresponding to the most neutron-deficient

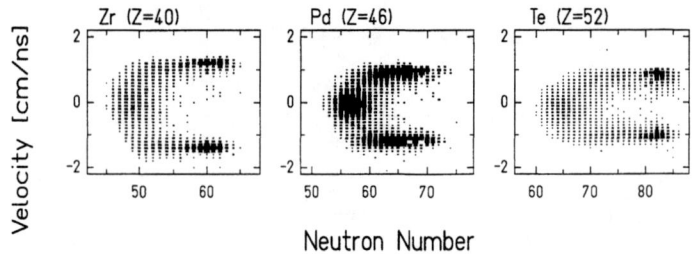

FIGURE 1. Two-dimensional cluster plots of the velocity distributions versus neutron number, given for a selected number of elements (Z = 40, 46, and 52) from the reaction of ^{238}U + ^{208}Pb. The velocities have been converted into the frame of the projectiles in the middle of the production target.

isotopes, peak close to the beam velocity, the fission products carry the recoil of the fission process. Due to the limited angular acceptance of the spectrometer, only those fission products which are emitted in forward or backward direction can be measured.

Figure 2 shows the isotopical distributions for two reactions with different projectile-target combinations.

In the reaction ^{238}U + ^{208}Pb, the most neutron-rich isotopes clearly show the double-humped structure characteristic for the low-energy fission of ^{238}U, while in the more neutron-deficient isotopes the dip at symmetric fission is filled. These different features can be traced back to the different excitation energies induced in either electromagnetic or nuclear interactions. The fragmentation products populate a corridor, half way between the valley of stability and the proton drip line, from the projectile down to lowest charges.

In the reaction ^{197}Au + H$_2$, the highest production cross sections result from the fragmentation process. The sharp decrease in the production cross sections for fragments below Z = 58 reflects to the limited excitation energy induced in the spallation process. The velocity distributions allow to clearly attribute the island of production below Z = 54 to fission fragments. These fragments are situated close to the evaporation corridor, indicating that the excitation energy induced in the spallation process was sufficiently high to reduce the A/Z ratio of fissioning nucleus.

V DISCUSSION

The experiments on spallation reactions performed in inverse kinematics have shown to be very powerful for determining spallation-residue cross sections. Important processes of the spallation mechanisms are revealed by the full isotopical distributions. Comparing the experimental values to the expectations of theoretical models, it will be possible to improve the understanding and the description of

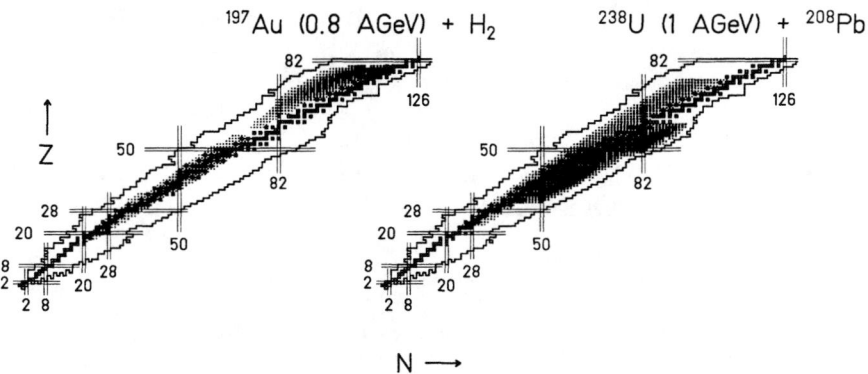

FIGURE 2. Two-dimensional cluster plots of the isotopic cross sections on a chart of the nuclides for reactions ^{238}U + ^{208}Pb (right) and ^{197}Au + H$_2$ (left). Open squares correspond to stable isotopes.

spallation reactions considerably as well for the nucleon-nucleon cascade as for the particle-evaporation process.

REFERENCES

1. K.-H. Schmidt et al., Nucl. Phys. **A452** (1992) 699.
2. K.-H. Schmidt et al., Phys. Lett. **B300** (1993) 313.
3. E. Hanelt et al., Z. Phys. **A346** (1993) 43.
4. M. de Jong et al., Nucl. Phys. **A628** (1998) 479.
5. J. Benlliure et al., Nucl. Phys. **A628** (1998) 458.
6. J. Benlliure et al., Submitted to European Phys. Journal.
7. Th. Rubehn et al., Z. Phys. **A353** (1995) 197.
8. M. Hesse et al., Z. Phys. **A355** (1996) 69.
9. P. Armbruster et al., Z. Phys. **A355** (1996) 191.
10. C. Donzaud et al., Eur. Phys. J. **A1** (1998) 407.
11. A.R. Junghans et al., Nucl. Phys. **A629** (1998) 635.
12. A.V. Ignatyuk et al., Nucl. Phys. **A593** (1995) 519.
13. Th. Rubehn et al., Phys. Rev. **C53** (1996) 3143.
14. W. Schwab et al., Submitted to European Phys. Journal.
15. C.D. Bowman et al., Nucl. Instrum. Methods **B320** (1992) 336.
16. C. Rubbia et al., CERN-LHC 96-011-ETT.
17. H. Geissel et al., Nucl. Instrum. Methods **B70** (1992) 286.
18. B. Voss et al., Nucl. Instrum. Methods **A364** (1992) 150.
19. M. Pfützner et al., Nucl. Instrum. Methods **B86** (1994) 213.
20. A. Junghans et al., Nucl. Instrum. Methods **A370** (1996) 312.

Neutron-rich nuclei produced in intermediate energy fission

P. Dendooven, S. Hankonen, A. Honkanen, M. Huhta, A. Jokinen,
G. Lhersonneau, M. Oinonen, H. Penttilä, V.A. Rubchenya*, J.C. Wang,
J. Äystö

Department of Physics, University of Jyväskylä, P.O. Box 35, 40351 Jyväskylä, Finland
*V.G. Khlopin Radium Institute, Prospekt Shvernika 28, 194021 St.-Petersburg, Russia

Abstract. For more than 10 years, neutron-rich nuclei produced in light-ion induced fission have been studied at the Ion Guide Isotope Separator On-Line (IGISOL) facility of the Department of Physics, University of Jyväskylä. The present performance of the system is briefly discussed. Some recent results concerning superasymmetric fission and β-delayed neutron emission are presented. The plans for producing isotopically pure beams are summarized.

THE FISSION ION GUIDE

For more than 10 years, neutron-rich nuclei have been produced at the IGISOL facility of the Accelerator Laboratory of the University of Jyväskylä using light-ion induced fission. Most often, a 25 MeV proton beam and a ^{238}U target have been used. In the ion guide method (1), reaction products recoil out of the target and are stopped in a helium-filled stopping chamber. Some of these products survive long enough as singly-charged ions to be evacuated from the stopping chamber by the helium gas flow and can be injected directly into the acceleration stage of the isotope separator. The present fission ion guide design is shown in figure 1 and described in greater detail in (2,3). Because the fission fragments are emitted more or less isotropically, the target can be tilted with respect to the primary beam, thus increasing the effective target thickness (in our design 8 times larger than the actual target thickness). Because fission fragments are emitted with high energy, the target and stopping chambers can be separated by a thin foil. This prevents the passage of the cyclotron beam through the stopping chamber. The resulting lower ionisation of the gas in the stopping chamber increases the ion survival time. A skimmer is used to prevent most of the helium gas from entering the isotope separator. The transmission of ions through the skimmer is enhanced by applying a negative voltage on the skimmer relative to the ion guide (typically -300 to -500 V). Since moving the IGISOL facility to the K130 Cyclotron Laboratory in early 1994, mass-separated yields of fission products have increased 100-fold due to increases in cyclotron beam intensity and ion guide efficiency. Presently, mass-separated beam intensities of 3×10^3 ions/s per mb cross section are available.

CP455, *ENAM98: Exotic Nuclei and Atomic Masses*
edited by B. M. Sherrill, D. J. Morrissey, and Cary N. Davids
© 1998 The American Institute of Physics   1-56396-804-5/98/$15.00

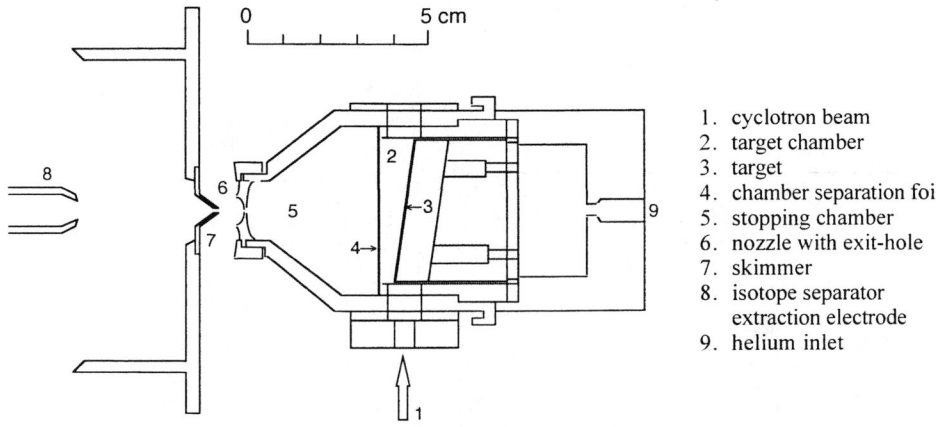

FIGURE 1. Schematic drawing of the fission ion guide. The gray areas are filled with helium

SUPERASYMMETRIC FISSION

Information on superasymmetric fission of heavy nuclei, where the light fragment mass is smaller than 80, has been scarce, mainly due to the very low yields, especially in spontaneous and thermal neutron-induced fission. However, the production rates can differ dramatically from one beam/target combination to the other. Therefore, for planning experiments to investigate neutron-rich exotic nuclei around Z=28, it is important to characterize the properties of the superasymmetric fission mode for different beam/target combinations.

Isotopic yields for the light fragments ($71 \leq A \leq 82$, $28 \leq Z \leq 32$) of 25 MeV proton-induced fission of ^{238}U were measured at the IGISOL facility (4). A substantial enhancement in cross-section relative to thermal-neutron-induced fission of ^{235}U was observed (see figure 2). In order to get a quicker overview of superasymmetric fission at intermediate energy, mass yields were measured using the HENDES detector array at Jyväskylä (5). Time-of-flight (TOF) measurements of fission fragments, measured together with neutrons, from ^{238}U(p,f) at E_p = 20, 35, 50 and 60 MeV, ^{232}Th(p,f) at E_p = 50 and 60 MeV and ^{238}U(d,f) at E_d = 30 MeV were performed. The mass distribution of the fission fragments was extracted from the original TOF data. The masses of the two fragments were assumed to add up to the mass of the compound system after pre-fission neutron evaporation. The results were analyzed in the framework of a model proposed earlier (6,7) and used here in a more advanced version (4). In the model, the superasymmetric division is determined by the nuclear shells at Z=28 and N=50 and by the tails of the second standard asymmetric component and the symmetric component, whose contributions are changing with excitation energy of the fissioning nucleus. In the more advanced version of the model, odd-even effects in charge and mass distributions are introduced and the model parameters are changed slightly taking into account the experimental data on very asymmetric fission obtained at IGISOL (4). Smoothed pre-neutron fission fragment mass distributions are

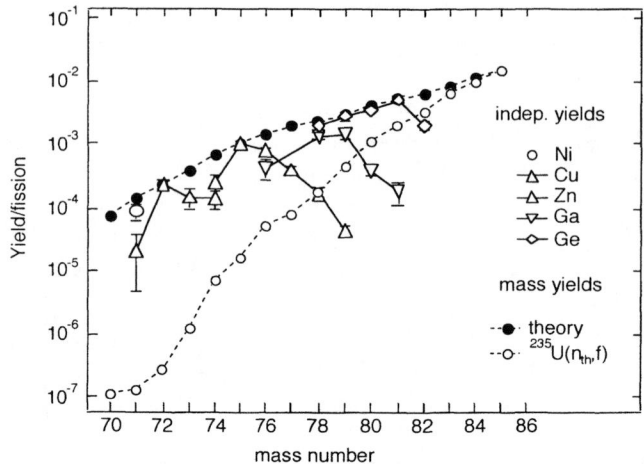

FIGURE 2. Measured independent yields (points with error bars) and calculated mass yields (closed circles) for 25 MeV proton-induced fission of ^{238}U. For comparison, the mass yields in thermal neutron induced fission of ^{235}U are shown (open circles).

approximated by the superposition of seven Gaussian distributions: one symmetric and three pairs of asymmetric components (each pair consisting of Gaussian distributions for both the heavy and light fragments). The asymmetric components are connected with:
- the $Z=50$ and $N=82$ nuclear shells in the heavy fragments;
- the "deformed" nuclear shell at $N=86$-90; and,
- the $N=28$ and $N=50$ nuclear shells in the light fragments.

The competition between fission modes is determined by fission dynamics and nuclear shells in the fission fragments. The HENDES measurements confirm the enhancement of fission fragment yields in the far asymmetric region. More details about the HENDES measurements and the theoretical model are given in (8,9).

Our results indicate that fission of heavy nuclei at intermediate excitation energy may become the tool of choice for the production of neutron-rich nuclei with mass number below 80.

NEW β-DELAYED NEUTRON EMITTERS

The interest in studying very neutron-rich nuclei is due to indications of new nuclear structure phenomena when moving towards the neutron-drip line (10,11) and the strong connections with astrophysics (12). Because of isobaric contaminants (produced with higher cross-sections), β-delayed neutron spectroscopy gives the highest sensitivity for identifying new neutron-rich nuclei. In collaboration with K.-L. Kratz, B. Pfeiffer and co-workers from the Institut für Kernchemie, University of Mainz, and using a neutron long counter, 14 new neutron emitters were discovered (see figure 3). For 6 of these, the half-life was measured for the first time. We refer to (13,14) for more details.

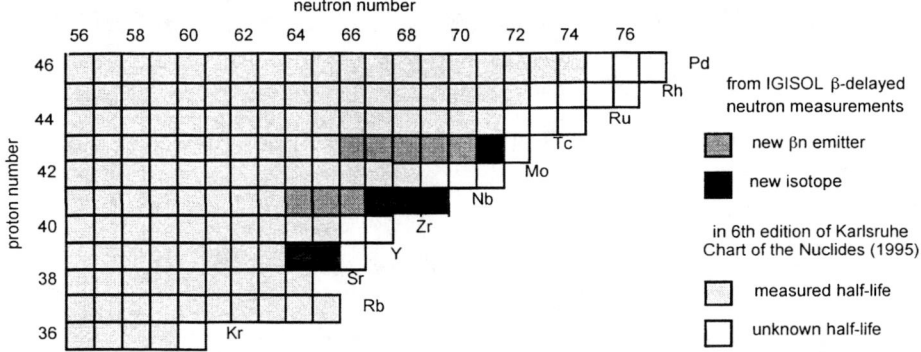

FIGURE 3. New β-delayed neutron emitters discovered at the Jyväskylä IGISOL facility. The new isotopes are those for which the β-decay half-life was measured for the first time.

ISOTOPICALLY PURE BEAMS

The study of new neutron-rich nuclei is hindered by the presence of contaminants in the mass-separated beam: isobaric isotopes of larger atomic number, isotopes of neighbouring masses and doubly charged ions (problematic at masses ≤ 80). All these problems can only be solved satisfactorily by performing isobaric separation of the beam. We are developing an ion trap for this purpose. However, the energy spread of the IGISOL beam (~100 eV) must be reduced before trapping is possible. Therefore, we will install a helium-filled RF quadrupole as ion beam cooler before the ion trap. This cooler/ion trap project is discussed in more detail in a separate contribution to this conference (15).

REFERENCES

1. P. Dendooven, Nucl. Instrum. Meth. Phys. Res. B 126 (1997) 182, and references therein
2. H. Penttilä et al., Nucl. Instrum. Meth. Phys. Res. B 126 (1997) 213
3. P. Dendooven et al., *Proc. 2nd Int. Workshop on Nuclear Fission and Fission-Product Spectroscopy*, April 22-25, 1998, Seyssins, France (American Institute of Physics, to be published)
4. M. Huhta et al., Phys. Lett. B 405 (1997) 230
5. W.H. Trzaska et al., In: *Application of Accelerators in Research and Industry*, eds. Duggen and I.L. Morgan, AIP Press (1997) p. 1059
6. E. Karttunen et al., Nucl. Sci. Engin. 109 (1991) 350
7. P.P. Jauho et al., Phys. Rev. C49 (1994) 2036
8. J. Äystö et al., *Proceedings International Conference on Fission and Properties of Neutron-Rich Nuclei*, Nov. 10-15, 1997, Sanibel Island, Florida (to be published)
9. V.A. Rubchenya et al., *Proc. 2nd Int. Workshop on Nuclear Fission and Fission-Product Spectroscopy*, April 22-25, 1998, Seyssins, France (American Institute of Physics, to be published)
10. J. Dobaczewski et al., Phys. Rev. Lett. 72 (1994) 981
11. N. Fukunishi et al., Phys. Lett. B 296 (1992) 279
12. K.-L. Kratz, in *Nuclei in the Cosmos III*, AIP Conf. Proc. No. 327 (AIP, New York, 1995), p. 113
13. T. Mehren et al., Phys. Rev. Lett. 77 (1996) 458
14. J.C. Wang et al., Department of Physics, University of Jyväskylä 1997 Annual Report, p. 23
15. A. Jokinen et al., contribution to these proceedings

Identification of μs isomers in fission products

J. Genevey*, J.A. Pinston*, H. Faust‡, T. Friedrichs‡, M. Gross‡, F. Ibrahim*, T. Larqué*, S. Oberstedt‡

*Institut des Sciences Nucléaires, 38026 Grenoble, France
‡Institut Laue Langevin, 38042 Grenoble, France

Abstract. Several μs isomers have been observed in the fission products of ^{241}Pu and ^{239}Pu. The detection is based on the time correlation between reaction products selected by the LOHENGRIN spectrometer and the γ-rays depopulating the isomers. New isomers have been observed in ^{96}Rb, ^{106}Nb, ^{127}Sn, and ^{130}Te and the others have been confirmed in ^{94}Y, ^{126}Sn, ^{129}Sn and ^{129}Sb.

INTRODUCTION

The study of excited states of neutron-rich nuclei is very important to test the nuclear models far from stability. Study of isomeric states of exotic nuclei is a way to obtain information on such excited states. These nuclei can be obtained from fission induced by thermal neutron. If there is an isomeric state in the μs range and if the fission fragments (FF) are selected with a spectrometer, it is then possible to measure the correlation between the FF and the γ-lines which depopulates the isomer. The LOHENGRIN spectrometer at ILL [1] has been used to separate the FF recoiling from thin targets of ^{241}Pu or ^{239}Pu. The neutron flux on the fission target is 5.10^{14} n.cm^{-2}.s^{-1}.

EXPERIMENTAL

A schematic view of the experimental set-up used for the detection is shown in fig.1. The FF are selected by LOHENGRIN according to their A/q ratio; they enter an ionisation chamber and are stopped in a single Si detector or in an array of 5 pin diodes detectors. The time correlation between the incoming FF and the γ-rays is measured in the 0-40 μs range. Details of the experimental procedure have been explained elsewhere [2] [3].
Independent isomeric yields down to 10^{-6} can be detected.

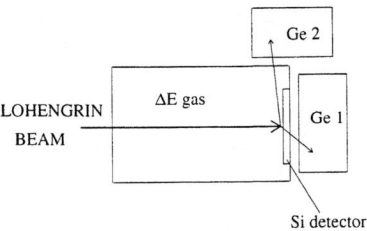

FIGURE 1. Schematic view of the experimental set-up

RESULTS

Many µs isomers in the mass range 88-109 have been observed [2] [3]. The preliminary results in the mass range 126-132 are also reported here. Fig.2. shows examples of decay curves.

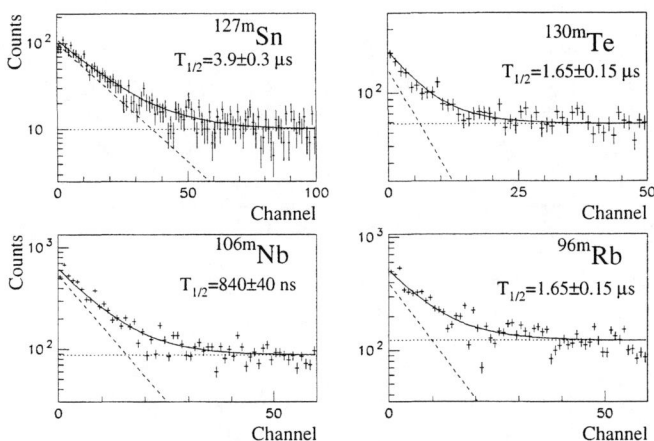

FIGURE 2. Decay curves of the isomers observed in this experiment. *Solid line=fitted curve, dotted line=background, dashed line=background subtracted result.*

The results are compiled in table 1. Our values of half-lives ($T_{1/2}^{exp}$), measured using the γ-rays indicated in column 1, are compared with the ones already known ($T_{1/2}^{lit}$). Our values agree generally with the ones previously reported, except in case of ^{107}Mo [3]. The isomer observed in the A = 127 chain is attributed to ^{127}Sn because all the lines deexciting this isomer have been already observed in the decay of ^{127}In ($T_{1/2}$=1.09 s) [24] ; the isomer observed in the A = 129 chain is attributed to ^{129}Sn for the same reason, all the lines deexciting this isomer have been already

TABLE 1. Isomers observed in this work

Nucleus	$E_\gamma(keV)$	$T_{1/2}^{exp}(\mu s)$	$T_{1/2}^{lit}(\mu s)$	$Y_{iso}\times 10^3$	$Y_{nucl}\times 10^3$	Ref.
^{88}Br	111	5.1 (4)	5.4 (7)	1.3(3)	4.5	5
^{94}Y	432	1.35 (2)	unpublished	1.8(3)	2.2	4
^{96}Rb	300+240+461	1.65 (15)	unknown	0.08(3)	1.3	
^{97}Sr	140	0.43 (3)	0.515+0.170	1.0(3)	14	6-7
^{98}Y	204	7.2 (1)	8.0 (2)	3(1)	6	8
^{99}Y	126	11 (2)	8.6 (8)	0.3(1)	2	9
^{99}Zr	130	0.40 (5)	0.293 (10)	2.6(8)	26	10
^{100}Nb	173+185	13 (1)	12	0.9(4)	4.2	7-11
^{104}Tc	69	5 (2)	3.5 (3	0.3(1)	2.5)	12
^{106}Nb	95+107+202+205	0.84 (4)	unknown	0.24(8)	17	
^{107}Mo	66	0.47 (3)	0.238 (7)	16(5)	17	13
^{109}Ru	96	0.68 (3)	0.780 (150)	8(3)	6	13
^{109}Rh	226	-	1.66	0.02(1)	0.2	14
^{126}Sn	112+909+1141	[9.0(4)]	6.6(14)	0.18(4)	2.3	15
^{126}Sn	269	7.6(4)	(6.5) unpublis.	0.21(4)	2.3	16
^{127}Sb	806+825	-	11(1)	0.12(2)	0.27	17
^{127}Sn	567+716	3.9(3)	unknown	0.22(5)	(3.8)	
^{128}Sn	79+321	3.4(1)	2.69(23)	0.5(1)	(6.7)	18
^{129}Sb	732+1128	2.2(2)	≥ 2	0.2(1)	(4.5)	19
^{129}Sn	382+570+1135+1325	3.6(1)	(3)	0.6(2)	(9.3)	20
^{130}Sn	97+391	1.6(1)	1.61(15)	0.5(2)	(10)	18
^{130}Te	502+833	1.65(15)	unknown	0.30(15)	2.1	
^{130}Te	182+331+793+839	[2.0(3)]	0.115		2.1	21
^{132}Te	776+926	3.7(2)	3.9(3)	3.0(15)	30	22
^{132}Te	150	-	28.1(15)		30	22

The independent fission yield of ^{241}Pu for A = 88-109 and ^{239}Pu for A = 126-133 are taken from JEF2 [23].

In ^{126}Sn and in ^{130}Te cumulative values are reported for the half-lives (noted [ ]).

observed in the decay of ^{129}In (T$_{1/2}$=0.61 s) [24]. Previously [20], it was attributed to ^{129}Sb. The ^{126m}Sn isomer (T$_{1/2}$=7.6(4)μs) was first reported in the Annual report of Argonne Lab. [16]. It has been confirmed by the present experiment. Its half-life value has been improved upon. Preliminary isomeric yields, when available, are also reported.

Fig.3 shows the systematics of even-even Te nuclei. By analogy with the case of ^{132m}Te, the new ^{130m}Te isomer (1.65 μs) at 2750 keV is probably the 10$^+$ state of the $\nu(h_{11/2})^{-2}$ configuration. It decays via two transitions of 102 and 502 keV to the 7$^-$ level which is interpreted as $\nu(h_{11/2})^{-1}$, $\nu(d_{3/2})^{-1}$. The others levels at 0$^+$, 2$^+$, 4$^+$, 6$^+$ correspond to the configuration $\pi(g_{7/2})^2$. It should be noted that the B(E2) values for the 10$^+ \to$ 8$^+$ transition in the two isotones ^{128m}Sn and ^{130m}Te are identical : 11.6 e^2.fm^4 for ^{128m}Sn and 12.0 e^2.fm^4 for ^{130m}Te.

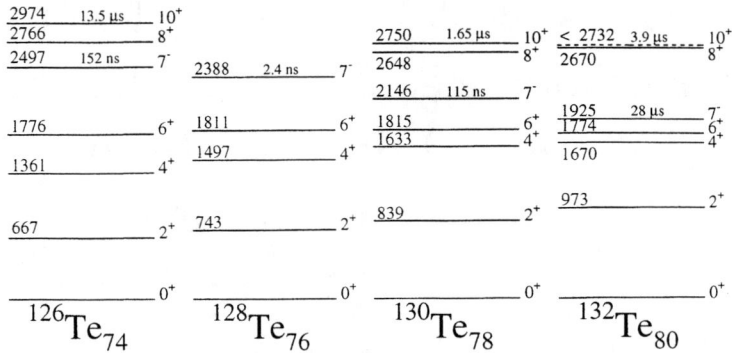

FIGURE 3. Systematics of the even-even Te nuclei

REFERENCES

1. E. Moll et al., Kerntechnik **19**, 374 (1977).
2. F. Ibrahim, J. Genevey, J.A. Pinston, et al., Phys. Rev. **C** , to be published
3. J. Genevey, J.A. Pinston, et al., Workshop on Nucl. Fission, Seyssins, France, 1998
4. R. Sellam, Thesis Univ. Grenoble 1976.
5. H. W. Müller, Nucl. Data Sheets **54**, 1 (1988).
6. R. G. Clark, et al., Int. At. En. Agency, Vienna, Vol2, p.221 (1974).
7. E. Monnand, et al., Ann. Rept. CEA Grenoble CEA-N-2176, p.20 (1980).
8. J. W. Grüter et al., Phys. Lett. **33B**, 474 (1970) and Rept. Jülich Jül-879-NP (1972).
9. R. A. Meyer, et al. Nucl. Phys. **A439**, 510 (1985).
10. K. Sistemich, G. Lhersonneau, R. A. Meyer, Jülich Jül-Spez-344, p.28 (1986).
11. G. Lhersonneau and K. Sistemich, Jülich Jül-Spez-344, p.29 (1986).
12. G. Tittel, et al., Univ. Mainz, Ann. Rept., 1980, p. 34 (1981).
13. E. Cheifetz et al., NFFS Cargese, CERN-76-13, 471 (1976) and Refs. therein.
14. N. Kaffrell et al., Nucl. Phys. **A470**, 141 (1987).
15. B. Fogelberg, P. Carlé, Nucl. Phys. **A323**, 205 (1979)
16. P.M. Carpenter et al., Ann. Rept. ANL, 95/14, 43.
17. K.E. Apt et al., Phys. Rev, **C9**, 310 (1974).
18. B. Fogelberg, K. Heyde, J. Sau, Nucl. Phys.,**A352**, 157 (1981).
19. C.A. Stone, W.B. Walters, Z. Phys, **A328**, 257 (1987).
20. K. Heyde et al., Phys. Rev, **C16**, 2437 (1977).
21. A. Kerek, P. Carlé, J. McDonald, Nucl. Phys., **A198**, 466 (1972).
22. K. Sistemich et al., Z. Phys., **A292**, 145 (1979).
23. JEF2, a computed file of fission data maintened by NEA.
24. L.E. De Geer, G. B. Holm, Phys. Rev, **C22**, 2163, (1980).

Systematics of Calculated Cold-Fusion Barriers for Reactions Leading to Compound Systems from $Z = 104$ to $Z = 126$

P. Möller[1],* P. Armbruster[†], S. Hofmann[†], and G. Münzenberg[†]

Theoretical Division, Los Alamos National Laboratory, Los Alamos, NM 87545, USA,
[†]*Gesellschaft für Schwerionenforschung, Planckstrasse 1, D64291 Darmstadt, Germany*

Abstract. We have previously shown that just as the decay properties of nuclei in the heavy region depend strongly on shell structure, shell structure also dramatically affects the fusion entrance channel. The six most recently discovered new elements were all formed in cold-fusion reactions. We discuss here the effect of the doubly-magic structure of the target in cold-fusion reactions on the fusion barrier and present a systematic study of cold-fusion heavy-ion reaction barriers for elements from Rf to $Z = 126$. We find that the systematics of the optimum reaction energy may change near $Z = 112$, because here the highest point on the interaction barrier shifts in location from near the touching distance at $r/R_0 \approx 1.5$ to $r/R_0 \approx 1.0$, which is a shape configuration just slightly more deformed than the ground state.

INTRODUCTION

The six heaviest-known elements were all produced in cold-fusion reactions with doubly-magic $^{208}_{82}\text{Pb}_{126}$ or near doubly-magic $^{209}_{83}\text{Bi}_{126}$ targets [1–6]. The cold-fusion reaction has long been thought to enhance heavy-element evaporation-residue cross sections primarily because it leads to compound nuclei of low excitation energy, which enhances de-excitation by neutron emission relative to fission. Higher excitation energies would lead to higher fission probabilities. However, the evaporation-residue cross section is the product of the cross section for compound-nucleus formation and the probability for de-excitation by neutron emission. One may therefore

---

[1]) Permanent address:*P. Moller Scientific Computing and Graphics, Inc., P. O. Box 1440, Los Alamos, New Mexico 87544, USA*

CP455, ENAM98: Exotic Nuclei and Atomic Masses
edited by B. M. Sherrill, D. J. Morrissey, and Cary N. Davids
© 1998 The American Institute of Physics   1-56396-804-5/98/$15.00

ask if cold fusion *also* enhances the cross section for compound-nucleus formation. Because of the low excitation energies in the entrance channel, the large negative shell correction associated with target nuclei near the doubly-magic ^{208}Pb should be almost fully manifested at touching and for an appreciable distance inside touching, in analogy with cold fission of nuclei near ^{258}Fm, an analogy which was proposed more than 10 years ago [7–10].

We have recently investigated this proposed analogy. We showed [11] that fusion-barriers calculated for cold-fusion heavy-ion reactions in the macroscopic-microscopic Finite-Range Droplet Model [FRDM (1992)] are very different from those obtained in a purely macroscopic model. The model has been described in sufficient detail elsewhere [12–16]. For reactions on targets near ^{208}Pb it was shown that shell effects in the entrance channel result in fusion-barrier energies at the touching point that are only about 5 MeV higher than the ground-state for compound systems near $Z = 110$. Moreover, the entrance-channel effects remain far inside the touching point, to configurations only slightly more elongated than the deformed ground-state shape. In particular, we showed that the magic gaps associated with the target were present in compound-nucleus single-particle level diagrams from the touching configuration to about $r/R_0 = 1.15$, where r is the distance between the centers of mass of the two parts of the system and R_0 is the spherical radius of the compound system. Because level crossings at the Fermi surface only start to occur far inside the touching point in these cold-fusion heavy-ion reactions it was concluded that dissipation sets in late in the collision process.

Here we will study the *systematics* of the cold-fusion heavy-ion reaction barrier for elements from Rf to $Z = 126$.

I THE COLD-FUSION ENTRANCE CHANNEL FOR $Z = 104 - 126$

We show in Figs. 1 and 2 calculated fusion barriers for the two elements just below $Z = 110$, namely Hs and Mt, for $Z = 114$, and for $Z = 120$. The fusion barriers for the merging systems are calculated for intersecting spheres. Just inside the peak in the fusion barrier at about $r/R_0 = 1.0$ we have switched from the intersecting-sphere parameterization to Nilsson's perturbed-spheroid ϵ parameterization so that we accurately obtain the energy of the ground state. The dotted line shows the calculated fission barrier, for which we considered ϵ_2, ϵ_4, and ϵ_6 deformations, except for the Mt calculation, which was done earlier, where ϵ_6 is omitted. The fusion barrier in the macroscopic FRDM is given by the dotted line. The thin, long-dashed line gives the r/R_0 location where the target and projectile just touch. In the right part of the Fig. 1 the thicker, long-dashed line gives the calculated adiabatic macroscopic-microscopic barrier without any specialization energy. The specialization energy due to the odd particle in ^{209}Bi is included in the solid "Total-fusion" curve. Despite the fairly high spin 9/2 of the ^{209}Bi ground state, the specialization energy is quite low. The adiabatic curve is shown only from touching

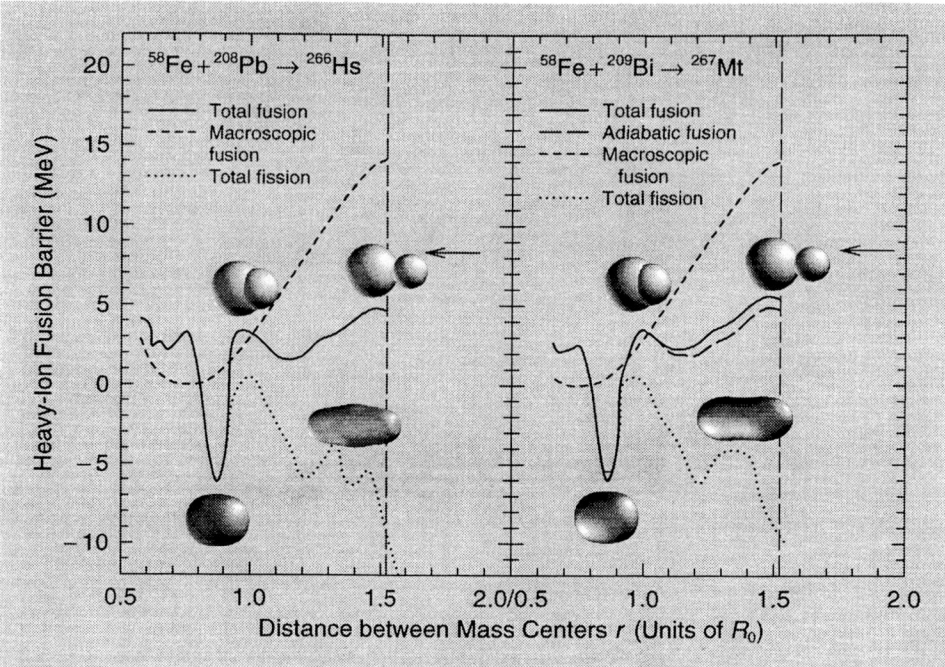

FIGURE 1. Total, adiabatic, and macroscopic fusion barriers for the reactions ^{58}Fe + ^{208}Pb → ^{266}Hs and ^{58}Fe + ^{209}Bi → ^{267}Mt. Also shown are the fission barriers corresponding to spontaneous fission from the ground state. The arrows indicate the incident energies corresponding to the maximum evaporation-residue cross sections.

to about $r/R_0 = 1.0$. However, the specialization energy is quite low also between $r/R_0 = 1.0$ and the ground-state shape.

Four shape configurations are also shown in Figs. 1 and 2. Two correspond to shapes along the Total-fusion barrier, namely the touching configuration and the shape at the peak of the fusion barrier. The shape at the minimum corresponds to the calculated ground-state shape, which except for Mt includes ϵ_6. The shape at $r/R_0 = 1.4$ is the shape along the fission barrier at this elongation. The shapes are rotated 30 degrees from a perpendicular orientation so that the right polar regions become visible.

In Fig. 3 we show the systematic behavior of various fusion-barrier and fission-barrier saddle points from Rf to $Z = 126$. The ground-state microscopic correction is also plotted, with a reverse sign, and is labeled "Enhancement to binding". It is by now well-known that for very heavy elements this quantity is a very good approximation to the maximum fission-barrier height. This is well born out by the results shown in Fig. 3.

One should observe that the fusion reactions presented here correspond to the most neutron-rich stable projectile available. For compound systems beyond $Z =$

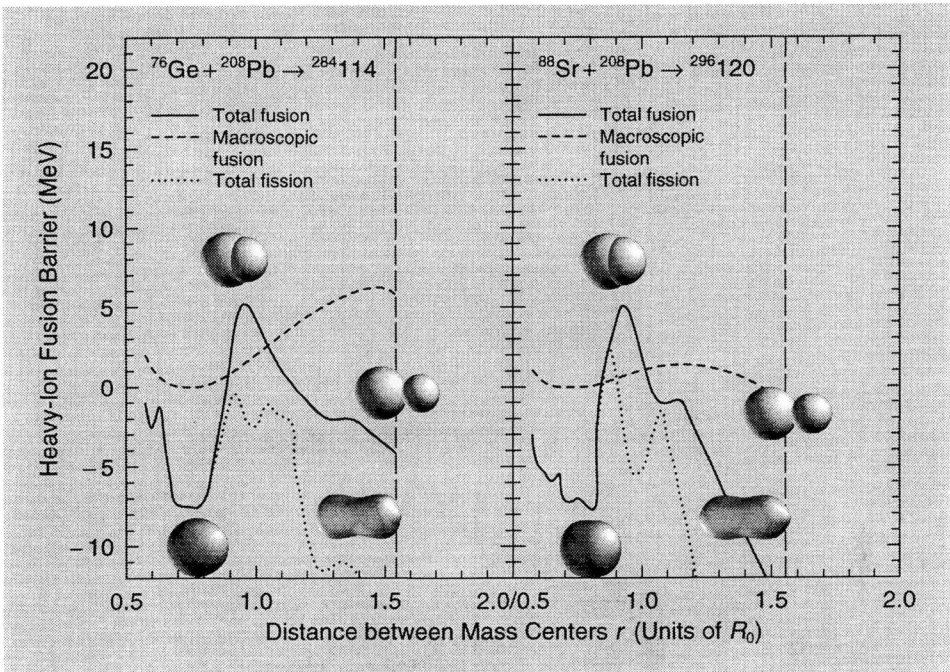

FIGURE 2. Total and macroscopic fusion barriers for the reactions ^{76}Ge + ^{208}Pb → ^{284}114 and ^{88}Sr + ^{208}Pb → ^{296}120. Also shown are the fission barriers corresponding to spontaneous fission from the ground state.

120 this choice no longer leads to the most favorable ground-state shell correction, so for these reactions *slightly less* neutron-rich projectiles should probably be chosen.

At $Z = 110$ the total fusion barrier at touching becomes lower than the fusion-barrier height at $r/R_0 \approx 1.0$. It is known experimentally that the optimum 1n cross section for evaporation-residue formation in cold-fusion heavy-ion reactions decreases from about 15 MeV excitation energy relative the ground state of the compound system for ^{258}Rf to only about 10 MeV for ^{278}112. This correlates well with the behavior of the calculated fusion-barrier maximum, which until about $Z = 110$ occurs at touching and gradually decreases in energy relative the compound-system ground state as the charge of the compound system increases from 104 to 110. When the energy for the optimum evaporation-residue cross section is correlated to the fusion barriers calculated here, one should recall that the fusion barrier is determined for a specific, "frozen", merging-sphere configuration. This configuration is expected to be quite realistic at touching and inside, until $r/R_0 \approx 1.15$ where the target magic gaps disappear in the calculated single-particle level diagrams corresponding to this frozen configuration. Inside $r/R_0 \approx 1.15$ one must

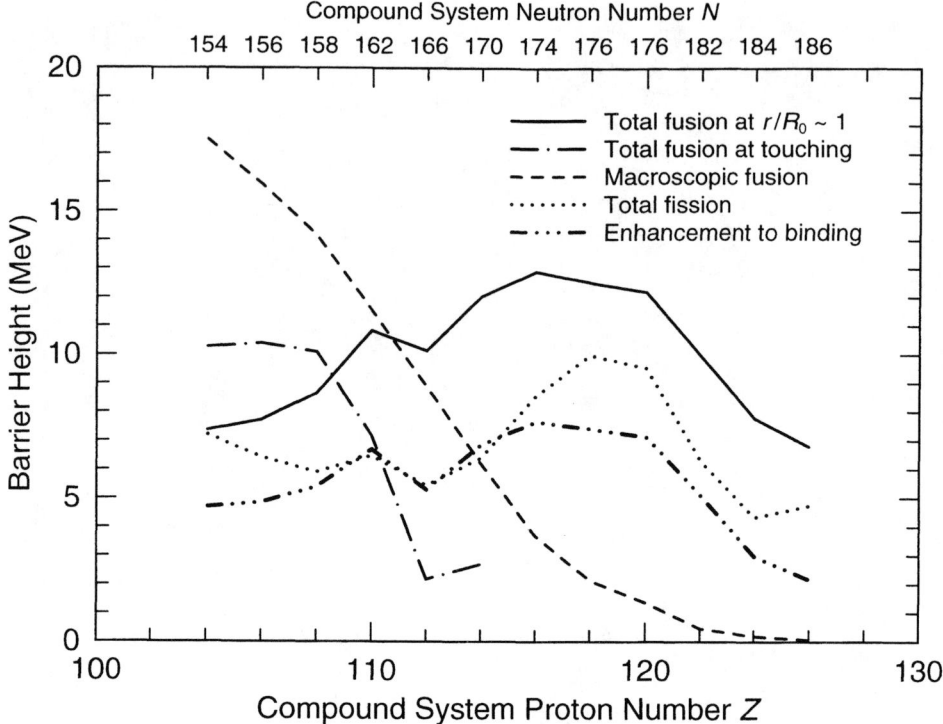

FIGURE 3. Potential-energy-surface barrier and ground-state systematics for cold-fusion reactions leading to compound systems from Rf to $Z = 126$.

therefore expect that it becomes profitable for the system to deviate from the merging-sphere configuration. Consequently, we expect that the proton number where the true inner fusion barrier height becomes higher than at touching occurs at $Z = 112$ or $Z = 114$ rather than at $Z = 110$.

Because the calculated fusion-barrier maximum shifts from the touching point to an elongation just beyond the deformed ground state at $Z = 110 - 112$, the calculated maximum fusion-barrier height will no longer systematically decrease as the charge of the compound system increases. The maximum fusion barrier is therefore no longer related to the barrier energy at the touching configuration. Thus, one can expect a break in the behavior of the systematics of the optimum energy for evaporation-residue formation at about $Z = 112$. This observation may obviously be of importance in the design of experiments to reach new elements beyond $Z = 112$.

References

1) G. Münzenberg, S. Hofmann, F. P. Heßberger, W. Reisdorf, K.-H. Schmidt, J. R. H. Schneider, P. Armbruster, C.-C. Sahm, and B. Thuma, Z. Phys. **A300** (1981) 7.
2) G. Münzenberg, P. Armbruster, F. P. Heßberger, S. Hofmann, K. Poppensieker, W. Reisdorf, J. R. H. Schneider, W. F. W. Schneider, K.-H. Schmidt, C.-C. Sahm, and D. Vermeulen, Z. Phys. **A309** (1982) 89.
3) G. Münzenberg, P. Armbruster, H. Folger, F. P. Heßberger, S. Hofmann, J. Keller, K. Poppensieker, W. Reisdorf, K.-H. Schmidt, H. J. Schött, M. E. Leino, and R. Hingmann, Z. Phys. **A317** (1984) 235.
4) S. Hofmann, V. Ninov, F. P. Heßberger, P. Armbruster, H. Folger, G. Münzenberg, H. J. Schött, A. G. Popeko, A. V. Yeremin, A. N. Andreyev, S. Saro, R. Janik, and M. Leino, Z. Phys. **A350** (1995) 277.
5) S. Hofmann, V. Ninov, F. P. Heßberger, P. Armbruster, H. Folger, G. Münzenberg, H. J. Schött, A. G. Popeko, A. V. Yeremin, A. N. Andreyev, S. Saro, R. Janik, and M. Leino, Z. Phys. **A350** (1995) 281.
6) S. Hofmann, V. Ninov, F. P. Heßberger, P. Armbruster, H. Folger, G. Münzenberg, H. J. Schött, A. G. Popeko, A. V. Yeremin, S. Saro, R. Janik, and M. Leino, Z. Phys. **A354** (1996) 229.
7) P. Armbruster, Proc. Int. School-Seminar on heavy ion physics, Dubna, USSR, 1986, JINR Report No. D7-87-68 (1987) p. 82.
8) P. Möller, J. R. Nix, and W. J. Swiatecki, Proc. Int. School-Seminar on heavy ion physics, Dubna, USSR, 1986, JINR Report No. D7-87-68 (1987) p. 167.
9) P. Möller, J. R. Nix, and W. J. Swiatecki, Nucl. Phys. **A469** (1987) 1.
10) P. Armbruster, J. Phys. Soc. Jpn. **58** Suppl. (1989) 232.
11) P. Möller, J. R. Nix, P. Armbruster, S. Hofmann, G. Münzenberg, Z. Physik, **A359** (1997) 251.
12) M. Bolsterli, E. O. Fiset, J. R. Nix, and J. L. Norton, Phys. Rev. **C5** (1972) 1050.
13) P. Möller and J. R. Nix, Nucl. Phys. **A229** (1974) 269.
14) P. Möller and J. R. Nix, Nucl. Phys. **A361** (1981) 117.
15) P. Möller and J. R. Nix, Atomic Data Nucl. Data Tables **26** (1981) 165.
16) P. Möller, J. R. Nix, W. D. Myers, and W. J. Swiatecki, Atomic Data Nucl. Data Tables **59** (1995) 185.

The Berkeley Gas-Filled Separator

V. Ninov, K.E. Gregorich, C.A. McGrath

Nuclear Science Division at Lawrence Berkeley National Laboratory
1 Cyclotron Rd, Berkeley, California, 94720

Abstract. The BGS is being constructed at the 88-Inch Cyclotron at LBNL in Berkeley. The magnetic configuration of the BGS will allow a large angular acceptance and good suppression of primary beam particles. BGS operates as a mass spectrometer with a $A/\Delta A \approx 200$ and as a gas filled separator at pressures between 0.1-50 hPa. The reaction products recoiling off a thin target will be collected with efficiencies from 10-80 % at the focal plane. A Monte Carlo simulation program of the ion transport through the gas-filled magnets in combination of 3-dimensional TOSCA field maps has been developed and reproduces closely the experimental behavior of BGS.

Introduction

Magnetic separators [1,2,3,4] have been used for the separation of compound nucleus products making use of the momentum imparted to the compound nucleus in the nuclear reaction. However, in heavy ion reactions, the low-energy recoils show broad distributions of velocity and ionic charge. The resulting magnetic rigidities are spread over a wide range, leading to a broad spatial distribution of the recoils at the focal plane of a separator. An alternative method applied in the Berkeley Gas-filled Separator (BGS) uses a magnetic deflection system filled with gas at low pressure. In the gas the recoils undergo atomic collisions, capturing or losing electrons. They take on a well-defined average charge state, which is nearly proportional to the velocity, resulting in a high efficiency, small image size, and large energy/momentum/velocity acceptance.

The Spectrometer

The BGS spectrometer is installed in Cave 1 at the 88" cyclotron at LBNL. The magnetic configuration consists of three independent magnets. The first is a vertically focussing quadrupole with a 15 cm aperture radius and a 60 cm yoke length located 25 cm downstream of the target. The maximum gradient is 8.5 T/m.
The second magnet is a large dipole modified to produce a strong horizontally focussing gradient and a bending angle of 25°. The length of the hyperbolic yokes is 91 cm. A three dimensional finite element calculation with TOSCA is shown in Figure 1.

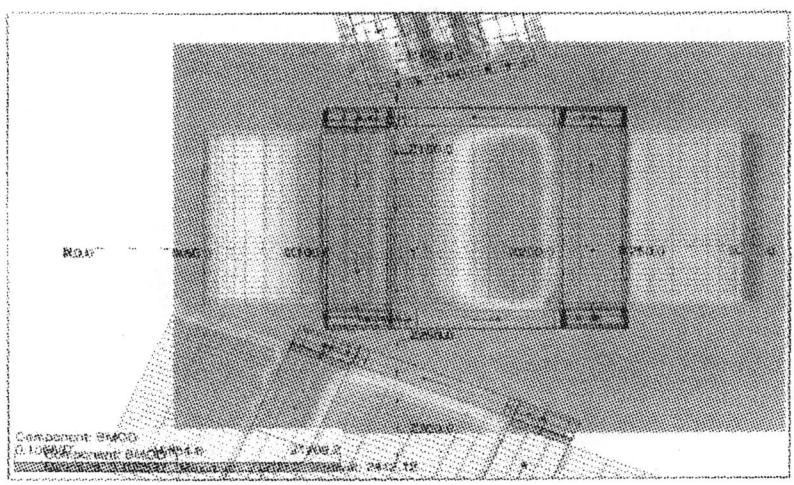

Figure 1: The gradient magnet was designed using TOSCA.

The third magnet is an H Magnet with a vertical gap of ± 6.5 cm and an yoke length of 91 cm. The bending angle of the last dipole is 45°. The total path length is approximately 4 meters. The maximum rigidity of the separator is 2.5 Tm and the totale bending angle 70°. The final 3D design of BGS is shown in Figure 2.

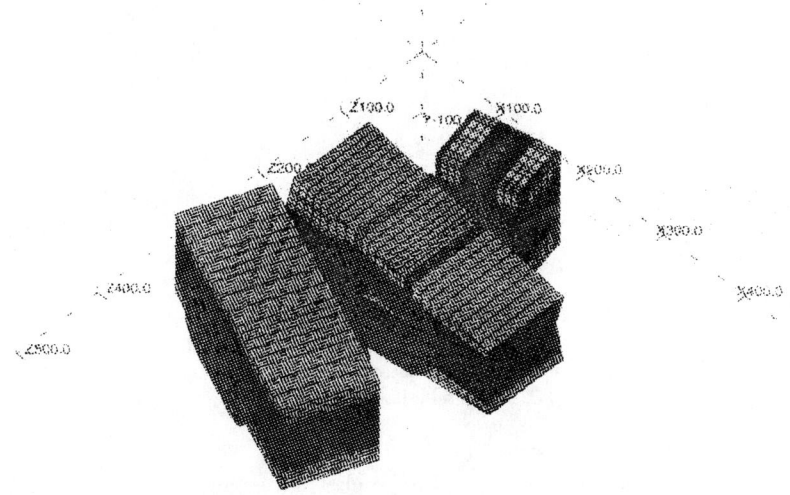

Figure 2: The final design of BGS. The recoils are entering the separator from the left. The focal plane is approximately 4 meters downstream of the target at the right.

With the present configuration of BGS a maximum angular acceptance of 45 msr can be achieved. The recoils arriving at the focal plane are dispersed by 2 cm / % of their Bρ.

Monte Carlo Simulation of Ions in BGS

A semi-microscopic model [5,6] for calculating the trajectories of heavy-ions in gas-filled magnetic systems has been developed. We consider charge-exchange processes [7,8], multiple small-angle scattering [9], and kinetic energy-loss of the ions passing through the gaseous media combined with three dimensional field maps calculated with TOSCA.

The trajectory of an ion is calculated by integration of the equation of motion determined by the Lorentz force exerted on the ions in the magnetic field. The interaction with the gas is incorporated in the calculations by introducing the mean free path λ between two collisions. The probability ω for an ion to survive a flight path s without any collisions is $\omega(s)=exp(-s/\lambda)$. The probability ω is sampled with random numbers and at each vertex an interaction with a neutral gas-atom is chosen. The energy and position spectra are accumulated collecting the coordinates $(x_i, y_i, \alpha_i, \beta_i,)$ of the ions reaching the focal plane. An initial result of the calculations is shown in figure 3. The solid line show the measured spatial distribution of α-particles at the focal plane. The accumulated spectrum is the result of a Monte Carlo simulation.

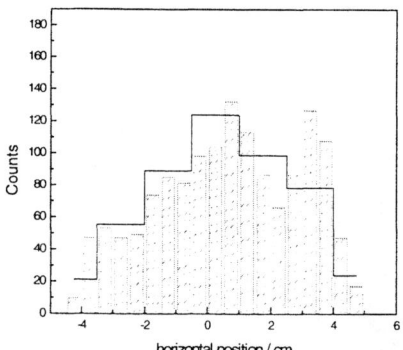
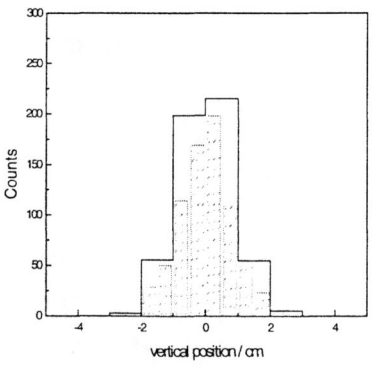

<u>Figure3:</u> Position spectra as calculated as described in the text for α-particles at the focal plane of BGS. The hatched areas denotes the Monte Carlo simulation. The solid line represents the experimentally obtained distribution. The size of the source is 6 mm in diameter.

The presented calculation was carried out under vacuum. The satellite peak in the horizontal distribution might be a result of image aberrations and saturation effects in the high field side of the gradient magnet.

Further calculations, shown in figure 4, have been carried out to determine the applicability of BGS as a mass spectrometer with large angular acceptance. By measuring the matrix element (x,α) and (y,β) a mass resolution of $A/\Delta A \approx 200$ can be achieved.

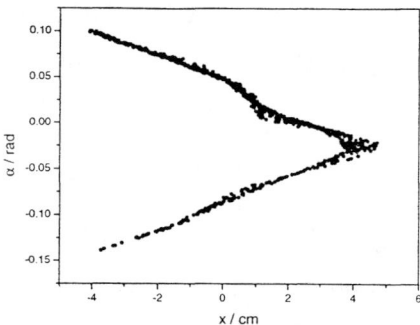 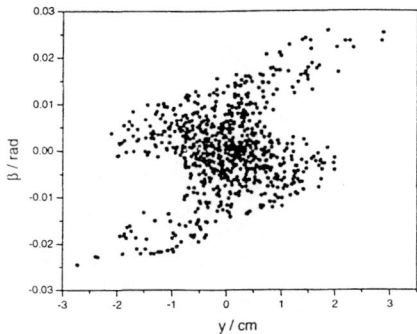

Figure 4: Plot of the matrix elements (x,α), (y,β) in the focal plane. Applying ray-tracing the mass resolution of BGS is $A/\Delta A \approx 200$.

The mass resolution in the gas filled mode is very poor and is determined mainly by the relative Bρ-width, which depends strongly on the initial phase space governed by the recoils after the nuclear reaction.

Conclusion

The BGS design is very versatile and allows a wide angular and momentum acceptance almost four times larger than existing separators. This makes the study of very exotic reactions far off stability possible. In addition BGS used as mass spectrometer can be combined with a wide variety of detectors.

References

[1] P.Armbruster et al. Nucl. Instr. and Meth. 91, (1971), 504
[2] G. Münzenberg et al., Nucl. Instr. & Meth., 161, (1979), 65
[3] A.V. Yeremin et al., Nucl. Instr. & Meth., A274, (1989), 528
[4] M. Leino et al. Nucl. Instr. and Meth. B 99, (1995), 653
[5] M. Paul et al, Nucl. Instr. and Meth. A277, (1989), 418
[6] V. Ninov et al. Nucl. Instr. and Meth. A 351, (1995), 486
[7] H.D. Betz, Rev. Mod. Phys., 44, (1972), 465
[8] H. Knudsen, Phys. Rev. A23, (1981), 600
[9] B. Efken et al. Nucl. Instr. and Meth. 129, (1975), 227

Fragment Angular Momentum and Descent Dynamics in ^{252}Cf Spontaneous Fission

G.S. Popeko[1], G.M. Ter-Akopian[1,2,3], J.H. Hamilton[2], J. Kormicki[2,a],
A.V. Daniel[1,2,3], Yu.Ts. Oganessian[1], A.V. Ramayya[2], J.K. Hwang[2],
A. Sandulescu[2-5], A. Florescu[2,3,5], W. Greiner[2,3,4], J. Kliman[1,6],
M. Morhac[6], J.O. Rasmussen[7], M.A. Stoyer[8], J.D. Cole[9], and GANDS95 Collaboration

[1]*JINR, Dubna,141980, Russia,* [2]*Vanderbilt University, Nashville, TN 37235 USA,* [3]*JIHIR, Oak Ridge, TN 37831, USA,* [4]*ITP, J.W. Goethe University, D-60054, Frankfurt am Main, Germany,* [5]*IAP, Bucharest, P.O. Box MG-6, Rumania,* [6]*IP SASc, Bratislava, Slovak Republic,* [7]*LBNL, Berkeley, CA 94720, U.S.A.,* [8]*LLNL, Livermore CA 94550, U.S.A.,* [9]*INEL, Idaho Falls, ID 83415 USA*

Abstract. Fragment angular momenta as a function of neutron multiplicity were extracted for the first time for the Mo-Ba and Zr-Ce charge splits of ^{252}Cf by studying prompt coincident γ-rays. The obtained primary fragment angular momenta do not continuously rise with the increase in the number of neutrons evaporated. In frame of the scission point bending oscillation model such regularity is explained due the decrease of the bending temperature. Adiabatic bending oscillations (T=0) are obtained at large (v_{tot}>5) and small (v_{tot}=0) scission point elongation. These oscillations are excited to the temperature of 2-3 MeV for the most probable scission configurations indicating a weak coupling between collective and internal degrees of freedom. A strong coupling between the collective bending and dipole oscillations was found.

The study of the fragment angular momentum can provide insights into the fission dynamics. We obtained new results on the fragment angular momenta of ^{252}Cf from the analysis of coincident prompt γ-rays detected using the Gammasphere facility. We measured the intensities of the γ transitions in the yrast bands of individual secondary even-even fragments (A_1, Z_1) when these fragments were obtained in pairs with specific complementary fragments (A_2, Z_2) such that $A_2+A_1+v_{tot}=252$, $Z_1+Z_2=98$. Transition intensities in function of the total number of neutrons, v_{tot}, evaporated from both primary fragments were obtained. From these transition intensities we determined the level populations, and calculated average spin values for secondary fragments. The primary fragment angular momenta were obtained from the average spin values assuming that the evaporation of one neutron reduces the primary fragment angular momentum by 0.5ħ. The mean numbers of neutrons evaporated by the light and heavy fragments v_l and v_h were calculated from the results of our paper [1].

More than 70 angular momentum values were found for primary fragments for the Ba-Mo and Ce-Zr pairs. Some results are presented in Fig.1. Analyzing the whole set of obtained results we arrived at the general conclusion that the primary fragment

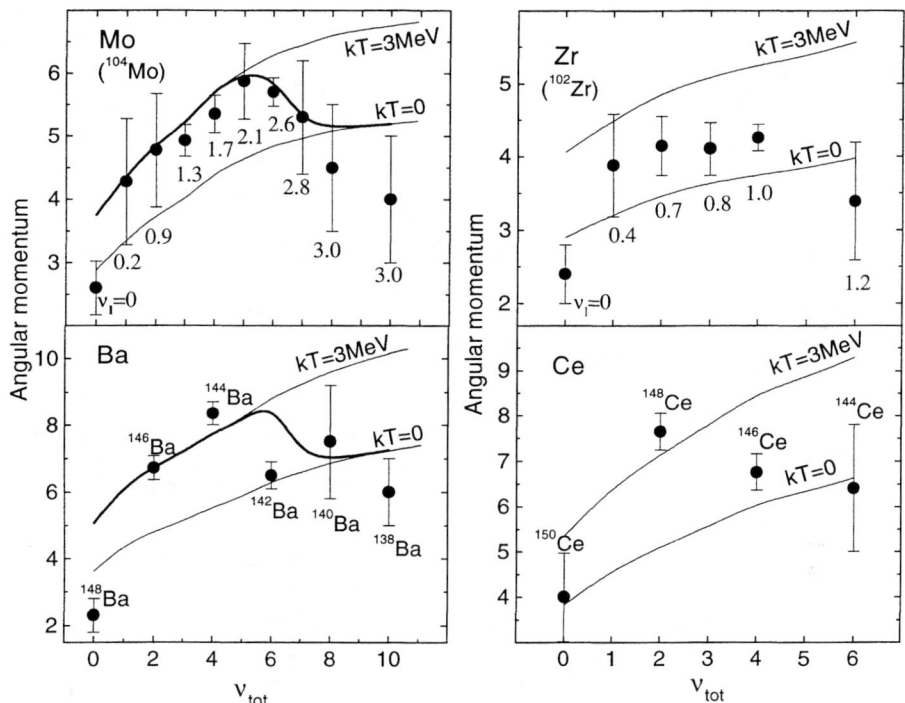

FIGURE 1 Circles show the average angular momenta of Mo and Zr (upper panels) and complementary Ba and Ce (lower panels) primary fragments in function of the total number of neutrons v_{tot} when the secondary pairs involve ^{104}Mo and ^{102}Zr, respectively, together with different Ba and Ce isotopes. Numbers given below the points of light fragments are the mean neutron numbers v_l evaporated from the primary light fragments. See other explanations in text.

angular momentum does not continuously increase with the increase in the number of neutrons evaporated. To understand the situation we carried out evaluations of the angular momentum values using the results of Ref. [2], where the scission point configuration was approximated as a system of two coupled deformed fragments performing the bending oscillation. Thin solid lines in Fig.1 show calculated results for two values of the bending temperatures kT=0 and kT=3 MeV. The comparison of experimental points with the calculations shows that the observed behavior of the angular momentum with an increase of the primary fragment deformation is explained as due to the decrease of the bending temperature.

In the Ba-Mo split, the existence of two different fission modes [1] could be the cause of this behavior of the fragment angular momentum. The second, peculiar fission mode presents an enormously large fragment deformation at the scission point. Evaluating the contributions of each of the two fission modes and assuming that the first, common fission mode has the bending temperature of kT=3 MeV, whereas the

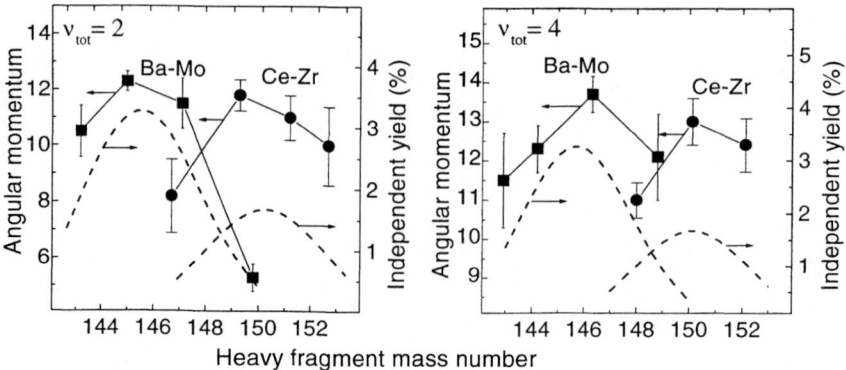

FIGURE 2 The sums of the angular momenta of the primary Ba-Mo (squares) and Ce-Zr (circles) fragments as a function of the heavy fragment mass number. The data are presented in two panels for the primary pairs that evaporated two and four neutrons. Dashed lines show the primary fragment yields.

second, peculiar mode is adiabatic we calculated the fragment angular momenta in a good agreement with the experimental values. In Fig. 1 thick solid lines show the results of these calculations. Thus, the second mode corresponds to the cold deformed fission.

For the single fission mode, as it is the case for the Ce-Zr split, one can suggest a more general picture. Following the idea of Ref. [3], we assume that, if the transverse oscillations are excited at the descent along the fission valley, so that their excitation energy approaches the height of the barrier dividing two valleys, the nucleus penetrates into the fusion valley. With the increase of the elongation, the excitation energy that is necessary for the barrier penetration decreases, and therefore decreases the fragment angular momentum.

We observed also an interesting correlation between the yields of the fragment pairs and fragment angular momenta. Fig. 2 shows the correlation between the sums of the angular momenta of the primary Ba-Mo and Ce-Zr pairs as a function of the mass number of the heavy fragment and the yields of the primary pairs obtained from [1]. The primary fragment pairs lying on the left and right wings of the Gaussian yield distributions are associated with the dipole oscillation that occurs at the descent to scission. The observed correlation gives a clear indication that there is a strong coupling between the two collective degrees of freedom, i.e. the dipole and bending oscillations.

[a] Also at UNISOR, ORISE. Oak Ridge, TN 37831; on leave from Institute of Nuclear Physics, Cracow, Poland.

REFERENCES
1. Ter-Akopian G.M. et al., Phys. Rev. **C55**, 1146 (1997).
2. Zielinska-Pfabe M. and. Dietrich K., Phys. Lett. **B49**, 123 (1974)
3. Berger J.F. et al., Nucl. Phys. **A502**, 85c (1989).

ANGULAR MOMENTUM EFFECTS IN MULTIMODAL FISSION OF ^{226}Th

G. G. Chubarian[1], B. J. Hurst[1], D. O'Kelly[1], R. P. Schmitt[1],
M.G. Itkis[2], N. A. Kondratiev[2], E. M. Kozulin[2], Yu. Ts. Oganessian[2],
V. V. Pashkevich[2], I.V. Pokrovsky[2], V. S. Salamatin[2], A.Ya. Rusanov[3],
L. Calabretta[4], C. Maiolino[4], K. Lukashin[4], C. Agodi[4], G. Bellia[4],
F. Hanappe[5], E. Liatard[6], A. Huck[7], L. Stuttgé[7]

[1]Cyclotron Institute, Texas A&M University, TX 77843, USA; [2]FLNR, JINR, Dubna, Russia 141980; [3]INP, Alma-Ata, Kazakhstan; [4]LNS-INFN, 57 Corso Italia, I-95100, Catania, Italy; [5]ULB, PNTPM CP229, av. F.D. Roosevelt, B1050, Brussels, Belgium; [6]ISN, 53 avenue des Martyrs, 38026, Grenoble, Cedex, France; [7]CRN, 23, rue du Loess, 67037 Strasbourg, Cedex 2, France

Abstract. The γ-rays from the multimodal fission of the ^{226}Th formed in ^{18}O+^{208}Pb was investigated at the near- and sub-barrier energies. The corresponding excitation energies at the saddle point, E^*_{sp}, ranged from 23 to 26 MeV. The average γ-ray multiplicities and relative γ-ray energies as a function of the mass of the fission fragments exhibits a complex structure and strong variations. Such strong variations have never been previously observed in heavy ion-induced fusion-fission reactions. Obtained results may be explained with the influence of shell effects on the properties of the fission fragments. Present work is the one in series of investigation of the multimodal fission phenomena in At-Th region.

INTRUDUCTION

It is well known that dramatic changes occur in the mass and energy distributions of certain spontaneously fissioning Fermium isotopes and some other transuranic elements [1,2]. For the isotopes ^{258,259}Fm and ^{260}Md, fission produces narrow symmetric mass distributions and fragments with relatively high kinetic energies. Only small changes in the nucleonic composition radically alter this picture.

Dramatic changes in the symmetric/asymmetric character of the fission probabilities can be accounted for by theories, which incorporate a micro-macroscopic approach including the effects of nuclear shell structure on the formation of the fragments. Detailed calculations of the nuclear potential energy surface indicate the possibility of a co-existence of various (at least two) distinct fission modes [3]. Multimodal fission phenomena have also been observed in the Pb-At region in proton and alpha induced reactions [4], as earlier predicted [5].

Unfortunately, with light charged particles it is not possible to extend this work into the largely unstudied At-Th region due to the absence of suitable targets. In this region dramatic changes are expected to occur in the fission properties as a function of the nucleonic composition and the excitation energy. The transition form symmetric to asymmetric fission along with the accompanying changes in the barrier height and the saddle to scission times can be expected to strongly influence the fission mass and energy distributions, the fission cross sections, the fragment γ-ray multiplicities as well as the pre- and post-fission neutron multiplicities.

A new strategy for probing the At-Th region was developed several years ago utilizing heavy ion beams in near- and sub-barrier fusion-fission reactions [6]. Similar experiments have been performed at GSI to investigate Coulomb fission reactions of secondary radioactive beams [7].

The current paper focuses on the results of studies of the γ-ray multiplicities from the fission fragments of the neutron deficient isotope ^{226}Th formed in reactions 78 and 75 MeV ^{18}O+^{208}Pb.

RESULTS AND ANALYSIS

The main problem in studying low energy fission (20<E^*<30 MeV) in the above mentioned transition region using heavy ion beams is the very small fusion-fission cross section near the Coulomb barrier. By choosing the appropriate target-projectile combination, one can achieve the lowest possible excitation energy.

In our experiments fission fragments were detected using a high efficiency time-of-flight spectrometer consisting of an array of position sensitive PPACS's. The γ-ray multiplicities were measured with six 63x63 mm NaI detectors. Detailed description of experimental set-up and procedures are presented in [6].

The characteristics of the reactions studied are summarized in Tab. 1. This table lists the beam energies, E_i, the excitation energies, E^*, the fission barriers, E_f, the saddle point excitation energies, E^*_{sp}, the average γ-ray multiplicities <M_γ>, and the standard deviation of the γ-ray multiplicities <$\sigma_{M\gamma}$>.

Figure 1(a) shows relative average γ-ray energies, <E_γ> (top, solid circles), <M_γ>, (middle, open circles) as a function of the mass of fragments for the reaction ^{208}Pb(^{18}O,f) at E_i=78 MeV. The mass distribution of the fission fragments is presented in the bottom of the figure. The variation of the <M_γ> with fragment mass is dramatic. Near symmetry <M_γ> is about 12. As one deviates from symmetry, <M_γ> stays approximately constant over a small mass interval and then plunges almost

TABLE 1. Parameters For the Fission of ^{226}Th.

Reaction	E_i, MeV	σ_f mb	E^* MeV	E_f MeV	E^*_{sp} MeV	< M_γ >	< $\sigma_{M\gamma}$ >
^{18}O+^{208}Pb	78	8±0.4	26.1	6.9	19.2	9.81±0.03	4.78±0.07
	75	0.042±0.007	23.3		16.4	9.56±0.60	5.26±1.35

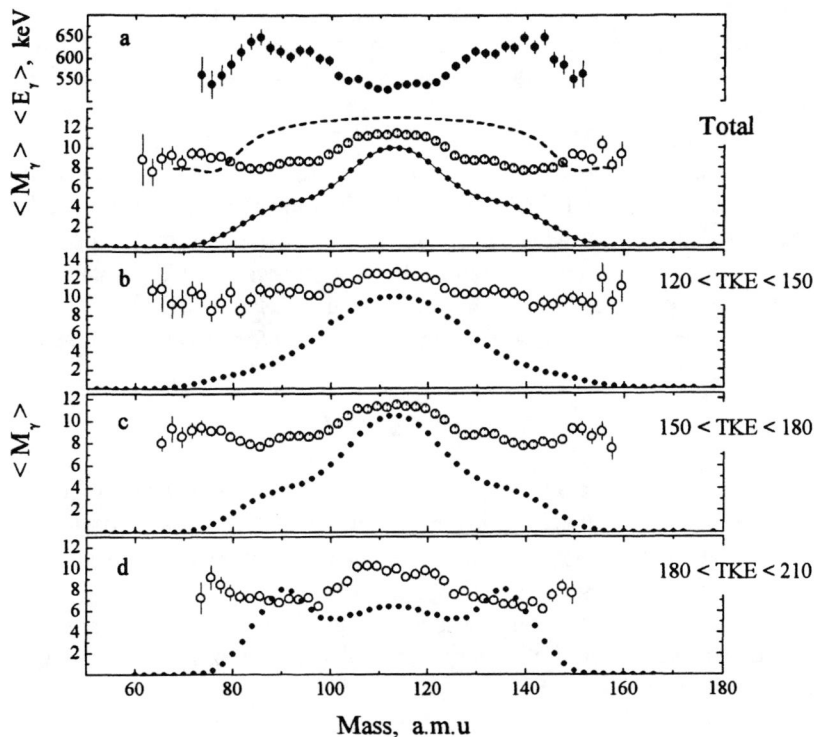

FIGURE 1. $\langle M_\gamma \rangle$ and $\langle E_\gamma \rangle$ as a function of the fission fragment mass. See details in text.

precipitously. Around $A \approx 94$ and 132, small shoulders are observed and then $\langle M_\gamma \rangle$ falls to a minimum value of about 8 around $A \approx 86$ and 142. At larger mass asymmetries, the multiplicity rises and then appears to fall.

Such a structure and strong variation in $\langle M_\gamma \rangle$ have never been previously observed in heavy-ion induced fission. The dependence of $\langle E_\gamma \rangle$ on mass also exhibits a strong variation. Near symmetry it has wide minimum and increases sharply while approaching the range of asymmetry in mass distribution. Around masses mentioned above, $\langle E_\gamma \rangle$ has well pronounced local maximums. For the largest masses, approximately in the range where $\langle M_\gamma \rangle$ rises again, $\langle E_\gamma \rangle$ appears to fall. The total variation of $\langle E_\gamma \rangle$ is about 30%. It seems likely that the marked structure in $\langle M_\gamma \rangle$ and $\langle E_\gamma \rangle$ is due to strong shell effects in the cold fission fragments, where predominantly rare, but high-energy γ-ray transitions are possible.

It is interesting to compare the present results with liquid drop calculations [8]. The dashed curve in fig.1(a) shows the results of such calculations assuming total fragment spin $\ell = 15\hbar$ and $S_T = 2(\langle M_\gamma \rangle - 6)$. The calculations predict a maximum at symmetry, however, the calculated values are larger than the experimental values. This is not particularly surprising since the assumed orbital angular momentum is a rough estimate. Furthermore, no corrections have been made for the spin removed by evaporated neutrons. The calculations do not include the influence of shell effects, which should have a significant impact on the fragments shape and the relevant temperatures.

As it were shown in [1,2,4,6], TKE may be a convenient tool for separation of distinct fission modes. Figure 1(b-d) shows $<M_\gamma>$ as a function of the fragment mass for three selected ranges of TKE. Mass distributions for the same TKEs' are also presented. For low TKEs', where the input of the asymmetric masses is small, $<M_\gamma>$ reaches its highest values, in the mean time the variation of $<M_\gamma>$ is the smallest. On the contrary, at the highest TKE where the yield of asymmetric fission fragments is dominating, variation of the $<M_\gamma>$ is the largest, but has the lowest absolute value.

SUMMARY

The properties of γ-rays from the fission fragments of the neutron deficient isotope of ^{226}Th have been investigated using near- and sub-barrier fusion-fission reactions. The data gathered in this work provide a detailed view of the evolution of the $<M_\gamma>$ and $<E_\gamma>$ with the mass of the fission fragment.

Complex structure and strong variation of the $<M_\gamma>$ and $<E_\gamma>$ as a function of mass may be explained with the influence of shell structure of the fission fragments on the yield and energy of the γ-rays.

For accurate fit of $<M_\gamma>$ dependencies one needs shell corrected theoretical calculations of the shapes and temperatures of the fission fragments at the scission point.

Results of the present work are in a good agreement and are supplementing the conclusions made in [6].

This was supported by the Russian Fundamental Research Foundation, under grant 96-02-17743A, and the US Department of Energy, under grant DE-FG03-93ER40773, and the US National Science Foundation, under grant PHY96-03143.

REFERENCES

1. Hullet E.K. et al., *Phys. Rev.* **C 40**, 779 (1989).
2. Hoffman D., Proc. of the 4th Int. Symposium on the Phys. and Chem. Of Fission, Julich, Germany, 1979, **1**, 275 (IAEA, Vienna, 1980).
3. Möller P., Nix R.J. and Swiatecki W.J, *Nucl. Phys.* **A 492**, 349 (1989);
   Pashkevich V.V., *Nucl. Phys.* **A 477**, 1 (1988).
4. Itkis M.G., Okolovich V.N., Smirenkin G.N., *Nucl. Phys.* **A 502**, 243 (1989) 243.
5. Pshkevich V.V., *Nucl. Phys.* **A 169**, 275 (1971).
6. Itkis M.G., Proc. Of the Tours Symposium on Nuclear Physics III, Tours, France, 1997, 189 (AIP, Woodbury, New York, 1997); Itkis M.G. et al., *Nucl. Phys. A,* in press.
7. Shmidt K.-H. et al., *Phys. Lett.* **B325**, 313 (1994).
8. Schmitt R.P. et.al.,*Z. Phys.* **A 321**, 411 (1985).

Early and Current Stability Predictions for Nuclei Near Z=110 and N=162

P. Möller [1,*]

Theoretical Division, Los Alamos National Laboratory, Los Alamos, NM 87545, USA

Abstract. We discuss why early super-heavy element calculations made no mention of the now observed island of deformed superheavy elements in the vicinity of proton number Z=110 and neutron number N=162.

HEAVY-ELEMENT STABILITY PREDICTIONS IN 1969 AND NOW

During session D1 Friday morning of this conference there was some reference, both in the presentations by S. Hofmann and by A. Soiczewski and in the discussions following the presentations, to what was obtained in early calculations on the stability of superheavy elements in the neutron-deficient *deformed* region near Z=110 and N=162. Results in the well-known 1969 paper (1) by Nilsson et al. were particularly discussed.

An unfortunate, common misconception, which also surfaced during this session, is that early studies of superheavy elements missed the deformed superheavy island because hexadecapole (ϵ_4) deformation were not taken into account at all, or not to correct order, or that the hexadecapole deformation space was only incompletely searched for a minimum. This is not the case. The Nilsson 1969 paper fully explored the hexadecapole shape-degree of freedom both at ground-state and fission saddle-pint distortions as is clear from figures 1, 12b, 13, 17a, and 17b, for example.

There are mainly tow reasons why the deformed superheavy regions of relatively stable nuclei was not identified in early calculations. First, nuclei in this region were not studied, since they are deformed nuclei close to the proton dripline and only spherical nuclei near β stability were thought to be of interest. Second, the models were not sufficiently developed to predict correctly this region of increased stability. These aspects of the early calculations are well illustrated in figure 16 of the Nilsson paper, a figure discussed briefly by S. Hofmann in his talk. The figure not included in the study. For the two chains, Sg and Hs, where N=162 is included the behavior in the vicinity of N=162 is incorrect. For Sg the shell correction is slightly higher fo r268Sg than for 260Sg, whereas our current mass model FRDM (1992) (2) predicts that the latter shell correction is about 1 MeV lower than the former. In the figure 270Hs is

---

[1] Permanent address: P. Moller Scientific Computing and Graphics, Inc. P.O. Box 1440, Los Alamos, New Mexico 87544, USA

shown to have a higher shell correction than 268Hs, whereas the former is 0.59 MeV deeper than the latter in our current model.

The realistic shell corrections obtained in our current FRDM (1992) are the results of a systematic development of the theory over more than 20 years, for example: a switch from the Nilsson modified-oscillator single-particle potential to the folded-Yukawa single-particle potential, an improved choice for the spin-orbit and diffuseness parameter of the potential, a switch from the liquid-drop macroscopic model to the finite-range droplet mode, the inclusion of ε_3 and ε_6 in the minimization of the ground-state energy, an improved paring model and a global adjustment of the model constants to ground-state masses and to fission-barrier heights. A more extensive discussion of model improvements is found in Ref. (3).

REFERENCES

1. S.G. Nilsson, C.F. Tsang, A. Sobiczewski, Z. Szymanski, S. Wycech, C. Gustafson, I.-L. Lamm, P. Möller, and B. Nilsson, Nucl. Phys. **A131** (1969) 1.
2. P. Möller, J.R. Nix, W.D. Meyers, and W.J. Swaitecki, Atomic Data Nucl. Data Tables **59** (1995) 185.
3. P. Möller, J.R. Nix, K.-L. Kratz, A. Wöhr, and F.-K. Thielemann, Proc. First Symp. on Nuclear Physics in the Universe, Oak Ridge, 1992 (IOP Publishing, Bristol, 1993) p. 433.

BETA DECAY AND FUNDAMENTAL MEASUREMENTS

Limits on Physics Beyond the Standard Model using Exotic Beams

A. García

University of Notre Dame
Department of Physics
Notre Dame, IN 46556

Abstract. The standard model of electroweak interactions contains purely vector and axial-vector currents. However, extensions of the standard model, such as super-symmetric theories and leptoquarks, predict in addition non-vector currents. Here we review the limits on possible tensor and scalar contributions to the weak interaction obtained from nuclear β decay. Future possibilities are discussed.

INTRODUCTION

The standard model prescribes purely vector and axial-vector currents for nuclear β decay mediated by the exchange of W bosons. However, extensions of the standard model, such as super-symmetric theories with more than one charged Higgs doublet, or leptoquarks, naturally predict scalar or tensor currents [1]. The most general effective Hamiltonian one can write, respecting Lorentz invariance is [2]:

$$H = (\bar{\psi}_p \gamma_\mu \psi_n)(C_V \bar{\psi}_e \gamma_\mu \psi_\nu + C'_V \bar{\psi}_e \gamma_\mu \gamma_5 \psi_\nu) +$$
$$(\bar{\psi}_p \gamma_\mu \gamma_5 \psi_n)(C_A \bar{\psi}_e \gamma_\mu \gamma_5 \psi_\nu + C'_A \bar{\psi}_e \gamma_\mu \psi_\nu) +$$
$$\frac{1}{2}(\bar{\psi}_p \sigma_{\lambda\mu} \psi_n)(C_T \bar{\psi}_e \sigma_{\lambda\mu} \psi_\nu + C'_T \bar{\psi}_e \sigma_{\lambda\mu} \gamma_5 \psi_\nu) +$$
$$(\bar{\psi}_p \psi_n)(C_S \bar{\psi}_e \psi_\nu + C'_S \bar{\psi}_e \gamma_5 \psi_\nu) + \text{Hermitian conj.} \qquad (1)$$

where a term proportional to $(\bar{\psi}_p \gamma_5 \psi_n)$ has been neglected because nucleons move slow. The standard model prescribes $C_V = C'_V$, $C_A = C'_A$, and $C_S = C'_S = C_T = C'_T = 0$. The two last terms are called scalar and tensor currents, respectively. For simplicity, we assume $C_V = C'_V$, $C_A = C'_A$, and present limits on the other couplings. From this Hamiltonian Jackson et al. [2] computed the consequences for the nuclear-β-decay rate, which, in the absence of orientation of the initial state and without detection of the lepton's helicity reads:

$$dW = dW_0(1 + a\frac{\mathbf{p}_e \cdot \mathbf{p}_\nu}{E_e E_\nu} + b\frac{m_e}{E_e}) \qquad (2)$$

where

$$a\xi = |M_F|^2(|C_V|^2 + |C_V'|^2 - |C_S|^2 - |C_S'|^2)$$
$$-\frac{1}{3}|M_{GT}|^2(|C_A|^2 + |C_A'|^2 - |C_T|^2 - |C_T'|^2) \qquad (3)$$
$$b\xi = \pm 2\gamma \text{Re}[|M_F|^2(C_V^* C_S + C_V'^* C_S')$$
$$+|M_{GT}|^2(C_A^* C_T + C_A'^* C_T')] \qquad (4)$$

and

$$\xi = |M_F|^2(|C_V|^2 + |C_V'|^2 + |C_S|^2 + |C_S'|^2)$$
$$+|M_{GT}|^2(|C_A|^2 + |C_A'|^2 + |C_T|^2 + |C_T'|^2) \qquad (5)$$

To make expressions simpler we have neglected the recoil-order corrections, the small Coulomb effects, and the possibility of orientation of the daughter in GT decays. However, these effects are taken into account in our plots below.

It is clear that precise measurements of the e, ν correlation coefficient, a, and of the Fierz-interference term, b, can potentially render information about physics beyond the standard model. The expressions above are complicated and depend on nuclear physics information through M_F and M_{GT}. **However, if one looks at either pure Fermi or pure GT transitions the expressions are much simpler and the M_F and M_{GT} factors cancel out.**

The fact that the correlation coefficient renders information on the nature of the couplings can be understood more easily with an example. In a $0^+ \to 0^+$ β^- decay the two leptons have to couple to zero angular momentum. Because the e^- is a particle and the $\bar{\nu}$ an anti-particle, the standard model prescribes they should tend to be emitted parallel to each other. However, if scalar currents dominated the β decay, then the leptons would be emitted anti-parallel to each other, because one of the leptons would have the 'wrong' helicity. This can be seen by writing the vector and scalar Hamiltonians of Eq. 1 in the following form:

$$H_V = (\bar{\psi}_p \gamma_\mu \psi_n)(C_V \bar{\psi}_e^L \gamma_\mu \psi_\nu^L)$$
$$H_S = (\bar{\psi}_p \psi_n)(\bar{\psi}_e^L (C_S - C_S')\psi_\nu^R + \bar{\psi}_e^R (C_S + C_S')\psi_\nu^L) \qquad (6)$$

where $\psi_R^L = (1 \pm \gamma_5)/2\psi$, $\bar{\psi}_R^L = (1 \mp \gamma_5)/2\bar{\psi}$. It is clear that vector currents conserve the chirality while scalar currents flip it. The same is true for Axial-vector currents (which do not flip chirality) and tensor currents (which do flip it).

The Fierz-interference term can also be understood looking at Eq. 6 and noticing that, by the time one makes projections onto helicity states, because the mass of the electron is finite, a fraction of the second term in H_S will look like the one in H_V and they interfere. So one expects a term proportional to $\frac{m_e}{E_e}(C_S + C_S')$ just as Eq. 4 dictates.

There is a long history of measurements that were performed to establish the $V - A$ character of the weak interaction. Today our aim is to look for small S or T components that could hint physics beyond the standard model. Boothroyd,

Markey, and Vogel [3] have put constraints on scalar and tensor currents using a compilation of all nuclear-β-decay experiments. One problem with this kind of compilation is that results from different experiments are averaged together. But some of them measure the same quantity in similar ways, so they may have common systematic errors. In addition, part of the information Boothroyd et al. used came from mixed transitions which, as explained above, do not address a single issue at a time. We therefore believe that those constraints should be taken with a grain of salt.

LIMITS ON TENSOR CURRENTS

The present limits on tensor currents come principally from two experiments. Wenninger, Stiewe, and Leutz [4] measured the shape of the β^+ spectrum from ^{22}Na decay, and obtained $b_{GT} = (0.8 \pm 2.8) \times 10^{-3}$. On the other hand, Johnson, Pleasonton, and Carlson [5] deduced the $e^-, \bar{\nu}$ correlation coefficient from ^{6}He β^- decay by measuring the distribution of momenta of the recoiling ^{6}Li ions. Their result, $\tilde{a} = -0.3343 \pm 0.0030$, [6] has been recently updated to $\tilde{a} = -0.331 \pm 0.003$ by Glück and Tóth [7] who pointed out that radiative corrections have to be properly taken into account. The limits from these two experiments (assuming all couplings are real) are shown in Fig. 1 and compared to the ones from Boothroyd et al. There is additional independent information from measurements of the helicity of

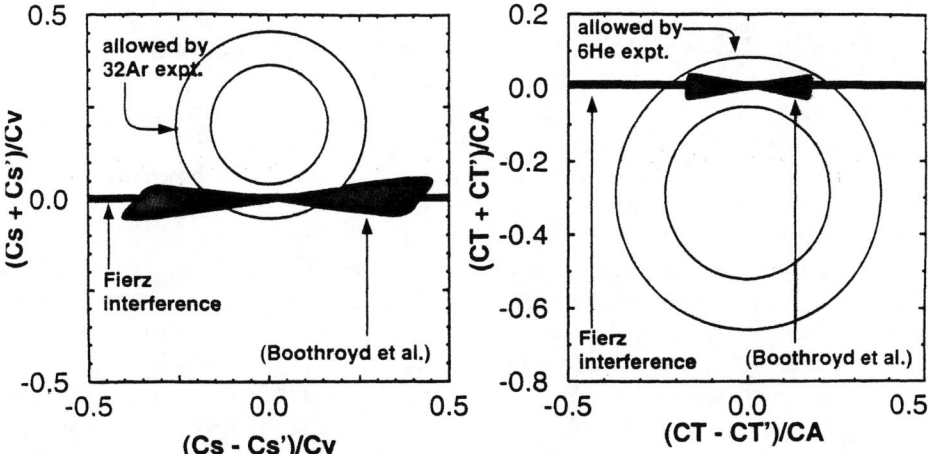

FIGURE 1. Limits on scalar and tensor couplings assuming them to be real. Limits indicate $\Delta \chi^2 \leq 4$. The open circles are the constraints from the e, ν correlation experiments. The horizontal bars come from Fierz-interference measurements.

β's in pure GT transitions, but these kind of experiments are mainly sensitive to

TABLE 1. Results of regressions using $\mathcal{F}t$ values from $0^+ \to 0^+$ transitions to find limits on b_F.

Assumed Unitarity?	Function used	Results fit	χ^2/ν
No	$\mathcal{F}t = \mathcal{F}t_0$	$\mathcal{F}t_0 = 3072.1 \pm 0.9$	1.3
No	$\mathcal{F}t = \mathcal{F}t_0(1 + b\langle \frac{m_e}{E_e}\rangle)$	$\mathcal{F}t_0 = 3075.4 \pm 2.5$ $b_F = (-4.1 \pm 2.9) \times 10^{-3}$	1.2
Yes	$\mathcal{F}t = \mathcal{F}t_0(1 + aZ)$	$\mathcal{F}t_0 = 3063.7$ $a = (1.6 \pm 0.2) \times 10^{-3}$	1.1
Yes	$\mathcal{F}t = \mathcal{F}t_0(1 + b\langle \frac{m_e}{E_e}\rangle)$	$\mathcal{F}t_0 = 3063.7$ $b_F = (7.5 \pm 1.1) \times 10^{-3}$	4.9
Yes	$\mathcal{F}t = \mathcal{F}t_0(1 + aZ + b\langle \frac{m_e}{E_e}\rangle)$	$\mathcal{F}t_0 = 3063.7$ $a = (1.3 \pm 0.3) \times 10^{-4}$ $b_F = (2.2 \pm 1.7) \times 10^{-3}$	0.8

$1-(C_T+C_T')^2$ and their error bars (not better than 2%) make them not competitive with the Fierz-interference limits.

LIMITS ON SCALAR CURRENTS

Limits on the Fierz-interference coefficient, b_F, can be obtained without the need to measure the shape of the β spectrum –usually prone to systematic errors– noticing that Eq. 2 above implies that $0^+ \to 0^+$ $\mathcal{F}t$ values should have a linear dependence on $\langle \frac{m_e}{E_e}\rangle$ (average over the full β spectrum) if $b_F \neq 0$. We have used the most updated values published by Hagberg et al. [8], and added the value from Fujikawa et al. [9] to extract limits on b_F under different assumptions, which we show in Table 1. As shown by Hardy [10] the $\mathcal{F}t$ values from Fermi transitions, with the state-of-the-art radiative and isospin-mixing corrections, imply a breakdown of the unitarity of the Kobayashi-Maskawa matrix. It has been shown that one can force unitarity allowing the $\mathcal{F}t$ values to have a linear or quadratic dependence on Z, which could account for some defects in the corrections. Because there is a correlation between the endpoints and Z (the larger Z, the smaller the value of $\langle \frac{m_e}{E_e}\rangle$) one may wonder as to the possibility that scalar currents and not defects in the corrections, are responsible for the apparent breaking of unitarity. Except for the fourth row in Table 1, which presents a huge value of χ^2/ν, the other tests show $|b|$ consistent with zero within 2σ's. We adopt $|b_F| \leq 0.007$ from this observable.

Until 1993 there had been no measurements on a (e,ν) correlation from pure Fermi transitions. Schardt and Riisager [11] noted that the β-delayed proton group emitted after the superallowed decay of ^{32}Ar showed clear evidence for a 'Doppler' broadening that originated in the (e^+,ν) correlation, i.e. they detected the momentum of the recoiling ^{32}Cl via the 'Doppler' broadening of the proton line. Adelberger [12] pointed out that one could use that information to tighten constraints on scalar currents. One problem with the experiment of Schardt and

Riisager is that e^+'s may distort the shape of the proton peak through summing. An experiment has been recently performed at Isolde [13] in which the summing with positrons was avoided by submerging the detection setup in a 3.5 Tesla field. The proton energy was measured using PIN diode detectors, cooled and stabilized in temperature, which yielded a record energy resolution of ≈ 4.5 keV (FWHM). Preliminary analysis yields $\tilde{a} = 0.998 \pm 0.005$ (statistical error only). However, this result is sensitive to the endpoint energy: $d\tilde{a}/dE \approx 1.2 \times 10^{-3}/\text{keV}$, and it turns out that the mass of ^{32}Ar is not well known. Fortunately, all 4 other members of the T=2 isospin quintuplet are known with precision and they follow the well-tested Isobaric Mass Multiplet Equation [14]. From here we deduce an endpoint with ≈ 3 keV uncertainty. It is unexpected to find a breaking of the IMME in the well-bound ^{32}Ar, but it would be nice to have a direct measurement of the mass of ^{32}Ar.

Fig. 1 shows the limits from the two sources discussed above, and compares them to the limits from Boothroyd *et al.*

DISCUSSION

We have pointed out the importance of using either pure GT or pure Fermi transitions to determine limits on scalar or tensor couplings.

The plots in Fig. 1 show that the limits from the absence of Fierz interference place very stringent constraints on the sum $(C+C')$ of the couplings (for both scalar and tensor currents). While for scalar couplings this comes from a variety of experiments and very careful determinations of possible systematic errors performed by several independent authors, for tensor couplings they come from measurements of the shape of the β spectrum in ^{22}Na, which is a very difficult measurement to perform without bias. It would be extremely nice to have a systematic search that could measure the shape of the β spectrum in a variety of pure Gamow-Teller transitions, all with the same apparatus and with sources prepared in similar ways. This enterprise could be feasible at an Isol-like facility.

The e,ν-correlation coefficient measurements complement the Fierz-interference ones putting constraints on the difference $C-C'$ of the couplings. The measurement from ^{6}He yields a 1% error in $\tilde{a}$ for pure GT transitions. This could be improved in the near future because there are other pure GT transitions where one could determine $\tilde{a}$.

The recent experiment on ^{32}Ar determined $\tilde{a}$ for pure Fermi transitions to $\approx 0.5\%$ (statistical). This result is based on the Isobaric Mass Multiplet Equation. Although there are no reasons to believe the IMME breaks down for ^{32}Ar, it would be nice to directly measure the mass of ^{32}Ar.

ACKNOWLEDGMENTS

A good part of the work presented here was done in collaboration with Eric G. Adelberger. I thank the NSF and the Warren Foundation for support.

REFERENCES

1. P. Herzeg, in *Precision Tests of the Standard Electroweak Model*, World Scientific, P. Langacker ed., (1995) page 786.
2. J.D. Jackson, S.B. Treiman, and H.W. Wyld Jr., Nucl. Phys. **4**, 206 (1957).
3. A.I. Boothroyd, J. Markey, and P. Vogel, Phys. Rev. C **29**, 603 (1984).
4. H. Wenninger, J. Stiewe, and H. Leutz, Nucl. Phys. **A109**, 561 (1968).
5. C.H. Johnson, F. Pleasanton and T.A. Carlson, Phys. Rev. **132**, 1149 (1963).
6. There is a subtlety on the translation from the measured correlation coefficient to limits on the coupling constants. Looking at Eq. 2 one realizes that the momentum distribution of the daughter will depend on

$$\frac{a}{(1+b\frac{m_e}{E_e})}$$

which depends on the electron energy. We here take the average value of $\frac{m_e}{E_e}$ and assume the e, ν correlation experiments put constraints on:

$$\tilde{a} = \frac{a}{(1+b\langle\frac{m_e}{E_e}\rangle)}.$$

This has the effect of making the limits on coupling constants from the e,ν-correlation experiments look like open circles. The effect is discussed in H. Paul, Nucl. Phys. **A154**, 160 (1970).
7. F Glück and K. Tóth, Phys. Rev. D **46**, 2090 (1992).
8. E. Hagberg, J.C. Hardy, V.T. Koslowsky, G. Savard, and I.S. Towner, nucl-ex/9609002.
9. B.K. Fujikawa et al., submitted to Phys. Rev. Lett.
10. J.C. Hardy, these proceedings.
11. D. Schardt and K. Riisager, Z. Phys. A **345**, 265 (1993).
12. E.G. Adelberger, Phys. Rev. Lett. **70**, 2856 (1993).
13. E.G. Adelberger, C. Ortiz, A. García, H.E. Swanson, M. Beck, O. Tengblad, M.J.G. Borge, I. Martel-Bravo, the ISOLDE collaboration, and H. Bichsel. The details of the experiment will soon be published elsewhere.
14. M. S. Antony, J. Britz, J. B. Bueb and A. Pape, At. Data Nucl. Data Tables **33**, 447 (1985).

Beta Strength Distribution in Neutron - Deficient Nuclei

Z. Janas[a], J. Agramunt[b], A. Algora[b], L. Batist[c], B.A. Brown[d]
D. Cano-Ott[b], R. Collatz[e], A. Gadea[b], M. Gierlik[a], M. Górska[e,a],
H. Grawe[e], A. Gulielmetti[e], M. Hellström[e], Z. Hu[e], M. Karny[a],
R. Kirchner[e], F. Moroz[c], A. Piechaczek[f] A. Płochocki[a],
M. Rejmund[e,a], E. Roeckl[e], B. Rubio[b], K. Rykaczewski[g,a],
M. Shibata[e] J. Szerypo[a], J.L. Tain[b], V. Wittmann[c], A. Wöhr[f]

[a] Institute of Experimental Physics, University of Warsaw, PL-00681 Warsaw, Poland
[b] Instituto de Física Corpuscular, C.S.I.C.-Univ. Valencia, E-46100 Burjassot, Spain
[c] St. Petersburg Nuclear Physics Institute, 188-350 Gatchina, Russia
[d] NSCL, Department of Physics and Astronomy, MSU, East Lansing, MI 48824-1321, USA
[e] Gesellschaft für Schwerionenforschung mbH, D-64291, Darmstadt, Germany
[f] Instituut voor Kern- en Stralingsfysica, University of Leuven, B-3001 Leuven, Belgium
[g] Oak Ridge National Laboratory, Physics Division, PO Box 2008, Oak Ridge, TN 37831, USA

Abstract. The results of recent studies of the Gamow-Teller β-decays of nuclei in the ^{100}Sn region are presented. Measurements performed with the use of the total absorption γ-ray spectrometer and the Cluster Cube array of germanium detectors revealed qualitatively new information on the Gamow-Teller strength distribution in the decays of ^{97}Ag and $^{103-107}$In. The shape of the measured β-strength distribution and the resulting total B_{GT} values are compared with the results of shell-model calculations.

Introduction

Understanding and reliable description of the β-strength distribution is of crucial importance for the complete characteristics of the nuclear β-decay. The β-strength function determines the gross properties of the decaying nuclei such as half-life or probability of β-delayed particle emission. Predictions and investigations of these basic characteristics of β-unstable nuclei are of primary interest in the decay studies far from stability. Data resulting from these studies provide an important nuclear physics input for the understanding of the element synthesis in the universe. The knowledge of the weak interaction rates in stellar matter evolution, in particular rates of the Gamow-Teller (GT) transitions for iron region nuclei, is crucial for the calculations of the electron capture (EC) rates during the presupernova core collapse of the massive stars [1]. High-precision measurements of superallowed $0^+ \rightarrow 0^+$ Fermi β-transitions rates allow one to verify the conserved vector current (CVC) hypothesis, to test the unitarity of the Kobayashi-Maskawa matrix and finally to set limits on extensions to the Standard Model [2]. Very sensitive probes for studying fundamental symmetries of electroweak interaction and properties of neutrino provide studies of double β-decay (2β) [3]. Reliable calculation of nuclear 2β-decay matrix elements is prerequisite for deduction of the neutrino mass from 2β-decay experiments. Matrix elements of the GT transitions are needed for the determination of the neutrino capture cross-section for (solar) neutrino detectors [4,5].

In this contribution we restrict our discussion to one of the most intriguing problems related to the β-strength distribution studies which concerns the question of the origin of the quenching observed for the strength of GT transitions. As it is shown by the analysis of the GT β-decays, the experimentally determined strengths appear to be systematically smaller than the calculated GT β-transition rates. Similar regularity is observed for the GT strengths extracted from the forward-angle intermediate-energy charge-exchange reactions [6,7].

The GT quenching can be quantitatively described in terms of the hindrance factor defined as the ratio of calculated and experimentally determined GT strength. The best description of nuclear wave functions and the GT matrix elements between nuclear states is provided by large-basis shell model calculations. The most complete calculations, based on the diagonalization of the effective hamiltonian within the full major oscillator shell, are feasible only for relatively light nuclei. Such calculations were used to determine hindrance factors for nuclei with A$\leq$50. A comparison of the GT decay rates measured and calculated within the full p-shell for A$\leq$18 nuclei revealed the quenching of the observed B_{GT} values by a factor of 1.49(3) [8]. The systematic analysis of GT strength for A=17-39 nuclei yielded the hindrance factor of 1.68(4) with respect to the complete sd-shell calculations [9]. The recent analysis of the GT decays in the mass range A=41-50 indicated that the agreement between the experimental data and the full fp-shell calculations demands the introduction of the average hindrance factor $h = 1.81(4)$ [10].

Two physically different mechanisms are usually considered to explain the observed quenching of the GT strength. The first one is a renormalization of the axial-vector coupling constant g_A in nuclear matter originating from non-nucleonic effects mediated mainly by the admixture of the $\Delta(1232)$-isobar nucleon-hole configurations to the GT states. The second mechanism responsible for the reduction of the observed GT strength is the higher-order nuclear configuration mixing arising from tensor correlations between nucleons. A recent estimate has shown that two thirds of the amplitude of the GT quenching originates from higher-order nuclear configuration mixing and one third from Δ-isobar admixtures [11]. A decisive experimental verification of this conjecture is, however, still lacking. In particular the role of the Δ particles contribution continues to be under discussion.

The question of the "missing" GT strength can be adequately addressed in β-decay studies of nuclei far from stability where the GT strength distribution can be investigated and confronted with theoretical predictions over a broad range of excitation energies. In this respect the ^{100}Sn region is of particular interest: Since the N=Z=50 shell closure occurs far from stability, isotopes in this region, especially non even-even ones, have relatively large Q_β-values. From simple single-particle model considerations one may expect that a substantial part of the total GT-strength resides within the Q_{EC} window and may thus be detected in β-decay measurements conducted with the proper experimental technique. As far as theoretical calculations are concerned, nuclei close to ^{100}Sn can be treated as closed-shell systems with a few valence particles only, which facilitates the model description.

GT strength distribution from β-decay measurements

The strength of the GT β-transition to the state at excitation energy E in the daughter nucleus can be derived from the measurements of the β feeding of this state $I(E)$, the decay energy Q_{EC} and the β-decay half-life $T_{1/2}$ according to the relation:

$$B_{GT}(E) = \frac{D \cdot I(E)}{(g_A/g_V)^2 \cdot f(Q_{EC} - E) \cdot T_{1/2}} \tag{1}$$

where D=6147(7) s is a constant, g_A/g_V =-1.262(4) is the ratio of axial-vector and vector coupling constants for the free-neutron decay and f is the statistical rate function.

Most frequently, β-feeding distribution has been derived from the detailed decay scheme established in conventional measurements employing standard-size Ge detectors. Such studies can often reveal a wealth of nuclear structure data on all individual levels in simple decay schemes and disclose information about the structure of low-lying states in complex decays. For complex decay schemes with high decay energy, however, such traditional measurements are generally unable, due to the low efficiency of detectors, to record all of the many weak γ transitions and hence to place them in the decay scheme. This is particularly true for energetic γ-rays depopulating states at high excitation energy in the decay product. As a consequence, the classical high-resolution γ-ray spectroscopy studies based on routinely applied Ge detectors usually overestimate the β-feeding intensity to low-lying states. Due to the very strong dependence of the β-decay rate function on the transition energy ($f \sim (Q_{EC} - E)^5$), the distortion of the β-feeding distribution has severe impact on the resulting B_{GT} distribution and causes underestimation of the apparent total GT strength.

A way to overcome the limitations of the standard discrete, high-resolution, low-efficiency γ-ray spectroscopy is a direct measurement of the distribution of β-decay feeding intensity. The ideal tool for this kind of measurements would be a γ-energy calorimeter with 100% full-energy peak efficiency for all γ-ray energies. In such a detector all members of each γ cascade depopulating an excited state would be summed to yield an output signal corresponding to the excitation energy of this state and provide an unambiguous signature for each β-feeding event to the given level.

Today the closest approach to such an ideal spectrometer represents an array of large scintillation detectors in a 4π geometry. The total absorption spectrometer (TAS) installed at the on-line mass separator at GSI consists of a large NaI crystal (ø 14"×14") for the detection of γ-rays [12]. A cylindrical well along the crystal's symmetry axis accommodates an assembly of auxiliary detectors. In the standard set-up it contains two Si counters for β particle detection, a high-resolution Ge X-ray detector and a "plug" NaI detector which restores the 4π geometry of the main crystal. The tape transport system is used to position mass separated radioactive sources in the center of the main crystal, between the two Si counters. By demanding coincidence with signals from the Si detectors, the positron component of the β^+/EC decay can be selected, whereas coincidences with characteristic X-rays recorded by the Ge detector can be used to select the EC events. The total γ-ray efficiency of TAS for monoenergetic photons in the energy range of 0.2 - 4.0 MeV exceeds 88%, and its full-energy peak efficiency is above 56%. The high efficiency values and their weak dependence on photon energy assure the operation of the detector as a satisfactory total absorption spectrometer. However, due to the apparent γ-efficiency loss effects, the determination of the β-feeding distribution from the experimental TAS spectra requires thorough knowledge of the detector response function and application of sophisticated deconvolution procedures. These problems have been elaborated by M. Karny et al. [13,14] and D. Cano-Ott et al. [15] who pointed to the importance of the high-resolution studies as a source of necessary input data for extracting the β-feeding distribution from the measured TAS spectra.

The genuine requirement for high-quality discrete spectroscopy data triggered the use of state-of-the-art Ge detectors for β-decay studies. In the experiments performed at the GSI on-line mass separator an array of 6 Euroball Cluster Ge detectors arranged to form a cube (Cluster Cube) has been used to complement the TAS measurements. The 42 Ge crystals of this array covered about 65% of the full solid angle with respect to a source positioned at the center of Cluster Cube, and allowed registration of 1.33 MeV γ-rays with a full-energy peak efficiency of about 19% and typical resolution of about 2.8 keV. The sensitivity of the Cluster Cube exceeds by far the sensitivity of any Ge detector set-up used in previous β-decay measurements and makes it indeed an excellent tool for spectroscopy studies of complicated decays. The usefulness of the Cluster Cube detector has been demonstrated in the measurements of ^{97}Ag and ^{150}Ho(2$^-$) decays performed at GSI on-line mass separator. More than 600 γ transitions (580 new) depopulating 150 levels were placed in the decay scheme of ^{97}Ag [16] An even more complex decay scheme has been established for ^{150}Ho(2$^-$) [17]. In both cases, the performance of the Cluster Cube pushed the measurements to the limit of the discrete γ-ray spectroscopy method, i.e. up to the excitation energies where deexcitation by statistical γ cascades deexcitation dominates. The comparison of β-feeding distribution for ^{150}Ho(2$^-$) determined from the Cluster Cube data and from the TAS measurement clearly illustrates the limitations of the discrete γ-ray spectroscopy technique for the investigation of β-strength function in complex decays. All in all, the high-resolution, discrete spectroscopy should not be considered as an alternative to the total absorption γ-ray studies but rather as a complementary method which, however, is indispensible for the reliable analysis of the TAS spectra and also needed to reveal the fine-structure of the β-strength distribution.

GT strength distribution for nuclei in the ^{100}Sn region

The GT decays of nuclei in the vicinity of ^{100}Sn proceeds via transformation of a $g_{9/2}$ proton into a $g_{7/2}$ neutron. On the basis of an extreme single-particle model one expects that this transformation will be dominated by the decay of the even-even core with the unpaired particles acting as spectators. In the decays of odd-odd

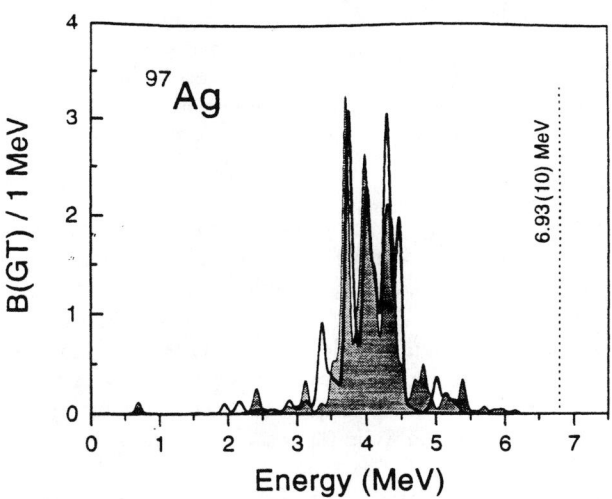

FIGURE 1. B_{GT} distribution for the decay of ^{97}Ag deduced from the Cluster Cube measurement (solid line) and obtained from the shell model calculations (shaded area), respectively. Both distributions were smoothed by folding with a Gaussian distribution of 60 keV FWHM. The theoretical data were normalized to the total B_{GT} value derived from the Cluster Cube measurement. The vertical line indicates the Q_{EC} value.

isotopes, this corresponds to the dominant population of four-quasiparticle states at excitation energies of 5-6 MeV in the daughter nucleus, whereas in the decays of odd-even nuclei one should observe substantial feeding of three-quasiparticle configurations at an excitation energy of 3-4 MeV. Due to the residual interactions, these configurations are expected to be spread over many levels of the daughter nucleus. Fig. 1 shows the preliminary B_{GT} distribution for the decay of ^{97}Ag resulting from the Cluster Cube measurement [16]. The global shape of this distribution is dominated by a resonance structure extending between 3 and 4.5 MeV, in general agreement with the results of the preliminary TAS data analysis [16]. The decay characteristics observed can be interpreted as the GT decay of the even-even core to three-quasiparticle configurations. As shown in Fig. 2, similar structure of the B_{GT} distribution has been obtained from the TAS measurements for $^{103-107}$In. As expected from the simple single-quasiparticle model, the experimental B_{GT} distribution is concentrated in resonances appearing at an excitation energy of about 3.5 and 5.5 MeV for odd-mass and even-mass indium isotopes, respectively. The full widths at half maximum of the distributions amount to about 1-1.5 MeV.

More realistic description of the GT strength distribution can be obtained in the shell model approach. Since the conventional shell model calculations within a full major oscillator shell are not feasible due to the very large number of configurations involved, the calculations were performed in the model space designated by SNB in ref. [18], where it was used to calculate β-decay properties of N=50 and N=51 nuclei near ^{100}Sn. In the full SNB model space the $1p_{1/2}$ and $0g_{9/2}$ proton orbitals are active, whereas neutron particles are allowed to occupy $0g_{7/2}$, $1d_{5/2}$, $1d_{3/2}$,

$2s_{1/2}$ and $0h_{11/2}$ orbitals. In Fig. 1, the GT strength function for ^{97}Ag resulting from these calculations is confronted with the experimental B_{GT} distribution. For a comparison of the *shape* of the distributions, the theoretical data displayed in Figs. 1 and 2 were normalized to the total observed B_{GT} (The hindrance factors with respect to the *absolute* B_{GT} values will be discussed below). As can be seen from Fig. 1, the SNB calculation fairly well reproduces the position and the fine structure of the GT resonance.

For $^{103-107}$In, the number of the final state configurations was much too large to handle in practical calculations and the SNB model space for neutrons had therefore to be restricted to $0g_{7/2}$ and $1d_{5/2}$ orbitals. The results of these calculations are compared to the experimental B_{GT} distributions in Fig. 2. For all indium isotopes studied in this work, the centroids and widths of the GT resonances are very well reproduced by the shell-model calculations performed in the restricted SNB model space. The shape divergence noticeable in the high-energy part of the B_{GT} distribution for ^{103}In most probably indicates the influence of the truncation of the full SNB model space [14].

The GT strength within the SNB model space arises only from the transformation of a $g_{9/2}$ proton into a $g_{7/2}$ neutron. The strength of this transition is maximized in ^{100}Sn, amounting to B_{GT}^0=17.78. In the general case, the total β^+/EC strength summed over all final states is $\Sigma B_{GT}=(N_{9/2}/10)\cdot(1-N_{7/2}/8)\cdot B_{GT}^0$, where $N_{9/2}$ and

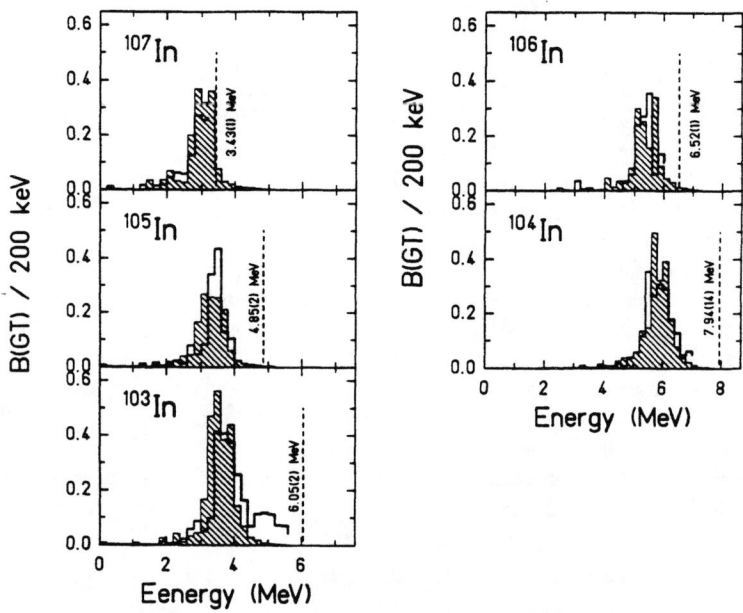

FIGURE 2. B_{GT} distribution for the decay of neutron-deficient indium isotope derived from the TAS measurements (blank histogram) and resulting from the shell model calculations (hatched histogram). Theoretical distributions were normalised to the total B_{GT} derived from the TAS measurements. The vertical lines indicate the Q_{EC} values.

$N_{7/2}$ are the initial state occupancies of the $\pi g_{9/2}$ and $\nu g_{7/2}$ orbitals, respectively.

Table 1 summarizes the total B_{GT} values predicted by the SNB calculations to lay within the Q_{EC} window and the ΣB_{GT} resulting from the TAS measurements. By comparing the theoretical and experimental ΣB_{GT} values the hindrance factors ranging from 5 to 8 were deduced indicating the importance of the $0\hbar\omega$ excitations beyond the SNB model space.

TABLE 1. Calculated and experimental ΣB_{GT} for isotopes studied by using the TAS. The respective half-lives ($T_{1/2}$), decay energies (Q_{EC}) and hindrance factors (h) are also given.

Nucleus	$T_{1/2}$(m)	Q_{EC}(MeV)	ΣB_{GT}^{th}	ΣB_{GT}^{exp}	h
^{97}Ag	0.422(5)	6.93(10)	12.8	2.63(37)	4.9(7)
^{103}In	1.08(2)	6.05(2)	12.7	2.47(25)	5.1(5)
^{104}In	1.80(3)	7.91(14)	12.6	2.04(14)	6.2(4)
^{105}In	5.07(7)	4.85(2)	10.9	1.4^{+6}_{-3}	$7.8^{+3.3}_{-1.7}$
^{106}In	6.2(1)	6.52(1)	10.7	1.43(8)	7.5(4)
^{107}In	32.4(3)	3.43(1)	8.0	1.15^{+30}_{-20}	$6.9^{+1.8}_{-1.2}$

Summary and conclusions

Measurements performed with the use of the TAS and the array of Euroball Cluster detectors have provided qualitatively new data on the GT strength distribution for nuclei near ^{100}Sn. For the first time, the complete β-strength in the domain of GT resonance has been mapped for such nuclei. The results obtained constitute a solid basis for a detailed comparison with theoretical predictions. The shell model calculations performed in the SNB model space fairly well reproduce the shape of the measured GT strength distributions. However, the total experimental GT strength was found to be a factor of 5 to 8 smaller than the shell model prediction. This discrepancy can be reduced if one assumes a universal character of the hindrance mechanisms observed in p-, sd- and pf-shells and adopts for the ^{100}Sn region the global hindrance factor of 1.81, which was introduced to account for the higher-order effects in the pf-shell. The possible source of the remaining disagreement must then be related to the truncation of the shell model configuration space. In fact, the importance of the intrashell configuration mixing has been demonstrated by the recent shell model Monte Carlo (SMMC) calculations performed for ^{94}Ru, ^{96}Pd, ^{96,98}Cd and ^{100}Sn within the complete g-d-s oscillator shell [19]. In particular, for ^{98}Cd the calculations in the SNB model space yield a free-nucleon value of 13.3 for the total B_{GT} [18], whereas the corresponding ΣB_{GT} obtained in the full $0\hbar\omega$ calculations amounts to 7.0 [19]. The latter value, when renormalized by the global hindrance factor of 1/1.81, is in excellent agreement with the experimental estimate of ΣB_{GT} equal $3.5^{+0.7}_{-0.6}$ [20]. It is worth noting that ^{98}Cd is the nucleus closest to ^{100}Sn for which the experimental B_{GT} can be confronted with the results of the complete $0\hbar\omega$ calculations. However, for the critical test of the theory, the extension of the SMMC calculations to non even-even isotopes in the vicinity of ^{100}Sn, for which reliable GT strength distributions have been determined from TAS measurements, is highly desirable.

Future experimental studies of GT decays should approach nuclei as close to ^{100}Sn as possible with the ultimate goal to measure the complete distribution of the GT strength of ^{100}Sn itself in a reliable manner, i. e. to improve the scarce data available for this decay. For the case of ^{102}In the TAS and Cluster Cube data are being analysed. Measurements for ^{100}In and ^{101}In decays are planned, and the reinvestigation of ^{98}Cd decay by means of the total absorption technique is considered. For the most exotic nuclei the γ-ray measurements have to be complemented by the investigation of β-delayed particle emission.

Acknowledgments

This work was supported in part by the Polish Committee of Scientific Research under grant KBN 2 P03B 039 13, by the Russian Fund for Basic Research and Deutsche Forsungsgemeinschaft under contract No. 436 RUS 113/201/0(R), by the U.S. N.S.F. under grant 9605207, by C.I.C.Y.T. (Spain) under project AEN96-1662, and by the EC Contract No. ERBFMGECT950083. M.K. would like to acknowledge financial support from the Foundation for Polish Science. B.A.B. wishes to thank the A. von Humoldt Foundation for support. K.R. would like to thank for a partial support from the U.S. D.O.E. under project DE-AC05-96OR22464 and ORNL managed by Lockheed Martin Energy Research Corporation.

References

1. Bethe, H.A., *Rev. Mod. Phys.* **64**, 491 (1992).
2. Hardy, J.C., Towner, I.S., *contribution to this conference*.
3. Ejiri, H., *Acta Physica Polonica* **B29** 47 (1998).
4. Trinder, W.,at al., *Phys. Lett.* **B349** , 267 (1995).
5. Liu, W., et al., *Z. Phys.* **A386** , 1 (1997).
6. Osterfeld, F., *Rev. of Mod. Phys.* **64**, 491 (1992).
7. Koonin, S.E., et al., *Annu. Rev. Nucl. Part. Sci.* **47**, 463 (1997).
8. Chou, W.T., Warburton, E.K., Brown, B.A., *Phys. Rev.* **C47**, 163 (1993).
9. Brown, B.A., Wildenthal, B.H., *At. Data Nucl. Data Tables* **33**, 348 (1985).
10. Martínez-Pinedo, G., et al., *Phys. Rev.* **C53**, R2602 (1996).
11. Brown, B.A., *Annu. Rev. Nucl. Part. Sci.* **38**, 38 (1988).
12. Karny, M., et al., *Nucl. Instr. and Meth. Phys. Res.* **B126**, 320 (1997).
13. Karny, M., et al., *contribution to this conference*.
14. Karny, M., et al., *Nucl. Phys. A*, in print.
15. Cano-Ott, D., et al., *contribution to this conference*.
16. Hu, Z., et al., *contribution to this conference*.
17. Agramunt, J., et al., *contribution to this conference*.
18. Brown, B.A:, Rykaczewski, K., *Phys. Rev.* **C50**, 2270 (1994).
19. Dean, D.J., et al., *Phys. Lett.*, **B367**, 17 (1996).
20. Płochocki, A., et al., *Z. Phys.*, **A342**, 43 (1992).

Superallowed Fermi Beta Decay

J.C. Hardy[a] and I. S. Towner[b]

[a] *Cyclotron Institute, Texas A & M University,
College Station, TX 77843*
[b] *Physics department, Queen's University,
Kingston, Ontario K7L 3N6, Canada*

Abstract. Superallowed $0^+ \to 0^+$ nuclear beta decay provides a direct measure of the weak vector coupling constant, G_V. We survey current world data on the nine accurately determined transitions of this type, which range from the decay of ^{10}C to that of ^{54}Co, and demonstrate that the results confirm conservation of the weak vector current (CVC) but differ at the 98% confidence level from the unitarity condition for the Cabibbo-Kobayashi-Maskawa (CKM) matrix. We examine the reliability of the small calculated corrections that have been applied to the data, and assess the likelihood of even higher quality nuclear data becoming available to confirm or deny the discrepancy. Some of the required experiments depend upon the availability of intense radioactive beams. Others are possible today.

CURRENT STATUS OF WORLD DATA

Superallowed Fermi $0^+ \to 0^+$ nuclear beta decays [1,2] provide both the best test of the Conserved Vector Current (CVC) hypothesis in weak interactions and, together with the muon lifetime, the most accurate value for the up-down quark-mixing matrix element of the Cabibbo-Kobayashi-Maskawa (CKM) matrix, V_{ud}. At present, the value of V_{ud} deduced from nuclear beta decay is such that, with standard values [3] of the other elements of the CKM matrix, the unitarity test from the sum of the squares of the elements in the first row fails to meet unity by more than twice the estimated error.

According to CVC, the measured ft-values for Fermi decays closely reflect the value of the weak vector coupling constant, G_V, and are independent of nuclear structure, outside of small correction terms that are of order 1%. Specifically for an isospin-1 multiplet

$$\mathcal{F}t = ft(1 + \delta_R)(1 - \delta_C) = \frac{K}{2G_V'^2}, \tag{1}$$

where f is the statistical rate function, t the partial half-life for the transition, δ_R is the calculated nucleus-dependent radiative correction, δ_C the calculated isospin-breaking correction, and K is a known [1] constant. The effective coupling constant

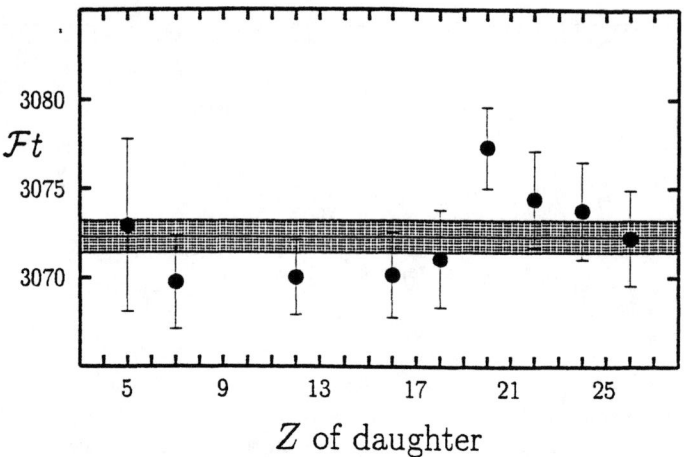

FIGURE 1. $\mathcal{F}t$-values for the nine precision data, and the best least-squares one-parameter fit

relates to the primitive one via $G'_V = G_V(1 + \Delta^V_R)^{1/2}$, where Δ^V_R is a calculated nucleus-independent radiative correction. For tests of the CVC hypothesis it is not necessary to consider this correction.

World data on Q-values, lifetimes and branching ratios were thoroughly surveyed [1] in 1989 and updated again [2] for the ENAM95 conference. Since then, there has been a new ^{10}C branching-ratio measurement [4] and a more precise ^{38m}K Q-value determination [5]. We have incorporated both measurements into our data base and extracted the $\mathcal{F}t$-values plotted in Fig. 1, which also uses the δ_R and δ_C corrections tabulated in our ENAM95 report [2]. It should be noted that those values of δ_C are, in fact, the averages of two independent calculations [6,7]. In a real sense, both experimentally and theoretically, Fig. 1 represents the totality of current world knowledge. The uncertainties shown reflect the experimental uncertainties and an estimate of the *relative* uncertainties in δ_C. There is no statistically significant evidence of inconsistencies in the data ($\chi^2/\nu = 1.1$), thus verifying the expectation of CVC at the level of 3×10^{-4}, the fractional uncertainty quoted on the average $\mathcal{F}t$-value (3072.3 ± 0.9 s).

In using the average $\mathcal{F}t$-value to determine V_{ud} and test CKM unitarity it is important to incorporate the 'systematic' uncertainty in δ_C that arises from the small systematic differences between the two independent model calculations [6,7]. The result is

$$\mathcal{F}t = 3072.3 \pm 2.0 \text{ s}. \qquad (2)$$

With this value, an estimate [8] of the nucleus-independent radiative correction of

$\Delta_R^V = (2.40 \pm 0.08)\%$, and the weak vector coupling constant [3] derived from muon decay, we obtain

$$V_{ud} = 0.9740 \pm 0.0005. \tag{3}$$

The quoted uncertainty is dominated by uncertainties in the theoretical corrections, Δ_R^V and δ_C. On adopting the values [3] of V_{us} and V_{ub} from the Particle Data Group, the sum of squares of the elements in the first row of the CKM matrix,

$$|V_{ud}|^2 + |V_{us}|^2 + |V_{ub}|^2 = 0.9968 \pm 0.0014, \tag{4}$$

differs from unity at the 98% confidence level.

To restore unitarity, the calculated radiative corrections would have to be shifted downwards by 0.3% (*i.e.* as much as one-quarter of their current value), or the calculated Coulomb correction shifted upwards by 0.3% (over one-half their value), or some combination of the two. In what follows, we discuss the accuracy of these two corrections and the direction of future research.

RADIATIVE CORRECTIONS

As mentioned, the radiative correction is conveniently divided into terms that are nucleus-dependent, δ_R, and terms that are not, Δ_R^V. These are written

$$\delta_R = \frac{\alpha}{2\pi}[\bar{g}(E_m) + \delta_2 + \delta_3 + 2C_{NS}]$$
$$\Delta_R^V = \frac{\alpha}{2\pi}[4\ln(m_Z/m_p) + \ln(m_p/m_A) + 2C_{\text{Born}}] + \cdots, \tag{5}$$

where the ellipses represent further small terms of order 0.1%. In these equations, E_m is the maximum electron energy in beta decay, m_Z the Z-boson mass, m_A the a_1-meson mass, and δ_2 and δ_3 the order $Z\alpha^2$ and $Z^2\alpha^3$ contributions. The electron-energy dependent function, $g(E_e, E_m)$ was derived by Sirlin [9]; it is here averaged over the electron spectrum to give $\bar{g}(E_m)$.

Typical values are

$$\delta_R \simeq 0.95 + 0.43 + 0.05 + (\alpha/\pi)C_{NS}\%, \tag{6}$$

where $(\alpha/\pi)C_{NS}$ is of order -0.3% for $T_z = -1$ beta emitters, ^{10}C and ^{14}O, and of order five times smaller for the $T_z = 0$ emitters, ranging from -0.09% to $+0.03\%$. Thus for $T_z = 0$ emitters $\delta_R \simeq 1.4\%$. If the failure to obtain unitarity in the CKM matrix with V_{ud} from nuclear beta decay is due to the value of δ_R, then δ_R must be reduced to 1.1%. This is not likely. The leading term, 0.95%, involves standard QED and is well verified. The order-$Z\alpha^2$ term, 0.43%, while less secure has been calculated twice [10,11] independently, with results in accord.

For the nucleus-independent term

$$\Delta_{\text{R}}^{\text{V}} = 2.12 - 0.03 + 0.20 + 0.1\% \simeq 2.4\% \tag{7}$$

of which the first term, the leading logarithm, is unambiguous. Again, to achieve unitarity of the CKM matrix, $\Delta_{\text{R}}^{\text{V}}$ would have to be reduced to 2.1%, *i.e.* all terms other than the leading logarithm summing to zero. This also seems unlikely.

COULOMB CORRECTIONS

Because the leading terms in the radiative corrections are well founded, attention has focussed more on the Coulomb correction. Although smaller than the radiative correction, the Coulomb correction is clearly sensitive to nuclear-structure issues. It comes about because Coulomb and charge-dependent nuclear forces destroy isospin symmetry between the initial and final states in superallowed beta-decay. The consequences are twofold: there are different degrees of configuration mixing in the two states, and, because their binding energies are not identical, their radial wave functions differ. Thus we accommodate both effects by writing $\delta_C = \delta_{C1} + \delta_{C2}$. Constraints can be placed on the calculation of δ_{C1} by insisting that the calculation reproduce the coefficients of the isobaric mass multiplet equation. Constraints on δ_{C2} follow by insisting that the asymptotic forms of the proton and neutron radial functions match known separation energies.

Recently Ormand and Brown (OB) [7] have recomputed their Hartree-Fock calculations with new results increasing δ_C over their earlier work [12] but still with values systematically smaller than the Saxon-Woods calculations of Towner, Hardy and Harvey (THH) [6] . Another recent work by Sagawa, van Giai and Suzuki [13] add RPA correlations to a Hartree-Fock calculation; these correlations, in essence, introduce a coupling to the isovector monopole giant resonance. This calculation, however, is not constrained to reproduce known separation energies. Finally a large shell-model calculation has been mounted for the $A = 10$ case by Navrátil, Barrett and Ormand [14] . Both of these two new works [13,14] have produced values of δ_C *smaller* than those used before, *i.e.* worsening rather than helping the unitarity problem.

The typical value of δ_C is of order 0.4%. If the unitarity problem is to be solved by improvements in δ_C, then δ_C has to be raised to around 0.7%. There is no evidence whatsoever for such a shift from recent works.

The δ_C calculations, as pointed out by OB [7], do predict that δ_C should be dramatically larger for nuclei in the fp-shell with $A \geq 62$. This is due to the increasing importance of the $1p$ orbital, which, with its extra node in the radial function compared to the $0f$ orbital, is much more sensitive to Coulomb effects. A similar effect was predicted earlier [6] for $T_z = -1$ nuclei in the middle of the sd-shell where the $1s$ orbital plays an equivalent role. Future experiments will test these predictions.

FUTURE PROSPECTS FOR EXPERIMENT

The nine superallowed transitions surveyed here have been the subject of intense scrutiny for at least the past three decades. All except ^{10}C have the special advantage that the superallowed branch from each is by far the dominant transition in its decay (> 99%). This means that the branching ratio for the superallowed transition can be determined to high precision from relatively imprecise measurements of the other weak transitions, which can simply be subtracted from 100%. Given the quantity of careful measurements already published, are there reasonable prospects for significant improvements in these decays in the near future? Given the uncertainty in the theoretical corrections, perhaps a more important question is whether there is any reason to seek experimental improvements at all.

If we begin by accepting that it is valuable for experiment to be at least a factor of two more precise than theory, then an examination of the world data shows that the Q-values for ^{10}C, ^{14}O, ^{26m}Al and ^{46}V, the half-lives of ^{10}C, ^{34}Cl and ^{38m}K, and the branching ratio for ^{10}C can all bear improvement. Such improvements will soon be feasible. The Q-values will reach the required level (and more) as mass measurements with new on-line Penning traps become possible; half-lives will likely yield to measurements with higher statistics as high-intensity beams of separated isotopes are developed for the new radioactive-beam facilities; and, finally, an improved branching-ratio measurement on ^{10}C has already been made with Gammasphere and simply awaits analysis [4].

Qualitative improvements will also come as we increase the number of superallowed emitters accessible to precision studies. The greatest attention recently has been paid to the $T_z = 0$ emitters with $A \geq 62$, since these nuclei are expected to be produced at new radioactive-beam facilities, and their calculated Coulomb corrections, δ_C, are predicted to be large [7,15]. They could then provide a valuable test of the accuracy of δ_C calculations. It is likely, though, that the required precision will not be attainable for some time to come. The decays of these nuclei will be of higher energy and each will therefore involve several allowed transitions of significant intensity in addition to the superallowed transition. Branching-ratio measurements will thus be very demanding, particularly with the limited intensities likely to be available initially for these rather exotic nuclei. Lifetime measurements will be similarly constrained by statistics.

More accessible in the short term will be the $T_z = -1$ superallowed emitters with $18 \leq A \leq 38$. There is good reason to explore them. For example, the calculated value [6] of δ_C for ^{30}S decay, though smaller than the δ_C's expected for the heavier nuclei, is actually 1.2% – about a factor of two larger than for any other case currently known – while ^{22}Mg has a very low value of 0.35%. If such large differences are confirmed by the measured ft-values, then it will do much to increase our confidence in the calculated Coulomb corrections. To be sure, these decays will provide a challenge, particularly in the measurement of their branching ratios, but the required precision should be achievable with isotope-separated beams that are currently available. In fact, such experiments are already in their early stages at the Texas A&M cyclotron.

CONCLUSIONS

The current world data on superallowed $0^+ \to 0^+$ beta decays lead to a self-consistent set of $\mathcal{F}t$-values that agree with CVC but differ provocatively, though not yet definitively, from the expectation of CKM unitarity. There are no evident defects in the calculated radiative and Coulomb corrections that could remove the problem, so, if any progress is to be made in firmly establishing (or eliminating) the discrepancy with unitarity, additional experiments are required. We have indicated what some relevant nuclear experiments might be.

Clearly, there is strong motivation to pursue them since, if firmly established, a discrepancy with unitarity would indicate the need for an extension of the three-generation Standard Model.

This work was supported in part by the U.S. Department of Energy under Grant number DE-FG05-93ER40773 and by the Robert A. Welch Foundation.

REFERENCES

1. J.C. Hardy, I.S. Towner, V.T. Koslowsky, E. Hagberg and H. Schmeing, Nucl. Phys. **A509**, 429 (1990)
2. I.S. Towner, E. Hagberg, J.C. Hardy, V. Koslowsky and G. Savard, *Proceedings of the Int. Conf. on Exotic Nuclei and Atomic Masses, Arles, France, June 1995*, eds. M. de Saint Simon and O. Sorlin (Edition Frontières, Gif-sur-Yvette, France, 1995) pp.711-721
3. C. Caso *et al.*, The European Physical Journal **C3**, 1 (1998)
4. B.K. Fujikawa *et al.*, APS e-print 1998may29_004
5. P.H. Barker, P.D. Harty and N.S. Bowden, private communication
6. I.S. Towner, J.C. Hardy and M. Harvey, Nucl. Phys. **A284**, 269 (1977)
7. W.E. Ormand and B.A. Brown, Phys. Rev. **C52**, 2455 (1995)
8. A. Sirlin, in *Precision Tests of the Standard Electroweak Model*, ed. P. Langacker (World-Scientific, Singapore, 1994)
9. A. Sirlin, Phys. Rev. **164**, 1767 (1967)
10. A. Sirlin, Phys. Rev. **D35**, 3423 (1987); A. Sirlin and R. Zucchini, Phys. Rev. Lett. **57**, 1994 (1986)
11. W. Jaus and G. Rasche, Phys. Rev. **D35**, 3420 (1987)
12. W.E. Ormand and B.A. Brown, Phys. Rev. Lett. **62**, 866 (1989); Nucl. Phys. **A440**, 274 (1985)
13. H. Sagawa, N. van Giai and T. Suzuki, Phys. Rev. **C53**, 2163 (1996)
14. P. Navrátil, B.R. Barrett and W.E. Ormand, Phys. Rev. **C56**, 2542 (1997)
15. J.C. Hardy, in *Proceedings of the Workshop on the Production and Use of Intense Radioactive Beams at the Isospin Laboratory*, ed. J.D. Garrett (Joint Institute for Heavy Ion Research, Oak Ridge, Tennessee: Conf. 9210121) pg. 51

Indication for superallowed Fermi decay from the N=Z nuclei ^{78}Y, ^{82}Nb, ^{86}Tc

Ph. Dessagne[a], C. Longour[a], J. Garcés Narro[b], D. Applebe[c],
L. Axelsson[d], B. Blank[e], A.M. Bruce[f], W.N. Catford[b],
C. Chandler[b], R.Clark[g], D. Cullen[c], S. Czajkowski[e], J.M. Daugas[h],
A. Fleury[e], L. Frankland[f], W. Gelletly[b], J. Giovinazzo[e],
B. Greenhalgh[i], R. Grzywacz[j], M. Harder[f], K.L. Jones[b],
N. Kelsall[i], T. Kszczot[j], M. Lewitowicz[h], Ch. Miehé[a], R.D. Page[c],
C.J. Pearson[b], A.T. Reed[c], P.H. Regan[b], O. Sorlin[k], R. Wadsworth[i]

[a] IReS Strasbourg, UMR7500, CNRS-IN2P3 et Université Louis Pasteur,
BP28, F-67037 Strasbourg Cedex 2, France
[b] Dept. of Physics, University of Surrey, Guildford, GU2 5XH, UK
[c] Dept. of Physics, Oliver Lodge Laboratory, University of Liverpool, Liverpool, L69 7ZE, UK
[d] Dept. of Physics, Chalmers University of Technology, S-412 96 Göteborg, Sweden
[e] CEN Bordeaux-Gradignan, Le Haut-Vigneau F-33175 Gradignan Cedex, France
[f] Cockcroft Building, University of Brighton, Brighton, BN2 4GJ, UK
[g] Nuclear Science Division, LBNL, Berkeley, CA 94720, USA
[h] GANIL, BP 5027, F-14021, Caen Cedex, France
[i] Dept. of Physics, University of York, Heslington, York, Y01 4DD
[j] Inst. of Expt. Phys., Warsaw University, Pl-00681, Warsaw, Poland
[k] Institut de Physique Nucléaire, 91406 Orsay, France

Abstract. The heavy odd-odd N = Z nuclei ^{78}Y, ^{82}Nb, ^{86}Tc have been produced at GANIL by fragmentation of a 60 MeV/u ^{92}Mo beam on a natural nickel target. Their lifetime have been measured for the first time, using the identification - implantation and subsequent time correlated β counting technique at the LISE spectrometer. The obtained half-lives for ^{78}Y, ^{82}Nb, ^{86}Tc are consistent with the microscopic predictions. These nuclei are good candidates to extend the mass range for the V_{ud} matrix element evaluation and our experimental results highlight the interest of further investigation of their excited states and β decay schemes as a function of isospin.

The study of odd-odd nuclei with equal numbers of protons and neutrons provides a unique laboratory for the study of pairing correlations and isospin symmetry in nuclear systems [1–4]. Charge independence suggests that for nuclei along the N=Z line, protons and neutrons fill the same shell, leading to a competition between $T = 0$ and $T = 1$ pairing in these nuclei. In particular, this competition determines the ground-state properties of odd-odd nuclei along the N=Z line [5]. For N=Z nuclei from the deuteron up to about mass 38, the $T = 0$ pairing seems to be more important leading to $T = 0$, $I \neq 0$ ground states. Heavier N=Z nuclei (with the exception of ^{58}Cu) have $I^\pi = 0^+$ ground states which demonstrates that the $T = 1$ pairing is responsible for the ground-state properties of these nuclei.

Fundamental aspects of the weak interaction can be tested with basic ingredients such as half-lives and β^+-decay energies. For example, Hardy et al. [6] used the superallowed $0^+ \rightarrow 0^+$ decay of nuclei from ^{14}O up to ^{54}Co to test the standard model via the Conserved Vector Current (CVC) hypothesis. For these studies, the extracted $log\ ft$ values must be corrected for radiative and Coulomb effects, the corrections being calculated using various theoretical approaches [7–10]. While these corrections are generally in good agreement with each other for nuclei where experimental data are available, for heavier nuclei, there are considerable differences between the predictions. A further question is whether the corrected $log\ ft$ values have an elemental (Z) dependence [6,8,11]. To a large extent, these questions can be addressed by precisely measuring $log\ ft$ values for Fermi superallowed β-decays in heavier nuclei and using the results to test the different predictions.

The nuclei of interest were produced following the fragmentation of a ^{92}Mo^{37+} beam on a natural nickel target of thickness 120 μm. The primary beam, provided by GANIL facility, was at an energy of 60 MeV per nucleon, with an average current of 200 enA. The fragmentation products were collected and separated using the LISE3 spectrometer [12]. An achromatic beryllium degrader of thickness 50 μm was placed between the two main dipole magnets and this, together with the velocity selection provided by the Wien filter allowed the selection of only the most exotic isotopes. At the final focus, the fragments were stopped in a three element telescope, the first element of which consisted of a 300 μm thick energy loss (ΔE) silicon detector, used to provide element (Z) identification. A twelve-strip segmented silicon detector of thickness 500 μm was situated behind the ΔE detector in which the ions of interest were stopped. A third silicon detector, also of thickness 500 μm, was placed behind the strip detector and was used in the off-line analysis to discriminate against any contaminating lighter ions reaching the final focus. A time-of-flight (TOF) for the fragments from the production target to the silicon telescope was measured by taking the time difference between a fast signal extracted from the ΔE detector and the cyclotron radiofrequency. This TOF together with the energy loss in the ΔE detector and the magnetic rigidity of the dipole magnets in the LISE3 spectrometer was used to obtain an unambiguous identification in Z and N for each fragment, using previously described techniques [13]. An array of four high-purity germanium detectors of 70% relative efficiency were packed in close geometry around the silicon stack with the aim of measuring discrete γ-rays following decays from isomeric states in the fragments arriving to the final focus, as used in references [14]. Decays from previously reported isomers in ^{76}Rb [15], ^{80}Y and ^{84}Nb [16] were identified, providing a validation of the experimental method and reference points for the particle identification spectra.

The β-decay half-lives of ^{74}Rb, ^{78}Y, ^{82}Nb and ^{86}Tc were determined in the following way. A hardware trigger was enabled such that, after implantation of a fragment of interest, the primary beam was immediately shut-off for one second. The conditions for the beam to be cut were that the detected fragment had to have $T_z \leq \frac{1}{2}$ and $43 \geq Z \geq 34$. Subsequent β^+-decays were then correlated with specific implants by looking for signals during this beam-off period in the same strip in which the fragment of interest was detected. To obtain a β^+ decay time spectrum, the time difference between the ion implantation and any subsequent decay signal in the same strip during the beam-off period was measured by two independent methods. Each event was time-stamped by a free-running external clock with a microsecond dispersion. Another clock, with millisecond time divisions, was reset each time a detected fragment caused the beam to be cut off. In the off-line analysis, software cuts were applied to a two-dimensional particle identification matrix of ΔE versus TOF. This allowed the radioactivity time distributions to be obtained for specific nuclear species.

The time spectra obtained using this method for the odd-odd, N=Z systems, ^{74}Rb, ^{78}Y, ^{82}Nb and ^{86}Tc are shown in figure 1. The background of radioactivity events from isotopes with half-lives longer than the beam-off period of one second contributed to a constant background level in the time spectra. The half-lives we have obtained for the four odd-odd N=Z nuclei are 60±10 ms for ^{74}Rb, 55±12 ms for ^{78}Y, 50±4 ms for ^{82}Nb, and 47±12 ms for ^{86}Tc. The half-life of ^{74}Rb has previously been measured by D'Auria et al. [17] to be (64.9±0.5) ms which is consistent with our result for this nucleus and thus provides a cross-check for the experimental method used in the current work.

Table 1 summarizes the experimentally deduced half-lives for the odd-odd N=Z nuclei ^{74}Rb, ^{78}Y, ^{82}Nb and ^{86}Tc together with the Q-values and the resulting *log ft* quantities. The *log ft* values for these β^+-decays were calculated assuming i) a branching ratio of 100% for the observed transition and ii) a ground-state to ground-state decay. Since, at the present time, no experimental data are available for the β^+-decay Q values of these reactions, we have taken the values as calculated from the mass evaluation of Audi and Wapstra [18]. Using these, we determine the statistical rate function f using a routine based on the formalism of Gove and Martin [19].

The calculated *log ft* values are all compatible within error bars with *log ft* = 3.5 which is indicative of a Fermi superallowed character for each of these decays. The daughter nuclei populated following the β^+ decay of these odd-odd, N=Z systems are even-even nuclei ($T_z = 1$), and thus have 0^+ ground states.

The question now arises whether the observed decays in these nuclei constitute the observation of 0^+, $T = 1$ ground states. There are a number of cases in odd-odd, N=Z systems, where a fast β^+ decaying $T = 1$, $I^\pi = 0^+$ state is observed to be an excited, isomeric state and not the ground state of this nucleus. For example,

TABLE 1. Half-lives, β-decay Q values and $\log ft$ values for the odd-odd, $T_z = 0$ nuclei ^{74}Rb, ^{78}Y, ^{82}Nb and ^{86}Tc

	$T_{1/2}$ (ms)	Q_{EC} (keV)	$\log ft$
^{74}Rb	60 ± 10	10440 ± 722	3.5 ± 0.3 [a]
^{78}Y	55 ± 12	10540 ± 400	3.4 ± 0.2
^{82}Nb	50 ± 4	11220 ± 592	3.5 ± 0.3
^{86}Tc	47 ± 12	11350 ± 532	3.5 ± 0.2

[a] Based on the half-life of D'Auria et al. of 64.9(5) ms, we obtain a $\log ft$ value of 3.5 ± 0.2

in the case of ^{26}Al, while the $T = 0$, $I^{\pi}=5^+$ ground state has a half-life of 7.4×10^5 years, the lowest-lying $T = 1$ level, a 0^+ state at an excitation energy of 228 keV, decays by a Fermi superallowed transition with a half-life of 6.3 s [20]. While the current work cannot definitely assign the observed decays as stemming from the 0^+, $T = 1$ *ground state* configuration, support for this assertion can be found by investigating the measured β^+ detection efficiencies.

Previous fragmentation studies have established that these reactions can populate high-spin isomeric states, with relatively large isomeric ratios of the order of 10-30 % [14]. This implies that the fragments are often produced in excited, high-spin states which subsequently decay by γ emission via a near-yrast cascade. In such a scenario, one would expect an yrast cascade to generally bypass a non-yrast 0^+ excited state, in favour of decaying in a cascade built upon a higher spin, $T = 0$, ground state. To test whether our results were consistent with the expectation of a 0^+ ground state, experimental ratios of the number of the observed β decays and the number of the implantations of each species were measured. The measured values of this ratio for the odd-odd, N=Z systems were (41 ± 11) % for ^{74}Rb, (21 ± 6) % for ^{78}Y, (50 ± 4) % for ^{82}Nb, and (47 ± 12) % for ^{86}Tc. The relatively high values observed for ^{74}Rb, ^{82}Nb and ^{86}Tc are consistent with most of the implants decaying by superallowed decays, suggesting 0^+ ground states, (higher-spin ground states would be expected to give rise to much slower, Gamow-Teller decays, thus reducing the values for this efficiency ratio). The smaller value for ^{78}Y can be explained by the recent report by Uusitalo et al. [21] of a Gamow-Teller decaying, $T = 0$, $I^{\pi} = 5^+$ excited state in $^{78}_{39}$Y. The measured half-life for this decay of (5.8 ± 0.6) s is long enough such that most of the nuclides where this high-spin isomeric state is populated would not decay in the one second time window after implantation used for the current work. The apparent reduction in β^+ efficiency observed for the Fermi superallowed decay in ^{78}Y can thus be explained by a substantial fraction of the fragments of this nucleus, being trapped in the *yrast* 5^+ state, consistent with our assumption of a general, near yrast population and decay scenario in fragmentation reactions. From these results the decays observed in our studies are presumably all of a $0^+ \rightarrow 0^+$ Fermi superallowed character.

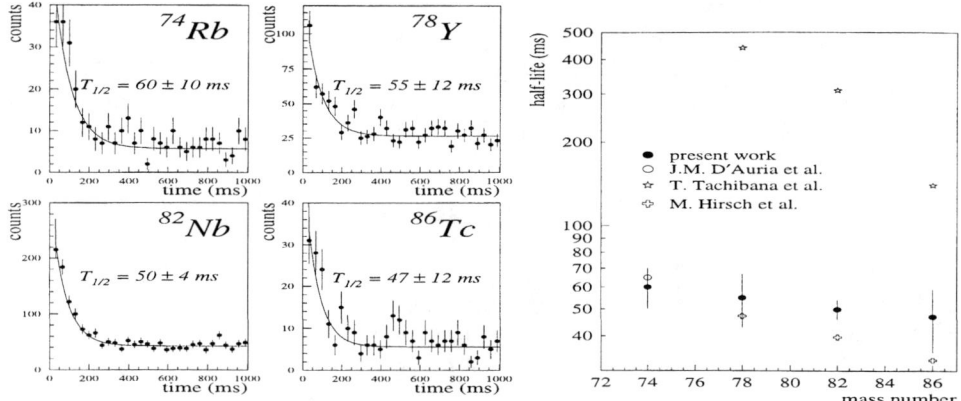

FIGURE 1. On the left part, the time spectra for the odd-odd N = Z nuclei as measured in the present work. The solid line is a least-square fit of an exponential curve with a constant level background to the data. On the right part, a comparison of the half-lives determined in the present work to values from the literature and from theoretical calculations. The full circles represent the results from the present work, the open circle shows the experimental result of D'Auria et al. [17]. The calculations of Hirsch et al. [23] are given by the crosses, whereas the results from the gross theory [22] are given by the stars.

Figure 1 shows a comparison of our results with theoretical predictions of the gross theory [22] and of microscopic calculations [23] using the mass formula of E.R. Hilf et al. [24]. The gross theory overestimates the half-lives by a factor of ten, because only Gamow-Teller transitions are taken into account. The pn-QRPA calculations of Hirsch et al. [23] are in good agreement with our experimental findings.

In conclusion, we have measured β-decay half-lives for the odd-odd, N=Z nuclei ^{74}Rb, ^{78}Y, ^{82}Nb and ^{86}Tc, the half-life of the last three nuclei being measured for the first time. The *log ft* values deduced assuming ground-state to ground-state transitions, 100% branching ratios and using β-decay Q values as determined from the Audi and Wapstra mass evaluation are indicative of Fermi superallowed ($0^+ \rightarrow 0^+$) β^+ decays. These are the heaviest N=Z nuclei for which Fermi superallowed decays have been established. The high detection efficiencies for these decays are suggestive of these 0^+ states forming the ground state configuration in these nuclei. Although in the current work the statistical precision of the data is not sufficient for a more detailed analysis, the half-life results suggest that these nuclei are excellent candidates to further extend the mass range over which the CVC hypothesis could be tested.

We would like to express our gratitude to the GANIL staff for providing an excellent primary beam and for a stable running of the equipment. This work was supported by IN2P3, CEA (France) and EPSRC (UK). CC, LF, BG, KLJ hold EPSRC Ph D research studentships. RDP acknowledges an EPSRC advanced fellowship. LA expresses his gratitude to CIES.

REFERENCES

1. Goodman, A.L., Adv. Nucl. Phys. **11**, 263 (1979)
2. Vincent, S.M. et al., to be submitted to Phys. Lett. B
3. Rudolph, D. et al., Phys. Rev. Lett. **76**, 376 (1996)
4. Dean, D.J. et al., Phys. Lett. **B399**, 1 (1997)
5. Van Isacker, P. and Warner, D.D., Phys. Rev. Lett. **78**, 3266 (1997)
6. Hardy, J.C. et al., Nucl. Phys. **A509**, 429 (1990)
7. Towner, I.S., Nucl. Phys. **A540**, 478 (1992)
8. Wilkinson, D.H., Nucl. Inst. Meth. Phys. Res. **A335**, 172 (1993)
9. Ormand, W.E. and Brown, B.A., Phys. Rev. **C52**, 2455 (1995)
10. Sagawa, H. et al., Phys. Rev. **C53**, 2163 (1996)
11. Savard, G. et al., Phys. Rev. Lett. **74**, 1521 (1995)
12. Mueller, A.C. and Anne, R., Nucl. Inst. Meth. Phys. Res. **B56/57**, 559 (1991)
13. Bazin, D. et al., Nucl. Phys. **A515**, 349 (1990); Blank, B. et al., Phys. Rev. Lett. **74**, 4611 (1995); Rykaczewski, K. et al., Phys. Rev. **C52**, R2310 (1995)
14. Grzywacz, R. et al., Phys. Lett. **355B**, 439 (1995); Chandler, C. et al., Phys. Rev. **C56**, R2924 (1997)
15. Hofmann, S. et al., Z. Phys. **A325**, 37 (1986)
16. Regan, P.H. et al., Acta Phys. Pol. **B28**, 431 (1997)
17. D'Auria, J.M. et al., Phys. Lett. **B66**, 233 (1977)
18. Audi, G. and Wapstra, A.H., Nucl. Phys. **A595**, 409 (1995)
19. Gove, N.B. and Martin, M.J., Nucl. Data Tables **10**, 205 (1971)
20. Firestone, R.B. and Shirley, V.S., *Table of Isotopes, 8th Edition, volume 1* (1996), John Wiley and Sons, New York
21. Uusitalo, J. et al., Phys. Rev. **C57**, 2259 (1998)
22. Tachibana, T. et al., Prog. Theor. Phys. **84**, 641 (1990)
23. Hirsch, M. et al., At. Data Nucl. Data Tables **53**, 165 (1993)
24. Hilf, E.R. et al., Proc. 3rd Int. Conf on Nuclei far from Stability, Cargèse, Corsica (France), CERN REPORT 76-13 (1976), p. 142

Beta-Decay Strength and Isospin Mixing Studies in the sd and fp-Shells

A. Jokinen, J. Äystö, P. Dendooven, A. Honkanen, P. Lipas, K. Peräjärvi, M. Oinonen and T. Siiskonen

Department of Physics, University of Jyväskylä, P.O. Box 35, FIN-40351 Jyväskylä, Finland
Ari.Jokinen@phys.jyu.fi

Abstract. We have studied beta decays of $M_T<0$ nuclei in sd and fp shells. The decay of ^{41}Ti shows a large, 10(8) %, isospin mixing of IAS and the Gamow-Teller strength is observed to be quenched by a factor of $q^2 = 0.64$. These results can be reproduced qualitatively in our shell model calculations. We have observed for the first time proton and gamma decay of the isobaric analogue state in ^{23}Mg. Our results on the isospin mixing of the isobaric analogue state agrees well with the shell model calculations. The obtained proton branch of the IAS is used to extract the transition strength for the reaction ^{22}Na(p,γ)^{23}Mg.

INTRODUCTION

Beta decays of proton-rich nuclei are associated with high Q_{EC}-values which result in an ideal possibility to probe the Gamow-Teller (GT) strength over a wide energy range, extending in some cases up to a Gamow-Teller resonance region. Another interesting feature of proton-rich nuclei deals with a Fermi-type beta decay to the isobaric analog state (IAS), whose proton decay is isospin forbidden. However, many of the $M_T=-3/2$ nuclei have a large proton decay branch from the IAS. To explain this observation the small isospin mixing of the IAS is proposed. In case of $0^+ \rightarrow 0^+$ transitions, where detailed decay data is available, the typical values of isospin mixing are of the order of 2 %.

Exceptionally high isospin-mixing values have been reported for T=3/2 isobaric analogue states in ^{23}Mg (1) and ^{41}Sc (2). To study these effects we have performed high-resolution decay experiments for T=3/2 nuclei ^{23}Al and ^{41}Ti (3,4). In this paper these results are reviewed and we discuss the perspectives to extend these studies up to $M_T=-5/2$ nuclei.

EXPERIMENTAL AND THEORETICAL METHODS

These studies have been performed in Jyväskylä applying light ion induced fusion reaction and the ion guide isotope separator (IGISOL) (5). In the ion guide technique the ions recoiling out from the target are stopped in a He-gas, where they reach 1+ charge state during thermalization. This method is rather independent of chemical properties of the ions wanted making it very suitable for isotopes of refractory elements like Ti. The mass-separated ion beam was implanted into the movable carbon-foil, which was surrounded by the recently developed gas-Si telescope (6) and gamma detector. In addition, conventional gamma and beta spectroscopy methods were applied to obtain complete picture of the decay properties. The implanted ion beam was pulsed and the periodic measurement was applied to check the half-life of the observed radiation.

LINKING sd AND fp SHELL BY THE DECAY OF ^{41}Ti

The experimental evaluation of the quenching of GT-strength suffers often from the lack of sensitivity for high-energy gammas from higher lying excited states. In this respect, the beta decay of ^{41}Ti is an adequate object for such a studies, since its beta-delayed proton branch is almost 100% and thus the feeding of the excited states can be determined from their proton decay. It is also worth to mention that the detailed study of the decay properties of ^{41}Ti paves the way to better understanding of the decay of ^{40}Ti. The latter one is important for the calibration of the solar neutrino detector ICARUS using ^{40}Ar as a target material.

The extracted beta-decay strength was compared to the shell-model calculations. The distribution of the GT-strength could be extracted up to an excitation energy of 8 MeV. The comparison of the experimental Gamow Teller strength to the calculated strength results in a quenching factor of about $q = 0.7$. This value compares well with generally accepted value of $q = 0.76(3)$ in sd-shell (7). We also extracted the experimental Fermi strength to the IAS by subtracting calculated GT-strength from the total beta decay strength. The obtained value $B(F) = 0.27$ implies 10(8) % isospin mixing. The earlier reported high isospin mixing of the IAS in ^{41}Sc was thus reproduced. The extracted value of mixing could be quantitatively understood by applying large-scale shell model calculations in the sdfp-space. According to calculations, most of the missing Fermi-strength goes to one excited state just below IAS.

STUDY OF βp DECAY OF ^{23}Al

Beta decay of ^{23}Al has been studied earlier by another group resulting in a very large isospin mixing estimation for the IAS. This value is in contradiction both with a shell model estimation and with general understanding of the magnitude of isospin mixing in this region of the nuclide chart. The transition strength of ^{22}Na(p,γ)^{23}Mg reaction extracted from the previous beta-decay study is also in contradiction with the recent reaction studies (8,9,10). The previous study was performed with unseparated sources and contaminations could not be fully excluded. For these reasons, the remeasurement of the beta decay of ^{23}Al was necessary.

The present study was performed at the IGISOL-facility with the same reaction, i.e. ^{24}Mg(p,2n), than in the previous study but applying the ion guide technique with mass separation. Additional improvement compared to previous study comes from the fact that in our experiment, both charged particle and gamma decay of IAS were observed simultaneously. The relative efficiency of the gamma and charged particle detection was calibrated by means of on-line sources of ^{24}Al and ^{20}Na. Thus, the relative gamma and proton intensity from the IAS in ^{23}Mg could be extracted with a good accuracy. Our result implies a much smaller proton branch from the IAS than obtained in the previous study. Our value is 0.42(3) %. Thus, the extremely high isospin mixing of IAS in ^{23}Mg could not reproduced in our study. On the contrary, the new value follows the general trend of relatively small isospin mixing. Using the gamma width of 3 meV from the mirror nucleus and the relative intensities of proton and gamma decay of IAS, we end with a proton width of 13(4) meV for the IAS. This value corresponds to 5.4(17) meV transition strength if the equation $\omega\gamma = (2J+1)\Gamma_p\Gamma_\gamma/14\Gamma_{tot}$ is applied. This value compares well with the recent reaction studies.

PERSPECTIVES TO M_T=-5/2 NUCLEI

The M_T=-5/2 series provides much higher Q_{EC} and thus the experimental and calculated GT-strength can be compared in wider energy range. This provides more accurate information on the quenching of GT-strength. Recently an experiment was performed at ISOLDE for ^{31}Ar (11) and at GANIL for ^{35}Ca (12). Although the energetics is favorable, these studies have to deal with additional complexity due to multiple decay channels, like beta-delayed two-proton emission (13). This fact makes it difficult to extract correctly the magnitude and distribution of the GT-strength. The key issue is to identify correctly charged particles and their emission order and angle. The first step towards better understanding of beta-delayed two proton decay is reported elsewhere in this conference showing the rich 2p decay pattern in the decay of ^{31}Ar (14).

ACKNOWLEDGMENTS

The support from the Academy of Finland is acknowledged.

REFERENCES

1. Tighe, R.J., Batchelder, J.C., Moltz, D.M., Ognibene, T.J., Rowe, M.W., Cerny, J., Brown, B.A., Phys. Rev. **C 52,** R2298-R2301 (1995).
2. Sextro, R., Gough, R.A. and Cerny, J., Nucl. Phys. **A234,**130-156 (1974).
3. Peräjärvi, K. et al., to be published, (1998).
4. Honkanen, A., Dendooven, P., Huhta, M., Lhersonneau, G., Lipas, P.O., Oinonen, M., Parmonen, J.-M., Penttilä, H., Peräjärvi, K., Siiskonen, T., Äystö, J., Nucl. Phys. **A621,** 689-705 (1997).
5. Penttilä, H., Dendooven, P., Honkanen, A., Huhta, M., Jauho, P.P., Jokinen, A., Lhersonneau, G., Oinonen, M., Parmonen, J.-M., Peräjärvi, K., Äystö, J., Nucl. Instr. Meth. Phys. Res. **B126** 213-217 (1997).
6. Honkanen, A. Oinonen, M., Eskola, K., Jokinen, A., Äystö, J., Nucl. Instr. Meth. Phys. Res. **A395,** 217-225 (1997).
7. Brown, B.A. and Wildenthal, B.H., At. Data Nucl. Data Tables **33,** 347-404 (1985).
8. Stegmüller, F., Rolfs, C., Schmidt, S., Schulte, W.H., Trautvetter, H.P., Kavanagh, R.W., Nucl. Phys. **A 601,** 168-180 (1996).
9. Schmidt, S., F., Rolfs, C., Schulte, W.H., Trautvetter, H.P., Kavanagh, R.W., Hategan, C., Faber, S., Valnion, B.D., Graw, G, Nucl.Phys. **A 591,** 227-243 (1995).
10. Seuthe, S., F., Rolfs, C., Schröder, U., Schulte, W.H., Somorjai, E., Trautvetter, H.P., Waanders, F.B., Kavanagh, R.W., Ravn, H., Arnould, M., Paulus, G., Nucl. Phys. **A 514,** 471-502 (1990).
11. Axelsson., L., Äystö, J., Borge, M.J.G., Fraile, L.M., Fynbo, H.O.U., Honkanen, A., Jokinen, A., Jonson, B., Lipas, P.O., Martel, I., Mukha, I., Nilsson, T., Nyman, G., Petersen, B., Riisager, K., Smedberg, M.H., Tengblad,O., the ISOLDE Collaboration, Nucl.Phys. **A635** 475-496 (1997).
12. Trinder, W. et al., to be published, (1998).
13. Axelsson., L., Äystö, J., Borge, M.J.G., Fraile, L.M., Fynbo, H.O.U., Honkanen, A., Jokinen, A., Jonson, B., Lipas, P.O., Martel, I., Mukha, I., Nilsson, T., Nyman, G., Petersen, B., Riisager, K., Smedberg, M.H., Tengblad,O., the ISOLDE Collaboration, Nucl.Phys. **A628** 345-362 (1997).
14. Borge, M.J.G. et al., this conference, (1998).

The mechanism of β-delayed two-proton emission in ^{31}Ar

M.J.G. Borge[1], L. Axelsson[2], J. Äystö[3,4], L.M. Fraile[1],
H.O.U. Fynbo[5], A. Honkanen[3], P. Hornshøj[5], A. Jokinen[3,4],
B. Jonson[2], I. Martel[1,4], I. Mukha[5], T. Nilsson[2], G. Nyman[2],
M. Oinonen[3], B. Petersen[5], K. Riisager[5], M.H. Smedberg[2],
O. Tengblad[1,4] and the ISOLDE Collaboration

1. Instituto de Estructura de la Materia, CSIC, E-28006 Madrid, Spain
2. Fysiska Institutionen, Chalmers Tekniska Högskola, S-41296 Göteborg, Sweden
3. Department of Physics, University of Jyväskylä, FIN-40351 Jyväskylä, Finland
4. PPE Divison, CERN, CH-1211 Genève 23, Switzerland
5. Institut for Fysik og Astronomi, Aarhus Universitet, DK-8000 Aarhus C, Denmark

Abstract. Beta-delayed two-proton decays, observed in several nuclei along the proton dripline, have only been studied in detail in very few cases. We have studied the decay of ^{31}Ar, as it was suspected that this nucleus has several β-2p branches with quite large branching ratios. We report here on two experiments done at ISOLDE-PSB (CERN) yielding to the most extensive investigation done so far on a β-delayed two-proton emitter. The analysis of the data have allow us to identify four β-2p branches connecting the IAS with ground and excited states in the two proton daughter ^{29}P. About ten β-2p peaks have been assigned to decays from excited states in ^{31}Cl at lower excitation energies than the IAS. In view of the energy and angular distribution of the individual protons the 2p-decay mechanisms are determined.

Introduction

Proton rich nuclei have always proven to be an excellent source of exotic decay modes. Beta delayed proton emission observed in almost 100 nuclei is considered an extremely sensitive nuclear probe. As experiments extend towards the proton drip line additional exotic decay modes as p-radioactivity, 2p-radioactivity, β-2p are predicted to dominate. The latter decay mode differ from βp-emission as it can proceed via three different mechanisms with differenciated experimental features: (i) Sequential emission via one or a few intermediate states. (ii) Simultaneous correlated emission where the two protons are emitted at the same time in a correlated way, often called "di-proton" emission. (iii) Simultaneous uncorrelated emission, also called "democratic" to emphasize that two-body resonances do not play a significant role.

Experimentally, only the first and last mechanisms have been identified. Most of the studies of the β-delayed 2p emitters show that sequential emission from the

IAS is the dominating decay mode. In the current statistics all but [22]Al [1] are consistent with transitions through only one intermediate state single out of the typically 50-100 states available in the energy region of the 1-proton daughter. On the other hand, the decay of the 2^+ excited state in [6]Be predominantly takes place directly to a three-particle final state [2].

In the following we describe the results of two experiments aiming to study in detail the decay of the $\beta 2p$ emitter [31]Ar as previous data [3] indicated that here the decay mechanism was more complex. Our first experiment [4] identified three β-2p branches from the IAS. This fragmentation of the strength reduces the statistics per branch making difficult to assess the mechanism. Therefore a second set-up was designed with higher granularity to look especifically to the β-2p and β-3p branches.

Experimental Set-up

Our first experiment on May 95 was dedicated to get a general view of the [31]Ar decay scheme with a thorough check to the different decay modes. The set-up is described in detail in ref. [4]. The Ar$^+$ beam passed through the central hole of an annular Si detector and stopped in a thin carbon foil tilted 45° with respect to the beam direction. A 70 % coaxial HPGe-γ-detector was located opposite to the annular detector, and in a plane perpendicular to the beam axis two telescopes were placed, one with Si front and back detectors and the other with a gas filled front detector and a Si back detector.

To disentangle the mechanism used in the 2p-emission, a second set-up was designed and used in June 97. The set-up consisted of a double sided Si strip detector (DSSSD) with 16×16 strips of 3×50 mm^2 and thickness of 276 μm and opposite a semi-sphere consisting of 15 standard 300 μm thick Si detectors with diameter 25.4 mm. The 60 KV Ar$^+$ from ISOLDE-PSB traverses the semi-sphere and it was collected in a thin C-foil placed in the center of the sphere. In order to be able to detect high energy protons a large area 700 μm thick Si detector with 7 cm diameter was placed behind the DSSSD. A factor of 13 for double hit and a factor of 50 for triple hit was gained with respect to the previous apparatus (May 95).

Results and Discussion

We first comment on the general features of the decay scheme. All main proton peaks reported in previous experiments [3,5–7] were observed; the new data are characterized by higher resolution and statistics and by covering a wider energy range. This has allowed us, for instance, to identify the βp from IAS to [29]P ground state. As normalization factor for the proton intensities, a weighted mean of the branching ratios [6,7] of the composite 2 MeV proton peak equal to 29(3) % was used. Combining the information from protons and gamma we get an estimate for the ground state beta feeding of [31]Cl, $B_{g.s.} = 23(8)$ % [4] in good agreement to the expected 25 % obtained in shell model calculation [7]. Four gammas were observed connected to the decay, none in [31]Cl.

From now on we concentrate on the beta delayed two proton branches. The two proton events were identified by proton coincidences between any pair of strips or detectors.

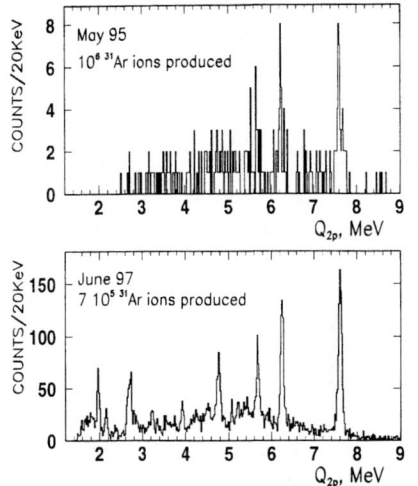

Figure 1 shows the recoil-corrected 2p-sum energy spectrum for protons observed in different detectors summed over all detector combinations. The energy of the missing recoil nucleus was deduced from overall momentum conservation and the finite solid angles were taken into account. The upper part of the figure shows the data from May 95 (cut-off at 1.3 MeV for the 2p-coincidences due to betas in the annular detector). The bottom part, corresponding to data taken on June 1997, shows the very high sensitivity of the second set-up. Both spectra correspond to approximate the same number of ^{31}Ar ions deposited in the foil .

A summary of the energies for the main peaks and respective branching ratios are given in table 1 for both experiments. From the observed Q-values, the excitation energies in ^{29}P [8] and the ground state mass excesses of ^{29}P and ^{31}Cl [9] we determine the excitation energy of the IAS to be E_{IAS}= 12313(60) keV, compatible but more precise than earlier determinations (see details in [4]).

TABLE 1. *Q-values and branching ratios for the resolved 2p-peaks*

	June 97		May 95		
N	Q_{2p} (KeV)	Branching Ratio (%)	Q_{2p} (KeV)	Branching Ratio (%)	E*(^{31}Cl)
1	7608(2)	0.99(11)	7635(25)	1.26(20)	IAS
2	6245(2)	0.75(8)	6230(20)	0.71(12)	IAS
3	5675(3)	0.51(6)	5680(20)	0.61(11)	IAS
4	5226(29)	0.19(2)			IAS
5	5069(5)	0.14(2)			9759(50)
6	4764(4)	0.50(6)			9455(50)
7	3927(6)	0.19(2)			8618(50)
8	3224(7)	0.17(2)			7915(51)
9	2803(8)	0.12(2)			7494(51)
10	2693(4)	0.42(5)			7384(50)
11	2532(11)	0.04(1)			7226(51)
12	2364(8)	0.08(1)			7054(51)
13	2150(4)	0.13(2)			6841(50)
14	1972(4)	0.30(4)			6662(50)

Three of the excited states in ^{31}Cl (N = 9,12,14 in table 1) also decay by 1-p with probabilities between a factor 3-10 higher as expected by phase space and barrier penetrabilities. The main contribution to the error in the determination of the excited state energies in ^{31}Cl is the uncertainty in the ground state mass excess [9].

The 2p-decay from IAS account for 2.44(15) % of the decay. Three other 1p-branch to the g.s. and first two excited states in ^{30}S has also been identified amounting to 0.41(6) %. The resulting observed beta feeding to the IAS accounts for 60 % of the value expected for a pure Fermi transition. Three other decay branches are energetically open to the IAS decay: β-3p, β-α, βpα. In [7] the observation of β-3p emission was reported with a branching ratio of 2.1(10) %. We have searched for these decays and strict upper limits for the different possible decay modes such as 2p-radioactivity ($6.0.10^{-4}$), β-3p (0.12 %), βpα (0.38 %) and β-α (0.03 %) are established (for details see [4] and H. Fynbo et al. in this Proceedings).

The individual proton energy distributions in the β-2p decay gives constrains in the decay mechanism. The data from May 95 could only identify three transitions from the IAS, the study of the energy and angular correlation of the individual protons ruled out the possibility of sequential decay through a few discrete states. A sequential decay fitted the data provided many intermediate states participate. The simultaneous uncorrelated emission was perfectly compatible with the data. The two mechanisms favour spin and parity for the IAS of $J^\pi = 5/2^-$.

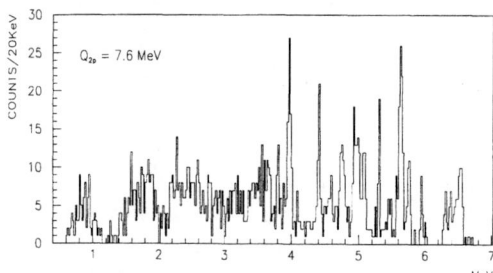
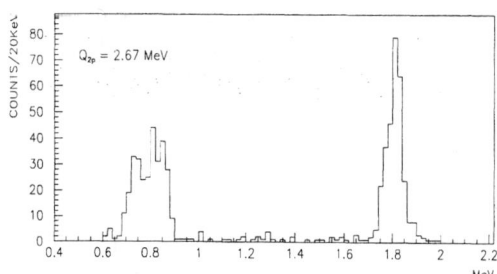

Fig. 2 shows the individual protons for the transitions with $Q_{2p} = 7.6$ MeV (upper) and 2.7 MeV (lower) connecting the IAS and the 7.38(5) MeV state in ^{31}Cl to the ground state in ^{29}P, respectively. In view of the different projections, we conclude that 2p-emission from the IAS very likely proceeds sequentially through many intermediate states in ^{30}S*. The high density of intermediate states available has blurred the signature with respect to the ^{22}Al decay. The 2p-emission from the other ^{31}Cl states proceed sequentially through very few levels. The proton feeding to an intermediate 5.2 MeV state in ^{30}S is suprisingly favoured, strong structural effects must be involved.

REFERENCES

1. M.D. Cable et al., Phys. Rev. **A30** (1984) 1276
2. O.V. Bochkarev et al., Nucl. Phys. **A505** (1989) 215
3. M.J.G. Borge et al., Nucl. Phys. **A515** (1990) 21
4. L. Axelsson et al., Nucl. Phys. **A628** (1998) 345; Nucl. Phys. **A634** (1998) 475
5. V. Borrel et al., Nucl. Phys. **A73** (1987) 331
6. V. Borrel et al., Nucl. Phys. **A531** (1991) 353
7. D. Bazin et al., Phys. Rev. **C45** (1992) 69
8. P.M. Endt, Nucl. Phys. **A521** (1990) 1
9. G. Audi and A.H. Wapstra, Nucl. Phys. **A595** (1995) 409

Spectroscopy of ^{22,23,24}Si and ^{22}Al

S. Czajkowski[a], S. Andriamonje[a], B. Blank[a], F. Boué[b],
R. Del Moral[a], J.P. Dufour[a], A. Fleury[a], E. Hanelt[b], N.A. Orr[c],
P. Pourre[a], M.S. Pravikoff[a], K.-H. Schmidt[d]

[a] *Centre d'Etudes Nucléaires de Bordeaux-Gradignan, F-33175 Gradignan Cedex, France*
[b] *Technische Hochschule, Schloßengartenstr. 9, D-64289 Darmstadt, Germany*
[c] *Laboratoire de Physique Corpusculaire, F-14050 Caen Cedex, France*
[d] *Gesellschaft für Schwerionenforschung, Planckstr. 1, D-64291 Darmstadt, Germany*

Abstract. β-delayed one- and two-proton and α spectroscopic studies of the neutron-deficient nuclei ^{22,23,24}Si and ^{22}Al produced in projectile fragmentation of a ^{36}Ar primary beam at 95 MeV/u have been performed at GANIL. Isotopes of interest were analyzed using the LISE3 spectrometer and were implanted in a telescope made of silicon detectors and a micro-strip gaseous counter where decay particles were detected.

I INTRODUCTION

β-delayed charged-particles are an important tool for spectroscopic studies of very neutron-deficient isotopes of light elements. Partial decay schemes obtained from decay energy spectra for nuclei close to the proton drip-line are often the only source of spectroscopic information. These nuclei may decay by exotic decay modes such as β-delayed one or several protons or β-delayed α emission. The understanding of these decay modes is important for nuclear-structure physics [1].

One can note a predicted $\beta\alpha$ mode for ^{22}Al [1], a known βp and β2p emitter, and β-2p,3p,pα for ^{22,23}Si. Only one transition was known for ^{24}Si [2].

II EXPERIMENTAL SETUP

The experiment has been performed in GANIL at the LISE3 facility which consists of an achromatic magnetic spectrometer followed by a velocity filter to improve the isotopic separation. [3]. The 95 A·MeV ^{36}Ar primary beam of 1 μA impinged on a ^{58}Ni target of 178 mg/cm^2 thickness located at the entrance focus of the spectrometer. The wanted isotope is transmitted to the exit focus with a few contaminants.

Ion identification was performed via ΔE-time of flight measurement. A telescope was located at the exit focus of the spectrometer to implant the selected ions and for spectroscopic studies. Before the implantation setup, a set of aluminium absorbers was used to adjust the residual range of each isotope to the wanted implantation depth.

The telescope consisted of two ΔE silicon detectors of 150 μm thickness and one of 300 μm where the ions were implanted and the emitted particles were detected. The last detector was located on the external face of the main cathode of a microstrip gas counter (MSGC) [4]. The beam was switched off for 100 ms after implantation of a selected isotope, except for ^{24}Si.

III HALF-LIFE AND SPECTROSCOPY OF ^{24}SI

For the ^{24}Si setting the beam structure was 100 ms pulses separated by a 100 ms or 600 ms beam pause. Contaminants are pure β^+-emitter (^{22}Mg) or have a very low βp branch at 832 keV (^{23}Al). All observed peaks above 1 MeV can be unambiguously attributed to β-delayed protons from ^{24}Si. In addition to the previously known proton decay of the IAS to the ground state of ^{23}Mg, seven new peaks have been observed: the decay of the IAS to the first excited state in ^{23}Mg, and several peaks from the decay of 1^+ states in ^{24}Al fed via GT-β-decay [5].

Branching ratios (BR) were determined from peak intensities relative to the number of ^{24}Si ions implanted in the last silicon detector. Both experimental results and theoretical values are reported on a partial decay scheme (Figure 1). There is a general good agreement with shell-model predictions. The BR values for the two GT transitions to the known proton-bound 1^+ states in ^{24}Al can be esimated by means of the β^--decay of the mirror nucleus 24Ne. By adding these values with the ones for the observed βp transitions one reaches nearly 100%, so that all important allowed transitions have been observed.

Time spectrum analysis was restricted to the decay events above 1.4 MeV in order to suppress the β-background contamination. The measured half-life (140±8) ms is in good agreement with the one predicted by Brown (148 ms) and slightly higher than the previous determination (103±42) ms.

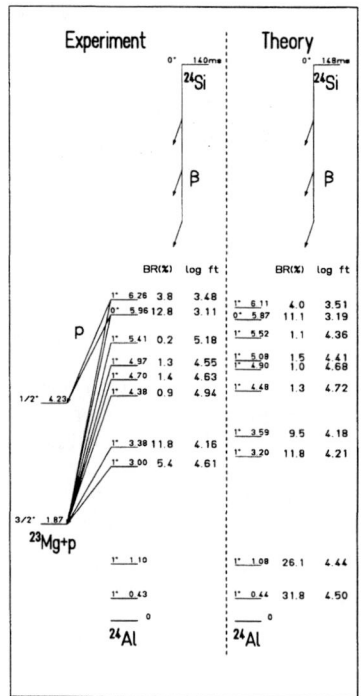

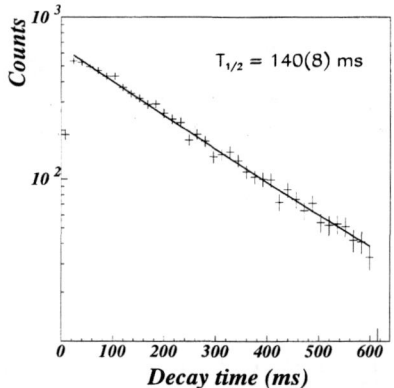

FIGURE 1. *left:* Partial decay scheme for ^{24}Si and theoretical values taken from a shell-model calculation performed by Brown. In addition to the proton decay of the IAS delayed protons from most of allowed transitions have been identified.

right: Time-spectrum for decay events above 1.4 MeV. The measured half-life agrees with but is slightly longer than the old value.

IV MASS MEASUREMENT OF ^{23}SI

β-delayed protons from ^{23}Si have been measured in both, the last silicon detector and the MSGC. Sixteen proton peaks have been observed. As no significant peak is observed where β-delayed α from ^{20}Na is expected one can neglect this contamination. Therefore all peaks can be attributed to β-delayed particles from ^{23}Si [6].

The half-life of ^{23}Si determined from the analysis of decay events above 1 MeV is $T_{1/2}=(42.3\pm0.4)$ ms, in agreement with shell-model predictions from Brown [1].

The main proton peaks are at (1.32 ± 0.04) MeV, (2.40 ± 0.04) MeV and (2.83 ± 0.06) MeV with respective branching ratio $(10\pm1)\%$, $(32\pm2)\%$ and $(14\pm1)\%$.

Two peaks are observed at higher energies (10.41 MeV and 11.66 MeV) that can be attributed to the decay of the IAS to the first excited state in ^{22}Mg at 1.25 MeV and to the ground state, respectively. We can deduce the excitation energy of the IAS in ^{23}Al: (11.78 ± 0.06) MeV, and its mass excess: (18.55 ± 0.06) MeV.

The IAS belongs to the same isobaric multiplet as the excited level at 19.586 MeV in ^{23}Na and as the ^{23}F ground state. Using their experimental mass excesses we determined the coefficients for the IMME and thus calculate the ground-state mass of ^{23}Si. We found a mass excess of m = (23.42 ± 0.10) MeV for ^{23}Si. This can be compared with the extrapolation from Audi and Wapstra which gives a value of (23.53 ± 0.30) MeV [7].

Events corresponding to β-2p decay can be detected simultaneously in the silicon and the MSGC. We identified two such peaks at sum energies of (6.18 ± 0.10) MeV and (5.86 ± 0.10) MeV, attributed to the two-proton emission from the IAS in ^{23}Al to the ^{21}Na ground state and to the first $\frac{7}{2}^+$ level at 320 keV, respectively. However, it is not possible to clearly separate the transitions in the intermediate nucleus ^{22}Mg. β3p and βpα emission are energetically possible for ^{23}Si. The branching ratios are expected to be small and they have not been observed in the present work.

V FIRST SPECTROSCOPIC STUDIES OF ^{22}SI

Several proton groups have been observed in the MSGC spectrum at (1.63 ± 0.05) MeV, (1.99 ± 0.05) MeV, (2.10 ± 0.05) MeV and (2.17 ± 0.05) MeV. Their branching ratios are $(6\pm2)\%$, $(20\pm2)\%$, $(4\pm2)\%$ and $(2\pm1)\%$, respectively. By comparison with the decay of the mirror nucleus, we can attribute these peaks to the proton decay of 1^+ excited states fed by GT transition [8].

Based on the decay events in the main proton groups, we measured the half-life of ^{22}Si. The value of (29 ± 2) ms is in excellent agreement with Brown's predictions [1].

Charged-particle activity has been measured at higher energies up to about 6 MeV. The activity observed below 5.6 MeV can belong to β-2p emission and the one below 3.4 MeV to β-3p emission mainly from the IAS but the low statistics do not allow to distinguish these events from β-delayed one-proton decay.

VI SPECTROSCOPY OF ^{22}AL

Protons can be distinguish from α-paricles in the MSGC by means of a drift time analysis since the range of α-particles is much shorter than the one of protons with similar energy. For this setting, the energy spectrum was dominated by α peaks from the decay of the contaminant ^{20}Na, but by taking into account only events correlated with an implantation in the MSGC one strongly reduces the intensity of all observed peaks except one at (3.27 ± 0.04) MeV that can be unambiguously attributed to β-delayed α decay of ^{22}Al [9]. This transition can be assigned to the α-emission from the IAS to the first 2^+ state in ^{18}Ne at 1.89 MeV. The branching ratio is $(0.31\pm0.09)\%$.

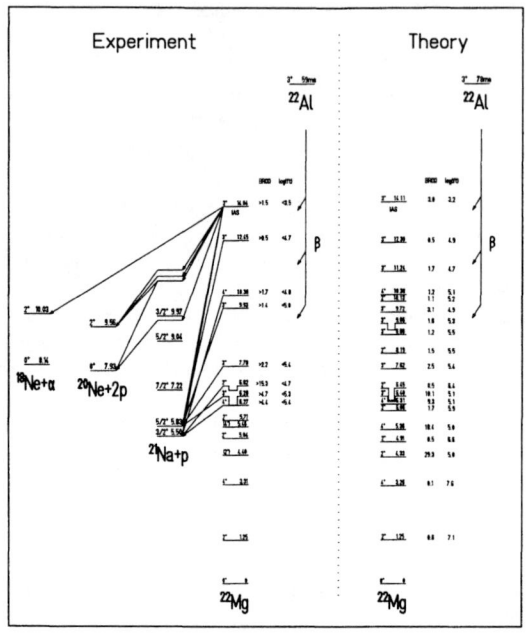

FIGURE 2. Partial decay scheme obtained for ^{22}Al.

In the decay-energy spectrum registered in the last silicon detector we observed several peaks corresponding to proton emission from excited states in ^{22}Mg fed by GT transitions. As above for ^{23}Si we identified several β-delayed two-proton groups. These transitions have been reported on the partial decay scheme shown in Figure 2. Relative intensities and J^π considerations allowed us to assign the different proton and two-proton lines to transitions between individual states. From this decay scheme the spin and parity of the ground state of ^{22}Al should be most likely 3^+ although its mirror nucleus has a 4^+ ground state [9].

REFERENCES

1. B.A. Brown, Phys. Rev. Lett. 65, 2753 (1990)
2. J. Äystö et al., Phys. Rev. C23, 879 (1981) and references therein
3. A.C. Müller and R. Anne, Nucl. Instrum. Methods B 56, 5597 (1991)
4. B. Blank et al., Nucl. Instrum. Methods A 330, 83 (1993)
5. S. Czajkowski et al., Nucl. Phys. A 628, 537 (1998)
6. B. Blank et al., Z. Phys. A 357, 247 (1997)
7. G. Audi and A.H. Wapstra, Nucl. Phys. A595, 409 (1995)
8. B. Blank et al., Phys. Rev. C 54, 572 (1996)
9. B. Blank et al., Nucl. Phys. A 615, 52 (1997)

Beta decay of neutron-rich cobalt and nickel isotopes

S. Franchoo*, B. Bruyneel*, M. Huyse*, U. Köster[†], K.-L. Kratz[‡],
K. Kruglov*, Y. Kudryavtsev*, W.F. Mueller*, B. Pfeiffer[‡],
R. Raabe*, I. Reusen*, P. Thirolf[||], P. Van Duppen*,
J. Van Roosbroeck*, L. Vermeeren*, W.B. Walters[¶], L. Weissman*,
and A. Wöhr*[1]

*Instituut voor Kern- en Stralingsfysica, University of Leuven, B-3001 Leuven, Belgium
[†]Physik-Department, Technical University of Munich, D-85748 Garching, Germany
[‡]Institut für Kernchemie, University of Mainz, D-55099 Mainz, Germany
[||]Sektion Physik, University of Munich, D-85748 Garching, Germany
[¶]Department of Chemistry, University of Maryland, College Park, Maryland 20742

Abstract. We report on the first β-γ spectroscopy measurements of the neutron-rich $^{68-70}$Co and $^{70-74}$Ni nuclei, produced in proton-induced fission of ^{238}U and ionized in a laser ion guide coupled to an on-line mass separator. Several γ lines from the decay of these nuclei have been identified, half-lives determined and production cross sections deduced. The derived level schemes for the copper and nickel isotopes show that the occupation of the $\nu(1g_{9/2})$ state has a strong influence on the structure of these neutron-rich nuclei. This may have a clear impact on the predicted structure and decay properties of doubly-magic ^{78}Ni.

The neutron-rich cobalt and nickel isotopes are of interest both for nuclear physics and astrophysics. From the nuclear physics point of view, the nickel isotopes are singly closed-shell nuclei on the way to ^{78}Ni. For astrophysics, the decay properties in this region of the nuclear chart can be crucial for understanding the nucleosynthesis r-process path. Multiparticle transfer studies have uncovered states in ^{68}Ni and ^{69}Cu [1], while the investigation of microsecond isomers produced in fragmentation reactions has revealed the first excited states in $^{68-70}$Ni and $^{71-72}$Cu [2]. However, the heaviest cobalt and nickel isotopes for which β-γ decay information was available remained ^{67}Co and ^{69}Ni [3].

The IGLIS Ion Guide Laser Ion Source development at the LISOL separator is especially devoted to the study of these isotopes by combining the high efficiency

---

[1]) Present address: Physics Department, University of Oxford, Oxford OX1 3PU, United Kingdom

TABLE 1. Measured half-lives for $^{68-74}$Ni, compared to literature values taken from [6–8] and QRPA predictions from [13].

	^{68}Ni	^{69}Ni	^{70}Ni	^{71}Ni	^{72}Ni	^{73}Ni	^{74}Ni
This work	29(2) s	11.2(9) s	6.0(3) s	2.56(3) s	1.57(5) s	0.84(3) s	0.9(2) s
Literature	19^{+3}_{-6} s	11.4(3) s	—	1.9(4) s	2.1(3) s	0.6(1) s	0.5(2) s
QRPA	28.4 s	18.9 s	6.9 s	2.3 s	1.75 s	0.74 s	0.77 s

and selectivity of resonant laser ionization with the fast evacuation times reached by the ion-guide technique [4]. A pulsed 30-MeV proton beam with an averaged intensity of 5 μA was directed onto the ^{238}U target foils. The laser-ionized reaction products were extracted through a Sextupole Ion Guide (SPIG) [5]) and mass separated. The nuclei were implanted into a movable tape around which the detection set-up was constructed. The set-up consisted of thin plastic ΔE detectors and two high-efficiency germanium detectors. The acquisition was triggered by β-γ or γ-γ coincidence events.

The $^{66-70}$Co and $^{68-74}$Ni isotopes were studied by comparing the β-delayed γ spectra with and without laser irradiation of the ion source. Coincidence requirements allowed for further identification of γ lines. Half-life information was extracted from the time behaviour of the γ-ray intensity, and was relied upon to discriminate between the mother nucleus and the daughter products. The results for the nickel isotopes are presented in Table 1. The half-lives measured in this work for $^{66-70}$Co are still under evaluation with the exception of ^{69}Co which is discussed below.

The achieved production rates for cobalt and nickel were deduced from the most intense γ ray following the decay of the isotope or its daughter. Branchings were adopted from literature [8], except at mass 70. The level scheme we constructed for ^{70}Zn differs considerably from the literature [9] and forced us to conclude that no groundstate feeding occurs in the decay of ^{70}Cu. For the 450-keV line of ^{73}Cu an intensity of 43(2)% was taken [10] and for the 606-keV transition in ^{74}Cu a value of 79(15)% [11]. The results are displayed in Figure 1.

Figure 1 also compares the experimental production rates with theoretical calculations [12]. For the nickel isotopes, the shape of the experimental curve is well reproduced by a gaussian fit with the maximum situated at mass 69.5(3), which is closer to stability than the theoretically predicted centroid at 71.1. For the cobalt isotopes only the slope of the production curve has been measured, since for ^{66}Co we could only deduce a lower limit. Therefore, we supposed a width equal to that of the experimental nickel fit and found a similar shift for the maximum. This maximum was found to be 66.7(3) for the experimental values, while the theoretical values maximize at 68.5.

Next to the nickel half-lives, we list, in Table 1, the QRPA calculations by one of the authors [13] that take into account locally known nuclear structure. These are in remarkable agreement with the data. For ^{68}Ni, the newly measured half-life corresponds much better to the expectations than the literature value. The

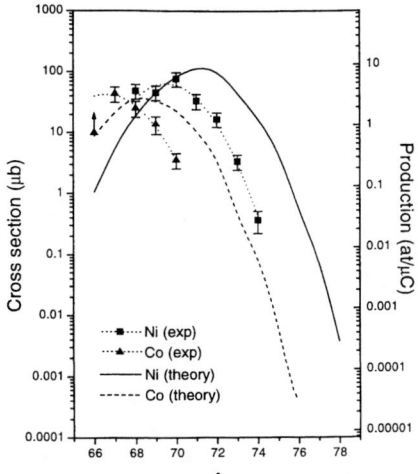

FIGURE 1. Comparison of experimental production yields for $^{68-74}$Ni and $^{66-70}$Co, versus cross section calculations for 30-MeV protons on ^{238}U [12].

theoretical model predicts a half-life of 210 ms for ^{78}Ni.

As an example of the nuclear structure information extracted, we discuss the ^{69}Co-^{69}Ni-^{69}Cu decay chain. In ref. [14] the β-gated γ spectra obtained at mass 69 with the laser on resonance for Co is shown. Two prominent γ-lines at 594 keV and 1296 keV are observed. Tuning the lasers to Ni resonance confirms that the 1296-keV line is from the decay of ^{69}Ni and the 594-keV transition is from ^{69}Co decay. In addition, from the time behaviour of the γ-intensities and the Z-identification, one concludes that the 1296-keV line can be attributed to the decay of an isomer in ^{69}Ni with $T_{1/2} = 3.5 \pm 0.5$ s while the 594-keV line belongs to the decay of ^{69}Co with $T_{1/2} = 216 \pm 9$ ms. Figure 2 shows part of the decay scheme of ^{69}Co-^{69}Ni-^{69}Cu displaying only the main decay branch. We also indicate the most probable shell-model configuration for the different states. The decay of ^{69}Co is governed by the $\nu(f_{5/2}) \to \pi(f_{7/2})$ Gamov-Teller decay while the subsequent decay of ^{69}Ni($1/2^-$) is determined by the $\nu(p_{1/2}) \to \pi(p_{3/2})$ Gamov-Teller decay. The latter feeds a state at 1296 keV in ^{69}Cu that has mainly a $\nu(p_{1/2})^{-2}(g_{9/2})^{+2}$ configuration. In the neighbouring even-even ^{68}Ni nucleus this (2p-2h) excitation accross the $N = 40$ subshell closure lies at an excitation energy of 1770 keV ($I^\pi = 0^+$) [15].

In conclusion, the laser ion-guide setup has been used to perform significantly improved measurements for the decay of very neutron-rich cobalt and nickel nuclei and determine new level structures for the heavy copper and nickel isotopes. The structure of these nuclei indicates a strong influence of the occupation of the $\nu(g_{9/2})$ orbital on the single-particle levels. The re-ordering of the single-particle levels could have profound effects on the shell structure for nuclei near and beyond the doubly magic ^{78}Ni.

We gratefully thank J. Gentens and P. Van den Bergh for running the LISOL

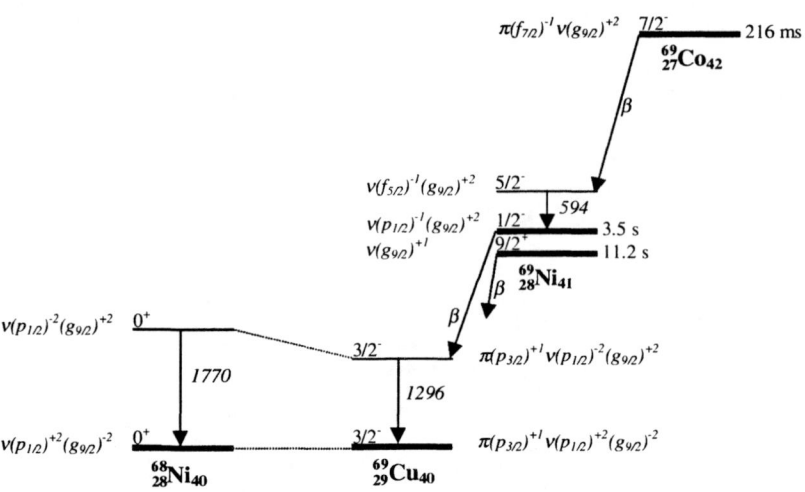

FIGURE 2. Partial decay scheme of ^{69}Co-^{69}Ni-^{69}Cu. The first excited state in ^{68}Ni is shown to the left of the scheme for comparison. Possible configurations are suggested.

separator. This work is supported by the Inter-University Attraction Poles (IUAP) Research Programme and by the Fund for Scientific Research – Flanders (FWO-Belgium). M.H. is Research Director, L.V. Postdoctoral Researcher, and S.F. Research Assistant of the FWO, Belgium.

REFERENCES

1. R. Broda et al., Phys. Rev. Lett. **74**, 868 (1995).
2. M. Pfützner et al., Nucl. Phys. **A626**, 259c (1997).
3. U. Bosch et al., Nucl. Phys. **A477**, 89 (1988).
4. Y. Kudryavtsev et al., Nucl. Instr. and Meth. B **114**, 350 (1996).
5. P. Van den Bergh et al., Nucl. Instr. and Meth. B **126**, 194 (1997).
6. M. Bernas et al., Z. Phys. A **336**, 41 (1990)
7. F. Ameil et al., Eur. Phys. J. A **1**, 275 (1998)
8. R. Firestone, Table of Isotopes, 8th Edition, John Wiley and Sons, New York (1996).
9. W. Reiter, W. Breunlich, and P. Hille, Nucl. Phys. **A249**, 166 (1975).
10. P. Mantica, private communication.
11. B. Fogelberg, private communication.
12. M. Huhta et al., Phys. Lett. B **405**, 230 (1997).
13. K.-L. Kratz et al., Z. Phys. A **332**, 419 (1989).
14. P. Van Duppen, this conference proceedings.
15. M. Girod et al., Phys. Rev. C **37**, 2600 (1988).

IS THERE A β3p BRANCH IN THE DECAY OF ^{31}Ar ?

H.O.U. Fynbo[1], J. Äystö[2], M.J.G. Borge[3], L.M. Fraile[3],
A. Honkanen[2], P. Hornshøj[1], Y. Jading[4], A. Jokinen[2], B. Jonson[5],
I. Martel[4], I. Mukha[1,†], G. Nyman[5], M. Oinonen[2], K. Riisager[1],
T. Siiskonen[2], O. Tengblad[3], F. Wenander[2] and the ISOLDE
collaboration[4]

1 *Institut for Fysik og Astronomi, Aarhus Universitet, Denmark*
2 *Department of Physics, University of Jyväskylä, Finland*
3 *Instituto Estructura de la Materia, CSIC, Madrid, Spain*
4 *PPE Division, CERN, Genève, Switzerland*
5 *Fysiska Institutionen, Chalmers Tekniska Högskola, Göteborg, Sweden*
† *on leave from Kurchatov Institute, Moscow, Russia*

Abstract. An improved limit on the beta-delayed 3p branch from ^{31}Ar is reported.

Beta-delayed multi-particle emission becomes increasingly important when approaching the drip lines [1]. On the proton rich side, where experimental studies are more feasible, the mechanism of two-proton (2p) emission from states fed in β-decay and in reactions has been studied intensively during the last 15 years, the question being whether the ^{1}S resonance in the proton-proton system will dominate the final state of the 2p-emission or whether the emission goes sequentially via intermediate states in the daughter nuclei. Some models suggest an enhancement of the angular distribution at small angles and the continuous energy distributions of the individual protons. In most cases studied experimentally the mechanism has been established to be sequential, see also [2,3]

In an experiment performed at GANIL 8 years ago the first observation of β-delayed three-proton emission was claimed in the decay of ^{31}Ar [4]. This observation was based partly on energetics, with the proposed β3p branch going through the isobaric analog state (IAS) in ^{31}Cl to the ground state in ^{28}Si and with an estimated branching ratio of 2(1) %. The detector used in this experiment consisted of effectively 8 segments enabling events with higher multiplicities to be detected, and the events assigned to the β3p branch had a multiplicity of at least 2.

In a recent experiment we have measured the β-decay of ^{31}Ar at ISOLDE with a setup designed to have a high efficiency for multi-particle decays. The ^{31}Ar beam,

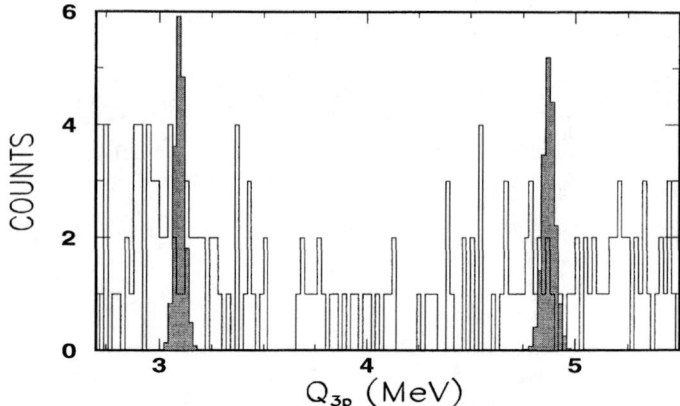

FIGURE 1. Q_{3p}-spectrum from multiplicity 3 events. The shaded peaks show the results of a Monte-Carlo simulation of the 3p emission.

with an intensity of 3 atoms/s from the ISOLDE separator, passed through the central hole of a specially designed semi-sphere consisting of 15 ordinary silicon (Si) surface barrier detectors, and was stopped in a carbon foil. On the other side of the foil was placed a double sided (16 × 16 strips) Si detector. With a total solid angle of 34 % of 4π divided into 271 segments this setup combines excellent efficiency with good angular resolution, both essential to this type of study. With it we can perform a very stringent test of the existence of the proposed $\beta 3p$ branch. For the results on the 2p branches from the same experiment see [2].

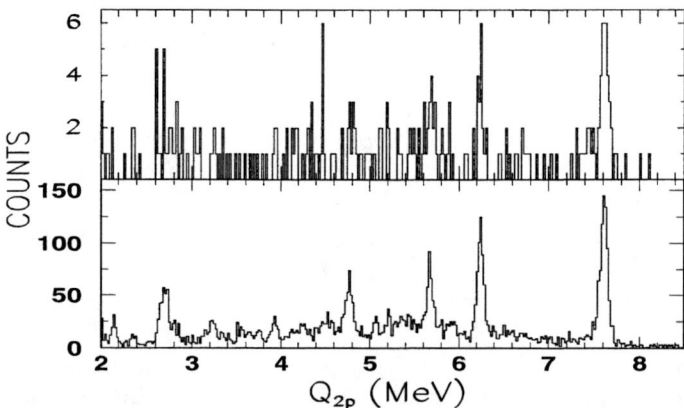

FIGURE 2. Top : Q_{2p}-spectrum obtained from the multiplicity 3 events assuming the signal with the lowest energy comes from a β-particle. Bottom : the Q_{2p} spectrum from multiplicity 2 events.

With the high segmentation of the setup, the analysis of the multi-particle events is in principle straight forward. From the energy and angles of the 3 detected particles we use momentum conservation to calculate the recoil energy of the daughter nucleus and thereby reconstruct the full decay energy (Q_{3p}) of the event. In Figure 1 is shown the Q_{3p} spectrum obtained in this way. Also shown in the figure is the result of a Monte-Carlo simulation of 3p-emission from the IAS to the ground state and first excited state of ^{28}Si. In this simulation the individual proton energies and angles are distributed according to phase space only. The energy resolution and geometry of the detectors is then taken into account leading to the peaks shown in the figure. First we notice that the data is not restricted to the kinematically allowed regions in the plot given by the simulation, meaning that there is a background of events of another type. This background is easily identified to be βpp events in the following way. For each multiplicity 3 event we assign the signal with the lowest energy as the β-particle and calculate the Q_{2p}-value of the remaining two signals assuming they correspond to protons. The result of this procedure is shown in Figure 2 where for comparison we also show the full Q_{2p} spectrum from the multiplicity 2 events. By comparing the two figures the hypothesis about the nature of the background in Figure 1 is clearly seen to be correct. By comparing the number of βp events with the number of p events and correcting for solid angle, the number of βpp events is found to be consistent.

In the analysis of the Q_{3p} spectrum we consider the βpp events to give a constant background in the kinematically allowed regions for the 3p events. This background is estimated from both sides of the relevant regions, which themselves are determined by the width of the peaks from the Monte-Carlo simulation. The problem of how to place upper limits in the case of Poisson statistics with constant background has been treated in the recent paper by Feldman et al. [5]. The result of this procedure is summarized in Table 1, where we also give limits corresponding to lower limits on the individual proton energies of 400 keV and 600 keV. We do not expect to see protons with energy less than 400 keV due to the Coulomb barrier. The limits given are at the 99% confidence level.

IAS → ground state in ^{28}Si : Q_{3p} = 4.781(50) MeV				
E_{TH}	COUNTS	BACKGROUND	UPPER LIMIT	BRANCH (%)
0 keV	15	8	19.81	$8.4 \cdot 10^{-2}$
400 keV	8	4	14.27	$6.1 \cdot 10^{-2}$
600 keV	4	2	10.23	$4.3 \cdot 10^{-2}$
IAS → 1. exited state in ^{28}Si : Q_{3p} = 3.092(50) MeV				
E_{TH}	COUNTS	BACKGROUND	UPPER LIMIT	BRANCH (%)
0 keV	21	6	29.5	$12.5 \cdot 10^{-2}$
400 keV	4	4	9.23	$3.9 \cdot 10^{-2}$
600 keV	0	2	3.26	$1.4 \cdot 10^{-2}$

TABLE 1. Upper limits for β3p branches in the decay of ^{31}Ar via the IAS in ^{31}Cl.

In both cases the introduction of low energy cuts lead to large reductions in the count numbers, which indicates that most of the multiplicity 3 events are βpp events.

The upper limits given in Table 1 are more than a factor of 25 below the intensity reported by Bazin et al. [4].

Turning now to a possible explanation of this discrepancy we first note that their observation was based partly on the sum energy in the detector system being the expected for a transition to the ^{28}Si ground state, and partly on the multiplicity being at least two. Secondly we note that with these criteria it is not possible to distinguish between a 3p-branch from the IAS and a 2p-branch from another state in ^{31}Cl that accidentally gives the same sum-energy. Thus we suggest that the 2p-peak seen in Figure 2 at 4.8 MeV explains the observation of Bazin et al. The intensity of this peak we find to be 0.6(1) %.

In conclusion, we have re-measured the β-decay of the drip line nucleus ^{31}Ar and found that the previous claim of the existence of a strong three-proton branch in this decay could not be verified.

REFERENCES

1. Jonson, B. and Nyman, G., in Nuclear Decay Modes, ed. Poenarn, D.N, Bristol: Inst. Pysics, 1990, p.102
2. Borge, M.J.G, These proceedings.
3. Axelsson, L et al., Nucl. Phys. **A628**, 345-362 (1998)
4. Bazin, D. et al., Phys. Rev. **C45** (1992) 69
5. Feldman, G.J and Cousins R.D Phys. Rev. **D57** (1998) 3873

Progresses in statistical analysis of β-delayed proton emission

J. Giovinazzo^a, Ph. Dessagne^b, Ch. Miehé^b
and the ISOLDE collaboration

^{a} Centre d'Etudes Nucléaires de Bordeaux-Gradignan, IN$_2$P$_3$, FRANCE
^{b} Institut de Recherches Subatomiques de Strasbourg, IN$_2$P$_3$, FRANCE

Abstract.
In the decay of the series of $T_Z = 1/2$ nuclei, ^{65}Ge, ^{69}Se, ^{73}Kr and ^{77}Sr, a statistical analysis was applied to the β delayed proton emission to obtain nuclear structure information. New results emerge from this study, in the light of improvements in the experimental techniques and developments of the analysis method. Experimental results are analysed by comparison with a statistical model, in order to extract information on level densities and partial transition widths of the proton emitting nuclei. Limitations in this comparison brought us to take into account more detailed descriptions of the processes involved in the decay. This concerns the proton transmission coefficients, the parity asymmetry in level densities and the evaluation of electromagnetic transition widths.

INTRODUCTION

The $A \simeq 70$ neutron deficient mass region is of special interest, since it is known to exhibit some structure effects like shape transitions or isomerism, or large deformations in the ground state A statistical analysis of delayed proton emission allows the characterization of level densities and partial transition widths in the $Q_{EC} - S_P$ window. New results could be obtained in that field, with the help of improvements in experimental techniques and analysis methods.

The β^+-EC decay of nuclei under consideration (^{65}Ge, ^{69}Se, ^{73}Kr, ^{77}Sr, with $T_Z = 1/2$) populates states in the $Q_{EC} - S_P$ window, that can decay neither via γ deexcitation, nor via proton emission. Due to the high level densities in this energy range, individual transitions are not accessible experimentally. The structure of proton energy distribution is then interpreted in terms of a statistical behaviour of levels spacing and partial decay widths [1].

In addition, the X-rays resulting from the K shell filling after an electron capture may be used as a clock for nuclear processes. In fact, the ratio R_X, defined as the ratio of intensities of X-rays emitted after or before the proton is, for an individual transition, directly related to nuclear lifetimes [2].

STATISTICAL ANALYSIS

In a first step, the experimental data are compared to a statistical model taking into account all processes of the decay [3]. The level densities are calculated with the *back shifted Fermi gas model*, with a free density parameter a; β^+/EC intensity is determined with a linear S_β strength function; γ deexcitation is evaluated using the asumption that $E1$ transitions are dominating the electromagnetic decay; proton emission probabilities are based on proton transmission coefficients T_ℓ calculated in the Gamow formalism. Correction factors are introduced on Γ_γ and T_ℓ, as free parameters of the model. In addition, statistical distributions are taken into account for level spacing (Wigner or Poisson law) and partial decay widths (Porter-Thomas law).

Parameters are estimated in order to reproduce the proton energy distribution and the ratio R_X (as defined below). The comparison of the global structure of proton spectra is based on the center and the width of the energy distribution, and the fine structure of experimental and simulated data is compared in terms of its normalised variance, that characterises the fluctuations observed in the spectra. We also expect the model to reproduce the feeding of excited states in the final nucleus.

The variance analysis represents a strong constraint that fixes the level density parameter for all nuclei under interest. In order to reproduce the experimental distributions, it is necessary to apply strong correction factors to T_ℓ coefficients (see table 1).

MODEL DEVELOPMENTS

This first approach brought us to take into account more detailed descriptions of the processes involved in the decay [4]. Due to correction factors applied on the T_ℓ coefficients, calculations have been performed in order to evaluate the influence of nuclear deformation. Moreover, since the first model is unable to reproduce the final states feeding rates (via proton emission) in the decay of ^{77}Sr, we introduced low energy shell structure, by considering an asymmetry of parities in level densities. Finally, we estimated the relative importance of $M1$ transitions in the γ deexcitation.

For proton transmission coefficients, several optical model calculations were performed, using a phenomenological (Wood-Saxon type) or a semi-microscopic [5] potential. In all cases, a correction factor is still required to reproduce experimental data. In the case of the phenomenological potential, calculations were performed in the frame of a deformed optical model, but the effect on T_ℓ is one order of magnitude lower than the correction factors.

Due to the low excitation energy of levels implied in the decay, the shell structure is not smoothed. This effect is taken into account with parity asymmetry of level density, calculated in the frame of a combinatory analysis [6] based on Hartree-Fock calculation of individual particle states and considering particle-hole and collective excitations. The parity asymmetry has a sensible effect on Γ_γ calculation, but doesn't strongly affect the results of the statistical analysis.

Since this asymmetry changes the distribution of levels with different parities, it may also change the relative importance of $M1$ and $E1$ transitions. $M1$ deexcitation has been considered [7], and it appears not to be neglectible compared to $E1$ transition, as it was generally admitted in this kind of analysis. For nuclei under study, $M1$ partial widths are reinforced if one considers the parity asymmetry.

CONCLUDING REMARKS

Those investigations, performed in the light of improved experimental data, lead to a characterization of low spin states in proton emitting nuclei. We could obtain the level density parameter a and an evaluation of proton and γ partial decay widths. Resulting values for a are much higher than the ones deduced in previous work (see table 1). In present work, this parameter is strongly constrained by the variance analysis that could be performed because of the improvements in energy resolution in proton spectra, and high statistics data.

The proton transmission coefficients deduced from optical model calculations still have to be corrected to reproduce the data. The potential parameters result from an extrapolation far from stability since no data is available in this mass region. This points out the importance of measures far from stability to extend the description of nuclear processes.

The parity asymmetry introduced in the model shows the importance of the shell structure at low excitation energy, due to the selection rules of the invloved processes. In addition, it tends to increase the $M1$ transition widths compared to $E1$ deexcitation.

emitter nucleus	$a\ (MeV^{-1})$ this work	J.C. Hardy [1]	γ correction	T_ℓ correction
^{65}Ga	9.5 (2)	6.9	0.3 (1)	15 (4)
^{69}As	10.0 (2)	8.0	0.3 (1)	8 (2)
^{73}Br	13.2 (2)	9.2	0.7 (2)	22 (4)
^{77}Rb	11.2 (2)		0.5 (2)	12 (3)

TABLE 1. Level density parameters and partial widths correction factors resulting from the statistical analysis.

REFERENCES

1. J.C. Hardy, 4th Int. Conf. on nuclei far from stability CERN Report 81-09 (1981) 217.
2. J.C. Hardy, E. Hagberg, Particle emission from nuclei, vol. 3 (CRC Press Inc., Florida, 1989) p. 99.
3. J. Giovinazzo, Ph. Dessagne, Ch. Miehe, The delayed proton emission in the $A \simeq 65 - 77$ mass region, statistical aspects and structure effects, submitted to Nucl. Phys. A (1998).
4. J. Giovinazzo, PhD thesis, University of Strasbourg - France (1997).
5. E. Bauge, *et al.*, A semi-phenomenological spherical optical model potential for elastic scattering of nucleons up to 200 MeV.
6. S. Hilaire, Microscopic Level Densities, Spec. Meet. Nucleon-Nucleus Opt. Model up to 200 MeV (1996).
7. P.M. Endt, Strengths of gamma-ray transitions in A = 45-90 nuclei Atomic Data and Nucl. Data Tab. 23 (1979) 547.

Beta-decay of ^{97}Ag: Evidence for the Gamow-Teller Resonance near ^{100}Sn

Z. Hu[1], L. Batist[2], J. Agramunt[3], A. Algora[3], B. A. Brown[4],
D. Cano-Ott[3], R. Collatz[1], A. Gadea[3], M. Gierlik[5], M. Górska[1],
H. Grawe[1], M. Hellström[1], Z. Janas[5], M. Karny[5], R. Kirchner[1],
F. Moroz[2], A. Plochocki[5], M. Rejmund[1], E. Roeckl[1], B. Rubio[3],
M. Shibata[1], J. Szerypo[5], J. L. Tain[3] and V. Wittmann[2]

[1]*Gesellschaft für Schwerionenforschung, D-64291 Darmstadt, Germany*
[2]*St. Petersburg Nuclear Physics Institute, 188-350 Gatchina, Russia*
[3]*Instituto de Fisica Corpuscular, Dr. Moliner 50, E-46100 Burjassot-Valencia, Spain*
[4]*Michigan State University, East Lansing, Michigan 48824, U. S. A.*
[5]*Institute of Experimental Physics, University of Warsaw, PL-00681 Warsaw, Poland*

Abstract. The ^{97}Ag → ^{97}Pd β-decay was investigated by using a total absorption spectrometer and an array of 6 Euroball-Cluster Ge detectors. A total of 603 γ-rays de-exciting 151 levels in ^{97}Pd have been assigned. The Gamow-Teller β-decay strength distributions from the experiment and a shell-model calculation are compared, revealing a dominant resonance around a ^{97}Pd excitation energy of 4 MeV with a width of about 1.5 MeV. An experimental quenching factor of about 4.9(7) for the total Gamow-Teller strength was obtained, which is close to the predicted theoretical hindrance factor.

Introduction

The problem of missing strength in Gamow-Teller (GT) β-decay has attracted considerable experimental and theoretical interest in recent years, especially concerning the region near the double-magic nucleus ^{100}Sn (see *e.g.* [1]). As part of an ongoing research program on β-decay near ^{100}Sn, we investigated ^{97}Ag. On the basis of an extreme single-particle model, one expects this decay to be dominated by the "core decay", *i.e.* to mainly populate, after breaking a $\pi g_{9/2}^2$ pair, *3qp* states ($\pi g_{9/2}^{-2} \nu g_{7/2}$) in ^{97}Pd at excitation energies around 4 MeV.

As complementary spectroscopic tools, we used a total absorption spectrometer (TAS) [2] and a cube-like array of 6 Euroball-Cluster Ge detectors (Cluster Cube) [3]. The "double strategy" of combining high- and low-resolution studies can serve to map the GT strength distribution even at high excitation energies of the daughter nucleus, which indeed represents a challenge in studying nuclei far from stability.

CP455, *ENAM98: Exotic Nuclei and Atomic Masses*
edited by B. M. Sherrill, D. J. Morrissey, and Cary N. Davids
© 1998 The American Institute of Physics   1-56396-804-5/98/$15.00

Experimental Techniques

The TAS is a highly efficient NaI detector which allows to measure the β-intensity distribution rather than the individual γ-rays. The Cluster Cube represents a compromise between high resolution and high efficiency. It surrounds the source with a solid angle of ~65% of 4π sr, the total efficiency for 1.33 MeV γ-rays being ~19%. For this γ-ray energy, the actual energy resolution of an individual capsule is ~2.6 keV.

The experiment was performed at the GSI on-line mass separator equipped with a FEBIAD ion source [4]. ^{97}Ag was produced by fusion-evaporation reactions induced by a ^{40}Ca beam from the UNILAC on an isotopically enriched ^{60}Ni target. The mass-separated A = 97 beam was implanted into a tape. After a selected collection period, which was optimized for the half-life of ^{97}Ag and thus suppressed longer-lived activity such as the isobaric contaminant ^{97}Pd, the resulting radioactive source was periodically transported to the center of either spectrometer.

Experimental Results

Based on the preliminary analysis of the Cluster Cube data, we have placed a total of 603 γ rays (578 new) depopulating 151 (132 new) levels in the β-decay scheme of ^{97}Ag. This scheme was applied for de-convoluting the TAS data with a so-called "peel-off" method [5]. From the preliminary evaluation of the TAS data, a Q_{EC} value of 6.93(10) MeV was determined in agreement with a systematic estimate of 7.0(5) MeV [6]. The β-intensity distributions obtained from de-convoluting the TAS spectra and

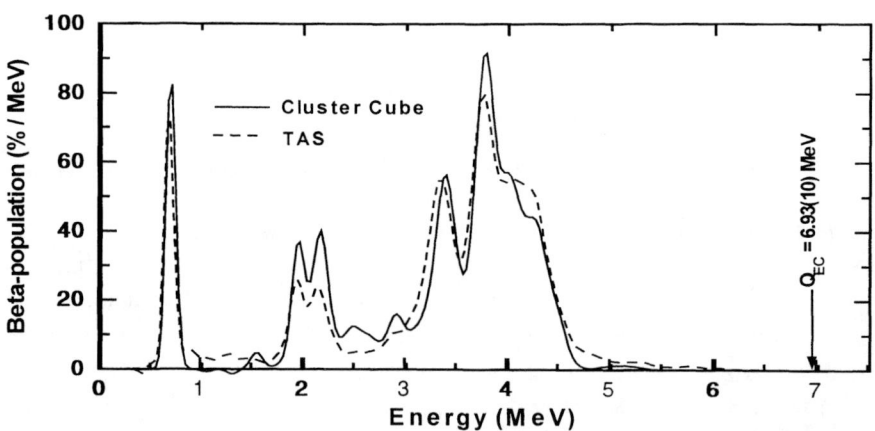

Figure 1. Beta-intensity distributions obtained from the TAS data (dashed line) and from the Cluster Cube data (solid line). The latter result was adapted to the TAS data by smoothing procedure.

from the γ-intensity balances based on the Cluster Cube data agree with respect to the overall shape which is dominated by a resonance between ~3 MeV and ~4.5 MeV (see Fig. 1). However, the Cluster Cube data show a little more β-feeding in the energy region below 4 MeV than the TAS data, while missing some β-feeding above that energy. According to the Cluster Cube data, the states above 4 MeV receive ~21% of the total decay intensity, compared to a value of ~26% from the TAS data. This ~5% difference can be interpreted by assuming that the Cluster Cube measurement has still missed some weak γ-rays emitted from high-lying ^{97}Pd levels.

Discussion and Conclusion

Using the Q_{EC} value obtained from the TAS data, the half-life value of 25.3(3) s determined from a previous work [7], and the β-intensity distributions deduced in this work, we have calculated the ^{97}Ag β-strength shown in Fig. 2. The global shapes of the distributions deduced from the TAS and Cluster Cube data, respectively, are in good agreement, showing a large GT resonance around 4 MeV with a width of about 1.5 MeV. However, as far as the total B(GT) values are concerned, we obtained a value of 2.14(38) from the Cluster Cube data in comparison to a value of 2.63(37) from the TAS data (We have excluded the B(GT)'s above 5.5 MeV due to large statistic uncertainties in the TAS result, as shown in Fig. 2). This discrepancy mainly results from the difference in the energy range above 4 MeV, where TAS yielded a summed

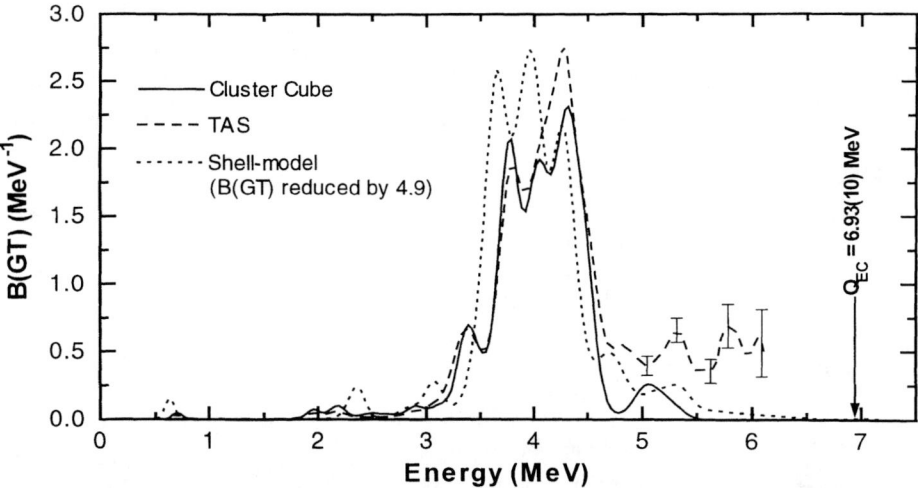

FIGURE 2. B(GT) distributions for the decay of ^{97}Ag from a shell-model calculation (dotted line) and from experiments performed with the TAS (dashed line) and the Cluster Cube (solid line). The B(GT) distribution from the shell-model calculation and the Cluster Cube measurement were adapted to the TAS data by a smoothing procedure. The error bars shown at the high-energy tail represent the statistic uncertainties of the TAS data in this region.

GT strength $\Sigma B(GT)$ of 1.65 compared to a value of 1.15 from the Cluster Cube data (excluding the B(GT)'s above 5.5 MeV). We conclude that, in comparison with the TAS data, the Cluster Cube data have missed 19(10)% of $\Sigma B(GT)$, which corresponds to the missing of ~5% of the β-feeding intensities above 4 MeV.

We have calculated the GT β-decay distribution of ^{97}Ag using the SNB model space, in which the interaction for protons is confined to the $1p_{1/2}$ and $0g_{9/2}$ orbits, and that for neutrons to the $0g_{7/2}$, $1d_{5/2}$, $1d_{3/2}$, $2s_{1/2}$ and $0h_{11/2}$ orbits [1]. As can be seen from Fig. 2, the shell-model calculation qualitatively reproduces the experimental results, especially concerning the centroid and width of the GT resonance. By comparing the $\Sigma B(GT)$ value deduced from the TAS data to the shell-model prediction, we obtained an experimental quenching factor of 4.9(7). This result is close to the value of 5.1±0.4 for ^{103}In [5], both experimental finding being in agreement with the hindrance factor of 4.4 expected from $0\hbar\omega$ excitations beyond the SNB space and from even higher-order configuration mixing (see refs. [5,8] for a detailed discussion). It is worth noting that ^{97}Ag and ^{103}In are the first two nuclei which are close enough to ^{100}Sn so that large space shell-model calculation can be performed and for which the complete GT resonance was observed experimentally.

Acknowledgments

This work was supported in part by the European Community under Contract No. ERBFMGECT950083, by the Polish Committee of Scientific Research under grant KBN 2 P03B 039 13, by the Russian Fund for Basic Research and Deutsche Forschungsgemeinschaft under contract No. 436 RUS113/201/0(R), by C.I.C.Y.T. (Spain) under project AEN96-1662, and by the U.S. National Science Foundation under grant 9605207. The authors would like to thank the German Euroball collaboration for making the Euroball Cluster detectors available for this experiment. These detectors were supported by the German BMBF, the KFA Jülich, GSI Darmstadt, and MPI-K Heidelberg. B.A.B wishes to thank the Alexander von Humboldt-Foundation for support.

References

1. B. A. Brown and K. Rykaczewski, Phys. Rev. C 50 (1994) 2270
2. M. Karny et al., Nucl. Instr. and Meth. in Phys. Res. B 126 (1997) 320
3. J. Ebert et al., Prog. Part. Nucl. Phys. 38 (1997) 29
4. R. Kirchner et al., Nucl. Instr. and Meth. in Phys. Res. A 234 (1985) 224
5. M. Karny et al., Nucl. Phys. A, in print
6. G. Audi et al., Nucl. Phys. A 595 (1995) 409
7. K. Schmidt et al., Nucl. Phys. A 624 (1997) 185
8. I. S. Towner, Nucl. Phys. A 444 (1985) 402

Charge-Exchange Reactions with a Radioactive Triton Beam

J. Jänecke[†]

University of Michigan, Ann Arbor, Michigan 48109-1120, U.S.A.

Abstract A high-resolution (t,^{3}He) test experiment has been performed recently by making use of a secondary triton beam produced by fragmentation of α-particles. The purpose of this charge-exchange experiment was to achieve good energy resolution in an (n,p)-type reaction at intermediate bombarding energies. The experiment was carried out with the K1200 cyclotron at the National Superconducting Cyclotron Laboratory using the A1200 beam-analysis system and the S800 magnetic spectrometer. The beam-analysis system was used to transport the energy-dispersed radioactive triton beam from the production target to the target position, and the magnetic spectrometer was used to focus the dispersion-matched ^{3}He particles from the (t,^{3}He) reaction at 0° onto the focal plane of the spectrometer. An energy resolution of 200–250 keV was achieved.

Charge-exchange reactions selectively excite isovector non-spinflip and spinflip states and giant resonances. The (p,n) and other (p,n)-like reactions, e.g. (^{3}He,t) or (^{6}Li,^{6}He), have been studied for many years and much information on spin-isospin excitations of nuclear states has become available for proton-rich nuclei. More recently, particularly the (^{3}He,t) reaction with its good energy resolution has added important data. Less information exists for neutron-rich nuclei from the (n,p) and (n,p)-like reactions. This is due to the fact that it is necessary to use a secondary neutron beam in (n,p) reactions, and the energy resolution is more limited. Other (n,p)-like reactions, e.g. (d,^{2}He), (t,^{3}He), (^{7}Li,^{7}Be), and (^{7}Li,^{7}Beγ), require coincidence measurements or a secondary beam as well. Whereas the (t,^{3}He) reaction does require a secondary beam, it has the potential of good energy resolution because of the expected good beam quality compared to neutron beams in conjunction with the use of magnetic spectrometers. A primary triton beam is in principle possible, though, but there is presently no accelerator facility in existence with a dedicated tritium ion source to accelerate ions to intermediate energies of >100 A·MeV.

Spin-isospin excitations in nuclei are of fundamental interest and have been studied for a long time both experimentally and theoretically. The nuclear responses for the operators $\sigma\tau$ and $\sigma\tau \mathrm{Y}_1$ have attracted special attention. These operators determine the Gamow-Teller and first-forbidden β-decay matrix elements. Gamow-Teller quenching is apparently still not fully understood. These matrix elements are also of basic interest in many astrophysical studies including supernovae explosions and in questions related to double β-decay and the detection of neutrinos.

(n,p)-like charge-exchange reactions permit the investigation of neutron-rich nuclei including light neutron-halo nuclei. Good energy resolution is especially desirable in most applications of basic nuclear physics and astrophysics. It is concluded from the present work that $(t,^3\text{He})$ charge-exchange reactions with good energy resolution will become competitive with (n,p) reactions as a tool to study spin-isospin excitations in neutron-rich nuclei including investigations of low-lying Gamow-Teller strength and the general characteristics of spin dipole, quadrupole and monopole resonances.

The recent successful attempt to use the $(t,^3\text{He})$ charge-exchange reaction to achieve high energy resolution (1) in an (n,p)-type reaction at intermediate bombarding energies and at 0° is a continuation of earlier work (2, 3) which was initiated a few years ago. Here, a secondary triton beam was produced by fragmentation of α-particles on a thick Be production target. The first stage of the A1200 beam-analysis system was used to direct the energy-dispersed triton beam to the target at an intermediate dispersive image. The second stage was used to focus the dispersion-matched ^{3}He particles from the $(t,^3\text{He})$ reaction onto the focal plane. An energy resolution of 780 keV was achieved (2,3). Experimental details are given in Ref. (2). The investigation of Gamow-Teller strengths from the $(t,^3\text{He})$ reaction on targts of ^{9}Be, ^{10}B, ^{11}B, ^{12}C, and ^{13}C has been described in Ref. (3).

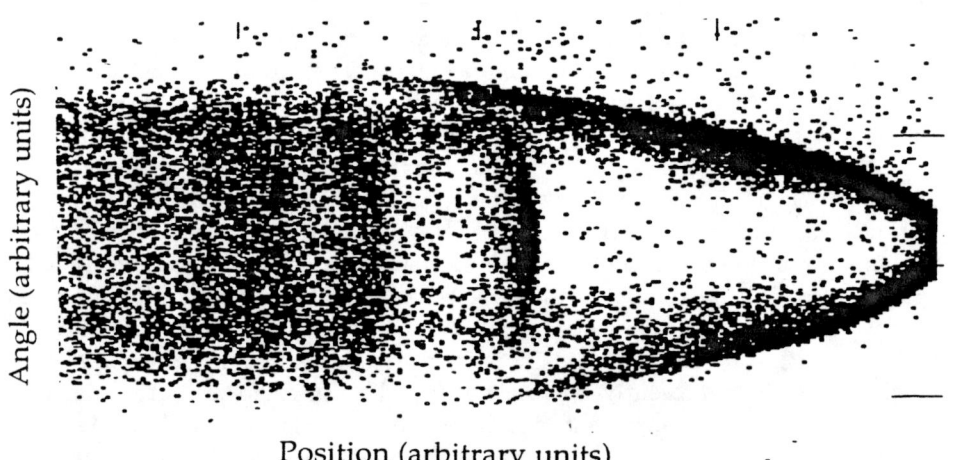

FIGURE 1. Plot of angle *versus* position for the polyethylene target (online). The events along the strongly curved line are from hydrogen. The essentially vertical lines are from ^{12}C.

The improved (t,^{3}He) test experiment (1) was performed by groups of researchers from the National Superconducting Cyclotron Laboratory, the Research Center for Nuclear Physics, the Indiana University Cyclotron Facility, Osaka University, the Kernfysisch Versneller Instituut, and the University of Michigan. It made use for the first time of the S800 high-resolution magnetic spectrometer. The triton beam was produced by fragmentation of 560-MeV α-particles, slightly lower in energy than in the previous experiments, on a thick Be production target (9.25 g/cm^2). A beam intensity of $(0.5-1.0) \times 10^6$ tritons/s was obtained. The energy-dispersed triton beam of median energy 350 MeV or 117 A·MeV, also slightly lower in energy than previously, was transported to the S800 target position using the A1200 beam analyzer. About 1% of the energy spread of the secondary triton beam was transmitted. This resultsd in a 5 cm tall beam spot at the target position. Dispersion matching was employed to obtain ^{3}He energy spectra with excellent energy resolution in the focal plane of the S800 magnetic spectrometer. The focal-plane detection system consisted of two cathode readout drift chambers (CRDC) and two scintillators. An energy resolution of 200–250 keV was achieved for the ground state of ^{12}B with a polystyrene target of thickness $\rho \Delta x \approx 5.5$ mg/cm^2. Reactions on the hydrogen component of the target did not interfere with the reactions on carbon. In fact, the intersection between the loci for p(t,^{3}He)n and ^{12}C(t,^{3}He)^{12}B$_{gs}$ in the angle *versus*

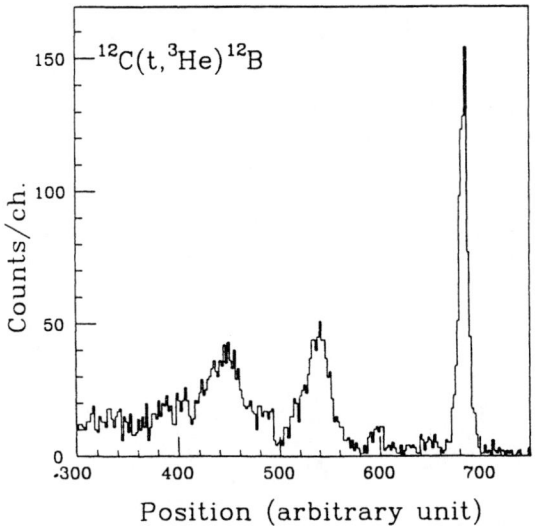

FIGURE 2. Spectrum for angles close to 0° for the ^{12}C(t,^{3}He)^{12}B reaction at 117 A·MeV (preliminary). The energy resolution for the transition to the ground state is 200–250 keV. The other two strong peaks are for the resonances at ~4.5 MeV and ~7.5 MeV, respectively.

position plane, as shown in Fig. 1, provides a convenient angle calibration. The weak curvature for the ^{12}B lines is easily removed in the offline analysis. Figure 2 displays a preliminary spectrum for angles close to 0°. The energy resolution for the ground state is <250 keV. The two other strong peaks are the known $2^-/4^-$ and $1^-/2^-$ resonances at ~4.5 and ~7.5, respectively. Additional weak states are also seen.

A nickel target was also bombarded in the present work to obtain initial information about low-lying Gamow-Teller strength. The GT_+ states are found in neutron-rich nuclei at low excitation energies due to the attractive contributions in the particle-particle interaction.

ACKNOWLEDGMENTS

This work was supported in part by the U.S. National Science Foundation, by the Ministry of Education, Science, Sports, and Culture of Japan (Monbusho), and by the Stichting voor Fundamenteel Onderzoek der Materie (FOM), the Netherlands.

REFERENCES

† For the collaboration.
1. Sherrill, B. M., Akimune, H., Alahari, N., Austin, Sam M., Bazin, D., van den Berg, A., Berg, G. P. A., Caggiano, J., Daito, I., Fujimura, H., Fujita, Y., Fujiwara, M., Harakeh, M. N., Jänecke, J., Roberts, and D. A., Steiner, M., (to be published).
2. Daito, I., Akimune, H., Austin, S. M., Bazin, D., Berg, G. P. A., Brown, J. A., Davids, B. S., Fujita, Y., Fujimura, H., Fujiwara, M., Hazama, R., Inomata, T., Ishibashi, K., Jänecke, J., Nakayama, S., Pham, K., Roberts, D. A., Sherrill, B. M., Steiner, M., Tamii, A., Tanaka, M., Toyokawa, H., and Yosoi, M., Nucl. Instr. Meth. **A 397**, 465-471 (1997).
3. Daito I., et al., Phys. Lett. **B418**, 27-33 (1998).

Beta–decay of ^{103}In studied by using a total absorption spectrometer

M. Karny[a,b], L. Batist[c], B.A. Brown[d], D. Cano–Ott[e], R. Collatz[f],
A. Gadea[e], R. Grzywacz[a], A. Guglielmetti[f], M. Hellström[f], Z. Hu[f],
Z. Janas[a], R. Kirchner[f], F. Moroz[c], A. Piechaczek[g], A. Płochocki[a],
E. Roeckl[f], B. Rubio[e], K. Rykaczewski[a,h], M. Shibata[f], J. Szerypo[a],
J. L. Tain[e], V. Wittmann[c], A. Wöhr[g].

[a] *Institute of Experimental Physics, University of Warsaw, PL–00681 Warsaw, Poland*
[b] *Joint Institute for Heavy Ion Research, Oak Ridge TN,37831 USA*
[c] *St. Petersburg Nuclear Physics Institute, 188–350 Gatchina, Russia,*
[d] *NSCL, Department of Physics and Astronomy, Michigan State University, East Lansing, MI 48824-1321, USA,*
[e] *Instituto de Física Corpuscular C.S.I.C.-Univ. Valencia, E-46100 Burjassot, Spain,*
[f] *Gesellschaft für Schwerionenforschung mbH, D-64291 Darmstadt, Germany,*
[g] *Instituut voor Kern- en Stralingsfysica, University of Leuven, B-3001 Leuven, Belgium,*
[h] *Oak Ridge National Laboratory, Physics Division, PO Box 2008, Oak Ridge, TN 37831, USA.*

Abstract. The β decay of the neutron–deficient isotope ^{103}In was investigated by using total absorption γ–ray spectrometry on mass–separated sources. The measurement reveals a high–lying resonance of the β–decay strength in striking disagreement with high–resolution γ–ray data. The result is discussed in comparison with shell–model predictions.

Within the last few years, the region of nuclei near the ^{100}Sn has been subject of intense experimental and theoretical investigations ([1–3] and references therein). Although ^{100}Sn was observed by using high–energy fragmentation reactions, its detailed spectroscopy appears to be still out of experimental range. For nuclei *near* ^{100}Sn, however, both in–beam and decay spectroscopy is already feasible today.

One of the particularly interesting features of decay studies in the ^{100}Sn region is the occurrence of fast β transitions related to the Gamow–Teller (GT) transformation of a $\pi g_{9/2}$ proton into a $\nu g_{7/2}$ neutron. A measurable quantity suited for comparison with theoretical predictions is the β strength defined as:

$$B_{GT}(E) = \frac{D \cdot I(E)}{f(Q_{EC} - E) \cdot T_{1/2}}, \qquad (1)$$

where $D = 3860(18)$ s is a constant corresponding to the value of the axial vector weak interaction coupling constant g_A for the decay of the free neutron, I the β intensity, E the excitation energy in the daughter nucleus, f the statistical rate function, Q_{EC} the total energy released in electron–capture (EC) decay, and $T_{1/2}$ the β–decay half–life. The $B_{GT}(E)$ distributions, deduced from measurements of $I(E)$, Q_{EC} and $T_{1/2}$, can be compared to the calculated square of the GT transition matrix element. The quenching of the experimental GT transition rates with reference to model predictions has been a puzzle for many years. A renormalization of g_A (or of the GT operator) has been applied [2,3] in order to account for the missing GT strength in the ^{100}Sn region. This led to a consistent picture for the decays of even–even nuclei, but the dramatic reduction of the shell–model GT strength remained to be explained for the decays of non even–even nuclei (see [4] for a recent example).

We report on a re–investigation of the β decay of ^{103}In, a five–quasiparticle configuration with respect to the ^{100}Sn core. In order to deduce B_{GT} we performed a measurement of $I(E)$ by means of total absorption spectrometer (TAS), and took the values of Q_{EC} (6.05(2) MeV and $T_{1/2}$ (60(1) s) from the literature. In order to estimate the systematic uncertainties involved in the evaluation of TAS data, we performed two independent evaluations. They differ in the applied Monte–Carlo simulation codes, and in the assumptions made for the de–excitation pattern of high–lying ^{103}Cd levels. Details of this work are presented in [6].

The TAS [7], installed at the on-line mass separator at GSI, consists of a large NaI crystal (ϕ 356 $mm \times$ 356 mm) surrounding the radioactive source, two small Si detectors (ϕ 16 $mm \times$ 450 μm) above ("top") and below ("bottom") the source, and one Ge detector (ϕ 16 $mm \times$ 10 mm) placed close in the center of the NaI crystal just above the "top" detector. By demanding coincidence with signals from the Si detectors, the β^+ decay component for the nucleus of interest is selected, whereas coincidences with characteristic $K_{\alpha,\beta}$ X–rays recorded by the Ge detector can be used to select the EC mode. The total γ–ray efficiency of TAS for monoenergetic photons between 0.2 and 4.0 MeV is above 88%, and its photopeak efficiency is above 56%.

The ^{103}In isotope was produced by means of ^{50}Cr(^{58}Ni,3p2n)^{103}In fusion–evaporation reactions. The energy and intensity of the ^{58}Ni beam on the ^{50}Cr (3.6 mg/cm^2) target amounted to 285 MeV and 40–50 particle nA, respectively. After ionization in a FEBIAD-B2-C ion source [8], and mass separation the ^{103}In beam was implanted into a transport tape and moved to the center of the TAS.

In order to deduce the β–intensity $I(E)$ as a function of the excitation energy E in the daughter nucleus from an experimental TAS spectrum $S(x)$, one has to solve the equation

$$S(x) = \sum_i R_i(x) \cdot I_i, \qquad (2)$$

where I_i is the β–feeding to level i. Any column $R_i(x)$ of the response matrix, transformed from energy into experimental spectrum channels (x), represents the "level response function" of TAS to the cascade of γ rays deexciting the level i. In the case of TAS measurements of exotic nuclei one usually faces the problem that, even though some β-delayed γ rays are known from high-resolution measurements, many of them have escaped from observation in these experiments. Correspondingly, the decay schemes obtained from high-resolution data are incomplete, and hence assumptions have to be made in deducing I_i. Such assumptions may introduce *unpredictable* systematical uncertainties. Therefore, we decided to carry out two independent data analyses (recursive folding (RF) and peel-off (PO) methods) and to confront their results in order to estimate the systematical uncertainties (for details see [6]). The B_{GT} distributions for ^{103}In, obtained from the RF and PO methods (see Fig. 1), agree in the *gross* features, i.e. the dominant resonance around E = 3.8 MeV with a full width at half maximum of the order of 2 MeV. A closer inspection shows, however, that the B_{GT} resonance is split into two components, and that a long tail extends towards high excitation energies. Furthermore, distinct differences occur between the results obtained by the two unfolding procedures. These differences, which are interpreted as representing the systematical uncertainties involved in these procedures, concern more the shape

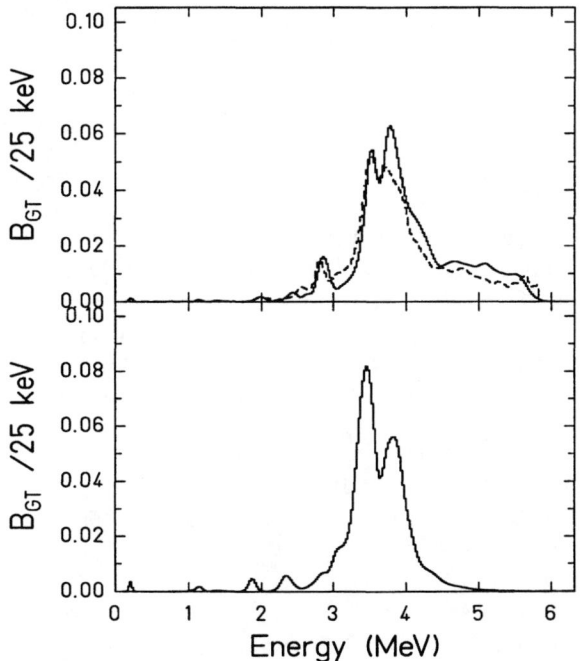

FIGURE 1. B_{GT} values for the ^{103}In decay as a function of the ^{103}Cd excitation energy. Upper panel: TAS data derived by the recursive-folding (RF) method (solid line) and the peel-off (PO) approximation (broken line); lower panel: results obtained by the shell-model calculations The shell-model B_{GT} values were divided by a factor of 5.1

than the summed B_{GT} values $\Sigma B_{GT}^{(exp)}$ of 2.54(25) and 2.40(25).

The model space and interaction we have used for the analysis of the ^{103}In GT β–decay is that denoted by SNB in [2], where it was introduced to calculate β–decay properties for nuclei near ^{100}Sn. A ^{103}In $\to$ ^{103}Cd GT-decay is calculated as $(\pi p_{1/2}, g_{9/2})^{11}; (\nu g_{7/2}, d_{5/2})^4 \to (\pi p_{1/2}, g_{9/2})^{10}; (\nu g_{7/2}, d_{5/2})^5$. The calculated B_{GT} distribution is shown in Fig. 1 in comparison with experiment. The centroid and width of the theoretical distribution is very close to that of the experiment. The main difference in the *shape* of the distributions is that the experiment yields a high–energy tail which is not present in our calculation.

The total experimental GT strength for ^{103}In, obtained from the average of the two analysis methods is 2.47±0.25, and the total theoretical strength from the SNB calculation is 12.7. Thus, h_{exp} is 5.1±0.5. This is the first time that one has been able to extract this factor for an odd-even nucleus close to ^{100}Sn with the confidence that most of the expected strength has been detected experimentally.

These results represent an important step in the understanding of the hindrance factors for β^+/EC decays near ^{100}Sn. It is the first time that the *complete* GT strength has been measured for a "non even-even" nucleus so close to ^{100}Sn that a comparison with full–space shell–model calculations becomes possible. Thus, the quality of the data obtained in this work goes beyond that reached in previous experiments for ^{100}In [9], ^{103}In [4], and ^{100}Ag [5].

Acknowledgements

This work was supported in part by the Polish Committee of Scientific Research under grant KBN 2 P03B 039 13, by the European Community under Contract No. ERBFMGECT950083, by the U. S. National Science Foundation under grant 9605207, and by the Russian Fund for Basic Research and Deutsche Forschungsgemeinschaft under contract No. 436 RUS 113/201/0(R), by C.I.C.Y.T. (Spain) under project AEN96-1662, by U.S. DOE under DE–AC05–96OR22464. M.K. would like to acknowledge financial support from the Foundation for Polish Science. B.A.B. wishes to thank the Alexander von Humboldt–Foundation for support. ORNL is managed by Lockheed Martin Energy Research Corporation.

REFERENCES

1. E. Roeckl, Nucl. Phys. News **7** (1997) 18.
2. B.A. Brown and K. Rykaczewski, Phys. Rev. **C 50** (1994) R2270.
3. I.S. Towner, Nucl. Phys. **A444** (1985) 402
4. J. Szerypo et al., Z. Phys. **A259** (1997) 117.
5. L. Batist et al., Z. Phys. **A351** (1995) 149.
6. M. Karny et al., GSI-Preprint-98-24, and Nucl. Phys. **A**, in print
7. M. Karny et al., Nucl. Instr. and Meth. Phys. Res. **B 126** (1997) 320.
8. R. Kirchner, Nucl. Instr. and Meth. Phys. Res. **B 26** (1987) 204.
9. J. Szerypo et al., Nucl. Phys. **A584** (1995) 221.

Decays of very neutron–rich fission products ^{113}Ru and ^{113}Rh

J. Kurpeta, A. Plochocki

Institute of Experimental Physics, Warsaw University, ul. Hoża 69, 00-861 Warszawa, Poland

G. Lhersonneau, J.C. Wang, P. Dendooven, A. Honkanen,
M. Huhta, M. Oinonen, H. Penttilä, K. Peräjärvi, J. R. Persson,
J. Äystö

Department of Physics, University of Jyväskylä, P.O.Box. 35, FIN-40351, Jyväskylä, Finland

Abstract. Exotic neutron–rich fission products obtained from IGISOL on–line mass separator for mass chain A=113 have been investigated by $\gamma\gamma$ coincidence and spectrum–multiscaling measurements. Gamma transitions following β decay of Tc to Ru have been identified and a 0.5 s isomeric state in Ru has been found. Decay schemes of Ru and Rh isotopes have been considerably extended. Large feedings to high–lying levels in Rh lead to revision of the B(GT) strength.

Introduction

Observation of new exotic activities and detailed investigations of the structure of their daughters provide stringend tests for nuclear models and in particular regarding their predictive power far from regions they have been designed to describe. This is especially important with respect to the efforts made currently to build facilities for the production of very exotic nuclei. In this work, we present decay data on the A=113 mass chain studied by γ-spectroscopy.

Experimental procedure

The experiment was carried out at the Ion Guide Isotope Separator On-Line Facility of the Jyväskylä Accelerator Laboratory (Finland) [1,2]. A target of ^{238}U was bombarded with 25 MeV protons from the K-130 cyclotron with intensities typically of 10 μA. Neutron–rich fission products were mass separated and implanted onto a movable collection tape in front of the detector system. Low energy γ–spectra (in the range of 14 to 610 keV) were measured with a LEGe detector and the higher range (40 to 2500 keV) with a Ge detector in anticoincidence with a BGO shield. The details of data evaluation, full list of observed γ–ray energies and intensities as well as complex decay schemes are presented in [3] and in a forthcoming paper [4].

Results

^{113}Tc decay and ^{113}Ru level scheme:

The nuclide ^{113}Tc was discovered in proton–induced fission of ^{238}U by Äystö et al. [5] who reported its half-life of $T_{1/2} = 130$ ms from the decay of Ru X-rays. This half-life is confirmed for X-rays and γ–lines of 99, 65 and 165 keV. The two–component decay pattern of the 99 keV line (a short one, 0.11 ± 0.03 s and the other of about 0.5 s) and prompt coincidences indicates that this transition is fed partially from an isomeric state with $T_{1/2} = 0.51 \pm 0.03$ s. Its energy is possibly too low to allow detection in our γ or X-ray spectra. The γ–rays following the decay of ^{113}Tc are listed in [6].

Decay of ^{113}Ru to ^{113}Rh: The decay scheme was known up to excitation energies of 1000 keV in ^{113}Rh [7]. We have introduced 74 lines and 29 levels on basis of X-γ and γ-γ coincidences. There are two groups of γ–lines connected with the 263 and 211 keV lines, with half–lives of 0.9 s and 0.6 s respectively. Levels connected with the 3/2$^+$ level at 263 keV are fed from the Ru ground state and those connected with the 9/2$^+$ level at 211 keV are fed from the ^{113}Ru isomeric state. This was not recognized in [7] because of the large statistical errors.

Decay of ^{113}Rh to ^{113}Pd: We have added 93 lines and 19 levels to the ^{113}Pd level scheme [3], [4]. The placement of a 84.5 keV line (compare [7]) creating a level with this energy is well supported by coincidence relationships.

Discussion

Features of A$\approx$113 region: A number of interesting phenomena are observed for neutron–rich nuclei around mass A = 100–110. These are shape coexistence [8], triaxiality [9], and existence of low-lying intruder states [10,11]. Predicted values of axially symmetric quadrupole deformation parameter β_2 [12] for $36 < Z < 48$, $61 < N < 73$ show a prolate to oblate shape transition in the vicinity of $N = 66$ whereas another calculation [13] predicts this transition at about $N = 60$. For nuclei in our region of interest potential energy minima for prolate and oblate shapes are separated by a low barrier and triaxial shapes are favoured. The relative depth of potential minima for prolate and oblate shapes can depend on excitation energy and in particular may be different for ground and isomeric state of the same nucleus. Nevertheless, experimental energies of first 2$^+$ excited states from Sr to

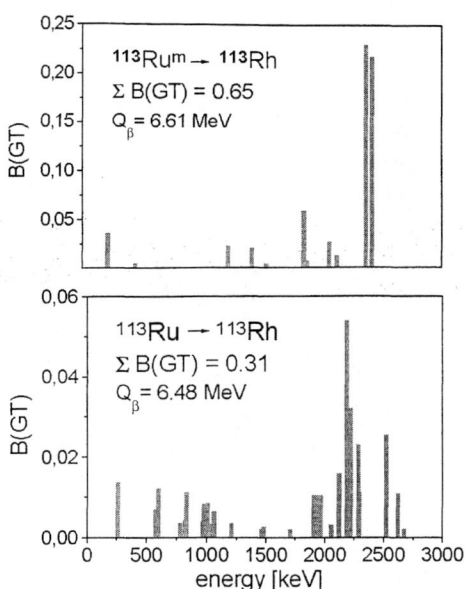
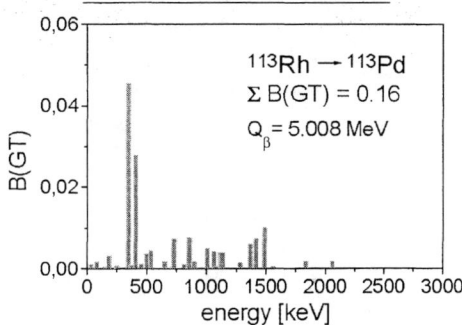

Table 1:

A	B(GT)	Ref.
108	0.34	[18]
110	0.26	[18]
111	0.22	[10]
112	0.23	[18]
113	0.31	this paper
113^m	0.65	this paper

FIGURE 1. Experimental B(GT) strength distribution for decay of ^{113m}Ru, ^{113}Ru and ^{113}Rh. In insert: Sums of GT strengths for Ru isotopes.

Cd behave in a rather smooth way if one excepts the region of shell closures near ^{96}Zr.

Systematics of $11/2^-$ states in odd–neutron nuclei from Ru to Sn for $61 < N < 73$ shows smooth tendency of lowering and then increasing their energy keeping isomeric character, except for Ru [14]. Thus, the 0.5 s state at 130 ± 30 keV in ^{113}Ru could be a $11/2^-$ isomeric state, although breaking the increasing energy trend of energies in Ru isotopes [6].

^{113}**Ru levels:** Systematics of the levels obtained in prompt fission experiments [15] [16] and from β-decay [7] for 107 – 111 Ru isotopes indicate sequences of $5/2^+$, $5/2^-$, $7/2^-$, $9/2^-$ and $11/2^-$ levels. In some cases, band structures and levels from β-decay can be merged. For ^{113}Ru, however, they create two separate structures and the 0.5 s isomer has not been reported in prompt fission. It seems that the $11/2^-$ level goes down and, being slightly above the 99 keV level, creates an isomeric state. Isomerism is well known to exist for all odd Pd isotopes (two protons more than Ru) with A = 109 – 115 where the $11/2^-$ isomeric state is observed as a second excited state with energy of 190 – 80 keV. It is probable that a band structure known from fission experiments is built on the isomer in ^{113}Ru.

^{113}**Rh levels:** As it was mentioned above we observe two sets of ^{113}Rh levels. One is fed by the ground state of ^{113}Ru and the other from the isomeric state.

Only the latter (211 and 444 keV states) are seen in ^{248}Cm prompt fission [15] creating a band structure. According to [7] there is no strong direct β–feeding to the ground state. From our data feeding to the states below 1.8 MeV is weak and the β–strength is concentrated around levels with 2.3 MeV excitation energy (see fig. 1) with log(ft) values as low as 4.2. This means that log(ft) values reported in [7] need a renormalisation. A very similar decay pattern is observed for ^{111}Ru decay but from the $5/2^+$ ground state [10], and in that case feeding to the Rh ground state is large. For both decays the odd 11/2 or 5/2 neutron is more or less a spectator and the decay has to involve a $g_{7/2}$ neutron inside of a pair, leading to a 3–quasi–particle final state. This mechanism also applies to the decays of the high–spin odd–odd Rh isotopes to Pd, where states near 2.5 MeV are strongly fed [17]. The sums of GT strength for the decay of Ru isotopes [18,10] in table 1 show similar values for other isotopes in this area in a case of ground state decays, but the isomeric decay of ^{113}Ru shows an exceptionally higher value.

^{113}Pd levels: In the decay scheme of ^{113}Rh to ^{113}Pd we have added several high–energy levels, resulting in modest changes of the β–strength distribution. Still the maximum is around 400 keV and with log(ft) = 4.7. Presumably the decay of the I=j-1 Rh ground state (j being the $g_{9/2}$ proton) populates final states coupling a core excitation with the remaining $g_{7/2}$ neutron. The GT strength for this decay is only 0.16, a half as for β–decay of Ru isotopes. The unknown feeding to the ground state was taken as neglectible but a branching of 60% has been derived recently [19]. Thus, the GT strength seems to be low in this decay.

REFERENCES

1. H. Penttilä et al., Nucl. Instr. Methods in Phys. Research B126, 213, (1997)
2. M. Huhta et al., Nucl. Instr. Methods in Phys. Research B126, 201, (1997)
3. J. Kurpeta, PhD Thesis, University of Warsaw, 1998 (unpublished)
4. J. Kurpeta, in preparation
5. J. Äystö et al., Phys. Rev. Lett 69, (1992) 1167.
6. J. Kurpeta et al., Eur. Phys: J. A1 (accepted for publication)
7. H. Penttilä, Ph. D. thesis, Univ. of Jyväskylä (1992)
8. G. Lhersonneau et al., Phys. Rev. C49, 1379 (1994)
9. J. Äystö et al., Nucl. Phys. A515, 365 (1990)
10. G. Lhersonneau et al., Eur. Phys. J. A1, 285 (1998)
11. G. Lhersonneau et al., Eur. Phys. J. A2, 25 (1998)
12. P. Möller et al.,At. Data Nucl. Data Tables 59, (1995)
13. J. Skalski et al.,Nucl. Phys. A617, (1997) 282
14. G. Audi et al.,Nucl. Phys. A624 (1997) No. 1
15. W. Urban, ^{248}Cm fission, private communication July 1997
16. K. Butler-Moore et al., Phys. Rev. C 52, (1995) 1339
17. J. Äystö et al., Nucl. Phys. A480, 104 (1988)
18. A. Jokinen et al., Z. Phys. A340, (1991) 21
19. J.C. Wang et al., (to be submitted)

Structure Studies of Nuclear Systems close to the Doubly-magic ^{132}Sn using Advanced β^--Spectroscopy

H. Mach [1], J. Blomqvist [2], B. Fogelberg [1], V. I. Isakov [3], L. Jacobsson [1],
A. Lindroth [1], K. A. Mezilev [3], M. Sanchez-Vega [1], and R. B. E. Taylor [1]

[1] *Department of Neutron Research, Uppsala University, S-61182 Nyköping, Sweden*
[2] *Department of Physics Frescati, Royal Institute of Technology, S-10 405 Stockholm, Sweden*
[3] *St. Petersburg Nuclear Physics Institute, Russian Academy of Sciences, 188350 Gatchina, Russia*

Abstract: A brief description of the OSIRIS fission product mass separator and an outline of the current program of research in the domain of nuclear structure at the doubly magic ^{132}Sn region are given. The most recent results on the study of the single particle states in ^{133}Sb are briefly summarized.

INTRODUCTION

The nuclear structure studies at the OSIRIS fission product mass separator at Studsvik in Sweden are largely focused on the simple nuclear systems at doubly magic ^{132}Sn. As of now the β^- spectroscopy offers the most detailed insight into the properties of these nuclei. More recently, however, complementary data on the Yrast states in these exotic nuclei became available from the γ-ray studies of prompt fission using multi-detector arrays (1). Our current research on nuclei at ^{132}Sn include:

- spectroscopy of the single proton nucleus of ^{133}Sb using high sensitivity singles and coincidence γ-rays measurements able to detect γ-rays having intensities down to about 10^{-5} per decay of the ^{133}Sn parent;
- high precision measurements of the transition rates in the two neutron-hole nucleus of ^{130}Sn using the advanced time delayed method;
- systematical theoretical investigation of the effective charges and transition rates at ^{132}Sn and comparison to the corresponding cases at ^{208}Pb [see a contribution by V.I. Isakov *et al.* to this Conference];
- systematic measurements of the Q_β values for the chains of nuclei A=131, 132, 133, 134 and 135 using high precision coincidence techniques. A strong effort has been made to resolve the Q_β puzzle at ^{132}Sn (1) [see a contribution to this Conference by B. Fogelberg *et al.*].
- systematical measurements by the OSIRIS-Oxford Collaboration of the static magnetic moments using the nuclear orientation techniques (2).

OSIRIS SEPARATOR

The OSIRIS on-line fission product isotope separator, located at the R2-0 1 MW nuclear reactor, represents a small but versatile facility. The movable reactor core allows for a flexible change of the neutron flux, and thus the intensity of the exotic beams, without the necessity to change the ion-source operational parameters. Furthermore, this small reactor is mostly dedicated to the separator, which allows also for varying the reactor power. Figure 1 illustrates the layout of the facility.

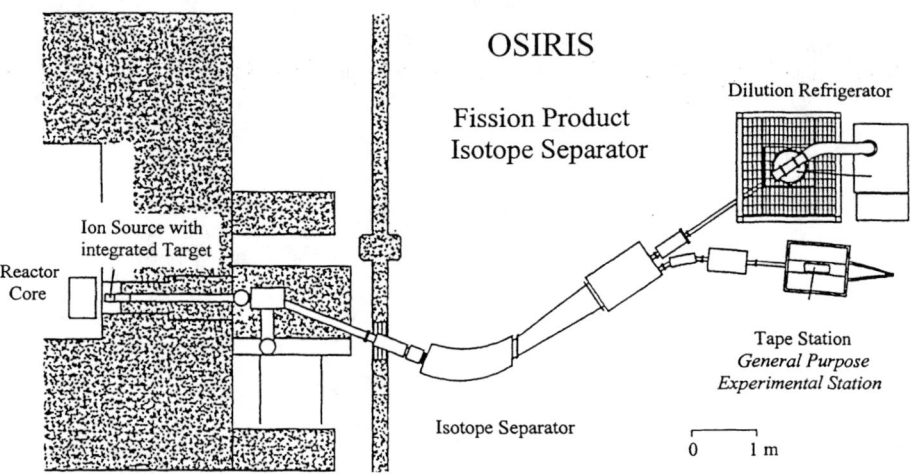

Figure 1. The layout of the OSIRIS facility illustrating the movable Reactor Core, the ANUBIS Ion Source with integrated Target, and two experimental stations: the general purpose Tape Station (used for the fast timing, Q_β and γ-spectroscopy measurements) and the Dilution Refrigerator station.

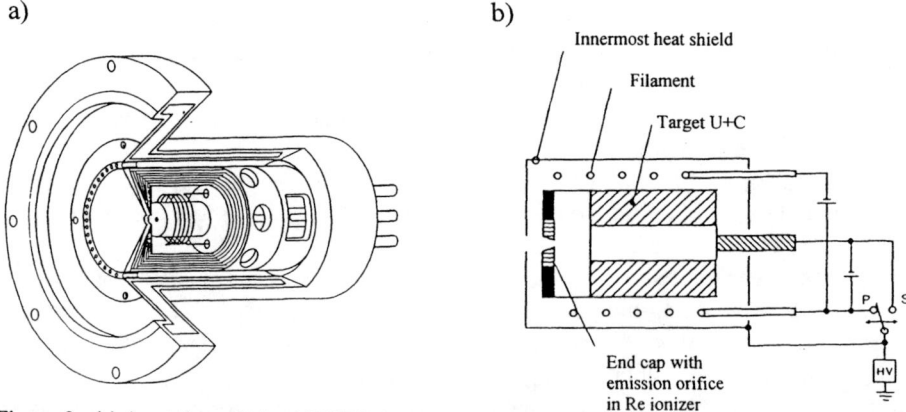

Figure 2. (a) A cut through the ANUBIS ion source showing the central W target chamber, surrounded by the electron bombardment filament and seven concentric heat shields. (b) A schematic picture of the electric power connections used for operation in the two different modes: of surface ionization and plasma ionization by electron impact. For details see Ref. (3).

At the heart of the facility is the versatile ANUBIS source providing exotic beams [a detailed description is given in Ref. (3)]. It is placed inside the reactor channel close to the reactor core. Typically it includes 0.25 to 1.0 g of the ^{235}U target. Its operational temperature range is 1800 – 2400 °C, with a maximum temperature for a short-term operation of about 2500 °C. Importantly, it can operate in two different modes: *pure thermal ionization* on a hot Re surface, which permits selective ionization of elements in the first three main groups of the periodic table, and in the *electron impact ionization*, which is non selective to the chemical species (3). One should note, that at OSIRIS there is a possibility (4) to obtain a higher production intensity of the very exotic nuclei and to reach by about 1 to 2 additional nucleons away from the line of stability, by using fast fission of ^{238}U rather than thermal fission of ^{235}U, and a substantial flux of fast neutrons available at the Studsvik's reactor.

EXPERIMENTAL TECHNIQUES

The ANUBIS ion source (3), described above, gives good intensities of the exotic beams at ^{132}Sn, which allows in turn to employ advanced spectroscopic tools. Although the beam intensities at the ISOLDE separator are higher, low radiation background and longer beam time available for specific measurements, make the OSIRIS facility very competitive. Further enhancement of the structure studies at ^{132}Sn is provided by three experimental techniques utilized at OSIRIS:

- measurements of the Q_β values using high precision $\beta\gamma$ coincidence technique, [see a contribution by B. Fogelberg *et al.* to this Conference]. This method, when applied to nuclei at closed shells, can give precision of the order of $\delta m/m \sim 3\times10^{-7}$;
- measurements of level lifetimes down to a few ps using fast timing $\beta\gamma\gamma(t)$ method. Presently, this is the only method available for the neutron-rich nuclei for the level lifetime measurements in the ps domain, for details see Ref. (5);
- measurements of the static magnetic moments using the nuclear orientation techniques (2).

Further enhancement of the experimental opportunities at OSIRIS, and specifically for the ^{132}Sn region, can be achieved with a modern compact array of efficient γ-detectors. The high quality $\gamma\gamma(\theta)$ angular correlation and γ-linear polarization measurements, that can be achieved with such a device, would be particularly useful.

DETAILED STRUCTURE STUDIES AT ^{132}Sn

In the last few years, we have performed a series of detailed spectroscopic studies at ^{132}Sn using the new capability of the ion-source and new experimental techniques. These studies include: ^{132}Sn (6), ^{132}Sb (7), ^{134}Te (8), and most recently ^{133}Sb (9).

^{132}**Sn:** The first excited negative parity state, the 3$^-$ level at 4351.9-keV, was shown to have a collective octupole character with B(E3) $\geq$ 7 W.u. Several particle-hole multiplets have been identified. The three positive parity states near 5.5 MeV were identified as members of the $\pi g_{7/2} g^{-1}_{9/2}$ multiplet, thus representing the first observation of proton p-h states in ^{132}Sn. More than about 20% of the theoretically estimated bound states of ^{132}Sn have been identified.

^{132}Sb: The structure of ^{132}Sb (7) provides a unique opportunity to study the coupling of a valence proton particle to a neutron hole outside the doubly magic ^{132}Sn core. In fact, it represents the best case in the Sn region from A=100 to 132, to study the T=0 matrix elements of the effective interaction. ^{132}Sb will be reinvestigated in more detail.

^{134}Te: This nucleus yields (8) an important information on the coupling of a pair of valence proton particles, and also some critical E2 and E3 transition strength.

^{133}Sb: The single proton states in ^{133}Sb have been investigated (9) via a high-sensitivity γ spectroscopic study of the β$^-$ decay of ^{133}Sn. The experiments included γγ coincidences and a series of γ-ray multi-spectra that provided accurate information on the energy and intensities of the transitions. The fast timing βγγ(t) method was used to extract the half-life of the single proton $h_{11/2}$ state. As a result the $d_{3/2}$ state has been identified at 2439.5 keV and its γ-ray branching to the $d_{5/2}$ and $g_{7/2}$ states was accurately determined. For the $h_{11/2}$ state at 2791.3 keV, a half-life of $T_{1/2}$ = 11.4(4.5) ps and a γ-ray branching to the $d_{5/2}$ were measured for the first time (9).

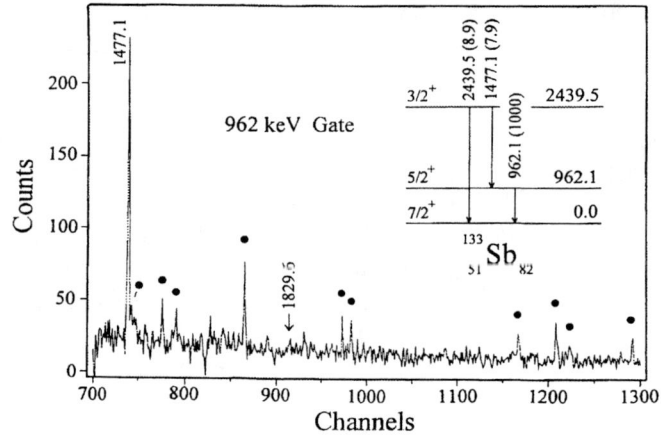

Figure 3. Coincidence spectrum with the 962-keV transition firmly defining a level at 2439.5 keV.

REFERENCES

1. Zhang, C. T., et al., Phys. Rev. Lett. **77**, 3743 (1996).
2. Stone, N. J., et al., Phys. Rev. Lett. **78**, 820 (1997).
3. Fogelberg, B. et al., Nucl. Instrum. Methods **B 70**, 137 (1992).
4. Fogelberg, B. et al., in *Research with Fission Fragments*, eds.: T. von Egidy et al., Singapore, World Scientific, 1997, pp. 69.
5. Mach, H., et al., Nucl. Phys. **A 523**, 197 (1991), and references therein.
6. Fogelberg, B. et al., Phys. Rev. Lett. **73**, 2413 (1994).
7. Mach, H., et al., Phys. Rev. **C 51**, 500 (1995).
8. Omtvedt, J. P., et al., Phys. Rev. Lett. **75**, 3090 (1995).
9. Sanchez-Vega, M., et al., Phys. Rev. Lett. **80**, 5402 (1998).

Deformation signature from the Gamow-Teller decay of N=Z nuclei

Ch. Miehé[a], J. Giovinazzo[a*], Ph. Dessagne[a], A. Huck[a],
A. Knipper[a], G. Marguier[a], C. Longour[a], V. Rauch[a]
M.J.G. Borge[b], I. Piqueras[b], O. Tengblad,[b] A. Jokinen,[c]
M. Ramdhane,[d]
and the ISOLDE collaboration[e]

[a] *Institut de Recherches Subatomiques, UMR7500, CNRS-IN2P3 et Université Louis Pasteur*
BP 28, 67037 Strasbourg Cedex 2, France
[b] *Instituto de Estructura de la Materia, Serrano 113bis,*
E 28006 Madrid, Spain
[c] *Department of Physics, Accelerator laboratory,*
University of Jyväskylä, FIN-40351, Finland
[d] *Institute of Physics, Dept. of Theoretical Physics, University of Constantine*
25000 Constantine, Algeria
[e] *CERN PPE division CH 1211 Geneva 23*
[*] *Centre d'Etudes Nucléaires de Bordeaux-Gradignan, 33170 Gradignan Cedex, France*

Abstract. The ^{76}Sr (N=Z=38) and the ^{72}Kr (N=Z=36) β^+ EC decay have been studied at the CERN/ISOLDE PSB facility where their beta-gamma and delayed particle decay modes have been investigated. The established decay schemes yield new information on the Gamow-Teller (GT) strength spread over the $J^\pi = 1^+$ states in the daughter nuclei. The delayed proton emission of an N=Z nucleus is observed for the first time in the case of ^{76}Sr. The experimental GT strength intensities and distributions are discussed in the light of the theoretical estimates for oblate and prolate deformations.

INTRODUCTION

The N=Z neutron deficient mass region above A=70 is currently motivating numerous theoretical and experimental investigations to answer questions about nuclear deformation, shape coexistence, shape transition, np pairing and isospin mixing. Among the most recent predictions, a noteworthy feature has been pointed out by I.Hamamoto and H.Sagawa [1] who found that, close to the drip lines, the main strength of the Gamow-Teller(GT) resonance might be located below the ground state of the mother nucleus. Further elaborate developments dedicated to the neutron deficient nuclei along the N=Z line have then been carried out, taking into account deformation and pair correlations [2]. According to those, the GT process is expected to bring in valuable information on nuclear deformation, as strong differences appear in the calculated GT strengths - total intensity and energy distribution - depending on the shape of the parent nucleus. Of special interest in this respect are the even-even N = Z nuclei in the A=70-80 mass region, where an oblate to prolate transition is predicted, and for which various deformation amplitudes have already been inferred from experimental results [3] [4].

I EXPERIMENTAL PROCEDURE AND RESULTS

The ^{76}Sr and the ^{72}Kr nuclei have been produced under exellent conditions at the CERN ISOLDE PSB facility, with production rates of 3.10^3 and 10^4 atoms/s respectively, by fragmentation of a thick Niobium target by the pulsed 1 GeV proton beam of the Proton Synchrotron Booster of $2\mu A$ mean intensity.

The mass selected beam is steered towards a tape transport system deserving two counting stations devoted respectively to gamma and delayed proton spectroscopy. The particle detection is ensured by a C_4H_{10} gas - silicon telescope covering a 10 % of 4π solid angle. It allows low particle rates to be observed in presence of the large amount of positrons, with an energy resolution of 60 keV FWHM. The measurements are performed in collection-measurement duty cycles managed by a microcomputer, permitting for the two stations independent data taking synchronized with the PSB supercycle.

A The ^{76}Sr decay

The registered data allowed to establish a decay scheme to bound states in ^{76}Rb carrying detailed information on the GT strength spread over the $J^\pi = 1^+$ states in the daughter nuclei, of major interest to be compared with theoretical predictions. The delayed proton emission is observed for the first time for an N=Z nuleus in the case of ^{76}Sr. The reconstructed proton energy distribution extends from 1.4 MeV up to 2.6 MeV. The relative intensity P_p of the delayed proton branch has been evaluated to be $3.6\pm0.7\ 10^{-5}$.

A comparison of our experimental results with the theoretical predictions of reference [2] may be attempted to obtain a signature of the nuclear shape. The Gamow-Teller strength expected to feed the bound and unbound states in ^{76}Rb in the case of oblate and prolate deformations are quoted in table 1, together with the corresponding experimental values. The measured $\Sigma B(GT)$ to the unbound states reflects only the delayed proton channel contribution to the strength. Indeed, mainly due to the spread of the γ decay over numerous open channels, the radiative deexcitation escapes the observation above the proton separation energy.

Given the weakness of the delayed proton branch, γ·proton coincidence were rendered unobservable and so proton emission to the ^{75}Kr ground state has been considered to evaluate the experimental B(GT) values.

TABLE 1. Experimental and theoretical Gamow-Teller strengths for ^{76}Sr.

Gamow-Teller strengths B(GT)=4139/ft	experiment	theory oblate	prolate
ΣB(GT) unbound states	0.07 - 0.22	0.4	4.5
Σ B(GT) bound states	0.6 - 0.8	2.9	1.4

Owing to the statistical analysis developed to obtain information on the level densities and partial widths of the $T_z = 3/2$ emitters [5], an estimate of the unobserved E1 radiative width in the $Q_{EC} - S_p$ window has been made, following J. C. Hardy et al. parametrization [6], with a proton penetrability ten times higher than in the spherical optical model evaluation, and a level density estimated in the Back shifted Fermi Gas Model, with a $=$ 11 MeV^{-1}. As a result, the delayed proton channel contributes for only 2% to ΣB(GT) in the $Q_{EC} - S_p$ window. Corrected for the gamma decay, the experimental ΣB(GT) value to unbound levels lies between 4 and 13, in the vicinity of the predicted strength for a prolate deformation. For the bound states, for which the radiative decay is known to be well detected, the corresponding strength is compatible with the prolate case when a quenching factor of 0.5 is taken into account. This result is in agreement with what is expected from the deformed Hatree Fock calculations with the SIII Skyrme type interaction, in terms of absolute minimum of the total potential energy curve [2].

B The ^{72}Kr decay

As far as the deformation of this nucleus is concerned, two different amplitudes of deformation have been inferred from in beam laser spectroscopy measurements ($\epsilon_2 = 0.3$) [3] and from in beam work ($\epsilon_2 = 0.4$) [4]. In the theoretical approaches, oblate deformation corresponding to the N=Z=36 subshell filling is predicted. From our data, a preliminary decay scheme has been established with the feeding of more than fifteen 1$^+$ bound states in ^{72}Br. No delayed proton emission is observed down to the level 10^{-6}. According to the theoretical predictions of I.Hamamoto et al, only weak differences are expected between the GT strength distributions for oblate and prolate deformations (table 2). The experimental value deduced from our measurements for the bound states is in the order of magnitude of both predictions, if a quenching factor of 0.5 is taken into account for the theoretical values. An upper limit of 0.0014 is inferred for the delayed proton channel. The missed E1 radiative decay in the $Q_{EC} - S_p$ window has been estimated, as described for ^{76}Sr, to reach 94% of the total strength - with a level density parameter $a = 10 MeV^{-1}$. Concerning the unbound states, an upper limit is set at 0.025 for the experimental Gamow-Teller strength. This value is more than one order of magnitude lower than the corresponding ones for both shapes.

TABLE 2. Experimental and theoretical Gamow-Teller strengths for ^{72}Kr.

Gamow-Teller strengths B(GT)=4139/ft	experiment	theory	
		oblate	prolate
ΣB(GT) unbound states	< 0.0014	1.4	0.7
Σ B(GT) bound states	0.5 - 0.7	1.5	1.0

II CONCLUDING REMARKS

This first tentative to sign the nuclear deformation of the parent nucleus on the basis of the Gamow-Teller strength distributions stresses the prominent part played by the delayed particle process to obtain valuable information in the upper part of the Q_{EC} window. From comparison with the theoretical predictions in the case of ^{76}Sr, strong indication is obtained for a prolate deformation from the measured GT strength distribution. One has to mention at this point that an evaluation of the missed radiative strength had to be made to overcome the limitations encountered on the experimental side. Our search for nuclear deformation signature via Gamow-Teller decay measurements will be extended to heavier nuclei and compiting investigations will be carried out with the help of a total absorption gamma ray spectrometer in a near future.

REFERENCES

1. Hamamoto, I. and Sagawa,H., Phys. Rev.C 48, 3, 2960 (1993)
2. Hamamoto, I. and Zhang, X. H., Z. Phys. A353, 145 (1995)
3. Gelletly, W. et al., Phys. Lett. B253, 287 (1991)
4. Lievens, P. et al., Cern report CERN-PPE/95-160
5. Giovinazzo, J., Thesis, Université Louis Pasteur, Strasbourg France (1997) and this conference PE5.
6. Hardy, J. C., Phys.Lett. 109B, 242 (1982)

Gamow-Teller strength in the $f_{7/2}$- nuclei ^{54}Co and ^{42}Sc studied through the beta decay of ^{54}Ni and ^{42}Ti

I. Reusen, A. Andreyev, J. Andrzejewski[1], N. Bijnens, B. Bruyneel,
S. Franchoo, M. Huyse, Y. A. Kudryavtsev, K. Kruglov, W. F. Mueller,
A. Piechaczek[2], R. Raabe, K. Rykaczewski[3], J.Szerypo[4], P. Van Duppen,
J. Van Roosbroeck, L. Vermeeren, J. Wauters[5], L. Weissman and A. Wöhr[6]

Instituut voor Kern- en Stralingsfysica,
University of Leuven, Celestijnenlaan 200 D, B-3001 Leuven, Belgium

Abstract. At the LISOL facility the neutron-deficient ^{54}Ni isotope was produced via a ^{3}He fusion-evaporation reaction on a ^{54}Fe target positioned in the element-selective Ion-Guide Laser Ion Source (IGLIS). The β decay of ^{54}Ni was observed for the first time. The Gamow-Teller β-decay strength associated with the ($J^\pi=1^+$) level at 937.2 keV in the daughter nucleus ^{54}Co was deduced. The Gamow-Teller β-decay strength is in agreement with B(GT) obtained from (p,n) reaction studies. For the first time an experiment was performed using a heavy-ion beam in combination with the IGLIS in order to produce ^{42}Ti.

INTRODUCTION

Experimental and theoretical studies of the β-decay of fp-shell nuclei are needed to search for a solution to the long-standing problem of the origin of B(GT) quenching. That is the overestimation of the B(GT) by shell-model calculations. The quenching of Gamow-Teller strength in the sd shell has been studied quite extensively [1]; however, in the fp shell there is still a lack of experimental information on B(GT), although much effort has been put into theoretical studies in the form of large-scale shell-model calculations [2] and shell-model Monte-Carlo calculations in this mass region [3].

The Gamow-Teller strength can be experimentally studied in β decay as well as in a (p,n) reaction if we assume isospin symmetry. Both methods are complementary. The reaction method has the advantage of not being limited by a Q-value window while β decay has the advantage of giving absolute numbers which can be used to normalize the

---
[1] Present address: University of Lodz, Poland
[2] Present address: Louisiana State University, Baton Rouge, LA, USA
[3] Present address: Oak Ridge National Laboratory, P.O. Box 2008, Oak Ridge, TN 37831-6371, USA
[4] Present address: Oak Ridge National Laboratory, P.O. Box 2008, Oak Ridge, TN 37831-6371, USA
[5] Present address: IMEC, Kapeldreef, B-3001 Leuven, Belgium
[6] Present address: Oxford University, Parks Rd., Oxford OX1 3PU, United Kingdom

CP455, ENAM98: Exotic Nuclei and Atomic Masses
edited by B. M. Sherrill, D. J. Morrissey, and Cary N. Davids
© 1998 The American Institute of Physics   1-56396-804-5/98/$15.00

B(GT) strength obtained from (p,n) reaction studies. Thus, the combination of both methods makes it possible to investigate the absolute B(GT) distribution.

Anderson et al. measured the GT strength in ^{54}Co using the ^{54}Fe(p,n) reaction at 135 MeV [4]. Up to now a β-decay experiment of ^{54}Ni, the mirror nucleus of ^{54}Fe, was not performed due to the chemical properties of nickel. Short-lived Ni isotopes are difficult to produce in conventional ion sources. Furthermore the ^{54}Ni beta decay will be dominated by ground state to ground state decay and the spectra will be overwhelmed by the decay of the ^{54m}Co ($T_{1/2}$=193.23 ms) and ^{54g}Co ($T_{1/2}$=1.48 m) isobars [5, 6]. The estimated half-life of ^{54}Ni is 140 ms if we assume pure Fermi decay from the ground state of ^{54}Ni towards the ^{54}Co ground state. From the B(GT) obtained by Anderson et al. we calculated the β-branching ratio towards the 1^+ level at 937.2 keV in the daughter nucleus ^{54}Co. This results in a β-branching ratio of 24% towards this level and a half-life for ^{54}Ni of 105 ms. In this article we present the β-decay study of ^{54}Ni produced in a ^{3}He fusion-evaporation reaction on an enriched ^{54}Fe target positioned in the element-selective Ion-Guide Laser Ion Source (IGLIS), based on resonant photo ionization in a gas cell [7, 8].

For the first time an experiment using an ^{16}O heavy ion beam in combination with the IGLIS was performed in order to produce a ^{42}Ti beam. The half life of ^{42}Ti was determined from in-beam experiments performed in the late-60s. The half life of ^{42}Ti is 199(6) ms but is the result of four scattered values [9, 10, 11]. The β-branching ratio towards the level at 611 keV in the daughter nucleus ^{42}Sc was determined by Aldridge et al. but shows a large uncertainty. In literature these values are used to calculate the B(GT) towards the level at 611 keV. The obtained B(GT) serves as a normalization for the B(GT) determined from (p,n) reaction studies on ^{42}Ca by Goodman et al.[12].

EXPERIMENTAL DETAILS

The ^{54}Ni nuclei were produced in a 45 MeV ^{3}He^{2+} fusion-evaporation reaction on an enriched ^{54}Fe target positioned in the Ion-Guide Laser Ion Source. The reaction products recoiling out of the target are thermalized and neutralized in 500 mbar He. The reaction products are transported by the gas flow towards the laser region where Ni atoms are ionized via a two-step resonant process [7, 8]. After ionization the charged nuclei are mass separated and transported to a tape system. The tape was used only to transport away long-lived activity. The mass-separated beam was pulsed (0.5 s ON – 1.0 s OFF) such that the half life of implanted nuclei could be measured. The implantation point was surrounded by a plastic ΔE-E detector and a H-P Germanium detector. An alternative set-up, consisting of two plastic ΔE detectors each followed by a H-P Germanium detector, was also used. In both experiments β-γ coincidences were recorded together with time information in order to deduce the half life of ^{54}Ni.

In the ^{42}Ti experiment we used an ^{16}O $^{5+}$ beam at incident energies of 40-60 MeV on a ^{28}Si target. The ^{42}Ti nuclei recoiling out of the target are thermalized in 400 mbar Ar.

EXPERIMENTAL RESULTS

The best production rate of ^{54}Ni was 10 at/μC. The selectivity, defined as the ratio of background subtracted counts during the implantation period while the lasers were ON relative to the lasers OFF was larger than 13. The half life of ^{54}Ni was determined in two ways. The first method consisted in measuring the half life of a 937.1 keV γ ray, which is only present when the lasers are tuned to nickel and which can be assigned to ^{54}Ni as the transition of the first 1^+ level in ^{54}Co to the ground state. Figure 1 shows the fit to the data yielding a half life of 103±22 ms. The second way is to subtract from the γ spectra, obtained with the lasers resonant to nickel, spectra obtained with the lasers resonant to cobalt by using the known γ lines of ^{54m}Co. The resulting γ spectra are dominated by the 511 keV annihilation radiation from the subsequent positron decay from ^{54}Ni to ^{54}Co to ^{54}Fe. Out of its time behavior a half life of 105±14 ms is obtained by using a mother-daughter fit. The weighted average of these two values results in a half life for ^{54}Ni of 104±12 ms. The branching ratio towards the 937.2 keV level is obtained by comparing the total amount of ^{54}Ni, determined from the intensity of the 511 keV annihilation radiation after correction for the contribution of ^{54}Co, and the emitted 937.1 keV γ rays. The efficiency of the 511 keV annihilation radiation and the 937.1 keV γ ray is needed. Extensive GEANT calculations have been performed in order to deduce the necessary efficiencies. This half-life and branching ratio result in a B(GT) of 0.69±0.16 to be compared with 0.736 from (p,n)-reaction studies [4].

While for mass 54, the ^{54}Ni γ line is clearly visible in the Ni on resonant γ spectrum, although dominated by the non-resonant ^{54}Co γ lines, at mass 42 the ^{42}Ti γ line at 611 keV only becomes visible in the γ spectrum dominated by the ^{42}Sc lines by special gating conditions. This is probably due to the combined effect of the strong survival probability of non-resonant Sc ions and the low laser ionization efficiency for Ti atoms. The measured ^{42}Sc production rate at 50 MeV ^{16}O on ^{28}Si was 8.8×10^4 at/μC, which results, by using the measured cross section of 120 mbarn by Dauk et al. [13], in an efficiency for guiding non-resonant ions through the gas cell of 0.44%. This survival probability is as high as in the heavy-ion ion guide from Jyväskylä [14]. The survival probability in our rather slow gas cell is probably so high due to the high purity level (ppb) of the used gas. If we use the ^{42}Sc production at 50 MeV as a normalization point, we notice that the experimental ^{42}Ti and ^{43}Ti production is two orders of magnitude smaller than what is predicted by HIVAP [6]. In this mass region Dauk et al. measured cross sections for a 25 MeV-50 MeV ^{16}O beam on ^{28}Si [13]. The measured cross sections are within a factor of 10 comparable to the cross sections predicted by HIVAP. This cannot explain the two orders of magnitude difference between our measured productions for ^{42}Ti and ^{43}Ti and the HIVAP predictions. A selectivity of only 7 was reached for ^{43}Ti. A possible explanation can be that Ti forms strongly-bound molecules which make laser ionization impossible and which cannot be dissociated.

CONCLUSION

We studied the β decay of ^{54}Ni. We measured the half-life of ^{54}Ni and the β-decay branching ratio towards the 937.2 keV level in the daughter nucleus ^{54}Co. Combining these two values a B(GT) strength associated with this level at 937.2 keV is obtained which is in agreement with B(GT) from (p,n) reaction studies [4].

For the first time the combination of an ^{16}O heavy-ion beam with the Ion-Guide Laser Ion Source was tested. From the production of ^{42}Sc we determined the efficiency of the IGLIS without laser ionization. The efficiency is comparable to the efficiency of the heavy-ion ion guide from Jyväskylä [14]. We observe two orders of magnitude less ^{42}Ti than what is predicted by HIVAP. Together with the low selectivity this indicates the possibility that Ti and trace amounts of ^{16}O are forming strongly-bound molecules which prevent these Ti atoms from being resonantly laser ionized and which cannot be dissociated.

We gratefully thank J. Gentens and P. Van den Bergh for running the LISOL separator. This work is supported by the Inter-University Attraction Poles (IUAP) Research Program and by the Fund for Scientific Research Flanders (FWO-Belgium). M.H. is Research Director, L.V. Postdoctoral Researcher, and S.F. Research Assistant of the FWO, Belgium.

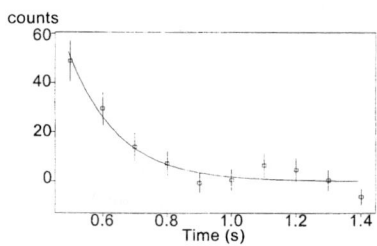

FIGURE 1 Decay curve of the β-delayed 937.1 keV γ rays.

REFERENCES

1. Brown B. A. and Wildenthal B. H., Ann. Rev. Nucl. Part. Sci. **38**, 29 (1988)
2. E. Caurier, G. Martínez-Pinedo, A. Poves, A. P. Zuker, Phys. Rev. **C52**, R1736 (1995)
3. K. Langanke *et al.*, Phys. Rev. **C52**, 718 (1995)
4. Anderson B. D. *et al.*, Phys. Rev. **C41**, 1474 (1990)
5. Junde H., Huibin S., Weizhong Z. and Qing Z., Nucl. Data Sheets **68**, 887 (1993)
6. Reisdorf et al., Z. Phys. **A300**, 227 (1981)
7. Vermeeren L. *et al.*, Phys. Rev. Lett. **73**, 1935 (1994)
8. Kudryavtsev Y. A. *et al.*, Nucl. Instr. and Meth. **B 114**, 350 (1996)
9. Aldridge A. M., Kemper K. W., Plendl H. S., Phys. Lett. **30B**, 165 (1969)
10. Gallman A., Aslanides E., Jundt F. and Jacobs E. , Phys. Rev. **136**, 186 (1969)
11. Nicholas F. M., *et al.*, Nucl. Phys. **A124**, 97 (1969)
12. Goodman C. D. *et al.* , Phys. Lett. **107B**, 406 (1981)
13. Dauk J., Lieb K. P. and Kleinfeld A. M., Nucl. Phys. **A241**, 170 (1975)
14. Dendooven *et al.*, to be published in Nucl. Instr. and Meth.

Statistical deliberations for exotic nuclei

K. Riisager

IFA, Aarhus Universitet, DK-8000 Aarhus C, Denmark

Abstract. An often encountered situation in experiments on nuclei far from stability is that the count number per channel in various differential spectra becomes small. This might require the use of not so well-known statistical methods. Some examples are given.

INTRODUCTION

The question of what statistical method to use in the analysis of experimental data has been considered in many textbooks, e.g. [1,2]. Experiments with nuclei far from stability nevertheless often result in situations where standard statistical methods do not immediately apply, partly due to the low count numbers that often are encountered. The aim of the present contribution is through three different illustrative examples to show how one then can proceed. The topics treated are biases in fit methods for Poisson distributed data, how to employ simple estimators in line shape analysis and efficient estimation of half-lifes.

LOW COUNT NUMBERS

When low count numbers occur one must use the Possion distribution explicitly rather than doing a Gaussian approximation. Since the Possion distribution is of exponential type the maximum likelihood will be an optimal method of analysis [2]. The often employed least squares minimization (χ^2-fit) gives a biased result. In two different variations of this method the count numbers or the theoretical values are used to estimate the error on the data points so that $\chi^2 = \sum_i (n_i - \theta_i)^2 / n_i$ or $\chi^2 = \sum_i (n_i - \theta_i)^2 / \theta_i$. Both methods gives a bias that asymptotically will be -1 and $1/2$, respectively [3]. For "experimental error bars" this follows since points below the "true value" will be attributed a too small error, conversely for points above. For "theoretical error bars" this follows since the denominator in χ^2 on its own would force θ_i to become large. A detailed derivation of the bias can be found in [3]. For a large number of channels N_0 with common "true value" μ the two variants of χ^2-minimization gives fit-parameters that are approximately $\mu^2/(\mu+1)$

and $\mu\sqrt{1+1/\mu}$, respectively. Results are also given in the limit of small N_0 and have in all cases been verified by means of Monte Carlo methods. Note that a maximum likelihood minimization not only gives the correct result for all N_0, but also gives the smallest variance of the estimate. It should therefore be used instead of least squares minimization whenever effects of order $1/n_i$ are judged important.

SIMPLE ESTIMATORS

Other methods than maximum likelihood and least squares are occasionally preferable, e.g. when the exact underlying distribution is not well established and one at first can be satisfied with obtaining only a few characteristics of the data. This situation might be encountered when the total statistics is limited and/or does not allow a detailed line shape analysis to be made, but one nevertheless would like to extract some main quantities. A way out is to resort to simple (distribution-free) methods [1,2]. As a practical example we can refer to data from a recent experiment at GANIL [4] where the aim is to obtain the width for the neutron longitudinal momentum distribution following break-up of ^{10}Be and ^{11}Be in a Be target.

Let the total number of counts be N. All summations should be understood to run over all counts, i.e. $\sum_i f_i$ should be interpreted as $\sum_{i=1}^{N} f_i$ or if the data are binned as $\sum_{i=\text{all bins}} n_i f_i$, where n_i is the number of counts in bin i ($\sum_i n_i = N$). For each count a "position variable" is determined as x_i, for binned data this is taken as the position in the middle of the bin. The data points are assumed to stem from some distribution that has mean value μ and variance V.

The location parameter can be estimated by $\hat{\mu} = \bar{x}$, the average value of x_i. The similar unbiased estimate of the variance is well-known to be:

$$\hat{V} = \frac{N}{N-1} \overline{(x-\bar{x})^2} = \frac{1}{N-1} \sum_i (x_i - \bar{x})^2 ,$$

and an error estimate for $\hat{\mu}$ is then given by $V(\hat{\mu}) = \hat{V}/N$. The corresponding procedure for obtaining an error estimate for the width [1] is less well known. A relatively simple derivation gives the following unbiased estimators:

$$\text{cov}(\hat{\mu}, \hat{V}) = \frac{1}{N-2} \frac{1}{N-1} \sum_i (x_i - \bar{x})^3 ,$$

$$V(\hat{V}) = \frac{N}{N^2 - 3N + 3} \frac{1}{N-1} \sum_i (x_i - \bar{x})^4$$

$$- \frac{1}{N} \left(3 \frac{2N-3}{N^2 - 3N + 3} + \frac{N-3}{N-1} \right) \left(\frac{1}{N-1} \sum_i (x_i - \bar{x})^2 \right)^2 ,$$

where also the covariance between $\hat{\mu}$ and $\hat{V}$ is included. The statistical efficiency of these simple estimators will of course depend on what the exact underlying distribution is (it might even asymptotically go to zero, cf. the discussion in section 8.7 in [2]), so if independent reliable knowledge on the distribution is available one should incorporate it by using more elaborate methods.

For the case in question the above formulas can be applied directly to extract parameters for the distribution from ^{10}Be (core) break-up. The position decreases and the width increases significantly as one goes to larger angles, i.e. larger transverse momenta. For break-up of ^{11}Be one might assume a two-component distribution: a contribution from the core (the one just extracted) and a contribution from the halo neutron. Using the ^{10}Be data a disentanglement can be done both for the position and for the width variables, but only the latter could be extracted with sufficient accuracy for the halo component. It is clearly smaller than the core width (the two variances differ by about a factor two) and also shows clear indications for an increase with angle. To draw more detailed physical conclusions one needs to introduce models for the process at this point.

HALF-LIFE DETERMINATIONS

Finally, the case of half-life determinations will be considered in some detail. The maximum likelihood is here again an optimal method of analysis also in the limit of low count numbers and can be used for finding parameters and their errors. As is well-known [1,2] the best estimator for the mean life τ is the average time $\bar{t}$ (multiply with $\ln 2$ to get the half life instead). If data only are taken up to an upper limit T one must iterate the equation

$$\tau = \bar{t} + \frac{T}{e^{T/\tau} - 1}$$

to obtain the mean life. If furthermore the data are binned with a bin width of Δt the equation to be iterated turns into

$$\tau = \bar{t} + \frac{T}{e^{T/\tau} - 1} + \left(\tau - \frac{\Delta t}{e^{\Delta t/\tau} - 1}\right).$$

The variance is similarly given by explicit formulas. It is τ^2/N (where N is the number of data points) in the simplest case and increases for finite T and for increasing Δt.

There are many powerful statistical methods for determining the fit-quality for an underlying exponential distribution [5]. As an example the Anderson-Darling statistics that is based on the empirical distribution function will be employed here to illustrate how the half life can be determined for ^{31}Ar from the observed time distribution of beta-delayed two-proton events [6]. The activity was produced at ISOLDE by a pulsed proton beam and diffuses gradually out of the target. The ions are collected while the decays are recorded and the goodness-of-fit test is therefore needed to ensure that one has reached a regime of "pure exponential decay" before a reliable value can be extracted. As illustrated in figure 1 this happens about 85 ms

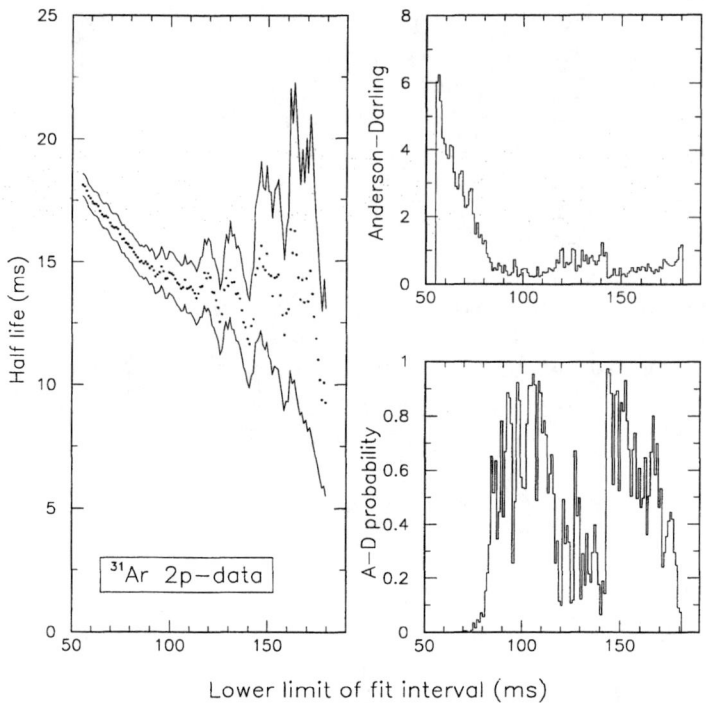

FIGURE 1. Half life of ^{31}Ar determined from two proton data for different fit-intervals as function of the starting point of the interval. Half life is found with maximum likelihood and goodness-of-fit is done with the Anderson-Darling statistics.

after start of collection (the "A-D probability" measures the agreement between the data and a single exponential distribution). Only after this point can the extracted half life values be trusted.

REFERENCES

1. Barlow, R.J., *Statistics, A Guide to the Use of Statistical Methods in the Physical Sciences*, Chichester: Wiley, 1989.
2. Eadie, W.T. et al., *Statiscial Methods in Experimental Physics*, Amsterdam: North-Holland, 1971.
3. Jading, Y., and Riisager, K., *Nucl. Instr. Methods* **A372**, 289–292 (1996).
4. Axelsson, L., Grévy, S. et al., in preparation.
5. Spurrier, J.D., *Commun. Statist.-Theor. Meth.* **13**, 1635–1654 (1984).
6. Borge, M.J.G. et al., these proceedings.

The Beta-Delayed Proton Decay of ^{23}Al

M. W. Rowe, D. M. Moltz, T. J. Ognibene, J. Powell and Joseph Cerny

*Department of Chemistry, University of California, Berkeley and
Nuclear Science Division, Lawrence Berkeley National Laboratory, Berkeley, California 94720*

Aluminum-23 was produced in two 40 MeV proton bombardments of Mg targets at LBNL's 88-Inch Cyclotron. Reaction products were transported by helium jet to a detection chamber. They were observed by two low-energy particle-identification telescopes; each consisted of two gas-ΔE detectors, a thin (< 70 μm) Si E detector and a 300 μm Si E-reject detector. Contrary to previous measurements, the lowest energy peak in the spectrum was observed at an energy of 246 ±20 keV with 33 ±3% of the intensity of the peak at 838 ±5 keV. Implications for isospin mixing and the proton-capture width of the ^{23}Al isobaric analog state in ^{23}Mg are discussed. Other weak beta-delayed proton-decay branches were also observed with energies up to ~2200 keV.

INTRODUCTION

Studies of light nuclei provide stringent tests of nuclear model predictions; this is particularly true for nuclides far from stability. Because relatively few nucleons are involved, it is possible to calculate realistic wavefunctions for nuclei in the $1s0d$-shell by including empirically-determined two-body interactions in the shell-model Hamiltonian (1). The operators for M1 and E2 gamma-ray emission and for Gamow-Teller beta decay are also relatively simple (2); thus, comparison between measurements of these decays and predictions generated by application of these operators to initial- and final-state wave functions provides a measure of the validity of the wave functions and the Hamiltonian which generated them.

The beta-delayed proton decay of ^{23}Al was first measured in 1972 (3); a single proton peak was observed at 832 ±30 keV. A recent measurement by this research group (4) used specially-designed low-energy particle-identification telescopes (5) to measure the spectrum to as low as ~180 keV. The primary goal of this measurement was to observe proton emission from the isobaric analog state (IAS) of ^{23}Al in ^{23}Mg; it was known from reaction work (6) to be unbound by only 215 ±6 keV. A proton peak at 223 ±20 keV (lab.) with an intensity 2.2 ±0.5 times that of the 832 keV group was assigned to this decay. Because this decay is isospin forbidden, it proceeds only to the extent that the initial and/or final states contain small admixtures of states with different isospin; when compared to the intensity predicted by Brown using an isospin-nonconserving (INC) Hamiltonian (7), the observed intensity suggested that the ^{23}Al IAS contains an unusually large T=1/2 component. The intensity also implied that the resonance strength ωγ of the IAS was significantly larger than the upper limits set by direct measurements (e.g., ref. 8) of the ^{22}Na(p,γ)^{23}Mg reaction at the energy corresponding to the IAS. This would make the above reaction a more important pathway in the NeNa cycle and/or the rp-process of explosive nucleosynthesis (9).

After publication of this measurement (4), it was discovered that ^{16}O recoils from the beta-delayed proton emitter ^{20}Na, which was produced simultaneously with ^{23}Al, could mimic very low-energy alpha decays if the recoils were detected in coincidence with the positron from the ^{20}Na beta decay (5). This caused concern that the proton peak at 223 keV might have been contaminated by these ^{16}O-recoil/beta events, particularly given the disparity between the measured and predicted intensity of the peak. We have remeasured this decay in order to address this concern.

EXPERIMENT

For each of two bombardments, a helium-jet system was used to collect the activity and transport it via an ~80 cm capillary to a detector chamber for counting. Activity exiting the capillary was deposited onto the rim of a slowly-rotating catcher wheel. Two telescopes viewed the activity deposition spot from above and below with solid angles of 1.6% and 1.7% of 4π sr. Four ~1 mg/cm^2 ^{nat}Mg targets were bombarded with ~200 mC of 40 MeV protons during the first measurement, which used a 0.9 mm i.d. capillary fed by ten smaller capillaries that collected the activity from behind the targets. A ~1.0 mg/cm^2 ^{24}Mg (99.9%) separated-isotope target was bombarded by ~100 mC of protons during the second run. A single 1.35 mm i.d. capillary collected the activity during this measurement. During both experiments, the beam was pulsed; data were acquired only during the beam-off period.

In order to eliminate the contamination from ^{16}O-recoil/beta coincidence events, modified detector telescopes were used for these measurements. Each consisted of two gas-ΔE detectors, backed by a thin Si E detector (28 and 63 μm in the upper and lower telescopes, respectively) and, behind this, a ~300 μm Si E_{reject} detector. The beta-delayed proton emitter ^{25}Si (10) provided calibration points from 386 keV to 5405 keV. In order to extend the calibration to the detector threshold, additional calibration spectra were taken with thin aluminum foils in front of the telescopes; these degraded the 386 keV proton group to as low as 203 ±3 keV, approximately the laboratory energy expected for proton from the IAS decay (206 ±6 keV).

RESULTS AND DISCUSSION

Figure 1 shows a spectrum resulting from these measurements; Table 1 lists the observed energies and relative intensities of the peaks, as well as possible decay assignments (11) and the corresponding log ft values. A more detailed discussion of

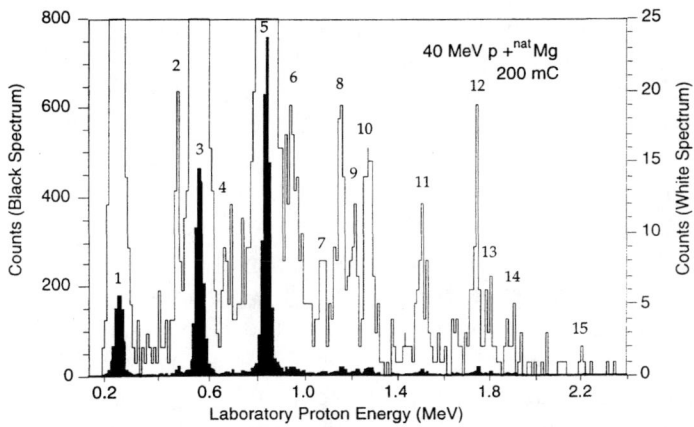

FIGURE 1. The beta-delayed proton spectrum of ^{23}Al.

these measurements may be found elsewhere (12). The lowest-energy peak in the spectrum at 246 ±20 keV is ~7 times less intense (relative to the 838 keV group) than the 223 ±20 keV peak observed in the previous measurment. Several other proton "peaks" are also indicated; some of the weaker groups require confirmation.

TABLE 1. Observed beta-delayed proton groups.

Peak #	Energy (lab.; keV)	Relative intensity (%)	Excitation in ^{23}Mg (keV) This work	Reference 11	log $f_A t$ [a]
1	246 ±20	33 ±3	7837 ±21	7795 ±6	---
				7852 ±6	5.82 ±0.04
			8420 ±21[b]	8420 ±6[b]	5.45 ±0.04
2	468 ±10	2.1 ±0.3	8070 ±10	8078 ±8	6.87 ±0.07
3	556 ±5	68 ±5	8161 ±5	8155 ±6	5.31 ±0.03
4	675 ±10	1.9 ±0.3	8285 ±10	8285 ±8	6.78 ±0.07
5	838 ±5	(100)	8456 ±5	8453 ±5	(4.94)
6	942 ±10	4.4 ±0.4	8564 ±10	8557 ±6	6.22 ±0.04
7	1075 ±10	1.2 ±0.2	8703 ±10	8758 ±6	6.68 ±0.08
8	1156 ±10	3.2 ±0.4	8788 ±10	8793 ±8	6.19 ±0.06
9	1215 ±10	2.1 ±0.3	8850 ±10	8870 ±8	6.32 ±0.07
10	1277 ±10	2.5 ±0.3	8914 ±10	8916 ±6	6.20 ±0.06
11	1505 ±10	1.9 ±0.3	9153 ±10	9138 ±6	6.11 ±0.07
12	1748 ±10	2.0 ±0.3	9407 ±10	9403 ±8	5.84 ±0.07
13	1797 ±10	0.9 ±0.2	9458 ±10	9465 ±6	6.13 ±0.11
14	1897 ±10	0.6 ±0.2	9563 ±10	9596 ±8	6.19 ±0.18
15	2201 ±10	0.2 ±0.1	9880 ±10	---	6.28 ±0.30

[a] Calculated assuming a 0.133% branch for the previously measured 838 keV group, as estimated from the measured (3) cross section for this decay (~220 nb) and a calculated total cross section of 165 μb.
[b] Assumes decay proceeds to the first-excited state of ^{22}Na at 583.03 keV excitation.

The energy of the 246 ±20 keV peak places its previous assignment to the IAS decay in doubt. Three possible assignments are listed in Table 1. Assuming it were the IAS decay, one may recalculate the isospin mixing and resonance strength. Using the same assumptions as Tighe, *et al.* (4), an isospin mixing of 1.7 ±1.1% and $\omega\gamma$ of 6.7 ±3.8 meV are obtained. The former is much closer to agreeing with 0.24% predicted by Brown (4,13); the latter is near the 2.6 meV limit set by Stegmüller, *et al.* (8). Additionally, it should be noted that these estimates are very sensitive to the assumptions made concerning the absolute decay branch and the gamma- and (unhindered) proton-decay widths for the IAS. If peak 1 is not from the IAS decay, the log ft values shown in Table 1 indicate it is probably just an allowed decay.

One may also calculate the Gamow-Teller (GT) strength at this excitation in ^{23}Mg. The total strength estimated from the log ft values shown in Table 1 is only about 10% of that predicted by Brown (13) for this energy region. However, since the log ft values have been determined from the relative intensity of the 838 keV peak and its estimated branching ratio, this result should not be taken too seriously. It is interesting to note that the *distribution* of GT-strength in this region is also quite different from what was predicted. Additional measurements (e.g., of absolute branching ratios and/or the gamma-decay width for the IAS) would allow much more reliable estimates of the GT strength, and the isospin mixing and resonance strength of the IAS, to be made.

ACKNOWLEDGMENTS

This work was supported by the Director, Office of Energy Research, Office of High Energy and Nuclear Physics, Division of Nuclear Physics, of the U.S. Department of Energy under Contract No. DE-AC03-76SF00098.

REFERENCES

1. Brown, B. A., Richter, W. A., Julies, R. E., and Wildenthal, B. H., *Annals of Physics* **182**, 191-236 (1988).
2. Brown, B. A., and Wildenthal, B. H., *Ann. Rev. Nucl. Part. Sci.* **38**, 29-66 (1988).
3. Gough, R. A., Sextro, R. G., and Cerny, J., *Phys. Rev. Lett.* **28**, 510-512 (1972).
4. Tighe, R. J., Batchelder, J. C., Moltz, D. M., Ognibene, T. J., Rowe, M. W., Cerny, J., and Brown, B. A., *Phys. Rev.* C **52**, R2298-R2301 (1995).
5. Rowe, M. W., Moltz, D. M., Batchelder, J. C., Ognibene, T. J., Tighe, R. J., and Cerny, J., *Nucl. Instrum. Meth. Phys. Res.* A **397**, 292-303 (1997).
6. Nann, H., Saha, A., and Wildenthal, B. H., *Phys. Rev.* C **23**, 606-615 (1981).
7. Ormand, W. E., and Brown, B. A., *Nucl. Phys.* **A491**, 1-23 (1989).
8. Stegmüller, F., Rolfs, C., Schmidt, S., Schulte, W. H., Trautvetter, H. P., and Kavanagh, R. W., *Nucl. Phys.* **A601**, 168-180 (1996).
9. Champagne, A. E., and Wiescher, M., *Ann. Rev. Nucl. Part. Sci.* **42**, 39-76 (1992).
10. Robertson, J. D., Moltz, D. M., Lang, T. F., Reiff, J. E., Cerny, J., and Wildenthal, B. H., *Phys. Rev.* C **47**, 1455-1461 (1993).
11. Endt, P. M., *Nucl. Phys.* **A521**, 1-830 (1990).
12. Rowe, M. W., Ph.D. Thesis, University of California, Berkeley (1998).
13. Brown, B. A., personal communication.

Delayed Neutron Emission in the Semi-Gross Theory of Nuclear β-Decay

Takahiro Tachibana and Masami Yamada

*Advanced Research Institute for Science and Engineering, Waseda University
3-4-1 Okubo, Shinjuku-ku, Tokyo 169-8555, JAPAN*

Abstract. The semi-gross theory of nuclear β-decay enables one to estimate some β-decay properties such as half-lives and delayed neutron emission probabilities (P_n-values) in the region far from the β-stability line. This theory has been obtained by refining the conventional gross theory to take into account some shell effects of the parent nucleus. In this theory, the one-particle energy-levels are assumed to be discrete and non-uniform, and the one-particle strength function depends on the orbital and total angular momenta of the decaying nucleon. The P_n-values obtained with use of the semi-gross theory are lower than the experimental values in many cases. In order to get more reasonable P_n-values, the strength of the semi-gross theory in each small energy interval is spread with a width depending on the excitation energy. Modified P_n-values thus obtained are compared with the experimental data as well as those estimated by the semi-gross theory.

SEMI-GROSS THEORY

The estimation of $T_{1/2}$ and P_n, which may be obtained from the β-decay strength function, is useful for the study of nuclear structure and, in the case of very neutron rich nuclei, necessary for the calculation of r-process nucleosynthesis. In our gross treatment, i.e., either in the semi-gross theory [1, 2] (referred to as SGT hereafter) or in the conventional gross theory [3] (referred to as GT2 hereafter) the β-decay strength function of the Ω–type (Fermi, Gamow-Teller and 1st forbidden) transition, $|M_\Omega(E)|^2$, is given by

$$|M_\Omega(E)|^2 = \int_{\varepsilon_{min}}^{\varepsilon_{max}} D_\Omega(E,\varepsilon) W(E,\varepsilon) \frac{dn_1}{d\varepsilon} d\varepsilon , \qquad (1)$$

where ε is the single-neutron (proton) energy for $\beta^-(\beta^+)$ -decay, E the transition energy measured from the parent state, $D_\Omega(E,\varepsilon)$ the one-particle strength function, $dn_1/d\varepsilon$ the one-particle energy distribution of a decaying nucleon, and $W(E,\varepsilon)$ a weight function to take into account the Pauli exclusion principle.

The one-particle energy distribution $dn_1/d\varepsilon$ used in GT2 is a continuous function obtained from the Fermi gas model with an effective mass. On the other hand, in SGT, it is a non-uniform and discrete function taking into account the shell effects and pairing effects in the parent nucleus. One-particle strength functions for the Gamow-Teller and 1st forbidden transitions have also been modified in SGT; each of them is assumed to be a superposition of several functions reflecting the effect of spin flip and change of oscillator quanta by the transitions.

In SGT, some part of the shell effects are correctly considered for the strength function around the ground state of the daughter nucleus. The accuracy and the predictive power of the semi-gross theory, in particular for the half-lives, have been examined in the whole nuclidic region, and it was found that they are, at least, comparable to the microscopic calculations QRPA [4]. (See Refs. [1] and [2] for more details)

DELAYED NEUTRON EMISSION

In the region far from the β-stability line, a nucleus has the possibility of delayed neutron emission. The emission rate of a delayed neutron P_n is given by

$$P_n = \frac{\lambda_n}{\lambda} , \qquad (2)$$

with

$$\lambda_n = \int_{-Q+S_n}^{0} S_\beta(E) f(-E) \frac{\Gamma_n}{\Gamma_n + \Gamma_\gamma} dE , \qquad (3)$$

and

$$\lambda = \int_{-Q}^{0} S_\beta(E) f(-E) dE . \qquad (4)$$

Here, Q represents the ground-state Q-value, S_n the neutron separation energy, and the total β-strength function S_β is the sum of allowed-equivalent strengths of all transitions $|M_\Omega(E)|^2$. In Eq.(3), Γ_n and Γ_γ represent the neutron and γ-radiation widths, respectively. In the following calculations, we put the competition factor $\Gamma_n/(\Gamma_n + \Gamma_\gamma)$ to unity for simplicity.

As discussed in Ref. [2], many P_n-values obtained from SGT are lower than the experimental values. This means that the strength functions estimated by SGT are smaller than the actual strengths in the delayed-neutron window. This underestimation gets worse if we introduce a competition factor with γ-ray emission.

In order to get more reasonable P_n-values with the use of SGT, we spread the strength in each small interval around the energy E_0 with a width depending on E_0. Namely, with a spreading function $G(E - E_0; E_0 + Q)$, the modified strength function is calculated by

$$S_\beta(E) = \int_{-Q}^{\infty} S_\beta^{SGT}(E_0)\ G(E-E_0; E_0+Q)\ dE_0\ , \quad (5)$$

where, S_β^{SGT} is the strength function of the semi-gross theory. For the function $G(E-E_0;\ E_0+Q)$, in this paper, we take a normalized Lorentzian function multiplied by Gaussian function.

$$G(E-E_0;\ E_0+Q) = \frac{C\ \exp(-[E-E_0]^2/2a_G^2[E_0+Q]^2)}{\sqrt{2\pi}a_G(E_0+Q)} \times \frac{a_L(E_0+Q)}{2\pi([E-E_0]^2 + a_L^2[E_0+Q]^2/4)}\ , \quad (6)$$

where C is the normalization constant, and the parameters (a_G and a_L), which define the absolute values of the widths, are fixed to reproduce reasonable half-lives and P_n-values in the whole nuclidic region. This spreading function G increases the strength function considerably in the delayed neutron emission window, but only slightly in the neighborhood of the ground state of the daughter nucleus. Since the function G is a δ-function at $E = -Q$, the strength function is not changed for the ground to ground transition. After some examinations, we have taken a_G=0.3 and a_L=0.2. An example of the modified strength function, Eq. (5), is shown in Fig. 1 (a) and (b).

The shape of the modified S_β in the giant resonance region is reproduced fairly well in comparison with the S_β^{SGT}. A slightly increased strength in the ground state region, however, gives shorter half-lives in comparison with those estimated by SGT

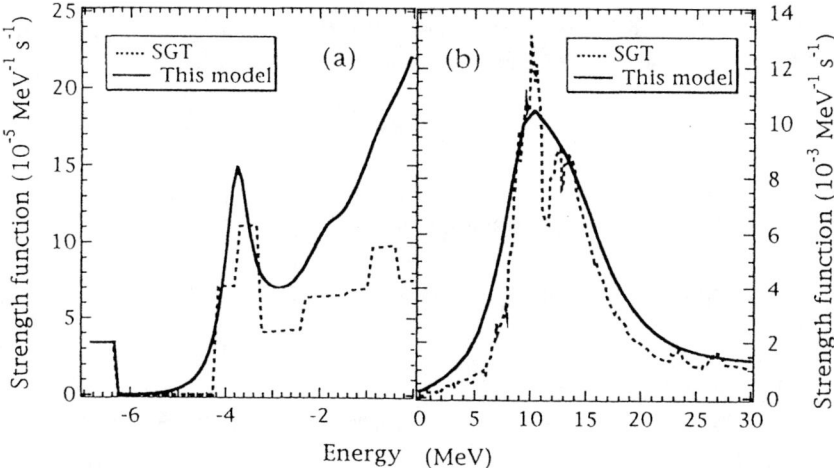

FIGURE 1. The Gamow-Teller strength functions of ^{87}Br in the β^- – decay region (a) and the giant resonance region (b). δ-functions in the strengths are drawn with a half-width of 0.5 MeV.

(about 80% on the average). We show the ratios between the calculated and experimental P_n-values [5] in Fig.2 (a); note that the experimental P_n-values are limited to those in the fission product region ($28 < Z < 40$ with $45 < N < 64$, and $46 < Z < 58$ with $73 < N < 94$). As seen in Fig.2 (a), this treatment modifies the calculated P_n-values mostly in the right direction, although the magnitude of the modification is not sufficient. In order to investigate the modified strength function in the delayed-neutron window, we also show the ratios between calculated and experimental λ_n-values (Eq. (3)) in Fig.2 (b). In this case, the effect of the modification is more clearly seen. We hope a further refinement of the spreading function G will improve the situation.

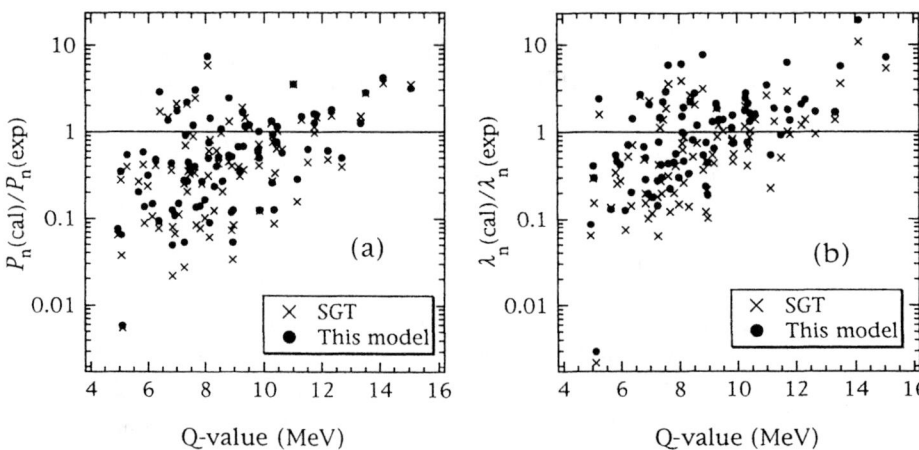

FIGURE 2. The ratios of calculated and experimental P_n-values (a) and λ_n-values (b). Evaluated nuclear masses in Ref. [6] and half-lives in Ref. [7] are adopted in the calculation.

This work was financially supported, in part, by the Grant for Special Research Projects, Waseda University.

REFERENCES

1. Nakata, H., Tachibana, T., and Yamada, M., Nuclear Physics, **A625**, 521-553 (1997).
2. Tachibana, T., and Yamada, M., *Tours Symposium on Nuclear Physics III*, IAP Conference Proceedings **425**, 495-504 (1998)
3. Tachibana, T., Yamada, M., and Yoshida, Y., *Progress of Theoretical Phys*ics, **84**, 641-657 (1990).
4. Homma, H., Bender, E., Hirsch, M., Muto, K., Klapdor-Kleingrothaus, H. V., and Oda, T., *Physical Review* **C54**, 2972-2985, (1996).
5. Rudstam, G., Aleklett, K., and Sihver, L., *Atomic Data and Nuclear Data Tables* **53**, 1-22 (1993).
6. Audi, G., and Wapstra, A. H., *Nuclear. Physics*, **A565**, 1-66, (1993) and **A595**, 409-480 (1995).
7. *Evaluated Nuclear Structure Data Files,* 1995 November version, communicated through Nuclear Data Center, Japan Atomic Energy Research Institute.

The GT Resonance revealed in β^+-Decay using new experimental techniques

J. Agramunt[a], A. Algora[a], L. Batist[d], R. Borcea[c], D. Cano-Ott[a],
R. Collatz[c], A. Gadea[a], J. Gerl[c], M. Gierlik[b], M. Górska[c],
O. Guilbaud[c], H. Grawe[c], M. Hellström[c], Z. Hu[c] Z. Janas[b],
M. Karny[b], R. Kirchner[c], P. Kleinheinz[f], W. Liu[c], T. Martinez[a],
F. Moroz[d], A. Płochocki[b], M. Rejmund[c], E. Roeckl[c], <u>B. Rubio</u>[a],
K. Rykaczewski[b], M. Shibata[c], J. Szerypo[b], J.L. Tain[a],
V. Wittmann[d] and the German Euroball Col.

[a] Instituto de Física Corpuscular, Dr. Moliner 50, E-46100 Burjassot-Valencia, Spain
[b] Institute of Experimental Physics, Warsaw University, PL-00-681 Warsaw, Poland
[c] Gesellschaft für Schwerionenforschung, D-64220 Darmstadt, Germany
[d] St. Petersburg Nuclear Physics Institute, 188-350 Gatchina, Russia
[f] Institut für Kernphysik, KFA Jülich, D-52425 Jülich, Germany

Abstract. The GT beta decay of ^{150}Ho has been studied with a Total Absorption Spectrometer (TAS), with an array of 6 Euroball CLUSTER Ge detectors (the CLUSTER CUBE), and with an alpha detector. The three techniques complement each other. The results provide the first observation of an extremely sharp resonance in GT beta decay.

Although the problem of the missing strength in Gamow-Teller decay is one of long-standing it remains unresolved. In light nuclei there is a great deal of experimental information, from both beta decay and charge-exchange reactions, which systematically indicates that ~40% of the strength is missing [1]. For heavier nuclei the beta decay information is sparse because of the difficulty of accessing nuclei with allowed decays. In this work we will report studies of Gamow-Teller beta-decay in heavier nuclei. In comparison with charge-exchange reactions they are reaction-model independent and free of background uncertainties. In addition they permit the study of nuclei far from stability. However, due to the selection rules, very few Gamow-Teller decays are allowed above the heaviest N~Z particle stable nuclei. This is because in general the required orbitals for allowed decays lie outside the beta-window. There are only two regions where the $\sigma\tau$ resonance is accessible in β-decay. In both of these regions a proton in a high J orbital decays into its spin-orbit partner neutron orbital with J-1, which is in general less bound than the J proton orbital and therefore empty. The nuclei we refer to lie just below ^{100}Sn and above ^{146}Gd.

The task of making precise β-strength measurements is far from trivial. Traditional high resolution techniques, based on a few Ge detectors to determine the β-feeding, often fail to detect significant but fragmented strength at high excitation energy. To tackle this problem we have performed β-decay measurements at the GSI On-line Mass-separator following two different approaches. First we have used a Total Absorption Spectrometer (TAS) [2] based on a large, cylindrical NaI crystal; see Fig. 1 (left). In order to separate events associated with the EC and β^+ processes we gated on the daughter X-rays or the positrons. This method is sensitive to the level population in the beta decay process rather than to the individual gamma rays. In a second set of experiments we measured the same decays with 6 Euroball CLUSTER detectors arranged to form a cube, see Fig. 1 (right).

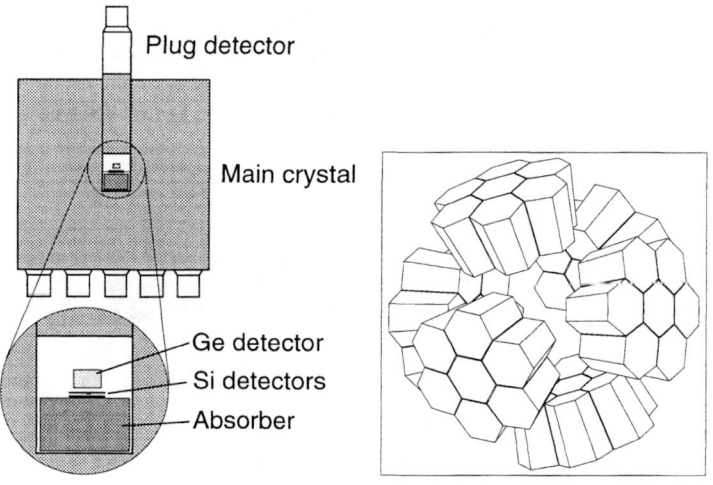

FIGURE 1. The two different experimental set-ups. Left: Total Absorption Spectrometer. Right: array of 6 Euroball CLUSTER Ge detectors (the CLUSTER CUBE)

Here we present the results for ^{150}Ho (2^- isomer) decay, which provides an ideal test of these methods since the strength is expected to be at high excitation energy, but still lie inside the Q_β window. From the TAS experiment we expect to extract the full β-strength distribution accessible in the decay, and from the CLUSTER CUBE experiment the fragmentation of the GT strength in a region of high level density. Moreover, the later experiment should provide the necessary input data to extract the β feeding from the TAS raw data. The comparison of the two independent results should reveal the limitation of the germanium detector techniques when the strength lies at high excitation energy.

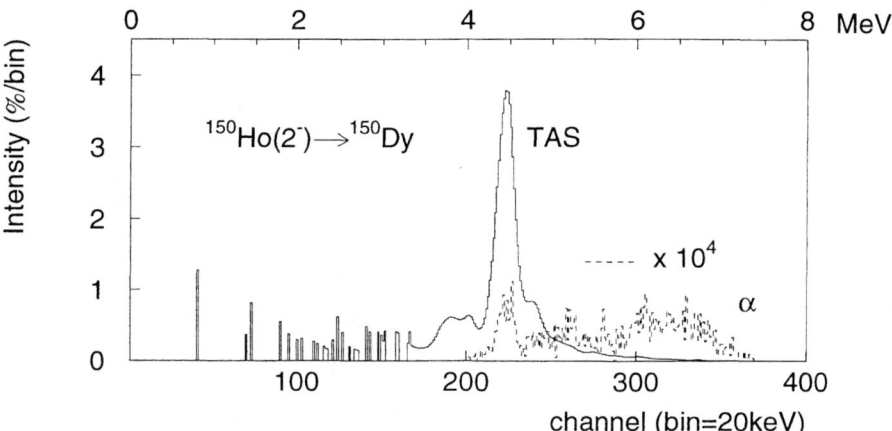

FIGURE 2. The ^{150}Ho (2^-) beta feeding followed by gamma de-excitation, measured with TAS (solid line), and by alpha emission (dashed line).

In the case of ^{150}Ho decay, β delayed alpha emission can compete with gamma de-excitation for sufficiently high excitation energy in the daughter nucleus. In order to check this possibility a third experiment was performed using both a Si ΔE-E particle telescope and a germanium detector at the same time. One limitation in the case of a TAS spectrometer is the impossibility of achieving 100% efficiency for all the gamma rays involved in the decay. The sophisticated procedure required to extract the β-feedings from the TAS raw spectra is explained in a separate contribution to this conference [3]. In Fig.2 we show the combined TAS result from the EC and β^+ decay of ^{150}Ho (solid line). An extremely sharp resonance of $\sim$240 keV width emerges at 4.5 MeV excitation energy. β-delayed alpha emission has been observed for the first time in this nucleus. The result is included in Fig.2 (dashed line). The ratio of alpha to gamma de-excitation is 6×10^{-5}.

The CLUSTER data gave a very rich ^{150}Ho 2^- decay scheme. In a preliminary analysis we have located 900 γ-transitions de-exciting 280 levels. The derived strength is represented by the solid line in Fig.3. In the same figure we show the TAS results. The CLUSTER CUBE data clearly demonstrate the correctness of the analysis method for the TAS experiments: Both experiments give a similar shape, but the CLUSTER CUBE experiment loses sensitivity towards higher energies. The total log ft obtained with the CLUSTER CUBE up to 5.45 MeV (the highest observed level) is 4.18. The log ft obtained with the TAS in the same energy range is 3.94, which represents 74% more strength. The integrated log ft up to the $Q_{EC} = 7400$ keV limit (from our data) amounts to 3.84.

In a very simple-minded approach, the 2^- isomer in ^{150}Ho can be viewed as $(\pi d_{3/2} \nu f_{7/2})_{2^-} (\pi^2)_{0^+}$. Thus, the only allowed decay is when the pair of protons

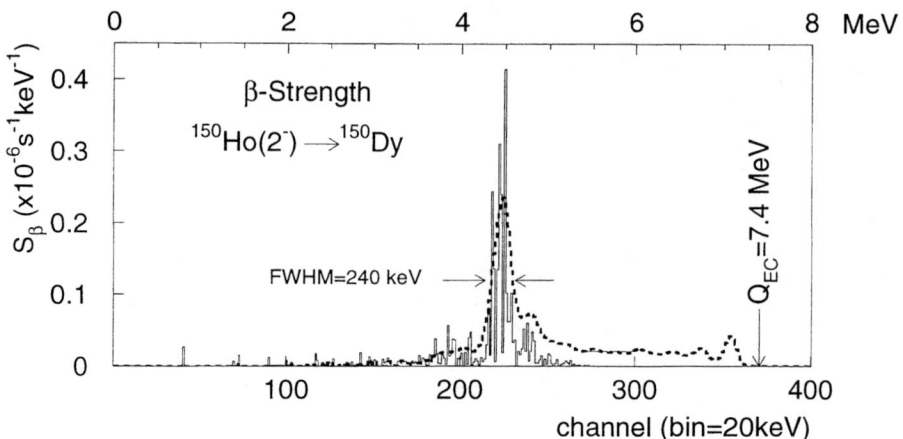

FIGURE 3. The TAS (dashed line) and CLUSTER CUBE (solid line) results for the ^{150}Ho 2^- β strength.

is in the $\pi h_{11/2}$ orbital, and the GT transition $(\pi h_{11/2})_{0^+} \to (\pi h_{11/2} \nu h_{9/2})_{1^+}$ can occur. The resulting decay will populate 4 qp states at ≈ 4 times the pairing gap energy (4-5 MeV). The very sharp resonance revealed at 4.5 MeV tells us that this picture is substantially correct, and that configuration mixing is very small here. The observed strength is only 0.15 of that derived from the extreme single particle picture where only two protons in the $\pi h_{11/2}$ orbit are considered and the $\nu h_{9/2}$ orbit is empty. Towner [4] has calculated the first order corrections to this simple picture. The predicted strength is still $\sim 30\%$ higher than the observed strength. Now that the GT resonance has been clearly seen for the first time, more elaborated calculations should also aim to reproduce the distribution of the strength.

Acknowledgments: Work supported by C.I.C.Y.T. (Spain) under project AEN96-1662, by E.C.C. contract ERBCIPD-CT-950083, by the Russian Fund for Basic Research and Deutsche Forschungsgemeinschaft contract 436-RUS-113/201/0(R) and by Polish Committee of Scientific Research grant KBN-2-P03B-039-13.

REFERENCES

1. B.A. Brown and B.H. Wildenthal, *At. Dat and Nucl. Dat. Tab.* **33**, 348 (1985)
2. M. Karny *et al*, *Nucl. Inst. Meth. B* **126**, 411 (1997)
3. D. Cano *et al*, *cont. to this conf.*
4. I.S. Towner *Nucl. Phys. A***444**, 402 (1985)

Gamow-Teller Decay of Even Isotopes ^{68}Ni to ^{78}Ni

Jan Żylicz*, Jacek Dobaczewski[+] and Zdzisław Szymański[+]

*Institute of Experimental Physics and [+]Insitute of Theoretical Physics,
University of Warsaw, 00-681 Warsaw, Poland

Abstract. Theoretical single-particle levels, obtained in the Hartree-Fock approach with four different Skyrme forces, and a schematic Gamow-Teller particle-hole interaction, have been used in the Tamm-Dancoff approximation to calculate the β strength distributions and half-lives for even isotopes ^{68}Ni to ^{78}Ni. For A ≤ 74, the best agreement with experimental $T_{1/2}$ data is found for the SkP force.

Far off the line of β stability, on the neutron-rich side, double shell closure has been established for $^{132}_{50}$Sn$_{82}$ [1], and a similar closure is expected for $^{78}_{28}$Ni$_{50}$. Some features of a doubly-magic system have been found for $^{68}_{28}$Ni$_{40}$ [2].

The β⁻ decay energy of ^{132}Sn is 3.1 MeV. The decay undergoes mainly via one Gamow-Teller (GT) transition [3]. This transition may be associated with a $\nu d_{3/2} \rightarrow \pi d_{5/2}$ transformation. However, with respect to the extreme single-particle shell model (ESPSM), it is retarded by a factor of 26. Blomqvist [4] showed that quenching of this transition can be understood in terms of shift of a large fraction of the β strength to the GT resonance. The estimate was based on the Tamm-Dancoff approach with experimental single-particle energies and a schematic GT particle-hole interaction

$$V_{ph} = \kappa (\sigma_1 \sigma_2)(\tau_1 \tau_2),$$

with κ = 23/A MeV (adjusted to reproduce the ^{208}Pb(p,n) reaction data).

This simple approach can be applied also to ^{78}Ni, but experimental single-particle energies have to be replaced by theoretical ones. Relevant formulae can be found in ref. [5]. To estimate the GT-strength distribution and half-life, we applied these formulae [6] to different sets of theoretical single-particle energies. The $f_{7/2}$, $p_{3/2}$, $f_{5/2}$, $p_{1/2}$ and $g_{9/2}$ neutron states below the Fermi surface and the $p_{3/2}$, $f_{5/2}$, $p_{1/2}$, $g_{9/2}$ and $g_{7/2}$ proton states above the Fermi surface were considered. With Q_β of about 10 MeV, a few GT transitions were predicted to have an essential contribution to the decay of ^{78}Ni.

In a recent experiment at GSI, Darmstadt, three atoms of ^{78}Ni were identified for the first time [7]. It is hoped that the next experiments will provide a half-life value and preliminary decay scheme of this isotope, and thus a chance to test our calculations.

In the present work, while waiting for new data on ^{78}Ni, we have used the same approach to the decay of lighter even isotopes of nickel down to ^{68}Ni. Single-particle energies are provided by the Hartree-Fock method with four Skyrme forces: SIII (Beiner et al. [8]), SkM* (Bartel et al. [9]), SkP (Dobaczewski et al. [10]) and SLy4 (Chabanat et al. [11]). The parameter of occupancy of the $g_{9/2}$ neutron level was taken as

TABLE 1. Energies of the 1^+ levels in ^{74}Cu relative to the ground state of ^{74}Ni and the GT strengths for the SkP force

$\pi \, \nu^{-1}$	$\kappa = 0$ $B_{sp}(GT)$	E_1+ (MeV)	$\kappa=23/A$ MeV ω_1+ (MeV)	$B_\omega(GT)$
$g_{7/2} \, g_{9/2}$	10.67	3.425	10.780	49.29
$f_{5/2} \, f_{7/2}$	13.71	0.986	2.342	1.35
$p_{1/2} \, p_{3/2}$	5.33	-2.819	-1.515	2.64
$g_{9/2} \, g_{9/2}$	7.33	-3.765	-3.191	0.33
$p_{3/2} \, p_{3/2}$	6.67	-4.227	-4.025	0.08
$p_{1/2} \, p_{1/2}$	0.67	-4.433	-4.419	0.01
$f_{5/2} \, f_{5/2}$	4.29	-5.085	-4.900	0.15
$p_{3/2} \, p_{1/2}$	5.33	-5.841	-5.642	0.15
$\Sigma B_{sp}(GT) = 54.00$			$\Sigma B_\omega(GT) = 54.00$	

$v^2 = n/10$, where $n = N-40$ is the number of neutrons in this orbit ($n=0$ for ^{68}Ni). It was assumed that the reduced beta strength is $B(GT) = B_{sp}(GT) \, v^2$, with

$$B_{sp}(GT) = 8 \, l \, (l+1)/(2l+1) = 160/9 \text{ and } (2l+2)(2l+3)/(2l+1) = 110/9$$

for the $\nu g_{9/2} \to \pi g_{7/2}$ and $\nu g_{9/2} \to \pi g_{9/2}$ transitions, respectively.

The results of the calculations are illustrated in Table 1 for ^{74}Ni. The ESPSM level energies correspond to the SkP force. All perturbed levels (energies ω_1+) are shifted upwards with respect to those unperturbed (E_1+). The perturbed GT strength is essentially redistributed. The highest level, at $\omega_1+ = 10.78$ MeV, takes 91% of the sum-rule value 3 (N-Z) = 54. For lower levels, the strength is reduced. Six levels with $\omega_1+ < 0$ can be fed in the β decay. A summing of decay constants over individual transitions gives $\lambda = 0.884$ s^{-1} and $T_{1/2} = 0.78$ s.

Results of half-life calculations are given in Table 2. Available experimental values as reported by the Leuven group [12] are given for comparison. The best agreement is obtained with the SkP force, and these results are suggested for extrapolation to ^{76}Ni and ^{78}Ni. Our HF/SkP results are close to those obtained in the RPA calculations by Kratz et al. [13]. However, the latter depend upon the choice of Q_β, while our calculations give the transition energies directly. The HF/SkP model is also the best for ^{132}Sn. For the GT decay of this nuclide, $T_{1/2}$ (SkP) = 30 s is rather close to the experimental value of 40 s, while half-lives calculated with the three other forces are essentially longer.

TABLE 2: Calculated and experimental half-lives (in seconds) of the nickel isotopes

Isotope	^{68}Ni	^{70}Ni	^{72}Ni	^{74}Ni	^{76}Ni	^{78}Ni
Force: SLy4	1340	78	17	5.7	2.2	0.95
SkM*	460	54	14	5.4	2.5	1.3
SIII	400	40	11	4.0	1.8	0.86
SkP	33	6.5	2.0	0.78	0.35	0.16
Exp. [12]	29(2)	6.0(3)	1.57(5)	0.9(2)	?	?

A more complete presentation of the calculations (with an account for the pairing effect) will be presented in a separate paper [14].

ACKNOWLEDGEMENTS

The authors are grateful to the Leuven colleagues for making available their experimental half-life data prior to publication. The work was supported in part by the Polish Committee for Scientific Research under the contracts KBN-2-PO3B-039-13 and KBN-2-P03B-040-14.

REFERENCES

1. B.Fogelberg et al., *Phys. Rev. Lett.* **73**, 2413 (1994).
2. R.Broda et al., *Phys. Rev. Lett.* **74**, 868-871 (1995).
3. C.A.Stone et al., *Phys.Rev.* **C39**, 1963 (1989).
4. J.Blomqvist, Lecture delivered at the XX Winter School, Zakopane 1985 (unpubl.) and priv. com. 1989.
5. Z.Szymański and J.Żylicz, "Predictions for the Gamow -Teller decay of ^{78}Ni", in *Proc. Int. Symp. on Nuclear Physics of Our Times*, Ed. A.V. Ramaya, World Scientific, Singapore 1993, pp. 100-103.
6. J.Dobaczewski, Z.Szymański and J.Żylicz, "Predictions for the ^{78}Ni → ^{78}Cu beta decay", in *Proc. of the Workshop on Nuclear Fission and Fission-Product Spectroscopy*, Seyssins (France), Eds. H.Faust and G.Fioni, Institut Max von Laue - Paul Langevin, 94FA05T, Grenoble 1994, pp. 190-195.
7. Ch.Engelmann et al., *Z.Phys.* **A352**, 351-352 (1995).
8. M.Beiner et al., *Nucl.Phys.* **A238**, 29-69 (1975).
9. J.Bartel et al., *Nucl.Phys.* **A386**, 79-100 (1982).
10. J.Dobaczewski et al., *Nucl.Phys.* **A422**, 103-139 (1984).
11. E.Chabanat et al., *Phys. Scripta* **T56**, 231-233 (1995).
12. S.Franchoo et al., *Ann. Rep. of the Instituut voor Kern- en Stralingsfysica*, Leuven 1997, see also these Proceedings .
13. K.-L.Kratz et al., *Z.Phys.* **A332**, 419-426 (1989)
14. J.Dobaczewski, Z.Szymański and J.Żylicz, to be published.

NUCLEAR ASTROPHYSICS

THE rp-PROCESS IN X-RAY BURSTS

M. Wiescher, A. Aprahamian, J. Döring, J. Görres, H. Schatz

University of Notre Dame, Department of Physics, Notre Dame, IN 46556, USA

Abstract. The rp-process was first suggested by Wallace and Woosley (1981) as the dominant nucleosynthesis process in explosive hydrogen burning at high temperature and density conditions. The process is characterized by a sequence of fast proton capture reactions and subsequent β-decays. The reaction path of the rp-process runs along the drip line up to Z$\approx$50. Within a sufficiently long time scale the associated nucleosynthesis produces N=Z even-even nuclei. Extended model calculations indicate a large production of in particular mass A=80 isotopes. The abundances are highly sensitive to the lifetimes and decay pattern of these nuclei. We present the implications of recent experimental results for the nucleosynthesis of mass 80 isotopes.

I INTRODUCTION

X-ray bursts have been suggested as possible sites for hot temperature hydrogen burning via the rp- and αp-process [1–4]. While X-ray bursts are frequently observed phenomena [5], the nucleosynthesis and the correlated nuclear energy generation, are not completely understood. Network simulations of the associated nucleosynthesis are still severely handicapped by the lack of knowledge about the nuclear structure and nuclear processes along the reaction path [6,7]. The standard models for type-I X-ray bursts are based on accretion processes in a close binary system. In the case of type-I X-ray bursts, accretion takes place from the filled Roche-Lobe of the extended companion star onto the surface of a neutron star. Typical predictions for the accretion rate vary from 10^{-10} to 10^{-9} M$_\odot$ y^{-1}. The accreted matter is continuously compressed by the freshly accreted material until it reaches sufficiently high pressure and temperature conditions which allow the thermonuclear ignition.

Nuclear burning is ignited at high density, $\rho \geq 10^6$ g/cm^3, in the accreted envelope, via the pp-chains and the hot CNO-cycles. Yet the limited energy release is insufficient for triggering the thermonuclear runaway at electron degenerate conditions near the bottom of the accreted layer. This requires the ignition of the triple α-process and the break-out from the hot CNO cycles via α capture on the CNO waiting points ^{14}O, ^{15}O, and ^{18}Ne. While the triple α reaction delivers the CNO-fuel, this break-out triggers the rp-process which in turn releases the required energy for initiating the thermonuclear runaway [2,3]. Peak temperatures up to $2 \cdot 10^9$ K can be reached before the degeneracy is completely lifted [8]. The time-scale for the thermal runaway and the subsequent cooling phase varies between 10 s up to 100 s [2] depending on the particular model parameters for the accretion process. Within this time-scale, the rp-process proceeds well up to ^{56}Ni [1] or even further up to ^{96}Cd [2,6,7]. In the following we will discuss the detailed characteristics of the

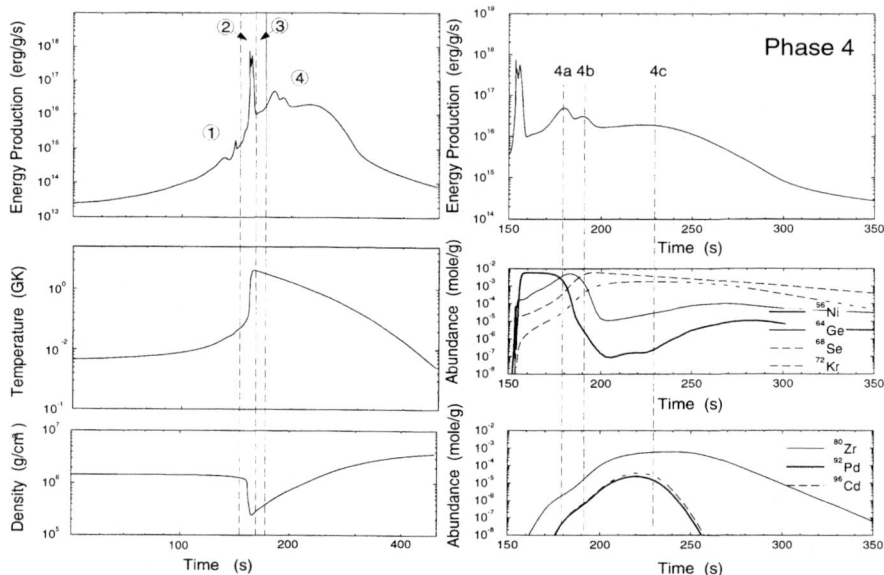

FIGURE 1. The left hand figure show the energy production (luminosity), the temperature, and the density as a function of time for the thermonuclear runaway in a simplified X-ray-burst model. Indicated are different phases in the thermonuclear runaway. Phases 1, 2, 4 correspond to periods of rapid nucleosynthesis, phase 3 correspond to a dormant period at the maximum of the temperature curve. The right hand figure shows the energy production (luminosity) and nucleosynthesis in the cooling phase (4) of an X-ray burst. The peaks in the luminosity correspond to the destruction of waiting point nuclei. For details see text.

rp-process in the framework of a simple one-zone model to investigate its influence on the time structure of the luminosity burst and its possible endpoint in the range of the highly deformed N=Z nuclei near A=80.

II THE RP-PROCESS IN THE THERMAL RUNAWAY OF THE X-RAY BURST

This section will focus on the rp-process characteristics calculated for temperature conditions during the thermonuclear runaway in an one mass zone X-ray burst model. For this model hydrostatic equilibrium is maintained.

Figure 1 shows the energy production, the temperature and the density in the accreted envelope (burning-zone) as a function of time over the duration of the X-ray burst. The structure of the energy production curve is directly correlated with the waiting point concept of the rp-process [7].

The energy production in the burst is characterized by four periods, the ignition phase of the burst (1), the peak of the burst (2), the dormant phase of the burst (3), and the cooling phase of the burst (4). These four different phases are characterized by different temperature and density conditions for the nucleosynthesis and energy generation.

Phase (1) is characterized by the hot CNO cycles triggered by proton capture reactions of the accreted hydrogen on the carbon, nitrogen, and oxygen isotopes which have not been destroyed by spallation processes in the outer atmosphere of the accreting neutron star [9]. The small two peaks in the energy production are caused by the conversion of the initial abundance of ^{12}C into ^{14}O and of ^{16}O into ^{15}O in the hot CNO cycles. Because of the slow decay of these two isotopes the CNO process is halted and the energy production drops. The main reaction flow is confined to the CNO cycles, but the temperature has already increased sufficiently to trigger phase (2) of the burst by the ignition of the triple α-reaction and the break-out from the CNO cycles.

Phase (2) is initiated at a temperature of $T \approx 2.4 \cdot 10^8$ K via the triple α-process. Parallel to that the ^{14}O and ^{15}O waiting point nuclei are rapidly depleted by the α-capture processes. The fine structure of the energy burst is characterized by the different waiting points along the process path. The burst is initiated by the conversion of the hot CNO waiting point isotopes ^{14}O, ^{15}O, and ^{18}Ne by the αp-process to ^{24}Si. After the decay of ^{24}Si the process is re-ignited at higher temperatures by proton and α capture reactions leading to the production of the next waiting point isotopes ^{29}S and ^{34}Ar. The rapid increase in temperature allows subsequent α capture to bridge these waiting points and leads to a rapid conversion of the waiting point isotopes to ^{56}Ni. At this time a continuous reaction flow converts the accreted He-abundance via the αp-process and the rp-process to ^{56}Ni. At this point, peak temperatures of $T \approx 1.5 \cdot 10^9$ K have been reached and further processing is halted by the ^{56}Ni(p,γ)-(γ,p) equilibrium [7]. The energy production drops rapidly while most of the initial heavy isotope abundances as well as a large fraction of the initial helium remains stored in the waiting point nucleus ^{56}Ni. The drop in energy production causes a slow-down in the temperature increase just before the peak temperature is reached.

At these peak temperature conditions the ^{56}Ni has a lifetime of approximately 100 s versus two-proton capture. However, with the decrease in temperature and the parallel increase in density the ^{56}Ni(p,γ)-(γ,p) falls out of equilibrium and the effective lifetime of ^{56}Ni decreases down to fractions of a second versus proton capture at temperatures $T \approx 1.0 - 1.5 \cdot 10^9$ K. In this temperature window, phase (4) of the energy burst is initiated which is characterized by nucleosynthesis via the rp-process beyond ^{56}Ni. Figure 1 shows the details of the burst structure in the cooling phase. Several peaks in the energy production are due to the depletion of ^{56}Ni and the further processing towards the N=Z waiting point nuclei ^{64}Ge and the subsequent nucleosynthesis towards ^{68}Se and ^{72}Kr. These waiting point nuclei have been suggested as termination points for the rp-process because of their rather long β-decay lifetime [10–12]. Yet, it has been shown [7,13] that the effective lifetime of these nuclei can be shortened by two sequential proton capture reaction which bridge the barely bound isotope ^{65}As [15] as well as the unbound isotopes ^{69}Br [11,12] and ^{73}Rb [12,14]. In the final phase ^{68}Se and ^{72}Kr are converted to heavier isotopes up to the mass 100 region. Figure 2 shows the reaction flow integrated over the duration of (4d). Notice that due to the lower temperature the αp-process is only dominant below sulfur, at higher masses the reaction path is characterized by the rp-process pattern leading up to ^{100}Sn. In the final phase most ($\geq$90%) of the initial helium as well as most of the other isotopes have been converted to heavy isotopes with masses A$\geq$72. The most abundant isotopes is the N=Z nucleus ^{80}Zr as indicated in figure 1.

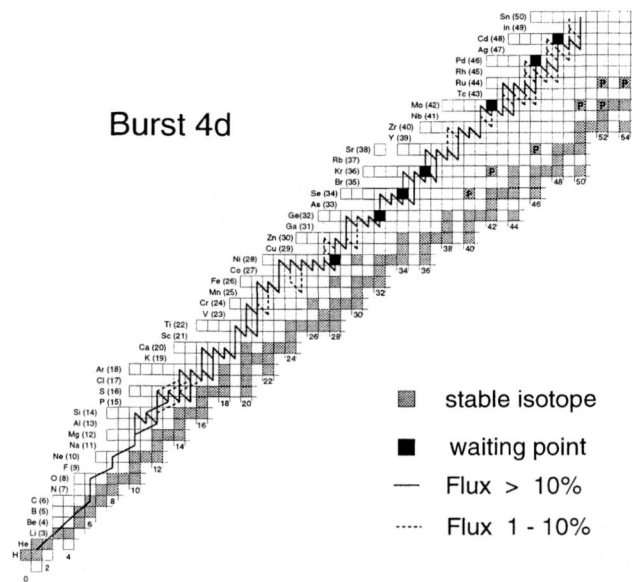

FIGURE 2. The reaction flux in the late phase (4d) of the x-ray burst. The overall flux is dominated by the rp-process. Only for very light Z nuclei (Z$\leq$14) is the Coulomb barrier low enough for α capture reactions to compete with the proton capture and the β-decay. Also indicated are the light p-nuclei which are possibly produced in the X-ray burst.

III NUCLEOSYNTHESIS IN THE MASS 80 RANGE

Experimental information on the nuclear reaction and nuclear decay processes along the rp-process path in the mass 76 to 100 range is very limited. The calculations are entirely based on model predictions for masses, life times and cross sections [7]. The nucleosynthesis predictions may therefore carry considerable uncertainties. The overall abundance conditions after the freeze out of the thermonuclear runaway are demonstrated in figure 3. The left part shows the abundance distribution of the material and the right part shows the comparison with the solar abundance distribution. While there is still an appreciable amount of hydrogen, the

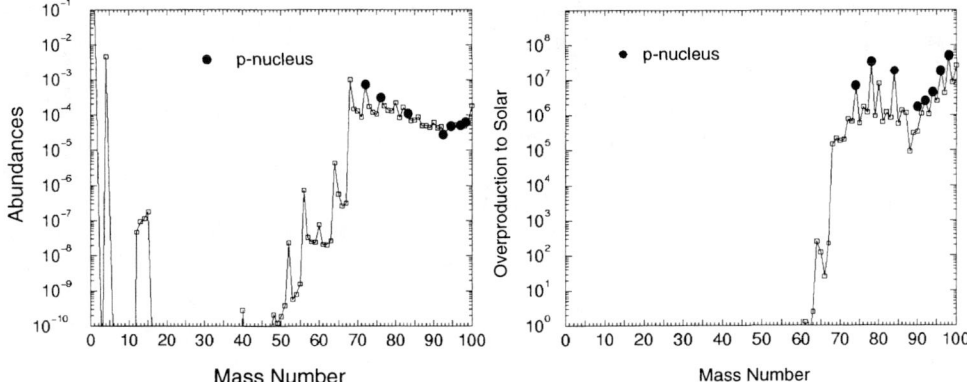

FIGURE 3. The abundance distribution after the thermonuclear runaway of a single X-ray burst. The left hand side shows the absolute abundances and the right hand side shows the abundances relative to the observed solar abundances.

bulk of the material has been converted to nuclei with masses A≥70. All isotopes above mass A=68 are enriched by more than four orders of magnitude compared to the solar abundances which have served as the initial distribution of the accreted material. Clearly peaking are the abundances of the light p-nuclei ^{74}Se, ^{78}Kr, ^{84}Sr, ^{92}Mo, ^{94}Mo, ^{96}Ru, and ^{98}Ru [16]. They are overproduced by up to seven orders of magnitudes compared to the solar abundances of these particular isotopes. This is in particular noticable because the relatively high observed abundances for these isotopes have not been explained yet in classical p-process scenarios [17-19] or in alternative neutrino-induced processes in type II supernovae [20]. If a suffiently high mass loss out of the gravitational potential of the neutron star is possible, X-ray bursts may be a potential source for these isotopes. However, ^{80}Kr also is largely overproduced by the decay of ^{80}Zr. This is clearly inconsistent since it is well known that ^{80}Kr is produced in the s-process [21].

IV   β-DECAY OF MASS A=80 NUCLEI AND THE INFLUENCE FOR A=80 ABUNDANCE

This high production of the progenitor ^{80}Zr in the rp-process model calculations results mainly from the long lifetime of $T_{1/2}$=6.85 s which has been predicted by QRPA model calculations for a prolate ground state deformation of ϵ_2=0.383 [7]. Recent attempts to measure the lifetime failed. The experiment did however allow the identification of an isomeric $J^\pi=1^-$ state in ^{80}Y at 228 keV excitation energy. This isomeric state β^+/EC decays to low lying states in ^{80}Sr with a half life of 4.7 s [22]and by emission of a 228 keV M3 γ-decay to the ground state in ^{80}Y. The half life of the ^{80}Y ground state was determined to be $T_{1/2}$=30.1 s.

The rp-process calculations previously only included the ground state β-decay of ^{80}Zr. However, calculations of β-decay processes in high temperature scenarios requires inclusion of the β-decay of thermally excited states as well [7]. These states have significantly different lifetimes from the ground state. For the highly deformed nucleus ^{80}Zr, the excitation energy of the first excited 2^+ state is rather low, $E_x \approx 290$ keV [23], therefore the decay through the thermally excited state is already initiated at temperatures above ≈ 1 GK. The β-decay will populate 1^+, 2^+ states (predicted for an excitation range $E_x \approx 300$-600 keV [22]) and the 2^+ 6 μs isomeric state at 312 keV in ^{80}Y [24] which decays by a 84 keV γ-emission to the 1^- isomeric state at 228 keV [22]. A schematic level diagram for such a decay is shown in figure 4. A QRPA calculation for such a decay suggests an effective half

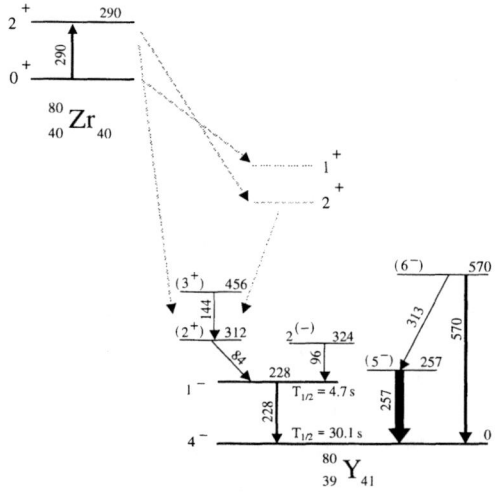

FIGURE 4. The level scheme of ^{80}Y. The gray lines indicate the proposed decay of ^{80}Zr at high temperature conditions.

life for ^{80}Zr of ≈ 3 s for the temperature conditions found in the cooling phase of the X-ray burst. Figure 5 compares the development of the abundances of the N=Z isotopes ^{56}Ni, ^{64}Ge, ^{68}Se, and ^{80}Zr for ground state decay of ^{80}Zr and fast decay by thermal excitation of ^{80}Zr through the isomeric states in ^{80}Y. The upper figure clearly shows a factor four reduction in the ^{80}Zr abundances while the abundances of the other isotopes remain unaffected.

For the formation of ^{80}Kr in the freeze out phase of the X-ray burst the decay of all A=80 neutron deficient isotopes have to be considered. Considering only the ground state decay, the abundances show a significant enrichment for ^{80}Y along the rp-process path due to its long half-life of 30.1 s. The feeding of the isomeric states

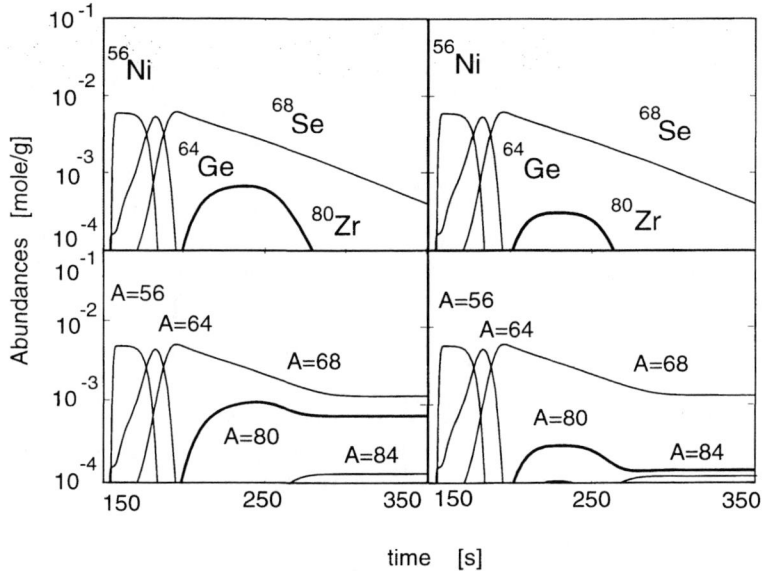

FIGURE 5. The development of the abundances for ^{56}Ni, ^{64}Ge, ^{68}Se, and ^{80}Zr as a function of time is shown in the upper part for ground state decay only (left) and after inclusion of the decay of excited states (right). The lower part shows the total abundances for mass A=56, 64, 68, 80, and 84 nuclei calculated for the same conditions.

in ^{80}Y by the decay of ^{80}Zr reduces the effective half life of ^{80}Y significantly and subsequently reduces its abundance during the cooling phase. Figure 5(lower part) compares the development of the 'total' A=80 abundance for both cases. That is, thermal excitation of ^{80}Zr with subsequent decay through the isomeric states of ^{80}Y reduces the A=80 abundance by nearly one order of magnitude compared to ground state decay only.

At the present state these calculations are rather speculative as neither the β-decay lifetime of ^{80}Zr nor the β-strength function has been measured yet. It remains to be confirmed that the β-decay of ^{80}Zr eventually populates the isomeric states in ^{80}Y.

V SUMMARY

It was shown in the framework of a simplified X-ray burst model that the rp-process is sufficiently fast to produce nuclei up to the mass A=100 range. However, the time structure of the nucleosynthesis and the associated energy production is highly structured and depends critically on the effective lifetimes of the waiting point nuclei along the rp-process path. These in turn are determined by the macroscopic temperature and cooling conditions in the burning zone. The main impedance for nucleosynthesis of nuclei beyond the mass A=90 range are the rather long effective lifetimes of ^{64}Ge, ^{68}Se, and ^{72}Kr. These are determined by the two-proton capture processes on these isotopes and the β-decay through their thermally excited states [7]. A further handicap for the reaction flow is the largely deformed N=Z nucleus ^{80}Zr. Two proton capture is not possible because ^{81}Nb is highly unbound [7]. It has been shown that the effective lifetime of ^{80}Zr can be considerably reduced by thermal population of the first excited state and by β-decay through the short-lived isomeric states in ^{80}Y. Reliable experimental information is needed on the β-decay pattern and the isomeric states in the mass 80 nuclei to perform a consistent study of rp-process nucleosynthesis in this mass range.

REFERENCES

1. Wallace, R.K., and Woosley, S.E., *Ap.J.Suppl.* **45**, 389 (1981)
2. Woosley, S.E., and Weaver, T.A. in *High Energy Transients in Astrophysics*, Vol. 115 of *AIP Conference Proceedings*, ed. S.E. Woosley (American Institute of Physics, 1984)
3. Taam, R., *Ann.Rev.Nucl.Sci.* **35**, 1 (1985)
4. Taam, R., Woosley, S.E., Weaver, T.A., and Lamb, D., *Ap.J.* **413**, 324 (1993)
5. Lewin, W., van Paradijs, J., and Taam, R., *Space Sci.Rev.* **62**, 233 (1993)
6. Van Wormer, L., Görres, J., Iliadis, C., Wiescher, M., and Thielemann, F.K., *Astrophys.J.* **432**, 326 (1994)
7. Schatz, H., et al, *Phys.Rep.* **294**, 167 (1997)
8. Bildsten, L., in *The many Faces of Neutron Stars*, eds. A. Alpar, L. Buccheri, J. van Paradijs (Dordrecht, 1998)
9. Bildsten, L., Salpeter, E.E., and Wassermann, I., *Ap.J.* **384**, 143 (1992)
10. Wallace, R.K., and Woosley, S.E., in *Proceedings of Accelerated Radioactive Beam Workshop, Parksville, Canada*, eds. L. Buchmann, J. D'Auria, (TRIUMF, Vancouver, 1985)
11. Blank, B., et al, *Phys.Rev.Lett* **74**, 4611, (1995)
12. Pfaff, R., et al, *Phys.Rev.*C **53**, 1753 (1996)
13. Wiescher, M., Schatz, H., and Champagne, A., *Phil.Trans.Roy.Soc.* **356**, 1 (1998)
14. Mohar, M.F., et al, *Phys.Rev.Lett* **66**, 1571 (1991)
15. Winger, J., et al, *Phys.Rev.*C **48**, 3097 (1993)
16. Lambert, D.L., *Astr.Astrophys.Rev.* **3**, 201 (1992)
17. Woosley, S.E., and Howard, E.M., *Ap.J.* **354**, L21 (1990)
18. Howard, E.M., Meyer, B.S., and Woosley, S.E., *Ap.J.* **373**, L5 (1991)
19. Rayet, M., et al, *Astr.& Astrophys.* **298**, 517 (1995)
20. Hoffman, R.D., Woosley, S.E., Fuller, G.M., and Meyer, B.S., *Ap.J.* **460**, 478 (1996)
21. Walter, G., Beer, H., Käppeler, F., Reffo,G., and Fabbri, F., *Astron. & Astrophys.* **167**, 186 (1986)
22. Döring, J., et al., *Phys.Rev.* C **57**, 1159 (1998)
23. Lister, C.J., et al., *Phys.Rev.Lett.* **59**, 1270 (1987)
24. Regan, P.H., et al, *Acta Phys.Pol.* B **28**, 431 (1997)

New information on r-process nuclei

Karl-Ludwig Kratz

Institut für Kernchemie, Universität Mainz, D-55128 Mainz, Germany

Abstract. The current status of nuclear-structure measurements on medium-heavy isotopes relevant to r-process nucleosynthesis is briefly reviewed. Here, the main focus is put on the region around the doubly-magic nucleus ^{132}Sn, where first experimental evidence for a weakening of the N=82 shell strength is presented. This new phenomenon leads to improved predictions of astrophysical observables like the solar-system isotopic r-abundances and the recent Ba – U elemental abundances of ultra-metal-poor neutron-capture-rich halo stars.

Introduction

Relative to other fields, astrophysics – and in particular explosive nucleosynthesis – is probably unique in its requirements that a very large number of physical environments be modeled to achieve a satisfactory description of the phenomena under study. On the nuclear level, cross sections, nuclear masses and decay properties are but a few examples of quantities that are of paramount importance in astrophysical models. Because nuclei of extreme N/Z composition, quite different from what has so far been studied on earth, exist in explosive stellar environments, an understanding of their nuclear-structure properties presents a continuing, stimulating challenge to the experimental and theoretical nuclear-structure community.

For example, only some 15 years ago, astrophysicists still believed that nuclear-structure information on isotopes in the r-process path would never become available from terrestrial experiments. However, already a few years later in 1986, a new area in nuclear astrophysics started with the identification of the first two classical, neutron-magic "waiting-point" isotopes [1] ^{80}Zn$_{50}$ and ^{130}Cd$_{82}$ [2–4].

Since then, considerable progress has been achieved in the study of neutron-rich medium- to heavy-mass nuclei at different laboratories (e.g. ILL/OSTIS, BNL/TRISTAN, Nyköping/OSIRIS, CERN/ISOLDE, GANIL/LISE, GSI/FRS, JYFL/IGISOL, LLN/LISOL, Mainz/SISAK, and the Eurogam and Gammasphere Spontaneous-Fission Collaborations), using various production, separation and detection methods. However, due to the generally very low production yields, the occurrance of isobar, multi-charged ion and molecular-ion contaminations from chemically non-selective ionization modes, or because of the application of non-selective detection methods, *direct* information on isotopes lying in the r-process path(s) could be obtained but in a few cases, like e.g. ^{79}Cu$_{50}$, ^{84}Ga$_{53}$ or ^{85}Ge$_{53}$ [5]. From the majority of investigations, originally dedicated to study nuclear-structure developments as a function of isospin, 'only' *indirect* – but nevertheless also important – information for r-process calculations can be deduced (see, e.g. [6–11]). In this context, we would like to remind the reader of the fascinating phenomenon of the *sudden onset and saturation of ground-state deformation at N=60* (see, e.g. [12–14]), which had in fact led to the first 'astrophysical request' of N=82 shell quenching [7].

Because of the severe experimental problems to identify r-process nuclei, recent progress has undoubtedly benefitted from the **selectivity** in their production and detection, by applying e.g. Z-selective laser ion-source (LIS) systems, isobar separation or multifold-coincidence techniques.

Recent experiments

There are mainly three mass regions, where at present nuclear-structure information is of particular astrophysical interest. The first is the *r-process seed* region which seems to be not well understood (see, e.g. [15,16]), involving very neutron-rich Fe-group isotopes up to the double-magic nucleus ^{78}Ni$_{50}$. Recent spectroscopic results can, for example, be found in [17–20]. The second region is that of far-unstable nuclei around $A\simeq 115$. Here, most r-process calculations show a pronounced *r-abundance trough*, which is believed to be due to nuclear-model deficiencies at and beyond N=72 mid-shell (see, e.g. [7,9,10]). Recent experimental information can, for example, be found in [14,21–24]. The third region of interest is that around the *double-magic nucleus* ^{132}Sn. Apart from astrophysical importance (formation of the $A\simeq 130$ peak of the solar-system r-abundance distribution ($N_{r,\odot}$) [7,11]), this region is of considerable shell-structure interest. The far-unstable isotope $^{132}_{50}$Sn$_{82}$ itself, together with the properties of the nearest-neighbour single-particle ($^{133}_{51}$Sb$_{82}$ and $^{133}_{50}$Sn$_{83}$) and single-hole ($^{131}_{50}$Sn$_{81}$ and $^{131}_{49}$In$_{82}$) nuclides are essential for tests of the shell model, and as input for any reliable future microscopic nuclear-structure calculations towards the neutron drip-line.

Because of this importance and the recent improvements in laser ion-source (LIS) technology to provide clean beams of these exotic nuclei, in the following, we will focus on new spectroscopic investigations around ^{132}Sn [25–32]. The bulk of the known data in this region has been obtained from β-decay spectroscopy at the mass-separator facilities OSIRIS and ISOLDE.

The structures of ^{131}Sn (ν-hole) and ^{133}Sb (π-particle) are fairly well known since more than a decade. More recently, the ν-particle states 2f$_{7/2}$ (g.s.), 3p$_{3/2}$ (854 keV), 1h$_{9/2}$ (1561 keV) and 2f$_{5/2}$ (2005 keV), as well as tentatively also the 3p$_{1/2}$-particle (1655 keV) and the 1h$_{11/2}$-hole ($\approx$3700 keV) states in ^{133}Sn$_{83}$ have been identified by measuring γ-rays in coincidence with delayed neutrons (d.n.) of

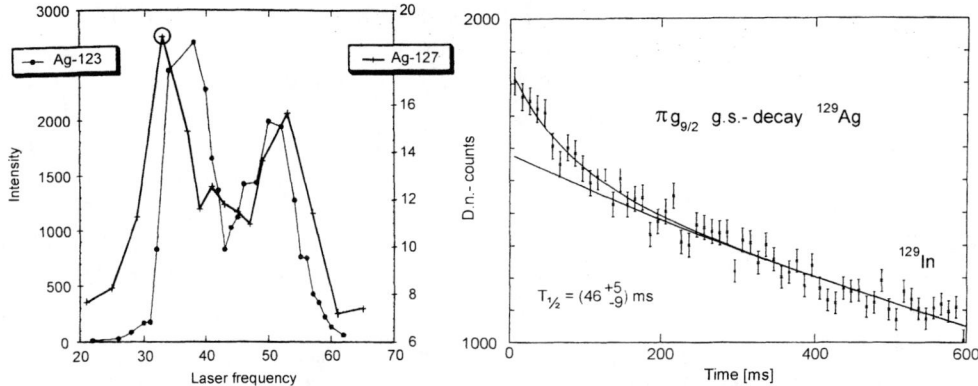

FIGURE 1. Broad-band laser frequency scans for ^{123}Ag and ^{127}Ag (left side) using d.n.-counting. Both distributions are consistent with the typical HF-splitting of a $\pi g_{9/2}$ configuration. As shown in the right part of this figure, a d.n.-decay measurement at an off-center frequency at A=129 (i.e. at maximum $\pi g_{9/2}$ yield; as indicated by the open circle in the ^{127}Ag curve) permitted the identification of laser-ionized ^{129}Ag $\pi g_{9/2}$ decay on top of the isobar activities from the two ^{129}In isomers which are due to surface ionization.

^{134}In decay [25]. From these data, valuable information on the spin-orbit splitting of the 2f-orbital and tentatively the 3p-state splitting was obtained. These results were compared to mean-field and HFB predictions, and it was found that none of the potentials currently used in *ab initio* shell-structure calculations was capable of properly reproducing the ordering and spacing of these states (see, e.g. [33], where also possible astrophysical consequences are given). Of particular interest in this context are the surprisingly low-lying $\nu p_{3/2}$- and $\nu p_{1/2}$-states in ^{133}Sn. According to the standard Nilsson model [34], for example, they are expected at 2.89 MeV and 4.36 MeV, respectively. Such lowering of the energies of low-j orbitals has recently been predicted as a *neutron-skin* phenomenon to occur only near the neutron drip-line (see, e.g. [35]), but certainly not yet in $(N_{mag}+1)$ ^{133}Sn which is still neutron-bound by $S_n \simeq 2.5$ MeV. Nevertheless, following the suggestion of Dobaczewski [35], we have modified the Nilsson potential by reducing the strength of the l^2-term, in order to study its effect on different orbitals. And indeed, this procedure led to the desired change in the position and even the ordering of the ν-particle states beyond N=82, and allowed at least a qualitative reproduction of the experimental observation of low-lying low-j and high-lying high-j orbitals in ^{133}Sn.

With these new data, the ^{132}Sn valence-nucleon region is nearly complete. The only missing information are the π-hole states in $^{131}_{49}$In, which can in principle be studied through β-decay of the very exotic nucleus ^{131}Cd$_{83}$. Recent LIS developments at Mainz [37] and CERN/ISOLDE, using a novel frequency-tripling technique for the Z-selective resonance ionization of Cd, indeed offer the possibility to perform such an experiment. First test measurements on neutron-rich Cd isotopes with this LIS system are already scheduled for Oct. 1998.

Another recent application of a chemically selective LIS at the ISOLDE facility are the spectroscopic measurements of neutron-rich Ag isotopes by a Mainz–Maryland–Troitzk–CERN collaboration [30–32]. Using an improved version of the LIS system described in [38] together with new microgating procedures [39], d.n.-counting and γ-ray spectroscopy could be performed up to 58-ms ^{128}Ag$_{81}$. This approach was of considerable assistance in minimizing the activities from (unavoidable) surface-ionized In and Cs isobars. Moreover, the identification of the long searched waiting-point isotope ^{129}Ag$_{82}$ was only possible by using the additional 'selectivity' of the spin- and moment-dependent hyperfine (HF) splitting to enhance the ionization of the $\pi g_{9/2}$ g.s.-decay (see Fig. 1).

The laser response to HF-splitting was calibrated on the stable $J^\pi=1/2^-$ ^{107}Ag whose magnetic moment is small ($\mu=-0.11$ n.m.) and on the radioactive $J^\pi=7/2^+$ isomer that has a large moment of $\mu=4.4$ n.m. (see [32,40]). Subsequently, the d.n.-activities were measured as a function of laser frequency for ^{122}Ag up to ^{127}Ag. As examples, the results for the two odd-mass isotopes ^{123}Ag and ^{127}Ag are shown in the left part of Fig. 1. Except ^{122}Ag, all other isotopes show – apart from the expected small isotope shift – the same pattern, where the moment of the odd $g_{9/2}$-proton is the dominant source of HF-splitting.

For ^{122}Ag, a different pattern was observed (see Fig. 2). Previous decay-studies of this isotope had indicated the presence of both low- and high-spin isomers [41]. And indeed, apart from the highly split peak that might be associated with a $(\pi g_{9/2}-\nu h_{11/2})_{9^-}$ configuration, another narrow peak shows up in the center that could arise from the $(\pi p_{1/2}-\nu d_{3/2})_{1^-}$ configuration. Subsequently, both d.n.- and γ-decay from ^{122}Ag was studied also as a function of laser frequency. As is evident from the partial γ-spectra shown in the lower part of Fig. 2, when the laser is set at the center of the frequency scale, γ-rays from both isomers can be observed. However, when the laser frequency is shifted away from the center (to 35 units), ionization of the low-spin isomer is suppressed, and the γ-spectrum indicates that only the decay of the high-spin isomer is oberved. Similarly, when following the d.n.-decay curves as a function of laser frequency, different half-lives are obtained which represent varying admixtures of the two isomers. From a preliminary analysis, we get an approximate half-life of the low-spin isomer of $T_{1/2} \simeq 550(50)$ ms and a value of $T_{1/2} \simeq 200(50)$ ms for the high-spin isomer of ^{122}Ag [42]. This experiment clearly demonstrates the sensitivity of ionization to laser frequency and marks the first observation of separation of short-lived isomers by this technique.

An important nuclear-physics quantity for r-process calculations (in particular for the build-up, the shape and the matter flow out of the A$\simeq$130 $N_{r,\odot}$-peak) is the β-decay half-life of the N=82 waiting-point isotope ^{129}Ag, sitting just below ^{130}Cd. QRPA calculations using different Q_β-values and single-particle wave functions resulted in decay rates between 15 ms and 170 ms [6,7,43]. This had left large uncertainties in the reproduction of the A$\simeq$130 peak when based on the simplistic assumption of $N_{r,\odot} \times \lambda_\beta \approx$const. (see, e.g. Fig. 4 in [44] and Fig. 1 in [11]).

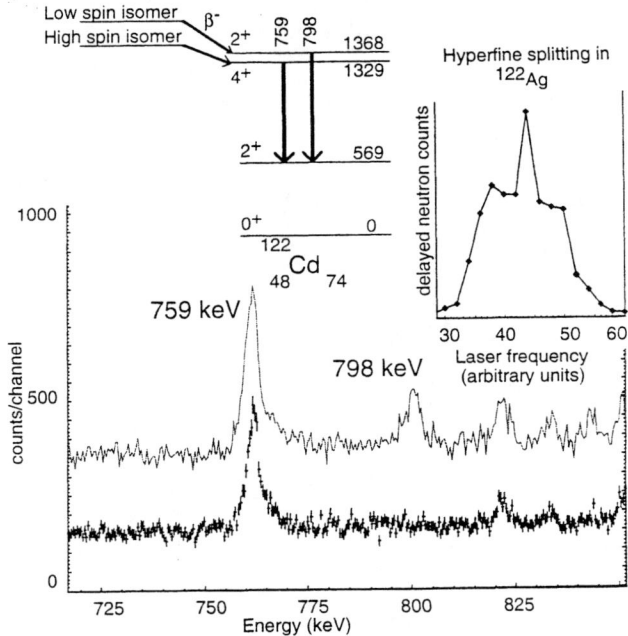

Figure 2. Portions of γ-ray spectra of the decay of ^{122}Ag isomers. The upper spectrum was taken with the laser frequency set at the center peak (44 units; see insert upper right part), and can be compared with the lower spectrum obtained at an off-center frequency (35 units). The γ-ray at 798 keV has been assigned to the $2_2^+ \to 2_1^+$ transition (see insert upper left part), whereas the line at 759 keV corresponds to the decay of the 4_1^+ level to the 2_1^+ state [41].

Previous attempts [30] to observe d.-n.-decay of ^{129}Ag had been performed with the (broad-band) lasers centered with respect to the mean frequency for the stable $\pi p_{1/2}$ ^{107}Ag. After the observation of enhanced ionization of the $\pi g_{9/2}$ level at an off-center frequency, a new experiment was performed with the laser setting at the peak value observed for ^{127}Ag (see left part of Fig. 1). This approach together with the microgating procedure [39], finally permitted the unambiguous identification of d.n.-decay from $\pi g_{9/2}$ ^{129g}Ag (see right part of Fig. 1). The half-life of 46^{+5}_{-9} ms is in very good agreement with the recent QRPA prediction of 47 ms [43], but is lower than the 'astrophysical requirement' of ≈130 ms [44]. However, that estimate is for an overall *average* half-life that would also include contributions from β-decay of a possible $\pi p_{1/2}$ isomer in ^{129}Ag. With the QRPA model of [45], it is possible to calculate the Gamow-Teller decay for such an isomer to $T_{1/2}\simeq 320$ ms. When including an estimate for the expected first-forbidden strength (extrapolated from the decay of the $J^\pi=1/2^-$ isomer in isotonic ^{131}In), a minimum value of $T_{1/2}\simeq 125$ ms is suggested. And, indeed, a careful reexamination and comparison of the A=129 d.n.-decay curve (taken with the laser at central frequency) with the pure ^{129}In curve from a laser-off run, gave a first indication of a 'longer-lived' ^{129}Ag d.n.-component with a half-life of roughly 160 ms. Now, it will be important to use isomer-specific laser ionization in combination with isobar separation at ISOLDE to ascertain the existence of this $\pi p_{1/2}$ isomer.

As an extension of our earlier γ-spectroscopic work [31], we have also studied lev-

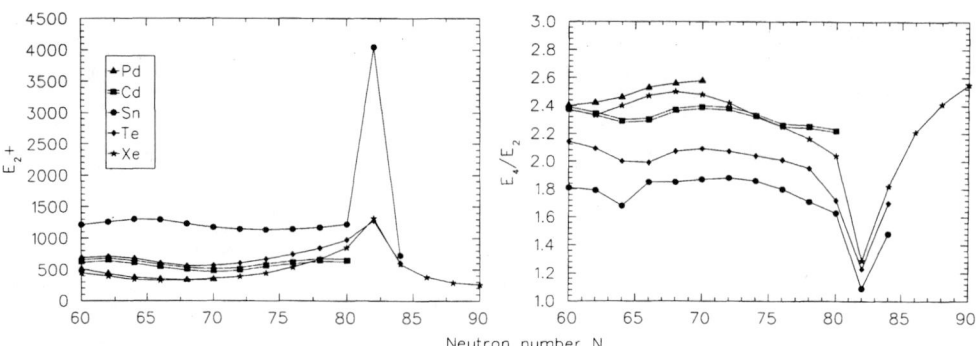

FIGURE 3. Systematics of the first 2^+ levels in neutron-rich $_{46}$Pd to $_{54}$Xe isotopes (left part); and $E(4^+)/E(2^+)$ ratios (right part).

els of even-even ^{126}Cd and ^{128}Cd from the decays of 95-ms ^{126}Ag and 58-ms ^{128}Ag. While for ^{126}Cd$_{78}$ still 8 out of 12 γ-lines were placed in a partial level scheme (see, e.g. Fig. 5 in [32]), for ^{128}Cd$_{80}$ so far only three lines could be assigned unambiguously, the two strongest corresponding to the $2^+\rightarrow 0^+$ and $4^+\rightarrow 2^+$ transitions. Our new results are shown in Fig. 3, together with known $E(2^+)$ and $E(4^+)/E(2^+)$ level systematics of neighbouring even-Z elements. The $E(4^+)/E(2^+)$ ratio for the 4n-hole nuclide ^{126}Cd is 2.25, a value almost unchanged relative to that for the 6n-hole isotope ^{124}Cd. Already this static ratio is in contrast to the larger reduction observed for the Z=52 isotones ^{128}Te and ^{130}Te. For the new 2n-hole nuclide ^{128}Cd, the energy of the first 2^+ state at 645 keV is even **lower** than the $E(2^+)$ of 657 keV in ^{126}Cd, and also the $E(4^+)/E(2^+)$ ratio of 2.22 is only insignificantly smaller than those in ^{124}Cd and ^{126}Cd. As can be seen from the right side of Fig. 3, this trend for the Z=48 isotopes clearly deviates from that of the Z=50 (Sn), 52 (Te) and 54 (Xe) isotones. This observation can be interpreted as a first direct experimental indication of a weakening of the N=82 shell strength below ^{132}Sn. A search for the $E(2^+)$ in N=82 ^{130}Cd is planned for the next Ag-run at ISOLDE.

Given this highly interesting result, one may have a look into existing data for additional 'hidden' or so far unrecognized signatures of shell quenching in this mass region. And indeed, first qualitative evidence for such a phenomenon comes from a comparison of the measured $T_{1/2} \simeq 195$ ms of ^{130}Cd$_{82}$ [4] with QRPA predictions. When applying Q_β-values from the 'unquenched' mass models FRDM [43] and ETFSI-1 [46], theoretical $T_{1/2}$ values of 1.12 s and 674 ms are obtained, respectively. However, with the 'quenched' mass formulae HFB/SkP [9] and ETFSI-Q [47], shorter values of 246 ms and 364 ms, respectively, are derived which are in better agreement with experiment. Another, more direct indication is given by the measured masses in the ^{132}Sn region [48,27]. As can be seen from Fig. 4, between $_{50}$Sn and $_{48}$Cd there is a significant change in the trend of the experimental and theoretical mass differences (normalized to the FRDM predictions, M_i-M_{FRDM}). Clearly, for the Cd isotopic chain the best agreement with the measured masses is

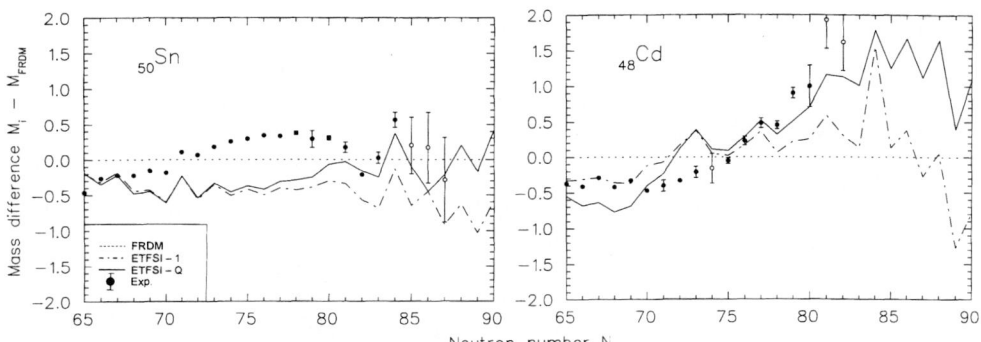

FIGURE 4. Experimental and theoretical mass differences for $_{50}$Sn and $_{48}$Cd isotopes normalized to FRDM values ($M_i - M_{FRDM}$).

obtained with the 'quenched' ETFSI-Q.

Reproduction of r-abundances

Mainly as a result of the improved nuclear-physics input, our site-independent r-process parameter studies have revealed that the entire isotopic $N_{r,\odot}$ pattern cannot be reproduced by assuming a global steady flow. Instead, even in a single astrophysical event (e.g., a SN II) requires a superposition of a multitude of components (minimum three) with different neutron densities or equivalenty different S_n's, characterizing different r-process paths [7,8]. In our approach, the weighing of the individual r-components is naturally given by the β-decay half-lives of the three waiting-point nuclei ^{80}Zn$_{50}$, ^{130}Cd$_{82}$ and ^{195}Tm$_{126}$ which represent the (main) progenitors of the stable isobars ^{80}Se, ^{130}Te, ^{195}Pt situated at the top of the respective $N_{r,\odot}$ peaks. That this approach has, indeed, a nearly one-to-one mapping to 'realistic' astrophysical environments has been discussed in some detail in [9,16].

In Fig. 5, we show global $N_{r,\odot}$-fits from a superpostion of sixteen components for the two versions of the ETFSI nuclear mass model (EFTSI-1 [46] and ETFSI-Q [47]). In both cases, identical conditions for the stellar parameters were used [10,11]. As is indicated by the dashed fit-curve, when using the original ETFSI-1, apart from pronounced A$\simeq$115 and A$\simeq$175 abundance troughs, too little r-process material is observed in the whole region beyond the A$\simeq$130 peak. These r-abundance deficiencies have been interpreted to arise from an overestimation of the N=82 and N=126 shell strengths below ^{132}Sn$_{82}$ and ^{208}Pb$_{126}$, respectively. As shown in the previous Sect., first experimental signatures of shell quenching are now, indeed, observed for N=82. This strongly justifies our further application of 'quenched' mass models in astrophysical calculations, like the spherical HFB/SkP or the deformed ETFSI-Q. The full curve in Fig. 5 shows the result of multicomponent fit with the latter mass model. When compared to the $N_{r,\odot}$-fit obtained with ETFSI-1, a considerable improvement of the overall fit is observed. In particular the prominent abundance troughs in the A≃115 and 175 regions are eliminated to a large extent.

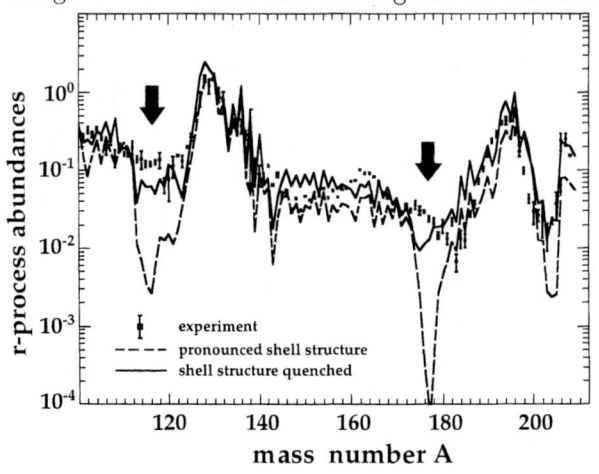

Figure 5. Global r-abundance fits for the ETFSI-1 mass model [46] with pronounced shell structure and the ETFSI-Q formula [47] with quenched shell gaps. For discussions, see text and [10,11].

Hence, a 'quenched' mass model for r-abundance calculations beyond A≃150 seems to be highly recommended, in particular if predictions for the A≥200 mass region are anticipated. Of special interest in this context is the good agreement for the region between ^{203}Tl and ^{209}Bi with the recent $N_{r,\odot}$ evaluation of [49]. For ^{209}Bi, for example, a nearly pure r-process origin of this isotope is thus indicated, which confirms that only a minor contribution is of s-origin. As a consequence, there is no further need for the so-called *strong s-process* component, which had been introduced initially to account for the Pb and Bi abundances.

This improvement with the ETFSI-Q masses should also make extrapolations up to the progenitors of the long-lived actinides ^{232}Th and ^{235,238}U more reliable, in particular since they are now for the first time based on an internally consistent, modern nuclear-physics input. Pfeiffer et al. [10] have shown that, only when using nuclear masses which correctly describe the entire *isotopic* $N_{r,\odot}$ pattern, consistent values for the Galactic age of 10-14 Gyr could be obtained. Recently, as a second r-observable *elemental* abundances between Ba and Th, U have been determined in several ultra-metal-poor, neutron-capture-rich halo stars, among them CS 22892–052 [50]. In Fig. 6, these values are compared to the solar r-element distribution and to our ETFSI-Q predictions [10,11], after adjustment to the solar metallicity. The excellent agreement clearly indicates, that the heavy elements in this star are of pure r-process origin. With the measured Th abundance and our calculated initial abundance, a 'decay age' of 13.0±4.6 Gyr [51] for CS 22892–052 can be deduced. This value overlaps with recent Galactic ages on the young side (see, e.g. [52]). Finally, the excellent agreement between the r-element abundances of this first-generation star and the global $N_{r,\odot}$ pattern, which represents a summation over the whole Galactic evolution, indicates that there is probably a *unique* scenario for the r-process (at least for A>130). In the meantime confirmed by further observations [51,53], this result seems to validate our site-independent approach of deducing astrophysical constraints on r-process conditions from fitting r-process observables, and thus strengthens once again the importance of future nuclear-structure investigations towards the neutron drip-line.

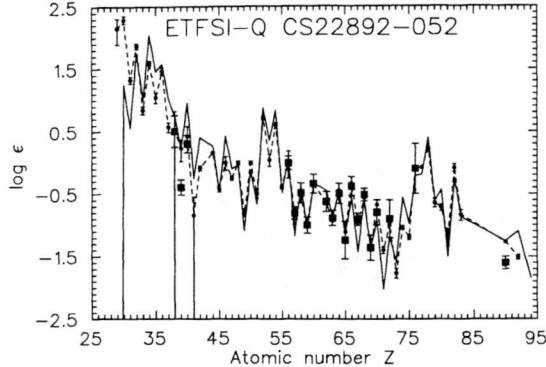

Figure 6. Elemental r-abundances calculated with ETFSI-Q are compared to the respective $N_{r,\odot}$ values (filled circles). Superimposed are measured abundances (filled squares) from the metal-poor halo star CS 22892–052 [50], which were scaled in log ϵ to the solar rare-earth values.

Acknowledgement

The author wishes to acknowledge collaboration with many colleagues, in particular B. Pfeiffer, G. Lhersonneau, W.B. Walters, P. Möller, F.-K. Thielemann, J. Dobaczewski, J.M. Pearson, V.N. Fedoseyev, V.I. Mishin, J. Äystö, H.L. Ravn and the ISOLDE Collaboration, and all present and former Mainz PhD students and postdocs. Support for this work was provided by various grants from the German BMBF, DFG, DAAD and FCI.

REFERENCES

1. Burbidge, E.M. et al., *Rev. Mod. Phys.* **29**, 547 (1957).
2. Lund, E. et al., *Phys. Scr.* **34**, 614 (1986).
3. Gill, R.L. et al., *Phys. Rev. Lett.* **56**, 1874 (1986).
4. Kratz, K.-L. et al., *Z. Phys.* **A325**, 489 (1986).
5. Kratz, K.-L. et al., *Z. Phys.* **A340**, 419 (1991).
6. Kratz, K.-L., *Rev. Mod. Astronomy* **1**, 184 (1988).
7. Kratz, K.-L. et al., *Ap. J.* **402**, 216 (1993).
8. Thielemann, F.-K. et al., *Nucl. Phys.* **A570**, 329c (1994).
9. Chen, B. et al., *Phys. Lett.* **B355**, 37 (1995).
10. Pfeiffer, B. et al., *Z. Phys.* **A357**, 235 (1997).
11. Kratz, K.-L. et al., *Nucl. Phys.* **A630**, 352c (1998).
12. "Nuclear Structure of the Zirconium Region", *Res. Rep. in Phys.*, Springer, ISBN 0-387-50120-7 (1988).
13. "Nuclei Far From Stability / Atomic Masses and Fundamental Constants 1992", *Inst. of Phys. Conf. Ser.* **132**, IOP, ISBN 0-7503-0262-3 (1993).
14. "Exotic Nuclei and Atomic Masses 1995", *ENAM95*, Edition Frontieres, ISBN 2-86332-186-2 (1995).

15. Takahashi, K. et al., *Astron. Astrophys.* **286**, 857 (1994).
16. Freiburghaus, C. et al., *Nucl. Phys.* **A621**, 405c (1997); and *Ap. J.*, in print.
17. Ameil, F. et al., *Eur. Phys. J.* **A1**, 275 (1998).
18. Franchoo, S. et al., Contrib. PE3 to *ENAM98*; and *Phys. Rev. Lett.*, in print.
19. Wöhr, A. et al., Contrib. C11 to *ENAM98*; and *Phys. Rev. Lett.*, in print.
20. Grzywacz, R. et al., Contrib. C7 to *ENAM98*; and *Phys. Rev. Lett.* **81**, 766 (1998).
21. Schoedder, S. et al., *Z. Phys.* **A352**, 237 (1995).
22. Mehren, T. et al., *Phys. Rev. Lett.* **77**, 458 (1996).
23. Lhersonneau, G. et al., *Eur. Phys. J.* **A1**, 285 (1998).
24. Kurpeta, J. et al., Contrib. PE10 to *ENAM98*; and *Eur. Phys. J.* **A2**, 241 (1998).
25. Hoff, P. et al., *Phys. Rev. Lett.* **77**, 1020 (1996).
26. Sanchez-Vega, M. et al., *Phys. Rev. Lett.* **80**, 5504 (1998).
27. Fogelberg, B. et al., Contrib. PC18, PC30 and PE11 to *ENAM98*.
28. Stone, N.J. et al., *Phys. Rev. Lett.* **78**, 820 (1997).
29. Zhang, C.T. et al., *Z. Phys.* **A358**, 9 (1997).
30. Fedoseyev, V.N. et al., *Z. Phys.* **A353**, 9 (1995).
31. Kautzsch, T. et al., *Phys. Rev.* **C54**, 2811 (1996).
32. Kratz, K.-L. et al., in: "Fission and Properties of Neutron-Rich Nuclei", Sanibel Island, Nov. 1997, World Scientific, in print.
33. Rauscher, T. et al., *Phys. Rev.* **C57**, 2031 (1998).
34. Ragnarsson, I. and Sheline, R.K., *Phys. Scr.* **29**, 385 (1984).
35. Dobaczewski, J. et al., *Phys. Rev. Lett.* **72**, 981 (1994).
36. Pfeiffer, B. et al., *Acta Physica Polonica* **B27**, 475 (1996).
37. Erdmann, N. et al., *Appl. Phys.* **B66**, 431 (1998).
38. Mishin, V.I. et al., *Nucl. Instr. and Meth.* **B73**, 550 (1993).
39. Jading, Y. et al., *Nucl. Instr. and Meth.* **B126**, 76 (1997).
40. Sebastian, V. et al., Contrib. PA15 to *ENAM98*.
41. Zamfir, N.V. et al., *Phys. Rev.* **C51**, 98 (1995).
42. Kautzsch, T., PhD Thesis, Univ. Mainz; and to be published.
43. Möller, P. et al., *At. Data Nucl. Data Tables* **59**, 185 (1995); and **66**, 131 (1997).
44. Kratz, K.-L., in: "Nuclei in the Cosmos III", *AIP Conf. Proc.* **327**, 113 (1995).
45. Möller. P. and Randrup, J., *Nucl. Phys.* **A514**, 1 (1990).
46. Aboussir, Y. et al., *At. Data Nucl. Data Tables* **61**, 127 (1995).
47. Pearson, J.M. et al., *Phys. Lett.* **B397**, 455 (1996).
48. Audi, G. et al., *Nucl. Phys.* **A624**, 1 (1997).
49. Beer, H. et al., *Ap. J.* **474**, 843 (1997).
50. Sneden, C. et al., *Ap. J.* **467**, 819 (1996).
51. Cowan, J.J. et al., *Ap. J.*, in print.
52. Salaris, M. and Weiss, A., *Astron. Astrophys.* **327**, 107 (1997).
53. McWilliam, A. et al., *Astron. J.* **109**, 2757 (1995).

Explosive Nucleosynthesis and the Astrophysical r-Process

F.-K. Thielemann[1], C. Freiburghaus[1], T. Rauscher[1], E. Kolbe[1],
B. Pfeiffer[2] K.-L. Kratz[2], and J.J. Cowan[3]

[1] *Departement für Physik und Astronomie, Univ. Basel, Klingelbergstr. 82, CH-4056 Basel, Switzerland*

[2] *Institut für Kernchemie, Univ. Mainz, Fritz-Strassmann-Weg 2, D-55099 Mainz, Germany*

[3] *Department of Physics and Astronomy, Univ. of Oklahoma, Norman, OK 73019, USA*

Abstract. We give an overview of chemical equilibria in explosive burning and the role which neutron and/or proton separation energies play. We focus then on the rapid neutron-capture process (r-process) which encounters unstable nuclei far from beta-stability with neutron separation energies in the range 1-4 MeV. Its observable features, like the abundances, witness nuclear structure as well as the conditions in the appropriate astrophysical environment. With the remaining lack of a full understanding of its astrophysical origin, parametrized calculations are still necessary. The classical approach is based on (constant) neutron number densities n_n and temperatures T over duration timescales τ. Recent investigations, motivated by the neutrino wind scenario from hot neutron stars after a supernova explosion, followed the expansion of matter with initial entropies S and electron fractions Y_e over expansion timescales τ. We compare the similarities and differences between the two approaches with respect to resulting abundance features and their relation to solar r-process abundances. Special emphasis is given to the questions (i) whether the same nuclear properties far from stability lead to similar abundance patterns and deficiencies in both approaches and (ii) whether some features can also provide clear constraints on the permitted astrophysical conditions.

INTRODUCTION

Hydrostatic burning stages in stellar evolution, like H, He, C, and Ne-burning with temperatures being essentially confined to $T<10^9$ K, are dominated by individual nuclear reactions and the precision of their cross sections.

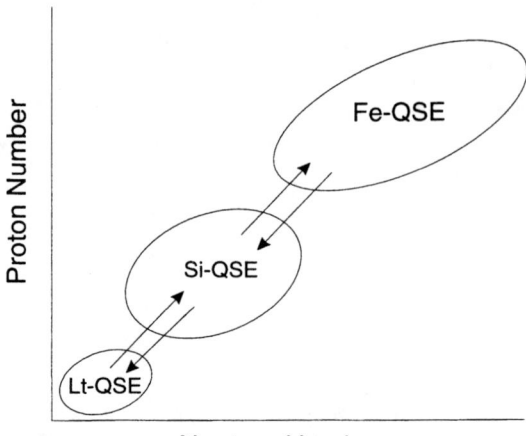

FIGURE 1. A schematic view of the nuclear chart with three quasi-equilibrium groups, (i) light nuclei including neutrons, protons and alpha partciles, (ii) the Si-group extending from Mg to the N=20 and Z=20 boundaries, (iii) the Fe-group containing nuclei above these boundaries.

The same is true for explosive events which never surpass critical temperatures. This is the case e.g. for explosive H-burning in novae which is limited to temperatures $T < 3 \times 10^8$ K or explosive burning stages up to Ne-burning in supernovae. Other explosive events attain high temperatures, but during their ignition stages individual reactions matter strongly. This is the case e.g. during the ignition stages of explosive H-burning in X-ray bursts, where break-out reactions from the hot CNO-cycle (bridging the gap between CNO nuclei and Ne) as well as other hot CNO-type cycles up to Ca or Ti are important. It also applies to central C-ignition in white dwarf progenitors of type Ia supernovae.

However, a whole variety of burning processes responsible for the abundances of intermediate and heavy nuclei, like e.g. hydrostatic Si and explosive O and Si-burning, are leading to partial or full equilibria of reactions. The abundances within an equilibrated reaction chain can be described by the chemical potentials of nuclei ($A + B \leftrightarrow C + D$ leads to $\mu_A + \mu_B = \mu_C + \mu_D$) which form a Boltzmann gas, being dependent on density, temperature, mass, and abundance of a nucleus. The abundance ratios are (besides thermodynamic environment properties) determined by mass differences, i.e. Q-values of reactions. Therefore, mass uncertainties matter, but uncertainties in cross sections do not enter abundance determinations [2,49]. Fig. 1 shows a schematic view of the nuclear chart with three such quasi-equilibrium (QSE) groups, as they occur in hydrostatic Si or explosive O and Si-burning. These groups are in equilibrium for neutron and proton captures. This implies equilibria also for other target nuclei as well as for mixed reactions like (p, α) etc. The reason for the equilibrium regions are fast forward and reverse reaction rates for such environment conditions, the non-equilbrium regions are identified by small cross sections and reaction rates. This is due to small Q-values for reactions out of the magic numbers N=20 and Z=20 and due to small level or resonance densities for light nuclei up to Ne. For sufficiently high temperatures, the Si and Fe-groups, or even all groups, can merge to a full or nuclear statistical equilibrium (NSE).

The chemical equilibrium for neutron or proton captures leads to abundance maxima at specific neutron or proton separation energies

$$\frac{Y(A+1)}{Y(A)} \approx 1 = n_{n,p}\frac{G(A+1)}{2G(A)}\left[\frac{A+1}{A}\right]^{3/2}\left[\frac{2\pi\hbar^2}{m_u kT}\right]^{3/2}\exp(S_{n,p}(A+1)/kT) \quad (1)$$

$$\frac{S_{n,p}(A+1)}{kT} = 24\ln\left[\left(\frac{A}{A+1}\right)^{3/2}\left(\frac{G(A)}{G(A+1)}\right)^{3/2}\left(\frac{T}{10^9 \text{K}}\right)^{3/2}\frac{N_A}{n_{n,p}/\text{cm}^3}\right]. \quad (2)$$

The equations are valid for neutron and/or proton captures, as indicated by the n,p subscripts. The Y's are abundances related to number densities n via $n = \rho N_A Y$. At the maximum in an isotopic or isotonic line we have $S_{n,p} \approx 24kT$, if the partition functions G are neglected (order unity). This is slightly modified by logarithmic dependences on density and temperature, i.e. for higher densities the equilibrium is pushed to smaller separation energies (closer to drip lines in extreme cases). When an equilibrium with neutrons and protons exists, the abundance maximum is found in the nuclear chart at the intersection of the relevant neutron and proton separation energies. The free neutron and proton densities reflect the total neutron/proton ratio in matter, which is determined by slow, weak interactions not in equilibrium. Thus, weak interactions affect abundances by changing Y_e, the total proton/nucleon ratio $Y_e = <Z/A> = \sum Z_i Y_i / \sum A_i Y_i = \sum Z_i Y_i$. $\sum A_i Y_i$ is the sum of all mass fractions (abundance × mass number) and thus equal to 1.

The understanding that explosive burning stages are governed by QSE or even NSE is growing [52,16,17,50]. This has been shown recently in calculations of type II supernova nucleosynthesis [55,20] with two libraries of nuclear reaction rates [53,47], differing by up to factors of 5-7. Of the nuclei produced in explosive O and Si-burning, i.e. most nuclei from Si to beyond Fe, only abundance differences were noticed in the transition region between the Si and Fe groups. Different types of QSE-groups can emerge in explosive burning. The high temperature phase of the rp-process in X-ray bursts witnesses isotonic lines in $(p,\gamma) - (\gamma,p)$ equilibrium, because neutrons are not available in hydrogen-rich layers [43,41] (see also Wiescher, this volume). The (slow) weak interactions are not in equilibrium and thus the β^+-decays (and also some connecting and competing (α,p) reactions) have to be followed explicitly. Another application is the r-process which should be the focus of the rest of this contribution. Here the QSE-groups are isotopic lines in $(n,\gamma) - (\gamma,n)$ equilibrium, and the connecting weak interactions are β^--decays. It should be mentioned that during the final stages when freeze-out from equilibria occurs due to the expansion of matter and temperature decline below equilibrium conditions, reaction rates count again. But the abundance differences will be smaller than expected from rate differences.

In general it should be pointed out that equilibria simplify the understanding of explosive nucleosynthesis processes and individual cross sections play a much less important role than reaction Q-values, i.e. mass differences. However, opposite to environments with neutron/proton ratios favoring nuclei close to stability, processes like the rp and r-process explore exotic nuclei close to the neutron or proton drip-lines where masses are not well known. Therefore the focus lies on nuclear masses and beta-decay properties and their experimental and theoretical investigations (see also Kratz and Wiescher, this volume).

R-PROCESS BASICS

The question whether we understand fully all astrophysical sites leading to an r-process is not a settled one. It is usually assigned type II supernovae (SNe II), the events accompanying the deaths of massive stars and formation of neutron stars (high entropy ejecta, see [56,46]). But galactic evolution timescales and the delayed emergence of r-process matter indicate that these can probably only be SNe II with long evolution timescales at the lower limit of supernova progenitor masses ($\sim$8-10 $M_\odot$) [7,28], while neutron star mergers or still other low entropy sites are not necessarily excluded [27,29,9]. Both these environments provide or can possibly provide high neutron densities and high temperatures which ensure an $(n,\gamma) - (\gamma,n)$ equilibrium in each isotopic chain before the decline of densities and temperatures during the explosion. However, recent observations shed some doubts on the supernova origin. On average type II supernovae produce Fe to intermediate mass elements in ratios within a factor of 3 of solar [49]. If they would also be responsible for the r-process, the same limits should apply. But the observed bulk r-process/Fe ratios vary widely in low metallicity stars. For example CS 22892–052 has an r/Fe ratio which is 30 times larger than solar [44]!

For the r-process source discussion ist is also helpful to ask the question which abundance pattern we need to explain. There are (i) the present r-process abundances in the solar system and (ii) elemental abundance observations of low metallicity stars. The latter are very old stars and display with their unchanged surface abundances the composition of the interstellar medium early in galactic evolution. The observations of low metallicity stars are all consistent with the solar r-abundance pattern of elements and the relative abundances among heavy elements do apparently not show any time evolution [44,6], at least for elements heavier than Ba. The analysis of Ba line profiles in several metal-poor stars showed that also the ratio of odd to even isoptopes agrees with a solar r-process distribution (making use of hyperfine splitting for odd isotopes [10]). This, plus the reproduction of the third

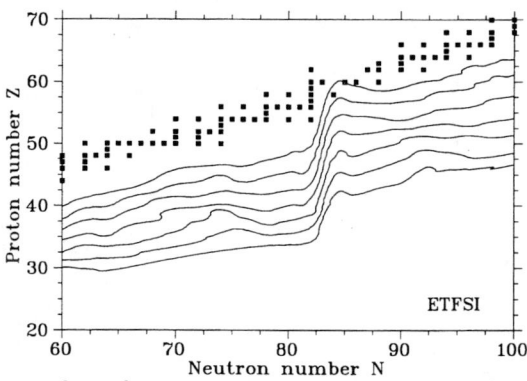

FIGURE 2. Contour plots of constant neutron separation energies S_n=1,2,3,4,5,6, and 7 MeV in the $80 \leq A \leq 140$ mass region for the ETFSI mass model [1]. Note the saddle point behavior caused by the shell closure at $N = 82$, also existing when using the FRDM masses [34], which leads to a deep trough before the peak at $A = 130$ (see dashed line in Fig. 3).

r-process peak (Os, Ir, Pt) suggests strongly that (at least beyond Ba) all contributing astrophysical events have a solar r-process composition, opposite to different claims [14]. Even if all events contributing to galactic evolution produce the same relative r-process abundances, a single astrophysical site will still have varying conditions in different ejected mass zones. This naturally guides towards a multi-event model (where all events are close to identical), however, each event displays a multi-component (i.e. superposition) behavior. A component is defined by a combination of neutron number density, temperature (defining the neutron separation energy of an r-process path) and duration time, or more physically for an adiabatic expansion, entropy, Y_e, and an expansion timescale. The physical conditions must vary smoothly, as expected from a single astrophysical site, opposote to other assumptions [13].

The site-independent classical analysis of Kratz et al. [23,24], based on n_n, T, and τ, led to the conclusion that the r-process experienced a fast drop from equilibrium (of the order of 0.05 s, at least for conditions producing the $A \simeq 80$ peak), in order not to wash out the odd-even staggering via slow freeze-out effects. A continuous superposition of components with neutron separation energies in the range 4-2 MeV (see Eq.(2), related to n_n and T) on timescales of 1 - 2.5 s, provided a good overall fit. The beta-decay properties along contour lines of constant S_n towards heavy nuclei [48] (see Fig. 2 for the region around the N=82 shell closure) are responsible for the resulting abundance pattern. These are predominantly nuclei not accessible in laboratory experiments to date. Exceptions exist in the $A = 80$ and 130 peaks [23,24] and continuous efforts are underway to extend experimental information in these regions of the closed shells N=50 and 82 with radioactive ion beam facilities (e.g. Kratz, this volume). Such classical r-process studies were extended to deduce necessary

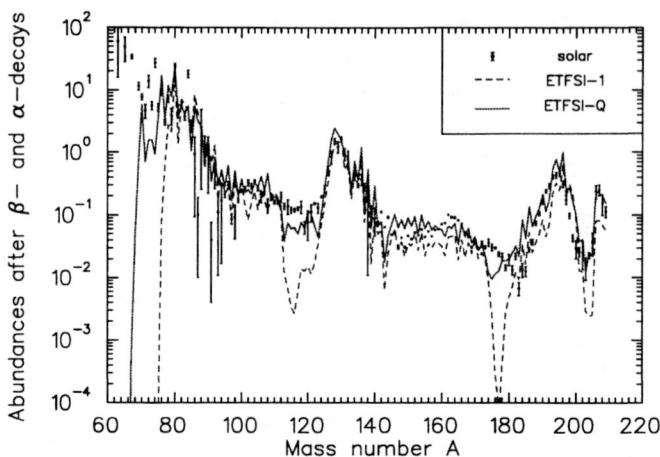

FIGURE 3. Fits to solar r-process abundances [22], obtained with a smooth superposition of 17 equidistant $S_n(n_n, T)$ components from 1 to 4 MeV. The dashed line presents the results for ETFSI masses [1] with half-lives $\tau_{1/2}$ and beta-delayed neutron emission P_n values from QRPA calculations [33]. For the solid line the ETFSI-Q mass model [36] was applied, which introduced a phenomenological quenching of shell effects. The quenching of the $N = 82$ shell gap leads to a filling of the abundance troughs. These results [37] are also the first ones which show a good fit to the r-process Pb and Bi contributions after following the decay chains of unstable heavier nuclei.

requirements for nuclear properties like masses, half-lives, and deformation [48,5,37,25]. One of the major conclusions was the quest for shell quenching far from stability in order to avoid abundance deficiencies, as seen in Fig. 3.

ENTROPY BASED CALCULATIONS

For the operation of an r-process 10 to 150 neutrons per seed nucleus (in the Fe-peak or somewhat beyond) have to be available to form all heavier r-process nuclei by neutron capture. For a composition of Fe-group nuclei and free neutrons that translates into a $Y_e = <Z/A> = 0.12$-0.3. Such a high neutron excess is not existing in typical stellar environments and only possible for high densities in neutron stars under beta equilibrium ($e^- + p \leftrightarrow n + \nu$) with neutrino escape and $\mu_e + \mu_p = \mu_n$, based on the high electron Fermi energies (chemical potentials) which are comparable to the neutron-proton mass difference [4,29].

Another option is a so-called extremely alpha-rich freeze-out in complete Si-burning with moderate $Y_e > 0.40$. After the freeze-out of charged particle reactions in matter which expands from high temperatures but relatively low densities, e.g. 90% of all matter can be locked into ^{4}He with $N=Z$, which leaves even for moderate Y_e's a large neutron/seed ratio for the few existing heavier nuclei. This corresponds to a freeze-out from QSE in Fig. 1 where

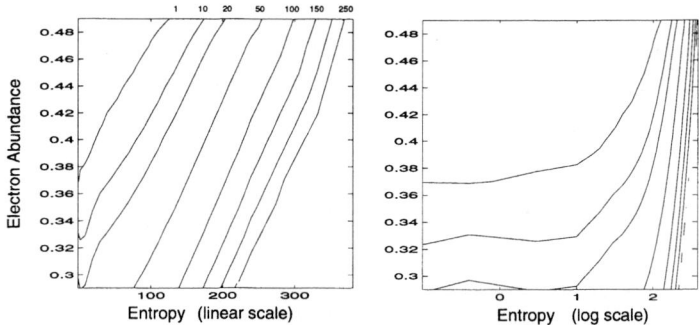

FIGURE 4. Y_n/Y_{seed} contour plots as a function of intial entropy S and Y_e for an expansion time scale of 0.05s. The left part shows how, for moderate Y_e-values, an increasing neutron/seed ratio, indicated by contour lines labeled with the respective Y_n/Y_{seed}, can be attained with increasing entropy which leads to a stronger alpha-rich freeze-out behavior. The results scale with Y_e, measuring the global neutron/nucleon ratio $1-Y_e$. The right part of the figure enhances on a logarithmic scale the low entropy behavior, where Y_n/Y_{seed} is only determined by the electron abundance Y_e. The contour lines are the same for both figures. A Y_n/Y_{seed} ratio of 150, which produces the heaviest r-process nuclei, would only be attained for a Y_e as small as 0.12.

groups 2 and 3 merged, but the connection between 1 and 2 is extremely weak due to low densities. The links accross the particle-unstable A=5 and 8 gaps is only possible via the three body reactions $\alpha\alpha\alpha$ and $\alpha\alpha n$ to ^{12}C and ^9Be, whose reaction rates show a quadratic density dependence. The entropy per gram of baryons in a radiation dominated gas is proportional to T^3/ρ and can be used as a measure of the ratio between the remaining He mass-fraction and heavy nuclei. $S\sim0.15$ k_B/nucleon represents the deviding line between a normal and alpha-rich freeze-out. Similarly, the ratio of neutrons to heavy nuclei (i.e. the neutron to seed ratio) is a function of entropy and permits for high entropies, with large remaining He and neutron abundances compared to small heavy seed abundances, neutron captures which proceed to form the heaviest r-process nuclei [54,32,46,56,18,19].

The behavior of these latter two environments, representing a normal and alpha-rich freeze-out, is summarized in Fig. 4. The available number of neutrons per heavy nucleus Y_n/Y_{seed} after charged particle freeze-out, when the lerge QSE-groups break up into isotopic lines, is shown as a function of entropy and initial Y_e. At low entropies the transition to a normal freeze-out occurs, indicated by the negligible entropy dependence. Recent r-process studies [56,46,38,19] have concentrated on the high entropy environment in the innermost ejecta of SNe II. These are the layers heated by neutrino emission and evaporating from the hot proto-neutron star after core collapse. Whether the entropies required for these conditions can really be attained in supernova explosions has still to be verified. In order to investigate the questions, whether

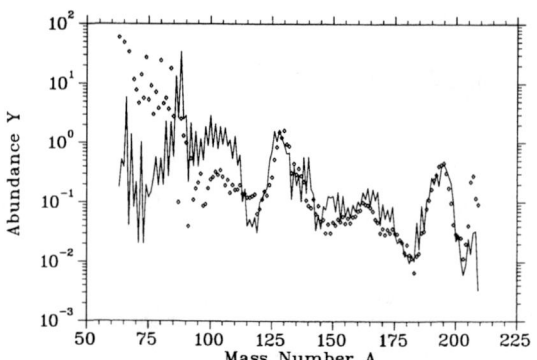

FIGURE 5. Similar to Fig. 3 with the ETFSI mass formula, making use of a superposition of entropies $g(S)$ to attain an overall good fit to solar r-process abundances from a high entropy neutrino wind in type II supernovae. These calculations were performed with $Y_e = 0.45$, but similar results are obtained in the range $0.30 - 0.49$, only requiring a scaling of entropy. The trough below $A = 130$ behaves similar to Fig. 3. This shows that a time dependent freeze-out (with a full treatment of neutron captures and photodisintegrations), resulting from a scenario based on entropy, can cause the same abundance deficiencies due to specific nuclear structure features as obtained in the site-independent classical approach with an instantaneous freeze-out from $(n,\gamma) - (\gamma,n)$-equilibrium.

and how explosion entropies can be translated into n_n and T (or S_n) of the site-independent classical approach, we performed a parameter study based on the entropy S and Y_e, in combination with an expansion timescale of typically 0.05 s [46], and varied nuclear properties (i.e. mass models). The neutron capture rates were calculated with the statistical model code SMOKER [40]. The β^--rates came from experimental data or QRPA calculations [33].

Following the reasons discussed above, we chose a smooth superposition with a weighting function $g(S_i) = x_1 e^{-x_2 S_i}$, where the index i counts the components. Fig. 5 shows a Y_e=0.45 sequence. Entropies from about 200 to about 350 give Y_n/Y_{seed}-ratios growing from approximately 30 to 150. The α-rich freeze-out always produces seed nuclei in the range $90 < A < 120$. This material can then be "r-processed", leading to a fully neutron dominated process, and the components have a very similar abundance pattern in the mass range A=110 - 200, as found in the classical approach. Thus, it is possible for this entropy range to establish a one-to-one correspondence for abundances obtained via entropy and expansion timescale (S, τ) in one type of calculation and neutron separation energy and timescale $(S_n(n_n, T), \tau)$ in classical calculations. The neutron separation energy S_n of the r-path is the one obtained during neutron capture freeze-out in the entropy based calculations.

Different mass models (in Fig. 5 only ETFSI [1] is shown) give fits of similar quality as in the site-independent classical approach. The discrepancies below the $A = 130$ r-process peak, in form of a pronounced trough, occur again for

FRDM [34] and ETFSI. Thus, our earlier conclusions can be translated from the classical analysis to realistic astrophysical applications in this mass region. The nuclear structure properties leading to agreement and deficiencies apply in the same way, due to the nature of a fast freeze-out, which preserves the abundances as they result from an initial (n,γ)-(γ,n)-equilibrium at high temperatures, even when neutron captures and photodisintegrations are followed individually [11]. There is possibly one difference to the conclusions given with Fig. 3. The calculations experiencing the highest entropies have the longest neutron freeze-out timescales and are responsible for the heaviest nuclei [11]. Our results show that the trough before the $A = 195$ peak, resulting in case of the ETFSI mass model and the classical approach, does not survive [48,5,3,37]. This region is changed by ongoing (non-equilibrium) neutron captures during the freeze-out and apparently does not directly witness nuclear properties far from stability at the $N = 126$ shell closure (Fig. 5). This outcome might change, however, back to the classical approach, as giant dipole resonance features based on neutron skins of neutron-rich nuclei far from stability could enhance neutron capture cross sections tremendously and ensure the existence of an $(n,\gamma) - (\gamma,n)$ equilibrium to lower temperatures [12].

There have been suggestions that neutrino-induced spallation of nuclei in the $A = 130$ peak, caused by a strong neutrino wind from the hot neutron star, could fill abundance troughs [39]. We refer to a more detailed discussion of this effect [11], including the requirements on neutrino luminosities and distances of ejecta from the neutron star at the time of the neutron freeze-out. We come to the conclusion in agreement with [39] that this effect can only be of importance for the low mass wings of the A$\sim$130 peak, but not for the mass region in the range 110-120. Nuclear structure effects (shell quenching far from stability) seem to be the best solution, especially as they are already observed experimentally and predicted theoretically [35,45,57,21,8,26]. For a general review whether neutrino interactions with matter actually support or hinder the occurance of an r-process see [31].

The correspondence between classical and entropy based calculations can only be established for entropies producing nuclei with $A >110$. Matter with $A<110$ results from lower entropies with a neutron-poor and alpha-rich freeze-out, where the abundances of heavy nuclei are dominated by those with alpha separation energies of $\approx$6 MeV [11]. This follows from an equation very similar to Eq.(2) for alpha separation energies. None of these entropies produces an abundance peak at charged particle freeze-out with $A < 80$, leaving a sufficient amount of neutrons for r-processing which would reproduce the typical neutron-induced abundance features in the range $A = 80 - 110$. A different choice of Y_e (e.g. 0.49) can influence that pattern somewhat in avoiding very large spikes for $A\approx 90$ and $N=50$ isotopes, but the overall features stay. Within an entropy based approach only conditions with lower Y_e, but also lower entropies, can lead to abundance features in agreement with the solar r-abundance pattern similar to the classical approach. Such features relate to

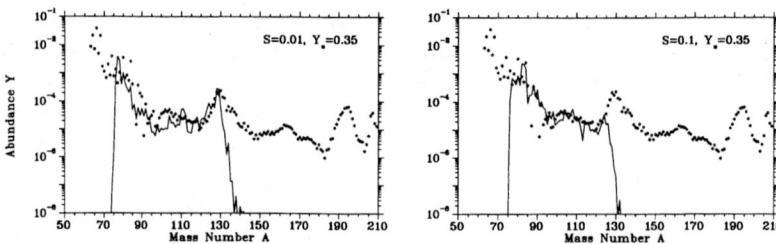

FIGURE 6. A low $Y_e = 0.35$ value and a high density $\rho > 10^7 \, \text{g cm}^{-3}$ cause a normal (low-entropy) freeze-out with a sufficient amount of neutrons left for the synthesis of $80 < A < 120$ nuclei. This calculation utilized a droplet model [15] which similar to ETFSI-Q avoids the trough for $110 < A < 120$. A further decrease in Y_e (see Fig. 4b) will increase the Y_n/Y_{seed}-ratio and could also account for the upper mass range of r-process nuclei. Such an environment is similar to low entropy scenarios related to neutron stars [7,30,42].

nuclei along contour lines of constant neutron (rather than alpha) separation energies (Fig. 6).

CONCLUSIONS

The r-process is a typical example of an explosive burning environment determined by reaction equilibria and particle separation energies. Its abundance features reflect the beta-decay half-lives along contour lines of constant neutron separation energy, which led to the classical site-independent approach based on a constant neutron number density n_n and temperature T over a process duration time τ. More realistic astrophysical environments are expected to follow the expansion of matter with initial entropies S on expansion timescales τ through the freeze-out of reactions with declining temperatures and densities. While this has been performed before by a number of investigators, it is the first time that different mass models have been explored in such investigations.

We find essentially a one-to-one mapping between the results of the classical and the entropy based approach for entropies producing nuclei with mass numbers $A > 110$. The neutron separation energy of the r-path at neutron freeze-out is the same as in the corresponding classical components with constant n_n and T. Concerning the influence of nuclear properties far from stability, the conclusions of the classical site-independent calculations of [24,5,37] remain valid for the $A \simeq 130$ region and the abundance troughs can be cured by quenching of shell effects.

However, for high entropy environments, as expected from a high entropy neutrino wind in supernovae, there exists no correspondence for contributions which cover the mass range $80 < A < 110$. They are the product of charged particle reactions and a neutron-poor, alpha-rich freeze-out. It is not meaningful to compare these contributions to classical r-process calculations and the resulting abundances do not fit the solar r-abundance pattern either. Low entropies $S < 0.1$ and low Y_e's of the order of ≤ 0.35, which lead to a normal (i.e. not alpha-rich) freeze-out, can provide neutron densities that resemble classical r-process studies and produce a good fit to the solar abundance pattern in the mass range of $80 < A < 120$. The latter findings seem to exclude high entropy supernova environments for producing the mass region $80 < A < 110 - 120$. This could point towards low entropy sources related to expanding neutron star matter.

Whether such an interpretation ($A < 130$ from low Y_e and S conditions, $A > 130$ from high S conditions) is a solution, might eventually be answered by observations. There seems to exist meteoritic evidence from the extinct radioactivities ^{107}Pd, ^{129}I, and ^{182}Hf [51] that the last r-process contributions to the solar system for $A > 130$ and $A < 130$ came at different times, i.e. from different types of events. It is highly desirable to have an independent verification from observations of low metallicity stars, which apparently show a completely solar r-process composition for nuclei with $A > 130$ [44,6].

ACKNOWLEDGMENT

This work was supported by the Swiss Nationalfonds (grant 20-47252.96 and 2000-053798.98), the German BMBF (grant 06Mz864) and DFG (grant Kr80615), the US NSF (grant AST 9618332) and the Austrian Academy of Sciencies.

REFERENCES

1. Aboussir, Y. et al. 1995, At. Data Nucl. Data Tables, 61, 127
2. Arnett, W.D. 1996, *Nucleosynthesis and Supernovae*, Princeton Univ. Press
3. Bouquelle, V. et al. 1996, A & A, 305, 1005
4. Cameron, A.G.W. 1989, AIP Conf. Proc. 183, 349
5. Chen, B. et al. 1995, Phys. Lett., B355, 37
6. Cowan, J.J., McWilliam, A., Sneden, C., Burris, D.L. 1997, Ap. J., 480, 246
7. Cowan, J. J., Thielemann, F.-K., Truran, J. W. 1991, Phy. Rep., 208, 267
8. Dobaczewski, J. et al. 1996, Phys. Rev. C53, 2809
9. Eichler, D., Livio, M., Piran, T., Schramm, D. N. 1989, Nature, 340, 126
10. Francois, P. & Gacquer, W. 1998, Nature, in press
11. Freiburghaus, C. et al. 1998, Ap. J., in press
12. Goriely, S. 1998, Phys. Lett. B, in press
13. Goriely, S., Arnould, M. 1996, A & A, 312, 327
14. Goriely, S., Arnould, M. 1997, A & A, 322, L29
15. Hilf, E. R., von Groote, H., & Takahashi, K. 1976, CERN-Rep 76-13, p.142
16. Hix, W.R., Thielemann F.-K. 1996, Ap. J. 460, 869
17. Hix, W.R., Thielemann F.-K. 1998, Ap. J., in press

18. Hoffman, R. D. et al. 1996, Ap. J., 460, 478
19. Hoffman, R. D., Woosley, S. E., Qian, Y.-Z. 1997, Ap. J., 482, 951
20. Hoffman, R. D., Woosley, Weaver, T.A., Rauscher, T. Thielemann, F.-K. 1998, Ap. J., submitted
21. Ibbotson, R.W. et al. 1998, Phys. Rev. Lett. 80, 2081
22. Käppeler, F., Beer, H., Wisshak, K. 1989, Rep. Prog. Phys. 52, 945
23. Kratz, K.-L. et al. 1988, J. Phys. G, 14, 331
24. Kratz, K.-L. et al. 1993, Ap. J. 402, 216
25. Kratz, K.-L., Pfeiffer, B., Thielemann, F.-K., 1998, Nucl. Phys. A630, 352c
26. Lalazissis, G.A., Vretenar, D., Ring, P. 1998, Phys. Rev. C57, 2294
27. Lattimer, J. M. et al. 1977, Ap. J., 213, 225
28. Mathews, G.J., Bazan, G., Cowan, J.J. 1992, Ap. J. 391, 719
29. Meyer, B. S. 1989, Ap. J., 343, 254
30. Meyer, B. S. 1994, Ann. Rev. Astron. Astrophys, 32, 153
31. Meyer, B.S., McLaughlin G.C., Fuller, G.M. 1998, Phys. Rev. C, in press
32. Meyer, B.S. et al. 1992, Ap. J 399, 656
33. Möller, P., Nix, J.R., Kratz, K.-L. 1997, At. Data Nucl. Data Tables 66, 131
34. Möller, P. et al. 1995, At. Data Nucl. Data Tables 59, 185
35. Orr, N. A. 1991, Phys. Lett. B258, 29
36. Pearson, J. M., Nayak, R. C., Goriely, S. 1996, Phys. Lett. B387, 455
37. Pfeiffer, B., Kratz, K.-L., Thielemann, F.-K. 1997, Z. Phys. A357, 235
38. Qian, Y.-Z., Woosley, S. E. 1996, Ap. J., 471, 331
39. Qian, Y.-Z., Haxton, W.C., Langanke, K., Vogel, P. 1996, Phys. Rev. C55, 1532
40. Rauscher, T., Thielemann, F.-K., Kratz, K.-L. 1997, Phys. Rev. C56, 1613
41. Rembges, F. et al. 1997, Ap. J. 484, 412
42. Rosswog, S. et al. 1998, A & A, in press
43. Schatz, H. et al. 1998, Phys. Rep. 294, 167
44. Sneden, C. et al. 1996, Ap. J. 467, 819
45. Sorlin, O. et al. 1993, Phys. Rev. C47, 2941
46. Takahashi, K., Witti, J., Janka, H.-T. 1994, A&A, 286, 857
47. Thielemann, F.-K., Arnould, M., Truran, J. W. 1987, in *Advances in Nuclear Astrophysics*, ed. E. Vangioni-Flam, Gif sur Yvette, Editions Frontière, p.525
48. Thielemann, F.-K. et al. 1994, Nucl. Phys. A570, 329c
49. Thielemann, F.-K. et al. 1998, in *Nuclear and Particle Astrophysics*, eds. J. Hirsch, D. Page, Cambridge Univ. Press, p.27
50. Wallerstein, G. et al. 1997, Rev. Mod. Phys. 69, 995
51. Wasserburg, G., Busso, M., Gallino, R. 1996, Ap. J. 466, L109
52. Woosley, S.E., Arnett, W.D., Clayton, D.D. 1973, Ap. J. Suppl. 26, 231
53. Woosley, S.E. et al. 1978, At. Data Nucl. Data Tables 22, 371
54. Woosley, S. E., Hoffman, R. D. 1992, Ap. J. 395, 202
55. Woosley, S.E. et al. 1997, Nucl. Phys. A621, 445c
56. Woosley, S. E. et al. 1994, Ap. J. 433, 229
57. Zhang, C.T. et al. 1997, Z. Phys. A358, 9

Review of Radioactive Beam Nuclear Astrophysics experiments

Jean Vervier

Institut de Physique Nucléaire, Louvain-la-Neuve, Belgium

Abstract. The use of low-energy radioactive beams to study problems of astrophysical interest is reviewed. The emphasis is placed on the direct measurements of the cross sections for nuclear reactions involving short halflifes radioactive nuclei in their entrance channels, and on the astrophysical implications of their results. Future perspectives of this field are also presented.

1. INTRODUCTION

Radioactive Beams can be produced by 2 methods [1]. In the ISOL method, large quantities of radioactive nuclei are generated by bombarding a thick (primary) target with high intensity primary stable beams produced by a first accelerator, or by irradiating a fissile element with high-flux neutrons from a reactor or from a target bombarded by deuteron beams. These nuclei are extracted from the target as atoms or molecules, transformed into ions by a suitable ion source, mass-separated by an Isotope Separator On Line and (post-) accelerated by a second accelerator. The energies of these secondary beams can be in the 0.2 to 25 MeV per nucleon range, depending on the postaccelerator, or in the tens to hundreds of keV range, when the ISOL ion source is raised at such high tensions. The resulting beams have good qualities (energy variability, energy spread, emittance) ; the range of their lifetimes is limited downwards to about 100 ms, due to the extraction time from the primary target and the transfer time between the target and the ion source. In the In Flight (sometimes called Fragmentation) method, a high-energy (from several tens to several hundreds of MeV per nucleon) primary heavy-ion beam is sent on a thin (primary) target, where it is broken into many different kinds of nuclei, either by fragmentation or by fission. The resulting fragments, most of which are radioactive, are mostly emitted in the forward direction, with about the same velocity as the primary beam. The desired Radioactive Beams, with energies comparable to the one of the primary beam, are separated from the latter and from the other fragments by suitable methods, which involve magnetic and electric fields, wedge-shaped energy degraders, velocity filters, time-of-flight measurements ... These secondary Radioactive Beams are then sent on a (secondary) target to study nuclear reactions, or are collected by a suitable catcher to investigate their spectroscopic properties. These beams have high energies (tens to hundreds of MeV per nucleon) and rather wide energy spreads and emittances ; the range of their lifetimes is limited downwards to a few hundreds of ns, due to their flight times in the separation devices. Their qualities can be improved by cooling and storage techniques, which however raise the lower limit of their lifetimes to a few seconds.

The main purpose of Nuclear Astrophysics is to understand the production of energy and the synthesis of elements in stars or during stellar events. Both processes occur through nuclear reactions, which often produce radioactive nuclei, as for example the ^{12}C(p,γ)^{13}N reaction where $T_{1/2}$ (^{13}N) = 10 minutes. In "quiet" stars like our sun, and in general during non-violent stellar events, the rate of these nuclear reactions is much smaller than the average decay rate of these radioactive nuclei ; as a consequence, these nuclei have ample time to decay before being involved in other nuclear reactions, so that mostly reactions between stable nuclei are then important. A notable exception is the ^{7}Be(p,γ)^{8}B reaction, which involves the radioactive nucleus ^{7}Be ($T_{1/2}$ = 33 days on earth) and which plays a role in the solar neutrino problem ; this reaction is extensively discussed in other contributions to this Conference [2] and will not be further considered here. It is believed that, during "violent" stellar events, example of which are given in other contributions to this Conference [3], the opposite situation prevails, i.e. the average time between successive nuclear reactions is much shorter than the average decay time of the radioactive nuclei : the latter do not have time to decay before being involved in new nuclear reactions, so that reactions in which at least one of the two partners is radioactive (the other one being often a proton, a neutron or an α-particle) then become very important. It is thus essential, in order to fully understand these types of events, to know the decay properties of the radioactive nuclei produced (halflifes, decay modes, binding energies ...), and the cross sections for nuclear reactions in which they are involved. It is the purpose of the present contribution to show how Radioactive Beams can be very useful to determine these nuclear reaction cross sections.

The uses of Radioactive Beams in Nuclear Astrophysics cover the following topics. Beams produced by the ISOL method in the 0.2 to 25 MeV per nucleon range can be used to measure directly the cross sections for nuclear reactions of astrophysical interest ; this will be the main topics of the present contribution (Sections 2 to 4). The tens to hundreds of keV beams, produced by Isotope Separators On Line, allow the study of the spectroscopic properties of very exotic nuclei, on the proton- and neutron-rich sides ; examples of such studies are given in another contribution to this Conference [4], and this topics will no longer be treated here. Nuclear reactions studied with high energy (tens to hundreds of MeV per nucleon) beams produced by the In Flight method often allow the indirect determination of the cross sections for other nuclear reactions which directly intervene in Nuclear Astrophysics ; this is developed in another contribution to this Conference [5], and will no longer be considered here.

To measure directly the cross section of a nuclear reaction in which one of the two partners is radioactive and the other one stable, two methods can be used : the "radioactive target" method, in which the radioactive partner is incorporated in a target which is bombarded by beams of the stable partner ; the "radioactive beam" method, in which a beam of the radioactive partner is produced and used to bombard a target containing the stable partner. It can be shown [6] that, for lifetimes of the radioactive partner shorter than about one hour, the radioactive beam method is more efficient, whereas above one hour, the radioactive target method is preferable. Most of the radioactive nuclei involved in stellar nuclear reactions belong the first category, so that the development of Radioactive Beams is very useful in Nuclear Astrophysics. It may sometimes be advantageous to use both methods, as illustrated by the $^7Be(p,\gamma)^8B$ reaction [2].

2. REACTIONS RELATED TO THE COLD AND HOT CNO CYCLES

The so-called "cold" and "hot" CNO cycles involve the following sequences of nuclear reactions and β-decays, respectively :

$$^{12}C(p,\gamma)^{13}N(\beta^+\nu)^{13}C(p,\gamma)^{14}N(p,\gamma)^{15}O(\beta^+\nu)^{15}N(p,\alpha)^{12}C \qquad (1)$$

$$^{12}C(p,\gamma)^{13}N(p,\gamma)^{14}O(\beta^+\nu)^{14}N(p,\gamma)^{15}O(\beta^+\nu)^{15}N(p,\alpha)^{12}C \qquad (2)$$

The cold CNO cycle, equation (1), is the dominant mode of energy production in heavy main-sequence stars, and is responsible for about 2 % of the energy of the sun [6]. The hot CNO cycle, equation (2), occurs in "violent" stellar events such as novae and X-ray bursts [7]. The dominance of either of these 2 cycles in stellar environments depends on the respective rates for the β^+-decay of ^{13}N and for the $^{13}N(p,\gamma)^{14}O$ reaction To evaluate the latter rate, it is necessary to know the cross section for this reaction.

The $^{13}N(p,\gamma)^{14}O$ reaction has been the subject of many recent studies. At Louvain-la-Neuve, its cross section has been measured directly in the astrophysically interesting energy region [8] using intense (a few 10^8 particles per second) ^{13}N Radioactive Beams produced by the ISOL method [9]. The same beams have been used to study the $^{13}N(p,p)^{13}N$ elastic scattering and $^{13}N(d,n)^{14}O$ proton transfer reactions [10,11]. This set of experiments have allowed to determine the parameters of the lower energy resonance which dominates the proton capture on ^{13}N in the astrophysically interesting energy region, energy E_R, total width Γ, radioactive width Γ_γ, resonance strength $\omega\gamma$ [6], as well as the direct capture contribution [12,13]. The $^{13}N(p,\gamma)^{14}O$ reaction has also been investigated indirectly, using the so-called Coulomb dissociation method [14,15], with Radioactive Beams produced by the In Flight method. Other useful information has also been deduced from branching ratio measurements dealing with the lower energy resonance [16,17] and from the energy and total width of this resonance [18].

All these data allow to determine the cross section for the $^{13}N(p,\gamma)^{14}O$ reaction in the whole energy range important in Nuclear Astrophysics, and to deduce, from it, the astrophysical conditions under which the cold, equation (1), and hot, equation (2), CNO cycles respectively dominate. The results are synthetized in the "phase diagramme" of Figure 1, in which the stellar density ρ multiplied by the hydrogen fraction in the star Y_p is plotted versus the stellar temperature T. The full line (and its uncertainty given by the dashed lines) delineates the regions where the cold and hot CNO cycles respectively dominate [13]. The density and temperature conditions which prevail in novae are $\rho = 10^3 - 10^5$ g/cm^3, T = 1 - 3.5 $\times$ 10^8 K. It is thus clear that the hot CNO cycle is dominant in novae. Other consequences of the measurements on the $^{13}N(p,\gamma)^{14}O$ reaction, in particular on the $^{13}C/^{12}C$ abundance ratio in novae ejecta, can also be drawn [19].

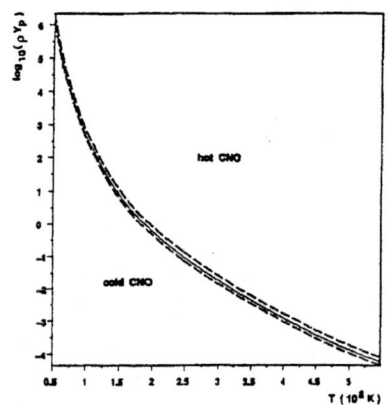

FIGURE 1. Phase diagramme of the stellar density ρ multiplied by the hydrogen fraction in the star Y_p <u>versus</u> the stellar temperature T, for the $^{13}N(p,\gamma)^{14}O$ reaction. For details, see text and ref. [13].

3. REACTIONS RELATED TO AN "ESCAPE" FROM THE CNO CYCLES

The nucleosynthesis in "violent" stellar events can "escape" from the CNO cycles, and proceed towards the production of nuclei with A $\geq$ 20, by various ways illustrated in Figure 2 [20]. One of them is the reaction sequence

$$^{15}O(\alpha,\gamma)^{19}Ne(p,\gamma)^{20}Na \tag{3}$$

If the nucleus ^{19}Ne is produced by this (and other) way(s), it can : either β^+-decay to ^{19}F with its halflife of 17 seconds, in which case the $^{19}F(p,\alpha)^{16}O$ reaction leads

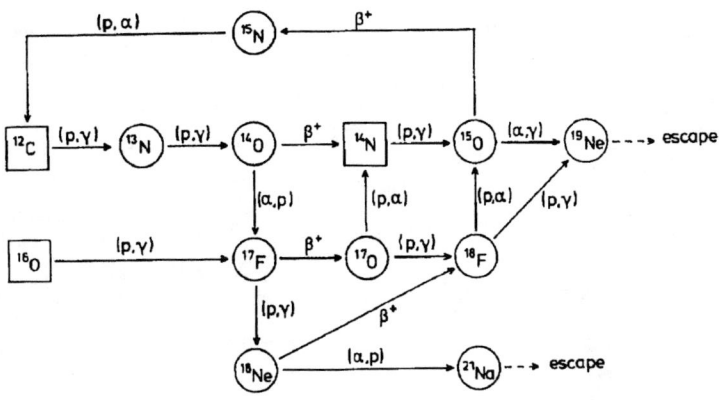

FIGURE 2. Set of reactions in the hot CNO cycles leading to a possible escape to the rp-process. For details, see text and ref [20].

back to CNO elements ; or capture a proton to produce ^{20}Na, which cannot be converted back to these elements, but rather leads to the synthesis of heavier A $\geq$ 20 nuclei. The cross section for the ^{19}Ne(p,γ)^{20}Na reaction is thus an important information to determine the stellar conditions under which such an escape from the CNO cycles occurs.

The ^{19}Ne(p,γ)^{15}O reaction has also been the subject of many recent studies. At Louvain-la-Neuve, its cross section has been measured directly using intense (a few 10^9 particles per second) ^{19}Ne Radioactive Beams produced by the ISOL method [21,22]. The same beams have been used to study the ^{19}Ne(p,p)^{19}Ne elastic scattering at several resonances in the astrophysically interesting energy region [23] and the ^{19}Ne(d,n)^{20}Na reaction to the bound levels of ^{20}Na [24]. This set of experiments have allowed to determine the parameters of these resonances (or upper limits to them), energies, total widths, resonance strengths, as well as to obtain some information on the direct capture contribution [24]. Other useful information on the levels of ^{20}Na above the ^{19}Ne + p threshold has also been obtained from studies of the ^{20}Ne(p,n)^{20}Na and ^{20}Ne(^{3}He,t)^{20}Na charge-exchange reactions, and from the β^+- and delayed proton decays of ^{20}Mg (see the references in [24]). All these data allow to determine the cross section for the ^{19}Ne(p,γ)^{20}Na reaction, or at least limits for it, in the energy range important in Nuclear Astrophysics, and to deduce, from it, the astrophysical conditions for an escape from the CNO cycle through the reaction sequence of equation (3). Figure 3 displays the phase diagramme of the stellar density ρ (for solar composition) as a function of the stellar temperature T and for the following reactions : ^{19}Ne(p,γ)^{20}Na (the shaded area corresponds to the uncertainties in the experimental data); ^{15}O(α,γ)^{19}Ne. It can be concluded, from these results : that, above T = 0.2 $\times$ 10^9 K, the rate of the ^{19}Ne(p,γ)^{15}O reaction dominates the β^+-decay rate of ^{19}Ne; that, for all temperatures, the "bottleneck" for the escape from the CNO cycles through the sequence of equation (3) is the $^{15}O(\alpha,\gamma)^{19}Ne$ reaction, which is always slower than the $^{19}Ne(p,\gamma)^{20}Na$ reaction.

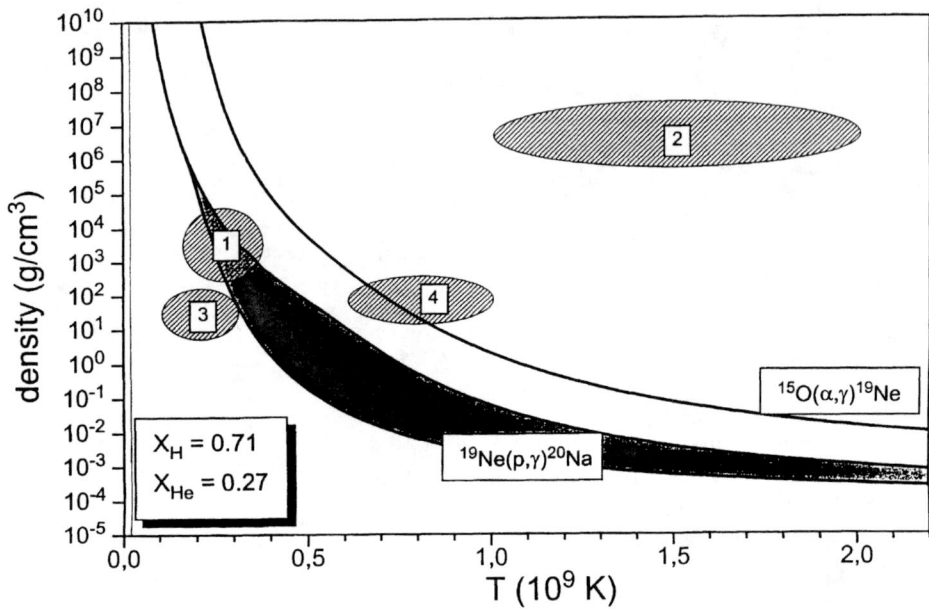

FIGURE 3. Phase diagramme of the stellar density ρ (for solar composition) *versus* the stellar temperature T for the $^{19}Ne(p,\gamma)^{20}Na$ and $^{15}O(\alpha,\gamma)^{19}Ne$ reactions. The hatched areas indicate typical values for the density and temperature in novae (1), X-ray bursts (2), type-II supernovae (3) and supermassive stars (4). For details, see text and ref. [24].

The $^{15}O(\alpha,\gamma)^{19}Ne$ reaction, whose rate has been estimated according to the results of ref. [25], will require a ^{15}O beam of about 10^{11} particles per second to be measured directly. Indirect methods can also be used to obtain information on this reaction, which involve radioactive beams.

Other paths to "escape" from the CNO cycles, included in Figure 2, involve the $^{18}F(p,\gamma)^{19}Ne$ reaction, which, in view of its Q-value, is in competition with the $^{18}F(p,\alpha)^{15}O$ reaction. These 2 reactions have been measured directly with ^{18}F beams : for $^{18}F(p,\alpha)^{15}O$ at a resonance at 638 keV in the $^{18}F + p$ system, both at Louvain-la-Neuve [26] and Argonne National Laboratory [27] ; for the same reaction at a resonance at 324 keV, at Louvain-la-Neuve [28] ; for $^{18}F(p,\gamma)^{19}Ne$ at the 638 keV resonance, at Argonne National Laboratory [29] (an upper limit was obtained for its cross section); for the $^{18}F(p,p)^{18}F$ elastic scattering at the 638 keV resonance, at Louvain-la-Neuve [26]. Further information on the ^{19}Ne levels reached by the $^{18}F + p$ interaction in the energy region of interest has been obtained through the $^{19}F(^3He,t)^{19}Ne$ reaction [30].

All these data allow the following astrophysical conclusions to be drawn. In the whole temperature region reached in novae and X-ray bursts, the ^{18}F β^+-decay halflife (100 minutes) is much longer than the one for the ^{18}F(p,α)^{15}O reaction. The latter reaction is in turn much faster than the ^{18}F(p,γ)^{19}Ne reaction, so that ^{18}F is converted into ^{15}O, and not into ^{19}Ne, in the "explosive" events just mentioned.

Still another way to "escape" from the CNO cycles, also shown in Figure 2, is through the ^{18}Ne(α,p)^{21}Na reaction, if ^{18}Ne is produced by the ^{16}O(p,γ)^{17}F(p,γ)^{18}Ne or ^{14}O(α,p)^{17}F(p,γ)^{18}Ne reaction sequences. A new method has been developed at Louvain-la-Neuve, in collaboration with scientists from Edinburgh, to study (α,p) reactions of astrophysical interest using radioactive beams, and has been tested successfully with the ^{13}N(α,p)^{16}O$_{g.s.}$ reaction whose inverse ^{16}O(p,α)^{13}N$_{g.s.}$ is well known [31].

4. THE ^{56}NI(D,P)^{57}NI REACTION AND THE RP-PROCESS

The nucleus ^{56}Ni is produced in the rp-process, as shown in other contributions to this Conference [3]. Since the ^{56}Ni(p,γ)^{57}Cu reaction has a low Q-value (0.695 MeV), the nucleus ^{56}Ni could be a "waiting point" for the rp-process, where a ^{56}Ni + p $\leftrightarrow$ ^{57}Cu + γ capture-photodissociation equilibrium could occur, in much the same way as the "waiting points" in the r-process [4]. In order to gain information on the ^{56}Ni(p,γ)^{57}Cu reaction, the ^{56}Ni(d,p)^{57}Ni reaction has been studied directly at Argonne National Laboratory with ^{56}Ni beams ($T_{1/2}$ = 6 days) of about 2.5 $\times$ 10^4 particles per second [32]. The coincidence detection of the ^{57}Ni recoils in the Fragment Mass Analyser and of the protons in a large Si-detector array allowed to study the ^{56}Ni(d,p)^{57}Ni reaction leading to the 3/2$^-$ ground state, 5/2$^-$ level at 0.768 MeV and 1/2$^-$ level at 1.113 MeV in ^{57}Ni, and to determine the corresponding spectroscopic factors, which turn out to be close to 0.9.

These data have been used to calculate the ^{56}Ni(p,γ)^{57}Cu reaction rate, assuming that the same spectroscopic factors are valid for the ^{56}Ni(d,n)^{57}Cu reaction as for ^{56}Ni(d,p)^{57}Ni, and considering a direct proton capture on ^{56}Ni. The resulting halflife of ^{56}Ni in explosive stellar events as a function of temperature T turns out to be very short, much shorter than previously calculated [33]. This suggests that ^{56}Ni is not so much a waiting point in the rp-process, and that the nucleosynthesis in the latter proceeds beyond A = 56 to higher mass number [34].

5. CONCLUSIONS AND PERSPECTIVES

The field of Nuclear Astrophysics with Radioactive Beams is likely to increase during the next few years. New Radioactive Beams will come into operation all over the world [1] in the adequate energy range, at Oak Ridge National laboratory (USA), TRIUMF-ISAC (Canada), REX-ISOLDE at CERN (Switzerland), SPIRAL-GANIL (France), EXCYT-CATANIA (Italy), INS-Tokyo (Japan). The

Louvain-la-Neuve facility will be upgraded by the commissioning of a new postaccelerating cyclotron CYCLONE44, which has been designed to cover the energy range most suitable for direct measurements of cross sections for nuclear reactions of astrophysical interest (0.2 - 0.8 MeV per nucleon) with an improved (by one order of magnitude) acceleration efficiency. New instruments have been or are being developed for such experiments : recoil separators (at Oak Ridge National Laboratory, TRIUMF-ISAC and Louvain-la-Neuve), most useful for (p,γ) reactions ; large Si-detector arrays for (p,α) reactions, such as the LEDA (Louvain-Edinburgh Detector Array) detector at Louvain-la-Neuve ; direct and indirect methods for (α,γ) reactions ; new technique for (α,p) reactions [31]. One can thus foresee that crucial data on Nuclear Astrophysics with Radioactive Beams will be presented at the ENAM 3 Conference.

REFERENCES

1. See the contributions to this Conference by A. Mueller, I. Tanihata and J. Nolen.
2. See the contributions to this Conference by G. Bogaert, M. Hass, R. Tribble, F.M. Nunes, T. Motobayashi and B. Davids.
3. See the contributions to this Conference by M. Wiescher, K.L. Kratz and F.K. Thielemann.
4. See the contribution to this Conference by K.L. Kratz.
5. See the contribution to this Conference by T. Motobayashi.
6. Rolfs C.E. and Rodney W.S., *Cauldrons in the Cosmos*, The University of Chicago Press, Chicago and London (1988).
7. See the contribution to this Conference by F.K. Thielemann.
8. Decrock P. et al., *Phys. Rev. Lett.* **67**, 808 (1991).
9. Darquennes D. et al., *Phys. Rev.* **C42**, R804 (1990).
10. Delbar Th. et al., *Nucl. Phys.* **A542**, 263 (1992).
11. Decrock et al. et al., *Phys. Rev.* **C48**, 2057 (1993).
12. Decrock et al. et al., *Phys.Lett.* **B304**, 50 (1993).
13. Delbar Th. et al., *Phys. Rev.* **C48**, 3088 (1993).
14. Motobayashi T. et al., *Phys. Lett.* **B264**, 259 (1991).
15. Kiener J. et al., *Nucl. Phys.* **A552**, 66 (1993).
16. Fernandez P.B. et al., *Phys. Rev.* **C40**, 1887 (1989).
17. Aguer P. et al., Proc. Int. Symp. Heavy Ion Physics and Nuclear Astrophysical Problems (S. Kubono, M. Ishihara and T. Nomura eds), *World Scientific, Singapore* p. 107 (1989).
18. Chupp T.C. et al., *Phys. Rev.* **C31**, 1023 (1985).
19. Arnould M. et al., *Astron. Astrophys.* **254**, L9 (1992).
20. Leleux P. et al., *Nucl. Phys.* **A621**, 183c (1997).
21. Page R.D.. et al., *Phys. Rev. Lett.* **73**, 3066 (1994).
22. Michotte C. et al., *Phys. Lett.* **B381**, 402 (1996).
23. Coszach R.. et al., *Phys. Rev..* **C50**, 1695 (1994).
24. Vancraeynest G. et al., *Phys. Rev.* **C57**, 2724 (1998), and references therein.

25. Mao Z.Q. et al., *Phys. Rev. Lett.* **74**, 3760 (1995).
26. Coszach R. et al., *Phys. Lett.* **B353**, 184 (1995).
27. Rehm K.E. et al., *Phys. Rev.* **C52**, R460 (1995); *Phys. Rev.* **C53**, 1950 (1996).
28. Graulich J.S. et al., *Nucl. Phys.* **A626**, 751 (1997).
29. Rehm K.E. et al., *Phys. Rev.* **C55**, R566 (1997).
30. Utku S. et al., *Phys. Rev.* **C57**, 2731 (1998).
31. Bradfield-Smith W. et al., . *Nucl. Instr. Meth. Phys. Res.* **A**, to be published.
32. Rehm K.E. et al., *Phys. Rev. Lett.* **80**, 676 (1998).
33. van Wormer L. et al., *Astrophys. J.* **432**, 326 (1994).
34. Rehm K.E., *Contribution to the Hirschegg workshop* January 1998.

Direct measurements of the $^7\text{Be}(p,\gamma)^8\text{B}$ reaction cross section

Gilles Bogaert*

*IN2P3-CNRS
Centre de Spectrométrie nucléaire et de Spectrométrie de Masse
91405 Orsay Campus, France

Abstract. The flux of high energy neutrinos produced in the decay of ^8B is proportional to the $^7\text{Be}(p,\gamma)^8\text{B}$ reaction cross section. Large uncertainties in this cross section limit our ability to test solar models. Direct measurements are underway to reduce the uncertainties to a level of 5 % at solar energies. The S_{17} value seems to converge towards a 19 eV b value instead of 22.4 eV b, which implies a significant reduction (15-20%) in the ^8B solar neutrino flux.

THE SOLAR NEUTRINO PUZZLE

From early times when Davis and Bahcall [1,2] wanted to detect solar neutrinos to check that nuclear reactions are the sun's fuel, to the present day, when four-five neutrino-detectors are in operation around the world, no agreement has ever been found between the measured neutrino flux and the theoretical prediction, although many efforts have been made to improve both solar models [3,4] and neutrino detection techniques : the measured flux has always been found lower than predicted. For years, two reasons have been put forward to account for this discrepancy : solar modeling and nuclear reaction rates. In order to reduce the uncertainties in the latter quantities all the key nuclear reactions have been investigated many times, with increasingly sophisticated methods and techniques, resulting in extremely good overall accuracy in all the reactions rates, except the $^7\text{Be}(p,\gamma)^8\text{B}$ reaction rate (see [6] for a recent review of all nuclear reactions involved in the sun modeling). The solar models have also been improved and for a few years now the helioseismology measurements have put severe constraints on sun modeling [5].

However the global deficit of neutrinos is only a piece of the puzzle. The results of the well calibrated experiments GALLEX and SAGE [7,8] reveal the disappearance of most or all ^7Be neutrinos on their path to Earth. Closer inspection shows that all the detection experiment results are inconsistent with each other. It has been shown that the set of detection results (Homestake, (Super-)Kamiokande and

GALLEX-SAGE) leaves no room for any nuclear or astrophysical solution to the total neutrino deficit [9]. The observed deficit of neutrinos has to be explained by particle physics, *ie* non standard neutrino properties. In contrast with the incompatibility of the detection results within the framework of standard physics, the MSW theory of neutrino oscillation constitutes an elegant explanation of all the results (Web-sites such as [10,11] are interesting sites devoted on the solar neutrino puzzle and implications).

The SNO detector will soon be operational, and thanks to neutral current detection, this detector will measure the total flux of solar neutrinos (ν_e and ν_x) of high energy (from ^{8}B beta decay) - independently of any solar model - and characterise the mechanism by which they oscillate (if they do).

Focusing now on astrophysics and tests of solar modeling, the total flux of ^{8}B neutrinos is found to be very sensitive to the sun's temperature : this flux is roughly proportional to S_{17} T^{24}, where S_{17} is the S factor at 0 keV of the ^{7}Be(p,γ)^{8}B reaction [12]. Knowing this cross section with accuracy allows the ^{8}B neutrino flux to become an accurate measurement of the nuclear burning zone temperature in the sun. Reducing the uncertainty on S_{17} to a very low value (5 per cent) is therefore a challenge for nuclear physicists.

$S_{17}(0)$ FACTOR DETERMINATION AND RECOMMENDED VALUE

Theoretical calculations of the cross section of the ^{7}Be(p,γ)^{8}B reaction use approximations and have not so far been able to provide reliable absolute values for the cross section (for recent calculations see [13,14], for A=8 exact calculations, see the reference [15]). The S factor at solar energy (15-18 keV) is then obtained by extrapolating experimental data measured at energies where the cross section is high enough (greater than a few picobarns). Thus the determination of S_{17} relies on experimental data and on calculated energy dependences that until now are found quite scattered and do not provide a well-established accurate result.

Regarding the extrapolation question, it was stressed [16] that calculations should not be trusted above 400 keV, due to manifold limitations . At lower energy, the reaction takes place at such great distances from the nucleus that the cross section behavior becomes independent of any ^{8}B structure assumption. In practice, theoretical models do not predict exactly the same energy dependence in the energy range 0-300 keV (disagreement by a few %), and the various curves diverge with increasing energy. This leads to the uncomfortable situation where the only data points permitting a model independent extrapolation are also the most difficult to obtain (for energies below 400 keV). However, all the experimental results for the ^{7}Be cross section show approximately the same behavior for the energy dependence, and measurements at higher energies are very useful.

The cross section at various energies (above 117 keV cm) has been measured 6 times from the pioneering work of Kavanagh in 1960 up to that of Filippone et al.

published in 1983 [17–22]. However the published absolute values differ by large amounts [23]. The common feature of all the experiments is that delayed alpha or beta particles were detected. Most experiments also rely on the ^{7}Li(d,p)^{8}Li normalization. This reaction was found to be very convenient for absolute normalization because the ^{7}Li content of the target originates from the ^{7}Be decay, and is thus directly connected to the number of ^{7}Be atoms over the beam spot, and because the same set-up may be used for detecting the delayed particles from the ^{8}Li decay. However, the cross section of the normalization reaction turned out to be rather difficult to measure correctly, and the published data for this reaction vary from 110 to 210 mb ! Recent compilations [6,24] converge to a value 147 mb ± 11 mb. (see also Strieder et al. [25]). In addition, a recent measurement at Rehovot gives a value of 155 ± 8 mb, and shows that a thick target backing of heavy material influences the measurements when delayed alpha detection is employed (see M. Haas's contribution at this conference [26], and [27]).

But even when normalized to the same value for the ^{7}Li(d,p)^{8}Li cross section, the experimental data divide into two groups with S_{17} values differing by 25 percent. The origin of this is not clear and may be due to poorly understood systematic errors. Due to these discrepancies the value for the cross section used in solar codes has never been the same and has always been disputed; some authors considering the world average value to be the right one ($S_{17}(0)$ = 24 eV b [28] or 22.4±2.2 eV b [16] or 21±2 eV b [24]) while others recommended [23] or adopted [4] a lower value. The recent reanalysis of all experimental works regarding the solar reaction rates [6] leads to a recommended value which relies on the Filippone et al. measurement $S_{17}(0) = 19 \pm^4_2$ eV b, that is 15% lower that the previous value recommended by Johnson et al. [16].

RECENT DEVELOPMENTS

In 1995 and 1996 two experiments were performed by the collaboration of three french laboratories (CSNSM, IPN, CENBG), [29] using the Van de Graaff at the CENBG (Bordeaux). The results of these two experiments performed with two targets at energies between 0.350 and 1.4 MeV are fully consistant. As in the Filippone experiment, the ^{7}Li(d,p)^{8}Li normalization was used in addition to the activity normalization. The latter was carried oot by measuring the ^{7}Be areal distribution on the target backing (through gamma ray detection), and by controlling the position of the beam spot on the target. In this experiment the two normalization procedures give the same final result (as in Filippone's experiment). Table 1 lists the sources of random and systematic errors obtained in this experiment whose authors claim a final accuracy of the order of 5 percent on the extrapolated S factor ($S_{17}(0) = 18.5 \pm 1.$ eV barn, using [32]). Taking into account the escape of ^{8}B nuclei due to backscattering on the platinum backing [26] slightly increases the result to 19. eV barn according to TRIM simulations [30].

For the purpose of extrapolation to zero energy, a set of theoretical predictions

[31,32,16,33,34] were fitted [29,35] to the data. Depending on the theoretical curve, the extrapolated values range from 18 to 21 eV b, with a reasonable or excellent χ^2 except when fitting the Csótó theoretical curve. The Hammache et al. results are found to be in excellent agreement with the Vaughn et al. and Filippone et al. data, and a global fit to these 3 sets of data gives the same value for S_{17} as for the Hammache data set alone.

A Student's test performed on all the world data sets shows clearly that the above set of data is incompatible with Kavanagh's and Parker's set of data (in the Student's test the assignment of error-bars to the data points does not play any role). This strong inconsistency was found to be the same whatever the theoretical curve used for the Student's test calculation. This indicates that the usual procedure for averaging all the world results is biased. As a consequence, only a restricted set of data should be taken into account for obtaining a reasonable S_{17} value as in [6,29]

Very good fits to the data are obtained whatever the data tange is taken into account for Descouvemont's, Barker's or Johnson's energy dependence. However the final values for S(0) depend slightly on the energy dependence, as stressed above [29,35]. New measurements below 300 keV with uncertainties better than 5% would be very valuable to distinguish between different nuclear models for the cross section.

Four experiments are, or will be, running soon. All of them avoid traditional normalization procedures by using more sophisticated techniques. At Orsay, an Orsay-Bordeaux collaboration aims to measure the energy dependence at energies lower than 240 keV. Improved accuracy in the data points should allow comparison with theoretical predictions of the behavior of the cross section in this crucial range of energy. The ^{7}Be nuclei are produced at the Bordeaux Van de Graaff, chemicaly separated and deposited on a backing at the IPN. The measurements are performed at the PAPAP electrostatic accelerator [36], using a superconductive solenoid (3 Teslas) [37] with a large cavity in which are positionned the target (140 mCi), 24 Si-detectors and 6 plastic detectors for $\alpha - \beta$ coincident detection. This experiment avoids the usual ^{7}Li normalization procedure and relies on the ^{9}Be(p,α)^{6}Li and ^{9}Be(p,d)^{8}Be reaction.

At higher energies, the Seattle-Vancouver [38], the Isolde-Rehovot [39] and the LUNA [40] collaborations avoid all ^{7}Li(d,p)^{8}Li normalization and measurement of the activity over the beam spot by using a rather small target spot uniformly swept by the beam. Provided the sweeping is uniform, and the beam size larger than the target size, the areal distribution of ^{7}Be does not matter in a first approximation. But for low-energy measurements, as planned at Seattle, inhomogeneities in the target thickness may induce a spreading of the effective energy [38].

Preliminary values for the cross section have already been obtained at Bochum and Rehovot, using low activity ^{7}Be targets, at energies above 800 keV. These results seem to be consistent with the Filippone et al., Vaughn et al., and Hammache et al. data. However, very low counting statistics have not yet allowed for any improvement in the S_{17} value.

It has been stressed many times that "experiments with ^{7}Be ions beams would be valuable. Such experiments avoid many of the systematic uncertainties that are important in interpreting measurements of proton capture on a ^{7}Be target" [6]. Such a beam (E=8 MeV) has been produced at the 3 MV tandem accelerator in Naples (Italy) by the NaBoNA collaboration [41] and accelerated into a proton gas target, at a level of 18 pA of ^{7}Be^{4+} nuclei on the target. Beam purification allows normalization on the Rutherford scattering cross section. Five counts where recorded in a ΔE-E telescope, which proved the ability of the setup to allow for compound nucleus detection. Results from all these experiments may be available within a couple of years.

CONCLUSION

The Adelberger et al. collaboration recommends the value 19^{+4}_{-2} eV b for the S factor. This value is in agreement with the recent result of Hammache et al., who reduced the error bars to the small value of 5% (19. ±1. eV b). However further experiments should be performed to establish a secure basis for assessing the best estimate for S_{17}. Currently, the S_{17} value seems to converge towards a value of 19 eV b instead of 22.4 eV b. This lower value implies a significant reduction (15-20%) in the ^{8}B solar neutrino flux.

TABLE 1. measurement uncertainties in the Hammache et al. experiment

energy cut	1%
energy loss	4. keV at 441 KeV
E_{cm}	1 keV at 350 keV
I beam	2 %
counting statistics	2.6 % - 5.5%
target area	4%
Detection efficiency	2%
activity	2%
Target inhomogeneity	5%

REFERENCES

1. Davis, R., Jr., *Phys. Rev. Lett.* **12**, 303 (1964).
2. Bahcall, J. N., *Phys. Rev. Lett.* **12**, 300 (1964).
3. Bahcall, J. N., et al., *Rev. Mod. Phys.* **67**, 781 (1995).
4. Turck-Chieze, S., and I. Lopes, *Ap. J.* **408**, 347 (1993) ; Dzitko, H., et al., *Ap. J.* **447**, 428 (1995)
5. Turck-Chieze, S., et al., *Solar Physics* **175.2**, 247 (1997).
6. Adelberger, E. G., et al., *Rev. Mod. Phys.*, to appear (October 1998).

7. Hampel, W., et al., *Phys Lett B* **388**, 384 (1996); Cribier, M., *proceed. of the 5th Int. Workshop, TAUP 97* (1997)
8. Abdurashitov, J. N., et al, *Phys. Lett. B* **328**, 234 (1994); *Phys. Lett.* **77**, 3720 (1996); hep-ph/9803418.
9. Berezinsky et al., *Phys. lett. B* **365**, 185 (1996); Hata, N., et al., *Phys. Rev. D* **49**, 3622 (1994); Heeger, K. H., and R. G. H. Robertson, *Phys. Rev. Lett.* **77**, 3720 (1996).
10. http://www.sns.ias.edu/ jnb/
11. http://dept.physics.upenn.edu/ www/neutrino/solar.html
12. Bahcall, J. N., and A. Ulmer, *Phys. Rev. D* **53**, 4202 (April 15, 1996).
13. Bennaceur, K., *J. of Phys. G* (July 1998), nucl-th/9802054
14. Zhukov et al. *Phys. Rev.* (1998)
15. Pieper, S. C., this conference.
16. Johnson, C. W., E. Kolbe, S. E. Koonin, and K. Langanke, *Ap. J.* **392**, 320 (1992).
17. Kavanagh, R. W., *Nucl. Phys.* **15**, 411 (1960).
18. Parker, P. D., *Phys. Rev.* **150**, 851 (1966) and *Ap. J. Lett.* **153**, 85 (1968).
19. Kavanagh, R. W., T. A. Tombrello, J. M. Moshe, and D. R. Goosman, *Bull. Am. Phys. Soc.* **14**, 1209 (1969).
20. Vaughn, F. J., R. A. Chalmers, D. Kohler, and L. F. Chase, Jr., *Phys. Rev. C* **2**, 1657 (1970).
21. Wiezorek, C., H. Krawinkel, R. Santo, and L. Wallek, *Z. Phys. A* **282**, 121 (1977).
22. Filippone, B. W., A. J. Elwyn, C. N. Davids, and D. D. Koetke, *Phys. Rev. Lett.* **50**, 412 (1983) and *Phys. Rev. C* **28**, 2222 (1983).
23. Barker, F.C., and R. H. Spear, *Ap. J.* **307**, 847 (1986).
24. Angulo, C., et al., *submitted to Nucl. Phys. A*
25. Strieder, F., et al., *Z. Phys. A* **355**, 209 (1996).
26. Weissman, L., et al., *Nucl. Phys. A* **630**, 678 (1998).
27. Strieder, F., et al., short note, *Eur. Phys. J. A*, accepted.
28. Fowler, W. A., G. R. Caughlan, and B. A. Zimmerman, *Ann. Rev. Astron. Astrophys.* **13**, 69 (1975).
29. Hammache, F., et al., *Phys. Rev. Lett.* **80**, 928 (1998).
30. TRIM, Ziegler, J. F., http://www.research.ibm.com/ionbeams/
31. Barker, F. C., *Nucl. Phys. A* **588**, 693 (1995).
32. Descouvemont, P., and D. Baye, *Nucl. Phys. A* **567**, 341 (1994).
33. Csótó, A., *Phys. Lett.* **394**, 247 (1997).
34. Nunes, F. M., R. Crespo, I. J. Thompson, nucl-th/9709070
35. Jennings, B. K., S. Karataglidis, and T. D. Shoppa, *Phys. Rev. C* **58**, 579 (1998)
36. Bogaert, G., et al., *Nucl. Instr. and Meth. B* **89**, 8 (1994)
37. Shapira, J. P., et al., *Nucl. Instr. and Meth.* **224**, 337 (1984).
38. *UW Nuclear Physics Lab Annual Report, University of Washington, Nuclear Physics Laboratory*, http://mist.npl.washington.edu/
39. Hass, M., *private commnication*
40. Strieder, F., *private communication.*
41. Campajola, L. , et al., *Z. Phys. A* **356**, 107 (1996).

Measurement of the ^{7}Be(p,γ)^{8}B cross-section with an implanted ^{7}Be target

M. Hass[1], C. Broude[2], V. Fedoseev[2], G. Goldring[1],
G. Huber[3], J. Lettry[4], V. Mishin[2], H.J. Ravn[4],
V. Sebastian[3] and L. Weissman[1]

[1] *Weizmann Institute of Science, Rehovot,* [2] *Institute of Spectroscopy, Troitzk,*
[3] *Johannes Gutenberg University, Institute of Physics, Mainz,*
[4] *ISOLDE, CERN, CH-1211 Geneva*
and the ISOLDE Collaboration

This is a report on the first result of a new measurement of the ^{7}Be(p,γ)^{8}B reaction cross section using an implanted target and a uniformly scanned particle beam larger than the target size. This experimental procedure overcomes some of the recognized experimental uncertainties of previous measurements[1]. We have reported elsewhere[2] on an earlier measurement of the ^{7}Li(d,p)^{8}Li reaction cross section with the same technique and the same apparatus. That measurement served *inter alia* as a test for the more involved ^{7}Be(p,γ)^{8}B measurement and one lesson learned was the realization that with a heavy target backing a significant fraction of the recoils (^{8}Li or ^{8}B) may be backscattered out of the target and escape detection. The ^{7}Be nuceli for the present experiment were therefore implanted in Cu for which, according to the simulations of ref. 2, the backscattering loss is insignificant. An ion-implantation machine at the Technion, Haifa, with a stable ^{9}Be beam at a similar energy as below, was used to investigate the implanted-target properties after a prolonged bombardment with an intense proton beam. Subsequent SIMS measurements of the irradiated target demonstarted that Be in Cu is stable under these conditions.

The ^{7}Be target was prepared at ISOLDE/CERN by direct implantation of a radioactive ^{7}Be beam into a copper substrate. The ^{7}Be ions are produced via proton induced reactions in a target which is connected to a laser ion source using stepwise resonant laser ionization inside a high temperature cavity. For beryllium a two step excitation scheme with laser light at a wavelength of λ=235 nm and λ=297 nm has been developed[3]. Using the 2p^2 ^{1}S$_0$ autoionizing state an ionization efficiency for beryllium of 3.4(7) was achieved. The yield of ^{7}Be from a standard 50 g/cm^2 uraniumcarbide/graphite target was 1.4 10^{10} atoms/s.μA and allowed, during a short test run, to implant 6 10^{14} ^{7}Be atoms into a copper substrate on a area of 2 mm diameter. The target intensity was rather low but sufficient to perform a measurement of the ^{7}Be(p,γ)^{8}B cross section at a relatively high proton beam energy of E$_p$ = 1.2 MeV.

The general scheme of the experiment is shown in Fig.1a. A proton beam out of the Weizmann Institute 3 MV Van de Graaff accelerator is raster scanned over a rectangle of 7 mm X 6 mm. The purpose of the scan is to obtain a beam of uniform areal density. The proton beam is collimated by a 3 mm diameter hole and impinges on the ^{7}Be target of 2 mm diameter. Under these conditions the reaction yield (no. of ^{8}B's) is given by:

$$Y = \sigma \frac{dn_b}{dS} n_t$$

where $\frac{dn_b}{dS}$ is the areal beam (p) density.

The target spot is aligned with a set of variable collimators downstream of the target. The target is mounted on an arm which is periodically moved out of the beam and in front of a 40 micron surface barrier detector registering the delayed α's following the β decay of ^{8}B. The detector was surrounded by a shroud to prevent scattered beam particles from reaching the detector. the time sequence of the whole cycle is: a.- 1.5 s beam-on-target; b. 100 ms rotation; c. 1.5 s target in the counting position; d. 100 ms rotation. In the counting position a gate signal from the control unit opens the ADC for α counting and the gated scaler for Faraday-cup beam monitoring. This sequence results in an efficiency factor for the α count of η(cycle)=0.400(1) (Fig.1b). A liquid nitrogen cold cryofinger is placed close to the target area to protect the target surface from contamination. The vacuum in the chamber was 8·10^{-7} torr.

The beam density $\frac{dn_b}{dS}$ was measured by collimating the beam by holes of known areas downstream from the target position, integrating the collimated beam in an electron-suppressed Faraday cup and counting the digitized counts in a gated scaler. The current digitizer and the scaler were checked during the experiment with a calibrated current source. The beam homogeneity was virtually insured by the nature of the raster operation: a low frequency triangular y scan and a high frequency triangular x scan, in small, digitally controlled steps in clock-fixed time intervals. The beam homogeneity was checked directly in two ways: 1. by measuring the areal density of x-rays from a tin foil induced by the scanned proton beam in a phosphor image plate[5], 2) by repeating the measurement with different downstream collimators of known apertures. The collimator hole areas where measured to an accuracy better than 1% by a microscope and by having an alpha source in front of the collimator-detector assembly. The number n_t of ^{7}Be's in the target was determined by registering γs from the β decay branch to the 478 keV state in ^{7}Li. The ^{7}Be mean life and the β branching ratio are known with high accuracy and the gamma activity was determined by comparison with calibrated ^{22}Na, ^{137}Cs and ^{133}Ba γ sources at a fixed distance from a Ge detector, shielded for

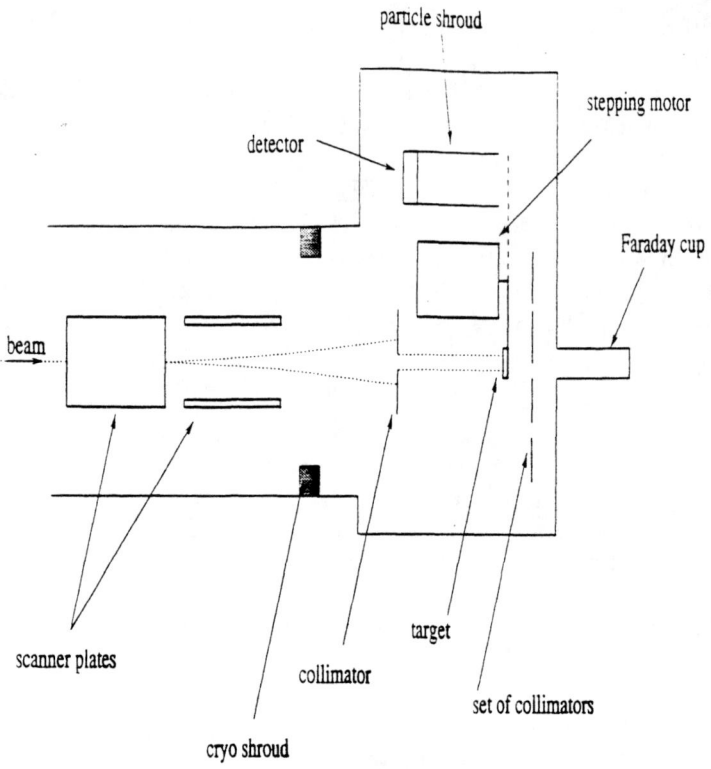

a)

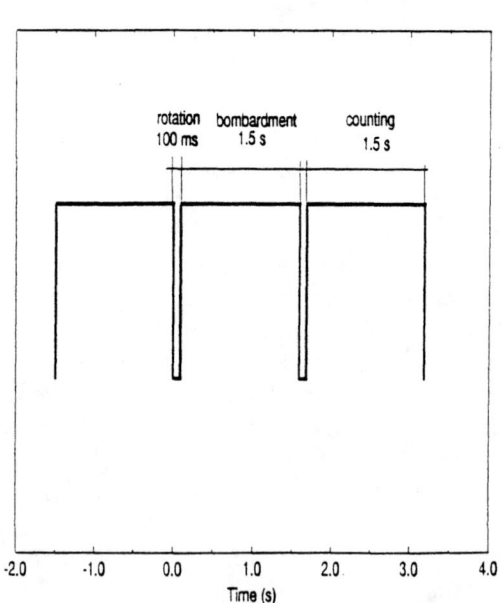

b)

low-background, at the γ-counting laboratory of the NRC-Soreq Nuclear Research Center.

More details about the various factors affecting the accuracy of the measurement can be found in ref. 2.

A weak target of implanted ^{7}Be with $n_t = $ 4-5 10^{14} was prepared in a recent test run at ISOLDE. A measurement of the ^{8}B production cross section was carried out at one energy of $E_p = 1.2$ MeV. The counting rate was rather low - about 10/h, and the α background, measured at intermittent intervals, was about 0.5/h. Because of the low counting rate the shape of the α spectrum could not be directly ascertained but the requisite information was obtained from the similar α spectrum from the ^{8}Li decay (of ref.2).

The results of the present measurement can be expressed as: $S_{17}(E_p=1.2$ MeV$)=22.5(2.5)$ eV. barn, in essential agreement with the compilation in [1]. Because of the low counting rate we consider this result as preliminary and expect to improve the measurement in the near future with more intense ^{7}Be targets which will be prepared at a forthcoming ISOLDE run.

We wish to thank Prof. R. Kalish for his help with the ion implantation and for helpful discussions.

REFERENCES

1. Adelberger et al., Rev. Mod. Phys. (in press); F. Hammache et al., Phys. Rev. Lett. **80**, 928 (1998).
2. L. Weissman et al., Nucl. Phys. **A630**, 678 (1998).
3. V. Sebastian et al., ENAM'98
4. J. Lettry et al., Rev. Sci. Instrum., 69, 761 (1998)
5. L. Weissman, M. Hass and V. Popov, Nucl. Inst. Meth. **A400**, 409 (1997).

Determination of $S_{17}(0)$ from Transfer Reactions

R.E. Tribble, A. Azhari, H.L. Clark, C.A. Gagliardi, Y.-W. Lui,
A.M. Mukhamedzhanov, A. Sattarov, L. Trache

Cyclotron Institute, Texas A&M University, College Station, Texas 77843

V. Burjan, J. Cejpek, V. Kroha, Š. Piskoř, J. Vincour

Institute for Nuclear Physics, Czech Academy of Sciences, Prague-Řež, Czech Republic

Abstract. The S-factor for the direct capture reaction $^7Be(p,\gamma)^8B$ can be found at astrophysical energies from the asymptotic normalization coefficients which provide the normalization of the tails of the overlap functions for $^8B \to {}^7Be+p$. Peripheral transfer reactions offer a technique to determine these asymptotic normalization coefficients. As a test of the technique, the $^{16}O(^3He,d)^{17}F$ reaction has been used to determine asymptotic normalization coefficients for transitions to the ground and first excited states of ^{17}F. The S-factors for $^{16}O(p,\gamma)^{17}F$ calculated from these $^{17}F \to {}^{16}O + p$ asymptotic normalization coefficients are found to be in very good agreement with recent measurements. Following the same technique, the $^{10}B(^7Be,{}^8B)^9Be$ reaction has been used to measure the asymptotic normalization coefficient for $^7Be(p,\gamma)^8B$. This result provides an indirect determination of $S_{17}(0)$.

INTRODUCTION

Nuclear capture reactions such as (p,γ) and (α,γ), play a major role in defining our universe. Until recently, the only reliable method to determine a reaction rate that is dominated by direct capture has been to measure it at laboratory energies with a low energy particle beam and extrapolate the result to astrophysical energies. Often the reaction of interest involves a radioactive target which makes measurements quite difficult. Hence it is important to develop alternative techniques to determine rates. Coulomb dissociation may provide an indirect method for obtaining this information, but it has not yet been subjected to a suitable reliability test.

Direct capture reactions of astrophysical interest usually involve systems where the binding energy of the captured proton is low. Hence at stellar energies, the capture proceeds through the tail of the nuclear overlap wave function. The shape of the wave function in this tail region is completely determined by the Coulomb interaction, so the amplitude of the wave function alone dictates the rate of the capture reaction. The $^7Be(p,\gamma)^8B$ reaction is an excellent example of such a direct capture process. Indeed recent calculations of the normalization constant have been used to predict the capture rate [1,2]. But new measurements are still needed as was underscored in a recent review of stellar reaction rates [3] which includes a detailed discussion of the uncertainties in our present knowledge of $S_{17}(0)$ and its importance to the solar neutrino problem.

The asymptotic normalization coefficient C for the system $A + p \leftrightarrow B$ specifies the amplitude of the single-proton tail of the wave function for nucleus B when the core A and the proton are separated by a distance large compared to the strong interaction radius. In previous communications [1,4], we have pointed out that astrophysical S-factors for peripheral direct radiative capture reactions can be determined through measurements of asymptotic normalization coefficients (ANC) using traditional nuclear reactions such as peripheral nucleon transfer. Direct capture S-factors derived with this technique are most reliable at the lowest incident energies in the capture reaction, precisely where capture cross sections are smallest and most difficult to measure directly. Of course it is extremely important to test the reliability of the technique in order to know the precision with which it can be applied. Determining the S-factors for $^{16}O(p,\gamma)^{17}F$ from its ANC's has been recognized as a suitable test for this method [3] because the results can be compared to existing direct measurements of the cross sections [5,6]. Furthermore, the $^{16}O(p,\gamma)^{17}F$ reaction has substantial similarities to the $^7Be(p,\gamma)^8B$ reaction. Below we briefly discuss the results of a measurement of the $^{16}O(^3He,d)^{17}F$ reaction to determine the ANC's for the ground and first excited states in ^{17}F. From these ANC's, we calculate S-factors for $^{16}O(p,\gamma)^{17}F$ and compare to experimental results. We then discuss our measurement of the $^{10}B(^7Be,^8B)^9Be$ reaction, the extraction of the ANC's for $^8B \to {}^7Be + p$ and our determination of $S_{17}(0)$.

A TEST CASE

The $^{16}O(^3He,d)^{17}F$ reaction was measured previously at a beam energy $E_{^3He}$ = 25 MeV [7]. We repeated the measurement at 29.75 MeV in order to obtain better angular coverage and to have a measurement at a second energy, both of which were necessary for extracting reliable ANC's. Data at laboratory scattering angles between 6.5° and 25° were obtained using Si solid state detectors and a ^3He beam from the U-120M isochronous cyclotron of the Nuclear Physics Institute of the Czech Academy of Sciences, while data at laboratory angles between 1° and 11° were obtained using the MDM magnetic spectrometer and a molecular $(^3He - d)^+$ beam from the Texas A&M University K500 superconducting cyclotron. More details of the experiments can be found in [8].

For a peripheral transfer reaction, ANC's are extracted from the angular distributions by comparison to a DWBA calculation. Consider the proton transfer

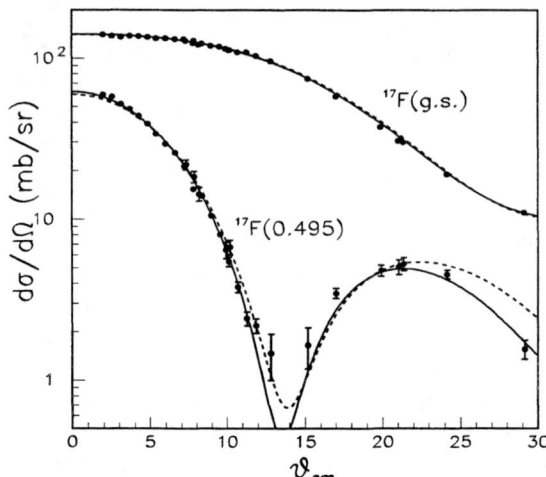

FIGURE 1. Angular distributions for the ground and first excited states of ^{17}F from the $^{16}O(^3He,d)^{17}F$ reaction. The curves are fits from DWBA calculations using two different optical potential sets.

reaction $a + A \to c + B$, where $a = c + p$ and $B = A + p$. The experimental cross section is related to the DWBA according to

$$\frac{d\sigma}{d\Omega} = \sum (C^B_{Ap})^2 (C^a_{cp})^2 R, \qquad (1)$$

where

$$R = \frac{\tilde{\sigma}^{DW}}{b^2_{Ap} b^2_{cp}}. \qquad (2)$$

$\tilde{\sigma}$ is the calculated DWBA cross section and the b's are the asymptotic normalization constants for the single particle orbitals used in the DWBA. The sum in Eq. (1) is taken over the allowed angular momentum couplings, and the C's are the ANC's for $B \to A + p$ and $a \to c + p$. The normalization of the DWBA cross section by the ANC's for the single particle orbitals makes the extraction of the ANC for $B \to A + p$ essentially independent of the parameters used in the single particle potential wells. See [9] for additional details.

The DWBA calculations were carried out with the finite range code PTOLEMY [10], using the full transition operator. The single particle orbitals were calculated in Woods-Saxon potentials with r_0 in the range $1.15 - 1.35$ fm and a_0 in the range $0.55 - 0.75$ fm, and indicated that the calculated DWBA cross sections are insensitive to assumptions about the ^{17}F wave functions in the nuclear interior. A range of optical model parameter sets were studied for both the entrance and exit channels, as detailed in [8], and the resulting fits to the ground and excited state angular distributions are shown in Fig. 1. Normalizing the DWBA calculations to the data and dividing by the ANC's for the single particle orbitals yields the product of the ANC's for the $^{17}F \rightarrow {}^{16}O + p$ and $^3He \rightarrow d + p$ systems. Dividing this product by the known ANC for $^3He \rightarrow d + p$ [11,12] provides C^2 for $^{17}F \rightarrow {}^{16}O + p$. Our final adopted ANC's are $C^2 = 1.08 \pm 0.10$ fm^{-1} for the ground state and $C^2 = 6490 \pm 680$ fm^{-1} for the first excited state.

The relation of the ANC's to the direct capture rate at low energies is straightforward. The cross section for the direct capture reaction $A + p \rightarrow B + \gamma$ can be written as

$$\sigma = \lambda |< I_{Ap}^B(\mathbf{r}) | \hat{O}(\mathbf{r}) | \psi_i^{(+)}(\mathbf{r}) >|^2, \qquad (3)$$

where λ contains kinematical factors, I_{Ap}^B is the overlap function for $B \rightarrow A + p$, $\hat{O}$ is the electromagnetic transition operator, and $\psi_i^{(+)}$ is the incident scattering wave. If the dominant contribution to the matrix element comes from outside the strong interaction radius, the overlap function may be replaced by

$$I_{Ap}^B(r) \approx C \frac{W_{-\eta, l+1/2}(2\kappa r)}{r}, \qquad (4)$$

where C defines the amplitude of the tail of the radial overlap function I_{Ap}^B, W is the Whittaker function, η is the Coulomb parameter for the bound state $B = A + p$, and κ is the bound state wave number. For $^{16}O(p,\gamma)^{17}F$, the required C's are just the ANC's found above from the transfer reaction. Thus, the direct capture cross sections are directly proportional to the squares of these ANC's.

Following the prescription outlined above, the S-factors describing the capture to both the ground and first excited states for $^{16}O(p,\gamma)^{17}F$ were calculated, with no additional normalization constants, using the standard definition of S [13]. The results are shown in Fig. 2 compared to the two previous measurements of $^{16}O(p,\gamma)^{17}F$ [5,6]. Both E1 and E2 contributions have been included in the calculations, but the E1 components dominate the results. The theoretical uncertainty in the S-factors is less than 2% for energies below 1 MeV. The agreement between the measured S-factors and those calculated from our $^{17}F \rightarrow {}^{16}O + p$ ANC's is quite good, especially for energies below 1 MeV where the approximation of ignoring contributions from the nuclear interior should be very reliable.

ANC FOR $^8B \rightarrow {}^7Be + p$

We have measured the $(^7Be, {}^8B)$ reaction on a 1.7 mg/cm^2 ^{10}B target in order to extract the ANC for $^8B \rightarrow {}^7Be + p$. The radioactive ^7Be beam was produced at 12 MeV/A by filtering reaction products from the $H(^7Li, {}^7Be)n$ reaction in the recoil spectrometer MARS using a primary ^7Li beam at 18.6 MeV/A from the TAMU K500 cyclotron incident on an H$_2$ cryogenic gas target cooled by LN$_2$. Reaction

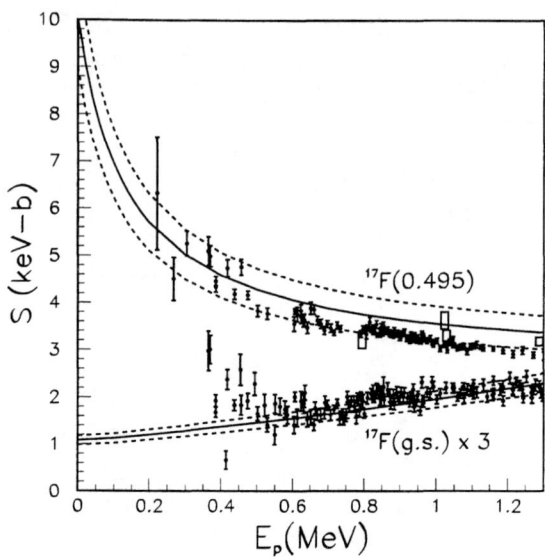

FIGURE 2. A comparison of the experimental S-factors to those determined from the ANC's found in $^{16}O(^3He,d)^{17}F$. The solid data points are from [5], and the open boxes are from [6]. The solid lines indicate our calculated S-factors, and the dashed lines indicate the $\pm 1\sigma$ error bands. Note that the experimental ground state S-factor may be contaminated by background [14] at energies below 500 keV.

products were measured by 5 cm × 5 cm Si detector telescopes consisting of a 100 μm ΔE strip detector, with 16 position sensitive strips, followed by a 1000 μm E counter. Two different detector configurations were used to take data. In one mode, two detector telescopes were mounted symmetrically about the beam axis. The angular coverage in this mode was from $\approx 3.2°$ to $\approx 17.8°$ in the laboratory frame for each telescope. In the second mode, a single detector telescope stack was placed at 0° and the beam was stopped just in front of the ΔE detector. The effective angular coverage in the lab for this mode was from $\approx 3.2°$ to $\approx 12.5°$.

A single 1000 μm Si strip detector was used for initial beam tuning. This detector, which was inserted at the ^{10}B target location, allowed us to optimize the beam shape and to normalize the ^7Be flux relative to a Faraday cup that measured the intensity of the primary ^7Li beam. Following optimization, the approximate ^7Be beam size was 6 mm × 3 mm (FWHM), the energy spread was ≈ 1.3 MeV, the angular spread was $\Delta\theta \approx 14$ mrad and $\Delta\phi \approx 24$ mrad, and the purity was $\geq 99.5\%$ ^7Be. Periodically during the data acquisition, the beam detector was inserted to check the stability of the secondary beam tune. The system was found to be quite stable over the course of the experiment with maximum changes in intensity observed to be less than 5%. The typical rate for ^7Be was ≈ 1.5 kHz/pnA of

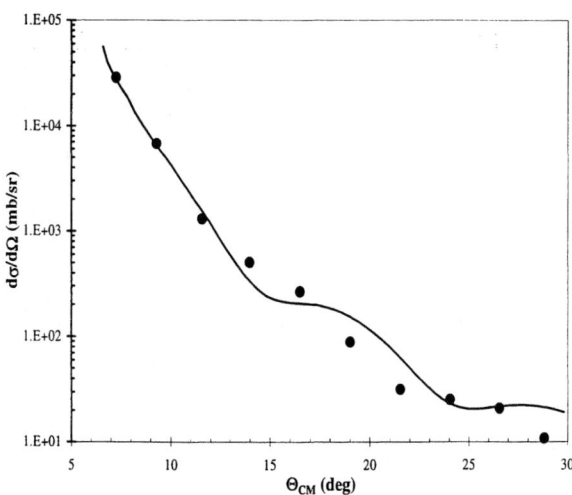

FIGURE 3. Angular distributions for elastic scattering. The curve is from optical model calculations smoothed over the angular acceptance of each bin.

primary beam on the production target. Primary beam intensities of up to 60 pnA were obtained on the gas cell target during the experiment.

Figure 3 shows the preliminary results for the elastic scattering angular distribution which includes contributions from three target components, B (86%), C(10%) and O(4%). A Monte Carlo simulation described below was used to generate the solid angle for each angular bin. The absolute cross section is then fixed by the target thickness, number of incident ^7Be, the yield in each bin, and the solid angle. The curve with the elastic scattering was found by adding together the cross sections from the three target components in the laboratory frame and then transforming the result to the center of mass assuming kinematics appropriate for the ^{10}B target.

The ^8B Q-value spectrum, shown in Fig. 4, was obtained by assuming a $^{10}B(^7Be,^8B)^9Be$ reaction and correcting the ^8B reaction products for kinematic energy shifts as a function of scattering angle. The two major contributions to the energy resolution seen in Fig. 4 are the beam energy spread and the nonuniformity of the target. Since the ground state of ^9Be is not cleanly separated from excited states, a Monte Carlo simulation of the experiment has been used to fix the lineshape and determine cross sections. The simulation, which was fine tuned to reproduce the resolution observed in elastic scattering, includes the geometry of the experimental setup, reaction kinematics, nonuniform energy loss in the target and the size, angular spread and energy spread of the beam. The three peaks shown in Fig. 4 correspond to the excitation of the ground and second excited states of ^9Be and the ground state of ^{15}N from the ^{16}O contamination in the target. They were obtained by including the predicted angular distributions for the states in the

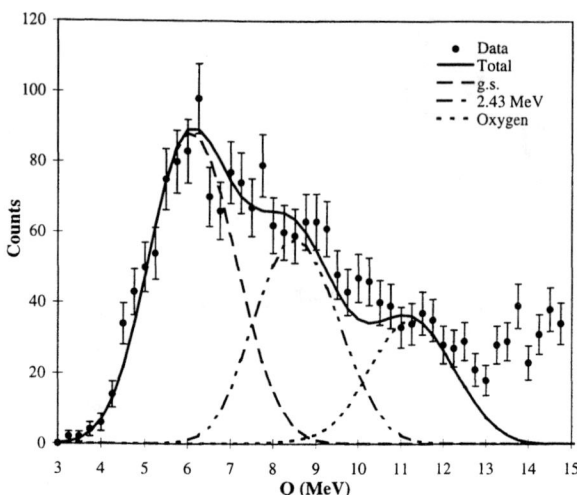

FIGURE 4. Q-value spectrum for ^8B reaction products. The three peaks correspond to the excitation of the ground state and second excited states of ^9Be and the ground state in ^{15}N from the ^{16}O contamination in the target.

Monte Carlo simulation and then extracting the associated Q-value spectrum. The normalization of the three peaks was done by a χ^2 minimization to the data.

The ANC for $^8B \rightarrow \, ^7Be + p$ was extracted based on the fit to the Q-value spectrum and the ANC for $^{10}B \rightarrow \, ^9Be + p$ [9] following the procedure outlined above in our test case. Two components, $1p_{1/2}$ and $1p_{3/2}$, contribute to the transfer reaction but the $1p_{3/2}$ dominates. In calculating the angular distribution, we used the ratio for the two orbitals as given by a microscopic description of the B ground state [1,4]. No optical model parameters are available for either the entrance or exit channel in this reaction. Consequently we have used parameters obtained from renormalized microscopic folding potentials using the M3Y effective NN interaction. The parameters used in the entrance (exit) channel were $V = 99(101.8)$ MeV, $W = 48.8(50.2)$ MeV, $R_V = R_W = 2.99(2.97)$ fm, $a_V = a_W = 0.83(0.85)$ fm. The entrance channel parameters were the same as those used in calculating the elastic scattering angular distribution for ^7Be on ^{10}B in Fig. 3. As part of this program, we have obtained elastic scattering data on several beam-target combinations in this mass and energy range and are working on developing appropriate optical model parameters that can be extended to ^7Be and ^8B projectiles [15]. We have checked the sensitivity of the calculations by varying optical model and well depth parameters. As in previous studies, the results are insensitive to single particle well parameters.

The S-factor for $^7Be(p, \gamma)^8B$ has been determined from the ANC which includes a 10% uncertainty for optical model parameters, a 12% uncertainty for experimental fits and normalization of the absolute cross section and the uncertainty in the

ANC for $^{10}B \to ^9Be + p$. The relative contribution of the two angular momentum couplings to the S-factor is straightforward to calculate and introduces a negligible additional uncertainty in our result [1,4]. The preliminary value that we find is $S_{17}(0) = 17.4 \pm 2.9$ eV b which is in good agreement with the recommended value of 19^{+4}_{-2} eV b [3].

One of the sources of uncertainty in the value quoted above for $S_{17}(0)$ is the optical model parameters that are used to predict the angular distribution. As indicated, we are working toward a set of global optical model parameters to be used for radioactive beams in this mass and energy region. Once this is complete, the calculation will be redone and a new ANC and hence $S_{17}(0)$ will be determined. In addition to the data on the ^{10}B target, we have recently measured the $^{14}N(^7Be,^8B)^{13}C$ reaction using the same ^7Be beam on a melamine target. Primary beam intensities of over 100 pnA were obtained during this recent experiment yielding more than 100 kHz of secondary ^7Be beam. Unlike the ^{10}B target, the melamine target was quite uniform. Thus the energy resolution in the ^8B spectrum is dominated by the beam energy spread and the differential energy loss in the target between the ^7Be beam and the outgoing ^8B. The transition to the ground state of ^{13}C is cleanly separated from transitions due to excited states or contaminants thus making the extraction of the angular distribution straightforward. We expect that the ANC derived from this more recent measurement will have a somewhat smaller uncertainty than the one that we have obtained to date from the analysis of the $^{10}B(^7Be,^8B)^9Be$ reaction.

This work was supported in part by the U.S. Department of Energy under Grant number DE-FG05-93ER40773 and by the Robert A. Welch Foundation.

REFERENCES

1. H.M. Xu et al., Phys. Rev. Lett. **73**, 2027 (1994).
2. L.V. Grigorenko et al., Phys. Rev. C **57**, R2099 (1998).
3. E.G. Adelberger et al., Rev. Mod. Phys. (in press).
4. A.M. Mukhamedzhanov and N.K. Timofeyuk, JETP Lett. **51**, 282 (1990).
5. R. Morlock et al., Phys. Rev. Lett. **79**, 3837 (1997).
6. H.C. Chow, G.M. Griffith and T.H. Hall, Can. J. Phys. **53**, 1672 (1975).
7. J. Vernotte et al., Nucl. Phys. **A571**, 1 (1994).
8. C.A. Gagliardi et al., Phys. Rev. Lett. (submitted).
9. A.M. Mukhamedzhanov et al., Phys. Rev. C **56**, 1302 (1997).
10. M. Rhoades-Brown, M. McFarlane and S. Pieper, Phys. Rev. C **21**, 2417 (1980); Phys. Rev. C **21**, 2436 (1980).
11. M. Kamimura and H. Kameyama, Nucl. Phys. **A508**, 17c (1990).
12. A.M. Mukhamedzhanov, R.E. Tribble and N.K. Timofeyuk, Phys. Rev. C **51**, 3472 (1995).
13. J.N. Bahcall, *Neutrino Astrophysics* (Cambridge University Press, Cambridge, 1989).
14. R. Morlock, private communication.
15. L. Trache et al., to be published.

Determining the Astrophysical S_{17} with transfer reactions

J.C. Fernandes*, R. Crespo*, F.M. Nunes* and I.J. Thompson[†]

* *Departamento de Física, Instituto Superior Técnico, 1696 Lisboa-codex, Portugal*
† *Physics Department University of Surrey, Guildford, Surrey GU2 5XH, U.K.*

Abstract.

The method of extracting the low energy proton capture cross section of ^{7}Be through transfer reactions is analysed. It is applied to the ^{7}Be $(d,n)^8$B reaction for which there is available data. We perform a systematic study of the uncertainty introduced by having to select from a set of optical potentials fitted to elastic scattering on nearby targets and at nearby energies.

Multi-step processes involving excited states of ^{7}Be are discussed within the DWBA framework. Our calculations show that these are not important.

INTRODUCTION

An accurate value for the ^{7}Be$(p,\gamma)^8$B capture cross section at the astrophysical energies is fundamental for the progress in resolving the so called *solar neutrino problem* [1,2]. Given the disagreement between the different sets of direct measurements [3,4], alternative methods have been suggested in order to settle the matter [5,6]. The transfer reaction ^{7}Be$(d,n)^8$B was measured with the purpose of extracting $S_{17}(0)$ [6], and we will base our studies on this example. Peripheral transfer reactions allow us to probe the tail of the ^{8}B wavefunction, determining the asymptotic normalisation constant (ANC) from which a $S_{17}(0)$ can be directly obtained [7]. Even if the experimental aspects are truly optimal, the analysis of transfer data has to take into account the entrance and exit nuclear interactions as well as the possibility of multistep processes. In this paper we estimate the magnitude of the uncertainties due to the optical potentials and evaluate the effect of multistep processes.

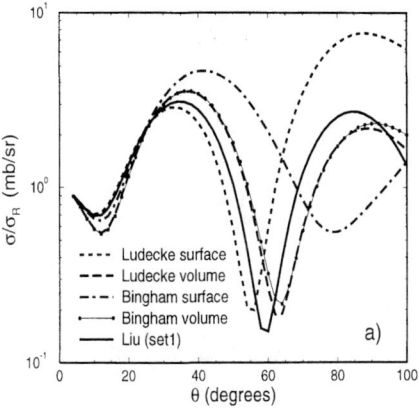
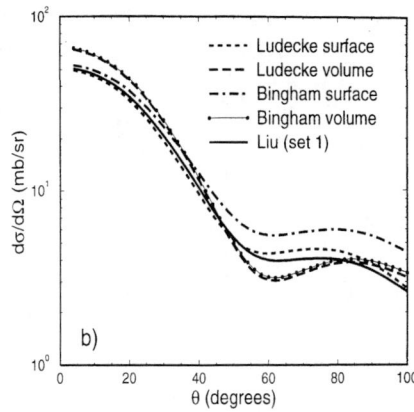

FIGURE 1. a) the d-^{7}Be elastic scattering at 7.5 MeV for a set of optical potentials taken from fits to 12 MeV data; b) the corresponding transfer reaction cross sections.

SENSITIVITY TO THE OPTICAL POTENTIALS

Our calculations are performed within finite range DWBA[1], for which entrance and exit channel optical potentials are required. Presently there are no elastic data available for the necessary targets, at the corresponding energies. We have done a systematic study of the predicted transfer cross section for a wide range of optical potentials in order to quantify the uncertanties introduced from ignoring the optical input .

The DWBA calculated differential cross section for the transfer reaction is very sensitive to the elastic optical potential inputs. The lack of measured observables to constrain these optical potentials induce uncertainties in the S_{17} extracted from the transfer studies, uncertainties which arise because one is using elastic scattering data either at a shifted energy or on a different target (or both).

We first evaluate the DWBA differential cross section [8] for a set of entrance channel optical parameters from [11,12](volume and surface), that fit the elastic scattering data on ^{7}Li at higher energies ($E_{lab} \simeq 12$ MeV) including the optical parameters used in [6]. We find that partial waves up to $l_{max} = 20$ and $R_{max} = 55$ for the T-matrix integrals is sufficient to obtain converged results. In fig.(1a) we show the d–^{7}Be elastic scattering at the correct energy ($E_{lab} = 7.5$ MeV) for this set of potentials. Even at small angles, there are differences in the results. Consequently the transfer cross section varies considerably (see fig.1b) producing up to 16% uncertainty in the extracted $S_{17}(0)$ (calculated relative to the mean

---

[1] It is important to note that in the zero range approximation one should not use D_0 as done in [6]. We have reproduced the zero range calculation in [6] and find a reduction of 17% on the $S_{17}(0)$ if a finite range DWBA calculation is performed.

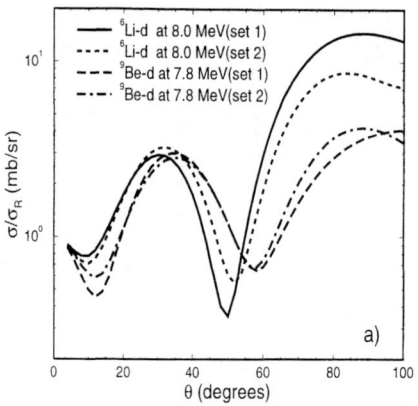 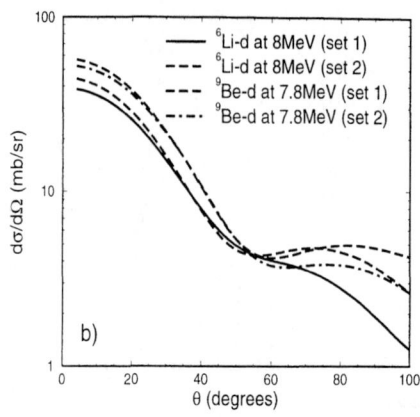

FIGURE 2. a) the d-^{7}Be elastic scattering at 7.5 MeV for a set of optical potentials taken at the correct energy but on different targets; b) the corresponding transfer reaction cross sections.

value). For these calculations we have used a n–^{9}Be potential from [16] for the exit channel and Kim's single potential model for ^{8}B [9].

Apart from the error introduced by extracting the optical potential from elastic scattering at the wrong energy, there is also an error due to the changes in the mass or charge of the target. We have analysed this effect by comparing the results obtained when using deuteron optical parameters from elastic scattering on ^{6}Li and ^{9}Be at the correct energy [12,13]. We still use the neutron potential of [16], and the ^{8}B structure of [9]. The calculated elastic scattering and transfer differential cross sections are shown in fig.(2). We find that the uncertainty on the $S_{17}(0)$ relative to its mean value is 17%.

Next, we concentrate on the effect of the exit channel n–^{8}B optical potential. Here it was not possible to distinguish energy effects and mass/charge effects, due to the reduced number of optical potentials available at the correct energy. We have studied their combined effect by testing a set of optical potentials available from elastic scattering at nearby energies and on nearby targets [6,15–18] and find an uncertainty of 26% for a deuteron potential from [12] and 27% for a deuteron potential from [13].

If we indiscriminately take the whole range of optical potentials for entrance and exit channels here mentioned, we obtain $S_{17} = 23.5^{+12.3}_{-7.9}$ eVb. These results are in very good agreement with the recent work by the Texas A&M group [19]. However one can minimise this uncertainty by using only potential parameters taken from elastic scattering at the correct energy. If we reduce our set to d-^{6}Li or d-^{9}Be optical potentials from [12,13] at the correct energy for the entrance channel and n-^{9}Be optical potentials from [16] at the correct energy for the exit channel, we obtain $S_{17} = 20.8^{+3.5}_{-3.3}$ eVb which is in disagreement with the result $S_{17} = 27.4 \pm 4.4$ eVb presented in [6].

CORE EXCITATION AND MULTISTEP EFFECTS

If a serious estimate of S_{17} is to be made from transfer reaction measurements one should make sure that the reaction mechanism is well understood. In the previous section we have only considered the first term of the DWBA expansion [20]. There are low energy excited states in ^{7}Be that can be excited by the deuteron, giving rise to multistep processes.

Let us then assume that the core (^{7}Be) is deformed. We consider quadrupole rotational nuclear couplings between its 3 bound states. In doing so, the transfer process will depend on couplings to and from the g.s. of ^{7}Be and the excited states, in addition to the transfer couplings themselves. A diagram of 1-step, 2-step and

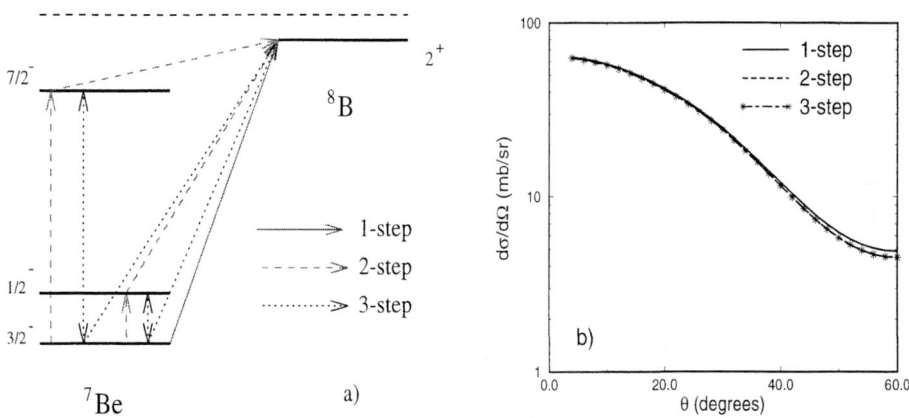

FIGURE 3. a) the coupling diagram for multistep processes; b) the transfer cross section for 1-step, 2-step and 3-step.

3-step paths is shown in fig.(3a). It is now necessary to use a ^{8}B bound state with known fractional parentage to all the states of ^{7}Be. We therefore use one of the coupled channel wavefunctions from the ^{8}B models developed in [21] where a proton is coupled to a deformed and excitable ^{7}Be.

We first verify that the inclusion of deformation and/or core excitation in the entrance channel optical potential (without transfer back coupling) does not alter the elastic scattering. The inclusion of both excited states of the core reduces the transfer cross section by 10% due to the additional admixture components in the ^{8}B g.s. After obtaining the experimental spectroscopic factor from the fit to the transfer data [6], one can rescale the S-factors obtained in the direct capture calculations presented in [21]. Further work on the application of deformed core models to transfer reactions for extracting the ANC will be discussed elsewhere.

In fig.(3b) we show the transfer cross section for 1-step, 2-step and 3-step calculations (here we have used model EXC1 from [21]). From these results it is clear that 2-step effects are negligible and only become visible for larger angles ($\simeq 60°$).

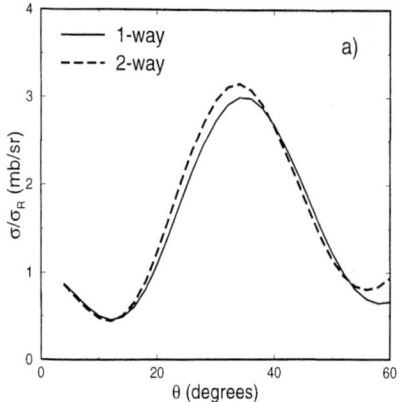 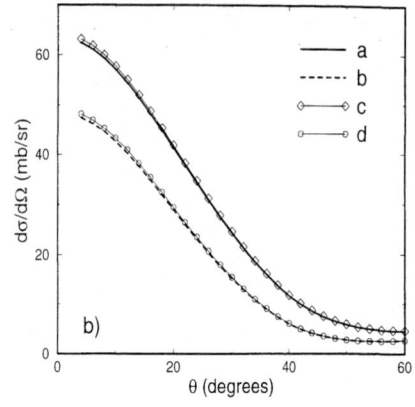

FIGURE 4. a) Comparing the d-^{7}Be elastic scattering when two-way transfer couplings are included and when only forward transfer is allowed; b) the transfer cross sections for different type of couplings.

As expected, 3-step effects are even smaller. This same result was confirmed for the other models from [21].

It is known that when introducing back couplings to the original DWBA problem, the optical potential fit to the scattering data may well be lost [22]. Our calculations above have included all couplings between the 3 states of ^{7}Be and transfer couplings between the 3 states of ^{7}Be and the g.s. of ^{8}B. We found that the elastic scattering cross sections obtained with and without down couplings between the ^{7}Be states is almost the same. However, when we include transfer back couplings, which in nature are present, we find that the various scattering components are changed considerably. As an example we show in fig.(4a) the effect of back transfer couplings on the elastic scattering channel.

In fig.(4b) we show the transfer cross section for 4 cases: a) all couplings between the core's states are included but no back transfer couplings; b) all couplings including back transfer couplings; c) no back transfer couplings and only up couplings for the core; and d) only up couplings for the core but two way transfer couplings. The couplings from the excited states to the g.s. of ^{7}Be reduces the transfer cross section by less than 2% whereas the back transfer couplings reduce it by 25%.

This large difference can be discounted, however, if we realise that when the couplings affect the elastic scattering, the given entrance-channel optical potential is no longer appropriate. When back couplings are present, one should use a bare optical potential that would not fit *by itself* the elastic scattering data. Alternatively we may assume that the back transfer couplings are effectively included in the optical potential already, and, to avoid double counting, should therefore be removed from the transfer calculations.

CONCLUSIONS

From the low energy transfer data of [6] we obtain a range for the S-factor based on using only potential parameters taken from elastic scattering at the correct energy: $S_{17}(0) = 20.8^{+3.5}_{-3.3}$ eVb. The uncertainty could be further reduced if elastic scattering data were available for the exact entrance and exit channel. We find multistep effects to be negligible as long as no back transfer couplings are included, which is the correct procedure if we have a entrance channel optical potential adjusted to the elastic scattering data.

Portuguese support from JNICT PRAXIS/PCEX/P/FIS/4/96 and UK support from the EP-SRC grant GR/J/95867 are acknowledged. One of the authors, F. Nunes, was supported by JNICT BIC 1481.

REFERENCES

1. Eric Aldelberger et al., *Solar Fusion Cross Sections*, to appear in Phys. Rep. (1998)
2. R. Crespo, A.M. Mourão and F.M. Nunes, *Nuclear Physics limits to the ^8B solar neutrino flux and the Kamiokande experiment*, Proc. Workshop on New Worlds in Astroparticle Physics, Springer-Verlag
3. W. Kavanagh et al., Bull. Amer. Phys. Soc. **14** (1969) 1209
4. B.W. Filippone et al., Phys. Rev. Lett. **50** (1983) 412
5. T. Motobayashi et al., Phys. Rev. Lett. **73** (1994) 2680; T. Kikuchi et al., Phys. Letts. **B391** (1997) 261
6. W. Liu *et al*, Phys. Rev. Lett. **77** (1996) 611.
7. A.M. Mukhamedzhanov, R.E. Tribble, N.K. Timofeyuk, Phys. Rev. **C 51** (1995) 3472
8. I.J. Thompson, Comp. Phys. Rep., **7** (1988) 167.
9. K.H. Kim, M.H. Park and B.T. Kim, Phys. Rev. **C35** (1987) 363.
10. S. Matsuki et al., J. Phys. Soc. Japan **26** (1969) 1344
11. H. Ludecke et al., Nucl. Phys. **A109** (1968) 676
12. H.G. Bingham et al., Nucl. Phys. **A173**(1971)265
13. G.R. Satchler, Private commmunication to P.E.Hodgson (1965)
14. M. El-Nadi et al., Ann. Phys. (Leipzig) **25** (1970) 1
15. H.F. Lutz and al., Nucl. Phys. **47**(1963) 521
16. J.H. Dave and C.R. Gould, Phys. Rev. **C28** (1983) 2212
17. M. Hyakutake et al., J. Nucl. Sci. Technol. **11** (1974) 407
18. A.J. Frasca et al., Phys. Rev. **144** (1966) 854
19. C.A. Gagliardi et al., Phys. Rev. Letts. **80** (1998) 421
20. T. Tamura and T. Udagawa, Phys. Rep. **65** (1980) 345
21. F.M. Nunes, R. Crespo and I.J. Thompson, Nucl. Phys. **A 615** (1997) 69; Nucl. Phys. **A 627** (1997) 747
22. G.R. Satchler, *Direct Nuclear Reactions*, Clarendon Press, Oxford 1983

High energy Coulomb breakup experiments for nuclear astrophysics

Tohru Motobayashi*

Department of Physics, Rikkyo University, Toshima, Tokyo 171, Japan

Abstract. The Coulomb breakup method has been applied to investigations of several radiative capture processes of astrophysical interest. The method has an advantage of high experimental efficiency when the incident energy is above several tens MeV/nucleon. This makes many experiments possible even with radioactive nuclear beams with rather weak intensities. Two examples of studies for resonance and continuum excitations are discussed: the reactions $^{13}N(p,\gamma)^{14}O$ and $^{7}Be(p,\gamma)^{8}B$. They are the key reactions in the hot CNO cycle in massive stars and the one related to the solar neutrino problem, respectively. Preliminary results of two new experiments on the $^{8}B(p,\gamma)^{9}C$ and $^{11}C(p,\gamma)^{12}N$ reactions are also reported.

INTRODUCTION

For thermonuclear burning processes in various astrophysical cites, nuclear reactions at low energies play important roles. Reactions on unstable nuclei can be faster than their competing β decays, when temperature and matter density are high. Such reactions can only be studied directly with low-energy radioactive (RI) beams, if the relevant nuclei are short-lived and cannot be served as targets. The first attempt has been made at Louvain la Neuve on the $^{13}N(p,\gamma)^{14}O$ reaction, a key in the hot CNO cycle hydrogen burning [1].

Another possibility is to use intermediate-energy RI beams, which are now available at many laboratories with variety of ions. Low-energy cross sections can be extracted indirectly from the data of Coulomb breakup (dissociation) at several tens MeV/nucleon. Targets commonly used in actual experiments are thicker by an order of 3 or 4 than those for low-energy direct measurements, where the thickness is limited by the energy loss of the incident beam. Due to the above advantage together with the enhancement of its cross sections, which will be discussed later, the Coulomb dissociation method may have easier access to the cases with lower cross sections than the $^{13}N+p \rightarrow {}^{14}O^{*}(1^{-}) \rightarrow {}^{14}O+\gamma$ process, which could be measured due to the relatively large cross section of the order of 100 μb.

COULOMB DISSOCIATION EXPERIMENTS

In order to study the capture reaction A(x,γ)B, the residual nucleus B bombards a high-Z target and is Coulomb excited to an unbound state that decays to the A+x channel. This Coulomb dissociation process is regarded as absorption of a virtual photon, *i.e.* B(γ,x)A, the inverse process of the radiative capture of interest [2]. Reviews on the method are given by Baur and Rebel [3]. As mentioned before, the Coulomb dissociation method enhances the original capture cross section by a large factor. This is due to intense virtual-photon flux and large gain due to the phase space factor of the photo absorption compared with the corresponding radiative capture. The two factors can be in the order of 100 or 1000 in some actual cases such as the ^{14}O and ^{8}B dissociation, which will be discussed later in this report.

Pioneering studies of the Coulomb dissociation were made for the stable Li isotopes, ^{6}Li$\rightarrow \alpha$+d and ^{7}Li$\rightarrow \alpha$+t, at around E_{in}=10 MeV/nucleon [4–7]. A measurement at a higher incident-energy of 26 MeV/nucleon was also performed for the ^{6}Li breakup [8].

The first Coulomb dissociation experiments with RI beams were made for the ^{208}Pb(^{14}O,^{13}N p)^{208}Pb reaction at yet higher incident energies of 87.5 MeV/nucleon [9] and 70 MeV/nucleon [10]. The results demonstrates the usefulness of the method, and triggered extension of the Coulomb dissociation method to ^{12}N$\rightarrow^{11}$C+p [11] and ^{8}B$\rightarrow^7$Be+p [12–16].

^{14}O BREAKUP AND HOT CNO CYCLE

When the ^{13}N(p,γ)^{14}O reaction becomes faster than the β^+ decay of ^{13}N, the regular CNO cycle is replaced by the hot CNO cycle. This is believed to occur in the core of super massive stars, novae or at the surface of neutron stars. However this key reaction ^{13}N(p,γ)^{14}O was difficult to be studied because the radioactive nucleus ^{13}N (10 min. half life) is involved. The experimental goal is to extract the E1 radiative-width Γ_γ of the first 1^- state in ^{14}O, because this resonant state dominates in the hot CNO burning. The experimental accuracy obtained in branching-ratio measurements was rather poor. We performed an experiment of Coulomb dissociation [9] hoping that a more accurate Γ_γ value might be obtained.

An ^{14}O beam was obtained by the projectile-fragmentation with the RIPS separator at RIKEN [17]. It bombarded a thick (350 mg/cm^2) ^{208}Pb target. Outgoing particles were detected in coincidence by a mosaic of 24 ΔE(Si)-E(CsI(Tl)) detector telescopes and two (x and y) planes of plastic scintillator hodoscopes.

The resultant radiative width Γ_γ is 3.1±0.6 eV, which agrees with the width obtained in the direct measurement of the ^{13}N(p,γ)^{14}O reaction [1] reported almost at the same time. Another Coulomb dissociation experiment at 70 MeV/nucleon at GANIL [10] gives a value consistent with these results within errors.

^{8}B BREAKUP AND SOLAR NEUTRINO PROBLEM

The ^{7}Be(p,γ)^{8}B reactions at low energies are responsible for producing high-energy neutrinos in the sun through the β^+ decay of ^{8}B. Many experiments has been and is being performed for the reaction with radioactive ^{7}Be targets. To study this important process by an independent method, the Coulomb dissociation of ^{8}B was measured at RIKEN. It should be noted that the astrophysical process is dominated by the E1 transition to the unbound continuum state of ^{8}B.

The first measurement is described in two articles [12,13], and a part of the second result is reported in Ref. [14]. A schematic view of the experimental setup is shown in Fig. 1. Beams of ^{8}B were produced by the ^{12}C+^{9}Be interaction at 92 MeV/nucleon. The ^{8}B energy in the center of the target, 50 mg/cm^2 ^{208}Pb, was approximately 50 MeV/nucleon. The outgoing particles were detected by a plastic hodoscope consisting of ΔE-E scintillator strips, which was set 3-5 m from the target. The time-of-flight (TOF) technique was employed to determine the energies of the fragments. The fragments' scattering angles are determined by the positions in the 10×16 segments formed by the ΔE and E scintillators. The p-^{7}Be relative energy spectrum could be constructed from the measured energies and angles, and was converted to the ^{7}Be(p,γ)^{8}B cross section with the help of a DWBA type calculations[1] together with a

---

[1] We employed the coupled channel code ECIS79. However, the higher order terms are very small, and hence the calculated results should be almost those of DWBA.

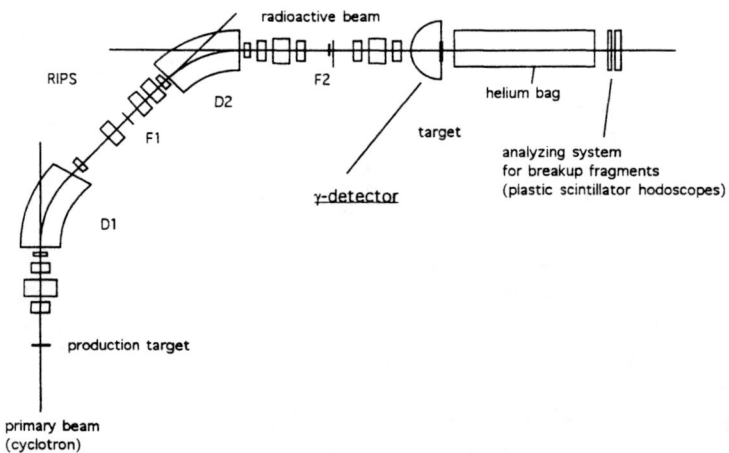

FIGURE 1. Experimental setup for the ^{8}B Coulomb dissociation experiment.

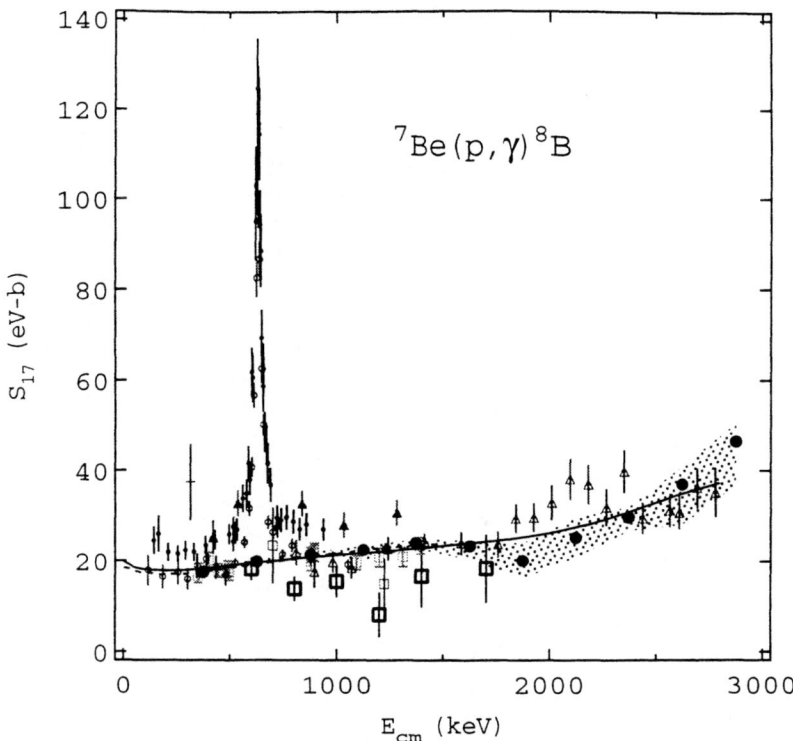

FIGURE 2. Astrophysical S_{17}-factors for the ^{7}Be(p,γ)^{8}B reaction extracted from the first (large squares) and second (large solid circles) experiments. Existing direct (p,γ) data are also shown.

Monte-Carlo calculation which simulates detection efficiency.

In the second experiment, the DALI setup [18] was used to measure the deexcitation γ rays from the first excited state of ^{7}Be at 429 keV ($1/2^-$) populated in the dissociation process. This might affect the core-excited component of the ground state of ^{8}B, i.e. ^{7}Be$^*(1/2^-)\otimes\pi(p_{3/2})$. The contribution from this process was measured to be about 5% of the Coulomb dissociation yield.

In Fig. 2 the astrophysical S-factors obtained in the experiments are shown together with the ones determined in direct (p,γ) measurements. Our Coulomb dissociation data are consistent within errors with the results by Filippone *et al.* [19] and Vaughn *et al.* [20]. The most recent measurement by Hammache *et al.* [21] reproduces the Filippone's data, and hence is consistent with the Coulomb dissociation results.

Two theoretical curves are normalized to the present data as shown in the figure. If the data points below 1.4 MeV are used for the fits, the S factor

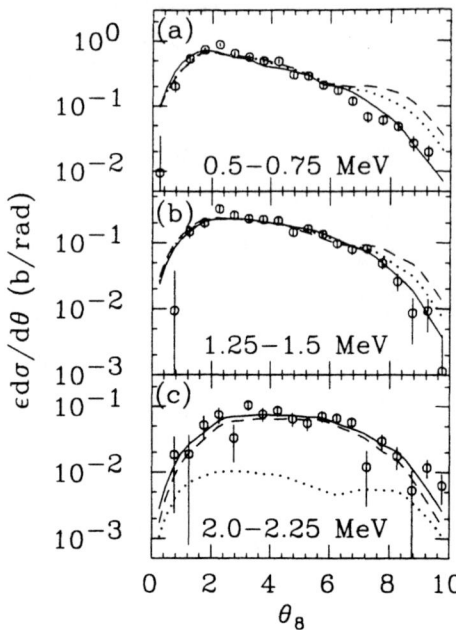

FIGURE 3. Angular distribution of the ^{8}B Coulomb dissociation reaction.

at zero-energy $S_{17}(0)$ is extrapolated to be 18.5±1.6 eV-b and 19.6±1.6 eV-b with the one-body potential model of Barker and Spear [22] and the p-^{3}He-α cluster model by Descouvemont and Baye [23], respectively. The values change slightly if the data for the fits are limited to a lower-energy region. By including these uncertainties in the extrapolation as well as the statistical and systematic errors in the measurement, our second ^{8}B dissociation experiment provides $S_{17}(0)$=18.9±1.8 eV-b. This is well within the errors of the most recent recommendation, 19^{+4}_{-2} eV-b [24].

Figure 3 shows the experimental angular distributions for the scattering angle θ_8, the angle of the center-of-mass for the p-^{7}Be system. They are fitted by calculated distributions. Two free parameters are introduced. They are the magnitudes of $\ell=1$ and $\ell=2$ components, where ℓ denotes the angular momentum transfer. Only the E1 transition is assumed for $\ell=1$ and nuclear contributions are neglected, because a microscopic calculation [25] predicts very small values. For $\ell=2$, both the E2 and nuclear $\ell=2$ transitions are taken into account. The transition form factor for nuclear excitation is constructed with the collective model assuming a deformation length common to the electromagnetic (E2) transition. As shown in the figures, the best fits are obtained with no or negligibly-small $\ell=2$ component. This supports the analysis on the first Coulomb dissociation data [12,13] assuming pure E1 transition.

In general, higher order processes including the post acceleration is less important at higher incident energy. We also performed a Coulomb dissociation experiment at GSI using ^{8}B beams of 254 MeV/nucleon [15]. Outgoing fragments were measured in coincidence by a magnetic spectrograph KaoS [26], which has a momentum acceptance large enough to allow for detecting protons and ^{7}Be that have quite different magnetic rigidities. Our preliminary analysis gives S_{17} factors quite similar to the Coulomb dissociation results obtained in the second RIKEN experiment, and therefore to the direct (p,γ) results by Filippone et al. [19], Vaughn et al. [20], and Hammache et al. [21]. The agreement between the two results obtained at the different energies (50 MeV/nucleon at RIKEN and 250 MeV/nucleon at GSI) might indicate smallness of the higher-order contribution.

All the quoted results support the analysis assuming a pure E1 transition for the ^{8}B dissociation data. Furthermore an inclusive ^{7}Be measurement for the ^{8}B dissociation at a low energy of 3.25 MeV/nucleon suggests an E2 mixture much smaller than any of the existing theoretical predictions. However, experimental results that may conflict to the above conclusion are reported. Kelley et al. [27] and Davids et al. [28] measured inclusive $p_{//}$ spectra of ^{7}Be produced by the ^{8}B+^{208}Pb interaction. Clear asymmetries were observed, and they are attributed to interference between E1 and E2 amplitudes. The amount of the asymmetries was roughly consistent with theoretical predictions. To study the effect more in detail, we are analyzing data of the angular-correlation, that is the fragment's angular distribution in the p-^{7}Be center-of-mass. It should be noted that the $p_{//}$ spectra corresponds to the energy-average of the angular correlation projected to the longitudinal axis. Detailed theoretical analyses including possible $\ell=2$ nuclear components are also desirable.

FURTHER APPLICATIONS OF COULOMB DISSOCIATION

Recently we are trying to extend the Coulomb dissociation method to several cases where the capture cross sections are not known. An example is the ^{8}B(p,γ)^{9}B reaction. In very high temperature and high pressure situation, this reaction becomes faster than the β^+ decay of ^{8}B, and the regular pp-chain nuclear burning is switched to the hot pp mode [29]. No experimental information is available so far for this reaction, and the cross sections are only estimated by theoretical calculations [30,31]. A Coulomb dissociation measurement of ^{9}C+^{208}Pb$\rightarrow$^{8}B+p+^{208}Pb has been performed with almost the same technique as for the ^{8}B dissociation experiments. Preliminary analyses give S_{18} factors which are consistent with those prediction by Descouvemont [31], but are smaller by a factor of three to four than the evaluation of Wiescher et al. [30].

The ^{12}N dissociation was also studied with a better $E_{\rm rel}$ resolution than the previous measurement at GANIL [11]. The extracted Γ_γ for the second excited state in ^{12}N is in between the two available theoretical predictions [30,32].

SUMMARY

The Coulomb dissociation method has been studied for several reactions of astrophysical interest. The ^{14}O and ^{8}B dissociation provide typical examples of resonant and non-resonant breakups. Their experimental results encourage new measurements of astrophysical reactions. Further experimental investigation to control the possible ambiguities as well as theoretical studies are desirable.

One of the question to be solved is for the amount of nuclear excitation especially for the angular momentum transfer $\ell=2$ as discussed before. Figure 4 shows a comparison between the Coulomb and nuclear cross sections for $\ell=2$. They are calculated quantum mechanically. Collective deformation model is used for constructing the transition form factors with a common deformation length. As shown in the figure, the nuclear contribution dominates for light nuclei, and Coulomb dominance starts at around $Z=10$. This is consistent with the considerations presented in the present symposium by Aumann for ^{8}He [33] and by Varner for ^{11}Be [34]. The Coulomb dominance for the high-Z region justifies the analysis for the excitation of ^{32}Mg [35], ^{56}Ni [36] and various neutron-rich nuclei around $A=40$ [37] where the pure E2 transitions are assumed. However in several theoretical works on the ^{8}B dissociation including the one presented by Thompson in this symposium [38], negligibly small nuclear contributions are predicted. Further investigation is necessary for these apparently contradicting conclusions.

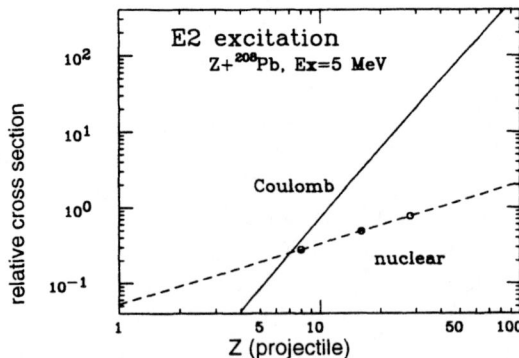

FIGURE 4. Coulomb and nuclear excitation cross sections as a function of the projectile charge Z.

REFERENCES

1. P. Decrock et al., *Phys. Rev. Lett.* **67**, 808 (1991).
2. G. Baur, C.A. Bertulani, and H. Rebel, *Nucl. Phys.* **A458**, 188 (1986).
3. G. Baur and H. Rebel, *J. Phys.* **G20**, 1 (1994); *Ann. Rev. Nucl. and Part. Sci.* **46**, 321 (1996).
4. A.C. Shotter et al., *J. Phys. G: Nucl. Phys.* **14**, L169 (1988).
5. H. Utsunomiya et al., *Nucl. Phys.* **A511**, 379 (1990).
6. J. Hesselbarth et al., *Z. Phys.* **A331**. 365 (1988).
7. S.B. Gazes et al., *Phys. Rev. Lett.* **68**, 150 (1992).
8. J. Kiener et al., *Phys. Rev.* **C44**, 2195 (1991).
9. T. Motobayashi et al., *Phys. Lett.* **B264**, 259 (1991).
10. J. Kiener et al., *Nucl. Phys.* **A552**, 66 (1993).
11. A. Lefebvre et al., *Nucl. Phys.* **A592**, 69 (1995).
12. T. Motobayashi et al., *Phys. Rev. Lett.* **73**, 2680 (1994).
13. N. Iwasa et al., *J. Phys. Soc. Jpn.* **65**, 1256 (1996).
14. T. Kikuchi et al., *Phys. Lett.* **B391**, 261 (1997).
15. N. Iwasa et al., *Proc. Tours Symp. Nucl. Phys. III, AIP Conference Proceedings* **425** (1998) p. 382.
16. J. Schwarzenberg et al., *Phys. Rev.* **C53**, R2598 (1996).
17. T. Kubo et al., *Nucl. Instr. Meth.* **B70**, 309 (1992).
18. T. Nishio et al., *RIKEN Accel. Prog. Rep.* **29**, 184 (1996).
19. B. Filippone et al., *Phys. Rev.* **C28**, 2222 (1983).
20. F.J. Vaughn et al., *Phys. Rev.* **C2**, 1657 (1970).
21. F. Hammache et al., *Phys. Rev. Lett.* **80**, 928 (1998).
22. F.C. Barker and R.H. Spear, *Astrophys. J.* **307**, 847 (1986).
23. P. Descouvemont and D. Baye, *Nucl. Phys.* **A567**, 341 (1994).
24. E.G. Adelberger et al., *Rev. Mod. Phys.* in print.
25. C.A. Bertulani, *Phys. Rev.* **C49**, 2688 (1994).
26. P. Senger et al., *Nucl. Instr. Meth.* **A327**, 393 (1993).
27. J.H. Kelley et al., *Phys. Rev. Lett.* **77**, 5020 (1996).
28. B. Davis et al., preprint; this symposium.
29. G.M. Fuller, S.E. Woosley and T.A. Weaver, *Astrophys. J.* **307**, 675 (1986); A. Jorrisen and M. Arnould, *Astron. Astrophys.* **221**, 161 (1989).
30. M. Wiescher et al., *Astrophys. J.* **343**, 352 (1989).
31. P. Descouvemont, *Astrophys. J.* **405**, 518 (1993).
32. P. Descouvemont and I. Baraffe, *Nucl. Phys.* **A514**, 66 (1990).
33. T. Aumann, this symposium.
34. R. Varner this symposium.
35. T. Motobayashi et al., *Phys. Lett.* **B346**, 9 (1995).
36. Y. Yanagisawa et al., this symposium.
37. H. Scheit et al., *Phys. Rev. Lett.* **77**, 3967 (1996); T. Glasmacher et al., *Phys. Lett.* **B395**, 163 (1997); R.W. Ibbotson et al., *Phys. Rev. Lett.* **80**, 2081 (1998).
38. I. Thompson, this symposium.

Measurement of E2 transitions in the Coulomb dissociation of ^{8}B

B. Davids[1,2], D.W. Anthony[1,3], Sam M. Austin[1,2], D. Bazin[1],
B. Blank[1,4], J.A. Caggiano[1,2], M. Chartier[1], H. Esbensen[5], P. Hui[1],
C.F. Powell[1,3], H. Scheit[1,2], B.M. Sherrill[1,2], M. Steiner[1], P. Thirolf[6],
J. Yurkon[1], A. Zeller[1]

[1] *National Superconducting Cyclotron Laboratory, Michigan State University, East Lansing, Michigan 48824*
[2] *Department of Physics and Astronomy, Michigan State University, East Lansing, Michigan 48824*
[3] *Department of Chemistry, Michigan State University, East Lansing, Michigan 48824*
[4] *Centre d'Etudes Nucléaires de Bordeaux-Gradignan, F-33175 Gradignan Cedex, France*
[5] *Physics Division, Argonne National Laboratory, Argonne, Illinois 60439*
[6] *Ludwig Maximilians Universität München, Am Coulombwall 1, D-85748 Garching, Germany*

Abstract. In an effort to understand the implications of Coulomb dissociation experiments for the determination of the ^{7}Be$(p,\gamma)^8$B reaction rate, longitudinal momentum distributions of ^{7}Be fragments produced in the Coulomb dissociation of 44 and 81 MeV/u ^{8}B beams on a Pb target were measured. These distributions are characterized by asymmetries interpreted as the result of interference between E1 and E2 transition amplitudes in the Coulomb breakup. At the lower beam energy, both the asymmetries and the measured cross sections are well reproduced by first order perturbation theory calculations, allowing a determination of the E2 strength. This measurement yields $S_{E2}/S_{E1} = 6.7 ^{+2.8}_{-1.9} \times 10^{-4}$ at the 0.63 MeV 1^+ resonance. If higher order dynamical effects are not negligible, then the E2 strength extracted here is a lower limit.

A recent study [1] of the Coulomb dissociation of ^{8}B has been performed in order to garner information on the rate of the inverse radiative capture reaction ^{7}Be$(p,\gamma)^8$B, which is critical for solar neutrino flux predictions. Although at solar energies the radiative capture reaction proceeds almost exclusively by E1 transitions, E2 photons can contribute significantly in Coulomb dissociation. The size of this E2 contribution must be known in order to infer the radiative capture rate from Coulomb dissociation data.

Measurements of the distribution of longitudinal momenta of ^{7}Be fragments resulting from the breakup of ^{8}B on heavy targets can be used to gauge the E2 strength in this reaction. First order perturbation theory calculations of the

Coulomb dissociation of ^{8}B predict that the distribution of the longitudinal momenta of the emitted ^{7}Be fragments will be asymmetric due to interference between E1 and E2 transition amplitudes [2]. The magnitude of this asymmetry depends on the beam energy, because the ratio of the number of virtual E1 photons to E2 photons increases with beam energy. Hence, an effective way to gauge the strength of E2 transitions in the Coulomb dissociation of ^{8}B is to measure the asymmetry of the longitudinal momentum distributions of the emitted fragments at different beam energies. A recent study of the longitudinal momentum distribution of ^{7}Be fragments from the Coulomb dissociation of 41 MeV/u ^{8}B on a gold target [3] found an asymmetry of roughly the predicted size, but poor statistics prevented a definitive conclusion.

We used the new S800 spectrometer at the National Superconducting Cyclotron Laboratory (NSCL) to carry out a much improved experiment. The large solid angle ($7° \times 10°$), high resolution, and large momentum acceptance (6%) of the S800 made possible a high precision, high statistics measurement. The beam energies were chosen to be 44 and 81 MeV/u. The lower energy is close to the energies of the earlier experiments of Refs. [1,3].

^{8}B nuclei were dissociated in a 28 mg cm^{-2} Pb target. The spectrometer was set at 0° to detect ^{7}Be fragments, and was operated in a dispersion matched mode, so that the momentum spread of the incident beam did not limit the momentum resolution of the spectrometer. The focal plane of the S800 was instrumented with two position sensitive cathode readout drift chambers (CRDCs) [4], a 16 segment ionization chamber, and 3 thick plastic stopping scintillators. Energy loss signals were provided by the ionization chamber, and the first scintillator was the source of total energy signals. Reaction products were unambiguously identified by comparing the energies and energy losses of the detected particles with those of a calibration beam of ^{7}Be having the same rigidity as the ^{8}B beam. The ion optics code COSY INFINITY [5] was used to calculate momenta and scattering angles for each event from the magnetic field settings and the two dimensional position signals provided by each CRDC.

First order perturbation theory calculations of the Coulomb dissociation of ^{8}B on Pb at the energies of this experiment based on the model of Ref. [2] have been performed. The model predicts that $S_{17}(0) = 18$ eV b, and that $S_{E2}/S_{E1} = 9.5 \times 10^{-4}$ at the 0.63 MeV 1^+ resonance. This E2 strength is smaller than that of the model of Kim et al. [6] by about a factor of 2. The total momentum distribution was calculated for several ^{7}Be scattering angle cuts. Projecting these events on the beam direction yielded the calculated longitudinal momentum distribution.

The measured longitudinal momentum distributions of ^{7}Be fragments produced in the Coulomb dissociation of ^{8}B on Pb at 44 MeV/u are shown in Fig. 1 (a) for three different angle cuts. Uncertainties in the target thickness and beam intensity resulted in systematic uncertainties in the measured cross sections of ± 9%. The predicted longitudinal momentum distributions, convoluted with the experimental resolution of 5 MeV/c, are superposed on the measured distributions. The description of the data with the original model was quite good, but not precise. Both the

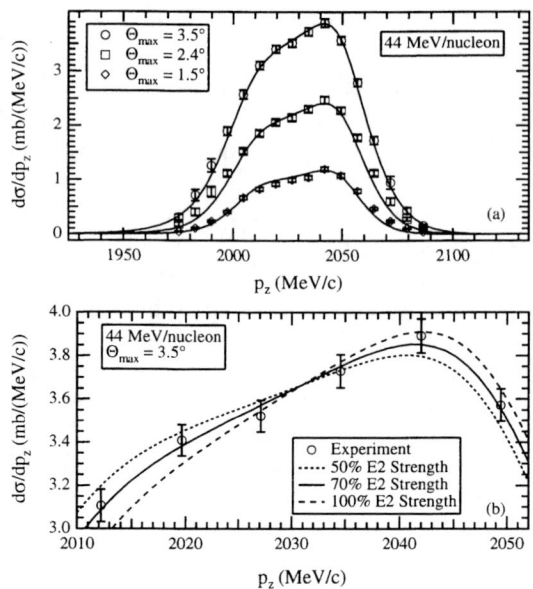

FIGURE 1. (a) Laboratory frame longitudinal momentum distributions of ^{7}Be fragments formed in the Coulomb dissociation of 44 MeV/u ^{8}B on Pb with maximum scattering angles of 1.5, 2.4, and 3.5°. The curves are the results of first order perturbation theory calculations convoluted with the experimental resolution of 5 MeV/c. The error bars indicate the relative uncertainties of the data points, which are dominated by statistical errors. (b) Central region of the 3.5° angle cut at the same beam energy. The curves are calculations performed with different E2 strengths, normalized to the center of the distribution.

value of $S_{17}(0)$ and the E2 strength are implicit in the structure model used, but are not robust predictions of such models. Since the predicted cross sections depend on $S_{17}(0)$ and the E2 strength, the normalization and E2 strength are effectively free parameters. We therefore adjusted the E2 strength and the overall normalization of the model to minimize χ^2 for the central 6 points of the 3.5° distribution. This yields a normalization factor of 1.22 and an E2 strength 0.7 times as large as the original value. Fig. 1 (b) shows the central region of the 44 MeV/u momentum distribution for the 3.5° angle cut. Also shown are three calculations with different E2 strengths, expressed as fractions of the original E2 strength of the model, normalized to the center of the distribution. The dependence of the calculated asymmetry on the E2 strength is apparent.

Good agreement between the observed and predicted shapes of the distributions implies that the shapes of the E1 and E2 responses predicted by the model of Ref. [2] are realistic, requiring only slight adjustments in absolute magnitude. Independent DWBA calculations [7,8] find that the cross section for nuclear breakup is less

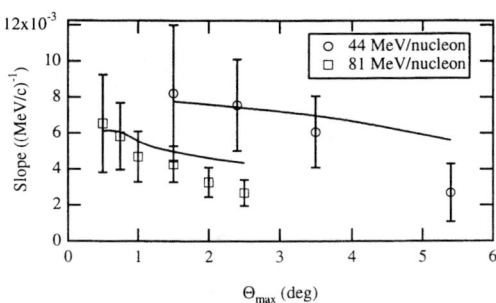

FIGURE 2. Comparison of slopes extracted from the central regions of the logarithms of the measured and theoretical longitudinal momentum distributions plotted versus maximum scattering angle. The agreement between the experimental results and the calculations is good at the smallest angle cuts. Deviations between the experiment and the theory (solid lines) are evident at large scattering angles, suggesting that nuclear breakup becomes important at these angles.

than 0.5% of the cross section for Coulomb breakup in the angular range of our 44 MeV/u measurement. Hence nuclear breakup is not expected to have a significant influence on either the shape or size of the 44 MeV/u longitudinal momentum distributions. The evident asymmetry of the distributions, characteristic of interference between $\ell=1$ and $\ell=2$ amplitudes, is therefore interpreted as the result of interference between E1 and E2 transition amplitudes.

The value of $S_{17}(0)$ corresponding to the best fit normalization factor is 22 eV b, well within the limits of the recommendation of the 1997 INT Workshop on Solar Fusion Cross Sections [9]. The value of the ratio S_{E2}/S_{E1} at the 0.63 MeV 1^+ resonance corresponding to this E2 strength is 6.7 $\times 10^{-4}$, which is consistent with the upper limit of 7 $\times 10^{-4}$ given in Ref. [10]. However, the E2 strength was extracted from the present experiment under the assumption of first order perturbation theory. If higher order dynamical effects are significant, then the E2 strength extracted here is a lower limit; for a given E2 strength, the predicted asymmetry is smaller when higher order effects are included than when they are neglected [2]. The results of this experiment apparently contradict those of two earlier experiments that found the E2 strength to be smaller than all published theoretical predictions [11,12].

In order to compare the asymmetries measured in the experiment with those predicted by the model, the slopes of the central regions of the longitudinal momentum distributions were extracted. However, the slopes of the theoretical distributions are proportional to the normalization factor by which the calculation has been multiplied. It is possible to eliminate this dependence on the normalization factor by taking the logarithm of the distributions before extracting a slope. Therefore, straight lines were fitted to the logarithms of the measured and theoretical distributions between 2020 and 2035 MeV/c for the lower energy, and between 2771 and 2791 MeV/c for the higher energy. The results of this comparison are shown in Fig. 2. The calculations at both energies were performed with the same optimal

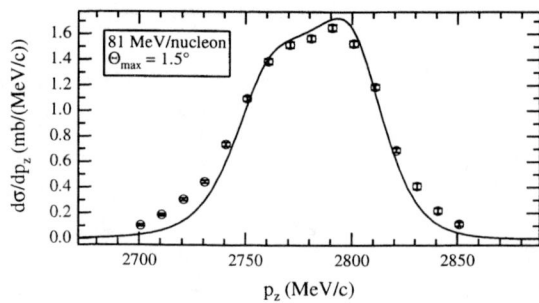

FIGURE 3. Measured longitudinal momentum distribution of ^{7}Be fragments with scattering angles less than 1.5° formed in the dissociation of 81 MeV/u ^{8}B, along with the prediction of the model. The slightly greater width of the experimental distribution indicates the presence of a broad component possibly due to nuclear breakup not accounted for in the calculation.

E2 strength described above. The systematically smaller slopes at the higher beam energy reflect the lesser relative importance of E2 transitions there. At both energies, the experimental slopes decrease more rapidly with angle than the Coulomb dissociation calculation predicts. This is interpreted as the result of nuclear induced breakup at angles approaching the grazing angle, which is 4.4° at the higher beam energy. Nuclear breakup results in a symmetric longitudinal momentum distribution [3], and could lead to smaller slopes.

Fig. 3 depicts a measured longitudinal momentum distribution of ^{7}Be fragments produced in the dissociation of ^{8}B on Pb at 81 MeV/u. Also shown is the prediction of the model with the same optimal E2 strength, again convoluted with the experimental resolution. The agreement between experiment and calculation at this energy is not as good as at 44 MeV/u. The main difference between the prediction of the model and the experimental measurement is the slightly greater width of the measured distribution. It is possible that nuclear breakup not accounted for by the model is responsible for broadening the measured distributions. The nuclear breakup cross sections are approximately equal at the two beam energies of this experiment [13], while the Coulomb excitation cross section at a given impact parameter is roughly proportional to the inverse square of the beam velocity. Hence nuclear breakup is relatively more important at 81 MeV/u. For this reason we utilized only the 44 MeV/u data in extracting the E2 strength. Nevertheless, including the 81 MeV/u data in the analysis would change the result little.

In summary, longitudinal momentum distributions of ^{7}Be fragments formed in the Coulomb dissociation of 44 and 81 MeV/u ^{8}B on Pb have been measured with high precision and statistics. At 44 MeV/u, the shapes of the measured distributions and the magnitudes of the measured cross sections agree with first order perturbation theory calculations based on a simple single-particle model of the structure of ^{8}B. The model predicts that interference between E1 and E2 transition amplitudes will produce observable asymmetries in the longitudinal momentum distributions of the fragments formed in the Coulomb dissociation of ^{8}B at the beam energies investigated in this experiment. These asymmetries were in fact observed. The 81 MeV/u distributions were not as well fit, perhaps because of a small nuclear component, so the 44 MeV/u distributions alone were used to extract the E2 strength. Nevertheless, the theoretical predictions for the slopes of the central regions of the momentum distributions based on this E2 strength agree well with the 81 MeV/u data at small angles, where nuclear effects are expected to be small. The measured distributions are consistent with a value of $6.7^{+2.8}_{-1.9} \times 10^{-4}$ for the ratio S_{E2}/S_{E1} at the 0.63 MeV 1^+ resonance. The calibration of the magnitude of the E1 response of the model made in order to reproduce the measured cross sections implies a value for $S_{17}(0)$ that agrees with the recommendation of Ref. [9]. However, the longitudinal momentum distribution of ^{7}Be fragments is not the most sensitive probe of this quantity. We hope that a careful measurement of the ^{8}B decay energy spectrum at low excitation energies, planned later this year, will allow a more precise determination of $S_{17}(0)$.

This work was supported by the U.S. National Science Foundation. One of us (H.E.) was supported by the U.S. Department of Energy, Nuclear Physics Division, under contract No. W-31-109-ENG-38.

REFERENCES

1. T. Motobayashi et al., *Phys. Rev. Lett.* **73**, 2680 (1994).
2. H. Esbensen and G.F. Bertsch, *Phys. Lett. B* **359**, 13 (1995); Nucl. Phys. A **600**, 37 (1996).
3. J.H. Kelley et al., *Phys. Rev. Lett.* **77** 5020 (1996).
4. J. Yurkon et al., *National Superconducting Cyclotron Laboratory Annual Report*, 207 (1996).
5. M. Berz et al., *Phys. Rev. C* **47**, 537 (1993).
6. K.H. Kim, M.H. Park, and B.T. Kim, *Phys. Rev. C* **35**, 363 (1987).
7. C.A. Bertulani, *Phys. Rev. C* **49**, 2688 (1994).
8. R. Shyam, I.J. Thompson, and A.K. Dutt-Mazumder, *Phys. Lett. B* **371**, 1 (1996).
9. INT Workshop on Solar Fusion Cross Sections, 1997, *Rev. Mod. Phys.* (to be published).
10. M. Gai and C.A. Bertulani, *Phys. Rev. C* **52**, 1706 (1995).
11. J. von Schwarzenberg et al., *Phys. Rev. C* **53**, R2598 (1996).
12. T. Kikuchi et al., *Phys. Lett. B* **391**, 261 (1997).
13. K. Hencken, G. Bertsch, and H. Esbensen, *Phys. Rev. C* **54**, 3043 (1996).

Asymptotic Normalization Coefficients for $^{14}N \to {}^{13}C + p$ and $^{10}B \to {}^{9}Be + p$

A. Azhari, H. L. Clark, C. A. Gagliardi, Y.-W. Lui, A. M. Mukhamedzhanov, L. Trache, R. E. Tribble, H. M. Xu, X. G. Zhou, V. Burjan[a], J. Cejpek[a], V. Kroha[a], F. Carstoiu[b]

Cyclotron Institute, Texas A&M University, College Station, TX 77843, U.S.A.
[a] *Institute for Nuclear Physics, Czech Academy of Sciences, Prague, Czech Republic*
[b] *Institute of Physics and Nuclear Engineering, Bucharest, Romania*

Abstract. The proton exchange reactions $^{13}C(^{14}N,^{13}C)^{14}N$ and $^{9}Be(^{10}B,^{9}Be)^{10}B$ were studied at energies where they are peripheral. Angular distributions for elastic scattering and proton transfer were measured, and the results were used to extract ANC's by comparison with modified DWBA calculations. These ANC's will be used together with proton transfer reactions with radioactive beams to determine astrophysical S-factors for radiative capture reactions, in particular $S_{17}(0)$ for $^{7}Be(p,\gamma)$.

Radiative proton capture reaction rates are one of the primary inputs needed in astrophysical calculations aimed at understanding elemental abundances in the universe. However, direct measurements of such reaction rates at astrophysical energies have proven difficult, if not impossible, in the laboratory, so indirect techniques have been sought out. Coulomb dissociation may provide an indirect technique to determine direct radiative capture cross sections at astrophysical energies, but it has not been put through a reliability test yet. Since most direct radiative capture reactions of astrophysical interest involve a loosely bound proton within the core's potential, the capture reactions tend to be peripheral and one need only to consider the contribution from the tail of the wave function. The asymptotic normalization coefficient (ANC) for a nuclear system specifies the amplitude of this tail. It can be measured in peripheral transfer reactions and can be used in subsequent calculations of direct capture rates [1,2].

In a proton transfer reaction, one determines the product of the ANC's for two vertices: the proton-core systems from which and to which the transfer occurs [3]. As part of our program to determine the astrophysical S-factor $S_{17}(0)$ for the proton radiative capture reaction $^{7}Be(p,\gamma)^{8}B$, using results from the proton transfer reactions $^{14}N(^{7}Be,^{8}B)^{13}C$ and $^{10}B(^{7}Be,^{8}B)^{9}Be$ with ^{7}Be radioactive beams, we have studied the proton exchange reactions $^{13}C(^{14}N,^{13}C)^{14}N$ at $E(^{14}N) = 162$ MeV and

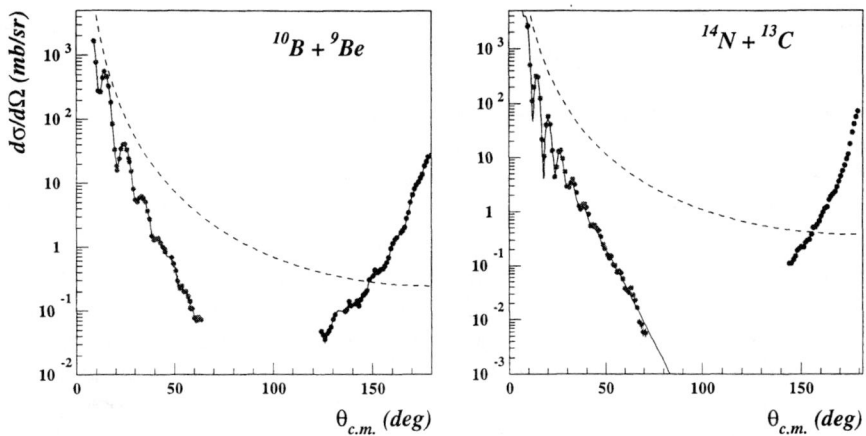

FIGURE 1. The angular distributions for elastic scattering of ^{10}B (100 MeV) on ^9Be and ^{14}N (162 MeV) on ^{13}C. The cross sections are in mb/sr. The dashed curve is the prediction for Rutherford scattering.

$^9Be(^{10}B,^9Be)^{10}B$ at $E(^{10}B) = 100$ MeV. These reactions determine the asymptotic normalization coefficients for the wave functions of the systems $^{14}N(g.s.) \to {}^{13}C+p$ and $^{10}B(g.s.) \to {}^9Be + p$, which represent "the other vertices" in the radioactive beam measurements mentioned above.

The experiments were performed using beams from the Texas A&M University K500 superconducting cyclotron. Reaction products from thin self supporting ^{13}C and ^9Be targets were observed in the Multipole Dipole Multipole magnetic spectrometer. The experimental setup was described in detail in Ref. [3]. The modified Oxford detector was used in the focal plane to do particle identification and determine the focal plane position and target angle using ray trace reconstruction. Elastic scattering was measured for the laboratory angular range $\theta_{lab} = 2° - 34°$, corresponding to the center-of-mass range $\theta_{cm} = 4° - 70°$, by detecting $^{14}N^{+7}$ ($^{10}B^{+5}$) in the focal plane of the spectrometer. Proton exchange reactions were measured by retuning the magnetic fields of the spectrometer for the rigidity of the outgoing $^{13}C^{+6}$ ($^9Be^{+4}$) at forward angles in the range $\theta_{lab} = -3°$ to $+18°$. In Fig. 1 one can see clearly that potential scattering at forward angles and elastic proton exchange at backward angles are independent. Five distinct families of optical model potentials were found for the ^{14}N experiment and three for ^{10}B from fits to elastic scattering data. The families had regularly increasing volume integrals, and the potentials providing the best fits in the two cases had very similar volume integrals. In addition to the elastic proton exchange, inelastic transfer to 4 excited states in ^{14}N and 3 in ^{10}B was measured.

We analyzed the proton exchange angular distributions in DWBA using the optical model potentials extracted above. The calculations were done with the code PTOLEMY [4]. In the general case the result contains the product of the spectroscopic factors for the initial and final states when using the standard DWBA or the product of the squares of the ANC's of the initial and final states in the modified DWBA we use [3]. In the particular case of an elastic proton exchange reaction, the entrance and exit channels are identical and the result is the square of the ground state spectroscopic factor or the fourth power of the ANC ($C_{g.s.}^4$). This allows us to extract the ANC from a single experiment. For the ground state of ^{14}N, we supposed a mixed configuration in which the last proton in either the $\pi 1p_{1/2}$ or $\pi 1p_{3/2}$ orbital is coupled to the ground state of the ^{13}C core. For the ground state of ^{10}B, only the $\pi 1p_{3/2}$ orbital participates and the fit is simpler. The calculations were done using a Woods-Saxon well plus spin-orbit term for the potential describing the movement of the last proton in the mean field of the ^{13}C (^9Be) core. In Fig. 2 we compare the stability of the extracted spectroscopic factors S_j and ANC's C_j^2 versus the ANC's of the single particle bound state orbitals. These orbitals were found with Woods-Saxon single particle potentials over a wide parameter range $r_0 = 1.0 - 1.3$ and $a = 0.5 - 0.7$ fm. The extracted spectroscopic factors S_j depend strongly on the choice of the potential – varying by more than a factor of 2 over the whole range, or 25% if we impose the additional condition that the potential well reproduces the experimental rms radius of the charge distribution within errors. At the same time, C_j^2 remains stable to better than $\pm 2\%$ on the whole range. We find the ground state ANC's to be $C_{p_{3/2}}^2 = 5.06(46)$ fm^{-1} for ^{10}B and $C_{p_{1/2}}^2 = 18.6(12)$ fm^{-1} and $C_{p_{3/2}}^2 = 0.93(14)$ fm^{-1} for ^{14}N. ANC's for the ^{10}B and ^{14}N excited states may be found in [3,5].

Further, we checked the dependence of the extracted ANC's on the ambiguities in the optical potentials used. For ^{14}N we found that the variations in the calculated values of the cross section for the elastic exchange are very small. The total contribution of optical model potential ambiguities to the uncertainty in C^2 is about 3% [5]. This is not surprising since the elastic scattering itself was studied for an angular range where the surface contributes most. In the case of ^{10}B we chose the shallowest two potentials and found that the extracted C^2 differed by only 9% [3].

The constancy of the extracted ANC's in Fig. 2 shows that the transfer takes place in the asymptotic region. Indeed, the calculations show that in the ^{14}N case the contribution to the transfer reaction amplitude comes from partial waves with $26 \leq l \leq 50$, with a peak around $l = 34$ [5]. This corresponds to a rather narrow window of distances between the two core nuclei, centered at $R = 6.8$ fm. This is larger than twice the radius of ^{13}C and certifies that the proton exchange reaction is peripheral.

In conclusion, we have measured elastic scattering and the elastic and inelastic proton exchange reactions $^{13}C(^{14}N,^{13}C)^{14}N$ and $^9Be(^{10}B,^9Be)^{10}B$. The measurement of the proton transfer reactions was used to extract the ANC's describing the tail of the wave function of the outer proton in the field of the ^{13}C and ^9Be cores. We used modified DWBA calculations with optical model potentials extracted from

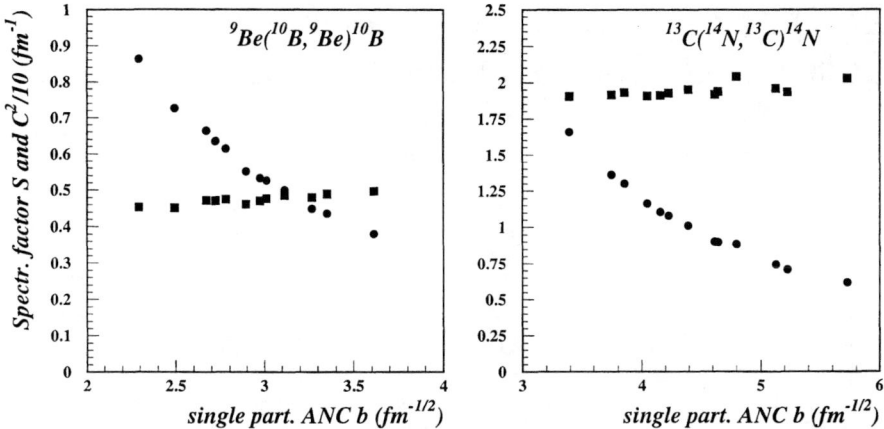

FIGURE 2. The comparison between the spectroscopic factors S (circles) and the squares of the asymptotic normalization factors C^2 (squares) extracted for the ground states of ^{10}B and ^{14}N from DWBA analyses in its standard use and the present modification, as functions of the ANC's of the proton-core single particle bound state orbitals.

the elastic scattering data. The results are stable for changes in the core-proton potential used in the DWBA and for different optical model potential parameters. This demonstrates the utility of a parametrization of peripheral transfer reactions in terms of an ANC rather than spectroscopic factors. The ANC's extracted here will be used to extract the ANC for the virtual decay of ^8B from the proton transfer reactions $^{14}N(^7Be,^8B)^{13}C$ and $^{10}B(^7Be,^8B)^9Be$ that we performed recently with ^7Be radioactive beams at the Texas A&M University Cyclotron Institute [1].

This work was supported in part by the U.S. Department of Energy under Grant number DE-FG05-93ER40773 and by the Robert A. Welch Foundation.

REFERENCES

1. R.E. Tribble et al., this Proceedings.
2. H.M. Xu, C.A. Gagliardi, R.E. Tribble, A.M. Mukhamedzhanov and N.K. Timofeyuk, Phys. Rev. Lett. **73**, 2027 (1994).
3. A.M. Mukhamedzhanov et al., Phys. Rev. C **56**, 1302 (1997).
4. M. Rhoades-Brown, M. McFarlane and S. Pieper, Phys. Rev. C **21**, 2417 (1980); **21**, 2436 (1980).
5. L. Trache, A. Azhari, H.L. Clark, C.A. Gagliardi, Y.-W. Lui, A.M. Mukhamedzhanov, R.E. Tribble, and F. Carstoiu, Phys. Rev. C (submitted).

Extrapolation of the astrophysical S factor for $^7\text{Be}(p,\gamma)^8\text{B}$ to solar energies

S. Karataglidis, B. K. Jennings, and T. D. Shoppa

TRIUMF, 4004 Wesbrook Mall, Vancouver, British Columbia, Canada, V6T 2A3

Abstract. We investigate the energy dependence of the astrophysical S factor for the reaction $^7\text{Be}(p,\gamma)^8\text{B}$, the primary source of high-energy solar neutrinos in the solar pp chain. Below 400 keV the energy dependence is well understood in terms of a subthreshold pole arising from the binding of the valence proton to the ^8B ground state.

The $^7\text{Be}(p,\gamma)^8\text{B}$ reaction, at energies of approximately 20 keV, plays an important role in the production of solar neutrinos [1]. The subsequent decay of the ^8B is the source of the high energy neutrinos to which many solar neutrino detectors are sensitive. The cross section for this reaction is conventionally expressed in terms of the S factor which is defined in terms of the cross section, σ, by:

$$S(E) = \sigma(E) E \exp\left[2\pi\eta(E)\right], \tag{1}$$

where $\eta(E) = Z_1 Z_2 \alpha \sqrt{\mu c^2 / 2E}$ is the Sommerfeld parameter, α is the fine structure constant, and μ is the reduced mass. The definition of the S factor eliminates from it most of the energy dependence due to Coulomb repulsion by factoring out the penetrability of a Coulomb potential. However, it does not make the S factor energy independent, as there are still energy dependences due to the structure of the final bound state and resonances. The reaction rate, obtained by folding the thermal distribution of nuclei in the stellar core with the cross section, peaks at approximately 20 keV. Because the cross section diminishes exponentially at low energies, the only method of obtaining information about the S factor at energies of astrophysical interest is to extrapolate data taken at experimentally accessible energies ($E > 100$ keV). To do the extrapolation reliably we must understand the physics associated with the S factor.

Most calculations of the S factor follow the pioneering work of Christy and Duck [2] to which we refer the reader for more details. Here we present a brief overview of the model. The S factor, for the $^7\text{Be}(p,\gamma)^8\text{B}$ reaction, may be written as

$$S = C(I_0^2 + 2I_2^2)E_\gamma^3 \left(J_{11}\beta_{11}^2 + J_{12}\beta_{12}^2\right) \frac{1}{1 - e^{-2\pi\eta}}, \tag{2}$$

where

$$I_L = \int_0^\infty r^2 dr \; r \; \psi_{iL}(r)\psi_f(r) \tag{3}$$

$$C = \frac{5\pi}{9}\frac{1}{(\hbar c)^3}(2\pi\eta k)e^2\mu^2 \left(\frac{Z_1}{M_1} - \frac{Z_2}{M_2}\right)^2. \tag{4}$$

In Eq. (2), J_{LS} is the spectroscopic factor for a given angular momentum, L, and channel spin, S, β_{LS} is the asymptotic normalization of the bound state wave function, E_γ is the photon energy, and k is the momentum of the incident proton. The extra factor of r in the integrand comes from the photon wave function, which is $E1$ for real photons. The final bound state wave function $\psi_f(r)$ is normalized in the asymptotic region to $\psi_f(r) = W_{\alpha,l}(\kappa r)/r$ while the initial wave function reduces to the regular Coulomb wave function divided by $kr\sqrt{2\pi\eta}/(e^{2\pi\eta}-1)$. The unusual choice of normalizations is just to simplify the mathematics and generate integrals that are well-behaved at threshold. The initial state has both Coulomb and nuclear distortions. The Coulomb distortions are large and give the penetration factor included in the definition of the S factor, Eq. (1). They are included in all calculations. The nuclear distortions are much smaller but they are important and introduce a significant model dependence into the calculations, as described in the next section.

The absolute magnitude of the S factor is determined primarily by the spectroscopic factor and the asymptotic normalization (see also Ref. [3]). The spectroscopic factor contains the many-body aspects of the problem and is calculable from standard shell model theory. The asymptotic normalization also depends on the many-body wave function, but is far more difficult to estimate from first principles: it requires detailed knowledge of how the 8-body wave function extends beyond the nuclear potential and its mapping to the Whittaker function in this region. This may be estimated crudely by approximating that behavior by using a suitably chosen Woods-Saxon wave function for the weakly bound proton. Instead we treat the overall factor, $A_n = J_{11}\beta_{11}^2 + J_{12}\beta_{12}^2$, as a free parameter, which is independent of energy, and determined by the S-factor data. For simplicity we will refer to this combination of asymptotic normalization and spectroscopic factor as the asymptotic strength.

To investigate the behavior of the integrals in Eq. (2), we first consider $\psi_f(r) = W_{\alpha,l}(\kappa r)/r$ for all radii and take $\psi_{i0}(r) = F_0(kr)/\{kr\sqrt{2\pi\eta}/(e^{2\pi\eta}-1)\}$. The s-wave integral then becomes

$$I_0 = \int_0^\infty dr \; r \frac{W_{\alpha,l}(\kappa r)F_0(kr)}{k\sqrt{2\pi\eta}}\left(e^{2\pi\eta}-1\right). \tag{5}$$

The integral is smooth as k passes through zero and diverges as $k \to i\kappa$ ($E \to -E_B$). The nature of the divergence is determined by the asymptotic forms of the Coulomb wave function and Whittaker function for large r. There the Whittaker function is proportional to $r^{-|\eta k|/\kappa}e^{-\kappa r}$ [2] (ηk is independent of k). Above threshold the Coulomb wave function oscillates at large radii, however below threshold it is exponentially growing and is proportional to $r^{|\eta|}e^{|k|r}$. Thus the behavior of the integrand at large radius is

$$r^{1-|\eta k|(1/\kappa - 1/|k|)} \exp[-(\kappa - |k|)r] \qquad (6)$$

and the integral diverges as

$$I_0 \sim 1/(\kappa - |k|)^2 \sim 1/(E_B + E)^2 = 1/E_\gamma^2 . \qquad (7)$$

The S factor is proportional to $I_0^2 E_\gamma^3$, and gives rise to a simple pole in S at $E_\gamma = 0$. Hence, the product $E_\gamma S$ should be a straight line. This was demonstrated in [4]. However, the first correction term is not simply $1/E_\gamma$ but rather of the form $(1 + c \log E_\gamma)/E_\gamma$, the logarithmic term coming from the $r^{-|\eta k|(1/\kappa - 1/|k|)}$ factor. Both the leading and first correction terms are both determined purely by the asymptotic behavior of the wave functions. The second correction term, of order E_γ^0, is not determined purely by the asymptotic value of wave function alone but also depends on the wave function at finite r.

The presence of the pole suggests the S factor may be parametrized as a Laurent series:

$$S = d_{-1} E_\gamma^{-1} + d_0 + d_1 E_\gamma + \ldots \qquad (8)$$

The coefficients of the first two terms, d_{-1} and d_0, are determined purely by the asymptotic forms of the wave functions while the third coefficient, d_1, is also dependent on the short range properties of the wave functions.

To account for nuclear distortions, which only affect the s wave contribution to the S factor [5], we construct a simple model where the initial state wave function is zero inside some radius, r_c, and a pure Coulomb wave outside. We impose the boundary condition that the wave function be zero at r_c. This generates a phase shift and is equivalent to having an infinitely repuslive potential with a radius r_c. The d-wave scattering state is taken to be an undistorted Coulomb wave function, and the bound state is assumed to be a pure Coulomb state, described by a Whittaker function, for all radii.

We use this model, which preserved the character of the pole, along with the others available, to analyse the data and extract an S factor. This has been done in Ref. [5], and the results of the fits is shown in Fig. 1. From fitting the energy range 0 to 400 keV we obtain $S(20) = 18.4 \pm 1.0 \pm 0.2$ eVb, or equivalently, $S(0) = 19.0 \pm 1.0 \pm 0.2$ eVb. The first error is experimental while the second is from model dependencies in the fit. Those model dependencies increase with energy, and so we also fit to 1.5 MeV, for which $S(20) = 17.6 \pm 0.7 \pm 0.4$ eVb and $S(0) = 18.1 \pm 0.7 \pm 0.4$ eVb.

In summary, we have determined the low energy behavior of the S factor for the $^7\text{Be}(p,\gamma)^8\text{B}$ to be dominated by a pole at $E_\gamma = 0$ which arises from the subthreshold ^8B ground state and the Coulomb interaction between the ^7Be and the proton and

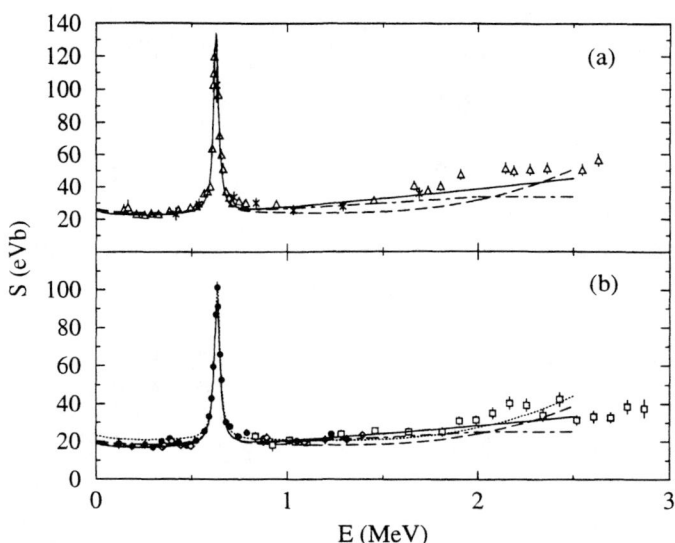

FIGURE 1. Fits to the data using various models of extrapolation [5]. The simple hard sphere model is indicated by the solid line in both cases.

large radii. That pole induces the upturn at threshold in the S factor which is observed in the calculations, and is also seen in the data for the S factor for the $^{16}O(p,\gamma)^{17}F$ (0.498 MeV) reaction [6].

REFERENCES

1. J.N. Bahcall and M. Pinsonneault, Rev. Mod. Phys. **67**, 885 (1992); J. N. Bahcall, S. Basu, and M.H. Pinsonneault, astro-ph/9805135.
2. R. F. Christy and I. Duck, Nucl. Phys. **24**, 89 (1961).
3. H. M. Xu, C. A. Gagliardi, R. E. Tribble, A. M. Mukhamedzhanov and N. K. Timofeyuk, Phys. Rev. Lett. **73**, 2027 (1994).
4. B. K. Jennings, S. Karataglidis, and T. D. Shoppa, Phys. Rev. C, in press (July, 1988).
5. B. K. Jennings, S. Karataglidis, and T. D. Shoppa, in preparation, to be submitted to Phys. Rev. C., and references therein.
6. R. Morlock, *et al.*, Phys. Rev. Lett. **79**, 3837 (1997).

Beta-decay of ^{40}Ti and Its Implication for Solar-Neutrino Detection

W. Liu[1], M. Hellström[1], R. Collatz[1], J. Benlliure[1], L. Chulkov[2],
D. Cortina Gil[1], F. Farget[1], H. Grawe[1], Z. Hu[1], N. Iwasa[1], M. Pfützner[1,3],
A. Piechaczek[4], R. Raabe[4], I. Reusen[4], E. Roeckl[1], G. Vancraeynest[4],
A. Wöhr[4]

[1]*Gesellschaft für Schwerionenforschung, D-64291 Darmstadt, Germany*
[2]*Kurchatov Institute, 123182 Moscow, Russia*
[3]*Institute of Experimental Physics, University of Warsaw, PL-00681 Warsaw, Poland*
[4]*Institut voor Kern-en Stralingsfysika, Katholieke Universiteit, B-3030 Leuven, Belgium*

Abstract. The β-decay of ^{40}Ti was studied by measuring the β-decayed proton- and γ-emission. The half-life for ^{40}Ti was determined to be 54(2) ms. The experimental β-decay strength distribution is compared with shell-model calculations and results from other experiments. Based on the experimental ^{40}Ti β-decay strength, the neutrino absorption cross section and induced neutrino event rates for ^{40}Ar were determined to be $14.3(3) \times 10^{-43}$ cm^2 and $9.4 \pm 0.2(\text{stat.})^{+1.3}_{-1.6}(\text{syst.})$ SNU, respectively.

INTRODUCTION

The proposed ^{40}Ar detector ICARUS [1] will be capable to detect neutrino oscillations, which is important in connection with the solar-neutrino puzzle. For a reliable evaluation of ICARUS data, the cross sections for the different interaction processes must be known very well. A possibility of calibrating ICARUS for the capture of electron-neutrinos is to use, under the assumption of isospin symmetry, the B(GT) values of the mirror β-decay of ^{40}Ti. The large energy release of the ^{40}Ti decay (Q_{EC} = 11680(160) keV [2]) enables one to extract information that is relevant for the Gamow-Teller (GT) contributions to the rate of solar neutrinos absorbed by ^{40}Ar.

EXPERIMENTAL TECHNIQUES

A ^{58}Ni beam of 500·A MeV with an intensity of 1×10^9 ions/s from the heavy-ion synchrotron SIS at GSI in Darmstadt was used to produce ^{40}Ti by fragmentation reactions in a 4 g/cm^2 thick ^{9}Be target. By using the projectile fragment separator FRS, an isotopically separated ^{40}Ti beam was produced. The intensity of the ^{40}Ti beam at the final focus of the FRS was measured to be about one atom per minute. During the experiment of 5 days, about 1.1×10^4 ^{40}Ti ions were produced. The ^{40}Ti ions identified by energy-loss (ΔE) and mass-to-charge ratio (A/q), were slowed down at the final focal plane from 240·A MeV to 60·A MeV by using a 0.7 g/cm^2 thick aluminum degrader.

This measure was taken in order to shift the implantation profile to the center of the silicon detector stack consisting of eight 300 μm thick, 30 mm diameter silicon detectors. The three central counters were used to measure positrons and β-delayed protons, whereas the outer ones served as veto detectors to reject unwanted particles as described below. About 90 % of the ^{40}Ti ions were implanted in the central three counters. An array of 14 large-volume NaI detectors were mounted close to the silicon detector stack to measure γ–rays emitted in the ^{40}Ti decay process. The photopeak efficiency was measured to be 15(2) % at a γ-energy of 1.33 MeV. The proton energy calibration was based on the β-delayed proton spectrum of ^{41}Ti, measured in a separate FRS experiment.

DATA ANALYSIS

The delayed-coincidence technique, used in the off-line data analysis of the ^{40}Ti data, was based on a time window of 200 ms. This window was opened by a ^{40}Ti event, selected according to conditions with respect to ΔE and A/q, and closed by the subsequent decay event. In this way, decay events from positron and/or β-delayed proton emitters (e.g. ^{38}Ca) were rejected. The proton energy spectrum, shown in Figure 1, was generated by an anti-coincidence condition with the front and rear silicon counters in order to eliminate heavy ions, which were stopped in the first counter or penetrated through all of them. The proton events, which were not completely stopped in a single detector, were rejected by an anti-coincidence condition with adjacent counters. To get the proton branching ratios for individual proton transitions, the proton peak intensities were corrected for the full-energy detection efficiency, and were normalized to the total number of implanted ^{40}Ti ions. The latter quantity was determined by selecting events recorded in coincidence between the adjacent counters. The sum of individual proton branching ratios for ^{40}Ti amounts to 101(5) %. The proton-γ coincidence data were used for identifying proton emission from ^{40}Sc levels to excited states of ^{39}Ca. This information was used

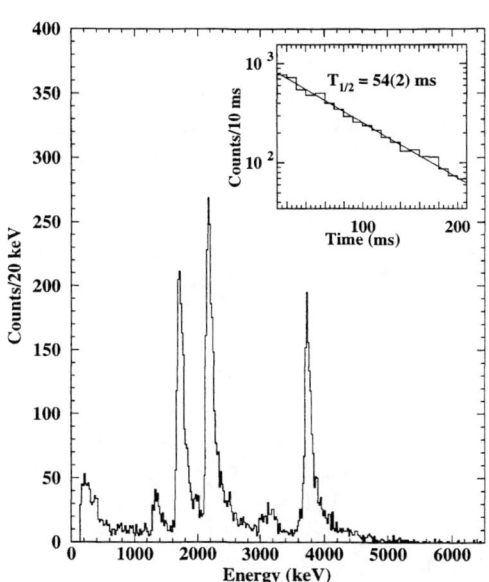

FIGURE 1. ^{40}Ti β-delayed proton energy spectrum, and time distribution of proton events with energies above 1.5 MeV.

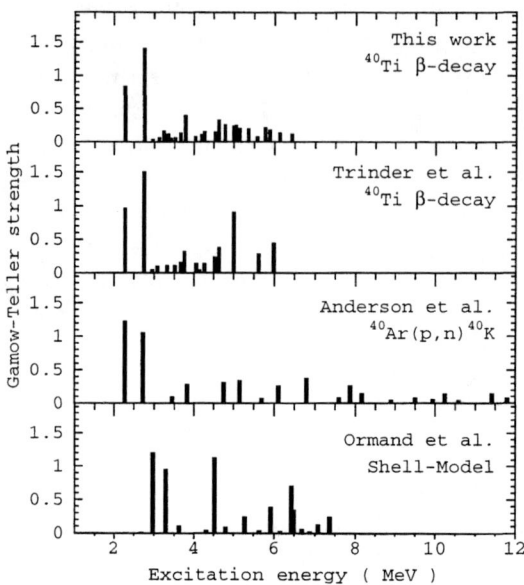

FIGURE 2. Experimental Gamow-Teller strength distributions for the ^{40}Ti β-decay and the ^{40}Ar(p,n)^{40}K reaction in comparison with a shell-model prediction [7].

for deducing the β-branching ratios from proton branching ratios. A ratio of 12(3) % was found between the number of γ-gated events in the 1322 keV proton peak and the number of events observed for this peak in the ungated spectrum. This value is in good agreement with the calibrated NaI photopeak efficiency of 13(3) % at 2.5 MeV. Correspondingly, the intensity of the 1322 keV proton line is assigned to a transition between the 4365 keV state (IAS) in ^{40}Sc and the 2649 keV state in ^{39}Ca. The remaining ^{40}Ti proton-γ coincidence data were interpreted in a similar way.

We used the revised ^{40}Ti Q_{EC} value of 11466(13) keV [5], to calculate the transition strengths for ^{40}Ti. The B(F) and B(GT) values from this work, which stem from an improved data evaluation [4] compared to a previous shortnote [3], are displayed in Figure 2 in comparison with results from the other recent β-decay measurement [5], a ^{40}Ar(p,n)^{40}K experiment [6], and a shell-model calculation [7].

DISCUSSION

The J^π assignment for ^{40}Sc levels populated by β-decay of ^{40}Ti is 1^+, except for the 4365 keV level which was identified as the 0^+, IAS. The latter interpretation is based on the observed β-strength of 4.01(31), which agrees with the theoretical value of |Z-N| = 4 for a pure Fermi transition. The integrated B(GT) value is 5.84(39), which can be compared with the shell-model result of 5.62 [5] that was obtained by the free-nucleon GT operator quenched by a factor of 0.775. However, we note that the shell-model overestimates the excitation energies of the first two excited states of ^{40}Sc by roughly one MeV, as can be seen from Figure 2.

Excellent agreement was found between the present results and those obtained by Trinder et al. [3] up to a ^{40}Sc-excitation energy of 5 MeV. At higher excitation energies, however, the GT strength from this work is more fragmented than that obtained by Trinder et al.. This discrepancy is not surprising, as the statistics of our data as well as of those obtained by Trinder et al. are too poor for proton energies above 5 MeV for an *unambiguous* decomposition into individual transitions. Our data agree with those of the (p,n) work [6]. From this comparison, we conclude that within the experimental uncertainties we did not observe any isospin asymmetry in the ^{40}Ti-^{40}Ar mirror pair.

Following the procedure suggested by Ormand et al. [7], we calculated electron-neutrino absorption rates in ^{40}Ar, based on the experimental ^{40}Ti β-decay strength from this work and the updated solar neutrino flux of $6.6^{+0.9}_{-1.1} \times 10^6$ cm^{-2}s^{-1}[8]. In this calculation, we used the experimental ^{40}K excitation energies whenever possible; otherwise the ^{40}Sc excitation energies from this work were taken. Based on the ^{40}Ti β-decay measurement of the present work and that of Trinder et al. [3], we derived a recommended value of $14.3(3) \times 10^{-43}$ cm^2 for the electron-neutrino absorption cross-sections of ^{40}Ar as the weighted average of the experimental values of $13.8(6) \times 10^{-43}$ and $14.5(4) \times 10^{-43}$ cm^2 [5], respectively. This result, together with the solar neutrino flux according to [8], yields a recommended value of the electron-neutrino induced event rates for the ICARUS detector of 9.4 ± 0.2(stat.)$^{+1.3}_{-1.6}$(syst.) SNU.

The summary of the results obtained in this work and in previous experiments, as well as the interplay between the reaction rates and the solar model prediction of the ^{8}B-neutrino flux, are shown in Figure 3. It can be clearly seen that the ^{40}Ar-absorption cross section appears to be accurate enough, whereas the solar model prediction for the ^{8}B-neutrino flux represents the major source of the systematical uncertainty in our prediction of the event rate in the ICARUS detector.

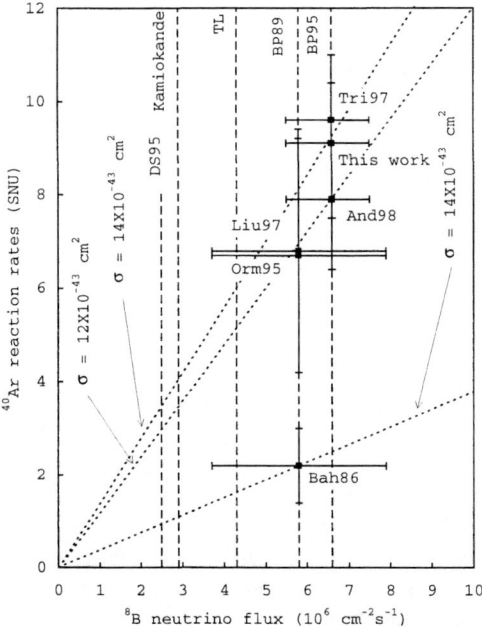

FIGURE 3. Experimental data and theoretical calculations for the ^{40}Ar neutrino-absorption rates in the ICARUS detector. The vertical dashed lines show the model prediction of ^{8}B neutrino fluxes; the dotted lines show the measured absorption cross-sections. See [9] for details.

REFERENCES

1. ICARUS Collaboration, ICARUS II proposal, LNGS Report 995/99-I (1993), see also http://www.aquila. infn.it:80/icarus/main.html.
2. Audi, G. et al., Nucl. Phys. A624 (1997) 1.
3. Liu, W. et al., Z. Phys. A386 (1997) 1.
4. Liu, W. et al., GSI Preprint 98-19 (1998), to be published.
5. Trinder, W. et al., Phys. Lett. B415 (1997) 211.
6. Anderson, B. D., private communication.
7. Ormand, W. E. et al., Phys. Lett. B345 (1995) 343.
8. Bahcall, J. N. et al., Rev. Mod. Phys. 67 (1995) 781.
9. Liu, W., Ph.D. Thesis (1998).

Study of the d(^{7}Be,^{8}B)n Reaction

C.F. Powell[*‡], D.J. Morrissey[*‡], D.W. Anthony[*‡], B. Davids[†‡],
M. Fauerbach[†‡], P.F. Mantica[*‡], B.M. Sherrill[†‡], and M. Steiner[‡]

Department of Chemistry
†*Department of Physics and Astronomy*
‡*National Superconducting Cyclotron Laboratory*
Michigan State University, East Lansing, MI 48824

Abstract. The (d,n) reaction of ^{7}Be to form the ground state of ^{8}B was studied in inverse kinematics using a 25 MeV/nucleon beam of ^{7}Be nuclei from the A1200 fragment separator at the NSCL. The recoiling ^{8}B nuclei were observed in a pair of position sensitive detector telescopes at lab angles between 1° and 5.5° (7° to 38° CMS). A large array of neutron detectors was placed at angles ranging from 158° to 104° (35° to 7° CMS) that measured the energy of coincident neutrons by their time-of-flight. The cross section for the d(^{7}Be,^{8}B)n reaction was determined from both the charged-particle singles and from the neutron coincidence data. In addition to measuring the transfer reaction, the quasi-elastic scattering of ^{7}Be from deuterium was measured. The cross sections of these reactions are compared with DWBA and coupled-channels calculations.

INTRODUCTION

The nuclear structure of the lightest boron isotope remains uncertain even thought its properties have been intensely studied. The odd proton in ^{8}B is loosely bound and potentially exists in a halo state. This nucleus has been studied with a variety of techniques with differing results for even the root-mean-squared radius. The total reaction cross section has been measured [1,2], the momenta of the ^{7}Be core nucleons have been measured several times in nuclear and Coulomb breakup reactions [3-5], the quasi-elastic scattering cross section with carbon has been fitted with a microscopic potential [6], and the quadrupole moment has been measured [7]. The root-mean-squared radii of ^{8}B derived from these measurements range from 2.21 to 2.98 fm. The one proton overlap intergral for ^{7}Be and the size of ^{8}B has been very recently discussed by Timofeyuk [8]. Given such large discrepancies among these values, and the uncertainty they bring to the underlying structure of ^{8}B, we performed a new experimental study: the (d,n) single nucleon transfer reaction. A similar study, but at an extremely low bombarding energy, has been recently reported [9].

The d(^{7}Be,^{8}B)n reaction is also important from a very different standpoint. Xu, et al. have shown that the asymptotic normalization coefficient (ANC) for the overlap of the ^{7}Be + p in ^{8}B extracted from an analysis of the transfer reaction can be used to extract the S factor for the proton capture reaction on beryllium [10,11]. This ^{7}Be(p,γ) reaction has itself been the subject of intense study as a possible solution to the solar neutrino puzzle.

The short half-life for ^{7}Be (53 days) makes the traditional approach of studying the valence proton state of ^{8}B via direct reactions with a ^{7}Be target very difficult. However, the recent availability of energetic beams of exotic nuclei allow the study of inverse reactions. In the present work a 25 MeV/u beam of ^{7}Be was reacted with a CD$_2$ target and the elastically scattered ^{7}Be beam and the ^{8}B reaction products were observed at forward angles. We also observed neutrons in kinematic coincidence with a range of ^{8}B products.

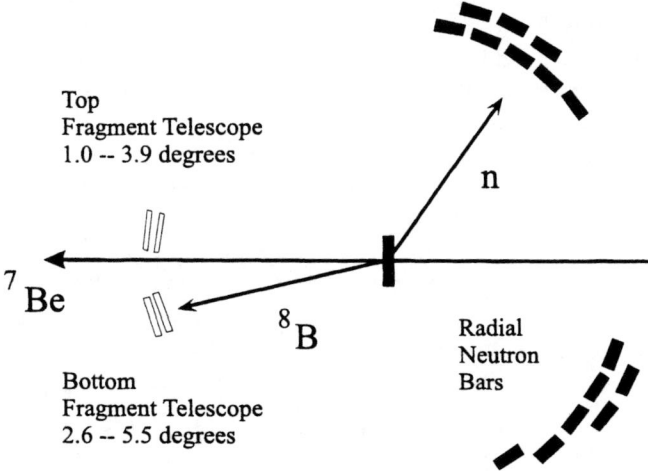

FIGURE 1. Schematic diagram of the detector system. The silicon telescopes were positioned above and below the plane of the beam and each neutron detector subtended an arc 90° by 4.4° at a 1 meter radius.

EXPERIMENTAL AND RESULTS

The present experiment was carried out at the National Superconducting Cyclotron Laboratory. The A1200 fragment separator provided a secondary beam of ^{7}Be nuclei (25 MeV/nucleon) from the reaction products from the interaction of a ^{12}C beam (60 MeV/nucleon) in a 587 mg/cm^2 beryllium target using a 425 mg/cm^2 aluminum wedge. The intensity of the ^{7}Be beam was carefully monitored and was approximately 335,000 particles per second. The secondary target was an 18 mg/cm^2 deuterated polyethylene foil. The contribution from reactions on the carbon nuclei were measured with a 31 mg/cm² carbon foil and subtracted from the data.

The scattered charged particles were identified and measured in a pair of silicon telescopes placed above and below the beam. The telescopes consisted of a double-sided silicon strip detector (5cm x 5cm x 300 or 500μm, 16x16 strips) and a silicon PIN diode detector (5cm x 5cm x 500μm) placed at lab angles that ranged from either 1.0° to 3.9° or 2.6° to 5.5°. Neutrons in kinematic coincidence with the scattered particles were detected in the NSCL neutron time-of-flight array [12]. The neutron array was calibrated by observing the beta-delayed neutrons from a ^{17}N source produced by the A1200 and implanted into a beta-detector at the target position [13]. A schematic layout of the two types of detectors is shown in Figure 1. The overall response of the detector system was checked by measuring the d(^{12}C,^{13}N)n reaction at 22 MeV/nucleon under identical, i.e., low intensity, conditions. These results are not presented here due to constraints on space.

The cross section for quasi-elastic scattering was obtained by grouping the charged particles identified in each of the 256 pixels in each silicon telescope according to laboratory angle. Note that the (only bound) excited state in this system could not be resolved from the ground state by the silicon detectors. The cross section for the transfer reaction was obtained by a similar grouping of the single-

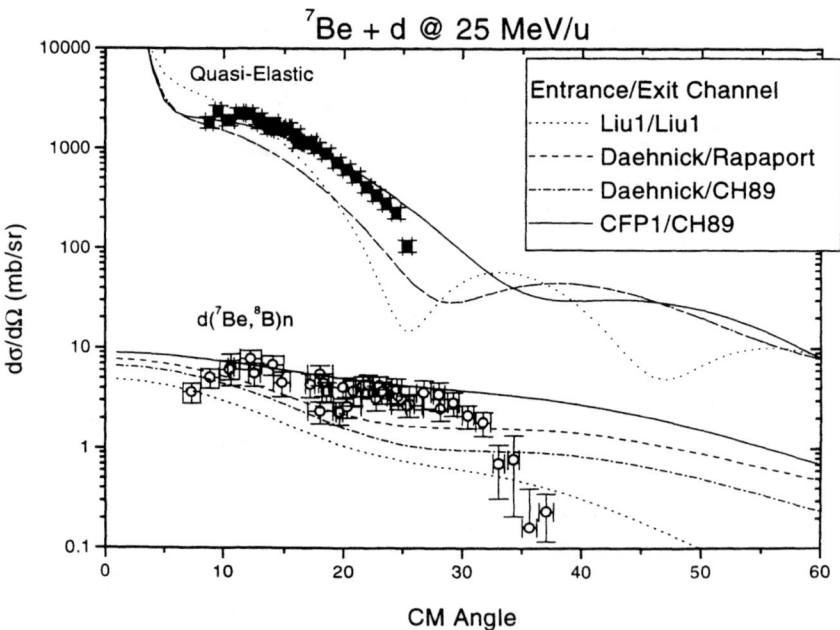

FIGURE 2. Results for the elastic scattering and transfer reaction are compared to DWBA calculations with several optical potentials from the literature, see the text for details.

particle inclusive data from the silicon telescopes and by the neutron coincidence data. The resulting angular distributions for both reactions are shown in Figure 2, the error bars indicate the combined uncertainties (ordinate) and the angular bin width (abscissa).

The measured angular distributions are compared with the predictions of finite-range DWBA calculations performed with the FRESCO [14] coupled-channels computer code in figure 2. All calculations were performed with the deuteron bound by a gaussian potential, and ^{8}B bound by the potential of Tombrello [15], a Woods-Saxon with V_0= 32.62 MeV, r_0=1.25 fm, a=0.65 fm, giving an ANC value of 0.764. The calculations were performed using several different sets of optical parameters from the literature. Global deuteron parameters from Daehnick et al. [16] were used for the ^{7}Be + d entrance channel, while the global parameters from Rapaport [17] and the CH89 set [18] are used for the ^{8}B + n exit channel. The entrance channel parameter set labeled as CFP1 was constructed from the ^{8}B real potential described by Tombrello and the ^{9}Be imaginary potential described by Mani et al. [19]. For completeness, calculations using the parameter sets determined by Liu et al. [9,11] for the same reaction at a much lower energy are also shown. Although reasonable agreement was obtained with one parameter set, further analysis is necessary.

This work was supported through the cooperative agreement PHY-95-82244 with the National Science Foundation. We would like to especially acknowledge the help and insight provided by J.A. Tostevin with the DWBA calculations.

REFERENCES

1. Tanihata, I., et al. *Phys. Lett.* **B 206**, 592-596 (1988).
2. Warner, R.E., et al. *Phys. Rev.* **C52**, R1166-R1170 (1995).
3. Schwab, W., et al. *Z. Phys.* **A 350**, 283-284 (1995).
4. Kelley, J.H., et al. *Phys. Rev. Lett.* **77**, 5020-5023 (1996).
5. Davids, B., et al. these proceedings and to be published, (1998).
6. Pecina, I., et al. *Phys. Rev.* **C52**, 191-198 (1995).
7. Minamisono, T., et al. *Phys. Rev. Lett.* **69**, 2058-2061 (1992).
8. Timofeyuk, N.K., *Nucl. Phys.* **A 632**, 19-38 (1998).
9. Liu, Weiping, et al. *Phys. Rev. Lett.* **77**, 611-614 (1996).
10. Xu, H.M., et al. *Phys. Rev. Lett.* **73**, 2027-2030 (1994).
11. Gagliardi, C.A., et al., *TAMU Prog. in Res.* I17-I20 (1997).
12. Harkewicz, R., et al., *Phys. Rev.* **C44**, 2365-2371 (1991).
13. Scheller, K.W., et al., *Phys. Rev.* **C49**, 46-50 (1994).
14. Thompson, I.J., *Comp. Phys. Reps.* **7**, 167-1xx (1988).
15. Tombrello, T.A., *Nucl. Phys.* **71**, 459-464 (1965).
16. Daehnick, W. W., et al., *Phys. Rev.* **C21** 2253-2274 (1980).
17. Rapaport, J., *Phys. Reps.* **87**, 25-75 (1982).
18. Varner, R.L., *Phys. Reps.* **201**, 57-119 (1991).
19. Mani, G.S., et al., *Nucl. Phys.* **A 165**, 145-151 (1971).

Status of the LOREX:Geochemical ^{205}Tl Solar Neutrino Experiment

K. M. Subotic and M. K. Pavicevic

Institute of Nuclear Sciences Vinca, Pob 522, 11001 Belgrade, Yugoslavia

Abstract: The LOREX-^{205}Tl solar neutrino experiment is designed to measure the solar neutrino flux integrated over the last few million of years using lorandite (TlAsS$_2$) minerals from Allchar mine (Macedonia), by measuring the ^{205}Pb content resulting from neutrino capture in ^{205}Tl. The mean depth of the sample has to be determined by measuring the depth dependent cosmic ray induced production of isotopes ^{10}Be, ^{26}Al, ^{36}Cl and ^{53}Mn to evaluate the background ^{205}Pb contributions. The basic idea is to accelerate the beams of above isotopes to energies required to obtain the fully stripped ions and separate them employing the resulting difference in q/m ratios.

The Sun, stars and the more distant constituents of the Universe are characterised mainly by studying their electromagnetic radiation. Many of the most fundamental processes, such as stellar energy generation take place in regions, which are not accessible by electromagnetic observations. Weakly interacting neutrinos emitted in nuclear reactions which generate the energy in the Sun are considered the powerful tool to probe these regions. Recent measurements of the flux of solar neutrinos have shown that the measured values are two to three times smaller than predicted by theory. The possible explanations of the solar neutrino puzzle are that the theories of stellar energy generation are incomplete or that the reduced flux of the solar neutrinos at terrestrials detectors results from the neutrino oscillations. Several experimental set-ups are nowadays tempting to resolve the aspects of Solar Neutrino Problem related to the accuracy of the available experimental information. They are mainly based on the use of the large volume detectors having the potential to record the signals of weakly interacting neutrinos coming from the Sun.

The more economical LOREX project, that has the lowest energy threshold ($Q = 52\,\text{keV}$) for the detection of the pp neutrinos was originally proposed by Freedman. (1). Namely, by absorbing a neutrino a stable ^{205}Tl nucleus via β- decay can be transformed into a ^{205}Pb nucleus. The difference in atomic mass between ^{205}Tl and ^{205}Pb is very small, ($52\,\text{keV}$). It is then expected that the observation of the reaction ^{205}Tl(ν,e-)^{205}Pb can provide the relevant information, related to the low-energy solar neutrinos from the hydrogen burning reaction p+p---d+e$^+$+430keV.

The leading idea in this experiments is *to determine the solar neutrino flux integrated over geological time, measuring the* ^{205}Pb *content resulting from neutrino capture in* ^{205}Tl *in 4.2 million years old samples of lorandite mineral (TlAsS$_2$) from Allchar mine.* The determination of the ^{205}Pb background contribution produced by cosmic radiation that is dependent on the mean depth of the lorandite mineral sample has been proposed to be made by measuring the erosion rates. This can be made using long-lived radioisotopes ^{10}Be, ^{26}Al, ^{36}Cl and ^{53}Mn produced by cosmic radiations in the collocated monitor minerals for example, quartz (SiO$_2$), realgar (As$_2$S$_2$) and marcasite (FeS$_2$). The expected concentrations of ^{205}Pb, ^{10}Be, ^{26}Al and ^{53}Mn, relative to their stable isotopes range between 10^{-12}-10^{-15} Tandem based AMS techniques were successfully applied in the measurement of ^{10}Be and ^{26}Al nuclei The principles of the possible use of cyclotron based AMS to measure the trace concentrations of ^{53}Mn and ^{205}Pb nuclei are considered here (alternatives are considered elsewhere (2)).

Cyclotrons are inherently high mass resolution devices. When tuned for a given atomic mass, they are detuned for isobars, due to the acceleration phase slip Φ given by: $\sin\Phi = 2\pi hndf/f = 2Rdf/f$, that arises from the lack of coincidence between time of a particle's arrival and the peaking of accelerating voltage at the dee's gap (3).

Device resolution is given by $R = m/\delta m = \pi nh$ where n is the number of the executed cyclotron turns and h is the applied harmonic of the RF tune, while $f = qB/m$ represents the frequency of the particle's orbit having mass m and charge q, in magnetic field B.

There are two basic possibilities in using the advantage of the extremely long effective particle's path and related AMS features to identify and count the intensity of the desired beam component.

1. To accelerate beams to energies required obtaining the fully stripped ions. Equilibrium charge attained at beam velocity v is given by: $q = Z[1-\exp(-137\beta/Z\gamma)]$ were Z is the atomic number, $\beta = v/c$, c is the velocity of light, and γ is approximately $2/3$. If it is fully stripped, the higher Z component has an effective charge, which is the one unit higher than the effective charge of the lower Z isobar beam component. In this way the extremely high mass resolution $R = m/\delta m$ required to resolve the interfering isobars is reduced to the requirement $R' = Z/\delta Z$ due to the fact that the interfering isobars after the passage through the stripping foil are characterised by different q/m ratios. Then they may be separated using standard mass separation techniques. If beam energy is not sufficiently high to obtain the fully stripped ions this method may still be applied in the cases where the differences in the energy losses for the interfering isobars are larger than widths of the isobar lines caused by the energy straggling (4).

2. Use of the time of flight (TOF) technique based on the different travelling times of interfering isobars along the cyclotron trajectories. The total time of flight is given by $\tau = nT$, where n is again the number of the executed particle turns, while T represents the period of the particle's orbit. Typical values for n and T are 300 turns and 300ns respectively, leading to the values of $\tau \approx 0.1$ms. At the resolving time $\delta\tau > 0.2$ ns the interfering isobars can not be separated with resolving power $R = \tau/\delta\tau = m/\delta m > 500000$ (5)

The possibility to use the NSCL ECR ion source-K1200/K500 cyclotron tandem-A1200 fragment separator system, to separate and identify the products of the neutrino capture in ^{205}Tl is discussed here. Anticipated measurements on the level of "single atom counting" require the highest possible efficiency of beam production and transmission. Expected consumption rate of the order of 25 μg/h/pμA as quoted for the case of ECR production of the expensive ^{48}Ca beam at the system transmission of 10% enables an efficiency of $5 \cdot 10^{-3}$, which suggests the sample sizes of the order of few μg (7). The superconducting cyclotron K1200 at NSCL at Michigan State University can accelerate the ^{53}Mn ions up to energies required to obtain the fully stripped ions, while the future coupling mode of K500/K1200 operation makes possible acceleration of ^{205}Pb and ^{205}Tl ions up to energies of 140A MeV enabling the equilibrium charge states $q = 80$ and $q = 79$ respectively.

1. **The ^{205}Pb case.** At flux and cross-section values corresponding to 200 SNU (SNU=10^{-36}), expected atom ratio of ^{205}Pb to ^{205}Tl nuclei produced in the neutrino capture reaction in ^{205}Tl in the 4.5 million years old lorandite samples is of the order of 10^{-19}. The estimated $^{208}Pb/^{205}Tl$ atom ratio in lorandite material is however of the order of 10^{-6}. Thus after chemical purification we expect to have the ^{205}Pb and ^{205}Tl components present in the ^{208}Pb beam delivered by processed lorandite probe, on the level 10^{-13}-10^{-12}, respectively. We expect a counting rate $> 10^{-5}$ for ^{205}Pb nuclei and the ^{208}Pb beam intensity $> 10^8$ p/s. At beam energy of 140A MeV the equilibrium charge state for ^{205}Pb nucleus will be $q = 80$ and that of the interfering ^{205}Tl isobar $q = 79$. It is expected that this difference in equilibrium charge state should result in the separation of particle response lines after analysis in the magnet separator.

2. **The ^{53}Mn case.** In order to determine the solar neutrino flux by counting ^{205}Pb nuclei produced by neutrino capture in ^{205}Tl, one has to determine the background contributions induced by cosmic radiation. These contributions depend on the mean depth of the measured sample that in turn may be obtained by determining erosion rates by measuring the depth-dependent cosmic ray-induced production of the long living radioisotopes ^{10}Be, ^{26}Al, and ^{53}Mn. These radioisotopes are mainly produced by the capture of negative muons. Their half-lives are evidently suitable for measurement of the ages ranging from 5-10 million of years.

Experimentally the Mn sample from FeS_2 (marcasite or pyrite) could be ionised in an ECR ion source to obtain an efficient production of Mn14+ beam component. The K1200 cyclotron tuned to accelerate $^{53}Mn^{14+}$ beam component can resolve the $^{55}Mn^{14+}$ beam component after the first 10 turns ($R = 26$). The energy of 50A MeV required to obtain the fully stripped isobars $^{53}Mn^{25+}$ and $^{53}Cr^{24+}$ can be easy attained in K1200 stand-alone operating mode. After passing the stripper foil the ^{53}Mn beam component may be separated and identified using the standard PID arrangement in A1200 separator.

3. **Anticipated experimental difficulties.** Recent attempts to use the cyclotron based AMS in experiments with low statistics experiment show that several

experimental difficulties have to be solved in order to attain the required confidence level for the measured data. The most important of these are noted here.

- Beam isobar impurities arising from the ECR construction materials are unavoidable contamination. In order to evaluate their presence in each particular case the "blank" run of the ECR ion source has to be performed at the given cyclotron RF tune.

- Systems that control the particle motion are designed to regulate the collective behaviour of the bunch of the ion species. This feature does not imply the same quality of the control for the single particle motion in a "few event" experiment that aims for the ^{205}Pb identification and counting. The analogue pilot beams like ^{41}Ca^{16+} in the case of ^{205}Pb^{80+} run have to be used to identify the quality of the response of the system to the single particle ^{205}Pb motion.

- Statistical laws can not be directly applied in the processing of the data in a "few event" experiment. Confidence levels should be carefully determined. One of the critical points is certainly the evaluation of confidence levels in the use of the statistical distributions at low statistics in determining the effective charge state, relevant for particle identification

- Background contributions from nuclear reactions induced in the stripper foils and detectors produce severe difficulties in identifying and counting the true events. These products are expected, however, to have different energies that may facilititate their rejection while identifying the true events (8).

The beta transitions from $J = 1/2+$ ground state of ^{205}Tl to low-lying states in ^{205}Pb may be considered the first forbidden beta transitions. *There are no experimental data on the transition probability for that process or inverse. Therefore, the theoretical calculations of the high precision are highly needed. Recently, P.Kienle proposed to solve this problem by measuring bound state beta decay* of completely stripped ^{205}Tl^{81+} to ^{205}Pb^{81+}. *that enables extraction of the relevant matrix* element for the neutrino induced reaction on ^{205}Tl.

REFERENCES

1. E.Nolte and M.Pavicevic ed., *Nucl.Instr.Meth.* **A271,** 237-342 (1988).
2. W.Henning, GSI Report 1993.
3. K.M.Subotic, *J.Phys. G*, **17,** S363 (1991).
4. D.Novkovic, and K.Subotic, *Nucl.Instr.Meth.B* **94,** 369-372 (1994).
5. W.Mittig,, Ganil Report, 1993.
6. Proposal for K500/K1200 coupled mode operation, MSU 1994.
7. V.Kutner, private communication.
8. W.Kutchera et al. Proc.Internat.Conf on AMS, Canbera1993, Nucl.Instr.Meth.(1993)

EXPERIMENTAL DEVELOPMENTS AND RADIOACTIVE BEAMS

New Results from Advances in ISOL Techniques

Piet Van Duppen

Instituut voor Kern- en Stralingsfysica, University of Leuven,
Celestijnenlaan 200 D, B-3001 Leuven, Belgium

Abstract. Recent developments at on-line isotope separators are discussed. Emphasis is put on laser ionisation, ion manipulation in traps and segmented gamma detection systems.

INTRODUCTION

On-line isotope separators (ISOL) have played an important role in the research related to nuclei that are situated far from the line of stability -- so called exotic nuclei. Over a period of more then 30 years, continuous developments in target-ion-source systems and mass separation (1) as well as in experimental techniques positioned at the end of the beam lines (decay-, laser and mass spectroscopy, on-line nuclear orientation,...) have revealed an impressive amount of data that improved our understanding of the manifestation of the strong, weak and electromagnetic interaction in the atomic nucleus substantially.

With the advent of several projects producing - or planning to produce - post-accelerated radioactive nuclei, the role of ISOL-systems as injectors of low-energy radioactive beams becomes increasingly important (2). However this is not the only reason why on-line isotope separators will continue to play an important role in the research field of exotic nuclei.

Recently several new developments have been introduced at ISOL-systems which enlarge their possibilities. Changes in the primary beam intensity and time structure, laser ion sources, ion- and atom traps, electronically segmented germanium detectors, and charge state breeders are only a few examples of developments that have been or will be implemented at ISOL-systems. In this contribution, we will discuss some of these developments. It is not the aim of this contribution to review the whole field of important experiments that are currently taken place at ISOL-systems. With a focus on the new developments we will try to present to the reader an outlook to the future of ISOL-systems.

As a guide throughout the paper, we summarise in a schematic way the different developments that are taking place or that are proposed (fig. 1). The central part of the figure shows an ISOL-system and the relevant developments are highlighted in different frames.

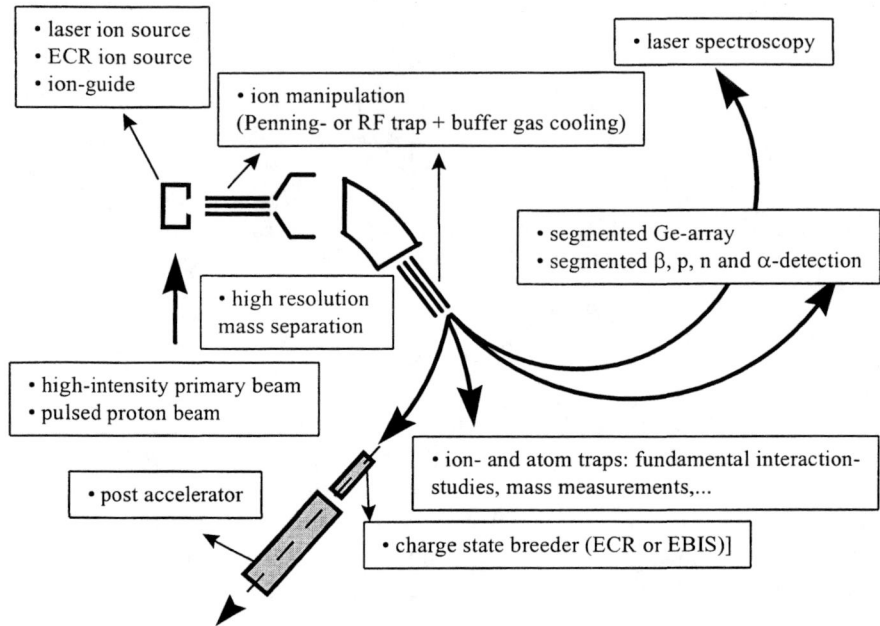

FIGURE 1. Schematic drawing of an on-line mass separator. The frames highlight recent developments and are discussed in the text.

RECENT DEVELOPMENTS

Over the last few years, recent developments have taken place at on-line isotope separators that have improved the beam intensity and quality and have created new experimental possibilities. In this chapter we will discuss some of these developments.

Primary Beam

One obvious development to improve the intensity of the mass-separated radioactive beams is an increase of the primary beam intensity. Typical primary beam intensities used at on-line isotope separators are 2µA for high-energy protons, <1 µA for heavy-ion beams, and thermal neutron fluxes of the order of 10^{11} /s. Several projects to increase these intensities are under construction. The ISAC project at TRIUMF aims at using a 100 µA current of 500 MeV protons (3) while the SPRIAL project at GANIL plans to use high-energy heavy ions (up to 96 MeV/amu) with intensities that correspond to a power deposition of 6 kW in the target (4). The design and construction of a target-ion source system that can cope with the power densities and radiation levels and that keeps the necessary properties of fast and efficient diffusion and ionisation is a technological challenge. A prototype tantalum target has been constructed at the Rutherford Appleton Laboratory in collaboration with ISOLDE-CERN (5,6). Projects to increase the thermal neutron flux up to the level of 0.3 10^{14} and 1.5 10^{14} neutrons/(cm²s) exist at the PIAFE (7) and the fission fragment separator of the FRM-II reactor (8) respectively. Finally an idea to use a high flux (100 µA) of energetic (energy peaked at 100 MeV) neutrons produced with a primary deuteron beam, has been proposed for Argonne National Laboratory's "Exotic Beam Facility" (9). All these projects will go through an interesting technical development phase in the coming years.

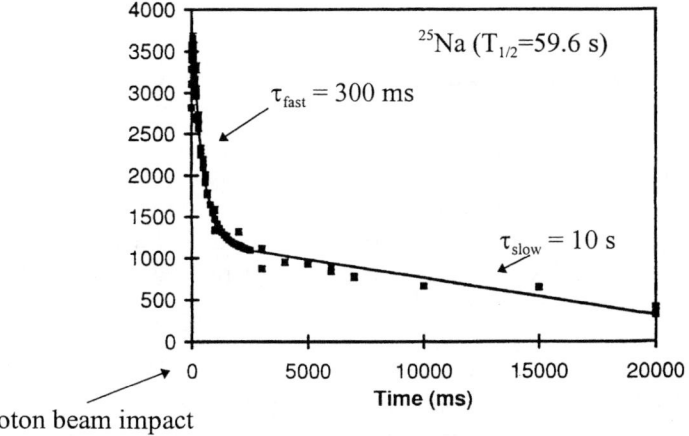

FIGURE 2 Release profile of ^{25}Na produced in a 1 GeV proton induced reaction on a tantalum foil target. The number of mass separated ^{25}Na nuclei as a function of the time elapsed after the proton beam impact is shown. The influence of the shock wave induced by the proton beam impact is clearly observed (10)

Apart from the intensity, the time structure of the primary beam can play an important role in the properties of the radioactive beams. For example, the proton beam from the PS-Booster used at the ISOLDE facility is pulsed (pulse width=2,4 µs, repetition rate = 0,4-0,8 Hz, average intensity = 2 µA, energy = 1 GeV). The high power density during the proton pulse gives rise to a thermal shock wave through the target. This phenomenon reduces the diffusion-effusion time as can be seen in figure 2. As a consequence for certain elements the production rate for the "far from the line of stability" isotopes with short half lives increases substantially. This has for example allowed detailed laser spectroscopy measurements of the neutron-rich Na isotopes up to ^{31}Na ($T_{1/2}$ = 17 ms) (11). Other consequences of the pulsed proton beam are an improvement of the peak to background ratio in certain experiments and the possibility to explore new regions of the chart of nuclei. In this way new neutron-rich isotopes around ^{208}Pb have recently been studied (12).

Target - Ion Source Systems

At the GSI on-line mass separator an elegant way to study diffusion and effusion properties of various elements has been developed and extensively used (13). The information extracted from these measurements are important for any high-temperature based target-ion source system.

The continous development of ion-guide systems (14) have led to higher secondary beam intensities allowing for the first on-line laser spectroscopy measurements on isotopes from refractory elements like the neutron-deficient Hf isotopes (15).

Electron Cyclotron Resonance (ECR) ion sources have been adapted for use at an on-line mass separator and development continues to utilise them as charge state breeders (16). The latter holds also for the Electron Beam Ion Sources (17). At present the use of ECR ion sources directly couple to a target is limited to gaseous elements since the connection between the target and the ion source is at room temperature.

An important development that took place over the last decade is the implementation of resonant photo ionisation in the ion source. High efficiency and ultimate element selectivity can in principle be reached with laser ionisation. With commercially available lasers almost all elements can be ionised which shows the universality of the principle. Laser ionisation has been implemented in two ways: in a hot cavity connected to a target and in a gas cell. A recent review of laser ion sources can be found in ref. (18).

In the hot cavity approach, developed at Gatchina (19) and ISOLDE-CERN (20, 21), a high temperature target is connected to a heated cavity or tube. The radioactive atoms of interest that are produced in the target diffuse from the target to the cavity where they are kept in gaseous form (Fig. 3). The light, from a Cu-vapor laser, enters the cavity through the exit hole. Several elements (Li, Be, Mg, Mn, Ni, Cu, Zn, Ag, Cu, Sn) have been laser ionised at ISOLDE with efficiencies between 3 and 20 %. Different experiments have been performed over the last few years making use of laser ionisation and are reported in these proceedings: decay study of neutron-rich Ag (up to ^{129}Ag) and Mn nuclei, decay study of neutron-deficient ^{58}Zn, production of a ^{7}Be target, and laser spectroscopy of ^{11}Be (22-25). The laser ion source can also be used for laser spectroscopy measurement (26,27). At the GSI-on-line mass separator a laser ion source has been used to study the β-delayed proton emission of neutron-deficient Sn nuclei (28).

Laser ionisation in a gas cell is based on photo ionisation of reaction products that are thermalised and neutralised in a buffer gas (29). A schematic drawing is given in fig. 3. This principle uses a thin target and is complementary to the above described ion source. Atoms of refractory-type elements with short half life, which are difficult in the hot-cavity approach, can be ionised in this way as they stay in the gas phase during the

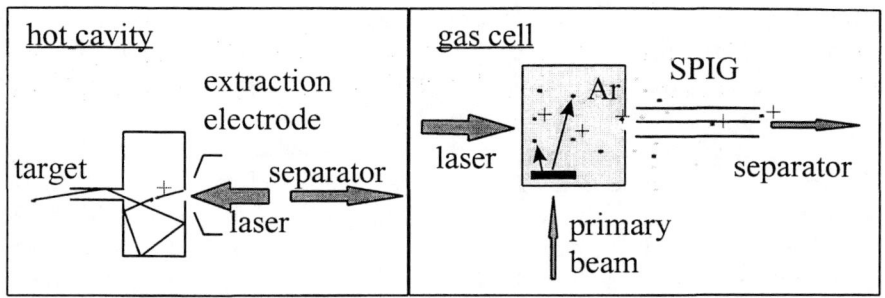

FIGURE 3. Schematic drawing of the two approaches of a laser ion source at ISOL systems. On the left hand side the hot-cavity approach is shown while on the right hand side the gas-cell approach is sketched.

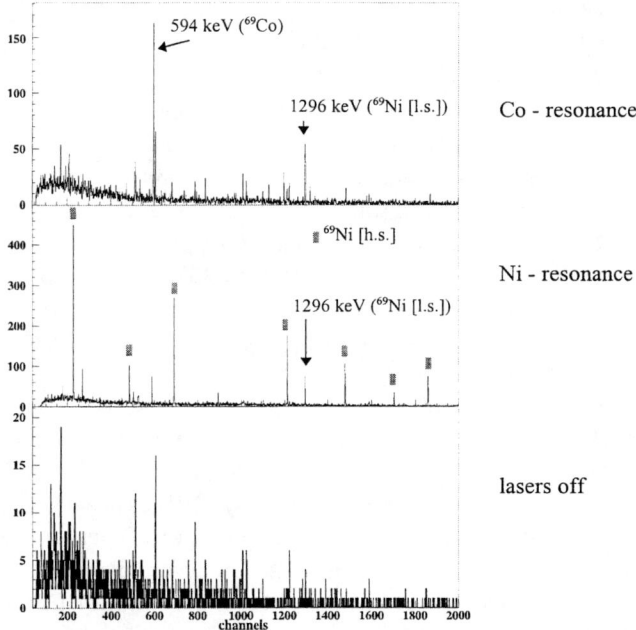

FIGURE 4. γ-spectra obtained at mass 69 with the LISOL laser ion source using a (30 MeV) proton induced fission reaction on ^{238}U. The lasers were tuned for Co resonance (top spectrum), Ni resonance (middle) and were switched off (bottom) (30). l.s. and h.s. stands for low-spin and high-spin respectively.

whole thermalisation, ionisation and extraction process. Results have been obtained with light ion fusion and fission reactions and the β-decay of $^{70-74}$Ni and $^{68-70}$Co has been studied (31).

To show the strength of resonant photo ionisation used in the ion source at an on-line mass separator, we show in fig. 4 the γ-spectra obtained at mass 69 using a proton induced fission reaction at the LISOL laser ion source. From these spectra one can immediately identify the γ-rays that belong to ^{69}Co or ^{69}Ni and at the same time extract information on the decay pattern of the ^{69}Co-^{69}Ni-^{69}Cu decay chain. More details will be given in ref. (30).

Laser ion sources will play an important role in the further exploration of the chart of nuclei. They will also be needed for the production of radioactive beams for post-acceleration. The mechanical construction of the target-ion source that is positioned in the high radiation zone can be kept as simple as possible while the laser system can be positioned in an area with free access all the time.

Ion- and Atom Traps

Using a combination of electric and/or magnetic fields, ions can be trapped for an extended period of time. This property opens up a whole series of possibilities for manipulating radioactive ions. The Penning trap coupled to the ISOLDE separator (so-called ISOLTRAP) has been operational for several years and precise mass measurements of exotic nuclei have been performed (32). An atom trap is installed at the TRIUMF isotope separator for weak interaction studies (33). This particular use of traps to store and cool radioactive ions in order to determine their properties will increase in the future. But ion traps can also be used to improve the beam quality of the secondary beams (Fig. 1). After the laser ion source at LISOL (see fig.3) a Sextupole Ion Guide (SPIG) (34) has been installed (35). This radiofrequency (RF) ion trap guides the ions from the high pressure zone to the low pressure zone for acceleration. The ions are cooled at the same time by buffer gas collisions. For example, after installation of the SPIG the mass resolving power of the separator increased by a factor of five (35). Using a trap for cooling of secondary beams after mass separation has also been proposed. A 60 keV radioactive ion beam will be injected into the ion trap. For this project a Penning trap (as for the REX-ISOLDE project) (32) as well as an RF-trap is proposed (36). Apart from the cooling property mass selectivity can also be achieved in these ion manipulators (32). This technique of radioactive ion trapping and cooling is very promising but requires much technical development. Therefore a group of research institutes and universities (JYFL-Jyväskylä, GSI-Darmstadt, CERN-Geneva, LMU-Munich, GANIL/LPC-Caen, KULeuven) have joined efforts in the so-called EXOTRAPS network aimed at dealing with the specific problems related to trapping and cooling of exotic nuclei (37).

New Detection Systems

A last point we would like to mention here is the development of new detection systems. With the advent of electronically segmented Ge detector and the existence of highly segmented Si detectors, a new generation of decay experiments will be performed at on-line isotope separators (Fig. 1). The combination of pure and spacially well-defined radioactive sources with the new generation of detectors will give rise to unique experiments. A first example of such experiments was performed at the GSI on-line mass separator where the Gamow-Teller strength in the β-decay of neutron deficient nuclei around Z=50 and around N=82 has been studied (38,39) using the so-called cluster cube. This detector set-up consisted of 6 Euroball cluster detectors that were positioned in close geometry with respect to the source. The results were complemented by total absorption measurements using the "Total Absorption Spectrometer": a highly efficient NaI-detector. The progress in the technology of electronically segmented Ge detectors, where high efficiency is combined with a small solid angle, will allow the detection of weakly produced radioactive nuclei and decay channels, and provide a way to correctly treat true summing effects that are of importance for determining exact branching ratios. At a later stage one might reach a level of sophistication where even γ-ray tracking would become possible and identification of a γ-transition can be based on only a few counts. The combination of these detectors with pure sources from laser sources and ion manipulators will be powerful. Groups like the Ge-Mini Ball, the EXOGAM and the GRETA collaboration aim at the realisation of such compact arrays of highly segmented and efficient detectors (40,41). Segmented particle detectors for the detection of higher muliplicity events are used for the study of exotic decay modes like for example two proton decay (42).

CONCLUSION

In this paper we discussed some recent developments at on-line isotope separators. Most important achievements and new developments are expected from the implementation of laser ion sources, ion traps and segmented detector arrays. The combination of these techniques will allow us a further unravelling of the structure of exotic nuclei. On-line isotope separators have played an important role in the "far from the line of stability" research. With the advent of different projects for post-accelerated radioactive ion beams they will stay on the scene as supplier of intense radioactive beams. But with the developments discussed in this paper it is clear that the latter is not the only reason for supporting ISOL systems. These developments, together with the ongoing research, guarantees important and unique contributions to the field of exotic nuclei using the low-energy beams of on-line isotope separators.

REFERENCES

1. Ravn, H.L. and Allardyce B.W., Treatise on Heavy-Ion Science (Ed. A. Bromley), Plenum Press, New York, 1989, vol. 8, pp. 363
2. Kubono, S., Kobayashi, T. and Tanihata, I., ed. of the proceedings of the 4th International Conference on Radioactive Nuclear Beams, Nucl. Phys. **A616** (1997)
3. Bricault, P.G., et al., Nucl. Instr. and Meth. **B126**, 231 (1997)
4. Villari, A.C.C., Nucl. Phys. **A616**, 21c (1997)
5. Bennett, J.R., Nucl. Instr. and Meth. **B126**, 105 (1997)

6. Drumm, P.V., Nucl. Instr. and Meth. **B126**, 121 (1997)
7. Pinston, J.A., Nucl. Instr. and Meth. **B126**, 22 (1997)
8. Habs, D., et al., Nucl. Phys. **A616**, 39c (1997)
9. Nolen, J.A., "A Target Concept for Intense Radioactive Beams in the ^{132}Sn Region", presented at the "The Third International Conference on Radioactive Nuclear Beams", East Lansing, Michigan, Ed. D.J. Morrissey, Editions Frontières, 111 (1993)
10. Lettry, J., et al., Nucl. Instr. and Meth. **B126**, 130 (1997)
11. Wilbert, S., et al, these proceedings (1998)
12. Van Duppen, P., et al., Nucl. Instr. and Meth. **B134**, 267 (1998)
13. Kirchner, R., Nucl. Instr. and Meth. **B126**, 135 (1997)
14. Dendooven, P., Nucl. Instr. and Meth. **B126**, 182 (1997)
15. Campbell, P., et al., these proceedings
16. Villari, A.C.C., Nucl. Instr. and Meth. **B126**, 35 (1997)
17. Habs, D., et al., Nucl. Instr. and Meth. **B126**, 218 (1997)
18. Van Duppen, P., Nucl. Instr. and Meth. **B126**, 66 (1997)
19. Alkazov, G.D., et al., Nucl. Instr. and Meth. **A306**, 400 (1997)
20. Fedoseyev, V.N., et al., Nucl. Instr. and Meth. **B126**, 88 (1997)
21. Lettry, J., et al., Rev. of Scien. Instr. **69**, 761 (1998)
22. Fedoseyev, V.N., et al., these proceedings
23. Keim, M., these proceedings
24. Köster, U., et al., these proceedings
25. Wöhr, A., et al., these proceedings
26. Sebastian, V., et al., these proceedings
27. Barzakh, A.E., et al., Eur. Phys. J. **A1**, 3 (1998)
28. Janas, Z., et al., Physica Scripta **T56**, 262 (1995)
29. Kudryavtsev, Y.A., et al., Nucl. Instr. and Meth. **B114**, 350 (1996)
30. Franchoo, S., et al., these proceedings
31. Franchoo, S., et al., submitted to Phys. Rev. Lett.
32. Bollen, G., Nucl. Phys. **A616**, 457c (1997)
33. D'Auria, J.M., et al., Nucl. Instr. and Meth. **B126**, 7 (1997)
34. Xu, H.J., et al., Nucl. Instr. and Meth. **A333**, 274 (1993)
35. Van den Bergh, P., et al., Nucl. Instr. and Meth. **B126**, 194 (1997)
36. Lunney, D. and Moore, B., private communication
37. http://www.jyu.fi/~armani/exotraps/frames.htm
38. Hu, Z., et al., GSI-annual report 1997, **98-1** 23 (1998)
39. Agramunt, J., et al., GSI-annual report 1997, **98-1** 29 (1998)
40. Habs, D., et al., Progr. Part. Nucl. Phys. **38**, 111 (1997)
41. http://www.ganil.fr/exogam/
42. Borge, M.J.G., et al., these proceedings

REXTRAP, an Ion Buncher for REX-ISOLDE

F. Ames[1], G. Bollen[2], G. Huber[1], P. Schmidt[1]
and the REX-ISOLDE collaboration[2]

1) Institut für Physik, J. Gutenberg - Universität, D-55099 Mainz, Germany
2) CERN, ISOLDE, CH-1211 Geneva 23, Switzerland

Abstract. Accelerating radioactive ions to several MeV/u opens up new interesting experimental fields in nuclear physics. To produce such beams at the ISOLDE facility at CERN REX-ISOLDE will post-accelerate the existing radioactive ion beams from 60 keV to up to 2.2 MeV/u. For an efficient use of the continuous ISOLDE beam an ion buncher system has been set up. It consists of a large gas filled Penning trap for accumulation and a pulsed release of the ions, allowing cooling and purification of the beam as well.

REX-ISOLDE

The aim of REX-ISOLDE (Radioactive beam EXperiment at ISOLDE) (1,2) is to post-accelerate light (A≤ 50) radioactive ions produced at the ISOLDE facility at CERN up to an energy of 0.8 - 2.2 MeV/u. This energy is well suited to study coulomb excitation and particle exchange reactions for light ions. First experiments planned are the investigation of nuclear structure near the magic neutron numbers 20 and 28. γ - spectroscopy following reactions caused by the collision of high energy radioactive ions with target atoms will be used. Besides these there are several experiments proposed, dealing with nuclear structure physics, nuclear astrophysics, atomic physics and solid state physics. They depend on several features REX-ISOLDE can provide, i.e. the high but also variable energy of the ions, the possibility of using ion bunches and the high charge state of the ions.

Figure 1 shows the layout of the post-accelerator system. A continuous beam of singly charged 60 keV ions produced by ISOLDE is decelerated and accumulated in REXTRAP, a large Penning trap located on a high voltage platform. Bunches of cooled ions are produced and reaccelerated to 60 keV. They are transported to an electron beam ion source (EBIS), which is also placed on a high voltage platform at a potential near to 60 kV. There their charge state is increased to reach a charge-to-mass ratio of $q/m \approx 1/4.5$. During the charge state breeding the platform potential can be lowered to 20 kV before extracting the ions. A magnetic q/m selector following the EBIS allows choosing the desired charge state for the injection into the accelerator. The latter consists of an RFQ to accelerate the ions to 300 keV/u, an IH-structure with an exit energy of 1.1 - 1.2 MeV/u and three 7-gap resonators to accelerate or decelerate the ions to their final energy of 0.8 to 2.2 MeV/u.

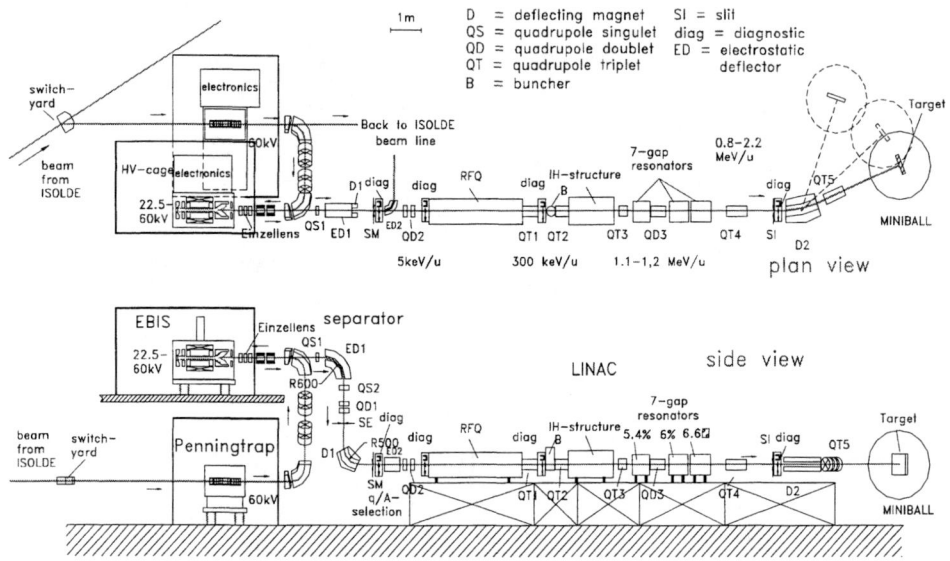

FIGURE 1. Layout of REX-ISOLDE

REXTRAP

For an efficient use of the EBIS and the accelerator system an accumulation, cooling and bunching of the ion beam is required. To insure the possibility to measure also short-lived isotopes the time separation of the bunches should be less than 20 ms. The transversal acceptance for ion injection into the EBIS will be $\varepsilon_{EBIS} \approx 5\ \pi$ mm mrad @ 60 keV. As ISOLDE emits ions with an emittance of $\varepsilon_{ISOLDE} = 30\ \pi$ mm mrad @ 60 keV cooling is required to insure a high total efficiency.

All these functions can be achieved by using a large gas filled Penning trap. The concept of bunching and cooling And also the possibility of a mass purification of the ions in such a trap has already been demonstrated in the ISOLTRAP experiment at ISOLDE (3).

The REXTRAP Principle

In a Penning trap ions are confined in radial direction by a strong magnetic field and in longitudinal direction by a quadrupolar shaped electrical field. This allows three eigen motions for the ions: one harmonic oscillation (ω_z) in axial direction and two oscillations in radial direction, the reduced cyclotron motion ω_+ and the magnetron motion ω_-. The sum of the radial frequencies gives the cyclotron frequency ω_c, which only depends on the mass of the ion and the magnetic field strength $\omega_c = q/m\ B$. Typical values for a singly charged ion with mass A = 30 amu in a magnetic field of B = 3 T, an electrical potential depth of 10 V and a characteristic trap dimension of d = 15 mm are v_z = 60 kHz, v_- = 1.2 k Hz and $v_+ \approx v_c$ = 1.5 MHz.

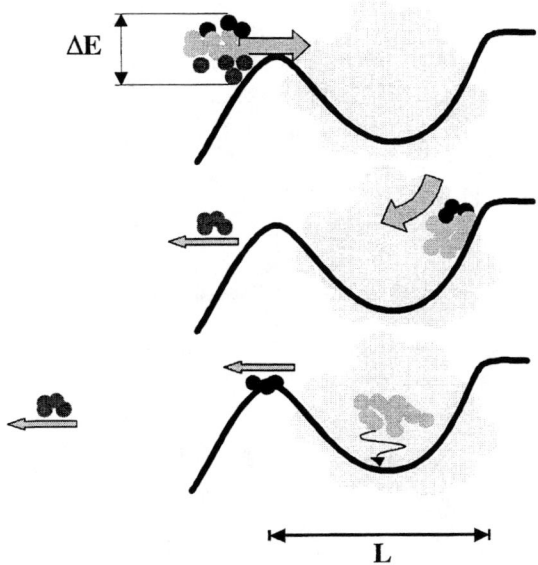

FIGURE 2. Principle of capturing ions in a gas filled trap

Capturing of externally injected ions in the potential depression of the trap requires an energy loss after entering the trap. This can be achieved by collisions with buffer gas atoms. The principle is illustrated in figure 2. Ions will enter the trap with an energy spread ΔE. All ions are captured if the energy loss during one oscillation in the trap is larger than ΔE and if the ions with lowest energy are just allowed to overcome the potential well at the trap entrance. ΔE is mainly determined by the pick-up of radial energy while the ions enter the high magnetic field region. In the case of REXTRAP this energy pickup is fully determined by the emittance of the ISOLDE beam of ε_{ISOLDE}=30 π mm mrad @ 60 keV and will be typically $\Delta E \leq 100$ eV. Figure 3 shows the energy loss calculated from the known ion mobility (4) of K^+ ions in helium or argon for a 1 m long flight through the gas and an initial energy of 200 eV. An energy loss of 100 eV in Argon can be achieved with a buffer gas pressure of p = 10^{-3} mbar.

The buffer gas inside the trap also provides cooling of the ion motion. The ions finally reach the temperature of the gas, which is room temperature in the case of REXTRAP. The lower part of the figure shows the decrease in energy for K^+ ions. For a pressure of p≥10^{-4} mbar Argon the time constant is less than 1 ms. Thus, the axial and reduced cyclotron oscillation is cooled very fast and their amplitudes decrease. However since the magnetron motion in a Penning trap is unstable, its amplitude increases with decreasing energy. This increase can be circumvented by a mass selective side band cooling technique. An application of an rf-field at the cyclotron frequency in addition to the damping provided by the buffer gas results in a simultaneous reduction of the amplitudes of both motions (5,6,7). This results in a centering of the ions in the trap.

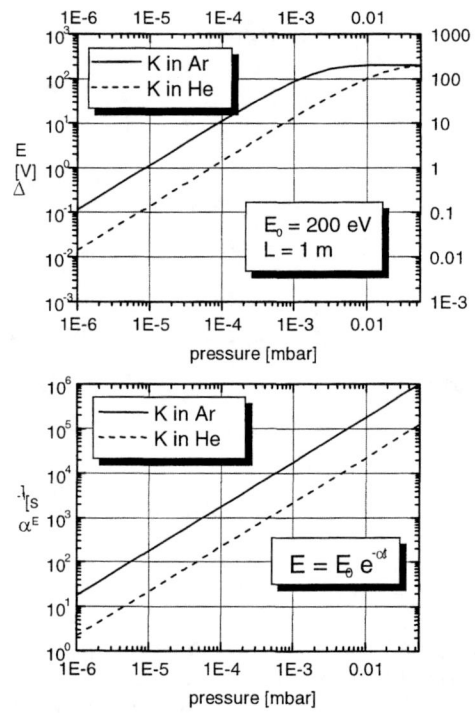

FIGURE 3. Energy loss of K^+ ions in Argon and Helium

The ejection of the ions is done by lowering the electrostatic potential well of the trap. Due to the small transverse spread of the ions inside the trap the expected transverse emittance of the beam will be reduced to $\varepsilon_{trap} \leq 3$ π mm mrad @ 60 keV. The longitudinal emittance depends mainly on the longitudinal dimension of the ion cloud in the trap and the acceleration fields. It will be less than 10 eVμs.

The REXTRAP Set-Up

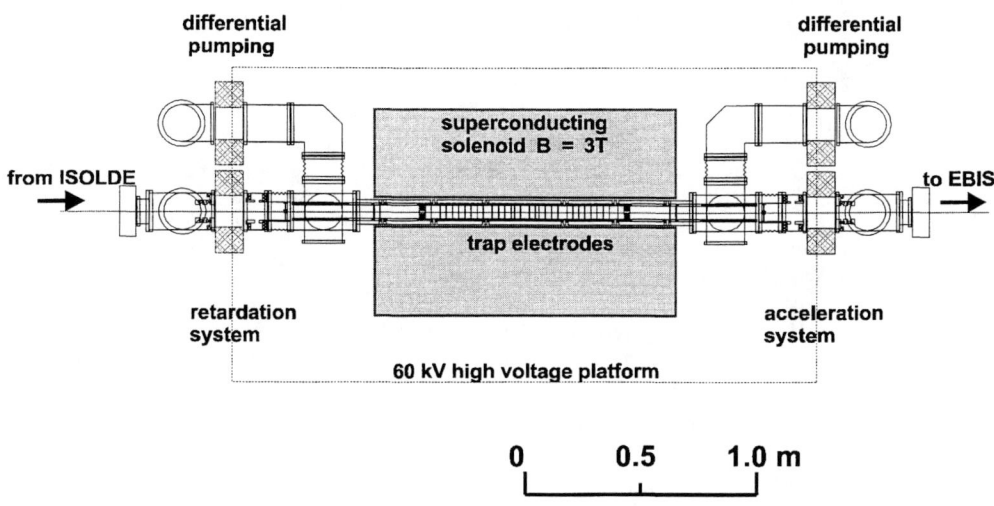

FIGURE 4. Experimental set-up

In figure 4 the experimental set-up is shown. The magnetic field for the penning trap is provided by a superconducting solenoid. It has a central field strength of $B = 3$ T over a length of 90 cm. A cylindrical electrode structure has been chosen to form the electrical field. It is made from gold plated copper rings insulated by ceramic spacers. To capture the 60 keV ions from ISOLDE they are first decelerated in two stages. This system was designed by using ion optical simulations to minimize the pick-up of radial energy when the ions enter the high magnetic field region.

The electrode structure to form the trap potential is divided into two sections separated to each other and to the outer part by diaphragms. Their dimensions are chosen in such a way that it results in the pressure distribution shown in figure 5. The first region with a buffer gas pressure of up to several 10^{-3} mbar provides the energy loss necessary to stop the ions. The second part which houses the harmonic part of the trap potential has about one order of magnitude less pressure. This gives good performance for the cooling technique described above and minimizes collisions with the gas while ejecting the ions. For the ion ejection the potential at the exit side is ramped down and after leaving the trap the ions are reaccelerated to 60 keV. The whole set-up including the electronics to regulate the trap voltages is placed on a high voltage platform at 60 kV. On both sides of the trap two differential pumping stages are installed to ensure good vacuum conditions in the beam lines going to and from the trap.

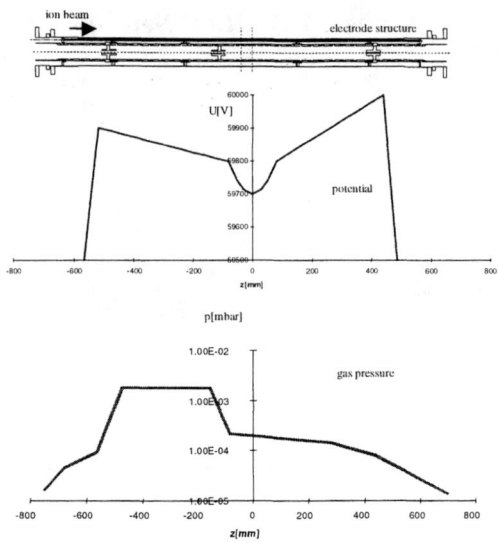

FIGURE 5. Electrode structure, voltage and pressure distribution inside the trap

RESULTS AND OUTLOOK

The system is nearly completely assembled and first measurements of the deceleration of ions and their injection into the magnetic field have already been carried out. A beam of $^{40}Ar^+$ ions from an ISOLDE plasma ion source was used and the ion current inside the high magnetic field region after deceleration and passing through the entrance diaphragm of the trap was measured. No beam losses could be observed. A full test of the REXTRAP performance with ISOLDE beam is planned in autumn 1998.

REFERENCES

1. Radioactive beam Experiment at ISOLDE: Coulomb excitation and neutron transfer reactions of exotic nuclei, proposal to the ISOLDE committee, CERN/ISC 94-25
2. D. Habs et al. Nucl. Instr. Meth. B 126 (1997) 218-223
3. H. Raimbault-Hartmann et al. Nucl. Instr. Meth. B 126 (1997) 378-382
4. L. A. Viehland and E. A. Mason, Atomic Data and Nuclear Data Tables 60, 37-95 (1995)
5. G. Bollen et al. J. Appl.. Phys. 68, (1990) 4355-4374
6. G. Savard et al. Phys. Lett. A 158, (1991) 247
7. M. König et al. Int. Journ. Mass Spectr. Ion Proc. 142, (1995) 95-116

Status of RNB Facilities in Europe

Alex C. Mueller

IN2P3-CNRS
R&D Accélérateurs, Institut de Physique Nucléaire
F-91406 Orsay, France

Abstract. The present paper discusses the status of the main European Radioactive Beam Facilities. Several facilities offer beams that are attracting large user communities, often European collaborations, but also scientists from abroad. At many laboratories, major upgradings are under way, new projects are planned, have entered construction or are even in completion. Prominent examples of this on-going effort will be described. Furthermore, the important R&D programs which often are pursued by European Networks under the auspices of the European Union are mentioned. They constitute the basis for future projects (production of radioactive beams through neutron induced fission, the development of new ultra-intense driver accelerators and of targets withstanding the highest possible charged-particle beam intensities, beam handling with emmitance improving and storage devices....). This is also the context for the work of the NuPECC study group for next generation Radioactive Beam Facilities.

INTRODUCTION

It is probably fair to state that the present proceedings, prolific in new results, demonstrate the intense interest of Nuclear Physicists in radioactive beams, including those from the old continent. If one compares to the preceding conference ENAM 95 (1), it is striking to see the wealth of new results that have been reported at today's ENAM 98 from experiments made at, e.g., CERN-ISOLDE, GANIL, GSI, Jyväskylä, Louvain-la-Neuve.

From the beginning, it may be interesting to note that these European facilities rely on very different production schemes and separation schemes. During the last years this complementarity, rather than exclusively generating a merci-less competition, has more and more developed into a source of particular strength: European networks constituted by groups from the different university laboratories and facilities are now tackling common scientific cases with various approaches and are performing the important R&D programs together.

MAKING RADIOACTIVE NUCLEAR BEAMS IN EUROPE

ISOL and In-Flight

Both techniques for the separation of Radioactive Nuclear Beams (see figure 1), ISOL ("Isotopic Separation On-Line") and In-Flight separation, are extensively used in Europe.

The ISOL method, was the first to be developed, the archetype for a radioactive beam facility, operating since more than a quarter century, is ISOLDE at CERN (2). In the ISOL technique the radioactive ions are produced by the beams of a primary accelerator or by the neutrons from a nuclear reactor bombarding a target that in general is sufficiently thick (else a catcher is used) to stop the recoiling reaction products. The latter are then transported (diffusion, jet transport...) into an ion source (surface ionization, plasma, laser...) providing element separation through chemical selection. After extraction (of the desired charge-state) the wanted mass is then obtained through electromagnetic separation allowing for experiments with low-energy beams (a few tens to a few hundred of keV). High-resolution laser spectroscopy (5) and injection into of traps (4) are examples underlining the remarkable potential of ISOL beams.

Information on the technically demanding, but also very promising R&D topics for ISOL techniques, can e.g. be found in the proceedings of the EMIS conference series (5). Some recent high-lights, note e.g. the progress of laser ion sources at ISOLDE and Louvain-la-Neuve, have been reported by van Duppen in the present proceedings (6).

Louvain-la-Neuve was the first facility in the world to *inject* ISOL beams into a second accelerator in order to obtain energies in the range 0.65-5 MeV/u. Other projects in Europe, of which will be given a few details below, will soon offer radioactive ion beams of a large variety of isotopes and/or with still higher energies. The description of projects in Asia and America has been presented at this conference by Tanihata (7) and Nolen (8).

High energy radioactive beams are presently made in Europe by the use of the heavy-ion accelerators at GANIL and GSI. The technique of in-flight separation by means of recoil spectrometers relies on the forward focusing which is present in peripheral (and also certain other) nuclear reactions. The concept of "fragment-separators" was originally pioneered with relativistic heavy ion-beams at Berkeley using beam-line elements, and at GANIL for intermediate-energy beam with a dedicate spectrometer, "LISE" (9). Consecutively, other large devices for fragment separation have been constructed and put into operation: "FRS" (GSI), "A1200" (MSU), "RIPS" (RIKEN) and "SISSI" (GANIL), see e.g. the review by Sherrill (10). Among the many highly attractive features of in-flight separated high-energy radioactive beams let us underline here the connection to a storage-ring opening completely novel experimental methods (11), or reaction studies like the hitherto unknown fission properties of many heavy nuclei (12), as reported by Geissel and Schmidt, respectively.

Yet the optical quality of secondary projectile-like fragment beams is somewhat limited, in particular while aiming at a high transmission, which means privileging the angular and momentum acceptance of the separator. The situation is best at high energy, where also contamination of incompletely-stripped charge states is minimal. However, the probability for "destroying" the wanted isotope by nuclear reactions in the various materials it passes (production target, Z-selective degrader, detector systems) increases naturally with the amount of matter to be passed.

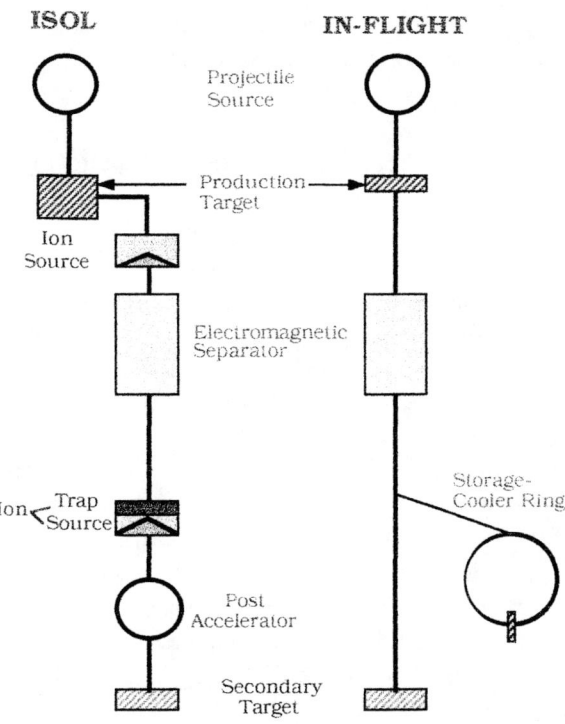

FIGURE 1. "ISOL" and "In-Flight", the two basic ways used at European radioactive beam facilities for production and separation (see text).

Up to now, the reaction experiments at the fragment separators have essentially been made down to a minimum energy of, say 25 MeV/u. Indeed, it is impossible to attain the energy band of 0-25 MeV/u by degrading the high-energy beam from a fragment separator through passage of matter and to simultaneously maintain reasonable optical properties and conserve the intensity. Cooling and decelerating by means of a storage cooler-ring is an elegant alternative, however still limited at present by transmission and, in particular for short-lived nuclei, by the necessary cooling time.

It appears today that development of techniques for the "handling" of radioactive beams (accelerating/decelerating cooling, trapping storing...) is a highly fascinating topic, with high rewards at stake for both the ISOL and the In-Flight technique. For exempla, the laboratories of Jyväskylä, Leuven, Munich, GSI Darmstadt, CERN-ISOLDE, GANIL & LPC Caen have formed EU-RTD ("European Union Research and Technical Development") Network EXOTRAP which aims at "developing techniques for cooling and trapping of radioactive nuclei", see also the contribution by van Duppen (6) for further information. Indeed, the European Union is now providing significant financial support for such common scientific endeavors where the partner institutions bring in their specific expertise for a common general goal.

Aspects Concerning Production Reactions

(In particular at high energy), numerous reactions channels may be populated. Thus, in addition to the wanted reaction product, a large amount of contaminants may be present and has to be filtered out by the carefully designed (fragment- or ISOL-) separator. Whereas the in-flight separation at GANIL or GSI separation roughly takes 1 μs, the time scales for the ISOL-method are extremely case- and design-dependent. The IGISOL method, used at the Jyväskylä facility (13), can e.g. produce beams of 100 μs isomers of quite refractory species.

Consequently, the final intensity may differ very strongly from the in-target production rate. The latter is determined by the reaction cross-section, the target thickness and the primary beam intensity. The product of the latter two quantities, the luminosity L is a characteristic for the primary ("driver") beams of a given facility.

With the exception of some early experiments various reactors (for an overview, see (14)) and recent work at Studsvik (15), predominantly a variety of charged-particle induced reactions has been used in Europe, including fragmentation/spallation, fusion, nucleon transfer, deep-inelastic or fission. The luminosities are very high for proton-induced reactions, because of the high intensities of proton accelerators and the larger possible target thickness compared to heavy ion reactions of similar energies. Thus the latter have basically lower in-target production rates and seem at disadvantage compared to proton-induced ones. Yet, the pioneering experiments mapping the drip-lines and studying magic species far-off stability like ^{28}O, ^{44}S, ^{78}Ni and ^{100}Sn (16) have been only feasible with heavy-ion reactions, because of the efficiency of the associated in-flight separation method.

A general limitation is given by the maximum heat-deposition in the target, due to the energy-loss in the slowing-down of the primary charged-particle beam. Extensive

R&D is under way at various places (CERN-ISOLDE, Rutherford Appleton Laboratory (RAL), IPN Orsay, GANIL-SPIRAL...) for the design of targets with minimized local thermal overstress (17). The SIRIUS-facility design study of the RAL relies, e.g. on this R&D.

In contrast to charged particles, neutrons, however, as primary beams, will heat the target only through the energy released by the "useful" nuclear reactions. Since the production cross sections for thermal neutrons are extremely large in the top of the distribution where one finds neutron-rich isotopes of high interest for post-acceleration, this originated the PIAFE project at high-flux reactor ILL (18) and is the basis for the newly planned facility at the FRM-II reactor at Munich (19).

Nolen (20) recently proposed the use of fast neutrons: Very large target thicknesses could e.g. be used for the production of fission fragments from a ^{238}U target, luminosites in the order of 10^{15} barn^{-1} sec^{-1} might be envisaged. The efficient release of the activity from such thick targets is a major R&D issue.

First promising experimental results have been recently obtained at Orsay. Here the deuteron beam from the 15 MeV tandem is used to produce neutron fluxes of more than 10^8 s-1 with an energy around 10 MeV. These neutrons impinge in turn on a ^{238}U target from which, by means of the device PARRNE, the produced radioactive noble gases are extracted and collected. So far, 20g of 238U have been used in form of uranium carbide (21), or molten uranium (22). For the latter, it is expected that variants containing several hundred grams of ^{238}U may soon get developed. An ISOL separator is presently installed at Orsay where PARNNE-type targets will be connected to different ion-sources of relevance for fission fragments, including laser ion sources.

A European Union RTD Network has now been constituted between GANIL, Jyväskylä, KVI Groningen, Louvain-la-Neuve and Orsay (23) in order to develop the PARRNE devices at various energies and thus to found the basis for strong fission-fragment beams at the GANIL/SPIRAL facility.

Discussing the future, "second generation", facilities and projecting experiments very far off stability, cross-section calculations for the various envisaged production reactions are very important. In this sense, the complementarity of different methods for the production of very neutron-rich species has been analyzed by Benlliure et al. at the present conference (24). Other cross section calculations for fast particle induced fission are presently made by Rubchenya (25) and Mirea (26); experimental data mainly come from GSI and Jyväskylä.

(SOME INFORMATION ON) THE EUROPEAN FACILITIES

The exact definition of a facility providing RNB on a regular basis (as compared to a set-up) is somewhat arbitrary. Here I take in principle the rather strict "large facility" criterion, as used by the European Union and apologize to all those who feel that the information given below is incomplete for this rapidly advancing field.

In-Flight Facilities

GSI Darmstadt, FR Germany, where the whole mass range of heavy ions is available with relativistic energies (up to 2 GeV/u) exploits very successfully the combination of the fragment separator FRS and the storage ring ESR, see, e.g. (11). The physics programme with exotic nuclei and beams will continue to benefit considerably by the substantial upgrade which has been scheduled since a few years. Around the year 2000, in particular due to a reconstruction of the injecting accelerator UNILAC (27), the heavy-ion synchrotron SIS can be filled up to its incoherent space charge limit. Consequently, the luminosities will have been increased by several orders of magnitude as compared to the original pioneering experiments.

Discussions related to a long-range up-grade are presently under the way at GSI (28). Very intense relativistic beams in low charge states could be extracted from a rapid-cycling synchrotron with a connection to a dedicated storage/cooler ring. The latter would also provide the possibility to operate in an electron/RNB collider mode for electron scattering experiments. To constitute a true second generation RNB facility would be the clear ambition of such a project.

In Russia, the **Flerov Laboratory** at Dubna operates two cyclotrons, U400 (K=400-540) and U400M (K=450-630), reputed for their high intensity. Physics with light radioactive nuclei from beam fragmentation can be made at the separators ACCULINNA (29) and COMBAS (30).

ISOL Facilities

Concerning ISOL+post-accelerator facilities, as already mentioned, the radioactive beam facility at **Louvain-la-Neuve**, Belgium, has been pioneering this method and is, as of today, the only facility based on this principle with a full scale experimental programme in the world. The target is connected to an ECR source. An important upgrade, called ARENAS3 (31) is presently under commissioning: A K=44 cyclotron with 25% acceleration efficiency (first beams were made in June 1998) will deliver beams for the astrophysically interesting energy range 0.2-0.8 MeV/u. For the radioactive beam production one may use, in addition to the (presently used) driver-accelerator CYCLONE 30, the cyclotron CYCLONE presently operating as post-accelerator. CYCLONE 30 delivers up to 500 µA of 30 MeV protons, whereas the

K=110 CYCLONE with its higher energy and diversity of particles may produce nuclei much further off stability, albeit at much lower intensity.

The **ISOLDE** facility at CERN, Geneva, Switzerland has from the beginning been operated as a large European collaboration. It has been outstanding in developing over a quarter-century the thick target concept for the production of radioactive species of very many isotopes (about 600 species), generally available in the 1+ or 1- charge state. Since a few years the PS-Booster (2 mA of 1GeV protons) acts as "driver"-accelerator for producing low-energy (60 keV total) mass-separated radioactive beams for a large user community. From 1999 on, ISOLDE will benefit from an energy increase of the PS booster to 1,4 GeV. Also next year the construction of a post-accelerator accepting the ISOLDE beams up to mass A=80 will be completed. Called REX-ISOLDE (32) post-acceleration up to 2 MeV/u will be available: Prior to injection into a linear structure (RFQ, interdigital H-type, linac) bunching and charge-state breeding is assured by a novel scheme based on a Penning trap and an EBIS source.

ISOL Beams are also available from the **Jyväskylä**, Finland K=130 cyclotron. There is no post-accelerator at present, but the facility has an very active low-energy RNB physics programme and contributes to the R&D on the ISOL method (13).

As already pointed out above, the neutrons from a high-flux reactor can be a very interesting driver concept for the production of fission fragments. Although the **PIAFE** project (18) is now officially discontinued, much of the invested R&D will be found back in the **Munich**, FR Germany project (19) in which fission fragments would be accelerated by a linear-structure post-accelerator to energies around the Coulomb bar. The construction of the Munich driver, the reactor FRM-II, is rapidly progressing.

At the Italian national Laboratory at **Legnaro**, the 16 MV tandem with its superconducting booster LINAC provides 20 MeV/n beams up to the mid-mass region and thus constitutes, similar to the ATLAS facility in the US, a post-accelerator of great potential (33). In the frame of the Italian long-range plan, studies for a possible RNB facility based on an intense LINAC driver are presently made.

Facilities using both In-Flight and ISOL

This short overview will finish by two facilities which (will) have the special feature of providing RNBs by both in-flight fragment separation as well as ISOL+post-acceleration. Thus very different production methods may be used and radioactive beams over a very wide energy range will be available at these places.

The super-conducting cyclotron of the **Catania** National Laboratory (K=800) in Italy delivers intermediate-energy heavy-ion beams which can be used in connection with the fragment separator ETNA or with the EXCYT (34) facility (both under construction). This latter project, connects an ISOL system to the existing 15MV tandem. Vital for the project are the most efficient injection and ejection for the cyclotron: It is based on the high-current high charge-state super-conducting source SERSE, and special care is taken for the critical electrostatic deflector in the ejection system. In the context of the Italian long range plan, a new 200 MeV proton driver is under consideration.

The **GANIL** facility at Caen, France (two coupled K=380 cyclotrons) has, since 1984, a broad RIB physics programme. It is based on In-Flight fragment separation of heavy-ion beams up to 95 MeV/u. A specialty are the high intensities available from ECR sources for very rare enriched isotopes. The new ISOL+postaccelerator SPIRAL (35) is based on the existing GANIL as a driver. The project relies on ECR techniques to inject highly-charged RNBs into the cyclotron CIME for post-acceleration to energies between 1.8-25 MeV/u.

Concerning the ISOL production scheme a vigorous R&D effort is under the way where production targets have been connected on-line to different ECR-sources (36). For radioactive isotopes of several gaseous elements produced in a "universal" projectile fragmentation (carbon) target (37), an originality of a heavy-ion driver accelerator, encouraging production rates for short-lived species, extracted in high charge states, have been observed. The construction of the cyclotron CIME, by a collaboration between GANIL and IPN Orsay, has well progressed during the last years. First (stable) test beams have been accelerated in December 1997, various connecting elements are presently installed. Thus the facility is expected to be finished by the end of 1998. The early operation will use the present experimental equipment while a dedicated new spectrometer, VAMOS and a dedicated Ge-array, EXOGAM, are both presently constructed by European Collaborations.

CONCLUDING REMARKS

Hopefully, this short article demonstrates the fantastic activity which is nowadays connected to RNBs. In lieu of an elaborated conclusion, I would like to point to a cross-disciplinary connection of this field where scientific progress is intimately linked to the luminosity progress of the used accelerators. Ultra-high current accelerators are presently developed for different, highly interesting applications: nuclear waste burning, hybrid reactors, neutron spallation sources, primary beams for muon colliders...... Design studies for high current cavities are presently under way at Legnaro, Italy (38), at CERN (39), where the re-use of the

LEP cavities is considered for this purpose, and in France, where the injector accelerator IPHI (0.1 A current!) is in construction (40).

Concerning the long term future of RNB facilities in Europe, the Nuclear Physics European Collaboration Committee, **NuPECC**, had since 1990 study groups reporting on this field. The "first generation" facilities which are now starting operation were subject of a comprehensive report (41). During 1996-1997 NuPECC initiated a review procedure of the whole field of nuclear physics. Based on the reports from the several study groups which interacted with the whole community through a series of town meetings, the new report "Nuclear Physics in Europe: Opportunities and Perspectives" has recently been published (42). The recommendations concerning the long-term future of Nuclear Structure state that the planning of European second generation facilities should now be initiated. To this end a new NuPECC group has now been formed investigating the main options for two facilities based on the two complementary methods ISOL and In-Flight. News from this ongoing work can be found on NuPECC's WWW server (http://www.e12.physik.tu-muenchen.de/nupecc/), the final conclusions are expected next year.

REFERENCES

1. *Proceedings of the International Conference on Exotic Nuclei and Atomic Masses 1995*, Arles, France, 19-23 June 1995, edited by M. de Saint-Simon and O. Sorlin (Editions Frontières, Gif-sur-Yvette), 1995, pp. 1-884
2. H.L. Ravn, *Nucl. Inst. Methods* **B70** 107-112 (1992)
3. see, e.g. M. Keim, "Recent Measurements of Nuclear Moments far from stability", and F. Leblanc et al., "Large Odd-Even Radius Staggering in the Very Light Platinum Isotopes from Laser Spectroscopy", this volume
4. G. Bollen, "Penning trap mass measurements at ISOLDE", this volume
5. See, e.g., *Proceedings of the 13th International Conference on Electromagnetic Isotope Separators and Techniques related to their Applications*, Bad Dürkheim, FR Germany, 23-27 September 1996, edited by G. Münzenberg, *Nucl. Inst. Methods* **B126** 1-440 (1996)
6. P. van Duppen, "New Results from Advances in ISOL Techniques", this volume
7. I. Tanihata, "Status of RNB facilities in ASIA", this volume
8. J. A. Nolen, "Status of RNB facilities in North America", this volume
9. Alex C. Mueller and Rémy Anne, *Nucl. Inst. Meth.* **B56/57** 559-563 (1991)
10. B. M. Sherrill, "Radioactive nuclear beam facilities based on projectile fragmentation", in *Proceedings of the 2nd Int. Conf. on Radioactive Nuclear Beams*, August 1991, Louvain-la-Neuve, Belgium, edited by Th. Delbar (Adam Hilger IOP Publishing, Bristol) 1991 pp. 3-20
11. H. Geissel et al. "Experiments with Stored Relativistic Exotic Nuclei at the FRS-ESR Facilities", this volume
12. K.H. Schmidt et al., "Fission Studies of Nuclei far from Stability", this volume
13. M. Huhta et al. in (5), pp. 201-205
14. U. Köster et al., "Comparison of Reactor Based Radioactive Beam Facilities", in *Proceedings of the International Workshop on Research with Fission Fragments*, 28-30 October 1996, Benediktbeuren, Germany, edited by T. von Egidy et al. (World Scientific, Singapore), 1997, pp. 29-40
15. B. Fogelberg, M. Hellström, L. Jacobsson et al., *Nucl. Inst. Methods* **B70** 137-140 (1992)

16. See, e.g., in reference(1) : M. Lewitowicz, pp. 427-436; O. Sorlin et al., p. 605-606 ; M. Bernas et al., pp. 481-490 ; F. Heine et al. pp. 565-569
17. See, e.g. in reference (5), J.R.J. Bennett et al., pp. 105-108; J.C. Putaux et al., pp. 113-116; J.R.J. Bennet et al., p.117-120; P.V. Drumm et al., p. 121
18. J.A. Pinston in (5), p.22-25
19. D. Habs, O. Kiester, P. Thirolf, K.E.G. Löbner et al., *Nucl. Phys.* **A616**, 39c-47c (1997)
20. J.A. Nolen in *3rd Int. Conf. on Radioactive Nuclear Beams*, 24-27 May 1993, East Lansing, Michigan, USA, edited by D.J. Morrissey (Editions Frontières, Gif-sur-Yvette) 1993, pp.111-115
21. F. Clapier, A.C. Mueller et al., *Phys. Rev. ST - Accelerators and Beams* **1** 013501 (1998)
22. Authors of reference (22), to be published
23. The SPIRAL Phase II Project, http://ganila.in2p3.fr/spiral2
24. J. Benlliure et al., "Possibilities for the Production of Neutron-Rich Isotopes", this volume
25. V. Rubchenya et al., private communication (1998) and to be published
26. F. Mirea et al., private communication (1998) and to be published
27. "Beam intensity up-grade of the GSI accelerator facility", Report GSI-95-05, Darmstadt, FR Germany, 1995
28. G. Münzenberg and H. Geissel, private communication (1998)
29. A.M. Rodin et al. in (5), pp. 236-239
30. A.G. Artukh et al. in (5), pp. 246-249
31. J. Vervier, *Nucl. Phys.* **A616** 97c-106c (1997)
32. D. Habs et al. in (5), p. 218-221
33. G. Fortuna, private communication (1998)
34. G. Ciavola, L. Calabretta, G. Cuttone, G. Di Bartolo et al., *Nucl. Phys.* **A616** 69c-75c (1997)
35. The SPIRAL Radioactive Beam Facility GANIL report R-94-02, Caen 1994
36. A.C.C. Villari in (5), pp. 35-39
37. J.C. Putaux et al. in (5), pp. 113-116
38. A. Pisent, private communication (1998)
39. G. Bollen and H. Haseroth, private communication (1998)
40. J.M. Lagniel, S. Joly, J.L. Lemaire, A.C. Mueller, "The IPHI project at Saclay", *Proc. 1997 Particle Accelerator Conference and International Conference on High-Energy Accelerators*, 12-16 May 1997, Vancouver, BC, Canada, in press
41. "European Radioactive Beam Facilities", Report of the NuPECC Study Group, Rolf H. Siemssen convenor, edited by E.G. Körner, NuPECC, (Munich, FR Germany) 1993
42. "Nuclear Physics in Europe: Highlights and Opportunities", NuPECC Report, edited by J. Vervier et al., (Munich, FR Germany), 1997, http://www.e12.physik.tu-muenchen.de/nupecc/

Status of RIB Facilities in Asia

Isao Tanihata

RIKEN, 2-1 Hirosawa, Wako, Saitama 351-0198, Japan

Abstract. Radioactive Ion Beam Facilities in Asia are presented. In China, in-flight separation type facilities are in operation at the Institute of Modern Physics in Lanzhou and the other at Tandem facility in China Institute of Atomic Energy in Beijing. The storage-ring facility is proposed and approved in Lanzhou. In India, the Variable Energy Cyclotron Facility in Calcutta start to construct an ISOL-type facility. In Japan, in-flight separation type facilities are working at Research Center for Nuclear Physics in Osaka, and at RIKEN. Also a separator start its operation in medical facility in Chiba. In RIKEN, the construction of RI Beam Factory has been started. An ISOL-type facility is proposed in the Japan Hadron Facility in KEK. Table I summarize these facilities.

1. IN CHINA

Two facilities are in operation now in China and two new facilities are proposed. These activities are at Institute of Modern Physics (IMP) in Lanzhou and at China Institute of Atomic Energy in Beijing.

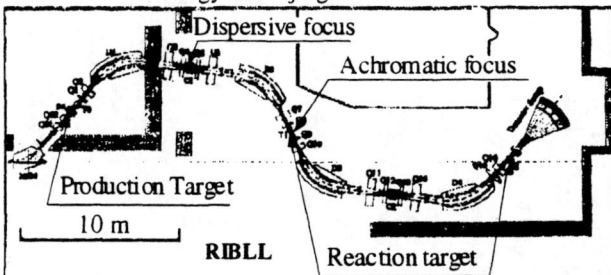

Figure 1. The fragment separator RIBLL at IMP.

The Heavy-Ion Research Facility in Lanzhou (HIRFL), which has a Separate-Sector-Cyclotron (K=450), completed the construction of a separator. This separator (RIBLL) can separate and produce the secondary radioactive beams up to 80A MeV. This separator (Fig. 1) has advanced layout; two achromatic separators in series. The second separator not only provides additional separation power but also provides a mean to tag each incident nucleus. In a single-stage separator, many unseparated nuclides are mixed at the first dispersive focus and often make it very difficult, if it is possible, to tag by the momentum of a nucleus. Separated beam includes much less nuclei at the second-stage dispersive focus and thus a high-resolution momentum tagging is possible. Therefore it provides a new mean for high-resolution studies with 'poor quality' beam. Progressing experiments include the production cross section of neutron rich nuclei, reaction cross sections with radioactive nuclei. Experiments are also planned to study proton halos in ^{17}F and ^{17}Ne.

A new storage ring facility has been proposed at HIRFL. It has been approved in Government and is waiting for funding. This facility consist of two rings called CSRm (Cooler-Synchrotron main Ring) and CSRe (Cooler Synchrotron experimental Ring).

Heavy-ion beam of energy $10A$ - $50A$ MeV from the cyclotron will be accumulated, cooled and accelerated to $900A$ MeV for C^{12+} and $400A$ MeV for U^{72+} in CSRm. The direct beam as well as the separated radioactive beams is injected into CSRe for internal target experiments. The electron cooling is applied for both of the rings.

Table I. Asian Facilities of Radioactive Ion Beams

In operation

Facility	Type	Properties	Ref.
RIBLL in IMP/Lanzhou	in-flight with K=450 Cyclotron	$\Omega=7$ msr, $\Delta p/p=10\%$, $B\rho=4.2$ Tm	
GIRAFFE in CIAE/Beijing	in-flight with HI-13 Tandem	$\Omega=1.8$ msr, $B\rho=1.4$ Tm	1
E-arena R&D in KEK/Tanashi	ISOL K=68 Cyclotron + Linac	< $1A$ MeV for A/q<30	2
Sec. Beam Line in NIRS/Chiba	in-flight with HI synchrotron	$\Omega=13$ msr, $\Delta p/p=2.5\%$, $B\rho=8.13$ Tm	3
FRS in RCNP/Osaka	in-flight with K=400 Cyclotron	$\Omega=2.8$ msr, $\Delta p/p=4\%$, $B\rho=3.2$ Tm	4
RIPS in RIKEN/Wako	in-flight with K=540 Cyclotron	$\Omega=5$ msr, $\Delta p/p=6\%$, $B\rho=5.76$ Tm	5

Facility	Type /Acc.	Properties
CSRm/CSRe in IMP Lanzhou	in-flight/Cooler synchrotron and exper. storage ring	< $900A$ MeV for light and <$400A$ MeV for heavy elements. Internal target operation in CSRe.
BRNBF in CIAE/Beijing	ISOL/Cyclotron +ISOL+ HI-13 Tandem + Superconducting linac	< $5A$ MeV up to A~140.
E-Arena/JHF in KEK/Tsukuba	ISOL/ 3 GeV, 10 µA proton synchrotron +ISOL + linac	< $1.05A$ MeV and $6.5A$ MeV stations.
RI Beam Factory in RIKEN/ Wako	K=1600 SRC + MUSES	< $400A$ MeV from SRC and $1.4A$ GeV in MUSES. DSR with electron+HI

1. X. X. Bai et al., Nucl. Phys. A588 (1995) 273c.
2. S. Kubono et al., Nucl. Phys. A616 (1997) 11c.
3. K. Sato et al., Nucl. Phys. A588 (1995) 229c.
4. S. Mitsuoka et al., Nucl. Inst. Meth. in Phys. Res. A 372 (1996) 489.
5. T. Kubo et al., Nucl. Inst. Meth. in Phys. Res. B70 (1992) 33.

In Tandem accelerator laboratory at China Institute of Atomic Energy (CIAE) has the separator GIRAFFE that provides low energy beams of radioactive nuclei. A clean beam of ^7Be are delivered. An astrophysical S17(0) factor of $^7Be(p, \gamma)^8B$ reaction has been determined from a measurement of $^7Be(d, n)^8B$ differential cross sections.[1]

An ISOL-type RNB facility has been proposed in CIAE that includes construction of 70 MeV proton cyclotron for production of radioactive nuclei and Superconducting linac for acceleration of RIB.

2. In INDIA

The construction proposal of an ISOL-type RNB facility at Variable Energy Cyclotron Center (VECC) has recently been approved. The present cyclotron will be upgraded to obtain 30 µA of p, ^{3}He and α particles in the energy range from 15 to 60 MeV. The separated radioactive nuclei will be accelerated up to $80A$ keV by an RFQ linac and subsequently to $1A$ MeV by a linac in the first phase of the project.

Reactions such as (beam, n) or (beam, xn) reactions will be used for production of radioactive nuclei. High temperature diffusion and He-jet transport system are under development. Recently, nuclei produced by ^{127}I(α, xn) reaction have been transported by the He-jet system that is combined with CsI target of 200° C. The transport efficiency was found to be 10^3 times higher than the thin target recoil transport system. Developments of ECR ion source and 35Mhz RFQ linac are in progress under the collaboration with RIKEN.

3. IN JAPAN

Three in-flight separators are in operation and two large facility projects exist. These facilities are summarized in Table I together with the other facilities in Asia.

3.1 E-Arena Project

The E-arena project is a part of the Japan Hadron Facility (JHF) at KEK, that aims to obtain high-intensity 50 GeV proton for hadron physics. (Fig. 2) The E-arena use 3 GeV proton beam from the booster synchrotron. The proton intensity of 10 µA is planned to be used in the E-arena for production of radioactive nuclei. Several different types of ion source such as a laser, a surface ionizing, and an ECR are being developed. After an ISOL of mass separation power of 9000, radioactive nuclei are accelerated to $6.5A$ MeV by three linacs; SCRFQ, IH1, and IH2. The design parameters are listed in Table II.

Table II Design parameters of E-arena

Primary Beam	3 GeV proton	10 µA
ISOL	M/ΔM	9000
	injection energy to linac	2A keV
Ion Source	Laser, Surface ionizer, ECR	
Linac	duty factor	30%
SCRFQ	frequency/ energy	25.5 MHz/ 170A keV
IH1	frequency/ energy	51 MHz/ 1.05A MeV
IH2	frequency/ energy	102 MHz/ 6.5A MeV

As a part of R&D project of the E-arena, a small size facility has been built. It uses 40 MeV proton beam from K=68 cyclotron. (Fig. 3) Beam of ^{19}Ne has been accelerated up to $465A$ keV by an SCRFQ and an IH linacs. Further improvement on the intensity is in progress. This facility intends to deliver the beam of nuclei within ±5 nucleons from the stability line with 10^9 /s of intensity and energy up to 1A MeV.

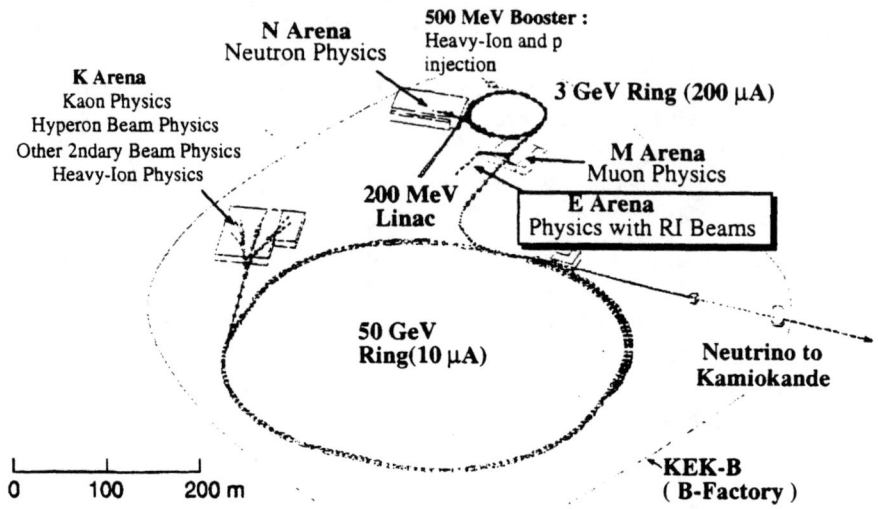

Figure 2. Layout of Japanese Hadron Facility. E-arena provides an ISOL type RNB facility of energy up to 6.5A MeV.

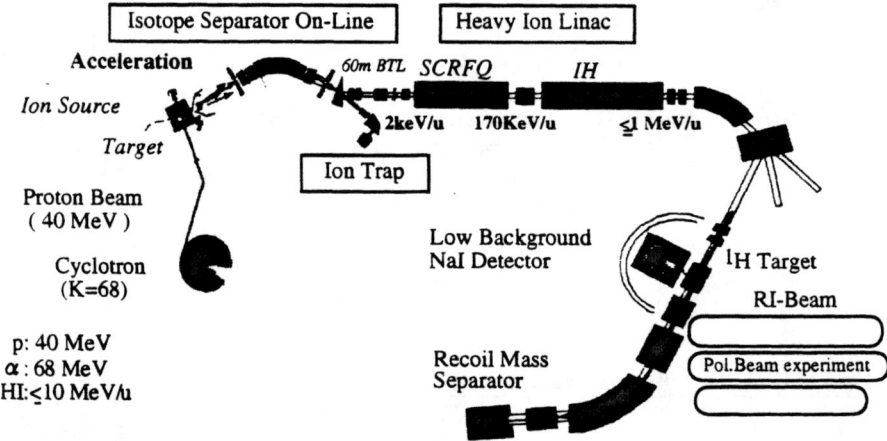

Figure 3. E-arena R&D project layout. Beam of ^{8}Li has been accelerated.

Mainly studies of reactions with astrophysical interests are planned. A low back ground NaI(Tl) detector and a recoil mass separator are under construction. A reaction related to big-bang nucleosynthesis, ^{8}Li(α, n)^{11}B, and a reaction related to hot CNO cycle and starting of rp process, ^{19}Ne(p, γ)^{20}Na, are planned in the first series of the experiment.

3.2 The RIPS and RI Beam Factory Project in RIKEN

3.2.1 Recent Development in the RIPS

The RIPS has been working since 1992 and operating as one of the pilot facility of RIB. Many experiments has been reported in these three years in variety of studies. They includes researches on new isotopes[2], nuclear moments[3], nuclear matter radii, elastic and inelastic scattering[4], Coulomb excitation[5], electromagnetic dissociation[6], (p, n) reaction to isobaric analog state of halo nuclei[7], quasi-free nucleon-nucleon scattering[8], and fusion[9], with radioactive nuclei.

In addition to nuclear physics studies, other field of studies also are going on;

1. The Mössbauer studies has been developed using radioactive beam. The use of radioactive beams enable the deep implantation of a Mössbauer element into any mother material. Also short lived isotopes can be used as Mössbauer source because the implantation and the Mössbauer measurement can be made simultaneously.[10]
2. Implantation imaging has also been developed using positron emitters and a PET.
3. The β-NMR technique that has been developed for determination of nuclear moments are also used to study internal fields of material.

Several new systems have been constructed recently. They are; i. injection system to HI linac, ii. a superconducting solenoid before the RIPS target, and iii. new detector systems. Brief descriptions are shown below.

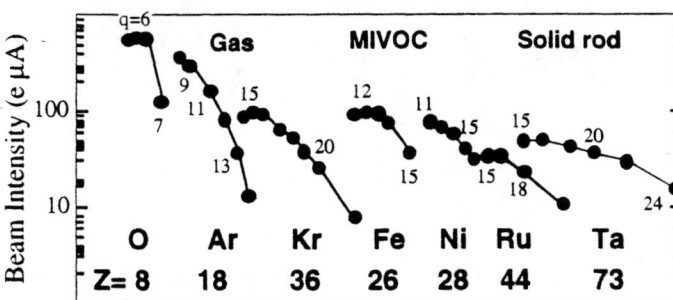

Figure 4. Beam intensity obtained by ECRIS-18.

A 18 GHz ECR ion source and a frequency variable RFQ linac have been constructed as new injection device for the variable energy linac(RILAC) that is used as an injector for the Ring Cyclotron (RRC). The new 18 GHz ECR ion source (ECRIS18) gives much higher intensity of higher charge-state ions than the present ECR ion source neomafios. The intensity of the extracted beams from ECRIS18 are shown in Fig. 4.[11] These intensity are also high enough to supply 1pµA of beams of 400A MeV at the RI beam factory. For metallic ions 'Metal Ion by Volatile Compound' (MIVOC) method has been applied using organic metallic compounds $Fe(C_5H_5)_2$, $Ni(C_5H_5)_2$, and $Ru(C_5H_5)_2$.

The variable frequency RFQ has folded-coaxial type rf resonator and thus is extremely compact (1.7 m in length) and provide energy variability of the facility.[12] Heavy ions are accelerated to 200 q·keV of energy. The transmission efficiency was measured to be 88 % at the beam intensity of 120 eμA.

A superconducting solenoid has been installed in front of the RIPS target. The field strength is 6.2 T and the beam spot size was reduced to be 0.4 mm (full width). This solenoid is also used to collect the fusion products in high efficiency. In this operation a gas target is set upstream of the solenoid and the fusion product are collected.

A new set of Li SSD spectrometer (RIKEN-telescope) for elastic and inelastic scatterings and transfer reactions has been constructed. (Fig. 5) It consists of 1 layer of 250 μm and 7 layers of 750 μm annular shapes SSD's. The first one has readout in both θ and φ using signals from both side of the detector. The other SSD's has only segmented readout in φ. The last SSD is usually used as veto counter. This telescope can be used together with the RIPS magnetic spectrometer and the neutron wall. The search experiment has been completed for an excited state of ^{7}He using p(^{8}He, d)^{7}He* reaction.[13] In the measurement, deuterons are detected by the telescope and forward going fragments, such as ^{6}He and ^{4}He, were detected by the magnetic spectrometer. A search for ^{7}H is also planned using p(^{8}He, ^{2}He)^{7}H reaction. A high segmentation of the detector enables us to detect two protons that formed a ^{2}He.

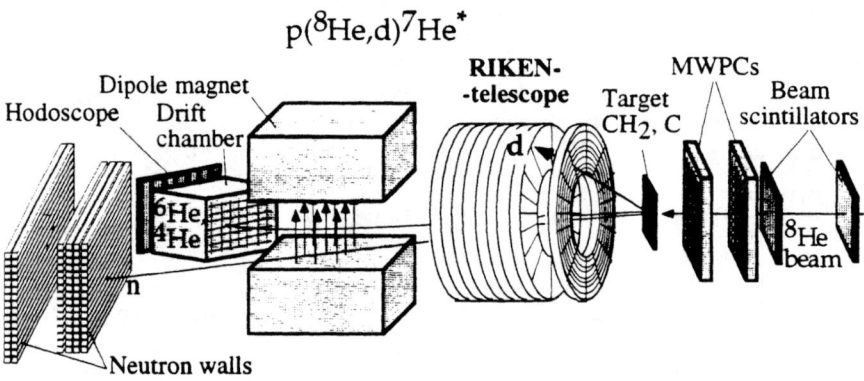

Figure 5. A newly constructed RIKEN telescope.

A detection system for quasi-free nucleon scattering has also been constructed recently. It consist of two 5" NaI(Tl) scintillation counters and drift chambers in front of them. The determination of the scattering angle is very crucial to the energy resolution. The present system gives about 1 MeV resolution for a hole state of excitation energy around 20 MeV. This system can also be combined with the forward magnetic spectrometer. It is very advantageous to use forward spectrometer for observation of a deep hole state. The deep hole state usually decays by emissions of particles. The invariant mass measurement of these final state provides a mean to restrict the excitation energy of the final state and thus can be used to eliminate the background from other region of the excitation energy. Hole states of neutron rich He and Li isotopes including ^{11}Li have been measured.[9]

3.2.2 RI Beam Factory at RIKEN

The RI beam factory has a configuration shown in Fig. 6. The accelerator system consists of a 4 sector ring cyclotron (IRC) and 6 sector (SRC) superconducting ring cyclotrons, and Multi-USe Experimental Storage rings(MUSES). A beam from the existing ring cyclotron (RRC) is injected into the IRC. Light nuclei are accelerated to $400A$ MeV and the heaviest nuclei are accelerated to $150A$ MeV with the highest intensity. Higher-energy beams of heaviest nuclei are available for a lower intensity. For example, $200A$ MeV is available if the beam is about 1/10 of the maximum intensity.

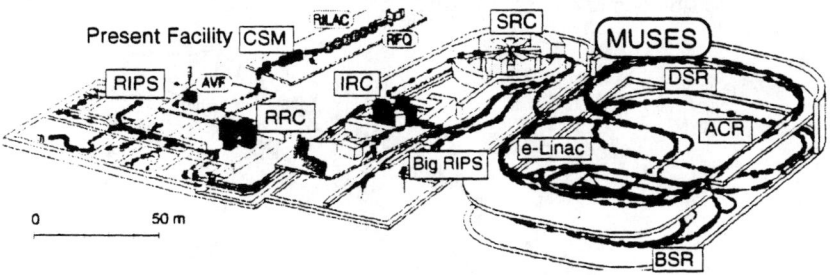

Figure 6. The layout of RI Beam Factory.

The heavy-ion beams obtained from the SRC will be converted to RI beams by a separator complex, called 'Big RIPS'. The Big RIPS consists of three fragment separators: RIPS-M, RIPS-II, and RIPS-III. The RIPS-M is a fragment separator of an ordinal energy-loss-achromat, which delivers beam to the MUSES. The RIPS-II and RIPS-III will be used in time sharing mode to deliver RI beams to experiments. Two different arbitrarily selected nuclides can be delivered to two beam lines, simultaneously. The nuclide at one line can be selected without disturbing the other.

The MUSES has four rings: Accumulator Cooler (ACR), Booster Synchrotron (BSR), and Double Storage (DSR). The ACR accumulates and cools radioactive nuclei separated by the RIPS-M. Internal-target experiments and low-energy electron-ion spectroscopies can be carried out. After cooling, ions are injected into the BSR and accelerated to $1400A$ MeV. Then they are brought into the DSR where various collision experiments can be performed. Electrons are also accelerated by the BSR to 2.5 GeV and brought into the DSR. Of course, any heavy-ions accelerated by the SRC can be used as well in all of the stages.

The available primary beam intensities and energies are shown in Fig. 7. A

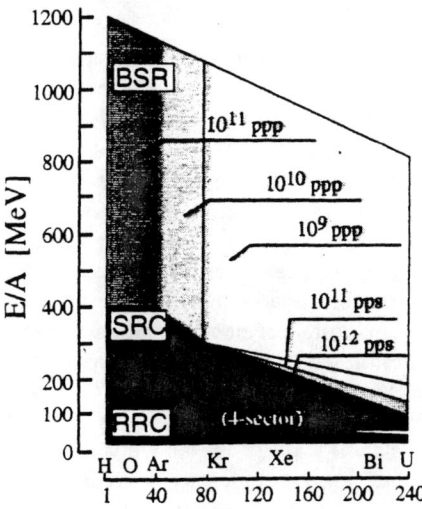

Figure 7. Energies and intensities of primary ions.

beam intensity higher than 1 particle µA (pµA) would be available for most of the elements. The intensities of the radioactive beams are shown in Fig. 8. A primary beam intensity of 1 pµA was assumed for the estimation, and the thickness of the production target and the charge states of secondary ions are selected so as to give the highest intensity. In addition to so far observed 2000 isotopes, many new isotopes are expected to be synthesized and be available as beams for experiments. A proton dripline up to Z~90 would be reached. For the neutron rich side, nuclei with A/Z<3 would also be available. For example, 10 particles per second of beam provides opportunities for measuring the interaction cross sections and momentum distributions of fragments. More than 2000 nuclide will be supplied at higher intensities than that.

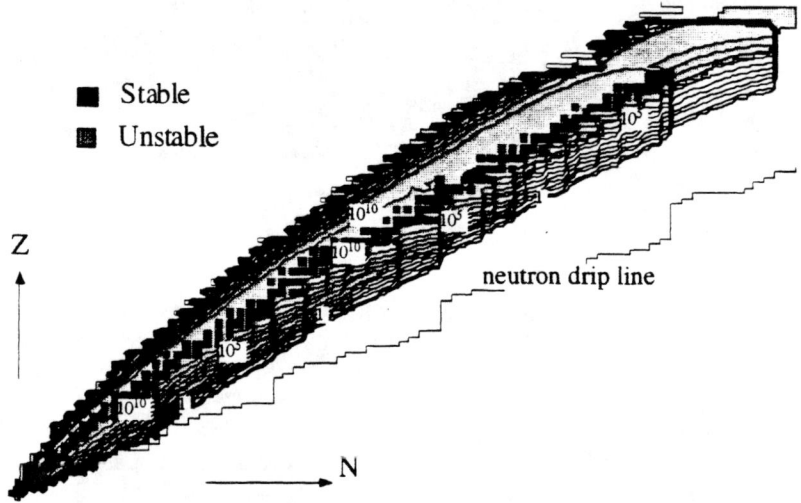

Figure 8. Intensity of radioactive nuclei separated by Big-RIPS.

The physics opportunities have already been presented in several places so that I do not present them here except for electron scattering. Interested readers are asked to refer the previously published articles.[14] Here I show an opportunity of using an electron beam in MUSES. The number of stored electrons amounts up to 2.7×10^{12} particles, which is limited by the longitudinal coupled-bunch instability. The typical colliding luminosity for the electrons and heavy ions is estimated to be 1×10^{32} /cm²s, provided that space charge limit of ions are stored and synchronously collided with electron bunches. In order to further improve the luminosity, the installation of a powerful pulsed heavy-ion source, e.g. a laser-ion source or a metal-vapor ion source are under consideration. The luminosity for electron and unstable nucleus collisions depend strongly on a nuclide. The estimated luminosity are shown in Fig. 9. The typical colliding luminosity for the electrons and RI beam is 5.6×10^{26} cm²/s when 1×10^{7} ions are stored. Yield estimations for elastic-scatterings are made for ^{39}Ca and ^{132}Sn which have luminosity of 3.9×10^{23} /cm²s and 1.4×10^{25}, respectively. Root-mean square radii of charge distribution would be determined if luminosity is larger than 10^{21}. Differential cross sections up to second minimum can be studied if luminosity is as large as 10^{25}.

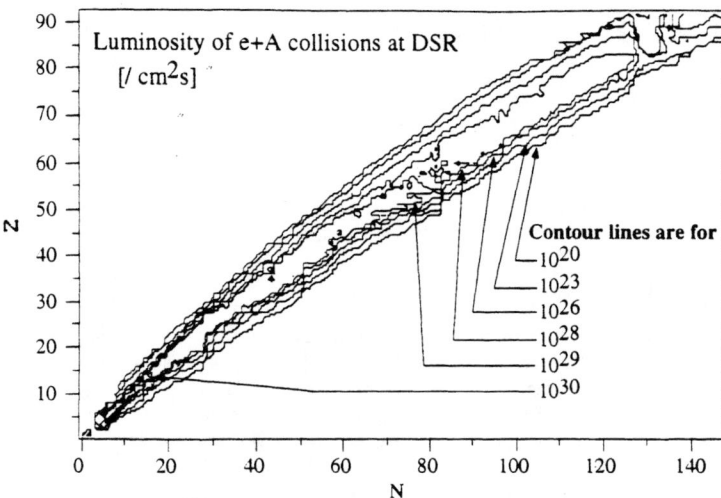

Figure 9. The estimated luminosity for electron-radioactive nucleus collisions in DSR of MUSES.

The MUSES also provides opportunities for studying a reaction with a large energy and a momentum transfer. In e+p reaction, an energy transfer up to 35 GeV and Q^2 up to $(60 \text{ GeV/c})^2$ would be available. Although the luminosity is not extremely high (10^{32} /cm^{2}s) some experiments concerning the color transparency and nucleon structure would be possible.

The construction of the first part (including SRC and Big RIPS) will be completed in 2002. The construction of MUSES will follow.

---

[1] W. Liu and X. X. Bai, Phys. Rev. Letters **77** (1996) 611.

[2] H.Sakurai et al., Phys.Rev.C54, R2802 (1996). H.Sakurai et al., Nucl.Phys.A616, 311c (1997).

[3] K. Matsuta et al.,Nuclear Physics A 588(1995)153c-156c. K. Matsuta et al., Hyperfine Interactions 97/98(1996) 519-526. T. Minamisono et al., Physics Letters B 420 (1998)31-36. H. Ueno et al., Phys. Rev. C 53 (1996) 2142. H. Izumi et al., Phys. Lett. B366 (1996) 51.

[4] A. A. Korcheninnikov et al., Phys. Rev. C **53** (1996) R537-R540. A. A. Korcheninnikov et al., Phys. Rev. Letters, **78** (1997) 2317 -2320.

[5] T. Motobayashi et al., Phys. Lett. B346 (1995) 9. T. Nakamura et al., Phys. Lett. B394 (1997) 11.

[6] N. Iwasa et al., J. Phys. Soc. Jpn **65** (1996) 1256., T. Kikuchi et al., Phys. Lett. B391 (1997) 261.

[7] T. Teranishi et al., Phys. Lett. B407 (1997) 110.

[8] T. Kobayashi et al., Nucl. Phys. **A616** (1997) 223c.

[9] A.Yoshida et al., Phys. Letters B389 (1996) 457. M. Petrascu et al., Phys. Letters 3(4) (1995) 214.

[10] Y. Kobayashi et al., Hyperfine Interactions, in press

[11] T. Nakagawa et al., Proceedings of 13th International Workshop on ECR Ion Source, 26-28 Feb. 1997, Texas A&M Univ. College Station, TX. USA, p10.

[12] O. Kamigaito et al., Proc. 18th Int. Linear Accelerator Conf., Geneva, Switzerland, pp. 863-865 (1996).

[13] A. A. Korcheninnikov et al., contribution to this conference.

[14] I. Tanihata, Nuclear Physics **A616** (1997) 56c-68c. Isao Tanihata, Nuclear Instruments and Methods in Physics Research B **126** (1997) 224-230.

Status of RNB Facilities in North America

Jerry A. Nolen

Physics Division, Argonne National Laboratory
Argonne, IL 60439, U.S.A.

Abstract. This paper presents the status of accelerator facilities in North America that are involved in research using radioactive nuclear beams (RNB), including existing and operating facilities, ones currently under construction or undergoing major upgrades, and ones being planned or proposed for the future. Existing RNB facilities are located at TRIUMF (TISOL) in Vancouver, B.C., the Holifield Radioactive Ion Beam Facility (HRIBF) at Oak Ridge National Laboratory, the Argonne Tandem Linear Accelerator System (ATLAS) at Argonne National Laboratory, the National Superconducting Cyclotron Laboratory (NSCL) at Michigan State University, the Nuclear Structure Laboratory at The University of Notre Dame, the 88" Cyclotron at Lawrence Berkeley National Laboratory, and the Cyclotron Institute at Texas A&M University. Currently, there are two major RNB facility upgrades in progress in North America, one at TRIUMF, the ISAC project, and one at NSCL, the Intensity Upgrade project. For the future, the U. S. Nuclear Science Advisory Committee has given high priority for an advanced RNB facility of the ISOL type. Concepts for such a facility, currently being developed at Argonne National Laboratory and Oak Ridge National Laboratory, are presented. Plans are also being developed in Canada at TRIUMF for a major upgrade of the ISAC facility.

INTRODUCTION

There are four qualitatively different types of radioactive beam facility: isotope-separator-on-line (ISOL), batch, in-flight, and fragmentation. ISOL facilities have existed at reactors and accelerators for over 30 years, but methods to accelerate the separated isotopes for nuclear physics research with energetic radioactive beams have only been developed recently. At ISOL facilities the production targets and ion sources are closely coupled so that isotopes are produced, extracted, ionized, separated, and delivered for research on time scales of milliseconds to seconds. In batch-mode facilities the production is physically separated from the ionization and acceleration so that isotopes must have half-lives typically in the minutes to hours range. The in-flight and fragmentation modes use nuclear reactions of energetic primary beams to produce secondary radioactive species that are delivered for research without stopping, on time scales of microseconds. In this paper the in-flight mode is distinguished from the fragmentation mode. The in-flight mode refers to the production of secondary beams via direct reactions with inverse reaction kinematics in the beam line. Typical reactions are inverse (p,n) or (d,n) using plastic foil or hydrogen gas cell targets. The primary heavy-ion beams are typically at energies somewhat above the Coulomb barrier. The fragmentation mode is similar except the primary beam energies are typically 100 MeV per nucleon or higher and the reaction is via an abrasion/ablation mechanism rather than a simple transfer or charge-exchange.

EXISTING FACILITIES

There are seven nuclear physics laboratories in North America, which currently do research with radioactive beams. They are listed briefly in the Table 1 below along with an indication of the production mechanisms used at each. They are separately discussed in separate sub-sections below, sorted according to the production mechanisms used.

TABLE 1. Summary of Existing Facilities

Location and name	Type
TRIUMF/TISOL	ISOL (not accelerated)
ORNL/HRIBF	ISOL (accelerated)
LBNL/88"/BEARS	Batch
ANL/ATLAS	Batch & In-flight
UND/TWINSOL (UND/UofM)	In-flight
TAMU/K500-MARS	In-flight
MSU/NSCL/K1200	Fragmentation

Current ISOL Facilities

There are two active ISOL facilities in North America, one at TRIUMF in Vancouver, BC, and one at Oak Ridge National Laboratory in Tennessee.

TRIUMF/TISOL

The TISOL facility (1) at TRIUMF is a traditional ISOL facility, where the radionuclides are mass separated at ion source energies of 60 keV, but not post accelerated. TISOL occupies one of several beam lines from the TRIUMF 500 MeV proton cyclotron and uses up to 1.5 microamperes of primary beam current. Typical isotopes produced and used recently in research include ^{9}C, ^{16}N, ^{17}Ne, and ^{37}K, with intensities up to about 10^9/s. Beta-decay studies related to nuclear astrophysics and fundamental symmetry studies in atom traps (TRINAT) are being pursued.

ORNL/HRIBF

The Holifield Radioactive Ion Beam Facility (2) at Oak Ridge was recently completed and is currently in the process of developing radioactive beams for research. Radionuclides are produced using beams such as p, d, or α's from the ORIC cyclotron at energies of 50 to 100 MeV and with currents of 10-20 microamperes. The production target/ion source complex produces negative ions for post acceleration by a 25 MV Tandem accelerator. Beams of ^{69,70}As have been developed and delivered to targets at intensities on the order of 10^5/s, and beams such as ^{17}F and ^{56}Ni are currently being developed with expected intensities of 10^5-10^8/s. Experimental apparatus includes the Daresbury Recoil Separator (DRS) and the Recoil Mass Separator (RMS) for nuclear astrophysics and nuclear structure research.

Current "Batch-Mode" Facilities

The superconducting heavy ion linac ATLAS at Argonne has been doing research with accelerated radioactive beams for three years, using both the batch mode and in-flight production mechanisms; they are described separately in this sub-section and the next one below. A group at the LBNL 88" cyclotron is currently developing a batch-mode radioactive beam capability described below in this sub-section.

ANL/ATLAS

In 1995 a ^{18}F ($T_{1/2}$=110 min.) beam was developed and used for nuclear astrophysics measurements at ATLAS. The radionuclide was produced and prepared for the ion source at the University of Wisconsin medical cyclotron and flown to Argonne where ^{18}F(p,α) cross section excitation functions were measured (3). More recently, a ^{56}Ni beam was developed and used to measure spectroscopic factors via the (d,p) reaction in inverse kinematics (4). The ^{56}Ni ($T_{1/2}$=6 days) was produced at the 50 MeV proton linac of the ANL pulsed neutron source (IPNS) and transferred to the ion source of the Tandem at ATLAS for ionization and acceleration. Beam intensities on target were from 3×10^6/s at 0.6 MeV/u for ^{18}F to 3×10^4/s at 5 MeV/u for ^{56}Ni. The experimental apparatus used at ATLAS included the split-pole magnetic spectrograph in the gas-filled mode and the Fragment Mass Analyzer (FMA) in conjunction with a large area silicon detector array near the target.

LBNL/BEARS

A project called Berkeley Experiments with Accelerated Radioactive Species (BEARS) has been initiated at the LBNL 88" cyclotron (5). The plan is to produce radioactive isotopes such as ^{11}C, ^{13}N, and ^{14}O at the LBNL medical cyclotron that is located 300 meters from the 88" cyclotron. The activity will be carried through a tube via a carrier gas from the medical cyclotron to the 88" cyclotron. Initial tests with activity created at the 88" cyclotron have demonstrated that ^{11}C and ^{14}O can be separated from the carrier gas in a liquid nitrogen trap and subsequently delivered slowly into the ECR ion source. Ionization and transport efficiencies of over 3% of the trapped ^{11}C into the 4+ charge state were measured after the ion source analyzing magnet (6). Once the two cyclotrons are coupled via the transfer line, accelerated beams in the 1-20 MeV/u energy range with intensities $\sim 10^4$-10^8/s are envisioned. The 8π-detector array, large area silicon detectors, and the new Berkeley Gas-filled Separator (BGS) are available for research.

Current In-Flight-Production Facilities

There are three nuclear physics laboratories in North America that are currently doing research with radioactive beams produced via direct reactions in inverse kinematics. Their facilities and activities are summarized below.

Notre Dame/University of Michigan

For several years a University of Notre Dame and University of Michigan collaboration has been doing research using superconducting magnetic solenoids with large solid angle to capture and refocus radionuclides produced via reactions in the beam line (7). Their current project is based on a pair of superconducting solenoids known as TwinSol (8). The results of first experiments with TwinSol using beams of ^{6}He and ^{8}Li are reported in paper G3 of this conference (9).

ANL/ATLAS

An in-flight production system has been developed at ATLAS during the past two years. The system consists of gas cells with thin Havar windows operating at about 600 Torr of H_2 or D_2. Inverse (p,n) or (d,n) reactions are induced by primary heavy ion beams with intensities of roughly 250 particle nanoamperes. The reaction products are collected and focussed by a superconducting solenoid. A superconducting resonator following the solenoid is used to debunch the time structure of the secondary beam to reduce its energy-spread (10). Excitation functions of the cross section for the fusion of ^{17}F+^{208}Pb have been measured (11) at beam energies of 87-100 MeV with beam intensities of about 3×10^6/s on target.

Texas A&M/MARS

Groups at the Texas A&M Cyclotron Institute have been using primary heavy ion beams from the K500 superconducting cyclotron to initiate reactions in gas targets in the beam line. Radioactive ^{7}Be beams have been used to study the (^{7}Be,^{8}B) reaction on ^{10}B and ^{14}N to determine parameters relevant to the ^{7}Be(p,γ) capture reaction rate (12). The experiments used the recoil mass separator, MARS, to achieve a 99.5% pure ^{7}Be beam at a rate of 10^5/s at the focal plane, where the secondary reaction targets were located. Other research involves the use of radioactive beams to study the isospin dependence of nuclear reactions.

Current Fragmentation Facility

The only fragmentation facility in North America is at the National Superconducting Cyclotron Laboratory (NSCL) at Michigan State University.

Michigan State/NSCL

The NSCL K1200 superconducting cyclotron (13) produces primary beams of heavy ions in the 100 to 200 MeV per nucleon energy range. Energetic radioactive beams produced via the fragmentation mechanism have been used for nuclear physics research at the NSCL for about 10 years. The primary beam, with currents up to approximately 100 particle-nanoamperes, interacts with a target at the front of a Projectile Fragment Separator, the A1200 (14, 15). Studies of reaction mechanisms, nuclear structure, and nuclear astrophysics are carried out using the A1200 itself or downstream in other apparatus. The instruments available include magnetic spectrographs (the old S320 or the new S800), recoil separators (the RPMS), large-acceptance charged particle and neutron detector arrays, the University of Michigan superconducting solenoid (BIGSOL), and general-purpose scattering chambers. A measurement using the S800 and a radioactive ^{8}B beam was done recently (16, 17).

FACILITIES UNDER CONSTRUCTION

TABLE 2. Facilities with Upgrades Currently in Progress

Location and name	Type	Construction status
TRIUMF/ISAC	ISOL	Commissioning in '99-'00
NSCL/K500⊗K1200	Fragmentation	Commissioning in '00-'01

TRIUMF/ISAC

The construction of ISAC (Isotope Separation and Acceleration) at TRIUMF is well underway (18,19). The new facility involves adding a new beam line from the TRIUMF cyclotron to deliver 500 MeV protons with currents up to 100 µA to a new target/ion source complex. Radioactive ions are mass analyzed and transported to the RNB accelerator. The accelerator comprises an 8-m long RFQ and an IH-type drift-tube linac. The accelerator runs CW and can accept ions with m/q up to 30. The first half of the RFQ has been assembled and tested successfully with stable beams, N$^+$ and N$_2^+$. It operated CW at the required voltage and the energy-gain and acceptance efficiency agreed with design calculations (19). The output energy of the completed accelerator will be continuously variable from 150 keV/u up to 1.5 MeV/u. The construction of a recoil-mass separator, DRAGON, optimized for measuring capture-reaction cross sections has begun. A windowless gas target apparatus to be used for (p,γ) and (α,γ) measurements is nearing completion.

Michigan State/K500⊗K1200

A major project involving the coupling of the K500 and K1200 superconducting cyclotrons at the NSCL is underway (20). It also involves replacing the present A1200 beam analysis system by an improved A1900 beam analysis system which will have more bending power and a higher capture efficiency to allow preparation of very neutron-rich secondary beams. Large gains in radioactive beam intensity, often by two to three orders of magnitude, and significant gains in maximum energy for very heavy beams will be achieved. Primary beam intensities of up to 1 particle microampere at energies of 200 MeV/u for light, N=Z, ions will be available. For the heaviest beams, such as ^{238}U, the intensities and energies will be approximately 10^9/s and 100 MeV/u, respectively. The project is scheduled for completion in the year 2000.

PROPOSALS FOR NEW FACILITIES OR FACILITY UPGRADES

In North America there are proposals being developed for ISOL-type facilities in Canada and the United States. A proposal to increase the radioactive beam energy and mass range at ISAC (ISAC2) is being developed at TRIUMF. In the U. S., the Nuclear Science Advisory Committee has given high priority for an advanced RNB facility of the ISOL type. Concepts for such a facility, based on addressing the scientific goals spelled out in a recent white paper (21), are currently being developed at Argonne National Laboratory and Oak Ridge National Laboratory, as indicated in Table 3 below. Either facility would provide instrumentation for research with RNB over a broad mass range and at the various energies appropriate to address the physics delineated in the white paper.

TABLE 3. Proposals for New or Upgraded Radioactive Beam Facilities in North America

Facility proposals	Time scale
TRIUMF/ISAC2	Next 5-year plan
U.S. DOE Advanced ISOL Facility	Begin construction ~FY 2003

Possible DOE sites	Facility concepts
Argonne:	Multi-beam driver + ATLAS
Oak Ridge, option 1:	Spallation neutron source + Post Accelerator
Oak Ridge, option 2:	HRIBF facility upgrade

TRIUMF/ISAC2

The ISAC2 proposal currently being developed at TRIUMF is to greatly extend the range of the radioactive beams available for research by increasing the maximum energy available to about 6.5 MeV/u and the mass capability up to roughly A=150 (22). For masses above 30 the RFQ requires higher charge states (m/q≤30), which would be accomplished either with ECR ion sources or some type of charge-state multiplier. A new drift-tube linac for the high-energy end would be either a room temperature IH-type or a superconducting linac based on short, independently phased resonators. A major building addition would provide space for new experimental apparatus.

ANL Concept

The Physics Division at Argonne National Laboratory is developing a concept for an advanced ISOL-type radioactive beam facility based on ATLAS (23,24). The ANL concept features a multi-beam driver and a radioactive beam accelerator with an injector capable of accelerating very heavy ions from 1+ ISOL-type ion sources. Preliminary work is currently in progress to design a 100-kW-beam-power, multi-beam driver based on a 200-MV CW superconducting linac (25). The low q/m RNB injector is based on the development of a 12-MHz CW RFQ. A full-scale prototype of this RFQ has been operated successfully at the voltage required to accelerate mass 132/1+ ions (26). This prototype will be tested with ^{132}Xe beam in the near future. The multi-beam driver concept gives the facility the flexibility to select the type and energy of primary beam appropriate for optimal production of specific radionuclides. Driver beams available from the 200-MV linac include protons and deuterons at any energy up to 200 MeV, secondary neutron beams (27), and heavy ions at energies up to 100 MeV/u. Proton beam energies above 200 MeV could be made available in the future by using the 200-MV CW driver as the injector for an isochronous cyclotron or higher energy superconducting linac booster. With the present ATLAS linac serving as the heart of the RNB accelerator very high quality radioactive beams over a broad mass range at energies up to 15 MeV/u would be available for research

ORNL Concepts

Two options for an advanced ISOL facility are being considered at Oak Ridge (2). One option is based on using a fraction of the 1-GeV proton beam from the Spallation Neutron Source (SNS) that is currently planned for construction at ORNL. About 100 µA of the SNS proton beam would be diverted to a new ISOL-type RNB facility. The second option is a major upgrade of the present HRIBF. A new high power driver linac would replace the present ORIC cyclotron and a booster linac would follow the 25 MV Tandem to provide RNB over a broader mass range and at higher energies.

LINKS TO RELATED WEB SITES

There are many sites on the World Wide Web with figures and more detailed descriptions of the material summarized in this summary paper. A few are listed in Table 4 below.

TABLE 4. Web Sites Related to Radioactive Beam Facilities in North America

Location and name	Web Site
TRIUMF/TISOL/ISAC	www.triumf.ca/isac/lothar/isac.html
ORNL/HRIBF	www.phy.ornl.gov/hribf/hribf.html
LBNL/88"/BEARS	www-nsd.lbl.gov/88_docs/whatsnew.html
ANL/ATLAS	www.phy.anl.gov/
UND-UofM /TWINSOL	www.physics.lsa.umich.edu/twinsol/
TAMU/K500-MARS	wwwcyc.tamu.edu/

MSU/NSCL   www.nscl.msu.edu/facility/home.html
Columbus White Paper   www.er.doe.gov/production/henp/isolpaper.pdf

REFERENCES

1. Domsky, M., *et al.*, " *Nucl. Instr. Meth.* **B70** 125 (1992).
2. Alton, G. D., and Beene, J. R., *J. Phys. G: Nucl. Part. Phys.* **24** 1347-1359 (1998).
3. Rehm, K. E., *et al.*, *Phys. Rev.* **C52** R460 (1995).
4. Rehm, K. E., *et al.*, *Phys. Rev. Lett.* **80** 676-679 (1998).
5. Powell, J., *et al.*, "BEARS: Radioactive Ion Beams at LBNL," paper PG12, ENAM98.
6. Xie, Z.-Q., private communication.
7. Becchetti, F. D., and Kolata, J. J., "Low-Energy Radioactive-Beam Experiments Using the UM-UND Solenoid RNB Apparatus at the UND Tandem: Past, Present, and Future," *Proceedings of the 14th International Conference on the Application of Accelerators in Research and Industry*, Denton, Texas, USA, November 1996, AIP conference proceedings No. 392, 369-375 (1997).
8. Lee, M. Y., *et al.*, "TwinSol: A Dual Superconducting Solenoid System for Low-Energy Radioactive Nuclear Beam Research," *ibid.*, 397-400.
9. Kolata, J. J., "First Results From the TwinSol RNB Facility," paper G3, ENAM98.
10. Pardo, R. C., et al., "Enhanced Transport and Control of In-Beam Produced Radioactive ^{17}F," *Proceedings of the XIX Intern. Linac Conf., Chicago, IL, August 23-28, 1998*, to be published.
11. Rehm, K. E., *et al.*, Phys. Rev. Lett., to be published.
12. Tribble, R. E., et al., "Determination of the ^{7}Be(p,γ)^{8}B rate, $S_{17}(0)$," paper F7, ENAM98.
13. Nolen, J. A., *et al.*, "Commissioning Experience with the NSCL K1200 Superconducting Cyclotron," *Proceedings of the 12th International Conference on Cyclotrons and Their Applications, Berlin, May 8-12, 1989*, World Scientific, 5-8 (1991).
14. Sherrill, B.M., *et al.*, *Nucl. Instr. and Meth.*, **B56/57** 1106 (1991).
15. Sherrill, B.M., *et al.*, "Initial Operating Experience with the A1200 Fragment Separator," *Nucl. Instr. and Meth.*, **B70** 298 (1992).
16. Davids, B., *et al.*, "Measurement of E2 Transitions in the Coulomb Dissociation of ^{8}B," paper F10, ENAM98.
17. Davids, B., *et al.*, *Phys. Rev. Lett.* **81** 2209-2212 (1998).
18. D'Auria, J. M., *et al.*, "The ISAC Radioactive Beams Facility in Canada: Progress and Plans," paper PG5, ENAM98.
19. Laxdall, R. E., et al., "Status of the ISAC Accelerator for Radioactive Beams," *Proceedings of the XIX Intern. Linac Conf., Chicago, IL, August 23-28, 1998*, to be published.
20. York, R. C., "The NSCL Coupled Cyclotron Project - Overview and Status," *Proceedings of the 15th Intern. Conf. on Cyclotrons and their Applications, Caen, June 14-19, 1998*, to be published.
21. "Scientific Opportunities with an Advanced ISOL Facility," November, 1997.
22. Baartman, R., Laxdal, R. E., and Root, L., "Long Range Plan Proposal for an Extension to ISAC," *Proceedings of the XIX Intern. Linac Conf., Chicago, IL, August 23-28, 1998*, to be published.
23. "Concept for an Advanced Exotic Beam Facility Based on ATLAS, a Working Paper," Physics Division, Argonne National Laboratory, February, 1995.
24. Nolen, J. A., "Accelerator Complex for a Radioactive Ion Beam Facility at ATLAS," Proceedings of the 1995 Particle Accelerator Conference, IEEE 95CH35843, 354-356 (1996).
25. Shepard, K. W., *et al.*, "Niobium Spoke Cavities for a Superconducting Light-ion Linac," *Proceedings of the XIX Intern. Linac Conf., Chicago, IL, August 23-28, 1998*, to be published.
26. Shepard, K. W., *et al.*, "A Low-charge-state CW RFQ," *Proceedings of the XIX Intern. Linac Conf., Chicago, IL, August 23-28, 1998*, to be published.
27. Nolen, J. A., "A Target Concept for Intense Radioactive Beams in the ^{132}Sn Region," *Proceedings of the Third International Conference on Radioactive Nuclear Beams, East Lansing, MI, May 24-27, 1993*, Ed. D. J. Morrissey, Editions Frontieres, pp. 111-114.

Possibilities for the production of neutron-rich isotopes

J. Benlliure[a,b], F. Farget[a], A.R. Junghans[a] and K.-H. Schmidt[a]

[a] *Gesellschaft für Schwerionenforschung, 64291 Darmstadt, Germany*
[b] *Univ. of Santiago de Compostela, 15706 Santiago de Compostela, Spain*

Abstract. The production of very neutron-rich isotopes is discussed in terms of the two possible reaction mechanisms leading to the formation of these nuclei, projectile fragmentation and fission. The reactions are described by means of an abrasion-ablation model including a decription of the fission channel and an analytical description of the cold fragmentation process. The results of the model calculations allow to conclude about future perspectives in the production of very neutron-rich isotopes.

INTRODUCTION

The possibility to extend the limits of the chart of the nuclides towards the neutron-rich side opens new opportunities for nuclear-structure investigations. Two reactions mechanisms are mainly used for the production of these isotopes, projectile fragmentation and fission. Different experimental techniques are also applied. Heavy-ion collisions in inverse kinematics together with in-flight separators allow an unambiguous identification of the reaction products that can be used as secondary beams of these neutron-rich nuclei. The ISOL method has been used mainly with proton or neutron induced reactions with ^{238}U targets to produce neutron-rich fission fragments. The advantage of this technique is the larger intensity of primary beams while the final production rates will depend on its chemical properties and half lives. The common point to all these production methods is the understanding of both, fission and fragmentation reaction mechanisms wich populate different regions in the chart of the nuclides.

Using a modern version of the abrasion-ablation model [1] which includes an analytical description of the cold fragmentation process and a detailed description of the fission process [2] we discuss the production of these neutron-rich isotopes and the expected yields for both, fragmentation and fission.

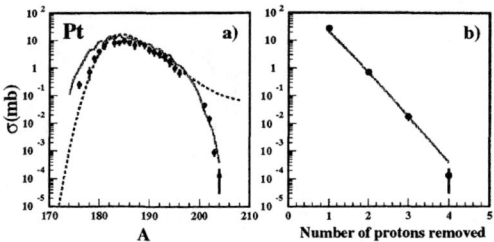

FIGURE 1. *Isotopic production cross sections obtained in the reaction Pb(1 A.GeV)+Cu (circles) from ref.[6]. a) Cross sections for Pt isotopes compared with the result of the abrasion–ablation model (solid line) and the EPAX formula (dashed line). b) Cross sections of the proton removal channel for the same reaction.*

THE MODEL

Heavy-ion reactions at relativistic energies can be described as a two-step process consisting of a fast interaction between the projectile and the target and the subsequent deexcitation of the reaction products. The initial stage can be described by a Glauber-type model [1,3]. At larger impact parameters without nuclear contact, electromagnetic excitations are considered.

The second stage of the reaction is treated in the framework of the statistical model. The emission probability of a particle is calculated using the Weisskopf formalism.

We use a simplified approach, similar to the one proposed in ref. [4], to estimate the production yields of very neutron-rich nuclei produced by cold fragmentation. Excited nuclei with large neutron excess almost exclusively deexcite by evaporating neutrons; they have a negligible probability to emit protons or other charged particles. In this way, the computing time could be reduced by many orders of magnitude, allowing to estimate very low production cross sections.

The fission decay width in the statistical deexcitation has been included according to the transition-state method of Bohr and Wheeler [5].

We introduced also a semiempirical description of the fission-fragment properties taking into account the influence of nuclear structure on the fission process. The model is supposed to be able to predict the magnitudes and widths of the different fission channels (symmetric and asymmetric), the even-odd fluctuations and the neutron-to-proton ratio of the fission-fragment distributions as a function of nuclear charge, mass and excitation energy of the fissioning nucleus in a global way. A more detailed description of the model can be found in ref. [2].

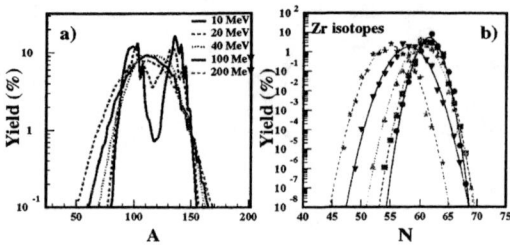

FIGURE 2. *Mass distribution of fission fragments (a) and N/Z ratio of Zr isotopes (b) produced from ^{238}U fission at different excitation energies. The steps in excitation energy are the same in both cases.*

I RESULTS ON PRODUCTION RATES BY FISSION AND FRAGMENTATION REACTIONS.

A complete calculation with this model allows us to predict the complex nuclide distribution resulting from peripheral relativistic heavy-ion collisions. In order to illustrate the different reaction mechanisms contributing to these distributions we have compared existing experimental data with our calculations.

In figure 1 we report the production cross sections of Pt isotopes (a) and the proton-removal cross sections (b) obtained in the reaction Pb(1 A.GeV)+Cu [6]. These data are compared with our model calculations and the EPAX formula [7]. The discrepancy for very neutron-rich isotopes between the data and the EPAX calculation shows that this formula does not correctly reproduce the cold fragmentation. In contrast, the abrasion-ablation model shows a fair agreement with the data even for the most neutron-rich isotopes. The predictions of the present model closely compare also with the cross sections of the proton-removal channels(b), indicating a realistic description of the cold fragmentation process which is dominant in the production of very neutron-rich isotopes.

Fig. 2 shows the variation of the total mass distribution and the N/Z ratio of Zr isotopes for fission of ^{238}U at different excitation energies. When increasing the temperature, the mass distribution broadens considerably, leading to drastically increased production rates of very light and heavy fission fragments. The width of the isotopic distribution grows also when increasing the excitation energy. However, due to neutron evaporation, the neutron-rich side is almost independent of energy up to about 100 MeV.

In order to compare the isotopic distributions produced by fission and fragmentation, in figure 3 we report the calculated Zr isotopic distribution in the induced fission of ^{238}U in a beryllium target (triangles and solid line) and in the projectile fragmentation of ^{136}Xe in an aluminium target. In this picture, the open squares represent experimental production cross sections of neutron-rich Zr fission fragments measured in the reaction ^{238}U+Be [9]. The fair agreement between the

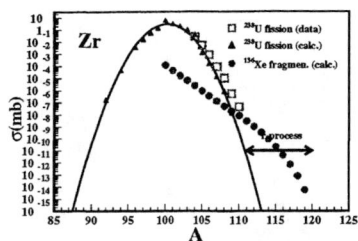

FIGURE 3. *Estimated production cross sections of Zr isotopes produced by fission of $^{238}U(750$ A.MEV) on a Be target (solid line) and fragmentation(full circles) in the reaction $^{136}Xe(1$ A.GeV)+Al. The results of these calculations are compared with experimental data obtained in the reaction $^{238}U(750$ A.MeV)+Be [8,9] (open squares).*

measured and calculated production cross sections of fission residues indicates that the proposed model gives a good description of the N/Z ratio of fission residues. As we have already seen, this ratio together with the width of the fission-fragment mass distribution determines the population of neutron-rich isotopes by fission. From this picture, we can conclude that the fission mechanism allows to produce high intensities of moderately neutron-rich isotopes. However, the fragmentation process dominates the production of very neutron-rich isotopes, in particular those involved in the r-process path, as indicated in the picture.

II FUTURE PERSPECTIVES ON THE PRODUCTION OF NEUTRON-RICH SECONDARY BEAMS.

Inverse-kinematic reactions at relativistic energies seem to be a very promising tool for the production of very neutron-rich isotopes. The choice of different projectiles allows to profit from the advantages of fission and fragmentation reaction mechanisms.

We have estimated the expected production rates of neutron-rich isotopes taking into account typical primary-beam intensities. The results of these calculations are shown in figure 4. In this picture, the different hatched areas represent different production rates of neutron-rich isotopes ranging from 200 ions/s till 10 ions/day, as indicated in the scale. Calculations were performed assuming a primary-beam intensity of $2\ 10^9$ ions/s and a 1.5 g/cm^2 thickness Al target for fragmentation and a Be target for fission, respectively.

The predictions of the proposed model show that the r-process path can be reached for a large number of nuclear species with production rates about 100 ions/week by using different projectiles. ^{238}U fragmentation seems to be the only chance to approach the r-process at the waiting point around A=195, while ^{176}Yb and ^{136}Xe fragmentation can compete with ^{238}U fission to approach the r-process at the waiting point around A=130.

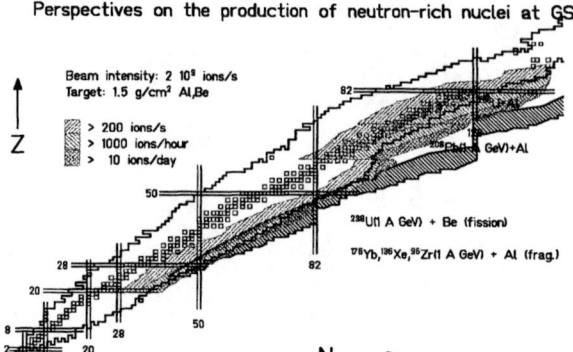

FIGURE 4. *Calculated production rates of neutron-rich isotopes obtained with primary beam intensities available after the intensity upgrade of GSI. The predictions are given in three different regions with different projectile-target combinations. The neutron-deficient limit in the calculations is determined by the maximum of the production rate.*

III CONCLUSION.

We conclude that fission is an efficient reaction type to produce high-intensities of moderately neutron-rich isotopes in the mass region of fission products. Projectile fragmentation reaches further to the neutron-rich side, however, with very low production cross sections. Nontheless, projectile fragmentation seems to be the most promising tool for extending our knowledge on nuclear properties when approaching the neutron drip line. The primary-beam intensities are decisive for the limits to be in reach.

REFERENCES

1. J.-J. Gaimard, K.-H. Schmidt, Nucl. Phys. A 531 (1991) 709
2. J. Benlliure et al., to be published in Nucl. Phys. A
3. K.-H. Schmidt et al., Phys. Lett. B 300 (1993) 313
4. K.-H. Schmidt et al., Nucl. Phys. A 542 (1992) 699
5. N. Bohr, J.A. Wheeler, Phys. Rev. 56 (1939) 426
6. M. de Jong et al, to be published in Nucl. Phys. A
7. K. Sümmerer et al., Phys. Rev. C42 (1990) 2546
8. M. Bernas et al., Phys. Lett. B 331 (1994) 19
9. M. Bernas et al., to be published in Phys. Lett. B

A Radio Frequency Quadrupole Ion Beam Buncher for ISOLTRAP

G. Bollen[a], J. Dilling[b], A.M. Ghalambor Dezfuli[c], S. Henry[d], F. Herfurth[b], A. Kellerbauer[c], T. Kim[c], H.-J. Kluge[b], A. Kohl[b], E. Lamour[b], D. Lunney[d], R.B. Moore[c], W. Quint[b], S. Schwarz[b], P. Varfalvy[c] and L. Vermeeren[b]

[a]*CERN, CH-1211 Geneva 23, Switzerland*
[b]*GSI, Planckstr.1, D-64291 Darmstadt, Germany*
[c]*McGill University, 3600 University St., Montreal, QC, H3A 2T8, Canada*
[d]*CSNSM-IN2P3-CNRS, F-91405 Orsay-Campus, France*

Abstract. ISOLTRAP is a Penning trap spectrometer at the on-line mass separator ISOLDE at CERN for the mass determination of radioisotopes. It consists of three electromagnetic traps in tandem; a Paul trap for ISOLDE beam collection, a Penning trap for cooling and purification and a high-precision Penning trap for the measurement of masses by cyclotron resonance. The Paul trap, which collects radionuclide ions using only electric fields and a noble buffer gas, has been essential for the masses of radionuclides that cannot be surface ionized. The success with this system has led to the present program to increase the collection efficiency by replacing the Paul trap by a radiofrequency quadrupole ion guide operating as a buncher. This system would also provide a DC ISOLDE beam of emittance approaching 1 π-mm-mrad.

INTRODUCTION

ISOLTRAP is a Penning trap spectrometer installed at the on-line mass separator ISOLDE at CERN for the mass determination of short-lived isotopes (1). The spectrometer consists presently of three electromagnetic ion traps in tandem, each serving different functions (Fig. 1). The first trap is a RFQ Paul trap which acts as an accumulator and buncher for the continuous 60 keV ISOLDE ion beam. The second is a Penning trap in which the ions are cooled by a mass selective buffer gas cooling technique (2) and the third trap is a high precision Penning trap from which the mass is deduced from cyclotron resonance, detected by the time of flight of ejected ions.

The configuration of ISOLTRAP is based on the engineering principle of separation of function, which allows optimization of the components of the system to their individual tasks. The capabilities of the resulting system are indicated by the determination, to date, of the mass of more than 100 radioactive isotopes.

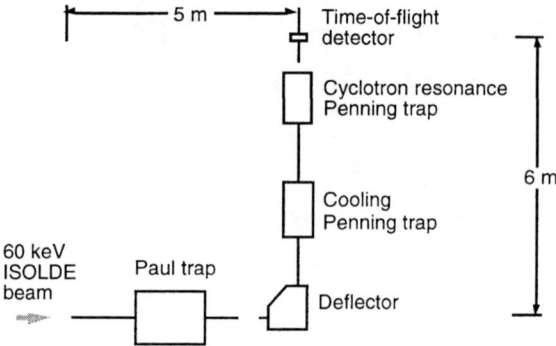

FIGURE 1. The present configuration of ISOLTRAP.

PRESENT COLLECTION AND COOLING SYSTEM

Essential for Penning trap mass measurements on short-lived radionuclei is fast and efficient transfer of their ions into the trap without restriction of species. In its earlier version ISOLTRAP was limited to surface ionizable elements, and so to increase its applicability a Paul trap system was installed to arrest, accumulate and bunch the ISOLDE beam without any interaction with material other than a noble gas as a buffer. In this system (Fig. 2) the ISOLDE beam is slowed by electrostatic retardation and injected into the Paul trap where a fraction of it is captured and cooled by collisions with the buffer gas. After ejection at low energy the collected bunch passes through a drift tube which is brought to near ground potential while the ions are still in it. In this way a low-energy ion bunch is prepared which can easily be captured in the subsequent Penning trap. This system was used for the first time for measurements in the isotopic chain of mercury, when masses were determined to an accuracy of $\delta m/m = 10^{-7}$ for neutron deficient isotopes down to ^{184}Hg.

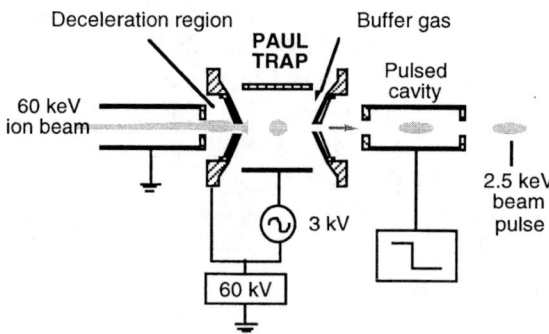

FIGURE 2. The present Paul trap ion buncher system.

THE NEW COLLECTION AND COOLING SYSTEM

The success of the on-line measurements on radionuclides that could not be surface ionized led to consideration of schemes to increase the efficiency by which ions could be arrested and captured in buffer gas. Such a scheme was presented by recent work (3) on the use of buffer gas in a radiofrequency quadrupole rod structure to capture and cool beams of 100 eV cesium ions. Essentially, this scheme involves segmenting the rod structure and applying dc potentials to the segments (Fig. 3). In this way a decelerating axial electric field can be applied, ending finally in axial potentials which form an electric trap. In this new scheme, the gas-filled four-rod structure replaces the Paul trap. By placing the trap close to the extraction end of the system it is easy to extract the collected and cooled ions. Extraction would also be eased by the much lower radiofrequency fields that are required in the linear trap system compared to the Paul trap.

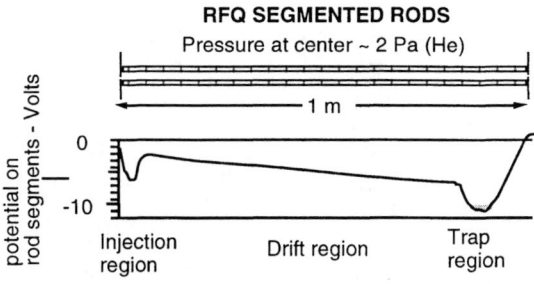

FIGURE 3. The new buncher system.

Numerical simulations of the ion trajectories for various decelerator systems have been carried out, resulting in a system as outlined in Fig. 4. The main decelerator electrode, a paraboloid, brings 60 keV ions to about 2 keV near its exit orifice.

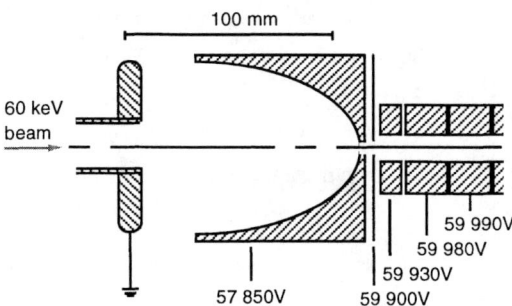

FIGURE 4. The deceleration and injection system into the 4-rod structure.

A retardation plate immediately following the main retardation electrode brings the ions to about 100 eV. At this energy the transverse radiofrequency quadrupole electric field can contain the radial motion while the potentials on the rod segments shown in Fig. 4 bring the captured ions to an axial energy of 10 eV. This remaining energy is then removed by the damping of the buffer gas in the rod structure.

More detailed calculations on the finalized design of the capture system have yielded the acceptance results shown in Fig. 5. It appears that the acceptance of the system will be sufficient for the full ISOLDE beam emittance of 35 π-mm-mrad. With an expected efficiency of 50% in the decelerator it would appear that overall collection and bunching efficiencies of several tens of percent can be expected with this new system.

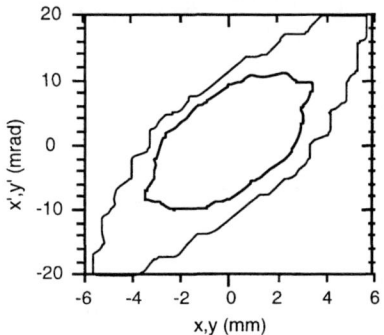

FIGURE 5. Calculated acceptance of the new buncher system. The inner envelope is for acceptance of incoming ions at all RF phases, the outer envelope being the limit for acceptance at any RF phase. The area of the inner envelope is about 30 π-mm-mrad. The RF amplitude is 300V at 980 kHz.

This cooler system would also be very effective as an emittance reduction device for an ISOLDE type beam. Indeed, as part of the tuning of the system it will be used in a dc mode, which is achieved by simply turning off the trapping potentials on the rod segments forming the trap and leaving the pulse-down electrode at ground potential. Work on this mode of operation (3) has shown that beams of up to 1 nA can be cooled to temperatures and sizes corresponding to emittances of about 1 π-mm-mrad for an ISOLDE beam of 60 keV.

REFERENCES

1. Bollen, G., Becker, St., Kluge, H.-J., König, M., Moore, R.B., Otto, Th., Raimbault-Hartmann, H., Savard, G., Schweikhard, L., Stolzenberg, H., *Nucl. Instr. and Meth.* **A 368**, 675-697 (1996).
2. Savard, G., Becker, St., Bollen, G., Kluge, H.-J., Moore, R.B., Otto, Th., Schweikhard, L., Stolzenberg, H., and Weiss, U., *Phys. Lett.* **A158**, 247-252 (1991).
3. Kim, T., *Buffer Gas Cooling of Ions in a Radio Frequency Quadrupole Ion Guide*, PhD Thesis, McGill University, 180 p (1998).

New Approach to the Analysis of Total Absorption Spectra

D. Cano-Ott[a], A. Gadea[a], B. Rubio[a], J.L. Tain[a], M. Karny[b],
Z. Janas[b], K. Rykaczewski[b], R. Kirchner[c], E. Roeckl[c], L. Batist[d],
F. Moroz[d], V. Wittmann[d]

[a] *Instituto de Física Corpuscular, Dr. Moliner 50, E-46100 Burjassot-Valencia, Spain*
[b] *Institute of Experimental Physics, University of Warsaw, PL-00681 Warsaw, Poland*
[c] *Gesellschaft für Schwerionenforschung, D-64291 Darmstadt, Germany*
[d] *St. Petersburg Nuclear Physics Institute, 188-350 Gatchina, Russia*

Abstract. Total Absorption Spectroscopy is a powerful technique to study the Gamow–Teller β-strength in nuclei far from stability. It has not been widely used however, because of the complexity of the data analysis. We present several new improvements related to the analysis problem: a) precise numerical correction method for electronic pulse pileup, b) a Monte Carlo simulation of the response of a NaI(Tl) total absorption spectrometer that considers the light yield non-proportionality, and c) three different methods to solve the inverse problem.

The determination of the β-decay strength distribution is of fundamental importance to our understanding of the nuclear structure. Experimentally the β-strength can be determined from the β intensity or feeding probability (per unit energy interval) I_β:

$$S_\beta(E_x) = \frac{I_\beta(E_x)}{f(Q_\beta - E_x)T_{1/2}} \quad (1)$$

where f stands for the statistical rate Fermi function of the available energy in the decay $Q_\beta - E_x$ and $T_{1/2}$ is the β-decay half-life. We are particularly concerned with the investigation of the problem of the missing Gamow-Teller (GT) strength: the consistent overestimation of the strength for the transitions governed by the GT spin-isospin operator $\vec{\sigma}\vec{\tau}$. We have shown [1] that even an array of 42 large Ge detectors closely packed missed a substantial fraction of the GT strength in the ^{150}Ho (2^-) decay ($Q_{EC} = 7.4$ MeV). A solution is given by the Total Absorption Gamma-ray Spectroscopy with high efficiency scintillation detectors [2]. An ideal Total Absorption Spectrometer (TAS) with 100 % peak efficiency sums up the energy of every γ-ray cascade de-exciting a fed level in the daughter nucleus and

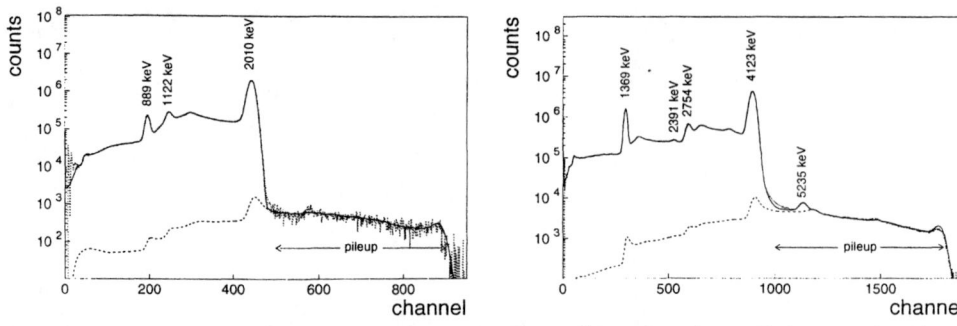

FIGURE 1. Left: experimental TAS spectrum of the ^{46}Sc decay (dotted) compared with the simulated response (solid) added to the calculated pileup distortion (dashed). Right: experimental TAS spectrum of the ^{24}Na decay (dotted) compared with the simulated response (solid) added to the calculated pileup distortion (dashed). The arrow marks the region where the pileup becomes a significant contribution in the spectra.

registers a count at the corresponding level energy. The spectrum obtained is proportional to the sought $I_\beta(E_x)$. A real TAS has always limited efficiency, and the beta intensity has to be extracted from the measured spectrum using the spectrometer response. The TAS used at the GSI (Darmstadt) on-line mass separator [3] for our experiments in the ^{146}Gd and ^{100}Sn regions consists of a ⌀35 cm×35 cm NaI(Tl) detector. The relation between a measured TAS spectrum **d** and the level feeding distribution **f** ($\equiv NI_\beta$, $N=$ number of decays) is best represented by the equation:

$$d_i = \sum_{j=0}^{j_{max}} R_{ij} f_j , \quad i = 1, i_{\max} \quad \text{or} \quad \mathbf{d} = \mathbf{R} \cdot \mathbf{f} \qquad (2)$$

Each column j of the response matrix **R** represents the average response of the spectrometer to the decay into the levels in the energy bin labeled by the channel j. **R** depends on the emitted radiation as well as on the apparatus and is constructed [5] in terms of the individual quanta responses intervening in the decay. The way to accomplish this task is based on Monte Carlo simulation. The Monte Carlo code GEANT3 [6] was chosen because of its powerful geometry package, which allowed an extremely detailed modelization of the geometry of the spectrometer. It was necessary to include the non proportional light yield of NaI(Tl) [7] to γ radiation in the simulation [5]. This non-proportional light production has the striking consequence that γ-ray sum peaks are displaced in the measured spectrum with respect to single γ-ray peaks. It is no longer possible to talk strictly about an energy calibration of a TAS spectrum since the counts in a certain channel correspond to different deposited energies. In Fig.1 we compare the experimental TAS spectra (dotted) for the ^{46}Sc ($Q_{\beta-} = 1176$ keV) –left– and ^{24}Na ($Q_{\beta-} = 5516$ keV) –right– decays with the corresponding simulated responses (solid) added to the calculated pileup distortion (dashed). The excellent overall agreement between the simulated and

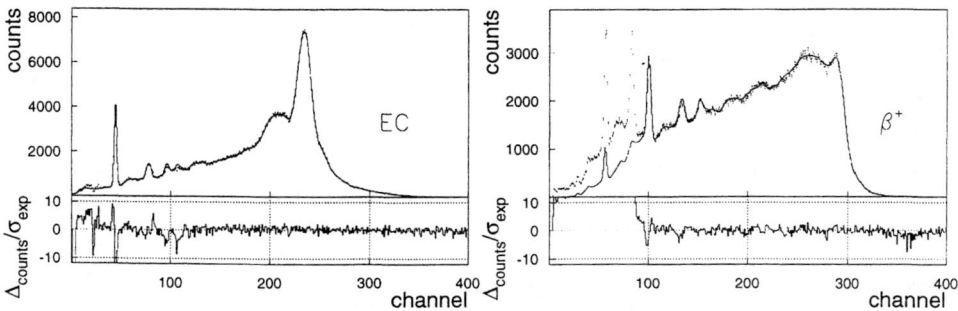

FIGURE 2. Experimental spectra for the ^{150}Ho(2^-) decay (dotted), corresponding reconstructed responses (solid) and difference Δ_{counts} over σ_{exp} (solid) below. Left: EC. Right: β^+.

experimental responses reflects the quality with which **R** is obtained. The pileup distortions present in both experimental spectra is also excellently reproduced by the calculated ones (dashed). Even for low counting rates, TAS spectra have to be corrected from electronic pulse pileup distortions. Pileup happens when two or more events occur so close in time that the corresponding electronic signals overlap and are processed as a single pulse carrying a wrong amplitude information. This effect has an influence on the accurate determination of the spectrum end-point and also introduces a large deviation on the deduced β-strength. We have developed a numerical method based on the true pulse shape of the energy signals [4] that allows a very precise calculation of the distortion.

The TAS employed in our experiments allows to distinguish between the competing EC and β^+ processes. Two Si(Li) detectors surrounding the source permit to establish a coincidence between the β particles and an energy signal in the NaI(Tl), while a small Ge detector placed close and above the source allows to make the same for the characteristic X-rays emitted after the EC process. Two spectra are obtained in this way, and its independent analysis guarantees the consistency of the results. The corresponding feedings **f** fulfilling Eq.(2) are then obtained by three different methods: Linear Regularization [8], Maximum Entropy [9] and Iterative Bayesian [10]. These methods require some a priori information on the de-excitation pattern in order to construct the response matrix R. This apparently dramatic problem is however solved by the high efficiency of our TAS and some knowledge on the low energy part of the decay scheme. Only assumptions on the branching ratios for the high energy part of the level scheme are needed. In Fig.2 we show the experimental (dotted) and reconstructed spectra (solid) corresponding to the EC –left– and β^+ –right– processes for the ^{150}Ho(2^-) decay. The difference Δ_{counts} between both experimental and simulated spectra relative to the experimental uncertainty are plotted below (solid) for each case indicating the accuracy obtained with the Iterative Bayesian method. The large deviations for β^+ case below channel 100 are due to a ^{150}Er contamination and have no influence in the analysis. Such a good agreement is also achieved with the two other methods. The

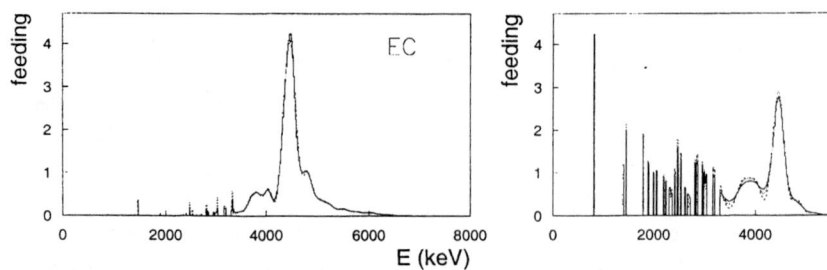

FIGURE 3. Feeding distributions normalized to 100 for the ^{150}Ho(2^-) decay obtained with the Iterative Bayesian (solid), Maximum Entropy (dotted) and Linear Regularization (dashed) methods. Left: EC. Right: β^+.

consistency of the results obtained by the three methods shown in Fig.3 guarantees the reliability of the extracted β-feeding.

Acknowledgments: Work supported by C.I.C.Y.T. (Spain) under project AEN96-1662, by E.C.C. contract ERBCIPD-CT-950083, by the Russian Fund for Basic Research and Deutsche Forschungsgemeinschaft contract 436-RUS-113/201/0(R) and by Polish Committee of Scientific Research grant KBN-2-P03B-039-13.

REFERENCES

1. Agramunt J., Algora A., Cano-Ott D., Gadea A., Rubio B.,Tain J.L., Gierlik M., Karny M., Janas Z., Plochocki A., Rykaczewski K., Szerypo J., Collatz R., Gerl J., Górska M., Grawe H., Hellström M., Hu Z., Kirchner R., Rejmund M., Roeckl E., Shibata M., Batist L., Moroz F., Wittmann V. and Kleinheinz P., *Proc. Int. Symp. New Facet of Spin Giant Resonances in Nuclei (Tokyo, 1997)* in press.
2. Duke C.L., Hansen P.G., Nielsen O.B., Rudstam G. and the ISOLDE Collaboration, *Nucl. Phys. A* **151** (1970) 609.
3. Karny M., Nitschke J.M., Archambault L. F., Burkard K., Cano-Ott D., Hellström M., Hüller W., Lewandowski S., Roeckl E. and Sulik A., *Nucl. Instr. Meth. B* **126** (1997) 411
4. Cano-Ott D., Taín J.L., Gadea A., Rubio B., Batist L., Karny M. and Roeckl E., submitted to *Nucl. Instr. Meth. A*
5. Cano-Ott D., Taín J.L., Gadea A., Rubio B., Batist L., Karny M. and Roeckl E., submitted to *Nucl. Instr. Meth. A*
6. *GEANT: Detector description and simulation tool*, CERN Program Library W5013, Geneve, 1994.
7. Engelkemeir D., *Rev. Sci. Instr.* **27** (1956) 589.
8. *Solutions of Ill-Possed Problems*, New York: Wiley, 1977
9. Collins D.M., *Nature* **298** (1982) 49
10. D'Agostini G., *Nucl. Instr. and Meth. A* **362** (1995) 487

The ISAC Radioactive Beams Facility in Canada: Progress and Plans

G. Ball[1], R. Bartmann[1], J. Behr[1] P. Bricault[1], L. Buchmann[1],
J.M. D'Auria[2], P. Delhaij[1], M. Dombsky[1], G. Dutto[1], R. Kiefl[3],
K.P. Jackson[1], R. Laxdal[1], J.M. Poutissou[1], P. Schmor[1], and G. Stanford[1]

[1] *TRIUMF, 4004 Wesbrook Mall, Vancouver, B.C., Canada V6T 2A3*
[2] *Simon Fraser University, 8888 University Way, Burnaby, B.C., Canada V5A 1S6*
[3] *University of British Columbia, 2329 West Mall, Vancouver, B.C., Canada V6T 1Z4*

Abstract. In 1995 the Governments of Canada and British Columbia jointly funded the installation of a high intensity, thick target, ISOL (Isotope Separator, On-Line) based, accelerated radioactive beams facility. This facility, named ISAC (Isotope Separator and ACcelerator) will take advantage of the high intensity (100 microamps), intermediate energy (500 MeV) proton beam available at TRIUMF to produce a wide range of radioisotopic ion beams. Beams with masses below 30 will be accelerated to energies from 0.15 to 1.5 MeV/u, an energy range optimal for studies important for understanding explosive nucleosynthesis in cataclysmic events in the universe. There will be two main experimental areas, one using only the mass separated, radioactive beam in studies such as fundamental interactions, material science, and nuclear structure, and the second using the accelerated beam for nuclear physics and astrophysics studies. Operations into the first area are expected to commence in November of this year. This report will provide a summary of the scientific program planned for the first years of operation along with a brief report of the facility.

INTRODUCTION

A new ISOL based, accelerated radioactive beams (RB) facility is being installed at the 500 MeV, intense proton cyclotron facility, TRIUMF located in Vancouver, Canada. ISAC received funding in 1995 and the first beam for experiments is expected in 1998. RB will be produced using the intense (<100 μ) proton beam to intercept a thick (< 19 cm) target; the products diffusing from the heated target (< 2500°C) will be ionized on-line, mass analysed, and directed either to an experimental systems or into a specially designed accelerator. Following acceleration using a LINAC, beams with q/A $\geq$ 1/30 with energies from 150 keV/u to 1.5 MeV/u will be available. Both accelerators are room temperature systems, operating cw.

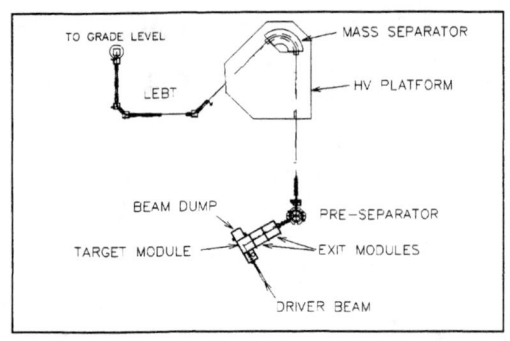

FIGURE 1. Schematic representation of the ISAC target and separator areas.

Protons of 480-500 MeV are extracted from the H-cyclotron into a 50 m long beamline (B1-2A) simultaneously with other operating beam lines at TRIUMF. This beam line was commissioned in April 1998.

The target system (Figure 1) is designed to satisfy the following specifications, namely containment of large amounts of mobile radioactivity, separation of radiation damageable material from locations near to the high radiation fields, high voltage isolation needed for ion beam extraction, precision alignment of target and ion beam elements, and routine, remote changing of the target. These issues were addressed by placing the target in a sealed self-contained module, 2 m in length, which can be transferred to a hot cell facility for maintenance. Access to such modules is done vertically with an overhead crane and all vacuum seals are located at the top of the module which itself is positioned inside of a vacuum chamber. This chamber contains 5 modules for the driver beam entrance, the target/ion source, the driver beam dump, and two for extraction of the ion beam. Sufficient shielding (iron and concrete) is packed around the chamber to allow use of proton beam intensities up to 100 µA although initial operations will be limited to < 10 µA.

Mass selection is achieved in two stages. The pre-separator magnet acts as a cleaning stage to limit contamination. The optics of the matching sections is designed to provide similar ion beam properties form each of two target stations. The mass separator was obtained from the Chalk River laboratory and is positioned on a high voltage platform to minimize contaminants in the mass selected, ion beam. The mass separation system will handle beams of A<238 with source extraction voltages from 12 to 60 kV. Commissioning of this system will be performed in the summer of 1998.

The ion beam from the mass-separator is transported vertically to experimental systems located on two different levels. The first level is where the TRINAT (TRIUMF Neutral Ion Trap) facility will be located to receive the first beam from ISAC. At the second level (at grade level) an electrostatic switchyard directs the beam either to the low energy experimental area or to the accelerator chain and high energy area. An off line ion source (OLIS) is also available at ground level to provide a beam of stable heavy ions to either the low or high energy areas for operation in the absence of the RB.

The post-acceleration system consists of an RFQ followed by a DTL. The RFQ is 8 m long, housing 19 split ring structures each feeding 40 cm lengths of modulated electrode; both water cooled. It is expected that 81 % of the beam will be accelerated to

150 keV/u in bunches with a time spread of 86 ns (11.7 MHz). A separated function, room temperature DTL will complete the acceleration to a final energy, variable up to 1.5 MeV/u. A stripping system between the two accelerators increases the q/A to > 1/6. Transmission of 100 % is expected with no longitudinal emittance growth over the entire energy range. Commissioning is scheduled for 2000.

THE EXPERIMENTAL PROGRAM

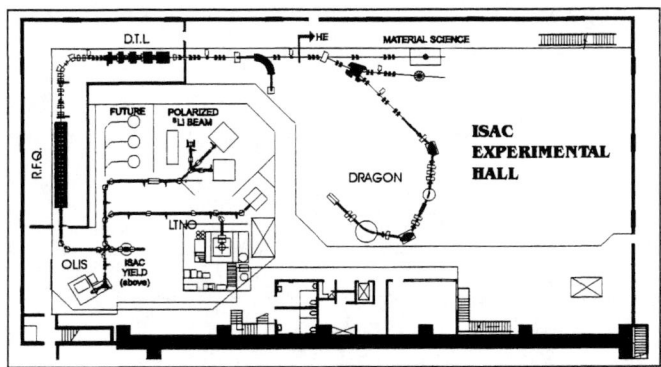

FIGURE 2. ISAC Experimental Hall

Figure 2 displays the planned layout of the ISAC Experimental Hall. The experimental program is expected to begin in 1998 with the delivery of radioactive beams of ^{37,38m}K (from a target of CaO thin pressed discs) to the TRINAT facility. In this study optically trapped atoms are confined in small volumes of space (mm^2) with negligible source thicknesses. ^{37}K and ^{38m}K each decay predominantly by a single superallowed transition. The β-ν correlation in the pure Fermi decay of ^{38m}K is sensitive to the exchange of hypothetical scalar bosons and a 1% measurement of the correlation factor, a, is attainable. The mixed Fermi-Gamow-Teller decay of ^{37}K is suitable for β-asymmetry experiments and positron longitudinal polarization measurements, which are sensitive to the presence of right-handed currents in the weak interactions. Studies have been initiated on the TISOL facility and will be relocated to utilize the more intense RB at ISAC.

A second experiment scheduled to receive these same beams is a high precision measurement of their half lives using an approach pioneered at the TASCC laboratory. Using essentially the same equipment including an in-air tape transport system and a 4π beta proportional counter, a precision of 1% is the goal. In 1999 an attempt to attain a similar precision for the half-life of ^{74}Rb will be made.

With the development of other beams, other experimental systems will be utilized. A beam of ^{8}Li will be polarized with a co-linear laser beam and used with a β-NMR system to perform thin layer studies of high temperature superconductors. A low temperature nuclear orientation (LTNO) system, obtained from ORNL, will be used to measure nuclear properties (magnetic and quadrupole moments) of a range of nuclei.

The accelerated beam will be used primarily for nuclear astrophysics studies. One of the first studies will use a ^{8}Li beam to perform an excitation function of the (α,n) reactions; the strength of this reaction could provide evidence of the important of the inhomogeneous big-bang model. The rates of a number of reactions considered important for explosive nucleosynthesis in nova and x-ray bursts will be measured at the new DRAGON (Detector of Recoils And Gammas Of Nuclear reactions) recoil mass separator facility. A windowless gas target (H_2 or He) will be used to study radiative proton and alpha capture reactions with inverse kinematics. Following commissioning in 2000, the first reaction to be studied will be $^{1}H(^{21}Na,^{19}Ne)\gamma$. In addition, a scattering chamber will be installed to perform studies of particle reactions and elastic scattering reactions.

FUTURE OPTIONS

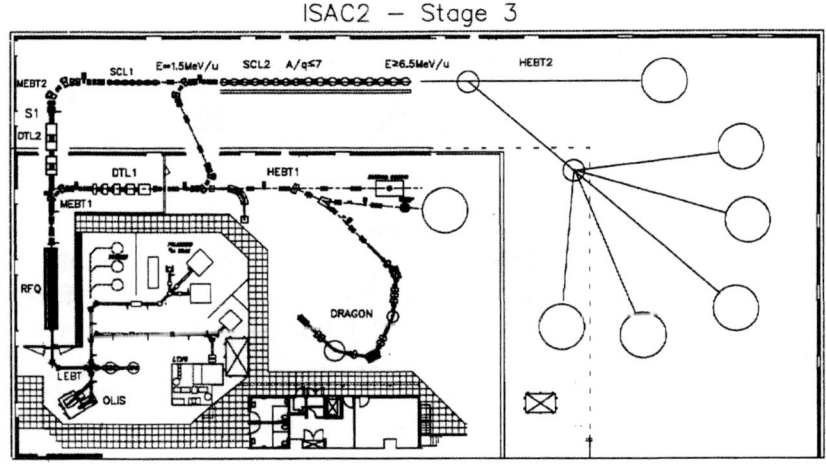

FIGURE 3. Proposed ISAC 2 Layout

TRIUMF is currently preparing a new five year plan requesting additional funding from the Canadian and British Columbia for the period beginning in April, 2000. A major part includes an upgrade of the ISAC facility to permit acceleration of RB with A<150 up to 6.5 MeV/u. Figure 3 displays one possible layout of this proposed facility, ISAC-2. The final proposal will be submitted by the end of 1998.

More information on any of the topics mentioned herein can be found on the TRIUMF WEB page (http://www.triumf.ca).

EXOGAM: A γ-ray Spectrometer for Exotic Beams

Gilles de FRANCE for the EXOGAM collaboration

GANIL, BP 5027, 14076 Caen Cedex 5, France

Abstract. The availability of radioactive ion beams will open up a vast range of new nuclear physics. The new SPIRAL facility at GANIL will be operational within a few months from now. This paper describes EXOGAM, a highly efficient and powerful γ-ray array for nuclear spectroscopy using the exotic radioactive beams from SPIRAL.

Radioactive beam spectroscopy requires a new compact, highly efficient, highly segmented, device for gamma-ray spectroscopy, designed in a flexible way in order to cope with different experimental set-ups and different ancillary devices. EXOGAM is a European collaboration to build such a device. Today, the collaboration includes Denmark, Finland, France, Germany, Hungary, Sweden, and United Kingdom.

DESIGN SPECIFICATIONS

The expected beam intensities, at least at the start up of SPIRAL, will be from 10 to 1000 times lower than those obtained with stable species. The design of the γ-ray array must therefore maximize the efficiency. This is also true for high γ-ray energies (i.e. up to 5-6 MeV) since such values will be observed for example in the Coulomb excitation of light nuclei or in β-decay studies.

Physics is always related to the study of "weak" phenomena not already observed or studied with the previous generation of detection systems. Interest will be in the study of the most exotic species and hence will require the weakest beams. Therefore, in the analysis of γ-ray spectra, it must be possible to extract tiny peaks from a possibly huge backround (corelated Compton background, background from target radioactivity, scattered beams,etc.) To push the limits further down, it is vital to have good signal-to-noise ratio and energy resolution.

There will be a large variety of nuclear reactions which will be used with radioactive ion beams. This variety will impose severe experimental conditions in terms of γ-ray energy (from x-rays of tens of keV to γ-ray energies up to 5-6 MeV), of multiplicity (from one to $\sim$ 15 coincident photons); of recoil velocity (from zero to ∼ 10 % of light velocity); and of kinematics of the reaction mechanism (from recoiling fusion products emitted at ∼ 0° or scattered particles between 0° and 180°). This means that the setup of the array must be adapted for each experiment. The radioactive nature of the beam is also a concern and shielding of the detectors becomes an important design criterion.

THE SEGMENTED CLOVER DETECTOR

One detector which is a good candidate for γ-ray spectroscopy with radioactive beams is the segmented clover detector [1]. The individual crystals for the EXOGAM clover will be 60 mm diameter and 90 mm long before shaping. The shape and the performances of this individual detector have been optimized using simulations calculations. The diameter allows for a compact design maximising the solid angle coverage with Ge. The length maintain a reasonable efficiency at high γ-ray energy. The calculated photopeak efficiency and peak-to-total as a function of gamma-ray energy and multiplicity are shown in Figure 1.

The calculated addback factor for this detector is ∼ 1.6. The granularity, and hence the Doppler broadening, will be improved by electronically splitting the crystal. Consequently, pile-up effects (arising from the simultaneous detection of more than one γ-ray in an individual element) are reasonable up to a multiplicity of M_γ ∼ 15 for a distance target to Ge of 11 cm. This limitation is compatible with the EXOGAM physics case which is focused on low and medium multiplicities.

The great advantage of the electrical segmentation is the significant improvement obtained on the energy resolution. Calculations have shown that the energy resolution is multiplied by 2 when going from the clover to the corresponding non-composite (monolithic) detector. Another factor of 2 is obtained between the segmented and the non-segmented clover. This improvement is fundamental when the background is large.

Another improvement on the energy resolution can be obtained by analysing the shape of the pulse delivered by the preamplifier. Indeed, this shape is characteristic

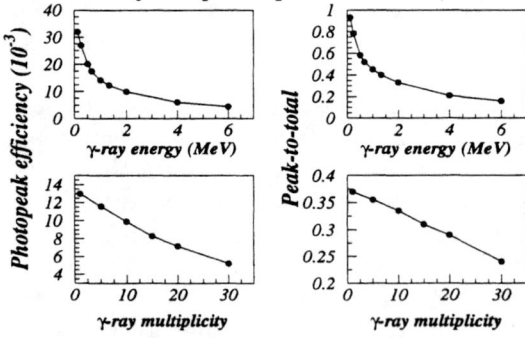

FIGURE 1. Calculated effective photopeak efficiency and unsuppressed peak to total as a function of gamma-ray energy and multiplicity at a distance of 11 cm for the EXOGAM clover.

of the distance between the interaction point and the anode of the Ge diode. A substantial gain will however be obtained only if the hit localisation is precise enough and tests have been performed which are still under analysis to estimate this gain. Information on the pulse shape for radial position determination will be included in the VXI cards on the signal from the central contact.

In addition, information on the pulse shape will improve the timing performance of the detector. This is extremely important in radioactive beam experiments since the uncorrelated background radiation can be greatly reduced by precisely relating the events in the detector to the beam. The use of the time relationship between the pulsed beam and germanium detectors is vital for the success of the experiments.

SUPPRESSION SHIELD

Each segmented clover Ge detector is surrounded by an escape suppression shield. The shield designed for the EXOGAM clover is based on a new concept in which the shield comprises several distinct elements, a backcatcher, a rear side element and a side shield, see Figure 2. Designing suppression shields from individual elements, creates greater flexibility for different configurations.

The shields will be operated in two configurations. The first is with the back catcher and rear-side element, configuration A, and the second with the additional side elements, configuration B. The only free space is left for the cold finger from the liquid nitrogen dewar of the Ge detector.

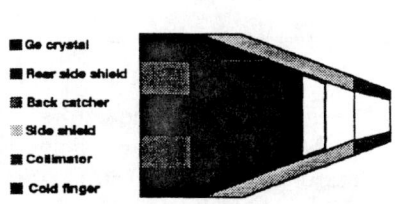

FIGURE 2. Left: the different elements of the suppression shield for the segmented clover Ge detectors (not to scale). Right: the full EXOGAM array in configuration A (see text). The suppression shields are not represented on this drawing.

The use of a back catcher prevents forward scattered events from escaping through the large angular section located behind the clover. Calculations have shown that the use of the back catcher will improve the peak to total by $\sim 10\%$. To maximise the solid angle coverage with Ge material (i.e. the photopeak efficiency), the side elements of the shield have to be in BGO. However, behind the Ge detector there is sufficient space for a scintillator with lower stopping power to be used, such as CsI(Tl).

SEGMENTED CLOVER ARRAYS.

The EXOGAM segmented clovers can be arranged in different geometries. Configuration A is the close packed geometry where the Ge detectors can essentially touch at the front. Configuration B is the pulled back geometry in which the detectors are further from the target to allow for the inclusion of the additional side suppression elements.

An array of 16 clover detectors can be arranged with 4 detectors at 135°, 8 at 90° and 4 at 45° to the beam direction. In configuration A the signals from adjacent Ge crystals can be summed to increase the efficiency. The calculated increase is $\sim 6\%$.

Simulation calculations have been also performed to calculate the whole array in a realistic way (see Table 1). The configurations considered in the calculations are the cube geometry with 4 clovers with full suppression shields and the 16 clover geometry in configurations A and B.

TABLE 1. Total photopeak efficiency and peak to total for EXOGAM.

	Photopeak efficiency (%)		Peak-to-total (%)	
	662 keV	1.3 MeV	662 keV	1.3 MeV
EXOGAM configuration A	28	20	57	47
EXOGAM configuration B	17	12	72	60
Gamma-Cube	15	10	72	60

It is envisaged that the array will be used with several ancillary detectors as their use in association with the gamma-ray array will be vital to obtain efficiently the data necessary to achieve the physics goals. A major ancillary detector is the proposed high resolution large acceptance spectrometer VAMOS (the VAriable MOde high acceptance Spectrometer). This spectrometer imposes additional constraints for EXOGAM since the array may be required to rotate.

REFERENCES

1. Beck F.A. et al., *Proceedings of the Workshop on Large Detector Arrays* **AECL 10613, Vol. 2 (1992)**.

Cooling, Bunching and Isobar Separation of Radioactive Ion Beams at IGISOL

A. Jokinen, J. Äystö, P. Dendooven, V.S. Kolhinen, J. Huikari,
A. Nieminen and K. Peräjärvi

*Department of Physics, University of Jyväskylä,
P.O. Box 35, FIN-40351 Jyväskylä, Finland
Ari.Jokinen@phys.jyu.fi*

Abstract. Universality of the ion guide technique allows the mass separation of isotopes of any element. However, this feature also means unselectivity in respect to chemical element number and results in severe isobaric background. Also the operation of the ion guide with the maximum yield increases the energy spread and the emittance of the ion beam. To overcome these problems, new means to cool, bunch and mass purify the ion beam will be developed. The weak radioactive ion beam will be cooled in a gas-filled radiofrequency quadrupole and further mass purified and bunched in a Penning-trap.

INTRODUCTION

Handling of the weak ion beams has become increasingly important in the recent years. One of the reasons is the advent of the radioactive ion beam facilities, but also extension of the nuclear structure studies further away from the stability. In the Department of Physics, University of Jyväskylä, a new project has been launched to improve the quality of radioactive ion beams produced at IGISOL-separator. This is achieved by a combination of buffer-gas cooling (1) in an RFQ and a mass-selective Penning trap. Their unique combination with the universality of the IGISOL-technique will provide clean radioactive ion beams of practically all elements for nuclear structure studies far from the stability. The principle of the project is illustrated in the figure 1.

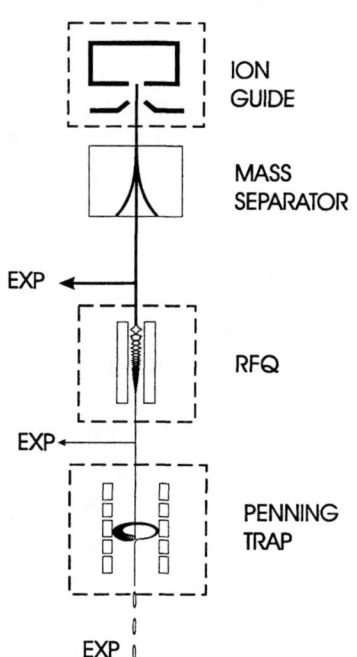

CP455, *ENAM98: Exotic Nuclei and Atomic Masses*
edited by B. M. Sherrill, D. J. Morrissey, and Cary N. Davids
© 1998 The American Institute of Physics   1-56396-804-5/98/$15.00

PROPERTIES OF IGISOL BEAMS

In the IGISOL-method, nuclei recoiling out of a target after nuclear reactions, are stopped in a He-gas, where they reach a 1+ charge state as a result of numerous collisions with the buffer-gas atoms. The ions are transported out from the target chamber by the continuous He-flow. Immediately after the exit of the target chamber the neutral He-gas and the ions are separated by means of differential pumping and an electric field. Ions are further accelerated to 40 keV and separated according to their mass in a dipole magnet with a mass resolving power of the order of 400.

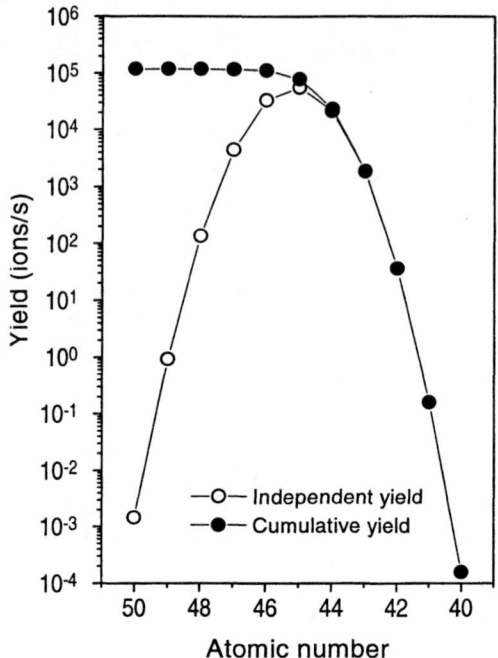

Presently IGISOL-technique can be used for fission and light ion and heavy-ion fusion reactions. It can handle all elements independent of their chemical or physical properties, but due to its chemical unselectivity, it delivers ion beams with strong isobaric contamination. This is illustrated in figure 2, which shows the independent and cumulative production rates of the A = 112 isobars. Due to subsequent stopping and extraction, no separate ionization phase is needed. This makes the IGISOL-method very fast. Due to collisions with He-gas atoms in the extraction/acceleration stage it has relatively wide energy spread.

The properties of the IGISOL-beams mentioned request for the reduction of energy spread and emittance. In addition, subtraction of isobaric contamination is a necessity to improve the spectroscopic sensitivity far from the stability. These improvements should be obtained without the loss of the speed of the system and universality in respect to element number.

ION BEAM COOLER

Before injecting the ion beam into the Penning trap its energy spread and transverse emittance is reduced by an ion beam cooler. The cooler consists of deceleration and acceleration section and a Radio Frequency Quadrupole (RFQ) (3) with buffer gas load. The oscillating quadrupole field in transverse plane compared to the optical axis of the RFQ constrains the ion motion in that plane. Collisions between buffer-gas atoms and ions tend to decrease the energy of the ions. Thus the ion motion is reduced in all directions. Simulations show that the emittance of a 40 keV IGISOL-beam can be reduced to less than 1 π×mm×mrad and its energy spread from initial 80 eV down to 0.5 eV. A low transverse emittance and low energy spread will allow for an efficient injection into the cylindrical Penning trap and provides higher accuracy for precision spectroscopies such as collinear laser spectroscopy (4).

PENNING TRAP

Cylindrical Penning trap (5) is a quadrupole trap which makes use of static electric and magnetic fields for three dimensional confinement of ions. The static quadrupole electric fields for axial trapping of ions are created by a set of ring electrodes at negative potential and two endcap electrodes at positive potential. The radial confinement can be achieved with a homogenous axial magnetic field of 3-6 T produced by the superconducting solenoid. With the combination of buffer gas cooling and an azimuthal quadrupole RF-field a mass dependent centering of ions can be achieved in the Penning trap (6,7). This allows for a mass selective ejection of ions in short time bunches.

OUTLOOK

Main application of the JYFL cooler-Penning trap system described here will be the preparation of pure isotopic beams of neutron-rich nuclei produced in intermediate energy fission. The collinear laser spectroscopy will be one the first experiments to gain from the project. Already reduction in energy spread and reduced spatial size after ion beam cooler will provide more than order of magnitude improvement for the laser spectroscopy. The bunching of the beam and mass purification in the Penning trap will further improve the experimental conditions not only for the laser spectroscopy, but for nuclear structure studies in general. Finally, one can see this project as a starting point for further studies, like post acceleration of the IGISOL-beams or high resolution spctroscopy of the refractory elements.

ACKNOWLEDGEMENTS

The support from the EU under the RTD contract ERBFMGECT950037 and support from exchange program between the Academy of Finland and Germany is acknowlegded.

REFERENCES

1. Itano, W.M., Bergquist, J.C., Bollinger, J.J., Wineland D.J., Phys. Scr. **T59** 106-120 (1994).
2. Penttilä, H., Dendooven, P., Honkanen, A., Huhta, M., Jauho, P.P., Jokinen, A., Lhersonneau, G., Oinonen, M., Parmonen, J.-M., Peräjärvi, K., Äystö, J., Nucl. Instr. Meth. Phys. Res. **B126** 213-217 (1997).
3. W. Paul, Rev. Mod. Phys. **62** 531-540 (1990).
4. Billowes,J., Campbell, P., cochrane, E.C.A., Cooke, J., Dendooven, P., Evans, D.E., Grant, I.S., Griffith J.A.R., Honkanen, A., Huhta, M., Levins, J.M.G., Liukkonen, E., Oinonen, M., Pearson, M.R., Penttilä, H., Persson, J.R., Richardson, D.S., Tungate, G., Wheeler, P., Zybert, L., Äystö, J., Nucl. Instr. Meth. Phys. Res. **B126** 416-418 (1997).
5. H. Raimbault-Hartmann, H., Beck, D., Bollen, G., König, M., Kluge, H.-J., Schark, E., Stein, J., Schwartz, S., Szerypo, J., Nucl. Instr. Meth. **B126** 378-382 (1997) 378.
6. Savard, G., Becker, St., Bollen, G., Kluge, H.-J., Moore, R.B., Otto, Th., Schweikhard, L., Stolzenberg, H., Wiess, U., Phys. Lett. **A158** 247-252 (1991).
7. König, M., Bollen, G., Kluge, H.-J., Otto, T., Szerypo, J., Int. J. Mass. Spec. Ion Proc. **142** 95-116 (1995).

ENHANCEMENT OF THE RADIATIVE TRANSITIONS BETWEEN THE GROUND AND THE 3.5-EV ISOMER STATES IN THE HYDROGEN-LIKE ^{229}TH^{89+} ION

F.F.Karpeshin[1,2], S. Wycech[3], I.M.Band[4], M.B.Trzhaskovskaya[4], M. Pfützner[5] and J. Żylicz[5]

[1] Universidade de Coimbra, Departamento de Física, Coimbra, Portugal
[2] St. Petersburg University, St. Petersburg, Russia
[3] Soltan Institute for Nuclear Studies, 00-681 Warsaw, Poland
[4] St. Petersburg Nuclear Physics Institute, St. Petersburg, Russia
[5] Institute of Experimental Physics, Hoża 69, Warsaw

Abstract. Lifetimes for the M1 transitions from the isomeric 3.5 eV 3/2+ state to the ground state are predicted to be enhanced in the hydrogen-like ^{229}Th89+ relative to the bare 229Th nucleus by several orders of magnitude. A possibility of experimental study of this phenomenon is discussed.

In ^{229}Th, a uniquely low lying nuclear level with spin and parity $3/2^+$ has been observed at the excitation energy $E_{isom} = 3.5 \pm 1.0$ eV with respect to the $5/2^+$ ground state [1]. This presents exceptional possibilities of discovering resonance effects in the atoms of ^{229}Th, which arise in an externally applied field of a laser ([2] and refs. cited therein). Another interesting possibility has been noted in ref. [3]. It was noted that due to an extremely low energy, the hyperfine interaction arising in the H-like ions of ^{229}Th can mix nuclear states of the same parity that differ by one unit of spin. This mixing effect is enhanced by the small energy separation between the $5/2^+$ and $3/2^+$ nuclear states. Until now, such nuclear-spin mixing due to the magnetic interaction has only been observed in muonic atoms [4].

In the case of ^{229}Th^{89+}, such mixing leads to drastic effect of acceleration of the nuclear deexcitation from the isomeric state [5]. In the present paper, we briefly remind the mechanism of acceleration, and dwell on prospects of experimental studying the effect.

As well known, in a hydrogen-like ion every nuclear level with spin I splits into two hyperfine components with full momenta $F \pm \frac{1}{2}$. Moreover, in $^{229}Th^{89+}$, additional interaction arises between the isomeric and ground-state components with the same full momentum F = 2. This interaction causes additional mutual repulsion of the levels, as illustrated in Fig. 1. This mixing reduces the lifetime of the upper component by orders of magnitude, from hours for the bare nucleus [2, 3] to fractions of second. The two other components with F = 1 and 3 remain unshifted. Experimental observation of the effect would allow one to precisely

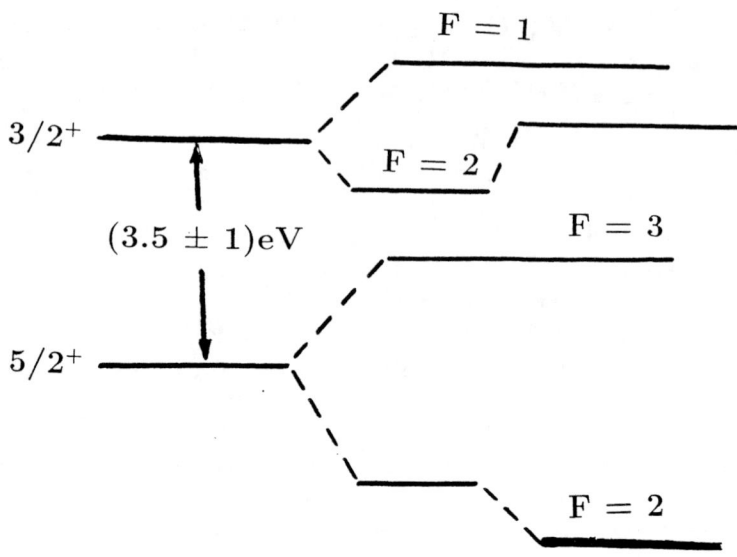

Figure 1: Scheme of the h.f.s. components cooresponding to the doublet nuclear states of the H-like ion of ^{229}Th.

determine full information about the spectroscopic properties of the doublet nuclear levels. Consider the question in finer detail.

The mixed states are denoted by $|F = 2i, M; mixed\rangle$, where the index i (= up, low) distinguishes one from the other:

$$|2\,up, M; mixed\rangle = \{|3/2\ 1/2; 2M\rangle + \beta|5/2\ 1/2; 2M\rangle\}\,N_\beta, \qquad (1)$$
$$|2\,low, M; mixed\rangle = \{-\beta|3/2\ 1/2; 2M\rangle + |5/2\ 1/2; 2M\rangle\}\,N_\beta,$$

where β is the mixing amplitude and $N_\beta = 1/\sqrt{\beta^2 + 1}$. Structure of the two non-mixed states with F=1 and 3 holds unchanged.

It follows from eq. (2) that mixing gives rise to the transitions between the levels which are diagonal with respect to the nuclear states, which are much faster than the transition between the isomeric and ground states of a bare nucleus. They occur via spin-flip of the orbital electron, just as transitions between normal components of hyperfine structure. To better realize the effect, one can imagine a neutral atom of ^{229m}Th with the excited nucleus. Its lifetime is hours, as aforementioned. If, however, all the electrons are stripped but one, the nucleus deexcites promptly! This mechanism of acceleration may be compared to acceleration of interband transitions arising due to coriolis mixing.

Calculation leads to the following expressions for the transition rates W:

$$W(2\,up \to 2\,low) = \frac{25}{18}\,W_0\,\beta^2\,N_\beta^4, \qquad (2)$$

$$W(3 \to 2\,low) = 5/9\ W_0\ N_\beta^2, \qquad W(1 \to 2\,low) = 5/6\ W_0\ \beta^2\ N_\beta^2,$$
$$W(2\,up \to 3) = 7/6\ W_0\ \beta^2\ N_\beta^2, \qquad W(1 \to 2\,up) = 5/6\ W_0\ N_\beta^2, \qquad (3)$$

where, for the Coulomb field of the atom,

$$W_0^{Coul} = \frac{\alpha}{m^2}\left(\frac{1+2\gamma}{3}\right)^2 \omega^3. \qquad (4)$$

Then, ω is the transition energy, m is the electron mass, α is the fine structure constant, with $\gamma = \sqrt{1-Z^2\alpha^2}$.

The results of calculation for the hfs energy levels and mean lifetimes are presented in Table 1. Since the most recent value of the isomer energy is 3.5 ± 1.0 eV [1], we have performed the calculations for $E_{isom} = 4.5$, 3.5, and 2.5 eV. As may be seen from Table 1, the nuclear-spin mixing in ^{229}Th^{89+} has drastic influence on the electromagnetic transitions between the pairs of the hfs components. The calculated rates of these transitions are enhanced by several orders of magnitude with respect to the isomeric nuclear γ transition.

The hfs energy levels and relevant transition energies, the mixing amplitude β and the decay rates are determined by the energy E_{isom}, the magnetic moments of the ground and isomeric states, and by the mixing matrix element. In ref. [3], the magnetic moments of the ground and isomeric states were adopted to be $0.45 \pm 0.06\mu_n$ and $-0.08 \pm 0.08\mu_n$, respectively. The mixing matrix element was found to be $\langle 2\,3/2|H_{magn}|5/2\,2\rangle = 0.59\pm0.06$ eV. It should be noted that the results presented in Table 1 have been obtained with no account for the experimental uncertainties.

The role of the uncertainties in magnetic-moments has been studied in ref. [5]. It has been found that the changes in magnetic moments of the ground and isomeric states within certain limits may cause rather significant changes in the level and transition energies, and in the mean lifetimes. For example, the order of the $F = 1$ and $F = 2up$ levels may be reversed. This is a result of the mixing and the related mutual repelling of the $F = 2\,up$ and $F = 2\,low$ states. By contrast, the changes in the mixing matrix element within its limits lead to a minor change ($\approx 4\%$) in the spacing of the $F = 3$ and $F = 2low$ levels.

A relevant experiment could be conceived at GSI, in Darmstadt, where hyperfine structure of ^{209}Bi has been recently measured [6]. To this end, hydrogen-like ions of heavy atoms can be produced as products of fragmentation reactions. In a fragmentation of ^{238}U nuclei on a Be target, the ground- and isomeric states of ^{229}Th are presumably produced with a comparable yield. In view of the long half-life of the isomer, in the beam of bare ^{229}Th^{90+} nuclei injected into the storage ring, both states would be present. A transformation of ^{229}Th^{90+} into hydrogen-like ^{229}Th^{89+} by an electron capture inside the ring would be followed by transitions with the lifetimes between 10 and 100 ms (see Table 1). Efforts should be made to detect the energies and lifetimes of the transitions. Then the four values: B(M1), E_{isom}, magnetic moments μ_1 and μ_2 in the ground and excited states, are unambiguously

Table 1: Calculated energies ω and mean lifetimes τ for radiative transitions between the four states of hfs in the ^{229}Th^{89+} ion.

E_{isom} [eV]	4.5		3.5		2.5	
mixing β^2 [%]	1.35		2.06		3.51	
Transition $F_i \to F_f$	ω [eV]	τ [s]	ω [eV]	τ [s]	ω [eV]	τ [s]
1 → 2up	0.12	24.0	0.10	41	0.08	82
1 → 2low	5.24	0.022	4.26	0.026	3.28	0.033
2up → 3	4.11	0.031	3.13	0.046	2.15	0.085
2up → 2low	5.12	0.013	4.16	0.017	3.20	0.023
3 → 2low	1.01	0.060	1.03	0.056	1.05	0.054

determined by experimentally observable lifetime of any of the transitions involving F = 2 component, and three energies of the excited hyper-fine components (with respect to the ground level 2low).

This work has been suported by the PRAXIS-XXI Program, Defense Special Weapons Agency (USA) under contract Nr. DSWA01-98-C-0040, by Grant No 96-02-18039-a of the Russian Fund for Fundamental Research, and by the Polish Committee of Scientific Research under Grant No 2-P03B-048-12 (S.W.) and BST (M.P. and J.L.Z.).

References

[1] Helmer, R.G., and Reich, C. W., *Phys. Rev.* **C49**, 1845 (1994).

[2] Karpeshin, F.F., Band, I.M., Trzhaskovskaya, M.B., and Listengarten, M.A. *Phys.Lett.* **B372**, 1 (1996).

[3] Wycech, S., and Żylicz, J., *Acta Physica Polonica* **B24**, 637 (1993).

[4] Kankeleit E., and Tomaselli, M. *Phys. Lett.* **32B**, 613 (1970).

[5] Karpeshin, F.F., Wycech, S., Band, I.M., Trzhaskovskaya, M.B., Pfützner, M., and Żylicz, J. *Phys. Rev.* **C57**, 3085 (1998).

[6] Finkbeiner, M., Fricke, B., and Kühl, T., *Phys. Lett.* **A176**, 113 (1993).

On-line separation of short-lived beryllium isotopes

Ulli Köster[a,1], James Barker[b,2], Richard Catherall[b],
Valentin N. Fedoseyev[c], Uwe Georg[b], Gerhard Huber[d], Ylva Jading[b],
Ove Jonsson[b], Mitsuo Koizumi[e], Karl-Ludwig Kratz[f],
Erich Kugler[b], Jacques Lettry[b], Viatcheslav I. Mishin[c], Helge Ravn[b],
Volker Sebastian[d], Claire Tamburella[b], Andreas Wöhr[g,3] and the
ISOLDE Collaboration[b]

[a] *Technische Universität München, Physik-Department, 85748 Garching, Bavaria*
[b] *ISOLDE, CERN, 1211 Genève 23, Switzerland*
[c] *Institute of Spectroscopy, Russian Academy of Sciences, 142092 Troitsk, Russia*
[d] *Johannes-Gutenberg Universität, Institut für Physik, 55099 Mainz, Germany*
[e] *Japan Atomic Energy Research Institute, 1233 Watanuki, Takasaki, 370-12, Japan*
[f] *Johannes-Gutenberg Universität, Institut für Kernchemie, 55099 Mainz, Germany*
[g] *KU Leuven, Instituut voor Kern- en Stralingsfysika, 3001 Leuven, Belgium*

Abstract. With the development of a new laser ionization scheme, it became possible to ionize beryllium efficiently in the hot cavity of the ISOLDE laser ion source. The high target and ion source temperatures enable the release of short-lived beryllium isotopes. Thus all particle-stable beryllium isotopes could be extracted from a standard uraniumcarbide/graphite target. For the first time the short-lived isotopes ^{12}Be and ^{14}Be could be identified at an ISOL facility, ^{14}Be being among the most short-lived isotopes separated so far at ISOLDE. The release time from the UC/graphite target was studied with several beryllium isotopes. Profiting from the element selectivity of laser ionization, the strong and isotopically pure beam of ^{12}Be allowed to determine the half-life to $T_{1/2} = 21.34(23)$ ms and the probability of beta-delayed neutron emission to $P_n = 0.48^{+0.12}_{-0.10}$ %.

Introduction

Beryllium is a "difficult" element for ISOL (isotope separation on-line) facilities, which is not or only very slowly released. With a vapor pressure of ≈ 10 mbar at

---

[1] Corresponding author: Tel. +41 22 767-9786, Fax -8990, Email: Ulli.Koster@cern.ch.
[2] Present address: University of Greenwich, Woolwich, London SE18 6PF, UK
[3] Present address: Oxford University, Department of Physics, Oxford OX1 3PU, UK

2000 K [1], about equal to tin, it is less a problem of evaporation than of adsorption to surfaces. To allow for a fast desorption the whole target-ion source assembly has to be kept at high temperatures. However, the traditional high temperature unit including a surface ion source cannot be applied due to the high ionization potential (9.32 eV) of beryllium. Off-line tests with ^{7}Be in a high temperature plasma ion source MK5 [2] gave an efficiency of 10^{-4} at best. About 95 % of the ^{7}Be got stuck to cold spots and did not reach the ionizer. It accumulated especially on the surface of the oxide insulators in the plasma source, forming very stable BeO.

For many metals of the groups IIA and IIIA the release speed can be improved considerably by adding CF_4 and separating the molecular side band of the monofluoride [3]. However on-line tests with ^{11}BeF$^+$ and a hot plasma source (all parts above 1800 °C) gave only an efficiency of $4 \cdot 10^{-4}$.

A recently presented ECR (electron cyclotron resonance) ion source for the production of ^{7}Be beams at TRIUMF [4] reaching an efficiency of 2 to 3 % might be an alternative for off-line produced ^{7}Be, but the source body is too cold to allow a rapid release of short-lived beryllium isotopes. Moreover the beam is mainly separated as BeF$^+$ and BeF$_2^+$. With a heavy target, huge contaminations of isobars like Ne$^+$, Na$^+$, Ar$^+$, K$^+$, AlF$^+$, etc. are unavoidable and will disturb nuclear spectroscopy experiments strongly due to their comparable lifetimes and decay modes (e.g. βn).

Resonance ionization laser ion source

A new way was opened with resonant photo ionization in a hot cavity laser ion source. For beryllium a new scheme had to be developed including frequency tripling of the dye laser light to reach the first excited state with UV light [5]. An off-line measurement gave an efficiency of 3 to 4 %, the on-line yields indicate an even higher efficiency.

Yields

With a standard[4] ISOLDE uraniumcarbide/graphite target (20 cm long, thickness 52 g/cm^2 of uranium and about 10 g/cm^2 of graphite) an ion implanted ^{7}Be target was produced. It was used for measurements of the astrophysical S_{17} factor [6]. The same UC/graphite target was used to produce a ^{11}Be beam for a collinear laser spectroscopy experiment, where the magnetic moment of ^{11}Be was measured with β-NMR [7,8]. In a target test also the short-lived isotopes ^{12}Be ($T_{1/2} = 21.3$ ms) and ^{14}Be ($T_{1/2} = 4.4$ ms) could be separated for the first time at an ISOL facility. ^{14}Be is among the most short-lived isotopes separated so far at

---

[4] Note that for the production of the long-lived isotopes ^{7}Be and ^{10}Be a "long" (40 cm length) and dense (73 g/cm^2) target from pure graphite was used, giving significantly higher yields when coming into saturation.

ISOLDE. The given yields (ions per s) are normalized to 1 µA of primary proton beam, the average proton current at ISOLDE is between 2 and 4 µA.

TABLE 1. Measured yields of beryllium isotopes from a standard ISOLDE UC/graphite target.

Isotope	Half-life	Yield (ions/µA·s)	Measurement method
^{7}Be	53 d	$1.4 \cdot 10^{10}$	Faraday cup
^{9}Be	stable	$1.7 \cdot 10^{10}$	Faraday cup
^{10}Be	$1.6 \cdot 10^6$ a	$5.9 \cdot 10^9$	Faraday cup
^{11}Be	13.8 s	$7 \cdot 10^6$	Faraday cup & β detector
^{12}Be	21.3 ms	1500	β detector
^{14}Be	4.4 ms	4	neutron detector

Release time

The release time structure was measured with ^{11}Be using the sampling technique (cf. ref. [9]), see fig. 1. This allows to access a time scale of about 0.2 to 30 s.

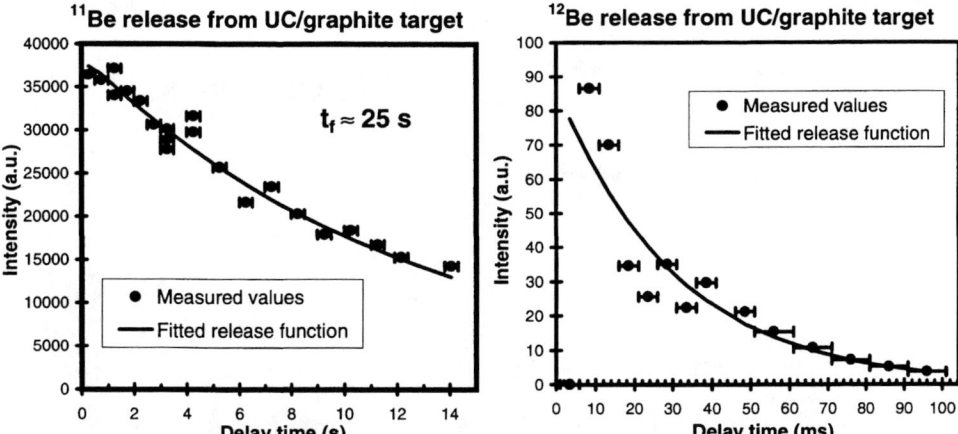

FIGURE 1. Release function of beryllium measured with ^{11}Be and ^{12}Be.

For the simplified release function (not taking into account the contribution of a possible slow component):

$$p(t) = A \cdot (1 - \exp(-t/t_r)) \cdot \exp(-t/t_f)$$

a fall time t_f of about 25 s can be deduced after correction for the radioactive decay of ^{11}Be. The very first part of the release function (up to about 100 ms) was monitored with ^{12}Be. Despite some fluctuations a rise time t_r in the order of 1 ms can be deduced. This extremely fast rise might indicate a small component which is released as a bunch due to the impact of the intense proton pulse. Such a "radiation enhanced release" has already been experienced before, e.g. for noble gases from oxide targets [10]. A slower tail in the minute range can be observed in a Faraday cup from the release of the long-lived radionuclides ^{7}Be and ^{10}Be after switching off the proton beam. A quantitative evaluation is however difficult, since the proton beam contributes to the target heating and to the just mentioned bunched release. During the whole run the target was kept at temperatures above 2000 °C and the tungsten ionizer cavity above 2300°C.

Beta-delayed neutron emission

^{14}Be was detected via its beta-delayed neutrons with a 4π neutron long counter. ^{3}He proportional counters are arranged in three concentric circles and embedded in a polyethylene matrix for moderation. The neutron detection efficiency is about 30 % in this arrangement. Figure 2 shows the release and decay curve of ^{14}Be.

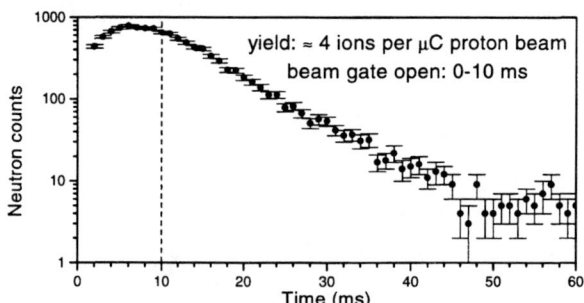

FIGURE 2. Grow-in and decay curve of ^{14}Be.

^{12}Be decay

^{12}Be has a Q_β value of 11.71 MeV. The channel for beta-delayed neutron emission can open at 3.37 MeV excitation energy, thus leaving a rather large window of 8.34 MeV for βn decay. Detecting simultaneously the betas with a scintillation detector and the beta-delayed neutrons allows to deduce the probability of beta-delayed neutron emission P_n.

The neutron spectrum can be fitted with a single exponential decay, since any neutron emission in the decay of the daughter ^{12}B is energetically excluded as well as a $\beta 2n$-decay of ^{12}Be. For the fit of the beta spectrum the half-live of the daughter ^{12}B, which is very precisely known (20.20(2) ms), was kept constant. No boron beam is released directly from the source (the surface ionization of boron is negligible with an ionization potential of 8.3 eV) but the ^{12}B produced by ^{12}Be decay in the grow-in phase has to be taken into account. The fits were done with the maximum likelihood method to account correctly for the Poisson distribution of the low statistics events [11].

The resulting half-life of 21.34(23) ms is shorter than the older values in literature [12–14], but in agreement with more recent values [15,16].

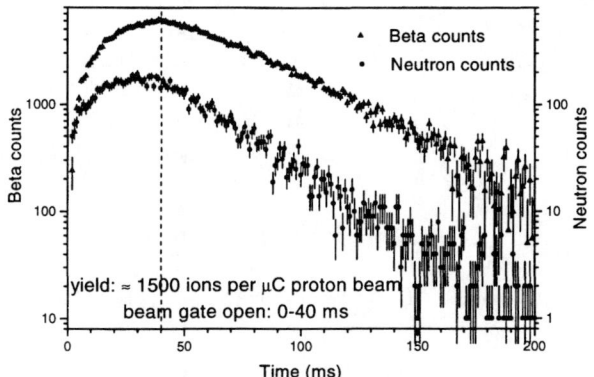

FIGURE 3. Grow-in and decay curve of ^{12}Be.

The efficiency ratios of the neutron and beta counters were calibrated on-line with several neutron emitters with known P_n values. Since the alignment of these weak ion beams cannot be checked directly with a scanner or Faraday cup, a systematic error arises due to differences in the effective solid angle of the detectors to a possibly shifted implantation position. While this effect is small for the neutron detector it can affect the efficiency for beta detection considerably. An error of about 25 % (relative) for the P_n value was derived from the different calibration measurements including the response to various beta and neutron energies of the calibration isotopes. This mainly systematic uncertainty could still be considerably reduced with an enlarged set of calibration data and a careful study of the effective detector efficiencies by Monte Carlo simulations.

The obtained value ($P_n = 0.48^{+0.12}_{-0.10}$ %), which is remarkably low compared to the large energy window available for neutron emission, is in agreement with the upper limits from [12,13] and the value reported in [16]. It disagrees with an older measurement [17] where the beam was probably contaminated with other neutron emitters.

Prospects

Already now new experiments with beryllium beams are scheduled (detailed nuclear spectroscopy of ^{11}Be, ^{12}Be and ^{14}Be [18]) or planned for the future (measurement of S_{17} in inverse kinematics at REX-ISOLDE [19]). In our set-up the neutron energy could not be determined accurately. With a different set-up (^{3}He ionization chambers or a time-of-flight arrangement) it should be possible to determine the populated neutron-emitting levels in ^{12}B. Together with a detailed $\beta\gamma$ spectroscopy this should allow to monitor the beta strength distribution over a rather large energy range and to compare it to theoretical predictions, e.g. [21,22]. The laser ion source can even be used to measure directly the isotope shift of the beryllium atoms, see [23].

To perform these experiments a further increase in yield would be very useful. This could be achieved by optimizing the target geometry for fastest release (more compact design). A tantalum foil target which can be heated to very high temperatures (well above 2400 °C) could be an alternative to UC/graphite. By adding CF_4 the release time may be improved, but the beryllium-fluoride molecules have to be dissociated in the ionizer tube before laser ionization can be applied.

Acknowledgements

Thanks for support by the "Beschleunigerlabor der TU und LMU München".

REFERENCES

1. R.E. Honig and D.A. Kramer, RCA-Review 30 (1969) 285.
2. S. Sundell, H. Ravn and the ISOLDE Collaboration, Nucl. Instr. Meth. B70 (1992) 160.
3. R. Kirchner, Nucl. Instr. Meth. B126 (1997) 135.
4. K. Jayamanna et al., Rev. Sci. Instr. 69 (1998) 756.
5. J. Lettry et al., Rev. Sci. Instr. 69 (1998) 761.
6. Michael Hass et al., contribution to these proceedings.
7. Matthias Keim, contribution to these proceedings.
8. Stefan Kappertz et al., contribution to these proceedings.
9. J. Lettry et al., Nucl. Instr. Meth. B126 (1997) 130.
10. H.L. Ravn et al., Nucl. Instr. Meth. B126 (1997) 176.
11. Y. Jading and K. Riisager, Nucl. Instr. Meth. A372 (1996) 289.
12. D.E. Alburger et al., Phys. Rev. C17 (1978) 1525.
13. J.P. Dufour et al., Z. Phys. A319 (1984) 237.
14. F. Ajzenberg-Selove, Nucl. Phys. A506 (1990) 1.
15. M.S. Curtin et al., Phys. Rev. Lett. 56 (1986) 34.
16. G. Audi, O. Bersillon, J. Blachot and A.H. Wapstra, Nucl. Phys. A624 (1997) 1.
17. A.M. Poskanzer, P.L. Reeder and I. Dostrovsky, Phys. Rev. 138 (1965) B18.
18. Beta-decay study of dripline isotopes of Be, ISOLDE Proposal P99, Spokesperson: G. Nyman
19. Measurement of the $^7Be(p,\gamma)^8B$ absolute cross section in inverse kinematics, ISOLDE Letter of Intent I20, Spokespersons: C. Rolfs and U. Greife
20. H. Keller et al., Z. Phys. A348 (1994) 61.
21. T. Suzuki and T. Otsuka, Phys. Rev. C56 (1997) 847.
22. T. Suzuki, J. Phys. G24 (1998) 1455.
23. Volker Sebastian et al., contribution to these proceedings.

MISTRAL: THE BEGINNING OF A NEW MASS MEASUREMENT PROGRAM AT ISOLDE

D. Lunney[1], C. Toader[1,2], M. de Saint Simon[1], G. Audi[1], C. Borcea[2], H. Doubre[1], M. Duma[2], M. Jacotin[1], S. Henry[1], J.-F. Képinski[1], G. Lebée[3], G. Le Scornet[1], C. Monsanglant[1], C. Thibault[1], and the ISOLDE collaboration[3]

[1] CSNSM-IN2P3-CNRS, F-91405 Orsay, France
[2] Inst. Atomic Physics, Bucharest, Romania
[3] CERN, EP Division, Geneva, Switzerland

Abstract. The *MISTRAL** experiment is now on-line at *ISOLDE*. Installed in May 1997, *MISTRAL* received its first stable beam in October and first radioactive beam in November 1997. These first tests, with a plasma ion source, resulted in excellent isobaric separation and reasonable transmission. Further testing and development enabled first data taking in July 1998 on neutron-rich Na isotopes having half-lives as short as 31 ms.

The atomic mass is a global property that provides information of interest to nuclear physics through the nuclear binding energy. The elucidation of new nuclear structure is thus possible by systematic and high precision mapping over several isotopes and isotones around a given point on the mass surface. The need for accurate measurements of atomic masses particularly far from stability is acute since model predictions vary quite drastically only a few isotopes beyond known values. The modeling of the nucleosynthesis *r*-process, for example, not only suffers from the eventual error of the predicted mass itself, but can be biased by the propagation of uncertainty in the calculation of other nuclear physics properties that are derived from a fit to known masses (in the case of macroscopic-microscopic models).

Given such physics motivation, a new mass measurement program has been undertaken at the *ISOLDE* isotope separator facility at *CERN* [1]. This program, called *MISTRAL** [2-3], complements the existing Penning trap spectrometer, *ISOLTRAP* [4] by concentrating on very short-lived nuclides. By two coherent, high-harmonic radiofrequency excitations of the ion orbit inside the spectrometer magnetic field, an accurate determination of its cyclotron frequency is accomplished and when compared to that of a reference ion, provides mass measurements of very high precision.

---

*Mass measurements at ISolde using a Transmission RAdiofrequency spectrometer on-Line

CP455, *ENAM98: Exotic Nuclei and Atomic Masses*
edited by B. M. Sherrill, D. J. Morrissey, and Cary N. Davids
© 1998 The American Institute of Physics   1-56396-804-5/98/$15.00

A schematic diagram of the spectrometer with its nominal trajectory is shown in figure 1. Ions injected at the full *ISOLDE* beam energy (60 kV) follow a two-turn helicoidal trajectory inside the annular, homogeneous magnetic field and are counted using a secondary electron multiplier (figure 1, inset center). With an injection slit size of 0.4 mm and orbit radius of 0.5 m, a mass resolution of 2500 is obtained using no radiofrequency. In order to make a measurement, a longitudinal kinetic energy modulation is effected using two symmetric electrode structures (figure 1, inset right) located at the one-half and three-half turn positions inside the magnetic field. This way the ions make one cyclotron orbit between the two modulators. A radiofrequency voltage is applied to the central modulator electrodes. Depending on the phase of this voltage when the ions traverse the structure, the resulting longitudinal acceleration produces a larger or smaller cyclotron radius than that of the nominal trajectory (since all the trajectories are isochronous). The ions are transmitted through the 0.4 mm exit slit when the net effect of the two modulations is zero. This happens when the radiofrequency voltage is an integer-plus-one-half multiple of the cyclotron frequency which means that during the second modulation the ions feel exactly the opposite of what they felt during the first. For high harmonic numbers (e.g. larger than 1000) and a radiofrequency voltage of about 200 V, this gives narrow transmission peaks having resolutions of up to 100,000 (figure 1, inset left).

A mass measurement is made when an unknown mass is alternately injected with a reference mass. These comparisons are done in rapid succession (seconds) in order to eliminate short-term drift in the magnetic field. Comparing masses in this way, without changing the magnetic field, requires changing not only the transport energy of the reference beam but the voltages of all electrostatic elements in the spectrometer (two triplets, eight pairs of steering plates, and two benders plus the injection switchyard bender). Since, for the moment, the reference ion source does not withstand more than 60 kV, we are obliged to use a reference mass that is heavier than the ISOLDE mass in order to operate ISOLDE and its transport system at the nominal voltage. A reference source upgrade later this year should avoid this limitation for future runs.

ISOLDE uses a pulsed proton beam extracted from a set of synchroton booster rings [5]. In the case of short-lived isotopes (as well as elements with very rapid release times from the target matrix, such as Na) it is impossible to scan the entire required frequency range in time after the impact of the proton pulse. In this case, a special acquisition mode is used (called, appropriately: point-by-point). For each radioactive beam pulse, the ion transmission signal is recorded for only one radiofrequency point (determined randomly) and the resonance peak is reconstructed at the end. Moreover, for each point, the ion signal is recorded with the radiofrequency switched off so that not just the intensity but the true transmission is measured. This mode also allows us to increase statistics in the peak.

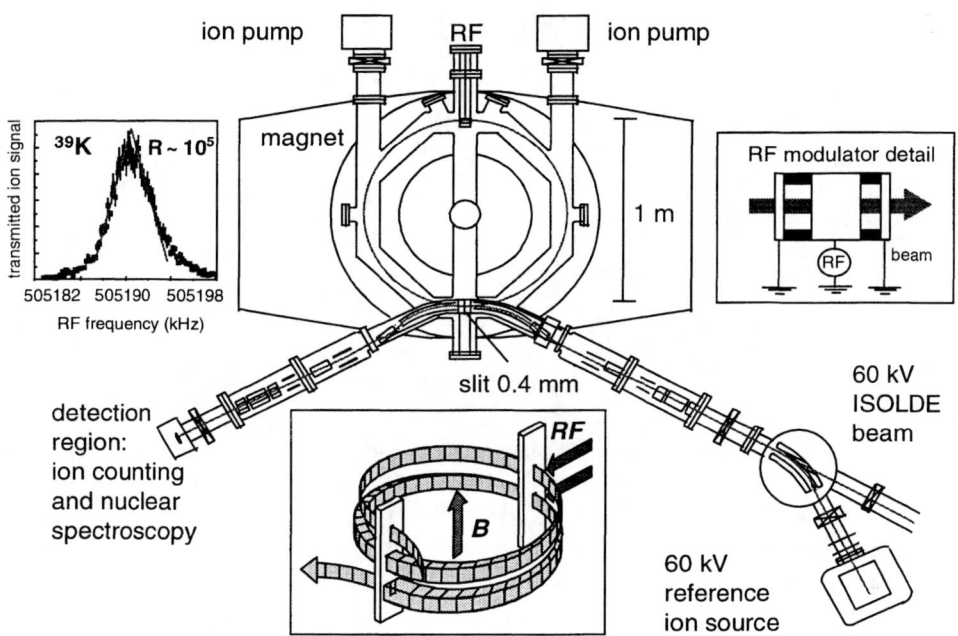

FIGURE 1. Layout of MISTRAL showing the nominal ion trajectory. Ions are injected from the *ISOLDE* beamline at the nominal transport voltage of 60 kV while the reference mass is alternately injected using the required transport energy for that mass in the magnetic field corresponding to the *ISOLDE* ion mass. Consequently, all electrostatic elements in the spectrometer have their voltages switched dynamically. Inset (right) shows the modulator electrode structure the geometry of which is selected depending on the mass range of operation. Inset (center) shows an isometric view of the trajectory envelope with the 0.4 mm injection slit followed by the first modulator at one-half turn, the phase-definition slit (up to 5 mm wide to incorporate the envelope of cyclotron radii), the second modulator at three-half turns and finally the exit slit. Inset (left) shows a transmitted ^{39}K ion signal as a function of radiofrequency for one harmonic number.

The MISTRAL spectrometer was installed in the new beam hall extension of ISOLDE in mid-1997 and a first test run using radioactive isotopes around $A = 27$ took place at the end of that year using a UC$_2$ target coupled with a plasma ion source. The spectrometer was able to cleanly separate the isobaric components with relatively good sensitivity and very encouraging indications for measurement precision.

The potential meaurement program at ISOLDE is quite rich. Though the calculated transmission of 1% has not yet been reached, there are some one hundred candidates for either new measurements or considerably reduced error. New developments with the ISOLDE laser ion source will also provide exciting possibilities [6]. Scheduled beamtimes for MISTRAL are July and November, 1998.

ACKNOWLEDGEMENTS

The authors would like to thank the following researchers who gave their advice and support to the original proposal for the *ISOLDE* experiments committee: G. Bollen, D. Guillemaud-Mueller, P.G. Hansen, B. Jonson, H.-J. Kluge, R.B. Moore, A.C. Mueller, G. Nyman and H. Wollnik. One of us (David Lunney) would also like to thank CERN for a one-year Scientific Associateship.

REFERENCES

1. E. Kugler *et al.*, Nucl. Instr. and Meth. B 70 (1992) 41
2. M.D. Lunney *et al.*, Hyperfine Interactions 99 (1996) 105
3. M. de Saint Simon *et al.*, Physica Scripta T59 (1995) 406
4. G. Bollen *et al.*, these proceedings
5. J. Lettry *et al.*, Nucl. Instr. and Meth. B 126 (1997) 130
6. V. Sebastian *et al.*, these proceedings

NOTE ADDED TO "THE PROOF"

Just two weeks after the end of ENAM98, the *MISTRAL* experiment recorded its first data on neutron-rich Na isotopes at *ISOLDE* using the UC_2 target and W surface ioniser. In a detailed and comprehensive mapping (for evaluation of the systematic error) the masses of the isotopes $^{23-30}$Na were measured. The last three isotopes of this chain have half-lives of 31, 45 and 48 ms, respectively.

BEARS: Radioactive Ion Beams at LBNL

J. Powell, F. Q. Guo, P. E. Haustein[†], R. Joosten, R.-M. Larimer,
C. Lyneis, D. M. Moltz, E. B. Norman, J. P. O'Neil, M. W. Rowe,
H. F. VanBrocklin, Z. Q. Xie, X. J. Xu and Joseph Cerny

Lawrence Berkeley National Laboratory, University of California, Berkeley, CA 94720, and
[†]*Chemistry Department, Brookhaven National Laboratory, Upton, NY 11973*

BEARS is an initiative to develop a radioactive ion-beam capability at Lawrence Berkeley National Laboratory. The aim is to produce isotopes at an existing medical cyclotron and to accelerate them at the 88" Cyclotron. To overcome the 300-meter physical separation of these two accelerators, a carrier-gas transport system will be used. At the terminus of the capillary, the carrier gas will be separated and the isotopes will be injected into the 88" Cyclotron's Electron Cyclotron Resonance (ECR) ion source. The first radioactive beams to be developed will include 20-min ^{11}C and 70-sec ^{14}O, produced by (p,n) and (p,α) reactions on low-Z targets. A test program is currently being conducted at the 88" Cyclotron to develop the parts of the BEARS system. Preliminary results of these tests lead to projections of initial ^{11}C beams of up to 2.5×10^7 ions/sec and ^{14}O beams of 3×10^5 ions/sec.

There is currently extensive world-wide activity in the development and construction of radioactive ion-beam facilities of various types. The availability of beams of unstable nuclei offers exciting new opportunities for research into nuclear structure and nuclear astrophysics. BEARS, or Berkeley Experiments with Accelerated Radioactive Species, is an initiative to develop a radioactive ion-beam capability at LBNL.

The basic concept for an initial BEARS system involves the coupling of isotope production at an existing medical cyclotron in building 56 of LBNL (see Fig. 1) with post-acceleration by the 88" Cyclotron. These accelerators are separated by a distance of about 300 m; isotopes will be transported between the two via a gas-jet capillary. Carrier gas would be pumped by a high-throughput Roots blower at the building 88 end. Preliminary tests have shown that a total transport time of less than a minute is easily achieved, and times as short as 10 sec may be possible.

After transport, the radioisotopes are to be injected into one of the ECR ion sources of the 88" Cyclotron. These sources can achieve good ionization efficiencies at high charge states; however, they require vacuums of the order of 10^{-6} torr to operate. Therefore the central technical challenge of BEARS is the coupling of a gas transport system to an ECR ion source. Two separate techniques are being explored.

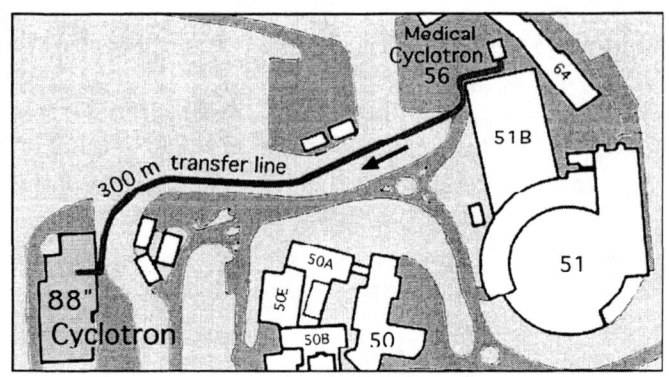

FIGURE 1. Map showing proposed transfer line between buildings 56 and 88.

The first involves transport of activity attached to small aerosol clusters suspended in the carrier gas (1). This is a standard technique that can achieve very high transport efficiencies. At the ECR, the vast majority of the carrier gas is skimmed away in a multi-stage differential-pumping system. The aerosol clusters, due to their high mass, pass though the skimming system into the high vacuum region. This technique has the potential for being broadly applicable to many isotopes, but does not work with volatile compounds.

The second technique involves separation of the radioactivity from the carrier gas by cryogenic trapping at liquid-nitrogen temperatures (1). After separation, the trapped gas is bled into the ECR plasma region. This method has the advantage of simplicity although it requires that the isotope of interest be in a suitable gaseous chemical form. It is expected that the two methods will complement each other.

The medical cyclotron produces a 10 MeV proton beam with typical intensities of 40 µA. Several light proton-rich isotopes can be produced, via (p,n) and (p,α) reactions on light-Z targets. Initially, we have focused on the production of ^{11}C ($t_{1/2}$=20 min) and ^{14}O ($t_{1/2}$=71 sec), both produced from ^{14}N. The target is either N_2, which doubles as the transport-system carrier gas, or a solid material such as boron nitride (BN).

Several tests of the various parts of the BEARS system have been carried out. As the 300 m transfer line has yet to be constructed, the 88" Cyclotron has been used to mimic the medical cyclotron, producing up to 5 µA of 10 MeV proton beams. Activity produced in our prototype target chamber is transported in one of the above described manners to one of two ECR ion sources (the other being used to supply the proton beam).

For the aerosol transport technique, a four stage differentially-pumped skimming system was constructed and coupled directly to the ECR. This is shown in Fig. 2. Aerosol clusters and carrier gas enter the first stage in a jet of near-sonic velocity. The

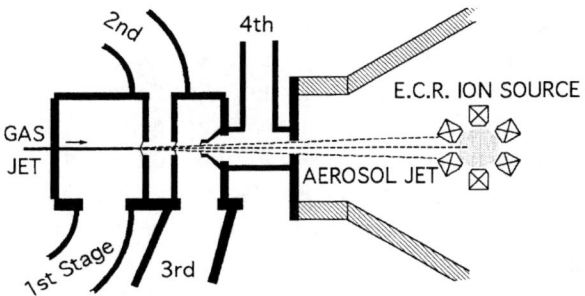

FIGURE 2. Skimming system for injecting aerosols into the ECR ion source.

heavier clusters exit in a narrower cone than the expanding gas, allowing then to pass through the small holes in the three skimmers. Once inside the ECR, the aerosols are caught on heated surfaces, to vaporize the activity. Tests have shown that, with a full gas load, ECR performance is not significantly degraded.

Unfortunately, it was found that this system failed to transport significant amounts of ^{11}C or ^{14}O, the initial BEARS production isotopes. This was traced to the majority of the activity forming gaseous compounds and thus not attaching to the aerosol clusters. The fraction of ^{11}C in a chemical form that could be transported was only on the order of 0.1-0.5%. However, this small amount of activity was successfully injected into the ECR, and an extracted beam of ^{11}C was identified, although at very low intensities.

The cryogenic trapping technique was found to be much more effective for transporting ^{11}C and ^{14}O; about 15% of the total produced activity being successfully trapped and then released into the ECR. The N_2 target/carrier gas was passed through a coil of 1/8" o. d. stainless steel tubing, about 1.5 m long, submerged in liquid nitrogen. After stopping the gas flow and allowing the remaining nitrogen to be pumped away, the trap was connected to the ECR. The liquid nitrogen was then replaced by an alcohol bath containing dry ice, quickly raising the temperature to 195 K. This

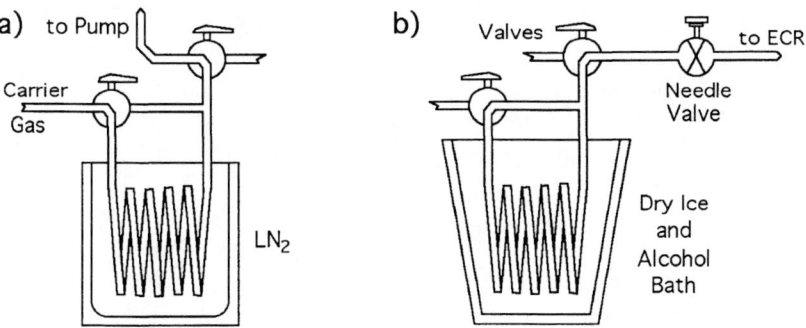

FIGURE 3. Cryogenic trapping system: (a) trapping and (b) release at dry ice temperatures into the ECR ion source.

temperature increase releases 90% of the trapped ^{11}C (^{11}CO$_2$) and about 40% of the trapped ^{14}O (believed to be N^{14}O or N$_2$^{14}O). The released gas was then bled directly into the ECR plasma region through an adjustable needle valve.

Care had to be taken to minimize the amount of non-radioactive gas that was trapped and introduced to the ECR along with the radioactivity. With too great a gas load, ionization efficiencies at high charge states are greatly reduced. By using the dry-ice/alcohol bath, rather than warming the coil to room temperature, reduced the resulting ECR gas load by 90%.

Beams of both ^{11}C and ^{14}O were successfully extracted from the ECR in charge states of up to 5+. The ionization efficiency for ^{11}C^{4+} was measured to be 0.7%. Similar results were found for 1+, 2+ and 3+ charge states, while the efficiency for ^{11}C^{5+} was about 0.1%, due to the difficulty in stripping an S-shell electron. Initial measurements of ^{14}O in 3+, 4+ and 5+ charge states indicate efficiencies in the range of 0.4 to 1%. Future developments, including the use of the 88" Cyclotron's second, advanced ECR ion-source (the AECR-U), are expected to increase these efficiencies to a range of 1 to 4%.

Beam transport efficiency through the 88" Cyclotron is estimated to be 10-25% for energies of 65 to 180 MeV (with ^{11}C$^{3+,4+}$ ions) and 5-15% for 11-50 MeV (with ^{11}C$^{1+,2+}$). With these and the above measured efficiencies, and starting with the known ^{11}C production at the medical cyclotron of 1.0×10^{11} atoms/sec (3), a ^{11}C beam intensity of 0.5 to 2.5×10^7 ions/sec can be projected for the completed BEARS system, with a reasonable expectation of greater than 10^8 ions/sec with expected future improvements in trapping and ECR efficiencies. For comparison, the Louvain-la-Neuve facility reports a current available ^{11}C^{1+} beam of 10^7 ions/sec in an energy range of 6.2-10 MeV (2). Similar projections for ^{14}O lead to initial beams of 3×10^5 ions/sec.

The transfer line between the two cyclotrons is to be built over the remainder of 1998. During this time, further tests will be conducted to maximize efficiencies and minimize transport times and ECR gas load. In addition, we will begin tests to develop other light-mass proton-rich ion beams, such as ^{13}N, ^{15}O, ^{17}F, ^{18}F, and possibly ^{10}C.

ACKNOWLEDGMENTS

This work supported by USDOE, under contracts DE-AC03-76SF00098 and DE-AC02-98CH10886. One of the authors, R. Joosten, would like to acknowledge the support of the German Academic Exchange Service (DAAD).

REFERENCES

1. MacFarlane, R. D. and McHarris, W. C., in *Nuclear Spectroscopy and Reactions, Part A*, Cerny, J. ed., New York: Academic Press, 1974, pp. 249-268.
2. Vervier, J., *Nucl. Phy.* **A616**, 97c-106c (1997).
3. Kitwanga, Sindano wa, et al., *Phys. Rev.* C **42**, 748-752 (1990).

The use of (d,xn) reactions: RIB production and energy generation

D. Ridikas and W. Mittig

GANIL, BP 5027, F-14076 Caen Cedex 5, France

Abstract

The interest in neutron production reactions has been recently renewed because they are the basis in the development of powerful neutron sources for various purposes like nuclear energy production and incineration of nuclear waste, material structure analysis, tritium production, etc. Another interest is related to the possibility of the production of radioactive nuclei and exotic beams by neutron induced fission. One of the crucial problems in this context is to determine the most efficient way to convert the beam energy into neutrons produced afterwards. By performing the calculations with the LAHET code system [1] we try to estimate whether protons or deuterons are more favourable for neutron production. Consequently, a few direct applications of our investigations are presented and compared with experimental data.

1. Introduction

The total energy is one of the essential parameters which defines the actual costs of particle acceleration. Deuterons seem to be more efficient projectiles for neutron production for all target materials [2] if compared to protons at the same incident energies as it is shown in Fig. 1. In the case of light targets like Be, neutron yield is higher for deuterons by a factor of 1.5-2.5 depending on energy, and the difference decreases with increase in incident energy. However, for heavy targets like U, neutron production by 200MeV protons and deuterons is almost the same. Increase in projectile energy makes deuterons more efficient only by 10% if compared to protons.

We also estimated [2] that for protons the angular distributions are relatively flat, while for deuteron induced reactions the neutron production is more concentrated at forward angles. On the other hand, both (p,xn) and (d,xn) reactions showed a some-what similar angular dependence at more backward angles; the cross section decreases considerably with the angle in the case of light and intermediate mass targets, while for the heavy ones neutrons are emitted in the entire scattering angle range.

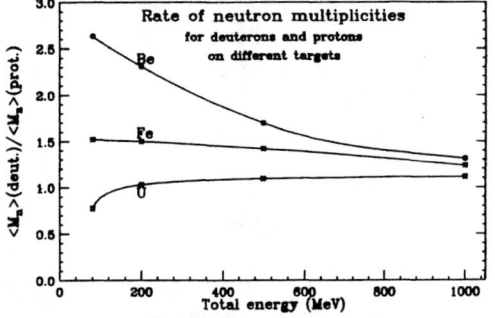

Figure 1: Ratio of neutron multiplicities from deuteron induced reactions over multiplicities from proton induced reactions on different targets (thick) as a function of total incident energy.

The energy distributions of neutrons emitted from (d,xn) and (p,xn) are quite different as well. We found that much more energetic neutrons are produced by light targets than by heavy ones. Moreover, deuterons (comparing to protons) both in light target and heavy target materials produce considerably more energetic neutrons in absolute value. We refer the reader to Ref. [2] for more detailed analysis.

2. Production of neutron-rich isotopes

In a charged particle induced cascade one can distinguish two qualitative physical regimes: a) a spallation driven, high energy phase and b) a neutron driven, fission dominated regime. Neutrons from the first phase are acting as a "source" for the second phase. The choice of the target materials for both of these physical regimes depends on the final purpose of such a device.

Our combined target system, as initially proposed by Argonne National Laboratory group [3], is presented in Fig. 2. It consists of a stopping Be cylinder (neutron source) and different U cylinders (production targets). In order to minimise the neutron loss from the system, and at the same time to provide it with some cooling conditions, the targets are surrounded by a cylindrical light water blanket.

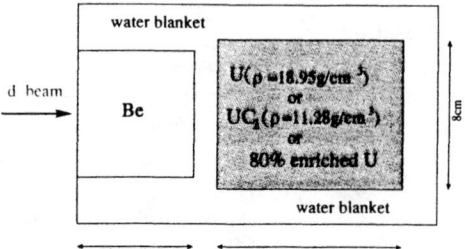

Figure 2: Combined target assembly for isotope production.

Neutron production in the Be target increases very sharply as a function of incident deuteron energy ranging from 50MeV up to 200MeV. A factor of 2 (or 4) increase in energy corresponds to a factor of 4 (or 10) increase in total neutron multiplicities. In this energy region, the beam intensity could not compensate neutron production at lower energies but the same beam power. Therefore, the highest 200MeV energy is suggested [4]. In the secondary target neutron multiplicities are proportional to the number of neutrons in the primary one; we found that approximately 35-45% of all neutrons produced in Be cross the surface of the secondary target, where neutrons are further multiplied by (n,xn) and (n,f) reactions. The energy spectrum of incoming neutrons is again very important in this process; more fissions and more (n,xn) reactions will occur for more energetic neutrons.

As expected, higher incident energies resulted in broader isotope distributions for an in-target isotope production [4]. Moreover, at the very peaks of the distributions, increase in incident energy from 50MeV to 100MeV (to 200MeV) increased the isotope production by a factor of 8 (or 15). On the neutron rich (or deficient) side these differences are similar or expected to be even higher (if extrapolated from presently calculated values) [4].

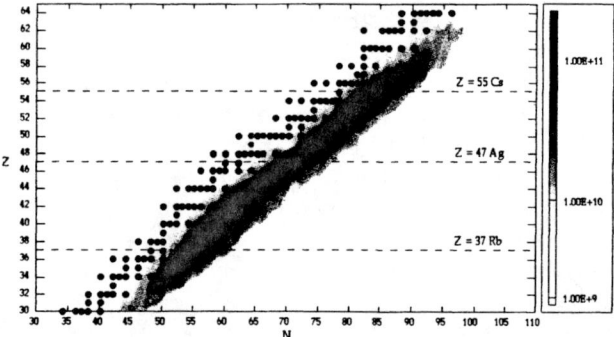

Figure 3: Isotope production in the system 200MeV $d+Be \rightarrow xn + {}^{238}U$.

In Fig. 3 we plot isotope production distribution from our combined Be+U target system with 200MeV deuterons. The isotope rates are normalised to $1.9*10^{14}$ d/s. For comparison we also included the fission yield distribution from the fission of 1g of the uranium ^{235}U by thermal neutrons for the neutron flux of $1.9*10^{14}$ n/s cm^2 [5] (see Fig. 4); it contains two well established peaks located at Z=36-38 and Z=54-56, between which at Z=44-46, the distribution has its local minimum.

We found that the mass distribution of fission products from the $n_{th}+^{235}$U reactions give higher yields by 1-2 orders of magnitude near the peaks. However, fission yields from high energy neutrons resulted in much broader isotope distributions, what makes these two different methods equally efficient on the neutron deficient side of the distributions. In addition, in the mass region with Z=40-50, where fission products from thermal neutron induced fissions exhibit their local minimum, high energy neutrons give higher fission yields by 2 (or higher) orders of magnitude* (compare Figs. 3 and 4).

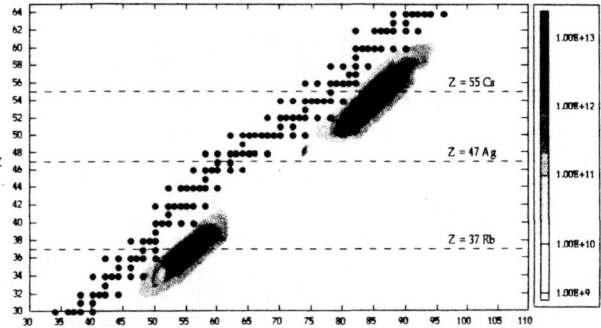

Figure 4: Isotope production in the system $n_{th}+^{235}$U [5].

The use of highly enriched uranium for production of isotopes also seems to be a very interesting possibility. 80% ^{235}U in the production target results in an effective neutron multiplication coefficient $k_{eff} \sim 0.9$ to be compared to $k_{eff} \sim 0.1$ for pure ^{238}U [4]. As a good approximation $N_{fiss} \sim k_{eff}/(1-k_{eff})$ [2]; consequently, one could obtain the increase in a number of fissions (and fission yield) by a factor of $N_{fiss}(80\%\ ^{235}U)/N_{fiss}(^{238}U) \sim 81$!

3. Nuclear energy generation

So far only high energy (0.8-1.6GeV) protons hitting a heavy target (uranium, thorium, lead) were considered as far as sub-critical hybrid systems are concerned [6]. The use of projectiles such as deuterons and light targets (lithium, beryllium) as converters to create the neutrons may have competitive, if not more efficient, features too. After each successive fission in ^{238}U or ^{235}U approximately 181MeV energy becomes available according to their fission Q-values. Therefore, the energy gain $G(N_{fiss})$ can be estimated by a very simple relation, i.e. $G(N_{fiss})=181 N_{fiss}/E_{beam}$, where E_{beam} is given in (MeV).

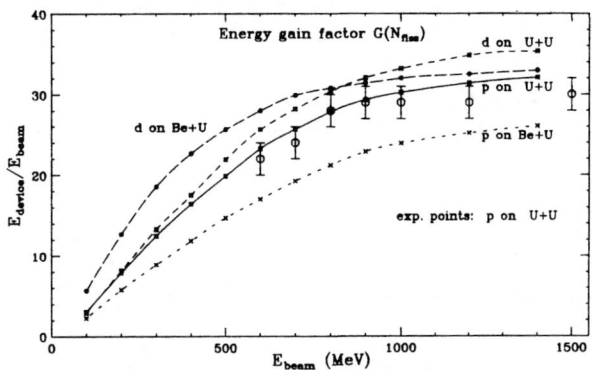

Figure 5: Calculated and measured [7] avarage energy gain for proton and deuteron induced reactions in the target assemblies Be+U and U+U.

Fig. 5 shows the calculated energy gain $G(N_{fiss})$ as a function of the incident beam energy and two different spallation targets (Be or U) inside the same fuel assembly (uranium and water mixture with $k_{eff} \sim 0.9$ [7]) for protons and deuterons. In the same Fig.

---

*Our present calculations do not include fission yields from fissions induced by neutrons bellow 20MeV, and the yields estimated represent the lower limits of a complete fission process.

we also plot the experimental points with error bars from [7], which in our notation should be compared to the "p on U+U" solid curve. Inspite of the simplifications we made in our geometry assumptions, we got a reasonable agreement with the experimental data. Our predicted values for deuterons on Be+U target combination (see "d on Be+U" curve) suggest that a practical energy amplifier could therefore be operated with deuteron beam energies of the order of 500MeV-800MeV [2], i.e. at lower incident energies by 400MeV than suggested in [7] for proton beam, and still give a comparable energy gain factor.

4. Conclusions

We have tried to estimate the isotope production and energy generation by neutron induced fissions in the combined target assemblies. The stopping beryllium converter interacting with intermediate energy deuterons served as an efficient energetic neutron source. The total incident energy of the chosen projectiles, the geometry, density and atomic contents of the combined target system were the major parameters for the complete optimisation of high number of fissions wanted.

We found that the isotope production by neutron induced fissions in uranium following the Be(d,xn) reactions is a very interesting possibility. Even with 6kW deuteron beam at 200MeV (compatible with the existing characteristics of GANIL) we expect a 2-3 orders of magnitude increase in the final secondary beam intensity estimated from present fragmentation of the target and/or the heavy ion beam. However, for a more realistic comparison it is very important to take into account the final release-delay-source efficiencies.

We also predicted that the optimum energy gain in a hybrid system may be reached at lower incident energies if deuterons are used instead of protons and light metal target is employed as a spallation target. This lower energy should result in higher beam intensities, lower costs of the system and facilitate radioprotection problems.

The present studies showed, that all characteristics examined are dependent on details of composition and geometry of the combined target. The present conclusions will hold, to our opinion, qualitatively for other configurations; however, one always must take into account the precise device for a realistic optimisation.

Acknowledgments

We wish to express our appreciation to R.E.Prael for providing us with the LAHET code system. We also would like to thank J.A.Nolen for fruitful discussions and comments at different stages of this contribution.

References

[1] R.E. Prael and H. Lichtenstein, "User Guide to LCS: The LAHET Code System," Los Alamos National Laboratory Report LA-UR-89-3014 (September 1989).

[2] D. Ridikas and W. Mittig, GANIL preprint **P 98 2**, GANIL, Caen (February 1998).

[3] J.A. Nolen, Proceedings of the 3rd International Conference on RNB, Gif-sur-Yvette, France, 24-27 May 1993, Ed. D.J. Morrissey, Editions Frontiers (1993) 111; Concept for Advanced Exotic Beam Facility Based on ATLAS, Argonne National Laboratory (February 1995).

[4] D. Ridikas and W. Mittig, GANIL preprint **P 98 ?**, GANIL, Caen (to be published).

[5] PIAFE Collaboration, Piafe Project: Physics Case, SARA/ISN, Institute des Sciences Nucléaires de Grenoble (June 1994).

[6] H. Nifenecker and M. Spiro, "Hybrid systems for waste incineration and/or energy production", Journées GEDEON, Cadarache (November 1997).

[7] S. Andriamonje *et al.*, Phys. Lett. **B 348** (1995) 697.

CONCLUDING REMARKS

Concluding Remarks

Dominique Guillemaud-Mueller

Institute de Physique Nucléaire
CNRS-IN2P3
F-91406 Orsay, France

INTRODUCTION

Before starting my conclusions, it is fair to confess that this summary will be a necessarily incomplete and even superficial panorama of the conference for the following reasons: a tremendous wealth of new results has been obtained, both in theory and experiment, since the previous ENAM95 in Arles; many new radioactive beam facilities are either projected or constructed. It is therefore impossible to cover all the physics presented in the different oral contributions and in the large numbers of posters.

This gives me the opportunity for two initial comments. First, the reading of these proceedings will be the best way to have a broad overview of the field at this precise time. Secondly, the large number of posters is undoubtedly the sign that our field is vast, alive, interesting and prosperous.

Furthermore, I got the impression that the striking progress made during the last decade and in particular since ENAM95 will open fantastic perspectives. Some will be in the near term future at newly starting facilities, others longer term at the second-generation facilities.

Another important remark, to my opinion, is the complementarity, in many cases, in methods developed in order to attack a physics question or to overcome a delicate experimental problem.

The competition between high-energy or intermediate-energy heavy-ion accelerators and ISOL-type facilities, which could have been viewed negatively at the beginning, proved to be a great richness in the developments of experiments and understanding of the structure of nuclei far from stability.

MASSES, MOMENTS AND RADII

One of the highlights of ENAM98 was the wealth of new experimental mass values obtained with a very high precision. Two dedicated methods have made this possible: the ISOL-TRAP leading to a precision of 0.1 ppm on long chains of isotopes and the ESR storage ring used to separate, store and cool the fragments from heavy-ion reactions leading to a determination of masses with a precision of $\delta m/m = 10^{-6}$. Operating the ESR in a second (isochronous) way allows to measure also masses of very short-lived nuclei like in the case of exotic Ni isotopes. Half-live determinations can now be performed for stored and cooled nuclei. It is further possible to study the influence of atomic charge states on beta decay half-lives. I would also like to mention the direct mass measurements using single-path time of flight techniques. Although less precise, they can give important information in some important areas of the nuclear chart. An example is the behavior of shell closure in the vicinity of N=28 for Si to Ar isotopes. The expected start-up of the MISTRAL experiment should provide new insight in the most exotic regions of the very short-lived nuclei.

As to the theory of masses, different methods have been reviewed from microscopic mass formulae to self-consistent methods showing the challenge to predict nuclear masses over large regions of the periodic table but also the progress made in reproducing the trends of the mass surface. The number of new results published since the Arles conference and the consecutive 1995 atomic mass evaluation AME'95 push their authors to prepare a new evaluation foreseen in 1999. Very impressive work has been done on isomers whose mistreatment would have important consequences on ground-state masses. NUBASE containing this and much more is a basic tool for everybody.

Coming to nuclear radii and moments, very interesting results have been reported using different complementary methods. The measurements of β-asymmetry through Nuclear Magnetic Resonance methods coupled or not to Level Mixing Techniques have proved to be very effective for polarized implanted species. As a highlight, I consider, e.g. the determination for the first time of the magnetic moment of the one-halo nucleus ^{11}Be with the newly developed ISOLDE laser ion source. Studying ground-state properties of daughter nuclei with the COMPLIS method gives access to long chains of Au, Pt and Ir isotopes. Improvements in theory allow us to reproduce most of the behavior of isotope shifts around magic numbers.

NUCLEI AT THE DRIP LINES

This session was the occasion to strengthen the advance made in the understanding of one-neutron and two-neutron halo nuclei. Progress has been made on the experimental point of view with more sophisticated, hence more exclusive, experiments as well as on the theoretical point of view with deeper insights into the nuclear structure itself and its interplay with the involved reaction mechanisms.

Complete kinematic secondary beam experiments, studies of different isotopes from He to C, of higher mass species like ^{19}C and on different aspects (size, interaction cross-sections, Coulomb breakup, momentum distributions, transfer reactions ...) have considerably broadened our knowledge of these weakly-bound nuclei. Once again, it was shown that using complementary techniques like reaction studies with secondary beams directly or in inverse kinematics, Coulomb dissociation as well as β-decay studies were crucial to disentangle the influence of the structure and the dynamics responsible for the observed phenomena. Important information is the reward as clearly illustrated in the case of the structure of the unbound nucleus ^{10}Li.

On theoretical aspects, the presentations and discussions reflected the need for theoreticians to set and to define the limits of the validity of their different approaches and to ask themselves what has to be taken into account in each approach. As examples, one can mention sudden approximations, final-state interactions in the projectile and in the target, multi-step processes ...

The newly developed method of Coulomb Excitation (at fragment separators like the A1200) for exotic beams proved to be a very efficient way to access nuclear parameters like the excitation energy of the lowest 2^+ state, B(E2) probability and hence to deduce deformation parameters for very exotic nuclei. This was clearly demonstrated in the regions of N=20 and N=28.

The proton drip-line was also inspected. Studies of the one-proton radioactivity have now become "adult". They have entered the "age of spectroscopy" in particular through the combination of two experimental techniques which are already by themselves very powerful: Recoil Decay Tagging and large γ-arrays like those used at the FMA and RITU. The search for 2p-radioactivity is still under the way and remains a considerable future challenge.

Concerning the neutron drip-line for very light elements, we heard about the recent discovery with RIPS of the stability of ^{31}F. Looking in the region, from Z=6 to Z=10 it is interesting to note the case of oxygen isotopes of which the last bound isotope is ^{24}O. With the instability of higher mass ^{26}O and ^{28}O, oxygen isotopes seem to be unable to minimize their energy by adopting strong deformations as done by their neighbors.

NUCLEAR STRUCTURE AND SHAPES

The "Nuclear Structure and Shapes" session was largely dedicated to state-of-the-art descriptions of the nuclear many body problem.

The large shell model calculations with an extended basis really allow important progress in understanding the behavior of shell closures when going away further from stability. This has been proven in the region of ^{44}S at the N=28 magic number. Here the shell closure, while still present, is definitely eroded for P, S and Cl.

Relativistic mean field theories are now available to reproduce quite nicely different nuclear properties such as neutron separation energies.

Monte Carlo shell model calculations are also progressing in the description of light exotic nuclei. One can mention as examples the excellent agreement for the ^{6}Li electromagnetic form factor using the variation Monte Carlo method, as well as the good description of the Gamow Teller strength in fp-shell nuclei using the KB3 interaction in the frame of shell model Monte Carlo, and also the calculations of the energy location of the first 2^+ and 4^+ states in the series of the Silicon isotopes around N=20 in agreement with experimental measurements.

Concerning the experiments, it is important to note that the combination of a large γ-array with decay recoil tagging techniques gives access to very fine spectroscopy even for exotic nuclei. This was illustrated in the case of high-spin spectroscopy in the A~60 region. It was possible, for the first time, for the N=Z ^{60}Zn to identify a superdeformed band and its link to the normal deformed ones.

Another striking example was the evidence of an octupole deformation in the neutron deficient isotope ^{226}U.

Combining also two techniques - the recoil fragment separator and γ-detection-, it was possible to produce and to identify many new microsecond isomers. This method proved to be very efficient for spectroscopic studies and complementary to in-beam spectroscopy. Let me state that this area, just now starting, will be one of the bright futures in nuclear spectroscopy.

HEAVY ELEMENTS, FISSON, CLUSTER RADIOACTIVITY

Another highlight of the conference was undoubtedly the discovery of new elements.

With the ever improving SHIP on its new expeditions, the nucleus ^{277}112 was synthesized via fusion reactions using 1n-deexcitation channels and lead and bismuth targets and identified though α-decay. The deduced half-life is 240 μs and the measured cross-section 1 pb. An attempt to synthesize the nucleus ^{279}113 was made but today only an upper limit of the cross-section of 0.6 pb can be given.

The outstanding intensities of the U400 cyclotron made it possible to use the nuclear fusion-fission process in order to synthesize the nucleus ^{286}112 in the reaction

^{238}U+^{48}Ca at an energy lower that the Coulomb barrier. The deduced half-live is around 120 s and the measured cross-section around 5 ± 2 pb.

Also very nice theoretical work has been shown on the properties of heavy elements such as deformation, mass, shell structure and half-lives with respect to α-decay and spontaneous fission. From these calculations, one finds a good agreement with the measured half-lives. The theories also predict a shell stabilization region higher up in mass for higher Z with elements of spherical shapes.

Up to now, the newly discovered elements are just situated at the transition from deformed to spherical shapes. It will be highly interesting to see if beams of neutron-rich nuclei can push forward this field in the longer term.

The investigation of low-energy fission of more than 70 neutron deficient isotopes ranging from ^{234}U down to ^{205}Ra was the second highlight of the session. The nuclei were produced by fragmentation of a 1 GeV/A ^{238}U projectile delivered by SIS on a beryllium target. Fission was induced in-flight by electromagnetic excitation in a secondary lead target in the FRS. From this experiment it was possible to measure nuclear charge distributions, total kinetic energies for long chains of isotopes and of course mass distributions, showing a nice transition around ^{226}Th from asymmetric to symmetric fission. From this study, one will get much new important information of fission such as the influence of shell effects and paring correlations on the dynamics on nuclear matter.

The use of the new mechanism at the FRS of U-projectile fission proved to be a very powerful method to produce very neutron-rich isotopes in a region difficult to access by other production mechanisms.

As an example, 66 new neutron-rich nuclei were identified, including ^{78}Ni, in the region from calcium to niobium and the distribution of isotope yields were studied. This production method will undoubtedly offer large possibilities of spectroscopic studies. One can think immediately about half-life measurements.

BETA DECAY AND FUNDAMENTAL MEASURMENTS

Important questions about fundamental interactions have been address through experiments using radioactive nuclear systems.

Among them, one can cite the limits on the possible tensor and scalar contributions to the weak interaction (providing a test of the standard model) obtained from β-decay of ^{32}Ar by measuring the correlation between the positron and the neutrino, and the quenching of the Gamow-Teller strength in β-decay near ^{100}Sn.

Measurement of the superallowed Fermi β-decay provides a good test of the CVC hypothesis and the unitarity of the Cabibbo-Kobayashi-Maskawa matrix. This has been illustrated through new measurements of super-allowed β-decay in the region of Ga to Tc on more exotic nuclei than the nine previously, extensively studied decays that range from ^{10}C to ^{54}Co.

As an outlook from discussions, I got the feeling that trap techniques will become a powerful tool in this domain.

NUCLEAR ASTROPHYSICS

The session dedicated to nuclear astrophysics gave the audience an opportunity to travel along the different processes and possible sites of nucleosynthesis: the rp-process in explosive hydrogen burning at high temperature, the r-process responsible for the abundances of elements higher than iron. This gave a flavor of the interdependence between nuclear astrophysics and nuclear structure. As an example, one can mention the possible implication of the newly established nuclear structure around ^{132}Sn on the build-up and shape of the A=130 r-process abundance peak and the r-matter flow to higher masses. Concerning the rp-process, half-life and mass determinations are of interest in some precise regions of the chart of nuclides. This is particularly the case around ^{80}Zn and ^{84}Mo.

Another major point addressed in this topic was the solar neutrino problem through the determination of the cross section of the ^{7}Be(p,γ)^{8}B reaction which governs the flux of energetic neutrinos produced in side the Sun. The different experiments (direct measurements, Coulomb dissociation, Asymptotic Normalization Coefficients ANC) already performed, running and planned were reviewed. Different methods have different sources of uncertainties and it remains difficult to obtain the precision (better than 5%) astrophysicists ask for. The various ways when applied to the extraction of the S_{17} astrophysical factor are now in relatively good agreement.

EXPERIMENTAL DEVELOPMENTS AND RADIOACTIVE BEAMS

The highlights of this conference were obtained either by in-flight separation of energetic secondary beams or by the ISOL-method showing the complementarity of the two techniques. Traps, storage rings, laser ion sources have lead to important experimental breakthroughs. The future of radioactive beams has been reviewed worldwide. It is very satisfying to note that several continents are participating in this endeavor: Europe, America and Asia. Some of the projects are under construction and will already give access in 1999 to a variety of post-accelerated beams such as SPIRAL in France, REX-ISOL at CERN and ARENAS 3 at Louvain la Neuve. They will soon be followed by the fragmentation beams from the MSU upgrade and the new RIKEN facility. It will be highly interesting to learn at ENAM2001, which will be hosted by our Finnish colleagues from the Jyväskylä facility about he progress of the other projected facilities.

ENAM98 Scientific Program

TUESDAY, June 23, 1998
A1 - Masses, Moments and Radii

8:50   *Welcome*

9:00   **G. Bollen** (ISOLDE-CERN, Geneva, Switzerland)
Penning trap mass measurements at ISOLDE

9:30   **H. Geissel** (GSI, Darmstadt, Germany)
Experiments with stored relativistic exotic nuclei at the FRS-ESR facilities

10:00   **G. Alkhazov** (Petersburg Nuclear Physics Institute, Gatchina, Russia)
Study of exotic nuclei by proton scattering in inverse kinematics

10:30   *Coffee break*

11:00   **A. Wapstra** (NIKHEF, Amsterdam, The Netherlands)
The 1999 status of atomic mass knowledge

11:30   **A. Zuker** (IReS, Strasbourg, France)
Towards a microscopic mass formula

11:50   **H. Savajols** (GANIL, Caen, France)
Magicity far from stability?

12:10   *Lunch*

A2 - Masses, Moments and Radii

2:00   **M. Keim** (ISOLDE-CERN, Geneva, Switzerland)
Recent measurements of nuclear moments far from stability

2:30   **G. Neyens** (University of Leuven, Belgium)
Nuclear moments of exotic nuclear states studied with level mixing techniques

3:00   **P. Mantica** (Michigan State University, E. Lansing, USA)
Isospin nonconservation effects in light, $T_Z = -3/2$ nuclei

3:20 **P. Campbell** (University of Manchester, UK)
On-line laser spectroscopy of refractory radioisotopes at the JYFL IGISOL facility

3:40 *Coffee break*

4:10 **F. Le Blanc** (Institut de Physique Nucléaire, Orsay, France)
Large odd-even radius staggering in the very light platinum isotopes from laser spectroscopy

4:30 **T. Suzuki** (Niigata University, Japan)
Nuclear radii of ^{14}Be, ^{17,19}B

B1 - Nuclei at the Drip Lines

4:50 **P. Woods** (University of Edinburgh, UK)
Exploration of the proton drip-line

5:30 **B. Jonson** (Chalmers University of Technology, Sweden)
Status of neutron drip-line nuclei

6:00 **I. Thompson** (University of Surrey, UK)
Probing halo structure with breakup reactions

6:30 **A. Korsheninnikov** (RIKEN, Japan)
Studies of excited states in ^{11}Li

7:30 *Dinner*

9:30 *Student session*

Wednesday, June 24, 1998

B2 - Nuclei at the Drip Lines

8:30 **J. Tostevin** (University of Surrey, UK)
Sizes of the He isotopes deduced from proton elastic scattering measurements

9:00 **T. Aumann** (GSI, Darmstadt, Germany)
Alignment of ^{5}He fragments from ^{6}He breakup

9:30	**T. Glasmacher** (Michigan State University, E. Lansing, USA) *Coulomb excitation of exotic nuclei*
10:00	**N. Alahari** (Michigan State University, E. Lansing, USA) *Search for proton halos in phosphorus isotopes*
10:20	**T. Nakamura** (University of Tokyo, Japan) *Coulomb dissociation of ^{19}C*
10:40	*Coffee break*
11:10	**M. Chartier** (Michigan State University, E. Lansing, USA) *New Evidence for parity inversion in ^{10}Li from ^{9}Li and γ-ray coincidences*
11:30:	**P. Ring** (Technische Universität München, Germany) *Self-consistent microscopic approaches to nuclear masses*
11:50	**H. Sakurai** (RIKEN, Japan) *New neutron-rich isotope ^{31}F and particle instability of ^{25}N and ^{28}O*
12:10	**S. Fortier** (Institut de Physique Nucléaire, Orsay, France) *Investigation of ^{11}Be microscopic structure by means of the $p(^{11}Be,^{10}Be)d$ reaction*
12:30	**R. Varner** (Oak Ridge National Laboratory, USA) *Measurement of the E1 strength function of ^{11}Be*
12:50	*Lunch, poster preparation*

C1 - Nuclear Structure and Shapes

3:00	**W. Nazarewicz** (University of Tennessee and ORNL, USA) *Exotic nuclei from a theoretical perspective*
3:30	**T. Otsuka** (University of Tokyo, Japan) *Monte Carlo shell model calculations of exotic nuclei*
4:00	*Poster session with refreshments*
7:30	*Transport to Schuss Mountain*
8:00	*Dinner at Schuss Mountain*
9:00	*Show - Schussy-Cats and The Thunderstorm*

Thursday, June 25, 1998

C2 - Nuclear Structure and Shapes

8:30 **S. Pieper** (Argonne National Laboratory, USA)
Quantum Monte Carlo calculations of light nuclei

9:00 **C. Svensson** (McMaster University, Hamilton, Canada)
High-spin spectroscopy in the A ~ 60 mass region: superdeformation and smooth band termination

9:30 **D. Seweryniak** (Argonne National Laboratory, USA)
Gamma-ray spectroscopy near ^{100}Sn

10:00 **M. Leino** (University of Jyväskylä, Finland)
In-beam studies of very neutron-deficient isotopes of heavy elements using recoil-decay tagging

10:30 *Coffee break*

11:00 **R. Grzywacz** (University of Tennessee, USA)
Microsecond isomer studies

11:30 **D. Dean** (Oak Ridge National Laboratory, USA)
Shell-model Monte Carlo studies of nuclei far from stability

11:50 **C. Gross** (Oak Ridge National Laboratory, USA)
Identification of gamma rays in the ground-state proton emitter ^{113}Cs

12:10 *Lunch*

1:30 *Outing*

8:00 *Dinner*

9:30 *Student session*

Friday, June 26, 1998

C3 - Nuclear Structure and Shapes

8:30 **S. Vincent** (University of Surrey, UK)
Competing T=0 and T=1 structures in the N=Z odd-odd nucleus ^{62}Ga

8:50 **A. Wöhr** (Oxford University, UK)
Decay of very neutron-rich Mn nuclides and deformation of heavy Fe isotopes

D1 - Heavy Elements, Fission, Cluster Radioactivity

9:10 **S. Hofmann** (GSI, Darmstadt, Germany)
New elements produced at GSI

9:40 **M. Itkis** (Flerov Laboratory of Nuclear Reactions, Dubna Russia)
FLNR JINR experiments on synthesis of superheavy nuclei with ^{48}Ca beam

10:00 **A. Sobiczewski** (Soltan Institute of Nuclear Problems, Warsaw, Poland)
Stability of the heaviest elements

10:30 *Coffee break*

11:00 **K.-H. Schmidt** (GSI, Darmstadt, Germany)
Fission studies of nuclei far from stability

11:30 **M. Bernas** (Institut de Physique Nucléaire, Orsay, France)
Fission process in ^{238}U collisions on Pb and Be targets

12:00 **A. Ramayya** (Vanderbilt University, Nashville, USA)
Structure of neutron-rich nuclei and rare processes in spontaneous fission of ^{252}Cf

12:30 **S. Kadmensky** (Voronezh State University, Russia)
Proton and cluster radioactivities and nucleus shapes

12:50 *Lunch*

Friday, June 26, 1998

E1 - Beta Decay and Fundamental Measurements

2:00 **S. Freedman** (University of California, Berkeley, USA)
Use of traps in fundamental measurements

2:30 **A. García** (University of Notre Dame, USA)
Using exotic nuclei in studies of fundamental symmetries

3:00 **Z. Janas** (University of Warsaw, Poland)
Beta strength distribution in neutron-deficient nuclei

3:30   **J. Hardy** (Texas A&M University, College Station, USA)
        Superallowed Fermi beta decay — status and future prospects

3:50   **Ph. Dessagne** (IReS Strasbourg, France)
        Indication for superallowed Fermi decay from the N=Z nuclei ^{78}Y, ^{82}Nb, ^{86}Tc

4:10   *Coffee break*

4:40   **A. Jokinen** (University of Jyväskylä, Finland)
        Beta-decay strength and isospin mixing studies in the sd and fp-shells

F1 - Nuclear Astrophysics

5:00   **M. Wiescher** (University of Notre Dame, USA)
        Recent measurements on rp-process nuclei

5:30   **K.-L. Kratz** (Johannes Gutenberg-Universität, Mainz, Germany)
        Recent measurements on r-process nuclei

6:00   **F.-K. Thielemann** (Universität Basel, Switzerland)
        Explosive nucleosynthesis

6:30   **M. Huyse** (University of Leuven, Belgium)
        Review of radioactive beam astrophysics experiments

7:30   *Conference Banquet*

Saturday, June 27, 1998

F2 - Nuclear Astrophysics

8:30   **G. Bogaert** (CSNSM Orsay, France)
        Direct measurement of the $^7Be(p,\gamma)^8B$ cross-section

9:00   **M. Hass** (Weizmann Institute of Science, Rehovot, Israel)
        Measurement of the $^7Be(p,\gamma)^8B$ cross section with an implanted ^7Be target

9:20   **R. Tribble** (Texas A&M University, College Station, USA)
        Using the Asymptotic Normalization Coefficient in nuclear astrophysics

9:50   **F. M. Nunes** (Instituto Superior Técnico, Lisbon, Portugal)
        Determining the astrophysical S_{17} with transfer reactions

10:10 **T. Motobayashi** (Rikkyo University, Tokyo, Japan)
High energy Coulomb breakup experiments for nuclear astrophysics

10:40 *Coffee break*

11:10 **B. Davids** (Michigan State University, E. Lansing, USA)
Measurement of E2 transitions in the Coulomb dissociation of ^8B

G1 - Experimental Developments and Radioactive Beams

11:30 **P. Van Duppen** (University of Leuven, Belgium)
New results from advances in ISOL Techniques

12:00 **F. Ames** (CERN-ISOLDE, Geneva, Switzerland)
REXTRAP, an ion buncher for REX-ISOLDE

12:20 **J. Kolata** (University of Notre Dame, USA)
First results from the TWINSOL RNB facility

12:40 *Lunch, Student session*

Saturday, June 27, 1998

G2 - Experimental Developments and Radioactive Beams

2:00 **A. Mueller** (Institut de Physique Nucléaire, Orsay, France)
Status of RNB facilities in Europe

2:40 **I. Tanihata** (RIKEN, Japan)
Status of RNB facilities in Asia

3:10 **J. Nolen** (Argonne National Laboratory, USA)
Status of RNB facilities in North America

3:40 *Announcement*

3:45 **D. Guillemaud-Mueller** (Institut de Physique Nucléaire, Orsay, France)
Concluding Remarks

4:30 *Conference close*

LIST OF PARTICIPANTS

Dr. Irshad Ahmad
Physics Division
Argonne National Laboratory
9700 S. Cass Ave.
Argonne, IL 60439-4843
USA

Dr. Navin Alahari
Michigan State University
S. Shaw Lane
East Lansing, MI 48824-1321
USA

Dr. Dmitri V. Aleksandrov
RRC, The Kurchatov Institute
Kurchatov Square 1
RU-123182 Moscow,
RUSSIA

Dr. Gueorgui D. Alkhazov
St. Petersburg Nuclear Physics Inst.
Leningrad District
RU-188350 Gatchina,
RUSSIA

Dr. Friedhelm Ames
EP-ISOLDE
CERN/J. Gutenberg Universität Mainz
CH-1211 Geneva 23,
SWITZERLAND

Dr. Alberto Andrighetto
Dipartimento di Fisica
"G. Galilei" via Marzolo 8
I-35100 Padova,
ITALY

Mr. Donald Anthony
Michigan State University
S. Shaw Lane
East Lansing, MI 48824-1321
USA

Dr. Nori Aoi
Dept. of Physics/Radiation Laboratory
Rikkyo University/RIKEN
2-1 Hirosawa, Wako-shi
351-0198 Saitama,
JAPAN

Prof. Ani Aprahamian
Dept. of Physics
University of Notre Dame
Notre Dame, IN 46556
USA

Dr. Thomas J. Aumann
KPII GSI
Planckstrasse 1
D-64291 Darmstadt,
GERMANY

Prof. Sam M. Austin
Michigan State University
S. Shaw Lane
East Lansing, MI 48824-1321
USA

Mr. Leif Axelsson
Dept. of Physics
Chalmers University of Technology
S-41296 Göteborg,
SWEDEN

Prof. Juha Äystö
Dept. of Physics
University of Jyväskylä
P.O. Box 35
FIN-40351 Jyväskylä,
FINLAND

Dr. Afshin Azhari
Cyclotron Institute
Texas A & M University
College Station, TX 77845-3366
USA

Ms. Jeanette Bakker
Executive Editor for Nuclear Physics A
Nuclear Physics A
Sara Burgerhartstraat 25
1055 KV Amsterdam,
THE NETHERLANDS

Prof. Robert C. Barber
Dept. of Physics
University of Manitoba
Winnipeg, Manitoba, R3T 2N2,
CANADA

Mr. Charles J. Barton
Wright Nuclear Structure Laboratory
Yale University/Clark University
P.O. Box 208124, 272 Whitney Ave.
New Haven, CT 06520-8124
USA

Dr. Jon Batchelder
JIHIR/ORNL, Bldg. 6008, MS6374
Oak Ridge Associated Universities
Oak Ridge, TN 37831-6374
USA

Mr. Thomas Baumann
GSI
Planckstrasse 1
D-64291 Darmstadt,
GERMANY

Dr. Daniel Bazin
NSCL
Michigan State University
S. Shaw Lane
East Lansing, MI 48824-1321
USA

Dr. Didier Beaumel
Institut de Physique Nucléaire, Orsay
B. P. No. 1, Bâtiment 100
F-91406 Orsay,
FRANCE

Dr. Fred D. Becchetti
Randall Laboratory
University of Michigan
500 E. University
Ann Arbor, MI 48109-1120
USA

Dr. Jose Benlliure
Departamento de Fisica de Particulas
Universidad de Santiago de Compostela
E-15706 Santiago de Compostela
SPAIN

Dr. Monique Bernas
Div. Recherche Expérimentale
Institut de Physique Nucléaire, Orsay
B. P. No. 1, Bâtiment 100
F-91406 Orsay Cedex,
FRANCE

Dr. Pallab K. Bhattacharyya
Dept. of Chemistry
Purdue University
1396 PHYS
West Lafayette, IN 47907
USA

Prof. Carrol Bingham
Dept. of Physics
University of Tennessee
401 A.H. Nielsen Physics Bldg.
Knoxville, TN 37996
USA

Dr. Bertram A. Blank
Centre d'Etudes Nucléaire de Bordeaux
Le Haut Vigneau, B.P. 120
F-33175 Gradignan cedex,
FRANCE

Dr. Yorick Blumenfeld
Institut de Physique Nucléaire, Orsay
B. P. No. 1, Bâtiment 100

F-91406 Orsay cedex,
FRANCE

Dr. Gilles Bogaert
CSNSM Orsay
Bâtiment 108
F-91405 Orsay cedex,
FRANCE

Dr. Georg Bollen
PPE-ISOLDE
CERN
CH-1211 Geneva 23,
SWITZERLAND

Prof. William H. Brantley
Dept. of Physics
Furman University
3300 Poinsett Hwy.
Greenville, SC 29613
USA

Prof. Daeg S. Brenner
Dept. of Chemistry
Clark University
950 Main St.
Worcester, MA 01610
USA

Dr. James A. Brown
Dept. of Physics
Allegheny College
Meadville, PA 16335
USA

Mr. Joseph Caggiano
NSCL
Michigan State University
S. Shaw Lane
East Lansing, MI 48824-1321
USA

Dr. Paul Campbell
Schuster Laboratory
University of Manchester
Brunswick Street
Manchester M13 9PL,
UNITED KINGDOM

Mr. Daniel Cano-Ott
IFIC - Fisica Teorica
CSIC
Avda. Dr. Moliner 50
E-46100 Burjasot (Valencia),
SPAIN

Dr. H. Kennon Carter
Bldg. 6008, MS 6374, UNIRIB/ORISE
Oak Ridge Associated Universities
P.O. Box 2008
Oak Ridge, TN 37831-6374
USA

Dr. Marielle Chartier
NSCL
Michigan State University
S. Shaw Lane
East Lansing, M I 48824-1321
USA

Mr. Luke Li-Ming Chen
NSCL
Michigan State University
S. Shaw Lane
East Lansing, MI 48824-1321
USA

Mr. Marcus J. Chromik
Section Physik
Ludwig Maximilans Universität
am Coulombwall 1
D-85748 Garching,
GERMANY

Dr. Greg Chubaryan
Cyclotron Institute
Texas A & M University
College Station, TX 77843-3366
USA

Dr. Tatjana V. Chuvilskaya
Institute of Nuclear Physics
Moscow State University
ul. Kossygina
RU-119899 Moscow,
RUSSIA

Dr. Jolie A. Cizewski
Dept. of Physics & Astronomy
Rutgers University
P.O. Box 849
Piscataway, NJ 08855
USA

Dr. Dolores Cortina
GSI
Postfach 11 05 52
D-64291 Darmstadt,
GERMANY

Prof. Paul D. Cottle
Dept. of Physics
Florida State University
211 Keen Building
Tallahassee, FL 32306-4350
USA

Dr. Gerard M. Crawley
Michigan State University
S. Shaw Lane
East Lansing, MI 48824-1321
USA

Dr. Serge Czajkowski
Centre d'Etudes Nucléaire de Bordeaux
Le Haut Vigneau, B.P. 120
F-33175 Gradignan,
FRANCE

Prof. John M. D'Auria
Dept. of Chemistry
Simon Fraser University
Burnaby, B.C. V5A 1S6,
CANADA

Prof. Patrick J. Daly
Dept. of Chemistry
Purdue University
West Lafayette, IN 47907
USA

Dr. Boris V. Danilin
RRC,
The Kurchatov Institute
Kurchatov Square 1
RU-123182 Moscow,
RUSSIA

Mr. Barry S. Davids
Michigan State University
S. Shaw Lane
East Lansing, MI 48824-1321
USA

Dr. Cary N. Davids
PHY203
Argonne National Laboratory
9700 S. Cass Ave.
Argonne, IL 60439-4843
USA

Dr. Gilles de France
GANIL
B.P. No. 5027
F-14076 Caen cedex, 5
FRANCE

Prof. Denis J. A. de Frenne
Vakgroep Subatomaire &
Stralingsfysica
Universiteit Gent
Proeftuinstraat 86
B-9000 Gent, BELGIUM

Mr. Robert C. de Haan
Dept. of Physics
University of Notre Dame
Notre Dame, IN 46556
USA

Dr. David J. Dean
Physics Division, Bldg. 6003, MS 6373
Oak Ridge National Laboratory
P.O. Box 2008
Oak Ridge, TN 37831-6372
USA

Dr. Peter Dendooven
Dept. of Physics
University of Jyväskylä
P.O. Box 35
FIN-40351 Jyväskylä,
FINLAND

Dr. Phillippe Dessagne
Institut de Recherches Subatomiques
IReS
B.P. 28, 23 rue du Loess
F-67037 Strasbourg cedex,
FRANCE

Dr. Claude Detraz
IN2P3
CNRS
3 rue Michel-Ange
F-75794 Paris cedex 16,
FRANCE

Dr. Hans Joachim Doering
Dept. of Physics
University of Notre Dame
Notre Dame, IN 46556
USA

Mr. George Drafta
Dept. of Physics
Vanderbilt University
Box 1807, Station B

Nashville, TN 37235
USA

Dr. Andre Emsallem
IN2P3, France
Institut de Physique Nucléaire
43, bd. du 11 Novembre
F-69622 Villeurbanne cedex,
FRANCE

Dr. Timo T. Enqvist
GSI
Planckstrasse 1
D-64291 Darmstadt,
GERMANY

Prof. Kari A. Eskola
Dept. of Physics
Helsinki University
P.O. Box 9 (Siltavuorenpenger 20D)
FIN-00014 Helsinki,
FINLAND

Dr. Friedrich Everling
Ringheide 24 f
D-21149 Hamburg,
GERMANY

Dr. Michael Fauerbach
Dept. of Physics
Florida State University
211 Keen Building
Tallahassee, FL 32306-4350
USA

Dr. Sergei Fayans
RRC,
The Kurchatov Institute
Kurchatov Square 1
RU-123182 Moscow,
RUSSIA

Dr. Simone Fortier
Institut de Physique Nucléaire, Orsay

B. P. No. 1, Bâtiment 100
F-91406 Orsay cedex,
FRANCE

Dr. Nikolaos Fotiadis
Dept. of Physics & Astronomy
Rutgers University
P.O. Box 849
Piscataway, NJ 08855-0849
USA

Mr. Serge Franchoo
Instituut voor Kern- en Stralingsfysica
University of Leuven
Celestijnenlaan 200 D
B-3001 Leuven,
BELGIUM

Dr. Stuart J. Freedman
Dept. of Physics
University of California, Berkeley
366 LeConte Hall
Berkeley, CA 94720
USA

Dr. Martin Freer
Dept. of Physics
University of Birmingham
Edgebaston, P.O. Box 363
Birmingham, B15 2TT
UNITED KINGDOM

Dr. Hans O. U. Fynbo
Institute of Physics and Astronomy
Aarhus Universitet
Ny Munkegade
DK-8000 Aarhus C,
DENMARK

Dr. Aaron Galonsky
Michigan State University
S. Shaw Lane
East Lansing, MI 48824-1321
USA

Prof. Alejandro Garcia
Dept. of Physics
University of Notre Dame
Notre Dame, IN 46556
USA

Dr. Maria J. Garcia Borge
Instituto de Estructura de la Materia
CSIC
Serrano 113 bis
E-28006 Madrid,
SPAIN

Dr. Hans Geissel
GSI
Planckstrasse 1
D-64291 Darmstadt,
GERMANY

Prof. Claus-Konrad Gelbke
Michigan State University
S. Shaw Lane
East Lansing, MI 48824-1321
USA

Dr. Janine Genevey
Institut des Sciences Nucléaire
53 ave. des Martyrs
F-38026 Grenoble cedex,
FRANCE

Mr. Thomas N. Ginter
Dept. of Physics
Vanderbilt University
133 West Farragut Road
Oak Ridge, TN 37830
USA

Dr. Jérôme Giovinazzo
Centre d'Etudes Nucléaire de Bordeaux
BP 120, Le Haut Vigneau
F33175 Gradignan cedex
FRANCE

Dr. Thomas Glasmacher
NSCL
Michigan State University
S. Shaw Lane
East Lansing, MI 48824-1321
USA

Dr. Stefan Gmuca
Institute of Physics
Slovak Academy of Sciences
Dubravska cesta 9
SK-842 28 Bratislava
SLOVAK REPUBLIC

Dr. Vladilen Z. Goldberg
RRC,
The Kurchatov Institute
Kurchatov Square, Building 1
RU-123182 Moscow,
RUSSIA

Dr. Gvirol Goldring
Dept. of Particle Physics
Weizmann Institute of Science
P.O. Box 26
76100 Rehovot,
ISRAEL

Dr. Sergey A. Goncharov
Skobeltsyn Institute of Nuclear Physics
Moscow State University
ul. Kossygina
RU-119899 Moscow,
RUSSIA

Dr. Kurt Govaert
NSCL
Michigan State University
S. Shaw Lane
East Lansing, MI 48824-1321
USA

Prof. Indra M. Govil
Physics Dept.

Panjab University, Chandigarh
Chandigarh -160014,
INDIA

Dr. Stephane J. F. Grévy
Laboratoire de Physique Corpusculaire
ISMRA-IN2P3
6, bd du Marechal Juin
F-14050 Caen cedex,
FRANCE

Prof. Konstantin A. Gridnev
Institute of Physics
St. Petersburg State University
Uljanovskaja 1, Peterhof
198904 St. Petersburg,
RUSSIA

Dr. Carl J. Gross
Physics Division, MS-6371
ORISE/Oak Ridge National Laboratory
P.O. Box 2008
Oak Ridge, TN 37830-6371
USA

Dr. Robert Grzywacz
Dept. of Physics
University of Tennessee/Warsaw Univ.
401 A. H. Nielsen Physics Bldg.
Knoxville, TN 37996-1200
USA

Dr. Daniel Guerreau
GANIL
B.P. No. 5027
F-14076 Caen cedex 5,
FRANCE

Dr. Dominique Guillemaud-Mueller
Div. Recherche Expérimentale
Institut de Physique Nucléaire, Orsay
B. P. No. 1, Bâtiment 100
F-91406 Orsay cedex,
FRANCE

Dr. Gregory S. Hackman
NSCL
Michigan State University
S. Shaw Lane
East Lansing, MI 48824-1321
USA

Prof. Joseph H. Hamilton
Dept. of Physics
Vanderbilt University
Box 1807, Sta. B
Nashville, TN 37235
USA

Prof. P. Gregers Hansen
NSCL
Michigan State University
S. Shaw Lane
East Lansing, MI 48824-1321
USA

Prof. John Hardy
Cyclotron Institute
Texas A & M University
College Station, TX 77845-3366
USA

Dr. Michael Hass
Dept. of Particle Physics
Weizmann Institute of Science
P.O. Box 26
76100 Rehovot,
ISRAEL

Dr. Karl Hauschild
MS-280
Lawrence Livermore National
Laboratory
P.O. Box 808
Livermore, CA 94551
USA

Dr. Peter E. Haustein
Dept. of Chemistry, Bldg. 555A
Brookhaven National Laboratory
20 Pennsylvania St.
Upton, NY 11973-9999
USA

Dr. Walter F. Henning
Physics Division
Argonne National Laboratory
9700 S. Cass Ave.
Argonne, IL 60439-4843
USA

Dr. Daisy Hirata
JASRI
323-3 Mihara, Mikazuki-cho
679-5198 Sayoj-gun, Hyogo,
JAPAN

Dr. Sigurd Hofmann
GSI
Postfach 11 0552
D-64220 Darmstadt,
GERMANY

Prof. Poul Hornshoj
Institute of Physics and Astronomy
Aarhus Universitet
Ny Munkegade
DK-8000 Aarhus C,
DENMARK

Prof. Gerhard Huber
Institut für Physik
Johannes Gutenberg-Universität Mainz
Staudinger Weg 7
D-55099 Mainz,
GERMANY

Prof. Dr. Mark L. Huyse
Instituut voor Kern- en Stralingsfysica
University of Leuven
Celestijnenlaan 200 D
B-3001 Leuven,
BELGIUM

Dr. Richard W. Ibbotson
Michigan State University
S. Shaw Lane
East Lansing, MI 48824-1321
USA

Dr. Shin-ichi Ichikawa
Nuclear Chemistry Laboratory
Japan Atomic Energy Research Inst.
(JAERI)
Tokai-mura
319-1195 Ibaraki,
JAPAN

Dr. Hideki Iimura
Nuclear Chemistry Lab.
Japan Atomic Energy Research Inst.
(JAERI)
Tokai-mura
319-1195 Ibaraki,
JAPAN

Prof. Mikhail G. Itkis
Laboratory of Nuclear Reactions
Joint Institute for Nuclear Research
Joliot Curie str. 8
141980 Dubna, Moscow region,
RUSSIA

Prof. Etienne T. C. Jacobs
Vakgroep Subatomaire &
Stralingsfysica
Universiteit Gent
Proeftuinstraat 86
B-9000 Gent,
BELGIUM

Dr. Ylva Jading
EP-division
CERN
CH-1211 Geneva 23,
SWITZERLAND

Dr. Zenon Janas
Institut of Experimental Physics
Warsaw University
ul. Hoza 69
PL-00 681 Warsaw,
POLAND

Prof. Joachim W. Janecke
Dept. of Physics, 2245
University of Michigan
Ann Arbor, MI 48109-1120

Docent Aksel S. Jensen
Institute of Physics and Astronomy
Aarhus Universitet
Ny Munkegade, Bygn. 520
DK-8000 Aarhus C,
DENMARK

Dr. Ari Jokinen
Dept. of Physics
University of Jyväskylä
P.O. Box 35
FIN-40351 Jyväskylä,
FINLAND

Prof. Björn N. G. Jonson
Dept. of Physics
Chalmers University of Technology
S-41296 Göteborg,
SWEDEN

Dr. Rainer Joosten
MS88
Lawrence Berkeley National Lab.
One Cyclotron Rd.
Berkeley, CA 94720
USA

Prof. Stanislav G. Kadmensky
Voronezh State University
Plekhanovskaya Str. 22 360, P/B 24
394030 Voronezh,
RUSSIA

Dr. Steven Karataglidis
TRIUMF
4004 Wesbrook Mall
Vancouver, B.C. V6T 2A3,
CANADA

Mr. Marek Karny
Nuclear Physics Division
Institut of Experimental Physics
ul. Hoza 69
PL-00 681 Warsaw,
POLAND

Dr. Matthias Keim
EP-IS
CERN
CH-1211 Geneva 23,
SWITZERLAND

Prof. Kirby W. Kemper
Dept. of Physics B-159
Florida State University
211 Keen Building
Tallahassee, FL 32306-4350
USA

Mr. Peter S. Kim
Dept. of Physics
University of Tennessee
401 A.H. Nielsen Physics Bldg.
Knoxville, TN 37996
USA

Prof. Michael W. Kirson
Dept. of Particle Physics
Weizmann Institute of Science
76100 Rehovot,
ISRAEL

Prof. Dr. H.-Jürgen Kluge
GSI
Planckstrasse 1
D-64291 Darmstadt,
GERMANY

Dr. Oleg M. Knyazkov
Physical Research Institute
St. Petersburg State University
Peterhof, Uljanovskaja 1
198904 St. Petersburg,
RUSSIA

Prof. James J. Kolata
Dept. of Physics
University of Notre Dame
Notre Dame, IN 46556
USA

Prof. Jan Kormicki
Dept. of Physics & Astronomy
Vanderbilt University
Box 1807, Sta. B
Nashville, TN 37235
USA

Dr. Gabriele-Elisabeth Körner
NuPECC c/o Physik Dept. E12
Technischen Universität München
James-Franck-Straße 1
D-85748 Garching,
GERMANY

Prof. Hans-Joachim Körner
Physik-Department E12
Technischen Universität München
James-Franck-Straße 1
D-85748 Garching,
GERMANY

Dr. Alexei A. Korsheninnikov
RIKEN
2-1 Hirosawa, Wako-shi
351-0198 Saitama,
JAPAN

Mr. Ulli Köster
Physik-Dept. E18
Technischen Universität München
James-Franck-Straße 1

D-85748 Garching
GERMANY

Mr. Hiroyuki Koura
Advanced Res. Institute Science & Eng.
Waseda University
3-4-1 Okubo, Shinjuku-ku
169 Tokyo,
JAPAN

Prof. Dr. Karl-Ludwig Kratz
Institut für Kernchemie
Johannes Gutenberg-Universität Mainz
Fritz-Strassmann-Weg 2
D-55218 Mainz,
GERMANY

Dr. Toshiyuki Kubo
Institute of Physical & Chemical Res.
RIKEN
2-1 Hirosawa, Wako-shi
351-0198 Saitama,
JAPAN

Dr. Theodore M. Lach
433 Roe Ct.
Downers Grove, IL 60516
USA

Dr. François LeBlanc
Div. Recherche Expérimentale
Institut de Physique Nucléaire, Orsay
B. P. No. 1, Bâtiment 104
F-91406 Orsay cedex,
FRANCE

Prof. Matti Leino
Dept. of Physics
University of Jyväskylä
P.O. Box 35
FIN-40351 Jyväskylä,
FINLAND

Dr. Alinka Lépine-Szily
Instituto de Fisica - IFUSP
Universidade de Sao Paulo
Caixa Postal 66318
05315-970 Sao Paulo, SP
BRAZIL

Dr. Marek Lewitowicz
GANIL
B.P. No. 5027
F-14076 Caen cedex 5,
FRANCE

Mr. Benyuan Liu
WNSL
Yale University
10 Pearl Street, 3rd floor
New Haven, CT 06511
USA

Mr. Patrick A. Lofy
NSCL
Michigan State University
S. Shaw Lane
East Lansing, MI 48824-1321
USA

Dr. David Lunney
CSNSM-CNRS
Bâtiment 108
F-91405 Orsay cedex,
FRANCE

Dr. Henryk A. Mach
Dept. of Neutron Research
Uppsala University
S-61182 Nykoping,
SWEDEN

Ms. Valentina Maddalena
Michigan State University
S. Shaw Lane
East Lansing, MI 48824-1321
USA

Dr. Paul F. Mantica
NSCL
Michigan State University
S. Shaw Lane
East Lansing, MI 48824-1321
USA

Dr. François Marechal
Dept. of Physics
Florida State University
211 Keen Building
Tallahassee, FL 32306-4350
USA

Mr. Karin G. Markenroth
Dept. of Physics
Chalmers University of Technology
S-41296 Göteborg,
SWEDEN

Dr. Christiane Miehe
Institut de Recherches Subatomiques
IReS
B.P. 28, 23 rue du Loess
F-67037 Strasbourg cedex,
FRANCE

Dr. John Millener
Dept. of Physics, Bldg. 510A
Brookhaven National Laboratory
20 Pennsylvania St.
Upton, NY 11973-5000
USA

Dr. Peter Moller
Theory Division T-2, MS-B243
Los Alamos National Laboratory
P.O. Box 1663
Los Alamos, NM 87545
USA

Prof. David J. Morrissey
Michigan State University
S. Shaw Lane
East Lansing, MI 48824-1321
USA

Prof. Tohru Motobayashi
Faculty of Science/Dept. of Physics
Rikkyo University
3-34-1 Nishi-Ikebukuro, Toshima-ku
171-8501 Tokyo,
JAPAN

Dr. Alex C. Mueller
Dept. Head, RDA/FE
Institut de Physique Nucléaire, Orsay
B. P. No. 1, Bâtiment 100
F-91406 Orsay cedex,
FRANCE

Dr. Ivan Mukha
Institute of Physics and Astronomy
Aarhus Universitet
Ny Munkegade
DK-8000 Aarhus C,
DENMARK

Prof. Takashi Nakamura
Dept. of Physics
University of Tokyo
7-3-1 Hongo, Bunkyo-ku
113-0033 Tokyo,
JAPAN

Prof. Witold Nazarewicz
Dept. of Physics & Astronomy
University of Tennessee
401 A. H. Nielsen Physics Bldg.
Knoxville, TN 37996
USA

Prof. Rainer Neugart
Institut für Physik
Johannes Gutenberg-Universität Mainz
Postfach 3980
D-55099 Mainz,
GERMANY

Dr. Gerda Neyens
Instituut voor Kern- en Stralingsfysica
University of Leuven
Celestijnenlaan 200 D
B-3001 Leuven,
BELGIUM

Dr. Victor Ninov
Nuclear Science, MS88
Lawrence Berkeley National Lab.
One Cyclotron Rd.
Berkeley, CA 94720
USA

Dr. Jerry A. Nolen
PHY203
Argonne National Laboratory
9700 S. Cass Ave.
Argonne, IL 60439-4843
USA

Dr. Masahiro Notani
The Institute of Physical & Chemical Res.
RIKEN
2-1 Hirosawa, Wako-shi
351-0198 Saitama,
JAPAN

Dr. Filomena M. Nunes
Instituto Superior Técnico
CENTRA
Av. Rovisco Pais, 1
1096 Lisboa Codex,
PORTUGAL

Prof. Göran H. Nyman
Dept. of Physics
Chalmers University of Technology
S-41296 Göteborg,
SWEDEN

Mr. Tom O'Donnell
Randall Laboratory
University of Michigan
500 E. University
Ann Arbor, MI 48109-1120
USA

Dr. Alexei A. Ogloblin
RRC,
The Kurchatov Institute
Kurchatov Square 1
RU-123182 Moscow,
RUSSIA

Dr. Ana-Maria D. Oros
Institut for Theoretical Physics
University of Gent
Proeftuinstraat 86
B-9000 Gent,
BELGIUM

Prof. Takaharu Otsuka
Dept. of Physics
University of Tokyo
7-3-1 Bunkyo-ku
113-8654 Tokyo,
JAPAN

Dr. Robert Page
Dept. of Physics
University of Liverpool
Oxford St., P.O. Box 147
Liverpool L69 3BX,
UNITED KINGDOM

Mr. Donald Peterson
Dept. of Physics
University of Notre Dame
South Bend, IN 46556
USA

Dr. Andreas Piechaczek
Bldg. 6000, Room 252
Louisiana State Univ./Oak Ridge Nat'l. Lab.
P.O. Box 2008
Oak Ridge, TN 37831
USA

Dr. Steven C. Pieper
PHY203
Argonne National Laboratory
9700 S. Cass Ave.
Argonne, IL 60439-4843
USA

Dr. Prof. Andrzej Plochocki
Institute for Experimental Physics
Warsaw University
ul. Hoza 69
PL-00 681 Warsaw,
POLAND

Mr. Chris Powell
NSCL
Michigan State University
S. Shaw Lane
East Lansing, MI 48824-1321
USA

Dr. James D. Powell
MS88-210
Lawrence Berkeley National Lab.
One Cyclotron Rd.
Berkeley, CA 94720
USA

Ms. Joann I. Prisciandaro
Michigan State University
S. Shaw Lane
East Lansing, MI 48824-1321
USA

Mr. Boris V. Pritychenko
Michigan State University
S. Shaw Lane
East Lansing, MI 48824-1321
USA

Dr. Philippe G. Quentin
Centre d'Etudes Nucl. de Bordeaux
Le Haut Vigneau, B.P. 120
F-33175 Gradignan,
FRANCE

Dr. Torsten Radon
GSI
Planckstrasse 1
D-64291 Darmstadt,
GERMANY

Prof. Akunuri Ramayya
Dept. of Physics & Astronomy
Vanderbilt University
Box 1807, Sta. B
Nashville, TN 37235
USA

Dr. Helge L. Ravn
ISOLDE
CERN
CH-1211 Geneva 23,
SWITZERLAND

Mr. Alan T. Reed
Dept. of Physics
University of Liverpool
P.O. Box 147, Oxford St.
Liverpool L69 3BX,
UNITED KINGDOM

Ms. Ils Reusen
Instituut voor Kern- en Stralingsfysica
University of Leuven
Celestijnenlaan 200 D
B-3001 Leuven,
BELGIUM

Prof. Lee L. Riedinger
Dept. of Physics & Astronomy
University of Tennessee
401 A. H. Nielsen Physics Bldg.
Knoxville, TN 37996-1200
USA

Dr. Karsten Riisager
Institute of Physics and Astronomy
Aarhus Universitet
Ny Munkegade
DK-8000 Aarhus C,
DENMARK

Prof. Peter Ring
Physik-Department T30
Technische Universität München
James-Franck-Straße 1
D-85748 Garching,
GERMANY

Dr. Alexander Rodin
Flerov Laboratory of Nuclear
Reactions
Joint Institute for Nuclear Research
Joliot Curie str. 8
141980 Dubna, Moscow region
RUSSIA

Dr. Ernst Roeckl
GSI
Planckstrasse 1
D-64291 Darmstadt,
GERMANY

Mr. Michael W. Rowe
Physics Dept., MS88-163
Lawrence Berkeley National Lab.
One Cyclotron Rd.
Berkeley, CA 94720
USA

Dr. Berta Rubio
Instituto de Fisica Corpuscular
CSIC
Dr. Moliner 50
E-46100 Burjassot (Valencia),
SPAIN

Dr. Krzysztof P. Rykaczewski
Bldg. 6000, MS-6371
Oak Ridge National Laboratory
P.O. Box 2008
Oak Ridge, TN 37831-6371
USA

Dr. Hiroyoshi Sakurai
Radiation Lab
RIKEN
2-1 Hirosawa, Wako-shi
351-0198 Saitama,
JAPAN

Prof. Stanislav B. Sakuta
RRC,
The Kurchatov Institute
Kurchatov Square 1
RU-123182 Moscow,
RUSSIA

Mr. Peter A. Santi
Dept. of Physics
University of Notre Dame
South Bend, IN 46556
USA

Dr. Domenico Santonocito
Institut de Physique Nucléaire, Orsay
B. P. No. 1, Bâtiment 100
F-91406 Orsay cedex,
FRANCE

Dr. Jocelyne Sauvage
Div. Recherche Expérimentale
Institut de Physique Nucléaire, Orsay
B. P. No. 1, Bâtiment 100
F-91406 Orsay cedex,
FRANCE

Dr. Hervé Savajols
GANIL
B.P. No. 5027
F-14076 Caen cedex 5,
FRANCE

Mr. Heiko Scheit
NSCL
Michigan State University
S. Shaw Lane
East Lansing, MI 48824-1321
USA

Dr. Karl-Heinz Schmidt
GSI
Planckstrasse 1
D-64291 Darmstadt,
GERMANY

Dr. Hans A. Schuessler
Dept. of Physics
Texas A & M University
College Station, TX 77843-4242
USA

Ms. Jazmin Schwartz
PHY203
Argonne National Laboratory
9700 S. Cass Ave.
Argonne, IL 60439-4843
USA

Dr. Ronald Schwengner
Institut für Kern- und Hadronenphysik
Forschungszentrum Rossendorf
FWKK/
Postfach 510119
D-01314 Dresden,
GERMANY

Mr. Volker Sebastian
Institut für Physik
Johannes Gutenberg-Universität Mainz
Staudinger Weg 7
D-55099 Mainz,
GERMANY

Prof. Paul B. Semmes
Dept. of Physics
Tennessee Technological University
Box 5051, TTU
Cookeville, TN 38505
USA

Dr. Dariusz Seweryniak
PHY203
Argonne National Laboratory
9700 S. Cass Ave.
Argonne, IL 60439-4843
USA

Prof. Kumar S. Sharma
Dept. of Physics, Fort Garry Campus
University of Manitoba
301 Allen Physics Laboratories
Winnipeg, Manitoba, R3T 2N2,
CANADA

Dr. Bradley M. Sherrill
NSCL
Michigan State University
S. Shaw Lane
East Lansing, MI 48824-1321
USA

Ms. Sunniva Siem
PHY203
Argonne National Laboratory
9700 S. Cass Ave.
Argonne, IL 60439-4843
USA

Dr. Adam Sobiczewski
Zaklad PVII
Instytut Problemów Jadrowych
ul. Hoza 69
PL-00 681 Warsaw,
POLAND

Dr. Alejandro Sonzogni
Physics Division
Argonne National Laboratory
9700 S. Cass Ave.
Argonne, IL 60439-4843
USA

Dr. Kruno Subotic
Institute of Nuclear Sciences Vinca
POB 522
11001 Belgrade,
YUGOSLAVIA

Dr. Sergey I. Sukhoruchkin
Neutron Research Dept.
St. Petersburg Nuclear Physics Inst.
Leningrad District
188350 Gatchina,
RUSSIA

Dr. Takeshi Suzuki
Dept. of Physics/Faculty of Science
Niigata University
8050, Ikarashi-2
950-21 Niigata,
JAPAN

Mr. Carl E. Svensson
Dept. of Physics & Astronomy
McMaster University

1280 Main St. West, ABB-241
Hamilton, ON, L8S 4M1,
CANADA

Dr. Jerzy Szerypo
JIHIR
Oak Ridge National Laboratory
P.O. Box 2008
Oak Ridge, TN 37831
USA

Dr. Takahiro Tachibana
Advanced Res. Ctr. for Science & Eng.
Waseda University
3-4-1 Okubo, Shinjuku-ku
169-8555 Tokyo,
JAPAN

Dr. Satoshi Takeuchi
Faculty of Science/Dept. of Physics
Rikkyo University
3-34-1 Nishi-Ikebukuro
171-8501 Toshima-ku, Tokyo
JAPAN

Mr. Xiao Tang
Cyclotron Institute
Texas A & M University
College Station, TX 77845-3366
USA

Prof. Isao Tanihata
RIKEN
2-1 Hirosawa, Wako-shi
351-0198 Saitama,
JAPAN

Dr. Olof Tengblad
Instituto de Estructura de la Materia
CSIC
Serrano 113 bis
E-28006 Madrid,
SPAIN

Ms. Stéphanie Teughels
Instituut voor Kern- en Stralingsfysica
University of Leuven
Celestijnenlaan 200 D
B-3001 Leuven,
BELGIUM

Dr. Catherine Thibault
CSNSM Orsay
Bâtiment 108
F-91405 Orsay cedex,
FRANCE

Prof. Friedrich-Karl Thielemann
Dept. of Physics & Astronomy
Universität Basel
Klingelbergstrasse 82
CH-4056 Basel,
SWITZERLAND

Dr. Michael Thoennessen
NSCL
Michigan State University
S. Shaw Lane
East Lansing, MI 48824-1321
USA

Dr. Ian Thompson
Dept. of Physics
University of Surrey
Guildford,
Surrey GU2 5XH,
UNITED KINGDOM

Mr. Christian-Florentin V. Toader
CSNSM Orsay
Bâtiment 108
F-91405 Orsay cedex,
FRANCE

Dr. Jeffrey A. Tostevin
School of Physical Science
University of Surrey
Guildford,
Surrey GU2 5XH,
UNITED KINGDOM

Dr. Kenneth L. Toth
Physics Division, Bldg. 6000, MS6371
Oak Ridge National Laboratory
P.O. Box 2008
Oak Ridge, TN 37831-6371
USA

Dr. Robert E. Tribble
Cyclotron Institute
Texas A & M University
College Station, TX 77843-3366
USA

Ms. Monica Trotta
Physics Dept.
Naples University
Via E. Nicolardi, Parco arcadia 5
I-80131 Naples,
ITALY

Mr. Erik Tryggestad
NSCL
Michigan State University
S. Shaw Lane
East Lansing, MI 48824-1321
USA

Dr. Kazuaki Tsukada
Advanced Science Research Center
Japan Atomic Energy Research Inst.
(JAERI) Tokai-mura
319-1195 Ibaraki,
JAPAN

Dr. Masahiro Uno
Elementary and Secondary Education
Bureau Ministry of Education, Science,
Sports & Culture
3-2-2-Kasumigaseki, Chiyoda-ku,
100 Tokyo,
JAPAN

Dr. Juha S. Uusitalo
Physics Division
Argonne National Laboratory
9700 S. Cass Ave.
Argonne, IL 60439-4843
USA

Prof. Dr. Jan S. Vaagen
Dept. of Physics
University of Bergen
Allégaten 55
N-5007 Bergen,
NORWAY

Dr. Piet Van Duppen
Instituut voor Kern- en Stralingsfysica
University of Leuven
Celestijnenlaan 200 D
B-3001 Leuven,
BELGIUM

Dr. Robert L. Varner
Physics Division, Bldg. 6000
Oak Ridge National Laboratory
P.O. Box 2008
Oak Ridge, TN 37831
USA

Dr. David J. Vieira
CST-11, MS J514
Los Alamos National Laboratory
P.O. Box 1663
Los Alamos, NM 87545
USA

Mr. Stuart M. Vincent
Dept. of Physics
University of Surrey
Guildford,
Surrey GU2 5XH,
UNITED KINGDOM

Prof. William B. Walters
Dept. of Chemistry & Biochemistry
University of Maryland
College Park, MD 20742
USA

Prof. Dr. Aaldert H. Wapstra
NIKHEF
P.O. Box 41882, Kruislaan 411
N-41882 DB Amsterdam,
THE NETHERLANDS

Dr. Christopher Wesselborg
Editorial Office
American Physical Society
P.O. Box 9000, One Research Rd.
Ridge, NY 11961
USA

Prof. Michael C. Wiescher
Dept. of Physics
University of Notre Dame
Notre Dame, IN 46556
USA

Dr. Jeff A. Winger
Dept. of Physics & Astronomy
Mississippi State University
Drawer 5167
Mississippi, MS 39762
USA

Prof. Andreas Woehr
Clarendon Laboratory
Oxford University
Parks Rd.
Oxford OX1 3PU,
UNITED KINGDOM

Prof. John L. Wood
School of Physics
Georgia Tech.
Atlanta, GA 30332-0430
USA

Dr. Philip J. Woods
Kings Bldgs./James Clerk Maxwell Bldg.
University of Edinburgh
Mayfield Road
Edinburgh EH9 3JZ,
UNITED KINGDOM

Dr. Yoshiyuki Yanagisawa
Faculty of Science/Dept. of Physics
Rikkyo University
3-34-1 Nishi-Ikebukuro
171-8501 Toshima-ku, Tokyo
JAPAN

Dr. Nicolae Victor Zamfir
WNSL
Yale University
P.O. Box 208124, 272 Whitney Ave.
New Haven, CT 06520-8124
USA

Prof. Nissan Zeldes
The Racah Institute of Physics
The Hebrew University of Jerusalem
Givat Ram Campus
91904 Jerusalem,
ISRAEL

Prof. Edward F. Zganjar
Dept. of Physics & Astronomy
Louisiana State University
Baton Rouge, LA 70803-4001
USA

Prof. Mikhail V. Zhukov
Dept. of Physics
Chalmers University of Technology
S-41296 Göteborg,
SWEDEN

Mr. James A. Zimmerman
Dept. of Chemistry/1500 Chemistry Bldg.
University of Michigan
930 N. University Ave.
Ann Arbor, MI 48109
USA

Dr. Andres Zuker
CNRS
IReS
Bâtiment 27, B.P. 28
F-67037 Strasbourg cedex 2,
FRANCE

Prof. Jan L. Zylicz
Institute of Experimental Physics
Warsaw University
ul. Hoza 69
PL-00 681 Warsaw,
POLAND

Author Index

A

Agodi, C., 711
Agramunt, J., 725, 769, 809
Ahmad, I., 470, 682
Akovali, Y. A., 444
Alahari, N., 209, 272
Alamanos, N., 478
Aleksandrov, D. V., 252
Algora, A., 725, 769, 809
Al-Khalili, J. S., 227
Alkhazov, G. D., 188
Allatt, R. G., 564, 570, 572
Ames, F., 3, 927
Amro, H., 486
Ando, Y., 371, 610
Andreyev, A. N., 363, 462, 486, 581, 793
Andriamonje, S., 753
Andrzejewski, J., 793
Angélique, J. C., 118, 314, 327, 494, 570, 572
Anne, R., 278, 327, 494, 568, 570, 572
Anthony, D. W., 66, 532, 890, 908
Antropov, A. E., 519
Aoi, N., 215, 233, 359
Applebe, D., 739
Aprahamian, A., 589, 819
Armbruster, P., 686, 698
Aryaeinejad, R., 523
Asai, M., 540
Asztalos, S. J., 523
Attallah, F., 11
Audi, G., 3, 30, 98, 118, 995
Auger, F., 478
Auger, G., 44
Aumann, T., 196, 256, 268
Austin, S. M., 890
Axelsson, L., 260, 268, 314, 327, 494, 739, 749
Äystö, J., 72, 581, 690, 745, 749, 761, 781, 981
Azhari, A., 245, 598, 868, 896

B

Bacher, A., 682
Back, B. B., 682
Bai, Y., 90
Baiborodin, D. S., 44, 568
Baktash, C., 444
Balabanski, D. L., 138
Ball, G., 973
Band, I. M., 985
Barber, R. C., 130
Bark, R., 450
Barker, J., 989
Barrett, B., 560
Bartmann, R., 973
Barton, C. J., 102, 466, 556, 614
Barzakh, A. E., 94
Batchelder, J. C., 264, 367, 375, 444, 474
Batist, L., 602, 725, 769, 777, 809, 969
Baumann, T., 268, 290
Baye, D., 363
Bazin, D., 209, 272, 598, 890
Beaumel, D., 233
Beausang, C. W., 556, 614
Beck, D., 3
Beckert, K., 11
Beene, J. R., 221, 245
Behr, J., 973
Bellia, G., 711
Belov, S. E., 519
Belozyorov, A. V., 44, 568
Benlliure, J., 276, 647, 686, 904, 960
Benoit, B., 314
Bentaleb, M., 470
Benton, D. M., 72
Béraud, R., 494
Berg, G. P. A., 682
Bergmann, U., 268, 314
Bernas, M., 664, 686
Beyer, C. J., 523
Bhattacharyya, P., 470
Bijnens, N., 462, 793
Billowes, J., 72
Bingham, C. R., 264, 367, 375, 444, 474, 486, 575

Blank, B., 221, 272, 276, 323, 739, 753, 890
Blazevic, A., 347
Blomqvist, J., 470, 785
Blumenfeld, Y., 239, 478, 490, 598
Blumenthal, D., 102, 450
Bochkarev, O., 84
Böckstiegel, C., 647, 686
Bogaert, G., 858
Bohlen, H. G., 347
Bollen, G., 3, 927, 965
Bolotov, D. V., 306
Boos, N., 585
Borcea, C., 44, 98, 278, 347, 494, 568, 570, 572, 995
Borcea, R., 809
Boretzky, K., 256
Borge, M. J. G., 260, 314, 351, 749, 761, 789
Bosch, F., 11
Boudard, A., 686
Boué, F., 276, 753
Bradfield-Smith, W., 363
Brenner, D. S., 102, 466, 556
Brenner, M., 519
Bricault, P., 973
Broda, R., 470
Broude, C., 864
Brown, B. A., 66, 142, 209, 602, 725, 769, 777
Brown, J. A., 598
Brown, L. T., 486
Bruce, A. M., 739
Bruyneel, B., 757, 793
Buchinger, F., 130
Buchmann, L., 973
Burjan, V., 572, 868, 896
Burtebaev, N., 618
Bush, J. E., 272
Butler, P. A., 422
Bykov, A., 602

C

Cabaret, L., 78, 585
Caggiano, J. A., 272, 890
Calabretta, L., 711
Campbell, P., 72
Cano-Ott, D., 725, 769, 777, 809, 969

Carjan, N., 282
Carlson, B. V., 527
Carpenter, M. P., 444, 450, 486, 506
Carstoiu, F., 278, 896
Casandjian, J. M., 118
Casten, R. F., 466, 556, 614
Cata-Danil, G., 556, 614
Catford, W. N., 239, 314, 572, 739
Catherall, R., 126, 989
Cejpek, J., 868, 896
Cerny, J., 801, 999
Chabannat, E., 494
Chandler, C., 739
Chapman, R., 239
Chappell, S. P. G., 239, 314
Chartier, M., 118, 221, 323, 890
Chasman, R. R., 682
Chen, L., 221, 272, 323
Chromik, M. J., 286, 490
Chubarian, G. G., 711
Chulkov, L. V., 84, 196, 904
Chuvilskay, T. V., 482
Ciurczak, D., 575
Cizewski, J. A., 474, 486, 506, 589
Clark, H. L., 868, 896
Clark, R., 739
Clarke, N. M., 239, 314
Clerc, H.-G., 647
Cocks, J. F. C., 422, 462, 486
Cole, J. D., 523, 708
Collatz, R., 276, 725, 769, 777, 809, 904
Conticchio, L., 575
Cooper, J. R., 556, 614
Cortina-Gil, D., 84, 138, 256, 268, 290, 904
Costa, G., 314
Cottle, P. D., 490, 498, 598
Coulier, N., 138
Coussement, R., 138
Cowan, J. J., 837
Crawford, J. E., 78, 130, 585
Crespo, R., 876
Cub, J., 256
Cullen, D., 739
Cunsolo, A., 118
Curtis, N., 239, 314
Czajkowski, S., 276, 686, 739, 753

D

Daly, P. J., 470
Daniel, A. V., 523, 708
Danilin, B. V., 294
D'Arrigo, A., 314
Daugas, J. M., 494, 739
D'Auria, J. M., 973
Davi, F., 276
Davids, B., 66, 209, 272, 890, 908
Davids, C. N., 102, 367, 375, 444, 450, 474, 486, 506, 575, 589
Davinson, T., 264, 363, 367, 375, 444
Deak, F., 323
Dean, D. J., 438
deAngelis, G., 594
DeBoer, J., 375
De Coster, C., 560
Decrock, P., 486
Decroix, B., 560
de FRANCE for the EXOGAM Collaboration, G., 977
de Goes Brennard, E., 314
de Haan, R., 575, 589
de Jong, M., 647, 686
Delhaij, P., 973
de Lima, A. P., 523
Del Moral, R., 276, 753
Dem'yanova, A. S., 510
Dendooven, P., 72, 690, 745, 781, 981
de Oliveira, F., 44, 138, 314
de Oliveira-Santos, F., 494
de Saint Simon, M., 3, 995
Dessagne, Ph., 739, 765, 789
de Vismes, A., 44, 239
Dilling, J., 965
Ding, K. Y., 486
Dlouhý, Z., 278, 568, 570, 572
Dobaczewski, J., 813
Dodder, R. S., 523
Dolinskiy, A., 11
Dombsky, M., 973
Donangelo, R., 523
Dönau, F., 594
Donzaud, C., 118, 570, 572
Döring, J., 819
Dorvaux, O., 314, 422
Dostal, W., 256
Doubre, H., 995

Duflo, J., 38
Dufour, J. P., 276, 686, 753
Duisebaev, A., 618
Duisebaev, B. A., 618
Duma, M., 995
Duong, H. T., 78, 585
Durell, J. L., 470
Dutto, G., 973

E

Eberlein, B., 256
Eberth, J., 594
Eickhoff, H., 11
Ellis, D., 444
Elze, Th. W., 256
Emling, H., 256
Enqvist, T., 564, 686
Erokhina, K. I., 552
Ershov, K. N., 519
Ershov, S. N., 294
Esbensen, H., 890
Eskola, K., 422, 462, 564
Ethvignot, Th., 494
Evans, D. E., 72
Evensen, A.-H., 581
Everling, F., 298

F

Falch, M., 11
Farget, F., 686, 904, 960
Farnea, E., 594
Fauerbach, M., 286, 498, 598, 908
Faust, H., 694
Fayans, S. A., 260, 302, 306, 310
Fedorov, D. V., 94, 335, 339, 343
Fedoseyev, V. N., 126, 456, 864, 989
Feng, X., 130
Fernandes, J. C., 876
Fischer, S. M., 486, 589
Fleury, A., 276, 739, 753
Florescu, A., 708
Fogelberg, B., 502, 552, 785
Fomitchev, A., 278
Forest, D. H., 72
Forkel-Wirth, D., 456
Fornal, B., 470

Fortier, S., 239
Foster, C. C., 682
Foti, A., 118
Fotiades, N., 486, 506
Foy, B. D., 102, 575
Fraile, L., 268
Fraile, L. M., 749, 761
Franchoo, S., 494, 757, 793
Franczak, B., 11
Frankland, L., 739
Franzke, B., 11
Frascaria, N., 478
Frauendorf, S., 594
Freer, M., 239, 314
Freiburghaus, C., 837
Friedrichs, T., 694
Fujita, Y., 11
Fujiwara, H., 610
Fukuda, N., 215, 233, 359
Fukutani, H., 130
Fulop, Z., 182
Fulton, B. R., 314
Fynbo, H. O. U., 351, 749, 761

G

Gadea, A., 594, 725, 769, 777, 809, 969
Gaelens, M., 363
Gagliardi, C. A., 868, 896
Galès, S., 239
Galindo-Uribarri, A., 444
Galonsky, A., 221, 323
Galster, W., 363
Gan, N., 221, 245
Garcés Narro, J., 739
García, A., 719
Gardina, G., 314
Garrido, E., 335, 339, 343
Geissel, H., 11, 84, 256, 268, 290
Geithner, W., 110, 142
Gelletly, W., 450, 739
Genevey, J., 78, 585, 694
Georg, U., 126, 142, 989
Georgiev, G., 66, 138
Gerl, J., 809
Gervais, G., 209
Ghalambor Dezfuli, A. M., 965
Gierlik, M., 725, 769, 809
Gill, R. L., 102, 466, 556

Gillibert, A., 44, 118, 478
Ginter, T. N., 264, 444
Giovinazzo, J., 739, 765, 789
Girod, M., 78, 585
Glasmacher, T., 204, 209, 272, 286, 478, 490, 498, 536, 598
Glogowski, M., 494
Gmuca, S., 106
Godwin, M., 478
Goldberg, V. Z., 260, 319
Goldring, G., 864
Golovkov, M. S., 182
Goncharov, S. A., 510
Gore, P. M., 523
Görres, J., 819
Górska, M., 725, 769, 809
Govaert, K., 209, 221, 272, 323
Govil, I. M., 515
Grabowski, Z. W., 470
Grant, I. S., 72
Grawe, H., 494, 725, 769, 809, 904
Greene, J. P., 682
Greenhalgh, B., 739
Greenlees, P. T., 422, 486, 564
Gregori, C., 314
Gregorich, K. E., 704
Greiner, W., 708
Grévy, S., 260, 314, 327, 570, 572
Grewe, A., 647
Gridnev, D. K., 519
Gridnev, K. A., 331, 519
Griffith, J. A. R., 72
Gross, C. J., 264, 444
Gross, M., 694
Grosse, E., 594
Grünschloß, A., 256
Grzywacz, R., 264, 430, 444, 602, 739, 777
Guglielmetti, A., 777
Guilbaud, O., 809
Guillemaud-Mueller, D., 260, 278, 314, 327, 351, 494, 568, 570, 572, 1009
Guimarães, V., 272, 347
Gulbekian, G. G., 633
Gulick, S., 130
Gulielmetti, A., 725
Guo, F. Q., 999

H

Hackman, G., 130
Halbert, M. L., 245
Hamilton, J. H., 523, 655, 708
Hanappe, F., 314, 711
Hanelt, E., 753
Hankonen, S., 690
Hannawald, M., 456
Hansen, P. G., 209, 221, 272, 323, 327
Harder, M., 739
Hardy, J. C., 130, 733
Hass, M., 864
Hausmann, M., 11
Haustein, P. E., 999
Heinz, A., 276, 647
Helariutta, K., 422, 462, 486
Hellström, M., 11, 84, 209, 256, 268, 276, 725, 769, 777, 809, 904
Henderson, D. H., 589
Henderson, D. J., 367, 375, 450
Henry, S., 965, 995
Herfurth, F., 3, 11, 965
Herman, M., 482
Heusch, B., 314
Heyde, K., 560
Higurashi, Y., 323
Hirai, M., 233, 371
Hirata, D., 118, 527
Hirzebruch, S. E., 490, 598
Hisanaga, I., 610
Hoff, P., 581
Hofman, D., 130
Hofmann, S., 625, 698
Hokanen, A., 781
Holeczek, J., 256
Holzmann, R., 256
Honkanen, A., 72, 690, 745, 749, 761
Hornshøj, P., 327, 351, 749, 761
Horvath, A., 323
Hu, Z., 276, 725, 769, 777, 809, 904
Huber, G., 78, 126, 585, 864, 927, 989
Huck, A., 711, 789
Huhta, M., 66, 532, 581, 690, 781
Hui, P., 890
Huikari, J., 981
Hurst, B. J., 711
Huyse, M., 363, 462, 486, 494, 581, 602, 757, 793
Hwang, J. K., 523, 655, 708

I

Ibbotson, R. W., 66, 209, 272, 286, 490, 498, 536
Ibrahim, F., 585, 694
Ichikawa, S., 540, 544
Ideguchi, E., 233, 359
Ieki, K., 323
Iimura, H., 544
Ikezoe, H., 548
Ikuta, T., 548
Ilievski, S., 256
Imai, N., 233
Irvine, R. J., 367, 474
Isakov, V. I., 502, 552, 785
Ishihara, M., 215, 233, 359, 371, 610
Ishii, T., 682
Itkis, M. G., 633, 678, 711
Ivanov, G. N., 618
Ivanov, M., 84, 268
Iwasa, N., 256, 268, 371, 904
Iwasaki, H., 215, 233, 359, 610
Iwata, Y., 323, 610

J

Jackson, K. P., 973
Jacobsson, L., 785
Jacotin, M., 995
Jading, Y., 126, 761, 989
Janas, Z., 264, 276, 494, 602, 725, 769, 777, 809, 969
Jänecke, J., 773
Janik, R., 84, 268
Janssens, R. V. F., 444, 450, 486, 506, 589
Jennings, B. K., 900
Jensen, A. S., 335, 339, 343
Jewell, J. K., 490, 598
Johnson, J. W., 444
Jokinen, A., 72, 581, 690, 745, 749, 761, 789, 981
Jones, E. F., 523
Jones, K. L., 239, 739
Jones, P., 422, 486, 564
Jonson, B., 166, 260, 268, 314, 327, 351, 749, 761
Jonsson, O., 126, 989
Joosten, R., 999

Julin, R., 422, 486, 564
Jungclaus, A., 594
Junghans, A. R., 647, 960
Juutinen, S., 422, 486

K

Kadmensky, S. G., 672
Källman, K.-M., 260
Kalpakchieva, R., 347, 568
Kankaanpää, A., 486
Kankaanpää, H., 422
Kappertz, S., 110, 142
Karataglidis, S., 900
Karnes, D., 272
Karny, M., 264, 276, 494, 581, 602, 725, 769, 777, 809, 969
Karpeshin, F. F., 985
Kartamishev, M. P., 519
Kaspar, M., 256
Katko, G., 110
Käubler, L., 594
Kautzsch, T., 456
Kawade, K., 540
Keim, M., 50, 110, 142
Keller, H., 602
Kellerbauer, A., 965
Kelly, G., 314
Kelsall, N., 739
Kemper, K. W., 490, 498, 598
Képinski, J.-F., 995
Kerscher, Th., 11
Kettunen, H., 422, 462, 486
Khan, E., 478
Khoo, T. L., 506
Kiefl, R., 973
Kikuchi, T., 371
Kim, S. H., 264
Kim, T., 965
Kimura, K., 84
Kirchner, R., 602, 725, 769, 777, 809, 969
Kiss, A., 323
Kleinböhl, A., 256
Kleinheinz, P., 809
Klepper, O., 11, 602
Kliman, J., 708
Kluge, H.-J., 3, 11, 965
Knipper, A., 789

Knyazkov, O. M., 306
Kobayashi, T., 84, 215
Kohl, A., 3, 965
Koizumi, M., 126, 456, 989
Kojima, Y., 540
Kolata, J. J., 272
Kolbe, E., 837
Kolhinen, V. S., 981
Kondratiev, N. A., 678, 711
Kormicki, J., 523, 708
Korsheninnikov, A. A., 84, 182
Köster, U., 126, 456, 757, 989
Kotrotsios, G., 110
Koura, H., 114
Kozhuharov, C., 11
Kozulin, E. M., 678, 711
Kratz, J. V., 256
Kratz, K.-L., 126, 456, 757, 827, 837, 989
Krieg, M., 78, 585
Kroha, V., 868, 896
Krouglov, I. V., 519
Krücken, R., 556, 614
Kruglov, K., 757, 793
Krupa, L., 678
Kruse, J., 221, 323
Kryger, R. A., 286
Kszczot, T., 739
Kubo, T., 215, 233, 359
Kuchtina, I. N., 306
Kudryavtsev, Y. A., 757, 793
Kugler, E., 126, 581, 989
Kulessa, R., 256
Kumagai, H., 233
Kurpeta, J., 581, 781
Kusaka, K., 182, 233
Kutner, V. B., 633
Kuusiniemi, P., 422, 462, 486, 564
Kuzmin, E. A., 182
Kuzumaki, T., 548

L

Labiche, M., 314
Lalleman, A., 44
Lamour, E., 965
Langevin-Joliot, H., 239
Lapoux, V., 347, 478
Larimer, R.-M., 999

Larqué, T., 694
Lassen, J., 150
Laurent, H., 239
Lauritsen, T., 470, 486, 506
Laxdal, R., 973
Lebée, G., 995
Le Blanc, F., 78, 585
Le Brun, C., 314
Leddy, M. J., 470
Lee, I. Y., 523
Lee, J. K. P., 78, 130, 585
Leenhardt, S., 314
Legrain, R., 686
Leifels, Y., 256
Leino, M., 422, 462, 486, 564
Leistenschneider, A., 256
Lenske, H., 268
Lépine-Szily, A., 118, 138, 347
Leray, S., 686
Le Scornet, G., 995
Leth, T., 351
Lettry, J., 78, 126, 456, 581, 864, 989
Levins, J. M. G., 72
Lewitowicz, M., 44, 118, 138, 260, 276, 278, 314, 327, 494, 568, 570, 572, 739
Lhenry, I., 239, 478
Lhersonneau, G., 690, 781
Liang, J. F., 444
Liatard, E., 711
Libert, J., 585
Lichtenthaler, R., 118, 347
Lieb, K. P., 594
Lievens, P., 110, 142
Lima, G. F., 118
Lindroth, A., 785
Lingk, C., 594
Lipas, P., 745
Lister, C. J., 444, 450, 506, 589
Liu, B., 556, 614
Liu, W., 809, 904
Lo Bianco, G., 594
Löbner, K. E. G., 11
Lofy, P. A., 66, 532
Longour, C., 494, 739, 789
Lönnroth, T., 260
Lopez-Jimenez, M. J., 494
Lougheed, R. W., 523
Lozowski, W. R., 682
Lu, J., 548
Lubkiewicz, E., 256, 470

Lui, W., 606
Lui, Y.-W., 868, 896
Lukashin, K., 711
Lukyanov, S. M., 44, 118, 233, 278, 359, 568, 570, 572
Lunney, D., 3, 78, 965, 995
Lyneis, C., 999

M

Ma, W. C., 523, 606
Macchiavelli, A. O., 506, 523
MacCormick, M., 118, 347
MacDonald, B. D., 264, 444
Mach, H., 502, 552, 785
Maddalena, V., 209, 221, 272, 323
Maidikov, V. Z., 568
Maier, H. J., 375
Maiolino, C., 711
Maison, J. M., 239
Manngård, P., 260
Mantica, P. F., 66, 532, 536, 598, 606, 908
Marchand, C., 276
Maréchal, F., 478, 490, 598
Marguier, G., 789
Markenroth, K., 260, 268, 314
Marqués, F. M., 314, 572
Martel, I., 260, 749, 761
Martinez, G., 572
Martinez, P., 130
Martinez, T., 809
Mas, J. F., 264, 444
McConnell, J. W., 264
McGrath, C. A., 704
McKenzie, J., 363
McLeod, R. W., 506
McNabb, D. P., 486
Mengoni, A., 215
Mezilev, K. A., 502, 785
Miehé, Ch., 739, 765, 789
Minemura, T., 610
Mishin, V. I., 126, 456, 864, 989
Mitsuoka, S., 548
Mittig, M., 138
Mittig, W., 44, 118, 1003
Miyaji, M., 544
Mizusaki, T., 391
Möller, P., 698, 715

Moltz, D. M., 801, 999
Monsanglant, C., 995
Moore, R. B., 3, 130, 965
Morhac, M., 708
Morimoto, K., 182
Moriya, S., 371
Moroz, F., 725, 769, 777, 809, 969
Morrissey, D. J., 118, 478, 490, 598, 908
Morss, L. R., 470, 682
Motobayashi, T., 371, 610, 882
Motta, M., 314
Mueller, A. C., 260, 278, 314, 327, 494, 568, 570, 572, 933
Mueller, W. F., 757, 793
Muikku, M., 422, 486
Mukha, I., 260, 351, 749, 761
Mukhamedzhanov, A. M., 868, 896
Müller, J., 647
Münzenberg, G., 11, 84, 256, 268, 698
Murakami, H., 371, 610
Murgatroyd, J. T., 314
Mustapha, B., 686
Musumara, A., 478

N

Nagame, Y., 540, 548
Nakamura, T., 215, 233, 359, 371, 610
Napoli, D. R., 594
Navratil, P., 560
Nayak, R. C., 122, 355
Nazarewicz, W., 381
Negoita, F., 278
Neugart, R., 110, 142
Neuroth, M., 142
Neyens, G., 58, 138
Nickel, F., 84, 268
Nieminen, A., 581, 981
Nilsson, T., 260, 268, 314, 327, 351, 749
Ninane, A., 314, 363
Ninov, V., 704
Nishinaka, I., 540, 548
Nishio, T., 371, 610
Nisius, D., 450, 486, 506, 589
Nolan, P. J., 572
Nolden, F., 11
Nolen, J. A., 952
Norman, E. B., 999
Notani, M., 215, 233, 359, 610

Novák, J., 572
Novak, J. R., 556, 614
Novatskii, B. G., 182, 252
Novikov, Yu., 11
Nowacki, F., 570
Nowak, A., 494
Nunes, F. M., 876
Nyman, G., 260, 314, 327, 351, 749, 761

O

Oberstedt, S., 694
Obert, J., 78, 585
Oganessian, Yu. Ts., 523, 568, 633, 678, 708, 711
Ogawa, H., 233, 359
Ogloblin, A. A., 84, 182, 510
Ognibene, T. J., 801
Ohtsuki, T., 548
Oinonen, M., 72, 690, 745, 749, 761, 781
O'Kelly, D., 711
O'Leary, C. D., 450
Oliveira, F., 347
Oliveira, Jr., J. M., 347
Oms, J., 78, 585
O'Neil, J. P., 999
Oros, A. M., 560
Orr, N. A., 44, 118, 239, 260, 314, 327, 347, 478, 494, 570, 572, 753
Osa, A., 540
Oshima, M., 544
Ostrowski, A. N., 118, 347
Otsu, H., 182, 215
Otsuka, T., 391
Ottini, S., 478
Oura, Y., 540
Ozawa, A., 84, 182, 268

P

Page, R. D., 422, 564, 570, 572, 739
Panteleev, V. N., 94
Pashkevich, V. V., 711
Patyk, Z., 11
Paul, S. D., 444
Pavicevic, M. K., 912
Pearson, C. J., 450, 739

Penionzhkevich, Yu. E., 44, 233, 278, 359, 568, 570, 572
Penttilä, H., 72, 575, 690, 781
Peräjärvi, K., 72, 745, 781, 981
Persson, J. R., 781
Péru, S., 78
Petersen, B., 749
Petrascu, H., 182
Pfeiffer, B., 456, 757, 837
Pfützner, M., 84, 276, 494, 602, 647, 904, 985
Phillips, S. K., 606
Phillips, W. R., 470
Piattelli, P., 478
Piechaczek, A., 264, 363, 444, 602, 725, 777, 793, 904
Pieper, S. C., 399
Piercey, R. B., 606
Pinard, J., 78, 585
Pinston, J. A., 694
Piqueras, I., 314, 789
Piskoř, S., 868
Pita, S., 239
Płochocki, A., 494, 581, 602, 725, 769, 777, 781, 809
Pokrovsky, I. V., 678, 711
Poli, G. L., 367
Pollacco, E. C., 478
Polyakov, A. N., 678
Ponomarenko, V. A., 678
Popeko, G. S., 708
Pougheon, F., 278, 327, 570, 572
Pourre, P., 753
Poutissou, J. M., 973
Powell, C. F., 890, 908
Powell, J., 801, 999
Prade, H., 594
Pravikoff, M. S., 276, 686, 753
Pribora, V., 84
Prisciandaro, J. I., 66, 532
Pritychenko, B., 209, 272, 498, 606
Prokhorova, E. V., 678
Prussin, S. G., 523
Pustylnik, B. I., 678
Putaux, J. C., 78, 585

Q

Quint, W., 11, 965

R

Raabe, R., 363, 757, 793, 904
Radford, D. C., 444
Radon, T., 11
Rahkila, P., 422, 486
Ramakrishnan, E., 245
Ramayya, A. V., 523, 655, 708
Ramdhane, M., 581, 789
Rasmussen, J. O., 523, 708
Rauch, V., 789
Rauscher, T., 837
Ravikumar, V., 606
Ravn, H. L., 126, 456, 581, 864, 989
Reed, A. T., 570, 572, 739
Regan, P. H., 450, 572, 739
Reich, H., 11
Reiter, P., 256, 486, 506, 589
Rejmund, M., 256, 725, 769, 809
Ressler, J. J., 264, 375, 575
Reusen, G., 602
Reusen, I., 757, 793, 904
Reviol, W., 444, 474
Richardson, D. S., 72
Richter, A., 268
Ridikas, D., 44, 1003
Riisager, K., 260, 268, 314, 327, 351, 749, 761, 797
Riley, L. A., 490, 598
Ring, P., 22
Roeckl, E., 276, 602, 725, 769, 777, 809, 904, 969
Rogatchev, G. V., 260
Rogde, T., 294
Rogers, W. F., 138
Ronningen, R. M., 532, 606
Roussel-Chomaz, P., 44, 138, 239, 347, 478
Roussière, B., 78, 585
Rowe, M. W., 801, 999
Roynette, J. C., 478
Rubchenya, V. A., 690
Rubio, B., 725, 769, 777, 809, 969
Rusanov, A. Ya., 678, 711
Rykaczewski, K., 264, 444, 494, 581, 602, 725, 777, 793, 809, 969

S

Saint-Laurent, M.-G., 260, 278, 314, 327, 494, 568, 570, 572
Saitoh, T., 450
Sakama, M., 540
Sakurai, H., 44, 215, 233, 359, 371, 610
Sakuta, S. B., 618
Salamatin, V. S., 711
Sanchez-Vega, M., 785
Sandulescu, A., 708
Santonocito, D., 478
Sarazin, F., 44, 314
Satpathy, L., 122
Sattarov, A., 868
Satula, W., 444
Sauvage, J., 78, 585
Sauvestre, J. E., 478, 494
Savajols, H., 44
Savard, G., 130
Savelius, A., 422, 486
Scarpaci, J. A., 478, 598
Schark, E., 3
Schatz, H., 819
Scheidenberger, C., 11, 256, 268
Scheit, H., 209, 272, 286, 490, 498, 536, 890
Schlegel, Ch., 256
Schlitt, B., 11
Schmidt, K.-H., 602, 647, 686, 753, 960
Schmidt, P., 927
Schmitt, R. P., 711
Schmitt, W., 682
Schmor, P., 973
Schnare, H., 594
Schrieder, G., 268, 351
Schuessler, H. A., 150
Schulz, N., 470
Schwab, W., 268, 570
Schwartz, J., 450, 589
Schwarz, S., 3, 965
Schwarzenberg, J., 602
Schwengner, R., 594
Sebastian, V., 78, 126, 456, 585, 864, 989
Seifert, H. L., 90
Sekine, T., 544
Seleznev, Yu. G., 482
Seliverstov, M. D., 94
Seres, Z., 323

Serikov, I. N., 260
Seweryniak, D., 102, 130, 367, 375, 415, 444, 450, 474, 486, 506, 575, 589
Shapira, D., 444, 466
Sharma, K. S., 130
Shawcross, M., 239
Sherrill, B. M., 118, 209, 272, 890, 908
Shibata, M., 276, 540, 725, 769, 777, 809
Shimoura, S., 215, 371, 610
Shirokova, A. A., 482
Shoppa, T. D., 900
Shotter, A. C., 363
Siiskonen, T., 745, 761
Simon, H., 84, 256, 268
Simpson, J., 450
Singer, S., 314
Sitar, B., 268
Sitár, B., 84
Skobelev, N. K., 278, 568
Skoda, S., 594
Slinger, R. C., 264, 375
Smedberg, M. H., 268, 327, 351, 749
Smith, A. G., 470
Smith, M., 239
Sobiczewski, A., 639
Sokol, E., 570, 572
Sonzogni, A. A., 367, 375
Sorlin, O., 260, 278, 314, 327, 494, 568, 572, 739
Soumijärvi, T., 490
Spohr, K., 239
Stadlmann, J., 11
Stanford, G., 973
Steck, M., 11
Steiner, M., 66, 209, 498, 532, 598, 606, 890, 908
Steinhäuser, S., 647
Stepanov, D. N., 252
Stéphan, C., 118, 686
Stephenson, E. J., 682
Stolla, Th., 347
Stoyer, M. A., 523, 708
Stracener, D. W., 245
Strmeň, P., 84, 268
Stroth, J., 256
Strottman, D., 282
Stuttgé, L., 314, 711
Subotic, K. M., 912
Sukhoruchkin, D. S., 134

Sukhoruchkin, S. I., 134
Sümmerer, K., 11, 84, 256, 268, 276, 290
Suomijärvi, T., 118, 239, 478, 598
Suzuki, T., 84, 268
Svensson, C. E., 407
Swider, J., 575
Szerypo, J., 3, 264, 581, 602, 725, 769, 777, 793, 809
Szymański, Z., 813

T

Tachibana, T., 114, 805
Taibin, B. Z., 519
Taieb, J., 686
Tain, J. L., 725, 769, 777, 809, 969
Takeuchi, S., 323, 371, 610
Talou, P., 282
Tamburella, C., 126, 989
Tanihata, I., 84, 182, 943
Tarasov, O. B., 568, 570, 572
Taroutina, T. V., 331, 519
Tassan-Got, L., 118, 686
Taylor, R. B. E., 785
Tengblad, O., 260, 351, 749, 761, 789
Ter-Akopian, G. M., 523, 708
Teranishi, T., 215, 233, 359, 371, 610
Ternier, S., 138
Teughels, S., 138
Thibault, C., 995
Thielemann, F.-K., 837
Thirolf, P. G., 245, 286, 757, 890
Thoennessen, M. R., 221, 245, 286, 323
Thompson, I. J., 174, 876
Toader, C., 995
Tokanai, F., 182
Tolokonnikov, S. V., 302
Toneev, V. D., 568
Törmänen, S., 450
Tostevin, J. A., 209, 227
Toth, K. S., 264, 444, 474
Towner, I. S., 733
Trache, L., 868, 896
Tribble, R. E., 868, 896
Trinder, W., 568, 570, 572
Trykov, E. L., 302
Trzaska, W. H., 422, 462
Trzhaskovskaya, M. B., 985

Tsukada, K., 540, 548
Tungate, G., 72

U

Uchibori, T., 371
Uno, M., 114
Ur, C. A., 594
Urban, W., 470
Utsuno, Y., 391
Uusitalo, J., 130, 367, 375, 422, 444, 474, 506, 564

V

Vaagen, J. S., 294
Vakatov, V. I., 678
VanBrocklin, H. F., 999
Vancraeynest, G., 363, 904
Van Duppen, P., 363, 462, 494, 602, 757, 793, 919
Van Roosbroeck, J., 757, 793
Varfalvy, P., 965
Varley, B. J., 470
Varner, R. L., 221, 245
Vermeeren, L., 11, 110, 142, 757, 793, 965
Vervier, J., 849
Vieira, D. J., 90, 118
Villari, A. C. C., 118
Vincent, S. M., 450, 572, 589
Vincour, J., 868
Volant, C., 686
Volkov, Yu. M., 94
von Oertzen, W., 347
Vyvey, K., 138

W

Wadsworth, R., 739
Wajda, E., 256
Walter, G., 581
Walters, W. B., 264, 367, 375, 456, 474, 532, 575, 757
Walus, W., 256
Wan, S., 256
Wang, J. C., 690, 781

Wapstra, A. H., 30
Warner, D. D., 102, 450
Watanabe, Y. X., 215, 233, 359, 371, 610
Watson, D. L., 314
Wauters, J., 474, 486, 575, 793
Weintraub, W., 264, 444
Weissman, L., 757, 793, 864
Wenander, F., 260, 761
Wiedenhover, I., 506
Wiescher, M., 819
Wilbert, S., 110, 142
Winfield, J. S., 239, 260, 347, 494, 570
Winger, J. A., 606
Winkler, M., 11, 84, 268
Winkler, Th., 11
Wittmann, V., 602, 725, 769, 777, 809, 969
Wlazlo, W., 686
Wöhr, A., 126, 363, 456, 581, 602, 725, 757, 777, 793, 904, 989
Wollnik, H., 11
Wolski, R., 252, 260
Wood, J. L., 474
Woods, P. J., 157, 264, 286, 367, 375, 444
Wouters, J. M., 90, 118
Wycech, S., 985
Wyss, R., 462, 560

X

Xie, Z. Q., 999
Xu, H. M., 896
Xu, X. J., 474, 999

Y

Ya. Chubukov, I., 94
Yamada, M., 114, 805
Yamanaka, T., 682
Yanagisawa, Y., 371, 610
Yeandle, G., 72
Yeremin, A. V., 633
Yokoyama, S., 245
Yoneda, K., 215, 233, 359
Yoshida, A., 233, 359
Yoshida, K., 84, 182
Younes, W., 486
Yousif, H. H., 606
Yu, C.-H., 264, 444
Yu. Nikolskii, E., 182, 252
Yurkon, J., 209, 890

Z

Zamfir, N. V., 102, 466, 556, 614
Zawischa, D., 302
Zeldes, N., 146
Zeller, A., 890
Zemlyanoi, S., 78, 585
Zganjar, E. F., 264, 444, 474
Zhang, C. T., 470
Zhang, X. Q., 523
Zhou, X. G., 896
Zhu, S. J., 523
Zhukov, M. V., 268
Zuker, A. P., 38
Zylicz, J., 602, 813, 985